THE BRACONID AND ICHNEUMONID PARASITOID WASPS

THE BRACONID AND ICHNEUMONID PARASITOID WASPS

BIOLOGY, SYSTEMATICS, EVOLUTION AND ECOLOGY

Donald L. J. Quicke

Faculty of Science, Chulalongkorn University, Bangkok, Thailand

WILEY Blackwell

Registered office: John Wiley & Sons, Ltd, The Atrium, Southern Gate, Chichester, West Sussex, PO19 8SQ, UK

Editorial offices: 9600 Garsington Road, Oxford, OX4 2DQ, UK
The Atrium, Southern Gate, Chichester, West Sussex, PO19 8SQ, UK
111 River Street, Hoboken, NJ 07030-5774, USA

For details of our global editorial offices, for customer services and for information about how to apply for permission to reuse the copyright material in this book please see our website at www.wiley.com/wiley-blackwell.

Library of Congress Cataloging-in-Publication Data

Quicke, Donald L. J.
The Braconid and Ichneumonid Parasitoid Wasps: Biology, Systematics, Evolution and Ecology / by Donald L.J. Quicke.
pages cm
Includes bibliographical references and index.
ISBN 978-1-118-90705-4 (cloth)
1. Parasitoids. 2. Parasitic wasps. 3. Braconidae. 4. Ichneumonidae. I. Title.
QL496.12.Q55 2015
595.7–dc23

2014015272

A catalogue record for this book is available from the British Library.

Wiley also publishes its books in a variety of electronic formats. Some content that appears in print may not be available in electronic books.

Cover image: © Donald L. J. Quike

Set in 9/11pt Photina by Laserwords Private Limited, Chennai, India

1 2015

To the memory of a wonderful dog called Mii,
who lived at Saphan Taksin pier.

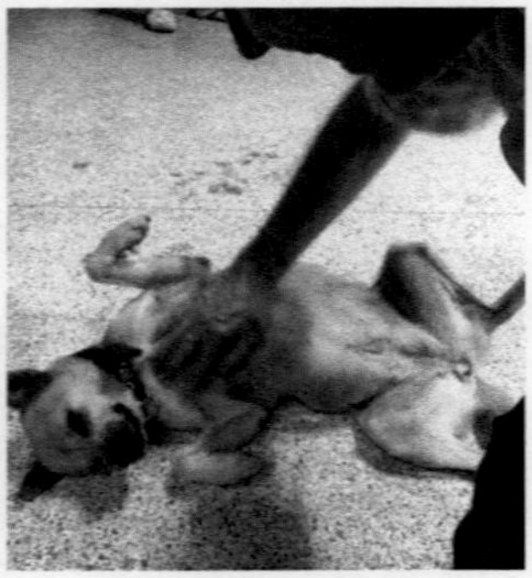

CONTENTS

Preface, xiii

Acknowledgements, xv

1 INTRODUCTION, 1

Life history, 5
Systematics, 6

PART 1 MORPHOLOGY AND BIOLOGY, 7

2 ADULT EXTERNAL MORPHOLOGY, 9

Head, 10
Antennal sensilla, 12
Antennal glands and tyloids, 14
Palps, 15
Mesosoma, 15
Legs, 17
Wings, wing venation and wing cells, 18
Confusing and sometimes erroneously applied vein names, 26
Wing flexion lines, 27
Metasoma, 29
Sexual dimorphism, 30
Male external genitalia, 32

3 THE OVIPOSITOR AND OVIPOSITOR SHEATHS, 35

The act of oviposition, 39
Functional morphology of wood-drillers, 41
Ovipositor stabilisation guides and buckling force, 43
Ovipositor notches and endoparasitism, 44
Ovipositor steering mechanisms, 44
Proposed evolutionary and related ovipositor transitions, 48
Number, position and possible functions of ovipositor valvilli, 50
Venom retention and delivery, 52
Ovipositor secretory pores, 53
Ovipositor sensilla, 54
Ovipositor sheaths, 55

4 INTERNAL AND REPRODUCTIVE ANATOMY, 57

Nervous system, 58
Digestive tract, 58
Female internal reproductive system, 59
Ovaries, 59
Time scale of egg maturation, 60
Spermatheca, 61
Common oviduct and vaginal gland, 62
Venom gland and reservoir, 63
Dufour's gland, 64
Cuticular hydrocarbons, 66
Sex pheromones, 67
Male internal reproductive system, 68
Sperm ultrastructure, 69
Spermatogeny index, 70

5 IMMATURE STAGES, 71

Eggs and oögenesis, 72
Hydropic and anhydropic eggs, 72
Embryogenesis, 73
Embryonic membranes, 75
Larva, 76
Larval feeding and nutrition, 81
Larval food consumption and dietary efficiency, 82
Lipid metabolism, 82
Respiration in endoparasitoids, 83
Larval secretions, 83
The pupal stage, 84
Cocoons, 84

6 IDIOBIONTS, KOINOBIONTS AND OTHER LIFE HISTORY TRAITS, 87

Parasitoidism, 88
Idiobiont and koinobiont strategies, 88
Generalists and specialists, 89
Ecto- and endoparasitism, 90
Permanent host paralysis, 91
Gregarious development, 91
Superparasitism, 92
Larval combat and physiological suppression, 93
Adaptive superparasitism, 95
Multiparasitism, 96
Obligate and preferential multiparasitism, 99
Hyperparasitism and pseudohyperparasitism, 99
Kleptoparasitism, 100
Evolution of life history strategies, 100

7 SEX, COURTSHIP AND MATING, 107

Sex determination, 108
Local mate competition and avoidance of inbreeding, 110
Sex allocation, 110
Protandry and virginity, 112
Thelytoky and cytoplasmic incompatibility, 113
Mate location, 117
Courtship, 119
Swarming and lekking, 120
Mating position, 121
Multiple mating and sperm competition, 121
Sex-related scent glands, 123
Genome size and recombination, 125
Cytogenetics, 125

8 HOST LOCATION, ASSOCIATIVE LEARNING AND HOST ASSESSMENT, 127

Tritrophic interactions, 129
Host acceptance, 130
Associative learning, 130
Biosensors, 134
Patch use, 134

9 OVERCOMING HOST IMMUNE REACTION AND PHYSIOLOGICAL INTERACTIONS WITH HOST, 137

Overcoming host immunity in endoparasitoids, 138
Passive evasion of encapsulation by parasitoid eggs, 139
Avoiding encapsulation by physical means, 139
Effect of host age and haemocyte number, 141
Other host defence mechanisms, 141
Venoms, 141
Neurophysiological venom actions, 143
Venom effects on host immune response, 144
Polydnaviruses, 145
Effects of polydnaviruses on hosts, 152
Other reproductive viruses, 155
Improving host quality, 156
Host castration and similar effects, 156
Teratocytes, 158
Intraspecific variation in resistance to parasitoids, 159
Effects on host moulting pattern, 160
Parasitoid-induced changes in host behaviour, 160

10 CONVERGENT ADAPTATIONS, 163

Antennal hammers and vibrational sounding, 164
Enlarged mandibles, 167
Chisel-like mandibles, 168
Concealed nectar extraction apparatus, 168
Reduced number of palpal segments, 169
'Facial' protruberances, 169
Frontal depressions, 170
Dorsal ridges on head or mesosoma, 170
Brachyptery and aptery, 170
Dorso-ventral flattening, 171
Postpectal carina, 173
Propodeal spines, 173
'Fossorial' legs, 173
Fore tibial spines, 174
Fore tibial apical tooth, 174
Expanded hind basitarsi, 174
Toothed hind femur, 174
Distitarsal scraper, 175
Pectinate claws and claws with angular basal lobes, 175
Glabrous wing patches and wing membrane scleromes, 176
Carapacisation, 177
Petiolate metasomas, 177
Modifications to the posterior metasomal margin, 178
Spermathecal colour, 179

Compression of apical part of metasoma, 179
The 'ophionoid facies', 179
White antennal stripes and tips, 180
White ovipositor sheath stripes and tips, 181
Number of larval instars, 182
Egg-larval parasitism, 182
Disc-like larval antennae, 182
Reduction of larval hypostomal spur, 183
Wide and heavily sclerotised larval epistoma, 184
Suspended cocoons, 184
Polyembryony, 184
Phytophagy and cecidogenesis, 184

PART 2 TAXONOMIC AND SYSTEMATIC TREATMENT, 187

11 OVERVIEW OF ICHNEUMONOIDEA: RELATIONSHIPS AND SYSTEMATICS, 189

Monophyly of Ichneumonoidea, Ichneumonidae and Braconidae, 190
Relationship of Ichneumonoidea to other Hymenoptera, 190
Fossil history and family-level phylogeny, 192
Brief history of classification, 194
Ancestral biology of Ichneumonoidea, 196
Separating ichneumonids from braconids, 197
Identifying specimens, 198

12 PHYLOGENY AND SYSTEMATICS OF THE BRACONIDAE, 201

Historical perspective, 202
Morphophylogenetic hypotheses, 202
Molecular phylogenetics, 204
Braconid classification, 205
  Eoichneumoninae†, 205
Trachypetiformes, 205
  Trachypetinae, 205
Cyclostomes *incertae sedis*, 209
  Protorhyssalinae et al., 209
  Apozyginae, 210
The aphidioid clade or 'Gondwanan' complex, 212
  Aphidiinae, 212
  Maxfischeriinae, 224
  Mesostoinae (including Canberreriini and Hydrangeocolini), 225
The remaining cyclostomes, 229
  Doryctinae (including Ypsistocerini), 231
  Pambolinae, 236
  Rhysipolinae, 237
  Rhyssalinae, 238
  Rogadinae s.l., Hormiinae, Lysiterminae, 243
  Betylobraconinae, 243
  Hormiinae, 243
  Lysiterminae, 245
  Rogadinae *sensu stricto*, 246
  Alysioid subcomplex, including Braconinae, 250
  Alysiinae and Opiinae, 250
    Alysiinae, 251
    General Alysiinae biology, 251
    Alysiini, 253
    Dacnusini, 255
    Opiinae, 256
  Braconinae, 260
  Exothecinae, 269
  Gnamptodontinae (= Gnaptodontinae), 270
  Telengaiinae, 271
The non-cyclostomes, 271
  Sigalphoid complex, 271
    Agathidinae, 272
    Sigalphinae, 275
  Helconoid complex, 278
    Helconinae, 279
    Helconoid group *incertae sedis*, 281
  Blacinae, 282
  Acampsohelconinae, 283
  Macrocentrine subcomplex, 284
  Macrocentrinae, 284
  Charmontiinae, 287
  Amicrocentrinae, 287
  Xiphozelinae, 288
  Homolobinae, 290
  Microtypinae, 292
  Orgilinae, 292
  Euphoroid complex, 294
    Euphorinae, 294
    Cenocoeliinae, 310
  The microgastroids, 311
    Cardiochilinae, 312
    Cheloninae (including Adeliini), 315
    Dirrhopinae, 319
    Ichneutinae, 320
    Khoikhoiinae, 322

Mendesellinae, 322
Microgastrinae, 322
Miracinae, 335
Unplaced subfamilies, 335
Masoninae, 335
Meteorideinae, 337

13 PHYLOGENY AND SYSTEMATICS OF THE ICHNEUMONIDAE, 341

History of ichneumonid classification, 342
Henry Townes (1913–90) and his idiosyncratic nomenclature, 344
The extinct subfamilies, 344
Tanychorinae†, 344
Palaeoichneumoninae†, 346
Labenopimplinae†, 348
Pherombinae†, 349
Townesitinae†, 349
The xoridiformes, 349
Xoridinae, 349
The labeniformes, 353
Labeninae, 353
Groteini, 355
Labenini, 355
Poecilocryptini, 356
The pimpliformes, 356
Acaenitinae, 356
Collyriinae, 359
Cylloceriinae, 360
Diacritinae, 360
Diplazontinae, 361
Orthocentrinae (= Helictinae), 366
Pimplinae, 367
Delomeristini, 369
Ephialtini (= Pimplini of Townes), 369
Polysphincta group, 371
Pimplini, 373
Poemeniinae (= Neoxoridinae), 378
Poemeniini, 378
Pseudorhyssini, 378
Rodrigamini, 378
Rhyssinae, 379
The ichneumoniformes, 383
Adelognathinae, 383
Agriotypinae, 385
Alomyinae, 387
Cryptinae, 388
Aptesini, 391
Cryptini, 391
Phygadeuontini, 393
Ichneumoninae, 394
The brachycyrtiformes, 398
Brachycyrtinae, 398
Claseinae (Clasinae), 398
Pedunculinae, 399
The orthopelmatiformes, 400
Orthopelmatinae, 400
The ophioniformes, 400
Lower ophioniformes, 402
Banchinae, 402
Lycorininae, 406
Sisyrostolinae, 407
Stilbopinae, 407
Tryphoninae, 411
Middle ophioniformes, 416
Ctenopelmatinae, 416
Mesochorinae, 421
Metopiinae, 422
Oxytorinae, 424
Tatogastrinae, 425
Tersilochinae (including Neorhacodinae and Phrudinae *s.s.*), 426
Higher ophioniformes, 430
Anomaloninae, 430
Campopleginae, 432
Cremastinae, 438
Hybrizontinae, 439
Nesomesochorinae, 442
Ophioninae, 442
Unplaced subfamilies, 445
Eucerotinae, 445
Microleptinae, 447

PART 3 ECOLOGY AND DIVERSITY, 451

14 ECOLOGY, 453

Adult diet, 454
Host-feeding, 454
Water, sugar and pollen feeding, 457
Fecundity, 460
Voltinism and seasonality, 462
Daily activity patterns, 462
Diapause, 463
Cold hardiness, hibernation and overwintering, 465
Coloration and thermoregulation, 467
Biological control, 467
Effect on host food consumption, 471

Artificial diets, 474
Artificial hosts, 475
Use of alternative hosts, 475
Hyperparasitism and kleptoparasitism, 476
Predation, 477
Pathogens, 477
Transmission of host pathogens, 479
Dispersal, 480
Coloration and mimetic rings, 480
Palatability and odours, 481
Competition, 482
Apparent competition, 482
Host ranges of parasitoids, 483
Parasitoid guilds and food webs, 484
Evolution of host ranges and speciation, 486

15 LOCAL AND GLOBAL PATTERNS IN DIVERSITY, 489

Field research in the tropics and anomalous diversity, 490
Estimation of global ichneumonoid species richness, 492
Distribution related to climate and latitude, 496
The nasty host hypothesis, 497
Biogeography, 503
Islands and their parasitoid faunas, 505
Species accumulation curves, 506
Altitudinal gradients, 507
Estimating local species diversity, 508
Ichneumonoidea as biodiversity indicators, 510
Conservation, 510
Effect of habitat degradation on ichneumonoid composition, 511
Significance of cryptic species, 511

16 COLLECTING AND REARING ICHNEUMONOIDEA, 513

Field collecting adults, 516
- Pan traps, 518
- Sweep netting, 519
- Light trapping, 521
- Canopy fogging, 521
- Malaise traps, 521

Rearings from wild-collected hosts, 523
- Rearing leaf rollers and tiers, 524
- Substrate rearings, 524

Culturing, 524
- Mating in captivity, 525

Mass rearing, 525
Mounting specimens for taxonomic study, 526
- Preparing specimens from alcohol storage, 526
- Direct pinning, 527
- Side gluing, 527
- Card rectangles and card points, 527
- Secondary staging, 528
- Labelling, 528

Preserving specimens for DNA analysis, 528
Packaging and posting specimens to other workers, 530

17 EPILOGUE, 533

Phylogenetic questions, 534
Host and parasitism questions, 534
Physiological questions, 535
Ecological questions, 536

Glossary, 539

References, 547

Author index, 633

General index, 653

Host index, 659

Ichneumonoid genus, tribe and subfamily index, 665

Ichneumonoidea species index, 677

COLOR PLATE SECTIONS ARE INSERTED BETWEEN PAGES NOTED BELOW

First 13-page colour plate section (between pages 112 and 113)
Second 13-page colour plate section (between pages 224 and 225)
Third 13-page colour plate section (between pages 336 and 337)
Fourth 13-page colour plate section (between pages 448 and 449)

PREFACE

The Ichneumonoidea is one of the largest and ecologically most important groups of insects, with approximately 41,000 valid described species and estimates of several times that number still awaiting to be described. They are responsible for regulating the populations of a large proportion of other insects, many of which would be harmful to both the natural and agro-forestry systems if their numbers were unchecked; ichneumonoids therefore have a large economic and aesthetic impact.

My previous book on parasitoid wasps (Quicke 1997) attempted to cover the biology and ecology of all the Hymenoptera with a parasitoid way of life and consequently did not go into any great detail about the life histories of particular groups. In this book, I am again covering almost all aspects of development, biology, physiology, behaviour and ecology and additionally devoting more space to morphology, systematics and phylogeny, biodiversity research and the roles of groups in biological control. By concentrating here on the Ichneumonoidea, the group with which I have most research experience, I am able to present a great deal more information within a phylogenetic context and also to discus what is known of the biologies of each of the subfamilies. Hence readers who are interested in only one group can find out something about them, their life histories, what research they have been involved in, something about their classification and also what there is still to be discovered about them.

The classification of the whole of the Ichneumonoidea, along with most other insect orders, has been plagued by typology, i.e. the notion that if a species possesses some particular character or combination of characters, it must belong to a particular group. Typology is a common disease of traditional taxonomy and although that is not to say that what constitutes traditional taxonomy is wrong or was necessarily a bad thing, when strong new evidence comes to light, we have to be willing to consider major changes. The new evidence that is now becoming available is provided by molecular data and we are only just beginning to see the implications. I want to take the opportunity in this book to push a molecular side of the argument because I think that creating phylogenies and classifications based only on morphological characters is particularly likely to be flawed. Of course, I do not believe that molecules are incapable of being misleading, they can certainly be that, but in general, and not necessarily under any very particular circumstances, molecular evidence is likely to provide our best insight into phylogeny. In recent years, it has become apparent that many morphological characteristics associated with basic life history features, specifically whether wasps are ecto- or endoparasitic, can be grossly misleading in terms of the phylogeny they suggest.

One of the key issues with understanding the evolutionary traits in the superfamily is that, despite a vast literature, most work on aspects of physiology or development is very patchily distributed and often in ways inadequate to control for phylogeny. Thus, what might appear to be an interesting biological association could just represent a single evolutionary origin, since critical taxa have not been studied.

Recent advances in taxonomy and systematics, the concomitant publication of better identification works and the realisation that many earlier host records (perhaps 50% or more) are wrong, and sometimes badly so, means that we are becoming better able to address some of the gaps and discover new, interesting evolutionary links. Progress is likely to continue at an even greater pace with better illustrated and more accessible web-based tools. In this book, I hope to draw attention to areas where particular types of research might be most effectively employed, including groups that need more work on host associations or general biology, and to follow up intriguing hints from preliminary data on other groups. In particular, new phylogenetic hypotheses have highlighted groups for which either additional biological information or, in numerous cases, any biological information at all might be particularly revealing for our understanding of the evolution of the group.

Donald L. J. Quicke
Bangkok, Thailand

ACKNOWLEDGEMENTS

Over the many years that I have been interested in parasitic wasps in general and ichneumonoids in particular, I have benefited from discussions with a very large number of people and to all of those I offer my thanks. However, I must make a few mentions in particular and hope that I have not missed anyone out who will be too upset by that. The following people, representing a number of fields from taxonomy and systematics, behaviour, functional morphology and ecology have been of particularly great help to me and have all, in various ways, contributed to the ideas in and the development of this book: Andy Austin, Sergey Belokobylskij, Ferdinando Bin, Gavin Broad, Mike Fitton, Charles Godfray, Vladimir Gokhman, Seraina Klopfstein, Nina Laurenne, Victoria Pook, Alexandr P. Rasnitsyn, A. M. [Guida] Santos, Mike Sharkey, Mark Shaw, Scott Shaw, Kees van Achterberg, Bob Wharton, Jim Whitfield and Alejandro Zaldivar-Riverón. I would also like to thank the many people who have kindly allowed me to use their published or unpublished photographs and micrographs: as they say, a picture speaks a thousand words. All graphs were produced using the statistical computing language R (R Development Core Team 2009).

Chapter 1

INTRODUCTION

The Braconid and Ichneumonid Parasitoid Wasps: Biology, Systematics, Evolution and Ecology, First Edition. Donald L. J. Quicke.

Although most people are blissfully unaware of them, the ichneumonoid wasps are one of the most diverse groups of insects, and in terms of their ecological role they are probably of enormous importance. No-one really has a good idea about how diverse they are and estimates vary widely. The total number of valid species described to date, 18,000 braconids and 23,000 ichneumonids[1], is certainly a great underestimate, but by how much is still anyone's guess. Many works cite estimates of 40,000 and 60,000, based upon expert opinion (Townes 1969, Gauld & Bolton 1988). Similar values have also been obtained by various objective estimation measures, but it seems likely that these too are underestimates, and narrowing the numbers down is not going to be easy for the reasons explained in Chapter 15.

Unfortunately, neither family has attracted a lot of attention from amateur entomologists, which seems to be a prerequisite for a good knowledge of a group's taxonomy, distribution and biology. This may be partly because many of the species are rather small and often dull coloured, although this does not seem to have deterred generations of amateur coleopterists. Probably the most important factor has been the dearth, until fairly recently, of reliable and accessible identification guides to the major groups (subfamilies), confounded by the fact that the subfamily-level classification is only now becoming fairly stable, largely as a result of much new molecular work. Problems have been compounded because numerous names were mis-applied by early workers and, as these errors were slowly discovered and corrected, many groups accumulated a historical backlog of alternative names. In many fields of science, the really old literature seldom has to be cited, but in zoology, a great deal of excellent work on anatomy and biology was carried out 50 to 100 or so years ago. As this may be the only detailed work on a given group, it is still relevant today and the reader therefore has to deal with the sometimes confusing or even misleading nomenclature.

Difficulties in the correct identification of specimens, and publications dealing with incorrectly identified specimens, have also been a major stumbling blocks. To quote Perkins (1959), 'It is perhaps, not surprising that keys to subfamilies are very imperfect, as exceptions can be found to almost all characters that have been used in defining any subfamily, even in the limited British fauna'. Partly because of overall improving taxonomic and systematic understanding, published research on both families is growing, that dealing with the braconids slightly more quickly than that for the ichneumonids (Fig. 1.1). There may be several reasons for this growth, not the least of which is that most researchers are now under great pressure to publish their findings quickly in bite-sized chunks and in high-impact journals, rather than presenting single, large tomes representing the results of many years of their work. The difference in the rate of publication between the families could well be due to the ease of identification – recognising subfamilies is generally easier for braconids and knowing what subfamily you are dealing with is the essential first step towards a proper identification.

The Ichneumonidae and Braconidae are each such large groups that few people since the early 20th century have attempted to work seriously on the whole of either one of them, so it is hardly surprising that in recent years almost no-one has attempted to tackle them both. This, of course, means that the similarities and differences between them may have been less well considered than they should have been. Superficially, it might seem that these two families essentially parallel one another, they are sister groups and they broadly occupy the same range of niches – they predominantly parasitise exposed and concealed moth and beetle larvae with a few incursions into attacking fly and Hymenoptera larvae, rarer ones into other insect groups and a few other ways of life such as spider egg predation and even a few instances of true phytophagy. However, things may not be as simple as they seem, because despite some remarkable parallels, they also show strong group differences in precisely what they do and in the types of adaptations they typically employ.

It should come as no surprise therefore, that ichneumonids and braconids do not 'behave' in the same way in so many aspects of their biology and morphology. If they did, it seems likely that one would have driven the other to extinction or pushed them a long way in that direction. That both groups are highly speciose seems very likely to indicate that they do not compete in a precise and consistent way, although many individual species no doubt do. Hence there are various sorts of adaptations that appear to evolve frequently in one family but not or only rarely in the other. For example, numerous braconids have evolved carapace-like metasomas where the basal 3 (or sometimes 4) metasomal terga are enlarged, frequently fused, or at least more or less immovably joined and conceal all more posterior ones (see Chapter 10, section *Carapacisation*). Only a very few ichneumonid groups have members

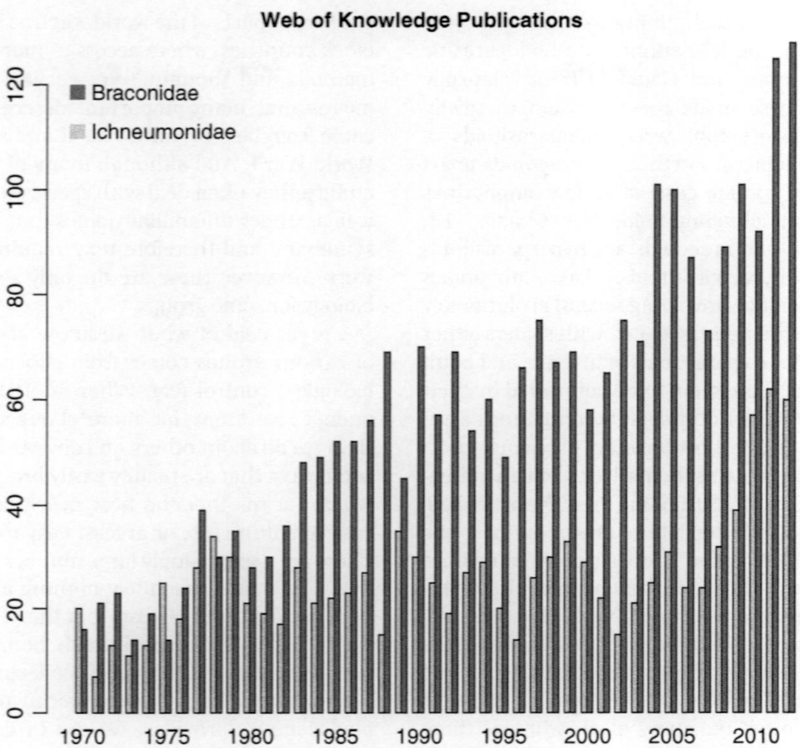

Fig. 1.1 Numbers of papers on Braconidae and Ichneumonidae published each year in *Science Citation Index* (SCI) journals from 1970 to 2012.

with carapaces and the numbers of species involved is very small. Is this associated with the difference in articulation between the second and third metasomal terga, which is one of the diagnostic features for separating the two families? Endoparasitoid larvae belonging to several different braconid lineages have apparently independently evolved an everted rectum forming a structure called an anal vesicle (see Fig. 5.1) that serves a variety of physiological roles, but this adaptation, as far as is known, has only evolved in two genera within the Ichneumonidae. Similarly, very elongate mouthparts (although variously involving the glossa, malar region or maxillary palps) have evolved on numerous independent occasions within those Braconidae dwelling in relatively arid habitats (see Chapter 10, section *Concealed nectar extraction apparatus*), but the number of such occurrences in the Ichneumonidae is small (e.g. *Rhynchobanchus*: Banchinae). These modified mouthparts, collectively referred to as a concealed nectar extraction apparatus, are an adaptation to obtain nectar from plants such as Asteraceae or Dipsaciaceae, which in turn are adapted to prevent their nectar from drying up in places where water is in short supply. In this case, it may be because braconids tend to comprise a relatively larger proportion of species in such habitats, but the data are not really available to test this.

Ichneumonids collectively utilise a somewhat different spectrum of hosts than braconids. They include many more taxa that are parasitoids of other Hymenoptera, including both endo- and ectoparasitism, in addition to acting as pseudohyperparasitoids of other ichneumonoids (see Fig. 13.1; *cf.* Fig. 12.2), and endoparasitism including developing as true hyperparasitoids within a host, as well as some being predators within aculeate wasp and bee nests. In the Braconidae, members of two tribes within the Euphorinae are endoparasitoids on adult Hymenoptera, a few ectoparasitoids attack leaf-mining sawflies and only a few members of the Ichneutinae are endoparasitic within sawfly larvae, and Gauld (1988a) plausibly suggested that these made the transition to sawfly hosts from ancestors that were endoparasitoids of leaf-mining Lepidoptera. Further, no braconids apart from the rather special case of a few euphorines parasitising adult ichneumonoids (see Chapter 12, section *Syntretini*), no braconids are hyperparasitoids or even pseudohyperparasitoids. Two subfamilies within Ichneumonidae, involving several evolutionary transitions, have become associated with spiders either as egg predators or as parasitoids of juvenile and adult individuals. All of these seem to be connected by their use of silk, or volatile or non-volatile compounds associated with silk, in the host location – because of its non-solubility, silk proteins themselves seem an incredibly unlikely source of host-finding cues. Nevertheless, at least some braconids do utilise cues from host silk trails (Ha et al. 2006), but it does not seem to have become an important part of their behavioural repertoire. Perhaps partly associated with this and the places where silk-cocooned hosts occur, ichneumonids appear to have evolved vibrational sounding (a sort of echolocation) as a host location tool on multiple occasions (and lost it on many also), whereas there is no evidence for this host location mode in the Braconidae (see Chapter 10, section *Antennal hammers and vibrational sounding*).

Another important question that we ought to consider is why the ichneumonoids and chalcidoids have not out-competed one another in one direction or another. Some niches occupied by chalcidoids are not available to ichneumonids; for example, egg parasitism, which necessitates body sizes smaller than or at least at the very bottom range of that which ichneumonoids (e.g. Miracinae or Cheloninae–Adeliini) have thus far achieved. Ichneumonids described to date are, in general, larger bodied than braconids (see Fig. 15.6), and this may correlate with some differences in host utilisation, since only braconids can attack small insect hosts such as psocids, aphids, plant bugs and tiny beetles (Čapek 1970).

It seems to me a very great shame that many traditional areas of study, such as those on comparative embryology and detailed descriptions of natural history, have suffered a serious decline in recent years and effectively have ceased in most Western universities. For a long time they have been largely restricted to workers in parts of the world, such as the former Soviet block countries, where access to more trendy modern methods and thought were perhaps restricted. This means that many important descriptions of biology come from before World War II and sometimes before World War I. And although many of these are of high quality, they often deal with species serendipitously, as well as under unfamiliar names that have been lost in synonymy and therefore may require some detective work. However, these are the only sources of detailed biology for some groups.

A great deal of what we know about the biologies of various groups comes from efforts to use them for biological control (e.g. Wharton 1984). As a consequence, we know far more about some subfamilies than we do about others and obviously we know more about taxa that are readily easily brought into culture, which means that the host nearly always has to be easy to culture too, or at least easy to find and collect. There are a surprisingly large number of subfamilies for which we know absolutely nothing about the biology, not even the order of hosts that they attack or whether they are ecto- or endoparasitoids. Some of them include fairly common and frequently collected species.

Now that we have the powerful tool that modern phylogenetics provides, we are in desperate need of more such studies to help test hypotheses about the adaptive natures of particular character states within a comparative framework. Although it is possible in some cases to go out and obtain the necessary taxa, there is a general mismatch of people skills. Many excellent physiologists and molecular geneticists carry out their work on taxa of real or potential economic importance and a considerable amount of their research receives funding because of this. By their nature, the hosts of potential biocontrol insects are generally easy to obtain and culture, although admittedly wood-borers may pose more of a logistic issue than say grass-feeding aphids or cotton-feeding moths. The parasitoids that are therefore best investigated are those which attack these hosts with the consequence that much work has been carried out on a relatively small subset of taxa, aphidiine and microgastrine braconids and campoplegine ichneumonids being prime examples. Many of the laboratory researchers would love to obtain some other parasitoid taxa into culture to study, but this involves setting up host cultures, obtaining the parasitoids and working out rearing techniques, all probably with less funding available. Although there are numerous exceptions, many laboratory-based experimentalists often do not have the field entomological or natural history backgrounds to facilitate the finding of some of the other taxa.

The other side of the skills mismatch is that there are, at least in many 'Western' and East Asian countries, excellent natural historians who are good at and enjoy going into the field in search of insects and rearing them, but they often do not necessarily know what more detailed pieces of information about an insect's biology are missing. Added to this is the problem that many taxonomically interesting taxa, which might well have particularly interesting biologies and associated physiology, biochemistry, etc., are simply rare, very local in distribution or attack hosts that are very difficult to obtain or bring into culture. There are multiple examples of all of these.

One of the aspects that really needs to be revisited, as soon as sufficient independent molecular phylogenies become available, is all the hypothesised evolutionary transitions and trends that have been based on purely morphological phylogenetic estimates. It is surprising how often the networks obtained from morphological and molecular analyses are similar, which is good, but the rooting is extremely different. Such different results may reflect either that the outgroups (if used) are too distant to provide much meaningful evidence of true ancestry or even that workers had a soft spot for an elegant biological story. It was certainly common practice in early cladistics studies to ignore characters that the worker 'knew' to be homoplastic. Currently available molecular data have provided a considerable number of new insights and reasonably well-supported big pictures for both families, but there are still many areas of the evolutionary tree where there is a real shortage of resolution and several taxa whose placements are far from certain. Quite possibly much of the radiation at subfamily level occurred subsequent to the Cretaceous period – there are few Cretaceous fossils that can be assigned to modern subfamilies with confidence, especially within the Ichneumonidae, yet the Eocene (53 to 33.7 Mya) fossil record contains many species that are fairly certainly recognisable to modern subfamilies and sometimes possibly to a genus.

The large size of both families mean that there are inevitably many scientific names which may seem daunting or confusing to beginners. Even when dealing with the relatively small number of frequently cultured species there are still many of them to get to grips with. I think it is certain that the nomenclatural aspect of work on this group has been off-putting, not aided by the fact that some workers have employed alternative systems [see Chapter 13, section *Henry Townes (1913–90) and his idiosyncratic nomenclature*, although it is not just Townes' work where confusion can arise].

It is always difficult in a book such as this to decide whether to start with morphology, biology or taxonomy. I have opted for the first, but in order to be able to make some sense of the features, it is necessary to refer to some aspects of each of the others in this section. I have therefore included below very brief outlines of some of the important biological concepts and systematics to facilitate understanding. I have also chosen to arrange things in rather a small number of chapters, each consequently with a fairly broad remit. Nevertheless, some topics have had to be shoe-horned in at places where they might seem slightly incongruous. It also seemed logical to include a few physiological aspects within more broadly morphological sections since the understanding of the morphology is sometimes intimately linked with other processes. As this will be used mainly as a reference book, some facts are repeated in two or more places. I hope that in the end, the structure more or less makes sense.

LIFE HISTORY

There are two important terms to be learnt here. The fairly obvious difference between parasitoids which (generally) lay their eggs within a host and whose larvae develop internally surrounded by wet host tissues, i.e. **endoparasitoids**, and those that lay eggs externally and whose larvae complete feeding from the outside, surrounded by air, i.e. **endoparasitoids**.

A second important distinction, – indeed, in many respects possibly more important – is between parasitoids whose hosts do not develop further after being parasitised, referred to as **idiobionts**, and those parasitoids whose hosts continue feeding and usually moulting after the parasitoid has oviposited (usually) in them, which are called **koinobionts**.

Ecto- and endoparasitism and idiobiont/koinobiont strategies both explain a great deal about other life history features and they are strongly correlated, although asymmetrically. Most koinobionts are endoparasitoids, but idiobionts can be either ecto- or endoparasitic, but endoparasitic idiobionts are almost entirely, within the Ichneumonoidea, parasitoids of host pupae and complete their development therein.

SYSTEMATICS

Necessarily, many subfamilies and genera have to be mentioned, and typically members of the same subfamily show very similar biologies and members of genera even more so, although there are some exceptions. Both Braconidae and Ichneumonidae include a large clade whose members are all koinobionts and, with the exception of the Tryphoninae within the Ichneumonidae, endoparasitoids. In both families, the sister group to the entirely koinobiont clade is a predominantly idiobiont ectoparasitoid lineage with numerous independent transitions to endoparasitism and koinobiosis. In the Braconidae, members of this lineage are called cyclostomes (see Fig. 2.1b) in reference to their mouthpart morphology, although some members of the cyclostome lineage have secondarily become non-cyclostomes. When I use the term 'cyclostome' in this book, unless specified otherwise, I am referring to the lineage rather than the condition. There is no equivalent term within the Ichneumonidae, although the biologically equivalent lineage of (predominantly) koinobiont endoparasitoids is dominated by a group informally referred to as the ophioniformes. The ending '-formes' is used throughout to indicate groupings of subfamilies that are believed to be monophyletic and usually have relatively consistent biologies.

Many readers will not know where a given taxon belongs, either within the above larger framework or to what subfamily it belongs. I have therefore very largely specified this as I go along, despite its clumsiness, because in that way the reader might most readily search for further information on other members of group of interest.

ENDNOTE

1. Over 60,000 species names have been published so some 19,000 are synonyms.

Part I

Morphology and Biology

Chapter 2

ADULT EXTERNAL MORPHOLOGY

The Braconid and Ichneumonid Parasitoid Wasps: Biology, Systematics, Evolution and Ecology, First Edition. Donald L. J. Quicke.

The pupal and adult bodies of apocritan (i.e. wasp-waisted taxa) Hymenoptera are divided into three tagmata, head, mesosoma and metasoma. The mesosoma comprises the three thoracic segments in addition to the first abdominal segment, in other words, the narrow waist that defines the apocrita is located between the first and second abdominal segments. The first abdominal segment at the posterior of the mesosoma is called the propodeum. The reason for the waist being there rather than between the thorax and abdomen is that in flying species, one of the large thoracic muscles associated with flight is attached to a posteriorly directed internal phragma of the metathorax. Readers should be aware that in the 19th century, this was not always recognised and so some works refer to the propodeum erroneously as the metathorax, which can be confusing, especially with descriptions of colour patterns.

Karlsson and Ronquist (2012) provide an excellent guide to the skeletal anatomy of two exemplar braconids, both members of the Opiinae, which deals with both external and internal chitinous structures. This work updates and, because of the use of scanning electron microscopy (SEM), is in greater detail than Alam's studies of the braconine, *Stenobracon deesae* (Alam, 1954, 1955). It serves as a basis for the terminology used here and also is almost entirely applicable to the Ichneumonidae.

HEAD

The head bears three pairs of multi-articulate appendages and also a pair of mandibles, eyes and three ocelli. Males tend to have larger eyes than females and the eyes and ocelli of crepuscular and nocturnal taxa, such as those showing the ophionoid facies, are usually relatively larger than those of related diurnal species. A few taxa that live largely concealed in termite nests can have very reduced eyes.

In the Ichneumonoidea, the head is virtually always orthognathous, with the line between the ocelli and mandibles being more or less perpendicular to the long axis of the body. The main exception is the definitely prognathous Masoninae (see Fig. 12.73). The top of the head comprises, from posterior to anterior, the occiput [which is often margined anteriorly with a carina (for occipital carina, see Figs 12.12e and 12.26)], the vertex, which is the dorsal part behind and somewhat lateral to the ocellar triangle, and the frons (that part between the ocelli and the antennal sockets). Presence, partial absence or complete absence of the occipital carina are often important characters in braconid identification. The three ocelli are usually on a distinctly differentiated, roughly triangular raised zone referred to as the stemmaticum. The front of the head (Fig. 2.1) from antennal sockets to the mandible insertion is comprised of the face and, below this, the clypeus, the two regions usually being clearly demarked by a carina or groove, although in a few groups (notably campoplegine and metopiine ichneumonids) they are effectively completely united in a uniform shield. At the border between clypeus and face are a pair of pits [anterior tentorial pits (Fig. 2.1a)], that are the external manifestation of a pair of internal cuticular protuberances (tentorial arms) which, together with posterior equivalents, form an internal supporting skeleton for the brain and pharynx. One of the most important features in braconid systematics involves the clypeus. In the majority of species in the cyclostome lineage, the lower part of the clypeus (called the hypoclypeus) is reflexed posteriorly and usually forms a dorsally convex arch exposing the labrum, which is, again usually, convex and glabrous (Fig. 2.1b). In the non-cyclostome lineage of braconids and nearly all ichneumonids, the labrum is hidden and, if exposed, clearly setose. Its partial exposure is also an important feature in the Ichneumonidae. The dorsal, clypeal margin of the hypoclypeal depression in cyclostome braconids usually bears a pair of clusters of small bristles, which are presumably sensillae (hypoclypeal hair brushes). The lower anterior clypeal margin nearly always bears a transverse row of strong setae and in the Ichneumonidae, in particular, their distribution and regularity of spacing are often an important taxonomic characters.

Between the ventral part of the compound eye and the base of the mandible articulation is a region called the malar space. Various measurements such as the relative lengths and widths of these parts are often important in species recognition, although there has not been complete consistency in how various measures are taken and care should be taken to understand what a particular author means by their use of the measurements or ratios they employ.

The top of the head comprises the frons, located between the three ocelli and the antennal sockets, and the occiput, that part which lies behind the ocelli. The back of the head is often completely or partially margined by a carina, the occipital carina, and this

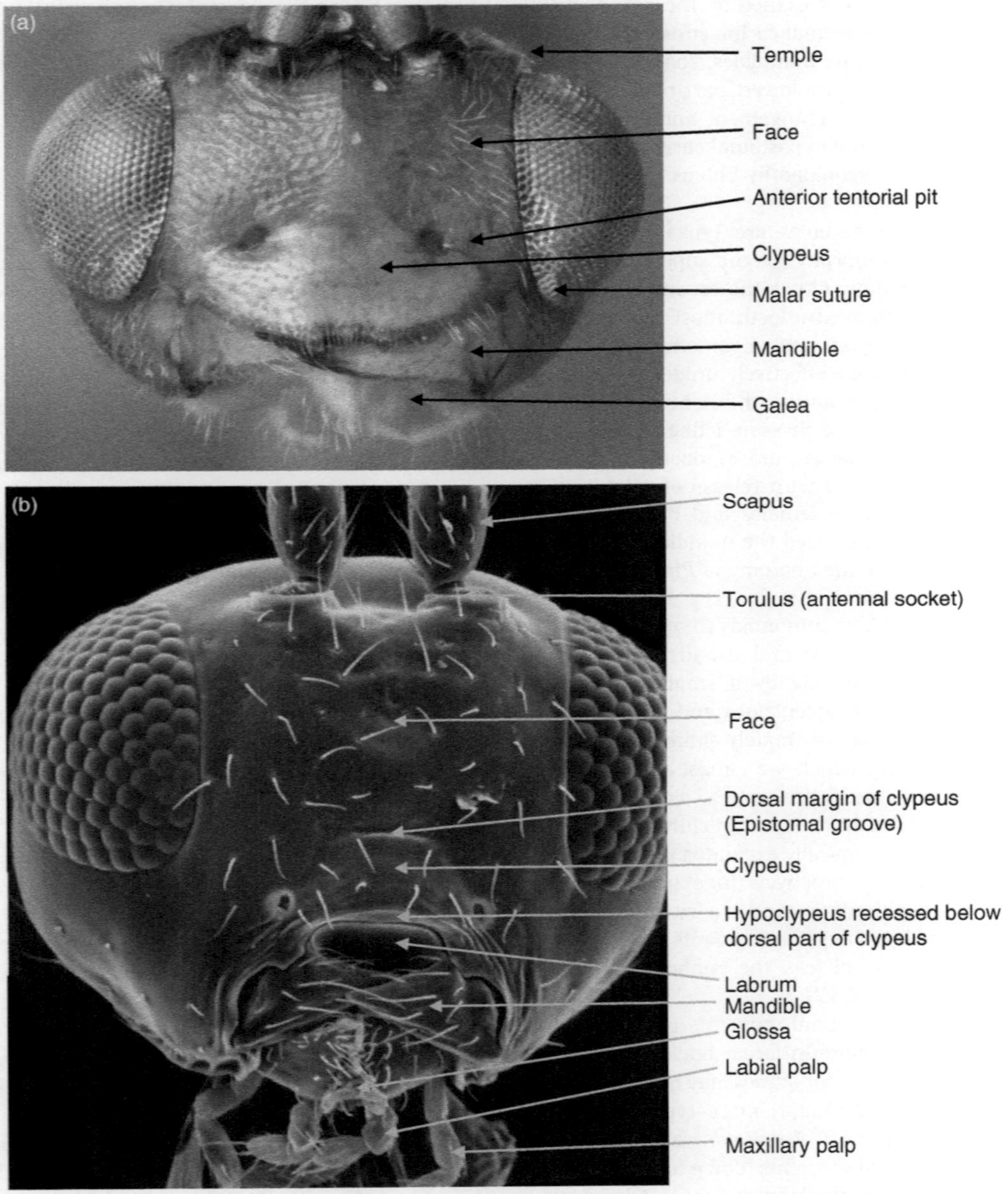

Fig. 2.1 Parts of head in frontal aspect. (a) *Phanerotoma behriae* (Cheloninae), a non-cyclostome braconid showing lack of impressed space between lower margin of clypeus and mandibles. (Source: Reproduced by permission of Rebecca Kittel.) (b) *Acrisis* sp. (Rhyssalinae), a cyclostome braconid showing hypoclypeal depression and concave, glabrous labrum.

is often of considerable taxonomic importance. The region where the head attaches to the prothorax (occipital foramen) is margined by the post-occipital carina, and the hypostomal carina runs from there to close to the base of the mandibles, sometimes fusing part of the way with the lowest part of the occipital carina. Details of the arrangement and development of the occipital and the hypostomal carinae have been surveyed in the Braconidae by Tobias and Potapova (1982).

Ichneumonoid mandibles are typically bidentate, which is a synapomorphy for the superfamily, but a number of taxa show either the secondary development of one or more extra teeth, most notably in the Alysiinae and Diplazontinae, or sometimes the reduction to one so they are effectively unidentate. Little is known about the distribution or function of mandibular glands, but those of rhyssine ichneumonids have been described by Davies and Madden (1985) and appear to be associated with release of volatiles that attract males (both by females and by males), and Shaumar (1966) illustrated the mandibular gland in his detailed study of the anatomy of *Pimpla rufipes* (as *instigator*).

The antennae of ichneumonoids comprise two basal segments, a rather large and usually bulbous scapus, followed immediately by a smaller pedicellus, although in some orthocentrines and hormiines the two structures are approximately similar in size. The following segments, which are almost always filiform with many segments, are referred to as the flagellum. No ichneumonoids have strongly clubbed antennae but they are slightly apically expanded in a few genera such as *Brachycyrtus* (Brachycyrtinae) (see Fig. 13.32) and *Hellwigia* (Ophioninae), and in a few rare instances the scapus and basal flagellar segments may be highly modified and raptorial (e.g. the euphorine braconid *Streblocera*; see Fig. 12.57).

The number of antennal segments is variable and there are typically more in larger bodied wasps. The largest braconines and doryctines may have more than 100 flagellomeres, although more typical numbers are between 20 and 60. A few genera have reduced numbers of antennal segments [down to about 12–14 flagellar segments in the ichneumonids *Adelognathus* (Adelognathinae) and *Pygmaeolus* (Tersilochinae, formerly Phrudinae) or 12 in miracine braconids and 9–12 in cedriine braconids (Pambolinae)]. Nearly all commonly encountered taxa have 15 or more flagellomeres. In the Microgastrinae, these often have two rings of placoid sensilla with a weak constriction between them, which can make them look superficially as if they have more segments than they do. Occasionally, the number of flagellar segments is fixed within a larger grouping, as in the braconid Microgastrinae (16 flagellomeres, with one exception) and Miracinae (12 flagellomeres in both sexes).

ANTENNAL SENSILLA

Numerous morphological studies, predominantly using SEM, have been carried out on the sensilla of ichneumonoids, predominantly on members of the Braconidae. Although all antennae have a variety of sensilla types, very little progress has been made in terms of identifying the individual roles of the different types. They fall into several morpho-categories, various types of setiform sensilla called *sensilla trichodea*, various types of *sensilla basiconica*, short sensilla *coeloconica* and elongate multiporous plate sensilla also called *sensilla placodea* or placode sensilla, the last being by far the most conspicuous and best studied.

The *sensilla placodea* often give the flagellomeres a longitudinally striated appearance and are sometimes referred to as rhinaria in reference to their likely olfactory function. They may occupy nearly the entire length of the flagellar segment or form two rings, one basally and one apically (notably in microgastrine braconids) or they may be arranged more irregularly. The *sensilla placodea* of ichneumonoids differ from those of chalcidoids in that they have the neural orifice entering it at its midlength rather than at the basal end and the tip of the sensillum is not free and protruding. The opening typically occupies about 25–60% of the length of the entire sensillum, but in most cyclostome braconids it is much smaller, only 15–20% (Barlin & Vinson 1981, Basibuyuk & Quicke 1999a). In the Braconidae, aphidiines (Borden et al. 1978b), a few Mesostoinae and the euphorine genus *Cosmophorus* lack the internal sensilla floor more or less completely and therefore the neural opening extends the entire length of the sensillum. What the significance is of these variations in size of the hole in the basal plate is can only be speculated upon. Maybe it is somehow related to the number of separate neurones innervating the placode and thus their relative discriminatory ability?

The ultrastructure of placode sensilla has been described in detail for the pimpline ichneumonids

Pimpla turionellae (Stepper et al. 1983) and *Itoplectis* (Borden et al. 1978a) and for the braconids *Aphidius smithi* (Borden et al. 1978b) and *Coeloides* (Richerson et al. 1972). The study by Stepper et al. additionally presented details of their morphogenesis through pupal developmental stages and provided evidence that they may ultimately have been derived from basiconic sensilla. They each develop from a typical cell set comprising two tormogen cells, two trichogen cells and two envelope cells.

Several other types of flagellar sensilla have been described in various ichneumonoids (Dweck & Gadallah 2007, L. Zhang et al. 2014) and those of various Braconidae have been more extensively studied, mostly just using SEM. About 6–13 morphotypes have been distinguished, although the exact roles that each type plays in host or mate recognition is far from clear and technical difficulties mean that most electrophysiological work has simply involved electroantennograms (e.g. Barbarossa *et al.* 1998, Trang & Dey 2013), which fail to distinguish the roles of particular receptor types.

Trichoid sensilla are slender, taper gradually to a point which may be somewhat bent, and they are longitudinally or spirally grooved. In the campoplegine *Eriborus applicitus*, L. Zhang et al. (2014) have shown that the curved-over tip actually partly shields a pre-apical excavation where pores open, but more studies are needed to know how general this is. They usually arise from a raised socket. Basiconic sensilla are shorter and usually thicker than trichoid sensilla and have pores either apically or along their shafts (Das et al. 2011), and they are blunt ended; they may or may not be longitudinally grooved. Coeloconic sensilla are the least abundant sensilla type, they are short and without pores, they originate from a pit and appear to be inflexible – they are often considered to be hygroreceptors or temperature receptors.

Trichoid sensilla tend to be orientated towards the antennal tip, whereas many basiconic sensilla stick out more or less perpendicularly. *Sensilla trichodea* are probably principally mechanosensory, but some may be porous and have chemosensory function too, whereas basiconic sensilla are probably largely or entirely chemoreceptors.

Some sex differences have been found; for example, Bénédet et al. (2002) discovered a multiparous sensillum type in the ichneumonine *Diadromus pulchellus* that is involved in detecting an insoluble host-associated kairomone (see Chapter 8, section *Host acceptance*). Males typically possess more *sensilla placodea* than females (e.g. Bleeker et al. 2004, Ahmed et al. 2013a, Trang & Dey 2013), but there are exceptions (Das et al. 2011). As *sensilla placodea* are chemosensory, it may indicate that mate location/recognition requires a different type of chemoreceptor than host location. However, in general, differences are slight, whereas Das and Fadamiro (2013) have recently shown very marked sex differences in the olfactory lobes of the brains in two genera of Microgastrinae, suggesting, as is sort of obvious, that neuronal signal processing is the most important thing underlying behavioural responses – such as why males generally search for females and not hosts. Das et al. (2011) interpreted interspecific differences in the abundance of chemosensory *sensilla placodea* and *sensilla basiconica* in two microgastrines in terms of their different needs for host discrimination, with larger numbers of both types occurring in the specialist *Microplitis croceipes* than in the generalist *Cotesia marginiventris*. Although such comparisons are interesting, it would probably be unwise to generalise until more, phylogenetically controlled, comparisons are made.

Elecroantennogram studies have been performed on a variety of species to compare responses to various host or host-plant related compounds and also to members of the opposite sex (e.g. Salom et al. 1992). Most of these have been fairly crude and do not permit the study of individual sensilla or nerve units. Nevertheless, Jewett and Carpenter (1998) found that electroantennogram responses of males of the cryptine *Diapetimorpha introita* to females or to extracts from females were greater than to males, although no differences were detected between body parts of the female. In this species males are attracted towards females but not vice versa. A few workers, either using more sophisticated equipment or careful experimentation, have managed to achieve some finer resolution. Barbarossa et al. (1998) found that the 'sensilla trichodea' of the opiine *Psyttalia* (as *Opius*) *concolor* each contained at least three chemosensory cells and were probably gustatory in function. Work by Ochieng et al. (2000) on *M. croceipes* looked at the responses of single sensory units, although the results are difficult to interpret from a behavioural perspective.

Very recently, L. Ma et al. (2014) investigated the expression of olfactory receptor genes in the microgastrine *Microplitis mediator* using expressed sequencing tags and found that out of 14 successfully cloned ones, three were highly expressed in female antennae and eight different ones highly expressed in male antennae, although some gene expression was found in a variety of other tissues. Hence it is hardly surprising that there are not more sex differences apparent in antennal receptors at a purely morphological level. It will be interesting to see what morphological sensilla types the genes are associated with and also how olfactory receptor expression might vary within apparently single sensilla morphotypes.

It is at present really impossible to see any 'big picture' in the available information as those taxa studied in detail have been so biased to a few taxononomic groups, particularly Aphidiinae, Microgastrinae and Pimplini, that are either important in biological control or easily cultured. It would seem likely that there might be overall differences between taxa that make direct antennal contact with their hosts and those that do not such as wood-borer parasitoids, but there have been very few detailed studies of species in the latter category. A good starting point would be a catalogue of sensilla morphotypes including understanding of any associated glands (see below), across a broad range of taxa.

ANTENNAL GLANDS AND TYLOIDS

The idea that parasitic wasp antennae are not only sensory but also often contain glands involved in either host acceptance or courtship has been slow to be appreciated, despite the first antennal glands, in the scape of the eulophid *Melittobia australica*, 30 years ago (Dahms, 1984). Chalcidoid, scelionoid and proctotrupoid antennae, being relatively shorter and more differentiated along their lengths, have been most investigated.

The presence of antennal glands in Ichneumonoidea and Aculeata, which almost always have longer filiform antennae, was not suspected for a long time, even though it was well known that the medial flagellomeres in some male ichneumonids bore distinct, often raised, structures with different sculpture, called tyloids (see Fig. 10.1a). These tyloids were widely used in identification keys and other taxonomic works as their precise distribution was often very species specific and they had previously been widely assumed to be sensory areas, and this was often boldly and erroneously stated. Because of their intimate role in courtship, it is not surprising that in several groups the distribution and shape of the male tyloids provide very useful taxonomic characters (e.g. Gauld 1984a,b, Jussila et al. 2010).

The first evidence that tyloids, or at least most structures called tyloids, were in fact glandular release sites came from a detailed electron microscopic investigation by Isidoro et al. (1996) and Bin et al. (1999a,b) of *Pimpla turionellae*. TEM revealed glandular cells, SEM of the outer surface revealed pores and SEM of the macerated internal cuticular surface revealed the internal glandular ducts leading to the pores. Not surprisingly, the tyloids of various other ichneumonids and other modified antennal regions are now being found to be glandular release areas.

Tyloids are found in members of the Adelognathinae, Cryptinae and Ichneumoninae (all Ichneumoniformes), Pimplinae, Poemeniinae (*Podoschistus*), Cylloceriinae (see Fig. 13.13b), Orthocentrinae and Diplazontinae (Pimpliformes) and some Ctenopelmatinae and Tryphoninae (but only in *Erromenus tarsator*; Kasparyan 1981) (Ophioniformes). Tyloids are only found in males, they are discrete, usually raised areas (but concave in the orthocentrine *Eusterinx*: Wahl & Gauld 1998) and are present on some of the medial flagellomeres (see Fig. 10.1a). Tyloid function has been studied in the Diplazontinae (Steiner et al. 2010; see also Klopfstein et al. 2010b) where their distribution on male flagellomeres corresponds to areas of contact with the female's antenna during courtship.

Gokhman and Krutov (1996) found that the tyloids of Ichneumoninae males were covered in trichoid and basiconic sensilla whereas those of other subfamilies lacked the sensilla and usually had pores suggesting a secretory role. SEM of opened and macerated flagellomeres of *Enclisis* (Cryptinae) showed the presence of internal glandular ducts in the tyloids and also external pores (Bordera & Hernández-Rodríguez 2003). Their conclusion that Ichneumoninae tyloids are not homologous with those of the other groups, though, is most likely to be incorrect as ichneumonines may well be either the sister group of the cryptines or even derived cryptines (Quicke et al. 2009), and therefore the differences may best be interpreted as a plesiomorphic tyloid form or an autapomorphic modification of a pre-existing tyloid structure.

Battaglia et al. (2002) showed that in the aphidiine braconid, *Aphidius ervi*, the first two flagellomeres, which typically have reduced numbers of elongate placoid sensilla compared with the rest of the flagellum, are the source of a secretion that is crucial for mate recognition and mating. Similar placoid sensilla-devoid basal segments are found in numerous other braconids, especially in basal cyclostomes such as the Mesostoinae (sister group of the Aphidiinae), Rhyssalinae and Braconinae (e.g. Dweck & Gadallah 2007), but none of these have been investigated. In the Ichneutinae, the genus *Paroligoneurus* has the placoid sensilla on the apical 0.75 of the antenna widely spaced ventrally with the area between them occupied by truncate-ended sensilla (Sharkey & Wharton 1994), and in the rogadine genus *Heterogamus* and the *Aleiodes coronarius* species group, the flagellomeres in the region of the white antennal stripe have a broad placode sensillum free zone (see fig. 4C,D in B.A. Butcher et al. 2012; see also Chapter 10, section *White antennal stripes and tips*), which seem likely to have a glandular function in the light of other findings. Males of some *Labena* (Labeninae) species appear to have a sub-apical antennal gland based on secretory products observed in pickled specimens (Gauld & Wahl 2000a).

PALPS

Insects have two pairs of palps, the maxillary and labial, which arise from the maxilla and the distal part of the labium (prementum), respectively. In contrast to the antennae, the maxillary palpal segments in ichneumonoids have intrinsic muscles in at least the basal three segments (Zaka-ur-Rab 1978). The plesiomorphic number of maxillary and labia palp segments is six and four, respectively, but some groups show moderate to extreme reductions (see Fig. 13.55b). The large African amicrocentrine braconids, for example, have both palps effectively one-segmented (van Achterberg 1979a) and the Australian *Praonopteron* (Mesostoinae) has only two and none, respectively (Tobias 1988).

In several rogadine braconids, e.g. *Colastomion*, *Cystomastax* and *Macrostomion*, the medial palpal segments in males (sometimes less so in females) are massively swollen, the function of which is yet to be determined, whereas in various taxa, the maxillary palps may be elongate and sometimes, adpressed with a central concavity forming a tube that is involved in sucking concealed nectar (see Fig. 60c; see CNEA type 5 in Table 10.1).

Little is known in general about the roles of palps; presumably they are involved in tasting food in most species, although in the alysiine *Polemochartus* they appear to be involved in host examination (Mook 1961), and Mudd et al. (1982) described briefly their use by *Venturia canescens* in examining host caterpillars and they appeared to be involved in the detection of the female's host marking pheromone. Navasero and Elzen (1991) described the sensilla on the palps of the microgastrine *Microplitis croceipes*, but there has been almost no other work on them, even simple excision-type experiments.

MESOSOMA

The mesosoma, also called the alitrunk by some workers, comprises four body segments, namely the three thoracic segments (pro-, meso- and metathorax) plus the first abdominal segments(called the propodeum). It is dominated in most taxa by the relatively larger mesothorax and propodeum (Fig. 2.2). As mentioned above, a lot of early workers did not realise this and so some references to the 'metathorax' are actually referring to the propodeum (Fig. 2.2: p). Therefore, when interpreting descriptions, it is necessary to determine how particular authors were interpreting the structures. The true metathorax is a rather narrow section dorsally (Fig. 2.2: mtn). A transverse groove, often concealed, at the anterior margin of the propodeum demarcates it from a very narrow sclerite called the metapostnotum.

The head is attached by a membranous connection at the front between the pronotum, which is a medium-sized, U-shaped structure, in dorsal view and triangular in lateral view (Fig. 2.2: pn) and the propleuron (Fig. 2.2: pp), which is a paired ventral structure. The prosternum is often virtually invisible, being located at the mid-posterior of the propleuron more or less between the fore coxae.

The largest part of the thorax is the second segment, comprising the (usually trilobed) mesoscutum dorsally and the (usually fused) mesopleuron/mesosternum. The three lobes of the mesoscutum (Fig. 2.2: m) are often demarked by grooves called notauli (Fig. 2.2: n) and are the insertion sites of the flight muscles. The anterior of the middle lobe of the mesoscutum is where the medial longitudinal flight muscle attaches to the cuticle and the lateral lobes are the dorsal origins of the dorso-ventral flight muscles. Behind these three lobes there is a usually a triangular cuticular plate called the scutellum (Fig. 2.2: s), at the anterior to which there is a transverse and often crenulate groove called the scutellar sulcus (Fig. 2.2: ss). The scutellar plate is sometimes separated from the mesoscutum,

Fig. 2.2 Major regions of the mesosoma illustrated in *Eadya* (Helconinae *s.l.*): (a, b), dorsal view; (c, d), lateral view. Abbreviations: mp, mesopleuron; ms, mesoscutum; mst, mesosternum; mtn, metanotum; mtp, metapleuron; n, notauli; pcs, precoxal sulcus; pn, pronotum; pp, propleuron; s, scutellum; ss, scutellar sulcus; t, tegula. For a colour version, see Plate 1.

anterior to the scutellar sulcus, by a flexible (maybe even narrowly membranous) trans-scutal suture, especially in smaller bodied taxa (absent in *Eadya*). This structure no doubt permits greater shortening of the mesosoma as the longitudinal flight muscles contract.

The mesopleuron (Fig. 2.2: mp) and mesosternum (Fig. 2.2: ms) are fused and often indistinguishable externally. Slightly posterior to the anterior margin there is often a transverse carina called the epicnemial or prepectal carina and its presence or absence is often of considerable taxonomic significance. Posteriorly, the mesopleuron/sternum abuts the metapleuron and posteriorly it bears the mid-coxal cavities. There is also sometimes a complete or partially complete transverse carina (postpectal carina) just anterior to the mid-coxal cavities, which again has taxonomic utility, although it also may be rather homoplastic, at least within the Braconidae (see Chapter 10, section *Postpectal carina*).

In many ichneumonoids, there is a longitudinal groove approximately at the mid-height or below it, which is often important taxonomically. The superficially similar structures in braconids and ichneumonids are not, however, homologous, although they no doubt play similar functional roles. The groove in ichneumonids is situated lower down the side and is called the sternaulus, while the groove in the Braconidae is usually referred to as the precoxal sulcus or suture or mesopleural furrow, but occasionally it is also called the sternaulus. The oriental opiine genus *Sternaulopius* (and another convergently similar opiinae) points to the differences between the two structures, in that it has two distinct grooves, one above the other. Following dissections to investigate the relationship of these to thoracic musculature, Wharton (2006) concluded that in *Sternaulopius*, the lower part of the groove corresponds to the true sternaulus of the Ichneumonidae, although given the derived position of this genus among the cyclostome braconids, its 'true' sternaulus can not be truly homologous to those of ichneumonids, instead representing an independent (homoplastic) acquisition of the same morphological structure.

The third thoracic segment comprises a short dorsal metanotum (Fig. 2.2: mtn), the middle of which is usually formed into a rounded and sometimes carinate mound, and the metapleuron (Fig. 2.2: mtp), which dorsally abuts the propodeum and ventrally the metasternum.

The large postero-dorsal part of the mesosoma is the propodeum, which is the true first abdominal segment. The wasp-waist of the apocritan Hymenoptera is thus situated between the first two abdominal segments, an arrangement that simultaneously allows a great deal of movement of the rest of the abdomen while allowing the mesosoma to house very large longitudinal flight muscles[1]. Many ichneumonids and some braconids have a pattern of transverse and longitudinal carinae on the propodeum which define, when the full complement is present, three medial and three lateral areas. Propodea with at least some of the medial areas defined are termed areolate and propodeal areolation is probably a synapomorphy for the superfamily as it is also apparent in at least some of the Early Cretaceous Eoichneumoninae† (Braconidae) (Rasnitsyn & Sharkey 1988) and Tanychorinae† (Ichneumonidae) (see Fig. 13.2a). The names of the carinae and areas of ichneumonids are given, for example, in Gauld (1991, 2002). However, propodeal carination has been completely lost in many taxa or only a few remnants may remain. In some taxa, a mid-longitudinal propodeal carina is present and in some of these it appears to have resulted from fusion of the lateromedian longitudinal carinae, but in others it may be of *de novo* origin. They have been a very useful source of taxonomic characters in the Ichneumonidae. Laterally, the propodeum may have a carina which runs just slightly below the spiracle or a carina above a groove or just a groove and these demark its separation from the metapleuron (see Fig. 13.28c). Sometimes, however, there may be no obvious external border or it is hidden by otherwise coarse uniform sculpturation.

LEGS

The legs are divided into five main sections, basally the coxa, a small trochanter, the femur, the tibia and a five-segmented tarsus, the last segment of which, the telotarsus, bears a pair of claws. The situation is slightly confused by a strong constriction at the base of the femur demarking a distinct but not articulated region called the trochantellus, which has the consequence, in taxonomic publications and identification keys that refer to femur length, of whether or not the author includes the trochantellus.

The apex of each tibia bears one (fore leg) or two (mid and hind legs) spurs. That of the fore leg is often referred to as the calcar, it is curved and bears a row of adpressed setae dorsally and opposes a notch in the basitarsus that has a lamellum and often a row of setae. The calcar and basitarsal notch together form the antennal cleaning apparatus (Basibuyuk & Quicke 1994, 1995, 1999b). During grooming, the wasp grasps the antennal flagellum near its base between the fore tibial spur and the basitarsal notch and pulls the antenna through the gap, often by a combination of head turning and leg movement (Basibuyuk & Quicke 1995, 1999b). Rarely, the fore tibia may have an ancillary, unmodified spur (only the genera *Rhamnura* and *Bicentra* in the Braconinae; van Achterberg 1981b) or the mid tibia may have only one or no spurs (in some Tryphoninae; *Metopius*: Metopiinae; some Anomaloninae) or the hind tibia only one spur (*Periope*: Metopiinae; *Sphinctus*: Tryphoninae) or no spurs (Exenterini: Tryphoninae). Little is understood about the function of spurs on the middle and hind legs, although it seems likely that they may help stabilise the tarsus when walking or standing; the hind legs may play a role in grooming other parts of the body and, of course, they may have sensory functions. Nearly all taxa have a dense brush of setae apicomedially on the hind tibia or in some there is an apico-medial comb formed of a line of adpressed setae; few have both and it seems that when one (probably the comb arrangement) evolves, the more basal patch becomes reduced. Both of these are involved in grooming.

The fore tibia of all Hymenoptera contains a subgenual organ, sometimes called an ear, but which detects substrate-borne vibrations. It is particularly well developed in females of wasps that use vibrational sounding to detect hosts and is indicated by a sex-specific swelling of the fore tibia (see Figs 10.2 and 13.6) (Vilhelmsen et al. 2001). The combination of a sexually dimorphic fore tibia (i.e. females having an enlarged subgenual organ compared with males) and modified female antennae was used by Broad and Quicke (2000) to infer which ichneumonids use vibrational sounding as a host detection mechanism (see Chapter 10, section *Antennal hammers and vibrational sounding*). Their conclusions seem to have been borne out by subsequent observations of host-locating females (e.g. Niemeyer & Flores-Prado 2012).

The tarsal claws are often important in taxonomy and their morphology is probably quite indicative of biology, although there have been no detailed studies, just some surmising. Sexual dimorphism in claw form, as shown by some Pimplinae (Gauld 1991), is probably indicative of a role in oviposition, but generally claws are similar in both sexes. Simple claws, with a small, rounded basal lobe and very few, often very small, pectin teeth near its base (see Figs 10.7a and 13.20a), are probably the groundplan; this form, and particularly when the claw is enlarged, is fairly characteristic of parasitoids that have to 'drill' through hard substrates with their ovipositors, such as most wood-borer parasitoids and some attacking hosts in seeds or galls, so presumably the large claw is important for maintaining a strong grip on the substrate. The Agriotypinae which search for hosts under fast-flowing water have particularly large claws (see Fig. 13.28) for holding on to stones. In the Labenini (Labeninae), which attack deeply concealed hosts in hard wood substrates, Gauld & Wahl (2000a) noted that in addition to having large, slender claws, they also have denticles or a dense patch of stout setae on the third tarsal article, and it seems likely, as suggested, that these act in opposition to the claw, providing a firm grip on the wood. The most common modifications of the claws are the expansion in size and extent of the basal pectin (see Fig. 10.7b) and the development of an angular basal lobe (see Fig. 10.7c). The likely roles of these are discussed further in Chapter 10 (section *Pectinate claws and claws with angular basal lobes*).

The ventral surface of the tarsal segments is often furnished with sensilla, and in those taxa that recognise hosts or host plants or trails left by hosts these are no doubt often important, but they have been almost completely neglected as an item of study. M.R. Shaw (2002b) noted that they are 'employed at length' by many *Aleiodes* (Rogadinae) species in the examination of potential host caterpillars prior to oviposition. There appear to have been only two SEM studies of tarsal sensilla, both dealing with microgastrines that attack exposed hosts:, by Navasero and Elzen (1991) on *Microplitis* and Obonyo (2011) on two species of *Cotesia*. In addition, the tarsal articles of a large number of taxa have a single longitudinal midventral row of enlarged, more or less erect, adpressed small spines. Nothing is know or has been postulated about their function.

WINGS, WING VENATION AND WING CELLS

Ichneumonids, in common with all Hymenoptera, have two pairs of wings, although in flight they are functionally dipterous, the posterior margin of the fore wing being narrowly recurved just beyond the middle, and the hind wing having a line of three or more strong, hooked or S-shaped structures on the corresponding part of its anterior margin, called hamuli. When the wings are extended, the hamuli interlock with the folded fore wing margin (Basibuyuk & Quicke 1997). Technically, these are the distal hamuli and, in some sawflies and wood wasps, there is also a row of similar hooks along the hind wing costal vein, called the basal hamuli, but in ichneumonoids these are replaced by so-called secondary hamuli that are more or less simple, stout setae. Instead of interlocking with the adjacent fore wing margin, the secondary hamuli reside underneath it in flight such that the fore wing margin cannot slip below the hind wing. The number of secondary (and probably distal) hamuli is largely related to body size, but sometimes it can have phylogenetic significance (Quicke 1982, Wahl 1993b).

The nomenclatural systems employed by various authors over the years for wing veins and cells in the Hymenoptera as a whole and also the Ichneumonoidea, have been various and are bound to

Table 2.1 Nomenclature systems used for wing veins in the Braconidae. Note that in some taxa with additional cross-veins, some of the terminologies presented would not be correct, particularly in terms of the use of numbers representing the abscissas of veins. The terminology employed in this book is that of Sharkey & Wharton (1997) and is in the shaded column.

Wing veins	Jurinian system; Rowher & Gahan (1916)	Abbreviations based on Jurinian system	van Achterberg (1976c, 1993a)	M.R. Shaw & Huddleston (1991)	Goulet & Huber (1993)	Sharkey & Wharton (1997)
Fore wing veins	Costa	–	C+SC+R	C and Sc+R+Rs	C+Sc+R	C+SC+R
	–	–	1-SR	–	Rs	1RS
	Metacarpus	met	1-R1	R	R1	R1
	Radius	r1	r	2r-rs	r-rs	r
	First intercubitus	cuqu1	2-SR	1Rs	Rs	2RS
	Radius	r2	3-SR	2Rs	Rs	3SRa
	Radius	r3	SR1	3Rs	Rs	3SRb
	Second intercubitus	cuqu2	1r-m	2rs-m	Rs	rs-m
	Cubitus	cu1	1-SR+M	Rs+M	Rs+M	(RS+M)a
	Cubitus	–	2-SR+M	–	–	(RS+M)b
	Basal vein	b	1-M	1M	M	1M
	Cubitus	cu2	2-M	2M	M	2M
	Cubitus	cu3	3-M	3M	M	3M
	Medius*	m	M+CU1	M+CU	M+Cu	M+CU
	Recurrent vein	nrec	m-cu	1m-cu	m-cu	1m-cu
	Discoideus	–	1-CU1	–	Cu	1CUa
	Discoideus	d1	2-CU1	1-Cu	Cu	1CUb
	Discoideus	d2	3-CU1	2Cu	Cu	2CUa
	Subdiscoideus	npar	CU1a	3Cu	Cu	2CUb
	Discoideus	d3	CU1b	2cu-a	2cu-a	2cu-a
	Nervulus	n	cu-a	1cu-a	1cu-a	1cu-a
	Submedius	–	1-1A or 1A+2A	1A	A	1-1A a,b,c
	Brachius	–	2-1A	1A	A	2-1A
	Axillus	–	2A	–	1a′	1a
	–	–	3A	–	–	2A
	–	–	a	–	2a′	2a
Hind wing veins	Costella	–	C+SC+R	–	C+Sc+R	C+Sc+R
	Costella	–	C	–	C	C
	Costella	–	1-SC+R	Sc+R+Rs	C&Sc+R	Sc+R
	–	–	2-SC+R	–	–	R
	Metacarpella	–	SC+R1	R	R1	R1
	Metacarpella	–	R1	R	R1	R1
	Radiella	–	SR or 1-SR and 2-SR	Rs	Rs	RS or 1RS and 2RS
	Basella	bas	1r-m	rs-m	1r-m	1r-m
	Mediella	med1	M+CU	M+Cu	M+Cu	M+CU
	Mediella	med2	1-M	1M	M	1M

(continued overleaf)

Table 2.1 *(continued)*

Wing veins	Jurinian system; Rowher & Gahan (1916)	Abbreviations based on Jurinian system	van Achterberg (1976c, 1993a)	M.R. Shaw & Huddleston (1991)	Goulet & Huber (1993)	Sharkey & Wharton (1997)
	Cubitella	cub	2-M	2M	M	2M or 2M and 3M
	Postnervellus	pnv	m-cu	–	–	m-cu
	Nervellus[†]	nv	cu-a	cu-a+Cu	Cu&cu-a	cu-a
	Nervellus	–	1-CU	–	Cu	CUa
	Discoidella	–	2-CU	–	–	CUb
	Submediella	–	1-1A	1A	A	1-1A
	Brachiella	–	2-1A	–	A	2-1A
	–	–	2A	–	a′	a
	Axiella	–	3A	–	2A	2A

*Also called the cubitus by some authors.
†When vein Cub is absent.

be confusing to newcomers trying to get to grips with identification keys. Indeed, one of the greatest annoyances in the ichneumonoid taxonomic literature is the use of multiple terminologies for wing veins. Further, braconid and ichneumonid researchers have tended to use different terminologies and there have even been arguments over the homologies of some veins in the two groups, and until recently there has been no attempt at consistency between the two families.

Wing venation terminology has had a complex history. Broadly two systems have been applied to ichneumonoid venation and cells, the very outdated and nowadays seldom used Jurinian system and a range of variants of the Comstock–Needham system (Comstock & Needham 1898, as interpreted for Ichneumonoidea by, for example, Riegel 1948 and Eady 1974). Use of the Jurinian system hung on for longer in the Ichneumonoidea than it did in many other groups of Hymenoptera (Sigwalt 1977) and its use was widespread until the mid-1970s and held on in some parts of the world, until more recently, especially for the Ichneumonidae, due in considerable part to the enormous influence of Henry Townes' work. The Jurinian names for veins and cells are given in Tables 2.1 and 2.2, but the rest of this discussion focuses on the various modifications of the Comstock–Needham system. The two systems differ fundamentally in the interpretation of what the parts middle-most longitudinal vein are and in particular the distal part is called the cubitus in the Jurinian system whereas it is (largely) the median vein in the Comstock–Needham system. In general, the breadth of wing venation variation shown among braconids is rather greater than that normally encountered in the Ichneumonidae, and this underlies the fact that many more variants of terminology are found in the former group.

The identities of the wing veins in relation to the standard insect pattern of, from the anterior margin of the wing working posteriorly and designated by capital letters, comprises eight primary longitudinal veins: costa (C), subcosta (SC or Sc)[2], radius (R), medius (M), cubitus (CU or Cu) and three anal veins (1A, 2A and 3A), although the last two of these are absent in most ichneumonoids. Almost all species have a conspicuous triangular pterostigma [a thickened structure about 0.6 of the distance from the base of the fore wing that provides a very useful landmark, although it is narrowed to the point of being almost indistinguishable from the following part of R1a in a few rather rare taxa, e.g. Mesostoinae: Hydrangeocolini (see Fig. 12.14a,b), Alysiinae: *Orthostigma* and Anomaloninae: *Ophionellus*]. Precisely what the pterostigma is derived from is unclear; it might be a thickened vein or a sclerotised wing cell. Functionally it has been proposed to be involved in inertial regulation during flight (Norberg 1972).

Vein identities are not immediately intuitive because in the Hymenoptera, especially in the Apocrita, these veins variously fuse, separate and even occasionally turn back on themselves. Careful study of fossils, vein development and tracheation, and in some cases,

Table 2.2 Nomenclature systems used for wing veins in the Ichneumonidae. Note that in some taxa with additional cross-veins, some of the terminologies presented would not be correct, particularly in terms of the use of numbers representing the abscissas of veins. The terminology employed in this book is in the shaded column. To facilitate comparison with the Braconidae, the vein at the base of the areolet is termed 2RS even though 1RS is not differentiated in extant ichneumonids.

Wing veins	Townes (1969) (based on Jurine)	Ross (1936)	Gauld (1991)	Gauld & Bolton (1988)	Goulet & Huber (1993)	Based on Sharkey & Wharton (1997)
Fore wing veins	Costa and subcosta	C+Sc+R	–	C and Sc+R+Rs	C+Sc+R	C+SC+R
	Metacarpus	R1	R_1	R	R1	R
	First abscissa of radius	r-rs	Rs+2r	Rs+2r-rs	r-rs	r-rs
	Second intercubitus	Rs	2rs-m	2rs-m	Rs	2RS
	Third intercubitus	3rs-m	3rs-m	3rs-m	r-m	rs-m
	Second abscissa of radius	Rs	Rs	Rs	1/Rs	3RSa
	Third abscissa of radius	Rs	Rs	Rs	2/Rs	3RSb
	Basal	M becoming Rs	Rs&M	Rs+M	Rs+M	RS&M
	Discocubitus*	1m-cu becoming Rs+M	1m-cu	1m-cu becoming M	M&1m-cu	1m-cu&(RS+M)b
	Second abscissa of cubitus	M	M	M	1/M	1M
	Third abscissa of cubitus	M	M	M	2/M	2M
	Fourth abscissa of cubitus	M	M	M	3/M	3M
	Medius	M+Cu	M+Cu	M+Cu	M+Cu	M+CU
	Second recurrent	2m-cu	2m-cu	2m-cu	2m-cu	2m-cu
	First abscissa of Discoideus	Cu	Cu_1	Cu	1/Cu	1CUa and 1CUb
	Second abscissa of discoideus or postnervulus	Cu	Cu_1	Cu	2/Cu	2CU
	Third abscissa of discoideus or postnervulus	2cu-a	Cu_{1b}	Cu2	2cu-a	2cu-a
	First abscissa of subdiscoideus	Cu	Cu_{1a}	Cu1	3/Cu	3CU
	Second abscissa of subdiscoideus	Cu	Cu_{1a}	Cu1	4/Cu	4CU
	Nervulus	1cu-a	cu-a	cu-a	1cu-a	1cu-a
	Submedius	1A	1A	1A	1/A	1-1A
	Brachius	1A	1A	1A	2/A	2-1A
Hind wing veins	–	C	–	–	C&Sc+R	C+Sc+R
	Costella	C	–	C	C	C
	Subcostella	R	–	R+Rs	Sc+R	Sc+R
	Subcostella	R1	R1	R	R1	–
	Metacarpella	R1	R1	R	R1	R1a
	First abscissa of radiella	Rs	Rs	Rs	1/Rs	RSa
	Second abscissa of radiella	Rs	Rs	Rs	2/Rs	RSb
	Intercubitella	rs-m	–	rs-m	r-m	r-m
	Mediella	M+Cu	M+Cu	M+Cu	M+Cu	M+CU

(continued overleaf)

Table 2.2 *(continued)*

Wing veins	Townes (1969) (based on Jurine)	Ross (1936)	Gauld (1991)	Gauld & Bolton (1988)	Goulet & Huber (1993)	Based on Sharkey & Wharton (1997)
	First abscissa of cubitella	M	M	M	1/M	1M
	Second abscissa of cubitella	M	M	M	2/M	2M
	Nervellus	1/Cu	Cu_1	–	1/Cu	CU1a
	Discoidella	Cu	Cu_1	Cu	2/Cu	CU1b
	Nervellus	cu-a	cu-a	cu-a	Cu&cu-a	CU&cu-a
	Submediella	1A	1A	1A	1/A	1-1A
	Brachiella	1A	1A	1A	2/A	2-1A
	Axillus	3A	–	–	–	2A

*If a ramulus is present on the discocubitus or it is otherwise discernible as two sections, Townes differentiated the posterobasal part as the first recurrent vein and the distal part as the first abscissa of cubitus.

reference to teratological or freak specimens (e.g. Riegel 1948), has led to what is now a fairly generally accepted view of their relationships. The situation is further complicated, however, because some major veins split into two. The radial vein splits into two just before the pterostigma, an anterior part which retains the name radius (R), and a posterior part called the radial sector (RS, Rs or SR), and the cubitus (CU or Cu) similarly splits in some other insects, although whether it does so in the ichneumonoids is not agreed upon by all workers. Thus, at the apex of the subdiscal cell, the vein closing it (when it is closed) may be considered the posterior part of a split vein (Figs 2.3a and 2.4a) or as a second cross vein between a single CU and the anal vein (Figs 2.3b and 2.4b). In general, longitudinal veins are indicated by capital letters and cross veins by lower-case letters separated by a hyphen indicating which two longitudinal veins the cross vein connects. Where two veins merge indistinguishably into one another because of the absence of another vein that would normally separate them into two distinct veins, an ampersand (&) is widely used to indicate which veins are involved, and a plus sign (+) is used where two longitudinal veins fuse for a part of their lengths.

The various versions of the Comstock–Needham system that have been applied vary in their orthography (Latinised versus English vein names and abbreviations, complete or partial use of capital letters) and in how the various abscissas of longitudinal veins are identified. The system employed for braconids by van Achterberg (1976c, 1993a) became widely but not universally adopted, and more recently another modification, proposed by Sharkey and Wharton (1997), based on a proposal originally put forward by Sharkey in *Ichnews*, in their New World braconid manual, has been gaining a lot acceptance and is the one used here with slight modification for consistency. The Sharkey and Wharton (1997) system attempts to be a more internally consistent modification of the Comstock–Needham system, with the abscissas of longitudinal veins formed by intersection with cross-veins running anteriorly from them are distinguished from abscissa formed by intersections with posterior cross veins. A number before the vein abbreviation is used for the former, and a lower-case letter after the vein abbreviation is used for the latter. Tables 2.1 and 2.2 compare the van Achterberg and the Sharkey and Wharton versions along with other widely used systems for the Braconida and Ichneumonidae, respectively. Because the van Achterberg and the Sharkey and Wharton systems have both been widely used in publications on the Braconidae, they are both illustrated in Fig. 2.3. Regarding the Ichneumonidae, a fairly widely used variant of the Comstock–Needham system (Fig. 2.4a) is compared with my adaptation of the Sharkey and Wharton system to this family. The two also indicate the main differences in opinion about vein homologies which are discussed below.

The advantage of the Sharkey and Wharton modification is that it more or less allows the topology of the wing venation to be reconstructed more accurately from just knowing the names of the veins that are present. A problem with all these systems, however, is that if a cross vein is interstitial, for example fore wing vein cu-a is often directly opposite RS&M in ichneumonids or 1M in braconids, so the long vein forming

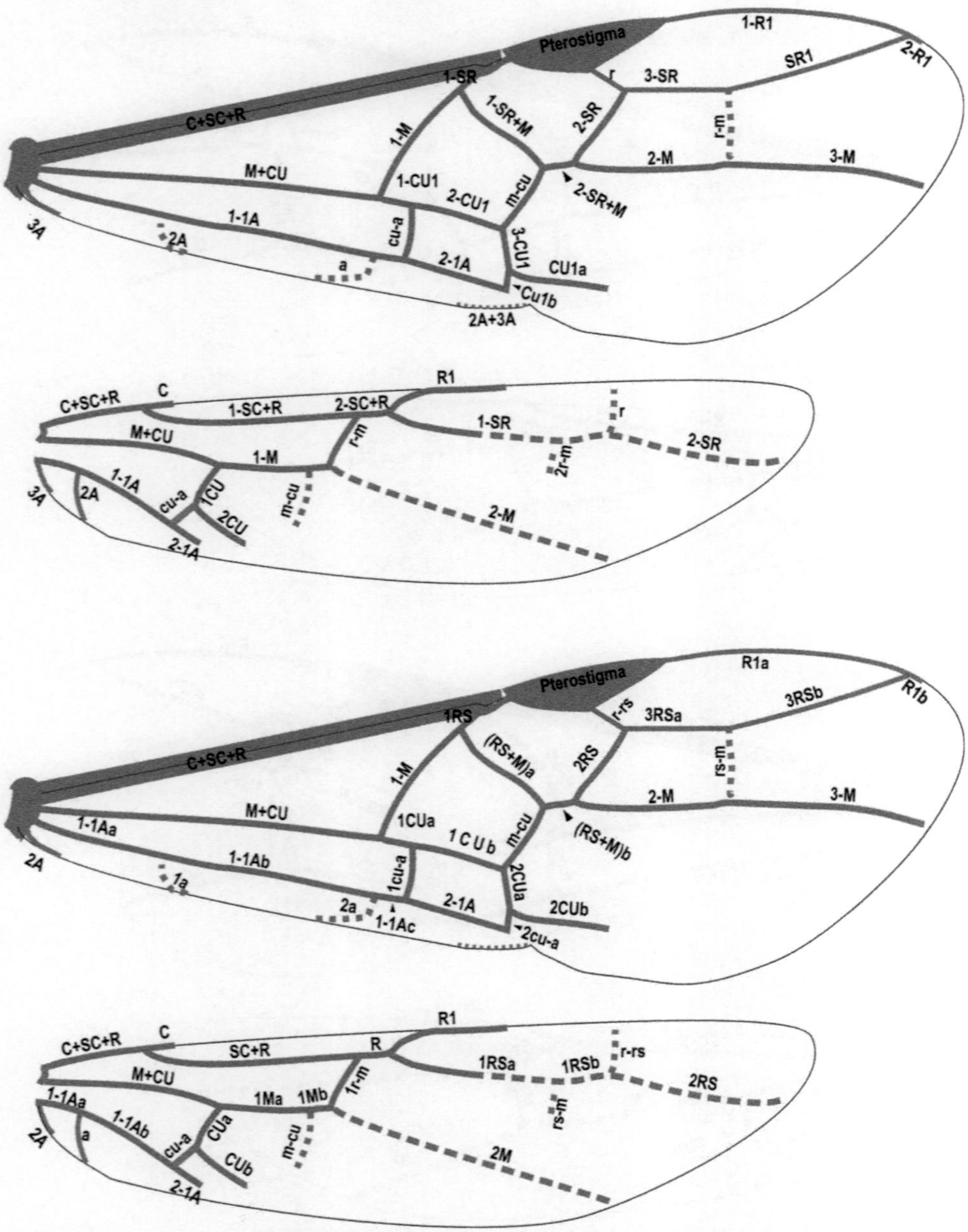

Fig. 2.3 Terminology for venation in hypothetical braconid wasp fore and hind wings: (a) based on Comstock–Needham system as applied by van Achterberg (1979c, 1993a) and (b) by Sharkey & Wharton (1997) with the longitudinal vein elements indicated by capital letters (C, costa; SC, subcostal; R, radius; SR, Sector radialis; M, medius; CU, cubitus; A, anal), cross-veins indicated in lower case, abscissas of vein numbered from base at the beginning, and branches of longitudinal veins indicated by number at end. For a colour version, see Plate 2.

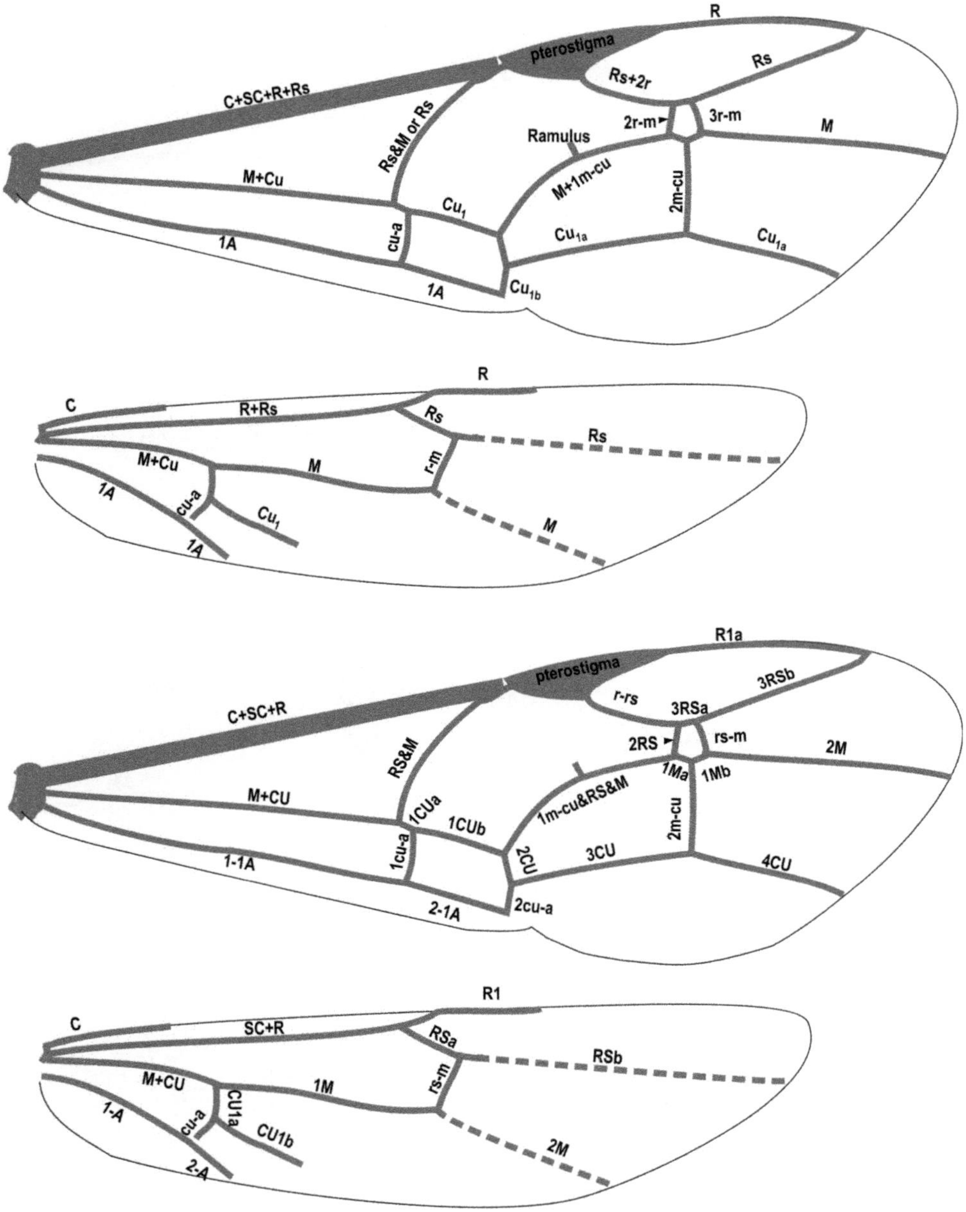

Fig. 2.4 Terminology for venation in hypothetical ichneumonid wasp fore and hind wings: (a) based on Comstock–Needham system and (b) applying the system of Sharkey & Wharton (1997). Longitudinal vein elements indicated by capital letters (C, costa; SC, subcostal; R, radius; SR, Sector radialis; M, medius; CU, cubitus; A, anal), cross-veins indicated in lower case, abscissas of vein numbered from base at the beginning, and branches of longitudinal veins indicated by number at end. See Table 2.2 for the vein names used by Henry Townes and others. For a colour version, see Plate 3.

the anterior margin of the subdiscal cell in these will be called 1CU, whereas if it is postfurcal, i.e. intersecting the CU vein some distance after M+CU has split, then the abscissa of CU forming the anterior margin of the subdiscal cell will be 2CU.

Another proposal to standardise braconid wing vein terminology is that of Goulet and Huber (1993), also using a modified form of the Comstock–Needham system worked out so as to be internally consistent and capable of accommodating any vein. This system is slowly being adopted by Western workers, but it seems probable that there will remain a multiplicity of systems, and students of the group will just have to cope with this.

One of the most important items of controversy about vein homologies concerns what veins form the basal and distal ends of the second submarginal cell (called the areolet by most ichneumonid workers and also sometimes by braconidologists working on microgastrines in which the cell may also be very small) and whether they are homologous in the Braconidae and Ichneumonidae. In the extant ichneumonids, this cell when present (i.e. closed basally by $2RS^3$ and distally by rs-m) is small and either rhombic or pentagonal (Fig. 2.5b) and its basal vein is rather short. In contrast, in most braconids it is a much more elongate cell and in particular in many of the basal cyclostome lineages may be very elongate (see Fig. 12.15). These differences led braconid workers generally to consider that the second submarginal cell is bounded basally by 2RS and distally by an rs-m vein probably homologous to that in ichneumonids, but many ichneumonid workers considered 2RS lost and that the basal cross-vein is actually an rs-m vein and was generally called 2r-m (see Fig. 2.4a), making the distal vein closing the areolet 3rs-m. Others considered these technically to be 1rs-m and 2rs-m. In either case, the second submarginal cells were not treated as homologous in the two families. However, I strongly suspect that they are homologous and at the very least the basalmost vein of the cell is the same. In the Lower Cretaceous taxa assigned (reasonably) to the Ichneumonidae, such as the Tanychorinae† (see Fig. 13.2), and also in some extant labenines, the second submarginal cell is far more elongate and vein 2RS slopes strongly and originates basally at the junction of (RS+M)a (which is complete in these basal taxa) (Quicke et al. 1999a). This strongly suggests that the basal vein of the cell is 2RS and that in most extant ichneumonids it has migrated distally. Assuming

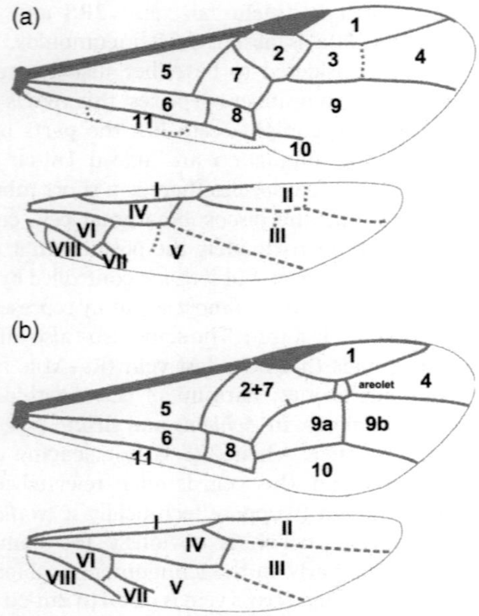

Fig. 2.5 Names of wing cells: (a) Braconidae; (b) Ichneumonidae. Fore wing key: 1, marginal cell; 2, first submarginal cell; 2+7, discosubmarginal cell; 3, second submarginal cell; 4, third submarginal cell; 5, basal cell; 6, sub-basal cell; 7, discal cell; 8, first subdiscal cell; 9, 9a, 9b, second discal cell (in ichneumonids technically the second and third discal cells are seldom referred to separately); 10, second subdiscal cell; 11, anal cell. Hind wing key: I, costal cell; II, marginal cell; III, submarginal cell; IV, basal cell; V, subdiscal; VI, sub-basal cell; VII, anal cell.

the ancestral ichneumonoid to have retained four submarginal cells as in the Praeichnemonidae† (see Fig. 11.1), it is possible that the distal vein of the 2ns submarginal cell in ichneumonids is not homologous with that of the braconids, especially as many of the putatively basal aphidioid clade have very elongate second submarginal cells, but I do not think that this can be ascertained in the absence of intermediate fossils. The second submarginal cell is not especially elongate in Early Cretaceous fossil braconids (Rasnitsyn & Sharkey 1988, Perrichot et al. 2009), but these are generally fairly small wasps with rather broad wings which might change the perception of the position of the distal cross-vein – it is difficult to know what reference frame should be applied.

Since most braconids have fore wing vein (Rs+M)a, the veins it joins distally are clearly 1m-cu and

(Rs+M)b, or if it is antefurcal, veins 2RS and 2M. But since (Rs+M)a is absent in ichneumonids, this vein is often considered to be either just 1m-cu or 1m-cu&Rs+M. In a number of species, this vein is sort of angled and so it might seem like the parts basal and distal to the angulation are indeed 1m-cu and (Rs+M)b and in a few species there is a short tubular vein spur entering the discosubmarginal cell, called the ramellus, at approximately the position that vein (Rs+M)a would connect, but if this is controlled by the same genes, then it would almost certainly represent a re-acquisition of a lost trait. The same issue also affects various braconids that have lost vein (Rs+M)a, such as various euphorines, Earinini of the Agathidinae and some aphidiines. In *Aphidius* and *Aclitus* (see Fig. 12.7c), for example, when 2RS is lost, leaving only the rs-m cross vein, this vein is often referred to as M&m-cu, although I suppose technically it would be m-cu&Rs+M&2M, which is obviously too cumbersome to use. Similarly, in the ichneumonid subfamily Ophioninae, the single cross vein is distal to 2m-cu and in many Anomaloninae and some Cremastinae (e.g. *Pimplomorpha, Sustenus*) it is more or less interstitial. A few workers have considered that the ophionine condition had arisen through the loss of the basal cross vein (i.e. 2RS in Fig. 2.4b), although I think from the intermediate conditions shown by the others that it has probably evolved by distal migration of the basal cross vein – it is to be hoped that some fossils showing transitional states will be found at some point.

Not all workers agree on the identities of a few other veins. In particular, the short cross vein closing the fore wing first subdiscal cell (cell 8 in Fig. 2.5) is considered by van Achterberg (1993a) and many ichneumonid workers to be the posterior arm of a bifurcating CU vein and have termed it CU1b or equivalents accordingly (Figs 2.3a and 2.4a), although many others, including Sharkey & Wharton (1997), treat it as a second cu-a cross-vein, i.e. 2cu-a (Figs 2.3b and 2.4b), which I suspect is more probable. Then there are the spur-like veins that originate from fore wing vein 1-1A in some braconids. Van Achterberg (1976c, 1993) considered that in braconids where a sub-basal spur arises from 1A, it is actually a remnant of 2A (Fig. 2.3a), whereas others have taken it to be a cross vein, 1a (Fig. 2.3b).

There is some lack of consistency in what the fore wing vein running from the antero-distal part of the first subdiscal cell to the postero-basal part of the areolet is called in the Ichneumonidae. In braconids, vein m-cu normally meets (RS+M)a distally, separating it from (RS+M)b, but (RS+M)a is absent in all extant ichneumonids and vein m-cu&RS+M often curves gradually before the areolet (which here I take to be bordered basally by vein 2RS, which is exactly the same as 2RS in Tanychorinae† and the Braconidae). The distal part of this vein is probably homologous to (RS+M)b, but the junction between them cannot be seen, although in some ichneumonids there is a distinct angulation and in others a short spur called the ramellus, both of which may indicate the junction between them. The presence of a 'ramellus' in the lower Cretaceous Palaeoichneumoninae† (see Fig. 13.3) must surely represent a retained distal remnant of (RS+M)a, although the ramellus, as it occurs in extant ichneumonids, might either be atavistic or something completely new – and thus can be cautiously referred to as (RS+M)a′. Various workers have referred to 1m-cu&Rs+M as either just 1m-cu or M+1m-cu (=M&1m-cu) or sometimes seem to have steered clear of giving it any name at all.

CONFUSING AND SOMETIMES ERRONEOUSLY APPLIED VEIN NAMES

Some vein terminology may be confusing, especially to beginners, and this might not be helped by some workers, myself included, who on particular occasions apply a technically incorrect terminology to describe a situation. There are two particularly common examples of this involving braconid veins. In the fore wing of most braconids, there is a short, longitudinal section of vein Rs+M distal to the m-cu cross-vein which is correctly termed (Rs+M)b (Fig. 2.3) and vein m-cu is sometimes said to be *antefurcal* [meaning before the fork or splitting on (Rs+M)a]. However, this vein is sometimes missing, in that vein (Rs+M)a divides into 1-SR and 2-M (i.e. postero-basal corner of the second submarginal cell) at virtually the same point that the 1m-cu cross-vein emerges – in these cases 1m-cu is usually said to be *interstitial*. In the more problematic cases terminologically, vein 1m-cu may leave after the split (i.e. it is postfurcal), but instead of arising somewhere along the longitudinal part of 2-M, 2-M more or less forms a right-angle at its junction with 1m-cu and so its basal portion runs transversely from where (Rs+M)a splits (see Fig. 12.22). Rather than referring to this short vertical section correctly as 2-M, the second abscissa of vein M when it is not fused with either CU or Rs and then the subsequent parts as 3-M and 4-M, workers usually call this part 2-SR+M, i.e. (Rs+M)b, and say that it is transverse rather than longitudinal. In that way, the vein bordering the posterior part of the second submarginal cell remains called 2-M and then comparing ratios is easier.

An equivalent situation occurs with vein R (= 2-SC+R in van Achterberg's terminology) in the hind wing of Braconidae, which normally runs longitudinally, but in a few appears transverse. Again, this is due to the migration of the anterior junction of cross-vein rs-m beyond the split of R and SR (which means that it should then properly be termed 1-SRa). The distal position of this intersection in the Ichneumonidae gives rise to their elongated basal cell, but when it occurs in Braconidae, the short abscissa of SR runs more or less transversely and may even be reclivous.

Also in the hind wing when vein CU1b is present (i.e. in most Ichneumonidae, but in the Braconidae principally in the extinct Protorhyssalinae and in the extant Apozyginae, Acampsohelconinae, Agathidinae, Meteorideinae, Sigalphinae and Trachypetinae), the vein issuing posteriorly from the apex of M+CU is deemed to be vein 1CU and cross-vein cu-a runs from this to the flexion line just anterior to vein 1A. However, in those braconids without 2-CU, the whole of the transverse vein from M+CU to 1-A is termed cu-a. In a few labenine ichneumonids (various *Apechoneura* species: see fig. 3d,e in Quicke et al. 1999a), the origin of vein 1CU has shifted to distal to the cu-a cross-vein, meaning that there are in reality two abscissas of M+CU which ought to be termed 1-M+CU and 2-M+CU.

Many cyclostome braconids have at least a trace of a transverse vein issuing from the distal part of hind wing vein 1M or sometimes from the proximal part of 2M. As with those occurrences of an antefurcal fore wing vein cu-a, the general practice has not been to regard these as creating a third abscissa of vein M. Similarly, a number of braconids, mostly non-cyclostomes, but occasionally among cyclostomes, have a distinct, although not normally tubular, cross-vein 2rs-m and sometimes also a cross-vein r-rs arising from hind wing vein 1RSa. If these were considered to divide the longitudinal vein into two or three separately numbered abscissae, that would again cause confusion and is not done in practice.

The naming of cells is if anything even more confusing and, in general, one sometimes has to work it out from context. I have tried to summarise the major systems in use in the Ichneumonoidea in Table 2.3. Two systems are largely at odds with one another and are particularly likely to cause confusion, namely that used by Goulet and Huber (1993) and the Jurinian system, which was followed in various forms principally by Russian and some American authors, since the former named cells after the longitudinal vein bordering their posterior margin and the latter named cells after the anterior vein. The system employed here for braconids and ichneumonids is shown in Fig. 2.5.

WING FLEXION LINES

In flight, insect wings do not remain permanently flat structures but their surface contours respond to changes in air pressure on their surfaces. Flexion lines, i.e. natural folds in the membrane, and associated bending points of wing veins shape their flight dynamics. They can usually be seen easily in transmitted light (Fig. 2.6), but in small taxa where wings are fairly unpigmented they may require more careful observation of surface contours and often the directions of setae. Although there is a broadly similar pattern in most ichneumonoids, there are also important differences.

In both families, the junction at the base of the pterostigma is flexible, perhaps particularly so in ichneumonids, and a major fold runs postero-distally from here. In braconids, there is usually a distinct break (desclerotisation; bulla) of the veins around the postero-basal part of the second submarginal cell, where this fold passes (Fig. 2.6a,b) and then tends to divide one arm going distally the other posteriorly. In ichneumonids, the fold passes through a bulla in vein 1m-cu&RS+M into the second discal cell and the longitudinal fold divides to form a Y-shape, either more or less at or distal to vein 2m-cu (Fig. 2.6c,d), in which case that cross vein has a single bulla, or basal to it, in which case the vein has two separate bullae (Fig. 2.6e,f), and this is an important systematic character that is almost invariably encountered in subfamily-level identification keys. The bulla in ichneumonid vein 1m-cu&RS+M is always distal (anterior) to the ramulus (ramellus) or any sharp angulation and may be adjacent to the base of the areolet, suggesting that it is in the equivalent of vein (RS+M)b, which in braconids always has a bulla or is desclerotised. The fold running into the discal cell also divides, sometimes near its base, and tends to form a weak transverse fold running towards the

Table 2.3 Wing cell nomenclature for Braconidae and Ichneumonidae (see labelled diagrams in Fig. 2.5). The names in the shaded column are employed here except that in the Ichneumonidae the areolet and discosubmarginal cells are referred to.

Wing	Jurinian	Russian and some American authors	Goulet & Huber (1993)	van Achterberg (1976c, 1993a)	Gauld, Wahl, etc. for Ichneumonidae
Fore wing					
1	Radial	Radial	First radial 2	Marginal	Marginal
2	First cubital	First radiomedial	First radial 1	First submarginal	NA
3	Second cubital	Second radiomedial	First + second radial sector	Second submarginal	Areolet
4	Third cubital	–	Third radial sector	Third submarginal	Submarginal
5	Medial	Medial	Radial	Basal	Basal
6	Submedial	Submedial	First cubital	Sub–basal	Sub-basal
7	First discoidal	Discoidal	First medial	First discal	NA
8	First brachial	Brachial	Second cubital	First subdiscal	First subdiscal
2 + 7	Discocubital	–	–	–	Discomedial, discosub-marginal or discal cell
9a	Second discoidal	–	Second medial	Second discal	Second discal cell
9b	Third discoidal	–	Third medial	Third discal	Third discal cell
10	Second brachial	–	Third cubital	Second subdiscal	Second subdiscal
11	Anal	Anal	First anal	First plical	Anal
Hind wing					
I	Costellan	–	Costal	Costal	Costal
II	Radiellan	Radial	First radial	Marginal	Marginal
III	Cubitellan	–	Radial sector	Submarginal	Discal
IV	Mediellan	Medial	Radial	Basal	Basal
V	Discoidellan	–	Medial	Discal + subdiscal	Subdiscal
VI	Submediellan	Submedial	Cubital	Sub-basal	Sub-basal
VII	Brachiellan	–	–	–	–
VIII	Anallan	–	First (and second) anal	First (and second) plical	Anal

junction between 2CU and 3CU. In both families, some taxa have vein rs-m with one (Fig. 2.6a,b) or two bullae (Fig. 2.6e,f) and well-developed folds in the third submarginal cell, or the rs-m vein may be largely desclerotised or absent. In the Ophioninae, which have the discosubmarginal cell closed distal to 2m-cu, the vein has no bullae (Fig. 2.6c,d), consistent with it being homologous to 2RS rather than rs-m, and consequently the distalmost submarginal cell (cell 4 in Fig. 2.5) has only weak longitudinal folds. In contrast, in those taxa with distinct bullae in the rs-m vein, there can be very strong longitudinal folds (Fig. 2.6e,f).

In both families, a second major fold runs longitudinally just anterior to the anal vein (1A), with always clear breaks at the posterior of vein 1cu-a, and runs between vein 2cu-a and 2–1A. A presumably segmentally homologous flexion line occurs in the hind wing and helps to differentiate the situation in braconids where vein CU1b is present (flexion line posterior to the vein) to when vein 2-1A is there (flexion line anterior to the vein spur). The hind wing of most ichneumonids has an additional flexion line running along the basal cell and causing a break near the posterior end of vein rs-m.

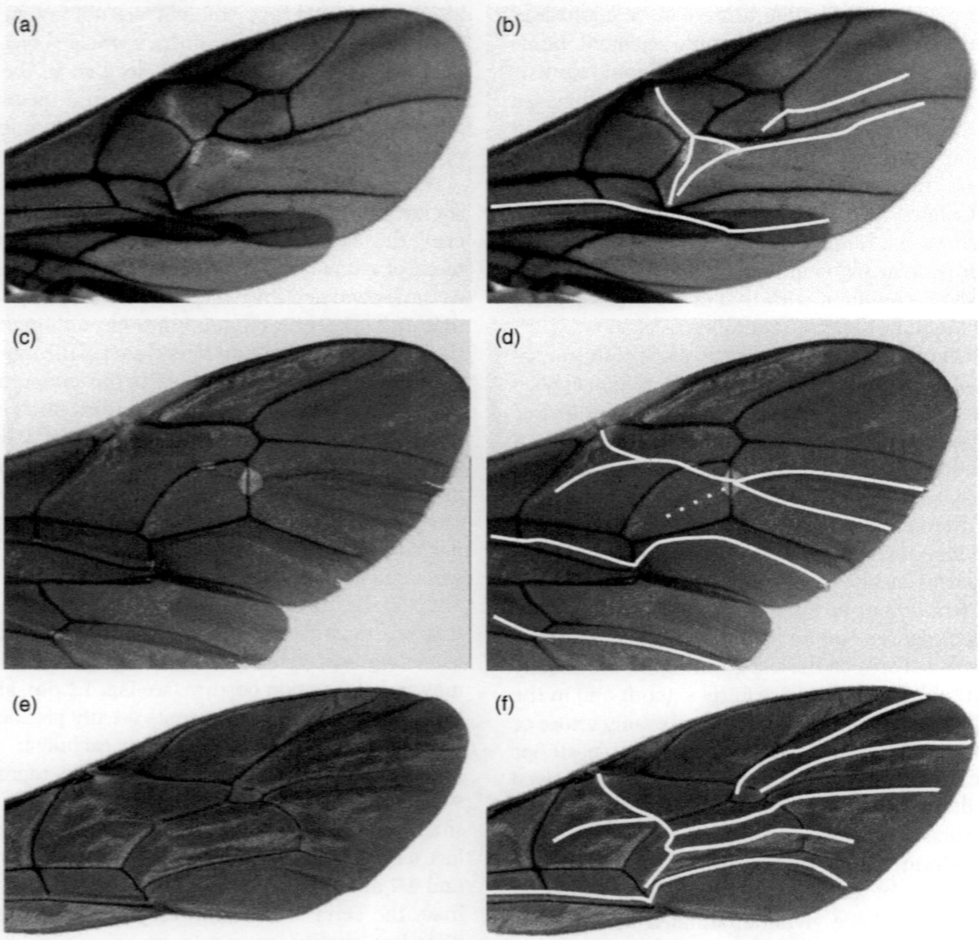

Fig. 2.6 Wings and with major flexion lines depicted in right-hand panels. (a, b) *Proterops* sp. (Ichneutinae); (c, d) *Hellwigiella* sp. (Ophioninae); (e, f) *Clydonium* sp. (Pimplinae: Ephialtini). (Source: a–d, reproduced by permission, © C. van Achterberg, Naturalis Biodiversity Centre; e, f, reproduced by permission of Ilari Sääksjärvi.)

There is a trend towards increasing flexibility of the distal parts of the wings, especially in small bodied taxa in both families. Within the Ichneumonidae, we see that while fore wing vein 2m-cu is generally retained, it has either one or two bullae. The number of bullae is widely used in identification keys, although it is variable in many subfamilies. When there are two bulli, it indicates that the flexion line splits into two closer to the base of the wing. In the hind wing, the shift in vein r-m basal to the split between R and RS that I presume to have occurred in the Braconidae (excluding trachypetines) frees the distal part of the wing for greater flexibility and the flexion line that runs through rs-m in the ichneumonids is lost.

Despite all the work on insect flight dynamics, there have been no studies of wing folding and flexion in ichneumonoids in flight, so it is not possible to be sure how these fold systems work. Surely this would be an interesting topic for investigation.

METASOMA

The metasomas of ichneumonids and braconids comprise 10 segments. Sometimes the term gaster is used to mean metasoma. The literature can be confusing, as some workers refer to abdominal segment numbers and others, the majority, to metasomal segment number, not always specifying which system they are using. The first metasomal segment, i.e. the second abdominal segment, often has important taxonomic characters. It is at least somewhat narrowed basally and articulates with the propodeum usually between or just above the hind coxae, and is often termed the petiole or petiolar segment. Usually the foramen of the hind coxae is confluent with that of the first segment, but that might be a derived condition.

The term tergum applies to all the parts of the dorsal sclerite of a metasomal segment and is not applied completely consistently, which can be another source of confusion. Each tergum can comprise three parts, a medial tergite and a pair of laterotergites, although the latter may be highly reduced or absent or not differentiated. Laterotergites may be fairly wide and pendulous or sometimes just a narrow strip or completely absent. The spiracles are normally located in the tergum or median tergite, but in some Braconidae, the spiracles of the second tergite are in the laterotergite. The metasomal tergites between the third and eighth are usually unremarkable; often they are fairly smooth and in the Braconidae may be either more or less evenly setose or have setae restricted to a single submarginal, posterior line, the latter condition being thought to be derived (van Achterberg 1988b).

Of particular importance in the Ichneumonidae is the position of the spiracle along its length and the possession of an elongate baso-lateral to lateral impression called the glymma, situated between the dorso-lateral and lateral carinae (i.e. below the level of the spiracle). Braconids also can have baso-lateral pits, situated either between the dorso-lateral and dorsal carinae when they are called dorsopes or below the dorso-lateral carina when they are called laterope (van Achterberg 1974). Although they may not be homologous, the ichneumonid glymmae and the lateropes of braconids, and also the dorsopes of the latter, probably serve the same function of strengthening that part of the tergite and probably also reflect muscle attachment sites. In the Ichneumonidae, when glymmae are present the anterior sclerotised part of the first sternite is virtually always separated from the tergum and when they are absent the tergum and sternum are always fused. Therefore, sternal/tergal fusion or the development of glymmae seem to be alternative ways of strengthening the base of the metasoma, although exactly how and why are not known. In the Braconidae, the position of the spiracle is also important, particularly whether it is located in the dorsal, more heavily sclerotised notum or in a membranous lateral part, which is characteristic of various members of the microgastroid complex of subfamilies.

The second and third tergites are both modified, but somewhat differently in the two families. Not considering those taxa with a carapace, in the braconids the terga of second and third segments (often referred to as the second and third tergites) are fused edge-to-edge to form a syntergum that is inflexible, although strictly there is often a pendulous lateral part of the tergites, the laterotergites, that are *not fused*. In the ichneumonidae, the joint between the second and third terga of metasomal segments 2 and 3, although not telescoped, remains flexible. The fourth and subsequent metasomal tergites are telescoped underneath the posterior margins of the previous tergum to which they are connected by a flexible arthrodial membrane. It is unclear what the flexible connection present between terga 2 and 3 in the braconid subfamily Aphidiinae represents. It seems most likely to be a reversal, i.e. a secondary loosening, associated with their highly mobile metasomas and oviposition posture (see Fig. 12.6a), although the possibility that it represents a truly plesiomorphic state cannot, I think, be completely excluded.

The postero-most segments bear the reproductive structures (ovipositor, male genitalia) and a pair of small cerci (also called pygostyles) and, just behind the last tergum, the anus. Frequently, abdominal terga 9 and 10 are fused to form a syntergite. In the Aphidiinae, the cerci are usually very enlarged (relatively) plate-like structures and in many genera are fused with the ninth abdominal tergum because the tenth tergum is reduced (Tremblay, 1979, 1981, 1983), and in some other groups they appear to arise from just behind the ninth tergum. In some pimpliformes, the ninth abdominal tergite is not only divided but also the two halves are widely separated, forming plates lateral to the tenth tergite (Peck 1937a,b, Gauld et al. 2002b).

Two features are of immense importance in determining the biology of parasitic wasps, namely their ovipositors and their venom systems, and both warrant large sections to themselves (see Chapters 3, 4 and 9).

SEXUAL DIMORPHISM

Male and female ichneumonoids are generally rather similar apart from primary and secondary sexual characters such as body and metasomal size, sex-related glands and numbers of antennal segments and their numbers of various classes of sensilla. Males also often have larger compound eyes and longer antennae, presumably involved in mate location and selection. Males not infrequently display features rather more weakly than females, sometimes associated with generally smaller body size, and this can, in many cases, make males much more difficult to identify. For example, Gauld (1991, p. 140) strongly advised anyone starting to study the Pimplinae to ignore males and start working only with females, and to slot the males in later once more familiar with the group. A few groups display fairly marked sexual colour differences. In the Braconinae there is a tendency for males to have darker heads and a darker apex to the metasoma; however, colour dimorphism (dichromatism) is most apparent in some of the larger Ichneumoninae (e.g. various *Ichneumon* species; see Perkins 1959, 1960, Gauld 1984a). For example, in Europe, the common *Ichneumon sarcitorius* provides a good example, with females being largely black and red whereas males, which are more commonly observed, have a slightly more 'wasp-like' yellow and black striped body (see Fig. 13.31b,c). In most species in which there is a pale antennal stripe, this is brighter and usually more extensive in females than males and is sometimes absent in males altogether (see Chapter 10, section *White antennal stripes and tips*).

In general, body size dimorphism is more pronounced in idiobionts than in koinbionts since the size of the host is limiting at the time of parasitisation in the former, and this has important consequences for sex allocation (see Chapter 6, section *Idiobiont and koinobiont strategies* and Chapter 7, section *Sex allocation*). Many koinobionts can regulate the size that their host reaches and only start to develop when there are sufficient resources. For example, Sato et al. (1986) showed that for two gregarious microgastrine parasitoids of the armyworm, *Pseudaletia*[4] *separata* (Noctuidae), adult wasp size was independent of the size and instar of the host when the eggs were laid and also those hosts that were more heavily parasitised consumed more food and reached a larger size than lightly parasitised ones, resulting in there being no difference in the size of the emerging wasps. In contrast, the males of many idiobiont rhyssines and pimplines can vary in length several-fold depending on host size, with females preferentially laying male eggs on small hosts.

Hurlbutt (1987) surveyed sexual differences in body size across a range of parasitic Hymenoptera and found that females were generally larger than males, although it was indicated that this was not so true in the Ichneumonidae and in particular many species in the subfamily Ichneumoninae (Godfray 1994). This suggestion received a quick repost from ichneumonid workers (Gauld & Fitton 1987; see also Teder 2005), who pointed out that the sample was highly biased in terms of the taxa included and that, in many ichneumonids and particularly in the Rhyssinae and Pimplinae (idiobiont taxa), there were often really quite dwarf males. This all fits into the context that body size has a more profound effect on fitness for females than it does for males (M.E. Visser 1994), although in some species, smaller males have been demonstrated to have lower mating success when 'paired' with large conspecific females in some cases, e.g. the opiines *Diachasmimorpha longicaudata* and *Psyttalia incisi* (Ramadan et al. 1991), although this does not appear to be the case with the braconine *Habrobracon hebetor* (Cook et al. 1994) or in the opiines *Psyttalia fletcheri* and *Fopius* (as *Biosteres*) *vandenboschi* (Ramadan et al. 1991). However, no-one appears to have tested this in any of those ichneumonoids in which males show a vast size range, such as Rhyssinae. Hence, particularly among idiobionts where host size is limiting at the time of attack, females tend to lay male eggs on smaller hosts and female eggs on larger hosts. In the Ichneumoninae, copulation usually follows a physical struggle between a male and a female (see also Hilpert 1992) and therefore selection would favour larger male size and strength if it increased mating success with reluctant females. Teder (2005) investigated four ichneumonine species (*Chasmias paludator*, *Exephanes occupator*, *Spilichneumon limnophilus* and *Vulgichneumon saturatorius*) which differed in their degree of sexual size dimorphism. Males of all species except *E. occupator* were larger than their conspecific females based on hind tibia length. In terms of live wet weight, *S. limnophilus* males were significantly heavier than their conspecific females, but *E. occupator* had significantly smaller males and no difference between the sexes was found in the other two. Examining mating success by males of these four species in relation to individual body size, age and dietary status, Teder showed that the fittest males were larger or in better condition and this meant that they were able to withstand a female's physical rejections in the pre-copulatory struggles, thus enduring longer struggles, and this resulted in a higher proportion eventually achieving copulation.

Not only do males, especially of idiobionts, tend to be smaller than conspecific females, but also their development time is often shorter and, as they typically emerge as adults only slightly (e.g. 2–3 days) before their females (protandry), there is sometimes a compensatory extension of the pupal phase. For koinobionts the picture is slightly different. Harvey and Strand (2003) presented data on life-history parameters of an interesting possible exception, the solitary microgastrine *Microplitis mediator*. In this species, males are significantly larger than females, but despite this, their overall development time is shorter. The longer developmental duration of the female largely resulted from an extended pupal stage. Males almost always have markedly shorter adult life spans than females, even when provided with ample food. The reasons for this are unclear.

MALE EXTERNAL GENITALIA

A great deal of use has been made of variations in male genitalia in the Braconidae for subfamily-level phylogenetics, whereas the Ichneumonidae show comparatively little subfamily-level differences. In both families, the intromittant organ, the aedeagus, is located between volsellar plates and lateral to these a pair of parameres (or claspers) (Fig. 2.7). The volselli on either side of the aedeagus each bear a pair of articulated, usually toothed structures, the digitus, and in addition, lateral to the digitus, some may have a fixed or sometimes articulated lobe that may also in some groups bear teeth (Fig. 2.7a). It seems likely, although there are no observations in ichneumonoids, that the digitus is involved in grasping the inside of the female's copulatory orifice so as to reduce the risk of the male becoming dislodged before he has transferred all of his sperm. When present, the articulated digital process appears positioned in such a way that it could work together with the adjacent digitus as a pair of pincers. This genital capsule is attached anteriorly to a basal, ring-like sclerite that may be open or closed, short or elongate, called the basal ring. Most ichneumonids and all non-cyclostome braconids have a short basal ring that is medially often very narrow or absent, but it can be very elongate in some cyclostomes (Fig. 2.7c) and is somewhat elongate in rhyssine ichneumonids, although in a slightly different way (Fig. 2.7e).

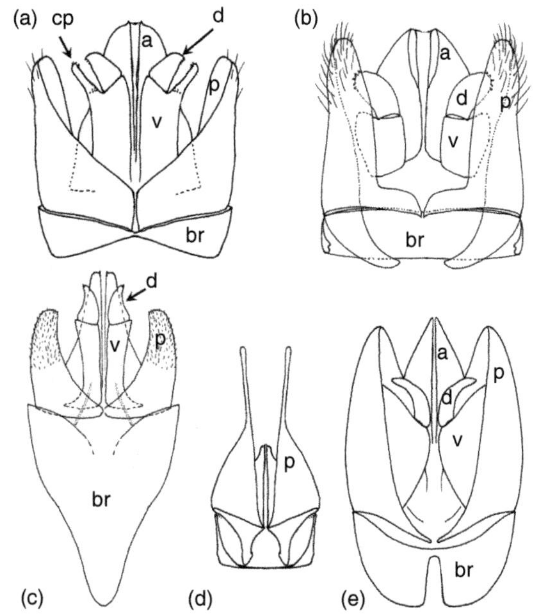

Fig. 2.7 Diagrams of selected male genitalia. (a) *Histeromerus mystacinus* (Rhysalinae) showing very short basal ring and partially articulated and apically toothed cuspidal process arising from volsella lateral to digitus. (b) *Agathis agilis* (Agathidinae). (Source: adapted from Brajković et al. 2010a.). (c) *Iphiaulax* sp. (Braconinae) showing very elongate basal ring. (Source: adapted from Quicke & van Achterberg 1990.). (d) *Mesochorus* sp. (Mesochorinae) showing almost unique needle-shaped parameres. (Source: adapted from O. Peck 1937a.). (e) *Megarhyssa lunator* (Rhyssinae) showing fairly large, bilobed basal ring. (Source: adapted from O. Peck 1937a.). Abbreviations: a, aedeagus; br, basal ring; cp, cuspidal process; d, digitus; p, paramere; v, volsella.

O. Peck (1937a,b) made a detailed study of male genitalia in the Ichneumonidae and attempted to determine the homologies of the various structures to the basic insect leg from which the genital parts are derived. Other major studies on braconid male genitalia are by Pratt (1939)[5] on a subset of Ichneumonidae (referred to by him as Ichneumoninae but corresponding to a group comprising Xoridinae, Labeninae, larger Pimpliformes and Banchinae), by Tobias (1967) various braconids, Tremblay (1979, 1981, 1983) for the Aphidiinae, Quicke (1988b) on the Braconinae, Maetô (1987) for the Microgastrinae and Brajković et al. (2010a) and Žikić et al. (2011) for the Agathidinae.

In both families, the male hypopygium may be more or less rectangular (see Fig. 103b) or the mid-anterior part of the sclerite may be strongly produced into a process (Figs 69 & 70 in Quicke & van Achterberg 1990). The function of this modification is unknown although it seems to me likely to be related to the orientation the genitalia assume during copulation and that the process is the site of muscle attachment.

ENDNOTES

1. The internal chitinous phragma to which the longitudinal flight muscles attach posteriorly projects far back under the propodeum.
2. It is also possible that the costa or the subcosta *per se* does not occur in the Ichneumonoidea, as there only ever appears to be one longitudinal division running along the 'C+SC+R' vein.
3. I make an exception to the strict application of the Sharkey and Wharton system for extant ichneumonids in order to make comparisons more clearly with the Braconidae, that is, I use 2RS for the basalmost vein of the areolet even though it is technically 1RS because the lack of (RS+M)a in extant ichnemonids means that 1RS as it would be in braconids is indistinguishable.
4. Sometimes treated as synonym of *Mythimna*
5. The species referred to as *Ichneumon irritator* by Pratt is what we now call *Dolichomitus irritator* and thus belongs to the Pimplinae and not Ichneumoninae, hence the now misleading title of his paper.

Chapter 3

THE OVIPOSITOR AND OVIPOSITOR SHEATHS

The Braconid and Ichneumonid Parasitoid Wasps: Biology, Systematics, Evolution and Ecology, First Edition. Donald L. J. Quicke.

Ovipositors in the Ichneumonoidea range from barely visible to extremely long organs, with the longest being up to approximately eight times the length of the body, for example, in the braconine braconid *Euurobracon yokahamae* (see Fig. 12.32a) and the pimpline ichneumonid *Dolichomitus hypermeces*. As these are rather large wasps, their ovipositors measure up to 18 cm (Townes 1975). Strangely, the '*Iphiaulax*? sp' (Braconinae) that Townes listed as having an ovipositor 14 times the length of the body is actually a member of a genus undescribed at that time, now named *Pheloura*, because the 'ovipositor' seen by Townes was in fact a pseudo-ovipositor comprised of three long, presumably mimetic, protuberances from the wasp's anal region, the true ovipositor being much more normal in proportion (van Achterberg 1989). Despite their great actual length, the record in the Hymenoptera for the longest ovipositor relative to body length goes to some parasitic chalcidoid fig-wasps.

In common with those of all other Hymenoptera, the ichneumonoid ovipositor system comprises a pair of sheaths that protect the ovipositor proper when it is not in use (and which may also have a sensory role) and the ovipositor itself, which comprises a single upper part (termed a valve) and a pair of lower valves. The more anatomical literature often uses technical terms for these parts based on their limb origins on the eighth and ninth abdominal segments, but in order to facilitate understanding, I have attempted to standardise on a simple terminology in accordance with Quicke et al. (1994); to allow this to be related to terms used in various other works, Table 3.1 provides a list of synonyms. Detailed discussions of anatomy can be found in Oeser (1961), Scudder (1961) and E.L. Smith (1968, 1969, 1970), and there are also distillations of these in various general entomology textbooks. There are still, however, despite the great deal of morphological work that has been done, some disagreements about how the apparatus works (see Quicke et al. 2000a).

The lower valves extend from an approximately triangular cuticular sclerite called the gonangulum or first valvifer (Fig. 3.1) and are technically the gonapophyses of the eighth abdominal segment. The upper valve and the ovipositor sheaths originate from opposite ends of gonocoxite 9. The gonangulum is connected closely with the edge of the ninth tergite and pivots against gonocoxite 9 such that as the angle it makes with the latter changes so the lower valves are moved longitudinally relative to the upper valve. The single upper valve is the result of the fusion of the paired appendages of the ninth abdominal segment, although it is secondarily largely re-separated except at the apex in some members of the ichneumonid ophioniformes group of subfamilies (see the section *Ovipositor steering mechanisms* below). The ventral surface of the upper valve interlocks with each of the lower valves by means of a sublateral, longitudinal ridge called the rhachis, which is T-shaped in cross-section and lies within a T-shaped groove called the aulax along the dorsal surface of each lower valve (Fig. 3.2). The combined rhachis and aulax mechanism which holds the valves together while allowing the three valves to slide independently relative to one another is called the olistheter. The upper and lower valves are all 'hollow', the lumen of

Table 3.1 Comparison of terms employed by different workers for the major parts of the ovipositor system.

Quicke et al. (1994) and this book	Elsewhere	Abdominal segment
Ovipositor sheaths	Gonoplac; gonosylus; third valvula	9
Upper valve	Gonapophyses; second valvula; stylet	9
Lower valves	Gonapophyses; first valvula; lancet	8

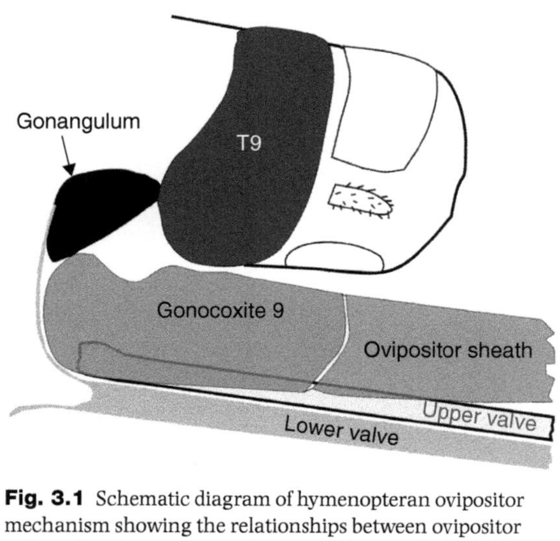

Fig. 3.1 Schematic diagram of hymenopteran ovipositor mechanism showing the relationships between ovipositor valves, sheath, gonocoxite 9, gonangulum and tergum 9. (Source: adapted from Quicke et al. 2000a.). For a colour version, see Plate 4.

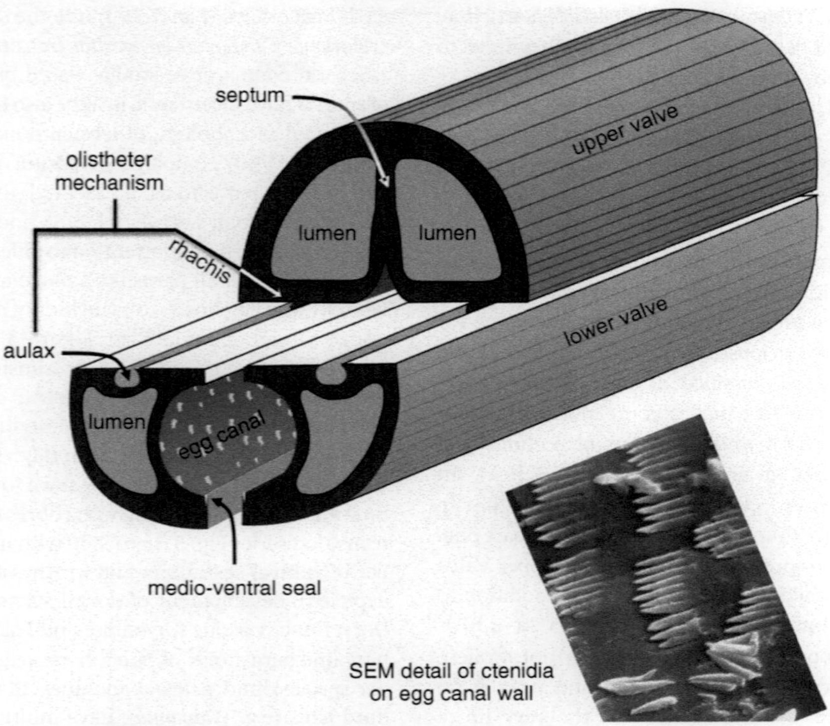

Fig. 3.2 Diagram illustrating the major features of medial part of an ichneumonoid ovipositor and inset showing microsculpture of ovipositor wall of a braconine. (Source: redrawn after Rahman et al. 1998a.). For a colour version, see Plate 5.

each containing haemolymph, tracheae, nerve fibres and other living tissues, although no muscles.

The ovipositor serves a similar set of functions in nearly all parasitic wasps. These are:

- navigating or penetrating the substrate (if the host is concealed);
- locating the host;
- assessing the host;
- piercing the host;
- staying in the host long enough;
- injecting venom;
- laying an egg.

Different aspects of the ovipositor are involved in each. They have to be long enough and on reaching the host may require 'drilling' through a substrate, which involves external serrations at the tip (see the section *Functional morphology of wood-drillers* below). Ovipositors may have to steer within, for example, galleries in wood or within leaf rolls and there are various associated modifcations near the tip (see the section *Ovipositor steering mechanisms* below). Having contacted a host, sense organs near the tip presumably often provide feedback about (a) having found a potential host, (b) whether it is an acceptable species and (c) whether it is acceptable for oviposition (age, health, etc.). This probably often involves penetrating it with the upper, the lower or both sets of valves. Hosts usually respond with strong movements when they are pricked and if an egg is going to be lain internally, a pre-apical dorsal notch of some species has evolved to help retain the ovipositor in the wriggling host (see the section *Ovipositor notches and endoparasitism* below). Depending upon species, a temporary paralysing venom might be injected at this stage. For idiobiont ectoparasitoids, the ovipositor nearly always has to deliver a paralysing venom at this stage. Envenomation requires both a means of preventing the venom from seeping out before the target and a way of creating the force to push it along an often narrow egg canal (see the section *Number, position and possible functions of ovipositor valvilli* below). Also, of course, they have to lay an egg in, on or, rarely, near the host.

Unlike in all other parasitic wasps, the most serrated valve in the ichneumonoids is the lower pair (Quicke et al. 1999a, Belshaw et al. 2003) and, when a pre-apical protrusion (nodus) or a notch is present, it is on the upper valve.

Despite its importance for understanding ichneumonoid evolution, very little detailed information is available on the modes of substrate or host penetration by parasitoid ovipositors. Broadly, ichneumonoid ovipositors can be classified as one of three main forms (although with numerous exceptions): (i) simple and needle-like without obvious projections and with very reduced lower valve teeth (Fig. 3.3e,k), (ii) typical wood-borer ectoparasitoid type with the tip deeper and wider than the shaft and with at least one, and sometimes more, distinct pre-apical upper valve protuberances called nodi (*sing.* nodus), in addition to well-developed lower valve serrations (Fig. 3.3a–d,h) or (iii) typical endoparasitoid type with the tip not swollen, with a pre-apical upper valve notch and usually no more than three weak serrations at the very tip of each lower valve (Fig. 3.3i–l) (see the section *Proposed evolutionary and related ovipositor transitions* below).

Several comparative studies based on small subsets of species have given some insight into the comparative functional morphology of ichneumonoid ovipositors. Lyngnes (1960) compared ovipositor gross morphology in three parasitoids, a braconid, an ichneumonid and a chalcidoid; LeRalec (1993) and LeRalec et al. (1996) paid particular attention to differences between ovipositors in small parasitoids associated with aphid, whitefly and egg hosts, thus including only Aphidiinae out of the Ichneumonoidea. Figure 3.3 depicts what I consider to be the most likely transitions in overall ovipositor shape from an ancestral ectoparasitising morphology with a simple dorsal nodus and a moderate number of apicoventral teeth (Fig. 3.3a). Specialist wood-borers which necessarily have to expend a lot of time and energy disrupting wood fibres occur along the transition series Fig. 3.3a to 3.3c with increasing number of ventral serrations and in Rhyssinae (Fig. 3.3c, depicted), development of dorsal serrations or, in the Doryctinae, various Rhyssinae and Labeninae, several pimpline parasitoids of pupal hosts (e.g. several *Theronia* species) and a few banchines that oviposit into hard galls (e.g. *Wahlamia*), have multiple dorsal nodi,

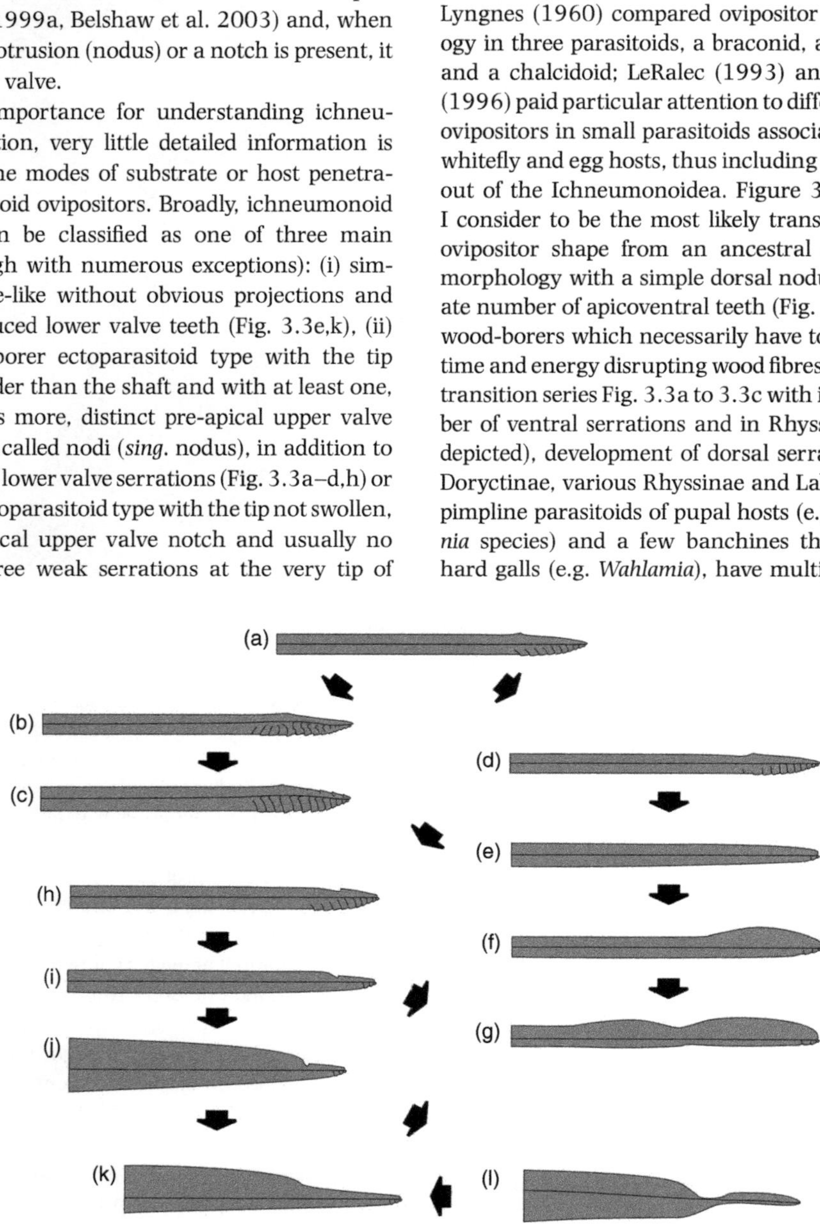

Fig. 3.3 Diagram illustrating possible series of transitions between ovipositors of different morphology and function.

suggesting an important role in penetrating hard hosts. A second nodus is also present in members of the braconid subfamily Doryctinae, for which it seems to be a largely reliable synapomorphy, although its function in these biologically rather diverse ectoparasitoids is not known (Quicke et al. 1992a).

Also derived from these 'drilling' forms is an evolutionary series in which the ovipositor is probably more flexible, the dorsal nodus and serrations become reduced although defined now more by a dorso-ventral narrowing behind it (Fig. 3.3d), and while still interacting with surrounding woody material, these ovipositors principally thread their way through pre-existing cracks in the host substrate (Eggleton 1989), and those which encounter hosts that are relatively close to the surface in more heavily fractured substrate (e.g. insects in dry twigs and narrow branches in xeric environments) lose the dorsal nodus entirely and the ventral serrations are usually reduced to just three small ones (Fig. 3.3e). Good examples of this occur in various *Iphiaulax*, *Digonogastra* and *Megalommum* (among many others) in the Braconinae. Sometimes these ovipositors become very thickened (robust) and may be moderately laterally compressed. In these, the upper valve almost certainly does not penetrate the host, and similarly it seems unlikely to in those or the apex of the upper valve is distinctly apically swollen (Fig. 3.3f; e.g. *Nesaulax* and *Sylvibracon*).

The morphologically simplest type is typical of taxa that oviposit in eggs (e.g. Cheloninae, some Ctenopelmatinae) or very small larval hosts, although in many cases those that attack very small hosts have an ovipositor that is fairly wide at the base but narrows dramatically to form a sharp, needle-like point (Fig. 3.3k) (Brajković et al. 2010b). Simple ovipositors are also typical of some species that lay eggs in spider egg masses (Fitton et al. 1987, 1988, Schwartz & Shaw 2000, Broad et al. 2004), although a very highly modified ovipositor with large recurved, tongue-like serrations is found in the ephialtine pimpline *Zaglyptus* with the same biology. The ovipositors of small idiobiont ectoparasitoid braconids, such as *Histeromerus*, *Hormius* and small *Bracon* spp., that use them only to inject venom into the host larva and do not have to penetrate any substrate to do so, often have a very reduced nodus and ventral serrations and may be completely smooth and simple.

The basic functional aspects of the pre-apically notched ovipositors typical of a large majority of endoparasitic taxa within both the Ichneumonidae and Braconidae has been explored in detail by Boring et al. (2009). Although in ichneumonoids the notch does not seem ever to be as relatively deep as in various eucoilid (Cynipoidea) ovipositors described by van Lenteren et al. (1998), its function appears to be analogous in that in restricts penetration of the host cuticle by the upper valve and observations on both *Banchus* and *Homolobus* (van Veen 1982, Boring et al. 2009) confirm the assumptions of Belshaw et al. (2003). As with many aspects of ichneumonoid biology, however, very few detailed observations have been published (see the section *Ovipositor notches and endoparasitism* below).

The way in which the ovipositors of some wood-borer parasitoids 'drill' through plant tissue and how others with long ovipositors steer them only by the differential pulling of the ovipositor valves at their bases have resulted in the ichneumonoid ovipositor attracting interest from several groups working on biomimetics and, in particular, work has been on-going to adapt the features to create useful, steerable, medical and other probes (e.g. Ko et al. 2011, Frasson et al. 2012).

THE ACT OF OVIPOSITION

The egg canal always has a posteriorly directed ctenidia-like microsculpture (Fig. 3.2), which helps drag, through the to-and-fro motions of the ovipositor valves, the chorion of the egg down it (Austin & Browning 1981), and this is also true for those of taxa such as aphidiines where the eggs are very small relative to the egg canal. There is a marked difference in the amount of variation in egg canal microsculpture between braconids and ichneumonids; the latter are, as far as we know, almost entirely uniform in having more or less wide lines or curved patterns of ctenidia (see the SEM images in Nénon et al. 1995), whereas in the braconids there are many more patterns and types of protuberance (Rahman et al. 1998a), and, sculpture within the Doryctinae is particularly diverse (Rahman et al. 1998b). Boring et al. (2009), noting that many aculeates also have ctenidia in the sting canal, along which no egg passes, suggested that they almost certainly have other functions than just dragging the egg chorion along during oviposition and suggested that the small gaps between them and the egg canal wall might act as miniscule reservoirs for lubricants that might otherwise be 'scraped away'.

Eggs, apart from the very smallest, become distorted as they pass along the narrow egg canal, and this act

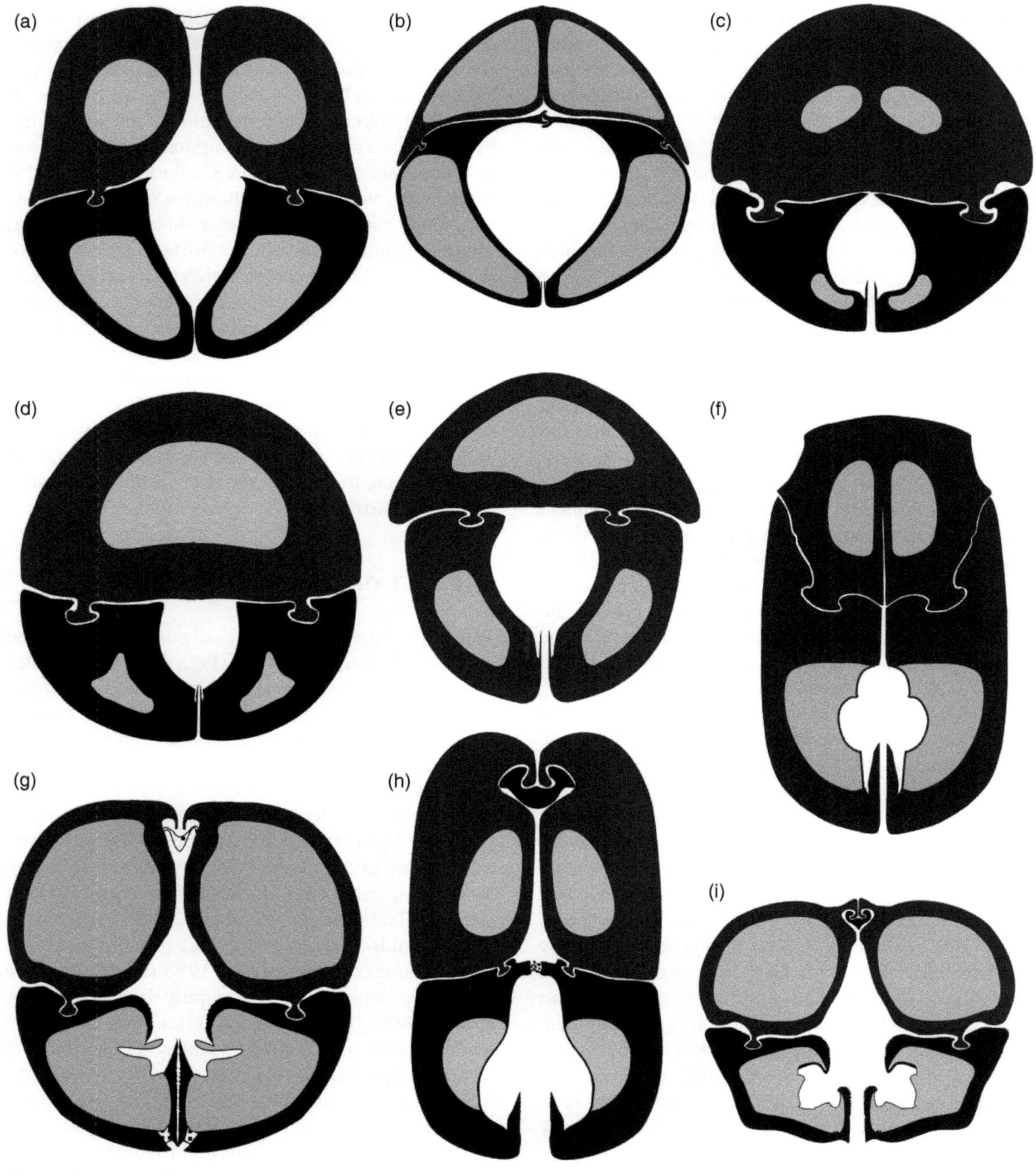

Fig. 3.4 Diagrams of transverse sections through the median part of an ovipositor for selected braconids and ichneumonids. (a) *Thoracoplites* (Rhyssalinae); (b) *Aleiodes* (Rogadinae); (c) *Zombrus* (Doryctinae); (d) *Cenocoelius* (Cenocoeliinae); (e) *Austrozele* (Macrocentrinae); (f) *Arotes* (Acaenitinae); (g) *Campoplex* (Campopleginae); (h) *Lycorina* (Lycorininae); (i) *Glypta* (Banchinae).

of distortion is frequently the trigger for meiosis to complete and embryogenesis to proceed (Amy 1961) (see Chapter 5, section *Embryogenesis*). In the case of the braconine *Habrobracon*, for example, the ovipositor egg canal is only one-twelfth the diameter of the mature ovarian egg and apparently the distortion of the egg during oviposition seems to shift the nucleus from a dorsal to a ventral position (B.R. Speicher 1936). Even in taxa with moderately small eggs such as most endoparasitoids, the egg is still stretched during its passage along the egg canal and, as Boring et al. (2009) noted, the chorion stores the elastic strain and, once the egg starts to emerge near the ovipositor tip, it will effectively help itself to escape as it resumes its original configuration, rather like a balloon being forced through a hole. Idiobiont parasitoids, especially those of deeply concealed hosts, typically have eggs with a long, narrow posterior tail (see Fig. 12.33b) (Abbott 1934, Iwata 1960, Skinner & Thompson 1960) which enters the egg canal first and probably occupies a great deal or all of the length of the ovipositor's egg canal during oviposition, thus giving the ctenidia (Fig. 3.2) a great deal of purchase on the egg's chorion. The egg canal wall of many Labeninae, pimpliformes, ichneumoniformes and even some basal ophioniformes (e.g. *Euryproctus*, *Xenoschesis*: Ctenopelmatinae) have a longitudinal ridge (Quicke et al. 1994, 2009), which probably serves to help grip the egg and prevent it from forcing its way out between the lower valves as it passes along the ovipositor. Although koinobiont endoparasitoids, many ctenopelmatines have rather large eggs with tough chorions (see Fig. 13.44d,f) which probably explains why they have possibly retained this feature. No similar longitudinal ridge occurs in any braconids with otherwise similar ovipositors and biologies, and its function might be replaced by the inwardly protruding medio-ventral margin of the lower valves.

Those ichneumonid taxa with the upper valve completely divided except at the apex (i.e. the majority of ophioniformes with long ovipositors) and with the halves interlocked by the aulaciform rod (Quicke et al. 1994; Fig. 3.4h,i), nearly always have a large articulated flap along the ventro-medial edge of the lower valves. This combination was postulated by Quicke et al. to enable the egg canal to be greatly enlarged while still maintaining a sealed tube for retention of egg and venom. Unlike xoridines, rhyssines and pimplines, for example, which often 'drill' through solid wood, these endoparasitoids mostly pass their ovipositors through pre-existing gaps into leaf-rolls and sometimes through cracks in wood. Thus the ovipositor is not constrained to remain narrow by its surroundings, which would restrain the egg to remain internal. The ability to expand the egg canal may therefore be a system enabling them to oviposit fairly large eggs without them forcing the valves apart in such a way that they could pop out before reaching the tip.

FUNCTIONAL MORPHOLOGY OF WOOD-DRILLERS

Le Lannic and Nénon (1999) presented a detailed study of the remarkably long ovipositor of the rhyssine ichneumonid *Megarhyssa atrata* and addressed the problem of how it can penetrate solid wood to reach its siricid wood wasp hosts. Since wood is a hard substrate, or can be, the parasitoid ovipositor has either to reach a host by navigating it largely or entirely through pre-existing cracks, fissures or borings made previously by other insects or it has to break the wood fibres. Different species definitely employ both techniques. Figure 3.5 presents a hypothetical mechanism by which the tip of a typical wood 'drilling'-type ovipositor may penetrate the substrate, although for clarity the amount of movement of the lower valves has been greatly exaggerated (see below). The system works somewhat like a ratchet; the upper valve is pushed into the hole as far as it will go (Fig. 3.5a) and the slightly wider apical part of the lower valves is then pushed level with the upper valve, and in so doing they force the nodus into the surrounding wood, which locks the upper valve in place (Fig. 3.5b). Further pushing of the lower valve crushes wood fibres and makes headway. When they have reached their limit, they withdraw (Fig. 3.5c,d), leaving a small opened cavity ahead of the ovipositor tip, into which the upper valve, now able to disengage from the wood, can push (Fig. 3.5e). The process is then repeated (Fig. 3.5f–h). Crucial to this, and perhaps rather counter intuitive, the first stage of penetration must be to locate a narrow pre-existing gap that allows the nodus on the upper valve to gain purchase and the process to start. When wasps are observed making several starts and aborting them, it may be that the substrate is actually too soft for them to get going and a new attempt has to be made (J.F.V. Vincent, personal communication).

The function of the lower valve teeth is not known for certain. It might be that they do rasp broken fibres out of the way, but when ichneumonoids are

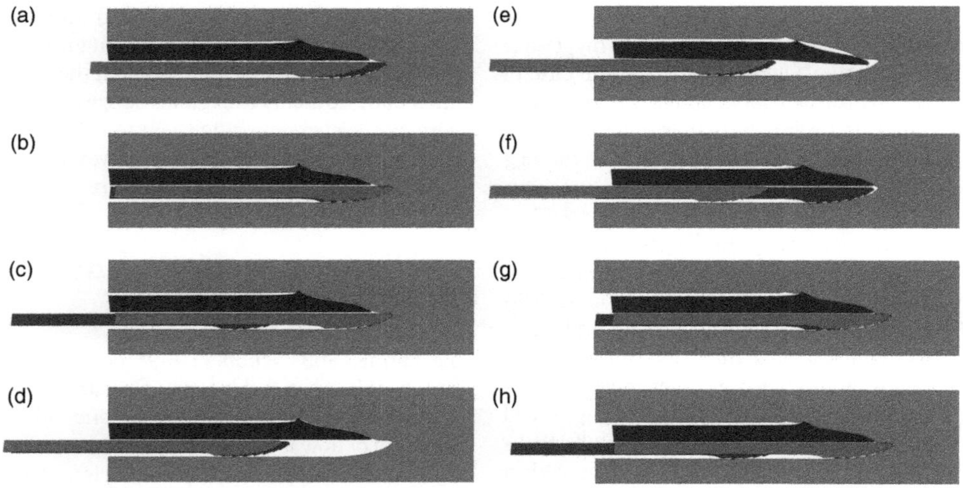

Fig. 3.5 Diagrams depicting a hypothesised mechanism of ovipositor penetration into wood with the dorsal nodus acting as a locking device enabling lower valves to make headway by crushing wood fibres and then retracting to allow further progression of the upper valve.

observed 'drilling' into wood, there is never any sawdust produced. There is also an almost complete lack of information on the speed and extent of the movements of the ovipositor valves during wood penetration. An interesting observation was made by Lyngnes (1960), who noticed that when the ichneumonid *Dolichomitus extensor* (as *Ephialtes*) and the doryctine *Spathius exarator* were dying after having been placed in dilute ethanol solution, the ovipositor valves would move rhythmically to-and-fro, the lower valves moving alternately. This suggested that the trauma had initiated a 'fixed action pattern' that most likely is similar to that occurring in substrate penetration, although without the concomitant sensory feedback. Lyngnes noted that in *E. extensor* the lower valves moved forwards and backwards about the length of two of their serrations beyond and then behind the tip of the upper valve, and in *S. exarator* the distance moved was equivalent to about three of the serrations and in this case the lower valves were moved alternately. With modern technology such as laser vibrometry or simply videoing valves in action with a minute spot of marker paint, it ought to be possible to obtain a far more detailed understanding of what the wasps do. The very considerable variation between taxa of the number, depth and spacing of lower valve serrations might suggest that more than one mode of action has evolved. Gauld and Wahl (2000a) suggested, and I strongly suspect that it is correct, that the presence of many closely spaced serrations on the lower and sometimes also the upper valve is indicative of 'drilling' through strongly lignified, i.e. hard, woody plant tissue, and this sort of modification is characteristic of many Rhyssinae and some Labenini (Labeninae). Such fine teeth are much more uncommon among braconids, but they do occur in the Southeast Asian braconine genus *Trullabracon* (van Achterberg & Quicke 1991).

Many parts of insect cuticle that are required to be particularly hard, strong or wear resistant are composed not of chitin but largely of a protein–metal ion complex. In the case of parasitic wasps, the ovipositor teeth (and also mandible cutting edges) can be hardened with either a zinc or manganese complex, and with ichneumonoids, the metal involved in the ovipositor teeth is usually manganese (Quicke et al. 1998). In one cryptine in particular, high levels of calcium were found associated with the cutting edges of the lower valve teeth (Quicke et al. 2004a), again almost certainly involved in hardening them, but perhaps not as a protein–metal complex. It is also obvious from SEM that the teeth of the ovipositor do experience some wear-and-tear, but there have been no studies relating this to metals or substrate or whether it ever becomes limiting. One of the diagnostic features displayed by most members of the braconid subfamily Doryctinae is that the apex of the upper and lower valves is very nearly black, indicating heavy sclerotisation and presumably greatly increased hardness and/or abrasion resistance, in addition to having manganese-based hardening of the ovipositor teeth (Quicke et al. 1992a) and a similar apical darkening is found in those *Bracon* species that oviposit into hard seeds (Flores et al. 2005). Although this presumed hardening is largely conserved among the doryctines, a few species that attack more easily accessible hosts have lost it, suggesting that some cost is involved.

The chitin of parasitoid ovipositors is a highly ordered composite. As in all chitinous cuticle, crystals of chitin molecules are surrounded by an orderly array of globular proteins (see Vincent & Wegst 2004) and the strength or flexibility of the cuticle in any given direction is determined by crystal size, length and orientation. However, instead of having all crystals orientated in the same direction, cuticle is considerably more sophisticated, with different layers with different crystal orientations, forming a complex structure. The long, thin ovipositor of *Megarhyssa* was used as an early model for understanding insect cuticle using X-ray diffraction (Blackwell & Weih 1980), and chitin crystal structure has also been studied in the ovipositor of *Rhyssa* using diffraction contrast TEM (Giraud-Guille et al. 1990).

Ultrastructural investigation of ichneumonoid ovipositors is in fact hampered by the very features that make them good at their job; the chitin–protein matrix, as seen under ultrahigh-resolution TEM, has almost no gaps in it and therefore it is virtually impossible for embedding resins to penetrate it. Therefore, when it comes to cutting ultrathin sectionsm the cuticle often breaks away from the embedding resin, making it difficult to obtain good-quality sections.

OVIPOSITOR STABILISATION GUIDES AND BUCKLING FORCE

The mechanics of ovipositor 'drilling' have been explored by Vincent and King (1996). Critical for the penetration of any substrate is that the cutting part at the tip must be able to deliver enough force to the substrate to break it, for example, by crushing or cutting the wood cells. If the wasp applies a large force to try to drive a long, thin ovipositor into the substrate, the ovipositor is likely to buckle, and once it has buckled no further force can be delivered to the substrate. Hence, in simple terms, for successful substrate penetration

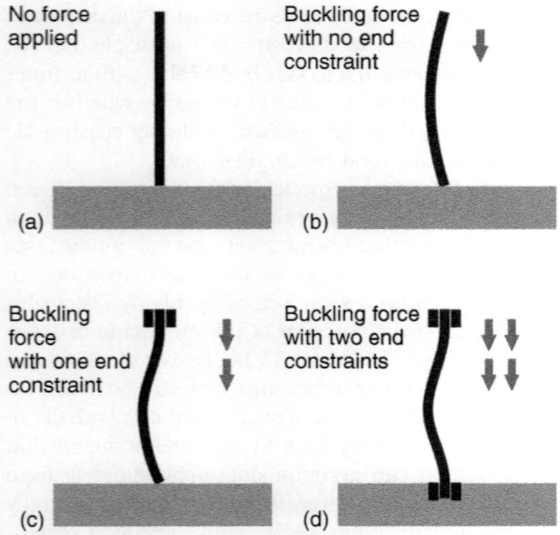

Fig. 3.6 Diagram illustrating how the amount of force that can be applied to an ovipositor before it buckles and is then no longer able to exert force depends on its end constraints. (Source: modified after Quicke et al. 2000a).

the force required to break the wood fibres must le less than the buckling force.

The buckling force depends on the stiffness of the ovipositor, the exposed length of the ovipositor (the longer the free length, the lower is the buckling force) and the end constraints (Fig. 3.6). The minimum Euler buckling force, P_{crit}, can be approximated by the equation

$$P_{crit} = \frac{\beta\pi^2 EI}{L^2} \tag{3.1}$$

where L is the exposed length of the ovipositor, E is Young's modulus of the ovipositor cuticle, I is the second moment of the cross-section of the cuticle and β is a constant related to the end constraints (Vincent & King 1996). β is 1 in the situation shown in Fig. 3.6a, 2 in Fig. 3.6c and 4 in Fig. 3.6d. For long ovipositors, the effective length can be reduced by it being clamped or stabilised by various adaptations of the metasomal sternites (Rhyssinae: see Fig. 10.5d) or grooves on inner face of the hind coxae as in many Labeninae (see Fig. 13.8b, arrow) (Gauld & Wahl 2000a) and some Cenocoeliinae (Pitz & Sharkey 2007), both of which grasp and hold straight the ovipositor at approximately half its exposed length.

The force that needs to be delivered to the ovipositor mechanism (by the Bowden cable principle, i.e. the same as happens in a bicycle brake cable with its inner wire and external sheath) is the same whether the ovipositor teeth make sawdust or simply crush aside the walls of the wood cells in their way.

Some *Megarhyssa* species with ovipositors much longer than the body are still able to stabilise them during oviposition using their sternal guides (see Figs 10.5d and 13.25), because the ovipositor can be looped within a sack formed by highly distensible membrane at the posterior of the metasoma between tergites 6 and 7 (Abbott 1934, Le Lannic & Nénon 1999). Since the membranous sack can be distended by the ovipositor to form a translucent disk with diameter of approximately 2 cm in *M. atrata*, it means that is adaptation can accommodate approximately 6 cm of of ovipositor. However, that adaptation is virtually unique and in other species with very long ovipositors, including other rhyssines, the ovipositor is or directed posteriorly into the substrate at an angle (van Achterberg 1986) (see Fig. 13.6b) or sometimes, but much ore rarely, anteriorly between the legs (see Fig. 13.30b). As some species have ovipositors several times the body length (e.g. various *Dolichomitus*, *Euurobracon* and *Shelfordia*), this must necessarily mean that once the female has located a host she has to walk some distance away from it before starting 'drilling'. Also, the ovipositor clearly cannot be supported, although Fig. 13.6b seems to show that the hind legs might play some sort of stabilising role. Since a long, thin, free ovipositor cannot withstand much pressure before buckling, the long ovipositors must penetrate the substrate using a ratcheting means, such that the tip is effectively pulling the ovipositor into the wood and only small advances need to be made in each cycle of upper/lower valve movement.

OVIPOSITOR NOTCHES AND ENDOPARASITISM

Keys to the subfamilies of Ichneumonidae very frequently make use of the presence of a distinct pre-apical notch on the dorsal ovipositor as a character to help differentiate a large group of subfamilies within the Ophioniformes valve (see Figs 3.3, 13.44d,e and 13.57c) (see e.g. Townes 1969, Gauld 1985, Wahl 1991). Surprisingly, therefore, although the same character state occurs in many braconid groups, this has not been used in keys to their subfamilies – this make me wonder whether it is used in the Ichneumonidae almost out of desperation. Anyway, pre-apical notches are most conspicuous in endoparasitic koinobiont taxa with longish ovipositors, although they also occur in short-ovipositored species. The notch is of considerable functional importance in that it limits the penetration of the upper ovipositor valve into the host and maintains it with just the tip penetrated. Although direct observations are lacking, a very detailed scenario for how it works was presented by Boring et al. (2009) based on observations of it in the braconid *Homolobus truncator*, but the proposed mechanism is almost certainly the same in other taxa with a pre-apical notch.

The sequence of largely hypothesised events in host penetration based on Boring et al. (2009) is shown in Fig. 3.7. The essential features proposed are that the initial penetration of host cuticle is by the lower valves, either singly or together, and, having made a hole (Fig. 3.7a–c), these may withdraw slightly (Fig. 3.7d,e) to allow the blunter tip of the upper valve to enter the notch to engage with the host's integument (Fig. 3.7f,g). The two lower valves then force their way further in (Fig. 3.7h,i) and subsequently, by a combination of ratcheting their ctenidea against the egg chorion and/or increasing venom pressure on the pumping action of the valvilli, cause the egg (Fig. 3.7j,k) and then the venom (Fig. 3.7l) to enter the host. The ovipositor then withdraws by retracting the two lower valves (Fig. 3.7m,n), which relieves the lateral pressure on the upper valve. allowing its notch to disengage from the host cuticle and the upper valve to be withdrawn (Fig. 3.7o). This process causes a minimum of damage to the host, the small cuticular prick being able to heal itself easily.

OVIPOSITOR STEERING MECHANISMS

The hymenopteran ovipositor is devoid of intrinsic musculature (although some aphidiine braconids may have some; A. LeRalec, personal communication), yet many members are capable of actively bending their ovipositors as an adaptation to reaching their hosts, and to achieve this several different steering mechanisms have evolved. The mechanism that appears to have evolved most often involves the formation of the distal part of the ovipositor into one or more arch-like regions. This type of modification is found in a number of braconines (Fig. 3.8b), opiines, tersilochines and cremastines (see Fig. 13.54) and also in the Mendesellinae

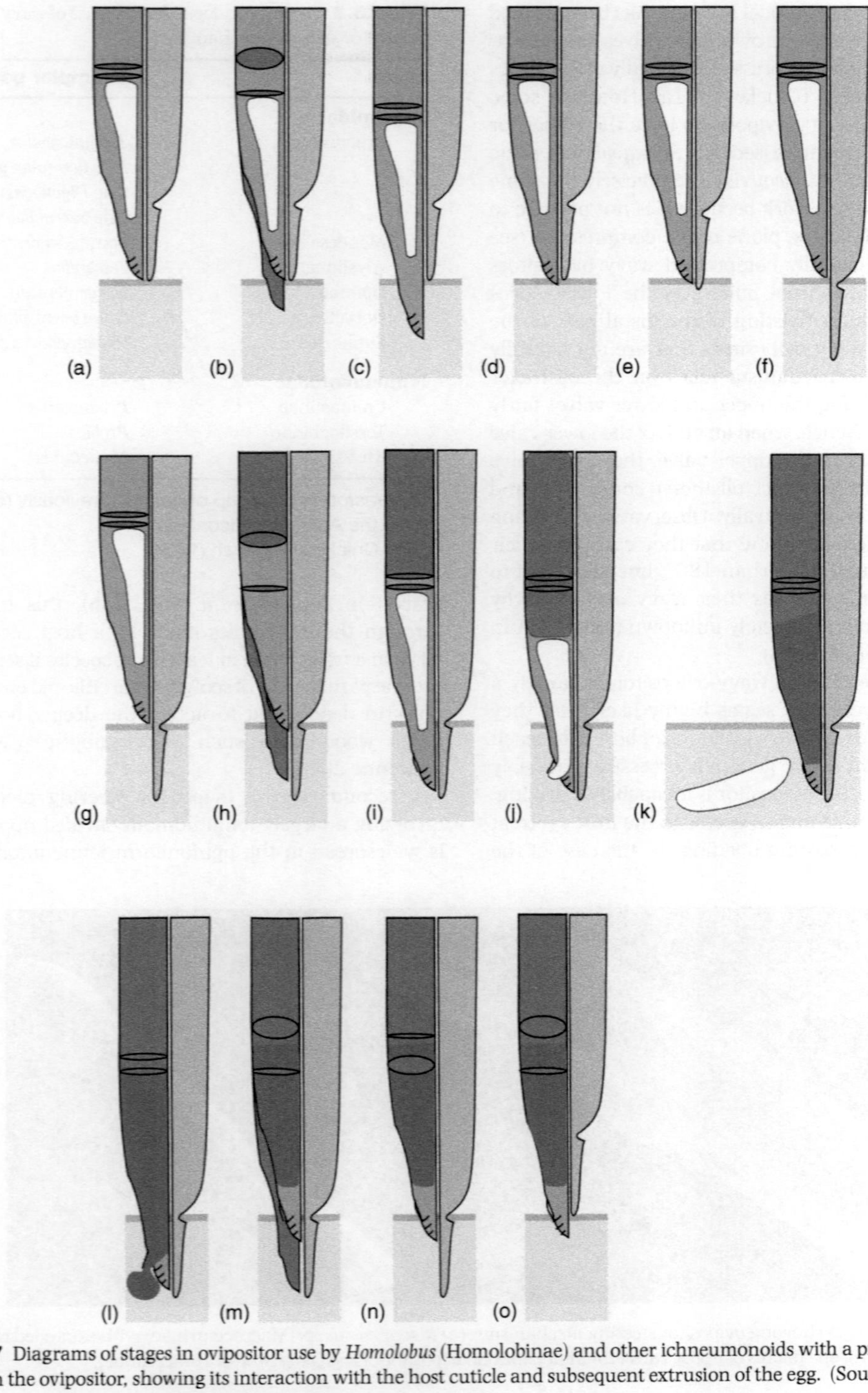

Fig. 3.7 Diagrams of stages in ovipositor use by *Homolobus* (Homolobinae) and other ichneumonoids with a pre-apical dorsal notch on the ovipositor, showing its interaction with the host cuticle and subsequent extrusion of the egg. (Source: adapted from Boring et al. 2009.). For a colour version, see Plate 6.

(Table 3.2). The mechanical principle has been worked out for the case where the ovipositor valves are more or less dorso-ventrally compressed, especially at the places between the arches (Quicke 1991a). However, some of the taxa with wavy ovipositors have the ovipositor strongly laterally compressed (e.g. *Sinuatophorus*, some *Zaglyptogastra* and *Myllenyxis*) and precisely the same mechanism will not work because it is not possible to bend the cuticle in the plane of the compression (see below). These laterally compressed wavy ovipositors probably use differential pulling of the lower valves in order to effect a twisting of the distal part of the ovipositor. For wavy ovipositors that are not laterally compressed, the mechanism relies on the olistheter mechanism holding the upper and lower valves fairly firmly together. Hence, when an arch of the lower valve is moved relative to the upper valve, the upper valve has to maintain its intact olistheter connection and to do so must bend ventrally. Observations of living *Zaglyptogastra* females show that they can bend their ovipositor through more than 180°, but the extent to which other taxa can flex their wavy ovipositors by this mechanism is completely unknown (see fig. 2A in Quicke & Laurenne 2005).

The exact use of these wavy ovipositors is largely a matter of conjecture. It seems highly likely that they allow the wasp to explore a leaf mine or boring to reach a host that might not be generally accessible. Probably in all these cases the ovipositor is incapable of 'drilling' through a substrate and so access to the host's retreat is through a pre-existing opening. In the case of the braconine *Zaglyptogastra* (Fig. 3.8b), this might be through the frass holes made by a host cerambycid larva in a twig, but in at least some species it seems that the wasps make use of scolytine curculionid emergence holes in dead wood to access the deeper borings of larger wood-borers such as cerambycids (Quicke & Laurenne 2005).

A second type of ovipositor steering mechanism involving a largely longitudinally divided upper valve is widespread in the ophioniform ichneumonids and

Table 3.2 Proposed independent origins of wavy-type ovipositor steering mechanism.

Taxon	Exemplar genera
Braconidae	
Braconinae	*Zaglyptogastra*
	Megalommum group*
	One *Digonogastra* sp.†
	One *Serrundabracon* sp.
Mendesellinae	Some *Mendesella* spp.
Alysiinae	*Terebrebus*
Opiinae	*Doryctobracon*
Doryctinae	One *Heterospilus* sp.†
Euphorinae	*Sinuatophorus* spp.
Ichneumonidae	
Cremastinae	*Pristomerus*
Tersilochinae	*Probles*
Rhyssinae	*Myllenyxis*

**Aphrastobracon* group of genera (previously referred to as the Aphrastobraconina).
†See Quicke and Marsh (1992).

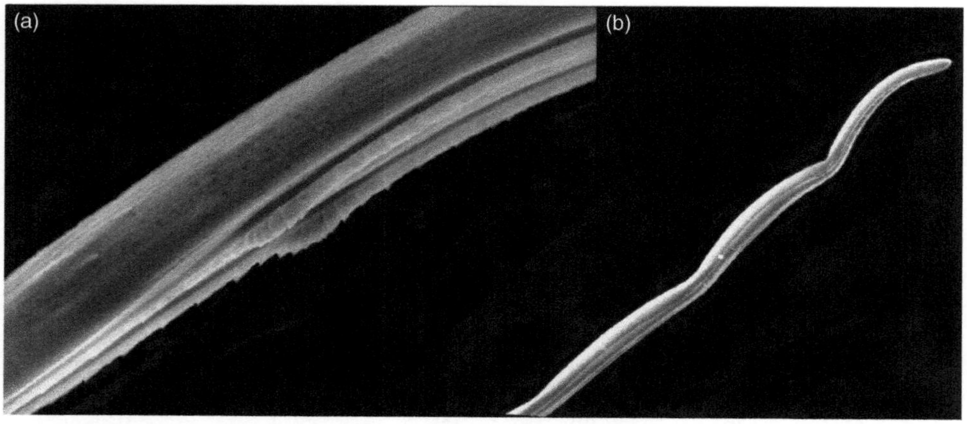

Fig. 3.8 SEM images showing ovipositor steering mechanisms. (a) *Macrocentrus* sp. (Macrocentrinae) with expanded region with coarser scaling on rhachis of upper valve; (b) intact apex of ovipositor of *Zaglyptogastra* sp. (Braconinae).

probably represents a basal (or near basal) adaptation of the clade, although it has been subsequently lost completely in some taxa, especially those with very shortened ovipositors. Upper valves with this mechanism fully developed are divided into two completely separate halves for almost their entire length, the pieces fusing just at the tip. The two halves held together by a third narrow cuticular, U-shaped strip termed either a fifth piece or aulaciform rod (Fig. 3.4h,i) (Quicke et al. 1994). This arrangement is associated with a widened basal piece (Fig. 3.9) and, although no detailed observations of its actions have been made in living wasps, it seems highly likely that differential pulling of the left and right halves of the basal plate will cause, as with a bimetal strip in a thermostat, bending left or right. Observations of living wasps with this mechanism show that they can bend their ovipositor both dorso-ventrally and laterally, being able to transcribe a full circle with the tip. Whether all the three-dimensional movement is associated with the split upper valve or whether some other additional mechanisms might be present to achieve bending in the dorsal/ventral plane is not known. Some ophioniformes have the two upper valve halves connected along their length by a thin cuticle membrane rather than a mechanical locking with an aulaciform rod. It seems likely that these are still capable of some lateral bending, although perhaps not as much.

Taxa with this mechanism fully developed include the Stilbopinae *s.l.* (i.e. *Stilbops* and *Panteles*), long ovipositored banchines (e.g. *Lissonota*, *Glypta*), Cremastinae, Lycorininae, Tersilochinae, phytodietine Tryphoninae (i.e. *Netelia* and *Phytodietus*) and Sisyrostolinae (=Brachycleromatinae). Several of these are part of a grade (along with the Ctenopelmatinae and other Tryphoninae) leading to the 'higher' ophioniformes based on combined morphological and molecular analyses (see Fig. 13.3) (Quicke et al. 2009), and therefore it seems likely that with an evolutionary shift on to exposed hosts and associated shortening of the ovipositor, the need for such a steering-associated adaptation was lost and the base of the ovipositor subsequently simplified. It is not known whether any of the few ctenopelmatines (i.e. Chrionotini) with a long ovipositor may have retained/re-expressed this mechanism.

Three other steering mechanisms have been discovered in the Braconidae (Quicke et al. 1995). Several members of the Macrocentrinae subgroup (*Amicrocentrum*, *Charmon* and various *Macrocentrus*) have the rhachis of the upper ovipositor valve strongly swollen over a short distance pre-apically and with coarser scale-like sculpture (Fig. 3.8a). The aulax is presumably correspondingly widened at the same distance from the ovipositor tip and it is proposed that it works again like a bimetal strip, such that when the lower valve is retracted (effectively shortening the lower half of the combined ovipositor), it causes it to bend downwards, and protraction of the lower valve causes upward bending (Fig. 3.10a). Members of both *Charmon* and *Macrocentrus* species have been observed to be able to bend their ovipositors actively both up and down when handled, although not side to side to any visible extent. The related Xiphozelinae and Amicrocentrinae have not been observed in life and it is not known whether the long ovipositored *Distiriella* (Xiphozelinae) has this mechanism. At least some mild thickening and coarser scaling are also found in some members of the Helconini and it may be that this

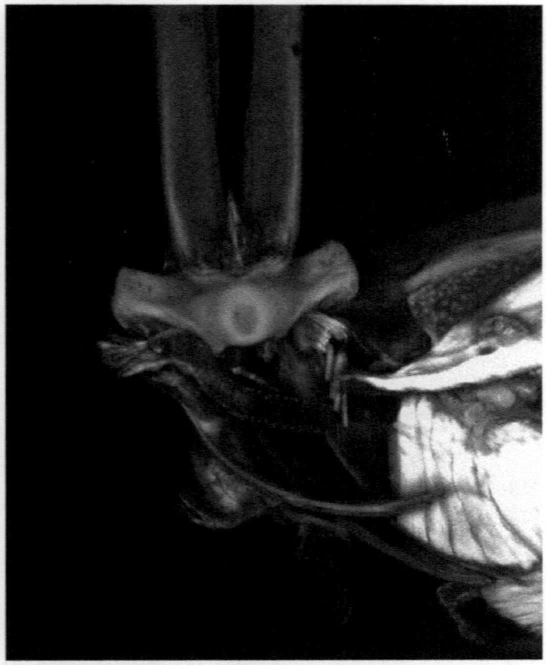

Fig. 3.9 Laser confocal scanning microscope images of the base of the dorsal ovipositor valve of *Glypta* sp. from Korea, showing the transverse basal piece and completely separated halves of the upper valve; the aulaciform rod is visible just to the right of the left-hand valve (Source: reproduced by permission of Istvan Miko.) For a colour version, see Plate 7.

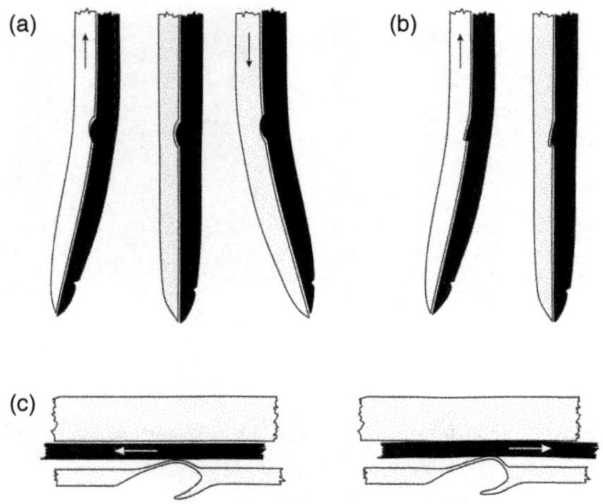

Fig. 3.10 Diagrams showing how three different ovipositor steering mechanisms found in the Braconidae work, with the upper ovipositor valve parts depicted in black. (a) System involving swollen, pre-apical part of the rhachis, with it located centrally within the corresponding part of aulax at rest and showing how retracting and extending the lower valves will cause the ovipositor to bend downwards and upwards, respectively. (b) System with pre-apical, scarped butt on the outer side of the upper valve and corresponding notch on the lower valve, which only permits bending ventrally. (c) Flexible constriction of lateral side of aulax that is believed to be dragged by the rhachis when the lower valve is retracted, allowing some ventral bending. (Source: adapted from Quicke et al. 1995.)

steering mechanism evolved basal to the helconiformes group (see Sharanowski 2011; Fig. 12.2).

Long-ovipositored agathidines such as *Agathis* and *Braunsia* spp. and at least some *Orgilus* (Orgilinae) have a railway buffer-type system with a strongly scarped lateral projection pre-apically on the dorsal ovipositor valve and a corresponding scarped notch on the lower valve (Fig. 3.10b). This mechanism works on exactly the same principle as the preceding one except that it only allows active downward bending.

A third mechanism has been found so far only in the Doryctinae and involves the ancillary tooth or teeth on the lower valves next to the thinned outer wall of the aulax (Fig. 3.10c). The mechanics seem clear, in that retraction of the lower valve, especially if the ancillary tooth is pushed medially, will cause the outer wall of the aulax to push against and restrain/lock the rhachis within the aulax at this point and thus cause the ovipositor to bend ventrally. This has been observed in hand-held individuals of *Heterospilus prosopidis*, so it clearly does not require that the ancillary tooth is being pressed medially by some substrate, but conceivably, during 'drilling' or probing in the host substrate, this might happen.

Finally, I have noted that some museum specimens of *Myllenyxis* have died with one or both lower valves protruding fairly far beyond the apex of the upper valve. In these cases, the lower valve apex curves fairly strongly contralaterally. This is similar to the condition that many non-pollinating fig wasp ovipositors display in dead specimens. I suspect that these also indicate a steering mechanism in which, if one lower valve is pushed into the substrate or into a tunnel first, and then the upper valve, the latter will tend to follow the curved course of that lower valve. Thus left or right steering could be achieved depending upon which lower valve was pushed out first.

In summary, the different steering mechanisms seem to be based on just three physical mechanisms. Those involving apical bosses or apically swollen regions of the rhachis, along with the split upper valve of many ophioniformes, all work in the same way as a bimetal strip, in that shortening of one 'side' will cause the ovipositor to bend in the direction of the shorter side. Wavy ovipositors, at least those that are not strongly laterally compressed, work by the rhachis and aulax having to stay united as one lower valve is pushed posteriorly, which means that the upper valve rhachis has to follow the downward curve set by the aulax and so the ovipositor will curve ventrally. In *Zaglyptogastra* with up to four separate arched regions, each node contributes more or less evenly to the bending, so their effects are additive, which allows such a strong curve to be achieved. Those with the lower valve tips having a strong innate contralateral bend work by the other valve having to follow the course set by the first valve to be pushed posteriorly.

PROPOSED EVOLUTIONARY AND RELATED OVIPOSITOR TRANSITIONS

The earliest fossil ichneumonids and braconids all had ovipositors with lengths that are somewhere between as long as the body and slightly shorter than the mesosoma (see Figs 13.2c and 13.3a), as does the extant braconid genus *Apozyx* (Apozyginae), which seems likely to represent a very basal offshoot (possibly one that would be best classified together with some Cretaceous taxa, notably *Aenigmobracon*†). This suggests that members of the superfamily originally attacked relatively weakly concealed hosts, and this is taken to be the starting point for considering evolutionary transitions (see Fig. 6.9). It also seems most probable that these early ichneumonoids would have been ectoparasitoids or at least their common ancestor at some point on the tree. The ovipositors of these taxa would probably have been rather generalised, just as they are in taxa with that biology today, having some apicoventral serrations and a weak dorsal nodus as in Fig. 3.3a. From this way of life ichneumonoids are taken to have evolved to attack both exposed and deeply concealed hosts, and, on numerous occasions, also evolving the endoparasitoid way of life.

The transition to becoming endoparasitic probably occurred with hosts that were weakly concealed for part of their development but which moved either to new shelters or to some other site to pupate (M.R. Shaw 1983, Gauld 1988a). The biology of the ectoparasitic koinobiont *Rhysipolis* is widely considered to indicate one possible scenario by which endoparasitism could have evolved (M.R. Shaw 1981, 1983). In this wasp, the egg is glued firmly to an intersegmental membrane of the host caterpillar and may hatch partly while its host is still feeding, but remains attached until the caterpillar has moved, prematurely, to a new site to construct its cocoon, at which point the parasitoid starts active feeding. It is easy to see how internal deposition of the wasp egg, if it can overcome the host immune defences and respire, etc., would reduce the risk of becoming dislodged from the still mobile host and also reduce risk of predation that an external egg/larval stage must normally entail. With very few exceptions (some Adelognathinae, some Tryphoninae and *Rhysipolis*), parasitism of exposed hosts is associated with endoparasitism and the laying a rather smaller egg, and endoparasitism is strongly associated with koinobiosis. In Shaw's scenario, the latter came first, but once a wasp is a koinobiont endoparasitoid, there is no reason why the hosts should not be deeply concealed at some point in their life, although attacking them may take place when they are still relatively accessible. Similarly, the ancestral ectoparasitoid attacking weakly concealed hosts can evolve to attack more deeply concealed ones.

Figure 3.3 shows my proposal for how the different morphological types of ovipositor might have evolved based on the biological transitions proposed in Fig. 6.9. Thus, from a relatively unspecialised ectoparasitoid form with a small preapical dorsal nodus and moderate number of lower valve serrations (Fig. 3.3a), the nodus and serrations may become stronger and/or more numerous (Fig. 3.3b,c) as adaptations to 'drilling' through hard substrates, and such features are found in groups such as various Doryctinae, Pimplinae (Ephialtini), Labeninae, Rhyssinae and Xoridinae with deeply concealed xylophagous hosts. Still dealing with ectoparasitoids, if the host is weakly concealed and there is no need for adaptations to cut through a substrate, simplification of the upper and lower valves occurs and, in at least some and possibly many cases, envenomation of the host is achieved by the lower valves alone which completely enclose the egg canal and may be interlocked by a tongue-and-groove system medio-dorsally, an adaptation to preventing venom leaking out from the egg canal dorsally as the lower valves alone penetrate the host (as in Fig. 3.4h; see also figures in Quicke 1991a, Quicke et al. 1994 and Dweck et al. 2008). From these simplified forms, some taxa have evolved blunt-tipped (Fig. 3.3d) and wavy-tipped (Fig. 3.3g) ovipositors that form a steering mechanism and are clearly not capable of 'drilling' through any substrate (see the section *Ovipositor steering mechanisms* above).

Going back to ectoparasitoids with a slightly reduced upper valve nodus (Fig. 3.3b), this often hardly protrudes and anterior to it there is usually a distinct notch which may help retain the ovipositor tip within the host, perhaps to deliver sufficient venom, especially if the host is fairly large (Fig. 3.3h). This arrangement makes an excellent preadaptation for endoparasitism, with the pre-apical dorsal notch being a widespread feature in endoparasitoid braconids and ichneumonids that attack larval hosts. Further development of the notch and narrowing of the tip beyond the notch then act as a feature limiting the penetration of the host by the upper valve (Fig. 3.3i) allowing the latter to be fairly large relative to the fine penetrating tip (Fig. 3.3j). In braconids at least, the apical part of the lower valve in these endoparasitoids becomes modified for releasing the egg just before the tip (see the section *Ovipositor notches and endoparasitism* above) and the egg canal wall of the upper valve near its tip evolves a pre-apical, mid-longitudinal ridge called a sperone

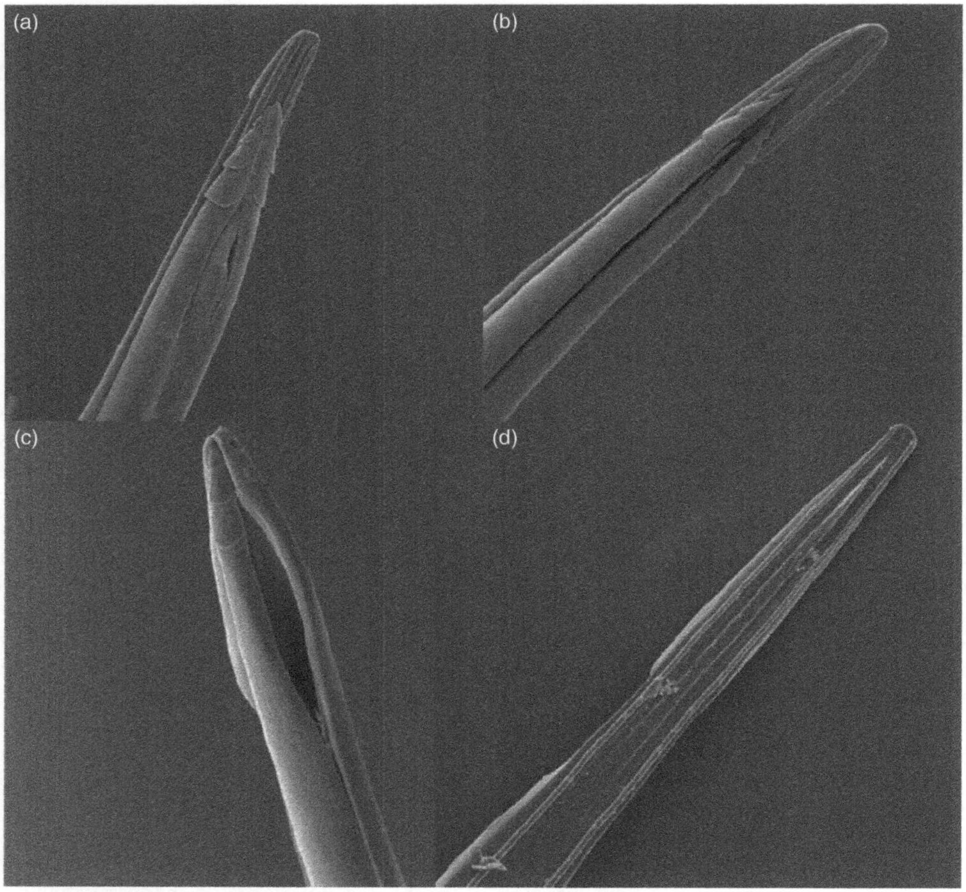

Fig. 3.11 SEM images illustrating features of non-cyclostome braconid wasp ovipositor apices. (a, b, c) Ventral views of intact ovipositors of *Streblocera* sp. (Euphorinae), *Eubazus* (Helconinae) and *Meteorus* (Euphorinae), respectively, the last having one lower valve removed; (d) ventral view of upper valve of *Homolobus truncator* (Homolobinae). All of these wasps show a small, feathery subapical flap at the base of the part where the egg will emerge and all have a medial apical ridge on the apex of the upper valve, called a sperone, which may serve to help push the egg out before it reaches the apex. (Source: Boring et al. 2009. Reproduced by permission of C. Andrew Boring.)

(Fig. 3.11a,b,d), which is believed to help guide or push the egg ventrally out of the ovipositor between the lower valves. From the condition in Fig. 3.3j, several groups show that either the notch may elongate (e.g. Fig. 3.3l) or disappear altogether (Fig. 3.3m). These needle-like tips are often associated with species that attack very early instar hosts, host eggs or, as in the hyperparasitic Mesochorinae, host parasitic wasp larvae within the body of a caterpillar. A few larval endoparasitoids have also evolved wavy ovipositor tips that are used for steering, but these have retained the pre-apical, dorsal notch (e.g. the cremastine *Pristomerus*; see Fig. 13.54).

NUMBER, POSITION AND POSSIBLE FUNCTIONS OF OVIPOSITOR VALVILLI

Valvilli are roughly circular, cuticular flaps that protrude into the egg canal from the wall of the lower ovipositor valves and are present in most ichneumonoids. Although they have been known for many

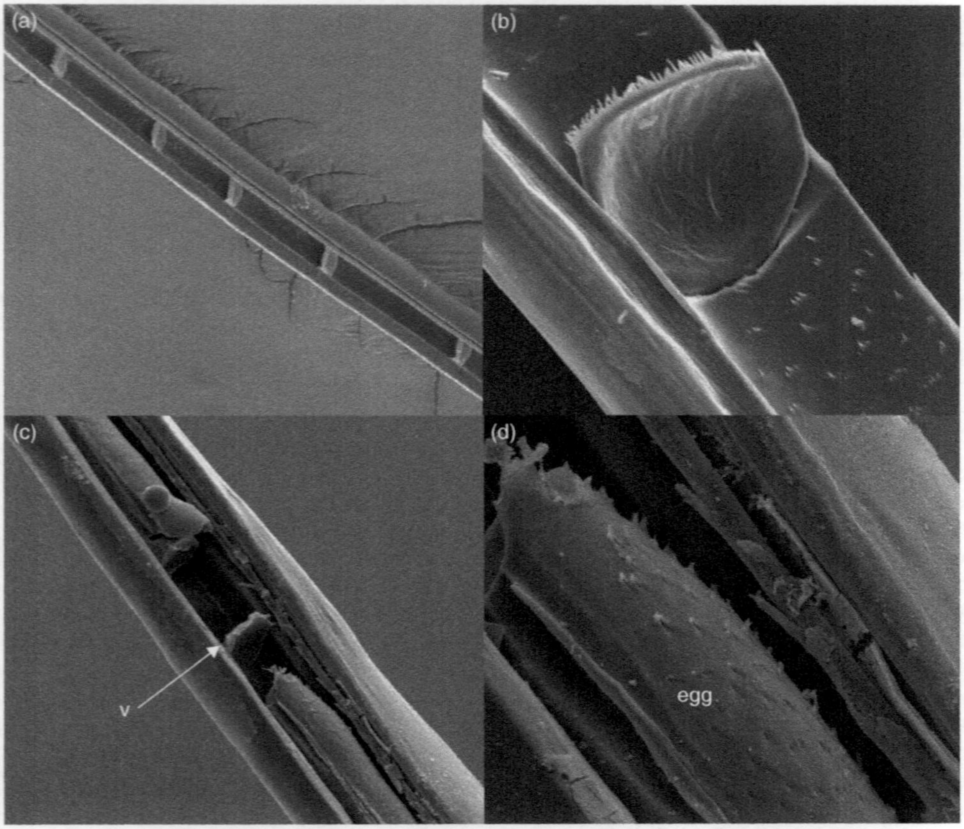

Fig. 3.12 SEM images illustrating features of the non-cyclostome braconid wasp ovipositor internal structure and egg. (a) Multiple valvilli of *Wroughtonia* sp. (Helconinae); (b) detail of one of the two valvilli of *Homolobus truncator* (Homolobinae); (c, d) *H. truncator* egg in position within the egg canal just anterior to the basalmost valvillus and detail of egg chorion showing spines. Also note the posteriorly directed scales on the edge of the aulax and groups of small ctenidial spines on the wall of the egg canal. (Source: Boring et al. 2009. Reproduced by permission of C. Andrew Boring.)

years (Oerser 1961), they have attracted little attention apart from having been considered as a synapomorphy for Ichneumonoidea+Aculeata (Quicke et al. 1992c, 1999a), although they have clearly been secondarily lost in many aculeates and in a few braconids. Recent molecular and morphological studies, however, have provided strong evidence that the ichneumonoids and aculeates are not in fact sister groups, leading to the conclusion that their valvilli must represent independent acquisitions (Sharkey et al. 2012).

In those aculeates that have valvilli, they are always present in a fairly closely opposed pair (rarely three) on each valve (Quicke et al. 1992c), but when present in ichneumonoids there may be a single one per valve or a more widely separated pair (Fig. 3.12c) or there may be up to seven or eight distributed more or less evenly along a long length of the egg canal (Fig. 3.12a). In the aculeates, it seems very certain that their primary role is as part of a venom pumping system and that they operate in a way similar to the piston in a force pump. The morphological differences between them in the two groups could indicate that they serve different functions.

An explicit hypothesis of the function of the pair of valvilli in *Homolobus* (Braconidae: Homolobinae) was presented by Boring et al. (2009), who proposed, although basing their argument on the assumption that valvilli were homologous in the aculeates and ichneumonoids, that they also acted as a venom pump, with the additional effect that the venom would help to push the egg out of the ovipositor. However, it is likely that even if the function is the same, the exact mode of operation might be different in those taxa with several pairs of valvilli along the length of their rather thin ovipositors, such as in the ctenopelmatine egg-larval parasitoid *Pion* and some Helconini, which have rather small eggs and rather long ovipositors (Fig. 3.12a). One possibility is that they help to prevent back-flow of venom and other secretions. The way in which valvilli are curved differs between major taxonomic groups, especially in the Braconidae. In those of the endoparasitic non-cylostomes (e.g. Fig. 3.12b) the margin of the valvillus is fringed and curves towards the ovipositor apex, whereas in ectoparasitic cyclostomes it is recurved and not fringed. Those of the endoparasitic rogadine cyclostomes are fringed but recurved. This suggests that these features are probably functional.

Observations on several taxa have shown that a liquid or a jelly-like substance is expelled from the tip of the ovipositor immediately before the egg appears. In the case of the endoparasitic ichneumonid *Banchus femoralis*, van Veen (1982) surmised that this is from the venom gland; in rhyssines, the jelly like substance also contains fragments of the symbiotic *Amylosterium* fungus that will attack the host tree's cambium (Skinner & Thompson 1960) and therefore probably originates from some other gland associated with the common oviduct. Clearly, in all idiobiont ectoparasitoids that paralyse their hosts and also in numerous koinobiont endoparasitoids that induce temporary host paralysis, the venom is injected into the host prior to egg deposition, sometimes involving a separate sting, so Boring et al.'s proposition that the valvilli pumping venom in *Homolobus* act to expel the egg cannot be true in all cases. I suspect that the primary function of valvilli varies between taxa and they may even have multiple roles in the same species. They almost certainly play a role in venom pumping but, in many idiobiont ichneumonoids with long ovipositors that lay a large egg, there is just a single pre-apical valvillus (Quicke et al. 1992c) and, as the egg is being extruded, this must either lie flat against the egg canal wall as the egg passes by or it might direct the egg to emerge from the ovipositor. Rogers (1972) noted that the campoplegine *Venturia canescens* 'carries a single egg in readiness for oviposition in a cavity right at the tip of the long ovipositor', i.e. just distal to the single valvillus. The loading of the egg into this cavity was called 'cocking' by Rogers and was a prerequisite for oviposition.

Valvilli have apparently been lost in most Aphidiinae, although a basal pair is present in at least some *Pausia* species. They have also been lost or are highly reduced in a few members of the Doryctinae, but their function appears to have been taken over by a highly modified ctenidium (Rahman et al. 1998b).

VENOM RETENTION AND DELIVERY

Venom must pass along the egg canal and only be released at or near the tip. At least in cyclostome braconids, which generally have a heavily muscularised, Type 1, venom reservoir (see Chapter 4, section *Venom gland and reservoir*), and similarly in xoridine ichneumonids, envenomation almost certainly involves contraction of reservoir muscles, forcing the venom down an often narrow and long lumen. This is probably true also of most other taxa, although the degree of reservoir muscularisation is somewhat less and also the reservoir muscles are not innervated to the same extent (Edson & Vinson 1979, Edson et al. 1982), perhaps suggesting that very quick action is not involved. Especially in those taxa in which venom appears to be forced along the egg canal under some pressure, there need to be adaptations to prevent the lower valves from simply being forced apart and the venom leaking out early. Nearly all Apocrita have one such adaptation, namely the orientation of the olistheter mechanism, which is usually either directly vertical or more usually ventrally converging, so that the lower valves are not so free to rotate laterally (Quicke et al. 1994). This contrasts with the majority of sawfly ovipositors in which the olistheter mechanism diverges ventrally, perhaps as an adaptation to allow distortion during passage of the egg.

Many wood-borer parasitoids have robust ovipositors with fairly thick chitin walls, meaning that distortion is restrained (Fig. 3.4c,d,f), but some, even among the cyclostome braconids, have flimsier looking ovipositors. Most braconids have a thin, recessed flap on the medial ventral part where they abut, and at rest these flaps are touching. Quicke et al. (1994) proposed that these flaps form a seal and that when venom pressure within the egg canal increases, it will act on the lateral side (i.e. the side next to the recess) of each flap, so pushing them medially and maintaining the seal. This will only work well when the lower valves are constrained not to be able to separate too far. A similar thin seal is present in some aculeates. Most chalcidoids, proctotrupoids and platygastroids have asymmetric, overlapping lower valves, such that when forced wider apart by venom/egg pressure, the overlapping parts are forced together more firmly, but the only ichneumonoid found with a similar mechanism is the Aphidiine *Ephedrus* (see fig. 13.32 in Quicke et al. 1994). Many ophioniformes Ichneumonidae have large, symmetric, medio-ventral, articulated flaps originating from the medio-ventral edge of each lower valve (Fig. 3.4g–i), and these are hypothesised to allow, in combination with the medio-dorsally weakly connected or divided upper valves, a great deal of expansion of the egg canal during oviposition/envenomation while, in a way similar to the braconid flaps, maintaining a seal. With the egg canal thus expanded, the pressure of the venom would be reduced.

At least some members of the cyclostome subfamily Rhyssalinae (including *Histeromerus* and *Thoracoplites*, but less so in *Rhyssalus*) have the upper valve halves connected medio-dorsally by a membrane (Fig. 3.4a), which might allow them to separate more when egg canal pressure increases and possibly thereby increasing the ventral convergence of the olistheter mechanism and the approximation of the lower valves. In *Histeromerus* in particular (see fig. 160 in Quicke et al. 1994), the ovipositor appears rather flimsy and ill-adapted to retain venom under pressure and, indeed, in these the large egg is thought to emerge from the ovipositor near its base. Their lower valves are very deep, which gives the egg canal a large surface area and large capacity. Further, rhyssalines do not have the tyical heavily muscularised venom reservoirs typical of other cyclostomes (see fig. 12.53 in Quicke & van Achterberg 1990), so actually forcing venom into the host might be accomplished primarily by the pumping action of their valvilli without the need to withstand high venom pressure within the egg canal over its whole length.

In many idiobiont ectoparasitoids, although it has not been extensively investigated, the lower valves abut both medio-ventrally and medio-dorsally near the tip, the dorsal part having a dovetail interlocking arrangement connecting the two valves (see fig. 2.3 in Quicke 1991a and figs 13.23, 13.27 and 13.29 in Quicke et al. 1994), which will prevent venom from escaping dorsally when the two lower valves together pierce the host and inject it. This is also true at least of some endoparasitic Rogadinae (Fig. 3.4b), which, unlike many endoparasitoids, have fairly large eggs, so this might also be associated with increasing resistance to distortion during oviposition.

OVIPOSITOR SECRETORY PORES

Some authors have claimed that there are external secretory pores, usually near the apex of the ovipositor, in several parasitic wasps, based on SEM observations (notably Nénon et al. 1995, 1997, Shah 2012, Ahmed et al. 2013b). To date there have been no definitive TEM studies showing glands, largely because of the aforementioned difficulty of cutting good sections through this material and the possibly sparse distribution of the putative pores. Most convincing are the micrographs of the ovipositor of the rhyssine *Megarhyssa* by Nénon et al. (1997), which show a possible secretory product oozing from pore-like structures along the length of the shaft. Although some TEM images were also presented that show some narrow holes running through the chitin from the lumen of the valve, they do not appear to originate from glandular tissue. As the lumens have to accommodate trachea and numerous nerves, any gland tissue would have to be rather small and also, given the constraints posed by diffusion, the amount of secretory product that they could release would likely also not be very voluminous. I find it difficult to imagine that any contribution of any enzymes released by such pores to a probably almost dry substrate could be significant. First, in terms of wood penetration, the postulated gland pores are apparently located along the length of the shaft and the place that ought to be softened might be anticipated to be near the tip. Second, it seems unlikely that sufficient haemolymph could be passed along the lumens of very long, thin ovipositor valves, such as those of rhyssines (see Fig. 13.25), to produce a useful amount of any lubricant secretion. Third, the proposal that the trans-cuticular pores observed by Nénon et al. (1997) might just connect to the intra-lumimal (i.e. haemolymph-filled) space seems highly unlikely – pores leaking haemolymph directly to the outside world would both open the way to infection and become blocked by the normal anti-bleeding properties of haemolymph. For a long ovipositor penetrating in a dry(ish) substrate, such as when *Megarhyssa* is 'drilling', adding liquids with enzymes that could soften what is already 'drilled'-through wood could even be positively disadvantageous, with the liquid likely to make the boring more sticky and therefore harder to slide the ovipositor valves within.

Given that the egg canal is joined by glandular ducts from Dufour's gland, the venom gland, the oviducts themselves and, in some taxa, including wood-borer parasitoids belonging to the Xoridine, Labeninae and Rhyssine, vaginal glands, it seems more probable that any lubricants or even specific wood-softening or mud-softening enzymes that might be released during 'drilling' would be produced by these rather than being released through any tiny (and not very abundant) pores on the ovipositor. A critical aspect to look at would be the water content of the 'typical' wood or other substrate being 'drilled' through by the various species. With *Pimpla* this is probably often a dry (even crispy) twig or grass stem or similar, and it not be soggy because that would probably negate their echolocation of prey – indeed, experimental research on them has mimicked the host substrate with paper or cardboard. They have no difficulty penetrating that with no lubrication.

Polidori et al. (2011) noted that as the ovipositor of the cryptine *Acrorictus seductor* penetrated the nest of its mud-dauber host (Sphecidae), the penetration site showed a dark halo that persisted for a few minutes after the ovipositor had been withdrawn, suggesting that the ovipositor had released a secretion to help soften this particularly hard substrate. Their SEM images, however, showed no evidence of pores and it seems almost certain that the fluid softening agent simply flows down the egg canal, especially as a fairly large amount of fluid must have been involved.

A very important aspect of parasitoid ovipositors that has never been researched is that their internal surfaces must be highly hydrophobic (even towards the presumably partly surfactant venom/Dufour secretions) because, if not, and given the incredible power of capillary attraction, once any liquid had passed down them in their first envenomation/oviposition, it would stay in the ovipositor and would be difficult to drag back completely into the oviduct by suction. SEM images of fresh ovipositor parts, including those from wasps that have already oviposited, do not normally show large amounts of matter adhering to the inner surfaces of the valves, suggesting that when not in use the ovipositor egg canals of many species are probably air-filled. Again, although it would seem an easy aspect to explore, no data have been reported.

OVIPOSITOR SENSILLA

In all parasitic wasps studied thus far, the upper and lower valves all bear some sensilla. There have been a number of moderately detailed SEM and, less frequently, TEM studies of these (Table 3.3) and it is possible to make some generalisations.

The commonest sensilla near the ovipositor tip are coeloconics, i.e. peg-like structures, lacking obvious pores, sitting in a small pit, and which most probably are hygroreceptors in this situation. Styloconic and campaniform sensilla are also common. LeRalec et al. (1996) found that in parasitoids of deeply concealed hosts, with consequently relatively long ovipositors,

Table 3.3 Summary of studies of ovipositor sensilla morphology.

Family/subfamily	Genus, species	SEM	TEM	Ref.
Braconidae				
Braconinae	*Habrobracon hebetor*	+		Dweck et al. (2008)
Doryctinae	*Spathius agrili*	+		X.-Y. Wang et al. (2010)
Macrocentrinae	*Macrocentrus cingulum*			Ahmed et al. (2013b)
Microgastrinae	*Cotesia flavies, sesamiae*	+		Obonyo (2011)
Opiinae	*Diachasmimorpha* (as *Biosteres*) *longicaudata*		+	Greany et al. (1977)
Orgilinae	*Orgilus lepidus*		+	Hawke et al. (1973)
Rogadinae	*Aleiodes esenbeckii*	+		Y.-P. Wang et al. (2008)
Ichneumonidae				
Campopleginae	*Eriborus applicitus*			L. Zhang et al. 2014
	Venturia canescens	+	+	Ganesalingam (1972), Shah (2012)
Rhyssinae	*Megarhyssa atrata*	+	–	Nénon et al. (1995, 1997)
				LeRalec et al. (1996)

sensilla are usually distributed along much of the length of the ovipositor, whereas in taxa attacking exposed hosts, they tend to be concentrated at the ovipositor tip. Nevertheless, whereas putative mechanosensory sensilla are usually distributed along the length of the ovipositor, chemosensory ones are concentrated at the apex, as would be expected since they are involved primarily with host assessment. Nearly all the ovipositor chemosensory sensilla appear to have a single pore, although Hawke et al. (1973) found that the sensilla of *Orgilus lepidus* (Orgilinae) had more than one chemosensory cell, and also a mechanosensory cell associated with an internal cuticular projection. The surface of the ovipositor of *Megarhyssa* has many small, asymmetric 'cuticular formations' (Nénon et al. 1997), but it is not known whether these are sensory or just have a tribological role, i.e. associated with their frictional interaction with substrate.

Towards the base of the upper valve there is often a row of trichoid sensilla that are believed to provide sensory feedback about the valve's movements and, at least in aphidiines, which have fairly thin-walled ovipositors, there are campaniform sensilla with distinct tubular bodies within the cuticle and their associated scolapidial neurones are clearly demonstrated by TEM, although they are usually not visible externally (Quicke et al. 2000a). Effectively nothing is known about the numbers or taxonomic or physical distribution of chordotonal organs in the Ichneumonoidea, so no generalisations can be made.

OVIPOSITOR SHEATHS

When a wasp is not ovipositing, the ovipositor is usually completely concealed by a pair of sheaths, which are concave medially. The outer surface is almost always densely clothed with setae, and the inner surface of the ovipositor sheath is always densely adorned with setiform or comb-like microsculpture. Vilhelmsen (2003) discussed the evolution of a flexible ovipositor sheath in the Hymenoptera, which is indicated by transverse ridges. Within the Ichneumonoidea, all those species with long ovipositors have at least a substantial part of the sheath flexible, but in groups with short ovipositors, such as the Aphidiinae, these may take the form of an inwardly concave triangular plate. A few taxa display an intermediate condition with the basal part of the sheath transversely ridged and flexible, but with the apical one-third or so forming a rigid plate. In some wood-borer parasitoids, they appear to help support the ovipositor during 'drilling' (see Fig. 13.25), thus reducing the likelihood of buckling or perhaps keeping it on course, although this is by no means always the case (see Figs 13.6a and 13.19d).

The apex of the sheaths in parasitoids of an exposed host in particular are often furnished with sensilla, although very little work has been carried out on their possible functions in host location or acceptance. In the Microgastrinae, concentration of the sheath sensilla (and indeed all setae) at the apex of the sheathes is one of the characters that define the 'macrolepidoptera suite' (W.R.M. Mason 1991, A.K. Walker et al. 1990a) and suggests that they are probably important in sensing external cues from potential hosts. Many microgastrines have peg-like or campaniform sensilla at the apex. Species of the microgastrine genera *Diolcogaster* and *Buluka* have a large spatulate sensillum at the apex of each sheath (Austin 1989); these wasps attack Immidae, although it is not known at what stage, and the gross morphology of this sensillum suggests that it may be gustatory. Labeyrie (1960) found in a simple experiment that cutting off the sheath tips in the ichneumonine *Diadromus* disrupted host selection. That they are probably important in sensing features of hosts seems likely for all taxa attacking exposed hosts. Whether the sheathes have a sensory function in groups that do not attack exposed larvae is unknown.

Chapter 4

INTERNAL AND REPRODUCTIVE ANATOMY

The Braconid and Ichneumonid Parasitoid Wasps: Biology, Systematics, Evolution and Ecology, First Edition. Donald L. J. Quicke.

Given the size of the group, there is an amazing lack of detailed studies of adult ichneumonoid internal non-reproductive anatomy, a fact noted by Given back in 1944. It has hardly changed since then and one remarkably good comparative study on ichneumonid internal metasomal anatomy published 100 years ago by Pampel (1914) is still useful today. Given (1944) presented a detailed description of the external and internal anatomy of the ichneumonine *Diadromus collaris*, although he did not describe any musculature. Other important studies are those of Alam (1954, 1955) on the Asian braconine *Stenobracon deesae*, Shaumar (1966) on various *Pimpla* and *Itoplectis* species (Pimplinae: Pimplini) concentrating on the nervous system but also including considerable information on the muscles, Vinson (1969) on the gut and internal reproductive organs of *Toxoneuron* (as *Cardiochiles*) *nigriceps* and Benham (1972) on the campoplegine *Eriborus molestae*.

NERVOUS SYSTEM

As with all insects (and indeed protostomes), the brain comprises a ring of ganglia that encircle the oesophagus in the head and a paired ventral nerve cord with segmental ganglia running from it to the posterior of the body. Shaumar's detailed morphological study of the nervous system of pimplines can provide a basis for study of the ganglia and nerves in other species (Shaumar 1966). There are three separate thoracic ganglia and six abdominal ones, the most posterior clearly being the result of fusion of the last few.

In gross terms, the brain comprises the dorsal cerebrum which receives nerves from the ocelli, the lateral paired optic lobes, the paired deuterocerebrum which receives input from the antennae and also the labrum and ventrally the suboesophageal ganglion with nerves running to the pharynx, mandibles and maxillary and labial palps.

Detailed morphological and functional-morphological studies of the brain have really been started only recently, in order to understand the neural bases underlying interspecific behavioural differences, especially in the microgastrine braconids. Important features are the olfactory bulbs that receive input from the chemoreceptor neurons in the antennae, with their complex groups of glomeruli each corresponding to inputs from receptors for particular molecules. Work on these structures has been greatly facilitated by advances in laser confocal scanning microscopy, with two excellent and detailed studies to date, the first by Smid et al. (2003), who compared one generalist/specialist pair of *Cotesia* species, and a more recent one by Das and Fadamiro (2013), who demonstrated both interspecific and marked sex-linked variations in the glomeruli in two other microgastrines representing the genera *Cotesia* and *Microplitis*, again, one a generalist the other a specialist (see Chapter 8, section *Associative learning*). Another important brain structure in parasitoid wasps, the mushroom bodies, are involved in associative learning, and it is significant that they are far more developed in parasitoid wasps than the phytophagous sawflies (Farris & Schulmeister 2011). However, almost nothing is known about their detailed organisation and role in the Ichneumonoidea apart from the *Ophion* sp. (Ophioninae) in the aforementioned study, which had deep calyces subdividing the lobes into a number of sub-compartments, similar to those found in aculeates.

DIGESTIVE TRACT

From the above mentioned work, the digestive tract can be separated into six regions, named in sequence largely like the parts of the gut of vertebrates. Anteriorly, the paired salivary gland ducts unite into a fairly long common duct that opens into the oral chamber. Unlike larval salivary (labial) glands, those of adults, which are located in the mesothoracic region, are branched (ampulate). The fore gut comprises a simple, narrow oesophagus which expands posteriorly to form a crop, followed by a short, narrow proventriculus connecting it to the ventriculus or mid-gut. The proventriculus acts as a one-way valve, being a narrow, posteriorly directed tube with strong opposing longitudinal and radial muscles and triradial symmetry. The ventriculus or mid-gut is endodermal and is where food absorption takes place, and correspondingly the cells lining it possess villi (brush border). It has hardly ever been studied in detail in the Ichneumonoidea, although Gonçalves et al. (2013) have shown that in the campoplegine *Campoletis flavicincta* there are differences in the cells from anterior to posterior along it, suggesting a gradient in absorptive functionality. Cells secreted at its anterior fold are principally responsible for secreting the thin, chitinous, peritrophic membranes and these membranes serve to separate 'raw' food substances from the villi-lined absorptive/digestive cells of the ventriculus wall. The ventriculus posteriorly connects with the hind gut or proctodaeum, which being of ectodermal origin is chitin-lined and comprises two parts, the first being sac-like and distensible and the second the rectum proper, which has from two to many rectal pads. The Malpighian tubules empty into the anterior part of the hind gut.

The histology of the rectal pads has been described by Alam (1954) and Vinson (1969). In the Braconidae, there are nearly always four pads arranged circumferentially; they may be more or less circular or very elongate (see fig. 12.38 in Quicke & van Achterberg 1990) and only a very few taxa have other numbers: the known exceptions are *Aphaereta minuta* (Alysiinae), *Monitoriella* (Doryctinae) and *Histeromerus* (Rhyssalinae) (see fig. 12.35 in Quicke & van Achterberg 1990), both with only two, and *Euurobracon* (Braconinae), which have many (>20) small circular ones of various sizes. In the Ichneumonidae there is far greater variation in both pad number and their arrangement. The majority of species have six, typically arranged in two off-set circumferential rings of three (Pampel 1914, Quicke et al. 2009). Numbers >10 characterise the Ophioninae and are also found in *Coleocentrus* (Acaenitinae), *Metopius* (Metopiinae) and *Dolichomitus* (Pimplinae) (Quicke et al. 1999a). In general, very small-bodied taxa are the ones most likely to have reduced numbers whereas the aforementioned ichneumonids with large numbers are nearly all fairly large wasps.

Salkeld (1967) studied the excretory system of larvae, pupae and recently eclosed adults of the alysiine braconid *Aphaereta pallipes* using light microscopy. In this species, prior to excretion, urates were found to be stored in the larval fat body and other inorganic waste salts in cells of the mid-gut, and these were mobilised ultimately to be excreted in the meconium at adult egression. The paucity of detailed anatomical studies of the gut means that it is impossible to know whether features such as the relative sizes of structures or the numbers and sizes of Malpighian tubules correlate with adult diet, such as whether or not the wasps feed on sugars, proteins, etc.

FEMALE INTERNAL REPRODUCTIVE SYSTEM

The female internal reproductive system in ichneumonids was first described in an excellent anatomical paper by Pampel (1914), which revealed considerable variation between the systems of the 52 species examined. This sort of comparative study was continued in a series of papers by Iwata (1958, 1960, 1966a), also on Ichneumonidae, and Iwata (1959) on the Braconidae. Cummins et al. (2011) provided a useful commentary on the history of the use of ovary features in ichneumonid phylogeny.

OVARIES

There is considerable and important variation in the ovaries in different ichneumonoid taxa that largely reflect biology rather than taxonomy. Each of the pairs of ovaries is composed on one or more ovarioles and the anterior ends of the ovarioles are connected to suspensory ligaments. At the anterior part of each ovariole is the germarium, where clusters of germline cells differentiate to form the nurse cell and oöcyte units that pass posteriorly as the egg matures, with the nutrients passed to it from the nurse cells, also sometimes called trophic cells or trophocytes. This more posterior part of the ovariole is sometimes called the vitellarium. Nurse cell differentiation was described in detail by Cassidy and King (1972) for the braconine *Habrobracon hebetor* (as *juglandis*) and by Sedlag (1957) for the Aphidiinae, and, using more modern protocols, by F. Huang et al. (2008) for the campoplegne *Diadegma semiclausum*. The degree of activity of the germarium in adult wasps varies greatly depending upon whether egg maturation is completed before emergence or continues for at least a part of adult life. These strategies were called pro-ovigenic and synovigenic, respectively, by S.E. Flanders (1950), although there is a spectrum of intermediates between the two (see the section *Time scale of egg maturation* below). In strictly pro-ovigenic species, the germariums may be hardly distinguishable (Fig. 4.1b), whereas in synovigenic species, there may be a large, even somewhat swollen, anterior germarium where cell division to produce new eggs is still taking place (Fig. 4.1a). Posteriorly, the ovarioles on either side unite to form a pair of lateral oviducts of variable length (see Figs 4.1 and 13.12; *cf.* Fig. 13.34) where mature eggs are stored in most species. The walls of these in microgastroid braconids and both campoplegine and banchine ichneumonids are swollen to form the calyx glands, which are the site of polydnavirus or virus-like particle production (see Chapter 9, section *Polydnaviruses*, and

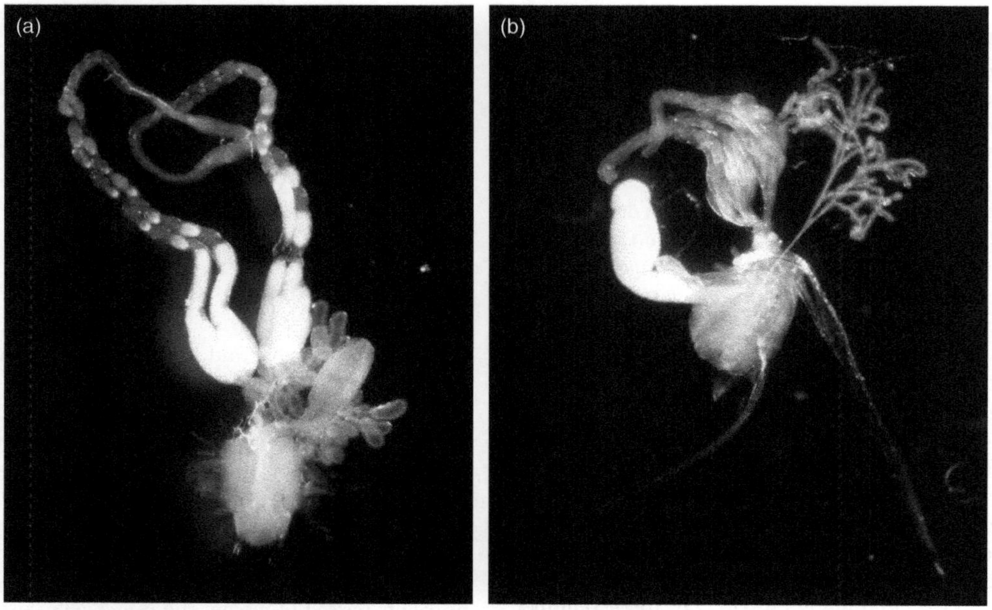

Fig. 4.1 Dissected female reproductive systems of cyclostome and non-cyclostome braconids. (a) *Aleiodes* sp. (Rogadinae), showing two pairs of synovigenic ovarioles and (on the right) the highly muscular Type 1 venom reservoir with branched globular venom glands; (b) *Meteorus* sp. (Euphorinae) showing multiple ovarioles per ovary, branched tubular venom gland on the right and venom reservoir and Dufour's gland on the left. (Source: photographs by Marie Frayssinet, reproduced with permission.)

Fig. 13.34). In addition to particles, their secretions have been found to contain other physiologically active compounds (e.g. Beck et al. 2000a).

TIME SCALE OF EGG MATURATION

As a general rule, idiobionts are more synovigenic that koinobionts and have relatively few fully mature eggs at any one time. However, many, perhaps the majority of koinobionts show some egg maturation after emergence, but this is normally nearly complete after 2–3 days (Fig. 4.2) (e.g. Khatri et al. 2009). In the case of the campoplegine *Bathyplectes curculionis*, females emerge with completely non-functional ovaries with individual ovarioles not visible and no eggs apparent, and even the calyx region is hardly distinguishable (Dowell 1978), but after only 1 day, the ovarioles have differentiated and egg production is happening at a rate of approximately 40 per day. Jervis et al. (2001) introduced the term 'ovigeny index' to express the proportion of mature eggs that are available to be used soon after eclosion and defined it as the number of

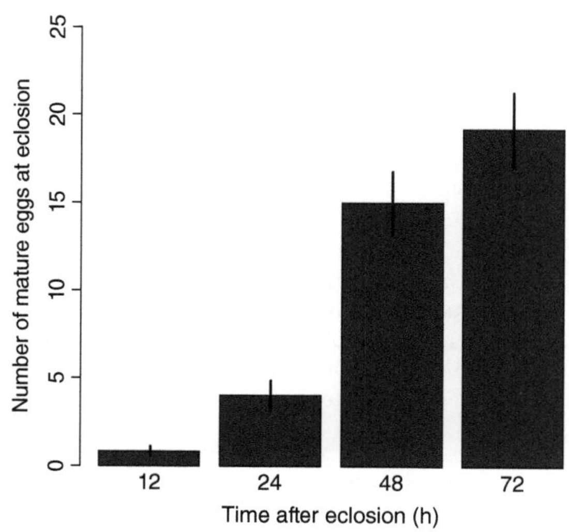

Fig. 4.2 Maturation of eggs (±SE) in the campoplegine *Diadegma semiclausum* in relation to time after eclosion. (Source: data from Khatri et al. 2009).

mature eggs at eclosion divided by the total lifetime potential fecundity. As expected, there was a continuum of values ranging from zero (i.e. no mature eggs at emergence) to one (all eggs that are ever going to be matured are mature at eclosion) (Table 4.1). Very protracted preoviposition periods are known for a few species, including both idiobionts and koinobionts. Females of one species of the doyctine genus *Spathius* take approximately 14 days to mature their eggs (Ayyar & Narayanaswami 1940), the tryphonines *Phytodietus obscurus* (as *pulcherrimus*) 15–21 days (Simmonds 1947b), and some *Netelia* species (Vance 1927) and the pimpline *Exeristes ruficollis* require nearly 1 month (Juillet 1959).

Although it is expected that idiobionts will have a low ovigeny index because of the reasons explained in Chapter 6 (section *Idiobiont and koinobiont strategies*), what is surprising is that the index is fairly or very low in many koinobionts (Table 4.1). The issue here, however, is really whether, as defined, the ovigeny index is a very informative measure. Egg maturation in many of these koinobionts does not require that the adult female does any feeding, relying only on redistributing resources already in her body, and in any case, is usually more or less completed after 2–3 days (Fig. 4.2) (e.g. Salkeld 1959, Khatri et al. 2009), whereas in many idiobionts, egg maturation continues for weeks or months. A more informative value might be something like a 2-day ovigeny index, where the number of mature eggs present after 2 days is compared with lifetime fecundity. Very few adult ichneumonoids have such short lives that this would be compromised (although see *Agriotypus*). Of course, total lifetime fecundity is often influenced by a wasp's access to food, be that carbohydrate sources or from host feeding. Rather counterintuitively, Hegazi et al. (2013) found that females of the microgastrine *Microplitis rufiventris*, although having lower lifetime fecundities, if not provided with a carbohydrate food source actually matured eggs more rapidly than fed wasps. Most likely, this is an adaptive response to potentially unfavourable environmental conditions in which the wasp might not have a long life expectancy and therefore it would be advantageous to be able to lay more eggs quickly before their ensuing demise.

From Table 4.1 and much additional data, it seems that although koinobionts are more pro-ovigenic than idiobionts, very few have an ovigeny index of 1.0. Ellers and Jervis (2004) asked the question of why so few parasitoids are pro-ovigenic. After modelling the effects of a number of parasitoid and habitat factors, they concluded that strict pro-ovigeny should be favoured when one or more of three conditions are met, viz. small body size relative to travel costs, low variation in the rate of host encounter and when eggs are large relative to other resources. Another aspect, however, might be that it takes a day or a few days to mature all the eggs that a female is going to produce in her adult lifetime, but if she does not need them all in one go, there might be an advantage to eclosing earlier with fewer than the maximum number of mature eggs available for oviposition so they can be used on those hosts that are initially available.

SPERMATHECA

The spermatheca in ichneumonoids tends to be a rather small, simple, spherical structure with a pair of associated glands, each with a small separate reservoir and a short primary or main duct connecting it to the anterior end of the vagina. An exception occurs in the aphidiine genus *Lysiphlebus*, in which only a single gland is present (De Marzo 2000). Spermathecal structure and function have received very little attention and I could find only two detailed studies. The largest survey is that of Chumakova (1968), who illustrated the anatomy of the system in a variety of Hymenoptera, including three species of *Cotesia* (as *Apanteles*) (Microgastrinae), *Habrobracon hebetor* (Braconinae), *Diadegma* (as *Horogenes*) *armillatum* (Campopleginae), *Gelis* sp. (as *Hemiteles*) (Cryptinae), *Ctenichneumon* (as *Amblyteles*) *panzeri* (Ichneumoninae) and *Pimpla turionellae* (as *examinator*) (Pimplinae). The microgastrines were most different from the other taxa in the degree of ramification of the gland ducts and, in some cases, the way in which their gland ducts joined the base of the reservoir separately and without an intervening short segment of primary duct. Also, the microgastrines do not appear, from Chumakova's illustrations, to have any reservoir in the glands themselves. Gerling and Rotary (1974) described the entry of sperm into the reservoir for the small braconid *H. hebetor* and Rojas-Rousse and Palevody (1981) presented some observations for the ichneumonid *Diadromus pulchellus*. In *Diadromus*, the spherical sperm reservoir is attached by muscles to the spermathecal duct, which runs from the spermathecal gland to the oviduct, and contraction of these may cause a kink in the short, narrow duct leading to the reservoir, thus precluding

Table 4.1 Ovigeny index*.

Species	Subfamily	Idiobiont/koinobiont	Ovigeny index
Ichneumonidae			
Bathyplectes anurus	Campopleginae	K	0.00
Bathyplectes curculionis	Campopleginae	K	0.00
Pimpla turionellae	Pimplinae	I	0.00
Braconidae			
Aphidius picipes	Aphidiinae	K	0.65
Aphidius rhopalosiphi	Aphidiinae	K	0.70
Asobara tabida	Alysiinae	K	0.3
Biosteres arisanus	Opiinae	K	0.33
Biosteres vandenboschi	Opiinae	K	0.06
Cotesia flavipes	Microgastrinae	K	1.00
Cotesia plutellae	Microgastrinae	K	0.32
Microconus hyperodae	Euphorinae	K	0.00
Microplitis croceipes	Microgastrinae	K	0.37
Orgilus obscurator	Orgilinae	K	0.62
Trioxys complanatus	Aphidiinae	K	0.75

*The Ichneumonoidea data used by Jervis et al. (2001), kindly supplied by Mark Jervis.

outflow of sperm. In *Habrobracon*, Gerling and Rotary found gland reservoirs to be empty just after copulation and it seems likely that their contents were mixed with the spermatozoa in the sperm reservoir; after a few days, the gland reservoirs were seen to have refilled. In the aphidiines, there is very marked variation in the degree of muscularisation of the main duct (De Marzo 2000). Gerling and Rotary observed changes in the sperm reservoir wall as females aged and suggested, probably correctly, that its function becomes impaired, which would agree with the common observation that the proportion of male progeny increases as female parasitoids get older.

Spermathecal colour varies considerably in both families, being whitish/pale yellow or very dark red-brown/black with a few intermediates, and shows considerable phylogenetic constraint. The significance of this colour variation is not clear, but it is noteworthy that in the Braconidae the cyclostomes, including the basal Rhyssalinae, have pale spermathecae (Quicke & van Achterberg 1990, Belshaw & Quicke 1997) and in the Ichneumonidae it is pale in the ichneumoniformes and most of the 'higher' ophioniformes, although with rather more exceptions than in the braconids. This distribution does seem to rule out the idea that dark pigment might provide some sort of UV protection in those taxa with longer adult life-spans and therefore more exposure to UV radiation, but of course, no-one has examined the UV absorbance/reflectance of the pigments so it could be the other way around. If the wasps have pale exoskeletons, another possibility might be that the darker spermathecae warm up more quickly, but any such effect would be very small and unlikely to affect sperm activity or survival significantly.

COMMON OVIDUCT AND VAGINAL GLAND

The oviduct itself is usually rather thin walled and not strongly muscularised and is generally unremarkable, as a medical doctor might put it. The anterior of the common oviduct in some ichneumonids (principally Xoridinae, Labeninae, Pimplinae, Poemeniinae, Rhyssinae and Orthopelmatinae) has a large single- or, more usually, bi-lobed pouch that appears to be glandular and is termed variously the vaginal, uterine or collaterial gland (see Poemeniinae) and whose functions are only partially understood, and indeed, only for the endoparasitic Pimplini. Osman (1978) showed that the product of these glands in *Pimpla turionellae* had anti-encapsulation properties on nylon pieces injected into host pupae and also inhibited the formation of pseudopodia in various classes of host haemocyte, and in the same species, Führer (1973b) found that both the anterior of the common oviduct and in particular the vaginal glands produce copious amounts of an Alcian Blue-positive, mucopolysaccharide which coats the wasp's eggs just before they are laid, and thought that it seems likely that these protect the egg from the host's immune response. The ultrastructural characteristics of the gland change as the female wasp matures and are fully developed after about 3 weeks (Blass & Ruthmann 1989). Four distinct cell types were apparent. The distribution of developmental states of the ichneumonidae is surveyed in the data matrix of Quicke et al. (2009). Although somewhat reduced, a distinct single-lobed gland is also present in some basal ophioniformes such as Stilbopinae, Lycorininae and some Banchinae. Its secretory contents were analysed by Osman and Führer (1979) in *Pimpla turionellae* and were found to contain a mixture of hyaluronic acid and lipoprotein. Such compounds might be well suited to act as lubricants, so it would be interesting to know whether the same sort of chemical profile occurred in the glands of ectoparasitoids. Other possible roles might be to produce a glue to help ectoparasitoids stick their eggs on their hosts, and it is worth noting that these glands are particularly massive in Adelognathinae, which do glue their large eggs on to the outside of a non-paralysed host. It would be well worth revisiting the secretions of these glands using more up-to-date analytical methods. There is no equivalent gland known in the Braconidae.

Posteriorly, where the oviduct approaches the base of the ovipositor (where it ends at the base of the egg canal), its inner surface becomes furnished with strong, posteriorly directed spinose which presumably aids in getting the egg to move posteriorly and prevent it from moving antigrade.

VENOM GLAND AND RESERVOIR

Ichneumonoid venom apparatus varies considerably in gross morphology, but essentially comprises a tubular or globular, usually branched, gland region that usually opens directly into a reservoir, which then connects via a primary venom duct to the base of the ovipositor In at least some aphidiines, the primary duct actually runs along part of the length of the upper ovipositor valve before opening directly into the egg canal (Quicke et al. 2000a). In some cyclostome braconids (Braconinae, Doryctinae, some Rogadinae and Rhyssalinae), the venom glands join the primary venom duct directly and venom must presumably flow retrograde to fill the reservoir. In many other taxa, including most non-cyclostome braconids and most ichneumonids, the glands open into the anterior end of the reservoir (Togashi 1963, Quicke & van Achterberg 1990, Quicke et al. 2009), which seems likely to be the plesiomorphic state. Some members of the braconid subfamily Rogadinae have a conical arrangement of interiorly projecting setum-like spines within the part of the secondary venom duct just before it joins the reservoir (Zaldivar-Riverón et al. 2004), which would appear likely to function as a one-way valve, preventing back-flow of venom from the reservoir to the gland during envenomation. Why only this group should have such a mechanism is unknown, and it is possible that other, soft-tissue systems have been overlooked.

Gross morphology appears to contain considerable phylogenetic information and has been surveyed in several groups. For the Ichneumonidae, important papers are a major survey by Togashi (1963), with considerable additional features and taxa scored and illustrated by Quicke et al. (2009). For the Braconidae, Edson and Vinson (1979) carried out an important initial survey which revealed two major reservoir forms, their type 1, which was heavily muscularised with many both circular and longitudinal muscle fibres and a relatively thick chitinous intima (Fig. 4.1a), and their type 2, which had relatively few and scattered muscle fibres and a thinner intima. Type 1 muscles are innervated and type 2 muscles are not (Edson et al. 1982), indicating perhaps that in the former the wasp had to have precise and rapid control of venom, whereas in the latter its expulsion might be brought about more gradually, perhaps aided in part by general metasomal movements during oviposition, although presumably the reservoir muscles are also somehow stimulated to contract when needed. The type 1 apparatus characterises the cyclostomes and type 2 apparatus the non-cyclostomes, with both types being present in the Aphidiinae. They also noted some differences between the types in the nature of the gland filaments, but these differences broke down when many more taxa were investigated. Sharkey (1993) postulated that the thin muscles of the type 2 apparatus of non-cyclostome braconids might also mean that these wasps do not have to inject much venom into a host at any one oviposition and noted that as these taxa are koinobionts, they often attack hosts at an early stage of development and thus are physically small. Nevertheless, as reported by Edson and Vinson (1979) and Quicke and van Achterberg (1990), a few non-cyclostomes, including members of the genus *Homolobus*, do have a rather strongly muscularised venom reservoir (although without the spiral pattern of intima thickening, probably indicating that they need rapidly to inject a fair volume of venom when they oviposit (Wharton 1993a); as suggested by Boring et al. (2009), perhaps it also help to push out the egg in this subfamily. In addition to the previously mentioned surveys, considerably more cyclostome and non-cyclostome taxa were examined and illustrated by Quicke and van Achterberg (1990), and more detailed treatments for various groups of cyclostomes were provided by Jiménez Peydró et al. (1996) and Barbalho and Penteado-Dias (1997), for Doryctinae by Quicke et al. (1992e), for Opiinae and Alysiinae by Quicke et al. (1997a), for Rogadinae by Zaldivar-Riverón et al. (2004) and for Aphidiinae by de Marzo (2000). Within the Ichneumonidae, there is a less clear-cut distinction between type 1 and type 2 and the only taxon with a definitive type 1 apparatus is the Xoridinae (Togashi 1963).

Detailed histological and TEM studies have been carried out on the glands of the ichneumonids *Pimpla turionellae* (Güven & Yel 1990, Uçkan 1999), *Diadromus collaris* (W.D. Li et al. 2006a) and *Diadegma semiclausum* (Pan & Chen 2003) and the braconids *Habrobracon* (as *Microbracon*) *hebetor* (van Marle 1977), *Opius caricivorae* (Wan et al. 2006), *Cotesia plutellae* (Pan & Chen 2003) and *Asobara* species (Moreau et al. 2009). Essentially all show the same sort of gland cell structure. The secretory dermal gland cells are of type III (Noirot & Quennedey 1974) and contain a vesicular organelle with a large number of microvilli. They often have a large nucleus, abundant mitochondria and extensive rough endoplasmic reticulum (RER), indicating that they are producing large quantities of protein. The cells of the venom glands are rich in actin filaments (Fig. 4.3c,d) suggesting a high level of intracellular transport.

Venom gland ultrastructural changes have been described for *D. collaris*; the cells develop more secretory apparatus over the first few days after eclosion and later start to degrade through autolysis at about 2 weeks. Frequently, secretory cells are also present in the reservoir (Wan et al. 2006), but in far lower densities and, in at least some cyclostome braconids, however, despite the lower density of glandular cells on the reservoir of *Opius caricivorae*, they had larger numbers of mitochondria than those of the gland proper. Glandular units may also be present on the primary venom duct (Fig. 4.4a). In the Braconinae and some Rogadinae, the primary venom duct is completely covered in chitin-lined globular vesicles (Fig. 4.4b,c), indicating that it plays a major part in the production of venom components, and it is tempting to speculate that the active components of their paralysing venoms may result from mixing of products made in the venom gland and the primary duct. An extremely similar state, presumably convergent, with the primary duct completely covered in globular chitin-lined vesicles is found in the ichneumonid subfamily Xoridinae. Unfortunately, although I have made some TEM images from these vesicles in the Braconinae (Fig. 4.4c), I am unsure as to precisely what their structure is. Each vesicle appears to contain a membrane-bound structure that is clearly extracellular, but whether this is an artefact or not requires further work. It is tempting to speculate that these primary duct secretory units must produce compounds that interact with the rest of the venom from the reservoir, perhaps activating it. Maybe fully active venom in these wasps would be too dangerous to store in the reservoir.

DUFOUR'S GLAND

The Dufour's or alkaline gland is a tubular, unpaired structure at the posterior of the metasoma that opens into the oviduct near to the base of the ovipositor near where the primary venom duct normally also joins it (Vinson 1969). It is also known as the alkaline gland because of its histological properties, the secretory cells being highly vacuolated. Its role in parasitic wasps appears to be primarily in producing hydrophobic compounds, probably most of them being at least partly volatile, rather than proteins, and is involved in host marking (e.g. *Toxoneuron nigriceps*, Vinson & Guillot 1971, Guillot et al. 1974; *Microplitis croceipes*, Vinson & Guillot 1971; *Venturia canescens*, Harrison et al. 1985) and, in some taxa, courtship, i.e. intraspecific communication, as is its role in the many aculeates in which it has been far better studied. The suggestion by Weseloh (1976) that it was the source of the sex pheromone in the microgastrine *Cotesia melanoscelus* was subsequently refuted (Weseloh 1980). The role of the Dufour's gland in the ichneumonoids is therefore probably largely host marking.

The chemistries of its products have been studied in only a few taxa: *Venturia canescens* by Mudd et al. (1982), *Toxoneuron nigriceps* (as *Cardiochiles*) by Syvertsen et al. (1995), the braconine *Habrobracon hebetor* by Howard *et al.* (2003) and, although mixed with venom

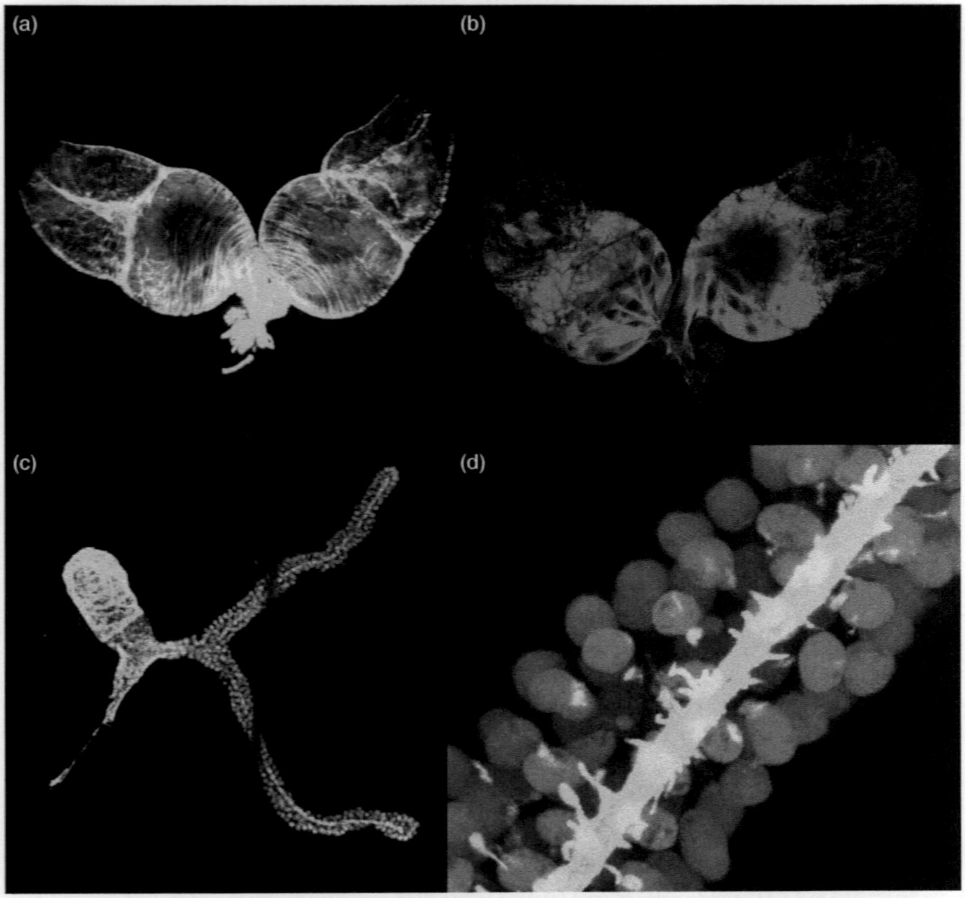

Fig. 4.3 Fluorescence microscopy images of the female internal reproductive system of the microgastrine *Cotesia congregata* showing DNA (both wasp nuclear and polydnavirus) fluorescing blue and actin, stained with phalloidin, fluorescing green. (a, b) Ovaries and calyx gland; note the intense blue fluorescence surrounding eggs in the calyx due to high concentration of polydnavirus particles; (c) entire venom apparatus; (d) detail of venom gland filament. (Source: reproduced by permission of Germain Chevignon). For a colour version, see Plate 8.

gland, Fukushima et al. (1990), and two species of *Bracon* used in the biocontrol of *Cephus cinctus* by J.E Baker et al. (2005) and therefore it is impossible to draw any firm conclusions. Howard et al. (2003) reported novel diterpenoids, β-springene, a homo-β-springene and a homo-geranyllinalool, as had Fukushima et al. (1990) previously, viz. (*E,E*)-11-ethyl-7,15-dimethyl-3-methylenehexadeca-1,6,10,14-tetraene (tentatively named homo-β-springene), the first authors apparently being unaware of the earlier findings. In the case of *H. hebetor*, it is not clear which if any of the components detected are pheromones. Howard et al. did show, however, that the Dufour's gland terpenoids were not used to coat the wasp's eggs.

Mudd et al. (1982) found a mixture of C_{21}, C_{23} and C_{25} saturated and mono-unsaturated hydrocarbons with (*Z*)-10-tricosene as the main component in extracts from the campoplegine *Venturia canescens* Dufour's gland, and Harrison et al. (1985) showed that the gland's secretions were involved in host discrimination. However, of the four major compounds detected by Mudd et al., only one, heneicosane, had any effect on wasp behaviour. Contact with this compound, either

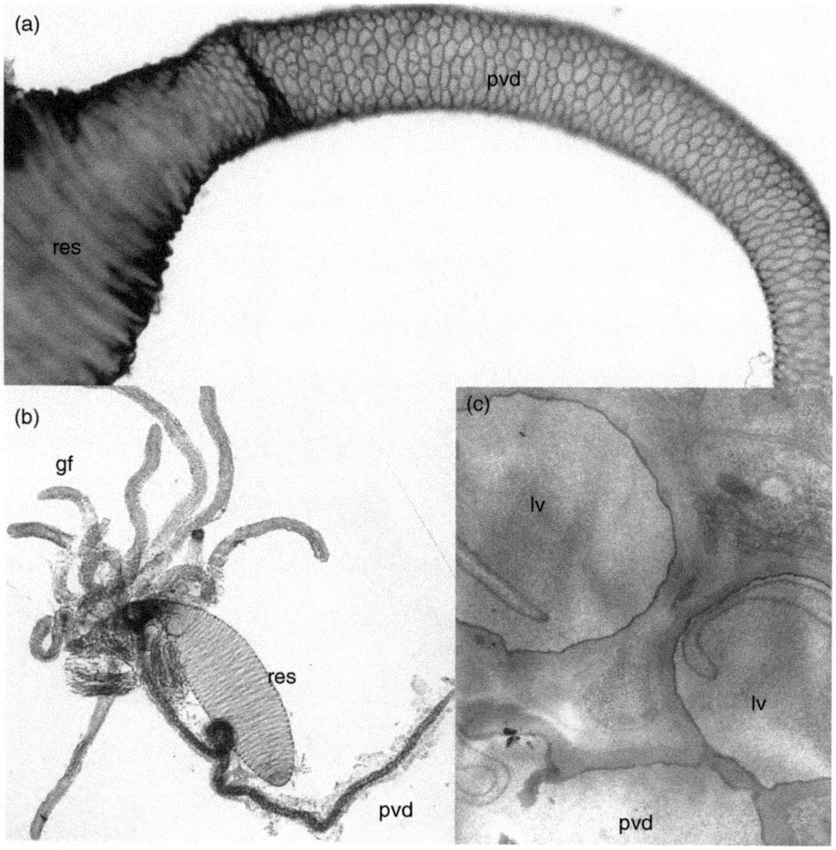

Fig. 4.4 Features of the venom apparatus in some cyclostome braconids. (a) *Aleiodes* sp. (Rogadinae), macerated intima of distal part of venom reservoir (res) and primary venom duct (pvd) showing the latter divided into many small polygonal areas, each with a secretory ductile. (b) *Bracon* sp. (Braconinae) macerated venom apparatus, showing insertion of all tubular venom gland filaments (gf) just posterior to the ovoid, thick-walled and highly sculptured venom reservoir (res) and the mass of near spherical clitin walled vesicles along the length of the primary venom duct (pvd). (c) TEM section of part of the edge of the primary venom duct showing two of the connected lateral vesicles (lv), apparently full of secretion and displaying an apparently membrane-bound structure of unknown structure/function.

alone or when marked artificially on a host *Ephestia* larva, caused the wasp to become very agitated and often elicited a flight response, which in the wild might be expected to cause the wasp to leave a patch that had already been utilised by a conspecific.

The morphology of the gland has hardly been investigated. Togashi (1963) distinguished three forms in the Ichneumonidae, single long and straight, coiling and branched, the last being found only in *Anomalon* and not in the other anomalonine genus investigated.

CUTICULAR HYDROCARBONS

Compared with social hymenopterans, where they play important roles in colony and species recognition, there have been relatively few studies of cuticular hydrocarbons in the Ichneumonoidea. Liepert and Dettner (1996) studied those of the aphidiine *Lysiphlebus cardui*, which attacks aphids in ant-attended (*Lasius niger*) colonies of the black bean aphid, *Aphis fabae cirsiiacanthoidis*, and found that the cuticular hydrocarbons of both comprised exclusively *n*-alkanes and mono-, di- and trimethylalkanes. In contrast, *Trioxys angelicae* hydrocarbons were very different [dominated by (Z)-11-alkenes]. Whereas the ants largely ignored the *Lysiphlebus*, they vigorously attacked *Trioxys* (see Fig. 12.6b). Collectively, these results suggest that the *Lysiphlebus* cuticular hydrocarbon composition has evolved to mimic those of the ants that attend its hosts. In another ant–aphidiine–aphid association, Takada and Hashimoto (1985) found that whether or not the aphidiine *Paralipsis eikoae* was attacked by an ant (*L. niger*) depended on what colony they had come from, being attacked by ants from a different colony. However, if the aphidiine managed to get on top of an ant from a different colony (riding it) and rubbed her legs and body on it, she soon became acceptable to ants of that colony. Hence they concluded that the wasps acquired some of their cuticular hydrocarbons through direct physical association with the ants they live with, thus enhancing their chemical camouflage.

Howard and Baker (2003) presented detailed analyses of the cuticular hydrocarbons and wax esters present in *Habrobracon hebetor* through pupal development to the adult stage. The main components found had carbon numbers from C_{21} to C_{45}, consisting of a homologous series of *n*-alkanes, 11-, 13- and 15-methylalkanes, 13,17-dimethylalkanes and (Z)-5-, (Z)-7- and (Z)-9-alkenes. Espelie and Brown (1990) examined cuticular hydrocarbons in a tetratrophic system (apple, codling moth, *Cydia pomonella*, the cheloinine primary parasitoid *Ascogaster quadridentata* and the hyperparasitoid *Perilampus fulvicornis*). The profile from *Ascogaster* was remarkably similar to that of adults of its host and female *Ascogaster* also shared one dominant component (pentacosene) with its hyperparasitoid, but what this means in evolutionary terms is unclear.

SEX PHEROMONES

Female ichneumonoids usually produce volatile compounds or mixtures to attract males and/or to elicit courtship, mounting and copulation by males. It seems likely that different compounds or mixtures are probably involved in long-range mate attraction and in the subsequent stimulation of male courtship through to copulation.

Several workers have investigated the location of the female sex pheromone gland in the microgastrine genus *Cotesia* (*C. glomerata*, Obara & Kitano 1974, Tagawa 1977, 1983; *C. melanoscelus*, Weseloh 1980; *C. rubecula*, Field & Keller 1994), and all found that the source was a pair of glands in the posterior of the metasoma associated with the female 'genital capsule'. Field and Keller (1994) examined the male response of the *Cotesia rubecula* to washed separated female body parts and found the female genital capsule, which comprises the ninth tergum, the ovipositor, hypopygium and second valvifer (= gonocoxite 9), to be the most attractive, and this elicited courtship song from them; however, rather than attempting to copulate with the genital capsule, they did so preferentially to the anterior of the metasoma, suggesting that having been lured to the female chemically, they may use vision to orientate for copulation. Two possible glandular sources of the sex pheromone were identified through a combination of SEM and dissection, but they failed to locate its exact source. One site was a group of pores located more or less coincident with the mechanosensory 'hair plate' at the anterodorsal part of the second valvifer. The second candidate site was a pair of glands located next to the dorsal membrane running medially between the outer edge of the nase of the hypopygium and the body. Failure of the authors to use an explicit anatomical terminology does not help interpretation of their findings and, of course, the pheromone might comprise a mixture of products from the two glands. Tagawa (1983) found that males and females of *C. glomerata* both possessed a patch of setae on the ninth tergite, but only females had an associated patch of columnar cells, which he described as being typical class 1 secretory cells (Noirot & Quennedey 1974), and these are certainly possible sex pheromone sources.

There have been few studies on the chemistry of the sex pheromones in Ichneumonoidea, partly because of the chemical difficulty of identifying compounds that may be present in only minute quantities [see, for example, Shu and Jones (1993) on the campoplegine *Eriborus* and Danci et al. (2006) on the microgastrine *Glyptapanteles*]. It is therefore hard to make any broad summaries, although probably most involve multiple components and the ratios between these are important (Shu & Jones 1993, Danci et al. 2006). Only a few actual compounds have been identified, with studies of the braconid *Macrocentrus* being particularly thorough and showing that the female 'sex pheromone' is actually a mixture of compounds (Swedenborg & Jones 1992a,b). In this wasp, females release a series of (Z,Z)-9,13-dienes that are air oxidised to form the pheromonally active compound

(Z)-4-tridecenal, and in addition they also release a range of dimethyl-6-(methylethyl)-3,4,5,6-tetrahydropyran-2-ones (Swedenborg et al. 1993, 1994). Eller et al. (1984) identified ethyl (Z)-9-hexadecenoate, a sex pheromone of the ctenopelmatine *Syndipnus rubiginosus*. Kainoh et al. (1991) found (Z)-9-hexadecenal in the chelonine braconid *Ascogaster reticulatus* and DeLury et al. (1999b) identified the similar (Z,Z)-9,12-octadecadienal as the sex pheromone that attracted males of *A. quadridentata*, although two related compounds elicited responses in the male antenna. Neral [(Z)-3,7-dimethyl-2,6-octadienal] and geranial [(*E*)-3,7-dimethyl-2,6-octadienal], two isomers of citral, are involved in the sex pheromone of the pimpline ichneumonid *Itoplectis* (Robacker & Hendry 1977). In terms of short-range effect compounds, Syvertsen et al. (1995) showed that female *Toxoneuron* (as *Cardiochiles*) *nigriceps* release (Z,Z)-7,13-heptacosadiene plus at least one other compound to initiate male antennation and mounting.

MALE INTERNAL REPRODUCTIVE SYSTEM

The musculature of the male genitalia is very similar to that of other Hymenoptera, as shown by O. Peck (1937a) for the rhyssine ichneumonid *Megarhyssa lunator*. Rojas-Rousse (1972) studied male gross internal anatomy and histology in the ichneumonine *Diadromus pulchellus*. As with virtually all ichneumonids, the testes are fused above the gut, i.e. they are enclosed in a tough common envelop (Fig. 4.5a) (e.g. Benham 1972), although a small proportion (~0.5%) of individuals were found to have separate testes (see also Wharton et al. 1989). The fused or separate testes are attached via a suspensory ligament that runs anteriorly to the base of the first tergite. Each testis comprises many follicles and these connect ultimately to form the vas deferens. The vas deferens is always somewhat convoluted next to the testes and then runs as a long narrow tube on either side to join the accessory glands (Rojas-Rousse 1972, Maetô 1987). The accessory glands secrete acidic mucopolysaccharides, which mix with the spermatozoa before they are ejaculated and their forceful secretion may aid the expulsion of sperm in part by providing volume.

The disposition of the testes, gut and vas deferens in the Braconidae was surveyed by Maetô (1987), who recognised three groups of subfamilies. The testes may be fused above the gut with the vas deferens attaching posteriorly or medially on the accessory glands (as in Fig. 4.5a), the presumed plesiomorphic state based on outgroup comparison with ichneumonids, or they may be fused below the gut or they may be separate and closely connected to the accessory glands but at the apical end. Occasionally, as in ichneumonids, some cyclostomes may have the testes separated (Fig. 4.5b),

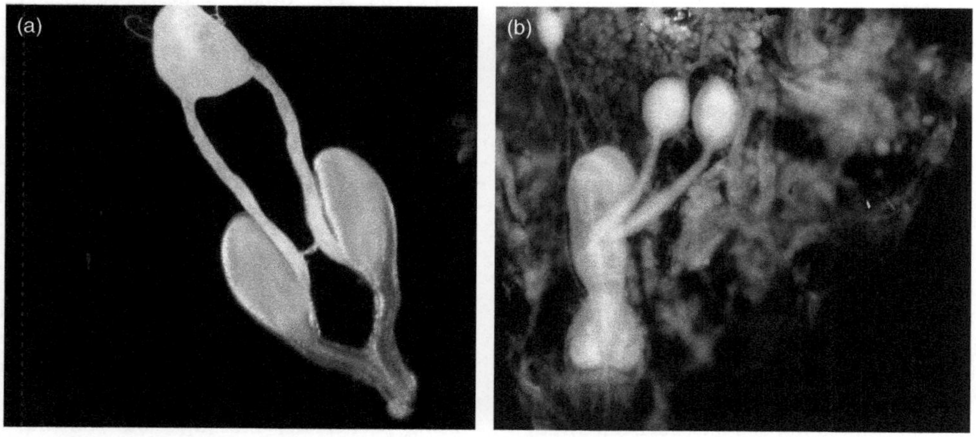

Fig. 4.5 Testes in dissected wasps. (a) *Hyposoter didymator* (Campopleginae) showing testes fused in single envelope and vas deferens inserting posteriorly on accessory glands. (Source: photograph by Marie Frayssinet, reproduced with permission.) (b) *Digonogastra* sp. (Braconinae) showing atypical separation of testes as found in occasional species and individuals within the cyclostome lineage.

which can be due either to intraspecific variation or occasionally a species-specific character. In addition to Maetô's data, various other taxa were scored and illustrated by Quicke and van Achterberg (1990) and here I can add that the Mesostoinae, based on *Hydrangeocola* sp., has the typical cyclostome condition. The plesiomorphic condition is shown by the great majority of members of the cyclostome lineage, but excluding Opiinae and Alysiinae and also the Aphidiinae, which show some variations. The separated state with anterior attachment to the accessory glands appears to be a synapomorphy for the non-cyclostomes (see also Trachypetinae), but is found also in the aphidiine genera *Praon* and *Lysiphlebus*, while the separate state with posterior attachment or fusion below gut is characteristic of Opiinae and Alysiinae. Given what we know now of braconid phylogeny, it appears that testes separation occurred independently in the Opiinae+Alysiinae but with secondary fusion below the gut in some Opiinae. It is interesting that separate testes have also been observed in a few members of the Braconinae, which seem to belong to the same lineage. It would undoubtedly be interesting to survey more members of that lineage and also of Aphidiinae to try to determine where transitions occurred.

The structure and function of the seminal vesicle, which is a convoluted part of the vas deferens located adjacent to the accessory glands, were studied in *Habrobracon hebetor* by Gerling and Rotary (1974) and it is the primary location of sperm storage. The accessory glands vary from being ovoid to very elongate and may be separate or joined along their medial interface (see fig. 1 in Gerling & Rotary 1974). In transverse section they show five internally projecting lobes comprised of tall, columnar secretory cells (Rojas-Rousse 1972; see fig. 6.2a in Quicke 1997a) and many microvilli (F. Huang et al. 2007). It seems highly probable that they are involved in spermatophore production, at least in some species.

SPERM ULTRASTRUCTURE

In all ichneumonids investigated to date, the sperm are isolated in the vas deferens and not formed into spermatodesmata as they are in sawflies and some aculeates. Enormous variations are apparent in sperm structure and ultrastructure between subfamilies in the Braconidae (Quicke et al. 1992b). Sperm of the cyclostome subfamilies are unmodified (plesiomorphic), with an elongate, rather straight nucleus ranging from approximately 7 to 20 μm long and a flagellum from 30 to 150 μm long, with the shortest ones being found in aphidiines and alysiines. However, all non-cyclostomes investigated have highly apomorphic sperm. Those of the euphoroid, helconoid and microgastroid complexes are very short, often with a bent/curved nucleus and with a very short flagellum, 3–11 μm, and very reduced mitochondrial derivatives, which are involved in providing energy to the axoneme. Agathidines, in contrast, have an enormously long flagellum and spermatozoa extracted from the seminal vesicles form a gelatinous mass and are difficult to separate; their axonemes appear to have only one or no central singlet microtubules, which might suggest reduced motility, but this has not been investigated. More work is required, but there is a hint that sperm are longer in those wasps, principally ectoparasitic ones, that have larger mature eggs.

Unfortunately, there are fewer detailed published studies on ichneumonid sperm, but straightforward examination of light microscope smears suggests that they are rather uniform in gross morphology and similar to those of cyclostome braconids. Detailed studies have been presented for the ichneumonine *Diadromus pulchellus* (Chauvin et al. 1988) and for the cryptines *Lymeon dieloceri* and a *Pachysomoides* species (Moreira et al. 2010). TEM suggests that many have a moderately complex extracellular matrix forming an acrosomal cap around the acrosome and extending posteriorly over the anterior part of the nucleus (Fig. 4.6). Moreira et al. showed that the dense 'soup' of filaments that fills the surrounding seminal fluid (Fig. 4.6c) peels off from the acrosomal cap, which might also be visible in the section through an *Ophion* sperm (Fig. 4.6b). The transverse sections through the acrosomal cap shown in Fig. 4.6(c) of the metopiine *Colpotrochia* sp. shows a complex organisation. The outer casing at the anterior end (middle section) shows 20 longitudinal, rather thick rods, the slightly more posterior section (left) has slightly more less-well-formed structures and the even more posterior section (right) appears to show that these have fragmented to form filaments that appear very similar to those in the fluid surrounding the sperm.

A very few studies have indicated that sperm are delivered into the female genital tract in the form of a spermatophore. This was first reported for the braconid *Macrocentrus* (S.E. Flanders 1945, Finney et al.

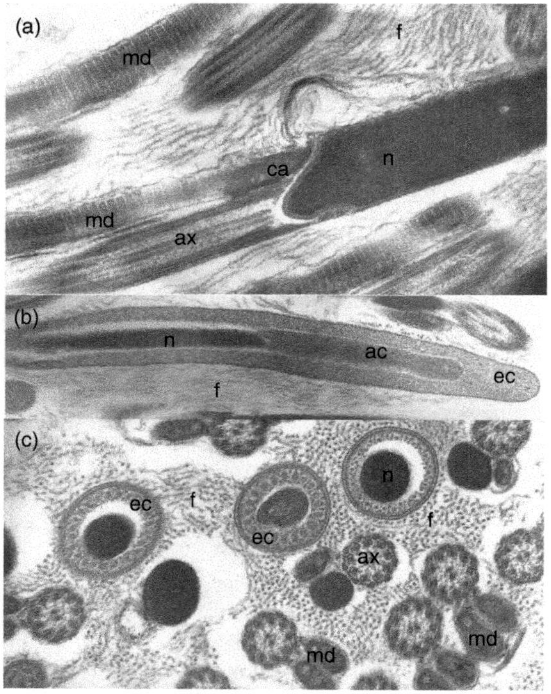

Fig. 4.6 Features of mature spermatozoa and seminal fluid of Ichneumonidae. (a, b) *Ophion* sp. (Ophioninae) showing, respectively, the posterior end of the nucleus with the centriolar adjunct interposed between the nucleus and one of the mitochondrial derivatives and longitudinal section through the anterior of the nucleus and acrosome, the whole being embedded within a long, tubular, extracellular acrosomal cap; (c) *Colpotrochia* sp. (Metopiinae), transverse sections of acrosome and acrosomal cap (upper middle), nucleus and acrosomal cap, naked nuclei, tail regions with axonemes at various levels and the dense fibrous strands that originate from the acrosomal caps filling the semen. Abbreviations: ac, acrosomal cap; ax, axoneme; ca, centriolar adjunct; f, (extracellular) filaments; md, mitochondrial derivative; n, nucleus.

1947) and other instances include campoplegines (*Diadegma semiclausum*, Madel et al. 1975; *Bathyplectes anurus*, Gordh & Hendrickson) and the ichneumonine *Diadromus pulchellus* (El Agoze et al. 1995). All ichneumonoids have large accessory glands, and it is almost certain that they could produce far more secretion than eventually ends up in the female spermatheca along with the sperm, so presumably they are often involved in the production of either a spermatophore or have some other role.

SPERMATOGENY INDEX

Some ichneumonoids complete spermatogenesis before emerging as adults and can mate very soon after eclosion, others may need a day or so to complete sperm maturation and some species appear to carry on producing sperm through much of their adult life and so are able to replenish stocks as needed (Boivin et al. 2005). The stages of spermatogenesis have been described in some detail only in the rogadine *Aleiodes coxalis* (Newman & Quicke 1998) and are certainly complex and show some differences from those of other insects.

ENDNOTE

1. I use the term vaginal gland here since 'uterus' implies a place where something remains for a while and matures and 'anterior common oviduct gland' is just too cumbersome.

Chapter 5

IMMATURE STAGES

The Braconid and Ichneumonid Parasitoid Wasps: Biology, Systematics, Evolution and Ecology, First Edition. Donald L. J. Quicke.

EGGS AND OÖGENESIS

Egg production starts in follicles at the anterior end of the germarium at the anterior end of each ovariole, where oögonia are produced. Mitotic divisions of these give rise to a cluster of cells, the most posterior of which will form the oöcyte, and the others, all interconnected with each other and the oöcyte (King & Cassidy 1973; see fig. 4.7a in Quicke 1997a), called nurse cells, are responsible for passing nutrients to the egg cyctoplasm and in lecithal eggs this mainly consists of yolk precursors. Nurse cells are connected to the developing oöcyte by large nutritive pores or intercellular bridges. The whole complex is surrounded by a thin layer of follicle cells.

Iwata (1958, 1960, 1966a) published a very detailed survey of the form of mature ovarian eggs in the Ichneumonidae and, in general, the eggs of ichneumonids show a greater range of morphological variation than do those of braconids. The eggs of most endoparasitic ichneumonoids are fairly nondescript, weakly asymmetric, weakly curved ovoids, whereas those of many of the larger ectoparasitoids have a long, narrow tail. Outside these basic forms, the eggs of some groups, particularly among the Ichneumonidae, show remarkable adaptations, and in particular several groups have evolved eggs with anchors, something hardly known within the Braconidae.

The outer coverings of ichneumonoid eggs have been little reported, apart from broad comments on colour and texture, and in particular, an outer fibrous layer appears to be important in avoiding host immune defence in several taxa. D.H. Davies et al. (1986), using TEM and histochemical techniques, investigated the production of the chorion in the cardiochiline *Toxoneuron* (as *Cardiochiles*) *nigriceps* in considerable detail. It is secreted around the oöcyte by the surrounding layer of follicle cells and in this species was found to comprise five morphologically distinct layers, from inside to outside referred to as 'vitelline envelope, endochorion, an electron-dense "irregular layer", papillary layer and fibrous layer'. The outer layer had neutral glyco- or mucoproteins and no acidic ones, which is important because host haemocytes are unable to encapsulate negatively or neutral charged surfaces such as Sephadex C50 cation-exchange beads but readily encapsulate positively charged surfaces such as Sephadex A50 beads (Vinson 1974). Further, the chorion may not always be inert, as Nguyen et al. (2013) seem to have found that it possesses enolase activity in the aphidiine *Aphidius ervi*. The egg of *Venturia* (as *Nemeritis*) *canescens* (Campopleginae) was studied using electron microscopy by Rotheram (1967, 1973a), who showed that the inner layer (vitelline envelope which lies between the chorion proper and the external plasma membrane of the oöcyte) was rather thin, and associated this with its hydropic nature and need to take up water and nutrients from the surrounding host haemolymph. The external surface of the chorion in this species was densely covered with long, perpendicular projections, which appeared to be important in helping the virus-like particles produced in the calyx region of the oviduct of this wasp adhere to the egg (see Chapter 9, section *Polydnaviruses*). The micropyle, through whose pores the sperm gain access to the inside of the egg, is located at the anterior end (Bronskill 1959, Cassidy & King 1972).

The process of oögenesis has been described, largely based on electron microscopy investigations, for a range of ecto- and endoparasitoid taxa whose eggs differ most conspicuously in the amount of yolk they contain. Major studies of egg development include those of Meng (1968) for *Pimpla turionellae*, Henschen (1928) for the idiobiont braconine *Habrobracon hebetor* (as *juglandis*), Cassidy and King (1972) and King and Cassidy (1973) for *H. hebetor* and King et al. (1971) and LeRalec (1995) for the koinobiont microgastrine *Cotesia glomerata* and the aphidiine *Ephedrus plagiator*, respectively.

The egg maturation rate in at least one synovigenic species, *Habrobracon hebetor*, has been shown to depend on the lighting conditions: with females kept under a continuous light regime, eggs mature at twice the rate of those kept in a normal light–dark day regime and even more so than those kept completely in the dark (Lum & Flaherty 1973).

HYDROPIC AND ANHYDROPIC EGGS

All ectoparasitoid species produce anhydropic, yolk-rich eggs with stiff chorions; however, most but not all koinobiont endoparasitoids have what are called hydropic eggs. These are small when laid but can absorb a great deal of water and nutrients from the host haemolymph and tissues during development (Ferkovich & Dillard 1986), such that by the time they are ready to hatch they may be from several hundred to 1000 or so times larger in terms of volume than when they were first laid (e.g. O.J. Smith 1952). Eggs of a few endoparasitic ichneumonids such as many ctenopelmatines, banchines and *Panteles* have much thicker, dark brown to black chorions, and are clearly anhydropic (see Figs 13.41b and 13.44f) (Cummins et al. 2011). Although not dark, similarly anhydropic eggs are found in the endoparasitic Rogadinae. Minimally, highly hydropic eggs appear to have evolved on six separate occasions, viz. in the non-cyclostome braconids, in the Opiinae, at least twice in the Alysiinae, in the ophioniformes and in the Aphidiinae, although most likely it has happened on considerably more occasions. Both lack of detailed biological studies and lack of robust phylogenies within groups prevent much further analysis.

Hydropic eggs contain no, or virtually no, yolk, only a small number of lipid globules and no protein resource (King et al. 1971, LeRalec 1995), whereas anhydropic eggs are larger and contain a large yolk reserve comprising lipoid and protein 'bodies'. There is necessarily a marked difference in the stretchability of the chorions of anhydropic versus hydropic eggs, those of the latter being far thicker and possibly less convoluted. Yolk proteins in most insects are synthesised initially in the female fat body as the large precursor protein vitellogenin, which is then cleaved into two or more smaller protein chains and secreted into the haemolymph, from where it is then taken up by ovarian cells, although there is some variation; in the honey bee, *Apis mellifera*, for example, the vitellogenin is secreted into the haemolymph uncleaved. Nose et al. (1997) therefore investigated the vitellogenin of a parasitic wasp, *Pimpla nipponica* (Pimplinae), to determine whether it conformed to the standard insect pattern or showed features characteristic of the derived honey bee condition, and found it most similar to the former. Nothing seems to be known about vitellogenin genes in those ichneumonoids that produce hydropic eggs, i.e. whether they are functional at all or lost.

Parasitoid wasp taxa that produce large, highly lecithal eggs are typically idiobiont ectoparasitoids and those that produce small, alecithal eggs are all koinobiont endoparasitoids. These two extreme egg types are exactly what would be predicted from considerations of the idiobiont–koinobiont spectrum. Large, yolky eggs give the first instar parasitoid larva a good head start and enable it to survive before it can bite through the host cuticle and commence feeding – tiny alecithal eggs just would not survive outside a host.

Several groups have independently evolved eggs with part of the chorion being modified to form an anchor to attach the egg either through the host's external cuticle (Tryphoninae) or to internal organs of the host [some Anomaloninae, Ctenopelmatinae (Westwoodiini), Stilbopinae (*Panteles*), some Tersilochinae]. Two other uncommon groups have eggs with long, thin anchors, the Australian Maxfischeriinae (Boring et al. 2011) and the Lycorininae (Iwata 1958, Coronado-Rivera et al. 2004, M.R. Shaw 2004b). The lycorine ichneumonids, which probably, based on the observation that one species, oviposits through the anus of its host caterpillar (M.R. Shaw 2004b), may well anchor the egg inside the host's rectum before the parasitoid larva chews its way through into the host's body cavity, as happens in some Metopiinae, although this is pure conjecture as the location of the egg within the host has not actually been observed. Nothing is known about the biology of Maxfischeriinae.

Eggs of the ichneumonid subfamily Eucerotinae, a biologically very aberrant hyperparasitic group of ichneumonids, are extremely small and stalked (see Fig. 13.58c). The females lay very large numbers of them around the edges of leaves near where her secondary host caterpillars are feeding. Each mature egg in the ovary has a short posterior sucker-like extension that Tripp (1961) suggested is continuous with the chorion, but it appears to be a specialised extrachorionic layer. During oviposition the sucker is attached to a leaf and the rest of the egg complex is drawn out to form a thin supporting strand with the egg proper at the tip, with the support strand apparently hardening quickly, so they appear very like the eggs of many lacewings (Neuroptera), although much smaller (Tripp 1961, Varley 1965). For some while, ichneumonidologists considered the anchor of Eucerotinae eggs to be homologous to that of tryphonines, but their stalks are stated in the literature not to be chorionic, with the resemblance being purely superficial. However, given the published diagrams, I find it hard to understand why the stalk apparatus does not have the same origin as the chorion, i.e. from follicle cell secretions, and it certainly is not a simple glue stalk. The exact nature and origin of their stalks are, I believe, still very uncertain and require further investigation, although even if it turns out to be chorionic, that does not mean that the structure is homologous to that in tryphonines.

EMBRYOGENESIS

The first meiotic division occurs after the egg (oöcyte) has fully matured, or nearly so, but often before it has been laid. In the idiobiont ectoparasitic braconine *Habrobracon hebetor* (as *juglandis*), B.R. Speicher (1936) reported that the 'first maturation division [i.e. first meiotic division] of the egg is initiated before oviposition but is then discontinued until after the egg is deposited' and that eggs in early stages of this division are present in the 'uterus' with no attached nurse cells. Meiosis is often suspended at this stage with the second meiotic division occurring only after oviposition (B.R. Speicher 1936), and the plane of cell division is roughly perpendicular to the egg surface. In *Habrobracon*, the four nuclear meiotic products have different fates. One of the polar nuclei (polar nucleus 1) remains close to the cell's chorion, while the other two remain close to one another and fuse to form a 'polar fusion nucleus'. Polar nucleus 1 usually disintegrates and never forms a nuclear membrane. If the egg is fertilised, the sperm nucleus fuses with the oötid nucleus and the first and second egg cleavages occur within approximately the next 15 min.

In many taxa with large eggs, cleavage in unfertilised eggs (i.e. ones destined to become males) and also, it is presumed, in fertilised eggs has been shown to be initiated by the mechanical stress and disruption experienced during oviposition. This has been shown for the ichneumonids *Campoletis sonorensis* (Vinson & Jang 1987), *Pimpla turionellae* (Went & Krause 1974) and *Diadegma* (R. Butcher, personal communication). In *Pimpla*, embryogenesis could be initiated by passing mature, unfertilised eggs, dissected out of the female oviducts, through a fine capillary, simulating ovipositional distortion, and in many other Hymenoptera, exposure to stress caused by immersion in hypotonic saline can achieve the same effect. However, it seems likely that in some taxa with very small eggs and relatively wide ovipositors, mechanical stress during oviposition would be far less and chemical factors on contact with host haemolymph might be the triggers, and chemical triggering has been demonstrated for a microgastrine (Ferkovich et al. 1991). The mechanism behind initiation due to stress appears to be the intracellular release of Ca^{2+} ions, as Wolf and Wolf (1988) were able to demonstrate by activating eggs by injecting them with a calcium ionophore, although there appear to have been no similar studies of alecithal eggs.

In some tryphonine ichneumonids, embryogenesis can actually commence within the female internal reproductive tract (intrauterine development) or sometimes slightly later while the anchored eggs are held at the base of the ovipositor (Kasparyan 1981). Although intrauterine development may sometimes be the result of a lack of availability of hosts, it is of more regular occurrence in some *Netelia* species. Although a female is usually able to dump eggs that are developing too far within her, in extreme cases the egg may actually hatch within the female's body, the first instar then entering her body by tearing its way through her oviduct and killing her (Shevyrev 1912).

The eggs of ichneumonoids appear to follow Hallez's law in that the embryonic head develops at the end of the egg that is antero-most in the ovary (Hallez 1886). There have been few studies, however, on how the anterior–posterior or dorso-ventral orientations of embryonic development are brought about. Achtelig and Krause (1971) used a blunt needle to 'destroy' the RNA-rich oösome[1] material that occupies the posterior 10% of the length of a *Pimpla* egg and found that the remaining undamaged part of the egg gives rise to a fully developed larva including gonads. Despite their detailed experiments, they were unable to come up with a completely satisfactory interpretation of the oösome apart from that it gives rise to pole cells, although these did not appear to be essential for embryonic development or gonad formation. This is unusual as in most insects the oösome is essential for production of the germline cells and thus gonads in the adults (Bilinski 1991b). Working on egg production and embryogenesis in the ectoparasitic tryphonine *Cosmoconus meridionator*, Bilinski et al. (1993) found it likely that the oösome produces factors that will be important for subsequent differentiation. Both anterior–posterior and dorso-ventral polarity in the eggs of this wasp start to become apparent in the so-called stage three just prior to the deposition of yolk (vitellogenesis) and are marked by migration of the oöcyte nucleus. Meng (1968) obtained similar results for the pimpline *Pimpla turionellae*. The oöplasm (oöcyte cytoplasm) of *C. meridionator* and other ichneumonoids is also unusual because it includes organelles referred to as accessory nuclei, which originate from the oösome nucleus and have a double membrane, but are far smaller and 'and comprise polymorphic inclusions that are termed "pseudonucleoli"'. Previous studies on them in *C. glomerata* and *Ophion luteus* indicated that they are formed by the folding of lamellae that originate at the surface of the nuclear envelope and that they contain RNA (King & Fordy 1970). Other suggestions for the role of accessory nuclei include that they are involved in membrane formation or they may be involved in the production of the vitelline membrane, although working with the sawfly *Tenthredo*, Bilinski (1991a) presented some evidence that they are most likely involved in gradient formation.

There have been relatively few detailed studies of embryogenesis in the superfamily. To most scientists currently working on the Ichneumonoidea, embryology is usually not their idea of a trendy topic, although in fact with the huge variations found within the Ichneumonoidea, and even within the Braconidae, several quite different modes of early development occur (Grbic & Strand 1998, Strand 1999, Harvey & Strand 2002). Insects, being protostomes, show largely deterministic development, that is, the future fate of cell lines following each embryonic cell division is determined by the ancestry of that cell line rather than its relative position within the developing embryo, and, at least during early development, chemical gradients play little or no role. It is therefore possible, just as in *Drosophila*, to construct fate maps, and Petters (1977) did this for *Habrobracon hebetor* (as *juglandis*) carrying the *ebony* mutation, which both acted as a marker and was associated with an increased incidence of gynandromorphism. Using information from 1211 haploid–diploid gynandromorphs he was able to recognise which groups of adult cuticular structures were derived from which individual cells formed during embryogenesis.

A light micrographic atlas, including Nomarski interference images, of the egg and embryonic development in the campoplegine *Venturia canescens* was published by Amos and Salt (1974), although with only descriptive detail. After just under one day, a layer of cells just beneath the chorion becomes visible and these form the serosal membrane, with the rest of the embryonic development occurring below this. Segmentation starts to become visible after approximately 36 h and the first signs of movement can be seen after about 66 h. Bronskill (1959, 1964) described the embryogenesis of *Pimpla turionellae* (Pimplinae) and *Mesoleius tenthredinidis* (Ctenopelmatinae) in great detail and compared it with that in the only other two ichneumonids that had been studied up until that time, *Banchus* (Banchinae) and *Diadegma* (Campopleginae). At about the same time, Ivanova-Kasas (1956) described the embryogenesis in the aphidiine genera *Aphidius* and *Ephedrus*. Although the first two are not at closely related, they were deemed to be more similar to one another than to either of the other two and to display the more plesiomorphic development.

Whereas the braconine *Habrobracon* (Amy 1961), and presumably others that lay large yolky eggs, form a syncytium and then develop as typical long germ band, insects show segmentation occurring more or less simultaneously along the length of the egg (hence 'long') during mid-embryogenesis. The first few syncytial nuclear divisions occur within the central yolk region and after the tenth cleavage they migrate to the periphery where, following two further nuclear cleavages, a cellular blastoderm is formed. In contrast, some other taxa display holoblastic cleavage (e.g. Aphidiinae), as can be demonstrated by injection at various stages of fluorescent dyes (Grbic & Strand 1998), and show signs of being a short germ band insect, i.e. only that part of the shorter germ band forming the head and thorax is initially specified, with the abdominal segments developing sequentially. This variation is remarkable since the mode of embryogenesis is normally fixed in far larger taxonomic groupings even at ordinal level. The key factors that led to this were probably the loss, in some endoparasitic taxa, of the need to have a hard chorion which normally serves to prevent desiccation, and of the need to provide the embryo with a large yolk resource, since taxa such as aphidiines can obtain nutrient direct from the haemolymph of their host.

EMBRYONIC MEMBRANES

The formation of the membrane that surrounds the embryo and in some taxa forms a trophamnion or dissociates to form teratocytes varies considerably. In the ichneumonids *Pimpla*, *Mesoleius* and *Diadegma*, it forms from a dorsal patch of the blastoderm that spreads out to surround the embryo. In contrast, in *Mesochorus*, the outer layer of blastoderm cells separates from the layers below (delamination) to form a complete surrounding membrane. Considerable work carried out by Tremblay and Calvert (1971) on the embryology of the aphidiine braconids revealed major differences between different taxa. Members of the putatively basal Ephedrini (*Ephedrus*) and Praini (*Praon*) form the membrane through delamination as in microgastrines, euphorines and *Mesochorus*, but in the derived clades Aphidiini (*Aphidius*, *Diaretiella*, *Lysiphlebus*) and Trioxini (*Trioxys*) and also in *Macrocentrus* (Macrocentrinae) there is total cleavage, and in *Trioxys* some nuclei from the second cleavage (polar nuclei) fuse with polar bodies and form the membrane that will eventually disassociate to form the teratocytes (see Fig. 12.52a). This means that teratocytes formed from the serosal membrane share all their genes with the developing embryo, whereas those from polar body-derived cells are necessarily less closely related.

LARVA

Although there are many peculiar larval forms found throughout the parasitic Hymenoptera, especially when it comes to the first larval instar (Quicke 1997a), the larvae of ichneumonoids, with very few exceptions, are rather uniform and generally of a maggot-like – so-called 'hymenopteriform' – type. The plesiomorphic number of larval instars in both families appears to be five, with reductions occurring in various, principally endoparasitoid, taxa (see Table 10.3; see also van Achterberg 1988b), although determining the number can be difficult and sometimes there are disagreements between workers even about the same species (O'Donnell 1987, Rojas-Rousse & Benoit 1977). In most cases, body segmentation (three thoracic and 10 abdominal) is clearly visible in the first instar at eclosion.

Ichneumonoid larvae, in common with other parasitic wasps, do not have a complete gut; instead, the mid-gut ends blindly although is attached to the rectum and a through connection does not occur until the prepupal stage just prior to pupation, and the meconium is normally voided at this stage. This is likely to be an adaptation to not 'fouling' the parasitoid's environment but is also testament to their high food conversion efficiency such that there is very little waste (especially solid waste) to void. A few groups that pupate within the host remains do not void the meconium until the adult ecloses (Ichneumoninae, H.D. Smith 1932; Alysiinae and Opiinae, Salkeld 1959), although there seems to be no information about whether the gut becomes completely open earlier.

Early instar larvae of endoparasitoid taxa often lack a tracheal system (e.g. Salkeld 1959, Narayanan et al. 1961) or, if one is present, the spiracular apparatus is not connected to the tracheal system present. When, in middle instars, a tracheal system is present and connected to the spiracles, the spiracles are often cited as being closed or non-functional. Ectoparasitoids nearly always have fully functional spiracles and tracheal system even from their first instar, although in some Tryphoninae it has been reported that early instars do not have a full complement of spiracles (Cushman 1926b).

Early instars tend to vary in the degree of development of the head capsule and mandibles and, in the case of endoparasitoids, whether they are caudate (i.e. with an extended tail, or both. The caudal appendage and its possible functions have been studied in several taxa, in the campoplegine *Venturia canescens* by Ullyett (1944), who concluded that its main role was probably rupturing the chorionic membrane, enabling the young larva fully to eclose and that it probably also plays a role in absorbing nutrients from the host haemolymph and maybe in 'balancing'. Thorpe (1932) had earlier investigated whether it was involved in respiration, as had been suggested by some earlier workers, in the cremastine *Cremastus interruptor* and the campoplegines *Campoplex* (as *Omorgus*) *mutabilis*, *Diadegma* (as *Angitia*) *nana* and *Sinophorus* (as *Eulimneria*) *cassifemur*, using the chemotactic flagellate protist *Polytoma* (Chlorophyceae), which form bands/shells of individuals in zones of optimal oxygen concentration. He found that the protists first aggregated along the ventral surface of the ichneumonid larvae, but not along the tail. Then, as the oxygen concentration fell below optimal, they formed a shell surrounding a largely protist-free zone around the body of the wasp larva, indicating that oxygen uptake and CO_2 release were restricted to the body of the larva, which also made sense since the tail did not contain either abundant tracheae or any circulatory organ. Although this may seem an antiquated technique, there is no reason to be doubtful about the results. More recent work has shown that the caudal appendage is used for moving around the host and swimming (O'Donnell 1989, Harvey et al. 2013), and in the ovo-larval chelonines it is used to enable the newly eclosed larva to penetrate the haemocoel of the late embryonic host caterpillar (Hawlitzky 1972, Kaeslin et al. 2005a).

Early instar larvae of several endoparasitic braconid subfamilies [Microgastrinae, Cardiochilinae, Cheloninae, Helconinae (Brachistini), Macrocentrinae, Orgilinae] and also two ichneumonids, the banchine *Banchus* and, to a lesser extent, the metopiine *Triclistus* (Gerig 1960), have a different posterior modification, an everted rectum called an anal vesicle (sometimes also with a small pointed caudal appendage present) (Fig. 5.1). McConnell (1938) reported a first instar anal vesicle also in the ctenopelmatine *Glyptorhaestus*,

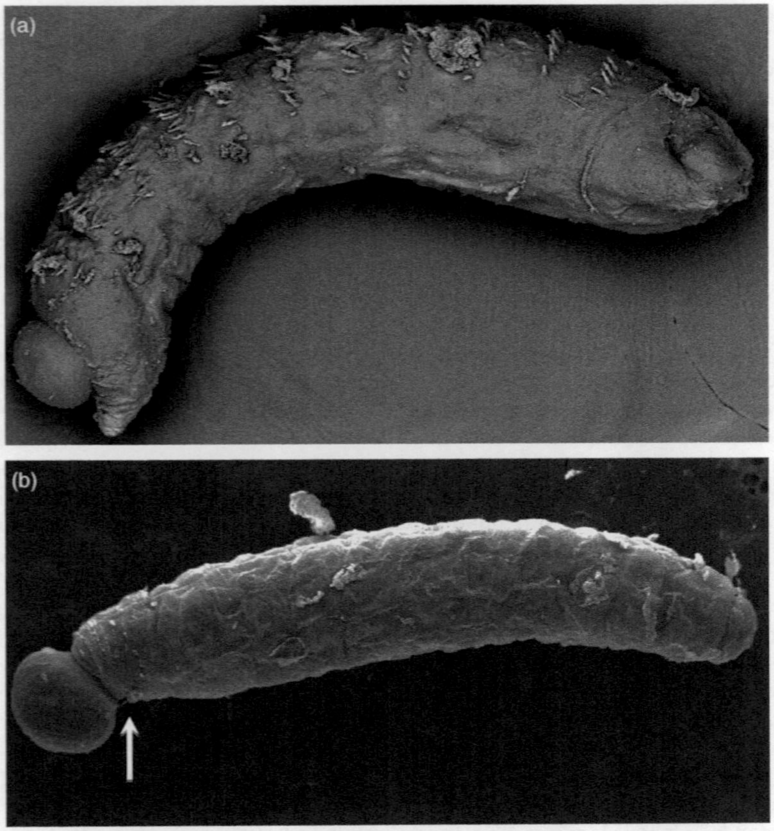

Fig. 5.1 SEM images of the first and third instar larvae of *Cotesia flavipes* (Microgastrinae); arrow indicating greatly reduced 'tail' of later instar. (Source: reproduced by permission of Daniel Russ Solis.)

but nothing is known about its structure. Diagrams of histological sections of the anal vesicle of several braconids and of *Banchus* were presented by Thorpe (1932), and Edson et al. (1977) reported a detailed TEM study of the vesicle of the microgastrine *Microplitis croceipes* and Kaeslin et al. (2006) follow its development in the chelonine *Chelonus inanitus*. In the ichneumonid genus *Banchus*, the entirety of the vesicle appears to be closely approximated to the mid-gut, whereas in the microgastroid braconids, the larger part of the vesicle is in close proximity to an expanded posterior part of the aorta with a smaller region next to the mid-gut, although the structure is rather different in *Orgilus*, in which it is composed of a thick proctodeal membrane with no large internal cavity. Thorpe emphasised that the structural diversity present should mean that it would be unwise to generalise function across all these taxa.

Using the same technique with *Polytoma* outlined above, Thorpe (1932) showed that the anal vesicle of microgastroid (Cheloninae and Microgastrinae) larvae plays a large respiratory role consistent with its large contact with the aorta, whereas that of *Orgilus* had very little detectable respiratory role. The anal vesicle has also been shown to be involved in both nutrient absorption and excretion (Edson & Vinson 1976, 1977, Kaeslin et al. 2006) and the epithelial cells on its surface are densely covered with microvilli. In *C. inanitus*, during the second instar, the anal vesicle starts to show signs of degeneration and it becomes relatively smaller and concomitantly the mid-gut cells develop a fully formed brush-border, indicating that by this stage, nutrient uptake is now predominantly through the gut.

Apart from the caudal modifications, ichneumonoid larvae tend to be rather maggot-like (hymenopteriform) with varying degrees of cuticular microsculpture, and in early instars this often takes the form of transverse rows of posteriorly directed spines, especially dorsally (Fig. 5.1), that no doubt assist them in moving around in or on hosts. Polypodiform early instars with ventral clawed protuberances occur in the Agathidinae+Sigalphinae clade (M.R. Shaw & Quicke 2000) (see Fig. 12.40d) and Clausen (1940) reported them in some Aphidiinae and they may be present in the enigmatic helconoid genus *Eadya*.

Ectoparasitoid larvae tend to be similar through all their instars whereas endoparasitoids often have rather different forms in their first, middle and final instars (Grossniklaus-Bürgin et al. 1994). Also, ectoparasitoid larvae almost always have papilliform antennae whereas those of endoparasitoids are nearly always disc-like without any protruding part or with only a small central papilla (but see Chapter 10, section *Disc-like larval antennae*, and also Chapter 12, section *Gnamptodontinae*) and they nearly always have a fully developed tracheal system with open spiracles from the earliest instar. In endoparasitoid larvae, the palps appear to be the dominant sense organs, although there are additional sensilla like structures on the head in some (e.g. aphidiines; see Couchman & King 1977).

Larval salivary glands (= labial glands; = silk glands) are usually present throughout development and are likely to have multiple functions, the most conspicuous being the production of silk for cocoon construction and also for pre-eclosion anchors that enable the larvae of some species to gain purchase to push themselves through the host integument (Nakamatsu et al. 2007). They are long and tubular and in most taxa comprise two branches on either side, one branch typically running close to the dorsal vessel, the other more laterally (e.g. Dowden 1938, Ahmad 1944, Thireau et al. 1990), although those of aphidiines appear to have only one branch on each side (Wheeler 1923, Schlinger 1961), and only a single branch was also illustrated for the doryctine *Heterospilus cephi* by Hill & Smith (1931). Also in aphidiines, they are absent from the earlier instar larvae (Wheeler 1923). Whether the different branches have different functions is unknown. Anteriorly, the glands from either side unite to form a common duct which opens at a slit in the larval head capsule, marked by a chitinised structure called the silk press (Fig. 5.2). The shape of the salivary orifice (N.B. not the silk press, see Wahl 1990) appears to be of considerable phylogenetic significance in the Ichneumonidae. In addition to salivary glands, Couchman and King (1977) reported, but did not illustrate, a small gland '[o]pening onto the mandible' in first but not subsequent instars of the aphidiine *Diaretiella rapae*, whose function was postulated to be to produce a cytolytic enzyme. Adult ichneumonoids (or at least those large pimpliformes ichneumonids which seem to be the only ones studied in detail) also have a gland associated with the mandible.

The salivary glands have been a relatively neglected organ system and the main work on their ultrastructure, functioning and other features was carried out by Baccetti (1958), who investigated them in the campoplegine *Rhimphoctona megacephalus* (as *Bathyplectes corvina*), the diplazontine *Diplazon tarsatorius* and the microgastrine *Protapanteles* (as *Apanteles*) *fraternus*. Baccetti's paper, together with Vance's study of *Chelonus* (Vance 1932a), are about the most detailed studies of general larval internal anatomy in any ichneumonoids. The glands produce both protein and glycoprotein products. In the ectoparasitic pimpline *Exeristes roborator*, they appear to produce a phenoloxidase which causes melanisation of the artificial culture medium (S.N. Thompson 1980), although how this might help the parasitoid larva is unclear. In endoparasitoids, they produce various secretions, but little is known about their precise functions as it is difficult to isolate them from possible secretions from the rectum/anal vesicle. The secretory cells of the salivary glands appear to conform to the Class I type of Noirot and Quennedey (1974), lacking the chitinous end apparatus that characterises Class III exocrine glands such as are found in the venom apparatus (Sutherland et al. 2010). The salivary gland cells have highly anastomosed nuclei in sawflies and presumably in ichneumonoid larvae, probably indicating a high level of endo-polyploidy.

The final larval instar head capsule is weakly sclerotised, except for a number of more or less distinct skeletal elements, and is widely variable in both the Braconidae and Ichneumonidae (Figs 5.2, 5.3f and 13.39f) and for this reason, unlike most other groups of parasitic wasps, it has been a source of a large number of characters for phylogenetic inference, although it is probably true to say that many workers have underestimated the extent to which they may show convergence as a result of selection

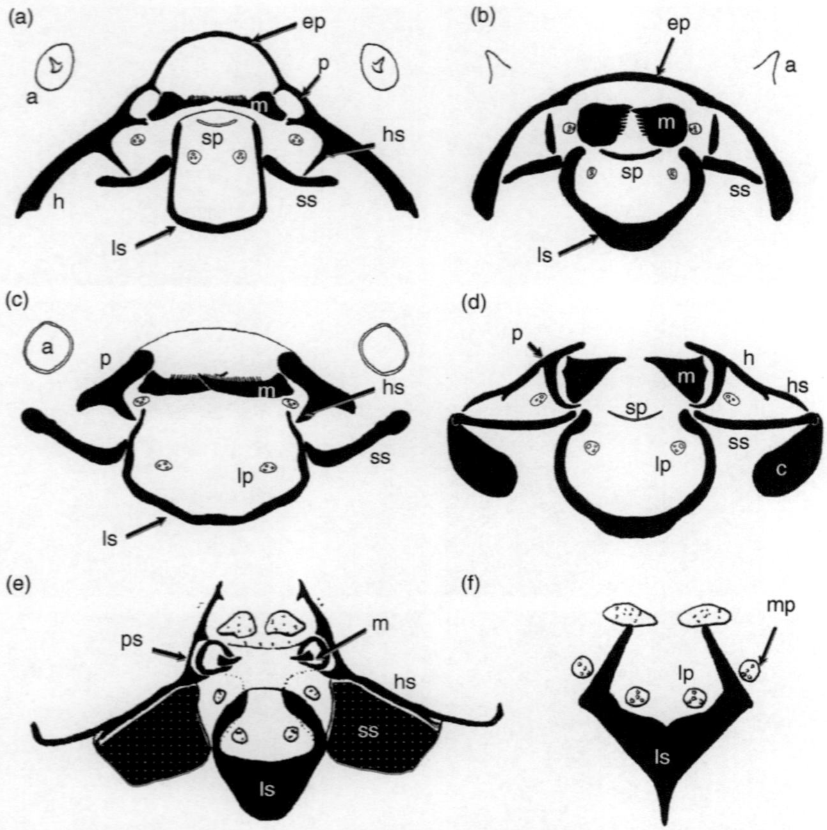

Fig. 5.2 Drawings showing features and some of the variations in ichneumonoid final larval instar head capsule features. (a) Generalised cryptine showing fairly plesiomorphic condition with complete epistoma, processes from either end of the pleurostoma towards mandible articulation points (also called superior and inferior mandibular processes), long narrow hypostoma and hypostomal spur, relatively narrow labial sclerite, toothed mandibles and papilliform antenna; (b) *Coeloides* (Braconinae) showing hypostomal spur reduced and detached from hypostoma; (c) *Phanerotoma* (Cheloninae), showing reduced epistoma and hypostoma parts, disc-shaped non-papilliform antennae and long, crossing and finely toothed mandibles; (d) *Acanthocaudis* (Aphidiinae), showing reduced epistoma and hypostoma, toothless mandibles and large, paddle-shaped expansion at outer end of stipital sclerite that is probably homologous to cardo in most braconids; (e) *Arotes amoenus* (Acaenitinae), showing reduced mandibles with only apex and a basal ring sclerotised and fusion of hypostoma and stipital sclerite forming a hypostomal–stipital plate; (f) *Syrphoctonus* sp. nr *columbiensis* (Diplazontinae), showing complete loss of most features except for palps and labial sclerite, the latter being produced into a long, ventral process. Abbreviations: a, antenna; ep, epistoma; h, hypostoma; hs, hypostomal spur; lp, labial palp; ls, labial sclerite; m, mandible; mp, maxillary palp; p, pleurostoma; sp, silk press; ss, stipital sclerite. (Source: b, c, adapted from Čapek 1970; d, adapted from Finlayson 1990; e, f, adapted from Wahl 1990).

pressure. There have been numerous detailed studies and surveys of the sclerotised or skeletal parts, particularly for the Ichneumonidae (Beirne 1941, J.R.T. Short 1952, 1959, 1970, 1977, 1978, 1982, J.R.T. Short & Nakanishi 1981, Finlayson 1967, 1975). Final instar larval head capsule features have been widely used to try to understand relationships because they were seen as providing both additional features that were independent of the vagaries of adult morphology and, even more important, because they represent an earlier developmental stage. They have played a particularly important role in ichneumonid systematics,

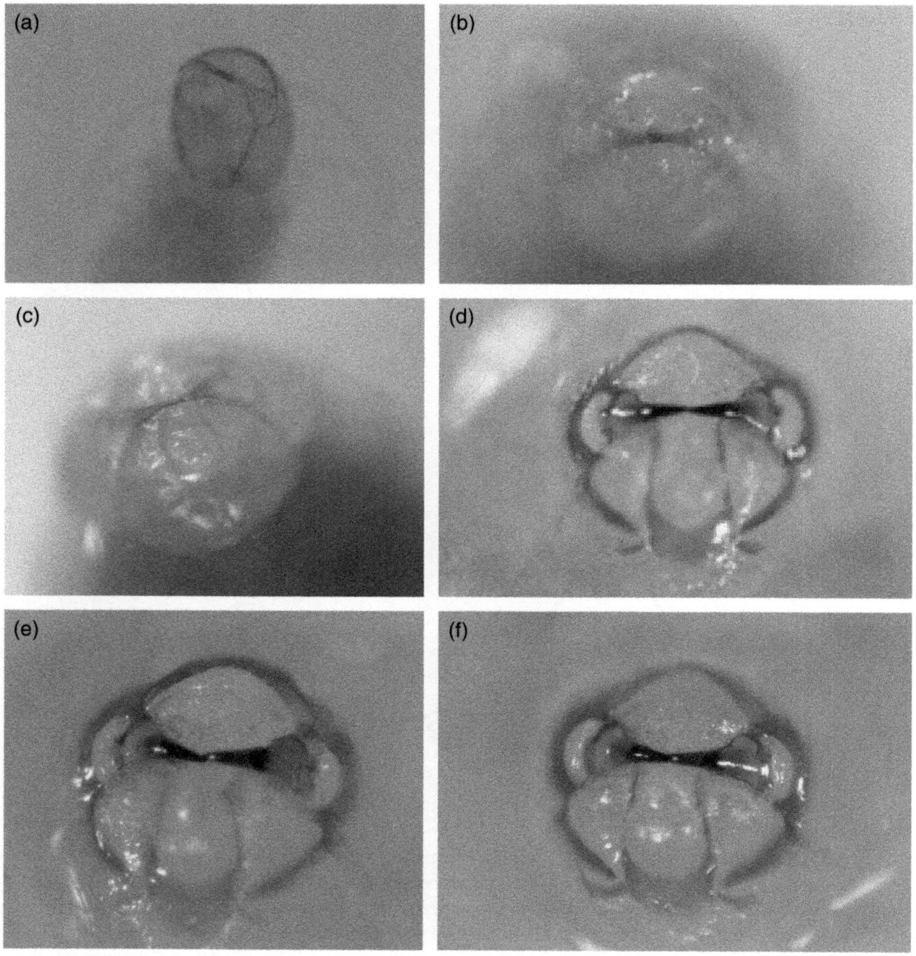

Fig. 5.3 Larval head development in *Itoplectis melanocephala* (Pimplinae). (a, b) two views of first instar head; (c–f) cephalic strictures of living second–fifth instars (not to scale). (Source: reproduced by permission of Ayşegül Özdan.) For a colour version, see Plate 9.

as the above references suggest. Finlayson (1967, 1975, 1990) used final instar cephalic structures to obtain hypotheses for relationships within the 'Pimplinae', Campopleginae and Aphidiinae, respectively. Detailed papers by Wahl have extended these studies for various groups of Ichneumonidae, provided some corrections to previous interpretations and placed many characters in a clear phylogenetic context (Wahl 1988, 1990, 1991).

Čapek (1970) similarly revealed a great deal of variation in the Braconidae, much of which appeared to offer phylogenetic clues, basing a lot of his phylogenetic scheme on what he found (see Fig. 12.1b), and subsequently provided a key to the final instar larvae of Braconidae (Čapek 1973). Why Čapek's work has been followed up so little by braconid workers, apart from occasional descriptions (e.g. Huddleston & Short 1978, van Achterberg 1979a, Quicke & Huddleston 1989) is a mystery.

The putatively basal, ectophagous lineages of both families have relatively complete sets of cephalic sclerites (Fig. 5.2a). The following description is based largely on Short (1978). There is a wide sclerotised arch running from the sides and dorsally comprising the hypostoma, pleurostoma and epistoma, the last passing through the anterior tentorial pits. Protruding medially from the anterior and posterior ends of the pleurostoma are the anterior (= superior) and posterior (= inferior) mandibular processes that articulate with the base of the mandible. The hypostoma usually has a distinct posterior spur, usually originating near to the pleurostoma, and usually abuts and is articulated with (or may be fused to) a transverse sclerotised bar called the stipital scleite. The latter, together with the lateral cardo (when present) and the membranous area (stipes) between them and the hypostoma, comprise the maxilla and the maxillary palp is located within the stipes. Medially to the stipital sclerite is the roughly U-shaped labial scleite which surrounds the generally membranous prelabium and which bears the labial palps. The silk press and salivary orifice are located at the anterior margin of the prelabium. An additional, often Y-shaped sclerite (prelabial scerite, not shown) is sometimes present below the silk press, particularly in taxa that spin a robust cocoon (notably Campopleginae). The hypostomal spurs may be a synapomorphy for the Ichneumonoidea and serve to maintain rigidity of the maxillae during silk extrusion and cocoon spinning (Short 1978). Ectoparasitoids and many endoparasitoids have toothed mandibles (Fig. 5.2a–c), but toothless ones (Fig. 5.2d,e) tend to be found only in endoparasitoids and also in a few gall-forming taxa [e.g. *Mesostoa* (Mesostoinae); Quicke & Huddleston 1989]. When teeth do occur in endoparasitoids, they tend to be finer even if the larva shows a final ectoparasitic feeding phase, whereas in ectoparasitoids, the teeth are usually fewer and more robust.

In addition to cephalic structures, larval exuviae also preserve the spiracles and adjacent spiracle closing apparatus. The external opening of the spiracle is connected to a small, subspherical chamber called the atrium and then leads to a special structure called the spiracular closing apparatus. Particular attention has been paid to this in the Ichneumonidae, in which most endoparasitoids have the atrium and closing apparatus completely adjacent to one another, whereas in most ectoparasitoids there is a distinct segment of trachea between them.

Despite all this work and the occasional phylogenetic insight it has given, there can be little doubt that the idea that larval head capsule features might be good sources of phylogenetic characters is substantially marred because we know so little of the actual functions of the many, easily observed, parts of the larval mouthparts and we almost certainly lack an appreciation of the different features that will be selected for in response to particular types of feeding, egress and cocoon-spinning needs. Some inferences have been made from a few observations of head and mouthpart movements. For example, in the Diplazontinae and Orthocentrinae, the mandibles are often reduced or even completely lost (Fig. 5.2e) (Wahl 1986, 1990), and in the Diplazontinae, the labial sclerite protrudes ventrally (Fig. 5.2e), and observations of head movements in these suggest that the latter structure takes on a similar functional role of breaking up host tissue (see below).

Almost nothing is known about the endocrine system of ichneumonoid larvae. The main studies are rather old, by Vagina (1982, 1987), who investigated the neuroendocrine cells of larvae of the braconid *Alysia manducator* and in particular the accumulation and release of apparent neuro-hormones during and after diapause. It seems surprising that despite a vast quantity of work having been carried out on the interactions of parasitoid larvae and their hosts, little has been done on the internal mechanisms that trigger features of their development. Further, nothing seems to be known about the precise mechanisms by which endoparasitoid larvae respond to changes in host hormonal levels, that is, do they detect and monitor concentrations through external sense organs or respond as a result of direct absorption through their gut or cuticle?

LARVAL FEEDING AND NUTRITION

Again, little work has been carried out on the actual mode of feeding by parasitoid larvae. A number of endoparasitic taxa have intermediary instars that lack or have greatly reduced mandibles and essentially drink host haemolymph. Ectoparasitic species bite through their host cuticle and imbibe haemolymph, and later in their development they may chew all of the host tissue except usually the head capsule. Slobodchikoff (1967) made careful observations of feeding by larvae of the labenine *Grotea californica*, placed with larvae of the host *Ceratina* bee. They rapidly pierced the posterior of the bee larva using their mandibles and then inserted their stipital sclerites into the hole they had made, and Slobodchikoff proposed that this served to keep the puncture wound open so that the parasitoid larva could continue feeding on the exuding haemolymph.

The way in which larvae of endoparasitoids feed in their early and middle stages of development may differ markedly from the way in which ectoparasitoids feed, but it is, of course, harder to study. This is indicated to a small extent by the often large differences in their head capsule and mouthpart features, and it had been suggested a long time ago that maybe first instar parasitoid larvae may absorb some nutrients directly through their cuticle, very much as hydropic eggs do. Recent applications of more sophisticated techniques have now shown conclusively for the larvae of aphidiine braconids that many nutrients (sugars and amino acids) are indeed directly absorbed from the host haemolymph via the cuticle (de Eguileor et al. 2001, Giordana et al. 2003). According to Corbet (1968), first instar larvae of the campoplegine *Venturia canescens* release 'cytolytic' enzymes from their salivary glands and fore gut which help to digest host tissues. Whether this is a general feature is unknown. Larvae of the ichneumonid subfamily Diplazontinae and related pimpliformes taxa have mandibles that are markedly reduced or are even completely absent (Fig. 5.2f), but they have a large ventral spur on the labial sclerite and Wahl (1990) suggested that in these, the latter structure may be used to disrupt host internal tissues and noted that J.R.T. Short and Nakanishi (1981) observed that *Diplazon laetatorius* third instars moved their heads up and down in a way that could be involved in breaking up host tissues with the labial spur.

Chudek et al. (1996) employed NMR spectroscopy to study the feeding done within the living host by the koinobiont campoplegine *Venturia canescens*. They found that the parasitoid first ate the host's fat body and digestive system and stated that the parasitoid larva allowed '... the host to continue to grow and only progressed to the vital organs when its own development had neared pupation'. Although it is true that the host is allowed to continue to grow, it is difficult to imagine that the host's digestive system is not somehow also important for its growth. It would be most interesting if more NMR studies could be carried out as the costs of using such facilities must presumably gradually come down and resolution increase.

Many endoparasitic ichneumonoids that egress from their hosts to pupate externally have an obligate external feeding phase in the last larval instar. Although lost in many microgastrines (but see Fig. 12.66), this stage might be construed as a remnant of past completely ectophagous biology. Tobias (1989) considered this to be an atavistic feature reflecting the ancestral totally ectoparasitic biology, but being a reversal to the ancestral biology rather than it having been retained as a symplesiomorphy.

Baker and Fabrick (2000) also showed, using a range of specific proteinase inhibitors, that the main protein digestion system in the larval gut of the ectoparasitic braconine *Habrobracon hebetor* probably involved a serine protease. There appear to have been no equivalent studies in any other parasitoids, so generalisation should not be attempted.

LARVAL FOOD CONSUMPTION AND DIETARY EFFICIENCY

Most people probably consider that carnivory is exceedingly energy inefficient in that only about 10% of the food energy in a plant becomes converted to the meat of a warm-blooded herbivore. Parasitoids, however, can show remarkably high conversion rates, often in excess of 50% dry weight (Howell & Fisher 1977). Rojas-Rousse and Kalmes (1978) have shown for the ichneumonid pupal parasitoid *Diadromus pulchellus* that between 80 and 84% of the energy potentially available to the host is obtained by the parasitoid. Similarly, Harvey et al. (2006, 2009) showed that this is not uncommon, with a high percentage of host nitrogen being utilised by the parasitoid, and this goes a long way towards explaining why some parasitoid systems can extend to a fifth trophic level. Perhaps because they have to move around rather more, dietary efficiency in egg predators may be lower, and Edgar (1971) reported efficiencies, depending on season, of only 31.2 and 16.2% for the cryptine *Hidryta sordidus* feeding on wolf-spider eggs.

LIPID METABOLISM

Parasitoid Hymenoptera, including all ichneumonoids investigated, are unable to synthesize their own lipids *de novo* from carbohydrate sources (B. Visser & Ellers 2008), hence they must obtain all fats from their food source. Different taxa vary in their ability to modify biochemically the fats they ingest. D. Jones et al. (1982) showed that the ability of parasitoids to synthesize their own glycerides (i.e. a molecule with a glycerol backbone and one, two or three ester links to an acyl group, i.e. a fatty acid) rather than just acylating endogenous mono- and diacylglycerides, was reflected in whether the parasitoid's adult lipid profile closely resembled that of the host or not. Some species have the ability to synthesize *de novo* and some do not. In their study, they isolated lipid microsomes from adults of a chalcidid, three ichneumonids and a braconid [*Brachymeria*, Chalcididae; *Aphaereta pallipes* (Alysiinae), *Exeristes roborator* and *Itoplectis conquisitor* (Pimplinae) and *Hyposoter exigua* (Campopleginae)]. Although most of these had lipid profiles quite dissimilar from that of their hosts, indicating both *de novo* synthesis and acylation of existing non-triglycerides, the two pimplines had very similar lipid profiles to their hosts. Working with another polyphagous *Exeristes* species, *Exeristes comstockii*, Bracken & Barlow (1967) found that the wasp had no characteristic fatty acids of its own but had a fatty acid profile that duplicated those of its hosts, and S.N. Thompson and Barlow (1970) showed the same for *Itoplectis*. However, as *Itoplectis* and *Exeristes* both belong to the Pimplinae (although in tribes Pimplini and Ephialtini, respectively), the above findings do not tell us whether there is a general principle involved here or just that the Pimplinae are different. That it might be a more general phenomenon was hinted at by Norullahoğlu et al. (2004), who found that the fatty acid profiles of the koinobiont microgastrine *Apanteles galleriae* was qualitatively similar to that the pyralid host *Achroia grisella* on which it was investigated, but they did not make a comparative study with other hosts.

Lipids are stored primarily in the insect's fat body and provide the adult with a long-term energy reserve. In the braconine *Habrobracon hebetor*, lipid reserves accumulated in the larval stage provide the wasp with energy for approximately 10 days, after which they are exhausted (Işṭvan et al. 2011). Ellers (1996) showed that lipid reserves were depleted linearly with age in the alysiine *Asobara tabida* and therefore fat reserves at egress are a good predictor of longevity. Further, different strains of *A. tabida* showed different body lipid levels and those with more lipid reserves had longer life expectancies.

RESPIRATION IN ENDOPARASITOIDS

Obtaining enough oxygen is clearly a prerequisite for successful development and one of the ways in which encapsulation of parasitoid eggs or early larvae kills them might be that it starves them of oxygen. Oxygen uptake in early stages is by diffusion from surrounding host haemolymph through the very thin cuticle, since first instars never have a fully functioning tracheal system, usually do not possess spiracles and, even if they do, then they are not connected to the tracheal system. Respiration may also be aided in some groups by their anal vesicles (see above). However, at least one species has a behavioural adaptation that maximises its exposure to oxygenated haemolymph. The cardiochiline *Toxoneuron nigriceps* oviposits into the haemocoel of its host *Heliothis virescens* (Noctuidae) caterpillar almost anywhere, but the first instar larva soon moves itself to the eighth abdominal segment of the host, which is where there is a specialised tracheole tuft free in the haemocoel that is forms part of a system for providing oxygen to the host's haemocytes (Rao et al. 2009). It seems highly improbable that *T. nigriceps* is unique in this behaviour and it would be very interesting to know how widespread the occurrence of this caterpillar oxygenating organ is and also whether larvae in other groups of endoparasitoids display the same tropism (see also Chapter 5, section *Meteorini*).

LARVAL SECRETIONS

A number of species, and probably the majority of endoparasitic ones, are known to release factors into their hosts that may interact with host physiology, or, in those that produce teratocytes, the teratocytes do so as well. Having said that, there have still been relatively few studies on larval secretions. Führer and Willers (1986) showed that an anal secretion produced by larvae of *Pimpla turionellae* was important in inhibiting the melanisation process involved in host encapsulation response.

The combined effects of parasitoid secretions into the host affect the physiology of the latter to the benefit of the parasitoid. *Spodoptera* caterpillars parasitised by the chelonine *Chelonus inanitus* produce more lipids and glycogen and their haemolymph contains less protein than in controls, and this seems to be essential for the successful completion of development by the wasp larva (Kaeslin et al. 2005b). Larvae of *Ascogaster reticulatus* secrete ecdysteroids into artificial medium, suggesting that the larva may be regulating host development directly (Brown et al. 1993).

THE PUPAL STAGE

Hymenoptera pupae are exarate, that is, the wings, antennae and limbs are separate from the body. As such, they tend to be rather delicate and are easily damaged if they are handled. Pupal development goes through a similar pattern of pigmentation development in all taxa, the dark eye pigments (ommochromes) often being visible even in the prepupal stage. Development of colour tends to start anteriorly with the mesosoma and finally the posterior of the metasoma (Fig. 5.4). The different stages in colour development allow pupal development to be divided into a number of discrete stages against which the development of internal structures and physiological changes can be correlated [e.g. Albrecht et al. (1994) recognised six stages for pupal development in *Chelonus*]. In their study of ecdysteroids of *Chelonus inanitus*, Marti et al. (2003) divided Albrecht et al.'s stage 1 into two finer subdivisions. Leibenguth (1967) followed the development of the ommochromes in the pupae of the braconine *Habrobracon hebetor* (as *juglandis*) and made additional inferences about the chemical pathways involved by analysing the pigments of various eye colour mutant trains which lacked various of the specific enzymes involved in the ommochrome synthesis pathway.

In taxa with long ovipositors, the pupal ovipositor is almost always curved over the postero-dorsal surface of the metasoma (Fig. 5.4b–d), but in the xoridine *Odontocolon* (see Fig. 13.7) and the doryctine, *Spathius agrili* (Z.Q. Yang et al. 2010), it projects posteriorly, although the reasons for this are unknown and, indeed, how this posterior projection is achieved at the moult from prepupal stage has not been described.

COCOONS

Although cocoons are not technically part of the wasp, they are made by them and play a crucial role in their survival and, because they are usually rather robust structures, the parasitoid's pupal stage is generally very fragile with thin cuticle and easily damaged if disturbed. In addition, cocoons have been shown to help prevent hyperparasitism and desiccation and no doubt protect against pathogens and predators. They

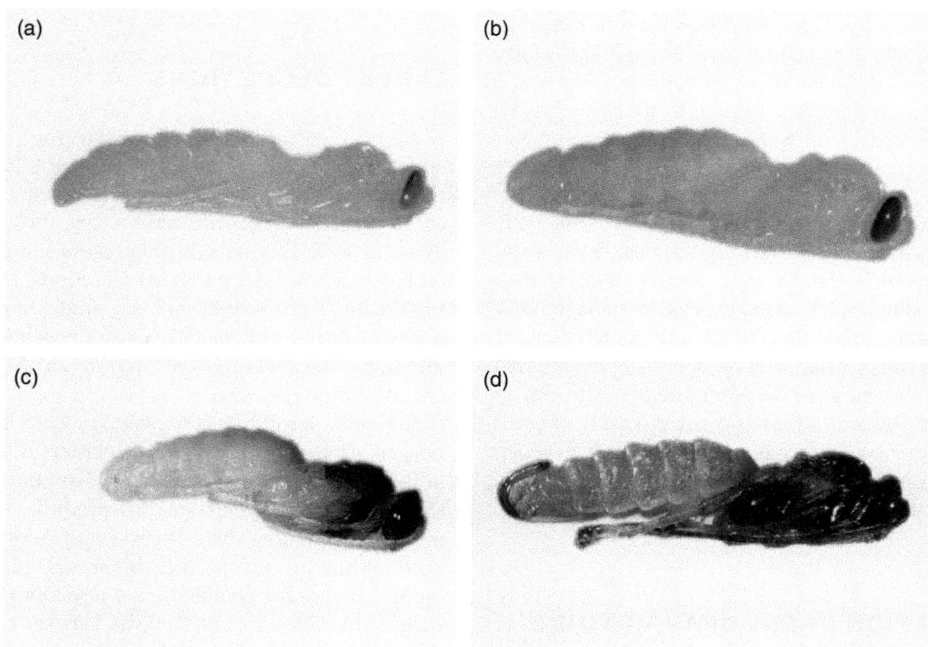

Fig. 5.4 Pupae of *Itoplectis melanocephala* (Pimplinae). (a) Male, just moulted; (b–d) female at three stages of darkening; note the ovipositor is curved over the back. (Source: reproduced by permission of Ayşegül Özdan.) For a colour version, see Plate 10.

are also important in protecting host and parasitoid when they are submerged by heavy rain or floods (Heron 1960).

Unlike most other parasitic wasps, nearly all ichneumonoids construct a cocoon in which the pupal phase is spent, and this is true even of many taxa that pupate within hard host remains such as Lepidoptera pupae (e.g. Anomaloninae, Metopiinae), fly puparia (e.g. Alysiinae) or mummified host larvae (e.g. Rogadinae), although in these latter cases the cocoon is often rather insubstantial it still serves to increase the structural strength of the pupation site (Caudri 1941). The major exception is the braconid subfamily Opiinae, most of which form no cocoon at all within the puparium of the fly hosts (Shaw & Huddleston 1991, R.A. Wharton, personal communication). There are a few other exceptions, e.g. the ichneumonine *Tricyphus respinozae*, where cocoon construction has been lost entirely, although its host sphingid pupa is a particularly robust environment in which to pupate (Ward & Gauld 1987) and naked pupae also occur in some orthocentrines and are formed within the thin cocoon made by their mycetophilid hosts. The highly aberrant *Hybrizon* (Hybrizontinae), an endoparasitoid of worker ants, makes no cocoon at all and its completely naked pupae can be found in the nest of its host ant (Donisthorpe & Wilkinson 1930, as *Paxylomma*). It is likely that other hybrizontines do the same, although no data are available for members of any of the other genera. Absence of a cocoon has also been reported in some *Heterospilus* (Doryctinae) that attack sphecid nests, but this requires confirmation (Marsh & Melo 1999).

Typically, the silken cocoon is formed of three, more or less distinct, major layers: a loosely woven outer layer, a dense middle layer and a thin, nearly transparent inner lining of very fine fibres. The middle and inner layers are also sometimes distinctly multi-layered. Many members of the ophioniformes group of subfamilies make cocoons with a distinctly paler equatorial region that I denote the central equatorial band (CEB). There appear to be no available data on the nature of the dark, usually red–brown to almost black, material that colours the sub-equatorial regions, or on how the larva regulates their disposition while constructing the cocoon layers.

In some species, cocoon silk accounts for a staggeringly large proportion of the dry weight of the final instar larva, which indicates the importance that protection from physical damage, predators, hyperparasitoids and desiccation has on parasitoid survival (Tagawa 1996). In temperate region bivoltine species attacking more or less exposed hosts, usually there is a seasonal dimorphism in the structure and robustness of the cocoon, with the overwintering one thicker, darker and tougher than the summer one (the overwintering one probably on average has to last longer and is likely to be exposed to harsher weather conditions) (*Sphecophaga burra*, Schmeider 1939; *Bathyplectes curculionis*, Cross & Simpson 1972; *Bracon mellitor*, Adams et al. 1969; *Microplitis* species, see also Shaw & Huddleston 1991). The cocoon of the cardiochiline *Hansonia chavarriai*, which is constructed in the leaf litter of the dry forest in Costa Rica and endures a 6-month dry season there, is extremely hard with a wall 2 mm thick and whose inner surface is 'extremely hard and polished, almost appearing to be gold foil' (Dangerfield et al. 1996). Similarly, cocoons of the ophionine *Enicospilus lebophagus*, which enters a very protracted diapause, have been noted to have a very dense, nacreous inner wall, which presumably affords considerable protection against desiccation and penetration by microorganisms (Gauld 1991), but its composition has not been investigated. Those ichneumonoid parasitoids that make very weak cocoons are almost always those that pupate in an already robust casing such as a host pupa, as is the case with rogadine and meteorideine braconids and anomalonine, metopiine and some ichneumonine ichneumonids.

Ichneumonoid cocoons are composed of silk fibres, i.e. fibrous proteins of high molecular weight that can only be dissolved with difficulty (i.e. in calcium chloride in ethanol at low temperature, in calcium nitrate in methanol, or in *N*-methylmorpholine *N*-oxide), covered in or glued together by various other, lower molecular weight, proteins and glycoproteins. The silk is manufactured in the labial glands (also called silk or salivary glands), which switch over to this from likely earlier production of secretions involved in either digestion or host immune response control. There are no known instances of anal (i.e. Malpighian tubule-derived) silk production, in contrast to some chalcidoids (Quicke 1997a). Silk is a functional term describing large, insoluble fibrous molecules and can have several molecular configurations, and its chemistry can vary fairly widely within the constraint that the amino acids forming the regular α-helices, parallel and cross-β sheets and collagen-type folding patterns are ones with short side chains (Rudall & Kenchington 1971). A survey of silk amino acid composition based on gross chemical analysis was presented by

Quicke & Shaw (2004) for a range of non-cyclostome braconids and Quicke et al. (2004b) for a few other ichneumonoids as part of a study to examine whether the availability of host amino acids (as used by the host Lepidoptera larva to form its own cocoon) may have led to the evolution of particular amino acid use by the parasitoid, for which weak evidence was found.

Within the microgastrine braconids, Quicke & Shaw (2004) revealed that a major compositional shift in silk chemistry occurred in the almost certainly monophyletic group that excludes *Microgaster* (and presumably its very close relatives). This silk turned out to have a novel structure (Yamada et al. 2003, 2004) and provides a strong synapomorphy for the remaining genera of the subfamily. Although far from as clear-cut, considerable differences were found between members of other ichneumonoid subfamilies, nevertheless indicating that there is probably substantial variation in silk chemistry.

Several braconids have been noted as 'paying particular attention' to the silk covering what will be the anterior end of the cocoon once the wasp larva pupates. In the Meteorini (Euphorinae: both *Meteorus* and *Xele*), the parasitoid larva spins a 'soft' internal silk cap and it has been suggested that this may provide protection for the head of the fragile pupal stage (see Fig. 12.51c) (Barrantes et al. 2011). Similarly, extra layers were noted in the cocoon of *Macrocentrus ancylivorus* (Fink 1926).

During cocoon construction, the parasitic wasp larva rotates many times (e.g. Boling & Pitre 1970, Cross & Simpson 1972), which must make it very difficult if not impossible for it to remember which way around it is, and this could be very important for its survival. Salt (1977) considered this problem in the case of *Venturia canescens*, which constructs its cocoon within that of its host *Anagasta* (= *Ephestia*) caterpillar. The *Anagasta* cocoon has a definite orientation, with one end destined for the emergence of the adult moth, and it is towards this end that the *Venturia* ideally needs to exit from its own cocoon. Salt found that the inside of the *Venturia* cocoon, although generally smooth and constructed with fine silk fibres glued together by other oral secretions, has a distinctly textured transverse band which he postulated provides the final, pre-pupation larva with an orientation cue so that it does not pupate in an orientation that would make emergence difficult. My own observations support Salt's in that a distinct rough zone of various conformations can be discerned in most if not all ichneumonoid cocoons.

ENDNOTE

1. Comprising the polar granules, and they are more or less the only conspicuous morphological structures in the oöplasm and are located at the posterior pole of the egg.

Chapter 6

IDIOBIONTS, KOINOBIONTS AND OTHER LIFE HISTORY TRAITS

The Braconid and Ichneumonid Parasitoid Wasps: Biology, Systematics, Evolution and Ecology, First Edition. Donald L. J. Quicke.

PARASITOIDISM

Most ichneumonoids are parasitoids, that is, they complete their whole larval development on a single host, which is almost invariably killed as a result, although there appear to be some very rare instances in which parasitised hosts may go on to complete development and maybe reproduce (Principi et al. 1979, R.E. Ryan & Milligan 2009). When aphidiines attack late instar aphids, the host may also moult to an adult and produce some offspring before being killed. Because the hosts are killed, parasitoids are frequently important in regulating the populations of their hosts, and this has been put to practical use in a large number of biological control programmes.

IDIOBIONT AND KOINOBIONT STRATEGIES

The major biological distinction recognised within the parasitic Hymenoptera for a very long time was whether or not the species was ecto- or endoparasitic. Haeselbarth (1979) first drew some attention to another very important feature, namely whether the parasitoid allows the host to continue to develop after the wasp has oviposited into it or whether further development is prevented, which he termed koinophyte and idiophyte strategies, respectively. This was elaborated on considerably by Askew and Shaw (1986), who provided far more explicit argumentation and introduced the slightly more appropriate but still far from euphonious terms koinobiont and idiobiont into the general language of parasitoid workers. Godfray (1994, Figs 8.5 and 8.6) expressed the differing webs of selection pressures affecting idiobionts and koinobionts graphically, which emphasises the complex nature of the interactions.

The major correlated life history features of idiobionts and koinobionts are summarised in Table 6.1, which is adapted from Gauld and Hanson (1995) and Quicke (1997a). It must be emphasised that these terms represent ideal ends to a spectrum of biologies and, across a large range of parasitoid wasp taxa, some life history traits show higher degrees of correlation with these strategies than others. Statistical assessment by Mayhew and Blackburn (1999) found some support for some correlations but not for others, but given the taxonomic breadth of their data set, this is perhaps not surprising, as across large taxonomic gaps many other features will vary and, in addition, their data set was limited to taxa whose biologies were well known. These features will be considered below.

Although the larval stage of koinobionts is commonly very long, it is often the case that the bulk of this is accounted for by a very protracted first larval instar, which can in some cases last 9 months (Tothill 1922, Clausen 1940). Within the koinobiont category, the development time may also reflect the host ecology, with parasitoids of exposed hosts sacrificing size for more rapid development and parasitoids of concealed (and thus less vulnerable to predation) hosts spending longer to develop in order to maximise size, although the data for this are currently rather limited (Harvey & Strand 2002). Most koinobiont ichneumonoids that attack exposed Lepidoptera larvae do not kill the host until it has completed making its cocoon or other shelter, the main exceptions being some members of the Campopleginae and Microgastrinae, Meteorini (Euphorinae) and Rogadinae. Some of these campoplegines and *Meteorus* (Meteorini) species spin pendulous cocoons (see Figs 12.51 and 13.51a,b) and rogadines mummify their host caterpillars more or less *in situ* (see Fig. 12.51). As Jerman and Gauld (1988) noted, these always have some special adaptations for protecting their own cocoons from predation (Shirai & Maetô 2009, Harvey et al. 2011) or attack by pseudohyperparasitoids. Among the braconids, a very large number of microgastrines and most Meterorini (Euphorinae) complete feeding on their hosts before the latter have entered their pupation retreats or spun cocoons, thus often leaving a considerable amount of unconsumed host tissue, so that is not limiting.

Host quality can only decline after having been attacked by an idiobiont, so there is an evolutionary advantage to completing feeding on a host in as short a period as possible, and in some *Bracon* and *Habrobracon* species the whole developmental period can be as short as 6 days (C.G. Jackson & Butler 1984) and Yang et al.'s study of the doryctine *Spathius agrili* showed that the first four instars last on average little more than 1 day each and the final one only 5–10 days depending on temperature (Z.Q. Yang et al. 2010). Even in the large idiobiont endoparasitoid *Pimpla rufipes*, the second to fourth instars last only 1 day each (Rojas-Rousse & Benoit 1977). This is achieved in part by them laying large, yolk-rich (lecithal) eggs such that upon egress, the first instar larva is already fairly large. The drawback is that if a wasp lays large eggs, she cannot carry as many of them with her at any one time (see

Table 6.1 Principle differences between idio- and koinobiont strategies in Ichneumonoidea, although all are trends with exceptions.

Idiobionts: host does not develop after parasitisation	**Koinobionts: host allowed to continue development after parasitisation**
Ectoparasitoids or endoparasitoids of pupae	Mostly endoparasitoids
Hosts generally concealed	Hosts generally exposed
Often generalists	Often specialists
Larval hosts permanently paralysed	Larval hosts not or only temporarily paralysed
Larval development rapid	Larval development usually protracted, often with long-duration first instar
Eggs large with much yolk	Eggs usually small with little or no yolk
Females with few mature eggs at any one time	Females usually carry many mature eggs at any one time
Synovigenic	Pro-ovigenic, or at least usually with large ovigeny index
Host feeding more common	Host feeding less common
Long adult life span	Short adult life span
Host stage attacked larger than wasp	Host stage attacked often smaller or much smaller than wasp
Male eggs often preferentially laid on small hosts	Offspring sex not normally strongly influenced by host size
Sexual size dimorphism often strong	Sexual size dimorphism absent or less pronounced
Mostly diurnal	Diurnal or nocturnal

Fig. 4.1a; *cf.* Fig. 4.1b). This does not mean that lifetime fecundity is that constrained to the number of eggs a female can carry, since many may have fat and protein reserves to mature more (synovigeny) as some are lain and some wasps can consume protein-rich food as adults and this can result in the maturation of additional eggs (see Chapter 14, section *Host feeding*). Nevertheless, the lifetime fecundities of idiobionts are generally much lower than those of otherwise comparable koinobionts. Within the Ichneumonidae, relatively large eggs for endoparasitic koinobionts are laid by various Banchinae and Ctenopelmatinae (see Fig. 13.44d). The latter may be considered rather basal in the koinobiont ophioniformes group, hence their large eggs could be an evolutionary hangover, but they might also indicate a different sort of adaptation to that life style. As Wharton (1993a) pointed out, much of what is written about synovigeny is surmise based on correlations with egg size and ovariole number (see also Chapter 4, *Time scale of egg maturation*).

Across a spectrum of parasitoid Hymenoptera, Blackburn (1991) found that when body size is controlled for, koinobionts have a generally longer pupal duration and pre-adult lifespan than idiobionts. The picture is complicated, as one might expect, and taking simple measures such as larval development time misses the fact that in many koinobionts this period is protracted owing to a very long first instar larval duration, during which the parasitoid hardly feeds on its host and often does not influence its biology much, waiting until it has constructed a cocoon, at which point their larval development might be fairly rapid.

GENERALISTS AND SPECIALISTS

The idiobiont/koinobiont spectrum, more usually treated as if it were a strict dichotomy, is often used as a surrogate for whether a parasitoid of unknown biology is a generalist or specialist, because even though the biology of the individual species is likely to be unknown, whether or not a taxon is idio- or koinobiont or ecto- or endoparasitoid, is almost always conserved at tribe level[1]. Although there are numerous exceptions, the generality that idiobonts have broader host ranges than koinobionts is well accepted and borne out by analyses. See Chapter 14, section *Host ranges of parasitoids*, for a discussion of how a host range should most meaningfully be defined.

Althoff (2003), based on data provided in Wharton et al. (1997), showed that a significantly higher proportion of koinobiont than idiobiont braconids in the New World fauna attacked hosts belonging to a single family (Fig. 6.1). There are good and bad points about these data. The authors of the work are experts and so the information given is considerably cleaned up from erroneous records. However, the data presented are based on genus rather than species host ranges and there is no control for phylogeny. Of particular interest, I think, are some of the extreme exceptions such as some *Charmon* (Charmontinae) and *Meteorus* (Euphorinae) species, which, despite being koinobiont endoparasitoids, have host ranges including many host families.

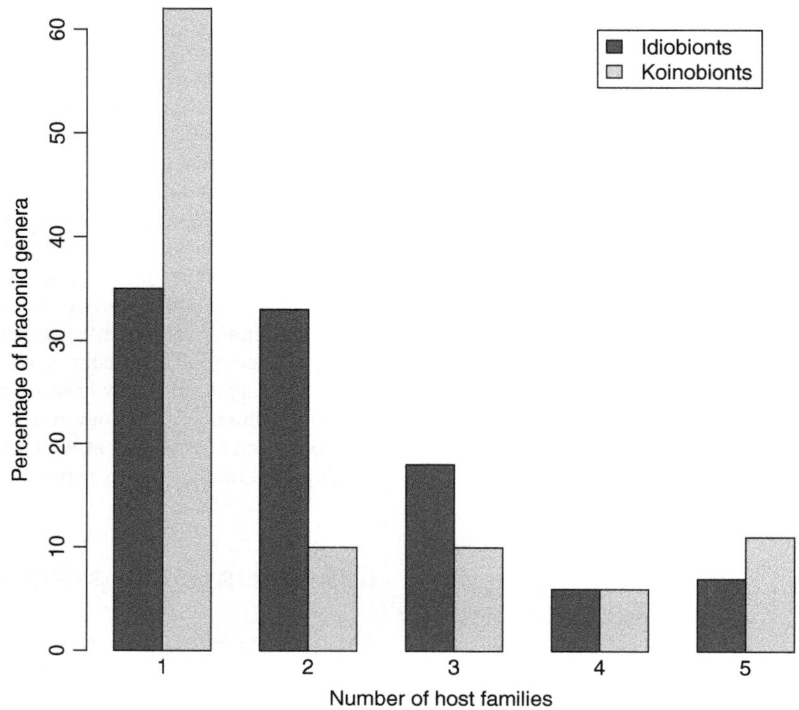

Fig. 6.1 Host range, represented as number of families attacked, of koinobiont and idiobiont braconids showing that a higher percentage of genera in the former are restricted to a single host family. (Source: data from Althoff 2003.)

ECTO- AND ENDOPARASITISM

With few exceptions, ectoparasitic ichneumonoids are idiobionts and attack concealed hosts because in this situation, the host can be permanently paralysed and fed on externally without normally increasing its risk of being predated and without the risk of the host being able to crush or dislodge the parasitoid egg/larva by its movements (e.g. Glover 1939). The exceptions, i.e. koinobiont ectoparasitoids (Tryphoninae, Adelognathinae and the *Polysphincta* group of Pimplinae among the ichneumonids[2] and the Rhysipolinae in the Braconidae), each have different sets of adaptations that enable them to overcome the normal increased risks that could be expected of being ectoparasitic on an active host, such as avoiding the host physically being able to dislodge the parasitoid larva through precise egg placement on inaccessible body regions.

Considering first the far more common ectoparasitic idiobiont strategy of attacking and paralysing concealed endophytic insects, this life history is widely considered to represent the ancestral biology of both ichneumonids and braconids, although I suspect that attacking deeply concealed hosts such as wood-borers represents a more recent evolutionary development (Gauld 1988a, Whitfield 1998). Whereas for the Ichneumonidae it still seems to be true that the most basal lineages (e.g. Xoridinae, many Labeninae) are ectoparasitoids of concealed hosts, the most recent phylogenies for the Braconidae suggest that the basal-most clade comprises a group taxa with very different biologies, including endoparasitism in Aphidiinae, cecidogeny in Mesostoinae and, based on ovipositor morphology, probable endoparasitism in Trachypetinae (Belshaw et al. 2003). Although I would not suggest that they show reversals from endo- to ectoparasitism, they do appear to have undergone shifts to endoparasitism earlier than the ichneumonids.

Without doubt there have been multiple independent evolutionary shifts from ecto- to endoparasitism within both families. The reverse transition seems much more unlikely on the grounds that endoparasitoid eggs, and more particularly larvae, have often lost features that would need to be re-evolved to cope with external conditions, such as yolk, early instar spiracles and tracheal system and thinned cuticle. Within the Braconidae, endoparasitism seems reliably to have evolved on eight occasions, of which four have led to major radiations, viz. at the base of the non-cyclostomes, in the Aphidiinae, in the Rogadinae and in the Opiinae+Alysiinae (Whitfield 1992, Quicke 1993, Zaldivar- Riverón et al. 2006). Additionally, a few species of Braconinae (i.e. the Aspidobraconina and possibly also in *Acrocerillia*), one putative doryctine (*Sericobracon*, Shaw & Edgerley 1986) and one lysitermine genus (*Katytermus*, van Achterberg & Steiner 1996) also show this biology. Among the ichneumonids, endoparasitism occurs in all members of the ophioniformes+orthopelmatiformes (except the koinobiont ectoparasitic Tryphoninae), in the Ichneumoninae, in several subfamilies of pimpliformes and separately in the tribe Pimplini of the Pimplinae and also in a few species of stilpnine Cryptinae.

The koinobiont endoparasitoids appear generally to be in a very favourable position to render their hosts maximally palatable, because they generally do not have to make their major growth spurt until after the host has reached a suitable state of development and they can usurp their host's physiology in the meanwhile, perhaps to produce more suitable proteins for parasitoid consumption and reducing the input that the host might invest in unnecessary structures such as gonads and other imaginal bud tissues (see Chapter 9, section *Host castration and similar effects*).

PERMANENT HOST PARALYSIS

There is a strong association between the idiobiont life history strategy, ectoparasitism and permanent host paralysis. Quicke et al. (2000a) considered how paralysing the host has advantages and disadvantages depending on where the host is feeding and where the parasitoid's egg is laid (see Fig. 6.6). Ectoparasitoids are vulnerable in a number of ways. The host may try to remove them, either using its mandibles or potentially by scraping them off or crushing them against some substrate. They might, if the host was exposed, render it and themselves more conspicuous to predators. The host's direct defence can be eliminated if the host is permanently paralysed, and this is what idiobiont ectoparasitoids of concealed hosts do, but obviously that would be too risky a strategy with exposed hosts, as they would no longer be able to defend themselves and would inevitably become more conspicuous to predators. The few ectoparasitic koinobionts have adaptations to minimise these additional risks. Egg placement by ectoparasitic tryphonines is nearly always in locations that the host cannot reach with its jaws, such as between the legs or on the dorsal surface just behind the head. With *Polysphincta* group pimplines, the egg and larva develop between the cephalothorax and abdomen, out of the host spider's reach. Tryphonines, if they attack exposed hosts, do not develop beyond the first larval instar and do not actually kill their hosts until they have constructed their cocoon or entered some other retreat, so when the host and parasitoid would become especially vulnerable to predator attack, they are fairly safely out of harm's way.

Permanent host paralysis makes sense for ectoparasitoids of concealed hosts (see Fig. 12.33b–d), because such hosts could easily protect themselves if they were mobile and could (even inadvertently) crush the juvenile ectoparasitoid. In these cases, the parasitoid has to live with the 'cost' of being an idiobiont, in that the host cannot gain more resources and just declines in quality as soon as it is paralysed.

GREGARIOUS DEVELOPMENT

Gregarious parasitoidism, that is, where more than one larva develops on a single host, is widespread and occurs among both ecto- and endoparasitoids (Table 6.2). Sometimes among endoparasitic taxa, very large numbers of individuals (>100) may be reared from a single, large host, as with many *Cotesia* and other Microgastrinae species, some euphorines and meteorines, the rogadine genus *Macrostomion* and kin (M.R. Shaw 2002a, Maetô and Arakaki 2005) and the Meteorideinae (Ghosh & Abdurahiman 1984). Gregarious development is common among endoparasitoids within the Braconidae occuring in numerous groups [Agathidinae, Alysiinae, Braconinae (*Aspidobracon* group), Macrocentrinae, Meteorideinae, Microgastrinae, Rogadinae, Sigalphinae and some Euphorinae]. Further, in many of these it has undoubtedly been evolved on numerous occasions. In marked contrast, endoparasitic ichneumonids are only rarely encountered (M.R. Shaw 1999a). Known gregarious endoparasitic ichneumonids include the west Palaearctic campoplegine *Olesicampe clandestina* and

Table 6.2 Distribution of gregarious koinobiont* endoparasitoids in Ichneumonoidea with particular reference to those groups that are seldom gregarious.

Taxon	Gregarious representatives
Braconidae	
Aphidiinae	*Ephedrus californicus* (facultatively)
Macrocentrinae	*Macrocentrus*, many species
Euphorinae	Several species of *Meteorus, Microctonus, Perilitus* and *Syntretus*
Microgastrinae	Nearly all genera for which biology is known include at least some gregarious species
Meteorideinae	All known species of *Meteoridea*
Sigalphinae	*Sigalphus bicolor*
Rogadinae	A few species of *Aleiodes* and a few species of the unrelated genera *Triraphis, Colastomion* and *Macrostomium*
Ichneumonidae	
Campopleginae	*Olesicampe clandestina*
Banchinae	*Lissonota mutator*

*Some gregarious Lysiterminae may be koinobionts (van Achterberg & Steiner 1996).

three species of banchine in two genera, *Lissonota* and *Diradops* (M.R. Shaw 1999a, Gauld et al. 2002a). *O. clandestina* produced broods of between 20 and 40 individuals, but in the case of the banchines, the gregariousness is unusual in that brood size is small, from two to nine individuals emerging from the host (see below about the significance of small gregarious broods). All the broods of the Palaearctic *L. mutator* were either all male or all female, leading Shaw to speculate on the possibility of polyembryony in this species.

Small gregarious broods are known in some microgastrines (e.g. M.R. Shaw 2012) and a few other cases (see Chapter 12, section *Rogadinae sensu stricto*, and particularly what is written about *Colastomion*) and some of these are facultatively gregarious (e.g. Quicke et al. 2012b). Facultative gregariousness involves species that attack hosts of a considerable range of sizes, with sometimes only one developing in the smallest host species, for example, in some *Cotesia* species (M.R. Shaw et al. 2009) and the tropical Rogdinae genus *Colastomion* (Quicke et al. 2012b), and even rarely in aphidiines (Mackauer & Chou 2012). Such species are of considerable interest from the point of view of behaviour as it requires a change from the typically aggressive nature of solitary parasitoids that will seek, fight and kill rivals in a given host, to ones that are tolerant of other individuals (siblings) (Mayhew & van Alphen 1999, Mayhew & Glaizot 2001, Guinnee et al. 2005). Mayhew and van Alphen (1999) considered the evolutionary problem of how gregariousness might evolve, with particular relevance to the Alysiinae, in which there is strong selection for mechanisms to kill superparasitoids, be they siblings or the offspring of an unrelated female. They found that first instars of both a solitary and a gregarious species of *Aphaereta* possessed sharp mandibles, but a simple behaviour change to non-aggressive behaviour in the gregarious *A. pallipes* might have been all that was needed to permit the evolution of gregariousness. Interestingly, polyembryony is also involved in several instances of gregarious endoparasitism, and in these cases it might be that genetically identical individuals do not recognise other individuals as targets for elimination.

Smart and Mayhew (2009) examined sex ratios of gregarious broods across a wide range of braconids and ichneumonids, but failed to find the same negative relationship with clutch size as had been found earlier in bethylids; instead, the data from a total of 31 species showed a wide scatter and no pattern. The reasons for this are probably numerous, but likely important ones are the wide differences in biology (ecto- and endoparasitoids), sex determination system and variability in the occurrence of local mate competition.

Gregariousness in endoparasitoids might be a means of ensuring that the contents of the host are virtually fully consumed so as not to form a rotting mess that could be a source of bacterial or fungal infection. In the case or rogadine braconids such as *Aleiodes* and *Clinocentrus*, which are normally solitary, but pupate within host remains before the host's internal contents have been fully consumed, the final instar larvae bite a hole through the host's cuticle so as both to glue them to the substrate and also probably to rid themselves of surrounding, potentially fouling, liquors in addition to securing their pupation site in a good position. There is also evidence in some species that larger broods are better able to utilise host resources (e.g. Gu et al. 2003), but whether this was a reason for the evolution of large broods or the result of a later reduction in the physiological ability of single larvae is unknown.

SUPERPARASITISM

Superparasitism is when a wasp lays an egg or eggs in (or on) a host that was previously parasitised by a member of the same species. Sometimes it may be the same individual female wasp, in which case it is referred to as self-superparasitism. In many cases, when a usual host is only large enough to support the production of a single individual wasp, selection has normally led to strong behavioural avoidance of superparasitism (Bragg 1974, van Alphen & Nell 1982, Labeyrie & Rojas-Rousse 1985), although there are numerous cases in which such avoidance does not occur and in some situations may be adaptive (van Alphen & Visser 1990, Gu et al. 2003). In the case of the aphid parasitoid *Aphidius ervi*, females can discriminate conspecifically parasitised hosts using internal cues and avoid ovipositing in these if alternative unparasitised hosts were available. However, if they were only presented with previously parasitised hosts, they would superparasitise ones that had been oviposited into within the past 24 h, which are ones in which their larvae had a reasonable chance of surviving, but reject ones parasitised a longer time ago (Bai 1991). In this case, the outcome of the contest seemed to depend on both larval fighting and probably also on physiological suppression. The chelonine braconid *Ascogaster reticulatus* discriminates previously parasitised host eggs using the presence of conspecific venom as the cue. This was neatly demonstrated by Yamaguchi (1987), who compared eggs laid into unparasitised hosts eggs, unparasitised eggs with venom injected and artificial eggs with extract of the wasp's alimentary canal added. Unparasitised eggs with venom added were strongly avoided.

Females of the idiobiont, pimpline pupal endoparasitoid *Itoplectis naranyae* are able to detect self-superparasitised hosts externally for approximately 30 min after the initial parasitisation event and show strong avoidance of self-superparasitism. However, they do not discriminate against pupae containing eggs from other conspecific individuals (Ueno 1994) and also preferentially host feed on pupae parasitised 40 h or more previously (Ueno 1998). The use of an external oviposition-deterring pheromone appears to be demonstrated in the opiine fruitfly parasitoid *Diachasma alloeum*, which attacks various *Rhagoletis* species, including *R. pomenella* in both apple (*Mallus*) and hawthorn (*Crategus*); females reared on hosts in either fruit were able to detect and reject oviposition into hosts that had previously been attacked in either fruit (Stelinski et al. 2007, 2010).

Not only are many species of ichneumonoid able to discriminate hosts already parasitised by conspecifics, some can detect patches where themselves or conspecifics have already searched, although little is known about what cues might be involved. Females of the spider egg predator *Gelis infestans* (Cryptinae), studied by Van Baarlen et al. (1996), detect their specific host spiders from the silk trails they leave, but when searching leaves for them, they spent significantly less time on ones that had been searched previously.

LARVAL COMBAT AND PHYSIOLOGICAL SUPPRESSION

The first instar larvae of many ichneumonoids have a morphology called 'mandibulate', meaning that they have relatively large mandibles and well sclerotised head capsules, which are adaptations for eliminating competitors (see Figs 6.2 and 118). As they are usually very active, first instars may actively seek out any egg or larva of a competitor and use their mandibles to kill it. This was elegantly demonstrated by Vinson and Mourad (2000), who used a miniature, aquatic Y-ditch perfused with insect tissue culture medium and restrained one first instar larva at the end of one of the arms of the Y and released another at the base. Significantly more of the second larvae crawled towards the arm containing the restrained one than into the control arm.

The high mobility of first instar larvae means that the interval between the different superparasitism events can be critical for determining the survival probability of the first- or second-laid parasitoid progeny and often the second-laid offspring may have the advantage over the less mobile second instar, which never has large falcate mandibles (Fig. 6.2b) (Tillman & Powell 1992a). Chau and Maetô (2009) showed this very nicely with parasitism of *Spodoptera litura* by the polyphagous euphorine braconid *Meteorus pulchricornis*. When superparasitism occurred within approximately 30 min, the contest between the two hatching larvae was symmetrical, with no distinct advantage to either based on age. When the time between ovipositions was roughly 1 day, the larger, older first instar parasitoid larva tended to kill the younger one; however, if the interval between ovipositions was sufficient that the older larva had moulted into its amandibulate second instar, the younger mandibulate one always won. Similar results were obtained by Marris and Casperd (1996), who investigated the importance of the interval between

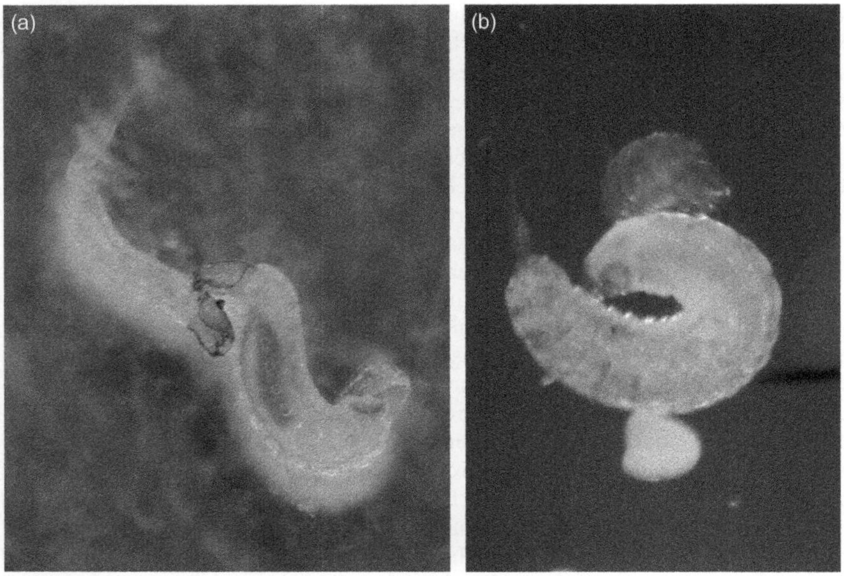

Fig. 6.2 Lethal combat between endoparasitoid larvae. (a) Intraspecific fight between two larvae of *Meteorus pulchricornis* (Euphorinae), the winner biting the loser behind the head, where the damage will expose the immunologically active soft tissue to further attack by host defence mechanisms. (Source: reproduced by permission of Kaoru Maetô.) (b) Interspecific contest between loser, *Hyposoter horticola* (Campopleginae), and winner, the much smaller larva of *Cotesia melitaearum* (Microgastrinae), parasitoids of the Glanville fritillary, *Melitaea cinxia* (Nymphalidae). (Source: reproduced by permission of Saskya van Nouhuys.) For a colour version, see Plate 11.

superparasitism oviposition events to the outcome of parasitoid larval competition in the campoplegine ichneumonid *Venturia canescens*.

Physical combat using mandibles is not the only means by which one parasitoid offspring can combat a rival, and it is clear that other physiological processes can be involved. Supernumerary larvae in the campoplegine *Diadegma* (as *Horogenes*) *chrysostictos* may eliminate one another either through physical combat if both are first instars or by humoral inhibition if one is a second instar; in both cases the older individual wins (Fisher 1959), so in this species the more mobile first instar does not have the advantage (*cf.* Fig. 6.2). The same result was found for another campoplegine, *Hyposoter didymator*, by Tillman and Powell (1992a). Fisher (1963) found that eggs and first instar larvae of *Venturia canescens* were particularly susceptible to low oxygen stress and suggested that physiological suppression might involve either release of chemical compounds or a reduction in oxygen level within the host. This idea was also put forward by Starý (1966) to explain the death of the second-laid *Aphidius* larva in a host that already contained a more mature larva. Timberlake (1910) came to the conclusion that elimination of supernumeraries in the aphidiine *Praon simulans* probably involved physiological suppression as '[they] were apparently already dead, being sometimes somewhat disintegrated although bearing no marks of violence'. He was able to exclude starvation as the cause since at the time of dissecting the aphid host, there was still a lot of unconsumed tissue available. With modern technology, it ought to be possible to obtain precise measurements of oxygen partial pressures inside hosts to test these hypotheses more thoroughly.

Two species of opiine parasitoids of tephritid fruit-flies studied by X. Wang and Messing (2003) showed completely different modes of elimination of supernumeraries resulting from superparasitism. In *Diachasmimorpha tryoni*, physical attack was involved, but in *Fopius arisanus*, the mechanism appeared to be physiological suppression and the same mechanism seemed to work interspecifically. Lawrence (1988) showed that larvae of another species of *Diachasmimorpha* (as *Biosteres*), *D. longicaudata*, release allelochemical toxins *in vitro* that suppress the development of their competitor (and therefore presumably also do so *in vivo*) in addition to physical combat. However, the eggs themselves were shown not to release any such active compounds. In the stilbopine ichneumonid *Panteles*, a koinobiont larval parasitoid of the incurvariid moth *Lampronia fuscatella*, most wild-collected, parasitised hosts contained multiple parasitoid eggs, but only ever one parasitoid larva, and it was apparent that only one of the eggs had ever hatched (*cf.* Fig. 183c,d) (Quicke 2005). Unfortunately, the local nature of the host, which forms galls on young birch (*Betula*) trees in heathland in Europe, together with its strictly annual life cycle, means that the mechanisms involved will probably never be elucidated, at least through physiological experimentation.

Physiological suppression might be fairly widespread as a mechanism of eliminating supernumeraries, but it is less easily demonstrated than physical attack where often one might see the damaged and perhaps encapsulated remains of the loser. There is also evidence that non-combative suppression of competitor eggs occurs in rogadine braconids (Dowden 1938), and when superparasitism occurs in the banchine *Lissonota dubia* (Jørgensen 1975, as *Pimplopterus*), the first larva to hatch migrates to the head end of its host caterpillar and this apparently gives it some sort of advantage as it is always the winner, although the exact mode by which its physical location aids its physiological suppression of competitors is unclear. It seems likely that physiological suppression is particularly important in various groups of ophioniformes, including most ctenopelmatines, some banchines and the aforementioned stilbopine *Panteles*. In many of these, the first instar larvae have only tiny mandibles that seem unsuitable for combat.

ADAPTIVE SUPERPARASITISM

Based on observations of the ichneumonid *Diadromus pulchellus*, a solitary idiobiont endoparasitoid of the leek moth pupae, Labeyrie and Rojas-Rousse (1985) argued that when superparasitism occurs it may actually be adaptive. When presented with multiple hosts, the female *Diadromus* concentrated oviposition in a subset of those available, as was also the case with the cryptine *Pleolophus bazonicus* (K.J. Griffiths 1971). Similarly, there is evidence that superparasitism might be adaptive in a number of systems involving microgastrine braconids. Sallam et al. (2002) found that *Cotesia sesamiae* could produce up to three offspring per host as a response to multiple oviposition events, suggesting that normally only a single egg is laid in a host. Gu et al. (2003) further showed that in the case of the gregarious *C. glomerata*, the weights of females that emerged from superparasitised caterpillars were significantly greater than those from singly parasitised hosts, which suggests that the extra stress caused by increased competition were offset, at least to some extent, by an increased availability of host resources for consumption, perhaps additive effects of venom/polydnavirus or maybe larval secretions. In a very interesting study, F. Schneider (1950), who was trying to investigate whether the parasitoid or its hoverfly host was in control of voltinism, attempted to get the diplazontine *Syrphoctonus* (as *Diplazon*) *fissorius* to parasitise syrphid species other than its regular host, *Episerphus balteatus*. These attempts failed because of a strong immune response of *E. balteatus* against its novel parasitoid. However, by forcing the wasp to superparasitise the novel host, successful rearing was achieved. In the gregarious ectoparasitic tryphonine *Netelia perfecta*, attacking *Helicoverpa* hosts, there is an optimum number of eggs per host and although females will sometimes massively superparasitise, this leads to the death of the host and all the parasitoid larvae (Fig. 6.3) (Johns & Whitehouse 2004).

Sometimes, when a host is sufficiently large, selection has not always led to superparasitism avoidance. J.-H. Zhang et al. (2010) showed that the normally solitary campoplegine ichneumonid *Campoletis chlorideae* can easily be made to superparasitise an acceptable, large host, *Mythimna separata*, and that although there is some avoidance of freshly parasitised hosts, females did not distinguish unparasitised hosts from those parasitised 24 h previously. Khafagi and Hegazi (2008) showed that superparasitism by the solitary parasitoid *Microplitis rufiventris* of *Spodoptera littoralis* was beneficial in older and unsuitable hosts. Later instar (fourth to sixth instars) were poor or even completely unsuitable for the solitary development of this wasp, but became suitable when doubly or triply superparasitised. When early instars were superparasitised, larvicide led to only one developing to adulthood, but larvicide did not seem to occur in larger superparasitised hosts and gregarious development ensued. Nevertheless, with superparasitism, parasitoid larvae were not successful in egressing from the host, perhaps owing to its greater than normal cuticular thickness or other requirements

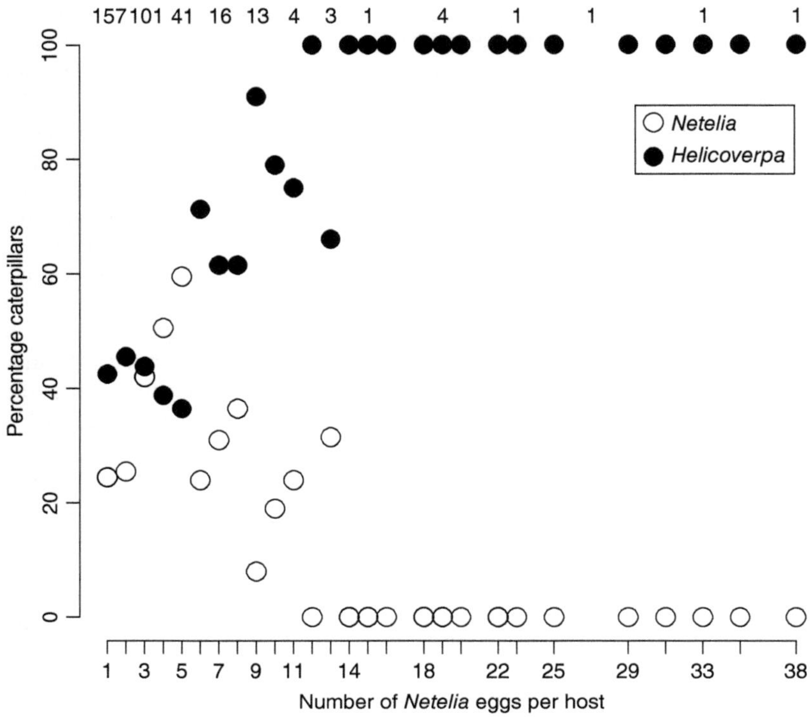

Fig. 6.3 Outcomes of progressively increasing superparasitism on the survival of *Netelia producta* (Tryphoninae) and its host moth, *Helicoverpa armigera* (Noctuidae), showing that maximum production per caterpillar occurred at brood sizes of around five and that high egg loads lead to the death of both parties. Numbers of caterpillars, approximately every other egg load, are given along the top; not shown for clarity are the caterpillars that died without yielding moth or wasp. (Source: data redrawn from Johns & Whitehouse 2004.)

not present in such a late host stage (Hegazi & Khafagi 2005, Khafagi & Hegazi 2008).

MULTIPARASITISM

Although in general, parasitic wasps can discriminate between hosts that have not been attacked and those that have been attacked by a conspecific (or even themselves) and so avoid superparasitism (Salt 1961), they do not, as a rule, discriminate between unparasitised hosts and those that have already been parasitised by another species (e.g. Wallner et al. 1982, Kenis 1997, Muli et al. 2006), and multiparasitism must be fairly common in Nature. Indeed, modelling of the *Drosophila–Asobara tabida–Leptopolina heterotoma*[3] system, in which both parasitoids show conspecific superparasitism avoidance but fail to discriminate allospecifically parasitised hosts, shows that as long as the females are not egg-limited there will be no selective advantage for avoiding multiparasitism even under fairly extreme scenarios (Turlings et al. 1985).

However, avoidance of multiparasitism is known in numerous instances and has been known for some while (Godfray 1994), the first reported case probably being that by Lloyd (1940), who found that the campoplegine *Diadegma semiclausum* (as *Angitia eucerophaga*) avoided ovipositing in diamond back moth hosts containing the microgastrine *Cotesia* (as *Apanteles*) *plutellae*. In the case of another campoplegine-microgastrine pair, *Campoletis sonorensis* and *Cotesia marginiventris*, parasitising *Spodoptera frugiperda*, Tamò et al. (2006) showed that *C. sonoriensis* was the superior competitor in cases of multiparasitism and that each species readily accepted hosts parasitised by the other, but that in an olfactometer experiment the *Cotesia* strongly avoided the odour of the *Campoletis*, which agreed with predictions that, since the parasitoids are unlikely to be egg-limited, they should only reject previously parasitised hosts if it will save them time, and the time saving is likely to be greatest in avoiding patches that have been visited by the other species rather than rejecting individual hosts once encountered. That both of these examples, which are some of the few showing innate interspecific host discrimination, involve the same two subfamilies is potentially interesting, but whether it reflects something about the strength of selection, the likelihood of finding previously attacked hosts, the suitability of the microgastrine-infected host and the probable outcome of multiparasitism is not clear. Microgastrines tend to attack hosts as early instars, so that could explain why the campoplegine has evolved the avoidance behaviour and not the other way around.

Similarly, avoidance of patches where another parasitoid species has been foraging was demonstrated by Price (1970) for cocoon parasitoids of the diprionid sawfly *Neodiprion*, all belonging to the Cryptinae (viz. species of *Endasys*, *Mastrus* and *Pleolophus* species). Using an arena, Price found that females avoided areas that had been visited previously by one of the other species in addition to avoiding areas of ground that they themselves had previously searched. With unrelated parasitoids of the same host, K.J. Griffiths (1976) found that the cryptine *P. bazonicus* strongly avoided ovipositing into *N. sertifer* cocoons in which the larva had previously been parasitised by the endoparasitic ctenopelmatine *Lathyroplectus luteator*. When multiparasitism did occur, the cryptine was always the winner, as would be expected since it is an ectoparasitoid, but the individuals that emerged were always smaller (and presumably less fit) than ones that attacked unparasitised hosts. Because this is a regular association and the cryptine consistently wins, it is not surprising that natural selection has resulted in interspecific discrimination.

Whether parasitoids can discriminate hosts attacked by another species may depend on how closely related the wasps are (Marris et al. 1996, Moser et al. 2008), possibly due to having preadaptations for the detection of their oviposition markers, although there are exceptions (Bai 1991). Vet et al. (1984) showed that two species of *Asobara* (Braconidae: Alysiinae) could distinguish allospecifically parasitised *Drosophila* larvae, McBrien & Mackauer (1990) showed that two closely related species of *Aphidius* could avoid multiparasitising aphids attacked by the other and Aluja et al. (2013) found that the opiine *Utetes anastrephae* discriminated against hosts already parasitised by another opiine, *Doryctobracon areolatus*, with which it routinely competes. S. Liu et al. (2001) showed that although the ichneumonine *Diadromus collaris* attacked host *Plutella xylostella* pupae that had been parasitised previously by the eulophid *Oomyzus sokolowskii*, they did not lay eggs in them.

As with superparasitism, the outcome of multiparasitism may be species specific or depend on the order in which the host is parasitised, although in some cases it is the first to attack that tends to survive and in others it is the second, especially if the earlier parasitoid has moulted to a less mobile and aggressive second instar (Fig. 6.4) (Ueno 1999b). In an interesting study by Magdaraog et al. (2012) in which armyworm (*Mythimna separata*) caterpillars were experimentally multiparasitised by a solitary *Meteorus* and two gregarious *Cotesia* microgastrines in all pairwise combinations, the *Meteorus* nearly always won, but occasionally, when both *Cotesia* attacked the same host in rapid succession, mixed broods of adult wasps sometimes emerged. Intrageneric multiparasitism was studied by Kalmes et al. (1983) using two obligately solitary species of *Diadromus* (Ichneumoninae), both of which can successfully parasitise *Acrolepiopsis assectella* pupae (Yponomeutidae), *D. collaris* and *D. pulchellus*. In this case, neither species seemed to have a significant advantage over the other. In the system studied by Tian et al. (2008), there was a clear advantage to being the first to oviposit. When the microgastrine *Microplitis mediator* was first, they eliminated the larvae (although not the eggs) of the competing campoplegine, *Campoletis chlorideae*, by physical combat using their large mandibles, the vanquished nearly always showing clear signs of injury. However, when the parasitism order was reversed, the *Campoletis*, whose first instar does not have large mandibles, won but apparently through a physiological process. Interspecific physiological suppression of competitors has also been reported for another campoplegine, *Hyposoter horticola*, against eggs and early larvae of its competitor, the microgastrine *Cotesia melitaearum* in one of the latter's generations (van Nouhuys & Punju 2010). Thus campoplegine larvae often appear to rely on physiological suppression of competitors under conditions of both superparasitism and multiparasitism.

In cases where oviposition of the two parasitoids is effectively simultaneous, the species whose egg is

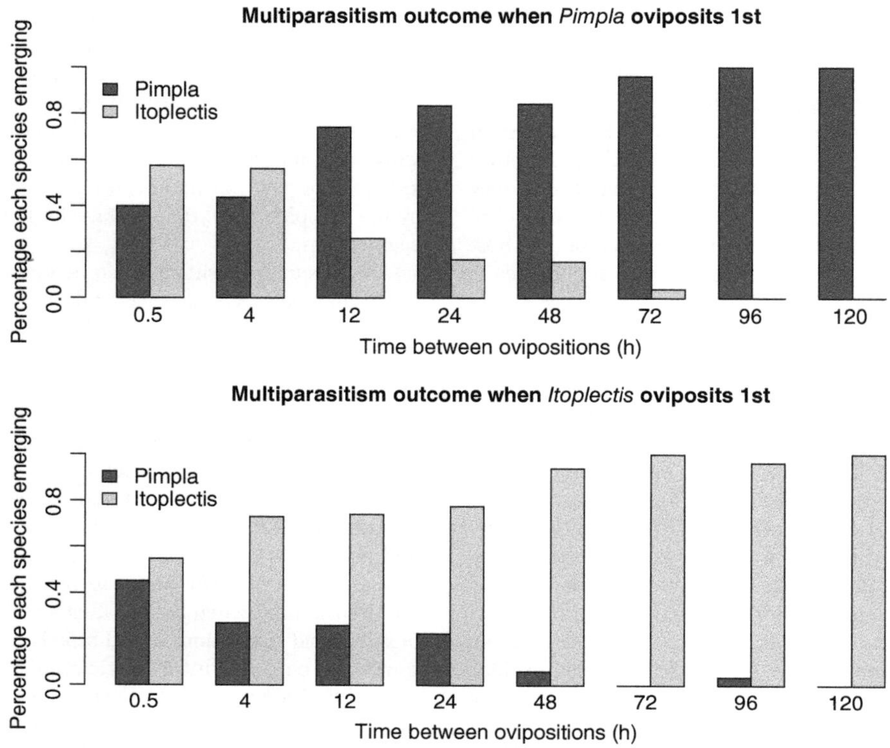

Fig. 6.4 Outcomes of multiparasitism of wax moth pupae (*Galleria mellonella*) by the pimpline idiobiont endoparasitoids *Pimpla nipponica* and *Itoplectis naranyae* depending on order and time interval between ovipositions. (Source: data from Ueno 1999.)

first to hatch sometimes has the advantage (Tillman & Powell 1992b). The pimplines *Apechthis ontario* and *Itoplectis quadricingulatus* will both attack wax moth (*Galleria*) pupae, and the former is the superior competitor when ovipositions by the two are separated by only a few minutes (R.B. Ryan 1971), perhaps because its eggs hatch earlier; however, when the time interval before multiparasitism events was increased to 3 days, the species first to oviposit had the advantage. In this system, *A. ontario* tended to avoid multiparasitism when host pupae contained a developing parasitoid larva, but the *Itoplectis* did not.

Whether a parasitoid avoids multiparasitism may also be influenced by prior experience. Naïve females of both *Pimpla disparis* (an introduced parasitoid for biological control of gypsy moth) and *Itoplectis conquisitor*, both generalist Lepidoptera parasitoids, were reluctant to oviposit in pupae previously attacked by the other species, but became more willing to multiparasitise after having previously oviposited (Moser et al. 2008).

In another study involving pupal parasitoids, the pimpline *Xanthopimpla stemmator* and the ichneumonine *Dentichasmias busseolae*, Muli et al. (2006) found that the *Xanthopimpla* was an intrinsically superior competitor if it oviposited first or at the same time as the ichneumonine, but that there was a strong time interval effect on the winner if the *Dentichasmias* oviposited first, with increasing success the longer the interval.

Whilst one might think that the majority of multiparasitisms involving the Ichneumonoidea involve either other ichneumonoids or tachinid Diptera, analogous situations occur when hosts have been infected previously by either pathogenic viruses, bacteria, fungi or nematodes. Lacey et al. (2003) showed that two ichneumonid parasitoids of the codling moth, *Cydia pomonella*, viz. the cryptine *Mastrus ridibundus* and the pimpline *Liotryphon caudatus*, both avoided ovipositing on hosts that were infected by the nematode *Steinernema carpocapsae*.

OBLIGATE AND PREFERENTIAL MULTIPARASITISM

As many parasitoid venoms have a suppressive effect on the immune capability of the host, it ought in principle to render the host more favourable to a second parasitoid, perhaps one that was not so adapted to it that it would have a high chance of survival in a healthy host with uncompromised immunity. Hence it is perhaps surprising that this niche, i.e. hosts housing a previous parasitoid of another species, has not apparently happened very often. Indeed, perhaps it does and might explain why it is sometimes not easy to culture parasitoids. As it must certainly happen in the wild, this might also be a mechanism that facilitates the acquisition of new hosts to a parasitoid's host range, allowing some survival in a new host which natural selection could then more easily improve upon.

When various species of the campoplegine genus *Hyposoter* attack the white-marked tussock moth, *Orgyia leucostigma*, their eggs are encapsulated by this unsuitable host; however, Guzo and Stoltz (1985) showed that if the *Orgyia* had been previously parasitised by the microgastrine *Cotesia melanoscela*, then all three of the *Hyposoter* species investigated could avoid encapsulation and develop to maturity. By injecting caterpillars with *Cotesia* venom and polydnavirus (calyx gland extract) (see Chapter 9, section *Polydnaviruses*) both separately and combined, they showed that a combination of the two was essential to enable the *Hyposoter*s to survive. Further, the host's immune response against *Hyposoter* larvae (but not eggs) could be blocked by chemically covalently cross-linking the polydnavirus DNA, indicating that active expression of polydnavirus genes was important to suppress the immune response completely. The *Drosophila* parasitoid *Asobara tabida* (Alysiinae) attacks both *D. melanogaster*, which is one of its normal hosts, and also, even in choice experiments, *D. simulans*, in which it cannot develop (A.R. Kraaijveld 1999). However, this behaviour is not necessarily disadvantageous as it can develop in *D. simulans* if this host has also been parasitised by the eucoilid *Leptopilina boulardi*, which commonly co-occurs with *A. tabida* in the field.

In another possible example, Arthur et al. (1964) showed that the cremastine *Temelucha interruptor* preferentially oviposited in larvae of the pest pine shoot moth, *Rhyacionia buoliana*, that had previously been parasitised by the braconid *Orgilus obscurator*, but what, if any, implication this had for survival of the *Temelucha* was not investigated, and it seems from their experimental results that *Temelucha* females were simply not very good at locating unparasitised hosts, the inference being that ones attacked by *Orgilus* might have been marked by some pheromone, rendering them more 'conspicuous'. Schröder (1974) subsequently demonstrated that not only *T. interruptor* showed a kleptoparasitic interaction but so too did two other ichneumonid parasitoids, the cremastine *Pristomerus* sp. nr *orbitalis* and the campoplegine *Sinophorus rufifemur*. Hence there is clearly something rather special going on in this *Orgilus*/*Rhyacionia* system. Nevertheless, this may have had serious implications for the biological control of the moth and the authors suggested that the importation and release of additional numbers of *Temelucha* into the United States to control the moth may have been unwise because of this parasitoid's negative impact on the *Orgilus*. Such kleptoparasitism also occurs at the expense of the ichneumonoid.

HYPERPARASITISM AND PSEUDOHYPERPARASITISM

The great majority of ichneumonoids are primary parasitoids, but a few are obligate hyperparasitoids, that is, they necessarily parasitise the larvae of other parasitic wasps while they are feeding on or in their primary host. Only two groups of ichneumonoids are obligate hyperparasitoids, the Eucerotinae and Mesochorinae, and no braconids have this biology. However, these two groups of ichneumonids differ enormously in their biologies.

Pseudohyperparasitism is a term given to situations where the host is a primary parasitoid, but it is not attacked until it has completed feeding on its host and is almost always in its own cocoon (or rarely mummified host or the adult). Hence the location of the host parasitic wasp may effectively be independent of the latter's host. Only a couple of instances of pseudohyperparasitism are known within the Braconidae, the euphorine genus *Syntretus*, some members of which attack adult parasitic wasps (of both Ichneumonidae and Braconidae) (Cole 1959a, van Achterberg & Haeselbarth 2003), and one species of the alysiine genus *Aphaereta* that has been recorded as attacking a tachinid (Wharton 1984). However, within the Ichneumonidae, pseudohyperparasitism has evolved on several occasions, notably in the Cryptinae with many species of *Gelis* and *Lysibia*, and also members of other genera, attacking other cocooned ichneumonoids, especially Microgastrinae as ectoparasitoids (Wieber et al. 1995) and some *Itoplectis* species (Pimplinae) being specialist pupal endoparasitoids of cocooned ichneumonids (M.R. Shaw 2009b), while others may be facultatively so (Evenhuis & Vlug 1983). Other slightly different instances include some cryptine parasitoids of solitary aculeates, for example, *Hoplocryptus* can develop not only on various Apidae, Crabonidae and Vespidae, but also on gasteruptiid and chrysidid primary parasitoids of these (Schwartz 2007), and then there is the rhyssine ichneumonid *Rhyssa persuasoria*, which can facultatively be an hyperparasitoid of Ibaliidae primary parasitoids of their siricid hosts (Hanson 1939). Probably such instances are commoner than currently realised as they would normally require careful observation of host remains and would probably not be detected in simple substrate rearings.

KLEPTOPARASITISM

The above example of *Temelucha* and *Orgilus* might also therefore be regarded more as an issue of kleptoparasitism, with attack by the latter rendering the host more easy to locate. The best known example however, involves the poemeniine *Pseudorhyssa* which waits until a rhyssine has oviposited on a host and then inserts her ovipositor down the same boring and lays her own egg. Given that 'drilling' or ovipositing female parasitoids in conspicuous situations such as those attacking bark-borers is a clear indication that there is likely to be a host beneath them, it is perhaps surprising that there are not more known instances of this sort of interspecific kleptoparasitism – after all, the first parasitoid has done a great deal of the work in locating the host already. Some wood-borer parasitoids, e.g. the braconines *Coeloides* and *Monilobracon*, display aggressive behaviour towards potential conspecific and allospecific competitors (Mills 1991, Quicke et al. 2005b) to protect their oviposition site. Only in the second case did encounters seem to be deliberately targeted at other drilling females, while in the first they seemed to be random.

EVOLUTION OF LIFE HISTORY STRATEGIES

Only now that some sort of phylogenetic hypotheses are emerging for both the Ichneumonoidea as a whole and for some particular groups in detail can we start to examine the evolution of host shifts. One of the most obvious features in both families is that the great majority of transitions are from ecto- to endoparasitism (see Figs 12.2 and 12.73) and this is very much the dogma. The 'Tryphoninae' which show an intermediate biology, being koinobiont ectoparasitoids but with the anchor of the egg placed internally within the host, appear close to the base of the endoparasitic ophioniformes (admittedly not as a monophylectic group) in both morphological, molecular and combined analyses (Quicke et al. 2000b, 2009), although never exactly at the base. This might suggest that they are true intermediates or, if they represent a reversal or reversals to ectoparasitism, it would have happened fairly soon after the basal members of the ophioniformes had evolved endoparasitism, and it is worth noting that several endoparasitoid tribes and subfamilies also recovered near the base of the ophioniformes have strongly sclerotised robust eggs [e.g. some Ctenopelmatinae, see Fig. 13.44d,f; Stilbopinae (*Panteles*), see Fig. 13.41a–d; some Banchinae] and at least some of these taxa retain papilliform larval antennae (e.g. *Stilbops*; Short 1970).

Both Gauld (1988a) and Quicke et al. (2000a) discussed the probable constraints on the evolutionary transitions between being an ectoparasitoid of a concealed host and an endoparasitoid of an exposed/concealed host and possible transitional stages. The latter authors concentrated on the joint issues of ecto- versus endoparasitism, host concealment, host paralysis and the evolution of koinobiosis. The overall picture might be summarised as follows. Koinobiont endoparasitism of different orders is not easy to evolve and may require some pre-adaptations. On the assumption that the ancestral ichneumonoid was an idiobiont ectoparasitoid of a (weakly) concealed host, Fig. 6.5 illustrates how if the host is exposed, there is distinct advantage to paralysing the host if the egg and subsequent larval development are external or else the host might squash them against the wall of its confines or otherwise dislodge them. There is no advantage to paralysing the host if the egg is laid internally and endoparasitism offers no advantage *per se* in this situation, except if the host may be attacked at an early stage of development when it is accessible if later it were to be either inaccessible or having to move to a new location, perhaps for pupation.

In contrast, for parasitoids of exposed hosts, Fig. 6.6 illustrates the distinct advantage of being an

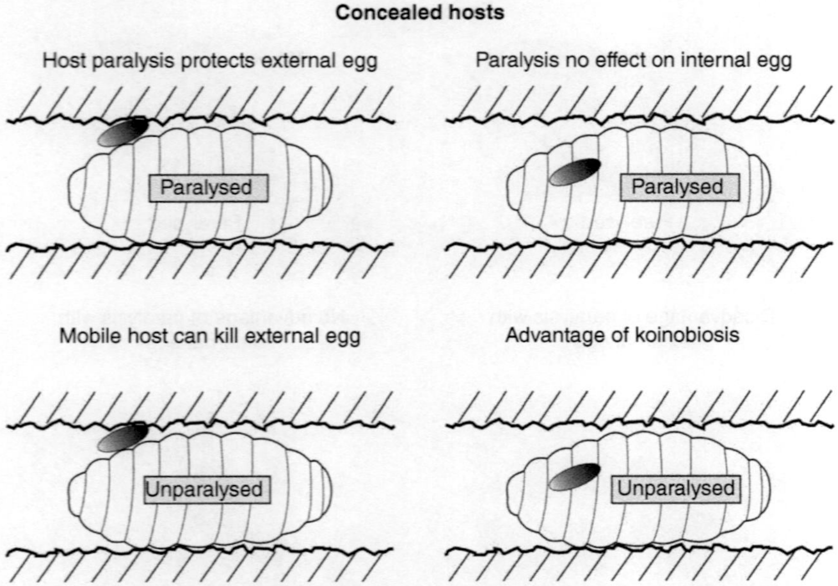

Fig. 6.5 Cartoons illustrating the advantages and disadvantages of four combinations of ecto- and endoparasitism of a paralysed or unparalysed concealed host. (Source: redrawn from Quicke et al. 2000a.)

endoparasitoid of an unparalysed host, that is, a koinobiont. If an exposed host were permanently paralysed, it is obvious that it would be vulnerable to attack in addition to not being able to continue feeding or growing. If the parasitoid's development were not internal, the host could potentially dislodge or kill it. Nevertheless, at least four origins of koinobiont ectoparasitism[4] have resulted in successful strategies, viz. the Rhysipolinae among the braconids, although technically biology has only been studied for European *Rhysipolis*, and in the Adelognathinae and Tryphoninae among the ichneumonids. Of these there are very few species of either *Rhysipolis* or *Adelognathus* and most of their hosts live somewhat concealed for part of their lives. Only the Tryphoninae and *Polysphincta* group of Pimplinae have capitalised on this way of life through the evolution of highly adapted larvae and larval strategies to remain attached to the host when it moults. Even then, development of the parasitoid is either delayed until the host has constructed its pupal chamber/cocoon (Tryphoninae) on the host, having reached sufficient size, has its behaviour modified so that final parasitoid development can continue in an appropriate site. It should be mentioned that some Metopiinae, and potentially Lycorininae, technically oviposit externally but within the moist confines of the host rectum and the parasitoid larva then has to bore into the host's body before continuing development, but these have highly specialised, and possibly species-specific, strategies, not much removed from, and undoubtedly having evolved from, true endoparasitism. Outside the Ichneumonoidea, only some Eulophidae successfully act as koinobiont ectoparasitoids (of exposed caterpillars) and their larval development is very rapid once the host becomes unable to feed and defend itself.

In relation to the above, M.R. Shaw (1983) drew attention to the biological similarities between two cyclostome braconid genera, the rogadine *Clinocentrus*, which is a koinobiont endoparasitoid, and *Rhysipolis* (Rhysipolinae), which is a koinobiont ectoparasitoid, both of weakly concealed Lepidoptera caterpillars. The former, atypically for both koinobionts and endoparasitoids, lays rather large, yolky eggs just below the host cuticle (see Fig. 7.2a), whereas *Rhysipolis* lays on to the host's exterior. Hence it was easy to imagine a scenario where a *Rhysipolis*-like wasp might evolve to deposit its egg internally, thus perhaps gaining some protection from possible dislodging actions of its host and allowing the host greater freedom that might be mutually beneficial, and Gauld (1988a) presented

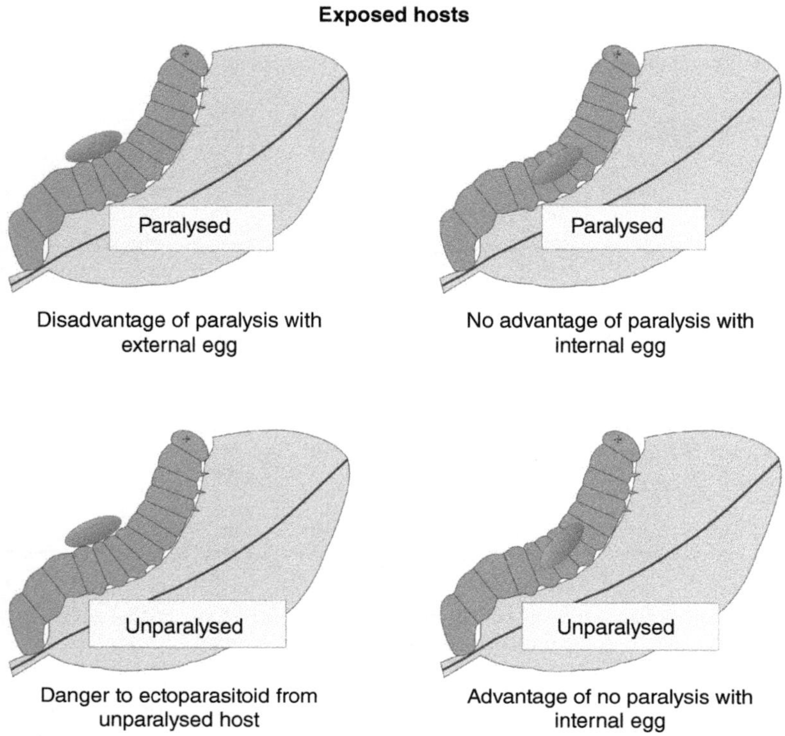

Fig. 6.6 Cartoons illustrating the advantages and disadvantages of four combinations of ecto- and endoparasitism of a paralysed or unparalysed exposed host. (Source: adapted from Quicke et al. 2000a.) For a colour version, see Plate 12.

a hypothetical set of plausible steps that would lead from an idiobiont ectoparasitoid of a concealed host to being a *Rhysipolis*-like koinobiont ectoparasitoid of an exposed host that subsequently seeks a secluded retreat for pupation. At the time, it seemed possible that these were in fact transitional stages leading to endoparasitism in the Rogadinae, although molecular phylogenies indicate that the two subfamilies are not in fact that closely related.

Several genera, such as the cryptine ichneumonids *Bathythrix* and *Gelis*, attack a number of taxonomically diverse hosts, but all are united by their association with silks. For example, each genus includes species that are predators within spider egg sacs, pseudohyperparasitoids of parasitic wasp cocoons and primary parasitoids of cocooned sawfly larvae (Fitton et al. 1987, Gauld 1988a). As a slight twist, some *Bathythrix* species are ectoparasitoids of *Aleiodes* (Rogadinae) prepupae or pupae within the mummified caterpillar skin, which has minimal associated silk, but the switch was probably made because in other respects these are cocoon-like structures. Gauld postulated the following sequence of host relationships as an example of how host silk might have become intrinsically linked with host location and acceptance in these groups (Fig. 6.7). Then, paying particular attention to the evolution of Pimplini, although it also applies to some Cryptines, this was extended and refined to accommodate the attack of exposed Lepidoptera pupae, which are the hosts of many members of the tribe Pimplini, a group of endoparasitic idiobionts (Fig. 6.8). In this scenario, the thinning of the host-provided protection for its own pupa would have provided an advantage for the parasitoid if it evolved to feed internally where it might be more protected from hyperparasitoids and predators. An example that might show how this could evolve involves the cryptine *Glabridorsum stokesii*, which belongs to a group that is normally

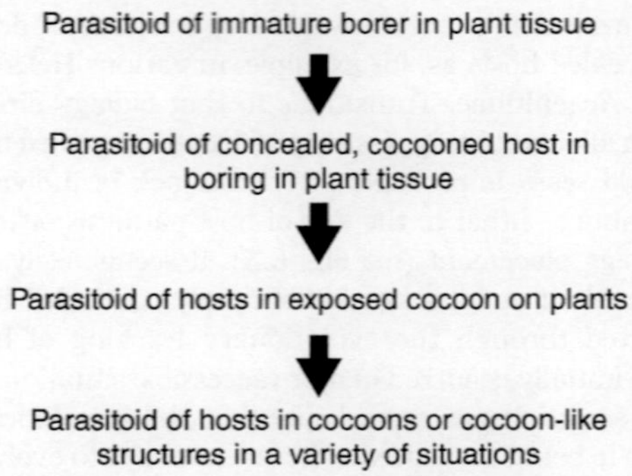

Fig. 6.7 Gauld's (1988a) proposed evolutionary pathway in ichneumonids stemming from utilisation of silk as a host location cue.

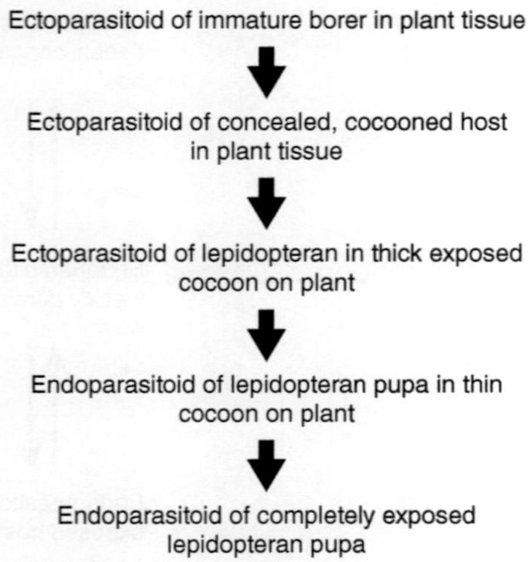

Fig. 6.8 Gauld's (1988a) proposed evolutionary pathway in ichneumonids from ectoparasitism of concealed hosts to endoparasitism of exposed pupae.

ectoparasitic. In this species, the female wasp attacks a pupal stage host and oviposits externally but the first instar wasp larva starts feeding below the host's wing case before burrowing through the host cuticle to become fully endoparasitic. It is interesting that within the Cryptinae, there appear to have been several shifts to endoparasitism rather than a single radiation as in the Pimplinae, that is, in the tribe Pimplini.

In the case of the polysphinctine group of pimplines, which are koinobiont parasitoids of young and adult spiders, the shift to this dangerous class of hosts has been hypothesised to have been via attacking their egg sacks as egg predators – or sometimes as spiderling predators (Townes & Townes 1960, Townes 1969). Potential intermediate evolutionary stages displayed by extant taxa were discussed by Fitton et al. (1987) and include examples wherein the parasitic wasp kills the adult spider guarding its egg sack or the parasitoid larva feeds both on the adult spider and on her eggs (Nielsen 1935). For example, in the case of *Zaglyptus varipes*, which is a normally a gregarious egg predator of *Cheiracanthium* spiders, the female stings and kills the spider guarding her egg sac and the parasitoid larvae consume not only the eggs but also the adult spider, and will do so even if the spider has not laid any eggs at the time of being attacked. A cautionary note needs to be added here because, as Fitton et al. (1988) pointed out, Nielsen's observations were fairly limited and it would be very good to have more detailed studies of the biologies of these spider egg sac predators, especially *Zaglyptus* and *Clistopyga*. Nevertheless, although morphological phylogenetic analysis supports this attractive hypothesis, no close relationship between the *Polysphincta* group and spider egg parasitising pimplines was found in effectively purely molecular analysis by Quicke et al. (2009).

A major difference between the two families concerns the occurrence of the supposed ancestral biology of being an idiobiont parasitoid of concealed hosts. Gauld (1988a) supposed that the ancestral condition is displayed by various, somewhat disparate, basal ichneumonids, alluding to groups such as labenines, xoridines and pimpliformes, whereas in the Braconidae, wood-borer ectoparasitoids are nested within the cyclostomes, being principally within the Braconinae and Doryctinae. Gauld concluded that probably the ancestral idiobiont ectoparasitoid lineages (attacking quite concealed hosts) in the Braconidae must have gone extinct. However, given the relatively short ovipositors of the earliest fossil ichneumonoids, it is at least as plausible to consider the labenines and xoridines as displaying rather derived biologies in their specialisations for attacking quite deeply concealed, wood-borer hosts.

I believe that the overall set of transitions that have occurred in the Ichneumonoidea can be summarised as in Fig. 6.9. Starting from an idiobiont ectoparasitoid of a weakly concealed host (as indicated by, for example,

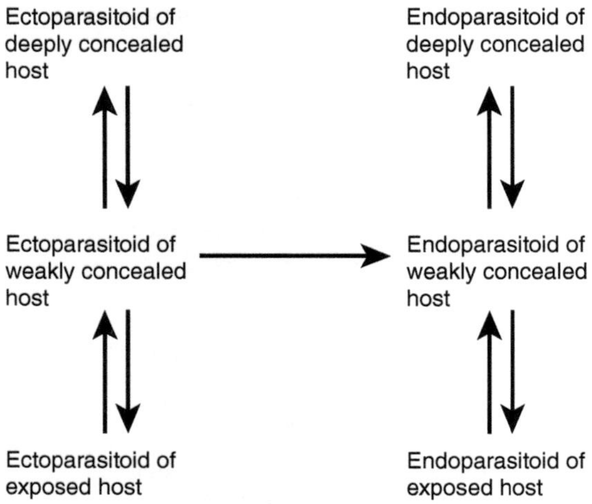

Fig. 6.9 Proposed most likely evolutionary pathways between various combinations of ecto- and endoparasitism and whether the host is concealed weakly, deeply or not at all, assuming only one feature can change at a time. (Source: modified after Quicke et al. 2000a.)

the small body size and the ovipositor being only about as long as the body or less in early ichneumonoid fossils), adaptations of the ovipositor could enable them to gain access to deeply concealed, e.g. xylophagous, hosts and there would, in that case, be no advantage to evolving endoparasitism. Similarly, from a weakly concealed host as a starting point, the biologies of taxa such as rhysipoline braconids and tryphonine and adelognathine ichneumonids must have evolved, each due to the particular circumstances of their host's biology, and in particular the host being available for parasitism earlier in its development but not suitable for the parasitoid to complete feeding upon until later, after it has constructed its cocoon elsewhere. According to Gauld's (1988a) scenario (Fig. 6.8), I suspect the transition to endoparasitoidism probably occurred on weakly concealed hosts, probably through a mixed strategy with external oviposition and subsequent endophagy of the larva, followed by many adaptations of the egg and early instar parasitoid larvae, losing those features associated with ectoparasitoid development such as papilliform antennae, but gaining many more features such as hydropic eggs, trophamnia and teratocytes. The transition to endoparasitoidism in a weakly concealed host space will, of course, enable the full endoparasitic koiniobiont attack of exposed hosts and also, although with far fewer examples, the evolution of koinobiont endoparasitism of deeply concealed hosts as, for example, in various Helconini and Acaenitinae. Transitions to that biology directly from idiobiont ectoparasitism of deeply concealed hosts would seem improbable given the lack of individual advantage either to the loss of host paralysis or internal egg placement (see Fig. 6.5). It seems likely that koinobiont endoparasitoids of deeply concealed hosts evolved through the evolutionary tracking of hosts that initially occurred in more accessible situations.

It seems to me remarkable that the pimpliformes, despite being very diverse, never managed to evolve to be successful endoparasitoids of caterpillars, and their sister group, the ichneumoniformes only do it poorly in the sense that only a very few species of Ichneumoninae have succeeded in attacking, as koinobionts, early instanr caterpillars, and these do not belong to particular species clades. Similarly, in the Braconidae, most non-cyclostomes are koinobiont endoparasitoids of Lepidoptera, with a few tribes/subfamilies attacking principally Coleoptera. Only once have the cyclostomes done so with many species (Rogadinae), and they are not very well adapted to being koinobionts despite *Aleiodes* and *Triraphis* being very speciose. It has been suggested that the relatively large eggs of Rogadinae (see Figs 7.2a and 4.1a) mean that they have only relatively recently evolved into the Lepidoptera koinobiont endoparasitoid niche, yet there are species in Baltic amber, some 40 million years old, that appear to be Rogadinae, and 40 million years can hardly be considered a short time.

Within the Ichneumonidae, koinobiont endoparasitism of Diptera larvae is exhibited by several pimpliformes subfamilies (Cylloceriinae, Diplazontinae, Orthocentrinae), and current phylogenetic hypotheses suggest that these form a monophyletic group (see Fig. 13.10) and also in a single lineage, the Stilpnina, within the Cryptinae. Only the Opiinae+Alysiinae clade do so in the braconids and there is no evidence of any members of these attacking other host orders, suggesting that the adaptations required to attack Diptera prevent the successful attack of other groups and host shifts on to them – a sort of large-scale evolutionary dead end. It seems unlikely that this has anything to do with host location, as at least two groups (Braconinae, Exothecinae) of idiobiont ectoparasitoids, even intragenerically, can attack Diptera as well as members of other orders when they co-occur. The same might be true, perhaps to a lesser extent, of koinobiont endoparasitoids of Lepidoptera.

Several very large subfamilies of both ichneumonids and braconids specialise on these and only a very few species, or perhaps genera, occasionally make the transition on to members of another host order such as Hymenoptera or Coleoptera.

Within the vast subfamily Microgastrinae, with an estimated 17,000–46,000 species (Rodriguez et al. 2013), only one species is known to attack a member of another 'traditional order', the Trichoptera, but these are in any case close to the Lepidoptera phylogenetically. The Campopleginae are slightly more 'adventurous', with a few genera specialising on beetles and a few other orders. Although the phylogenetic hypotheses for campoplegine genera are far from robust (Miah & Bhuyia 2001, Quicke et al. 2009; see Fig. 13.52), they do suggest that there have been several transitions on to the Coleoptera within this subfamily. Unfortunately, the genus *Pyracmon* which attacks Elateridae (Coleoptera) has not yet been sequenced and therefore it is not possible to check whether it is closely related to other beetle parasitoids within the subfamily such as *Rhimphoctona* and *Bathyplectes*. Cremastines and anomalonines have also made transitions (presumably) from Lepidoptera to Coleoptera. In the case of anomalonines, the beetle-parasitising genus *Anomalon* is generally recovered as the sister group to the remaining genera; it has also apparently occurred at least once in the Ophioninae. Further, the *Belesica* group, currently placed in Cremastinae, although perhaps representing a separate lineage, also attack Coleoptera.

Evidence for co-speciation between hymenopteran parasitoids and their hosts is virtually non-existent and most species can successfully attack more than one species of host. The reason for this is that parasitic wasps do make 'mistakes' and, although the majority of these fail, that is, the parasitoid fails to develop successfully on the accidentally attacked host, some of these mistakes are successful over evolutionary time. Otherwise, host ranges would simply not evolve. Unfortunately, there have been very few studies of the mistakes that parasitoids make (see, for example, D.C. Griffiths 1960, Johnson 1985). M.R. Shaw (2002b) noted that 'from experimental manipulations it is clear that some *Aleiodes* [Rogadinae] species show a willingness to oviposit into unsuitable hosts', and this was more likely when the hosts were either phylogenetically close to the normal host(s) or were physically or behaviourally similar to them.

ENDNOTES

1. Stilpnina among the cryptine ichneumonids are one of the few exceptions, showing very atypical biology for the group.
2. Also see Lycorininae, whose larvae might possibly technically be ectoparasitic at least for a while, but if so, would be concealed within the host caterpillar's rectum.
3. *Leptopilina* is a member of the Eucoilidae.
4. A fifth example could be considered the Eucerotinae, but their overall, hyperparasitoid biology is so different that it is difficult to see how it fits into any of the schemes discussed here.

Chapter 7

SEX, COURTSHIP AND MATING

The Braconid and Ichneumonid Parasitoid Wasps: Biology, Systematics, Evolution and Ecology, First Edition. Donald L. J. Quicke.

SEX DETERMINATION

In common with all other Hymenoptera, ichneumonoids have haplodiploid sex determination with daughters arising from fertilised eggs and males generally from unfertilised eggs, a condition known as arrhenotoky (Aubert 1982). The underlying mechanism, which applies to numerous hymenopterans, was first worked out by Whiting (1943 and many earlier papers) using the braconine *Habrobracon hebetor* and is called complementary sex determination (CSD). CSD may involve only one sex-determining gene locus, in which case it is called single-locus complementary sex determination (slCSD), or it may involve multiple loci, but in the latter case it is harder to detect and to investigate. In CSD, individuals possessing at least two different alleles at a sex-determining locus (or loci) will develop into females and those with only one form of the allele become males. As unfertilised and therefore haploid eggs can only have one form of the allele (or alleles), these become males, but fertilised diploid eggs will usually (depending on the alleles present in the mother and father) develop as daughters, although when the diversity of sex-determining alleles is low in the population, as often becomes the case in rearing programmes due to inbreeding, some fertilised eggs will not contain multiple allelic forms and so also develop into males – in these cases, diploid males – with greatly reduced or zero genetic fitness.

There is considerably inter-specific variation in whether diploid males survive to adulthood, and this can be marked even between closely related species (Holloway et al. 1999). Thus the CSD locus can be considered as displaying strong overdominance or heterozygous advantage and, as with other such cases, this helps to maintain a high diversity of alleles in the population. Apart from that, very few details are known about the CSD locus. The number of different alleles present in laboratory and 'wild' populations of *Habrobracon* have been estimated as between three and nine, with up to 12 being present across its geographic range (Heimple et al. 1999). Whiting (1961) had localised the sex locus in the *H. hebetor* linkage map using the colour and morphological markers available at the time and Antolin et al. (1996) and Holloway et al. (2000), using RAPDs[1], mapped it to one small linkage group; it is not clear whether some (small) differences observed between the studies represent specific differences, chromosomal rearrangements or possibly algorithmal features. Both molecular studies revealed 10 linkage groups, corresponding to the 10 haploid chromosomes (B.R. Speicher 1936, Gokhman 2009) whereas Whiting only found eight, but the latter's findings are easily explained by the distribution of marker genes across the chromosomes; unfortunately, especially considering the amount of research work involved, Whiting's original data are not available for re-analysis (Holloway et al. 2000).

Haplodiploid sex determination means that in most species, females can determine the sex of their offspring by choosing whether or not to fertilise a given egg as it is laid. In those that inject a large number of small eggs more or less simultaneously into a host, probably the female can still regulate to some extent the amount of sperm released to contact them and so affect the sex ratio in that way. This has enormous pay-offs from an evolutionary point of view, particularly for idiobionts because females can attempt to optimise future fitness by preferentially laying male-producing eggs on smaller hosts because, in general, smaller males are not at such a fitness disadvantage as equivalently smaller daughters, since they can still produce enough sperm to fertilise many eggs, assuming they can get a mating. In contrast, a small daughter may not be able to produce as many eggs (M.E. Visser 1994). Of course, the picture is more complicated, because in synovigenic species, a small female may, over time, still be able to produce as many eggs as a larger individual, but if adult lifespan is limiting rather than available hosts, one would still expect smaller individuals to have lower fitness. Antolin et al. (1995) only found weak evidence for size-dependent fecundity in the idiobiont braconine *Habrobracon* (as *Bracon*) *hebetor*, but in the alysiine *Asobara tabida*, which is time-limited, there is a strong correlation, with larger females having more eggs and greater longevity (Ellers et al. 1998). Thus in *A. tabida*, larger females also have a greater chance of locating and utilising new host patches.

Diploid males often show reduced viability and in many cases do not make it to adulthood (Petters & Mettus 1980), and those that do survive to adulthood are generally sterile and in that case if a female mates with one she will only be able to produce male offspring, a phenomenon called pseudovirginity. As many parasitic wasp females mate only once in their life, a high proportion of diploid males in a population could have a very adverse effect on sex ratio also through causing females to lay unfertilised eggs. The ichneumonine

Diadromus pulchellus and the microgastrine *Cotesia vestalis* are exceptions in that diploid males are viable, produce diploid sperm, mate at the same frequency as haploid males and their sperm even penetrates the female eggs, occasionally giving rise to triploid offspring (El Agoze et al. 1994, De Boer et al. 2007b).

Knowledge of which taxa have CSD and, of these, which may have slCSD, is very unevenly spread taxonomically (Asplen et al. 2009). It has been predicted that, within groups where it is known to occur, it ought to be commoner among solitary species rather than gregarious ones, because of the detrimental effects of inbreeding (de Boer et al. 2007a). In the Braconidae, it has been demonstrated for approximately nine species of the cyclostome lineage, of which four belong to the genus *Habrobracon*, and five species of Microgastrinae, of which three belong to the genus *Cotesia*, and within the Ichneumonidae all but one tested case is a member of the Campopleginae. Tagawa and Kitano (1981) found that in the field, approximately 60% of matings in *C. glomerata* were between siblings before the females subsequently dispersed, despite its slCSD (see below). slCSD has been confirmed for all but one of the ichneumonids investigated, viz. *D. pulchellus* (Periquet et al. 1993), and the campoplegine ichneumonids *Diadegma chrysostictos* (R.D.G. Butcher et al. 2000a,b) and *Venturia canescens* (Beukeboom 2001). Within the Braconidae, slCSD is definite in only a few cases, e.g. *Habrobracon hebetor* [Whiting 1939, 1943, B.R. Speicher & Speicher 1940 (as *brevicornis*), Clark & Rubin 1961 (as *Bracon serinopae*)] and in the microgastrine braconids *Microplitis croceipes* (W.W.M. Steiner & Teig 1989) and *Cotesia glomerata* (Y. Zhou et al. 2006). Strong evidence for slCSD, i.e. a very rapid increase in male production in small inbred (sib-mated) colonies, is also found in *Hyposoter didymator* (M.I. Schneider & Viñuela 2007).

Duplication of the slCSD locus can lead to multilocus complementary sex determination (mlCSD), which to date has been demonstrated in two species of microgastrine, *Coltesia vestalis* and *C. rubecula* (De Boer et al. 2007a, 2008, 2012).

However, two ichneumonoids have been shown definitely to lack CSD, the microgastrines *C. flavipes* and *C. sesamiae* (Niyibigira et al. 2004a,b), and slCSD definitely does not occur in the doryctine braconid *Heterospilus prosopidis* (Wu et al. 2005), the microgastrine *C. vestalis* and alysiine braconids *Asobara tabida* and *Alysia manducator* (Beukeboom et al. 2000), and W.-J. Ma et al. (2013) ruled out any form of CSD (up to a system involving at least 10 separate loci) in the case of *Asobara* (four species having been tested: *A. citri, tabida, pleuralis* and *japonica*). The almost certain occurrence of sib–sib mating in the basal, cyclostome braconid *Histeromerus mystacinus* (Rhyssalinae), together with its female-biased sex ratio (M.R. Shaw 1995), means that it too must also lack CSD.

All forms of CSD will be selected against in taxa with a high level of inbreeding, such as those displaying local mate competition, where brothers compete for mating with their own sisters, because this will lead to loss of genetic diversity including that of the alleles at the CSD loci. With less allelic diversity, the proportion of diploid males carrying two identical CSD alleles in slCSD species will increase and, although with more loci in an mlCSD system the increase in the production of diploids that are homozygous at all the CSD loci will be lower with any given level of inbreeding, it will nevertheless occur.

Many species with CSD probably have strategies to reduce inbreeding and its likely negative effects, especially in gregarious species or in ones that tend to live in dense patches. Early male emergence (protandry) may lead to males dispersing away from their natal patch before sib-mating can occur (e.g. Ode et al. 1995, Harvey & Strand 2003) or females may have a premating period after egress, again increasing the likelihood that close relatives will have dispersed further afield before they mate. At least in *Habrobracon*, females seem to be able to detect close relatives and avoid mating with them; for example, Petters et al. (1985) found that females mate faster with sons than with brothers and are therefore more likely to do so. However, given this, there may still be some local mate competition because females lay higher sex ratios when other mated females are present, although this could be a hang-over from a previous time when perhaps a different breeding system was in effect (Ode et al. 1997) (see also the sections *Mate location* and *Courtship*, below).

Several alternatives to CSD in haplodiploid Hymenoptera have been hypothesised, but so far only one alternative has been demonstrated, viz. maternal control sex determination in the pteromalid *Nasonia vitriipennis* (W.-J. Ma et al. 2013), and there is no evidence as to whether this might also be the case in the two *Cotesia* species that lack CSD or whether they may have some other mechanism. What is interesting from Ma et al.'s study and the distribution of instances of CSD and non-CSD in the Ichneumonoidea is that it must be evolutionarily fairly easy to change from one system to another and therefore it is likely that some of the same genes may be involved but functioning in a different way.

LOCAL MATE COMPETITION AND AVOIDANCE OF INBREEDING

In contrast to the Chalcidoidea, local mate competition (LMC) does not appear to be common in the Ichneumonoidea, but M.R. Shaw (1995) provided evidence that it may be routine in the genus *Histeromerus* (Rhyssalinae), which is a gregarious ectoparasitoid of wood-borer beetles, with males being reluctant to leave the cocoon mass and sex ratios always being very female biased, indicating probable routine sib–sib mating, and the same probably applies to the gregarious doryctine *Parallorhogas pyralophagus* (Bernal et al. 2001) and similarly to the gregarious rogadine *Macrostomion sumatranum* (Maetô & Arakaki 2005), in which the proportion of males increased with clutch size. LMC generally leads to highly female-biased sex ratios. Some species show distinct behaviours that reduce the probability of sib–sib mating, such as not mating until after dispersal (best studied in the braconine *Habrobracon hebetor*; Antolin & Strand 1992), whereas others that develop gregariously may show a mixed mating behaviour with most individuals dispersing but a few mating on the natal patch (e.g. the microgastrine *Cotesia glomerata*; Tagawa & Kitano 1981). The latter case could potentially be a bet-hedging strategy if there is a risk of finding no mate at all after dispersal, but the underlying mechanism behind the decision has not been investigated.

The mating system of the gregarious braconine parasitoid of store-product pests, *Habrobracon* (as *Bracon*) *hebetor*, was described in some detail by Antolin and Strand (1992) and by Ode et al. (1995). This wasp has slCSD, which means that inbreeding will lead to reduced fecundity through the production of increasing numbers of diploid males. Individuals of both sexes are reluctant to mate soon after emergence, which means that most will have dispersed away from their natal patch before copulating, thus reducing the risk of sib–sib mating. Further, using eye colour mutants, Ode et al. showed that females of *H. hebetor* avoid mating with brothers if they had developed on the same host individual, but willingly mated with them if they had developed on a different host, obviously using chemical cues to mediate this discrimination which lasted for up to approximately 5 days post-eclosion.

SEX ALLOCATION

One of the major consequences of haplodiploid sex determination is that hymenopterans have evolved the ability to control whether or not eggs are fertilised when they are laid and thus can often control the sexes of individual offspring or, at least the proportion of eggs that are fertilised. There are two important reasons why this is a potential advantage. First, if host quality can be assessed, it allows for females preferentially to lay daughters selectively on those hosts, usually the larger ones (Fig. 7.1) that are of the highest quality. This maximises fitness because of the normal differential effect of body size on potential fecundity between males and females (M.E. Visser 1994, and see above). Second, in situations where sons and daughters are likely to mate with one another (see the section *Local mate competition and avoidance of inbreeding*, above) it allows maximisation of fitness by increasing the sex ratio (i.e. males per female) if other females have also oviposited in the patch (see Godfray 1994 and Hardy 1994 for a detailed discussion). That LMC is not common among ichneumonoids is perhaps largely due to their widespread CSD system.

Many idiobionts attack concealed hosts so usually they cannot assess host size visually or by walking over them as is done by a few wasps. However, some are still able to assess the size or stage of the host, presumably via chemical or perhaps physical cues experienced by the ovipositor tip. Joyce et al. (2002) presented the solitary ectoparasitic idiobiont doryctine braconid *Syngaster lepidus* with concealed larvae of its wood-borer beetle hosts *Phoracantha recurva* and *P. semipunctata* of known ages, where age was an accurate proxy measure for host size. With the smallest hosts, only male wasps were produced and the proportion of females reared increased to 80% with the oldest host categories. Interestingly, the sizes of the wasps that emerged also correlated with host age, older hosts producing larger individuals of both sexes of parasitoid (Joyce et al. 2002, Paine et al. 2004).

Although wasps towards the far koinobiont end of the idiobiont–koinobiont spectrum have a great deal of control over when they kill the host (e.g. Elzinga et al. 2003), and therefore host size at oviposition is far less relevant to sex ratio decisions, others that perhaps

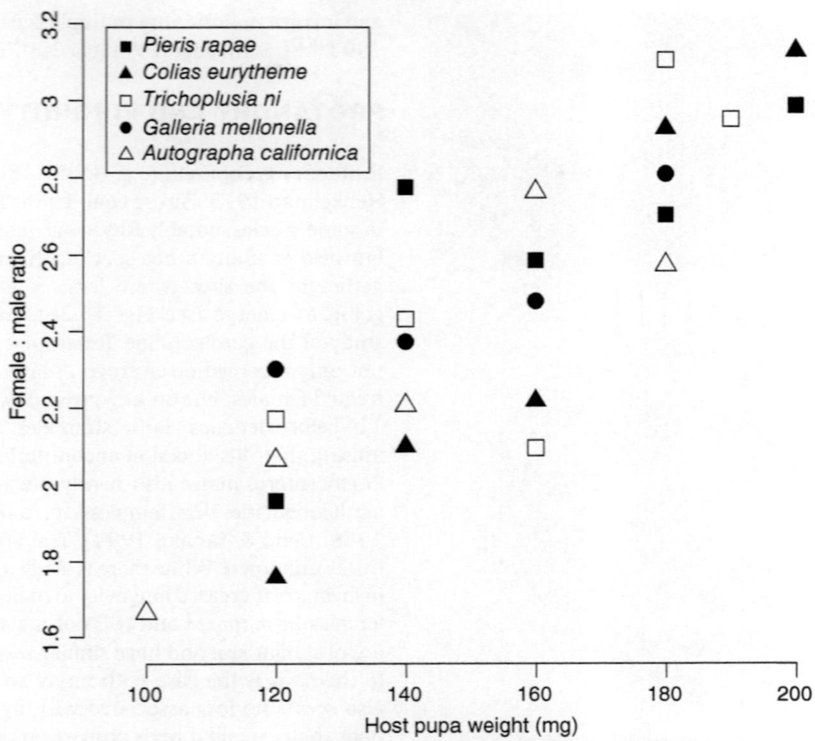

Fig. 7.1 Female to male ratio of the pupal parasitoid *Pimpla turionellae* (Pimplinae) reared batches of 20 pupae of each of five host species, showing strong positive correlation with host pupal weight both overall and within host species. (Source: data from Sandlan 1979.)

show fewer of the koinobiont adaptations may still adjust the sex ratio of their offspring in accordance with host size when it is attacked. For example, the rogadine *Clinocentrus cuncator*, which lays rather large, yolky eggs (Fig. 7.2a), appears to lay more male eggs into earlier instar hosts, again without being able to inspect them visually (M.R. Shaw 1983). In this case it is no doubt because the parasitoid causes premature metamorphosis, as can be seen by the development of pupal spines on the cuticle developing underneath the integument of a pseudoparasitised penultimate instar host (Fig. 7.2b) (M.R. Shaw 1981). The *Clinocentrus* then completes its own development after approximately the same duration of parasitism when either final or penultimate instar hosts are attacked, and therefore the potential size of and resources available to the emerging wasp may be affected by the host stage attacked. A similar host size effect was observed in the koinobiont opiine *Fopius* (as *Opius*) *concolor* by Avilla and Albajes (1984). Perhaps the parasitoids in these these particular host–koinobiont parasitoid associations have less control over host development than some others, and perhaps this is also a more widespread phenomenon than generally assumed.

The doryctine braconid *Parallorhogas pyralophagus*, lays precisely one male egg to four female eggs on each host larva – typical of local mate competition – irrespective of the size of its host, the pyralid moth *Eoreuma loftini* (Bernal et al. 2001). This sort of precise brood size and sex ratio is very unusual, but what is even more surprising is that a proportion of the females remain unfertilised (nearly 10%) and the number increases in larger broods. It seems unlikely that this reflects lack of sperm and maybe indicates that the single male does not have time to mate with all the females before they disperse (see the section *Protandry and virginity*, below).

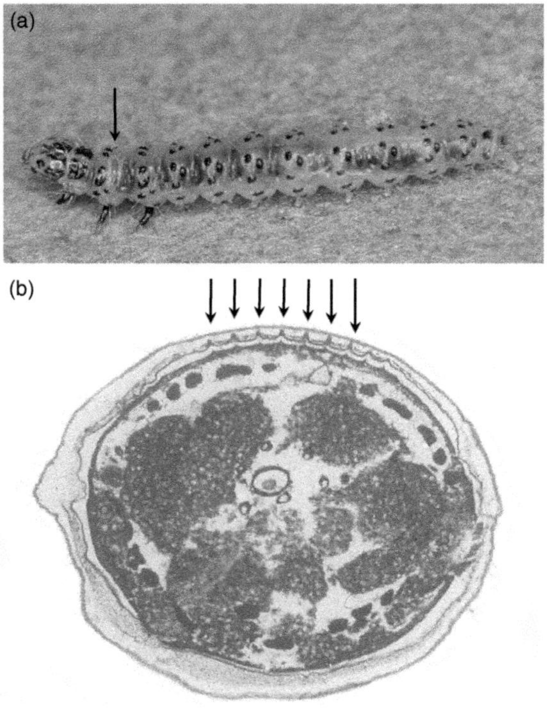

Fig. 7.2 Host of *Clinocentrus cunctator* (Rogadinae). (a) Larva of the choreutid *Anthophila fabriciana* host showing parasitoid egg located transversely just below the cuticle of the second thoracic segment; (b) transverse light microscope section through pre-penultimate instar of host envenomated but not oviposited in (pseudoparasitised), showing the prematurely developed dorsal pupal cuticle with spines that would normally form one instar later. (Source: reproduced by permission of Mark Shaw from M.R. Shaw 1981.) For a colour version, see Plate 13.

Increasing sex ratio (i.e. increasing proportion of males) is commonly observed in breeding programmes and, although not the only possible explanation, CSD and in particular slCSD are likely candidates when this happens (e.g. Johns & Whitehouse 2004). All forms of CSD, but particularly slCSD, have profound consequences for maintaining cultures for research or for mass rearing, as colony fitness declines rapidly as CSD alleles get lost and the proportion of sterile diploid males increases or, in the case of these being non-viable, general lower production of females. Breeding systems that minimise the negative effects have been devised and involve maintaining multiple separate lines (Cook 1993, M.I. Schneider & Viñuela 2007).

PROTANDRY AND VIRGINITY

Protandry is common (e.g. Gothilf 1969, Mackauer & Henkelman 1975, Hirose et al. 1988, Teder 2005) and in some species, notably Rhyssinae and some Pimplini, but also in many other species, this allows males to gather at the sites where females are subsequently going to emerge (see Fig. 13.26). In Hirose et al.'s study of the cardiochiline *Toxoneuron nigriceps*, males not only emerged on average 2 days before similarly treated females, but on any given day, emerged about 1 h before females, both strategies obviously maximising their likelihood of encountering virgin mates. Furthermore, males also nearly always have shorter adult longevities than females (e.g. Juillet 1959, Gothilf 1969, Ueno & Tanaka 1994). The precise reason for this is unknown. While there is likely to be no selection in favour of increased longevity in males if effectively all females have mated and will not mate again, if males are of similar size and have similar resources available to them, as is the case with many koinobionts, there also seems no loss associated with living longer. Perhaps they expend a high proportion of some resource in their early days searching for mates. No comparative study has been carried out on whether males might live significantly longer relative to females in koinobionts versus idiobionts – most research having been concentrated on female longevity, for obvious reasons.

Godfray (1994) examined the issue of virginity in parasitoids. Although producing males only is not going to be an optimal sex allocation strategy, the potential for female hymenopterans to produce sons even if not fertilised means that should no male be available (as could happen if a female's emergence was delayed beyond when protandrus males had mostly died), she does not necessarily have zero fitness. In many species, females are reluctant to oviposit until after they have mated or do so only at a low rate, although Narayanan et al. (1961) noted that females of the chelonine *Chelonus narayani* will very 'happily' oviposit in host eggs shortly after emergence, even if they are still virgins. Similarly in agathidines, *Agathis unicolor*, for example, readily oviposits irrespective of mating status (Fig. 7.3) (Odebiyi & Oatman 1977). In the case of the braconines *Habrobracon hebetor*, some females (approximately 10%) emerging from cocoons

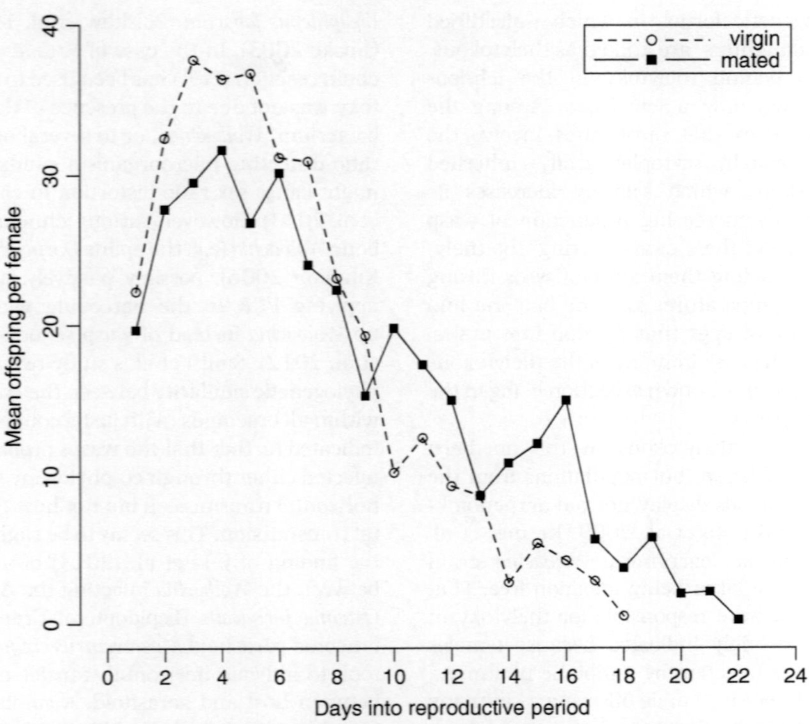

Fig. 7.3 Mean offspring production by mated and virgin females of *Agathis unicolor* (Agathidinae) on caterpillars of the potato tuberworm (*Phthorimaea operculella*: Gelechiidae). (Source: data from Odebiyi & Oatman 1977.)

in grain stores do not bother to crawl to the surface where mate-finding occurs, but simply go about the business of finding *Plodia* or similar hosts, remaining virgin even when males are present (as mating takes place on the surface), and this seems to be part of a so-called 'split sex ratio' strategy (Guertin et al. 1996). This might be a consequence of the fact that despite *H. hebetor* being an outbreeding species without or with very little local mate competition, it still has a female-biased sex ratio which these virgin females might be exploiting. In contrast, in another braconine, *Coeloides dendroctoni,* which parasitises bark beetles, unmated females completely refuse to oviposit (De Leon 1935). An intermediate situation is seen in the gregarious microgastrine *Cotesia glomerata* (Tagawa 1987). Virgin females will lay eggs into hosts, but these all-male broods are significantly smaller than the mixed-sex broods laid by mated females. However, when a virgin that has already oviposited is subsequently allowed to mate, her next broods increase in size. Dissections show that virgins have the same number of mature eggs available, so they appear to be making a decision to optimise their fitness by reserving some of their eggs in case a mating opportunity arises. Some species can only lay male eggs for a while after mating perhaps because of sperm behaviour (Genieys 1925, Flanders & Oatman 1982).

THELYTOKY AND CYTOPLASMIC INCOMPATIBILITY

Although most ichneumonoid species are fully sexual (though with males produced by arrenotoky), several well-studied species of are known entirely or almost entirely from females, and a number of others have both sexual and asexual strains which are frequently geographically isolated from sexual ones, possibly indicating that it can arise through a founder effect, although there are other possible explanations for this.

These parthenogenetic forms, in which unfertilised eggs give rise to daughters, are known as thelytokous. The mechanism behind thelytoky in the Ichneumonoidea is known only a few cases. Among the Chalcidoidea, it seems that most cases involve the intracellular, maternally (cytoplasmically) inherited bacterium *Wolbachia*, which selfishly increases its own propagation by increasing production of wasp daughters. In some of these cases, 'curing' the thelytokous female by feeding them antibiotics or raising them at elevated temperatures kills the bacteria and leads to the laying of eggs that develop into males. However, this not the case in many of the thelytokous ichneumonoids, the only known exception being in the alysiine *Asobara japonica*.

Asobara japonica is thelytokous in the northern mainland islands of Japan, but populations from the tropical southern islands display normal arrhenotokous reproduction (Murata et al. 2009, Kremer et al. 2009), with the former carrying a *Wolbachia* strain called *w*Ajap and the latter being infection free. That the *w*Ajap bacteria were responsible for thelytoky in this instance is strongly indicated because females can be cured of the infection by antibiotic treatment, and these go on to produce male offspring as in many thelytokous chalcidoids. Females of the thelytokous strain actually do occasionally produce male offspring and these vary in their *w*Ajap load. When these males are mated with females of the arrhenotokous strain, the sex ratio of their offspring is significantly more male biased than when they mate with normal, uninfected males. Possible explanations for this include that the males from the thelytokous strain are inheritantly less fertile, but K. Kraaijveld et al. (2011) showed that this effect was strongest with males having a relatively high *w*Ajap load, which suggests that the *Wolbachia* might be rendering their sperm less capable of fertilising an egg due to a cytoplasmic incompatibility effect, the uninfected females being unable to 'repair' the male's sperm after egg penetration. Unfortunately, the obvious experiment to test this, namely mating a male from a thelytokous mother with another thelytokous female, is not possible in this case since females of the thelytokous strain refuse to mate.

Thelytoky in all other ichneumonoids in which it has been investigated, which is not very many, appears to involve direct chromosomal or genetic changes in the wasps. The cytogenetics of thelytokous reproduction have been described in *Venturia canescens* (Beukeboom & Pijnacker 2000) and in the aphidiine braconid *Lysiphlebus fabarum* (Belshaw et al. 1999, Belshaw & Quicke 2003). In the case of *V. canescens*, polymerase chain reaction (PCR) has been used to show that thelytoky was not due to the presence of the *Rickettsia*-type bacterium, *Wolbachia*[2], or to several other possible sex ratio-distorting microorganism candidates that do or might cause sex ratio distortion in chalcidoids (Foray et al. 2013). However, various ichneumonoids do harbour *Wolbachia* (e.g. the opiine *Doryctobracon*: Cônsoli & Kitajima 2006), possibly passively, and occasionally applying PCR to the barcoding gene region turns up *Wolbachia* instead of wasp sequences (M.A. Smith et al. 2012). Smith et al.'s study revealed very strong phylogenetic similarity between the *Wolbachia*s present within all braconids (with just a couple of outliers) and indicated further that the wasps probably had become infected either through co-phylogeny or wasp-to-wasp horizontal transmission but not host-to-wasp horizontal transmission. This seems to be slightly at odds with the finding of J. Li et al. (2013) of a close similarity between the *Wolbachia* infecting the Asian corn borer, *Ostrinia furnacalis* (Lepidoptera: Crambidae), and its braconid parasitoid *Macrocentrus cingulum*, which they took to indicate horizontal transfer of the bacterium between host and parasitoid. A similar situation was found by Vavre et al. (1999), looking at *Wolbachia* in *Drosophila* parasitoids and their hosts. However, S.A. West et al. (1998), looking at *Wolbachia* phylogenies in two systems, also found no evidence of horizontal transfer from host to parasitoid. There would certainly seem to be cases where horizontal transmission occurs and others where it does not, and it would be interesting to know whether this was influenced by taxonomy or some aspect of the nature of the association. The report by Starý (1999) that thelytoky in the aphidiine *Lysiphlebus fabarum*, which was only because of the PCR detection of the bacterium, has been shown to be an overstatement and thelytoky in *L. fabarum* is not related to any known symbiotic bacterium (R Stouthamer, personal communication to R Belshaw, cited in Belshaw & Quicke 2003).

In both the microgastrine, *Cotesia sesamiae*, and the alysiine *Asobara tabida*, *Wolbachia* causes cytoplasmic incompatibility (Mochiah et al. 2002, Dedeine et al. 2004) and further, in the *A. tabida* system, one of the three *Wolbachia* strains that the wasps harbour, wAtab3, is essential for oögenesis, although in other *Asobara* species surveyed by Dedeine et al. (2005) *Wolbachia* were not always present and in none of them were they apparently essential for oögenesis.

The genetics underlying thelytoky in *L. fabarum* is central fusion automixis, in which the products of the second meiotic division reunite, restoring diploidy before the second anaphase. This mechanism does not prohibit occasional sex and recombination if a sperm nucleus were to fuse to one of the groups of univalent chromosomes from the second meiotic division before both products were to fuse again to form a diploid restitution nucleus. It also allows for some heterozygosity (at loci close to the centromere), features observed through DNA sequence polymorphisms (Belshaw et al. 1999). However, Tsutsui et al. (2014) have recently described the only known case of apomictic parthenogenesis in an ichneumonoid. It involves a thelytokous strain of the meteorine braconid *Meteorus pulchricornis*, which completely lack meiosis. The only other case in which the cytogenetics of thelytoky have been studied within the Ichneumonoidea is in *V. canescens*, which shows a similar system again with a restitution metaphase (Beukeboom & Pijnacker 2000). In contrast, when *Wolbachia* symbionts cause thelytoky, diploidy is restored at a later stage following the formation of four pronuclei, thus leading to homozygosity (Cook & Butcher 1999).

A summary of the best known cases of thelytoky in the Ichneumonoidea is given in Table 7.1, which is based largely on Stouthamer (2003). This table is inevitably biased in several ways and in particular it is heavily influenced by those taxa that have been investigated for biological control purposes, and this is probably the reason why there are far more cases known in the Braconidae than in the Ichneumonidae. Nevertheless, it shows what is almost certainly a genuine strong phylogenetic signal. Although both families show a broad scattering of examples across subfamilies and there are not enough known examples among ichneumonids to draw conclusions, within the Braconidae there are many examples, spread across multiple genera and tribes within both the Aphidiinae and the Euphorinae (including the Meteorini), and these clearly suggest that members of these two groups have some predisposition to becoming thelytokous. Whether this has anything to do with their similar biologies (i.e. largely attacking hemimetabolous hosts) is open to question and I cannot think of an obvious link, especially as the basic biological features of the hosts are otherwise so different. Another interesting feature is that all the known examples within the Microgastrinae are in the genus *Apanteles* with none among the far more widely studied *Cotesia* species. There is also an indication of bias towards endoparasitoids, but this may be an artefact of the applied sample rather than indicating anything biological.

In some species with both sexual and thelytokous forms it has been found that they differ in host preference (e.g. Day & Hedlung 1988), and they often have different or only partially overlapping geographic ranges (e.g. Kfir 1997, Tsutsi et al. 2014).

K.G. Speicher (1934) found that when a cross was made between two mutant strains (*tapering* and *spread*) of the braconine *Habrobracon hebetor* (as *juglandis*), in order to test for linkage, virgin F1 daughters produced a number of female offspring and this deuterotoky was maintained into the F2 generation. Tardieux and Rabasse (1988) reported a similar observation that thelytoky can be induced in the aphidiine *Aphidius colemani* by attempted crossing of males from one locality with females from another, specifically males from Reunion with females from Brazil or males from Brazil with females from France. The females apparently rebuffed the male's mating attempts, but thelytoky did not result if courting was artificially interrupted before mating. An apparently intraspecific population level polymorphism in chromosome number has been reported in another aphidiine, *Aphidius ervi*, and the extra pair of small, heterochromatin-rich, acrocentrics in some populations have been suggested possibly to represent selfish elements that could be responsible for thelytoky (Gokhman & Westendorff 2003). Although most *A. ervi* individuals have $n = 5$ and $2n = 10$, some females have an additional pair of small acrocentric chromosomes composed entirely of heterochromatin. If these were to mate with normal $n = 5$ males, one would expect $n = 11$ females, but the authors found neither these nor $n = 6$ males. Therefore, it seems most probable that these extra chromosomes are the cause of thelytoky in this species.

Also, although not thelytoky in its normal sense, Inaba (1940) showed that in *Habrobracon pectinophorae* (Braconinae), impaternate daughters could be produced by treating mothers just prior to eclosion with a 0.05% colchicine solution for about 5 h. His cytological observations indicated that there were 16 or more chromosomes[3] present at the second meiotic metaphase and concluded that the offspring were probably tetraploids, but with CSD, these would still have two different alleles at the CSD locus.

There have been a number of studies on the amount of genetic variation present among individuals of thelytokous strains compared with sexual strains.

Table 7.1 Occurrence of thelytoky in the Ichneumonoidea.

Family and subfamily	Species	Males known
Ichneumonidae		
Campopleginae	*Bathyplectes* (as *Biolysia*) *tristis*	Y
	Venturia cansescens	Y
Cremastinae	*Trathala flavoorbitalis*	Y (2%)
Cryptinae	*Gelis tenellus*	N
	Gelis cinctus	Y
	Hoplocryptus melanocephalus	N
	Sphecophaga burra	Y
	Sphecophaga vesparum	Y
Diplazontinae	*Diplazon laetatorius*	Y
Ichneumoninae	*Diadromus collaris*	Y
Mesochorinae	*Mesochorus nigripes*	Y (in Europe)
Metopiinae	*Triclistus epermeniae, pygmaeus* and other species	Y (c. 1%)
Pimplinae	*Polysphincta pallipes*	
Tersilochinae	*Stethantyx* (as *Tersilochus*) *parkeri*	Y
Braconidae		
Agathidinae	*Agathis stigmaterus*	Y (1.5%)
	Bassus rufipes (as *diversus*)	Y
Alysiinae	*Asobara japonica*	Y
Aphidiinae	*Aphidius colemani*	Y
	Ephedrus persicae	Y
	Lysiphlebus ambiguus	Y
	Lysiphlebus cardui	N
	Lysiphlebus confusus	Y
	Lysiphlebus fabarum	Y
	Lysiphlebus testaceipes	Y
Braconinae	*Habrobracon hebetor*	Y
Euphorinae	*Centistes ater* (as *excrucians*)	
	Dinocampus coccinellae	Y
	Meteorus pulchricornis (= *japonicus*)	Y
	Microctonus brevicollis	Y*
	Microctonus hyperodae	?N
	Microctonus sp.	
	Microctonus vittatae	Y (very rare)
	Microctonus nitidulidis	
	Peristenus howardi	Y
	Pygostolus falcatus	Y
	Pygostolus otiorhynchi	Y
Cheloninae	*Chelonus blackburni*	N
	Chelonus (*Microchelonus*) *sulcata*	Y
Microgastrinae	*Apanteles cerialis*	?
	Apanteles circumscriptus	
	Pholetesor bicolor (= *Apanteles pedias*)	Y
	Apanteles thompsoni	N
Rogadinae	*Aleiodes* (as *Rogas*) *unicolor*	Y

*When the parasitoid egg is laid into a host larva, only female progeny are obtained, but if into adult beetles, occasional males are produced (Künckel d'Herculais & Langlois 1891).
Source: citations to original reports for most are given in Stouthamer (2003); additional records from D.L. Jackson (1928) and Murata et al. (2009) (see also Hung et al. 1988).

In *Mesochorus nigripes* (Mesochorinae), Hung et al. (1988), using allozymes, found no significant difference in allelic diversity, although there was a trend towards greater diversity in the sexual form and Mateo Leach et al. (2012) also found more genetic variation in sexual strain of *Venturia canescens* using microsatellite markers. Just because a parasitic wasp is thelytokous does not necessarily mean that it is devoid of variation, and while Slobodchikov (1983) found that a laboratory strain of *Venturia canescens* could be distinguished morphometrically with 99% confidence from similarly thelytokous wild types, isofemale lines of the laboratory strain could not be distinguished because of a great deal of intra-line variability. Subsequently, Beck et al. (1999, 2000b, 2001) showed that two different thelytokous strains of this wasp (designated RP and RM) had different ovarian phenotypes and different reproductive strategies. The ovarian phenotype was shown to correlate with two different alleles of a locus encoding a protein of the wasp's virus-like particles (Beck et al. 1999) and that the two strains differed in egg maturation and embryonic development, one of them hatching more slowly as larvae, and that this had important consequences for the winner when superparasitism occurred. If females of the two strains were allowed to compete for hosts at the same time, then the RM strain had a marked advantage (Fig. 7.4a), whereas if either strain was allowed a 7 h head start, then the member of the RP strain always had the advantage (Fig. 7.4b,c).

Several workers have considered the importance of thelytoky in relation to biological pest control, notably because of the twofold cost of sex (e.g. Aeschlimann 1990, Stouthamer 2003). Stouthamer concluded that thelytokous races/species may be more effective control agents under some circumstances, especially when pest populations are high such that the increased female production gives its greatest advantage.

MATE LOCATION

Most insects, including nearly all species of ichneumonoid, are protandrous, indicating the selective advantage of being first on the scene when virgin females start to become available. This is especially true for most ichneumonoids in which females will mate only once or sometimes, only during a short time window after eclosing. Godfray (1994) discussed several different mate location strategies: females may attract males using pheromones, which appears to be by far the commonest strategy in the Ichneumonoidea; males may simply remain at the emergence site where siblings or perhaps other conspecifics are expected to emerge; males may actively search for female emergence sites (see, for example, Chapter 13, section *Rhyssinae*); males may search for oviposition sites where females will already be or will soon be frequenting; males may search for females at their feeding sites; or males may form leks or swarms (e.g. Blacinae, some Ctenopelmatinae, see Fig. 7.6) or aggregations (e.g. *Habrobracon hebetor*; Antolin & Strand 1992). In addition to these, it appears that individual males of some species (e.g. the euphorines *Napo* and *Cosmophorus*) may actively attract females using pheromones (see below).

If males remain at their emergence site where females are also about to emerge without searching for mates further afield, there is a high risk of them being siblings. A definite tendency to disperse away from the emergence site before mating is particularly important for species with slCSD and whose hosts live in aggregations or dense patches as known in some species such as the grain pest parasitoids of the genus *Habrobracon* (Braconinae) (Antolin & Strand 1992, Ode et al. 1995; see the section *Local mate competition and avoidance of inbreeding*, above), and it is likely that the obligate post-emergence premating period of some species has evolved to decrease the probability of mating with a close relative, through enforcing a period during which dispersal can occur. This is less likely to be the case for parasitoids whose hosts are widely dispersed, or at least will be by the time the parasitoid completes development, as with many koinobiont parasitoids of caterpillars that roam considerable distances before choosing a pupation site. Hence males of various species have to go in search of mates, either using innate or learnt cues to find locations/habitats from which other conspecifics are likely to be about to emerge. Males of the opiine fruit fly parasitoid *Psyttalia concolor* have been shown to be attracted to volatiles produced by fruit fly-infested fruit, which would likely be a good indicator of a site where freshly emerged females might soon be found, and Benelli and Canale (2013a) suggested that these cues could be considered as sexual kairomones. Once there, long- or medium-range sex pheromones generally play a major role and these can come from either the male or the female. In some species, males patrol a suitable habitat and detect freshly emerged

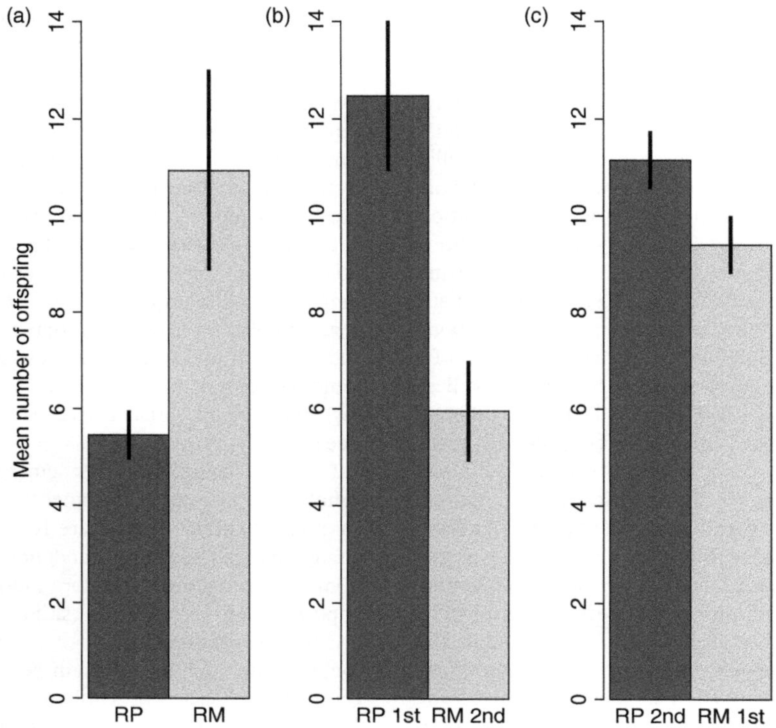

Fig. 7.4 Differences in competitive advantage (±SE) between genetically distinct, thelytokous strains (RP and RM) of the campoplegine wasp *Venturia canescens* in relation to timing of attack. (a) Wasps compete synchronously on a batch of 25 hosts; (b, c) wasp of one strain allowed access 7 h before the other as indicated. (Source: data from fig. 3 of Beck et al. 2000b.)

females by their sex pheromones, which sometimes can attract many males to her quickly, followed by a scramble competition between her suitors which swarm around her (e.g. the agathidine braconid *Alabagrus texanus*; Goh & Morse 2010). In this case, males have a short-term memory for a sight where a virgin female emerged, and if a second female is presented in the same place soon after a first one, those males which had swarmed to try to mate with the first female and then dispersed do not tend to come back, whereas if an hour is left between presenting the two virgins, then they will return (Fig. 7.5). Goh and Morse interpreted this as having evolved to prevent males wasting time searching for the same spot where a second female is very unlikely to emerge on the same day, given that they tend to be widely dispersed. Ashley Adams and Morse (2014) subsequently showed, using tethered virgin females of *A. texanus*, that they can apparently be fairly choosy about which males they will mate with, with approximately equal numbers either accepting the first male to come along, accepting the second or not mating. A different use of memory is shown by the pimpline ichneumonid *Pimpla disparsis*, in which the protandrous males search for host pupae from which females might emerge, memorise their locations and keep returning on consecutive days to the same place awaiting the eclosion of a mate (Danci et al. 2013).

Rhyssine males, whose potential mates egress from within wood, often aggregate in large numbers directly around a spot on the tree trunk from which (hopefully) a female is due to egress (see Fig. 13.26) (see Chapter 13, section *Rhyssinae*, for more details) and the same has been observed for some pimplines.

Mating at feeding sites has been observed in *Agathis* species visiting flowers (Belokobylskij & Jervis 1998). This strategy is most likely to evolve in synovigenic species that have no mature eggs ready to lay immediately after eclosion and may additionally have to feed

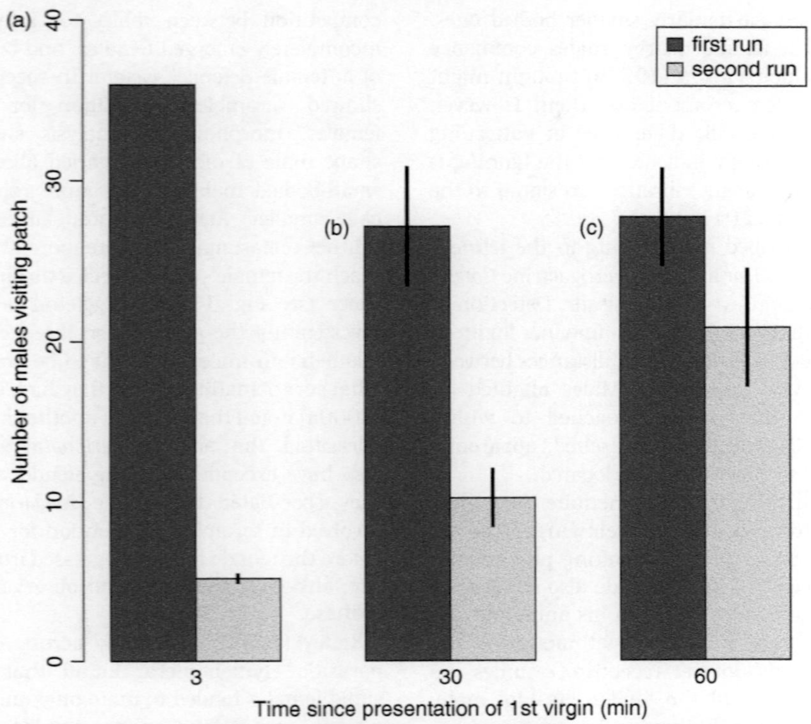

Fig. 7.5 Effect of time interval after presenting bagged virgin females of the agathidine braconid *Alabagrus texanus* on visits to a 1 m^2 site, showing that males have a short-term memory for a site where a virgin female was detected and do not return quickly to the same site as the probability of it containing an unmated female is low (±SE). (Source: data from Goh & Morse 2010.)

on hosts or at carbohydrate sources to mature eggs. How frequently this occurs in the Ichneumonoidea is at present almost completely unknown, largely because most workers collecting parasitoids at flowers are unaware of its significance.

Among flightless (i.e. brachypterous, micropterous and apterous) species, there are numerous instances of sexual dimorphism, although it varies as to whether the male or the female is brachypterous (see Table 10.2). It seems likely that in those ichneumonoids with brachypterous males the latter probably 'call' pheromonally to attract mates, and this has been observed in the euphorine *Cosmophorus cembrae* (Seitner & Nötzl 1925, M.R. Shaw 2009a). Buckingham and Sharkey (1988) further found that male *Cosmophorus* have large metasomal tergal gland reservoirs, but the chemistry of their contents has not been studied. Males also probably release pheromones to attract mates in another euphorine, *Napo townsendi*, and in this species appear to hold territories on leaves of a given species of tree seedling (S.R. Shaw 2012).

COURTSHIP

Compared with the Chalcidoidea and aculeates, comparatively little has been published on courtship and mating in the Ichneumonoidea (Matthews 1974). Males of most species probably use a combination of vision and smell to locate mates and then touch. In a simple experiment, Grosch (1947) showed that males of *H. hebetor* need their antennae in order to find or become attracted to females, but when the antennae were left intact and the eyes blackened, mating was unaffected. Short-range female pheromones are important in many cases and may trigger courtship behaviour or lower the male's threshold for courtship (Vinson 1972b).

In many species, particularly smaller bodied ones, having detected a female nearby, males commence wing-fanning, which Vinson (1972b) thought might serve to waft female pheromone over them. However, the consistent interspecific differences in patterning of the behaviour strongly indicate that the fanning is to do with the male using vibrations to signal to the female (Villagra et al. 2011).

Cole (1970) described mate finding in the ichneumonid *Phaeogenes invisor* and the microgastrine *Cotesia medicaginis* (as *Apanteles*) in moving air. Detection of virgin female pheromone elicited upwind flight in both species, which was detected at distances between 0.9 and 2 m away, respectively. Males alighted on vegetation when they had approached to within approximately 10 cm and they searched, apparently aimlessly, on foot until a female was located.

In *Banchus* (Banchinae), males orientate themselves to females head-to-head, fan with their wings and often raise their metasomas, probably wafting pheromones over her (van Veen 1982). The male also strokes the female's thorax and front legs with his antennae and will often make several attempts at mounting the female, but if she is not yet receptive, resumes the courtship activities, until she finally lets him mate. In *Habrobracon hebetor* (Braconinae), males perform short periods of wing vibration but then jump on to potential mates who seem generally to be receptive (Sudheendrakumar et al. 1978, as *brevicornis*).

There is considerable variation in whether wasps are ready to mate immediately upon eclosure or require a premating period of a day or so. For example, *Macrocentrus ancylivorus* individuals will normally copulate straight after emergence, although some individuals require a few hours before they will mate (Fink 1926). Short premating periods of about 1 h have been noted in a number of species [e.g. *H. hebetor* (as *brevicornis*), Sudheendrakumar et al. 1978; *Diadegma semiclausum*, Khatri et al. 2009], but some taxa require considerably longer (e.g. Hagen 1953). In the braconid *Orgilus jenniae*, Flanders and Oatman (1982) were not sure whether females had a pre-mating period, but noted that they only produced male offspring within the first 6 h after either emergence or mating, which might indicate that sperm need to be activated within the female in some way, or perhaps that further development of her reproductive system has to take place before she can fertilise eggs.

Eggleton (1991) described three different mating strategies in rhyssine ichneumonids: scramble competition between males for fully emerged and incompletely emerged females, and before emergence of a 'female defence' system. In species where males showed scramble competition for pre-emergence females, morphometric analysis showed that the shape male of metasoma varied allometrically, with small-bodied males having more robust (but physically smaller) metasomas and larger males having thinner metasomas that were more likely to be able to reach the female's genital orifice through the available space (see Fig. 13.26b). Eggleton proposed that by investigating the presence or absence of intraspecific allometry in males, it ought to be possible to predict what sort of mating system they had. Gauld and Wahl (2000a) noted that males of another large wood-borer parasitoid, the labenine *Certonotus vestigator* group, also have exceptionally long slender metasomas and they speculated that as in *Megarhyssa*, these are involved in scramble competition for a mate possibly before the female has fully egressed from her pupation site, although there were no observations of mating in these.

Ridley (1993), in a study across a wide range of parasitic Hymenoptera, found that solitary parasitoid females tended to mate only once (monoandry) whereas gregarious species tended to be multiple maters, and proposed that this may be due to the risk of inbreeding if sib–sib mating occurs in the gregarious species. This is likely to be especially important in species with slCSD, although if gregarious species also tend to have greater lifetime fecundities and utilise multiple hosts, then the prospect of sperm depletion also has to be considered.

SWARMING AND LEKKING

Males of a small number of species form swarms (leks), usually associated with some focal object such as a pile of dung or ant hill. These are best known in several European and North American species of *Blacus* (R.B. Benson 1944, Stelfox 1944, Arnaud 1952, Southwood 1957, Syrjämäki 1976, van Achterberg 1977), but swarming has also been reported in the microgastrine *Dolichogenidea* (as *Apanteles*) *coniferae* (Whitfield 1987), the opiine *Fopius arisanus* (Quimio & Walter 2000), a chelonine (S.R. Shaw 1991), a metopiine (*Leurus caeruliventris*) in Costa Rica (Eberhard 2007), the diplazontine *Diplazon pectoratorius* (Rotheray 1981c) and some ctenopelmatines, such as *Olesicampe*

Fig. 7.6 Swarming by *Xenoschesis fulvipes* (Ctenopelmatinae), a parasitoid of the sawfly *Cephalcia arvensis*. (Source: reproduced by permission of Tommaso Branzanti.)

monticola on *Cephalcia lariciphila* (Billany et al. 1985) and *Hadrodactylus* (Idar 1981). Although it is not certain whether it shows a single-sex or mixed-sex swarm or not, Fig. 7.6 shows a remarkable incidence of swarming by another ctenopelmatine ichneumonid, *Xenoschesis fulvipes*, during an outbreak of the spruce webspinning sawfly, *C. arvensis*, in Italy and, given the observations on *Olesicampe* and *Hadrodactylus*, it seems reasonable to assume that these too are all males.

The function of all-male swarms may be supposed to be similar to that of a lek: in having many males aggregated in one place, females may more readily be able to locate a mate because it seems unlikely that the male swarms rely purely on females locating them by sight, although there are as yet no known pheromones produced by males of most of the taxa. Having located a male swarm, the female is then presented with a choice of possible mates. Tobias (2002) suggested that the swarms may also provide feedback about population density (and sex ratio) in an area and stimulate males physiologically, and although the latter can not be discounted, the former begs the question of who would benefit. What it is that predisposes some species and, with the exception of *Blacus*, mostly isolated ones within their genera, to evolve male swarming behaviour is not clear. Biologically, the above-mentioned wasps are quite heterogeneous.

MATING POSITION

In most species, copulation takes place with the male atop the female and facing in the same direction (e.g. Steiner et al. 2010). A few exceptions are known. In some campoplegines and cremastines, for example, mating is strophandrous, i.e. with male and female joined end to end in coitus and facing in opposite directions (e.g. Khatri et al. 2009), whereas in another campoplegine, *Hyposoter fugitivus*, male and female stand side by side, just joined posteriorly (Barrows 1976), and in yet another, *Bathyplectes curculionis*, copulation is with the male in on the on-top position (Dowell & Horn 1978). Mating nearly always takes place in a settled position, but some anomalonines will fly *in copulo* (Camarao & Morallo-Rejesus 2003). Peck (1937a) illustrated the positions of the male and female genitalia of *Diadegma* (as *Angitia*) *fenestralis in copulo*.

MULTIPLE MATING AND SPERM COMPETITION

In many species, the female will mate only once, that is, they are monoandrous, at least under normal circumstances (e.g. Yeargan & Braman 1986), although little is known about whether in longer-lived species, females might mate again once initial sperm stocks become depleted. Males are typically polygynous and, indeed, I have found only one report of a monogamous species, by Gordh and Hendrickson (1976), stating that males of the campoplegine *Bathyplectes anurus*, 'that had copulated were not observed to show any subsequent interest in virgin females', and this might be due to the need to produce more spermatophores.

Males may also become more efficient at courtship and mating after a first experience, as X.-J. He and Wang (2008) showed for the aphidiine *Aphidius ervi*.

Multiple mating by females (polyandry) is generally uncommon in ichneumonoids; in the majority of species, females once mated refuse other male mating attempts (e.g. H.L. Parker 1931, Oatman & Platner 1974), although there may be a brief window when a second mating can occur (e.g. Broodryk 1969a), or a persistent male may overcome female rejection within a short period after her first mating (H.L. Parker 1931). In some species, after a female has mated she will rebuff future mating attempts fairly forcibly, as in *Habrobracon hebetor* (as *brevicornis*) (Genieys 1925) and *Perilitus rutilus* (Jackson 1928). Multiple female mating is, however, common in some species, including the poemeniine *Pseudorhyssa sternata* (Spradbery 1969), the campoplegine *Bathyplectes curculionis* (Dowell & Horn 1978), the pimpline genus *Theronia* (Shevirev 1913, cited in Jackson 1928) and rhyssines (Nuttall 1973), and has been observed in the campoplegines *Venturia canescens* (Vinson 1972b) and a few other species (see below). In another campoplegine, *Diadegma semiclausum*, multiple mating may occur when a female is paired with the same male for 12 h (Khatri et al. 2009). In a large laboratory culture of colour mutants, 11% of females of *Habrobracon hebetor* were found to have mated multiply (Dyson, cited by Antolin & Strand 1992). In this species, Antolin & Strand found that a considerable proportion of the females on the surface of the corn pile, where mating but not host searching takes place, had little or no sperm in their spermathecae and conjectured that they might actually be seeking further matings, having become sperm depleted. It appears that culture or ambient conditions might influence its occurrence and Matthews (1974) pointed out markedly contradictory reports obtained for the same species of microgastrine, *Cotesia* (as *Apanteles*) *flavipes*, by various authors. Dhillon and Sharma (2011) reported an interesting observation concerning multiple mating in the campoplegine *Campoletis chlorideae*, in which females that mated twice were significantly less successful at parasitising hosts than those that mated only once (20% of females versus 87%), and produced a higher proportion of male offspring.

In *V. canescens*, both virgin and mated females remain attractive to males, although multiply mated females were 'less active at eliciting male courtship', and in *D. semiclausum* the duration of the copulation increased significantly with subsequent rematings (Fig. 7.7) (Khatri et al. 2009). The reason for this is not known, but it might be because of the need of one of sexes, probably the male, to produce more secretions. In the alysiine braconids, Salkeld (1959) noted that some females would mate with several males in quick succession over a period of a few minutes, and also noted that it 'occurred frequently throughout the oviposition period', and Rappaport and Page (1985) similarly noted a period during which females of the banchine *Glypta fumiferanae* would permit several copulations. Multiple mating by males is common (e.g. Salkeld 1959). As Wharton (1984) noted, confining females with males may lead to occasional multiple female matings, but this is an artificial situation, and in the wild, females of these species probably do not mate more than once.

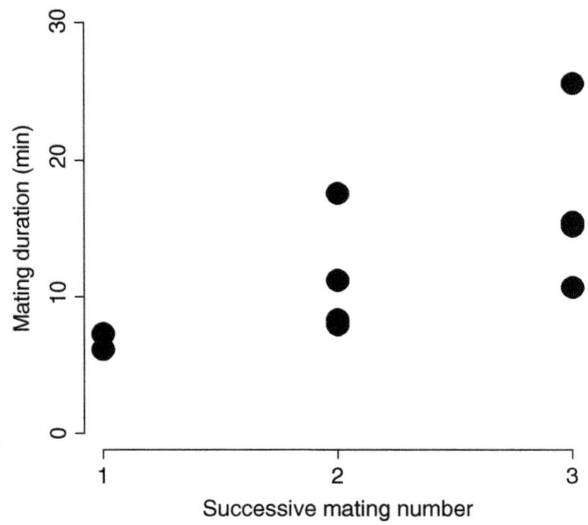

Fig. 7.7 Effect of number of matings on mating duration in the campoplegine *Diadegma semiclausum*. (Source: data from Khatri et al. 2009.)

In species where females mate with multiple males, there is the issue of sperm competition. As ichneumonoids have rather simple, usually spherical, spermathecae, that are separated from the vagina by a narrow duct, there is little structural complexity for evolution to work on. In *Diadromus pulchellus*, it has been shown, using two differently coloured strains, that the first male to mate gets the majority of the parentage (El Agoze et al. 1995), perhaps simply because one he has mated the female wasp's spermatheca is full and there is little or no room for sperm from a subsequently mated male. Given that easily recognisable genetic strains are known in numerous ichneumonoids, it is surprising that so little follow-up work has been done on this. The ready availability of colour variants of *H. hebetor* in which a proportion of females are known to mate multiply (Dyson, cited by Antolin & Strand 1992) would be a very suitable system for investigation. In the opiine *Diachasmimorpha longicaudata*, Martínez-Martínez et al. (1993) found that if a female mated with two genetically distinct males in rapid succession (<6 min), there was little sign of one male having precedence over another (a slight preference for the second to mate), but when matings were separated by a day or more, there was far stronger second male precedence. Several workers have noted that sperm in the spermatheca in dissected females appear very active (e.g. Gerling & Rotary 1974), so it is difficult to see how such preferential fertilisation could occur unless observed movements are an artefact.

SEX-RELATED SCENT GLANDS

The sources of sex pheromones have mostly only been postulated, often based on the discovery of female-specific glandular or putative glandular structures. Kainoh and Oishi (1993), using a series of progressive dissections of *Ascogaster* females finally localised the source of the female sex pheromone as the hind tibia. Tagawa (1983) and Field and Keller (1994) localised the sex pheromone-secreting gland in the microgastrine *Cotesia* to the female genital capsule.

Male parasitic wasps may also produce sex pheromones, and these are probably involved in close contact with the female and in getting her to accept him as a mate. However, there have been almost no studies of the compounds involved, even though male-specific glandular structures are extremely widespread in the Ichneumonoidea and often have conspicuous evaporation areas. Buckingham and Sharkey (1988) surveyed a range of braconids for male exocrine glands, which are often located at the anterior margin of the posterior metasomal tergites and are not normally exposed, but presumably (because there are very few relevant observations of living wasps) they are during either mate attraction or courtship. The best known are the Hagen's glands[4], which open in a series of pores along the anterior margin of metasomal tergite 8[5] in several braconid subfamilies (notably Opiinae, Alysiinae, Gnamptodontinae and Rhysipolinae), intertergal glands principally between metasomal tergites 5 and 6 and 7 in the Agathidinae, which are associated with brush-like evapatoria (Fig. 7.8d) (Buckingham & Sharkey 1988), and intertergal glands principally between the seventh and eighth metasomal tergites in addition to separate ones opening antero-laterally on the eighth metasomal tergite in many Braconinae (Quicke 1991b). Additionally, large tergal glands were found in males of the euphorine *Cosmophorus* and two genera of Ichneutinae. *Cosmophorus* is one of the few groups where males are known to attract females pheromonally (Seitner & Nötzl 1925). There are numerous other examples, such as the setose mid-tergal pits of many male *Heterogamus* and the medial, setose pores on metasomal tergites 4–6 of male *Aleiodes* (*Hemigyromeuron*) (Areekul Butcher & Quicke 2011), a group closely related to *Heterogamus*, although whether, despite being in the middle of the tergum of different metasomal tergites, these are genetically related needs further investigation.

Only in the fruitfly parasitising Opiinae is anything known about the chemistry of the products. H.J. Williams et al. (1988) identified (at least tentatively in some cases) the gland products in *Diachasmimorpha longicaudata*, *D. tryoni* and *Fopius arisanus*, all at that time placed in the genus *Biosteres*. Each included a unique set of products and there was no similarity even between the compounds of the two *Diachasmimorpha* species. Hagen (1953) and Buckingham (1968) had discussed possible roles in mating and courtship. Haramoto (1957), working on *Fopius arisanus* (as *Opius oophilus*), had suggested a defensive role and Buckingham (1975) provided some experimental evidence for this in *D. juglandis*, *Psyttalia concolor* (as *Opius concolor siculus*) and *Alysia alticola*. Spiders often released captured males quickly and seemed bothered by their contact with them, whereas ants, although readily consuming males from which the metasoma had been removed, hardly ever killed intact males. Given the large differences between their products in different taxa, the fact that the glands are only present in males and that the compounds produced have a generally pleasant odour to humans, a role in courtship seems almost certain, and a defensive function against predators also seems well demonstrated.

In comparison with the Braconidae, there have been very few studies of male (or female) scent glands in the Ichneumonidae, although apparently a considerable diversity exists (M.J. Sharkey, personal

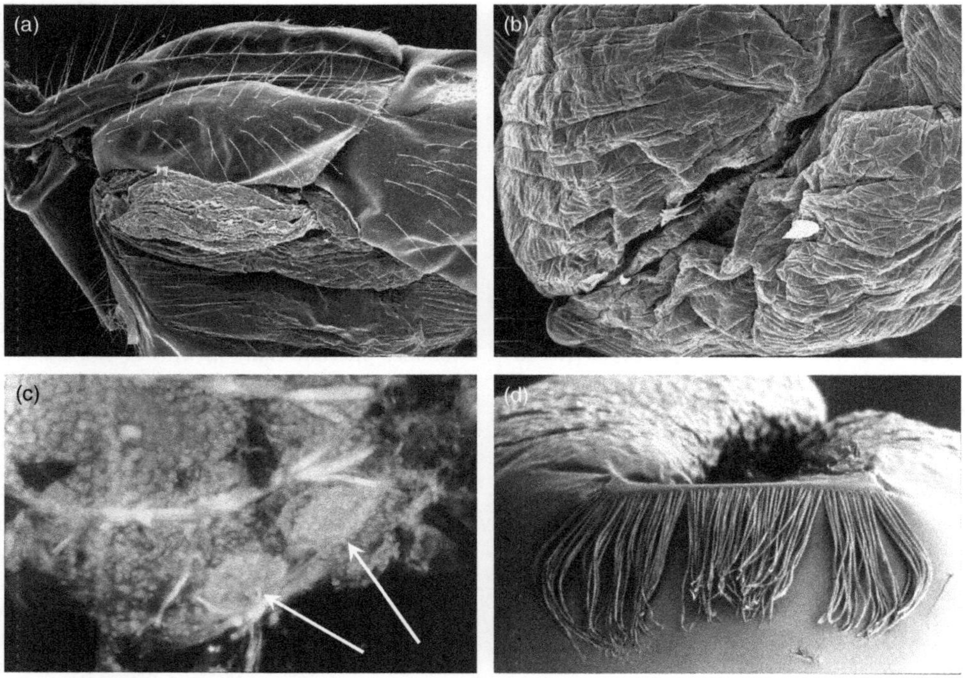

Fig. 7.8 Features of metasomal scent glands and their evaporatoria in the Braconidae. (a–c) Anterolateral abdominal glands in the braconine *Digonogastra kimballi* (Braconinae); (a, b) SEM images showing partially and largely everted corrugated external surface; (c) dissected anterior or metasoma showing internal right-hand pouches (arrows); (d) SEM image of *Agathis malvaceorum* tergum with brush-like anterior border which is normally concealed and large pouch-like glandular reservoirs. (Source: reproduced by permission of Mike Sharkey.)

communication.). The best known, and actually still not very well known, is a tergal gland near the anus of male Rhyssinae (Matthews et al. 1979, Eggleton 2008). Male specific glands are not always associated with the metasoma and may occur on the antennae or legs and possibly wings, although in almost all cases there is only the sex-linked correlation as evidence for roles in courtship. Some other potential examples include males of the north American braconine *Vipio monoleimae*, which have a swollen hind tibia with a setae filled groove (Quicke et al. 1998), as do those of one Neotropical species of the labenine genus *Apechoneura* (Gauld & Wahl 2000a), and in the latter case a glandular function is supported by the fact that specimens killed in alcohol often have the seat of the groove matted with large amounts of 'an emulsified oil-like substance'. Males of many doryctine braconids have a large, ventrally densely setose, 'pseudostigma' on the anterobasal margin of the hind wing, although unfortunately no histological/TEM studies have been carried out to date to determine whether these are indeed the sites of production of sex pheromones. Indeed, no-one appears to have carried out amputation experiments on easily cultured species such as *Heterospilus prosopidis* to see whether mating success is adversely affected if a male has the pseudostigma removed or perhaps blocked with something like wax. Of course, if the structure has evolved for purely mechanical purposes, amputation might well have an adverse effect on courtship but not because of pheromone loss, hence a dual approach is needed.

Even less is known about the chemistry of male sex pheromones than in females. Cossé et al. (2012) identified three male-sex pheromones, dodecanal, (4*R*,11*E*)-tetradecen-4-olide and (*Z*)-10-heptadecen-2-one, from the doryctine *Spathius agrili*.

GENOME SIZE AND RECOMBINATION

Sperm provide a good tissue for measuring wasp genome size since they are (virtually) always haploid, whereas many somatic tissues of adult ichneumonids are polyploid, especially those involved in very active metabolic processes. Methodologically, sperm should be air-dried from physiological saline to help prevent fragmentation and DNA quantified with an appropriate quantitative stain such as fuchsin or DAPI, along with a standard such as boar semen. Measurements for ichneumonoids vary considerably; unpublished data for the tryphonine genus *Netelia* range from approximately 3.5 to 7.3×10^{10} Da (João Lima, University of Guelph, personal communication) while 10×10^{10} Da was determined for two species of *Habrobracon* by Rasch et al. (1977). More recently Ardila-Garcia et al. (2010) reported values of 0.1–0.16 pg for more species of braconid and 0.26 pg for the ichneumonid *Neotheronia bicincta*, variously using flow cytometry (FCM) using brain tissue compared against *Drosophila melanogaster* or Feulgen image analysis densitometry. Because of the way in which the data are presented by various authors, it is not always straightforward to compare them, and when converting one measure to another there will be a slight effect due to the base composition differences between taxa and especially between the ichneumonids and braconids. However, roughly speaking, 1 pg is equivalent to 0.96×10^9 base pairs and one base pair is approximately equivalent to 660 Da. Therefore, Ardila-Garcia et al.'s braconid measurements correspond to approximately 6.3 to 10×10^{10} Da, and their value for *Neotheronia* equates to 16×10^{10} Da. The generally larger values found for ichneumonids may be related to the fact that they also tend to have larger numbers of haploid chromosomes (see the section *Cytogenetics*, below). Mateo Leach et al. (2012) estimated a considerably lower genome size for *Venturia canescens* of about 0.275×10^9 base pairs.

The recombination rate has only been determined for *Habrobracon hebetor* and *H.* sp. nr. *hebetor* with values around 3.2–4.8 3 cM Mb^{-1} (centimorgans per megabase), which is broadly within the range for other insects, although perhaps slightly on the high side.

In metazoans, there is a general correlation between nuclear DNA content and metabolic activity, which begs the question of how male ichneumonoids and other Hymenoptera cope. The answer appears to be that in metabolically active tissue such as flight muscles, the cells have doubled (endoreplicated) their DNA content so they are effectively diploid (Rasch et al. 1977), a 'trick' shown by all Hymenoptera except for the basal xyelid sawflies (Aron et al. 2005). Other highly active tissues such as Malpighian tubule cells show higher levels of endopolyploidy with up to eight times the haploid (1C) amount.

CYTOGENETICS

Largely due to the efforts of Vladimir Gokhman (Gokhman 2001, 2003, 2004, 2006, 2009, Gokhman & Westendorff 2003, Gokhman & Mikhailenko 2008), we now know a great deal more about the chromosomes of the Ichneumonoidea and of parasitic wasps in general than we did even 20 years ago (Gokhman & Quicke 1995). Within the superfamily, the haploid chromosome number (n) ranges from 3 to 23 (Fig. 7.9) and there is very considerable variation in the karyotypic morphology, with some species showing a more or less continuous range of chromosomal sizes and others having a few either markedly larger or smaller ones. In general, chromosome sets are dominated by metacentrics; however, karyotypes of some campoplegine Ichneumonidae (e.g. *Venturia canescens*) contain only acrocentric chromosomes.

The ancestral value of n for the parasitic Hymenoptera is believed to be around 14–17 (Gokhman 2003). In most Ichneumonidae, n is unimodal around 11, representing a moderate reduction from the groundplan, but braconids show a far more dispersed pattern, with many more taxa having greatly reduced haploid numbers (especially in Aphidiinae, Charmontinae, Cheloninae, Exothecinae and Homolobinae). Mapping these numbers on to a phylogeny shows that there have been multiple large changes in n in the family (fig. 4 in Gokhman 2004). Reduction in chromosome number appears to occur mainly through tandem fusions and, less commonly, centric fusions. Tandem fusions leading to fewer bi-armed chromosomes are a widespread feature among the Braconidae. Gokhman and Mikhailenko (2012) noted the great similarity in karyotype between members of two subfamilies within the macrocentroid complex *Charmon* (Charmontiinae) having $n = 5$ or 6 and *Homolobus* also having $n = 6$, whereas the Macrocentrinae have $n = 14$. This at first sight seem to suggest a sister group relationship between the former, as indeed has sometimes been proposed; however, molecular evidence now suggests fairly strongly that they are not sister groups and

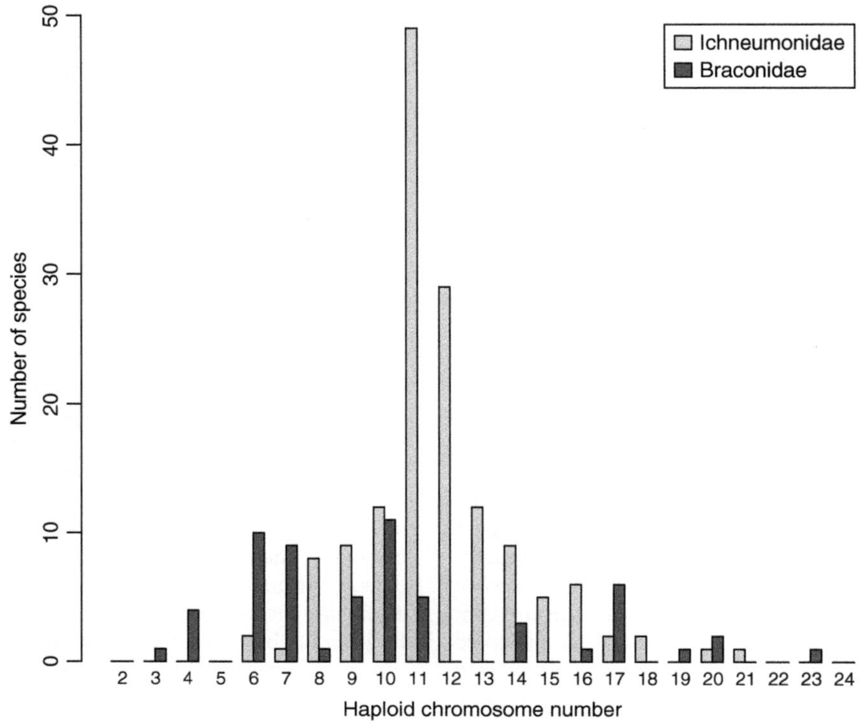

Fig. 7.9 Haploid chromosome numbers in the Braconidae and Ichneumonidae. (Source: data from Gokhman 2009.)

that their karyological similarity reflects homoplastic chromosomal fusions (Sharanowski et al. 2011).

Members of a few groups show consistently increased chromosomal numbers, notably in the pimpliformes subfamilies of the Ichneumonidae and the cyclostome Braconidae.

Chromosomal analyses have also revealed a number of cases where different individuals of a putative morphospecies have different chromosome numbers. Notable among these are the braconid *Charmon cruentatus* and several ichneumonids, especially in the Ichneumoninae (Gokhman 2006, Gokhman & Mikhailenko 2012). Most likely these observations indicate the presence of cryptic sibling species and it is noteworthy that at least *C. cruentatus* has a very large host range. The converse is not always the case and fairly often there are no apparent chromosomal differences among known cryptic species pairs.

ENDNOTES

1. RAPDs are randomly amplified polymorphic DNA markers.
2. It has been shown that parthenogenesis in some insects can be caused by bacteria other than *Wolbachia* (Mateo Leach et al. 2004), but these have also been excluded as being responsible for the *V. canescens* thelytoky.
3. *Habrobracon* have $n = 10$ as Inaba's later studies showed.
4. Originally called pygidial glands by their discoverer, but renamed by Buckingham and Sharkey because they are located on the penultimate tergum rather than the pygidium, i.e. the tenth abdominal tergite.
5. Or syntergum 8 + 9 (= abdominal 9 + 10) when these are fused.

Chapter 8

HOST LOCATION, ASSOCIATIVE LEARNING AND HOST ASSESSMENT

The Braconid and Ichneumonid Parasitoid Wasps: Biology, Systematics, Evolution and Ecology, First Edition. Donald L. J. Quicke.

This chapter considers three aspects of ichneumonoid wasp biology: how the wasps find their hosts, the role that learning plays in improving this and, having found a host, the subsequent decisions they make. Vinson (1976) dissected the basic pattern of parasitoid host location into a number of what are probably fairly distinct stages. He proposed a hierarchical system in which the habitat is located first (often the parasitoid will emerge in the relevant habitat, but migrating ones, for example, will have to locate a suitable habitat in which to find hosts). Vet et al. (1991) added the nuance that cues that are highly reliable may not be the most detectable: the reliability–detectability problem. Hosts have undoubtedly evolved for the most part to give out the fewest cues that natural enemies might use to locate them. Specific cues that emanate directly from the host itself will be the most reliable, but may be difficult to detect. Emission of some cues, however, is inevitable as they are intimately associated with necessary life functions. Defaecation results in the release of volatiles both directly from the macerated food and from the actions of the host, bacteria and fungi on it and, as with chemical signals emitted by host plants in response to the feeding of a herbivore, the host can do little to eliminate them (see the section *Tritrophic interactions*, below). The feeding noises created by endophagous, especially xylophagous, hosts are an inevitable consequence of their development (Wäckers et al. 1998), as are the volatiles emitted by wood–fungal associations of siricid and xiphydriid woodwasps which have symbiotic relationships with the fungi that help them to gain nutrition from the wood upon which they feed (Spradberry 1970b). Host sex pheromones are also obligate for hosts and likely to be highly reliable sources of information for parasitoids should they be emitted where a suitable host stage can be found – although this is mostly important only for parasitoids of adult hosts such as aphidiine and euphorine braconids.

Vinson (1998) expanded on his earlier thoughts and Vet et al.'s to emphasise the importance of learning for host location. Learning at a given place and time may greatly increase the reliability of cues used, in the sense that, across the complete range of its host, perhaps associated with different host plants or time of year, no single cue may be very useful, but locally associated cues may be very useful (see the section *Associative learning*, below). Vinson concluded that probably, at least when it comes to parasitoids of exophagous caterpillars, synomones which are generally highly detectable can nevertheless be just as reliable as generally less detectable kairomones. Many species of ichneumonoid, especially those with exophytic hosts, although this may be an artefact of research effort, learn odours and other cues that are present when they encounter and parasitise a suitable host. These are typically odours emanating from the host itself, the plant and the host's frass (e.g. Hendry et al. 1975, Dmoch et al. 1985, Nordlund & Lewis 1985, W. J. Lewis et al. 1988).

Different taxa require different stimuli to locate and subsequently attack hosts. Initial location of sites such as caterpillar- or aphid-infected plants or dead wood with beetle larvae often involves chemical cues. Living plants give off complex volatile mixtures when they are being damaged by herbivorous insects (see the section *Tritrophic interactions*, below) and parasitoids of those insects may have innate responses to these, in addition to improving their host location subsequently as a result of learning experience. Visual cues may also be important. Females of the microgastrine *Cotesia glomerata* are attracted to cabbage plants with holes in their leaves suggesting feeding damage, even when these holes have been created artificially (Vet 1999). This wasp can learn to distinguish the large holes made by its preferred host, *Pieris brassicae*, from the smaller ones made by the less preferred *P. rapae*. Parasitoids of wood-borers often rely on the vibrations produced by their hosts chewing; in fact, the braconid *Coeloides brunneri* can be induced to start 'drilling' through a piece of bark simply by scratching on the underside to simulate the host activity (R.B. Ryan & Rudinski 1962).

S. Fischer et al. (2004) pointed out that the location of pupal-stage hosts presents a different set of problems for parasitoids. Whereas larval hosts give out either sound signals if they are moving or feeding inside wood or leaf mines or lead through their feeding to the production of many volatiles, either induced by their feeding from the host plant or emanating from their frass, pupae are generally silent to both sensory modalities. Thus, for exposed pupae, visual cues are likely to play an important role and for hosts such as leaf-miners and rollers, the previous feeding damage to the plants may well be apparent visually as discoloured parts to leaves. Experiments conducted by Fischer et al. on the pimpline pupal parasitoid *Pimpla turrionellae* indicated that it has colour vision and uses visual contrasts in both hue and intensity in its host searching, but apparently does not respond to particular colours.

Considerably more comparative work in this area is needed.

From relatively few studies, it seems that parasitoids of active leaf-miner and wood-borer host larvae rely primarily on vibrations created by the hosts as they move or chew food and orientate to these vibrations (e.g. Sugimoto et al. 1988, Joyce et al. 2011); however, this might not always be the case. An earlier suggestion by Richerson & Borden (1972) that the braconine bark beetle parasitoid *Coeloides* locates its hosts through infrared signals resulting from their metabolism has since been effectively refuted by experiments conducted by Mills et al. (1991) (admittedly on a different species), but the latter authors also found no evidence that host-generated vibrations were used by the parasitoid either, suggesting that volatile chemicals probably provide the cues. If this is true, it is surprising that they provide sufficiently accurate information; however, the scolytine curculionid hosts probably do not move around much as they chew their galleries under the bark, so perhaps local concentrations of volatiles might be greater immediately above them.

Even from a distance, *Habrobracon hebetor* females can distinguish the odours emanating from host food alone, host food plus (unpreferred) young 10-day-old larvae, host food with larval frass and host food with (preferred) 30-day-old hosts (Darwish et al. 2003).

The way in which parasitoids of xylophagous hosts locate infected trees or parts of trees from a distance is also largely due to differences in volatiles emitted. Pettersson et al. (2001) found that the braconine *Coeloides bostrichorum*, a parasitoid of *Ips* beetles, can distinguish uninfested bark from infested bark. Then, by chemical analyses of volatiles associated with both types, they were able to synthesize a synthetic bait of volatiles associated with infested bark, which when added to the volatiles from uninfested bark rendered it attractive to the parasitoids. From my observations of large braconine parasitoids of wood-borers in dead trees in the tropics, it was nearly always the case that they were only active when sunlight was falling on to the infested part of the tree trunk or log. I strongly suspect that this has something to do with the warming effect of the Sun increasing the emission of attractive volatiles. At least in the case of the volatiles given off by wood–woodwasp combinations, xoridine and other parasitoids are attracted by compounds produced by wood–fungal interactions, although in this case the cues are reliable since the woodwasps co-transmit a symbiotic fungus into the host tree at the same time as they oviposit.

In general, host-seeking female parasitoids show arrestment and an increased rate of turning when they encounter the correct host kairomones that have been deposited on a substrate (Strand & Vinson 1982), and this behavioural change provides a ready bioassay for identifying kairomones.

TRITROPHIC INTERACTIONS

Long-range location by parasitoids of herbivorous hosts is often mediated through specific mixes of compounds released by the plant being attacked, and there is a large literature on these tritrophic interactions (e.g., Turlings et al. 1991, Whitman & Eller 1990, Geervliet et al. 1994, Mattiacci & Dicke 1995, Du et al. 1996, Powell et al. 1998, Guerrieri et al. 1999). Early work mostly concerned a few plants of great economic importance, but the release of kairomones that are attractive to parasitoids is now known to be a very widespread phenomenon. Young corn plants attacked by beet armyworm (*Spodoptera exigua*: Noctuidae) were found consistently to emit a mixture of 11 volatiles, and that naïve females of the microgastrine armyworm parasitoid *Cotesia marginiventris* were strongly attracted to an equivalent synthetic mixture of these, although slightly less so than they were to the odours from naturally damaged plants, presumably indicating that some other less abundant odour components were also important but perhaps below the analysable threshold (Turlings et al. 1991). Plants not only release different volatile profiles when attacked by a herbivore compared with being mechanically damaged by an investigator, but also may release different mixes depending on what particular herbivore is involved. In some cases, herbivore damage can be mimicked by applying caterpillar oral secretions to mechanically damaged plant tissue.

It has been a topic of considerable debate whether the plants release these chemical signals of herbivore attack as a result of natural selection or simply as a consequence of physiological disruption. It appears that the former is likely to be true, at least in some cases, as was demonstrated by Fritzsche-Hoballah & Turlings (2001) using plots of maize infected with *Spodoptera littoralis* and parasitised by the campoplegine *Campoletis sonorensis* and the microgastrine *Cotesia marginiventris*. Attraction of parasitoids to the attacked host plants led to them suffering significantly less caterpillar feeding damage and producing significantly more seeds. How such a signalling mechanism came about is not fully clear, but within a plant or perhaps within a plant kin group, signalling function, whereby the plant detects volatiles emitted from insect damage and responds in a way so as to reduce the palatability of other parts of the plant, may have been the original role, with parasitoids co-opting these signals as reliable indicators of the presence of hosts (Rodriguez-Saona & Frost 2010).

HOST ACCEPTANCE

After having located a host of a potentially suitable species, usually by a combination of olfactory and visual cues, the host still might not be acceptable. The ichneumonine *Diadromus pulchellus* has been a fairly intensively studied laboratory wasp in this respect. This species oviposits into the pupal stage of its weakly cocooned yponomeutid leek moth host (*Acrolepiopsis assectella*), which it attacks early after pupation. The monophagous nature of this species is, however, not physiological as females can be tricked into ovipositing into pupae of non-host species (e.g. *Plodia interpunctella*) if the latter are transferred to *A. assectella* cocoons shortly after having spun their own cocoons (Lecomte & Thibout 1993). Using this trick, it was possible to show that the response of females to frass-related volatiles of its natural host was innate with post-emergence refinement following experience rather than depending on pre-emergence factors. Bekkaoui and Thibout (1993) showed that this wasp responded to a water-soluble contact kairomone in the host's cocoon and that the response was host specific, in that females did not respond to cocoon extracts from three other cocoon-forming moths [*Bombyx mori*, *Anagasta* (= *Ephestia*) *kuehniella* and *Cacoecimorpha pronubana*], although none of these belong to the same family as the natural host of *D. pulchellus*. Bénédet et al. (2002) found that acceptance by the wasp is dependent upon a protein in the host cocoon silk, and as proteins are not volatile, the detection must be gustatory, i.e. the protein has to be solubilised before it will be able to contact the antennal receptors of the wasp. The actual compounds involved turned out to be soluble sericin-like polypeptides (referred to as B1 and B2) that were also strongly cross-linked with cysteine bridges such that they lost their effect when denatured (Gauthier et al. 2004). Not only is host cocoon silk important for host recognition by this parasitoid, but also female wasps encountering the silk trails left by host caterpillars changed direction to follow them and thus locate their host cocoons (Thibout 2005). Bénédet et al. (2002) showed that receptor cells of one, female-specific, multiporous, antennal sensillum type gave a positive electrophysiological response to the kairomone proteins, but how the wasps precisely detect these non-volatiles is not known, but it probably involves the release of secretions from antennal glands, perhaps associated with this sensillum.

ASSOCIATIVE LEARNING

It has been known for a long time that female parasitoids learn cues, principally chemical, about the environments where hosts are likely to be found either directly at emergence or as a more or less continuous process when each new and suitable host is encountered (Arthur 1971, Vinson et al. 1977, Wardle & Borden 1985), and may be particularly important in polyphagous species for which the commonest or best host available at any given time or place might vary from one wasp generation to the next (Wardle & Borden 1990). Learning plays a very important role in the location of hosts in most parasitoid species and is associated with highly developed mushroom bodies in the adult brain (Farris & Schulmeister 2011). Their complex brains also allow them greater flexibility in host patch utilisation, decision-making and utilisation of new resources. Numerous studies have shown that experience of different host-associated cues, be they natural or artificial, with different reliabilities, have different effects on the strength or duration of the learning response. Inexperienced parasitoids often show a lesser response to host kairomones than do experienced females (e.g. Strand & Vinson 1982). Vinson et al. (1976, 1977) accidentally found that the braconine boll weevil parasitoid *Bracon mellitor* learned, under culture conditions, to associate methyl parasept (methyl *para*-hydroxybenzoate), an antifungal agent used to keep the host's semi-synthetic diet infection free, with hosts and they would actively probe items infiltrated with the compound with their ovipositors. In an equally artificial experimental situation, when the braconid *Microplitis croceipes* was exposed to host frass in the presence of vanilla odour, the wasps orientated in wind tunnel experiments to vanilla odour alone for about 30 min following the initial experience

(Takasu & Lewis 2003). However, if they were allowed to oviposit into a host, then they continued to respond to the vanilla cue for 24 h.

However, learning is not necessarily always that important. For example, some species of parasitoid are fairly host taxon specific, but their hosts may feed on a very wide range of plants and so cues present at emergence or even at a host encounter may not be particularly good indicators about where future hosts will be found. In these situations, we would expect the parasitoid to have a strong innate response to compounds released by their hosts, and this is indeed what Sime (2002) concluded from a study of the responses of the ichneumonine *Trogus pennator*, which attacks papilionid butterflies on a wide variety of plants. Absence of odour-learning was demonstrated in the cereal stem-borer parasitoid *Cotesia sesamiae* by Potting et al. (1997), although why this species should not benefit from learning is unclear.

Some cues associated with a host are available for learning at the time of the parasitoid's eclosure, but others are only available once the wasp has located a host. A good example is provided by a campoplegine wasp, *Lemophagus pulcher*, that parasitises the European lily leaf beetle (Schaffner & Müller 2001). Larvae of the beetle, *Lilioceris lilii*, retain their exuviae/frass as a shield which they hold over their backs. Chemicals in the shield are attractive to the parasitoid and will stimulate ovipositor probing in naïve females. However, when attacked, the host larva produces an offensive oral discharge as part of its defensive repertoire and this secretion is attractive only to experienced wasps – a naïve eclosing female wasp would simply not have any contact with the relevant chemicals in the oral secretion whereas it would have experience of chemicals associated with the shield, and indeed it is these volatiles that are most important for its host location.

Not only are odours associated with a host important, but visual cues can be also, and Wäckers & Lewis (1994) showed that both odour and visual cues were important learned components of host site location by *Microplitis croceipes*. Another microgastrine, *Cotesia glomerata*, is rather unusual in that it is a very rapid learner, its long-term memory of host–host plant complex being consolidated after only a single host experience, in contrast to the closely related *C. rubecula*, which requires multiple experiences with host–host plant complexes (Fig. 8.1) (Smid et al. 2007). Smid et al. also knocked out long-term memory formation by feeding the wasps either anisomycin or actinomycin D, which inhibit protein translation and RNA transcription, respectively, and so prevent the manufacture of proteins essential for long term memory formation. Figures 8.2 and 8.3 compare the effects of LTM blocking in the two species.

Anisomycin has no effect on short-term memory in either species as the learned preference of *Nasturtium* plants used in the training is unaffected after 1 h. However, after 4 h, *C. glomerata* fed anisomycin has forgotten the learning experience and shows a greatly reduced level of preference for *Nasturtium*, in fact now down to the level shown by naïve females, whereas control wasps fed sucrose only retain their learnt preference almost unchanged for more than 5 days (Fig. 8.2). With *C. rubecula*, the action of actinomycin shows a very much more protracted time course (Fig. 8.3) and plateaus out (down) only after 72 h, demonstrating that the consolidation of long-term memory in this species is a slower process, with the necessary protein synthesis happening over a longer time period and with the wasp's short-term memory also lasting longer. The reason for this marked difference probably lies in the nature of their hosts. That of *C. glomerata*, the large white butterfly, *Pieris brassicae*, is gregarious, so when the female wasp has encountered a cluster of recently hatched hosts she has effectively found an enormous potential resource and will oviposit into most or all of the hosts there. In contrast, the host of *C. rubecula*, the small white butterfly, *P. rapae*, is solitary, and may be distributed potentially over several different Brassicaceae host plants in an area, hence fixing a long-term memory for a given host–host plant combination based on only the first encounter could easily be maladaptive, as the majority of potential hosts might in fact be on a different food plant and thus would be missed. *C. glomerata* does need to remember, though, because the hosts viciously try to defend themselves, biting the wasp and regurgitating on it, so the wasp has to leave the cluster of hosts to groom and clean up before returning to the same patch, to continue ovipositing. Kruidhof et al. (2012) have also shown that the magnitude of the reward is important for the consolidation of memory in *C. glomerata*, with long-term memory consolidation being effected after a single trial when the first experience is on the high-reward, gregarious *P. brassicae* and not on the lower reward, solitary *P. rapae*.

Responses to host–food plant complex volatiles have been shown to have both genetic and learned components. The best studied systems involve microgastrine braconids. Prévost and Lewis (1990) used

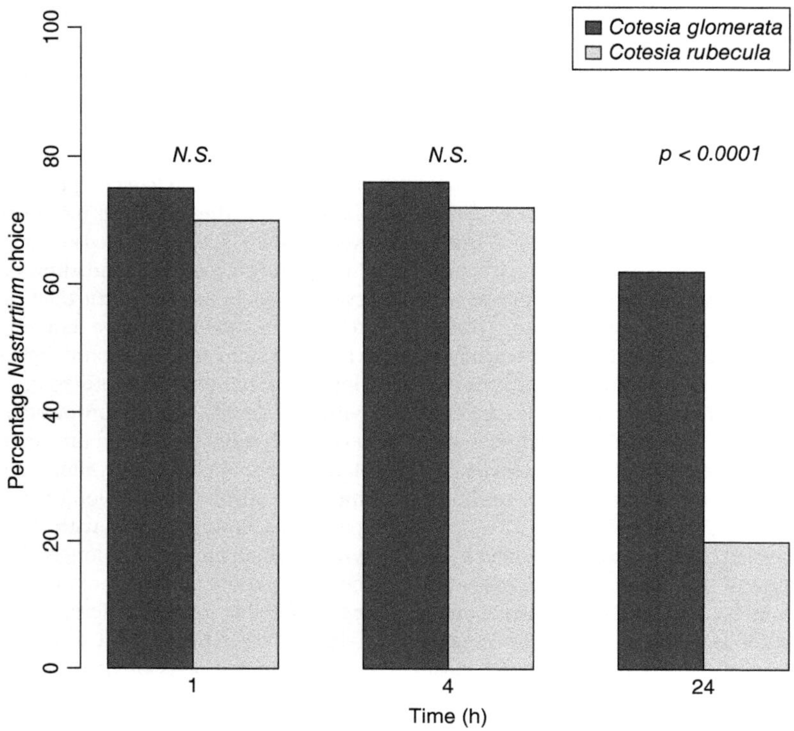

Fig. 8.1 Comparison of the microgastrines *Cotesia glomerata* and *C. rubecula* memory retention (learning) after a single presentation of a host on a *Nasturtium* plant, a normally unpreferred host plant, showing that *rubecula* females do not consolidate their memory after a single experience and show no learned preference for *Nasturtium* after 24 h. (Source: data from fig. 2c in Smid et al. 2007.)

isofemale lines to investigate genetic aspects of plant host-specific volatiles in *Microplitis croceipes* using experienced females and Gu and Dorn (2000) carried out similar experiments with *Cotesia glomerata*. In both cases, inherited differences were found between lines. More recently, Q. Wang et al. (2003) performed a bidirectional selection experiment on *C. glomerata*, examining two aspects of the wasp's response to plant host-specific volatiles, flight orientation and landing success, and found that they had high heritabilities of 0.248 and 0.216, respectively. This is in addition to learned effects, so both strains were significantly better at flight orientation and landing after prior exposure to the semiochemicals.

It has been proposed that endoparasitoids may be able to learn host-associated cues also prior to eclosion, possibly due to chemicals passed from the larval stage to the pupal stage, a process known as Hopkins' host selection principle. Working with the eulophid *Hyssopus pallidus*, Gandolfi et al. (2003) apparently demonstrated that early-stage larvae were able to learn host cues and pass this information to the adult whose subsequent behaviour it affected. Evidence for such preimaginal learning in the Ichneumonoidea is sparse and inconclusive. Technically, it can be difficult to exclude completely the possibility of chemical contamination (Godfray 1994). In addition, several studies have failed to show any effect; for example, Arthur (1965) demonstrated that it did not occur in the pimpline pupal parasitoid *Itoplectis conquisitor* when reared in two hosts, and Lecomte and Thibout (1993), working with the ichneumonine *Diadromus*

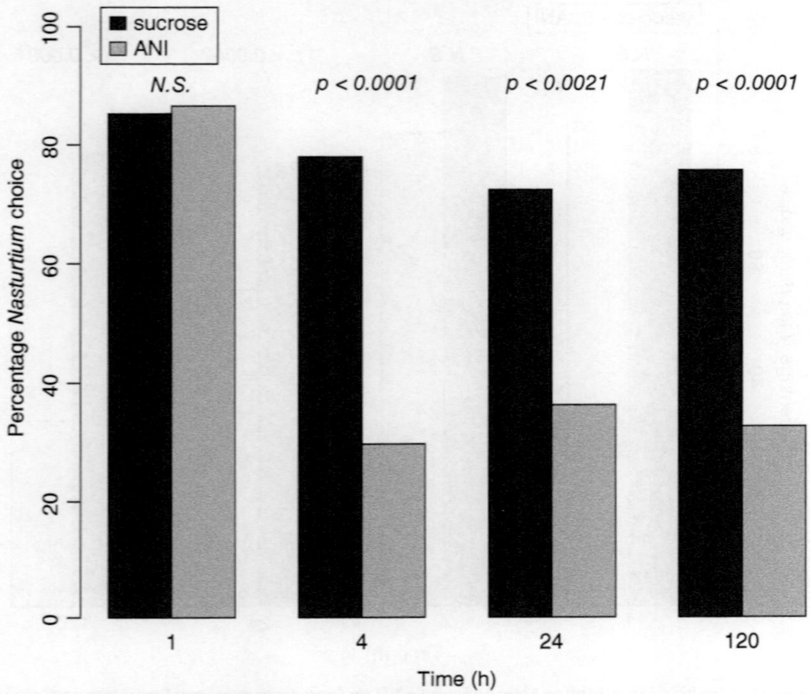

Fig. 8.2 Effect of the RNA translation inhibitor actinomycin (ANI) on long-term memory formation in *Cotesia glomerata* trained with three host–host plant (*Nasturtium*) experiences at 10 min intervals at the start of the experiment and showing that short-term memory is consolidated into long-term memory within 4 h of the oviposition experience. (Source: data from fig. 3a in Smid et al. 2007.)

pulchellus, similarly failed to detect any pre-imaginal experience effect. Van Emden et al. (1996), using a series of host-plant transfers and excision of aphidiine (*Aphidius rhopalosiphi*) pupae from mummies prior to olfactometer testing, showed that the preference of naturally emerging wasps for larval food plant odours was the result of food plant odour compounds having been absorbed by the mummy, so the learning was post-eclosion rather than larval/pupal. Similar findings that aphid mummies carry a chemical cue learned by emerging aphidiines were demonstrated by van Emden et al. (2008). However, Gutierrez-Ibanez et al. (2007), in a reasonably controlled experiment, found that the congener *A. ervi*, during the period when they bite through the host aphid cuticle to attach it to the host plant with silk, appeared to be able to learn vanilla odour.

Although it is not impossible that in some systems information that might be available to the sensory system of the juvenile ichneumonoid parasitoid could be translated into information utilisable be the adult wasp post-metamorphosis, proof would definitely require very carefully controlled experiments. The experiment of Gutierrez-Ibanez et al. (2007), for example, cannot fully rule out that part of the aphidiine larval cuticle had absorbed some vanilla compounds and that these were then smelt by the emerging adult. Indeed, even if the final larval instar skin were removed from the system, it would be incredibly difficult to eliminate the possibility that some highly potent molecules had not passed through it during the period of exposure and impregnated themselves on the developing pupal cuticle such that they could later be detected. I think only neurophysiological work will ever be able

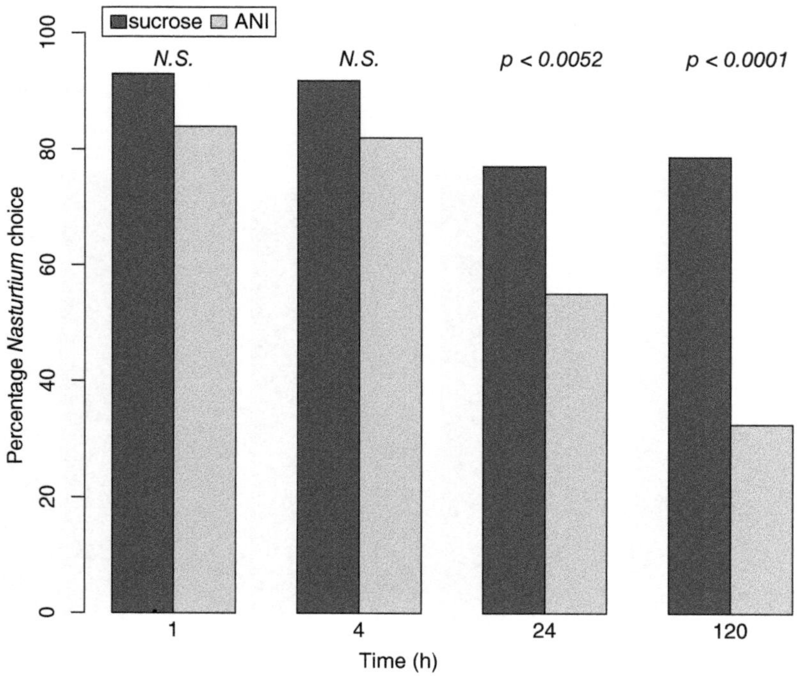

Fig. 8.3 Effect of the RNA translation inhibitor actinomycin (ANI) on long-term memory formation in *Cotesia rubecula* trained with three host–host plant (*Nasturtium*) experiences at 10 min intervals at the start of the experiment and showing that short-term memory is only consolidated into long-term memory over a period of days. (Source: data from fig. 3b in Smid et al. 2007.)

to prove this one way or another, and even if one or more systems do not show 'larval' learning, it will not prove that it is impossible.

BIOSENSORS

The considerable sensory ability of parasitoids to locate hosts, together with their learning capabilities, has led to some interesting research into potential applications, such as detecting land mines or drugs (reviewed by Tomberlin et al. 2008). The bulk of this work has concentrated on small but easily cultured and trained (associative learning) microgastroids, which are nevertheless within the size range for potentially attaching mini-aerials that could be detected using modern technologies, and could lead to a far better understanding of host (or mate or food)-finding strategies of ichneumonoids in the field in the near future.

PATCH USE

In a large number of cases, the hosts of a parasitic wasp are likely to be heterogeneously distributed in their environment, with some local patches containing several to many hosts and others none or relatively few. Probably also in many cases, the parasitoid cannot assess how many potential hosts a newly discovered patch might contain and it will therefore be faced with making a decision about when to leave the patch in search of others that are likely to yield more oviposition opportunities. This topic has been explored in great depth both theoretically and experimentally, many studies using ichneumonoid models (Godfray 1994), and I will not go into any great detail here. Particularly well investigated braconid taxa are *Asobara tabida* (Alysiinae), *Aphidius* , *Diaretiella* and *Lysiphlebus* species in the Aphidiinae, and several *Cotesia* (Microgastrinae, species. In the Ichneumonidae, most studies have involved the campoplegines *Diadegma semiclausum* and *Venturia canescens*.

Measurements in the field of patch use parameters are far scarcer than laboratory studies. Völkl (1994) studied patch use by the aphidiine *Aphidius rosae* on rose bushes at three spatial scales (bush, shoot, leaf) and found that the wasp was very efficient at discriminating aphid-bearing shoots from uninfected ones, thus wasting very little time searching where there were no hosts. The wasps searched for hosts on shoots and leaves by walking, but always flew between shoots. However, this flight time between patches constituted less than 1% of their time spent searching a bush. In contrast, I have observed females of the large Afrotropical wood-borer parasitoid *Merinotus* sp. in Uganda 'drilling' on an ebony (*Diospyros*) branch for more than 2 days, but when I subsequently dissected the wood, I found no trace of any suitable host and, given the biology of the whole group, it is virtually inconceivable that the host could be far smaller than the wasp.

Chapter 9

OVERCOMING HOST IMMUNE REACTION AND PHYSIOLOGICAL INTERACTIONS WITH HOST

The Braconid and Ichneumonid Parasitoid Wasps: Biology, Systematics, Evolution and Ecology, First Edition. Donald L. J. Quicke.

In addition to an egg or eggs, endoparasitoids inject into hosts a complex mixture of compounds and entities that have a diverse array of effects. Probably the most obvious are a venom produced by the venom gland (often called the acid gland in earlier work) and its associated reservoir and ducts, all of which have secretory units. However, in addition there are secretory products from the ovary or oviduct (so-called ovarian secretions), in some cases from the associated vaginal gland (e.g. in Banchinae). Members of three clades inject a slurry of polydnaviruses or virus-like particles, formed in the calyx gland cells, and an increasingly large number of ichneumonoids are being found to liberate other viruses from cells either in their oviducts or in their venom glands or the venom reservoir (e.g. Krell & Stoltz 1979). Although not known to affect host physiology, hydrophobic components from the Dufour's gland, which are probably only involved in host marking, might also be released. Further, host physiology may be affected by secretions from the serosal membrane around the embryo or first instar larva (Lawrence 1990), teratocytes formed by dissociation and swelling of serosal cells, secretions from the parasitoid, the larval salivary (labial glands) and larval anal vesicle (e.g. Führer & Willers 1986), hence the study of the effects of the various wasp products on the host can be complicated and very hard to dissect. A great deal has been achieved, especially in the past 30 years, but I am sure that more modern genetic and gene expression approaches will soon be leading to great advances. The complexity leads to an additional negative feature when it comes to understanding the evolution of the various physiological aspects of ichneumonoid host–parasitoid interactions, namely that the limited time and resources of researchers mean that nothing at all is known about some subfamilies, although a considerable amount is known about some others. Asgari (2006) provided a brief review, reasonably up-to-date at the time.

Insects in general have efficient immune systems which they employ against various foreign agents ranging from bacteria and fungi to parasitoids. Larger foreign objects are usually attacked by a process known as encapsulation (Salt 1965, 1970), and encapsulation of injected latex beads or bits of nylon can be used to assay the efficiency of a host's cellular immune defence. The way in which many parasitoids actively avoid encapsulation varies but often involves disrupting the host haemocytes from being able to spread (Prévost et al. 1990), or involves chemical camouflage or mimicry, and some parasitoids may employ a combination of these at different stages.

The eggs of insects have no encapsulation ability *per se* as they do not have free haemocytes, and the defences of the late embryonic and early first instars are similarly typically rather limited and the absolute numbers of defensive haemocytes increases greatly through the early instars of unparasitised hosts. Further, there is some evidence that the pupal stage of endopterygotes, which undergoes radical cellular reorganisation, might have a somewhat reduced immune capacity and certainly the realised and potential host ranges of many Pimplini, which are specialist idiobiont pupal parasitoids, principally of Lepidoptera, can be huge.

OVERCOMING HOST IMMUNITY IN ENDOPARASITOIDS

All endoparasitoids are potentially vulnerable to host immune defences and have evolved many mechanisms to overcome their host's actions. Within the Hymenoptera, nearly all work on host–parasitoid immune interactions is based on a few examples from within the Ichneumonoidea – probably because ichneumonoids are generally larger than other parasitic wasps that have been brought into culture. Although the best studied examples involve members of seven subfamilies (Ichneumonidae: Campopleginae and Braconidae: Alysiinae, Aphidiinae, Cardiochilinae, Cheloninae, Microgastrinae and, to a lesser extent, Opiinae), even this might give an exaggerated view of coverage, because the Cardiochilinae, Cheloninae and Microgastrinae all belong to the 'microgastroid' complex whereas the Alysiinae and Opiinae are probably sister groups or the former is derived from within the latter. Indeed, details of biology are almost unbelievably unknown for even some rather large and common groups and in particular the Helconinae (*Helcon* group), Cenocoeliinae and Lysiterminae – this is largely to do with host niche; we tend to know a considerable amount about groups that attack exposed hosts such as Lepidoptera larvae but much less about those that parasitised concealed hosts such as those dwelling in woody substrates.

PASSIVE EVASION OF ENCAPSULATION BY PARASITOID EGGS

Encapsulation of an endoparasitoid egg can only take place if it is exposed to haemolymph, so those that deposit eggs directly in the host haemocoel must obtain protection either through immune suppression (Strand & Pech 1995), by passive disguise, i.e. the egg surface not eliciting a response from otherwise fully competent immune system, or by hiding it from the host's immune system. Osman (1974) provided the first evidence that the protective coating of a parasitoid egg (*Pimpla turionellae*) was composed of mucopolysaccharide, which, when removed by enzyme treatment, rendered the egg susceptible to haemocyte attack. It turns out in this case that secretion from the vaginal gland which coats the egg is probably largely responsible. J. Hu et al. (2003) demonstrated apparent passive immunity mediated by the surface coating of *Macrocentrus* eggs, since the host immune response to other challenges was unaffected by parasitisation. In the microgastrine *Cotesia rubecula*, Asgari and Schmidt (1994) found that a protein secreted by the calyx gland attaches to the surface of the egg and helps to disguise it from the host's immune system. This only provides protection in the short term before the polydnavirus infection of host cells and subsequent suppression of host immune response take effect.

However, the best-studied system is also one of the more peculiar. Salt (1965, 1966, 1970) found that protection of eggs from encapsulation in the campoplegine *Venturia canesens* depended on a coating they received as they passed through the calyx gland region at the anterior of each lateral oviduct, which adhered to the surface of the eggs and rendered them immune from haemocyte attack. Salt, however, considered it most likely that the secretion was a fluid (G. Salt, personal communication to S.B. Vinson). Rotheram (1973a) examined the surface of the *V. canescens* egg and found its outer layer to be composed of many projecting fibres, and showed (Rotheram 1973b) that the small virus-like particles (VLPs) produced in the calyx adhered to these fibrils, and it is these VLPs that render the surface of the egg immune to the host haemocytic defence system. Feddersen et al. (1986) raised antibodies against the VLPs and demonstrated that they had similar antigenic determinants to the cells of the host *Ephestia kuehniella* caterpillar and then, by treating VLP-coated eggs with antibodies, showed that these determinants were responsible for avoiding encapsulation, although this might not be as clear cut as it seems since the protection of eggs also depends on the charge on the surface molecules, and this could have been changed by the antibodies.

The VLPs of *V. canescens* are in many respects unique in that unlike the virus particles produced by all other investigated members of its subfamily, the *V. canescens* VLPs do not contain any DNA (see the section *Polydnaviruses*, below). It should also be noted that within the genus *Venturia*, only that one species has been studied, so it is not known whether a similar purely particulate defence occurs in any other species. However, in addition to the role of polydnavirus gene expression in the targeted host cells (see below), the envelope protein of at least one of them, the virus of the microgastrine *Cotesia kariyai*, appears to protect the parasitoid's eggs from the host immune system (Hayakawa & Yazaki 1997) in a way analogous to that of the virus-like particles in *Venturia canescens*, and a similar protective coating effect was also found by Asgari and Schmidt (1994) for *C. rubecula* eggs. It would be very interesting to know how widespread this form of passive egg protection is. A fibrous outer layer of the chorion has also been implicated in passive encapsulation avoidance in *Macrocentrus cingulum* (Macrocentrinae) (J. Hu et al. 2003), but macrocentrines are not known to produce any virus-like particles.

AVOIDING ENCAPSULATION BY PHYSICAL MEANS

A number of ichneumonoids have evolved oviposition strategies to place their eggs in a part of the host where haemocytes have no access and thus where encapsulation cannot take place. Oviposition into host ganglia occurs in many agathidines, the sigalphine braconid *Acampsis alternipes* (see Fig. 12.41c) (M.R. Shaw & Quicke 2000), the aphidiine *Monoctonus paludum* (D.C. Griffiths 1961), the ctenopelmatine genus *Homaspis* (Eichhorn 1988) and some metopiine ichneumonids (Aeshlimann 1974a).

Partial but still effective defence against encapsulation can be achieved if only part of the parasitoid egg is exposed to host haemocytes, a strategy used by members of the alysiine braconid genus *Asobara*

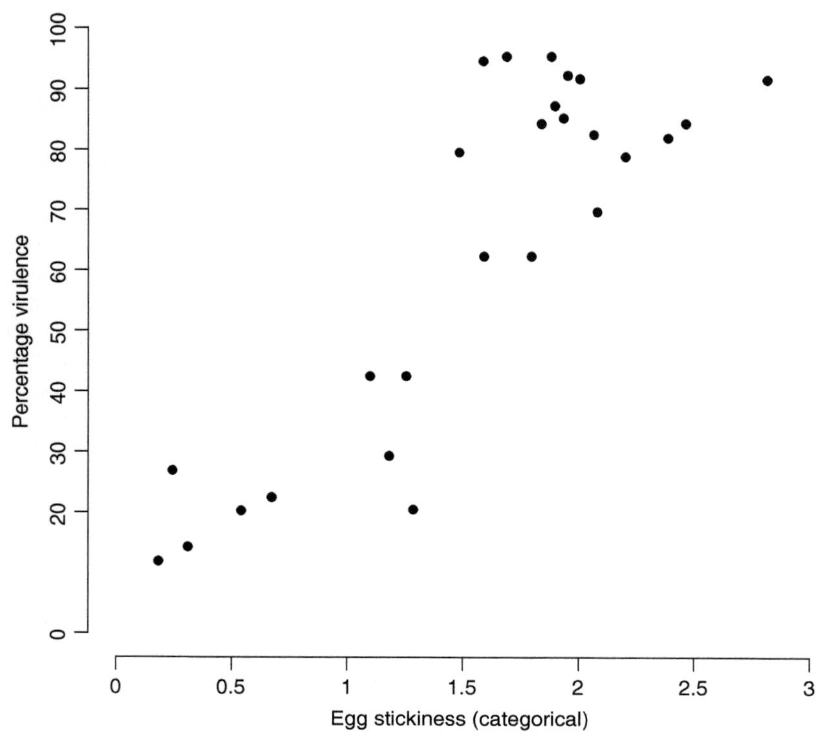

Fig. 9.1 Showing how the stickiness of eggs of the alysiine braconid *Asobara tabida* towards host, *Drosophila*, tissues affects their virulence, with the most sticky ones being less likely to succumb to lethal encapsulation. (Source: data from A.R. Kraaijeveld 1994.)

that attack *Drosophila* larvae, in which the surface of the eggs can be sticky and when inside the host they attach and embed themselves among host fat body tissue (A.R. Kraaijeveld & van Alphen 1994, Eslin et al. 1996). Sometimes virtually all the egg may be found adhering to host tissue and only that part of it which is exposed to haemocytes becomes 'encapsulated' and melanised, but the fact that part of the egg is protected means that the developing parasitoid embryo can still obtain enough oxygen and perhaps other nutrients and hatch successfully. The venom of *A. tabida* has no disruptive effect on the host larva's encapsulation ability, as shown by the host's continued ability to be able to encapsulate foreign objects such as paraffin droplets (Prévost et al. 2005). *A. tabida* shows inherited variation in stickiness, i.e. their tendency to adhere to host tissues. A.R. Kraaijeveld (1994) scored egg stickiness on a four-point categorical scale and showed that the virulence of the parasitoid (i.e. its ability to avoid death from being encapsulated) increased significantly with the stickiness of the egg's outer coating (Fig. 9.1). Different populations of *A. tabida* were found to display different levels of egg stickiness, which correlated with the encapsulation ability of the local host *Drosophila* population. However, the ability to the *Drosophila* host to encapsulate eggs of a given level of stickiness is also variable and under genetic control, so it is possible to select artificially for flies with increased encapsulation ability (A.R. Kraaijeveld & Godfray 1997). What is particularly interesting about this study is that there is a cost associated with possessing higher encapsulation ability, such that flies selected to be more resistant to the parasitoid fare less well in competition experiments when conditions become harsh (in this case, when larval food supply is restricted). Interestingly, the strategy used by another *Asobara* species to overcome the immune system of the same host is different. The Afrotropical species *A. citri* does not possess sticky eggs and instead parasitisation by it leads to reduced host haemolymph phenoloxidase activity, with reduced numbers of circulating haemocytes and an associated reduced encapsulation potential (Moreau et al. 2003).

In an aphid–aphidiine–fungal pathogen system, Ferrari et al. (2001) found that different clones of the host pea aphid had significantly different levels of resistance to two parasitoids and to a pathogenic fungus, but in this case there was no trade-off between the ability to protect against one of the natural enemies and the other and, indeed, there was a significant positive correlation for the ability to defend against the two wasp species. Presumably there is a trade-off against some other fitness parameter in their ability to mount a defence, otherwise it would be hard to understand why all aphid populations were not maximally efficient at defending themselves.

EFFECT OF HOST AGE AND HAEMOCYTE NUMBER

Eggs *per se* have no immune system, in that they lack haemocytes, although they may in some cases be able to protect themselves from attack. However, once the young host larva has a haemocoel and some haemocytes, the potential for defence is increased. Several studies have shown that the immune capability of larval holometabolous insects increases with age. Hazell et al. (2005) found that the predacious hoverfly *Episyrphus balteatus* has zero effective immunity against its ichneumonid parasitoid, *Diplazon laetatorius*, as a first instar, but kills 40% of parasitoids in the second instar and 100% in a third instar. To a considerable extent, such differences reflect the number of haemocytes that the host has at its disposal as it develops. The importance of haemocyte number in a host's ability to encapsulate parasitoids has also been shown in the *Drosophila*–*Asobara tabida* system (Enslin & Prévost 1996). Two species of *Drosophila*, *melanogaster* and *simulans*, differ markedly in their general ability to encapsulate *A. tabida*, the former usually failing to encapsulate and the latter nearly always doing so. Correlated with this, most *D. simulans* were found to have an approximately five times higher concentration of haemocytes than *D. melanogaster* larvae. However, those *simulans* that succumbed to parasitisation had equivalent haemocyte levels to those of *melanogaster*.

OTHER HOST DEFENCE MECHANISMS

In addition to or instead of encapsulation, a few hosts have other means of defence against ichneumonoid eggs. It has been known for a long time that nearly all cases in which aphids successfully defend themselves against aphidiine braconid attack do so by means of some sort of humoral response, with the parasitoid egg 'simply' disappearing after a few days. One report of encapsulation of an aphidiine egg, by D.C. Griffiths (1961), has not recently been verified and it is unclear whether it involved attachment of haemocytes or some humoral effect. Some plusiine noctuid caterpillars (genera *Trichoplusia* and *Autographa*), which include a number of pest species, sometimes, particularly as earlier instars, exclude a parasitoid egg by a process called cuticular encystment. This was first described by Arthur & Ewen (1975) working on *Banchus* (Banchinae) and subsequently shown by Namba et al. (2004) as a response to parasitisation by the microgastrine *Cotesia*, and other instances were investigated experimentally by Namba et al. (2008), this time working with the campoplegine *Campoletis*.

VENOMS

Venoms being complicated mixtures and employed by both ecto- and endoparasitic taxa, in addition to those attacking hosts at all stages from egg to adult, are necessarily diverse and here I will separate what is known of their actions in terms of interactions with the host nervous system (temporary or permanent paralysis), with the host's immune system and direct actions on the host's physiology (development and nutritional value).

Ichneumonoid venoms and those of other parasitic wasps generally lack the small peptides[1] that are present in most aculeate wasp and bee venoms (Leluk et al. 1989) and instead are dominated by acidic proteins. Nevertheless, these authors detected immunological similarities between venom/venom gland components of *Chelonus* (Braconidae: Cheloninae) and those of various aculeates, suggesting a level of phylogenetic conservation that might not have been expected given both the time between the separation of the groups and the different functions that one would expect the venom proteins to display. A summary of

Table 9.1 Functional enzymes currently known to be ichneumonoid venom components*. Enzymes are indicated by letters and subfamily level taxonomic groups are differentiated by shading.

	Enzyme†																		
	A	**B**	**C**	**D**	**E**	**F**	**G**	**H**	**I**	**J**	**K**	**L**	**M**	**N**	**O**	**P**	**Q**	**R**	**S**
Aphidius ervi							+												
Asobara tabida			+																
Chelonus sp. near *curvimaculatus*					+														
Chelonus inanitus	+	+		+	+	+			+				+						
Cotesia rubecula													+						
Microctonus aethiops (=*aethiopoides*)					+													+	
Microctonus hyperodae					+					+								+	+
Hyposoter didymator																+	+		
Pimpla hypochondriaca								+	+		+		+	+	+				
Pimpla turionellae												+							

*Mostly from Colinet et al. (2013); additional information from A.M. Crawford et al. (2008).

†A, α-*N*-acetylglucosaminidase; B, angiotensin-converting enzyme; C, aspartylglucosaminidase; D, C1A protease; E, chitinase; F, esterase/lipase; G, γ-glutamyltranspeptidase; H, laccase; I, metalloprotease; J, neprilysin; K, phenoloxidase; L, phospholipase B; M, serine proteases and serine protease homologues; N, trehalase; O, acid phosphatase; P, hyaluronidase; Q, arginine kinase; R, lysosomal thiol reductase; S, neutral endopeptidase.

parasitic wasp venom components with known actions was provided by Colinet et al. (2013), and that part referring to enzymes in Ichneumonoidea is given in Table 9.1. Given that the venom gland reservoirs are chitin lined, as indeed are the ducts from the glands (see Fig. 4.4), it is surprising that at least the braconid genus *Chelonus* venom contains an active chitinase (*Chelonus* sp. near *curvimaculatus*: Krishnan et al. 1994, *C. inanitus*: B. Vincent et al. 2010); what its function within the largely chitin-free (at least in the embryonic stage) egg might be is completely unknown, as is how it is safely stored within the wasp's own chitin-lined venom reservoir.

NEUROPHYSIOLOGICAL VENOM ACTIONS

Many parasitoids belonging to a wide range of subfamilies induce either temporary or permanent host paralysis through injection of venom. Nearly all work has been concentrated on those that induce permanent paralysis and essentially nothing is known of the chemistry or mode of action involved in temporary paralysis.

Of the permanently paralysing venoms, only that of the braconine braconid *Habrobracon hebetor* (also called *H. juglandis* in this literature) has been particularly studied. There appear to be three active proteins that presynaptically block the excitatory glutamatergic neurotransmission at the insect nerve–muscle junction[2] by blocking exocytosis of presynaptic vesicles (Rathmayer & Walther 1976, Walther & Reinecke 1983, Pennacchio & Strand 2006). The mode of action is not completely clear but presynaptic vesicles, while moving to and sticking to the presynaptic membrane, cannot fuse with it to release their neurotransmitter contents. El-Sawaf and Zohdy (1976) also showed that *H. hebetor* venom caused a reduction in host cholinesterase activity of about 40%, suggesting effects on the CNS where acetylcholine is a neurotransmitter. Additional *Habrobracon* venom components affect the host endocrine system and metabolism (Masler & Kovaleva 1999, Chanda et al. 2002). When *H. hebetor* envenomates a host *Ephestia* (=*Anagasta*) caterpillar that has already been parasitised by the endophagous ichneumonid *V. canescens*, i.e. multiparasitises it, the development of the latter is arrested but the *Venturia* larva can still continue moving so the neuroactive components of the *Habrobracon* venom clearly cannot gain access to the inside of the *Venturia*, at least not in an active form (Petters & Stefanelli 1983).

Many koinobionts have venoms that cause only temporary host paralysis, presumably as an adaptation to allowing the female parasitoid more accurately to deposit her egg(s), although fairly often there is only one ovipositor insertion. M.R. Shaw (1983) described the oviposition behaviour of five species of the tryphonine genus *Netelia* which lay an anchored egg through the cuticle of their caterpillar hosts. Some species oviposited in the host with a single ovipositor insertion and caused no host paralysis, whereas others used an initial ovipositor sting to cause weak temporary host paralysis before using a second ovipositor insertion to lay their eggs. Desneux et al. (2009) investigated the temporary paralysis, which lasts between 4 and 8 min, caused by two species of aphidiine braconid, both in the genus *Binodoxys*, against *Aphis glycine* and other aphids, and concluded that in addition to facilitating oviposition, it might also serve to reduce the probability of superparasitism occurring since paralysed aphids were significantly less likely to be attacked than active ones. However, if selection is indeed acting in that way it seems most likely that the primary role is to facilitate oviposition and that immobility is just a convenient by-product cue enabling other females to discriminate recently already parasitised hosts.

Selection will never be acting in the wild on any single venom component but on the combined effect of them on the host under natural conditions, which means that egg, larval and possibly teratocyte secretions will also be present. Venom from the alysiine *Asobara japonica*, a koinobiont parasitoid of Diptera larvae, has been shown to be lethal if delivered to the host by itself, even at a dosage that would normally occur during envenomation (Mabiala-Moundoungou et al. 2010, Prévost et al. 2012). As this wasp needs the host larva to continue to develop after parasitisation, the deadly effect of the venom is mitigated by an ovarian 'antidote'. This antidote was found to be a secretion originating from the lateral oviducts (Furihata et al. 2013) that enters the host soon after envenomation. The authors also showed that the venom of *A. japonica* is far more toxic than that of the congener *A. rossica*, and suggested that this might because the former has a far broader host range than the latter and will therefore have been selected to produce venom proteins that have sufficient potency against many different systems, which as a consequence could mean that they are more than potent enough by themselves in some hosts.

Several alysiines cause temporary host paralysis and it was thought for a while that the enzyme aspartylglucosaminidase, which is abundant in *A. tabida* venom (Vinchon et al. 2010), might be responsible as its action would increase aspartate concentration in the host and aspartate is an agonist of the excitatory glutamatergic neuromuscular transmission system in insects. However, all other tested species of *Asobara* also cause temporary host paralysis, but lack this venom component (Prévost et al. 2012). Another reason for this being an unlikely candidate for causing host paralysis is that the paralysing effect of the venom is very rapid.

VENOM EFFECTS ON HOST IMMUNE RESPONSE

The venom of *Pimpla rufipes* (as *hypochondriaca*) (an idiobiont endoparasitoid of Lepidoptera pupae) is chemically one of the best studied venoms, probably because the wasp itself is easily cultured on the tomato moth, *Lacanobia oleracea*, and is rather large (up to 15 mm), meaning that a single female can yield a substantial quantity of venom. Co-injection of Sephadex beads and venom into host pupae prevented the former from being encapsulated, thus showing that the effects include a generalised reduction in host immune competence. Similar findings were reported by Osman (1978) for *P. turionellae* venom and pieces of nylon. Although *Pimpla* is a pupal endoparasitoid, several of the studies on *P. rufipes* venom have explored its actions on caterpillar haemocytes, probably because of the ease of obtaining them (Richards & Parkinson 2000). It contains more than 20 components ranging in size from 5 to 100 kDa, many of which inhibit the motility of host haemocytes and haemocyte aggregation (Parkinson et al. 2002a, 2003, Parkinson & Weaver 1999, Richards & Dani 2008), and one of the anti-aggregation proteins has now been cloned (Dani & Richards 2009). Although the precise functions of most of its venom components has yet to be worked out, one 25.3 kDa protein showed some similarity to a serine protease, and others include a phosphatase, six hydrolases (acid phosphatase, β-glucosidase, esterase, β-galactosidase, esterase lipase and lipase; Dani et al. 2005). In the microgastine *Cotesia rubecula*, a serine protease identified by Asgari et al. (2003b) inhibits melanisation of host haemolymph. At sublethal concentrations, *Pimpla* venom causes a variety of changes in haemocyte morphology, including plasma membrane blebbing, degranulation and the formation of cytoplasmic vacuoles. Rivers et al. (2009) subsequently demonstrated that the apoptosis of haemocytes involves Ca^{2+}-dependent cell death pathways and Richards et al. (2013) showed that a recombinant venom protein (rVPR1) binds to the surface of some haemocytes and disrupts their cytoskeleton. This in turn reduces the haemocyte's ability to form pseudopodia, leading to reduced motility, phagocytosis and aggregation. Not surprisingly, this attack on the host's immune system has consequences and increases their susceptibility to the bacterial and fungal pathogens *Bacillus cereus* and *Beauveria bassiana*, respectively (Dani et al. 2004). A similar effect on host haemocytes, i.e. inhibition of spreading and phagocytosis, was found by Furihata et al. (2013) for the action of venom from the alysiine *Asobara japonica*.

A particularly interesting aspect of the study by Parkinson et al. (2002a,b) is that the proteins extracted from the venom reservoir were missing a few terminal amino acids encoded by cDNAs made from mRNA extracted from the venom apparatus, which suggests that some post-transcriptional modification of venom components might be taking place, possibly within the venom reservoir. Similar storage of inactive precursors of venom components has also bee found in the alysiine genus *Asobara* (Moreau et al. 2004, Vinchon et al. 2010). The complex structure of most hymenopteran venom glands with a diversity of secretory units on various parts of them suggests that a whole mixture of different products is being secreted into the reservoir and even into the primary venom duct. Another component group of *Pimpla* venom is phenoloxidases (Parkinson et al. 2001). These enzymes are normally associated with the immune responses of hosts to parasitoids as they are involved in the melanisation of the encapsulation process, catalysing the first stages in the pathway leading to the production of melanin, important for cuticle hardening, injury repair and encapsulation of parasitoids and other foreign bodies (Hartzer et al. 2005). However, the study of phenoloxidases is difficult as they are labile, hence much of the work on them involves the investigation of the more stable zymogen, prophenoloxidase. Parkinson et al. postulated that the *Pimpla* venom phenoloxidase might form products that bind to the host's haemocyte plasma membrane and disrupt their function.

In addition to all the work on *P. rufipes* venom, its congener *P. turionellae* has also been investigated in some detail (Uçkan et al. 2004, 2006, Er et al. 2010). As with *P. rufipes* venom, it is potently cytotoxic and cytolytic in addition to causing paralysis of its experimental host, *Galleria*, and it contains a mixture of biogenic amines and also noradrenalin, phospholipase B, peptides and proteins. The amounts of noradrenalin, serotonin and histamine (a minor component) were highest in the venom of newly emerged wasps and decreased with age. Histamine and noradrenaline, also present in bee venom, might be involved at least in part in protective stinging, although these wasps are not usually able to inflict much of a sting in humans.

Immunity to parasitoidism, at least in *Drosophila*, has been shown to have strong temperature sensitivity (Fellowes et al. 1999), but this is not a general effect. No temperature effect is shown, for example, in the immunity of *Hypera* beetles against the campoplegine *Bathyplectes* (Berberet 1986), whereas Blumberg and Ferkovich (1994) found neutral, positive and negative effects depending on host lepidopteran species against the microgastrine *Microplitis croceipes*. Fellowes et al. (1999) further showed that strains of *Drosophila melanogaster* selected for increased resistance against either the alysiine *Asobara tabida* or the eucoilid *Leptopilina boulardi* showed various, asymmetric, increases in resistance to the other parasitoid, although the opposite effect is also possible, depending on the details of what are being selected.

In addition to the general immunosuppressant effect of *Pimpla* venom discussed above, similar effects have been found for venom from the ichneumonine *Diadromus* (W.D. Li et al. 2006b) and the alysiine braconid *Asobara japonica* (Prévost et al. 2012). Such effects can be due to disruption of haemocyte spreading ability, direct induction of host haemocyte death through induced apoptosis or destruction of the tissues that produce haemocytes. Suzuki and Tanaka (2006) showed that haemocyte apoptosis caused by *Meteorus* venom was actually due to virus-like particles in it. Where host immunity has been generally suppressed, it would seem likely that they would be rendered more accessible to multiparasites. In the case of *Asobara*, it probably explains the cross-resistance to *Leptopilina* (Eucoilidae) found by Fellowes et al. (1999). Conversely, if a host has evolved a highly active immune system as part of its defence against one parasitoid, it may also be protected against other species of parasitoid whose numbers are not normally sufficient to create considerable selection pressure. Such cross-resistance has been demonstrated in lines of *Drosophila* selected for increased resistance to parasitism in the *Asobara* (Braconidae)/*Leptopilina* (Eucoilidae) system in artificially selected lines (Fellowes et al. 1999).

Surprisingly, venom from the idiobiont pupal endoparasitoid *P. rufipes* (Ichneumonidae: Pimplinae) causes lethal paralysis of not only its host's pupa, but also host larvae and adults (Parkinson & Weaver 1999), the factor involved apparently being a protein with approximate size 27 kDa. Whether the paralysing action of its venom is a hang-over from ectoparasitic ancestors with biologies similar to those of their sister group, the Ephialtini, is unknown and it is possible that the causative component also has some other function since host paralysis is unlikely to be directly important to this wasp.

Rather less is known about the individual components of venoms of wasps that have polydnaviruses, although there have been some investigations of whether they are necessary and have additive or synergistic effects. Asgari et al. (2003a) characterised a small venom protein (Vn4.6) from the microgastrine *Cotesia rubecula*, which interferes with the activation of the host hemolymph prophenoloxidase and therefore is likely to help prevent a full encapsulation response, although eggs of this species are passively protected in any case.

POLYDNAVIRUSES

A good short summary of the discovery of protective egg coatings and of polydnaviruses (pronounced puh-LID-nuh-VI-rus) was provided by Vinson (2012) and a more detailed exposition of the various milestones in polydnavirus research by Stoltz and Krell (2012). What is perhaps a coincidence is that the two taxonomic groups being studied intensively at the time, campoplegines such as *Venturia* and *Campoletis* and the cardiochiline *Toxoneuron nigriceps*, referred to as *Cardiochiles nigriceps* in all the early work, were taxa in which particulate products from the calyx gland played a major role in avoiding and overcoming host immune defence. Salt (1965) demonstrated that passage of the egg of the campoplegine *Venturia canescens* through the calyx region of the ovary conferred resistance to the host immune system. The importance of the calyx gland product from *T. nigriceps* in regulating host development and the successful parasitoid development was demonstrated by Vinson (1972c) and, through experimentally inactivating gland extracts using heat, ethanol or the addition of protease, the critical substance seemed likely to be a protein, but at that time the viral nature of the product was not recognised. Rotheram (1973a,b) showed that in *Venturia canescens* it was particles from the calyx adhering to the egg surface that were responsible for giving them protection from the host immune system. However, in this species, these particles are not viruses as they do not contain DNA and it appears to have been pure chance that this atypical system was the first to have been worked out. Working with Vinson, Norton's group carried out an electron microscope study of the production of what are true polydnaviruses in the calyx gland of the campoplegine *Campoletis sonorensis* (Norton et al. 1974), and the mode of their production, being seen budding through the nuclear membrane, gave the first clue that they might contain DNA, which was soon followed up by work on the calyx particles of the cardiochiline *Toxoneuron nigriceps*. These particles were shown to contain DNA by DNAase treatment which inactivated them and thus their 'viral' nature was confirmed (Vinson & Scott 1975). Soon afterwards, the virologist Stoltz became involved and started working on the calyx viruses from the gypsy moth parasitoid *Cotesia melanoscelus*, which at that time was classified in the same subfamily as *Toxoneuron*, still working on the assumption that the viruses were some form of *Baculovirus* (Stoltz & Krell 2012). Thus, the discovery that the particles contained multiple separate circular DNA molecules of various sizes came as a surprise, as that was not what baculoviruses have (Krell & Stoltz 1979). Further work demonstrated that these so-called polydisperse DNA molecules were characteristic of all these calyx gland viruses, although at that time numerous workers were sceptical and proof only really came with the use of electron microscopy, which actually visualised the individual circular DNA molecules of different sizes. At about that time, the term polydnavirus started to be used and it was discovered that the polydnavirus DNA was actually expressed within the host caterpillar (Fleming et al. 1983). It took a few years before it was finally confirmed that the DNA of the particles was actually integrated into the wasp's genome (Fleming & Summers 1986, 1991). Because of these multiple DNA copies, these 'viruses' are classified in the Polydnaviridae but despite this, the commonly used pronounciation defies the orginal etymology. In all cases, these 'viruses' contain multiple circular DNA molecules that are collectively encoded by a range of genes and each virus particle contains a mixture of these circular genes.

Polydnaviruses have now been demonstrated in numerous taxa (Table 9.2). What is striking about the taxa in Table 9.2, although not really surprising, is the highly clustered host list with a large percentage of studied cases being associated with a relatively few pest genera and species, notably *Choristoneura* species, the gypsy moths, *Lymantria* spp. (Lymantriidae), and armyworms, *Spodoptera* (Noctuidae). This ought to afford an excellent opportunity for examining whether horizontal gene transfer might occur, but perhaps means that rather less can be determined about the effect of systematic differences in host immune defence systems. It is unfortunate that whereas campoplegines attack Lepidoptera, sawflies and beetles, the microgastroids are effectively entirely restricted to Lepidoptera as hosts. Do the switches to other host orders by campoplegines indicate some additional feature of their virulence? Given that probably every microgastrine species has an associated polydnavirus 'species' and there may be a staggeringly large number of microgastrine species (16,000–40,000; Rodriguez et al. 2013), this would make the Polydnaviridae the most speciose virus family known.

Three groups of ichneumonoids, and potentially four, representing three (or four) independent evolutionary events, have evolved these extremely similar mechanisms involving completely unrelated polydnaviruses, to deal with host immunity, and in each case their genes are expressed only in the host. In all three cases, these virus particles contain multiple circular DNA molecules which are copied from proviral versions encoded in the wasp's genome. Within the Braconidae there appears to have been a single acquisition of these polydnaviruses in the 'microgastroid' lineage[3] and they are known in all members of the Microgastrinae, Cardiochilinae, Miracinae and Cheloninae that have been investigated (Wharton & Sittertz-Bhatkar 2002). These braconid polydnaviruses are termed bracoviruses and are classified as the virus genus *Bracovirus*, and those of ichneumonids are currently all placed in the genus *Ichnovirus*. However, unlike in the braconids, there are almost certainly two independent origins of polydnaviruses in the Ichneumonidae, one in the Banchinae (in the lower ophioniformes) (Fig. 9.2) and one in the Campopleginae (in the higher ophioniformes) (see Fig. 9.4). The name 'Banchovirus' has been suggested for the latter, and although I do

Table 9.2 Known polydnaviruses.

Parasitoid	Main host
Braconidae	
Cardiochilinae	
Toxoneuron nigriceps	*Heliothis virescens*
Cheloninae	
Ascogaster argentifrons	*Choristoneura fumiferanae*
Ascogaster quadridentata	*Cydia pomonella*
Chelonus altitudinis	*Homoeosoma electellum*
Chelonus blackburni	*Pectinophora gossypiella*
Chelonus nr. *curvimaculatus*	*Trichoplusia ni*
Chelonus inanitus	*Spodoptera littoralis*
Chelonus insularis	*Spodoptera frugiperda*
Phanerotoma leucobasis (as *flavitestacea*)	*Ephestia kuehniella*, *Ectomyelois ceratoniae*
Microgastrinae	
Apanteles crassicornis	Unknown
Apanteles fumiferanae	*Choristoneura* spp.
Cotesia congregata	*Manduca sexta*
Cotesia flavipes	*Diatraea saccharalis*
Cotesia kariyai	*Pseudaletia separata*
Cotesia marginiventris	*Manduca sexta*
Cotesia melanoscela	*Lymantria dispar*
Cotesia sesamiae	*Busseola fusca*
Cotesia rubecula	*Pieris rapae*
Cotesia ruficrus	*Pseudaletia separata*
Cotesia schaeferi	*Lymantria dispar*
Cotesia vestalis	*Plutella xylostella*
Diolcogaster facetosa	*Plathypena scabra*
Glyptapanteles flavicoxis	*Lymantria obfuscata*
Glyptapanteles indiensis	*Lymantria dispar*
Glyptapanteles liparidis	*Lymantria dispar*
Hypomicrogaster ecdytolophae	*Ecdytolopha insiticiana*
Microgaster canadensis	*Archips argyrospira*
Microplitis bicoloratus	*Spodoptera litura*
Microplitis croceipes	*Heliothis virescens*
Microplitis demolitor	*Chrysodeixis* (= *Pseudoplusia*) *includens*
Microplitis rufiventris	*Spodoptera littofolis*
Pholetesor ornigis	*Phyllonorycter* spp, *Camararia* spp., *Tischeria* spp., *Leucanthiza* spp.
Protapanteles palaecritae	*Paleacrita*
Miracinae	
Mirax sp.	Various leaf-miners
Ichneumonidae	
Banchinae	
Apophua simplicipes	*Choristoneura rosaceana*
Glypta fumiferanae	*Choristoneura fumiferanae*
Glypta sp.	Unknown
Lissonota sp.	Unknown

(continued overleaf)

Table 9.2 *(continued)*

Parasitoid	Main host
Campopleginae	
Bathyplectes anurus	*Hypera postica*
Campoletis aprilis	Noctuidae, especially Hadeninae
Campoletis chlorideae	*Helicoverpa armigera*
Campoletis flavicincta	*Spodoptera frugiperda*
Campoletis sonorensis	*Heliothis virescens*
Campoletis sp.	Unknown
Casinaria sp.	Unknown
Diadegma acronycta	Various Tortricidae
Diadegma terebrans	*Sesamia nonagrioides*
Dusona sp.	Unknown
Eriborus terebrans	*Ostrinia nubilalis*
Hyposoter annulipes	*Catocala* and *Pyrrhia* spp.
Hyposoter didymator	*Spodoptera frugiperda*
Hyposoter exiguae	*Trichoplusia ni*, several Geometridae
Hyposoter fugitivus	Various Saturniidae, Lasiocampidae
Hyposoter lymantrae	*Lymantria dispar*, *L. concolor*
Hyposoter rivalis	Various Noctuidae
Olesicampe benefactor	*Pristiphora erichsonii*
Olesicampe geniculatae	*Pristiphora geniculata*
Tranosema tenuifemur	*Choristoneura fumiferana*
Tranosema rostrale	*Choristoneura fumiferana*
Tranosema sp.	Unknown
Tranosemella interrupta	Various microlepidoptera, especially Pterophoridae

Source: modified after Stoltz & Whitfield 1992 and F. Huang et al. 2008.

not think it has been formally described, its usage is growing – it is unfortunate that one cannot change the name of the others to 'Campovirus'.

In all three wasp clades, these viruses are produced in epithelial cells of enlarged calyx glands at the distal part of the lateral oviduct (Stoltz 1990) and these swollen calyx glands are apparent from the diagrams of ichneumonid female reproductive apparatus dissections made by Iwata (1960, 1961). Interestingly, Cummins et al. (2011) noted that the ctenopelmatine genus *Barytarbes* also has enlarged calyx glands and suggested that these might also harbour polydnaviruses, but to date no-one seems to have checked, although Kroemer & Webb (2004) indicated that they also found polydnaviruses in the Ctenopelmatinae. If they were found to occur in ctenopelmatines in addition to banchines and campoplegines, it would almost certainly represent a third origin within the Ichneumonidae since all three are distantly related (see phylogenetic tree, Fig. 13.1). It has been proposed that one can tell whether a wasp has polydnavirus-producing calyx glands by dissecting adult females under saline: those with polydnaviruses have the lumens of the calyx gland appearing almost iridescent pale blue (technically Tyndall blue[4]) due to the optical effect of the vast number of virions. An alternative and easy approach to detecting polydnaviruses was presented by Ketseoglou and Bouwer (2008) and simply involves using Feulgen DNA staining, in which case the calyx lumen stains intensely because of all the viral DNA. It is hoped that with these techniques it will soon be possible to have the distribution of polydnaviruses, particularly among the Ichneumonidae, more thoroughly surveyed.

Not only are the bracoviruses and ichnoviruses (at least those found in Campopleginae) similar in having multiple circular DNA molecules (i.e. they have segmented genomes), sequence analyses have also shown that other genetic organisations have evolved in parallel (Fig. 9.3) (Kroemer & Webb 2004, Dupuy et al. 2006). Both campoplegine and microgastrine polydnavirus genomes lack replication genes, are segmented and have numerous families of virulence genes which

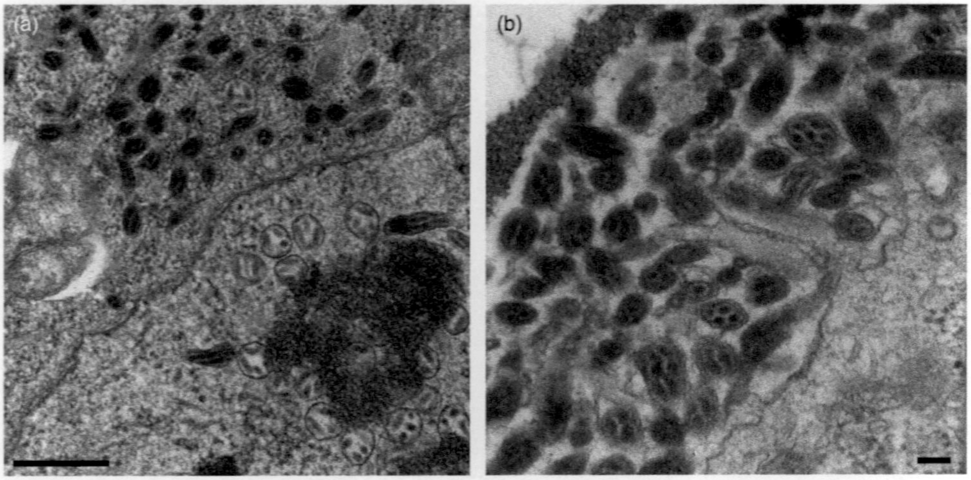

Fig. 9.2 TEM images of *Apophua simplicipes* (Banchinae) polydnavirus production (scale bar: a = 500 nm, b = 100 nm). (Source: reproduced by permission Michel Cusson.)

have replicated and diversified. However, they share few virulence genes and therefore Webb et al. (2006) concluded that they have evolved different strategies for making hosts suitable for, in their words, causing *disease* in the caterpillars.

Bracovirus particles have a cylindrical nucleocapsid or multiple nucleocapsids surrounded by a single membrane (Stoltz & Krell 2012) and are released by cell lysis, although the size of the virion varies considerably both within and between species (Albrecht et al. 1994). *Ichnovirus* particles from campoplegines form in the nucleus and at that stage they develop a single external membrane (Fig. 9.4). A second, outer membrane is acquired as they bud out through the calyx cell's plasma membrane so the nucleocapsids are surrounded by two membranes when injected into hosts (Fig. 9.4c). *Bracovirus* ontogeny has been described in detail for *Cotesia congregata* (de Buron & Beckage 1992) and *Chelonus inanitus* (Wyler & Lanzrein 2003, Marti et al. 2003) and *Ichnovirus* development has been described in detail by Norton et al. (1974) for *Campoletis sonorensis* and F. Huang et al. (2008) for *Diadegma semiclausum*.

Some molecular data are now available for the viruses of approximately 10 species and complete genomes are currently available for eight ichneumonoids, those of the braconids *Cotesia congregata, Glyptapanteles indiensis, G. flavicoxis* and *M. demolitor* and the ichneumonids *Campoletis sonorensis, Hyposoter fugitivus* and *Glypta fumiferanae*, which collectively represent three different subfamilies (Bézier et al. 2008, 2009), and the number will no doubt increase rapidly. Both *Bracovirus* and *Ichnovirus* genomes show little resemblance to those of viruses and mostly represent wasp virulence genes that have been shifted into the replicated viral gene region. Common gene families in bracoviruses include cystatins, which include cysteine protease inhibitors and protein kinases, while K. Tanaka et al. (2007) found that the ichnovirus genomes of the campoplegine ichneumonids *Campoletis sonorensis, Hyposoter fugitivus* and *Tranosema rostrale* all contained six conserved gene families: repeat element, cysteine motif, viral innexin, viral ankyrin, N-family and a newly defined putative family, the polar-residue-rich proteins. Provost et al. (2004) found that the same gene family, encoding for protein tyrosine phosphatases, was present in the bracoviruses in both cardiochilines and microgastrines. Thus, despite enormous differences in the gene complements of different bracoviruses, this suggests that these genes represent an ancient component of the bracovirus genome and that they probably play a central role in the mode of polydnavirus action. In vertebrates, protein tyrosine phosphatases are important in regulating signal transduction pathways and it is likely that they do the same when expressed in host tissues. In some microgastrine polydnavirus genomes, the protein tyrosine phosphatases have undergone an amazing amount of gene duplication and radiation such that

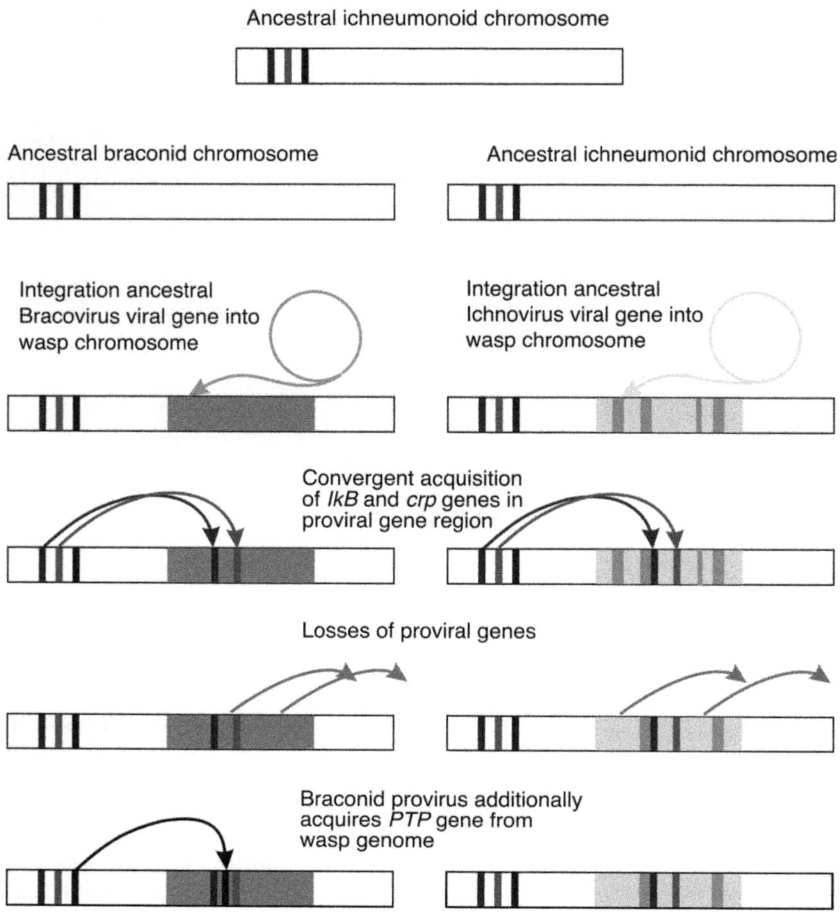

Fig. 9.3 Diagram of early evolutionary events (from top to bottom) leading to bracoviruses (left) and ichnoviruses (right) showing probably convergent evolutionary events in the two separate lineages prior to their subsequent radiation into the forms we know today. (Source: redrawn after part of fig. 3 in Dupuy et al. 2006.) For a colour version, see Plate 14.

there are as many as 42 different copies in some taxa. Serbielle et al. (2012) explored the mechanism by which these gene duplications had occurred and found that both tandem duplications (as commonly occur in genomes) and dispersed duplications were involved and proposed a novel mechanism by which this could occur. The genes have been generally conserved, but positive selection was evident from the sequences of recently produced copies that appeared to involve sites concerned in substrate recognition. Serbielle et al. (2008) had previously shown that the cystatin gene family of *Cotesia congregata* had also been under strong positive selection.

The presence of polydnaviruses in at least some Banchinae was only relatively recently confirmed, although examination of the drawings of ichneumonid female internal reproductive systems by Iwata (1960) clearly shows the calyx part of the lateral ovipositor ducts to be swollen in these almost as much as in the Campopleginae and Microgastrinae. The first record of a banchine polydnavirus was from a *Lissonota* sp. (Stoltz & Whitfield 1992), but the first studied in any detail

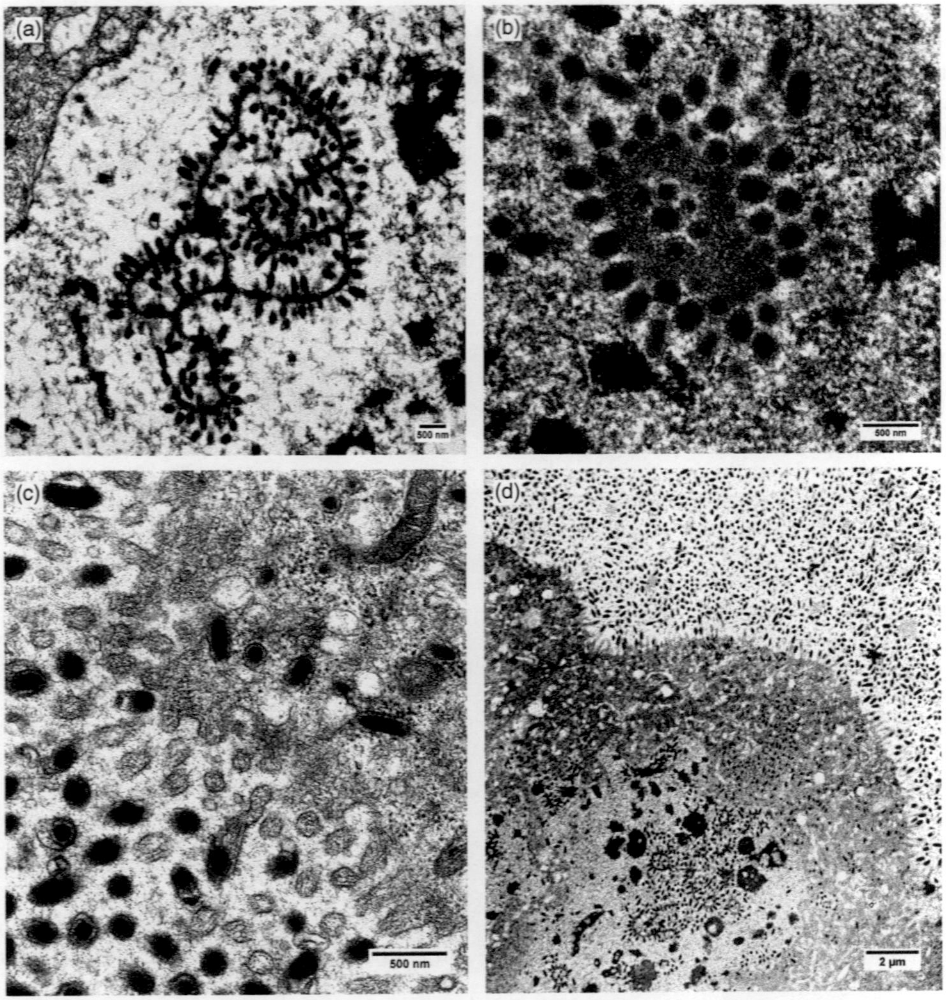

Fig. 9.4 TEM images of *Hyposoter didymator* (Campopleginae) polydnavirus production and calyx gland. (a, b) Virogenic stroma in nucleus of calyx gland cell; (c) budding of virions into the oviduct lumen; (d) view of the calyx with Ivs in nucleus, cytoplasm and oviduct. (Source: micrographs by Marc Ravallec; reproduced by permission Nathalie Volkhoff.)

was from *Glypta fumiferanae* (Lapointe et al. 2007) and a second instance has recently been documented for another glyptine, *Apophua simplicipes* (Djoumad et al. 2013). These viruses are currently all classified within the viral family Polydnaviridae, although this is certainly polyphyletic, and following general viral systematics it would seem sensible to treat each independent origin as representing a separate family or, if the precursor normal virus is clearly identifiable, then logically they should be considered as specialised genera within each of their respective viral progenitor families, thus creating monophyletic viral taxa.

Identifying the progenitors of these elements is, however, far from simple as not only have they lost a great deal of their original viral gene complement, but also they have occasionally picked up transposable elements and genes from other viruses that probably at some point infected parasitised hosts (Gundersen-Rindal et al. 2013). It seems highly likely now that bracoviruses originated from nudiviruses and not from baculoviruses, as had been thought earlier (Drezen et al. 2006). Bracoviruses encode 10 genes that are only known to occur in Nudiviridae (Bézier et al. 2009), and Herniou et al. (2013) provided a very clear summary of what is known about genes shared between nudiviruses and bracoviruses. Thézé et al. (2011) went on to use the dates of actual braconid fossils to calibrate a molecular tree of bracoviruses, nudiviruses and baculoviruses and concluded that the nudivirus clade originated approximately 310 Mya in the Carboniferous period. Whitfield and Asgari (2003) proposed a scenario in which polydnaviruses might have arisen from ascoviruses, based partly on the fact that the ichneumonine *Diadromus pulchellus* has an obligate mutualism with an ascovirus which helps suppress host immunity (Bigot et al. 1997). Subsequently, Bigot et al. (2008) presented evidence that ichnoviruses originated from ascoviruses; however, that hypothesis does not seem to be particularly supported (Volkoff et al. 2010, Cusson et al. 2012) and indeed it is not possible to say what its progenitor virus was – perhaps it has yet to be discovered (Stoltz & Krell 2012). Volkoff et al. (2010) identified genes of possibly viral origin and which are involved in *Ichnovirus* production that are clustered in the wasp's genome along with polydnavirus genes, but which are not packaged into the virus particle virions.

Despite the almost certain original incorporation of a true viral genome being responsible for all ichneumonoid polydnaviruses, they no longer behave as a virus and are more like a specialized nuclear secretory organelle of the wasp (Turnbull & Webb 2002, Federici & Bigot 2003, 2009, Whitfield & Asgari 2003). Indeed, whether polydnaviruses should be regarded as viruses proper is somewhat of a moot point because although they encode virulence genes they lack the core genes typically involved in particle production, i.e. they do not replicate themselves, and indeed, none encode 'pure' viral genes. Although the 'virus' particles are passed to the host during oviposition, enter various host cells and express their DNA within them, which results either in their death or a modification of its physiology, their lack of target-cell dependent replication renders them different from all other standard transmissible viruses. Selection is therefore only acting upon them via the survival of their 'host' wasps' offspring. Despite almost certainly three, independent origins, and some marked morphological differences, such as the number of surrounding membranes, their molecular convergence is extraordinary.

EFFECTS OF POLYDNAVIRUSES ON HOSTS

Two main types of polydnavirus effect have been found. A general suppression of the host immune response appears to be one of the primary roles of polydnaviruses in most cases and many also have profound effects on host development. In addition, some play a passive role in protecting parasitoid eggs from encapsulation (see the section *Passive evasion of encapsulation by parasitoid eggs*, above). They vary in the cells they infect and also in the timing of the effects that they induce. The dosages of polydnaviruses delivered to a host vary considerably and in some cases are rather constrained by host size. Egg-larval parasitoids such as chelonines deliver a total viral DNA load of about 0.1 ng per host egg, compared with 2.4 and 6.5 ng per host caterpillar for the solitary microgastrine *Microplitis demolitor* and the campoplegine *Campoletis sonorensis* (Lanzrein et al. 2001), and in the latter type of association, considerably more may be injected in cases of superparasitism.

A great deal of what we know about polydnavirus function comes from experiments in which either calyx extracts, eggs or venom are injected into hosts in various combinations, or, through use of X-ray or γ-ray sterilised wasps to pseudoparasitise hosts, such that effects of secretions from eggs or larvae can be excluded (e.g. D. Jones et al. 1986, Lanzrein et al. 2001, Tillinger et al. 2004). In the *Chelonus* sp./*Trichoplusia ni* system, approximately 10% of hosts that have been attacked by the parasitoid have no parasitoid eggs or larvae in them (D. Jones et al. 1986), so presumably the wasp failed to deliver an egg or the egg died soon afterwards leaving no trace. The hosts nevertheless continue development for a while, then prematurely spin a cocoon and enter a metamorphosis to a prepupal stage, making them a good subject for dissecting out the possible additional effects caused by secretions or activities of the parasitoid larva. However, these pseudoparasitised hosts will contain a presumably normal balance of venom, calyx and possibly other parasitoid products and working out the certainly different effects of venom and polydnaviruses requires finer analyses.

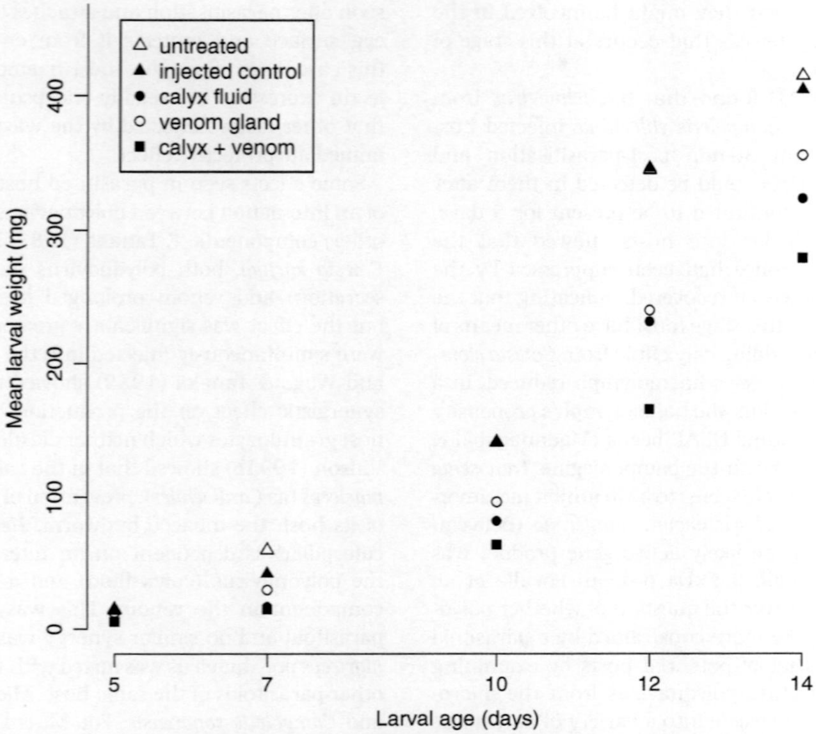

Fig. 9.5 Effect of injection of eggs of *Heliothis virescens* (Noctuidae) with extracts of the calyx gland, venom gland or the two combined from *Chelonus insularis* (Cheloninae) on subsequent host growth (*Spodoptera littoralis*; Noctuidae), showing that each separately retards host development. (Source: data from Ables & Vinson 1981.)

Injection experiments have shown that venom and polydnaviruses have an approximately additive effect on the retardation of host larval growth in the case of *S. ornithogalli* and *Heliothis virescens* parasitised by *Chelonus insularis* when injected into host eggs (Fig. 9.5) (Ables & Vinson 1981). When 5-day-old larvae were injected, the effects were similar, although the effect of combined calyx and venom gland extract was lost shortly before host pupation. However, although I believe the results, it is unfortunate that an extra control was not done, with an extract of some other wasp tissue, in case some of the effect was due to some sort of non-specific action.

Although all the bracoviruses seem to have a single common origin, 100 Mya or more (Murphy et al. 2008), i.e. pushed back from the 74 Mya estimated by Whitfield (2002), their effects on host development differ markedly between the egg-larval parasitoid chelonines and the larval parasitoid microgastrines[5]. Both have effects on the production of host juvenile hormone and juvenile hormone esterase production, but in opposite directions (Lanzrein et al. 2001). In the microgastrines that attack late host instars, they seem to be involved in up-regulating juvenile hormone and down-regulating the esterase, with the effect that the last larval instar is prolonged, preventing pupation, presumably giving the parasitoid larvae sufficient time to develop fully. In contrast, the opposite happens with *Chelonus*, resulting in premature metamorphosis to a final instar and the initiation of retreat construction, the chelonine finally completing development once the host has made its cocoon. Johner and Lanzrein (2002) cloned two polydnavirus genes from the chelonine *Chelonus inanitus* and showed that they coded for one moderate-sized protein and peptide. Interestingly, their expression in the host remained low until the host started to construct its pupation cell when it increased five- and 1000-fold, respectively, and it was postulated that they might be involved in the reduction of ecdysteroids that occurs at this stage of host development.

Yin et al. (2003) found that the *Ichnovirus* from the campoplegine *Campoletis chlorideae* infected host haemocytes within 30 min post-parasitisation and that viral transcripts could be detected in them after about 1 day and continued to be present for 5 days. Injection of Sephadex into hosts showed that the encapsulation response had been suppressed by the virus, but after 4 days it recovered, indicating that the parasitoid larva by this stage must have other means of defence. Similarly, adding calyx fluid from *Cotesia glomerata* to host, *Pieris rapae*, haemolymph reduced, in a dose-dependent fashion, the haemolymph's propensity to encapsulate cationic DEAE beads (Madanagopal & Kim 2006). However, in the campoplegine *Tranosema rostrale*, their main role seems to be to inhibit metamorphosis of its host, *Choristoneura fumiferana* (Béliveau et al. 2000), and the likely active gene product was identified as a small, 9.3 kDa protein. Lovallo et al. (2002) tried to answer the question of whether polydnavirus effects were more constrained by a parasitoid relationship or that of potential hosts by examining the effect of injecting polydnavirus from the microgastrine *Cotesia congregata* into a variety of non-hosts. Some non-hosts responded to the virus with effects that occurred in the natural hosts, although other hosts did not, and there was no clear phylogenetic pattern. I believe that this is a very promising avenue of research and that such studies in the future should make an effort to investigate a phylogenetically better sampled array of non-hosts and also compare between species of microgastrine with different degrees of relatedness rather than just investigating a few hosts of biological control interest that happened to be readily available.

A viral lectin is encoded in the bracovirus from *Cotesia plutellae* and can be found expressed in the haemlymph of parasitised hosts (S. Lee et al. 2008). When host haemolymph was extracted from parasitised and unparasitised hosts and fractionated into cellular and lymphal components, the addition of the lymph from parasitised hosts to haemocytes from unparasitised hosts caused a significant reduction in their ability to bind bacterial targets, indicating a general immunosuppressive effect, and this could be replicated by expressing the cloned gene region in a bacterial system and applying that to haemocytes in a similar fashion. The viral lectin gene is expressed in the host soon after parasitisation and attaches to the parasitoid egg surface and protects it from encapsulation. In this case, eggs exposed to host haemolymph prior to lectin expression are readily encapsulated, indicating that other factors injected by the wasp must have an immediate protective effect.

Some effects seen in parasitised hosts are the result of an interaction between polydnavirus and venom (or other) components. T. Tanaka (1987) found that with *Cotesia kariyai*, both polydnavirus (technically calyx secretion) and venom prolonged host development, but the effect was significantly greater when the two were simultaneously injected into the host caterpillar and Wago & Tanaka (1989) showed that they had a synergistic effect on the production of 'filopodia' by host granulocytes which neither did alone. T. Tanaka & Vinson (1991b) showed that in the case of *Toxoneuron nigriceps* (as *Cardiochiles*), prevention of metamorphosis of its host, the tobacco budworm, *Heliothis virescens*, caterpillars is dependent on an interaction between the polydnavirus (calyx fluid) and a 66 kDa protein component in the venom. This was specific to this parasitoid and no similar synergy was found when *T. nigriceps* polydnavirus was mixed with venom from two other parasitoids of the same host, *Microplitis croceipes* and *Campoletis sonorensis*. For *M. croceipes*, the combination of polydnavirus and venom led to only a few days' prolongation of the host larval stage rather than blocking pupation altogether. With *C. sonorensis*, parasitisation can be successful even if the venom glands have been removed (Webb & Luckhart 1994), showing that there is no obligate interaction of venom and polydnavirus in this system, although the venom does play some role in the suppression of the host immune response in the early stages of parasitisation (Asgari & Rivers 2011). Similarly, successful parasitisation of its noctuid host (*Spodoptera frugiperda*) by the campoplegine *Hyposoter didymator* is not reliant on venom despite the wasp having a well-developed venom gland and reservoir (as in all Campopleginae), polydnavirus alone being sufficient to protect it from encapsulation (Dorémus et al. 2013b). The latter authors suggested that the venom may have become redundant; however, this seems extremely unlikely to be the whole story, as when something as costly as a venom apparatus and venom is no longer important for an animals fitness, evolution rapidly leads to its degeneration and loss. More likely, the venom has become largely or entirely redundant in this particular host association but may still be important in other natural hosts and evolutionary time has not been sufficient for its loss in this particular case.

Although most of the published work on polydnaviruses is essentially pure biological research, there can be no doubt that many grant applications and grant-giving agency decisions have been based at least in part on their potential applied roles. One of the few papers to address this is by Rodríguez-Pérez et al. (2011), who described how tobacco hornworm caterpillars, *Manduca sexta* (Sphingidae), are normally refractory to *Autographa californica* multicapsid nuclear polyhedrovirus (AcMNPV), but when injected with their parasitoid, *Cotesia congregata*'s polydnavirus, they become far more susceptible (Washburn et al. 2000), the implication being that engineering the baculovirus to express the relevant polydnavirus genes could greatly improve their effectiveness as control agents.

OTHER REPRODUCTIVE VIRUSES

Although the polydnaviruses are the best known and most intensively studied, a number of other viruses have been found in association with ichneumonoid female reproductive systems, and although some appear to be just commensals, perhaps using the wasp as a vector, others have been shown to be essential for overcoming the host's immune defence against the wasp egg/larva (Krell & Stoltz 1979).

Particularly well studied of the commensal non-polydnaviruses are those of the ichneumonine *Diadromus pulchellus*, a pupal endoparasitoid of leek moth. Rabouille et al. (1994) described the presence of a member of the Reoviridae (DpRV) from cells, predominantly of the gut but also the venom gland, in both wild and laboratory populations. Differences in the number of genome segments were found between viruses of male and female wasps, with males having 10 and females an extra (supernumerary) one. Of the several known symbiotic viruses in this wasp, Renault et al. (2004) stated that 'The fourth, DpRV-1, is a commensal reovirus detected in most field populations of the wasp, *Diadromus pulchellus*. This reovirus is always found associated with an ascovirus, DpAV-4a, which is indispensable for host immune suppression'. Hence it seems that DpAV-4a represents a stage of virus–parasitoid association which is rather between that of a pathogen and the 'symbiotic' polydnaviruses (Stasiak et al. 2005). Some individuals of *D. pulchellus* from France have been found to lack DpRV-1 and DpAV-4a and instead harbour a *Cypovirus* (DpRV-2), which is transmitted to the host pupae during oviposition and reproduces in the pupal gut cells and also inhibits the host pupa's immune defence mechanisms. Rabouille et al. (1994) showed that *D. pulchellus* has a part of its associated Reovirus's genome integrated into the wasp DNA, which suggests that fully integrated polydnaviruses can arise easily.

Virus-like particles appear widespread in the venom gland filament cells of the Opiinae (Jacas et al. 1997, Wan et al. 2006). The venom gland filaments of the opiine braconid *Diachasmimorpha longicaudatus* were shown to contain virus-like particles by Lawrence and Akin (1990) and subsequently these have been identified as an *Entomopoxvirus* (Lawrence 2002), and they may have a mutualistic role with the parasitoid as they appear to lower the host's immunological defence.

Several taxa that have polydnaviruses have also been shown to carry true viruses associated with their reproductive systems. Stoltz et al. (1988) found that calyx cells of *Cotesia melanoscela* produce a virus called CmV2, which is transmitted matrilinearly and can replicate in host caterpillars, although it is non-pathogenic to them. Another microgastrine, *C. congregata*, in addition to its polydnavirus, has a filamentous virus-like particle associated with some calyx cells that do not produce polydnavirus (de Buron & Beckage 1992). These particles have a single nucleocapsid per envelope and are released into the oviduct in small numbers. Their role, if any, is not known. Among the polydnavirus-carrying ichneumonids, most individuals in one established colony of the campoplegine *Hyposoter exiguae* also has an associated Reovirus (HeRV) which was formed in a variety of tissues including the calyx gland and ovarioles, oviducts, midgut cells, Malpighian tubules and accessory glands associated with the male reproductive system, although it did not appear to have any pathogenic effect on the wasp and appeared to be commensal (Stoltz & Makkay 2000).

Several virus-like particles have been found in the reproductive and venom gland system of the euphorine braconid *Meteorus pulchripes*. One, termed MpVLP, which is found in the wasp venom, when injected into caterpillars of the noctuid host, *Pseualetia separata*, caused a reduction in the number of host haemocytes, although this was only a temporary effect compared with the persistent effect observed in parasitised hosts, indicating that other factors are also involved in host immune suppression (Suzuki & Tanaka 2007). Suzuki et al. (2008) narrowed the effect of the MpVLPs down to inhibition of the protrusion and spreading of filopodia and lamellopodia by host granulocytes and plasmatocytes and the granulocytes were prevented from forming focal complexes and adhesions, which are a critical first step in the recognition and encapsulation of foreign bodies. The exact nature of this virus does not seem certain at present and F. Huang et al. (2008) included it in their list of polydnaviruses, although I have seen no evidence that either its DNA is integrated into the host wasp's genome or that its genome comprises multiple circular DNA molecules.

The condition found in all females of the ctenopelmatine *Mesoleius tenthredinidis* by Stolz (1981) seems to be informative about how wasp–virus associations and partnerships evolved. Cells of the calyx region of this wasp produce what appears to be a typical *Baculovirus* and the nucleocapsids of these were observed to be uncoated as they exited the nuclear envelope, suggesting that the viruses might therefore be able to re-infect the same host individual. That all females examined had these viruses suggests that they must be transmitted vertically, either with high fidelity through co-injection into hosts along with the wasp's eggs or perhaps via some intracellular mechanism.

IMPROVING HOST QUALITY

Parasitism causes many changes in host physiology. In addition to the obvious necessities of avoiding encapsulation or other host defences, many species induce changes in host physiology that are beneficial to parasitoid development. The best studied are changes in the host's metabolism so that it may make more of compounds that are of greater benefit to the parasitoid.

Changes in the protein composition and other biochemical features of host haemolymph as a result of parasitism by ichneumonoids have been widely documented for both ecto- and endoparasitoids (Guryanova et al. 1977, Guerra et al. 1993a, Shelby & Webb 1997, Rahbé et al., 2002, Cônsoli et al. 2005, Salvador & Cônsoli 2008), although it is not always apparent and J.E. Baker & Fabrick (2000) were unable to detect any bulk changes due to parasitisation by the ectoparasitic braconine *Habrobracon hebetor* on host Indian meal moth (*Plodia interpunctella*: Pyralidae) haemolymph proteins between unparasitised and 72 h post-parasitisation caterpillars. Envenomation alone of the related *Bracon mellitor* caused two peaks in host whole-body extract free amino acid levels, each corresponding to a decrease in total soluble protein, suggesting increased protease activity. One peak occurs very rapidly and is over after 2 h and the second starts after approximately 4 days and would probably correlate with when the parasitoid larvae are starting to develop.

In the case of endoparasitoids with alecithal, hydropic eggs, there is some evidence that parasitoid-induced changes in host haemolymph composition can occur very soon after parasitisation in a way that facilitates the uptake of nutrients by the egg, thus permitting development of the embryo. For example, host haemolymph concentrations of particular free amino acids in *Heliothis virescens* caterpillars parasitised by *Toxoneuron nigriceps*, after a brief initial drop, increased significantly between 2 and 10 h post-parasitisation (Cônsoli & Vinson 2004). Some of the changes caused by polydnaviruses can be highly specific. Shelby & Webb (1997) showed that *Campoletis sonorensis* polydnavirus specifically blocks the synthesis of specific host, *H. virescens*, haemolymph proteins and further, since the numbers of mRNA transcript levels of these were unaffected, the block must be in translation of the mRNA to protein. It seems likely that up- or down-regulation of specific host protein production is something that polydnaviruses can achieve more readily than perhaps venom proteins because they can target not only specific host cell types but also a wide range specific biochemical pathways.

HOST CASTRATION AND SIMILAR EFFECTS

In some cases, the host is castrated, which means that resources that the host would normally put into its developing gonadial tissues can be diverted to produce chemicals or tissues that the parasitoid can consume. Because the host does not need its gonads for normal survival, castration has no negative effects for the parasitoid, whereas consumption of many other tissues might reduce the host's, and therefore the parasitoid's, chances of survival to the correct stage for the parasitoid to kill it. Host castration is probably best known in the Aphidiinae and has been demonstrated in numerous species (e.g. Soldán & Starý 1981). In the case of *Aphidius ervi*, it has been shown that a dimeric protein in the venom causes death of the germarium (the part of an insect's ovary comprising clusters of undifferentiated germ cells) and apoptosis of the ovarian cells

Table 9.3 Host castration or negative impacts of parasitisation on host testes development.

Wasp	Host	Ref.
Microctonus sp. (Euphorinae)	*Hypera postica* (Curculionidae)	Drea (1968)
Orgilus obscurator (Orgilinae)	*Rhyacionia buoliana* (Tortricidae)	Syme & Green (1972)
Choeras (as *Apanteles*) *tedellae* (Microgastrinae) and to a lesser extent *Lissonota dubia* (Banchinae)	*Epiblema tedella* (Tortricidae)	Führer (1973a)
Cotesia congregata (Microgastrinae)	*Manduca sexta* (Sphingidae)	Reed & Beckage (1997)
C. plutellae (Microgastrinae) and to a lesser extent *Diadegma semiclausum* (Campopleginae)	*Plutella xylostella*	Cai et al. (2006)
C. kariyae	*Pseudaletia separata*	Tagashira & Tanaka (1998)
C. glomerata	*Pieris brassicae* (Pieridae)	Junnikkala (1985)
Ascogaster sp. (Cheloninae)	*Cydia pomonella*	Reed-Larsen & Brown (1990), Brown & Kainoh (1992)
Asobara citri (Alysiinae)	*Drosophila* (Drosophilidae)	A.R. Kraaijeveld et al. (1999)
Aleiodes testaceus (Rogadinae)	*Earias* spp.	Ahmad (1944) (as *Rhogas*)

(*Aphidius*: Digilio et al. 2000, Falabella et al. 2007). It is not just the ovaries that are affected, but also the developing embryos within the host aphid, with their growth slowed and tissue degeneration being caused by physiological actions, although when older hosts are attacked, their largest embryos might escape the effects, and Polaszek (1986) suggested that this may be because by this time they have developed a resistant cuticle that prevents entry of damaging chemicals. The factor responsible for inducing ovarian apoptosis in *A. ervi* appears to be the enzyme γ-glutamyl transpeptidase, which is present in the female wasp's venom (Falabella et al. 2007). Imagobiont euphorines are in many ways similar to aphidiines in that adult hosts (assuming they are female) often have their egg production curtailed or eliminated (e.g. Loan & Holdaway 1961). There is also evidence for host castration or major testes development retardation in various other systems (Table 9.3).

The degree of host castration (i.e. disruption of spermiogenesis) may depend on the host stage parasitised. Tagashira and Tanaka (1998) showed that if the microgastrine *Cotesia kariya*, attacked early fourth instar *Pseudaletia separata* hosts, spermiogenesis was completely inhibited and testes cells displayed abnormal karyology, whereas if sixth instars were parasitised, sperm cell elongation was present as in unparasitised hosts. Experimental manipulation of oviposition by the very gregarious *C. glomerata* on *Pieris brassicae* to cause artificially small parasitoid broods had no effect on the degree of testis growth retardation (Junnikkala 1985). That so many of the known examples involve microgastrines and chelonines (microgastroid group), both of which employ polydnaviruses, could suggest that the viruses play an important role in this castration and Junnikkala's experiment shows that at least in that case it is unlikely to be due to secretions from the *Cotesia* larvae as there was no dose dependence. The effect of parasitism by another microgastrine, *Choeras*, on the testes and ovaries of its tortricid host was described in some detail by Führer (1973a). The blocking of gonad development occurs soon after parasitisation when the host has reached a late first or early second instar. Not only is the gonadal tissue itself damaged, but parasitism also affects the outer membrane of the gonads ('Peritonealhülle') and seems to cause it to transform into fat body tissue in the female.

In the case of the alysiine *Asobara citri*, A.R. Kraaijeveld et al. (1999) attempted to determine whether castration of male hosts might lead to a reduction in the size difference between male and female hosts (female *Drosophila* are the larger sex) but they detected no effect, probably because in insects the testes are not the source of important growth-regulating enzymes.

In a recently described case of parasitism of adult carabid beetles, dissected parasitised male hosts apparently had normal testes whereas a female host had no eggs, which was presumed to be the result of direct consumption by the euphorine braconid larvae (Sasakawa et al. 2011). In an analogous fashion to host castration, aphid parasitism by at least one aphidiine braconid, *Aphidius ervi*, inhibits or retards the development of wings in its host, the degree of disruption depending largely on the host instar attacked and also on whether the host was a male, a gynopara (i.e. a winged sexual female) or an alate virginopara (i.e. non-reproductive female) (Christiansen-Weniger & Hardie 1998), and Demmon et al. (2004) demonstrated that this allows the aphidiine to obtain more resources from its host aphid.

TERATOCYTES

In a number of braconid taxa, the membrane surrounding the embryo, whether it is strictly a serosal membrane or one formed by the polar bodies (see Chapter 5, section *Embryogenesis*), has evolved a new role. Instead of simply apoptosing, the cells of the membrane swell and separate and eventually become giant cells, which accordingly have been termed teratocytes (see Fig. 12.52a). Increases in teratocyte volume vary greatly from species to species, but there are several reports of those of euphorines swelling 1000-fold or more (e.g. Jackson 1928, O.J. Smith 1952, Waloff 1961), and in many cases they end up largely filling the host's haemocoele. Teratocytes do not divide in the host, but their ploidy level increases greatly, indicating that they are highly metabolically active synthesizing proteins. In those of the microgastrine *Microplitis demolitor*, DNA levels rise from 2–8C at the point of eclosion to 128–256C when fully grown (Strand & Wong 1991; see also Hotta et al. 2001). The surface of the teratocyte cells is typically densely covered in microvilli, presumably in order to increase their ability to absorb nutrients from the host's haemolymph.

Teratocytes are found in members of four braconid subfamilies, probably representing three separate evolutionary origins: Euphorinae (including Meteorini), Aphidiinae (Falabella et al. 2000), Cardiochilinae and Microgastrinae (Basio & Kim 2005). True teratocytes also occur in the Scelionidae. Given what we now know about euphorine phylogeny, we might expect them also to occur at least in taxa such as the euphorine tribes Mannokeriini, Planitorini and Ecnomiini that were previously treated as separate subfamilies or tribes in unrelated subfamilies. However, Poinar (2004) made no mention of them in his study of the adult ant parasitising Neoneurini.

No true teratocytes have been confirmed in the Ichneumonidae, although there is one instance of teratocyte-like cells and one old reference to something that might be teratocytes (see below). The first possible instance, is a report of what might be teratocytes in the campoplegine ichneumonid *Sinophorus* (as *Limnerium* = *Eulimneria*) by Timberlake (see Jackson 1928) and warrants further investigation. Cells of the serosal membrane of the ichneumonine *Diadromus pulchellus* dissociate within and carry on living within the haemolymph of the pupal *Acrolepiosis assectella* host, but unlike those of braconids, they do not swell up and so in themselves probably do not serve as an important food source for the parasitoid larva, although they could have some teratocyte-like functions (Rouleux-Bonnin et al. 1999).

Teratocytes have been studied in most detail in the Microgastrinae (Dahlman 1990, 1991) and Aphidiinae (Falabella et al. 2000). Probably the best studied are those of the microgastrine *Microplitis croceipes*, an important parasitoid of the tobacco budworm, *Heliothis virescens* (Vinson & Lewis 1973, D. Zhang et al. 1994, 1997, Dong et al. 1996, Schepers et al. 1998, Dahlman et al. 2003), those of the aphidiine *Aphidius ervi*, a parasitoid of the pea aphid, *Acyrthosiphon pisum* (Falabella et al. 2000, 2005), and those of the cardiochiline *Toxoneuron nigriceps* (Pennacchio et al. 1992b). Teratocyte ultrastructure in the euphorine *Dinocampus* (as *Perilitus*) *coccinellae* was described by Sluss and Leutenegger (1968) and the growth and development of those of *Cotesia kariyai* was described by Hotta et al. (2001). D. Zhang et al. (1994) showed how the plasma membrane of newly separated teratocytes is wrinkled and has few microvilli, but after 1 day they have swollen to become completely spherical and are by that time completely and densely covered in microvilli. After 3 days, the cells have increased in volume approximately fourfold, but still have the same apparent density of microvilli, indicating that as the cells enlarge additional microvilli are formed. TEM showed that the mature teratocytes contain abundant ribosomes and have a large, ramified nucleus, which probably indicates a high level of endopolyploidy.

Teratocytes have many roles. In some cases they may process host haemolymph and then be consumed by the growing parasitoid larva as in the case of the euphorine *Periltus rutilus* (Jackson 1928) in which they constitute the bulk of the parasitoid larval food during its rapid final feeding phase. In microgastrine braconids, they have been shown to be the source of a fungicidal compound (Führer & El-Sufty 1979), which may help prevent infection of the host with entomopathogens when its resistance is otherwise compromised, and El-Sufty and Führer (1981) went on to show that parasitism of *Pieris* larvae by *Cotesia* (as *Apanteles*) *glomerata* causes a pathology of the host cuticle which indeed renders it more susceptible to penetration by the entomopathogenic fungus *Beauveria brassiana*.

The teratocytes of *Microplitis mediator* have been shown to secrete collagenases (Qin et al. 2000). *M. croceipes* teratocytes secrete at least 15 different proteins in cultures, including a 14 kDa protein called TSP14 (Schepers et al. 1998, Hoy & Dahlman 2002) which is responsible for developmental arrest of the host and inhibiting its pupation. In *Cotesia kariyai* parasitising the noctuid *Pseudaletia*, Nakamatsu et al. (2002) showed that teratocytes enable the parasitoid larva to feed on the host fat body into the second instar. Development and consumption of the fat body during this instar were very rapid and Nakamatsu et al. found that many teratocytes had adhered themselves to the fat body surface and apparently released a collagenase-like enzyme that it was suggested made a hole in the outer covering and facilitated larval access to fat body resources, although the details are unclear. Teratocytes of the cardiochiline *Toxoneuron nigrceps*, based on a cDNA library, may secrete a chitinase (Rossi et al. 2012), which has been interpreted as possibly aiding weakening of the host cuticle to facilitate parasitoid larval egression. However, when expressed in the bacterium *Escherichia coli*, the product only bound to chitin rather than showing any enzymic activity.

Continued *in vitro* culture of *M. croceipes* teratocytes requires the presence of secretory products from host larvae (Hoy & Dahlman 2002). Secretory products from *in vitro*-cultured *M. croceipes* teratocytes when injected into the *Heliothis* host delay development of the caterpillar and induce other changes characteristic of normally parasitised hosts (Dahlman 1991). A role in suppressing host phenoloxidase activity was found for teratocyte secretions in the microgastrine *Cotesia* (as *Apanteles*) *kariyai* by T. Tanaka & Wago (1990), who showed that their presence was essential for the parasitoid larva to reach maturity. When host *Pseudoplusia includens* caterpillars were injected with approximately 600 *M. demolitor* teratocytes (the mean number in normally parasitised hosts being approximately 550 cells), they were caused to pupate almost immediately (Strand & Wong 1991), but when they were co-injected with calyx gland extract (i.e. polydnavirus), the developmental progression of the host much more closely resembled that of naturally parasitised larvae. Interestingly, the teratocytes are not themselves immune to encapsulation by this host but, as with the parasitoid larva, require the protection afforded by polydnavirus particles.

In aphidiines, specifically *Aphidius ervi*, venom and teratocytes act in a complementary way to increase food availability to the growing larva (Falabella et al. 2000). In aphids, endosymbiotic *Buchnera* bacteria are located in the bacteriocytes which are closely associated with the germarium. The bacteria synthesize and release amino acids that normally diffuse to and are taken up by germarium cells for the production of eggs and subsequently embryos. Venom components specifically kill the germarium cells, allowing more amino acids to be taken up by the teratocytes, many of which are also physically closely associated with the bacteriocytes. The teratocytes use these to synthesize and release into the host haemolymph large amounts of two proteins (P15 and P45). These proteins are then taken up by the developing parasitoid larva. In aposymbiotic aphids, the production of these proteins is greatly reduced and parasitoid growth similarly so. Caccia et al. (2012) showed that one of these is a fatty acid-binding protein (Ae-FABP), which, among others, shows a high affinity for C_{14}–C_{18} saturated fatty acids. The proteins appear to act as vectors, enabling the parasitoid larva to ingest large amounts of host fatty acids, which are probably released into the host haemocoel as a result of other parasitoid activities.

INTRASPECIFIC VARIATION IN RESISTANCE TO PARASITOIDS

Although it is obvious that there must be an 'arms race' between a parasitoid's ability to attack a host and the host's ability to defend itself by encapsulating the parasitoid's egg or early instar larva, many studies have found that the parasitoid is always or nearly always successful (in its natural host). Hence systems in which variation in the ability of a host species to encapsulate a given parasitoid are of interest. Sometimes, variation in encapsulation ability varies geographically, e.g. in *Bathyplectes curculionis* (Campopleginae) (Berberet et al. 1987). The best studied system is between *Drosophila* fly larvae and their alysiine braconid parasitoids of the genus *Asobara* (A.R. Kraaijeveld 1994, A.R. Kraaijeveld & van Alphen 1994). Using reference strains of either parasitoid or host against which to test the other, it was found that *A. tabida* from southern Europe were far more virulent against *D. melanogaster* than those from more northern countries such as the United Kingdom, The Netherlands and Sweden. This pattern was neatly explained because in the more northerly part of its range, *A. tabida* principally attacks a different host, *D. subobscura*, which is unable to encapsulate this and, indeed, various other parasitoids (Kraaijeveld & van der Wel 1994), whereas *D. melanogaster* is an encapsulator.

Aphids do not encapsulate their parasitoids but eliminate them by some poorly understood humoral mechanism. The ability to eliminate aphidiine parasitoids does vary intraspecifically and because aphids reproduce clonally, it makes them good experimental animals for the study of variation in resistance. S. Li et al. (2002) investigated the differences between a parasitism-susceptible green clone of *Acyrthosiphon pisum* and a partially resistant pink clone. In the case of the pink clone, *Aphidius ervi* could complete development if it attacked the aphid in its second instar but not if it attacked a fourth instar, and its development when it was successful was far slower. Normally, *A. ervi* teratocytes synthesize and release into the host haemocoel large quantities of a protein that is taken up and utilised by the parasitoid larva, but this production appeared to be inhibited in the pink clone and appeared to be the major mechanism in that clone's resistance. The resistance of aphids to aphidiines is also modulated by intraspecific differences in the secondary symbionts that they carry (see Chapter 12, section *Aphidiinae*).

EFFECTS ON HOST MOULTING PATTERN

Hosts parasitised by several endoparasitic ichneumonoids are caused to undergo precocious metamorphism, many examples having been reported for the hosts of Cheloninae (D. Jones et al. 1981) (see also the section *Effects of polydnaviruses on hosts*, above). In some cases, this effect is purely the result of the venom injected by the wasp, as it occurs irrespective of whether an egg is laid. In the case of *Spodopera* parasitised by the chelonine braconid *Chelonus inanitus*, it undergoes precocious metamorphosis forming a prepupal stage early, apparently through suppression of ecdysone production, and it is at this point that the *Chelonus* larva exit the host and commence external feeding (Grossniklaus-Burgin et al. 1994, 1998, Kaeslin et al. 2005a). On the other hand, Brown et al. (1993) found that larvae of another chelonine that causes premature host metamorphosis, *Ascogaster reticulatus*, actually secrete ecdysteroids, at least *in vitro*. In this system, however, precocious metamorphosis occurs only when the parasitoid larva is also present in the host, as shown by parasitoid larval removal and implantation experiments (Pfister-Wilhelm & Lanzrein 1996). The presence of a late first instar *C. inanitus* larva led to inactivation of the host's corpora allata, which consequently reduced the host caterpillar's juvenile hormone titre to nearly zero. Hochuli et al. (1999) show that the parasitoid larva releases proteins into the host that are essential for the change in host development and tentatively identified a 212 kDa candidate protein for this role. The cardiochiline *Toxoneuron nigriceps* prevents the host, *Heliothis virescens*, from pupating when in its final instar and compared with controls parasitised caterpillars have reduced haemolymph 20-hydroxyecdysone and higher juvenile hormone titres (S. Li et al. 2003).

Another interesting example is that larvae of *Anthophila fabriciana* (Choreutidae) parasitised the rogadine braconid *Clinocentrus cuncator* (as *gracilipes*) in their third instar, developed at the next moult the dorsally spiny cuticle of the final, fifth instar, i.e. one instar early (M.R. Shaw 1981) (Fig. 7.2b). This premature moult to a prepupal instar is induced entirely by the venom of the wasp. Shaw found that the venom of another cyclostome braconid, *Rhysipolis*, which while being an ectoparasitoid is also a koinobiont, has exactly the same effect on the host's endocrine system as that of *Clinocentrus*. It was postulated that this type of action evolved because of the need to modify the host's feeding and pupation biology and might actually have been an essential stage in the evolution of endoparasitism in the Rogadinae.

PARASITOID-INDUCED CHANGES IN HOST BEHAVIOUR

Many parasitoids are known to induce behavioural changes in their hosts in addition to influencing their development, such as changes in host pupation site, and this led to considerable speculation about which species, the parasitoid or the host, benefits from the change. Shapiro (1976) followed by Smith Trail (1980) put forward the idea that kin selection might favour, under certain conditions, evolution of host suicide, whereby a parasitised host such as a caterpillar changes its behaviour so as to render it more susceptible to predation, thereby killing its parasitoid also. The conditions required for this to occur would be where nearby individuals are likely to be close relatives, as in colonizing species where a new patch might be expected to be colonised perhaps by just one or two individuals, live in groups, and where the parasitoid emerging from them is likely to forage locally for hosts. The example cited was that of the North American nymphalid butterfly, *Chlosyne harrissii*, which is attacked by the microgastrine braconid *Cotesia* (as *Apanteles*) *euphydryidis*. Parasitised larvae move to a more conspicuous and exposed situation and do not defend themselves against further attack. Stamp (1981) and Godfray (1994) both suggested that this was highly unlikely to be suicidal behaviour as many other microgastrines induce climbing behaviour in their hosts before egress for other reasons (e.g. Sato et al. 1983). Stamp, looking at a system similar to that suggested by Smith Trail, argued that the behavioural changes in the hosts most probably benefited the parasitoid because it might lead to a reduction in hypereparasitism. Certainly in several other systems studied involving microgastrines manipulating host behaviour the wasp appears to be very much in charge (Harvey et al. 2008; see also Chapter 12, section *Microgastrinae*). However, none of this means that suicidal behaviour in parasitised hosts cannot evolve given the appropriate conditions and, as discussed by Godfray, the increased willingness of parasitised individuals of some aphids to drop from the host plant when disturbed and thus incur a high risk of death from other causes is certainly a possible example (McAllister & Roitberg 1987), and seems more plausible since aphids attacked at a stage where some future reproduction was still a possibility were found to be less likely to drop. Further, aphids live in clonal groups, which is an extreme form of kin grouping, and aphidiines that emerge within a group often go on to attack others in the vicinity. As Godfray pointed out, there could easily be other explanations for this behaviour and determining whether selection for host suicide had played a role would be very difficult to test conclusively, but it is certainly an interesting topic. Perhaps a fruitful system for examining this would be the interactions between neoneurine euphorines and their worker ant hosts, since worker ants within a nest are likely more than 50% and up to 75% genetically related to their sisters.

Some of the more sophisticated host behavioural modifications occur in spiders parasitised by polysphinctine pimpline ichneumonids, which lead to changes in the way in which they construct webs (e.g. Matsumoto 2008). The host spider's behavioural changes have been investigated in detail by Eberhard (2000a, 2001) for *Hymenoepimecis argyraphaga* and Eberhard (2010b) for *Polysphincta gutfreundi*. *Hymenoepimecis* stings its host, causing a temporary paralysis that lasts 10–15 min, during which time the parasitoid glues an egg to the anterior of the spider's abdomen. During its larval development, the *Hymenoepimecis* larva is snuggled between the spider's abdomen and cephalothorax and the spider continues to behave normally for up to 2 weeks, constructing normal webs and consuming the prey they catch. However, just before the spider is finally killed, it is induced to perform a unique construction activity, weaving a special 'cocoon web' which appears to be especially well suited to supporting the cocoon of the ichneumonid. Removal experiments show that the spider will sometimes carry out the same 'cocoon web' building in the absence of the parasitoid larva, suggesting that this stereotypical behaviour is induced by some chemical 'injected' by the parasitoid larva some time beforehand.

In a twist to the tale, caterpillars of the North American pest torticid moth, *Choristoneura rosaceana*, while not normally being cannibalistic, will readily consume conspecific larvae that have been attacked by the host-paralysing, ectoparasitoid, braconine *Habrobracon gelechiae* (Wang & Daane 2014), a behaviour that could have evolved through individual or kin selection, since, with fewer *H. gelechiae* around in the next generation, survival of that caterpillar's offspring might be markedly increased assuming that they tend to lay their eggs locally.

ENDNOTES

1. As there have been no detailed studies of those ichneumonoids whose stings cause considerable pain [e.g. *Enicospilus* and *Ophion* (Ophioninae) and *Netelia* (Tryphoninae)], and as several small aculeate sting peptides are involved in pain production, it is possible that some do.

2. In insects, glutamate is the excitatory neurotransmitter causing muscle contraction, whereas in chordates, glutamate is a neurotransmitter within the brain and central nervous system (CNS), and acetylcholine is a neurotransmitter within arthropod CNS and not at nerve–muscle junctions.
3. Polydnavirus status is not known for the Mendesellinae, Dirrhopiinae or Epsilogastrinae but seems likely given their phylogenetic relationship nested among the other microgastroid subfamilies.
4. John Tyndall in 1854 showed that the sky appears blue due to small particles that scatter shorter wavelengths of light (i.e. blue) more than longer ones.
5. Although a few microgastrines have secondarily evolved to be egg-larval parasitoids, they effectively are attacking very young caterpillars just prior to hatching.

Chapter 10

CONVERGENT ADAPTATIONS

The Braconid and Ichneumonid Parasitoid Wasps: Biology, Systematics, Evolution and Ecology, First Edition. Donald L. J. Quicke.

Why have no ichneumonids evolved to be cyclostome? Why are carapaces so rare among ichneumonids and yet so common in braconids? Why are facial horns almost entirely restricted to the Braconinae? Although individually these may seem to be trivial questions, and may remain unanswerable for the near future, I think they are actually rather important when considered in a broader context. If a feature has evolved on multiple occasions in a group but not in its almost equally sized sister group, this means that either the group lacked the necessary genetic variation or, if it had that variation, then either it offered no selective advantage directly or it was in conflict with other critically important features (perhaps through pleiotropy). For example, carapaces appear to have evolved at least 20 times in the Braconidae, but probably no more than four times in the Ichneumonidae.

One immediately odd thing here is that the above few examples all involve various Braconidae as having the derived state (with carapace, with facial horn, with cyclostome mouthparts), yet the Ichneumonidae have long been considered to show a high level of homoplasy, i.e. they include many examples of parallelisms (Gauld & Mound 1982). In addition to those discussed here, van Achterberg (1988b) noted a number of other parallel morphological changes seen among the Braconidae, but because of the absence of any biological hypotheses, not all are discussed, although no doubt they all do reflect some selective force.

ANTENNAL HAMMERS AND VIBRATIONAL SOUNDING

That parasitic wasps can detect concealed hosts through a combination of chemical cues and host-created vibrations has long been known. However, that some wasps might use a form of echolocation, called vibrational sounding, in which substrate-borne vibrations are produced by the wasps and whose reflections are detected and interpreted, was first discovered by Henaut & Guerdoux (1982) in the case of *Pimpla rufipes* (as *instigator*) (Ichneumonidae: Pimplinae), a common parasitoid of Lepidoptera pupae in concealed situations. This was a suitable subject for the study of host location in a species that attacks exposed hosts such as *Pieris brassicae* pupae, because it is easily reared and females will ovipositor probe even through paper tubes whose texture approximates that of the host's cuticle, and Halstead (quoted in Fitton et al. 1988) reported that one female even tried to oviposit into bean seeds within a packet that it 'no doubt perceived as cocooned pupae'. Henaut (1990) described how the antennae were not used as drumsticks and instead transmitted shockwaves made in the body to the substrate via vesicles at the tip of the terminal flagellomeres and stated that the vesicles are fluid-filled with a thin wall. However, SEM studies of several taxa with antennal hammers (Fig. 10.1a,b) suggest that in most cases these are probably solid cuticle as in the Orussidae (Vilhelmsen et al. 2001), although this is entirely based on SEM (Figs 10.1, 10.3).

Subsequently, members of some related genera have been demonstrated to show the same behaviour. This exotic mode of host detection, however, appears to have evolved on a number of separate occasions, both within the Hymenoptera and also within the Ichneumonidae. The association of hammer-like modifications of the terminal flagellomere and host location is indicated both by the antennal modifications being more pronounced in females than in males and by the accompanying relative enlargement of the fore tibia in females (Broad & Quicke 2000). This is because the subgenual (chordotonal) organ used for detecting substrate-borne vibrations in Hymenoptera is located in the fore tibia (Otten et al. 2002). Broad and Quicke (2000) carried out a comparative analysis of modifications of the antennal apex and sexual dimorphism in the fore tibia to assess the probable distribution of vibrational sounding in the Ichneumonidae, and concluded that it probably occurred in a wide variety of cryptines belonging to both the Cryptini and Phygadeuontini[1,2], labenine genus *Labena*[3], all xoridines, various Pimplini and the genus *Ecphysis* in the Claseinae. More detailed investigations within the Cryptinae indicated that this feature has been subject to multiple gains and losses (Laurenne et al. 2009).

A detailed structural and ultrastructural analysis of the equivalent hammering and vibration detection system in the distantly related, basal parasitic family Orussidae was presented by Vilhlemsen et al. (2001). Although no equivalent TEM analyses have been performed on the apparently vibrational sounding ichneumonids, it is likely that similar, if typically less extreme, modifications occur in these.

Cryptines that use antennal vibrational sounding move their angled antennae (Fig. 10.2) in smooth converging arcs, at the same time tapping them up and down with great rapidity. Those searching on

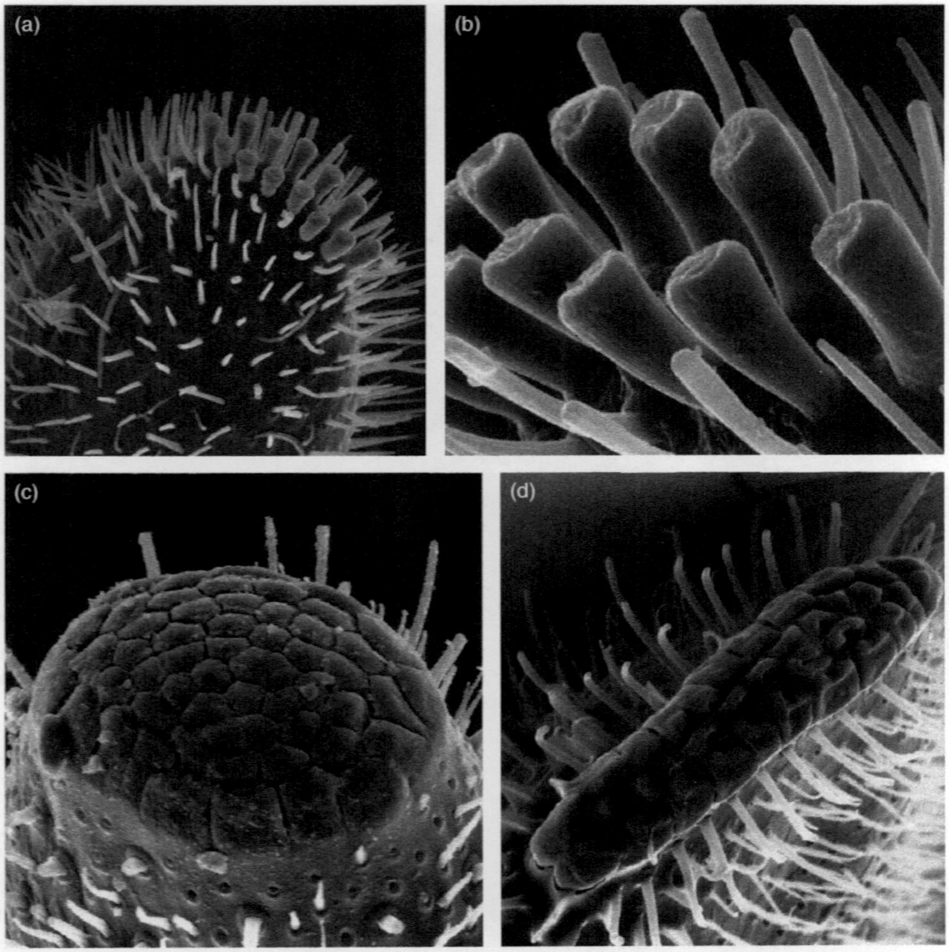

Fig. 10.1 Antennal hammers in Pimplinae and Cryptinae. (a, b) *Pimpla flavicoxis* (Pimplinae), terminal flagellomere and detail of hammer pegs; (c) *Schreineria* sp. (Cryptinae), flattened apex with grooves showing where pegs have fused; (d) *Eurycryptus* sp. (Cryptinae), showing narrow apical hammer.

barkless tree trunks typically walk to-and-fro doing this, homing in on to the sweet spot (Quicke et al. 2003, Niemeyer & Flores-Prado 2012). Those attacking cocooned hosts such as *Cryptus albitarsis* may spend a long time tapping the cocoon with the antennal tips (e.g. Simmonds 1947a, as *C. inornatus*). In the Cryptinae and Labeninae, the hammer modifications are at the apex of the terminal flagellomere, whereas in *Xorides* (Xoridinae), the hammers are simple or somewhat T-shaped pegs originating on the medial to subdistal, ventral antennal flagellomeres. These modifications are largely restricted to females.

A comparative SEM study of the antennal hammers across the Cryptinae by Laurenne and Quicke (2009) and Laurenne et al. (2009) suggested strongly that the hammer structures are derived from sensilla, and a range of degrees of modification from only slightly thickened setae, like sensilla (type 1), through to fully fused, solid hammer ends (type 5) was apparent. Highly modified hammers were present in numerous genera

Fig. 10.2 Female *Echthrus niger* in the process of vibrational sounding (echolocating) for a host. Note how the flagellum is sharply bent just distal to the white stripe such that the tips strike the substrate perpendicularly, and note the swollen (white) fore tibiae which house the enlarged subgenual organ. (Source: reproduced by permission of Stephen Luk.)

of Cryptini and Phygadeuontini, but were absent in the Hemigasterini, which more typically attack hosts in cocoons.

Many of the species with highly developed antennal hammers attack various aculeate Hymenoptera nests. Polidori et al. (2011) described the behaviours involved in host location and oviposition in the cryptine *Acroricnus seductor* attacking cells in the nest of the mud-dauber wasp, *Sceliphron caementarium* (Sphecidae). However, SEM of the antennal apex of *A. seductor* revealed that unlike the *Acroricnus* species examined by Laurenne et al. (2009), it had a highly modified (type 3) hammer and, conversely, the *Buathra* examined by Casiraghi et al. (2001) had a less well-developed hammer. Collectively, these results suggest that the degree of modification is rather plastic and can quickly evolve according the requirements of a given species.

Unlike most other types of sensory system involved in host location, the effectiveness of vibrational sounding would be expected to be temperature dependent because the production of the vibrations and the neural interpretation of the echoes might be expected to be differentially affected. This was explored in a series of studies on *Pimpla turionellae* (Kröder et al. 2006, 2007a,b) and it was found that *Xanthopimpla stemmator*, which originates from the tropics, has an higher optimal temperature for successful host location than the temperate *P. turionellae*. The success of the latter declined rapidly with increasing temperature whereas the former was accurate over a broad range of temperatures. Surprisingly, although temperature did not alter the basic vibration pulse train parameters, it inversely affected the vibration frequency within pulses. Also, larger wasps were better at echo-locating hosts, probably because they could cause stronger vibrations.

Although all the clear examples of the use of vibrational sounding in the superfamily involve ichneumonids, females of the euphorine braconid *Meteorus corax*, which attacks wood-boring beetle larvae, have a patch of 30 or so 'basiconic pegs' on one face of the penultimate flagellomeres (Maetô 1986) that are similar to the structures used for vibrational sounding in the genus *Xorides* (Xoridinae) (Fig. 10.3b,c).

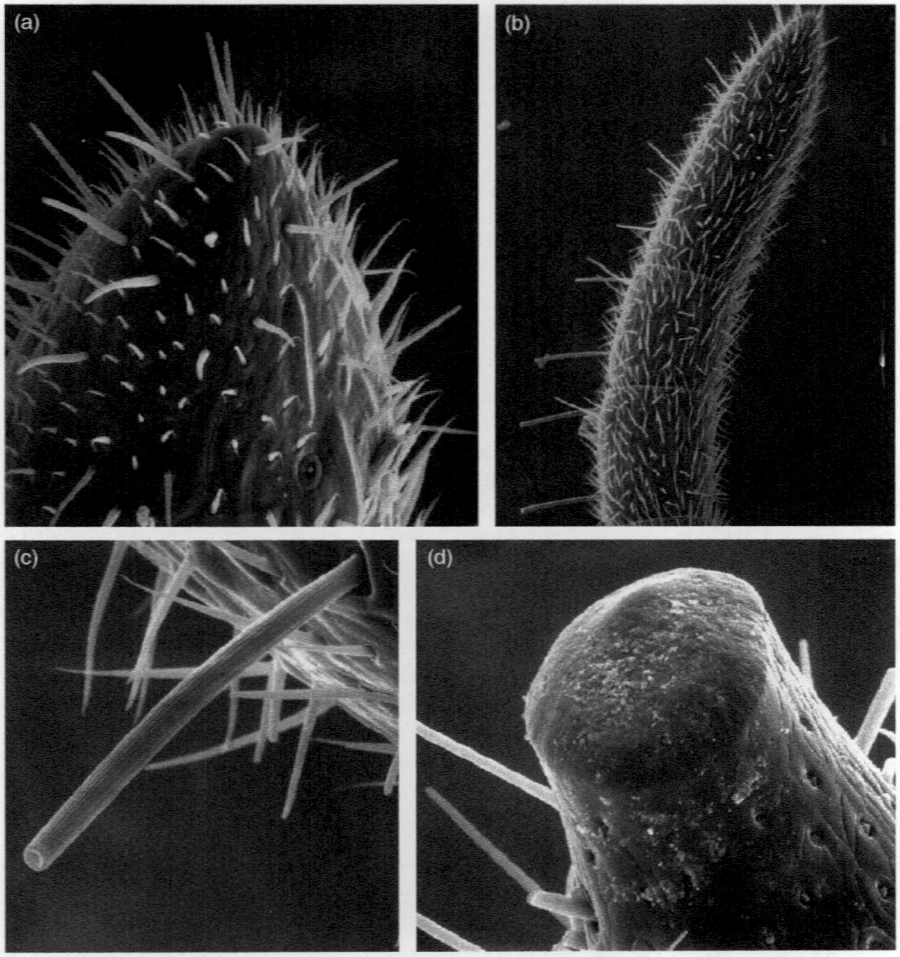

Fig. 10.3 Antennal hammers in Xoridinae and Labeninae. (a) *Xorides fulvicornis*, normal terminal flagellomere; (b, c) sub-apical flagellomeres showing thickened pegs on the anteroventral side and detail of a single peg; (d) *Labena schausi*, tip of terminal flagellomere.

Further, the braconine genus *Glyptomorpha* has the apex of the terminal flagellomere laterally compressed, blunt, smooth and devoid of sensilla (Quicke 1987a), again suggesting that it might have a possible hammering role.

ENLARGED MANDIBLES

A few ichneumonoids have remarkably developed, sickle-shaped mandibles, whose function is typically unknown, and given the biologies of the wasps involved, they must have different functions in different taxa. Really only in the euphorine Cosmophorini (see Fig. 12.53) is the function known, in that these endoparasitoids of scolytine beetle adults apparently use them to grab and restrain their hosts during ovipositioning. When similar such mandibles occur in ectophagous parasitoids of concealed hosts, such as in *Gnathobracon* (Quicke & Huddleston 1991) and *Pseudodicrogenium* in the Braconinae, it is far from clear what their function might be. They clearly cannot be involved in restraining the concealed host, so they might therefore be adaptations either for egressing from the pupation site – although that seems unlikely given that they have evolved on so few occasions – or in intraspecific interactions. They seem ill-suited for chewing their way out of solid and hard wood and I doubt, given their morphology, that they could be opened wide enough for that, but if the substrate were very soft this might still be a possibility. Unfortunately, no host records are available.

CHISEL-LIKE MANDIBLES

A large number of species, in several subfamilies, particularly those which pupate within host burrows within woody plant tissue, are modified to facilitate the wasp chewing its way out of the host substrate. Such mandibles are stocky, with the upper tooth or both teeth rather broad with a fairly straight edge and usually only a small gap between the upper and lower teeth. The mandibles also tend to be twisted such that in many species the upper tooth conceals the lower one. This arrangement is common, for example, in Rhyssinae, Labeninae, Poemeniinae and Ephialtini (Pimplinae) among the ichneumonids and in the Braconinae, Doryctinae and some Helconinae within the Braconidae.

Pucci and Sharkey (2004) also noted that unlike in other agathidines whose mandibles are rather slender and flattened and used to chew out of the parasitoid's and maybe also the host's cocoon, those of the genera *Agathirsia* and *Crassomicrodus* are far more robust and cylindrical in cross-section, suggesting that this is an adaptation to attacking hosts that spend the day constructing cocoons or pupation chambers under ground so the wasp has to use its mandibles either to dig down to reach them and/or dig its way out.

CONCEALED NECTAR EXTRACTION APPARATUS

Jervis (1998) detailed to a number of different adaptations that he collectively referred to as 'concealed nectar extraction apparatus' (CNEA) and recognised seven types within the Ichneumonoidea (Table 10.1). The commonest CNEA form involves elongation of the galea (Fig. 10.4a), which is commonly associated with at least some degree of elongation of the malar region (Figs 10.4a and 12.38e,f), elongation of the glossa (Fig. 10.4b) and elongation of the maxillary palps (Fig. 10.4c), which can be medially concave and form a drinking straw. In the agathidine genus *Agathirsia*, the glossa is furnished with hydrophilic scale-like setae

Table 10.1 Distribution of CNEAs in the Ichneumonoidea.

CNEA type	Description	Occurrence
1	Elongated glossa and galeae; glossa shallowly bi-lobed/forked but concealed in dry specimens	Agathidinae, Braconinae, Cheloninae, Cardiochilinae, Microgastrinae, Metopiinae, Ophioninae, Tersilochinae
2	Elongated glossa and galeae; glossa very deeply forked and glossa largely exposed in dry specimens	Cardiochilinae, Labeninae, Banchinae, Ophioninae
3	As in type 1 but additionally maxillary palps strongly elongated and adpressed forming a 'drinking straw'-like structure	Braconinae, Microgastrinae, Tersilochinae
4	Elongated glossa, galeae and labial palps; glossa exposed for much of its length in dried specimens; labial palps closely aligned with glossa and assist in nectar extraction/conduction	Agathidinae
5	Maxillary palps only, elongated	Braconinae, Cheloninae, Cryptinae
6	Elongated glossa and labial palps; inner surfaces of labial palps concave	Agathidinae, Cheloninae
7	Elongated labial palps enclosed within elongated maxillary palps; inner surfaces of maxillary palps concave	Agathidinae

Source: based on Jervis (1998).

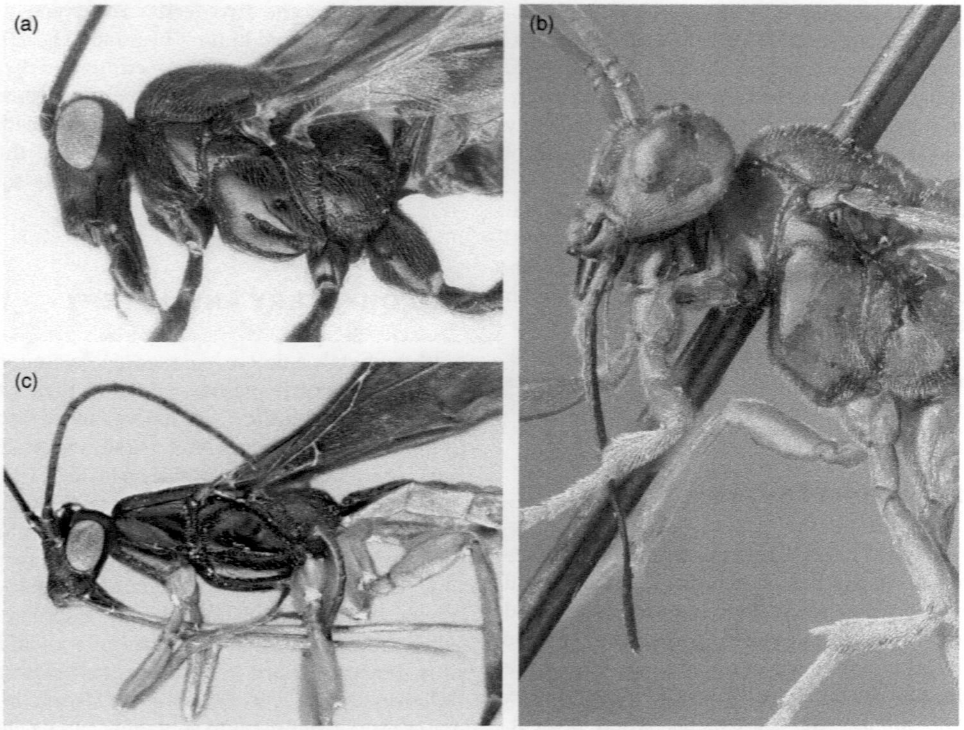

Fig. 10.4 Three examples of CNEA exhibited by members of the braconid subfamily Agathidinae. (a) *Agathis anglica*, showing elongation of malar region and of galea; (b) *Agathirsia proxima*, showing elongation of glossa (right); (c) *Agathis longipalpa*, showing great elongation of maxillary palps which are apposed and medially concave forming a tube. (Source: reproduced by permission of Michael Sharkey.)

externally and the inner surface is finely striated, both of which will cause fluid to spread towards the glossa by capillary action (Pucci & Sharkey 2004).

REDUCED NUMBER OF PALPAL SEGMENTS

The plesiomorphic condition is for the maxillary palp to be six-segmented and the labial palp four-segmented, and this condition is widely retained. Moderate reductions occur in several tribes and subfamilies and can be useful key characters. A few groups, however, show extreme reduction. For example, in the braconid subfamily Amicrocentrinae they are both effectively one-segmented (van Achterberg 1979a), in the mesostoine genus *Praonopterus* the palpal formula is 1 : 1 and in the ichneumonid Hybrizontinae they are also greatly reduced (see Fig. 13.55b). Van Achterberg (1988b) noted that this 'negative parallelism' has evolved on at least 17 separate occasions within the Braconidae alone, although it seems to be a less common trend among the Ichneumonidae.

'FACIAL' PROTRUBERANCES

This feature is largely restricted to the Braconinae, but is worthy of consideration here because of the number of separate occasions upon which they have evolved, and although the exact number of times is not yet certain, it is probably at least six, viz. *Plaxopsis*, *Zanzopsis*, *Deltaphyma*, *Malagopsis*, *Cervellus*, *Lasiophorus*, *Palabracon*, *Psittacibracon* and *Victoriviella*. Of these, the first two, and possibly the first four, may be closely related, but the form of the modification is so different that there may have been several origins of protuberances in the group. Similarly, both *Lasiophorus* and *Palabracon* have been shown using DNA sequence data to be derived within the huge genus *Digonogastra*, but again, they do not appear to be particularly closely related and probably represent independent origins of the modification.

All the taxa with them for which biology is known are parasitoids of moderately to deeply concealed hosts, from stem borers to wood-boring beetles, so perhaps they function to help the wasp egress, although it should be noted that these protuberances are frequently far less well developed in males and why they should not need them equally for this purpose is difficult to imagine.

FRONTAL DEPRESSIONS

Vilhelmsen (1997) drew attention to the importance of depressions on the face or frons in which are found in various parasitic wasps that pupate inside wood or within other hard substrates. They appear to have evolved to protect the base of the antenna (the scape and pedicellus) during egress from the pupation site. In ichneumonoids, the depressions, when present, are always on the frons and they vary in depth from shallow with smooth sides to deep, often with a sharp edge, and sometimes there is also a well-developed mid-longitudinal carina dividing the depression into two halves. The presence of such depressions seems to be a very reliable indicator of biology in this respect, although there are exceptions such as in some agathidines and in *Eadya*. They are well developed in various Rhyssinae and Poemeniinae in the Ichneumonidae and in several Braconinae, Doryctinae, Cenocoeliinae and Helconini in the Braconidae.

DORSAL RIDGES ON HEAD OR MESOSOMA

Eggleton (1989) drew attention to a syndrome of features associated with egress from pupation sites within wood which seem to be associated with allowing the wasp to gain sufficient purchase so that it can chew its way out – and of course there are also features of the mandibles which in these taxa tend to have chisel-shaped blades with little differentiation between the two teeth. Transversely striated mesonota are found in the Rhyssinae (Fig. 10.5a), the labenine genera *Certonotus*, *Asperellus* and *Apechoneura* (see Fig. 13,8b) and the Poemeniine genus, *Pseudorhyssa* (Fig. 10.5b), while the top of the head of many doryctine parasitoids of wood-borers and the temple of *Podoschistus* (Poemeniinae) are heavily sculptured (Fig. 10.5c).

BRACHYPTERY AND APTERY

Although only a tiny percentage of ichneumonoids are apterous or brachypterous, and even then sometimes only polymorphically so or in a sex-linked fashion, the morphological changes that result, especially of the mesosoma, can be very considerable (Reid 1941). With no need for flight muscles, the mesothorax becomes greatly reduced in size and no longer dominates the body region (see, e.g., Fig. 13.30c). These changes, and also the lack of wing venational characters (or major transformation in not fully apterous species), often make precise taxonomic assignment difficult and it is sometimes hard to tell what their closest winged relatives are (see, e.g., Wharton 2002), and in extreme cases even assignment to subfamily can be difficult, as with *Cuniculibracon* discovered in a rabbit burrow (see Fig. 12.42e).

In some species only males are apterous, in others it is only the females and in some it is both sexes (see Table 10.2). In the cryptine *Gelis corruptor* (see Fig. 13.30c), aptery is both sex linked and polymorphic within males, thus Salt (1952) referred to this species as trimorphic. He noted that in females there is no trace of wing discs (imaginal buds) even in the larval stage. However, the situation in this species is apparently more complicated because whether males develop wings depends on the quality/quantity of their larval food resource and their fate can be changed by transfer either from a small to a large host or by removing larvae on large hosts prematurely.

Whereas within the Ichneumonidae brachyptery appears to be largely restricted to the Cryptinae and a few northern *Stenomacrus* (Orthocentrinae) (see Table 10.2), in the Braconidae it has evolved on numerous occasions (e.g. Belokobylskij & Quicke 1999, Iqbal et al. 2003, Belokobylskij 2009, Belokobylskij & Kula 2012, Belokobylskij & Austin 2013) and occurs in a few members of virtually all subfamilies

Fig. 10.5 Features of ichneumonids associated with egress from and ovipositing into a hard wood substrate. (a, b) Transversely ridged mesoscutum of *Megarhyssa* sp. (Rhyssinae) and *Pseudorhyssa* sp. (Poemeniinae), respectively; (c) head of *Podoschistus alpensis* (Poemeniinae) showing dorsal sculpture; (d) ventral surface of anterior of metasoma of *Megarhyssa* sp. (Rhyssinae) showing medial groove and clasps of the ovipositor stabilising mechanism. (Source: reproduced by permission of Rikio Matsumoto.) For a colour version, see Plate 15.

within the cyclostome lineage and in four subfamilies of non-cyclostomes. The lack of detailed and independent phylogenies for nearly all the genera including brachypterous or apterous species makes it impossible to know precisely how many times it has evolved, but in the Braconidae it must minimally be 48 times and probably a lot more.

In both families aptery and brachyptery are concentrated, not surprisingly, among taxa that forage for prey either among ground litter or upon tree trunks and invariably involves physically small species. There may also be a tendency for them to evolve more on islands, somewhat like flightlessness among birds, and perhaps also in arid and arctic or high mountain environments where hosts may be exceedingly patchy or energetics limiting. Microptery is also found in a few slightly larger species such as in the summer morph of the cryptine *Sphecophaga vesparum*, which is slightly larger than most truly apterous and brachypterous species, and probably indicates a tendency to search among vespid nest cells on foot.

DORSO-VENTRAL FLATTENING

Several taxa of braconids and at least one ichneumonid that appear to develop and/or hunt for hosts in the narrow gap below loose bark are strongly dorso-ventrally compressed. In the Braconidae it is found in two unrelated genera of braconines (*Odontoscapus* (including *Chaoilta*) and in *Chartobracon*) and in some *Capitonius* (Pitz & Sharkey 2007). In the ichneumonidae, it is known in some *Aplomerus* (Xoridinae) (Watanabe & Matsumoto 2010).

Table 10.2 Incidence of brachyptery and aptery in the Ichneumonoidea*.

Family/subfamily	Subgroup	Biological notes
Braconidae		
Aphidiinae	All *Diaeretellus*, some *Trioxys*	Aphids in the far north but few instances
Mesostoinae	Males only of *Mesostoa* (see Fig. 12.12a,b)	Gall formers on *Banksia* plants
Rhyssalinae	*Achaiabracon*	From Greek islands
	One *Rhyssalus* and one *Acrisis* sp. *Metaspathius* sp.	
Doryctinae	Many *Spathius* spp., some *Heterospilus*, plus a few *Rhaconotus*, *Ecphylus*, *Nipponecphylus* and ~15 other genera	See Belokobylskij & Kula (1912), Belokobylskij & Austin (2013)
	Termitobracon males and *Ypsistocerus* males	Associated with termite nests
Pambolinae	Females of some *Pambolus* and *Dimeris* spp.; males of a few *Phaenodus* spp.	
Lysiterminae	*Neolysitermus*	Unknown
Hormiinae	One *Hormius* sp.	Unknown
Rogadinae	Female only is brachypterous in one *Aleiodes* sp.	From arid North Africa
Braconinae	One species tentatively assigned to *Bracon*	
Alysiinae	Members of several genera of Alysiini including *Aspilota* (some males), *Atopandrium*, *Chasmodon*, *Lodbrokia*, *Orthostigma*, *Pseudopezomachus* and *Pterusa*	Koinobiont endoparasitoids of Diptera, often associated with detritus/decaying matter/soil layer environment
Masoninae	*Masona* females of all four known species	Collected mostly in traps on the ground
Euphorinae	*Mannokeraia*	Unknown for *Mannokeraia* but, based on related euphorines, probably attacks nymphal or adult insects
	Cosmophorus, males of two species	Adult scolytine beetles, often found under bark, males probably call females pheromonally
	Meteorus, females of one species from New Zealand	Koinobiont parasitoid of moth caterpillars in grass layer in alpine zone.
Orgilinae	A very few *Orgilus* spp. are brachypterous	Koinobiont endoparasitoids
Helconinae *s.l.*	*Cuniculibracon* (see Fig. 12.42e)	Unknown, collected in rabbit latrine
Blacinae	A few *Blacus* spp.	
Ichneumonidae		
Pimplinae	A 'form' of *Sphecophaga vesparum*	Idiobiont ectoparasitoid of social vespids
Orthocentrinae	Some *Stenomacrus* spp.	Parasitoids of Diptera larvae
Cryptinae	Various spp. of *Gelis* (see Fig. 13.30c), *Thaumatogelis* and *Polyaulon* may be completely apterous and some members of various other genera are brachypterous; often only the females are apterous	See Salt (1952)
Ichneumoninae	A female 'form' of *Ichneumon latrator*	

*See Belokobylskij (2009) for more details.

POSTPECTAL CARINA

The postpectal carina is a transverse ridge running across the mesopleuron just anterior to the insertions of the middle coxae. It is variously developed and very variable in most ichneumonids but is absent from most braconids and its presence in some groups has been deemed an important aid to identification. Thus, one of the key features often used for recognising members of the Cheloninae is not just that they (excluding those taxa previously placed in the Adeliinae), in addition to possessing a carapace, also possess a well-developed postpectal carina (see keys by van Achterberg 1976c, 1993a). However, the situation is not so simple because a distinct and fairly well-developed postpectal carina also occurs in the Cenocoeliinae, which have a very odd metasoma that is inserted on the propodeum far above the hind coxae and is therefore superficially similar to that of evanioids (see Fig. 12.58). It is also present in two genera of Opiinae (*Coleopus* and *Bitomus*), both of which have a carapace, in the Acampsohelconinae (see Fig. 12.44) (van Achterberg 2003) and in at least one species of the sigalphine genus *Minanga*, which also has a carapace (Quicke et al. 2008). It seems extremely unlikely that this correlation with possession of a carapace or otherwise highly modified metasoma is coincidental. Rather, Quicke et al. (2008) proposed that these features are functionally related. Something about having to manipulate a carapaciform metasoma, or a very highly inserted one in the case of cenocoeliines, may lead to selection for strengthening part of the mesosoma.

PROPODEAL SPINES

Unlike in the Chalcidoidea, numerous ichneumonids have a pair or spine-like processes on the dorso-lateral part of the propodeum. These are most obvious in various ichneumonines and cryptines within the Ichneumonidae and in various braconines, pambolines (see Fig. 12.20), rogadines and chelonines within the Braconidae. In the Ichneumonidae they are usually referred to as propodeal apophyses. Their precise location varies between taxa. They are probably mainly involved in lateral stabilisation of the metasoma when it is raised during the early stages of penetration of the ovipositor into the substrate.

'FOSSORIAL' LEGS

Species that have to push their way through soil or similarly resilient media typically have short and very robust legs and often also rather protruding faces. Not only are the coxae and femurs and often the tibias swollen, but also the tarsal segments are usually relatively short and the telotarsus large, and the ovipositor is usually not or hardly exserted beyond the end of the metasoma. Although only a few ichneumonoids are known to crawl through the ground, such as *Alomya*, which searches for hepialid moth larvae feeding on grass roots (Hinz & Short 1983), several others reach hosts in situations such as leaf-rolls by physically forcing their way in. Many metopiine ichneumonids, especially smaller bodied species, show this behaviour and they have a typical fossorial appearance with robust legs and protruding faces (see Fig. 13.46d) and are believed to push their way into host leaf rolls and leaf ties. The enigmatic ichneumonid genus *Hyperacmus* (see Fig. 13.13c), which appears to belong to a clade including taxa that attack Diptera larvae and many Orthocentrinae, are strongly built, suggesting that instead of probing their substrate with a long ovipositor (which most do not have), they may actually crawl through the leaf litter or ground after them. Similarly, the tersilochine *Barycnemis blediator* attacks the burrowing dwelling larvae of the salt marsh rove beetle, *Bledius specabilis*, which tries to protects its young from the parasitoid (Wyatt & Foster 1989).

Numerous braconids have robust legs, particularly in the cyclostome lineage (van Achterberg 1995). Although never observed in life, notably taxa with very robust legs are classified in the Betylobraconinae *Betylobracon* itself and some other Rogadinae such as Yeliconini, *Bulborogas* and various *Aleiodes* (B.A. Butcher et al. 2012). Of these, only *Yelicones* has been reared and it attacks pyralid and crambid moth larvae which live in silk webs. Thus the swollen legs in these may well be associated with having to push their way through the silk to reach their hosts. In some members of the doryctine genus *Psenobolus*, the males are highly modified and live within fig syconia where they do battle, and their legs no doubt help them force their way between flower remains and in grappling with other males. Outside of the cyclostomes they are found in the Planitorini (Euphorinae) and enigmatic, prognathous Masoninae (see Fig. 12.73a,b), the latter seemingly being associated with the ground layer, so perhaps they use their legs for burrowing.

FORE TIBIAL SPINES

The outer face and apex of the fore tibia may possess a longitudinal or transverse apical row of peg-like structures or particularly robust setae, sometimes referred to as chaetobothria. They occur virtually exclusively in species that parasitise wood-boring hosts and it is likely that they play an important role in allowing egress of the parasitoid from its pupation site. In the Braconidae, these have been attributed some phylogenetic significance. They are especially well developed in the cyclostome subfamily Doryctinae, in which they usually form a single longitudinal row along most of the length of the tibia, whereas in some genera of the related Braconinae they are mostly restricted to the apex of the tibia where they form a transverse row, but in the braconids Histeromerinae and the *Doryctomorpha* (Mesostoinae *incertae sedis*; see Fig. 12.15b) and females of the ichnumonid genus *Alomya* (see Fig. 13.29), the whole of the anterior side of the fore tibia is more or less covered in strong spines. *Histeromerus* females presumably use these to tunnel through soft wood (see Fig. 12.22d) and *Alomya* through the soil, in search of hosts; the biology of *Doryctomorpha* is unknown.

Within the Ichneumonidae, similar peg-like structures are found on the fore tibia of many pimpliformes that attack wood-borer hosts. Very similar spines also occur on the mid-tibia of MesoAmerican species of the banchine genus *Procestus*, whose biology is unknown, but their long ovipositors suggest that they too attack at least concealed hosts.

FORE TIBIAL APICAL TOOTH

This feature is largely restricted to the Ichneumonidae, where it occurs in widely disparate subfamilies, although predominantly in the Ctenopelmatinae, and so unless it is a widely lost symplesiomorphy, it has probably (and most parsimoniously) evolved on numerous separate occasions. Its presence in both the Oxytorinae and Tatogastrinae, now that they appear to just be derived ctenopelmatines, is therefore unlikely to represent a separate origin. Within the Braconidae, a very similar structure is found only in the Afrotropical Pseudohelconina (Helconinae) (van Achterberg 1990a). Although it was always recognised as showing homoplasy, it was for a long time used as a character for recognising the subfamily Ctenopelmatinae. It occurs also in Xenothyrini (Labeninae) and the Chilean tersilochine (Phrudini) *Notophrudus*.

EXPANDED HIND BASITARSI

A number of groups have species with markedly expanded, usually laterally compressed, hind tarsi. Most notably, these occur in three genera of braconids, *Podorgilus* in the Orgilinae (see Fig. 12.49b), the Afrotropical genus *Cratodactyla* in the Rogadinae (although only the male is known) and the relatively species-rich genus *Hartemita* in the Cardiochilinae (see Fig. 12.60). Among the ichneumonids, a moderate degree of expansion is seen in a number of anomalonines (e.g. *Therion* and *Heteropelma*; Gauld 1976b) in which the hind tarsi are also rather pale (see Fig. 13.50c) and conspicuous in flight, and in males these have a depressed area that may be associated with a gland. In addition, the largest hybrizontines, *Eurypterna* spp., have conspicuously enlarged and flattened hind basitarsi (see Fig. 13.55a). The possibility that Hybrizontinae and Anomaloninae are closely related and that some aspect of this character is synapomorphic cannot be completely excluded.

TOOTHED HIND FEMUR

A small number of species/genera in rather disparate groups have evolved a flattened tooth on the mid-ventral margin of the hind femur. These include the braconid genera *Helconidea* and *Wroughtonia* in the Helconinae, *Euscelinus* in the Doryctinae, *Braconella* in the Braconinae, *Leptotrema* in the Alysiinae and in the Ichneumonidae, *Odontocolon* in the Xoridinae, *Dentifemura* and some *Jezarotes* in the Acaenitinae, *Lissopimpla* and *Odontopimpla* in the Pimplinae and *Pristomerus* (see Fig. 13.54) and some *Eiphosoma* in the Cremastinae. It is tempting to assume that this has something to do with egress from the cocoon or its environs, and several of these genera are associated with wood-borer or other concealed hosts, but no-one has observed this

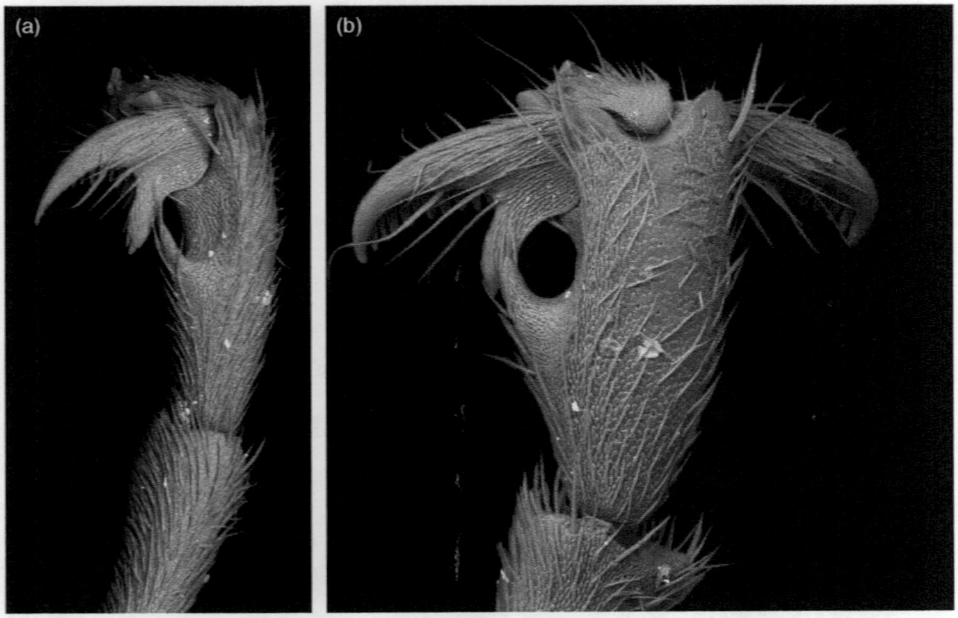

Fig. 10.6 Two views of distitarsal scraper of *Ophion zerus*. (Source: reproduced by permission of Gergely Várkonyi.)

feature in use. Another possibility for some instances is that they may play a role in ovipositor stabilisation, since the majority of the taxa with toothed femurs are associated with borer hosts and have ovipositors approximately the length of the body. However, this is not the case in the xoridine *Odontocolon* (see Fig. 13.7b), or in the case of *Pristomerus* (Cremastinae) whose steerable ovipositor does not have to penetrate anything very hard.

DISTITARSAL SCRAPER

W.R.M. Mason (1976a) was the first to draw attention to the presence in a small number of braconids of a modification of either the fore or hind telotarsus that has on its distal inner side a concavity bordered posteriorly by a tooth. He illustrated this structure in the hind telotarsus of the macrocentrine braconid *Austrozele* (as *Macrocentrus*) *uniformis* and noted that a similar structure was also present in another macrocentrine genus, *Dolichozele*, and a similar structure is also found on the fore telotarsus of various microgastrines. He proposed that it probably served a grooming function but no observations of it in action had ever been made. Gergely Várkonyi and I conducted a more extensive survey after finding an almost identical structure (Fig. 10.6) in an Australian ophionine wasp that we identified as *Ophion zerus* and also in members of several genera of microgastrines.

PECTINATE CLAWS AND CLAWS WITH ANGULAR BASAL LOBES

A common modification observed in many endoparasitoids of both families is increased development of the pecten spines such that they are of approximately equal length to the claw proper and extend right up to it, pectinate claws (Fig. 10.7b). These are particularly well developed in all or many members of groups such as the Ophioninae. Claws with a large, angularly produced basal lobe (Fig. 10.7c) occur in a wide range of taxa, sometimes being sexually dimorphic (notably in Pimplinae). Some Agathidinae show a sort of compromise condition with well-developed pecten basally

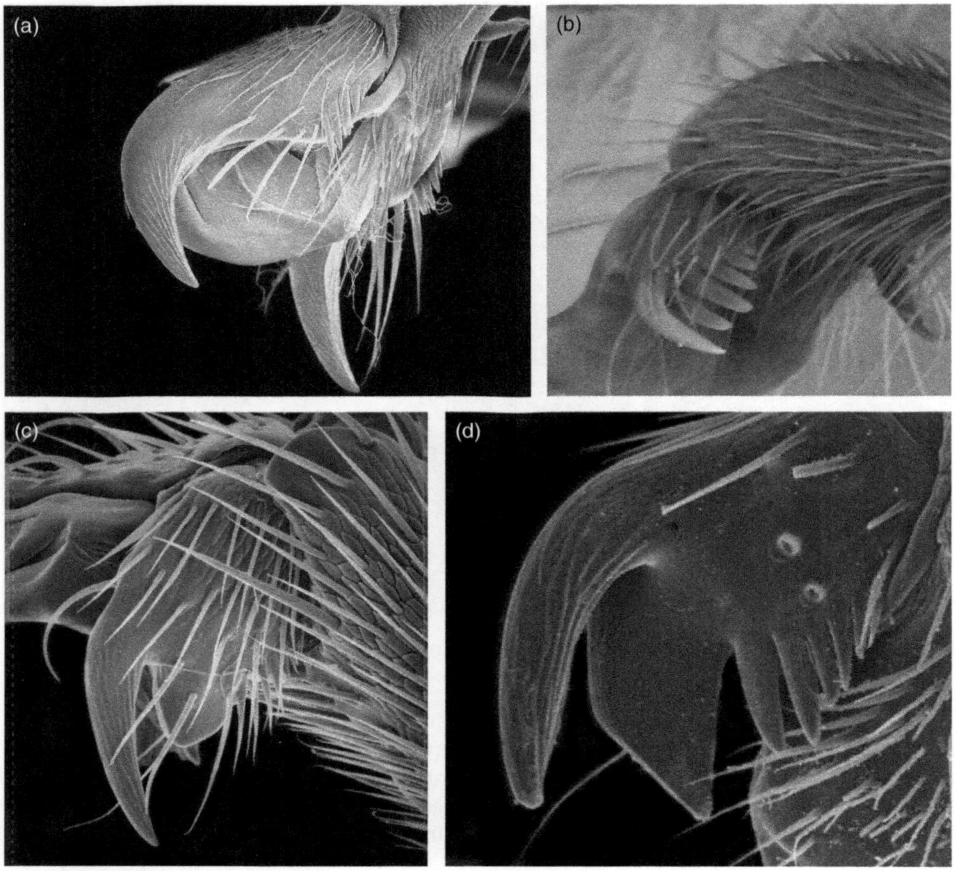

Fig. 10.7 SEM images of tarsal claws showing common different forms. (a) Large claw proper with small, rounded basal lobe and few peri-basal pectin teeth (Labeninae: *Labena grallator*); (b) with pectin highly developed, teeth of similar length to claw proper and extending to claw proper (Rogadinae: *Yelicones napo*); (c) basal lobe large and pointed (Agathidinae: *Austroearinus rufofemoratus*); (d) pectin modified with apical tooth broader, note how the apical tooth is offset towards the inner side of the claw (Agathidinae: *Cremnops* sp.). (Source: a, c and d reproduced by permission of Emily Tucker.)

and a square-ended large apical pectin tooth before the claw proper (Fig. 10.7d). Rarely, claws may be bifurcate with what might be derived either from the distal part of the basal lobe or a single, large, apicomedially situated pecten tooth.

Pectinate claws are found exclusively in koinobionts and probably in those that use their claws for holding on to, at least briefly, soft-bodied caterpillar type hosts and serve to increase contact and grip with the host cuticle while not piercing it (M.R. Shaw 2004b). The function of angularly produced basal lobes is unknown, as is that of bifurcate claws.

GLABROUS WING PATCHES AND WING MEMBRANE SCLEROMES

Although perhaps most widely known in the case of the anteriorly glabrous patch in the submarginal cell of the Ophioninae, which in the case of the common nocturnal genus *Enicospilus* (and a few other ophionines) possesses one or two conspicuous scleromes, many other, principally nocturnal or crepuscular, species, especially in the Braconidae, show highly modified wing cells with large glabrous patches and sometimes also with scleromes in them. Several cells can be involved. In the Rogadinae, some genera have the apical part of the fore wing sub-basal cell moderately [*Aleiodes* (*Hemigyroneuron*): see Quicke & Shaw 2005a, as *Pholichora*] or extremely expanded and surrounded by thickened veins (*Gyroneuron*). Molecular phylogenies suggest that these represent separate evolutionary events. Many *Megalommum* (see Quicke et al. 2000c, as *Curriea*) have the second subdiscal cell ovoid with a sclerome at the anterior of the longitudinal wing fold. The same types of modification are also observed in the hind wings of some species. The large nocturnal Afrotropical genus *Amicrocentrum* has the apex of the basal cell abruptly widened and glabrous (see Fig. 12.46b), and similarly the hind wing marginal cell of *Aulacocentrum* (Macrocentrinae) is widened basally and glabrous. The hind wing marginal cell of *Pseudoyelicones* (Rogadinae) is extremely modified, ovoid with a thickened curved vein 1RS posteriorly, and the wing membrane in this cell is corrugated.

It seems likely that all these variants have the same function or functions, associated somehow with nocturnality, but what it may be is not certain and there are no experimental data. Quicke and Shaw (2005a) postulated that the curved and thickened veins might be something to do with a generally lesser degree of sclerotisation and cuticular strength in these yellowish ophionoid taxa, although that would not explain the scleromes that some additionally have, or necessarily the glabrous membrane. A student suggested an alternative explanation that, at least in those with scleromes and a glabrous toughened wing membrane, these might be related to sound production or reflection, and as females of some nocturnal taxa such as ophionines and *Netelia* in the Tryphoninae can deliver painful stings, if there is a auditory aspect to these adaptations perhaps it is to warn bats.

CARAPACISATION

In many braconids, the first three tergites become fused and enlarged and can comprise the whole of the visible metasoma, although the reduced more posterior segments are still present, but concealed. This feature is typical of all of the subfamily Cheloninae and misled several early workers wrongly to classify several other taxa with carapaces in that group, for example, the braconine *Physaraia* and several genera of Helconinae: Brachistini (*Triaspis*, *Schizoprymnus*). In addition to these taxa, three-segmented complete carapaces occur in a number of Doryctinae (e.g. *Chelonodoryctes*), most members of the Lysiterminae, a few Rogadinae (e.g. *Aleiodes butcheri* and *A. darlingtonia*) and the microgastrine genera *Fornicia* and *Buluka* and in alysiine (*Gastralysia*).

Dudarenko (1974) made a detailed morphological study of the carapace in various groups of Braconidae, concentrating on the numbers of tergites involved and the distributions of spiracles, and concluded that carapaces had evolved on several occasions, although even before that detailed study, the error of placing all carapaced braconids in the cheloninae was realised. Van Achterberg (1988b) indicated that possession of a carapace was associated with oviposition either into a hard host or through a hard substrate, such as oviposition into plant galls or insect eggs (Cheloninae) (see also Wharton 1993a). Certainly these are usually hard substrates, but wood is also hard yet taxa that oviposit through wood tend to have a long, thin metasoma, so hardness cannot be the whole story.

In many cases the carapace is composed of just the first three metasomal tergites (Table 10.3), but some other robust carapace-like structures may include four or more tergites (e.g. in the *Spinaria* group of Rogadinae, see Fig. 12.27b), and there is obviously a continuum from normal metasomas to fully carapacised ones. A notable example of a four-segmented carapace occurs in a large number of New World *Aleiodes* species (which were formerly placed in their own genus, *Tetrasphaeropyx*). However, as their biology is only known to the level of host group, with no observations of oviposition having been made, it is impossible to interpret the function of the carapace.

Unlike the braconids, carapacisation has only rarely evolved within the Ichneumonidae. The most notable exceptions are two genera of Cryptinae (*Rothneya* and *Hemiteles*) and the aberrant banchine genus *Townesion* (see Fig. 13.38).

PETIOLATE METASOMAS

Basal narrowing of the first metasomal tergite is widespread and gives rise to the name petiole frequently used both in the Ichneumonoidea and elsewhere to refer to this segment. It is associated with increased mobility of the metasoma, principally in a dorso-ventral plain, that aids ovipositor manipulation. Once basally

Table 10.3 Taxonomic distribution of carapaces formed of just three or four terga and associated biology (when known).

Taxon	Biology and notes
Ichneumonidae	
Cryptinae: *Rothneyia*	Unknown
Banchinae s.l.: *Townesion*	Unknown
Braconidae	
Cheloninae: all genera although desclerotised in *Adelius* group (formerly Adeliinae)	Ovo-larval parasitoids of Lepidoptera, host eggs usually partly concealed
Rogadinae: *Aleiodes* (*Tetrasphaeropyx*), *A. butcheri*, *A. darlingtonia*	*Tetrasphaeropyx* is a parasitoid of Lepidoptera larvae, biology of the others unknown
Doryctinae: *Chelonodoryctes*	Unknown but surmised to be parasitoids of wood/twig-dwelling Coleoptera larvae
Braconinae: *Physaraia*	Reared from beetle larvae inhabiting plant galls
Lysiterminae: *Acanthormius*, *Lysitermus*, *Neolysitermus*, *Tritermus*, *Tetratermus*	
Exothecinae: *Aulosaphes*	Unknown
Alysiinae: *Hylcalosia*, *Trachionus*	Unknown but surmised to be parasitoids of Diptera larvae
Opiinae: *Coleopius*, *Bitomus*	Treated as synonyms by Wharton (1988) but major differences in venom apparatus and also adult morphology suggest separate origins
Microgastrinae: *Fornicia*, *Buluka*	The first attacks Limacodidae, which have tough integuments; the biology of *Buluka* is not known
Helconinae: Brachistini: *Triaspis*, *Schizoprymnus*	Egg-larval and possibly larval parasitoids of beetles living in wood or seeds
Basal Helconini group: *Spasskia*, *Chelonohelcon*†	
Acampsohelconinae: all genera	Only known for *Urosigalphus*, which attacks bruchine chrysomelid larvae in beans
Sigalphinae: all genera, complete fusion in *Minanga*	Not known for *Minanga*, but others attack early instar Lepidoptera larvae
Trachypetinae: *Trachypetus*	Unknown but inferred to be endoparasitoid (Belshaw et al. 2003)

narrowed, the first tergite has also evolved to become long and tubular, almost parallel-sided, in several separate subfamilies, notably in the Doryctinae (see Fig. 12.16b,e) and Euphorinae within the Braconidae and Tersilochinae (see Fig. 13.48) among the ichneumonids. In the first of these, it was a character used to define tribes and, indeed, a group including *Spathius* with this feature were sometimes regarded as constituting a separate subfamily and certainly a tribe. Molecular analysis has shown, however, that this feature is highly homoplastic and the genera treated in the Spathiini and morphologically similar Stephaniscini are not largely unrelated and not monophyletic (Zaldivar-Riverón et al. 2008).

MODIFICATIONS TO THE POSTERIOR METASOMAL MARGIN

These may take the form of a more or less sharp and deep emargination of the mid-posterior margin of the last heavily sclerotised and exposed metasomal tergite, and/or a pair or more of teeth or other protuberances. As with the formation of a carapace, modified posterior tergites are almost entirely restricted to the Braconidae and in particular to the Helconinae, Acampsohelconinae, Cheloninae, Lysiterminae, Rogadinae and Braconinae, but interestingly in the last of these, they are found also in a number of taxa which have relatively normal, although robust, metasomas. They can take the form of medial and/or submedial spines or emarginations and they are often present in those taxa that have a carapace, although they can also be found in taxa that have a more or less robust but not obviously carapacised metasoma. Within the Braconinae, they occur in *Soter*, *Spinadesha*, *Furcadesha*, *Trigastrotheca*, several Aspidobraconina, *Physaraia*, *Monocoila*, *Chelonogastra*, *Testudobracon*, *Pseudospinaria*, *Trispinaria*, *Rhytimorpha*, *Schiztobracon* and *Serrundabracon* (see figs 262–270 and 273 in Quicke 1987a) and these represent at least 10 separate origins! Unfortunately, biological data are available for only an handful of these. In contrast to the braconids, very few ichneumonids display spines or teeth on the posterior metasomal margin, and *Sachtlebania* and *Townesion* (Banchinae: Townesionini) have a pair of spines posterolaterally on the posterior of the carapace (see Fig. 13.38).

SPERMATHECAL COLOUR

Most parasitic Hymenoptera have unpigmented or pale-coloured spermathecae, but within both the Ichneumonidae and Braconidae there are taxa that have either dark red, red–brown or black spermathecae. Among the Ichneumonidae, dark spermathecae are found predominantly in the Labeninae, pimpliformes and some lower ophioniformes (character 131 in Quicke et al. 2009), in the Braconidae they are found in nearly all non-cyclostomes but only in a few cyclostomes. Thus, in the Ichneumonidae, dark spermathecae are associated with being ectoparasitic, whereas the opposite is the case in the Braconidae.

COMPRESSION OF APICAL PART OF METASOMA

Several groups in which the metasomal apex is used as part of the oviposition mechanism have the terminal abdominal segments strongly laterally compressed in order that they can be inserted into the host's hiding place, typically a gall or similar structure. This modification is found in the ctenopelmatine genus *Seleucus* (see Fig. 13.45), some diplazontines (e.g. *Phthorima*, strongly laterally compressed; *Campocraspedon*, strongly dorso-ventrally depressed), the orthocentrine *Neurateles*, the alysiines *Aspilota sylvaticae*, *Chorebus cultratus*, *Eucoelinidea compressa*, *Mesocrina mycetophila* and *Sarops* spp. which search for hosts between the gills of fungi, and the rogadine *Aleiodes compressor*, which inserts the apex of its metasoma into concealed places where it attacks a variety of different host groups (M.R. Shaw 2002b). In the ctenopelmatine *Homaspis*, the same functionality is achieved by its elongated and depressed (dorso-ventrally compressed metasoma) (see fig. 60 in Townes 1969).

THE 'OPHIONOID FACIES'

Gauld and Huddleston (1976) first proposed the existence of a suite of characters that are associated with being a nocturnal or crepuscular parasitic wasp. The most obvious of these were a general brown–yellow coloration, long antennae and large ocelli (see Figs 10.8, 12.46a and 12.47). Since then, a number of other features have been added to the list. The most interesting of these is that almost all species with this facies are koinobionts; indeed, at the time of Gauld and Huddleston, all were, although since then some braconines [members of the *Megalommum* (= *Aphrastobracon*) group of genera] are also crepuscular or nocturnal and display at least some of the typical characters associated with nocturnality. That they are koinobionts is because by being nocturnal they are able to access hosts that are not available (or not as available) during the day such as the many and various caterpillars that hide during the day but come out to feed at night. What the presumably idiobiont *Megalommum* group wasps are doing is less clear. The one available detailed host record is of *Megalommum* itself, which attacks the caterpillar of a *Eublemma* species that in turn is a predator of lac insects. Many members of the *Megalommum* group of genera have strangely modified ovipositors (Quicke 1987a); some species have the apex of the upper valve upturned, some have it spatula-shaped, some have the apical 0.25 sharply and strongly angled ventrally (as *Curriea*; Quicke et al. 2000c). All of these adaptations seem likely to be associated with inserting the ovipositor through a tight gap that maybe they have to prise open. If this is the case, we might imagine that the hosts of at least some species are hidden in portable cases, as for example are those of the pamboline genus *Notiopambolus* (Zaldivar- Riverón & Quicke 2002; see also p. 234).

Fig. 10.8 Four examples of ichneumonoids displaying the ophionoid facies with brown coloration, large eyes and ocelli and long antennae. (a) Macrocentrinae gen. sp.; (b) three *Netelia* sp. (Tryphoninae), at light sheet; (c) *Cystomastax* sp. (Rogadinae); (d) five *Ophion* sp. (Ophioninae) on underside of leaf, attracted to light. (Source: reproduced by permission of Kenji Nishida.) For a colour version, see Plate 16.

Although most species described as having an ophionoid facies are medium-sized to large wasps, light traps frequently attract considerable numbers of small, yellowish ichneumonoids such as members of the chelonine genus *Phanerotoma*, meteorideines and some euphorines. Although no details are known, meteorideines most probably attack larval lepidopteran hosts, but chelonines are all, so far as is known, egg-larval parasitoids and so it is not at all apparent what advantage these *Phanerotoma* would gain from being nocturnal and the possibility must exist that at least some attack host larvae as opposed to their eggs.

It should be noted that not all nocturnal ichneumonoids display this facies – for example, various braconids of the Charmontiinae, Blacinae, Euphorinae, Microgastrinae and Rhysipolinae are frequently taken in light traps but are small and dark (Lozan 2002).

WHITE ANTENNAL STRIPES AND TIPS

Many Ichneumonidae but only a few braconids have very conspicuously white-marked antennae (e.g. see Figs 10.2, 12.27a, 12.42c,d, 13.8a, 13.37b and 13.57a). In many ichneumonids, the stripes are found in both sexes, but when only present in one sex it is invariably in the female, e.g. *Ichneumon sarcitorius* (Ichneumoninae) (see Fig. 13.31b; cf. Fig. 13.31c), *Tycherus nigridens* (Ichneumoninae: H.D. Smith 1932), *Certonotus* (Labeninae: Gauld & Holloway 1986), many cryptines, rhyssines and xoridines. Sexually dimorphic antennal colour is also observed in some *Labena* species (Labeninae), but in this case the males have contrasting black-tipped antennae whereas those of the female are unicolorous (Gauld & Wahl 2000a).

In the Braconidae, white-striped antennae are less common but occur in some Helconinae s.s. and s.l.,

Rogadinae, Pambolinae (see Fig. 12.20), Doryctinae and white-tipped and -striped ones in a very few Braconinae. The Afrotropical braconine genus *Kimavu* has white-striped antennae in all known species.

In ichneumonids, the white antennal marking is usually a post-medial stripe, but among braconids the white (or buff) zone may be medial or apical. This pattern is found almost exclusively in tropical taxa, including various Mesostoinae (see Fig. 12.14a), Doryctinae (Wild et al. 2013), Pambolinae (see Fig. 12.20) and Alysiinae such as some *Phaenocarpa* and *Asobara* species (these are closely related genera in the *Phaenocarpa* group) and their relatives. In the Neotropical rogadine genus *Pseudoyelicones*, the terminal flagellomere is often whitish, contrasting with black penultimate antennomeres. Many Neotropical braconines also have white- or yellow-tipped antennae. Many of these species are found flying low over the leaf litter of moderately dense forests.

White-striped antennae seem to be widely distributed among biologies including idiobionts and koinobionts, and also taxa that attack deeply concealed hosts such as wood- or stem-borers, weakly concealed and even exposed hosts. M.R. Shaw (1980; referred to in Quicke 1984) suggested that they may be involved in providing the female wasp with visual feedback about the separation of her antennae, the example given being that of the cryptine *Mesostenidea obnoxious*, which is a pupal parasitoid of the burnet moth, *Zygaena filipendulae*. The female wasp moves her antennae along the sides of the host cocoon and seems to be using their separation as a guide to host size (perhaps to make a decision about sex allocation). As antennae have many articles, it is easy to imagine that summing up a lot of sensory feedback from stress receptors (if they occur) would not be a very accurate way of assessing separation, whereas the wasp can look directly at the white stripe. Of course, this cannot be the role in wood-borer parasitoids, but it is noticeable that in those ichneumonids that use vibrational sounding the antennae are frequently angled directly after the end of the white band (Fig. 10.2). Interpreting the function of white stripes is difficult as their distribution is probably confounded by mimicry, and white-tipped and -striped antennae are present in numerous small tropical Lepidoptera, Heteroptera and even Coleoptera species and the white fore tarsi of some tropical Micropezidae (Diptera).

Together with B. A. Butcher, I have found that a very high proportion of individuals, particularly males, of members of the *Aleiodes coronarius* group (Rogadinae) collected in Malaise trap samples had one or both antennae broken within the white stripe zone and therefore wonder whether this might indicate combat had taken place. In this group, there are fairly clear species distinctions in the position and extent of the white stripe and in some species, only the apical few flagellomeres of males are white, suggesting that the stripe might be used as part of a species recognition system, but nothing is known about the behaviour of these animals.

WHITE OVIPOSITOR SHEATH STRIPES AND TIPS

Some taxa that inhabit relatively shaded or dappled habitats and oviposit into deeply concealed hosts have the apical part of their ovipositor sheaths with conspicuous silvery-white setosity. This is very common among the Braconinae but occurs only rarely among ichneumonids, for example, in the labenine genus *Apechoneura*, the banchine *Lissonota lissolabis*. Hanson et al. (2008) noted that ovipositor-probing *Apechoneura* females hold their ovipositor sheaths straight out behind them and commented that Gauld (2000) and Gauld and Wahl (2000a) suggested that this might have evolved as an attack deflection device, drawing a potential avian predator's attention to the opposite end of the wasp as the more vulnerable white-striped antennae. Members of the Afrotropical braconine genus *Monilobracon* (see Fig. 12.32b), when ovipositor-probing into their lymexeliid beetle larval tunnels, hold their white-tipped ovipositor sheaths straight up rather like a flag (Quicke et al. 2005b). In this genus, individuals on a tree trunk or log frequently exhibit both inter- and intraspecific aggression, with the member of the larger species usually winning without actually fighting, so it is possible that the white tip serves as an easily measured signal related to size, and in these wasps the antennae are not white striped although they are apically yellow in some species (see Fig. 12.32b).

Quicke (1984) plotted the length of the white sheath apical region against the overall ovipositor length for a number of species of Braconinae, although without phylogenetic control, in order to test five hypotheses relating to the function of the coloration, including intra- and interspecific signalling, attack deflection, providing visual feedback during substrate penetration and warning coloration. The finding that those species with less rigid metasomas that probably bend them ventrally during ovipositioning (viz. *Bathyaulax* and *Zaglyptogastra*) had relatively shorter black basal parts to the sheathes supported the visual feedback hypothesis, while the slope of the increase in the length of black basal part against body length supported the attack deflection hypothesis. It is possible that this colour feature has several different functions. It would be interesting to extend this study across a far greater range of taxa with appropriate phylogenetic control.

NUMBER OF LARVAL INSTARS

The plesiomorphic number of larval instars for parasitic wasps is hard to be certain about, partly because of a dearth of evidence for some of the putatively basal taxa such as the stephanids and megalyrids, but dogma has it that it is around five or more, and S.R. Shaw (2004) considered five to be the plesiomorphic number for the Braconidae. All published records of numbers of instars have to be treated with caution, as it is very easy to be misled and numerous values in the literature are likely to be wrong (M.R. Shaw & Huddleston 1991). For example, the degree of sclerotisation of items like the mandibles, and hence perceived mandible size, can vary within an instar (O'Donnell 1987), or in those endoparasitoid braconids that leave their hosts either to continue feeding externally or simply to pupate, a moult might take place as the larva exits the host and the shed skin may be left 'plugging' the exit hole, as in the Microgastrinae (see Fig. 12.67). Further, some reports might have been influenced by dogma.

The number of larval instars is typically four or five in ectoparasitic ichneumonoids, four being observed only in the Ichneumonidae, irrespective of body size, but many endoparasitic species seem to have a reduced number, usually three or four, although five have been reported in some cases (Table 10.4). The reason for this reduction, if it is a true pattern, since no comparative analyses have been carried out, can easily be envisaged to result from endoparasitoid larval cuticles having fewer restraints since they do not have to prevent desiccation or penetration by pathogens and so may have greater elasticity. S.R. Shaw (2004) discussed the fact that whereas some (or perhaps all) *Meteorus* species (Euphorinae: Meteorini) have only three larval instars, which can be taken to be a derived state, the remaining euphorines have between three and five instars, suggesting that either they gained additional ones or that parallel reductions could have occurred in different lineages.

EGG-LARVAL PARASITISM

Although there are no egg-parasitic ichneumonoids, egg-larval parasitism is widespread, especially within the Braconidae (Clausen 1956), and has undoubtedly originated many times; in fact, even within the microgastrine genus *Cotesia*, it appears to have multiple origins. Egg-larval parasitism appears to be obligate to the Pionini (Ctenopelmatinae) and Collyriinae among the ichneumonidae and in the Cheloninae, Ichneutinae and probably all Helconinae (Brachistini) in the Braconidae. It has also been recorded in a few species of Opiinae and Alysiinae. Clausen (1956) reported it in the campoplegine ichneumonid *Campoletis perdistinctus*, although Gangrade (1964) indicated that this species oviposits into 2–4-day-old first instar hosts. However, it has now been shown to occur in odd species in two other campoplegine genera (*Hyposoter*, Castelo et al. 2010; *Melalophacharops*, Shimazaki et al. 2011) and in at least one species of *Ctenopelma* (Ctenopematinae: Ctenopelmatini) (Barron 1981). In all cases of ovo-larval parasitism in the superfamily, the host is not killed until it has virtually completed its larval development.

DISC-LIKE LARVAL ANTENNAE

Larval antennae range, in the Ichneumonoidea, from being distinctly segmented, protruding papilliform structures, through being flattened discs to being effectively indiscernible; it is possible that they are truly absent in some taxa. All ectoparasitoids, as far as I am aware, have papilliform antennae, but these are reduced in nearly all endoparasitoids, which often have just a flat disc or a disc with just a small central pair of sensilla, or rarely a small knob representing the papilla. Because of the strength of this correlation, it has been tempting to use larval antennal information to infer life history when the latter is incompletely known. Thus, for example, Gnamptodontinae have long been known from larval skins, but it is not known for certain whether they are ecto- or endoparasitoids of their leaf-miner hosts (Shaw & Huddleston 1991). Their final

Table 10.4 Numbers of larval instars and biology (dubious records are given in parentheses).

Ectoparasitoids		Endoparasitoids	
Taxon	**No. of larval instars**	**Taxon**	**No. of larval instars**
Braconidae			
Braconinae	5	Agathidinae	3
Doryctinae	4–5	Alysiinae	3–4
Exothecinae	5	Aphidiinae	3–4 (5)
Pambolinae	5	Cardiochilinae	4
		Cheloninae	3–4
		Euphorinae (Meteorini)	3
		Euphorinae (others)	3–5
		Macrocentrinae	5
		Microgastrinae	3–4
		Opiinae	4–5
		Orgilini	3
		Rogadinae	5
Ichneumonidae			
Cryptinae	4–5	Campopleginae	3–4 (5)
Pimplinae (Ephialtinae)	(3) 5	Collyriinae	?5
Rhyssalinae	4	Cremastinae	3–4
Tryphoninae	5	Ctenopelmatinae	5
Xoridinae	4	Ichneumoninae	4–5
		Metopiinae	5
		Ophioninae	3
		Pimplinae (Pimplini)	5
		Tersilochinae	4–5

instar larval antennae are distinctly papilliform (Čapek 1970), which suggests that they are likely to be ectoparasitoids, but considerable circumstantial evidence suggests that they must be endoparasitic. Indeed, papilliform antennae have been retained in some endoparasitic taxa and that is definitely the case with *Stilbops* (Stilbopinae) (Short 1978).

Another group for which we do not know the biology is the Brachycyrtinae. As Wahl et al. (2002) noted, its final instar larvae have several features typical of ectoparasitoids, including mandibles with denticulate dorsal margin, presence of a labral sclerite and spiracular closing mechanism separated from atrium by a length of trachea, but it would be good also to know whether their larvae had papilliform or disc-shaped antennae since no antennae were visible in any of the larval exuviae preparations available. Perhaps antennae are absent, in which case they are most likely endoparasitoids. Of course, there may be exceptions, as seems likely to be the case with Gnamptodoninae. Whether the disc-shaped antennae of some endoparasitoids in which no sensilla seem to be present have any sensory function is unknown, since they do not appear to bear any sensilla, and perhaps their role is taken over by the palps which always have distinct sensilla.

REDUCTION OF LARVAL HYPOSTOMAL SPUR

Gauld (1984b) noted that the hypostomal spur of the final instar larva is reduced in two ichneumonid subfamilies, the Anomaloninae and the Metopiinae, and postulated that this is because neither spins a large cocoon. Both anomalonines and metopiines are larval-pupal parasitoids and make only flimsy cocoons within the host pupa, which otherwise provides them with plenty of physical protection. Gauld argued that this is because the hypostomal spur represents the insertion point of the muscles needed to move the larval head from side to side, an important ability in cocoon spinning.

WIDE AND HEAVILY SCLEROTISED LARVAL EPISTOMA

This feature was noted by Wahl (1988) as a feature of the genus *Exetastes* in the Banchini and also of Anomaloninae, Metopiinae, Ichneumoninae and Pimplini, all of which (except *Exetastes*) pupate within the pupa of the host, so this character seems likely to be associated with that habit and possibly the fact that they spin little or no cocoon.

SUSPENDED COCOONS

Although cocoons are protective, they are easily susceptible to attack by pseudohyperparasitoids and possibly by other predators, so they are usually constructed somewhere concealed. Exceptions abound, however, and in those cases where ichneumonoids make their cocoons in exposed places they may well have protective adaptations even although these are not always either obvious or successful. Many are coloured with black and white transverse stripes which may be disruptive (see Fig. 13.51b,c). Very similar suspended and patterned cocoons have evolved among the campoplegine ichneumonids and in the meteorine braconids.

Sobczak et al. (2012) described how the suspended cocoon mass of an unidentified *Meteorus* species (Braconidae: Euphorinae) is nevertheless heavily pseudohyperparasitised (by a pteromalid chalcidoid) and Harvey et al. (2011) carried out an experiment to test whether suspended cocoons of *Meteorus pulchricornis* provided defence against the ichneumonid pseudohyperparasitoid *Gelis agilis*. In the latter case, cocoons glued directly to the plant suffered no more parasitism then suspended ones (Fig. 10.9). Similarly, the suspended mummies of Afrotopical members of the *Aleiodes buzurae* group (Braconidae: Rogadinae) are also heavily hyperparasitised by idiobiont chalcidoids and also *Mesochorus* (Mesochorinae), which would have attacked the *Aleiodes* larva earlier (Quicke et al. 2006, unpublished observations), so the protection afforded is wholly or largely likely to be avoidance of insect and avian predators rather than parasitoids, as Shirai and Maetô (2007), again with *Meteorus pulchricornis*, demonstrated in the case of ant predators.

Some species of the microgastrine genus *Pholetesor* also make suspended cocoons, but with the cocoon suspended between threads at both ends within their host's leaf mine. Whitfield (2006) postulated that this might serve to help protect them from avian predation, although I would guess that it would also be effective against pseudohyperparasitoids attempting to oviposit from outside the mine, as it would almost certainly make it harder for their ovipositors to gain purchase on the cocoon.

POLYEMBRYONY

Polyembryony, in which a single original 'embryo' divides to produce multiple (from two to many hundreds) of separate embryos is known in only a handful of parasitic wasps, most notably in the Encyrtidae (Quicke 1997a). However, it also occurs in at least one braconid, the genus *Macrocentrus* of the Macrocentrinae. It is possible that the few other genera of Macrocentrinae are also polyembryonic but they simply have not been studied. Gregarious *Macrocentrus* species such as the well-studied *M. cingulum* produce either all male, all female or mixed broods. In solitary species of *Macrocentrus* (although it has to be admitted that only one species has been studied), the zygote divides into two, giving rise to two or more morulae, but only one continues to develop to form an embryo (Daniel 1932, Finney et al. 1947). Circumstantial evidence suggests that at least a couple of other ichneumonoids might be polyembryonic. In Europe, *Lissonota mutator*, a banchine ichneumonid and one of the few gregarious members of the family, consistently produces single-sex broods of up to about nine wasps (M.R. Shaw 1999a), and the same is found in the Nearctic braconid *Sigalphus bicolor* with broods of up to 30 individuals (Clausen 1940).

PHYTOPHAGY AND CECIDOGENESIS

Whereas there are many transitions to secondary phytophagy, mostly through gall-forming or inquilinism, among the Chalcidoidea, it is far rarer in the Ichneumonoidea. Among the ichneumonids, there are no known gall-formers, but there is an instance of partial phytophagy in the case of the some Labeninae,

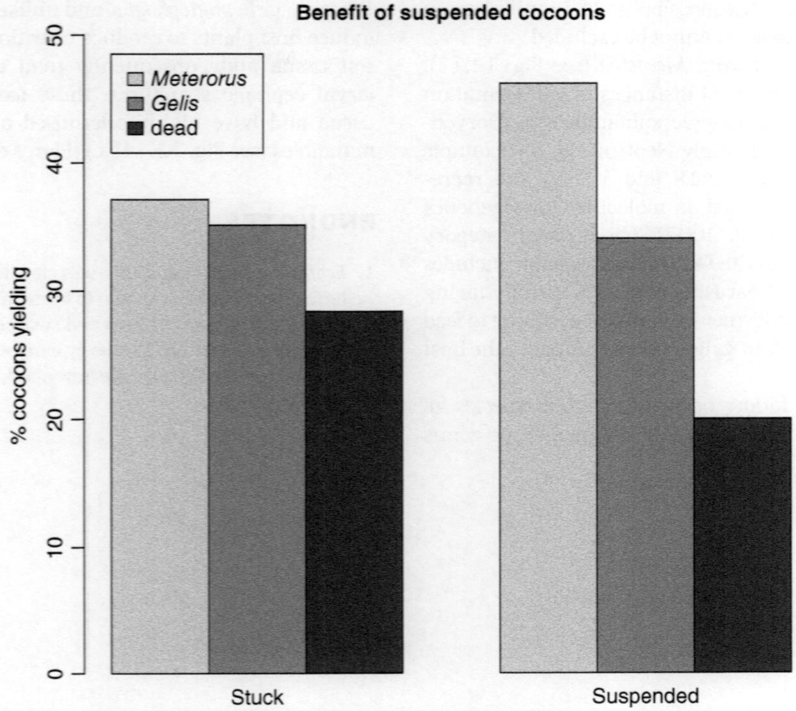

Fig. 10.9 Experimental test of effect of suspended cocoons of *Meteorus pulchricornis* (Euphorinae) on pseudoparasitism by the cryptine *Gelis agilis*, involving attaching cocoons directly back on the host plant and showing that that protection is not increased from parasitoidism by this hyperparasitoid. (Source: data from Harvey et al. 2011.)

which are parasitoid-come-predators of solitary bees, and in addition to consuming the 'host' bee larvae also eat the pollen cake and depending on the species this may involve the destruction of several cells (Slobodchikoff 1967). True cecidogeny occurs in several groups of braconids but is unknown within the Ichneumonidae.

Gall formation appears to have evolved on at least three occasions within the Braconidae and some evidence suggests that it might have arisen as many as six times (Wharton & Hanson 2004), although it was not known (or suspected) until 1989 when de Macêdo & Monteiro (1989) conclusively showed that a Neotropical species of the doryctine genus *Allorhogas* consumed plant tissue seeds of legume *Pithecellobium*. Although the behaviour was initially referred to as 'seed predation' (Wharton & Hanson 2004), it was subsequently shown that the wasp larva induces gall formation in the seeds and that it is the soft, nutritious gall tissue that the wasp larva consumes (de Macêdo et al. 1998). Other *Allorhogas* species have been reared from galls on *Quercus* (Fagaceae) and *Stryphnodendron* (Fabaceae) and have never been associated with any other gall-forming possible host (Marsh et al. 2000), although parasitism or inquilism cannot be ruled out for all species. In these the wasp lays her eggs in immature seeds with plentiful, soft and nutritious endosperm. In the case of *A. spermaphagus*, multiple individual wasps can develop on a single seed. Subsequently, seed-associated gall formation has been demonstrated for two other *Allorhogas* species and there is reasonable evidence that several other species may do the same, with some being associated with Bignoniaceae and Melastomataceae (Wharton & Hanson 2004). Various other *Allorhogas* species are known definitely to be idiobiont parasitoids of gall-forming cecidomyid flies in the Neotropics. Rearings from cynipid galls on oaks in the Nearctic might indicate that they can also be parasitoids of the cynipids, although the possibility of inquilinism in some of these situations cannot be excluded.

Interestingly, apart from *Mesostoa* (see Fig. 12.13), all of the other confirmed instances of gall formation involve members of a single subfamily (the Doryctinae) and all are exclusively Neotropical, for example *Monitoriella* (see Figs 12.18 and 12.19) and represent a single origin based on molecular phylogenetics (Zaldivar-Riverón et al. 2007). The broader category of gall association in the Doryctinae probably includes many that are strict parasitoids and some that, having devoured their gall-former host, might continue to feed on the nutritious plant gall tissue surrounding the host remains.

True seed predation occurs in a few species of neotropical braconine currently assigned to the genus *Bracon*, e.g. *B. phytophagus*, and unlike gall feeders that induce host plants to produce nutritious and relatively soft tissue and consequently tend to have reduced larval cephalic structures, these feed on hard seed tissue and have highly sclerotised and multidentate mandibles (see Fig. 12.34b,c) (Flores et al. 2005).

ENDNOTES

1. *Echthrus*, which at that time was classified in the Hemigasterini, has since been shown to be member of the Cryptini.
2. Indeed, the general habitus and swollen tibia led *Echthrus* to be classified within *Xorides* by early workers.
3. Members of the Groteini are now also known to use vibrational sounding.

Part II

Taxonomic and Systematic Treatment

Chapter 11

OVERVIEW OF ICHNEUMONOIDEA: RELATIONSHIPS AND SYSTEMATICS

The Braconid and Ichneumonid Parasitoid Wasps: Biology, Systematics, Evolution and Ecology, First Edition. Donald L. J. Quicke.

MONOPHYLY OF ICHNEUMONOIDEA, ICHNEUMONIDAE AND BRACONIDAE

Ichneumonoids are built to a fairly standard plan and, although many individual taxa show reductions to various parts, they collectively display many features that might be considered plesiomorphic within the Apocrita. However, morphological evidence for ichneumonoid monophyly is not particularly impressive (Sharkey & Wahl 1992). There are six putative synapomorphies for the superfamily: (i) adult mandible with two teeth, (ii) prepectus fused to posterior of pronotum, (iii) sternum of mestasomal segment one divided into a sclerotised anterior and a membranous posterior part (all from W.R.M. Mason 1981b, 1987), (iv) metasomal tergite one articulated to tergum 2+3 by a postero-lateral pair of condyles, (v) fore wing costal+subcostal and radial veins adjacent and adpressed (i.e. forming a more or less unified C+SC+R vein) and (vi) hind wing with only one r-m cross vein. Rasnitsyn (2002) additionally proposed the medially short pronotum immovably connected to the rest of the mesosoma and loss of medial mesoscutal sulcus and its internal apodeme, first tergite forming a heavily sclerotised plate and hind wing vein 1M straight as synapomorphies. The last of these is not true of various extant Ichneumonidae and interpretation of some of these is dependent upon the phylogeny, since Rasnitsyn depicted the ichneumonoids as the sister group to the aculeates, which is no longer tenable (see *Relationship of Ichneumonoidea to other Hymenoptera*, below). A long time ago, the Stephanidae, and sometimes also the Megalyridae, were regarded as belonging to the same group, but it has been realised for a very considerable while that they are not closely related to the Ichneumonoidea and are probably both more basal groups within the Apocrita (Sharkey & Wahl 1992, Wahl & Sharkey 1993).

Ichneumonoids tend to have long, many-segmented antennae and relatively complete wing venation in the groundplan. The loss of the fore wing costal cell with vein C running adjacent to or completely and indistinguishably fused to SC+R and combined vein being referred to as C+SC+R is probably the most instantly recognisable feature. There is a distinct but very narrow costal cell apically in the Cretaceous Praeichneumonidae† (Fig. 11.1), some Cretaceous Ichneumonidae (e.g. *Khasurtella* in the Tanychorinae†; see Fig. 13.2a), the Australian Maxfischeriinae and the large, enigmatic Australian Trachypetinae (tentatively placed in the Braconidae). Although the above morphological characters of the superfamily do not seem particularly strong, DNA sequence data support monophyly of its extant members. Having said that, however, phylogenetic reconstruction requires a root and that is usually the sister group of the taxon concerned, and determining the sister group of the Ichneumonoidea has been a major issue as so few characters seem to provide evidence (see below).

In modern usage, a few anomalous or fossil taxa have still been treated as separate families by various workers, notably the 'Agriotypidae' within the Ichneumonidae and 'Aphidiidae' and 'Apozygidae' within the braconids. Each of these is now generally regarded as constituting a subfamily, although in my opinion, the Apozyginae are likely to turn out to be very basal and part of the 'extinct' Eoichneumoninae† or Protorhyssalinae†, which are each probably paraphyletic.

Morphological phylogenies incorporating various extinct ichneumonoids have unavoidably been based on very limited numbers of characters, almost entirely of wing venation features, or, if including soft tissue features, have many missing entries for extinct representatives (e.g. Sharkey & Wahl 1992, Perrichot et al. 2009). The resulting trees therefore tend to lack much resolution (Fig. 11.2) and sometimes portray obviously erroneous relationships such as the recovery of the Hybrizontinae with the braconids by Perrichot et al., which was almost entirely a consequence of one character. It is unlikely, at least in the near future, that sufficient information will be forthcoming about these deep fossil groups to be able to define what genera form monophyletic groups, and therefore a strict, well-supported phylogenetic picture is unlikely to be obtainable and the best that might be expected is that a group of largely extinct taxa are accorded grade status and may most sensibly be referred to as the 'Eoichneumoninae†'.

RELATIONSHIP OF ICHNEUMONOIDEA TO OTHER HYMENOPTERA

For a long time, the Aculeata were taken to be the most probable candidate (e.g. Quicke et al. 1999a), largely because they share with the Ichneumonoidea specialised articulated flaps called valvilli (see Fig. 3.12a–c), usually one or two, on each lower valve of the ovipositor/stinger (W.R.M. Mason in widely circulated,

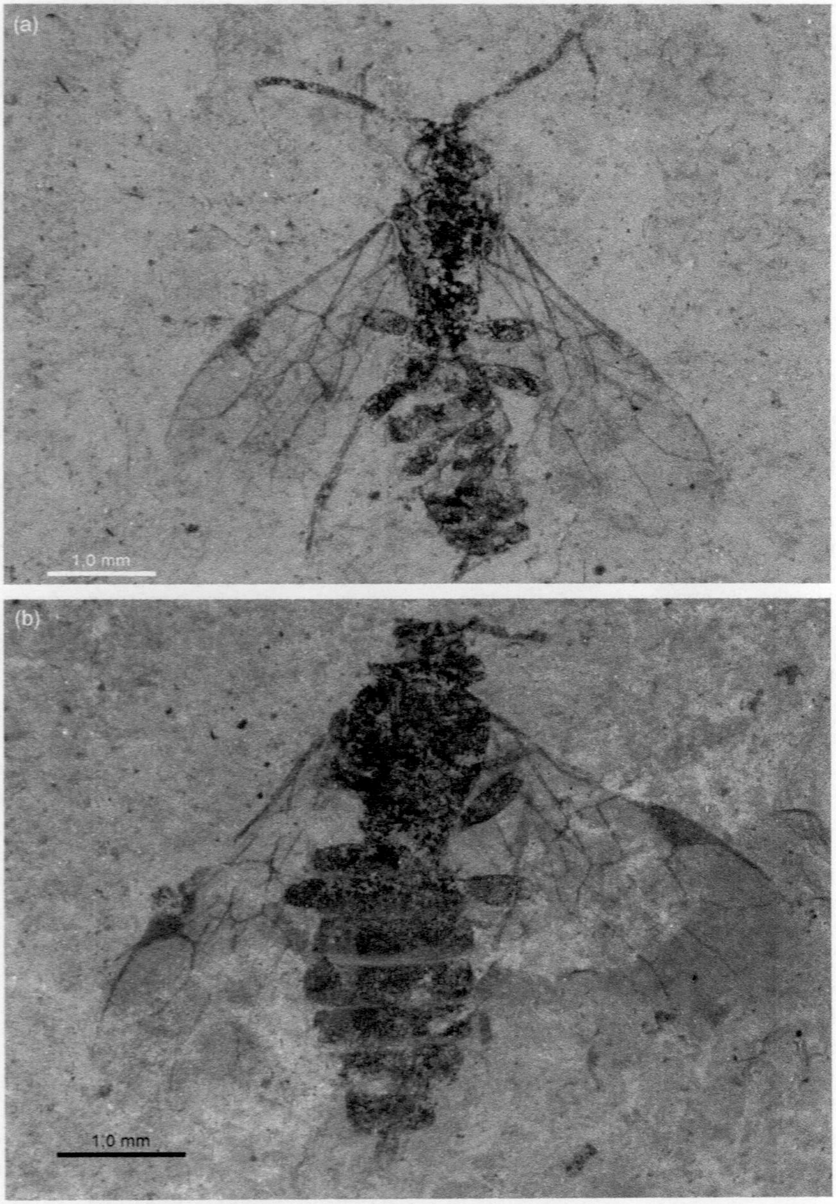

Fig. 11.1 Impression fossils of two recently discovered and better preserved species of the extinct family Praeichneumonidae†. (a) *Praeichneumon dzhidensis*†; (b) *P. khamardabanicus*. (Source: reproduced by permission of Dmitry S. Kopylov.) For a colour version, see Plate 17.

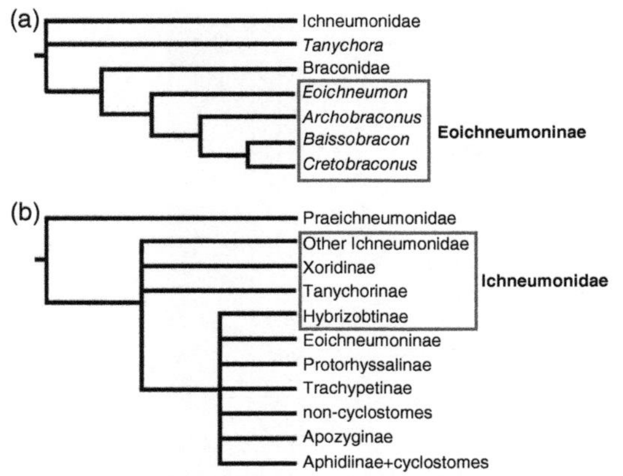

Fig. 11.2 Summaries of morphological phylogenies of the Ichneumonoidea. (a) Extant family summaries plus *Tanychora*† and the four genera assigned to Eoichneumonidae at the time (from Sharkey & Wahl 1992) that are now considered to be braconids, the tree shown being the most resolved of their maximum parsimony cladograms, the other two differing only in their relationships among the Eoichneumoninae genera; (b) summary of >50% bootstrap tree of Perrichot et al. (2009), in which Eoichneumoninae† represents the genera recognised by Sharkey and Wahl (as Eoichneumonidae†) plus *Aenigmobacon, Protorhyssalus* and *Protorhyssalodes*.

unpublished course notes). Comprehensive molecular phylogenies of recent years have not supported this and, in the most comprehensive sets of analyses to date, namely those of Sharkey et al. (2012) based upon 392 morphological characters and sequence data for four gene loci and of Klopfstein et al. (2013) based on two ribosomal and four protein-coding gene regions, the isolated position of the Ichneumonoidea is reaffirmed and, albeit with rather low support, they appear to be the sister group of the Proctotrupomorpha (comprising the Proctotrupoidea s.s., Chalcidoidea, Mymmaromatoidea, Platygastroidea, Diaprioidea and Cynipoidea) and not at all close to the aculeates.

FOSSIL HISTORY AND FAMILY-LEVEL PHYLOGENY

In addition to the two extant families of Ichneumonoidea, two other fossil families have been included, the Praeichneumonidae† (Rasnitsyn & Quicke 2002) and the Eoichneumonidae† (Jell & Duncan 1986). The latter is now considered to be a subfamily of Braconidae. Whether the Praeichneumonidae†, which is known only from rather few and often poorly preserved impression fossils from Lower Cretaceous of Mongolia and Russia (Transbaikalia) (Rasnitsyn 1983, 1990, Kopylov 2012a), is truly a member of the Ichneumonoidea is not totally certain, although in recent analyses it appears as the most basal known branch of the lineage. It was included within the Ichneumonoidea because of the reduced (but still distinct) space between C+SC and R (costal cell). The Praeichneumonidae have a weakly open costal cell, four submarginal cells in the fore wing (i.e. in addition to 2RS, two rs-m cross-veins are present), an areolated propodeum (Fig. 11.1) and, unlike all extant ichneumonids, their hind wing vein RS curves to and rejoins R on the wing margin far from the wing apex. No extant ichneumonids or braconids have four submarginal cells apart from some teratological specimens, especially in the latter. Kolarov's recent description of three new species based on better preserved fossils does nothing to contradict the idea that they are the sister group to (or possibly paraphyletic with respect to) the rest of the Ichneumonoidea.

The Eoichneumonidae† was originally based on a poorly preserved and fragile fossil from the Lower Cretaceous of Koonwarra fossil bed, Victoria, Australia (*Eoichneumon*) (Jell & Duncan 1986). These taxa appear well supported within the superfamily. A number of additional genera and species have been added based on impression fossils from Siberia and Mongolia (Rasnitsyn & Sharkey 1988) and subsequently, another poorly preserved impression fossil, *Purichneumon* was described from the British Duriston Bay beds near Swanage (Rasnitsyn et al. 1998).

Sharkey & Wahl (1992) provided a first formal cladistic attempt to define the group's monophyly and the relationships of the two currently recognised extant families along with the Eoichneumonidae†. In addition to refuting previous suggestions that Megalyridae and Stephanidae belonged with the ichneumoids, they treated the extant Hybrizontidae (as Paxylommatidae) and Apozygidae as members of the Ichneumonidae and Braconidae, respectively. However, they also excluded the Praeichneumonidae†, which were known at that time only from a few very poorly preserved fossils, because of a lack of clear evidence that they belonged to the superfamily, with their placement there relying heavily on only the very narrow fore wing costal cell. In Sharkey and Wahl's study, the Eoichneumonidae† were recovered as the sister group of the Braconidae but, as pointed out by Quicke et al. (1999a), their result is based on only two characters and one of these may have been misinterpreted. The data set used by Sharkey and Wahl was necessarily very minimalistic and has been criticised by Quicke et al. (1999a), who presented a more detailed study in which extant representatives were additionally scored for some soft tissue and molecular characters. The results of the new analyses were, however, still rather unsatisfactory.

The discovery of various amber fossils of three quite basal braconid genera, *Protorhyssalus* from the New Jersey Late Cretaceous and *Aenigmobracon* and *Protorhyssalodes* from the Mid Cretaceous of France, permitted a morphological cladistic analysis incorporating more characters, more fossil ichneumonoids in addition to extant taxa (Perrichot et al. 2009) and with the extant ichneumonids and braconids not treated as single terminal taxa. Although not all the relationships in the most parsimonious trees were likely to prove correct in the future, it was clear that the Eoichneumonidae† form a clade with the Braconidae (although with no bootstrap support), and because they are only distinguished from the latter by symplesiomorphies, Perrichot et al. duly synonymised them with the Braconidae constituting the subfamily Eoichneumoninae†.

Both the Ichneumonidae and Braconidae first appear in the fossil record in the very early Cretaceous, approximately 120 million years ago or possibly even at the latest Jurassic (H.C. Zhang & Rasnitsyn 2003), in fact, at the same time as the Praeichneumonidae†. The earliest known fossil braconids are listed in Table 11.1 and include various eoichneumonines, some of which are certainly from the very late Jurassic, and, in the modern sense, the genus *Eobraconus* from the lower Cretaceous (Rasnitsyn 1983 as *Eobracon*[1]). The placement of *Eobraconus* within the Braconidae is probably correct and is indicated by the apparent fusion of the second and third metasomal tergites and by the lack of fore wing vein 2m-cu. The poor state of preservation of the fossil makes further assessment difficult, although a few visible character states such as the origin of vein r-rs beyond the middle of the pterostigma may point to the plesiomorphic status for the family. Most of the early ichneumonid fossils belong to just three genera, *Tanychora*, *Tanychorella* and *Paratanychora*. Most of the ichneumonoid fossils are very uncommon and from the described material (although very little Braconidae from the Cretaceous has been worked on) they were not very morphologically diverse. However, it is not even absolutely certain that the above three genera are monophyletic with the Ichneumonidae in that several of the trees obtained by Quicke et al. (1999a) recovered *Tanychora* as the sister group of the Eoichneumonidae†, although I strongly suspect that they are stem ichneumonids.

The late Eocene Baltic amber, approximately 37–34 Mya (Rasnitsyn & Quicke 2002), contains many fossil ichneumonoids, nearly all of which, if well-enough preserved, can be placed with reasonable confidence in recent genera and include exclusively members of extant subfamilies (van Achterberg 1982a), although unfortunately many specimens were destroyed during the World War II and several private amber collections were broken up or lost at around that time, which has hampered study (van Achterberg 1982a). Baltic amber, and similarly aged, Late Eocene, Bembridge shale deposits (Antropov et al. 2014), contain a mixture of species, many clearly belonging to extant genera, but also some that are suffiently distinct to be recognised as representing probably extinct genera (see, for example, van Achterberg 1982a). However, most of the generic determinations of earlier workers cannot be taken at face value and many specimens are not sufficiently well preserved to permit completely certain identification. Nevertheless, Baltic amber braconids represent a very different fauna from any current one, being rich in euphorines, doryctines and various small cyclostomes that possibly belong to Cedriini or Chremylini, groups that are not commonly collected today and especially not in the Palaearctic. Many common groups in the present-day fauna, such as Braconinae and Microgastrinae, are extremely rare.

Studying amber inclusions of the size of some of the braconids is tricky. The traditional method of using very strong sucrose syrup of approximately the same refractive index as the amber itself and then placing a glass coverslip on top is still good, but recently workers have started using CAT scanning and synchrotron imaging both to obtain higher resolution external images and in some extremely well-preserved specimens, such as

Table 11.1 Early Cretaceous and extinct Cretaceous ichneumonoid fossil taxa.

Taxon	Provenance of fossils	Lagerstätten	Estimated age (Mya)	Ref.
Praeichneumonidae				
Praeichneumon	Mongolia	Lowermost Lower Cretaceous		Rasnitsyn 1983
Praeichneumon	Russia	Lowermost Lower Cretaceous		Rasnitsyn 1990
Praeichneumon	Russia	Lowermost Lower Cretaceous		Kopylov 2012b
Ichneumonidae				
Amplicella	Russia	Baissa	127–137	Townes 1973a
	Russia	Khasurty	130–145.5	Kopylov 2011
	China	Yixian	121–125	Zhang & Rasnitsyn 2003
Khasurtella	Russia	Khasurty	130–145.5	Kopylov 2011
	China	Laiyang	130–145.5	J.F. Zhang 1991
Megachora	Russia	Baissa	127–137	Kopylov 2010
Paratanychora	Mongolia	Anda-Khuduk	125–133	H.C. Zhang & Rasnitsyn 2003
Tanychora	Russia	Zara	130–145.5	Townes 1973a
	Mongolia	Bon-Tsagaan	113–125	Kopylov 2010a
	China	Yixian	121–125	H.C. Zhang & Rasnitsyn 2003
Tanychorella	Russia	Romanovka	130–145.5	Rasnitsyn 1975
	China	Yixian	121–125	H.C. Zhang & Rasnitsyn 2003
Eoichneumoninae				
Cretobraconus	Russia	Baissa	mostly 127–142	Rasnitsyn & Sharkey 1988
	Mongolia	Khotont	98.9–142	Rasnitsyn & Sharkey 1988
	Mongolia	Khutuliin-Khira	upper Jurassic	Rasnitsyn & Sharkey 1988
Eoichneumon	Australia	Koonwara, ?Aptian	113–125	Jell & Duncan 1986
Archobraconus	Russia	Baissa	127–137	Rasnitsyn & Sharkey 1988
Baissobracon	Russia	Baissa	127–137	Rasnitsyn & Sharkey 1988
Purichneumon	UK	Clements beds	139.8–145	Rasnitsyn et al. 1998
Protorhyssalinae				
Magadanobracon	Russia	Obeshchayushchiy	93.5–99.6	Belokobylskij 2012
Protorhyssalodes	France	Albian-Cenomanian amber	94–113	Perrichot et al. 2009
Protorhyssalus	USA	New Jersey amber	92	Basibuyuk et al. 1999
Protorhyssalopsis	Spain	Albian	105.5–113	Ortega-Blanco et al. 2011
incertae sedis				
Aenigmobracon	France	Albian-Cenomanian amber	94–113	Perrichot et al. 2009
Eobraconus	Mongolia	Khutel-Khara	125–145	Rasnitsyn 1983
Gen. sp. indet.*	China	Jiufotang	?129	H.C. Zhang & Rasnitsyn 2003

*This specimen was misidentified by Hong (1988).
Source: data from H.C. Zhang & Rasnitsyn (2003) and Menier et al. (2004).

the one shown in Fig. 11.3, even internal musculature, although the resolution obtained is limited by the size of the amber block and paring amber down to minimise block size sometimes results in catastrophic amber fracture. However, often modern scanning techniques allow, albeit again with limitations imposed by sample size, far greater resolution.

BRIEF HISTORY OF CLASSIFICATION

Although there will no doubt continue to be some shuffling of difficult-to-place taxa between subfamilies, and with increasing phylogenetic assuredness the erection of a few extra, small or monotypic subfamilies to allow a fully natural classification, the basic subfamily-level

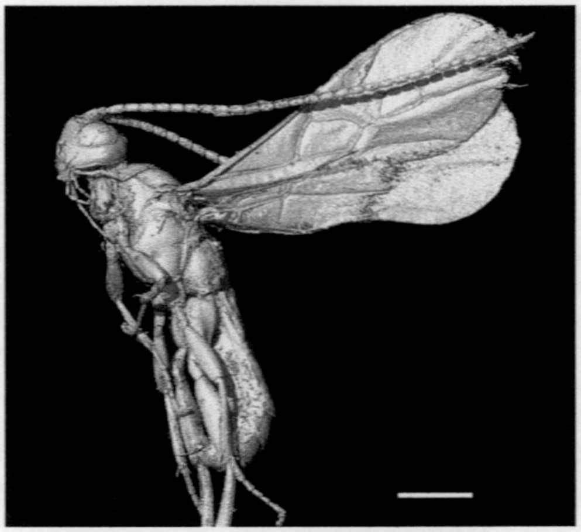

Fig. 11.3 Synchrotron image of Baltic amber braconid fossil. Scale bar: 0.5 mm.

classification of both Ichneumonidae and Braconidae is probably not going to change very drastically from now onwards. What we have now as a classificatory framework is the result of a large body of work over the past 100 or so years and is very different from what was used even 50 years ago.

The main reasons for giving a brief overview of the changes are so that the reader may get a feel about what of the older literature, which often contains an enormous amount of valuable and detailed observations, might be relevant today. In addition, it should serve to emphasise that using a typological approach, i.e. classifying taxa based on the possession or absence of what were perceived at the time to be 'good' diagnostic characters, can lead to all sorts of problems. Unfortunately, but excitingly, evolution does not always shape organisms to make classification or identification easy.

The close relationship of braconids and ichneumonids was almost always recognised in the early literature. Indeed, in works of the early part of the 19th century, the braconids were referred to as the 'Ichneumonides adsciti' in reference to the absence of fore wing vein 2m-cu. Currently, only two extant families are recognised, the Ichneumonidae and Braconidae, although it is possible that with additional phylogenetic understanding this could change slightly, especially if the placement of Trachypetinae shows that it belongs to stem Ichneumonidae rather than Braconidae, which I think is perfectly likely. Even fairly recently, a few groups whose placement now seems certain, either as braconids or ichneumonids, have been afforded separate family status and the reasons for the most important of these mistakes are outlined below. The groups involved are the Agryotypinae and Hybrizontinae in the Ichneumonidae and the Aphidiinae and Apozyginae in the Braconidae.

The braconid subfamily Aphidiinae, which many workers on aphid parasitoids have treated (and occasionally still do) as the Aphidiidae, have an unusual biology, often a very specialised morphology and in particular they have an articulation between the second and third metasomal tergites that is flexible (see Fig. 12.6a), whereas an inflexible fused syntergum is considered a synapomorphy for the Braconidae, although with some secondary reductions. The oviposition behaviour of most Aphidiinae, whereby the metasoma is very flexible and bent forward between the female's fore legs to achieve the act, is most likely to have led to a secondary loss, or at least a reduction, of the fused situation.

The Hybrizontinae, over which there is also a nomenclatural issue, was for many years referred to as the Paxylommatinae (van Achterberg 1976b, but see W.R.M. Mason 1981b for a discussion of why they should now be called Hybrizontinae), are a morphologically very peculiar group and all extant species have lost fore wing vein 2m-cu, which was considered a very important diagnostic feature of the Braconidae, but fossil taxa show its gradual reduction (Kasparyan 1988). A more important consideration is the hind wing venation, which has the rs-m cross-vein nearly interstitial to the split between R1 and RS, but as Sharkey and Wahl (1992) pointed out, vein rs-m is still distinctly distal to this (see Fig. 13.55d), and the confusion arises partly because RSa is not only short but also nearly transverse, although it is more distal in some fossil species (Kasparyan 1988, van Achterberg 1999). Another important character not discussed by Sharkey and Wahl is that hybrizontines have a distinct flexion line running through rs-m (Fig. 13.55d; indicated by a bulla), a character lost in all Braconidae. *Neorhacodes* (Tersilochinae: Neorhacodini), an unambiguous ichneumonid, does indeed have hind wing vein rs-m opposite the R1 and RS split, although not basal to it, and interestingly Neorhacodini also lack fore wing vein 2m-cu, suggesting that there is a functional connection between these features which most likely has to do with hovering flight and allowing the distal half of the wings greater capacity for distortion (see also Chapter 13, section *Anomaloninae*).

W.R.M. Mason (1981b) believed that the difference in hind wing venation between Ichneumonidae and Braconidae was that the distally located r-m cross-vein of the former was actually homologous to 2r-m of those sawflies that have two R to M cross-veins, whereas the that in the Braconidae was vein 1r-m, i.e. the two families had each retained a different r-m vein. He therefore concluded that Hybrizontinae, considering them not to have his 2r-m, shared no apomorphies with the Ichneumonidae and therefore he chose to regard them as a separate family. As Sharkey and Wahl (1992) pointed out, no extant Apocrita retain two hind wing r-m cross-veins, so it is implausible that the ancestral ichneumonoid did, thereby permitting the differential loss of r-m veins in the two extant families; rather, the vein is homologous in both. Based on the distal position of rs-m in the Praeichneumonidae (Fig. 11.1b, see the right-hand side) (Kopylov 2012a), it seems most probable that this vein has merely shifted to a more basal position in the Braconidae.

Of potentially greater interest is the extant Australian endemic braconid subfamily Trachypetinae, represented by a small number of large, nocturnal wasps from remote areas, which were not considered in the analysis by Sharkey and Wahl (1992). These large wasps have really always been considered to be braconids because of their loss of fore wing vein 2m-cu, and despite appearing in purely braconid-based analyses (Quicke & van Achterberg 1990) as derived members of the group, are almost certainly very basal and arguably should constitute a separate family. In relation to the above argument about hind wing vein rs-m, trachypetines definitely display the ichneumonid and thus putatively plesiomorphic state of having the vein located distal to the R1 and RS split and have the same GC to AT base composition of the 28S rDNA gene (Belshaw et al. 1998).

ANCESTRAL BIOLOGY OF ICHNEUMONOIDEA

In interpreting the polarity of characters and understanding the life-history shifts that may have occurred in each family, it is necessary to have some idea about what the group's ancestral biology was. Much has been written assuming that the ancestral ichneumonoid (or ichneumonid or braconid) was an ectoparasitoid of a moderately or even deeply concealed, probably xylophagous, host (e.g. Gauld 1988a). The logic is simple and appealing, but what I want to do in this section is just to show that it is essentially pure dogma with very little evidence in favour of it. I emphasise that I am not trying to suggest that the closest common ancestor of Ichneumonidae and Braconidae was not an ectoparasitoid, or that its host was not concealed, but just to lay bare the evidence behind this assumption.

Unfortunately, nothing is yet known about the biology of the Trachypetinae, which, despite appearing as derived braconids in morphological analyses (van Achterberg & Quicke 1990), are almost certainly basal. These large wasps are now almost entirely restricted to the arid inland of Australia, and interpreting the ancestral biology of the group poses an interesting problem, since consideration of their ovipositor morphology strongly suggests that they are endoparasitoids (Belshaw et al. 2003). Endoparasitism almost certainly evolved very early in the evolution of the Braconidae, possible even in the early Cretaceous if a recently described fossil is truly a member of the Aphidiinae (Ortega-Blanco et al. 2009).

For the most part, early fossil ichneumonoids such as the Praeichneumonidae, Tanychorinae†, Eoichneumoninae† and Protorhyssalinae†, although having exserted ovipositors, do not have spectacularly long ones and most of the species involved are just a few millimetres long, so it is unlikely that they were parasitoids of wood-boring hosts. Vilhelmsen and Turrisi (2011) instead proposed that parasitoidism of wood-borers in the Ichneumonoidea is secondary and that only Orussidae and Stephanidae retain the ancestral biology of the Apocrita.

The Ichneumonidae display a far stronger association with other Hymenoptera as hosts than do the braconids, and very often hymenoptera hosts appear to form the backbone from which various groups diversify on to other host orders. The pimpliformes includes many taxa that are endoparasitoids of Diptera or Hymenoptera or ectoparasitoids of Hymenoptera, with very few specialising on Lepidoptera, and of the latter, the Pimplini attack the pupal stage, which is relatively poorly defended physiologically. The large endoparasitoid orthopelmatiformes+ophioniformes lineage has largely specialist Hymenoptera parasitoids strongly clustered at its base, viz Orthopelmatinae as endoparasitoids of *Diplolepis* gall wasps, Tryphoninae largely as koinobiont ectoparasitoids of sawflies, Ctenopelmatinae as endoparasitoids of sawflies and giving rise to the Mesochorinae, which are hyperparasitoids predominantly of ichneumonoids. The diversity of the biologies of the putatively basal Aphidiinae+Mesostoinae lineage at the base of the predominantly ectoparasitoid cyclostome side of the braconid tree really means that nothing much can be said about their ancestral biology from that shown by extant taxa, especially as the extant cyclostomes and completely endoparasitic non-cyclostomes appear to be sister taxa. It has to be assumed that many important ancestral lineages are extinct.

SEPARATING ICHNEUMONIDS FROM BRACONIDS

It is always interesting to see how easily (in nearly all cases) experts can separate groups and how difficult it often is for beginners. Separating ichneumonids and braconids provides a good example. Probably this results from three things: greater familiarity with the characters, of course, but more importantly an understanding of the *Gestalt* of the group and memory of the individual taxa. Nevertheless, it is interesting that it is usually possible to distinguish ichneumonids and braconids in flight even from a distance with considerable accuracy, no doubt indicating that there are subtle differences in flight mode, and maybe this results from, or is associated with, features of wing venation and wing flexion lines (see Fig. 2.6). In this respect, it is worth noting that tersilochine (especially Neorhacodini) (see Fig. 13.49c) and hybrizontine ichneumonids have aberrant wing venation for ichneumonids, which in some ways more closely resembles that of braconids, and they are not infrequently misidentified as the latter.

Keys for separating members of the families have been provided by, among numerous others, Gauld and Bolton (1988), Shaw and Huddleston (1991) and Wahl and Sharkey (1993). Table 11.2 summarises the main characters that distinguish the two extant families. As with nearly all characters in very large taxa, there are nearly always at least a few exceptions to the rule. For most specimens (and species), two wing venation features will suffice to distinguish the two families, and these are illustrated in Figs 2.3 and 2.4. Ichneumonids nearly always have a second and more distal m-cu cross-vein (vein 2m-cu) in the fore wing running posteriorly from the second submarginal cell, which is usually referred to as the areolet (indicated at least basally by vein 2RS), whereas this is absent in all braconids except the largely extinct Cretaceous Eoichneumoninae†, some other basal fossils (see Fig. 12.4b), and the very rarely collected monotypic Chilean genus *Apozyx*, which might well belongs to that group also. This vein is, however, also absent in extant members of the relatively uncommon ichneumonid subfamily Hybrizontinae (see Fig. 13.55d), in the Neorhacodini (Tersilochinae) (see Fig. 13.49c), the anomalonine genus *Ophionellus* (see Fig. 13.51b), and also in a few other aberrant genera of Cryptinae.

The apical shape of the hind wing basal cell also differs in a very nearly consistent way, with its apex receding in the Braconidae but produced distally in the Ichneumonidae (see Fig. 2.3; cf. Fig. 2.4). More explicitly, hind wing vein r-m in ichneumonids originates

Table 11.2 Summary of morphological differences between Braconidae and Ichneumonidae'.

Character	Braconidae	Ichneumonidae
Fore wing cross-vein 2m-cu	Almost always absent, but present in Apozyginae and rarely teratologically in Histeromerinae, Rhyssalinae and Helconinae	Almost always present, but absent in Hybrizontinae and a few tersilochines, cryptines and anomalonines
Hind wing cross-vein r-m	Almost always antefurcal, but virtually interstitial in Trachypetinae	Almost always postfurcal, rarely nearly interstitial (Hybrizontinae, Tersilochinae: Neorhacodini)
Hind wing vein C+SC+R	Fused along all its length except for Trachypetinae	Veins C+SC and R separate but running in close parallel from near wing base
Junction between terga of metasomal segments 2 and 3	Inflexible with suture, if present, strongly sclerotised, but ichneumonid-like in Aphidiinae	Flexibly joined

anteriorly after the split between R and RS whereas in braconids it originates before the split. The only exceptions occur in the Braconidae, with the Trachypetinae having the cross vein distinctly arising from RS (i.e. it is technically rs-m in these) and a very few other taxa including a few rogadines (see, e.g. Quicke et al. 2012b), sigalphines (see Fig. 12.40b), amicrocentrines (see Fig. 12.46b) and a few braconines, although in these, this is clearly a taxonomically locally derived state. Of course, a few other 'problematic' taxa exist with what might at first appear to be rather intermediate conditions and, as luck would have it, these are also taxa that might be confused on other grounds also. Some ichneumonids, such as *Barycnemis* spp. (Tersilochinae: Tersilochini), *Scirtetes* (Campopleginae), *Neorhacodes* (Tersilochinae: Neorhacodini) and hybrizontines, also have their r-m cross-vein (rs-m) originating anteriorly only a very short distance after RS separates, but as far as I know, no ichneumonids display the typical braconid condition with the junction basal to the R and RS split.

The articulation of the second and third metasomal tergites (sometimes called the syntergum) differs between ichneumonids and braconids in a largely consistent way, with that of almost all braconids having the tergites fused and immovable, whereas in ichneumonids their junction is flexible. However, members of the braconid subfamily the Aphidiinae also have them flexibly joined, and a number of braconids in the cyclostome group have the tergites very largely membranous, and so in a sense also flexible. It seems most likely that the aphidiine character state represents a reversal.

IDENTIFYING SPECIMENS

In addition to the characters mentioned above, there are several easy-to-use, illustrated keys that will enable a worker to determine whether they are dealing with an ichneumonid or a braconid, which in most cases is very easy even if they do not have wings (e.g., Gauld & Bolton 1988, Goulet & Huber 1993 pp. 359–362, Hanson & Gauld 1995 p. 390). The next stage is to identify them to subfamily and that is considerably harder. Fortunately, changes in publishing technology and costs means that most recent keys are well illustrated. Some, although not many, subfamilies also have a restricted geographic distribution, which can make identification from some parts of the world slightly easier. It is worth considering briefly why identification keys are sometimes daunting. Evolution has resulted in much convergence between taxa; in other words, many characters show homoplasy, and there are very few subfamilies that can be unambiguously recognised on the basis of just one or two 'diagnostic' characters. Experts, in fact, hardly ever have to resort to keys at this level as they often have a feel for the *Gestalt* of a group or recognise genera, and, in a sort of reverse order, knowing the genus means that they then know the subfamily to which it belongs. Another issue is that taxonomists generally like their keys to be comprehensive, which means that they allow for all of the very many exceptions, some of which are so rare that most workers are never likely to encounter them. Using identification keys takes practice, and it is helpful to have some guidance either in the form of a local expert who can check, verify or correct identifications or perhaps access to some correctly identified reference material or through attending one of the few, but very useful, identification courses that are run approximately annually. It is also sensible when keys give extra ancillary bits of information such as 'only found in Australia' to take note of these – especially if you have collected the specimen in your back garden in Derbyshire or New Jersey. Although most subfamilies are cosmopolitan, that does not mean that the characters that enable their members to be recognised are universally applicable. I think it is fair to say that subfamily level identification is less problematic for braconids than for ichneumonids. The latter seem to have a larger number of groups lacking clear apomorphies such that even experts have sometimes made mistakes (see, e.g., Broad et al. 2004). On the other hand, it may be that the classification of some parts of the Braconidae, as reflected in keys, is misleading.

The main keys that I would recommend for Braconidae are those of Sharkey (1993, 1997d) and van Achterberg (1990d, 1993a, 1997b). Van Achterberg's keys are the more comprehensive and, importantly, treat separately several subfamilies within the cyclostomes that are lumped under Rogadinae in Sharkey (1993) or Rogadinae and Hormiinae (Sharkey 1997d). The key of van Achterberg (1993a) is also for world subfamilies whereas that of van Achterberg (1990d) is only for the Holarctic region and that of Sharkey (1997d) is only for those subfamilies found in the New World.

For Ichneumonidae, the key by Wahl (1993c) is good and global in scope, although exactly where a few aberrant genera will run is uncertain and sometimes a character's alternatives are not all that clear-cut – but it should nevertheless work for most species. The key of Gauld (1984a) for subfamilies that occur in Australia will work for many in that region, and that of Gauld (1991) permits the identification of Costa Rican subfamilies with the proviso that beyond couplet 43 it only works for females. Earlier keys, for example those of Perkins (1959) and Townes (1969), are difficult to use, partly because they try to key out each subfamily at just one place, which results in extremely difficult couplets. The unillustrated key of Chiu (1984) for Taiwanese ichneumonid subfamilies also includes keys to genera.

Although there have been some subfamily changes since even the more recent of the above, once a subfamily level identification has been reached, it should then be fairly easy for numerous groups to find genus-level keys. I have tried to add references where such keys are available after each subfamily entry, but in some cases there are really no recent, comprehensive publications.

The next problem, and one that is essential to overcome, is identification to species level. The literature on host relationships in the Ichneumonoidea is replete with errors. M.R. Shaw (1994) and Noyes (1994) both detail the level of the problem and its causes, not least of which is the misidentification of the parasitoid. Identifying an ichneumonoid to species correctly is not something that can be done easily or without substantial practice and effort. Anyone attempting this for the first time has many opportunities to make mistakes, getting the subfamily correct, the genus and the species, and from experience I can pass on that even experts sometimes make mistakes. I would strongly advise those who wish to put names to just one or a few species of parasitoid to contact relevant experts and make sure that their specimens are properly mounted (see Chapter 16, section *Mounting specimens for taxonomic study*). If one is dealing with material from known hosts, that will help to narrow down possibilities, but one must also be careful not to fall into the trap that the specimen *must* be one of those recorded from that host and trying to force it. If, however, you intend to do a lot of identification, especially of unreared material, then you will have to get to grips with the literature, which, in many cases, will be full of arcane terminology, sometimes difficult to access and possibly in an unfamiliar language. These are considerable impediments (Gauld 1986a).

Both families have experienced a number of nomenclatural issues that can undoubtedly be confusing, especially when delving into some of the early or mid-20th century literature in which a lot of important (and often the only) biological observations have been published. In the section *Henry Townes (1913–90) and his idiosyncratic nomenclature* in Chapter 13 I cover one particular potential source of name confusion that relates only to the Ichneumonidae, but within the braconids there have also been changes in the application of some names that can confuse the uninitiated. In particular, what are now called the Agathidinae were for a while referred to as the Braconinae, and the compositions of the Rogadinae, Hormiinae, Exothecinae and Doryctinae in particular have all undergone major shuffles. All these are discussed in the relevant sections. It should also be noted that some other issues may be traced back to Viereck (1914), because for much of the preceding time there was no generally accepted or applied type concept and therefore many authors of generic names did not specify what species was the type or indicate clearly what species they were based upon when more than one was included within a genus. To fix the meaning of generic concepts, Viereck designated a type species for all the generic names that he was aware of for which no type species had been specified previously. However, Viereck did this very much as a paper exercise, so when more than one species was included in a genus at its original description, he simply applied the rule of designating that which was first mentioned, i.e. 'page priority'. This means that many type species are atypical and maybe did not actually belong to the same group as the majority.

Several works are of great importance for identifying genera and species, and I will mention just a few here. For braconids, van Achterberg (see References) in particular has published generic revisions with well-illustrated keys for many subfamilies, and these often also provide keys to known species. For ichneumonids, the series of works in the *Ichneumonidae of Costa Rica* series (Gauld 1991, 1997, 2000, 2002) serves well for that country and neighbouring regions and covers most subfamilies, but not yet the large and difficult Campopleginae, Ichneumoninae or Cryptinae. Of course, beyond that country it will often be necessary to consult other sources such as the keys by Henry Townes (see References).

ENDNOTE

1. The name *Eobracon* was preoccupied.

Chapter 12

PHYLOGENY AND SYSTEMATICS OF THE BRACONIDAE

The Braconid and Ichneumonid Parasitoid Wasps: Biology, Systematics, Evolution and Ecology, First Edition. Donald L. J. Quicke.

HISTORICAL PERSPECTIVE

Braconid subfamily level classification has evolved in a similar fashion to that of ichneumonids, with 19th century workers recognising just a handful of major groups each more or less defined by presence or absence of a distinctive diagnostic character. Thus there were the 'Cheloni', 'Sigalphi', 'Microgasteres', Agathides', 'Bracones' and 'Bassi', all created by Nees von Esenbeck (1819, 1834) (see Wharton & van Achterberg 2000 for a more detailed summary of early groups). Of these, as explained by Wharton and van Achterberg, only the names Bracones and Bassi conformed to the notion of taxa comprising exclusive groups of other genera, and so of his names they are the only ones that can be regarded as the basis for modern family group names. However, for reasons explained by Wharton & van Achterberg, we can ignore von Esenbeck's Bassi.

The next major work, apart from the creation of a family group name based on *Alysia*, now the Alysiinae, by Leach (1815), was that of Haliday, who created three family group names which have passed directly to modern subfamily usage as the Agathidinae, Sigalphinae and Aphidiinae. Braconid classification then went off at a tangent with the widespread adoption of five informal terms. The *braconides exodontes* comprised the current Alysiinae while the *braconides endodontes* were split into cyclostomes, areolaires, polymorphes and cryptogastres (Wesmael 1835). Of these, the term cyclostome is still widely used, including here, for a large assemblage of subfamilies, although it also includes some that have secondarily lost the condition. Wesmael's terms, not being based on valid generic names, have no standing in formal nomenclature. Foerster (1862) produced a far more thorough classification which recognised 26 groups that are roughly equivalent to subfamilies, and in his system these were given the suffix '-oidae'. Van Achterberg (1976c) provided a more detailed discussion of the various historic groupings.

Throughout the next 80 years or so, valid family-level taxa were gradually added to the list, but it was a time also of confusion with the concepts of several groups changing considerably – misinterpretations of what other workers meant by particular names due to poor communication and inadequate descriptions played their part. Thus, as pointed out by Wharton and van Achterberg (2000), the concept of the Sigalphinae changed from one centred on what we now call Cheloninae together with those Helconinae *s.l.* that similarly possess a three-segmented carapace, rather than the smaller and less commonly collected group related to *Sigalphus*. Even after recognition that the Chelonoidae (= Cheloninae) formed a separate group, the Sigalphinae was still taken to refer primarily to the *Triaspis* group of Helconinae *s.l.* that is now called the Brachistini, and the matter was cleared up only in 1914 by Viereck, who designated type species for nearly all genera which did not explicitly state them in their original descriptions.

Although informal terms, the concept of the cyclostome braconids, with the remainder of the family referred to as non-cyclostomes, has been a useful, if imperfect, system. Monophyly of the majority at least of the traditional non-cyclostomes (i.e. excluding Aphidiinae, Opiinae and Alysiinae) seems to be supported by a couple of reasonably strong character: (a) their spermatozoa are highly apomorphic, being short with a reduced tail region with somewhat incomplete axonemal structure and reduced, mitochondrial derivatives, and also in having a large extracellular-like structure over the acrosome and nucleus (Quicke et al. 1992b), although this latter feature is shown by some ichneumonids (Fig. 4.6c) so might not in fact be derived; (b) their testes are separated and closely associated with the accessory glands (Maetô 1987). However, even these are not as clear-cut as could be liked because the sperm of the Agathidinae examined to date, although aberrant, appear to be so in a rather different way. They have very long tails (although still with modified axoneme) but study of them has been hampered by the fact that in the vas deferens at least, they are combined in a dense 'sticky' mass.

MORPHOPHYLOGENETIC HYPOTHESES

Phylogenetic analyses of the two families have not progressed apace. Reasonably explicit phylogenetic hypotheses for the Braconidae started appearing in the 1970s (Čapek 1970, Papp 1974, van Achterberg 1976c, 1984) (Fig. 12.1). Tobias (1967, 1987, 1989) presented a number of intuitive trees (Fig. 12.1a) in which he depicted the family as having had very rapid ancient radiation with considerable uncertainty about relationships and very few explicit synapomorphies, although with the cyclostomes generally appearing as a basal grade. Based principally on newly investigated final instar larval head capsule features, Čapek (1970) presented a rather attractive depiction of his view of

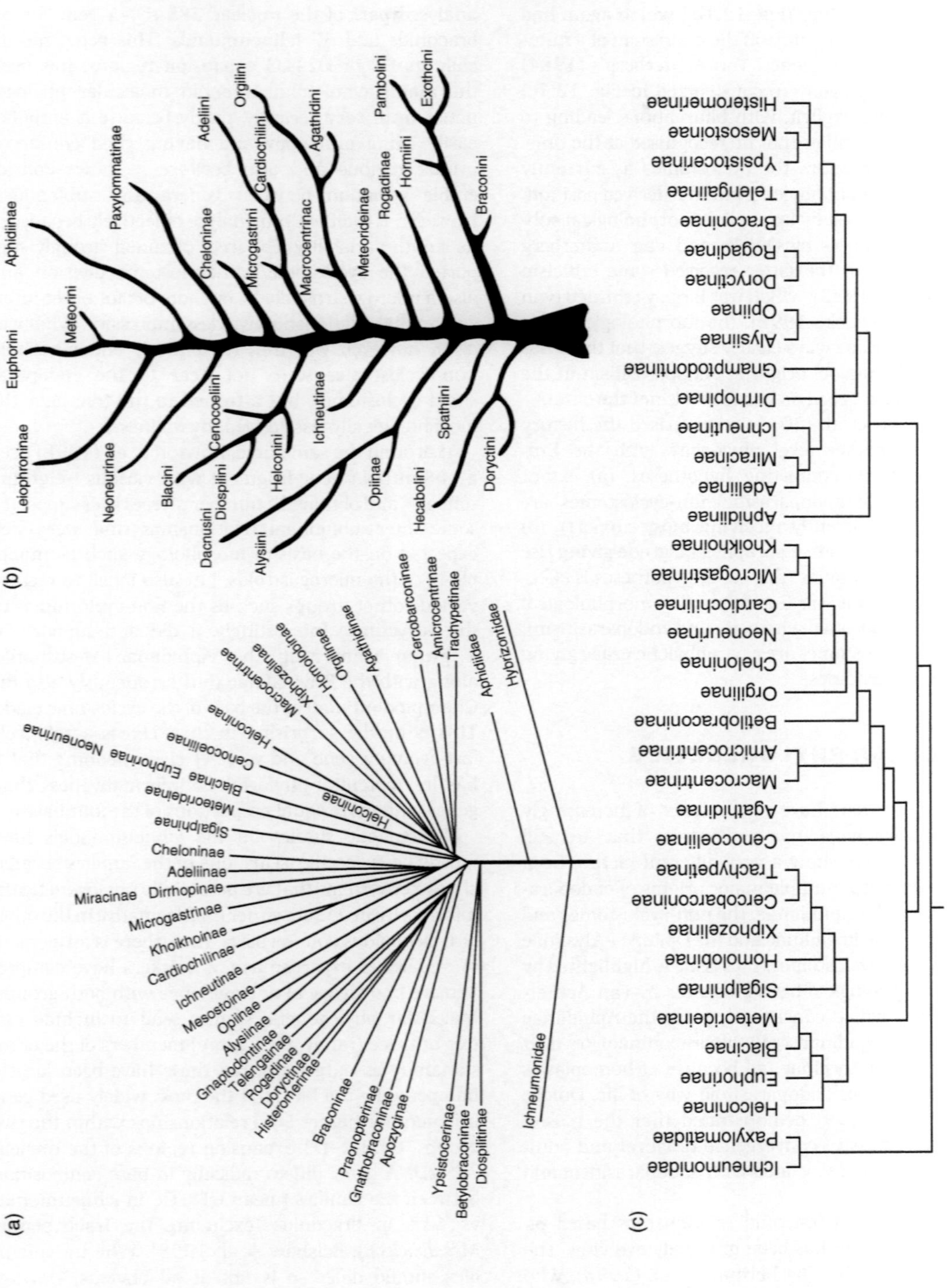

Fig. 12.1 Previous morphology-based phylogenetic hypotheses for the Braconidae. Sources: (a) Tree redrawn from Tobias (1989) with taxon names added; (b) tree redrawn from Čapek (1970); (c) tree proposed by van Achterberg (1994) assembled from subtrees.

subfamily relationships (Fig. 12.1b), which again had the cyclostomes branching off the main stem at a number of separate basal nodes. Van Achterberg's (1984) hand-drawn phylogeny (reconstructed in Fig. 12.1c) was much more explicit, with bifurcations leading to each of the subfamilies that he recognised at the time. In this reconstruction, the cyclostomes as currently recognised are monophyletic, but in a derived position. These were followed by an explicit morphological subfamily level analysis by Quicke and van Achterberg (1990). Although the latter received some criticism (Wharton et al. 1992), which was largely rebutted (van Achterberg & Quicke 1992), the morphological data analysed in various ways clearly suggest that there has been a single (major) origin of endoparasitism in the family (but please read on because it is not the case).

Sharanowski et al. (2011) summarised the history of proposed higher level phylogeny with the Braconidae as three competing hypotheses: (a) extant cyclostomes and endoparasitic non-cyclostomes are sister groups (currently receiving most support), (b) extant cyclostomes are a paraphyletic grade giving rise to the non-cyclostomes (this set of hypotheses is effectively rejected as it was due to parallel morphological adaptations associated with ecto- and endoparasitism); (c) the non-cyclostomes are a paraphyletic grade giving rise to the cyclostomes.

MOLECULAR PHYLOGENETICS

More recently, there have been a series of increasingly comprehensive molecular phylogenies that are still in progress. These show a very different picture from morphology, with multiple major origins of endoparasitism, viz. in the Aphidiinae, the non-cyclostomes and separately in the Rogadinae and the Opiinae+Alysiinae lineages of the cyclostomes. This issue is highlighted by a set of morphological trees presented by van Achterberg (1995), which consistently place the Aphidiinae next to the Rogadinae (±Betylobraconinae) or next to the Opiinae+Alysiinae, all because of homoplasies associated with an endoparasitoid way of life. Quicke and Belshaw (1999) demonstrated that the reason for this was massive convergence of larval and adult female characters associated with endoparasitism and koinobiosis.

Understanding of braconid relationships based on DNA sequence data has been gradually evolving. The first major study was by Belshaw et al. (1998), who analysed part of the nuclear 28S rDNA gene for 56 braconids and 57 ichneumonids. This gene, specifically its D2 or D2+D3 expansion regions, has been the mainstay of ichneumonoid molecular phylogenetics until very recently, partly because it amplifies easily being multicopy and having good conserved primer regions and also because it shows considerable variation between genera and subfamilies. However, it is difficult to analyse objectively because of its length variability. The trees obtained strongly supported the cyclostome/non-cyclostome division and also a microgastroid clade, but support for other inter-subfamilial relationships was less impressive. What was clear, however, was that the entirely endoparasitoid non-cyclostomes were not near to the endoparasitoid cyclostomes, but attached to the tree near the morphologically less derived rhyssalines.

At around the same time, Dowton et al. (1998) used a 16S mtDNA gene fragment with various weighting schemes and obtained a number of trees that supported some inter-subfamilial relationships that were well expected on the basis of morphology such as monophyly of the microgastroids, but also failed to recover various other groups such as the non-cyclostomes or the doryctines. Interestingly, it did find support for a group comprising the Aphidiinae+Mesostoinae along with the Rhyssalinae (but erroneously also the Gnamptodontinae) at the base of the cyclostome clade. This is hardly surprising in that 16S is a relatively fast-evolving gene and very AT rich, meaning that it has less potential phylogenetic informativeness than genes with more equal proportions of the four bases.

Phylogenetic studies on the ichneumonoids have mostly been specific to just one of the families because character systems that are potentially phylogenetically informative in one are either uninformative in the other or lack obvious counterparts. Also, there is an element of researcher expertise as few workers have comprehensive knowledge of or experience with both groups. Molecular phylogenetic studies tend to include just two or three (rarely a few more) members of the other subfamily as outgroups, but these have been largely hampered by the fact that the most widely used gene fragment for higher level relationships within the two groups, the D2–D3 expansion regions of the nuclear 28S rDNA gene, differs radically in base composition between the families [mean 61% GC in ichneumonids vs 43% in braconids (excluding the trachypetine, *Megalohelcon*); Belshaw et al. 1998]. Why these families should differ so is not at all obvious. Dowton and Austin (1997) also reported a base composition bias in the mitochondrial 16S gene between the two families, braconids having significantly more A and fewer G residues, probably attributable to stem sites where both G–T and A–T base pairing are permitted. The marked base composition shift between ichneumonids and braconids shown in the 28S gene provides a molecular synapomorphy for Braconidae excluding Trachypetinae.

Wei et al. (2010) presented the results of sequencing the entire mitochondrial genomes of seven, taxonomically widely separated braconids and compared them to the two available ichneumonid sequences. Their data revealed that three AT-rich regions were inverted in the braconids compared with the ichneumonids, suggesting that this might be a synapomorphy for the family, although Trachypetinae were not included in that study. Among their sampled taxa they found a considerable number of rearrangements involving transfer RNA genes, particularly among the non-cyclostomes. Gene order in the microgastrine *Cotesia vestalis* was very different from all the others with rearrangements and directional changes involving a considerable number of protein-coding genes also. It would be interesting to know if this was representative of all microgastrines or perhaps more microgastroids, although the chelonine *Phanerotoma flava*, a basal microgastroid, displayed a more or less unaltered ancestral pattern.

The most comprehensive molecular phylogenetic analyses of braconid relationships to date are those of Zaldivar-Riverón et al. (2004), which concentrated on the cyclostome subfamilies, and Sharanowski et al. (2011), which was predominantly concerned with non-cyclostomes, but nevertheless included a reasonable number of cyclostomes. The discussion below follows Sharanowski's phylogeny with just a few slight changes (Fig. 12.2).

BRACONID CLASSIFICATION

The subfamily level classification used in this book is given in Table 12.1. which also shows major informal groupings that are well supported, synonyms and some recent redistributions of taxa. Wharton (2000) presented a radical, perhaps tongue-in-cheek, proposal that would have reduced the number of recognised subfamilies from the 47 recognised here to just four or five, some blatantly paraphyletic, with the downgrading of most current subfamilies to tribes. No one has followed this and, if anything, with the desire to make all recognised groups monophyletic, the number is likely to grow. But Wharton's point was that having a huge number of names and inevitably long identification keys can have its drawbacks.

Eoichneumoninae†

This group, which dates from the Early Cretaceous, was originally described on the basis of a rather poorly preserved fossil from Australia and treated as a separate family within the Ichneumonoidea on the basis of primitive hind wing venation which was stated as having a second 'r-m' cross vein (Jell & Duncan 1986). However, the description of additional genera and species by Rasnitsyn and Sharkey (1988), along with re-examination of the type of *Eoichneumon* via high-resolution photographs, led these authors to conclude that 2rs-m of the hind wing is actually lacking. Nevertheless, these fossils show a taxon that probably represents the stem from which the Braconidae evolved. Rasnitsyn and Sharkey could find only one putative synapomorphy with the braconids, in the form of the anterior junction of hind wing vein r-m being close to split between veins R and RS. Eoichneumonine fossils often show the second and third metasomal segments have disarticulated *post mortem* (Rasnitsyn & Sharkey 1988), suggesting that the junction between them is rather weak as in the extant aphidiines, but they are clearly articulated sublaterally edge to edge.

In the phylogenetic analyses presented by Quicke et al. (1999a), depending on the protocol, they were recovered either unresolved at the base of the Ichneumonidae and Braconidae or as the sister group to the basal ichneumonid *Tanychora* or as a sister group to the Braconidae. With the discovery of additional early braconid fossils, the further analyses conducted by Perrichot et al. (2009) more clearly showed an association with the Braconidae (Fig. 11.2b) and they therefore synonymised the two families, the Eoichneumoninae† thus becoming a subfamily, although it lacks any obvious synapomorphy and might well be paraphyletic.

TRACHYPETIFORMES

Trachypetinae

These wasps, known only from Continental Australia, comprise three genera of rather large and spectacular insects (Fig. 12.3). *Trachypetus* is known from a

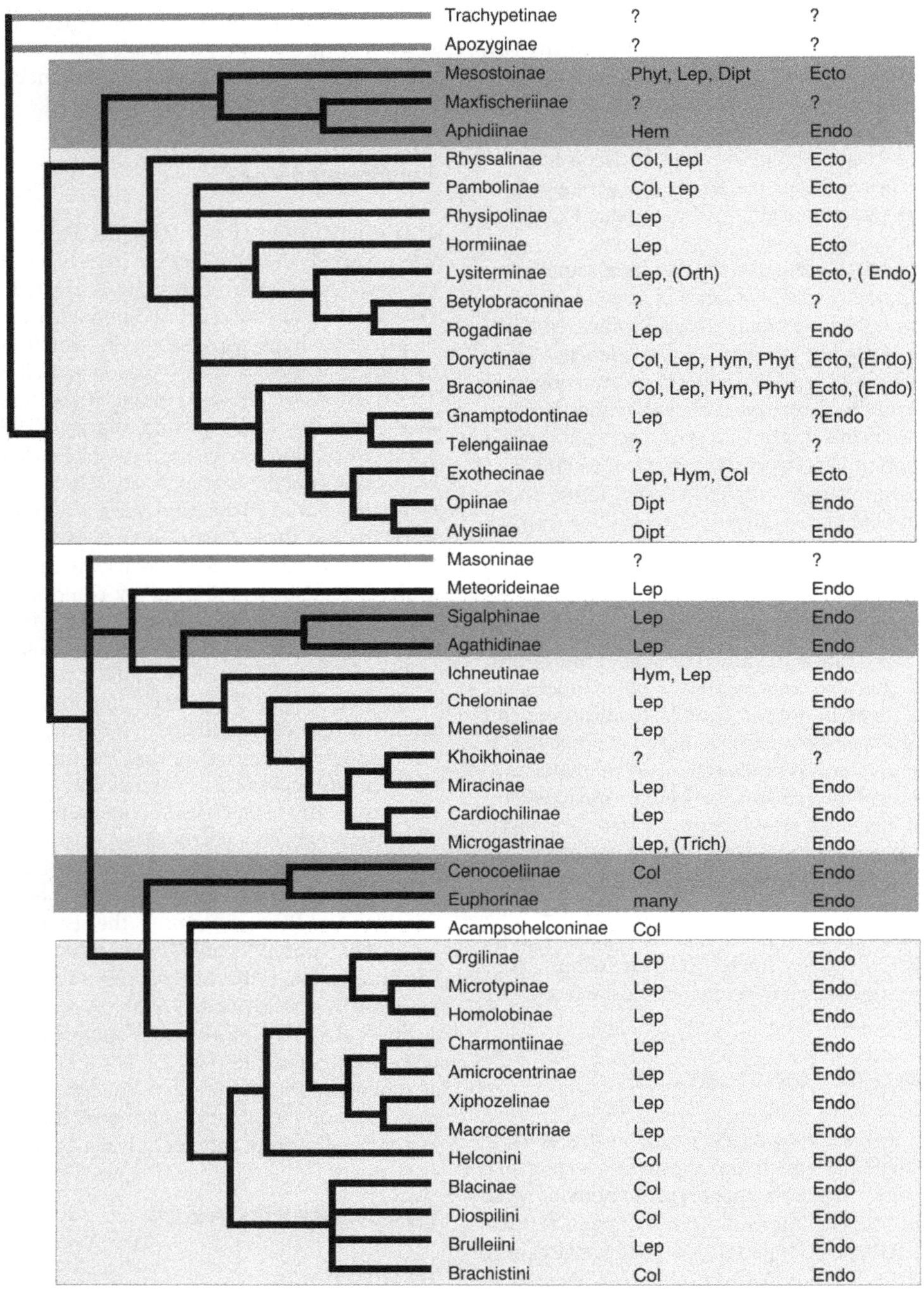

Fig. 12.2 Best-guess phylogenetic tree, major host groups and mode of parasitism for the subfamilies of Braconidae, based largely on the multi-gene analysis of Sharanowski et al. (2011) with some groups added according to Zaldivar-Riverón's (2006) cyclostome study, and Murphy et al.'s (2008) microgastroid study. Host orders: Col, Coleoptera; Hym, Hymenoptera; Dipt, Diptera; Lep, Lepidoptera; Neur, Neuroptera; Orth, Orthoptera; Trich, Trichoptera; Phyt, phytophagous. Mode of parasitism: Ecto, ectoparasitoid; Endo, endoparasitoid. Colour blocks from top to bottom indicate the informal group names following Sharanowski et al.; aphidioids, cyclostomes, sigalphoids, microgastroids, euphoroids and helconoids; non-coloured ones are isolated taxa of uncertain affinity. For a colour version, see Plate 18.

Table 12.1 Classification of Braconidae employed throughout this book, with synonyms.

Current name	Included groups	Synonyms, obsolete names, miss-spellings, etc.
trachypetiformes		
Trachypetinae		Cercobarconinae
apozygiformes		
Apozyginae		
'aphidioids'		
Aphidiinae		Aclitinae, Diospilitinae, Ephedrinae, Gynocryptinae, Incubinae, Praiinae, Trioxinae
Maxfischeriinae		
Mesostoinae	Canberriini	Praonopterinae, Hydrangeocolinae
cyclostomes		
Alysiinae		
Betylobraconinae		
Braconinae		Aphrastobraconinae, Vipiinae, Vipioninae, Gnathobraconinae, Pseudodicrogeniinae, Vaepellinae
Doryctinae	*Monitoriella*	Euspathiinae, Hecabolinae, Odontobraconinae, Spathiinae, Ypsistocerinae
Exothecinae		
Gnamptodontinae		Gnaptodontinae
Hormiinae		
Lysiterminae	Cedriinae	Acanthormiinae, Triterminae
Opiinae		
Pambolinae	*Chremylus*	
Rhysipolinae		
Rhyssalinae		
Rogadinae		Aleiodinae, Pelecystominae. Rhogadinae
Telengaiinae		
Non-cyclostomes		
'microgastroids'		
Cardiochilinae		Toxoneurinae
Cheloninae	Adeliinae	Acaeliinae*
Dirrhopinae		
Miracinae		
Khoikhoiinae		
Mendesellinae		
Microgastrinae		Apantelinae
Ichneutinae		Proteropinae
'helconoids'		
Acampsohelconinae		
Agathidinae		Aneurobraconinae, Bassinae, Eumicrodinae, Mesocoelinae
Amicrocentrinae		
Blacinae		
Cenocoeliinae		Capitoniinae

(continued overleaf)

Table 12.1 *(continued)*

Current name	Included groups	Synonyms, obsolete names, miss-spellings, etc.
Charmontiinae		
Euphorinae	Meteorinae, Neoneurinae, Ecnomiinae, Mannokeriini, Planitorini	Centistinae, Cosmophorinae, Elasmosomatinae, Helorimorphinae, Leiophroninae, Liophroninae, Perilitinae, Zelinae
Helconinae		Brachistinae, Calyptinae, Diospilinae, Triaspidinae
Homolobinae		
Macrocentrinae		
Masoninae		
Meteorideinae		
Microtypinae		
Orgilinae		Mimagathidinae, Stantoninae
Sigalphinae	Pselaphaninae	
Xiphozelinae		

*Mason (1985) explains why Adeliinae is the correct name, *Acaelius* being an unjustified emendation of the type genus *Adelius*.

†Due to misinterpretation of the name *Bassus*, Bassinae has also been used to refer to the ichneumonid subfamily Diplazontinae.

Fig. 12.3 Trachypetinae habitus photographs. (a) *Trachypetus petiolatus*. (Source: reproduced by permission Andrew Austin.) (b) *Megalohelcon* sp. (Source: reproduced by permission, © C. van Achterberg, Naturalis Biodiversity Centre). For a colour version, see Plate 19.

single species, *Trachypetus clavatus*, and was collected most frequently near the New South Wales coast close to Sydney, many from an area now occupied by the international airport. However, it was last collected more than 60 years ago and considerable efforts to find new localities and obtain fresh material for molecular analyses have proven fruitless (A.D. Austin, personal communication); sadly, it may be extinct, although I hope not. The other two genera, *Megalohelcon* and *Cercobarcon*, are both apparently restricted to very inhospitable, arid and barren areas in the central and north-western parts of the continent (Austin et al. 1993a).

Trachypetus and *Megalohelcon* were originally included within the Helconinae because they have fairly complete venation (with fore wing vein 1a present), whereas *Cercobarcon* was described in its own subfamily by Tobias (1979), who considered it to be related to the Helconinae, in agreement with most previous authors. Austin et al. (1993a) realised that all three genera were closely related and they synonymised Cercobarconinae with Trachypetinae. Among the synapomorphies they identified were the presence of a circular or oval 'plate' of sensilla on the outer face of the mandible which appears to be absent from all other Ichneumonoidea, although it is possible that this is also a release site for a secretion.

Austin et al. (1993a) discussed the relationships of the Trachypetinae at length but came to no firm conclusions. In the morphological analyses of Quicke and van Achterberg (1990), they tended to come out with the Agathidinae+Sigalphinae group, then thought to be highly derived within the family, largely because of the presence of hind wing vein 2-CU and a carapace-type metasoma, but neither of these is especially convincing. A very different picture emerged from molecular analyses. Although highly apomorphic in its own right, the 28S D2 rDNA sequence of *Megalohelcon* shows a strong similarity to those of the Ichneumonidae, with a G+C content of 59% compared with a mean of 43% in the Braconidae (Belshaw et al. 1998). Consequently, in rooted analyses, the Trachypetinae come out at the base of the Braconidae (Belshaw et al. 1998, Quicke et al. 1999a). This prompted a closer look at the morphology and, in addition to a previously noted ambiguous position of hind wing vein r-m (more or less opposite SC+R1), the Trachypetinae were found to possess an almost completely longitudinally divided hind wing vein C+SC+R, which is the condition found in the Ichneumonidae and not in any other Braconidae. hence both molecular and morphological evidence point to this subfamily being the most basal subfamily of the group and it is tempting to consider that they might best be regarded as representing a separate family. The venom reservoir is highly muscular and ovoid (see fig. 24 in Tobias 1979 for *Cercobarcon* and fig. 57 in Quicke & van Achterberg 1990 for *Megalohelcon*), closely resembling that of several groups of cyclostome braconid, but also the Xoridinae, a basal group of ichneumonids. Against this, however, is the disposition of the testes, which are separate and closely joined to the accessory glands (see fig. 101 in Quicke & van Achterberg 1990 for *Megalohelcon*), a condition that would appear to be a synapomorphy for the non-cyclostome braconids. It should be noted, however, that among the 'Symphyta' there is considerable variation in the arrangement of the testes and this is, after all, only one character. I strongly suspect that when sequence data for more genes and species of Trachypetinae, and also for Apozyginae, become available, the Trachypetinae will be found to warrant separate family status within the Ichneumonoidea.

Belshaw et al. (2003), considering the structure of the ovipositor in a phylogenetic context as a predictive tool for biology, concluded that trachypetines had a high probability of being endoparasitic. If this suggestion and a basal position of the subfamily within the Braconidae, are both correct, then it makes interpretation of the basal biology of the Braconidae far harder to fathom (see above).

CYCLOSTOMES *INCERTAE SEDIS*

Protorhyssalinae et al.

Described on the basis of a single species from the Late Cetaceous amber of New Jersey, USA, the Protorhyssalinae were thought likely to be cyclostome, although the mouth region of the available fossils was not sufficiently clearly visible to enable unambiguous interpretation (Basibuyuk et al. 1999). A cyclostome relationship was also suggested by the fore wing venation, which has a comparatively long second submarginal cell, a feature not seen among extant non-cyclostomes. To facilitate comparison between the early fossil braconid genera, important characters are listed in Table 12.2.

A second genus, *Protorhyssalodes*, which was described from French Cretaceous amber by Perrichot et al. (2009), bore some resemblance to *Protorhyssalus* and is clearly cyclostome (Fig. 12.4a). Recently, two additional genera (based on three species) have been tentatively added to the subfamily by Belokobylskij (2012), but as these are based upon impression fossils, their body structure is poorly preserved. Nevertheless, the specimen of *Magadanobracon rasnitsyni* is the first known female of this group and displays an exserted ovipositor with a total length approximately the same as the rather short metasoma. Also, this female displays an enlarged hind tibia, as seen for example in male *Rhyssalus*. In addition to the above taxa, Belokobylskij (2012) noted that '*Diospilus allani*' Brues from Canadian amber (Cedar Lake, 76.5–79.5 Mya) has similar venation and may well belong to this group rather than to the Helconinae.

Table 12.2 Summary of character states for early fossil braconids and the putatively basal extant Apozyginae.

Genus	Cyclostome	Fore wing vein C+SC+R completely fused	Fore wing vein 2m-cu present	Hind wing vein 2CU present
Aenigmobracon	Y	Y*	Y	Y
Apozyx	Y	Y	Y	Y
Magadanobracon	?	Y	N	?
Protorhyssalodes	?Y	Y	?N	Y
Protorhyssalopsis	Y	Y	N	Y
Protorhyssalus	?Y	Y*	N	Y
Eoichneumoninae	?	Y	Y	Y

*A very narrow short costal cell may be discernible apically but could be a preservation artefact.

In the phylogenetic analyses carried out by Perrichot et al. (2009), the Protorhyssalinae were not recovered as monophyletic and instead they formed a grade between the Eoichneumoninae and the extant Braconidae, including both *Apozyx* and Trachypetinae. However, although I suspect that the Protorhyssalines and *Anenigmobracon* are probably more or less correctly placed, I am not inclined to put much faith in the recovered relationships of either *Apozyx* or the Trachypetinae (see above). Molecular data for *Apozyx* are desperately needed, as also is more information on Trachypetinae.

Apozyginae

With its single species, *Apozyx penayi*, known only from a small region of Chile, this is an extraordinarily interesting taxon. When W.R.M. Mason (1978) originally described it, he placed it in its own family, the Apozygidae, and included this within the Ichneumonoidea because of its non-elbowed antennae with placoid sensilla, fused fore wing vein C+SC+R, presence of a pterostigma and various other features. The original description was based on a single male collected in *Nothofagus* forest, but subsequently a few more specimens, including the female (W.R.M. Mason 1987), have been collected and opinion about its relationships rapidly shifted. In almost all recent studies, *Apozyx* has been treated as belonging to the cyclostome lineage of Braconidae (Quicke & van Achterberg 1990, Sharkey & Wahl 1992, Quicke et al. 1999a), but unlike virtually all[1] other extant braconids *Apozyx* has fore wing vein 2m-cu, a putatively plesiomorphic feature present in the ichneumonids, although others have assumed it is just an atavism. Unfortunately, no *Apozyx* specimens have as yet become available for DNA sequencing, so interpretation of its relationships is only tentative, but its venation with 2m-cu in the fore wing, a well-developed hind wing vein CUb and the male genitalia with a large articulated and apically toothed cuspidal process suggest that it really occupies a very basal position. Like many Cretaceous fossil ichneumonoids it is rather small, with body length 2–3 mm. Other features that may have some phylogenetic significance include: placode sensilla with aperture extending the whole length (as in Aphidiinae and Mesostoinae and unlike other cyclostome braconids); hypoclypeus with long hypoclypeal setae; scuto-scutellar sulcus absent; the testes fused dorsal to the gut; three ovipositor valvilli (since there is only one known female specimen, this feature was ascertained by viewing the intact ovipositor with bright transmitted light and is not 100% certain); and parameres longer than digitus and densely setose.

The presence of both fore wing vein 2m-cu and of hind wing vein CUb could both be atavistic, but this seems unlikely as the venation of all known individuals is highly similar. Fore wing vein 2m-cu is absent in all other living braconids except atavistically in a few Rhyssalinae and the combination of being cyclostome and possessing hind wing vein CUb is also unique. Ortega-Blanco et al. (2011) noted correctly that the combination of characters in *Apozyx* is the same as that in the extinct Cretaceous genus *Aenigmobracon* (Fig. 12.4b) (Perrichot et al. 2009) which was not definitively assigned to any subfamily, would not prohibit its inclusion in the Apozyginae.

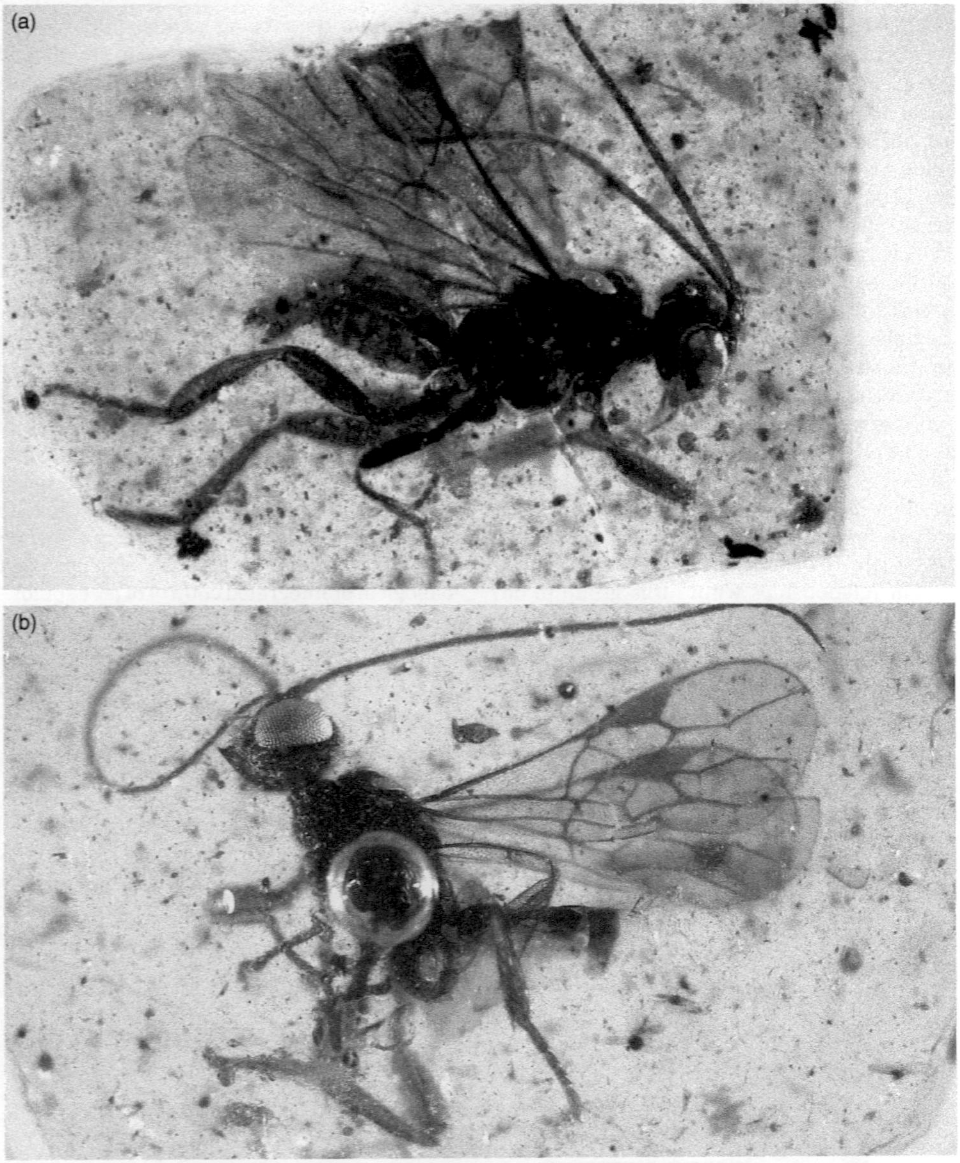

Fig. 12.4 Amber inclusions of a putative species of Protorhyssalinae and a candidate member of the Apozyginae from French Albian–Cenomanian Cretaceous deposits (ca 113–93.9 Mya). (a) *Protorhyssalodes arnaudi*; (b) *Aenigmobracon capdoliensis*. (Source: from Perrichot et al. 2009; reproduced by permission of Vincent Perrichot.).

THE APHIDIOID CLADE OR 'GONDWANAN' COMPLEX

Molecular evidence points firmly to the existence of a clade dominated in terms of number of extant species by the principally north temperate Aphidiinae, but also including several very diverse and almost entirely Gondwanan groups, the Australian Mesostoinae in its original sense, plus the South American Hydrangeocolini and Australian Canberriini which had been included in the Exothecinae (variously placed among Hormiinae *s.l.*), the New Zealand genus *Doryctomorpha* which was included in Doryctinae and, almost unbelievably, the Australian Maxfischeriinae, which were originally considered to belong to the Helconinae (Papp 1994). Two sets of recovered relationship are shown in Fig. 12.5, although one lacks *Maxfischeria* and one lacks *Doryctomorpha*. Some members of this clade are cyclostome, but most Aphidiinae and certainly the Maxfischeriinae are not. At present, for simplicity and because there is insufficiently strong molecular evidence about relationships within this group to do otherwise, they are all being treated as Mesostoinae *s.l.* Apart from these being mostly small wasps, they are morphologically very heterogeneous. It is likely that when additional data become available and a robust phylogeny is obtained, several of these tribes of Mesostoinae will need to be raised to subfamily level to create a classification composed of monophyletic groups. This would certainly lead to a more morphologically and biologically uniform set of subfamilies.

Aphidiinae

Aphidiines are all koinobiont endoparasitoids of aphids (Fig. 12.6a). Most can attack a range of instars of a given host, although a few specialise on attacking winged adults (Völkl & Mackauer 1996) and so are, in a sense, technically idiobionts or imagobionts.

This has been one of the most intensively studied groups of Braconidae from biological, taxonomic and phylogenetic points of view, largely motivated by the potential of some species to be effective agents in the biological or integrated control of pest aphids. They are

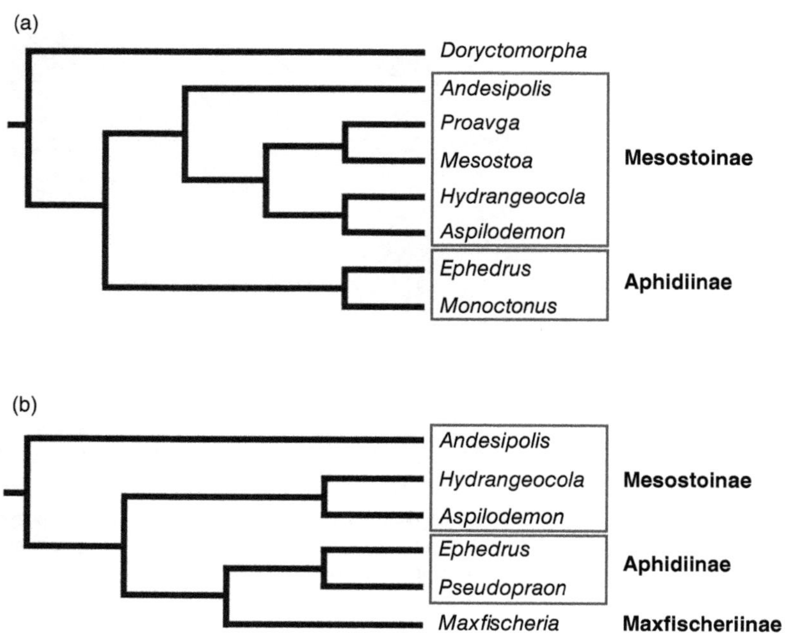

Fig. 12.5 Relationships of the Aphidiinae, Mesostoinae and related taxa. (a) Including *Doryctomorpha* as recovered by Zaldivar-Riverón et al. (2006) based on 28S rDNA D2+D3 region plus COI barcoding region plus morphology; (b) including Maxfischeriinae as recovered by Sharanowski et al. (2011) based on six fragments of four gene regions.

Fig. 12.6 Aphidiinae and their interactions. (a) *Pauesia unilachni* attacking *Schizolachnus pineti* on *Pinus sylvestris*; (b) ant, *Lasius niger*, attacking *Trioxys angelicae* among colony of *Aphis fabae* on *Cirsium arvense*. (Source: reproduced by permission of Wolfgang Völkl.). For a colour version, see Plate 20.

often very successful in controlling host populations, probably because the wasps have very high fecundities (see Chapter 14, section *Fecundity*) and many of the host aphids live in aggregations and have overlapping generations. More than 20 species have been deliberately released to help control exotic pests in classical biological control programmes (Hågvar & Hofsvang 1991) and approximately 20% of these have been successful (Hirose 2006).

The great majority of uses of aphidiines in pest control have been either in glass-house situations (e.g. *Praon volucre* against *Uroleucon ambrosiae*, a pest of a range of Asteraceae) or in warmer climates. In temperate regions, aphidiines often provide significant aphid control but are not normally very amenable to management, though are certainly more effective under some land management conditions than others. The presence of natural reserves of aphidiines close to the crops is a major advantage. Levie et al. (2005) described tests of using either mass release or strip reservoirs for parasitoids in the control of cereal aphids by *Aphidius rhopalosiphi* and found that both treatments were effectively equally efficient at reducing crop aphid numbers. *Binodoxys communis* was introduced from Asia to North America and has been released in several countries to control soybean aphid, *Aphis glycines*, a very important ecomonic pest. Similarly, *A. smithi* was introduced from India in 1958 to control the pea aphid, *Acyrthosiphon pisum*, and has been a large success having established and become its most important parasite over a large area (Halfhill et al. 1972), while potentially endemic[2] *A. ervi and P. pequodorum* also exert significant control and indeed *A. ervi* introduced from France in 1959 largely replaced *A. smithi* in the Eastern part of North America by the late 1960s (Angalet & Fuester 1977). *Trioxys pallidus* has been employed with great success to control walnut aphid, *Chromaphis juglandicola*, in California, although only after they were sourced from a country, Iran, with an appropriate climate (van den Bosch et al. 1982), however, introductions from cooler climatic regions were successful in the more northerly state of Oregon. Other important examples include the release, in 2000, of *Lipolexis oregmae* from Guam into Florida in an attempt to control brown citrus aphid, *Toxoptera citricola* (Persad et al. 2007). Even although parasitisation rates on the latter are currently rather low, it is hoped that natural selection will lead to more efficient control in the future. Several species are reared commercially for release against glass-house pest aphids. *A. ervi* is one such example and accepts several pest aphids as hosts, including pea aphid, *Acyrthosiphon pisum*, a major pest of alfalfa and other Fabaceae crops, the foxglove or glass-house potato aphid, *Aulacorthum solani*, which attacks a vast range of commercially grown plants, and the green-peach aphid, *Myzus persicae*.

Aphidiines are not always taxonomically easy and a lot of literature refers to *A. pulcher* being used in the biological control of *Acyrthosiphon pisum* in North America; however, Marsh (1977) showed that the real *Aphidius pulcher* is in fact a junior synonym of *A. nigripes* and does not attack pea aphid, and that what was being mass reared and released was something different.

Dispersal of the parasitoids is influenced by the host stage they attack. Thus, Olkowski et al. (1982) found that *Trioxys tenuicaudus* released in California to control elm aphid (*Tinocallis platani*), and which attacks almost entirely early instar, wingless, aphids, took several years to spread just a few blocks. Other species that attack late instars nymphs or winged adults, such as *Binodoxys communis*, have the potential to spread at far higher rates (Y. Zhang et al. 2012), although the earlier in development that the alatoid nymph is attacked the greater is the loss of subsequent flight potential and wing development (Christiansen-Weniger & Hardie 1998).

Aphidiines have often been treated as constituting a separate family, the Aphidiidae, and some workers persist with this practice despite a lot of morphological, and, perhaps more convincingly, molecular evidence that they are perfectly good braconids, albeit a rather basally derived clade. They are always fairly small and many have reduced wing venation (Fig. 12.7b–e), although *Ephedrus* and *Toxares* have a more or less complete set of veins (Fig. 12.7a). Morphological studies have long recognised four major groupings as tribes or subtribes (Mackauer 1961), the relatively derived Aphidiini including Trioxina and the Praini and Ephedrini, which are less derived, but a number of putatively basal genera that are not neatly accommodated by any of these four tribes and probably formal tribal classification will need modifying (see p. 224) and *Aclitus* is now often given tribal rank. Indeed, even the current composition of the Ephedrini, with *Toxares* included, is questionably monophyletic. In Finlayson's (1990) study of final instar larval head capsule, *Toxares* was found to differ considerably from *Ephedrus*.

Fecundity can be very high, with females carrying up to 800 eggs, although 150–200 or so is more typical, even although adult longevity is rather short, typically of the order of 1 week (e.g. van Steenis 1994, Latham & Mills 2010). Cohen and Makauer (1987) noted that one female of *Ephedrus californicus* had a realised lifetime fecundity of 1762. Along with their short generation times, these values mean that aphidiines can have very high intrinsic rates of increase with r up to 0.37 (see Chapter 14, section *Fecundity*). Many species are able to oviposit virtually immediately after eclosion (LeRalec 1993). Necessarily the eggs themselves are very small and contain virtually no resources and so have to develop rapidly. The eggs of Aphidiini (including Trioxina) are rather lemon-shaped (de Marzo 2000) and probably all members of this tribe are strongly pro-ovigenic. All aphidiine eggs are hydropic and become gradually rounder after oviposition as well as swelling. Embryonic development is rapid with cells of the serosal membrane differentiating early (Ivanova-Kasas 1956, Tremblay & Calvert 1971). These cells surround the embryo until the first instar larva is fully developed and they invade the surrounding host tissue where, by means of their dense covering of microvilli, they appear to divert food resources to the embryo, acting as a sort of placenta (Sabri et al. 2011). The serosal cells dissociate as soon as the first instar frees itself to live within the host haemocoel and continue swelling to become teratocytes and in *A. ervi* increase in diameter from about 40 to 200 μm diameter over a period of a couple of days, equivalent

Fig. 12.7 Aphidiinae. (a) *Toxares macrosiphagum*, male, showing near complete fore wing venation; (b) *Trioxys asiaticus*, female showing apival metasoma prongs; (c) *Aclitus* sp.; (d) *Lipolexis gracilis*; (e) *Monoctonia vesicarii*. (Source: reproduced by permission of Ehsan Rakhshani.).

to a volume increase of approximately 1000-fold. The ultrastructure of the teratocytes of *A. matricariae* was studied by Tremblay and Iaccarino (1971).

Larval anatomy and development have been described in detail probably more extensively than those of any other subfamily. Important studies are on *Aphidius colemani* (Hofsvang & Hågvar 1978), *A. ervi* (Pennacchio & Digilio 1990, de Eguileor et al. 2001). *A. nigripes* (Pare et al. 1979), *A. rhopalosiphi* (Muratori et al. 2004), *Diaretiella rapae* (Couchman & King 1977), *Ephedrus cerasicola* (Hofsvang & Hågvar 1978), *Lysiphlebus japonicus* (Ming et al. 2003), *Praon palitans* (Schlinger & Hall 1960), *P. pequodorum* (Chow & Sullivan 1984), *P. simulans* (Timberlake 1910),

P. volucre (Beirne 1942) and *Trioxys indicus* (Subba Rao & Sharma 1962) and *T. utilis* (Schlinger & Hall 1961). In addition, considerable inter-generic variation in first instar larval anatomy was described and illustrated by O'Donnell (1989) (see below).

The number of larval instars is variously stated in the literature as three, four or five (see Pennacchio & Diglio 1990) and reports of five instars [e.g. Hoek 1971 for *A. nigripes* (as *pulcher*)] are almost certainly erroneous and result from mistaken swelling and body shape change during the growth of a single instar as representing different ones. The study by Ming et al. (2003) indicated that in *L. japonicas*, the second and third instars are extremely similar under light microscopy and can only be distinguished by SEM due to the presence or absence of some small mandibular bristles. This suggests that little growth occurs between the two instars and so the moult might have some function other than growth. Schlinger and Hall (1961) also suggested that there was possibly a short-duration third instar between the amandibulate second and mandibulate final instar in *Trioxys*, but on the balance considered that it probably had onlt three instars.

First instar larvae are caudate and mandibulate and provide a number of potentially phylogenetically informative characters (O'Donnell 1989; see below). There appears to be variation in whether the first instar has a gut or not. Detailed anatomical studies on *Trioxys* (Subba Rao & Sharma 1962) and *Diaretiella rapae* (Couchman & King 1977) clearly illustrated and described one, whereas first instar *Aphidius ervi* completely lack a gut and also labial glands (de Eguileor et al. 2001). In this latter species, food absorption is through specially modified cuticle and epithelial cells which are furnished with microvilli and lie beneath a thin cuticle (de Eguileor et al. 2001, Giordana et al. 2003). The cuticle cells also show other classic features of absorptive epithelia including Na^+/K^+ ATPase and alkaline phosphatase activity. Whether this mode occurs also in other taxa, especially those with a gut, remains to be explored. The second instar, at least of some species of *Aphidius* (Chorney & Mackauer 1979) and *Trioxys* (Schlinger & Hall 1961), are amandibulate and of very short duration. The final instar larvae are hymenopteriform and their morphology has been surveyed by Finlayson (1990) with particular reference to the head capsule. An interesting difference between first and second instars in at least some species is that the former have well-developed salivary glands which are lost (greatly reduced) in the latter. These do not appear to be involved in digestion and it has been suggested that their secretions may be involved in inhibiting aphid immune defence mechanisms (Soldán & Starý 1981). The meconium is voided just before pupation (Couchman & King 1977).

First instar larval external morphology were surveyed across the subfamily by O'Donnell (1989), who found considerable variation in the development of the mandibles, dorsal spines and tail, which he suggested probably indicated different means of locomotion in the host in different genera (Table 12.3). However, the expectation therefore that *Praon* with its small mandibles ought to be inferior in interspecific competition against *Aphidius* was not borne out in the competition experiments of Chow and Mackauer (1984), which showed that *Praon* was superior. O'Donnell (1989) postulated that in these wasps the mandibles were unimportant for eliminating competitors and postulated that different taxa used combinations of tail movement and mandibular grabbing of host tissue to move around within the aphid host, although emphasising that more observations of living larvae were needed.

Table 12.3 Summary of first instar larval morphological features (adapted from O'Donnell 1989).

Genera	Mandibles	Dorsal spines	Tail
Aphidius, Diaretiella, Lysiphlebus, Pauesia (Aphidiini)	+	–	–
Praon, Dyscritulus (Praiini)	–	+	+
Ephedrus (Ephedrini)	–	–	+
Binodoxys, Trioxys (Trioxini)	–	±	±
Falciconius (Trioxini)	–	–	+
Monoctonus (Trioxini)	–	+	–

+, Well developed/large; –, small, unspecialised; ±, intermediate.

Aphidiines mummify their host aphids and most genera make a flimsy cocoon within the mummified and inflated host aphid skin, but Praini larvae (e.g. *Praon* and *Dyscritulus*) cut their way out of the mummified aphid skin through its ventral surface and construct a disc-shaped cocoon between the aphid mummy and the leaf. In a few taxa this behaviour is plastic, for example, in *Pseudopraon mindariphagum*, non-diapausing pupation occurs within the host and diapausing larvae construct an external cocoon under the host (Starý 1975) and both types of pupation occur in some *Areopraon* species (Mackauer & Finlayson 2012). In *Pseudopraon*, the variation is presumably adaptive and perhaps with cocoon attachment directly to the plant it is less likely to be washed away. In *Lipolexis*, diapausing cocoons are darker and separated further from the aphid mummy than non-diapausing cocoons (Shuja-Uddin 1977). Whether internal or external pupation is plesiomorphic for the Aphidiinae is not yet fully clear. Mackauer and Finlayson (2012) argued that internal pupation is plesiomorphic with external pupation adaptive whereas Tobias and Kyriac (1971) had argued that because the ancestral ichneumonoids were most likely ectoparasitoids with external cocoon construction, external pupation was a sympleiomorphy.

Aphidiines are largely pro-ovigenic, but most species continue to do some egg maturation for a couple of days after emergence; for example, in *Binodoxys communis*, a biocontrol agent against soybean aphid, females emerge with approximately 40 mature eggs, but during the following 24 h mature approximately 150 more (Dieckhoff & Heimpel 2010), but in *Trioxys utilis*, females emerge with approximately 75% of their eggs already matured (Schlinger & Hall 1961). Oosorption does not appear to occur in any aphidiines and eggs are stored in a thinly lined uterus expansion of the lateral oviducts (Schlinger & Hall 1961).

The adult parasitoid, if pupation occurred within the body of the mummified aphid, emerges either posteriorly below the level of the aphid siphunculi or dorsally, and the position is generally fixed within a given genus (Starý 1974). As with rogadines that mummify their host caterpillars, the mature aphidiine larva bites a small slit through the host's ventral cuticle and through this attaches the aphid skin to the substrate, but using silk threads rather than amorphous sticky liquid (Schlinger & Hall 1961, Gutierrez-Ibanez et al. 2007).

Most species are multivoltine, but a few are univoltine. In the case of the southern European *Monoctonia pistaciaecola*, it is restricted to being so because its host is a gall-forming aphid that is only available at the fundatrices because the others are too deeply protected by the gall tissue. There is, however, no obvious explanation for univoltine biology of the highly polyphagous *Pseudopauesia prunicola* (Polgár & Hardie 2000)

Internal anatomy, especially of the females, has been investigated in the Aphidiinae probably more intensively than in any other comparable group, with particular emphasis on the ovaries and venom apparatus (Sedlag 1957, Schlinger & Hall 1961, Tremblay 1964). The basal aphidiines (including Ephedrini and Praini) have two pairs of very elongate, meroistic ovarioles each with very swollen, egg-packed distal portions (sometimes called uteri). In contrast, the Aphidiina and Trioxina are unique among the Ichneumonoidea in having just a single pair of ovarioles which have a derived form with the anterior germarium wide and cap-like and the rest given over to being highly swollen reservoirs or 'uteri' for the very large numbers of lemon-shaped eggs produced (Sedlag 1957, Schlinger & Hall 1961, LeRalec 1993, De Marzo 2000). Unfortunately, ovarial structure is not yet known for many putatively basal genera, though most likely they will follow the plesiomorphic Praini/Ephedrini pattern. The primary venom duct is swollen distally in some species and enters the egg canal quite far down the length of the ovipositor. Aphidiines are the only subfamily known which display both major types of venom apparatus (Edson & Vinson 1979). Most have a highly muscularised venom reservoir with or without a spiral-lined cuticular reservoir intima, but *Praon* is described as lacking a reservoir (ampula) and only having a gland filament, but examination of the illustrations suggests to me that they actually lack gland filaments and have preserved a, presumably secretory, reservoir (Edson & Vinson 1979, De Marzo 2000). The Aphidiini (including Trioxina) also share highly derived embryology (Tremblay & Calvert 1971) in which the polar bodies fuse and produce a line of cells that go on to form an extraembryonic membrane that gives rise to the teratocytes.

Females of most species are monoandrous, although males often mate multiple times. Females of several species release sex pheromones that attract males. The role of pheromones in male attraction and courtship of *Praon volucre* and *Aphidius rhopalosiphi* was investigated by Decker et al. (1993), of *Aphidius nigripes* by McNeil and Brodeur (1995), of *A. ervi* by McClure et al. (2007) and for *Lysiphlebus testaceipes* by Lo Pinto et al. (2013). In *P. volucre* the pheromone appears to be released from the metasoma but is produced in such small quantities that it has thus far defied chemical identification (Nazzi et al. 1996). For *Lysiphlebus*, Lo Pinto et al. obtained evidence that non-polar, cuticular hydrocarbons, possibly of 27, 28 and 29 chain length, might be involved in eliciting male wing fanning. Female-baited water traps were highly effective at trapping males in the first of these studies, with catches markedly higher when traps were placed just below the top of the crop rather than above it. Courtship and mating have been shown to be dependent upon antennal contact between the sexes, the basal flagellar segments of virgin males being the release sites for a glandular secretion which is a contact pheromone required for mating acceptance by the female (Battaglia et al. 2002). Courtship success also depends on the male vibrating its wings to produce a courtship song. Villagra et al. (2011) presented sonograms and sound spectra of intact courting male *A. ervi* and of ones that had been half dealated (i.e. the apical half of each wing cut off) and completely dealated individuals, and showed that their artificial wing damage greatly reduced their singing ability and concomitantly reduced their mating success. Male mating behaviour has been described in some detail for three *Aphidius* species by Mackauer (1969) and also been shown to be affected by associative learning (Villagra et al. 2005). Females normally mate only once, but males mate multiply (e.g. Schlinger & Hall 1960).

The whole subfamily appears to be quite prone to the evolution of thelytokous lineages (Table 7.1) with it having evolved numerous times in the genus *Lysiphlebus* alone. Within the *L. fabarum* complex, the different asexual lineages show a high level of genetic diversity and, at least within the UK, the sexual and asexual forms attack different host aphid species. Belshaw et al. (1999) compared sequence data from a mitochondrial and a nuclear marker DNA marker for various populations and found that while the maternally inherited mitochondrial gene suggested that there were two asexual lineages in Western Europe which had diverged from sexual ancestors approximately 0.5 Mya ago, the nuclear marker varied independently and this was interpreted as indicating that the asexual lineages do occasionally have sex. Sandrock et al. (2011) found the same general pattern in a larger survey and concluded that the high genetic diversity of the asexual wasps could have resulted from rare sex, either with sympatric sexual wasps or with the occasional males produced by asexual lineages or 'contagious parthenogenesis' with rare males of the asexual strains infecting sexual females, resulting in the formation of new parthenogenetic lines. Sandrock et al. suspected that all three mechanisms probably play a role.

Host range is generally fairly narrow, with most species attacking only one or a few related genera of aphid (Fig. 12.8), although some species are far more polyphagous and can attack and develop in a taxonomically diverse set of aphids that co-occur in the same habitat (Mackauer & Starý 1967, Starý 1970, Starý & Rejmanek 1981). Probably a fairly high proportion of those that parasitise various southern hemisphere endemic hosts are quite host specific. Members of the genus *Pauesia* are specialists on Cinarini aphids on gymnosperms (Fig. 12.6a) and some aphidiine species are especially associated with ant-attended aphids, e.g. *Lysiphlebus cardui*, *L. fabarum*, *L. hirticornis*, *Paralipsis enervis*, *Pauesia picta* and *Protaphidius wissmannii* (Starý 1987). Others, although not completely specialised on ant-attended aphid colonies, nevertheless fare better on them; for example, *Pauesia pini* has been shown to oviposit more in an ant-attended colony of its host *Cinara* aphids (Völkl & Novak 1997). Failure of ants to attack these aphidiines results from both behavioural aspects of the parasitoids and also their cuticular hydrocarbon profiles, which in some cases are mimetic of those of the ants attending their hosts (Völkl & Mackauer 1993, Liepert & Dettner 1996, Dieckhoff & Heimpel 2010) (see Chapter 4, section *Cuticular hydrocarbons*). In other cases, probably the majority, ants defend their aphid colonies against aphidiine parasitism and will attack and kill aphidiines in addition to predating selectively upon mummified aphids (e.g. Wyckhuys et al. 2007). Ant mimicry is taken a stage further in some cases. *L. fabarum* might be unique in soliciting honeydew secretion from the aphid host, *Aphis fabae*, by antennating them in a fashion that is mimetic of their ant attenders (Rasekh et al. 2010). When the parasitoid was deprived of food for a period, it used the ant mimicry more often. Some root-feeding ant parasitoids apparently acquire part at least of their chemical camouflage by physical association with the ants in the nest where they emerge (Takada & Hashimoto 1985). The chemical and behavioural mimicry shown by *Paralipsis enervis* enables it to obtain food from its associated ants (*Lasius niger*) through trophylaxis (Völkl et al. 1996) and cannot survive outside of the ant association and it

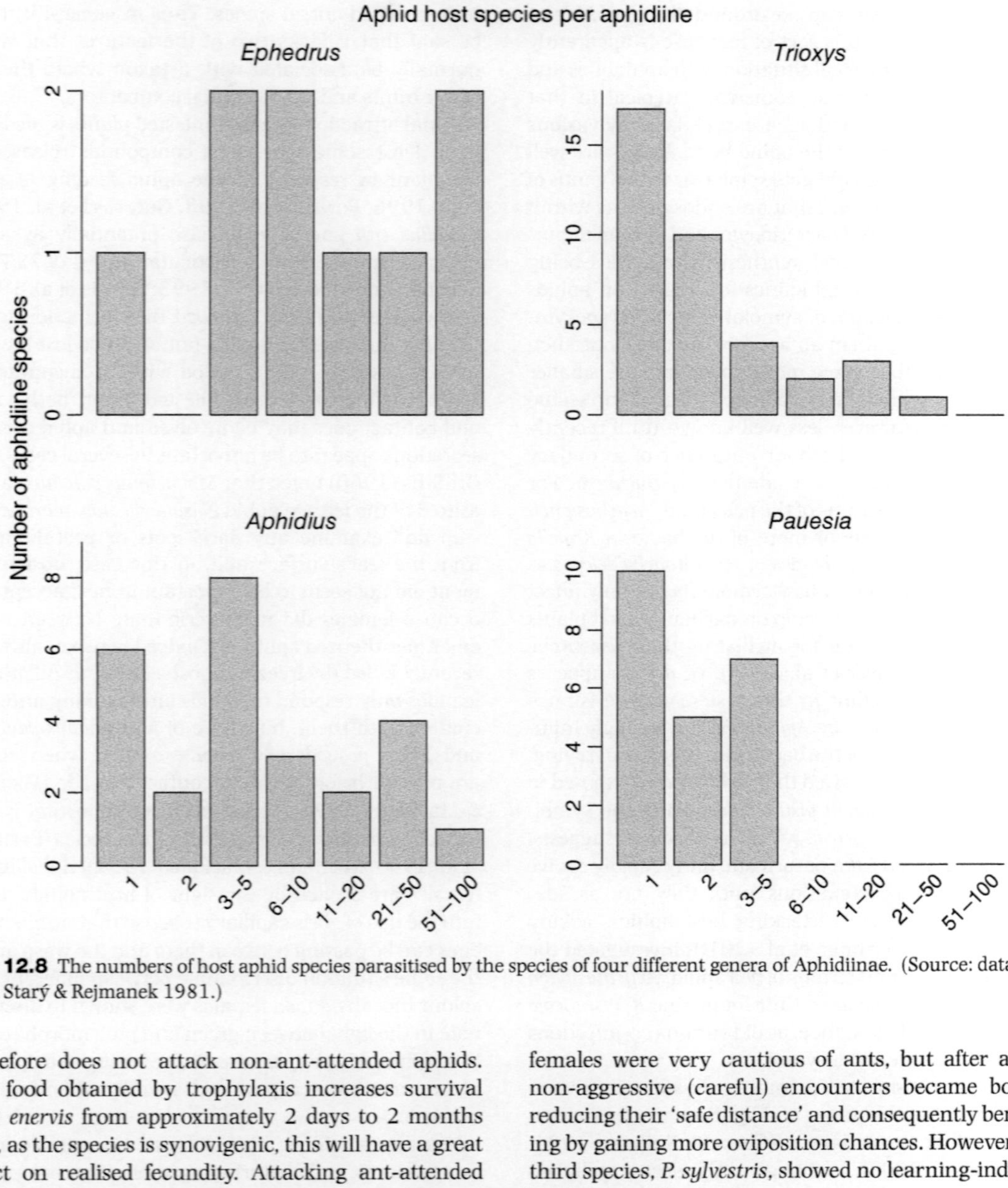

Fig. 12.8 The numbers of host aphid species parasitised by the species of four different genera of Aphidiinae. (Source: data from Starý & Rejmanek 1981.)

therefore does not attack non-ant-attended aphids. The food obtained by trophylaxis increases survival of *P. enervis* from approximately 2 days to 2 months and, as the species is synovigenic, this will have a great effect on realised fecundity. Attacking ant-attended aphids is dangerous, however, for non specialists and the ants having recognised an aphidiine will attack it and often kill it (Fig. 12.6b). Völkl (2001) investigated the behaviours of three *Pauesia* species that often have honeydew-collecting ants attending them. The behaviours of two of the species, *P. pinicollis* and *P. pictus*, changed with experience (i.e. learning). Naïve females were very cautious of ants, but after a few non-aggressive (careful) encounters became bolder, reducing their 'safe distance' and consequently benefiting by gaining more oviposition chances. However, the third species, *P. sylvestris*, showed no learning-induced behavioural changes.

Unlike most host insects, several, if not the majority, of aphids eliminate parasitoid eggs by a humoural mechanism, which makes them disappear, rather than by encapsulating them (Henter & Via 1995). Of four aphid hosts investigated, D.C. Griffiths (1961) found that *Aulacorthum circumflexum* formed a brown, haemolymph-derived capsule around the egg of *Monoctonus paludum*, but this sort of response is apparently rare. The immunological situation with aphidiines and their aphid hosts is also somewhat atypical in that aphidiine–host interactions are modulated by various bacterial symbionts of the aphid host. Aphids are well known for having an obligate symbiosis with a genus of bacteria called *Buchnera* that are endosymbiots within specialised cells called bacteriocytes and play an important role in amino acid synthesis, the aphids being deficient in these. If aphidiines are reared on aphids that have been cleared of symbiotic bacteria (aposymbiotic) by feeding them an appropriate antibiotic such as rifampicin, they grow more slowly and are smaller than those reared on symbiotic hosts (Pennacchio et al. 1999). However, less well known until recently is that many aphids harbour a number of secondary endosymbiont bacteria in addition to *Buchnera*. For example, various strains of the pea aphid, *Acyrhosiphon pisum*, may have one or more of the bacteria *Regiella insecticola*, *Serratia symbiotica* or *Hamiltonella defensa* as symbionts. These secondary symbionts not only affect the aphid's ability to develop on particular food plants but are also important for mediating their resistance to parasitoids (Oliver et al. 2005). *H. defensa* appears particularly important in this system and resistance of hosts to parasitism by *Aphidius ervi* is strongly influenced by the strain of the bacterium they are carrying. Dion et al. (2011) showed that *Aphidus ervi*, exposed to strains of *Acyrhosiphon pisum* harbouring the symbiotic bacterium *H. defensa*, which, as its name suggests, defends the aphid from parasitism, fairly rapidly evolve under laboratory conditions until they are as successful as parasitoids attacking host aphids lacking the bacterium. Nyabuga et al. (2010) investigated the resistance of various strains of pea aphid, *Acyrthosiphon pisum*, against *Aphidius ervi* and found that *R. insecticola* alone conferred resistance, as did various co-infections between the other endosymbionts.

Superparasitism is common, self-superparasitism probably being the norm, and fighting may be routine between first instar larvae (Chow & Sullivan 1984, Ming et al. 2003), although only occasionally do more than one individual survive to emerge from a single aphid mummy (Mackauer & Chow 2012). The majority of clades have first instar larvae with reduced/small mandibles (O'Donnell 1989), including those of *Ephedrus*, which is sometimes capable of gregarious development, and *Ephedrus* first instars also have reduced dorsal spines. Thus in general it might be said that it lacks two of the features that would normally be associated with a taxon where the first instar hunts and kills potential competitors.

Initial attraction to aphid-infested plants is mediated in at least some species by compounds released by the plant in response to the aphid feeding (e.g. Du et al. 1996, Powell et al. 1998, Guerrieri et al. 1999). *Aphidius ervi* and *A. eadyi* also potentially by aphid sex pheromones and in particular (+)-(4a*S*,7*S*,7a*R*)-nepetalactone (Powell et al. 1993; Hardie et al. 1994). Glinwood et al. (1999) showed that attraction to the sex pheromones was innate, but could be lost if wasps were exposed to it for a period without encountering hosts. Having located an infested plant, both visual and contact cues may be involved and aphid cornicle secretions appear to be important in several cases. D.C. Griffiths (1960) noted that *Monoctonus paludum*, a parasitoid of the lettuce aphid *Nasonovia ribis-nigri*, would stop and examine any dark spots or protuberances from the leaf's surface and, in this case, host movement did not seem to be important in host acceptance because females did not discriminate between active and anaesthetized aphids, or indeed between alive and recently killed (by freezing) hosts. *Ephedrus californicus* females only respond to aphids after making antennal contact with them, but those of *Monoctonus paulensis* and *Praon pequodorum* become excited when aphids are present before physical contact is made (Michaud & Mackauer 1995). Vision plays an important role in lost location and acceptance in some species (Battaglia et al. 1995, Michaud & Mackauer 1994). *Aphidius ervi* females are excited by the sight of host aphids sealed into the tips of glass capillary tubes so that no chemical cues can be passing between them and the wasp and in the same situation discriminate between different apid colour morphs. *Praon* females were shown to discriminate in the light between green and pink morphs of the host aphid, *Macrosiphon creelii*, preferring to oviposit in green ones, but accepted each equally when foraging in the dark.

Nearby aphidiines, having located a suitable aphid host, oviposit in it by bringing their telescopic metasoma under their mesosoma so that the ovipositor is pointing forwards between the wasp's middle or fore legs (Fig. 12.6a) and some genera have highly telescopically long metasomas that enable them to reach semi-concealed aphids under bark (e.g. *Lipolexis*, *Protaphidius*). Oviposition is typically brief but in the case of *Dyscritulus* and of *Monoctonus paludum planiceps* it may be protracted. In *D. planiceps* oviposition takes only 0.3–2.6 s although the probability of the female having successfully laid an egg while grappling the aphid was greater the longer the insertion, and insertions of less than 1.5 s did not result in successful oviposition. With *Monoctonus* and other Trioxini species, the female wasp grasps its aphid host with her fore legs (D.C. Griffiths 1960, Völkl & Mackauer 1996). *M. paludum* may insert her ovipositor for up to 15 s and lays its egg into the hosts ganglionic mass (D.C. Griffiths 1961), and *M. paulensis* similarly oviposits within host ganglia (Calvert & van den Bosch 1972). Whether some other species with long-duration oviposition behaviours also have to insert their eggs into specific tissues is unknown, but it seems likely that the ancillary host grasping apparatus in all Trioxini may be involved in precise oviposition since other taxa without such modifications manage to oviposit successfully. Völkl and Mackauer (2000) surveyed oviposition in 49 species representing 19 genera and classified the behaviours into six groups on the basis of speed of oviposition and whether the host is grasped. *Ephedrus* species show the least specialised oviposition behaviour, with no modifications for restraining the host and relatively long oviposition times. Females of some Troxina (*Acanthocaudus, Binodoxys, Trioxys*) have the hypopygium produced into a pair of straight or upcurved prongs that are used to restrain the aphid host during oviposition, while the Praiini, *Praon* and *Dyscritulus*, use their fore legs to grasp and help restrain the aphid (Volkl & Mackauer 2000). In these taxa, the ovipositor is more slender than in most other aphidiines and down-curved in the opposite direction to the prong. Very rapid oviposition (ca 0.5 s) is characteristic of *Aphidius* and a couple of other taxa.

Aptery is very rare but occurs in the genus *Diaeretellus*, with *D. ephippium* having the female apterous and the male fully winged and *D. svalbardicum* being extremely unusual in displaying wing polymorphism (macropterous and micropterous) in both males and females (Chaubet et al. 2013), and Gärdenfors (1990) described an apterous species of *Trioxys* from Ecuador. The only other apterous aphidiine is *Autriquella aptera* from bamboo forest mountains in Kenya in which both sexes are fully apterous; its biology is unknown but it is morphologically quite ant-like, leading to speculation that it might have an association with ant-attended aphids (Starý 1988).

Aphidius ervi is rapidly becoming one of the best studied of all ichneumonoids from a physiological perspective. Cloutier and Douglas (2003) showed that aphids parasitised by young first instar *A. ervi* produce more mycetocytes (the specialist cells containing the aphid's symbiotic *Buchnera* bacteria) and had lower embryo biomass, suggesting that the parasitoid has redirected host physiology, even at an early stage, towards making more of the tissue that is important in processing sap to make more insect friendly biochemical building blocks (Rahbé et al. 2002). The redirection of host resources usually entails castration with destruction of early embryos and of the germarium (Digilio et al. 2000) and disruption of host wing development (Demmon et al. 2004).

The nature of the hosts of aphidiines, usually living in dense colonies, has a number of impacts on the wasps. Some workers refer to them as quasi-gregarious, although what they do is not really very different from what some microgastrines such as *Cotesia glomerata*, do, i.e. utilise as many individuals in a colony as possible, progressively picking them off for oviposition one by one. Almohamad and Hance (2014) found that *A. ervi* avoids aphid colonies/patches where there were foraging larvae of two aphid predators, the hoverfly, *Episyrphus balteatus* (Syrphidae) and the multicoloured Asian ladybird, *Harmonia axyridis* (Coccinellidae), and also avoided areas with hoverfly larval tracks (Fig. 12.9). This all makes a great deal of sense, in that if they oviposit in aphids where either predator is present, there is a high chance that the aphid(s) containing their eggs will also be consumed. Further, syrphid larvae are likely to remain fairly close for a long time whereas ladybird larvae are more vagile.

Venom chemistry has been investigated by Tremblay et al. (1998) and Falabella et al. (2007). Several species induce transient host paralysis, e.g. *Monoctonus paulaensis* (Calvert & van den Bosch 1972) and *Binodoxys* (Desneux et al. 2009), and this may be a mechanism by which the wasps avoid both superparasitism and multiparasitism, although Calvert and van den Bosch suggested that it might just be a by-product of ovipositing into a host ganglion, and as Wharton (1993a) noted, 'The difference between temporary paralysis caused by venoms and behavioural immobility caused by the shock of oviposition is not always easy to distinguish'. Hågvar (1988) found that the venom of *Ephedrus cerasicola* is probably responsible for killing

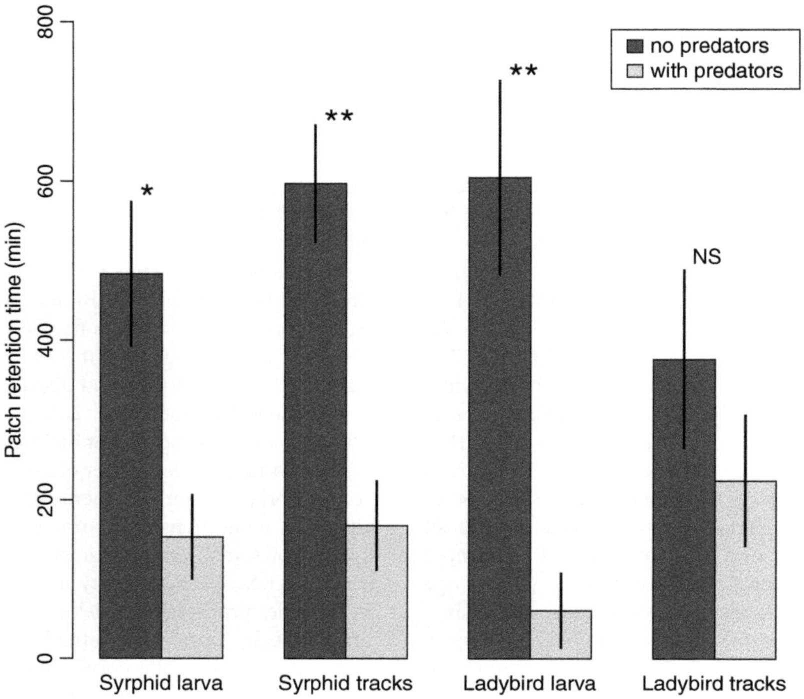

Fig. 12.9 Effects of the presence of aphid predator larvae or their trails, the hoverfly *Episyrphus balteatus* (Syrphidae) and the beetle *Harmonia axyridis* (Coccinellidae), on patch residence time (±SE) in the aphidiine *Aphidius ervi*. (* $p < 0.05$, ** $p < 0.01$.) (Source: data from Almohamad & Hance 2014.)

the egg of its competitor, *Aphidius matricariae*, if one is already present in the host aphid.

In addition to physiological changes, parasitism by aphidiines often leads to behavioural changes in the host aphid as its death approaches, and further, the change induced can depend upon the host's physiological state at the time. For example, Brodeur and McNeil (1990) reported that tomato aphids, *Macrosiphum euphorbiae*, parasitised by diapausing *Aphidius nigripes* larvae seek out darker (concealed) sites away from the host plant before they are mummified, while those hosting non-diapausing parasitoids tend to remain on the host plant and become mummified there.

Antennal sensilla have been studied in detail in *Aphidius rhopalosiphi* (Bourdaise et al. 2006) and *Lysiphlebus fabarum* (Xi et al. 2010). Six and seven types of sensilla were distinguished in the two species, respectively, using SEM and the numbers and distributions of three of these were found to differ between sexes in the former. Also, in the case of *A. rhopalosiphi*, it was demonstrated that cold storage, frequently used in the commercial production of large numbers of parasitoids, leads to a small increase in the development of abnormal sensilla.

Although the great majority of Aphidiinae are from the north temperate region where their hosts are certainly most abundant and speciose, there are a number of endemic Afrotropical, New Zealand and Chilean genera that presumably attack endemic southern genera of aphids, although host records are sadly lacking. The occurrence and endemicity in these southern regions of several genera (e.g. *Parephedrus*, *Psudephedrus*, *Vanhartenia*) displaying apparently plesiomorphic character states and probably representing very basal branches from the Aphidiinae tree have for a considerable time led workers to postulate that the group has a Gondwanan origin (Schlinger 1974; see also Gärdenfors 1986 for discussion).

Various informal hypotheses of tribe-level relationships were being postulated during the 1960s and 1970s based largely on appraisals of single character systems (e.g. Mackauer 1961, Tobias 1967, Edson & Vinson 1979; see P.T. Smith et al. 1999 for a summary). The first formal morphological analysis was by Chou (1984), who treated the Aphidiinae as a family, using the clustering UPGMA algorithm (which can be misleading in terms of phylogeny reconstruction because it unites groups on the basis of all similarities rather than just apomorphies). Three molecular phylogenies followed in fairly quick succession by a number of molecular phylogenies sung various gene fragments or combinations of fragments: Belshaw and Quicke (1997) (28S rDNA, elongation factor 1-α and cytochrome b), P.T. Smith et al. (1999) (NADH1 dehydrogenase), Sanchis et al. (2000) (18S rDNA), Kambhampati et al. (2000) (16S rDNA] and Shi and Chen (2005) (16S rDNA, 18S rDNA, ATPase 6]. All these studies had reasonable representation of each of the four major groups, they did not overlap well with the inclusion of several putatively basal taxa that might be expected to be important for rooting the phylogenies and they did not use the same gene regions. All agreed about monophyly of the Aphidiina+Trioxina, but in Smith et al.'s and Sanchis et al.'s studies they were not recovered as reciprocally monophyletic. The studies differed, however, in whether they place the Ephedrini or the Praini as the basalmost clade (Fig. 12.10) and, in

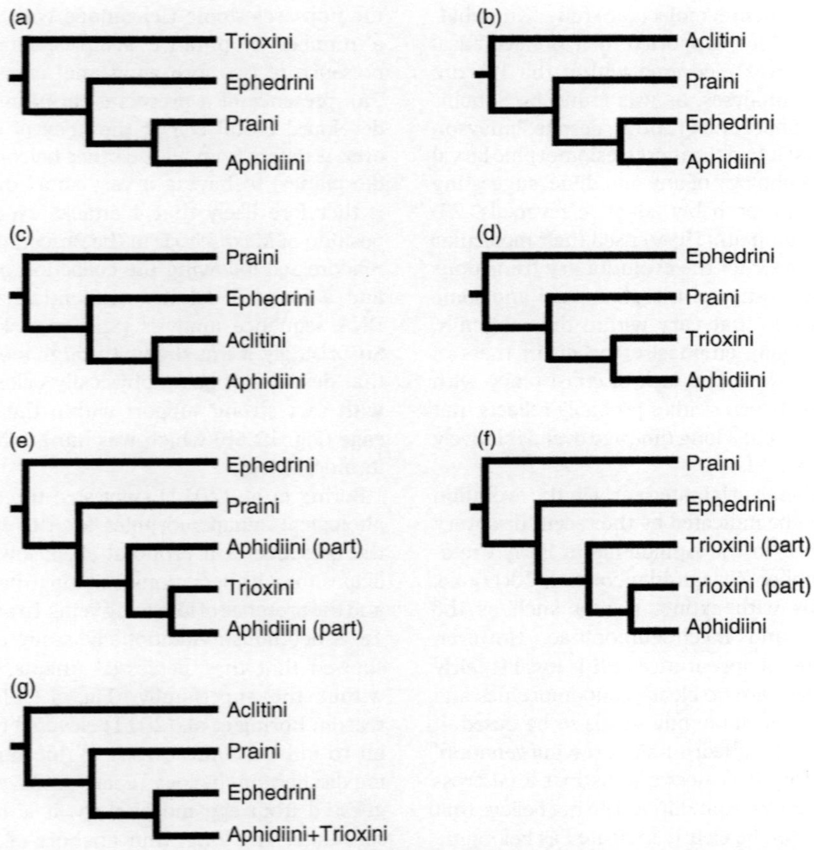

Fig. 12.10 Recovered relationships between the tribes of Aphidiinae from various studies. (a) From Finlayson (1990) based on final instar larval characters; (b) from Chou (1984) based on morphology and behaviour; (c) from Tobias (1967) and Edson and Vinson (1979) based on pupation habit and venom apparatus, respectively; (d–g), from Belshaw and Quicke (1997), Sanchis et al. (2000), P.T. Smith et al. (1999) and Kambhampati et al. (2000), respectively, based on various combinations of molecular markers (see text). Shi and Chen's (2005) tree was essentially the same as that of Sanchis et al.

those studies in which it was included, whether *Aclitus* was derived or basal (see Previous page). Male genitalia have been surveyed in detail by Tremblay (1979, 1981, 1983) and, through the presence of a cuspidal process, support the idea that Praiini are the most basal clade.

There is marked conflict in the placement of the uncommon or, at least, often hard to find genus *Aclitus*, members of which are associated with ant-attended root-feeding aphids. Morphological studies have placed it either in the Aphidiini or much more basally. Mackauer (1961) considered it a 'Tertiary relic' and separated it in its own subfamily, i.e. equivalent to tribes here as he considered the aphidiines as a separate family. However, he subsequently synonymised it with the Aphidiini (Mackauer 1968). Thus far they have only been included in one molecular study (Khambatpati et al. 2000), which supported their isolated basal position. *Lipolexis* was recovered within the Trioxini by the molecular analyses, as was found by Sanchis et al. (2000) and Shi & Chen (2005), despite Finlayson (1990) noting that it had the most plesiomorphic larval head capsule morphology of any aphidiine, suggesting that its features are probably adaptive reversals. P.T. Smith and Kambhampati (1999) used their molecular phylogeny to reconstruct the evolutionary transitions that occurred in a range of morphological and some biological characters that vary within the subfamily, and the results really emphasise the distinctness of the Aphidiini+Trioxini. The lack of consistency with rooting the tree between studies probably reflects that the basal branches split a long time ago over a relatively short geological period.

That the Aphidiinae originated early in the evolution of the family may be indicated by the recent discovery of a fossil assigned to the Aphidiinae in Early Cretaceous Spanish amber (Ortega-Blanco et al. 2009), i.e. contemporaneous with extinct groups such as the Protorhyssalinae and Eoichneumoninae. However, although the general appearance of this fossil is fairly aphidiine like, there are no clear synapomorphies and its assignment to the subfamily seems to be based in considerable part on 'Ephedrini-like forewing venation' despite the fact that it possesses a distinct anal cross vein absent in all extant Aphidiinae. I do not believe that at the moment it can be clearly indicated as belonging to this subfamily. However, if it truly is an aphidiine, it would, as Ortega-Blanco et al. pointed out, draw into question the Gondwanan origin of the subfamily that emerged from Belshaw et al.'s (2000) study.

Starý (1960) provided a key to the world genera (and subgenera) which is still useful even though a few new genera have been described since then and a few synonymised also. A key to the New World genera was provided by van Achterberg (1997c). Most of the species have been catalogued by Mackauer (1968). Ortega-Blanco et al. (2009) presented a key to the fossil genera which are well represented in the Eocene.

Maxfischeriinae

Papp (1994) described a new genus, *Maxfischeria* (Fig. 12.11a), based on a single species from Australia, which he placed in a new tribe, tentatively within the non-cyclostome Helconinae with which it shares a number of putative symplesiomorphies such as presence of two fore wing anal cross-veins (1a and 2a), presence of a prepectal carina and very slightly developed costal cell at the apex of C+SC+R. However, it differs from all the other helconines (including diospilines) in having a very short ovipositor, and it is therefore likely that it attacks exposed hosts. The position of *Maxfischeria* in the Braconidae was recently readdressed following the collection of fresh material and associated detailed anatomical study and also DNA sequence analysis (Sharanowski et al. 2011). Surprisingly, it was shown through molecular analysis that despite not being physically cylostome, it nested with very strong support within the cyclostome lineage (Fig. 12.5b) which was hard to believe based on its morphology.

Boring et al. (2011) suggested three external morphological synapomorphies for the Maxfischeriinae: the presence of a pronotal shelf; lower valve of the ovipositor with serrations from tip to base (Fig. 12.11e) and the presence of fore wing veins 1a and 2a (although 1a is nebulous). Additionally, study of fresh females showed that they possessed unique, anchored eggs within the superfamily (Fig. 12.11b–d). For this reason, Boring et al. (2011) elevated the Maxfischeriini to full subfamily status. As nothing is known of maxfischeriine biology apart from what might be guessed from egg morphology, it is only possible to speculate. The long, thin anchors of its eggs do not seem well adapted to being laid externally on a host, but instead it has been conjectured that they may be endoparasitoids with the anchor being used to secure the egg in some particular tissue or structure. They

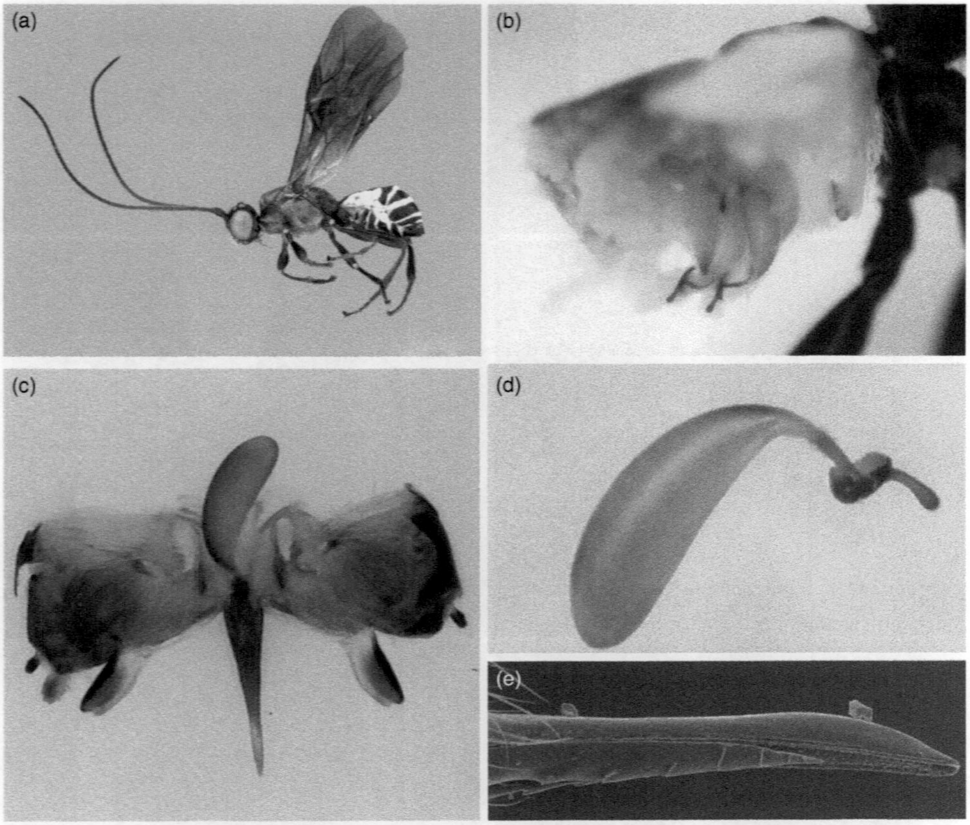

Fig. 12.11 Features of *Maxfischeria* spp. (Maxfischeriinae). (a) Habitus *of M. folkertsorum*; (b) eggs *in situ* in ovary of dissected female *M. ovumancora*; (c) dissection showing single egg with anchor in the posterior oviduct; (d) detail of single egg; (e) SEM image of ovipositor showing extensive lower valve serrations. (Source: reproduced by permission of Andrew Boring and Barb Sharanowski.) For a colour version, see Plate 21.

are slightly reminiscent of those of the ichneumonid subfamily Lycorininae in which they may be used to anchor the egg within a host's rectum.

It is interesting how the presence of the fore wing anal cross-veins was treated as a symplesiomorphy when *Maxfischeria* was considered to be a member of the helconoid group (many helconines have these two veins present), but as a synapomorphy when it was found to belong to a basal cyclostome clade, because these veins are usually absent in that group. However, a trace of 1a is present in some Rhyssalinae. Maybe its presence in *Maxfischeria* represents a true connection to ancestral braconids.

Mesostoinae (Including Canberreriini and Hydrangeocolini)

Van Achterberg (1975) erected the Mesostoinae for a new genus, *Mesostoa*, from Australia whose biology was then unknown and whose morphology gave little clue about relationships (Fig. 12.12). Independently, Tobias (1988) described another morphologically peculiar Australian genus, *Praonopterus*, and placed it in its own subfamily, the Praonopterinae. These two subfamilies were synonymised by Quicke and Huddleston (1989), who also reported for the first time that *Mesostoa* was associated with galls on the proteaceous

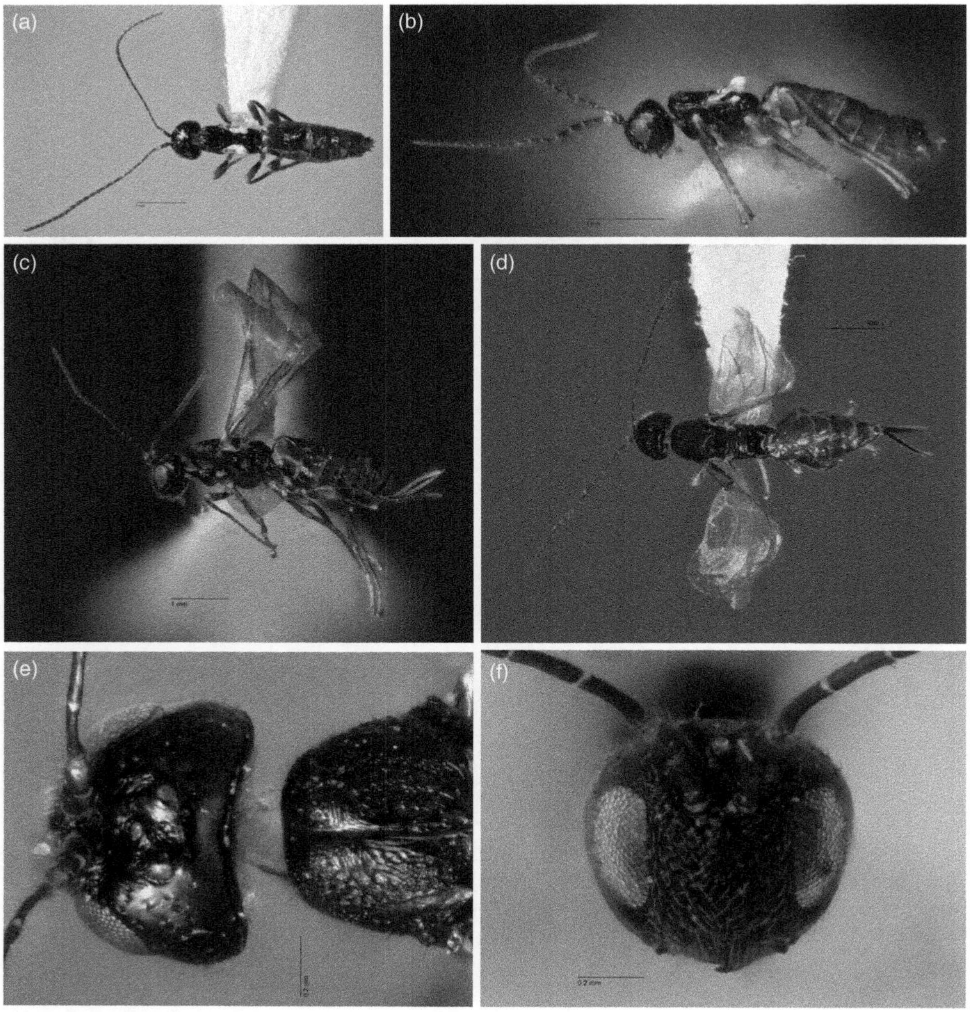

Fig. 12.12 *Mesostoa kerri* (Mesostoinae). (a, b) Micropterous male; (c, d) macropterous female; (e) close-up dorsal view of head and mesoscutum; (f) front view of face. (Source: reproduced by permission of Andrew Austin.). For a colour version, see Plate 22.

plant genus *Banksia*, although their assumption at that time was that *Mesostoa* was a parasitoid of a different gall-forming insect. That *Mesostoa* was actually the gall former on the *Banksia* host (Fig. 12.13) was discovered by Austin and Dangerfield (1998). The phytophagous nature of *Mesostoa* can now be seen to explain the confusing form of the final larval cephalic structures, described by Quicke and Huddleston, which showed a mixture of features associated with either ectoparasitism (papilliform antennae) and endoparasitism (toothless mandibles and loss of head sclerites). Since then, additional taxa have been discovered, but there is no complete revision and it should be emphasised that the biology is only known for a single species.

Molecular evidence strongly suggests that *Mesostoa* forms a monophyletic group with a few other very different looking groups from South America, Australia and New Zealand, viz. the Hydrangeocolini, Avgini and the enigmatic genera *Andesipolis* and *Doryctomorpha*. Both the Hydrangeocolini and Avgini

Fig. 12.13 Galls on *Banksia marginata* induced by *Mesostoa kerri* (Mesostoinae). (a) Herbarium specimen showing four large multilocular galls; (b) fresh gall cut sagitally showing *Mesostoa* cells, with larvae visible at left and right. (Source: reproduced by permission of Andrew Austin.)

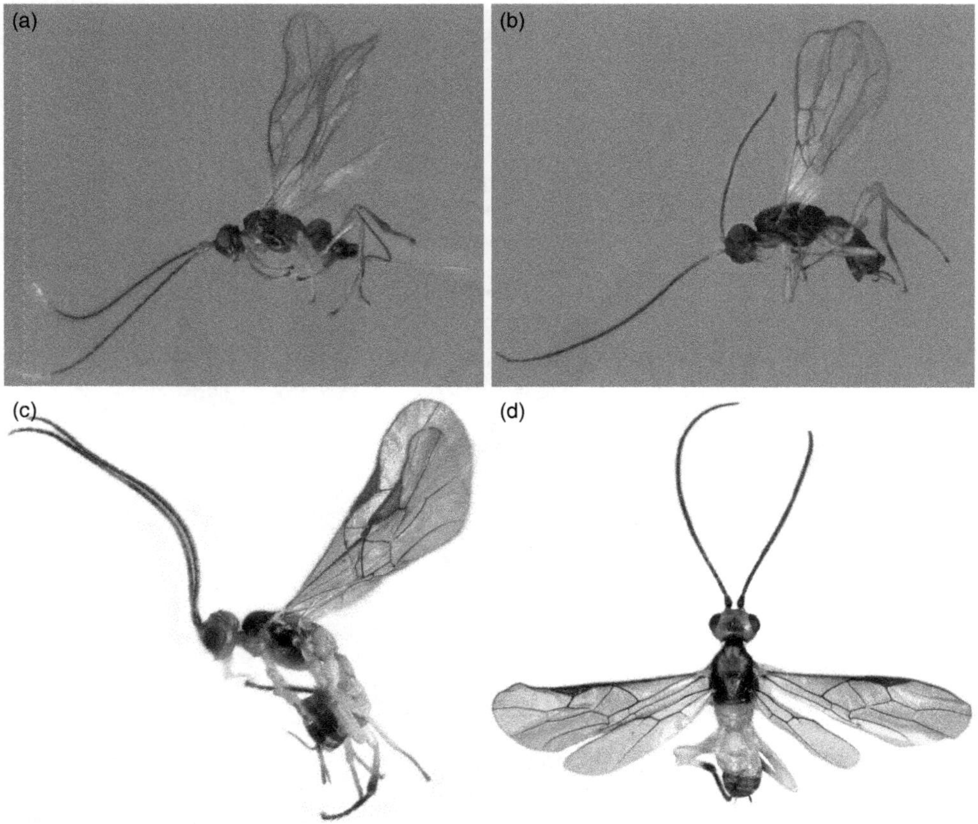

Fig. 12.14 Genera placed in Mesostoinae. (a) *Aspilodemon* sp.; (b) *Hydrangeocola* sp.; (c, d) *Andesipolis*. (Source: reproduced by permission of Angélica Penteado Dias.)

(as Canberreriini) were formally transferred to the Mesostoinae by Belshaw and Quicke (2002), who commented that it probably would be found to include other morphologically divergent taxa and that it would be unsatisfactory to elevate each of these small groups to separate subfamily level status. Wharton (1983) had previously recognised that they were probably rather basal and suggested an affinity with the Rhyssalinae. Zaldivar-Riverón et al. (2006) formally transferred *Andesipolis* to the Mesostoinae on the basis of DNA sequence data.

The Hydrangeocolini comprises three genera, *Aspilodemon* (Fig. 12.14a) and *Hydrangeocola* (Fig. 12.14b) from South America and *Opiopterus* from Australia (Wharton 1988). These all share a very narrow, virtually undifferentiated pterostigma and a very large second submarginal cell. *Hydrangeocola* was initially included in the Hormiinae but that was considered questionable by Wharton (1988), who could find no supporting symapomorphies, and he placed it tentatively, along with *Aspilodemon*, in the Rogadinae but refrained from according the group separate tribal status. Whitfield (1992) formally erected the subfamily Hydrangeocolinae. Wharton (1993b) considered the postfurcal vein m-cu which is similar to the condition found in Hormiini *s.s.* a result of convergence and instead considered them (specifically *Hydrangeocola*) probably to be related to Rhyssalinae. The biologies of *Aspilodemon* and *Hydrangeocola* are not fully known; both have been reared from galls, those of the former from cecidomyiid fly galls on Asteraceae (Oda et al. 2001), the latter from unidentified galls on *Hydrangea*.

The Avgini, which, according to van Achterberg (1995) is a senior synonym of both the Canberriini and

Austrohormiini, at least if the only sequenced genus, *Proavga*, which is thought to be closely related to *Canberria*, form another morphologically separate group that DNA suggests are closely related to *Mesostoa* (Belshaw & Quicke 2002). Although that relationship was not anticipated morphologically, Wharton (1993b) did note that it shared several features suggesting a relationship with *Opiopterus* (which he included in the Exothecinae). Nothing is known of the biology of any of the handful of, principally southern, included genera although *Avga* has been reported to parasitise Tortricidae, Gelechiidae and Limacodidae (van Achterberg 1995) although the reliability of these records is hard to assess. I suspect that further study will result in this group being found to be highly polyphyletic.

The aberrant South American genus *Andesipolis* (Fig. 12.14c,d) was not placed in any particular subfamily when it was first described (Whitfield et al. 2004) and its biology was unknown. Zaldivar-Riverón et al.'s (2006) placement of it within the Mesostoinae was disputed by Townsend & Shaw (2009) based on informal morphological consideration and its rearing as a gregarious koinobiont parasitoid of a pyralid moth larva. The *Andesipolis* larvae made cocoons external to the host remains although it was not discovered whether it was ecto- or endoparasitic. Hence Townsend and Shaw reasserted that the genus was best considered as a member of the Rhysipolinae. However, Sharanowski et al. (2011), in a multiple gene study, consistently recovered it with the clade Aphidiinae+Maxfischeriinae+Mesostoinae and never with the Rhysipolinae. Its placement in Sharanowski et al.'s analyses was not totally fixed and sometimes it was basal within the clade of these taxa and sometimes associated with the Maxfischeriinae, so they tentatively retained it within the Mesostoinae. As its biology seems unremarkable, it may well be a basal member of that group and simply displays plesiomorphic biology compared with the other derived or highly derived members of that group.

Doryctomorpha is a strange-looking animal (Fig. 12.15) which was described in the Doryctinae, although at a time when the limits of the group were quite vague. There is a single described species from New Zealand, and a second species from Baltic amber has also been placed in this genus. New Zealand specimens display considerable variation and there may in fact be quite a few undescribed ones in addition to the described one. Belokobylskij (2009) chose to discount the molecular placement of *Doryctomorpha* with the aphidioid group and instead transferred it to the Rhyssalinae (in tribe Doryctomorphini), on the basis of a few characters that are almost certainly associated with biology and are widely homoplastic. That placement is rejected here.

The described species of *Mesostoa* (Mesostoinae) are not cyclostome (Fig. 12.12f), although *Praonopterus* has a reduced but distinct depression (Tobias 1988). The other taxa treated here in the Mesostoinae are fully cyclostome and whilst the aphidiines are not generally considered cyclostome, some species do appear to be weakly cyclostome, so this feature appears to be moderately evolutionarily plastic.

THE REMAINING CYCLOSTOMES

This large group was recognised for a long time because of the modified lower clypeus and labrum of most of its members (see Fig. 1.2b). Variously treated as one 'subfamily' to up to about 13 subfamilies in recent work, the limits of this clade have been generally fixed for the last 50 years, although strict adherence to the idea that its members must be physically cyclostome has misled some. The majority of Opiinae, for example, are not cyclostomes (see Fig. 12.31b) and their close relatives the Alysiinae have such modified mouthparts that they cannot be classified that way (see Fig. 12.30a). It is clear that the cyclostome condition has been secondarily lost on a number of occasions within the group. We do not know with any certainty what the function of cyclostomicity is, but it is possible to deduce some sort of pattern. For example, the mandibles of those cyclostomes that attack wood-boring hosts (principally in the Doryctinae and Braconinae) are used to chew the freshly eclosed wasp's way out of their pupation site, through wood, are usually twisted such that the upper tooth is anterior and conceals the lower one in frontal aspect.

Phylogenetic relationships among the cyclostome subfamilies have been assessed in a number of studies and although they are improving in the sense that they are starting to reach some sort of consensus, they have undoubtedly been hampered by the lack of clarity over the borders or generic compositions of several subfamilies, most notably of the Rogadinae *s.l.*, Exothecinae and the nightmare grouping, the Hormiinae. Only recently have sequence data started to become available for a clearer picture to emerge. The main studies to date are those of Dowton (1999), Zaldivar-Riverón et al. (2004) and Sharanowski et al. (2011); although Sharanowski

Fig. 12.15 *Doryctomorpha antipoda* (probably Mesostoinae *s.l.*). (a) Head oblique view showing wide, dorsally weakly arched, hypoclypeal depression; (b) head frontal view, also showing densely spinose fore tibia; (c) female habitus side view. (Source: reproduced by permission of Darren Ward, Landcare Research, New Zealand.)

et al. (2011) did not include as many exemplars in their study, they did include a considerably larger number of gene fragments. As phylogenetic studies proceed, these subfamilies are being broken up such that most workers now recognise various groups (e.g. Pambolinae and Lysiterminae) as valid subfamilies.

Here I treat them in three groups of subfamilies. The first are a probably paraphyletic assemblage comprising Doryctinae. Pambolinae, Rhysipolinae and Rhyssalinae, the last of these very likely the most basal. Next a group that appears fairly closely related including Hormiinae, Lysiterminae, Betylobraconinae and Rogadinae. Finally is a group referred to as the alysioid complex by Sharanowski et al. (2011) within which I also include the Braconinae.

Identification of genera in the more fuzzy area of Hormiinae *s.l.* may require the use of several keys as there is incomplete overlap of groups treated. The main keys that cover a good proportion of the genera are those of van Achterberg (1995), which includes, along with members of some other groups, the Hormiinae and Lysiterminae, and Whitfield and Wharton (1997), which lumps into one key the New World genera of Exothecinae, Hormiinae, Lysiterminae, Mesostoinae:

Hydrangeocolini, Pambolinae, Rhysipolinae and Rhyssalinae, along with *Allobracon*, whose subfamily placement is uncertain, and the aberrant doryctine *Monitoriella*, which does not key readily to the Doryctinae.

Doryctinae (Including Ypsistocerini)

This is a very large, cosmopolitan subfamily which is particularly diverse in the Neotropical region – although past lack of detailed work on the tropical Old World fauna might be partly responsible for making this relatively neotropical taxonomic richness quite so apparent. The groups identity has remained more or less intact if one ignores some past decisions (e.g. Čapek 1970) to divide it into a small number of separate subfamilies (Doryctinae, Hecabolinae, Spathiinae), and this is perhaps rather surprising in that it has for the most part been defined by characters that are putatively plesiomorphic or simply associated with attacking wood-boring hosts (although this is by no means the only biology). Morphologically the group is rather diverse, although species do tend to have a certain *Gestalt* (Fig. 12.16) with a typically rather robust head. Past workers have defined the group usually on the basis of them having a longitudinal row of pegs (sometimes called chaetobothria), or at least thickened spines, along the fore tibia. These are typically more apparent in larger species and can be very difficult to make out in some small ones. It is also a feature associated with egress (Eggleton 1989) and similar pegs, sometimes in rows, are found in some braconines such as *Rhadinobracon* species (Braconinae). All parasitoid doryctine species, with the possible exception of *Sericobracon* (see below), are idiobiont ctoparasitoids associated with hosts that are concealed to various extents, ranging from leaf-miners and stem-borers to deeply concealed xylophagous beetles. Typical of wasps with this biology, their venom induces permanent flaccid host paralysis. Coleoptera probably constitute most hosts but a few species are parasitoids of members of other orders.

A relatively few species are involved in biological control, either through release or serendipitous utilisation of available harmful hosts. *Parallorhogas pyralophagus* is an important biological control agent of the pyralid rice and sugarcane pest *Eoreuma loftini* in Texas. *Heterospilus prosopidis* attacks bruchine chrysomelid beetle larvae feeding in beans and, because of its ease of laboratory culture, has been used as a model organism for studies of patch utilisation, metapopulations and sex ratio determination. Others involved in pest control are parasitoids of wood-boring cerambycid or buprestid pests, perhaps most importantly of the eucalypt cerambycid borers of the genus *Phoracantha*, which is host to the large and pretty *Syngaster lepidus*, and also several species of *Jarra* (Austin et al. 1994). In the wild in Australia, these parasitoids partition the host resource according to both host larval size and their depth within the wood (Paine et al. 2000). The gregarious *Jarra* species prefer larger hosts than the solitary *Syngaster*, while the latter utilises more deeply concealed hosts (Fig. 12.17). Efforts have been made in the USA to develop rearing techniques for *Syngaster* and *Jarra* species for augmentative release, including the development of new techniques for rearing subcortical beetle parasitoids (Millar et al. 2002). Very similar conclusions were reached by Urano & Hijii (1995) working with *Atanycolus initiator* (Braconinae) and *Spathius brevicaudis* (Doryctinae) attacking a variety of temperate subcortical hosts. More recently, with the invasion of North America by the emerald ash borer, *Agrilus plannipennis* (Buprestidae), from its native eastern Palaearctic, the doryctine *Spathius agrili*, originally from China, has been receiving considerable attention as a useful control agent, and was released in the USA in 2007 (X.-Y. Wang et al. 2010b). Short-range host location in this species appears to rely almost entirely on vibrations produced by the feeding activity of the host, and immobile ones, such as those already paralysed, are not attected.

Relationships within the subfamily are very poorly known and although a number of tribes and subtribes have been recognised, these are based very largely on the Palaearctic fauna and so are not necessarily appropriately defined if the world fauna is taken into consideration. Further, many of the features used by some authors in their tribal diagnoses are just 'trends' (e.g. Belokobylskij 1993) and typological ideas have been given importance. A formal maximum parsimony analysis of morphological data for all genera recognised at the time was presented by Belokobylskij et al. (2004), including many novel data on internal anatomy, and failed to recover as monophyletic most of the tribes, and some aberrant taxa were recovered among the outgroups. This was followed by a simultaneous molecular and morphological analysis (Zaldivar-Riverón et al. 2008), which again failed to recover the subfamily as monophyletic, a couple

Fig. 12.16 Various mounted specimens of Doryctinae showing some of the variation in body form. (a) *Iare* sp.; (b) *Sharkeyella* sp.; (c) *Histeromeroides onkotorebus*; (d) *Leptorhaconotus seyrigi*; (e) *Evaniodes spathiiformis*; (f) *Schlettereriella variegata*. (Source: reproduced by permission of Yves Braet.) For a colour version, see Plate 23.

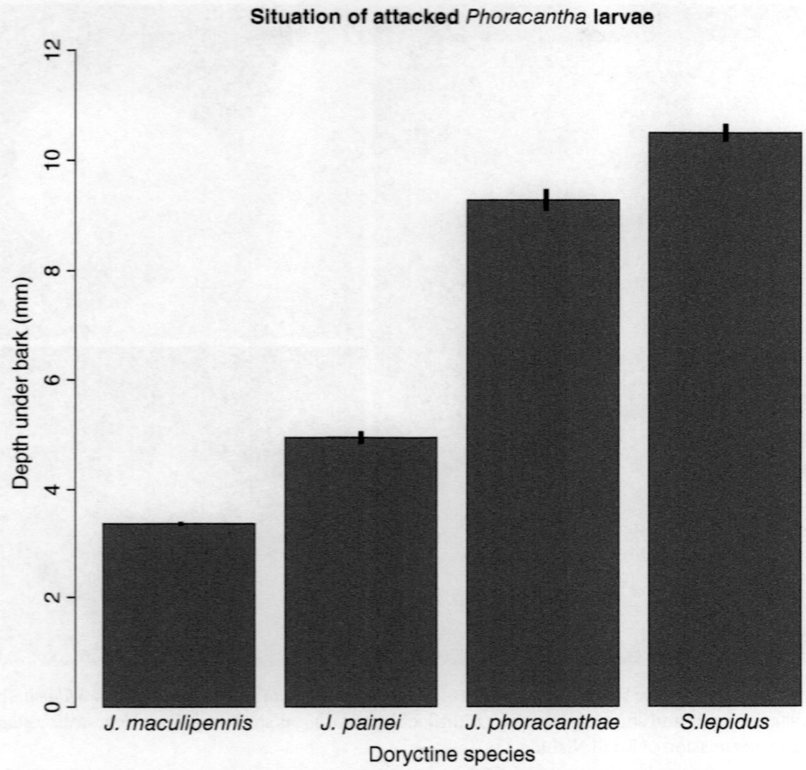

Fig. 12.17 Resource partitioning by four doryctine parasitoids of the cerambycid *Eucalyptus* borer, *Phoracantha semipunctata*, in its native range in Australia (mean ± SE). (Source: data from Paine *et al.* 2000.)

of less derived genera again associating separately with outgroups and therefore most likely reflecting lack of signal rather than true polyphyly. Nevertheless, they found a very strong biogeographic signal and also strong evidence for the polyphyly of several large cosmopolitan genera. The monophylies of both *Spathius* and *Doryctes* were strongly rejected statistically, indicating considerable convergence. Molecular dating estimates placed the separation of the African and New World clades at 53–59 Mya, considerably after the likely split of the land masses (93–110 Mya) indicating dispersal rather than vicariance as being responsible for current large-scale distribution.

Detailed descriptions of doryctine life history are sparse. Hill and Smith (1931) described the biology of *Heterostilus cephi*, a parasitoid of the wheat stem sawfly, *Cephus pygmaeus*, and Ayyar and Narayanaswami (1940) described the biology of *Spathius vulnificus*. This *Spathius* is interesting in that females have a very long preoviposition period lasting nearly 2 weeks. Z.Q. Yang et al. (2010) presented a particularly detailed study of *S. agrili*, a parasitoid of the emerald ash borer, a pest cerambycid, with careful control by isolating individual larvae and marking exuviae. As with most previous descriptions, Yang et al. found five larval instars, although Ayyar and Narayanaswami reported only four and Hill and Smith only three. All instars are hymenopteriform with an open spiracular system and well-developed papilliform antennae. Marsh and Melo (1999) described the biology of a group of species that are specialist parasitoids on solitary sphecid wasps. The egg of *Dendrosoter protuberans* is unusual in that it has a short, thick stalk that is bent back on itself, the purpose of which is unknown. Most temperate species are solitary but a few *Heterospilus*, *Spathius* and *Doryctes* species are gregarious (Hill & Smith 1931, Shaw & Huddleston 1991), but the biologies of most of the diverse tropical genera are unknown. There is

Fig. 12.18 Galls of cecidogenic New World Doryctinae. (a) Galls of *Monitoriella* sp. on *Philodendron wilburii*; (b) unilocular gall of *Monitoriella elongata* cut open and showing larva; (c) multilocular *Allorhogas* sp. fruit gall on *Conostegia xalapensis* cut open. (Source: reproduced by permission of Kenji Nishida.)

considerable variation in the strength of the cocoon and even within the, admittedly very speciose, genus *Heterospilus* they range from very thick and tough to very thin, and even may be absent in some Neotropical species attacking nests of pemphredonine Sphecidae, although as noted by Marsh & Melo (1999), this requires confirmation.

Some doryctines have evolved to become phytophagous either as seed predators or gall-formers (Fig. 12.18). The best known cases are members of the genera *Allorhogas* and *Monitoriella* (Fig. 12.19) (Infante et al. 1995). The enigmatic genus *Monitoriella* was originally described as a member of the Hormiinae and lacks the apparently diagnostic row of pegs on the fore tibia, although in its original description Hedqvist (1963) wrote that it fitted best in the Doryctinae. *Monitoriella* species are primary gall formers on *Philodendron* (Araceae). Not only has it now been shown to be a member of the Doryctinae using DNA sequence data, but it also belongs to a clade of other gall formers and gall-associated genera including *Donquickeia* and *Allorhogas* (Zaldivar-Riverón et al. 2007). Some *Allorhogas* species are primary gall formers (Marsh et al. 2000) whereas others are seed predators (de Macêdo & Monteiro 1989, Chavarría et al. 2009), and the response of the host plant to seed predation by one species has been described in detail (de Macêdo et al. 1998). More recently, other gall-associated New World genera have been shown to be phytophagous. Sometimes phytophagy might be in the form of inquilinism within galls formed by other insects and it can be hard to know the exact biology. Penteado-Dias (2000), for example, pointed out that the Neotropical genus *Donquickeia* is reared from cecidomyiid fly galls on a variety of plants, but its larval cephalic structures are far more like those of some other phytophagous taxa. Insect-induced galls are readily consumed from the inside and so gall feeders often have reduced mandibles and most other cephalic structures including the epistoma and enlarged labial sclerite. This contrasts with the larvae of the seed-feeding *Bracon* species such as *B. phytophaga*. Using an enlarged data set, Zaldivar-Riverón et al. (2014) have corroborated the earlier findings

Fig. 12.19 Mounted female of the gall-forming doryctine *Monitoriella costalimai*. (Source: reproduced by permission of Angelica Penteado Dias.)

that suggested a single origin of gall association within the subfamily though with *Allorhogas* being recovered as paraphyletic with respect to several other genera. Their results indicate that cecidogenesis arose via parasitoidism on other gall-forming hosts probably in the middle Miocene to early Oligocene, 16.33–30.55 Mya.

One member of the above, gall-associated group is the genus *Psenobolus*. It has been recognised as containing at least partly phytophagous species that live inside fig syconia and these show a remarkable degree of convergence morphologically and biologically to some fig wasps (Chalcidoidea: Agaonidae) such as *Idarnes* (Ramirez & Marsh 1996). The males of these *Psenobolus* are dwarfed and highly modified for fighting, just as in pollinating fig wasps.

Winglessness and brachyptery have evolved on several occasions within the subfamily, being found in a total of 16 genera (Kula 2011). These vary in whether the male has reduced wings while the female is macropterous (e.g. *Nipponecphylus*, some *Psenobolus* spp. including fig-inhabiting ones, see above), vice versa (e.g. *Australospathius pedestris*) or both have reduced wings (e.g. *Aptenobracon formicoides*, *Doryctopsis zealandicus*, *Echinodoryctes*, some *Heterospilus* spp.) or wing development polymorphic (e.g. some *Ecphylus* species) (data largely drawn from Kula 2011). The reason for the very high prevalence of rachyptery and aptery in the Doryctinae (40 species in some 20 genera) is not completely clear (Belokobylskij & Kula 2012, Belokobylskij & Austin 2013). It seems likely to be associated with host niche in most instances, with hosts dwelling in confined spaces that can only be reached by crawling through tunnels or substrate.

Males of a number of genera (predominantly Heterospilini, but also some Hecabolini and rarely Doryctini and Spathiini) possess a more or less complex pseudostigma in the hind wing, i.e. a large swollen area at the confluence of veins 1M, 1r-m and SC+R;

one genus is also known in which the female has an hind wing pseudostigma (Braga et al. 2002), but in this case the male is as yet unknown. The function of this structure is not known, although given that its lower surface is densely setose with wick-like groups of setae and is only seen in the females of one out of many species, it is tempting to speculate that it is a source of sex pheromones. Cossé et al. (2012) reported on a male-produced sex pheromone in *Spathius agrili* and out of seven male-specific compounds identified by GC–MS, three, viz. dodecanal, (4*R*,11*E*)-tetradecen-4-olide and (*Z*)-10-heptadecen-2-one, were the most important behaviourally active ones as assessed by flight tunnel assay.

The Ypsistocerinae was based on a highly aberrant and small braconid found only in termite (*Nasutitermes*) nests in Central and South America (Kistner et al. 2000). Originally including *Ypsistocerus*, the genus *Termitobracon* was included by Brues (1923). The only addition has been a single, morphologically rather less derived genus, *Embobracon*, whose biology is unknown and whose relationship with the others is not strongly supported. No ypsistocerines have been sequenced so interpretation of their relationships is necessarily based on features of their morphology. Although they do not possess any of the supposed ovipositor synapomorphies of the Doryctinae proposed by Quicke et al. (1992a), viz. a dark tip, double nodus and accessory teeth, the venom apparatus of *Termitobracon* (Quicke & van Achterberg 1990, Quicke et al. 1992e) is distinctly doryctine-like in having two separate insertions of the venom ducts on to the primary venom duct. Nothing is known of the biology of either of these termite-associated genera, i.e. it is not known whether they attack termites or some inquilines in their nests and whether or not they are ecto- or endoparasitoids. When placed in dishes with termites and inquilines they showed little response and although they occasionally mounted a termite no attempt at oviposition was observed (Kistner et al. 2000). They were inappropriately included within the Rogadinae by Sharkey (1993). Recently, two more and less morphologically bizarre genera of doryctines from Old world termite nests have been discovered (Belokobylskij 2002).

Sericobracon, a genus known from only two species and very few specimens from Trinidad, is an endoparasitoid of the embiopteran *Clothoda urichi*, the host being paralysed in its web and the final instar *Sericobracon* larva emerging to spin a cocoon (Shaw & Edgerly 1986). Neither species has been collected or reared since the original description and, although placed at the time in the Doryctinae, the authors noted that its subfamily placement is problematic and its endoparasitic (albeit idiobiont) biology would be unique within the Doryctinae. It shared with some other doryctines, especially *Heterospilus*, a posteriorly desclerotised fore wing vein 2SR and also possesses a ventral tubercle on the hind coxa, so it is possible that it is just a highly specialised doryctine, but molecular data, should more material become available, would be very welcome.

The New World genera can be identified using the key by Marsh (1997), although there have been a few additional taxa described or synonymised since then. There is no recent synthesis for the Old World fauna, except for the Russian 'Far East' (Belokobylskij 1998) and for Japan (Belokobylskij & Maeto 2009)[3].

Pambolinae

This group was afforded subfamily rank in the 19th century as the tribe Pambolini, but for a long time it was included within either the Hormiinae *s.l.* or the Rogadinae *s.l.*, in part due to the difficulty of defining natural groups within these entities (e.g. Wharton 1993b). With the desire to restrict the Rogadinae to a monophyletic group that mummify Lepidoptera larvae, the Pambolinae started to be treated as a separate subfamily (e.g. van Achterberg 1993a), and this is now widely accepted.

Morphologically it is a rather poorly defined group and its monophyly is far from certain. The group comprises the Pambolini, i.e. *Pambolus* (Fig. 12.20) and (putative) relatives, and the Chremylini (here restricted to just the genus *Chremylus*); however, Wharton (1993b) commented that the relationship between the two groups of genera (tribes) '... does not appear to be a close one'. In agreement with Wharton (1993b), van Achterberg (1995) had included *Cedria* and relatives (Cedriina) within the Chremylini as they also have few flagellomeres, although they differ from *Chremylus* in some other characters. The *Cedria* group, however, appear to belong to the Lysiterminae on the basis of DNA data (Zaldivar- Riverón et al. 2006), which is where Whitfield (1992) had included them; unfortunately, there are no molecular data available for any of the other taxa. Several *Pambolus* species and some *Chremylus* species are brachypterous (Belokobylskij & Kula 2012).

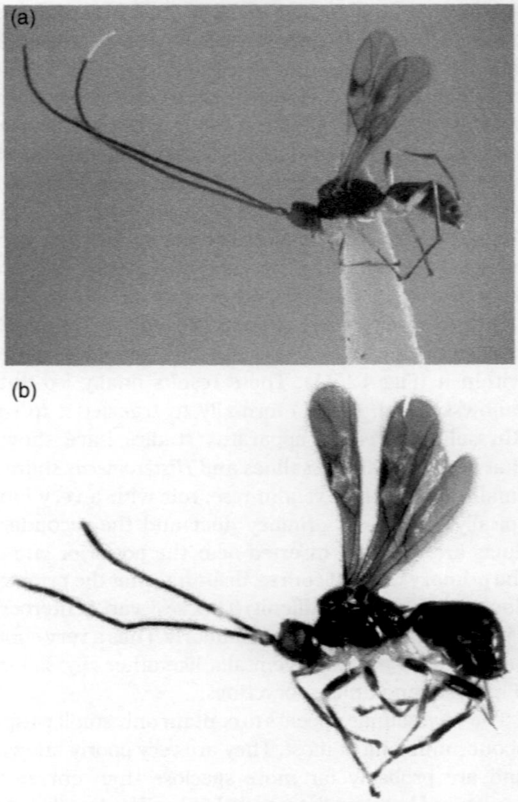

Fig. 12.20 *Pambolus* spp. (Pambolinae). (a) *P. albospina*; (b) *P. rastafari*. (Source: reproduced by permission of Juan Jose Martinez.)

Pambolus is common in many tropical faunas but very poorly known biologically. It has been reported as attacking Chrysomelidae, but details of the relationships are unknown. The Australian genus *Notiopambolus* has been found to be associated with a cryptocephaline chrysomelid beetle, whose larvae dwell in a psychid-type portable case feeding among *Eucalyptus* leaf litter (Zaldivar-Riverón & Quicke 2002), and it was proposed that the flattened ovipositor of *Notiopambolus* is probably used to gain access to the host by forcing it between the case and substrate. It is likely that *Notiopambolus* is ectoparasitic, but that is purely surmise. *Chremylus rubiginosus* has been reared from the case-bearing clothes moth, *Tinea pellionella*, in both Europe and North America (H.C. Mason 1948) and is reported to have five larval instars.

Rhysipolinae

This is a small group, based on the genus *Rhysipolis* (see Fig. 12.25d), and was recognised as comprising a distinct subfamily-level group by Belokobylskij (1984), and this was followed by van Achterberg (1995), but its definition, composition and status have been in almost perpetual flux, as summarised by Ortega-Blanco et al. (2012), and *Pseudorhysipolis* appears to be closely related to the Pambolinae. The rare neotropical genus *Andesipolis*, which was placed in the Rhysipolinae, was removed to the Mesostoinae on the basis of DNA data (Zaldivar-Riverón et al. 2006), subsequently transferred back to the Rhysipolinae by Townsend and Shaw (2009) on the basis of morphology and the newly discovered biology of a new species, and then re-transferred to the Mesostoinae by Sharanowski et al. (2011). All this leaves the subfamily comprising *Rhysipolis*, *Noserus*, *Cantharoctonus* (see Fig. 12.25c) and the extinct *Trichelyon*. In addition, the very distinctive New World genus *Allobracon* might well belong here on the basis of the DNA analyses performed to date (Zaldivar-Riverón et al. 2006). Van Achterberg (1995) and Belokobylskij & Maetô (2006) considered it to belong to the Avgaini, along with various other taxa that had variously been placed in Hormiinae *s.l.*, but this is not tenable given the molecular data.

Rhysipolines are small wasps but biologically of great interest because at least members of the genus *Rhysipolis* are koinobiont ectoparasitoids, although virtually nothing is known about the biologies of the other genera (Scatolini et al. 2002), so the generalisations often made may well be unfounded. *Rhysipolis* attack various leaf-mining caterpillars that construct multiple mines during their development or leave the mine to find a different place for pupation (M.R. Shaw 1983). The female wasp stings their host, causing temporary paralysis, which then gives her time to glue her large yolky egg firmly on to one of the host's intersegmental membranes. The host recovers for a while, continuing feeding, but the *Rhysipolis* venom causes it to cease normal moults even if it was not in its final instar, and instead the host is caused to construct its cocoon and (usually) enter a prepupal state prematurely, whereupon the wasp larva commences feeding properly.

This effect on host development, caused only by the wasp's venom, is remarkably similar to that shown by the rogadine *Clinocentrus*, and M.R. Shaw (1983) considered the biology of *Rhysipolis* as representing a stage in the development of truly koinobiont endoparasitoid biology in the Rogadinae. However, molecular phylogenetics does not support a direct link and it is best to consider the biology of *Rhysipolis* as representing one possible transitional stage. Of interest in Shaw's argument is that unlike in nearly all endoparasitic koinobionts, the eggs of rogadines are rather large and yolky, similar to those of *Rhysipolis*, suggesting that they might only recently have evolved to be endoparasitoids. However, given that there are apparently perfectly good rogadines in ca 40 Mya old Baltic amber, time may not be an important factor in the evolutionary strategy employed by that group.

The very few reared specimens of *Allobracon* are from leaf-mining Coloeoptera, probably Buprestidae and Hispidae, although really nothing can be said with great certainty (Wharton 1993b). They appear to be gregarious and make cocoons within the leaf mine outside the host remains and phylogenetic regression based on ovipositor measurements indicates that they are probably ectoparasitoids (Belshaw et al. 2003).

Numerous specimens have been collected at night and so they are probably principally nocturnal.

Rhyssalinae

The subfamily Rhyssalinae was erected by Quicke and van Achterberg (1990) to include *Rhyssalus* and a few related genera, its members having been placed variously in the Doryctinae, Rogadinae s.l., Hormiinae or Exothecinae. They are hard to define by morphological synapomorphies and thus the exact limits have not been constant, and DNA sequence data have shown that some taxa described therein are actually more closely related to the Rogadinae (e.g. *Anachyra*; Zaldivar-Riverón et al. 2006). Belokobylskij (2009) recognised and provided a key to four tribes in the subfamily, viz. Rhyssalini, Acrisidini, Achaiabraconini and Doryctomorphini but the last of these is considered, on the basis of molecular data, to belong to the basal, aphidioid clade and is treated here as Mesostoinae *sensu lato*. Their internal relationships as derived from molecular and combined analyses are shown in Fig. 12.21.

The highly apomorphic genus *Histeromerus* (Fig. 12.22), was often treated as a doryctine in the old broad concept of that subfamily until van Achterberg (1976c) transferred it to the Braconinae largely on the basis of it lacking an occipital carina. Although it is best known from the Holarctic, *Histeromerus* is now also known to occur also in South-East Asia and Australia. A broken (wingless) specimen was amazingly described as a new genus of ichneumonid, *Mithotynia*, by Hedqvist (1976). However, *Histeromerus* has nothing to do with either the Braconinae or Doryctinae, and van Achterberg (1984) re-instigated it as a subfamily in its own right. Consideration of its venom apparatus (Quicke et al. 1992e), male genitalia (Quicke & van Achterberg 1990) and all molecular studies that have included relevant taxa (Belshaw et al. 1998, Belshaw & Quicke 1999, Zaldivar-Riverón et al. 2006, Sharanowski et al. 2011) show it not only to be closely related to the Rhyssalinae but also probably nested within it (Fig. 12.21). These results finally led Sharanowski et al. (2011) formally to transfer it to the Rhyssalinae. Venom apparatus studies have shown that both typical rhyssalines and *Histeromerus* share a small unsculptured venom reservoir with a very long spirally sculptured primary duct and the secondary ducts are uniquely inserted near the posterior end of the primary duct (of course, defining what the primary duct is in this case is difficult) (Quicke & van Achterberg 1990, Quicke et al. 1992e). Similarly if has a very short basal ring to the male genitalia like other rhyssalines (Fig. 2.7a) and unlike doryctines.

The Rhyssalinae appears to contain only small wasps, about 5mm long at most. They are very poorly known, and are probably far more speciose than currently recognised because the tropical fauna has hardly been investigated apart from a few preliminary studies in Vietnam. Many occur frequently in Malaise trap samples from tropical moist forests. Given that rhyssalines, although small, are far from the tiniest of braconids and that they are fairly frequently encountered, it is surprising how little is known about their hosts and general biology. The realisation that rhyssalines probably occupy a fairly basal position within the family combined with the dogma that the ancestral parasitoids (and especially the ancestral ichneumonoids) are idiobiont ectoparasitoids of wood-boring hosts has led to the expectation that this is what rhyssalines will do. For the traditional members of the Rhyssalinae there are host records for only four genera. Shaw and Huddleston (1991) noted that *Oncophanes* appear to be mostly associated with arboreal leaf-miners, *Rhyssalus* with 'beetle-infested wood', and M.R. Shaw (1993b) reported a strange host association for *Dolopsidea*, which was reared from a scarabaeid beetle larva, probably Aphodiinae, but whose remains were, untypically, associated with dead wood, although other collecting

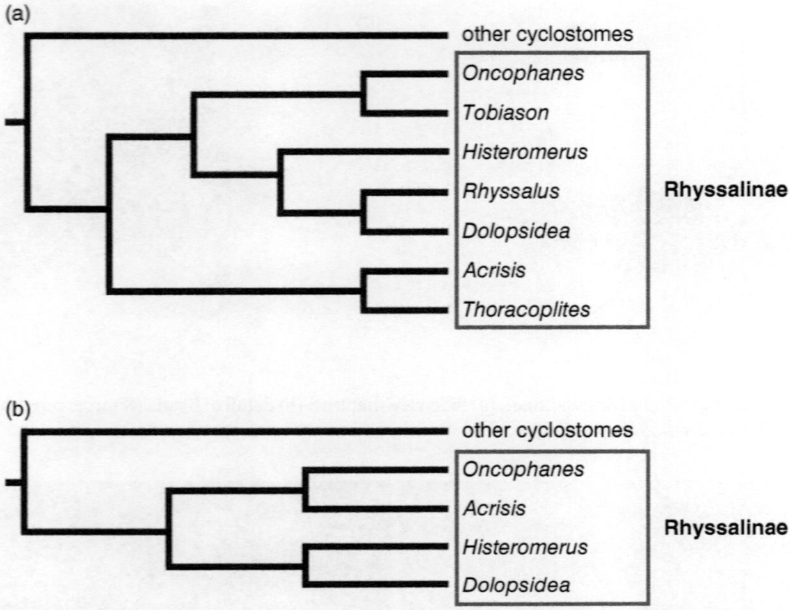

Fig. 12.21 Relationships of genera of Rhyssalinae and *Histeromerus*. (a) As recovered by Zaldivar-Riverón et al. (2006) based on 28S rDNA D2+D3 region plus COI barcoding region plus morphology; (b) as recovered by Sharanowski et al. (2011) based on six fragments of four gene regions.

evidence suggests that *Dolopsidea* may more normally be associated with old trees and there are host records from buprestids. *Pseudobathystomus* has been reared from oecophorid moth larvae. Clearly, much more information is needed before a full picture can be achieved.

The biology of *Histeromerus mystacinus* was described in detail by M.R. Shaw (1995), who showed that they are gregarious ectoparasitoids on larvae, prepupae and pupae of their wood-boring beetle hosts. Shaw and Huddleston (1991) discounted previous suggestions that it might be endoparasitic. In the studied case, broods ranged from seven to 46 larvae per host. The majority of host records are from relatively large hosts such as Buprestidae, Cerambycidae, Lucanidae, Lyctidae and possibly Elateridae (but see Shaw & Huddleston 1991). Indeed, as with many parasitoids of wood-borers, there is necessarily some degree of uncertainty associated with some of these. Van Achterberg (1992a) also recorded the Ptinidae as a host and if this is correct then while *H. mystacinus* is gregarious on larger hosts (Fig. 12.22a,b), it must almost certainly be a solitary parasitoid of the ptinid. Their modified heads and strong mandibles, together with their robust and highly spinose fore tibiae (Fig. 12.22), appear to enable them to dig their way through rotting, but still quite firm, wood to reach hosts in other burrows and M.R. Shaw (1995) reported that when some emerged in a corked glass tube in transit, they managed to bore their way out (Fig. 12.23d) – a feat few other insects could accomplish. The wood pieces bitten off as they burrow are passed ventrally beneath the wasp as she progresses. The host is paralysed by the female's venom and the female shows a degree of parental care, remaining with the developing brood, and in two instances where females deserted their broods, the eggs failed to hatch. The wasp is also interesting in that it probably shows local mate competition, with males routinely mating with siblings, and it correspondingly has a highly female-biased sex ratio. In addition, the females investigated by M.R. Shaw (1995) were reluctant to fly and this may account for their comparative scarcity in Malaise traps, although observations by coleopterists would suggest that they are not uncommon, a situation

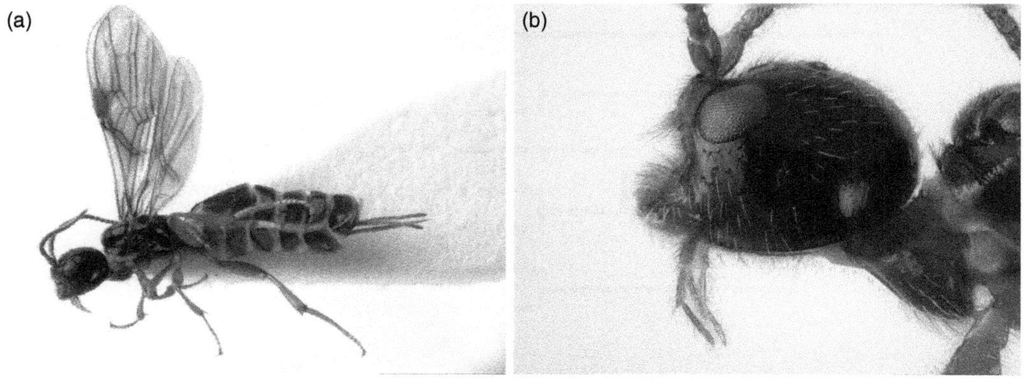

Fig. 12.22 *Histeromerus mystacinus* (Rhyssalinae). (a) Side view habitus; (b) detail of head. (Source: reproduced by permission, © C. van Achterberg, Naturalis Biodiversity Centre.)

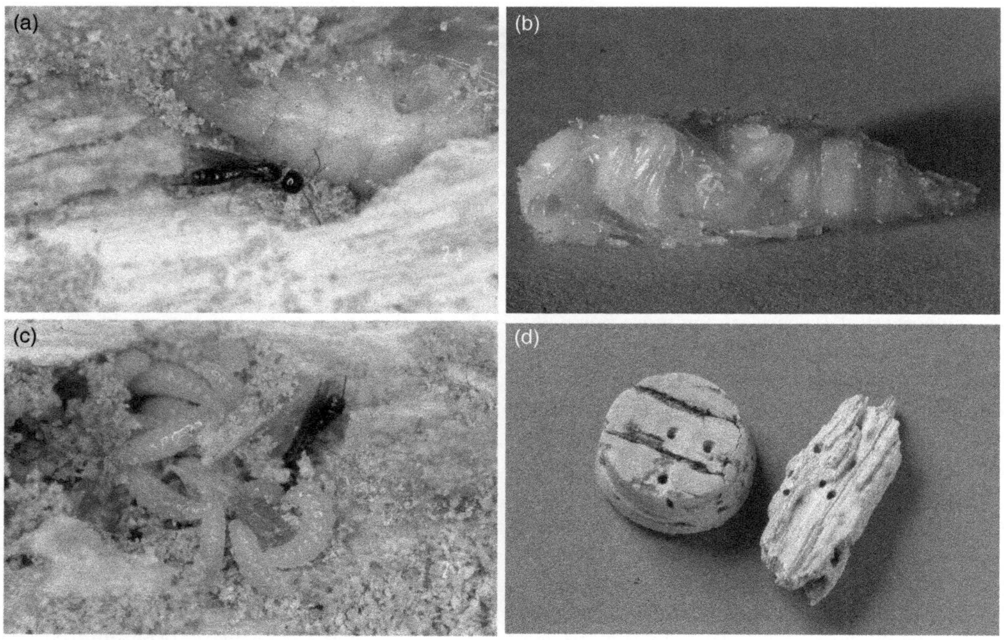

Fig. 12.23 Aspects of the biology of *Histeromerus mystacinus* (Rhyssalinae). (a) Female encountering pupal beetle host; (b) cluster of approximately eight eggs deposited on the side of the host just behind the legs; (c) mature larvae and female wasp still in attendance; (d) holes made in cork stopper of tube from which females escaped by chewing their way out, and in piece of wood substrate. (Source: reproduced by permission Mark R. Shaw.) For a colour version, see Plate 24.

very similar to *Cosmophorus* in the Euphorinae. *Histeromerus* ought to make a relatively easy system with to study the evolution of parental care.

Probably most species host-feed and Shaw & Huddleston (1991) suggested, I imagine because of their large yolky eggs, that it may be obligatory in many. It can be destructive or synchronous if the host is relatively large, as are those of some species of *Oncophanes* (M.R. Shaw 1983). The cocoons of *Dolopsidea* are 'spindle-shaped with sharply pointed ends' (M. R. Shaw 1993b).

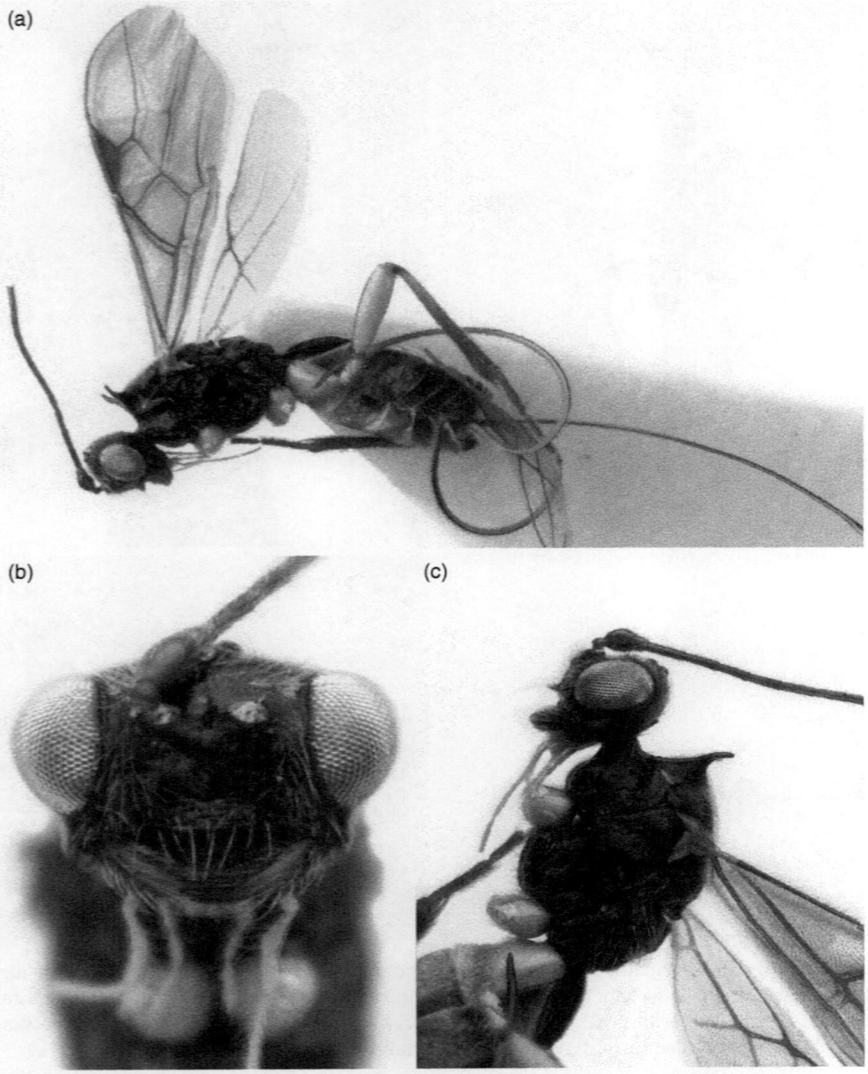

Fig. 12.24 *Thoracoplites bifurcatus* (Rhyssalinae). (a) Female habitus, (b) detail of face showing wide, shallow hypoclypeal depression, (c) head and mesosoma side view. (Source: reproduced by permission, © C. van Achterberg, Naturalis Biodiversity Centre.)

The Afrotropical genus *Thoracoplites* (Fig. 12.24) was described originally in the Opiinae (M. Fischer 1961), but that this was incorrect was noted by Wharton (1988), who placed it provisionally in the Doryctinae. However, that this placement was also likely to be erroneous became apparent after investigation of its venom apparatus (Quicke et al. 1992e), which suggested a relationship with the Rhyssalinae (including Histeromerinae) which share the long spirally sculptured primary venom duct intima and weak, unsculptured reservoir. Subsequently, this was further supported by molecular studies (Fig. 12.21a; Zaldivar-Riverón et al. 2006), although it appears most closely related to *Rhyssalus*

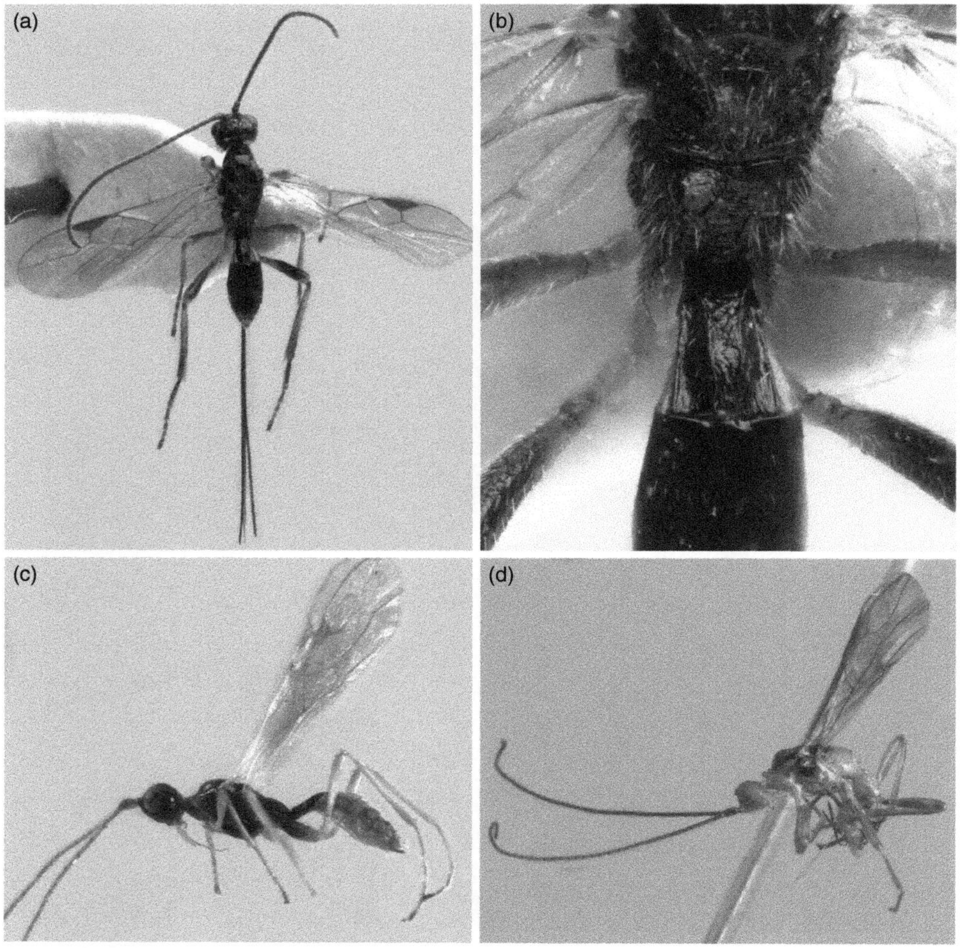

Fig. 12.25 Exemplar Rhyssalinae and Rhysipolinae. (a, b) *Dolopsidea indagator* (Rhyssalinae), the latter showing propodeal areolation and smooth metasomal terga after the first; (c) *Cantharoctonus jaragua* (possibly Rhyssalinae though currently classified in Rhysipolinae); (d) *Rhysipolis* sp. (Rhysipolinae), from Brazil. [Sources: (a, b) reproduced by permission, © C. van Achterberg, Naturalis Biodiversity Centre; (c, d) reproduced by permission of Angélica Penteado Dias.]

and *Dolopsidea* (Fig. 12.25a,b). Its biology is unknown but is likely atypical given the horn-like projection of the anterior of the middle lobe of the mesoscutum.

Acrisis, the type genus of the Acrisidini, is a small genus of very small wasps, only about 2 mm in length and long placed in either Hormiinae or Exothecinae with virtually nothing known about their biology, although there are records of them from Cecidomyiidae and Anobiidae – it seems unlikely that both are correct. Molecular data clearly place them within the Rhyssalinae (Zaldivar-Riverón et al. 2006).

The taxonomy of the Rhyssalinae is in a relatively poor state, with numerous widely dispersed publications and no unified treatment (even less surprisingly given that its exact composition is rapidly changing in the light of molecular results). The Asian genera of Rhyssalinae were keyed by Belokobylskij (2004), who reassigned some taxa. Although molecular studies support a monophyletic entity, it has no obvious synapomorphies. *Histeromerus* is instead a very easily recognised genus and has been treated by van Achterberg (1992a), with additional species now described from China, Japan and Taiwan and from Australia; *Histeromerus* is not yet known from the Afrotropical or Neotropical regions.

Rogadinae s.l., Hormiinae, Lysiterminae

While the Rogadinae in a fairly restricted sense is not hard to define biologically and to some extent morphologically, the remaining taxa from the old polyphyletic concept of the Rogadinae have been split into a number of still fairly ill-defined groups. One of these is the Hormiinae, but it is not at all obvious that even this is monophyletic. The systematic positions of several genera of Hormiinae *s.l.* are questionable (Wharton 1993b) and likely will not be resolved until molecular data become available for them.

Betylobraconinae

It is often the case that a species originally described as the type of a new higher group turns out to be extreme in its morphology, and this is certainly so with the Betylobraconinae. The type genus, *Betylobracon*, is a robust-looking animal from north-eastern Australia and it is non-cyclostome, although its wing venation gives the clue to its cyclostome origins. Tobias (1979) described this genus and placed it in its own subfamily and suggested that it might be related to the 'Braconinae phylogenetic branch', i.e. to the cyclostomes. Van Achterberg (1984), in his informal essay on braconid relationships, placed it deeply within the non-cyclostomes; however, he subsequently (van Achterberg 1995) recognised that it was closely related to a long-forgotten genus, *Mesocentrus*, also from Australia and New Guinea[4], and thus extended the circumscription of the subfamily since *Mesocentrus* are distinctly cyclostome. He also added two new tribes that extended its range into Asia and South America. However the key character of the group was the short fore tarsi, which, as van Achterberg's paper's title shows, was not exclusive to the group of taxa that he included within the subfamily. Not surprisingly, preliminary molecular data fail to recover all the tribes as a monophyletic group, although *Mesocentrus* and *Betylobracon* are clearly closely related and the former may be paraphyletic with respect to the latter as it lacks any obvious autapomorphy.

Van Achterberg's (1995) three tribes were the Betylobraconini from the Australian region and Oceania, Facitorini from the Indo-Australian, Neotropical and Nearctic regions and the Planitorini. Although van Achterberg placed *Planitorus*, the type genus of Planitorini, in the Betylobraconinae because of its similar habitus and robust legs, it has subsequently been removed to the Euphorinae (Sharanowski et al. 2011) and is clearly non-cyclostome. Belokobylskij and Long (2005) added a fourth tribe, the Aulosaphobraconini, to accommodate a new genus from Vietnam. Both Belokobylskij et al. (2008) and Zaldivar-Riverón et al. (2009) failed to recover the subfamily as monophyletic. *Aulosaphobracon* appeared as the sister group to the Rogadinae+Lysiterminae+Betylobraconini, and the Facitorini were recovered with significant support as part of the Yeliconini (Rogadinae), therefore Belokobylskij et al. synonymised it with the Yeliconini. The separate position of *Aulosaphobracon* was not strongly supported so it is currently regarded as Betylobraconinae *incertae sedis*. Its long ovipositor and some other differences suggest that it might actually belong to the Lysiterminae though it may be completely separate. Since the definition of the group is based largely on a widely homoplastic fore tarsal character, it is not surprising that they are polyphyletic.

Biology is not known for any of the included genera. *Mesocentrus* displays several morphological features that are quite similar to those present in some Rogadinae, and in collections the specimens are often mixed up. A species has recently been discovered Baltic amber (Quicke et al. in preparation). That no host records or biological details exist for any *Mesocentrus*, despite their wide distribution and numerous species, is somewhat of a mystery. Hence the possibility exists that the Betylobraconinae and Rogadinae may eventually be synonymised and indeed should be if the former were found to mummify their hosts, and we can only hope that their mode of parasitisation will soon be discovered.

Hormiinae

Even with the Hydrangeocolini having been transferred to a broad concept of the Mesostoinae and the Lysitermini raised to subfamily status, the rump Hormiinae is still a heterogeneous assemblage of small cyclostomes and may well be polyphyletic. Here I treat it as comprising the genus *Hormius* (Fig. 12.26) and close relatives, in agreement with van Achterberg (1995).

Little is known about the biology of this group, despite specimens being fairly common. Many are gregarious

Fig. 12.26 *Hormius* sp. (Hormiinae) from Papua New Guinea. (Source: reproduced by permission of Jan Hrcek.)

and hosts generally seem to be Lepidoptera larvae living in silken retreats (Basinger 1938, Shaw & Huddleston 1991). Recorded host groups are mainly Gelechiidae and Tortricidae and possibly some Coleophoridae and Pyralidae, although as with many groups some of these may be erroneous. *Hormisca tatianae* has been reared as a parasitoid of the pyralid, *Ancylosis* (*Heterographis*) *fulvobasella*. According to Basinger, the *Hormius* simply oviposits into the web containing the host tortricid larvae and her first instar offspring, which are apparently gregarious, attach themselves to the host caterpillar when the latter contacts them. Basinger then stated that 'The host soon surrounds itself and the attached parasites with a web and becomes inactive'. From this, it appears that their saliva has a paralysing or subduing effect on the host and indeed may induce a behaviour change, the web spinning, before this. Similar behavioural effects caused by parasitoid larval saliva are known in the *Polysphincta* group of Pimplinae.

One brachypterous species assigned to *Hormius* is known from North America, although in the original description Hedqvist (1963) writes that it '... differs from all species of the genus *Hormius* regarding shape of head and thorax, but I prefer to place it in this genus'. Given that statement, it might not be too surprising if future study showed it to belong elsewhere.

Wharton (1993b) considered the Hormiinae and Exothecinae to be synonymous (maintaining the concepts of Exothecini, Hormiini and Pentatermini) and argued that if one did treat them thus, against common usage, Hormiinae would be the appropriate group name. However, this synonymy merely resulted from a lack of any detailed phylogenetic investigation, partly no doubt because of their small body size. Molecular studies (Sharanowski et al. 2011) clearly place the traditional Exothecinae as a separate group, close to the Opiinae+Alysiinae, while the remaining 'Hormiinae' appear closer to the Doryctinae and Rogadinae but are almost certainly polyphyletic even with the exclusion of Lysiterminae, Pambolinae and Rhysipolinae.

The Austrohormiini are most likely not related to the genus *Hormius*, but rather to the Mesostoinae. Only one genus has a known host, *Nepticulihormius*, which as its name suggests is a parasitoid of nepticulid leaf-mining moths in New Zealand (van Achterberg & Berry 2004).

The Neotropical, gall-forming *Monitoriella*, originally treated as an hormiine (despite its original description stating that it was most like a doryctine) and retained there by Wharton (1993b) has now convincingly been shown to be a derived member of the Doryctinae though lacking the obvious morphological synapomorphies of the latter (Zaldivar-Riverón et al. 2007).

Lysiterminae

These are principally tropical and all rather small insects about which very little is known, despite the fact that they occur commonly in Malaise trap samples. Their taxonomy is still in a rather fluid state. Monophyly of the group is very far from certain. Indeed, monophyly of the type genus, *Lysitermus*, with a group of possibly monophyletic species that constitute the bulk of the subfamily as currently recognised is also far from certain. As noted by van Achterberg (1995), the carapace of *Lysitermus* is articulated between the first and second metasomal segments, whereas it is fused completely in the other genera. Misidentification of *Lysitermus* ensued because it is a rather rare genus and a similar organism with a completely fused carapace is properly called *Tritermus*. It is not clear to which genus some early generated DNA sequences refer.

Four tribes are recognised: Cedriini, Lysitermini, Pentatermini and Tetratermini. The first of these been relatively newly associated with the others based on the molecular study of Zaldivar- Riverón et al. (2006). *Cedria* and relatives in the tribe Cedriini, which are probably a generally rarer and are mostly tropical and Australasian group, although they occur commonly in Baltic amber. Van Achterberg (1995) had included several genera with very few flagellomeres in the Chremylini of the Pambolinae, including *Cedria* and *Chremylus*, and separated *Chremylus* in its own subtribe with the others placed in the Cedriina. However, monophyly of this group was questionable as it was based on a single, rather weak apomorphy. Here the two subtribes are regarded as belonging to two separate subfamilies, the Chremylina as a tribe within the Pambolinae and the Cedriina as a tribe in the Lysiterminae, since *Cedria* appears to belong there based on DNA data (Zaldivar- Riverón et al. 2006).

For the Cedriini, biology is only well known for one of the genera, *Cedria*, which attacks grain moth larvae. One species has caught the attention of biologists because it is one of the few parasitic wasps which display parental care and its biology is essentially similar to that of some bethylids which attack similar hosts and are also ectoparasitic. The female stays with her brood until they have completed development and emerged as adults (Beeson & Chatterjee 1935, Chu 1935, Mathur 1959). However, the adults are long-lived and, although apparently not feeding, can go on to attack up to four more hosts in the same fashion, with decreasing brood size each time (Chu 1935), although Beeson and Chatterjee suspected that normally only one batch would be laid. It is a pity that there have been no further studies on the biology of *Cedria* as it would make a very interesting and easy experimental system. The only other cedriine with a host record is the Neotropical *Carinichremylus peleopodae*, which has been reared from the oecophorid moth *Peleopoda* (van Achterberg 2000), but it is not known whether females show any parental care.

Because lysitermines have for a long time been associated with either the basal Rogadinae or the Hormiinae, they have usually been assumed to be ectoparasitoids of concealed hosts such as Lepidoptera or Coleoptera larvae, although there are very few records. Van Achterberg (1982b) and Wharton (1993b) noted that the more reliable records for *Lysitermus* indicate that hosts are larvae of Psychidae and Xyloryctidae (Lepidoptera). However, one South-East Asian species, *Katytermus palmicola*, currently classified within the tribe Tetratermini, has been reported to be a gregarious endoparasitoid of weakly concealed nymph of a katydid (Orthoptera: probably Gryllacrididae), with cocoon spinning and pupation externally after the 50 or so parasitoid larvae have emerged (van Achterberg & Steiner 1996; see also Chen & Whitfield 2002).

Pentatermus is widespread in the Old World tropics, but little is known of its biology. Hedqvist (1963) recorded the type species as being 'ex pupa of *Earias*' (Nolidae), but it could easily be that the label data referred to a cocoon rather than emerging from within the pupa and the extremely widespread (including China, Japan, South-East Asia, India, Madagascar, subsaharan Africa), *P. striatus*, has been reared from the hesperid *Parnara guttata*. Whether this is indeed a single widespread species or an aggregate of cryptic species remains to be seen.

The relationships of *Tetratermus* and *Pentatermus*, which, as their names suggest, have carapace-like metasomas with four and five segments visible, have

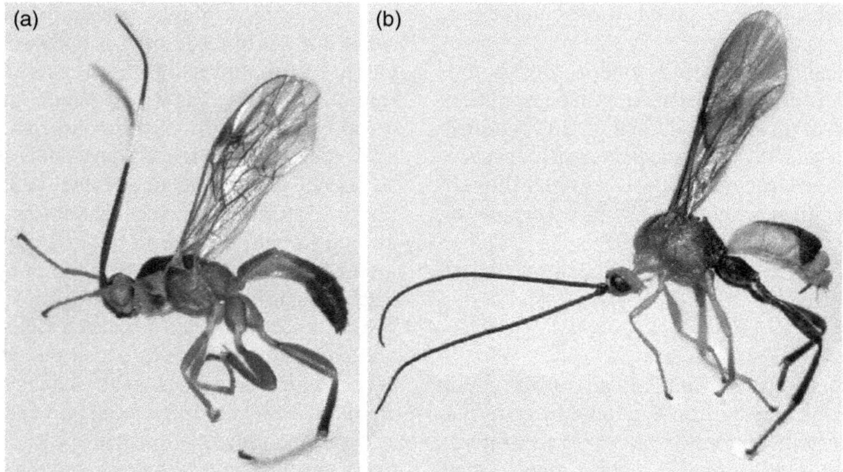

Fig. 12.27 Exemplar Rogadinae. (a) *Aleiodes* (*Arcaleiodes*) sp., from Nepal; (b) *Batothecoides yakushimensis*, from Japan, (Source: part b, reproduced by permission of Kaoru Maetô.) For a colour version, see Plate 25.

been virtually impossible to resolve using morphology alone, and even with the DNA sequence data available at present we cannot do much better. In many analyses they are close to the bulk of the other Lysiterminae, but sometimes rendering them into a grade taxon (Belokobylskij et al. 2007).

A key to the world genera was provided by Belokobylskij et al. (2007).

Rogadinae *sensu stricto*

Here I follow recent works that restrict this group to just those endoparasitic taxa that at completion of their growth, mummify their host to form a protection for the parasitoid pupa (Chen & He 1997, Zaldivar-Riverón et al. 2009). Probably all species are koinobiont parasitoids of Lepidoptera larvae, some ovipositing into early stages, others attacking later ones and some able to attack their host over a wide range of larval instars. One possible exceptional host record, of parasitism of a chrysopid lacewing larva by Johnson (1985), might be correct, but it has never been repeated and the parasitoid literature is replete with erroneous host records (Noyes 1994).

Few species have been involved in biological control. The European *Aleiodes* (as *Rhogas*)[5] *pallidator* (as *unicolor*) was released to control the satin moth, *Leucoma* (= *Stilpnotia*) *salicis*: Lymantriidae, which was accidentally introduced into North America in the 1920s (Dowden 1938). After some while it established, but it is interesting that whereas the source population was thelytokous, there is a high proportion of males in the material being reared from hosts in North America, leaving open the possibility that the North American wasps might actually belong to a cryptic sibling species (S.R. Shaw et al. 2013). Probably a number of other *Aleiodes* species are important in regulating pest numbers in their native ranges (S.R. Shaw 1997b, 2006).

Most species are nocturnal and typically brownish, with large eyes and ocelli and long antennae (Fig. 10.8c), but diurnal species can be brightly coloured (Fig. 12.27). These koinobionts are morphologically fairly uniform and most have a very distinctive metasomal sculpture with the dorsal carinae of the first tergite normally uniting to form a thin, mid-longitudinal carina, and the second and sometimes the third also having a mid-longitudinal carina. Five tribes have been recognised, Aleiodini, Clinocentrini, Rogadini, Stiropiini and Yeliconini; van Achterberg (1988b) added the Spinariini for a group largely with robust carapacised metasomas (Fig. 12.27b), although molecular analyses (Zaldivar-Riverón et al. 2009) suggested that these are just a derived group within the Rogadini. It has been proposed on the basis of biology that that the Clinocentrini, which have ovipositors about as long as the metasoma and used to attack host caterpillars concealed in webs, are the most basal branch (M.R. Shaw 1983). Apart from that, external morphology appears to provide few characters for phylogenetic analysis, although recent investigation of internal morphology and molecular data is proving useful for at least relationships within the Rogadini.

Clinocentrus can apparently assess the instar of its host just through the contact with it that its ovipositor makes, as they tend to lay more male eggs in younger hosts. Egg placement varies between the Clinocentrini and other Rogadinae (unknown for Stiropiini and Yeliconini), with the former placing the egg just below the host cuticle (Fig. 7.2a) and the latter placing it within the host's haemocoel. *Clinocentrus* and some *Aleiodes* have to induce temporary host paralysis with an initial sting and then withdraw their ovipositor and wait nearby until their venom has taken effect before approaching the host again to oviposit proper.

Either by interrupting this sequence or by studying the development of hosts which manage to encapsulate the parasitoid's egg, it has been possible show that different species vary in whether their venom has a long-term effect on host development. *Clinocentrus* venom, in the absence of a parasitoid larva, causes the host to enter a premature prepupal state irrespective of which instar is attacked (Fig. 7.2) (M.R. Shaw 1983), but this is not true of other investigated taxa. *Aleiodes testaceus* attacks second, third and fourth instar hosts. If it parasitises either of the first two of these, they are able to continue development and pass through one ore moult, but fourth instar larvae are prevented from moulting (Ahmad 1944), so death occurs at a roughly similar host size in all cases.

Given the size of the subfamily, there have been surprisingly few detailed studies of developmental stages. There appear to be five larval instars (Dowden 1938, Ahmad 1944), although only three have been reported in a few studies (Shaw & Huddleston 1991), although without very careful observation it might be easy to miss some (e.g. O'Donnell 1987). Dowden (1938) and Ahmad (1944) illustrated and described the larval stages of *Aleiodes* (as *Rhogas*) *pallidator* (as *unicolor*) and *A.* (as *Rhogas*) *testaceus*, respectively. The adults are synovigenic and the egg is fairly large (0.1 × 0.45 mm) given the size of the wasps, which are only 4.5 mm long, and most of its volume is occupied by yolk; the chorion is fairly strong and the egg is not noticeably hydropic. All five described instars are hymenopteriform. The first instar has normal-sized mandibles that are obviously not adapted for fighting, and its skin is described as 'smooth and delicate, without any cuticular structures' and thus lacking obvious adaptations for moving around the host quickly hunting for competitors, and Ahmad noted that they cannot move forwards or backwards and only make bending movements. First to early fifth instars feed only haemolymph and a pumping action can easily be observed in their pharyngeal apparatus. The final instar, which now occupies a large part of the host caterpillar's body, starts to consume small groups of fat cells and only towards the end of its development does it eat other host organs, including gut, gonads, muscles and nervous system. The tracheal system appears in the second instar, but functional spiracles only appear in the fifth instar.

Rogadine mummies are typically found on stems and twigs or sometimes on leaves, although the great majority of observations come from just one genus, the cosmopolitan *Aleiodes*. Probably the majority of species induce a behavioural change in the host shortly before finishing consuming it, to cause it to walk to a site that favours survival of the parasitoid, where, having consumed most (or at least all it needs to) of its host's soft tissues, usually bites a small slit antero-ventrally through the host cuticle usually between the mesothoracic legs (Fig. 12.28a) (Quicke et al. 2006). Through this slit, excess host fluids are pushed out, possibly together with parasitoid secretions, and these fix the host's remains *in situ*. In the case of all Aleiodini, Rogadini and Yeliconini, the parasitoid larva, which for most of its previous development has faced anteriorly, then rotates itself within the host remains so that when it pupates, it is facing posteriorly and subsequently chews its way out (Fig. 12.28b). Rotation of the larva is fairly time consuming and Ahmad noted that it takes approximately 2 h for that of *A. testaceus* to reorientate itself. Members of the genus *Aleioides* always emerge postero-dorsally and their mummies are highly tanned, often nearly black (Quicke & Shaw 2005b, Zaldivar-Riverón et al. 2009), presumably due to the release of enzymes and/or phenolic compounds by the parasitoid larva. Other taxa generally produce far less tanned cocoons and emergence may be from the anterior end (Clinocentrini, Stiropiini, some Yeliconini) and may be ventral, lateral or dorsal. Given the current phylogenetic hypotheses for the group, it seems likely that anterior emergence represents the plesiomorphic condition. Although it has been reported in the literature [e.g. in *Triraphis* (as *Pelecystoma*); O.J. Smith et al. 1955] that some species do not bite an incision

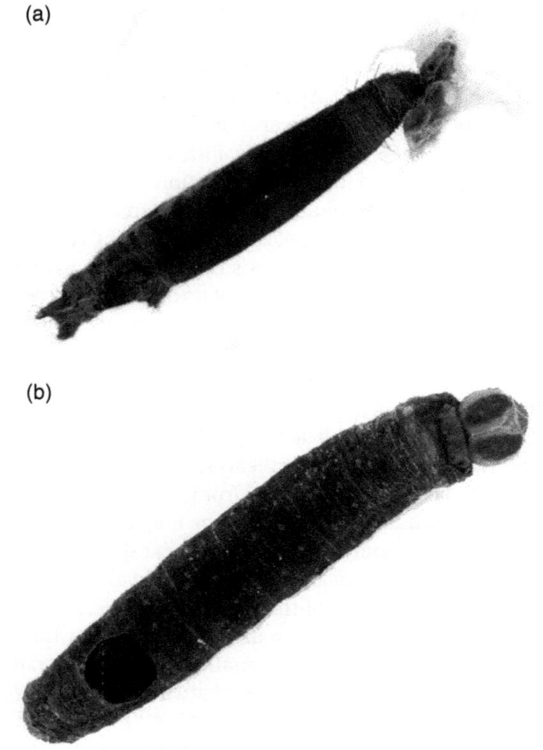

Fig. 12.28 Typical post-egress mummies (mummified host caterpillar skins) of *Aleiodes* spp. (Rogadinae). (a) Mummy of *A. aclydis*, showing the 'glue' (right) that congealed from liquids, possibly including parasitoid salivary compounds, that came out of antero-ventral cut in The host that adhered the mummy to the plant; (b) mummy of *A. arbitrium*, showing the postero-dorsal emergence hole made by parasitoids of this genus, and the dark, tanned nature of the host cuticle. (Source: reproduced by permission of Scott R. Shaw.)

through the host cuticle before pupation, this seems highly unlikely to be correct (Shaw & Huddleston 1991) as there would be a clear advantage to being able to expel unwanted host liquors, and in at least some reported cases subsequent detailed examination has revealed a cut. Nevertheless, at least some *Triraphis* do not glue their mummified hosts to the substrate as most other rogadines do (O.J. Smith et al. 1955, Kula et al. 2010a).

Most rogadines produce only a very flimsy, incomplete sort of cocoon within the mummified host remains and that would not protect the rogadine pupa from any liquefied, decaying host remains, but in some the silk lining appears to be more substantial and robust (Dowden 1938). More recently, a small clade of *Aleiodes*, including only a few described species, the *A. buzurae* group, has been found to cause its host to drop down from its host plant on a silk thread and then to spin a small silk mat between its legs and then it is mummified, suspended in mid-air (Quicke et al. 2006). This is just one of several probably adaptive manipulations of the host and may serve to reduce levels of predation and/or pseudohyperparasitism.

Most species are solitary parasitoids, but gregarious development is known to occur in two genera of the *Colastomion* group (viz. *Colastomion* and *Macrostomion*) (Maetô & Arakaki 2005, Shaw 2002a, Quicke et al. 2012b) and a very few *Aleiodes* (*A. pallescens*, *A. stigmator* and *A. leptocarina*). Interestingly, whereas *Macrostomion* attacks very large sphingid hosts and has brood sizes from about 20 to more than 100, some of the *Colastomion* species are facultatively gregarious and have brood sizes from one to four. As with solitary species, the gregarious *Aleiodes* orientate themselves prior to pupation with their head ends as postero-dorsal as possible so emergence takes place trough a series of holes along the mummy's back. Maetô & Arakaki (2005) reported a highly female-biased sex ratio, which would be consistent with predictions of local mate competition theory if there was significant sib–sib mating among the emerging parasitoids.

Despite their large eggs, their longevity means that realised fecundity can be fairly high. Dowden (1938) and Ahmad (1944) reported a maximum of 322 and 83 eggs, respectively, with approximately eight eggs being laid per day. Given that rogadines and almost all other cyclostome braconids have only two pairs of ovarioles (Fig. 4.1a), this suggests that each ovariole is capable of maturing two eggs per day, although Dowden found one female that was able to lay 88 eggs over a 2 day period, so presumably had a store of mature or nearly mature eggs ready. M.R. Shaw (1983) suggested that host-feeding is uncommon and probably facultative, so the two species whose biologies have been described in detail, both of which regularly host-feed, might be somewhat atypical. Shaw & Huddleston (1991) commented that egg–larval parasitism, although not yet known in the Rogadinae, might be expected to have evolved in some species. However, their large, yolky eggs would seem to make this unlikely.

Superparasitism in solitary species (i.e. most) appears to be fairly common, but it is reported that only one egg ever hatches to complete development, strongly suggesting that physiological suppression is involved, and certainly the first larval instars are not well equipped for combat (see above).

Many temperate *Aleiodes* species overwinter as first instar larvae within their hibernating hosts, especially plurivoltine species, but frequently complete development fairly quickly in the spring, hatch, and may then use an alternative host species for the next generation. Univoltine species may do the same (M.R. Shaw 2002b), but a larger proportion of these overwinter in their host mummies. A few species also hibernate as adults and in some cases, although not proven, adult wasp and host phenology seem to make this extremely likely, even though use of alternative hosts cannot always be excluded, e.g. *A. malacostomatos*, which attacks early instars of *Malacostoma* (Lasiocampidae) tent caterpillars in the spring, but adults emerge in late summer (S.R. Shaw 2006). Those that do overwinter as adults appear to do so in aerial sites such as in coniferous shrubs and trees (M.R. Shaw, personal communication). Many species cause their hosts to behave abnormally shortly before the parasitoid completes its development. *A. alternator* causes its host to climb high up stems from its normal feeding site, *A. dissector* causes precocious entry into soil litter by its *Orthosia* (Noctuidae) hosts and premature preparation of an aerial pupation site by penultimate instar larvae caused by parasitisation by *A. pallidator* (Quicke et al. 2006). As with members of the *A. buzurae* group that cause their hosts to descend from the host plant on a silk thread, nothing is known about how these behavioural changes are bought about, but at least in the cases of those that cause hosts to climb high up, it seems unlikely that they are just usurping a natural behaviour.

Courtship involves male wing fanning and the male mounting the female from behind. Copulation has not been described for many species; Ahmad (1944) reported that it lasts approximately 10 s. Females only very rarely mate more than once, usually rejecting subsequent attempts by males, whereas males can mate multiple times.

Detailed consideration of the venom apparatus revealed a number of new features to help understand rogadine relationships, including a new type of valve located at the insertion of the secondary venom duct on to the venom reservoir, which comprised a cone of filaments within the secondary duct (Zaldivar-Riverón et al. 2009). Given that rogadines have type 1 venom reservoirs with a thick muscular layer, the need for such a valve is uncertain. One possibility might be to prevent venom activating compounds secreted in the reservoir or primary duct, which have fairly thick chitinous linings, passing back into the more 'delicate' glandular system where active venom compounds might be harmful.

Molecular work has largely supported the inferences from the venom apparatus and suggests that the Rogadinae comprises two main groups of genera, the Aleiodini centred around the cosmopolitan *Aleiodes*, and the Rogadini centred around *Rogas* and including the Spinariina and a group of principally tropical genera related to *Colastomium* (Zaldivar-Riverón et al. 2009). The currently available molecular data additionally, although weakly, suggest that the Yeliconini are actually more closely related to the Aleiodini and are probably best considered as just a highly morphologically specialised group within the subfamily. Indeed, in collections the Neoptropical genus *Bulborogas* has sometimes been confused with *Yelicones* because it shows several features in common with the latter, including a rather short, robust habitus and strongly swollen femora and rather shortened fore tarsi the positions of Clinocentrini and Stiropiini are less certain.

Male-specific tergal glands are found in members of basal species of *Aleiodes* (including members of the subgenus *Hemigyroneuron*), which open via an internally setose pore on metasomal tergites 4–6 (Areekul Butcher & Quicke 2011). The reference in Buckingham and Sharkey (1988) to an obviously homologous structure in two *Rogas* species clearly refer to this group of *Aleiodes* and not to *Rogas* and not to *Triraphis*, a genus to which many New World species formerly described as *Rogas* belong, in which no male tergal glands have yet been discovered. As with so many groups and probably exacerbated by the predominantly nocturnal nature of these wasps, there are no observations of the male calling behaviour of these species.

The taxonomy of the group at genus level has been advanced considerably in recent years by the publication of revisions and keys to the New World, Afrotropical, Chinese and Thai faunas in particular (van Achterberg 1991a, S.R. Shaw 1997b, Chen & He 1997, B.A. Butcher et al. 2012), but there are still major difficulties with the Australian and Indo-Australian regions. The Stiropiini, with 3 genera, is restricted to the New World, the other tribes are cosmopolitan. The Rogadini is most generically diverse within the Old World tropics but with Triraphis, while most species rich in the New World also present in Europe, Africa and Asia. Zaldivar-Riverón et al. (2009), based on molecular data, showed that despite much past taxonomic uncertainty about generic limits, *Aleiodes* and *Heterogamus* formed a distinct group, well separated from other 'Rogadini', and recognised the Aleiodini for these. However, the Spinariini were not recovered as monophyletic based on DNA alone. Species-level phylogenies have been generated for North American *Aleiodes* by Fortier and Shaw (1999) based on morphology and for Thai *Aleiodes* based on CO1 barcoding gene fragment by B.A. Butcher et al. (2012). Unpublished molecular analyses of a very much larger number of species from all around the world (Quicke et al., in preparation) indicate a major split between New World and Old World faunas, with only a few apparent migrations between North America and the Palaearctic.

The Afrotropical genera were keyed by van Achterberg (1991a), the New World genera by S.R. Shaw (1997b) and the Chinese genera, which include most of those occurring in all of South-East Asia and Australia, by Chen and He (1997).

Alysioid Subcomplex, Including Braconinae

Sharanowski et al. (2011), in their proposed revised classification of the family, recognised a grouping of cyclostome subfamilies (Alysiinae, Exothecinae, Gnamptodontinae, Opiinae and Telengaiinae) that they termed the alysioid subcomplex. In their study, this group received high support, but was placed as the sister group to the Braconinae without statistical support (Fig. 12.29). Other studies have consistently recovered all these together with the Braconinae as the sister group to the rest (Belshaw et al. 1998, 2000, Dowton et al. 2002, Zaldivar-Riverón et al. 2006) and therefore the concept of this subcomplex is expanded here.

Alysiinae and Opiinae

These two groups are dealt with together here. A close relationship between these two subfamilies has long been recognised; members of both are exclusively koinobiont endoparasitoids of Diptera and molecular studies confirmed the group's combined monophyly (e.g. Gimeno et al. 1997, Wharton et al. 2006) except where only a very few exemplars are included (e.g.

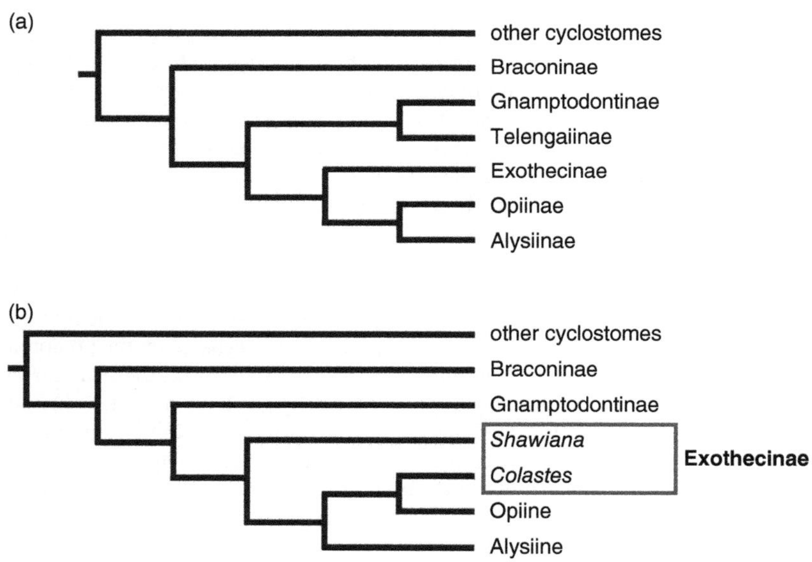

Fig. 12.29 Relationships of genera of Braconinae, Gnamptodontinae, Exothecinae, Opiinae and Alysiinae. (a) Including Telengaiinae as recovered by Zaldivar-Riverón et al. (2006) based on 28S rDNA D2+D3 region plus COI barcoding region plus morphology; (b) as recovered by Sharanowski et al. (2011) based on six fragments of four gene regions, although with low representation of genera, Exothecinae not recovered as monophyletic.

Sharanowsky et al. 2011). However, there have been no recent studies including markedly expanded taxonomic coverage and multiple gene regions. Having said that, there is some possibility that they may not be mutually monophyletic, as currently constituted. Gimeno et al.'s first study used rather quickly evolving genes, which was not ideal, and several genera currently in the Alysiinae, but somewhat questionably (e.g. *Parorthostigma, Glyphogaster, Oenonogastra* and *Gastralysia*: see Wharton 1980), have not yet been sequenced. Consideration of the female venom glands of both subfamilies (Quicke et al. 1997a), in which numerous species in various of these genera have an unsculptured, bulbous expansion at the anterior end of the venom reservoir, could indicate that some taxa will need moving to other groups once further evidence becomes available.

Shaw and Askew (2010), in their review of the attributes of hymenopterous parasitoids of Diptera, suggested that the latter have really only become major hosts for parasitic wasps relatively recently and wasps only attack those with particular types of larvae/larval habitat. Hence a large number of dipteran families are unaffected and no opiines, alysiines (or orthocentrines, cylloceriines or diplazontines in the Ichneumonidae) attack any of those Diptera with aquatic larvae.

Alysiinae

Alysiines are all koinobiont parasitoids of cyclorhaphous Diptera larvae which complete development in and emerge from the host puparium. This subfamily has for a long time been recognised as a natural group and has generally been treated as comprising two tribes, the Alysiini and the Dacnusini, although some workers have treated these as separate subfamilies. They are one of the easiest subfamilies to recognise as members always have exodont, multi-toothed mandibles (Fig. 12.30) and have no trace of an occipital carina. The exodont condition is found only in a very few other braconids (*Exodontiella* in the Gnamptodontinae and some species of *Paroligoneurus* in the Ichneutinae) and these are very rare. The function of the exodont mandibles was discussed by Griffiths (1964) and Wharton (1984), and appears principally to be associated with egress from within the host's puparium. Wharton's own observations of *Alysia* and *Alysiasta* showed that in these, the longer, second (middle) tooth was used to puncture the wall of the host puparium by the wasp pushing them outwards, and then further outward movement widened the tear that the initial punching had created. Depending on whether the initial puncture coincided with the weakened seam in the puparium of the cyclorhaphous dipteran host that the fly would otherwise push off, the break caused by the mandibles was either neat or a slightly more ragged. In addition to egress, some species use the mandibles to help burrow through soil to reach the surface and also possibly to dig for concealed hosts. Wharton noted that the outer/posterior face of field-collected alysiines often has material encrusted upon it, probably indicating that they were used for digging. No doubt associated with this action and the force it would require, alysiines typically have relatively large heads that are often described as 'cubical' or 'quadrate' in the older literature, and these presumably house large muscles for manipulating the mandibles.

The Dacnusini are largely characterised by the loss of fore wing vein r-m, although this is homoplastic within the Alysiini, having been lost independently in at least some members of three genera. As pointed out by Wharton (2002), monophyly of the Dacnusini seems fairly reliable, but the Alysiini are not known to be defined by any synapomorphy and hence they are probably paraphyletic with respect to the dacnusines. Indeed, features of the venom apparatus (Quicke et al. 1997a) and, to a lesser extent, the available DNA sequence data (Gimeno et al. 1997) suggest that dacnusines are probably closely related to a group of alysiine genera (including *Dapsilarthra* and *Oenonogastra*) that attack leaf-mining Agromyzidae flies.

A flurry of taxonomic descriptions of new genera means that the existing generic keys are all rather out-of-date. Tobias (1986a) provided a key to the genera of the former Soviet Union which works for most of the Palaearctic. Other workers mostly concentrated on either the Alysiini or the Dacnusini. Wharton (1980) provided a key to the Nearctic genera of Alysiini, but see also van Achterberg (1988a) for the *Aspilota* group of genera. G.C.D. Griffiths (1964) keyed the British genera of Dacnusini and Wharton and Austin (1991) to the rather impoverished Australian fauna of Dacnusini. Since dacnusines are largely absent from the tropics, the Palaearctic works are essentially world treatments.

General Alysiinae Biology

Although there are some biological aspects that differ between the two tribes and warrant some separate

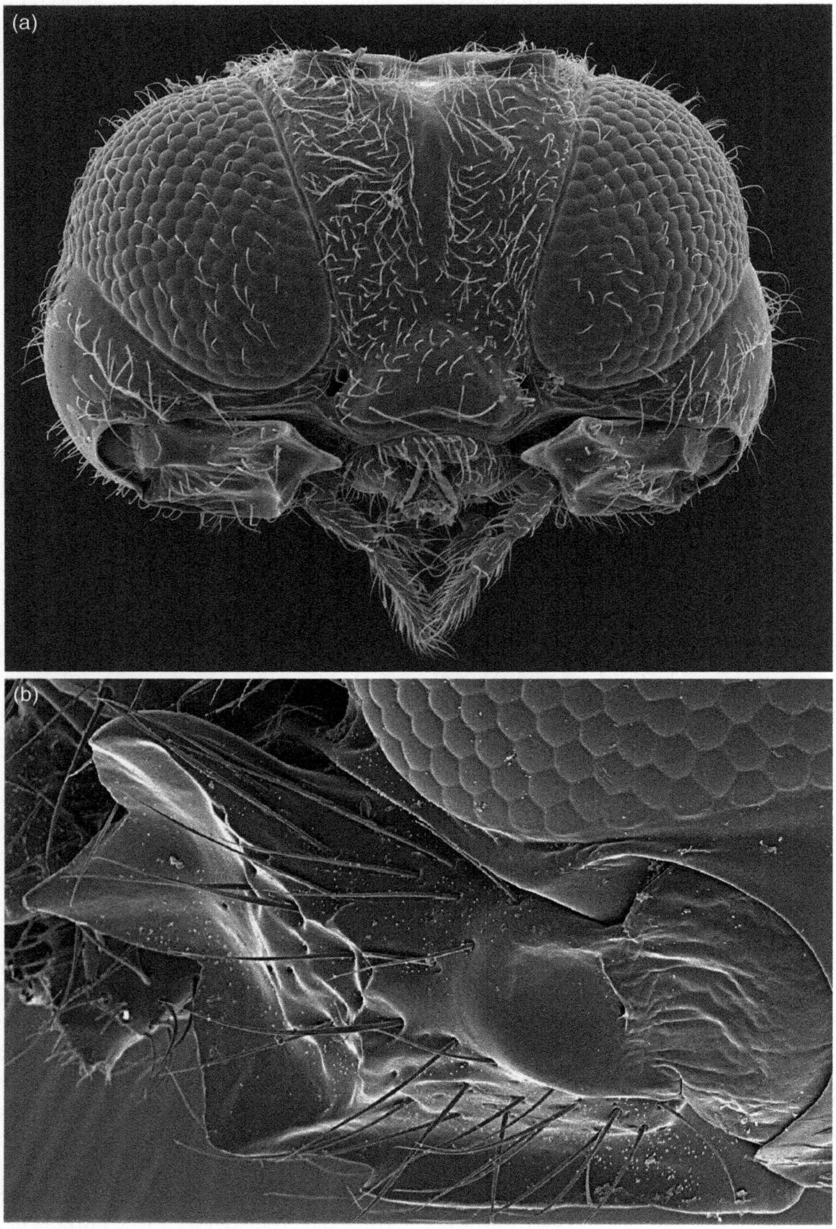

Fig. 12.30 Exodont mandibles of Alysiinae (Dacnusini). (a) *Chaenusa glabra*; (b) *Epimicta griffithsi*. (Source: reproduced by permission of Robert Kula.)

discussion, much of the biology is similar and so they are discussed together in this section.

The general biology of the Alysiini was summarised by Wharton (1984). Several species of both *Aphaereta* and *Asobara* are known to induce transient paralysis in their host larvae (Evans 1933, Salkeld 1959, Prince 1976, Moreau et al. 2002, 2003, 2009). Life history details have been described for a number of species (Guppy & Meloche 1987, Tormos et al. 2003). Early larval development occurs inside a trophamnion, i.e., a cellular membranous envelope derived from extra-embryonic serosal cells (Tormos et al. 2003). *D. sibirica*, a parasitoid of the chrysanthemum leaf-miner *Chromatomyia syngenesiae* (Agromyzidae), has been particularly well studied.

Detailed descriptions of egg and larval development have been provided for *Alysia* (Evans 1933, Caudri 1941), *Chorebus* (Tormos et al. 2003), *Dacnusa* (Haviland 1922, Guppy & Meloche 1987, Croft & Copland 1994), *Aspilota* (Evans 1933, M. Fischer et al. 2008) and *Aphaereta* (Evans 1933). Egg placement of alysiines varies considerably. Many taxa appear just to place their egg within the host haemocoel (Salkeld 1959, Guppy & Meloche 1987, M. Fischer et al. 2008), although some apparently attach it to the inner surface of the cuticle (e.g. *Alysia*; Evans 1933). The eggs of *Alysia* have a distinct knob at one end that Altson (1920) described as the micropyle being narrowly attached. It seems likely that this structure may function as an anchor, although Altson reported that the eggs are deposited in any part of the larval host. The eggs of some taxa of both tribes, such as *Aphaereta* and *Dacnusa*, are hydropic, increasing in volume at least fivefold within the host (Haviland 1922, Evans 1933, Salkeld 1959, Guppy & Meloche 1987). Salkeld reported that although laid free in the haemocoel, the eggs of *Aphaereta pallipes* often become embedded in host fat body (see also A.R. Kraaijeveld 1994) and are commonly found associated with host tracheoles.

Three larval instars have been reported for most species (Salkeld 1959, Croft & Copland 1994), although Caudri (1941) reported finding four in his study of *Alysia manducator*. It could be that larger-bodied taxa have more instars. The first instar larva of *D. sibirica* is caudate without any anal vesicle (Croft & Copland 1994, Tormos et al. 2003) and is initially surrounded by a trophamnion, and has a row of posteriorly directed spines postero-dorsally on the second and third thoracic segments and first eight abdominal segments. Development of the larval respiratory system varies between taxa. Evans noted that in *Alysia* it is well developed and open in the first instar but it is not in either *Aphaereta* or *Aspilota*. Final larval head capsule features have been described by Čapek (1970, 1973) and differ considerably between genera, with some such as *Alysia* and *Phaenocarpa* having a fairly complete and well-developed set of sclerites whereas *Aspilota* and various Dacnusini have considerably reduced structures (e.g. M. Fischer et al. 2008). It should be noted, however, as Fischer et al. pointed out, that the generic identities of some of the species included in earlier biological and developmental studies need careful appraisal as some may have been misidentified and others transferred to different genera with more recent taxonomic understanding.

From the available studies, alysiines are strictly larval or egg–larval parasitoids, because, although they emerge from the puparium, they complete development within the larva before pupation has occurred (Caudri 1941; R.A. Wharton, personal communication). At least in *Aphareta pallipes*, host movement is an essential prerequisite for oviposition (Salkeld 1959). The meconium is not voided until the adults emerge from the host puparium (Salkeld 1959). In *Alysia manducator*, superparasitism apparently leads to the death of the host larva and the parasitoid larvae within it (Altson 1920).

Most Alysiinae species are solitary but a few are gregarious, such as *Aphaereta pallipes*, a parasitoid of various species of Muscidae and Sarcophagidae (Wharton 1984), and one *Phaenocarpa* species is gregarious (van Achterberg 1998). It has also been noted that the gregarious species may generally have broader host ranges than the solitary species (Shaw & Huddleston 1991, Traynor & Mayhew 2005).

Courtship and mating have been described for rather few species. Wharton (1984) noted that members of *Alysia*, *Alysiasta* and *Aphaereta* that he studied all mated within 24 h of emergence. Males appeared to have to get very close to females (<5 mm) before becoming excited and starting wing-fanning, but an *Idiasta longicornis* observed in the field responded to females 2.5 cm away, possibly indicating that some pheromonal signal was detected. Stroking of the female's head and/or antennae by the male's antennae during courtship was observed in all these species and in others cited.

Alysiini

The hosts of Alysiini include members of more than 25 families of higher Diptera, and Wharton (1984) listed all reliable or acceptable host records up to then and also cautioned that many earlier references were unreliable and many incorrect due to the normal reasons of errors in association between parasitoid and supposed host or misidentification of the parasitoid, or both, and usefully provided a list of the records that can be discounted. Several species are reared regularly from farm filth where many maggots feed on manure or perhaps carcasses, and some have been used with mixed degrees of success in trying to control the flies, particularly Muscidae and Sarcophagidae, associated with intensive farming (Wharton 1980, 1984). Because of their hosts, many of which are synanthropic, some are potentially relevant to human health in that they attack disease vectors (e.g. *Coelalysia glossinophaga*, a parasite of tsetse flies, *Glossina* spp.), and others have relevance to forensics. *Aphaereta pallipes* has been investigated as a possible control agent for the blood-sucking hornfly *Haematobia irritans* in North America and *A. aotea* was introduced from New Zealand against nuisance-causing, dung-feeding bushflies (*Musca vetustissimma*; Muscidae) in Australia.

Although the Alysiinae as a whole seems to be predominantly boreal in terms of species richness, dacnusines are largely absent from the tropics (Wharton 1993a). An exception might be the Alysiini genus *Asobara*, which is specialist on Drosophilidae and has many undescribed tropical species Wharton (1993a).

Wharton (1980, 2002) recognised four major groups of genera within the Alysiini (the *Aspilota* group, the *Alysia* group, the *Phaenocarpa* group and the *Dapsilarthra* group), although he refrained from giving any of these formal subtribal rank because the relationships between them are really unknown and also because there are several genera which cannot safely be associated with any of these groups. One of the most problematic is the genus *Grandia*, which lacks fore wing vein r-m but has long been realised as not belonging to the Dacnusini, so the vein loss is a parallelism. Wharton (1984) noted that only 11 genera had more than a single reliable host record and therefore little could be said about host utilisation, although there was evidence that some genera may be restricted either to particular host fly families or major groups. At species level, many species attack hosts that are defined by microhabitat as well as taxonomy. For example, *Gnathopleura ridibunda* and *G. cariosa* attack Sarcophagidae and Muscidae feeding in dung and Sarcophagidae and Calliphoridae feeding on carcasses respectively. Most species however appear to be restricted to hosts of a single family.

Two genera of Alysiini, *Asobara* and *Aphaereta*, have been particularly important for increasing our understanding of parasitoid–host physiological interactions, the evolutionary arms race involved, and also the behavioural ecology of searching behaviour. As with so many things, it is hard to know how generalisable the findings are because both genera are specialist parasitoids of Drosophilidae and both belong to the *Phaenocarpa* group of genera. Particularly noteworthy is the study by M.E. Visser (1994) of the importance of female body size on a combination of fitness components in *Aphaereta minuta*, a result also found for the doryctine braconid *Parallorhogas pyralophagus* (Bernal et al. 2001). Mayhew and van Alphen (1999) compared the normally solitary *Aphaereta genevensis* and the gregarious *A. pallipes* to investigate the evolution of gregariousness in parasitic Hymenoptera and concluded that a reduction in larval aggressiveness is an essential precursor. This was followed by Pexton and Mayhew (2004), who compared larval motility between two *Aphaereta* species, the solitary *A. genevensis* and gregarious *pallipes* under multiparasitism, and confirmed the prediction that larvae of the solitary species would be found to be more active because of their need to locate and kill competitors. In this context, it would be very interesting to study in more detail the nature of larval combativeness among other taxa where there are both solitary and gregarious development (e.g. Microgastrinae, Aphidiinae).

Asobara tabida, a common parasitoid of *Drosophila*, shows genetic variation in the external egg coat (chorion) such that some individuals have a sticky coat and others do not (Fig. 9.1) (A.R. Kraaijeveld & van Alphen 1994). The sticky eggs stick to host tissues and become embedded in them, and this is the way they protect themselves from encapsulation. The part of the egg exposed to the host's haemolymph gets attacked by host haemocytes, but these protective cells do not have access to the part of the egg embedded in the host tissue and so avoid encapsulation. Since in these cases part of the egg covered by host tissue is not encapsulated and is still able to obtain oxygen, etc., the whole egg survives and hatches. After that, the host seems to have no further layer of defences. This stickiness of the egg coat is not universal within *Asobara*, however, and the Afrotropical *A. citri* overcomes host defences in a different way.

Despite some taxa, notably *Asobara* and *Aphaereta* species, having been employed extensively in behavioural ecology and evolutionary ecology studies, relatively little is currently known about their venom chemistry and most work has concentrated on an aspartylglucosaminidase-like protein (Moreau et al. 2004, Vinchon et al. 2010). Interestingly, although the 'enzyme' could be detected in the venom reservoir antigenically, no enzymic activity could be detected there, suggesting that activation occurred at a later stage, and I suspect that this involves the secretory units observed along the primary venom ducts of this wasp, as with many other cyclostomes (Quicke et al. 1997a). Venom apparatus gross morphology (of the chitinous intima) has been surveyed by Quicke et al. (1997a) and showed remarkably high diversity compared with many other groups of Ichneumonoidea.

Several species have been noted to cause their hosts to pupate prematurely, e.g. *Alysia manducator* (Salt 1932) and *Aphereata minuta* (Zinovjeva 1976), but others have disputed this or at least failed to find it (Wharton 1984). Larval diapause has been investigated in considerable detail in Holarctic *Alysia manducato*r. Myers (1927) found that in this plurivoltine species, juveniles may overwinter in the host and complete development and emerge the following spring, but that approximately two-thirds remained in diapause until the late summer. The cues for initiating and terminating diapause in this species have been investigated in detail and involve temperature, day length and host species (see Chapter 14, section *Diapause*).

Dacnusini

In contrast to the Alysiini, dacnusines are largely restricted to attacking just two host families, leaf-mining Agromyzidae and shore flies (Ephydridae), with a few attacking chloropids. This difference in host range and the range of host habits and sizes are likely to account for the considerable difference in genus level diversity in the two tribes, with more than two and a half times as many genera being recognised within the Alysiini (Wharton 1984). Within the Dacnusini there appear to be two major groups of genera, one dominated by *Dacnusa* and *Chorebus*, which are predominantly parasitoids of leaf-mining Agromyzidae, and the other (e.g. *Coelinius*, *Lepton* and *Polemochartus*[6]) that include parasitoids of Chloropidae. There are more than 500 described species distributed among approximately 20 genera, although the generic classification of the group is far from settled.

Several species may be economically important. Probably the best known of these is *Dacnusa sibirica*, which is used to control three agriomyzid pests: serpentine leafminer, *Liriomyxa trifolii*, tomato leafminer, *L. bryoniae*, and pea leafminer, *L. huidobrensis*. *D. sibirica* is marketed under the trade-name MINUSA (Koppert Biological Systems) primarily for low-level *Lyriomyza* infestations in greenhouses and is most effective against early larval instar hosts. Some species assigned to *Chaenusa* attack the beneficial members of the ephydrid fly genus *Hydrellia* which are used in classical biological control of various aquatic weeds such as *Hydrilla verticillata* and therefore potentially harmful and certainly undesirable. Other *Chaenusa* species are parasitoids of *Hydrellia* pests of rice and may be locally important.

Some dacnusines are egg–larval parasitoids e.g. *Polemochartus*, *Coelinidea* and *Trachionus* (= *Symphya*). *Polemochartus liparae* is a parasitoid of gall-forming *Lipara* flies (Chloropidae) and whereas nearly all alysiines (and cyclostomes for that matter) have only two pairs of ovarioles, *Polemochartus* has up to seven pairs, which might be correlated with the production of smaller eggs and possibly higher fecundity. Mook (1961) described the host searching and acceptance behaviour of *P. liparae*, which attacks the eggs of its host *Lipara* flies laid on the surface of reeds (*Phragmites*) before the fly larvae climb up the reed and then burrow down between the rolled-up leaves to the growing point, where they form characteristic galls. However, hosts parasitised by P. *liparae* metamorphose to the puparium stage prematurely (in August or September) rather than in the following March and April when unparasitised ones do so. Consequently, the galls from which *Polemochartus* emerge are typically smaller than those that will eventually produce flies.

G.C.D. Griffiths (1964) presented a partially resolved intuitive phylogeny of the genera of Dacnusini covering 16 genera and a number of separate species groups of *Chorebus*. Wharton (1984) divided the Alysiini into four genus groups but left a number of genera unplaced. Only two molecular phylogenetic analyses have been published and neither included many genera. The investigation by Gimeno et al. (1997) was based on three gene fragments and Wharton et al. (2006) used only 28S rDNA sequences but for many more genera, to resolve the placement of the exodont gnamptodontine genus *Exodontiella* (which turned out to be a gnamptodintine). Gimeno et al. recovered the Alysiinae paraphyletic with respect to the Opiinae and the Alysiini paraphyletic with respect to the Dacnusini. Wharton et al. recovered a monophyletic Alysiinae and a monophyletic *Phaenocarpa* group of genera with strong support; however, their Dacnusini and other Alysiini were muddled and with no significant bootstrap support for any of their relationships. The latter study seems to indicate that the venom reservoir similarities between some alysiines and opiines (Quicke et al. 1997a) may in fact represent a basal condition for the whole clade with subsequent reversals.

Opiinae

The opiines, as with the closely related alysiines, are composed, so far as is known, entirely of koinobiont endoparasitoids of Diptera larvae. Because most opiines are morphologically not or hardly cyclostome, in the old days they were often placed among the non-cyclostomes, but molecular and morphological data clearly place them close to the Exothecinae and Gnamptodontinae. Many opiines are superficially rather similar to alysiines but they lack the exodont condition of the mandibles (Fig. 12.31; *cf.* Fig. 12.29). A few are in fact distinctly cyclostome e.g. *Diachasma* and *Psyttalia* (Fig. 12.31a), but it is not clear whether this represents a retained plesiomorphic condition or is secondary. In any case, the anteroventral clypeal margin is at most weakly arched, unlike in most cyclostomes.

The majority of temperate species are parasitoids of leaf-mining Diptera (mostly Agromyzidae and Anthomyiidae) whereas the subtropical and tropical faunas have a much higher proportion of species associated with fruitflies (Tephritidae), and because of this, the Opiinae is one of the most important subfamilies from a biological control point of view, second only to the Microgastrinae in terms of the number of species involved, and introductions have a fairly high success rate (Gilstrap & Hart 1987, Wharton 1989). They are also one of the most taxonomically challenging. Generic concepts within the Opiinae have not been stable and creation of a taxonomy based on clear apomorphies has lacked behind other groups. Most of the larger bodied species were classified in the genus *Biosteres* and are frequently referred to as such in the literature. Considerable advances have been made by Wharton (1988, 1997) through the recognition of which morphological characters appear to be more reliable in defining monophyletic genera; Table 12.4 gives the current generic placements of those referred to in this book, and Wharton (1997) dealt with many others. The subfamily is dominated by the highly speciose and morphologically uniform genus *Opius*. All species are koinobiont endoparasitoids of Diptera and in particular of tephritid fruit flies, although a number of other families are attacked by particular genera. For example, *Ademon* species are aquatic and parasitise flies of the genus *Hydrellia*

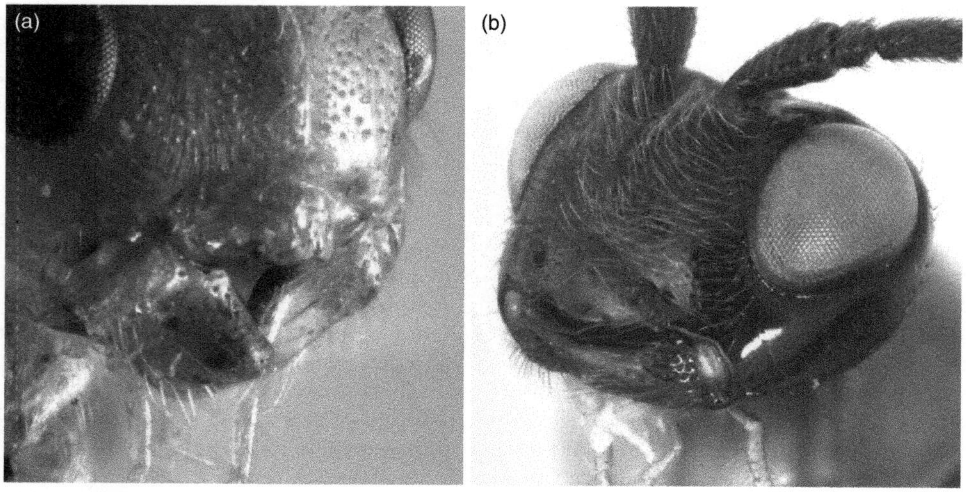

Fig. 12.31 Features of faces of Opiinae. (a) *Rhynchosteres* sp., showing distinct hypoclypeal depression with exposed labrum; (b) *Opius* (*Bellopius*) *hirtus*, showing absence of hypoclypeal depression. (Source: reproduced by permission of Robert A. Wharton.)

Table 12.4 Current generic placements of important fruitfly parasitising Opiinae that in much of the earlier literature are incorrectly treated under the generic names *Biosteres* or *Opius* (see Wharton 1997).

Species name	Current genus	Synonymy
anastrephae	*Utetes*	
areolatus	*Doryctobracon*	
arisanus	*Fopius*	
bellus	*Opius* (*Bellopius*)	
concolor	*Psyttalia*	
crawfordi	*Doryctobracon*	
juglandis	*Diachasmimorpha*	
incisus	*Utetes*	
longicaudata	*Diachasmimorpha*	
oophilus	*Fopius*	Junior synonym of *arisanus*
siculus	*Psyttalia*	Junior synonym of *concolor*
tryoni	*Diachasmimorpha*	
vandenboschi	*Fopius*	

(Ephydridae) in submerged aquatic plants (van Achterberg & Prinsloo 2012), and is reported to swim through the water using its relatively long and hairy legs (M.R. Shaw & Askew 2010).

Notable biocontrol species are *Diachasmimorpha longicaudata*, *D. tryoni* on *Anastrepha* species, *D. juglandis*, a specialist parasitoid of walnut (*Juglans*) feeding *Rhagoletis*, *Psyttalia concolor*, used to control olive fly (*Bactrocera oleae*), *Fopius arisanus*, which attacks a wide range of tephritid hosts (Ramadan et al. 1992), *Doryctobracon areolatus*, *D. crawfordi* and *Utetes anastrephae*, on *Anastrepha*, and *Opius hirtus*, principally on *Anastrepha* and also *Ceratitus*. In the case of the last of these species, Wharton (1983) noted interestingly and, from a biological control point of view, worryingly, that there were host-related differences in coloration and morphology. *Diachasmimorpha longicaudata* and *D. tryoni*, both introduced into Hawai'i to control fruit-infesting tephritids, subsequently also started attacking introduced and endemic gall-forming tephritids (Clancy et al. 1952, Duan & Messing 1997). Apart from those attacking tephritid hosts, a few are also important in the control of leaf-mining agromyzids; for example, *Opius caricivorae* is an important parasitoid of *Liriomyza sativae* (e.g. Chien & Chang 2012).

Opiines are either largely (e.g. Mohamed et al. 2003) or partly synovigenic with mature eggs ready at eclosion, although they continue to mature further for several days and, in the case of *F. arisanus*, access to hosts increases the rate of egg maturation. They also exhibit egg resorption and the rate of this is reduced when hosts are available (Ramadan et al. 1992, X. Wang & Messing 2003). Males of *F. arisanus* also need a period post-eclosion to produce mature sperm, and sperm were only found in the seminal vesicles 4 days after emergence (Hagen 1953, Quimio & Walter 2000).

Oviposition is usually into the early host larva, but ovo-larval parasitism is known in several species of *Fopius*, including *F. arisanus* (often referred to as *oophilus*), *F. caudatus* and *F. ceratitivorus*, although some of these will also occasionally oviposit into newly hatched host maggots. Wharton (1997), based on ovipositor morphology, i.e. with the tip being 'drawn out into a very fine point', additionally suggested that this biology might also be found in several other members of the genus, probably including all members of the *marangenses* group. Oviposition into eggs or early instar hosts can be viewed as an adaptation to hosts that rapidly become unavailable after hatching as they bore into their fruit substrate. *O. caricivorae* was found to accept all host instars with no effect on the successful development of the parasitoid (Chien & Chang 2012), and this species also showed non-destructive host feeding.

The eggs are normally rather simple, tapered ellipsoids, but in *Utetes* has a 'thumb-like' protuberance sub-posteriorly (Aluja et al. 2013). In larval parasitoids, the egg is deposited either just below the cuticle and sometimes attached to it: *Diachasmimorpha* (*Opius*) *mellea* deposits the egg just below the host cuticle (Lathrop & Newton 1933). Opiine eggs are slightly hydropic (Xu et al. 2007) and the serosal membrane cells of the developing embryo continue to surround the first instar larva (Pemberton & Willard 1918b, Lawrence 1990) but do not dissociate to form teratocytes. Some species cause temporary paralysis of their hosts at oviposition (Glas & Vet 1983). Movement of the host maggot appears to be a crucial stimulator for oviposition behaviour and Lathrop & Newton (1933) were able to persuade *D. mellea* females to show oviposition behaviour by simulating host larva movements under the skin of a blueberry using a pin or fine forceps, and Lawrence (1981a) showed that vibration was an essential cue to host location and oviposition in *Diachasmimorpha* (as *Biosteres*) *longicaudata*, with females showing no response to etherized or dead host larvae, but responding to artificial vibrations, and similarly, *Psyttalia concolor* females probe more into medium containing mobile than immobilised hosts, but also that the presence of medium itself was a significant factor (Canale & Raspi 2000, Canale 2003). Lawrence also found that the female's responsiveness to hosts was greatly influenced by the number of mature eggs in her ovaries at the time. Aggressive encounters between adult, host-searching female wasps have been observed in several species, although most notably in *D. longicaudata* (Lawrence 1981b, Jiménez-Pérez et al. 2007).

Host location has been investigated in several taxa involved in the biocontrol of fruitflies. Tephritid fly larvae induce various fruits to emit volatiles mixes that act as host-associated kairomones for female opiines (Benelli & Canale 2013a). Attraction to fruit (oranges) to *Diachasmimorpha longicaudata* was, however, found to be determined largely by more general habitat-associated cues, in particular volatiles released from damaged fruit due to fungal infection rather than those released as a result of host larval feeding (Segura et al. 2012). The presence of adult host flies themselves in an area may also indicate the presence of suitably aged hosts (i.e. young larvae) and Benelli et al. (2014) have gone on to show that *Psyttalia concolor* females respond to high concentrations of a host sex pheromone, such that they would find where male hosts were lekking, which in turn, is a good indicator of sites with many parasitisation opportunities. In an analogous system, Prokopy and Webster (1978) showed that marking pheromones used by the apple maggot, *Rhagoletis pomonella*, to deter oviposition by conspecifics, also serves as a kairomone for its opine parasitoid *Utetes canaliculatus* (correct name for *Opius lectus*).

Vision also plays a role in the host location in the case of some fruit fly-parasitising species. Rousse et al. (2007) found that naïve females of *Fopius arisanus* principally use the dark contrast of fruits and do not seem to detect or at least respond to short wavelengths and do not seem to show any hue discrimination. This response pattern was suggested perhaps to reflect the fairly generalist host relationships of this parasitoid where hosts might be encountered in many different fruit types. In the same species, Pérez et al. (2012) found that relative reflectance of the odour-baited coloured targets was important and that naïve females preferred dark-yellow targets, spherical targets more than other shapes and 10–12 cm diameter targets rather than smaller ones. Perhaps larger targets in the wild would correspond to older fruit that might be more likely to have been attacked by tephritids.

Opiines are reported normally to go through four larval instars (Lathrop & Newton 1933, Shaw & Huddleston 1991), but Xu et al. (2007) reported five larval instars in *Opius caricivorae*. Development beyond the first instar does not occur until the host maggot is about to pupate (Lawrence 1982, 1986). Detailed descriptions of developmental stages have been provided for *Diachasmimorpha longicaudata* (Carabajal-Palatino et al. 2010), *D. tryoni* (Pemberton & Willard 1918b, as *Diachasma*), *Fopius arisanus* (Rocha et al. 2004), *Opius ilicis* (E. Cameron 1941) and *O. caricivorae* (Xu et al. 2007). The eggs are rather lemon-shaped with a protuberance at either end (Pemberton & Willard 1918b, Iwata 1959). There appear to be four instars in most species (Pemberton & Willard 1918b) but Xu et al. reported five. The first instar is sub-hymenopteriform with a relatively large head but with rather small, sickle-shaped mandibles that are nevertheless capable of 'wide movement and quick action' (Pemberton & Willard 1918a), and there is a short tail segment. *D. tryoni* first instars also have a pair of ventral protuberances approximately as long as the first thoracic segment of uncertain function. The middle and final instar are unremarkably hymenopteriform. Development and pupation occur within the pupariumized larva before the host has moulted into a pupa as in the Alysiinae after the egg has hatched (Pemberton & Willard 1918b, Xu et al. 2007). The meconium is not voided at pupation as in most ichneumonoids, but retained in the body until emergence of the adult wasp, although it may pass from the mid-gut to the rectum during the pupal stage after fusion of the mid- and hind guts (Pemberton & Willard 1918b). Unlike nearly all other ichneumonoids, most opiines do not spin a cocoon or even, apparently, line their host puparium with any reinforcing silk (Shaw & Huddleston 1991), but a very few that attack hosts other than tephritids do line the inside of the puparium with a thin silk layer (R.A. Wharton, personal communication). What the function is of the labial glands in the larvae of those that do not produce any silk remains unknown.

Superparasitism is common but only one parasitoid ever completes development in a host, and this is most likely due to combat between first instar parasitoid larvae (Carabajal-Palatino et al. 2010, González et al. 2007).

Opiines and alysiines all have specialised glands in males that open at a number of pores or slits on the eighth metasomal tergum or near the anterior margin of the eighth plus ninth syntergum if the two are fused. These have generally been named Hagen's glands after their discoverer and have attracted considerable attention (Hagen 1953, Buckingham 1968, 1975, Buckingham and Sharkey 1988), and in some species they produce a distinct coconut-scented emission. Their secretory products have been analysed by H.J. Williams et al. (1988), who concluded that the variation observed between species indicated that they probably played a variety of functions, ranging from defence to sex pheromones (see Chapter 7, *Sex-related scent glands*). Detailed ultrastructual studies were carried out by Khoo & Lawrence (2002), who showed them to be typical Type III exocrine glands. The swarming behaviour observed by males of *F. arisanus* may be connected with the release of female-attracting volatiles (probably from the Hagen's glands), the effectiveness of this signal being increased by the local presence of many releasing males (Quimio & Walter 2000). At shorter distances, females also release a male-attracting pheromone.

Courtship behaviour has been investigated for the fruit fly parasitoids *Diachasmimorpha longicaudata* (Sivinski & Webb 1989), *D. kraussii* (Rungrojwanich & Walter 2000) and *Psyttalia concolor* (Canale et al. 2013). Males produce courtship songs through wing fanning/vibration. Interestingly, males produce a different song pattern when interacting with other males rather than trying to seduce females. Benelli & Canale (2013b) have shown that this might be because inexperienced males attempt to court members of both sexes, and this seems to be important practice and later courtship attempts by these more experienced males are initiated with a shorter latency, have more wing fanning and lasts longer, a very similar situation to that found in the pimpline *Itoplectis conquisitor* (Robacker et al. 1976). Boush and Baerwald (1967) presented some behavioural evidence for a female sex pheromone in *Diachama alloeum* (as *Opius alloeus*[7]) and noted that both sexes will mate multiply, at least during a short time window.

Viruses and virus-like particles have been reported from the venom apparatus of a number of species. Jacas et al. (1997) described coronavirus and cypo-like viruses in *Psyttalia* (as *Opius*) *concolor*, and Luo and Zeng (2010) reported a new rod-shaped virus from the venom gland filaments (which they termed 'accessory gland filaments') of *Diachasmimorpha longicaudata*, a species that previously has been shown to have entomopox virus (DlEPV) (Lawrence 2005, Hashimoto & Lawrence 2005) and a rhabdovirus (DlRhV) (Lawrence & Matos 2005) in its venom gland tissue. The rod-shaped virus and DlEPV replicate in host haemocytes and DlRhV in epidermal cells. The mode of transmission of these from one generation of wasp to another is uncertain and may involve multiple pathways, as they are found both underneath the chorion of mature eggs within the wasp and also within the guts of larvae feeding within the host. Also, interestingly, their report indicated that the same wasp can harbour different associated viruses in different parts of its range. Transmission of the virus to the wasp's offspring was examined ultrastructurally by Lawrence & Matos (2005), but despite careful study the exact pathway is unclear. Virions were not detectable using TEM in very recently lain eggs, but numerous virions were present in the sub-chorionic region of eggs that had been laid 30 h or so previously. Therefore, they probably formed as a result of unencapsidated RNA molecules being present in the pre-vitellogenic eggs through to the laid ones but not being detectable by TEM, but which were subsequently able to replicate and mature into mature virions. The alternative suggested, that somehow mature virions enter the egg through the chorionic membrane, seems very unlikely. They also found that within 24 h of the wasp larva hatching, the virions could be found in their gut lumen, suggesting that by this time they had infected and replicated in the host to a sufficient extent that they could be found in the parasitoid larva's diet.

Sense organs of the ovipositor of *Diachasmimorpha* (as *Biosteres*) *longicaudata* were described by Greany et al. (1977) and antennal sensilla, including electrophysiological responses, have been studied in *Psyttalia* (as *Opius*) *concolor* by Barbarossa et al. (1998). In a simple experimental system, Nishida (1956) showed that *P.* (as *Opius*) *fletcheri* females would pierce larvae of its house fly (*Musca domestica*) non-host when presented in a suitable medium but would not oviposit into them, hence the ovipositor sensilla enable it to discriminate between suitable fruit fly maggot hosts and unsuitable species.

There is considerable variation between genera and species in the morphology of the venom gland and reservoir and illustrations have been presented by a number of workers, including Genduso (1969) for *Psyttalia concolor* (as *Opius*), Wharton (1997) for various tephritid parasitising species and Quicke et al. (1997a) for a range of genera and species.

M. Fischer (1972) recognised four tribes (Ademonini, Desmiostomatini, Gnamptodontini and Opiini) and Wharton (1997) added Exodontiellini. However, Gnamptodontini were elevated to subfamily rank by van Achterberg (1983c) and Exodontiellini moved to Gnamptodontini by Wharton et al. (2006). Further, of the genera included by Fischer in the Ademonini, only *Ademon* remains, the others having been removed to Opiini, Rhyssalinae and Mesostoinae *s.l.*, while the type genus of Desmiostomatini was synonymised with *Opius* by Wharton (1988). This left the Opiinae with just two tribes, Ademonini and Opiini. The phylogeny of the group has only been approached formally in a molecular context twice, both times employing three gene fragments but with only partly overlapping gene coverage (Gimeno et al. 1997, X.-Y. Li et al. 2013). Node support for Gimeno et al.'s findings was generally low and Li et al. did not present any support measures but did raise several subgenera to full generic status on the basis of their results. Several species were also included in the molecular investigation by Wharton et al. (2006) of *Exodontiella*, which only recovered a monophyletic Opiinae under certain data analysis parameters, but found it roughly falling into two large genus groups, one centred around *Opius* and *Biosteres* and the other around *Doryctobracon*, *Diachasmimorpha* and *Fopius*. None of these studies included the putatively basal genus *Ademon*, but in the cyclostome tree of Zaldivar-Riverón et al. (2006) it was found derived among the small group of included genera despite it possessing a complete occipital carina. Similarly, *Ademon* has extremely reduced final instar larval head capsule features with only mandibles, prelabial sclerite and palps present (see fig. 40 in Čapek 1970). It seems likely that its possession of an occipital carina most probably represents a reversal rather than a retention of a plesiomorphic character.

Braconinae

This is one of the largest of braconid subfamilies, with 2800 described species in more than 185 genera (Shenefelt 1978, Quicke 1987a, D.S. Yu 2012). It is cosmopolitan and one of the morphologically most diverse subfamilies, and comprises species of a wide range of sizes (ca 1–30 mm body length) (Fig. 12.32), although it is dominated by the genus *Bracon*, an undoubtedly para- or polyphyletic group, which are predominantly small wasps (Belshaw et al. 2001). The temperate fauna is rather impoverished in terms of genera, with most species belonging to *Bracon* and far smaller numbers of species in genera such as *Coeloides*, *Cyanopterus*, *Digonogastra*, *Iphiaulax*, *Glyptomorpha*, *Atanycolus* and *Vipio*. Although the neotropics are rich in species, these dominated by members of just two genera, *Bracon* and *Digonogastra*, it is not rich in genera (1997b; cf. Quicke 1987a). In contrast, the Old World tropics, Africa and South-East Asia in particular, are far richer and include more than 170 valid genera (D.S. Yu 2012). Although future phylogenetic studies may reveal that some of these genera need to be synonymised, the difference between the Old and New World faunas is startling and seems to contrast strongly with the biologically very similar Doryctinae, which have a relatively far higher generic richness in the neotropics. Further, with the exception of *Bracon*, which is almost certainly polyphyletic and is not defined by any synapomorphies, there is a complete separation between genera occurring in Neotropics (and extending into the Nearctic) and the Old World genera.

Given the size of the subfamily, surprisingly few species have been of practical use in biological control, although a fair number do attack pest species including wood-borers, stem-borers, stored grain pests and a few seed predators, and may contribute to pest population regulation in a small way. The most noteworthy are *Habrobracon hebetor* (often called *Bracon* in the literature[8]), which is widespread and used in the control of a number of pest insects (X.F. Huang 1986), especially stored product pest Lepidoptera such as Mediterranean flour moth [*Anagasta* (= *Ephestia*) *kuehniella*: Pyralidae], Indian meal moth (*Plodia interpunctella*: Pyralidae) and the Angoumois grain moth (*Sitotroga cerealella*: Gelechiidae). *H. hebetor*

Fig. 12.32 Three long-ovipositored braconines. (a) *Euurobracon yokahamae* from Japan, body darkened post mortem, in life it is pale orange–yellow. (Source: reproduced by permission of Kyohei Watanabe.) (b) *Monilobracon* sp. from Uganda, searching for host beetle borings on barkless tree trunk. (c) *Glyptomorpha pectoralis*, a reasonably common species from the Mediterranean region. (Source: reproduced by permission of Cor Zonneveld.) For a colour version, see Plate 26.

may also be important in the control of cotton boll worm (*Helicoverpa armigera*: Noctuidae), millet head miner (*Heliocheilus albipunctella*: Noctuidae) and indeed, to a lesser extent, a number of other pest species. In many field situations, this parasitoid is used simultaneously with other treatments ranging from chemical pesticides and biological such as the chalcidoid wasp *Trichogramma*, the bacterium *Bacillus thuringiensis*, various entomopathogenic fungi such as *Beauveria bassiana* and *Metarhizium anisopliae* and even entomophagous *Heterorhabditis* nematodes. Other important braconines involved in biological control include *Bracon mellitor*, a parasitoid of the boll weevil (*Anthonomus grandis*) (Adams et al. 1969), *B. cephi* and *B. lissogaster* on the wheat stem sawfly *Cephus cinctus*, *B. nigricans* on the tomato leafminer (Biondi et al. 2013) and *Atanycolus* that attacks the emerald ash borer (*Agrilus plannipennis*: Buprestidae), which has recently become a major pest of *Fraxinus* trees in eastern North America following its introduction from Asia (Fig. 12.33). *B. celer* is currently being investigated as a possible biological control agent for olive fruit fly, *Bactrocera oleae* (Tephritidae), in the USA (Sime et al. 2006), although previous attempts to introduce it into southern Europe from its native range in South Africa have failed. Several species are important in their native ranges for controlling a variety of pests. *Coeloides bostrichorum* is an important control agent for *Ips* species (Curculionidae: Scolytinae) attacking spruce (*Picea*) in Europe. *Stenobracon deesae* is an important parasitoid of various lepidopterous borers in Asia, and members of the closely related genus *Euvipio* are frequently reared from the same sorts of hosts in Africa. Narayanan and Venkatraman (1952) described a technique for

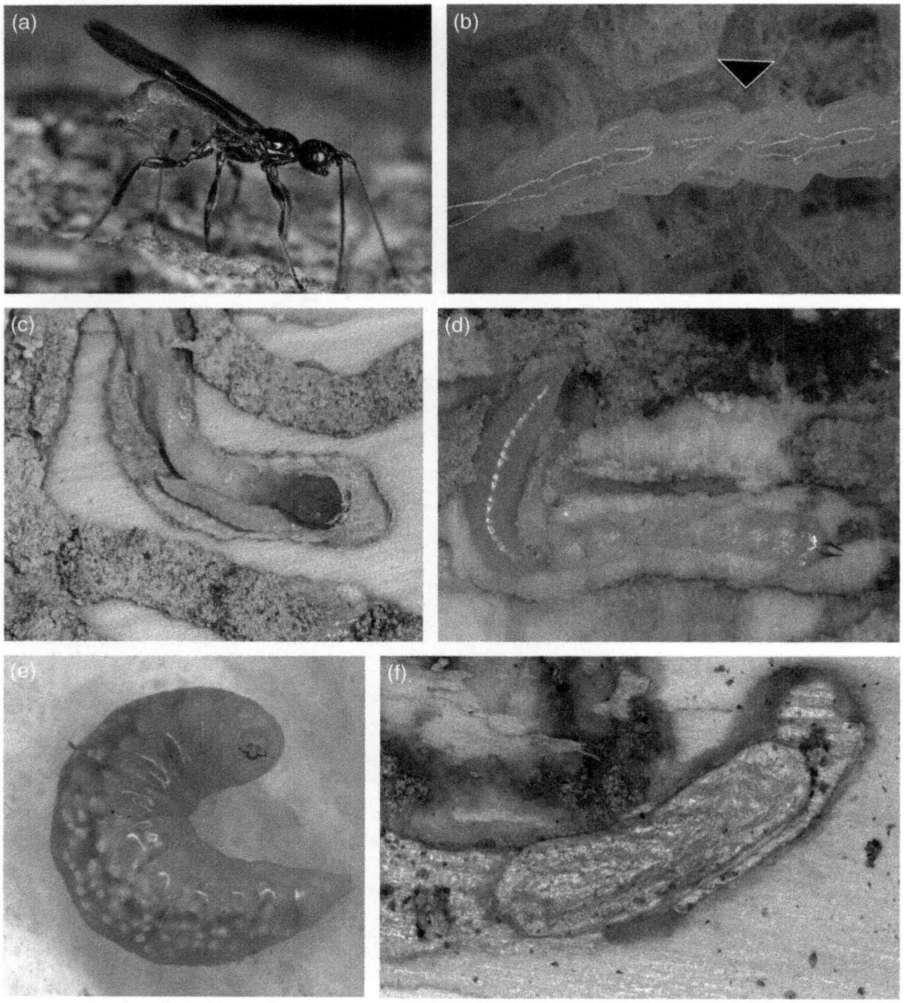

Fig. 12.33 Oviposition and development of the idiobiont, ectoparasitoid *Atanycolus cappaerti* (Braconinae) on larva of the emerald ash borer, *Agrilus planipennis* (Cerambycidae), an invasive pest of trees in North America, accidentally introduced from China. (a) Ovipositing female wasp; (b) egg on paralysed host larva (slightly to right of middle, arrow), note the long thin tail to the right of the egg body; (c, d) medium and nearly fully grown larva feeding on host; (e) isolated fully grown larva; (f) thin, papery parasitoid cocoon in host tunnel. (Source: reproduced by permission of David Cappaert, Michigan State University, Bugwood.org.) For a colour version, see Plate 27.

mass rearing the former for inundative release using alternative hosts.

Most braconines are idiobiont ectoparasitoids of weakly to deeply concealed larval hosts, particularly Lepidoptera and Coleoptera, but some also attack Diptera and Hymenoptera (Austin & Faulds 1989). In these cases the host is invariably paralysed by the parasitoid's venom, which is remarkably potent in this respect – a small female wasp may effect complete flaccid paralysis of even fairly large hosts. However, de Leon (1935) found that in the case of *Coeloides* parasitising the scolytine bark beetle, *Dendroctonus monticolae*, full host paralysis takes a day or so to develop, which would suggest that the parasitoid's eggs may be vulnerable to being squashed by the host, although this does not appear to happen. Host paralysis can then last up to several weeks (Beard 1952, Sláma & Lukáš 2011). The paralysing action of the venom is fairly generalised and injections into fictitious hosts effect almost the same degree of paralysis as in natural hosts. Given this, physiological interaction with hosts is unlikely to play a large role in determining host range. This is exemplified by the southern New World genus *Compsobraconoides*, which was first recorded as a parasitoid and egg predator of *Azteca* ants in Peru which live in domatia on the tree (D.W. Yu & Quicke 1997), but another species was found by Fortier and Nishida (2004) to show the more typical biology of being a parasitoid of a gall-forming weevil larva. The connection here would seem to be that the hosts live in swollen plant structures (galls, ant domatia) but can be taxonomically highly disparate. Parasitism of *Acacia* ants is also known in one species of *Trigastrotheca* from Africa (Quicke & Stanton 2005), but as these ants do not occur in South-East Asia, the *Trigastrotheca* species from there must have other host relationships. Predatory behaviour has also been observed for an Australian *Bracon* species that is associated with *Fergusonina* (Fergusoninidae) fly galls on *Melaleuca*, and Chadwick and Nikitin (1976) reported, but without detail, that an Australian *Bracon* sp. was reared from galls of the eriococcid hemipteran *Apiomorpha conica*, although as there have been no further reports it is possible that instead of feeding on the gall makers it might have been attacking an inquiline or another predator. Larvae of this species bore between chambers of the multilocular galls, consuming not just multiple host larvae and pupae, but also larvae and pupae of other associated parasitoids including the labenine *Poecilocryptus* and eurytomids and eupelmids (Goolsby et al. 2001). Particularly puzzling then, given the catholic nature of the venom action, is how the Old World tropics can support so many species and genera of wood-borer parasitoids, often with members of 5–10 species seen flying around and ovipositing in a single dead tree at any one time. However, nothing is known about how they might be partitioning up resources or competing.

Some species are associated with figs in both South-East Asia and Africa (F.X. Williams 1928, van Achterberg & Weiblen 2000, personal observations) and Williams reported that '*Microbracon*' *lendicivorus* was predatory on larvae of the cecidomyid *Asphotrophia fici* within figs in the Philippines. The body size of *Ficobracon* suggests that they may be at least partially predatory within the syconium, although no biological observations have been made. An apparently monophyletic group of genera, placed in the subtribe Aspidobraconina, are unusual in that they are endoparasitoids of butterflies with adults emerging from the host's pupa. Several of the aspidobraconine genera are gregarious, e.g. *Philomacroploea* and *Aspidobracon*, whereas some of the larger species may be solitary, e.g. *Hyboteles* (van Achterberg 1984b, Quicke 1987c).

At least two Neotropical species, at present attributed to the genus *Bracon* but much larger than typical members of that genus, are phytophagous in the seeds of *Protium* species (Burseraceae) (Flores et al. 2005, Perioto et al. 2011). The final instar larval, and probably earlier instars also, have strongly muscularised heads with very large, hard, multi-toothed mandibles (Fig. 12.34). The lower ovipositor valves are furnished with very strong serrations and the ovipositor tip is heavily sclerotised and black, more like those of the Doryctinae. Flores et al. (2005) speculated that this biology evolved from initial parasitisation of other seed predators such as bruchine chrysomelid beetles.

Juvenile development has been described for surprisingly few species, despite the fact that many have been cultured at various times. Good studies include those on *Bracon tachardiae* (Glover 1934), *Megalommum* (as *Aphrastobracon flavipennis*) (Glover 1939), *Stenobracon deesae* (Alam 1952) and *Habrobracon hebetor* (as *Bracon brevicornis*) (Sudheendrakumar et al. 1982). The eggs are nearly always ovoid, with a short or long stalk/tail that is usually at the posterior end and is better developed in those taxa with long, slender ovipositors that have to pass the egg a long way along a narrow egg canal (Fig. 12.33b) (Iwata 1959). The stalk of the egg of *Coeloides* is reported to have a segmented appearance (Russo 1938). There are five larval instars and all stages have functional tracheal system and are hymenopteriform. The first larval instar is often rather well endowed with cuticular spines that no doubt help it move over the paralysed host, and the mandibles are robust with a long blade and a basal group of teeth. The second to fourth instars are similar to the first but their cuticle is not as shiny (Fig. 12.33d,e). Nothing is known about the juvenile stages of the endoparasitic Aspidobraconina.

Courtship and mating have only been studied in a few, almost entirely smaller bodied, species, although Glover (1939) gave a short description of mating in *Aphrastobracon*. Sudheendrakumar et al. (1978) found that males of *Habrobracon hebetor* (as *Bracon brevicornis*)

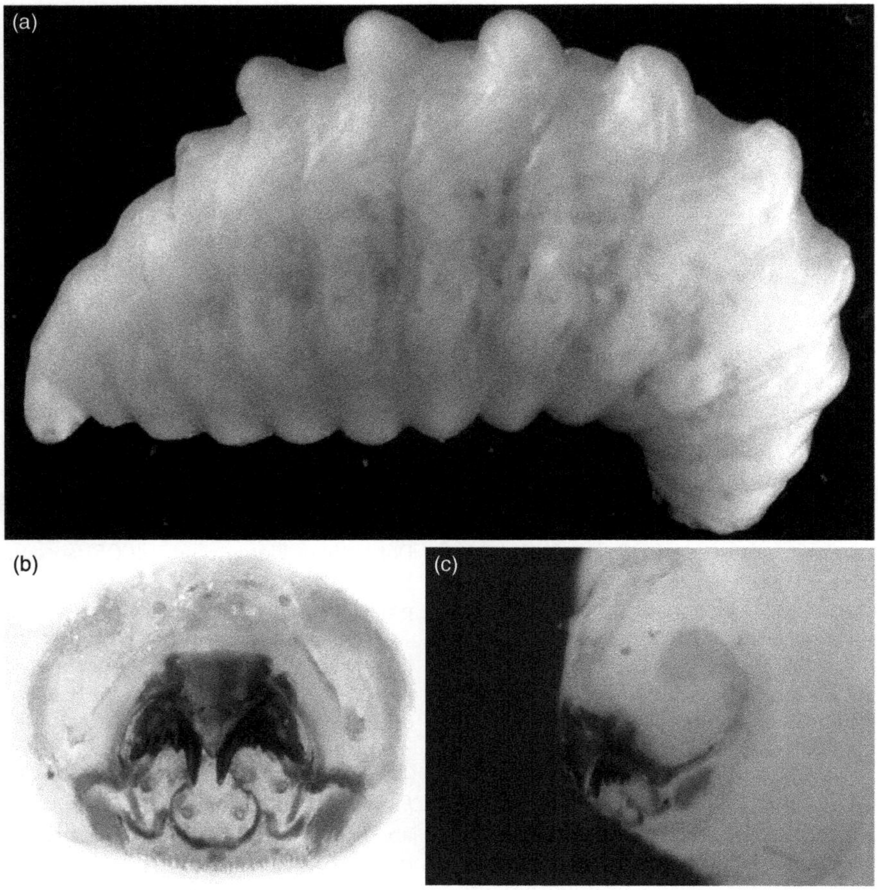

Fig. 12.34 Larval features of the phytophagous braconine *Bracon phytophagus*, which is a seed predator in *Protium tovarense*. (a) Whole larva side view; (b, c) larval cephalic structures, front and side view, respectively, showing large, heavily sclerotised and multidentate mandibles. Also note the small, papilliform antennae. (Source: from Flores et al. 2005, reproduced by permission of Taylor & Francis Ltd.)

have a short pre-mating period of approximately 1 h and freshly emerged females refused mating attempts by bending the metasoma ventrally. Males fanned their wings in bursts, as also seen in *H. hebetor* (Grosch 1947). Males of *hebetor* jump on to the backs of apparently receptive females, sometimes having to reorientate themselves if they land facing the wrong direction, and continue to buzz their wings until copulation starts, with both parties facing in the same direction and the male's metasomal tip curved down to enter the female's genital opening. Although wing buzzing stops, the mating male continues to stimulate the female by tapping her metathorax with his head, perhaps to keep her receptive until he has completed sperm transfer. Mating is relatively brief, 3–15 s. Antolin and Strand (1992) showed that in a 'natural' population infesting *Plodia interpunctella* in grain storage units males formed aggregations in high places and exhibited scramble competition for mates with polygyny. The problem of inbreeding with their slCSD sex determination system was largely avoided by the majority of females dispersing from their natal 'patch' before mating.

Regarding *H. hebetor*, there is an important taxonomic issue in that this incredibly widely used and published wasp shows considerable morphological variation, which is not surprising in an insect that has been bought repeatedly into culture and may show some founder effects. However, Heimple et al. (1997) showed that what people had been calling *H. hebetor* in at least two different culture systems were in fact completely genetically isolated cryptic species with different host range preferences, although they failed to go into the nomenclature and taxonomy to decide what the correct name might be for the second species. There are a number of potential names for it that have at various times been placed into synonymy with *hebetor* (Shenefelt 1978), but sorting them out correctly would certainly require very careful study of old type material and possibly necessitate neotype designations or descriptions of entirely new species.

As idiobiont ectoparasitoids which are essentially carnivores on protein- and lipid-rich host resource, they ought to be relatively easy to culture *in vitro*, but to date the only successful cases seem to be *Habrobracon hebetor* (Yazlovetsky et al. 1992, Xie et al. 1989, Magro & Parra 2004, Magro et al. 2006) and *B. mellitor* which have been reared from egg to adult on insect-free media (Guerra et al. 1993b).

Adult diet is not known for most genera. Nectar feeding is common and in Mediterranean environments in the Old World, and members of some genera are particularly common in these environments [e.g. *Glyptomorpha* (Fig. 12.32c) and *Vipio*, and also various *Bracon* spp. such as members of the *B. urinator* group] are frequent visitors to various Asteraceae (and also some Apiaceae) and display prominent CNEAs. *Habrobracon* is reported sometimes to construct feeding tubes (Genieys 1925), but there are no recent references and, basically, there are no detailed biological observations on the bulk of genera which are tropical.

Many of the larger species (and even some of the smaller ones) are brightly coloured and patterned attractive insects (Fig. 12.32) and they appear to be involved in a number of mimicry complexes, principally with other braconids but also with various other insect groups (Quicke 1986). The reason behind this conspicuous coloration has remained elusive and, although some other larger species can give a human handler a mild sting (e.g. Quicke & Laurenne 2005), the effect is not long lasting and nowhere near as acute as that given by some ophionines or *Netelia* (Tryphoninae). One remarkable species, placed in its own genus, *Pheloura*, takes the mimicry of other braconids to another level. *Pheloura* has only a short ovipositor, but issuing from its rectum are three very long thin strands (van Achterberg 1989) that look, as far as can be ascertained from pinned specimens since no one has observed it alive, very like the very long ovipositor and sheaths of several other ichneumonids and braconids. It is hard to see why this would be advantageous, as most ichneumonoids, and especially those with long, flexible ovipositors, are largely incapable of delivering a sting, although the tropical Old World genus *Zaglyptogastra* can (Quicke & Laurenne 2005). Given the obvious bright colours of many large tropical braconids, despite the general lack of ability to deliver a sting to a human and without obvious models among other conspicuously unpalatable or stinging Hymenoptera, it seems that the potential predators they encounter must be deterred by taste, although evidence for this is weak, predators such as anolid lizards consume them readily and numerous species do not taste unpleasant to humans (Quicke et al. 1997b).

Members of this subfamily collectively display an enormous range of ovipositor forms (Quicke 1987a), from a fairly typical form with apical dorsal nodus and ventral serrations typical of ectoparasitoids attacking concealed hosts to short and blade-like with virtually no ventral serrations (e.g. *Iphiaulax*, *Chelonogastra*), which are probably used to access concealed hosts through cracks in, for example, dry wood, dorsoventrally compressed forms upwardly curved ones (*Megalommum*, see Quicke et al. 2000c, as *Curriea*), as well as singly (*Cedilla*, one *Digonogastra* sp.) or multiply arched (*Undabracon*, *Zaglyptogastra*: Fig. 3.8b) forms that are capable of being steered (Quicke 1991a, Quicke et al. 1992e). Together this suggests that the diversity of Braconinae may be tied in part to a particular evolutionary plasticity in ovipositor manipulation.

Particularly puzzling then, given the catholic nature of the venom action, is how the Old World tropics can support so many species and genera of wood-borer parasitoids, often with members of 5–10 species seen flying around and ovipositing near the same place in a single dead tree at any one time.

Habrobracon hebetor, and probably other species with similar biology, practises ovicide if it encounters previously parasitised hosts and, in agreement with theoretical models, the likelihood of a female carrying out ovicide increased with the proportion of previously parasitised hosts encountered and the time interval between encountering potential hosts (Strand & Godfray 1989, Takasu et al. 1997).

The anatomy of the female reproductive system has been described in *Habrobracon hebetor* by Bender

(1943) and *Stenobracon deesae* by Alam (1954). They almost invariably have two pairs of ovarioles, although in *Habrobracon* some occasional variation occurs and can be induced by cold-shock to the larvae (Grosch et al. 1977). The venom gland system is almost entirely uniform throughout the subfamily, unlike in other cyclostome groups. The reservoir is undivided, ovoid and heavily muscularised, with numerous tubular, unbranched venom glands connecting in a ring at the junction between the reservoir and the primary venom duct (Fig. 4.4b), the latter being covered entirely by chitinous globular vesicles that are probably associated with additional glandular activity (Fig. 4.4c). The apparent double reservoir illustrated by Barbalho and Penteado-Dias (1997) is probably atavistic.

The characteristic smell, perhaps best described as slightly acrid, earthy and engine oil like, given off by nearly all braconines when they are disturbed or during agonistic interactions, and which may be detected by people up 1 m or more away (Quicke et al. 2005b), appears to be the result of compounds released by their anterolateral abdominal glands (ALAGs) located on the highly corrugated and eversible membrane between the tergite and sternite of the first metasomal segment (Fig. 7.8a–c). (Quicke et al. 1997b). These glands are well developed in most genera, although they may be reduced or absent in a few. A few tropical species also have subtly different odours.

Members of multiple species of the Afrotropical genus *Monilobracon* (Fig. 12.32b) that attack lymexylid beetle larvae in dead, barkless trees often seem to compete for the same host species, with some segregation of hosts according to size, very much as is seen in North America with *Megarhyssa* species. Both intra- and interspecific agonistic and aggressive interactions have been observed, with the outcome usually being determined by size and aggression only occurring when similarly sized conspecifics are involved (Quicke et al. 2005b).

In addition to the almost ubiquitous ALAGs, males of many species, probably the majority, have a pouch-like gland (reservoir) located between the sixth and seventh metasomal tergites (Quicke 1991b), similar to that found in some Gnamptodontinae (see fig. 110 in Quicke & van Achterberg 1990), and some members of the largely tropical *Atanycolus* group of genera have additional glands opening on either pore plates or through a collective duct anterolaterally on the seventh metasomal tergite (Quicke 1991b). Being male-specific, these are likely to be involved in courtship or perhaps mate location by females. Members of both sexes of some of the *Compsobracon* group of genera have a mid-ventral pouch-like and presumably gland-associated structure between the second and third metasomal sternites (Fig. 12.35), but since these are not sex linked, their likely function remains a mystery. Other possible pheromone gland structures include a deep, longitudinal groove on the hind tibia in the North American *Vipio moneilemae* (Quicke & Falco 1998) and a strangely swollen hind basitarsus in males of some Afrotropical *Rhadinobracon* species.

Another very strange and probably unique feature of males of quite a number of genera is the fusion of the dorsal surface of the rectum with a moderately to very large ovoid area on the seventh metasomal tergite (Fig. 12.36; Quicke et al. 1996). The chitinous wall of the rectum in this area is completely fused to the tergum and the latter appears to have pores, although the available micrographs are not completely clear on this issue. Light microscopic sections show the presence of some bacteria in the rectum near the fused zone, but whether there is any functional connection with the structure is unclear. Again, as this rectotergal fusion is only present in males and as these posterior tergites are often exposed during courtship or mating attempts, perhaps some volatiles produced in the gut or rectum are released and influence female behaviour. It is not yet clear what happens to the dorsal vessel in this region; presumably it is moved to the side.

Habrobracon hebetor (sometimes classified as *Bracon* or *Microbracon*), in addition to being an important control agent for several stored product pests, was for many years one of the 'laboratory rats' of the parasitoid world, with many hundreds of papers published on its biology (Shenefelt 1978). It is a small, gregarious ectoparasitoid capable of developing successfully on a large range of hosts (it is commonly reared in the laboratory on Mediterranean flour moth, *Ephestia* (= *Anagasta*) *kuehniella*, Indian meal moth, *Plodia interpunctella*, or greater wax moth, *Galleria mellonella*). Females generally have adult lifespans of about 40 days and lifetime fecundities of 300–500 (Antolin & Strand 1992), which involves maturing 10–15 eggs per day, while juvenile development may be as short as 6 days (Jackson & Butler 1984), making back-crossing relatively easy. For a while it was moderately important in the study of genetics, with several colour variants and other mutations (Whiting 1934, 1939, 1961). It was the first hymenopteran for which sex determination was shown to be determined by the segregation

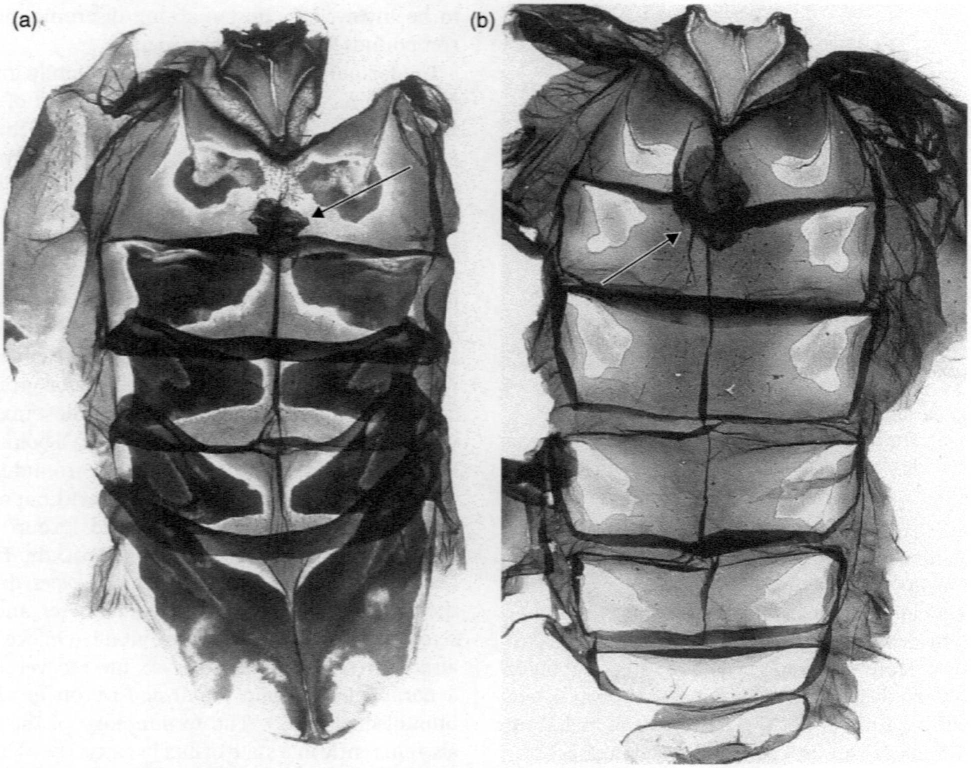

Fig. 12.35 Macerated metasomal sternites with arthrodial membrane stained with chlorazol black E showing a mid-ventral membranous pouch between the second and third metasomal sternites in two putatively related braconines. (a) *Virgulibracon vulsus*, female; (b) member of the *Compsobracon* group of genera, male.

of alleles at a single locus (Whiting 1943), and it and a closely related species have been extensively linkage mapped (Antolin et al. 1996, Holloway et al. 2000). Haplodiploidy makes it ideal for the generation of non-lethal mutants using either X-rays or chemical mutagens, and it was the first hymenopteran to be sent into space as part of the Biosatellite II experiment (von Borstel et al. 1968). J.F. Benson (1973) examined the effects of larval competition on survival, size and sex ratio. Taylor (1988) investigated its functional response and found that strong interference led to a type two response relationship, with number of hosts attacked plateauing out with increasing host density. Based on studies on *Habrobracon* species, braconines probably have single locus complementary sex determination. Whiting (1943) had concluded that *H. hebetor* has nine sex alleles and Heimple et al. (1999), using isofemale lines derived from populations collected from across the USA, found that the number of alleles per population ranged from three to nine (their methodology precluded detection of more than 10 alleles per population should any have had that many). Collectively, the populations appeared to have a total of 12 alleles.

Braconine venoms (excluding the few endoparasitic species) induce flaccid host paralysis and can be very potent, and some of the first studies on the chemistry and mode of action of parasitic wasp venoms involved *Habrobracon hebetor* (Spanjer et al. 1977, Piek et al. 1982, Piek 1990). Glover (1939), however, noted that the venom of *Aphrastobracon flavipennis* did not appear to be very toxic to tenth instar *Eublemma amabilis* (Noctuidae) hosts, 'which were in many instances still active after having been stung and oviposited upon', and this led to a high level of juvenile parasitoid mortality. Quistad et al. (1994) managed to isolate, with

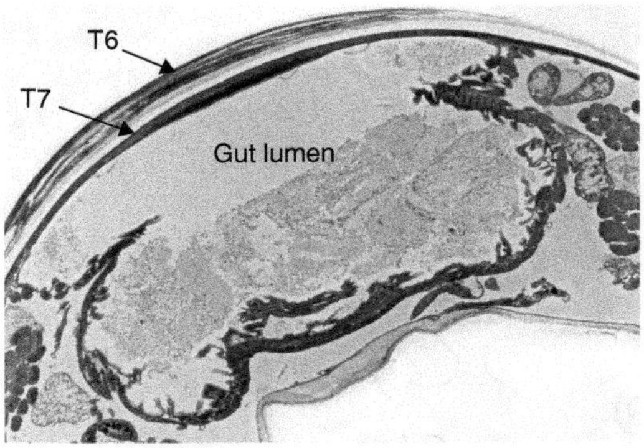

Fig. 12.36 Semi-thin transverse section through region of recto-tergal fusion (RTF) in a male of *Atanycolus ulmicola* (Braconinae). The posterior part of the sixth metasomal tergum is visible above the slightly thickened area of RTF on the seventh.

some difficulty, three proteins from *Habrobracon* venom, all large at approximately 73 kDa, two of which had high insecticidal potencies against *Heliothis* and other Lepidoptera larvae, with LD_{50} values ranging from 2 to 300 pmol/g. However, being large proteins (and fairly labile) they would be unsuitable for direct use as insecticides, although no doubt the possibility of including them in GM plants has not gone unconsidered.

Following a lag, new molecular tools are again making this an attractive choice of subject (J. E. Baker & Fabrick 2000, 2002, Hartzer et al. 2005). Kryukova et al. (2011) showed that *Habrobracon* venom induces a general reduction of immunity in its laboratory host *Galleria* and this is partly the result of suppression of the host's phenoloxidase system, which is involved in melanisation and encapsulation. Reducing this aspect of host immunity no doubt benefits the feeding *Habrobracon* larvae directly by reducing the host's ability to heal the feeding site; however, it also renders the host more susceptible to attack by other pathogens that may be present, such as entomopathogenic ascomycete fungi (Kryukov et al. 2013).

Dufour's gland chemistry has been investigated in *Habrobracon hebetor* (Fukushima et al. 1990, Howard et al. 2003) and in *Bracon cephi* and *B. lissogaster* by J.E. Baker et al. (2005). Technically, Fukushima et al. analysed a mixture of compounds from the venom gland, venom reservoir and Dufour's gland, but the nature of the compound they detected, a non-cyclic homoditerpene, strongly suggests that this was a component of the Dufour's gland. The gland is therefore more likely to be involved in host marking/discrimination than compounds from the venom glands.

The formal classification of the subfamily into tribes has most recently been largely the realm of Russian workers and based mostly on the fauna of the former Soviet Union, although a few tribes or synonymous subfamilies were erected around highly apomorphic taxa early in the 20th century, based mainly on the position of fore wing vein 1cu-a or highly enlarged mandibles. Not surprisingly, this approach has been far from satisfactory although, as it turns out, the three major groups recognised by the Russians hold out in the light of recent molecular studies. Thus we can now divide the subfamily into two major groups, one centred around *Bracon* and including most taxa with a ventrally short scapus and usually with a pointed basal lobe to the claw, and the other based around *Iphiaulax* with a ventrally elongate scapus and without a pointed basal lobe on the claws. The third 'group' are the Coeloidini, comprising just two genera, the Holarctic *Coeloides* and the neotropical *Megacoeloides*, defined by the distally expanded basal flagellomeres and highly setose parameres of the male genitalia (Quicke 1988b), although molecular studies so far recover these as a paraphyletic group (Zaldivar-Riverón et al. 2006, unpublished work). The terminology of the tribes is also currently in a state of flux because the oldest available family-group names stem from works by Ashmead based largely on the position of vein 1cu-a. Thus the group he called the Asphrastobraconini(-inae) and which for some time has been considered just to refer to a few taxa with an apomorphic ovalisation of the fore wing second subdiscal cell, now has to be treated as the whole of what used to be called the Iphiaulacini, i.e. the group largely characterised by a ventrally produced scapus. None of these characters are without homoplasy, of course, hence direct examination of a species will not necessarily allow it to be correctly placed within the framework.

There have been very few formal phylogenetic studies on the subfamily. A morphological phylogeny using maximum parsimony (conducted at a time when computational power and algorithms were not really up to the job of dealing with large data sets) was presented by Quicke (1988a). The only large molecular (28S rDNA) and combined molecular and morphological study was by Belshaw et al. (2001), which recovered in the combined analyses two large monophyletic groups, corresponding to the Braconini and Aphrastobraconini (i.e. Iphiaulacini of the authors).

Two genera, which may or may not be closely related, are of interest because they possess two fore tibial spurs, viz. *Rhamnura* and *Bicentra*. This is a feature otherwise almost completely absent from the apocritan Hymenoptera (the Ceraphronoidea being the exception) and therefore, when first discovered, this feature suggested that *Rhamnura* might be a very basal taxon within the Braconinae, and van Achterberg (1981b) correspondingly placed *Rhamnura* in its own tribe, the Rhamnurini. However, it is clear that for this to be truly a plesiomorphous feature, there would have to have been a huge number of parallel losses of one of the fore tibial spurs. Rather, the available data indicate that the two spurs are actually a synapomorphy for each genus separately and molecular data (unpublished) place *Rhamnura* deep within the Aphrastobraconina close to *Merinotus*; *Bicentra* morphologically is clearly closely related to another member of the subtribe, *Bathyaulax*. Thus, the presence of two fore tibial spurs in the Braconinae is almost certainly homoplastic.

A particularly interesting genus is *Argamania*, a group of small wasps from Israel and North Africa, which was originally described as an exothecine because unlike all other braconines it has a weak but complete occipital carina and a far postfurcal fore wing vein 1cu-a (Papp 1989). That it probably belongs to the Braconinae was noted by van Achterberg (1991b), who thought that undue weight had been given to its faint occipital carina and relatively long hind wing vein M+CU, and he created a new tribe for it, the Argamaniini, in the Braconinae. Another important feature pointed out by van Achterberg is the location of the second metasomal spiracle, which is in the laterotergite rather than the notum, and in this respect it differs from all other known braconines (but is the same as in exothecines). Wharton (1993b) was rather more circumspect and considered the placement as still tentative. The spiracular arrangement and the wing venation are probably best regarded as plesiomorphous and therefore may in fact be the most basal extant braconine. Although no material has become available for sequencing, van Achterberg's supposition is strongly supported by the venom apparatus of *Argamania*, which has a ring of venom glands inserting into the anterior end of the primary duct, a feature only known elsewhere in the Braconinae. However, the glands themselves are rather bulbous and not tubular as in other braconines. One particularly unusual feature of *Argamania* species is that they have a very distinct metallic appearance – black with a coppery sheen.

Another probably very basal braconine is the genus *Vaepellis*, which is thus far only from Ghana. Quicke (1987b) considered it so aberrant and lacking the braconine synapomorphy of a very short hind wing sub-basal cell that it was placed in its own subfamily, the Vaepellinae, although undoubtedly close to the Braconinae. Tobias (1988) considered it a derived braconine and his classification is followed here, although van Achterberg (1995) retained it as separate. This group is exceedingly rare in collections, no material has been available for molecular study and nothing is known of its biology. *Vaepellis* is interesting also because its body is densely covered with a fine pubescence, and it has very robust legs and very small eyes and ocelli, which suggests that it may spend much of its time in a concealed situation, perhaps in the ground.

The subfamily has a number of synonyms[9]. Gnathobraconinae was erected for a Neotropical wasp, *Gnathobracon babirussa*, with enormous mandibles (Szépligeti 1904), but this is just an highly derived species close to the genus *Digonogastra* and probably actually a derived member of that genus (Quicke & Huddleston 1991). A similar situation applies to *Pseudodicrogenium*, a very rare, monotypic genus from Africa also with enormous mandibles, upon which Fahringer based the Pseudodicrogeniinae, but is clearly a braconine and was synonymised by van Achterberg (1976c), although its relationships are far from certain. The Braconinae is also referred to as the Vipioninae principally by Russian authors.

The Old World genera were keyed by Quicke (1987a) and the New World genera by Quicke (1997b).

Exothecinae

The subfamily is treated here in a narrow sense equivalent to the tribe Exothecini *sensu* van Achterberg (1983a), which was then placed as a tribe within the Rogadinae or regarded as a group of Doryctinae, although molecular data in particular now indicate that they are not closely related and in fact Exothecinae is probably the sister group to the Opiinae+Alysiinae (Wharton et al. 2006, Zaldivar-Riverón et al. 2006, Sharanowski 2011), although there is a degree of lack of independence. In other recent treatments these have been included within the Hormiinae, indicating the lack of concrete morphological evidence for relationships.

Very little is known about exothecines apart from for the three genera found in the Palaearctic, viz. *Colastes*,

Shawiana and *Xenarcha*. It should be emphasised, though, that monophyly of the Exothecinae, even in this restricted sense, is far from certain, although they may well be closely related. *Colastes* is a relatively common parasitoid of a wide range of leafminers including Diptera, Lepidoptera, Coleoptera and also occasionally Hymenoptera (Tenthredinidae), with *C. braconius* having a particularly broad host range including members of all the above orders (Shaw & Huddleston 1991). Members of the other two genera are predominantly parasitoids of tenthredinid leaf miners (van Achterberg 1983a, Gorgiev 2006).

All species appear to be solitary parasitoids (Shaw & Huddleston 1991). Dowden (1941) gave a particularly detailed account of the biology of *Shawiana* (as *Phanomeris*) *phyllotomae*, a European parasitoid of the birch leaf-mining sawfly that was introduced into North America as a bio-control agent, but while it established it seldom kills more than 20% of hosts. There are apparently five instars and these are illustrated. Females are reportedly rather sluggish under laboratory conditions and when they locate a host mine they probe apparently randomly until they hit the host larva, which they partially, but incompletely, paralyse. Oviposition appears to take place upon an independent probe into the mine and the egg is most often laid away from the host towards the 'inner edge' of the mine. Upon hatching, the first instar larva, which is furnished with well-developed spines on the dorsum of most segments, searches and locates the still partially paralysed host and commences feeding externally. According to Dowden (1941), it seems that a large percentage die before becoming successfully attached to the host, although whether this is the case in the wild is less certain. As is typical of idiobionts, adult females are rather long-lived (up to 60 days) but larval development is rapid, only about 8–10 days from egg to cocoon construction. Their eggs are large and females carry only a small number of mature eggs (about six) at any one time. *S. phyllotomae* overwinters in its robust cocoon as a final instar larva and completes development shortly before emerging in late spring.

Colastes braconius development was illustrated by Beirne (1946), but some of the details given appear to be erroneous; in particular, M.R. Shaw (1983) showed that as with *Shawiana*, it is ectoparasitic and Beirne's assertion that it is endoparasitic appears to be a *lapsus*. The surprising thing here is that Beirne gave considerable detail of the five larval instars and the figures certainly seem to represent *Colastes*.

Fig. 12.37 Male of *Gnamptodon pumilio* (Gnamptodontinae). (Source: reproduced by permission, © C. van Achterberg, Naturalis Biodiversity Centre.

Gnamptodontinae (= Gnaptodontinae)

This is a small and rather poorly known subfamily of very small wasps, which were previously included within the Opiinae as a tribe until they were removed to their own subfamily (van Achterberg 1983c). Unlike opiines, these are not Diptera parasitoids but have been reared exclusively from leaf-mining nepticulid Lepidoptera. Gnamptodontines are fairly easily recognised by having a broad, smooth, raised area at the base of the second metasomal tergite (Fig. 12.37), together with their small size, typically being less than 1 mm long.

Apart from several host records and a strong indication that gnamptodontines are koinobiont, little else is known about them. The final instar larval head capsule shows prominent papilliform antennae, a putatively plesiomorphic feature generally associated with ectoparasitism, but detailed observations are lacking. Given their phylogenetic position close to the generally ectoparasitic Braconinae (with which some genera share diverging antero-lateral grooves on the third tergite) and the ectoparasitic Exothecinae, ectoparasitism would indeed seem the most likely possibility; however, other observations (or lack thereof) rather strongly suggest that they are endoparasitic (M.R. Shaw 1983) and, if this were so, it would be a corroboration of the prediction made by Belshaw et al. (2003) based on ovipositor characters using phylogenetic regression.

Whitfield (1992) carried out a morphological phylogenetic analysis of a selection of cyclostome braconids and recovered the Gamptodontinae as the sister group to the Opiinae+Alysiinae and noted, '… internal and external morphological evidence is mounting elsewhere (Buckingham & Sharkey 1988) that the Gnamptodontini are the sister group to the large Opiinae/Alysiinae lineage …'. The internal evidence referred to was the presence of Hagen's glands on their eighth metasomal tergite, which seemed like good evidence that they might be closely allied to the Opiinae+Alysiinae. However, there is now firm molecular evidence that the Gnamptodontini clade with the ectoparasitic Exothecinae or as sister group to the Braconinae (Wharton et al. 2006), so the Hagen's glands in them appear to be another example of misleading homoplasy, or perhaps they may be found among the Exothecinae when more taxa are examined.

Van Achterberg originally included four genera in the subfamily, *Gnamptodon*, *Pseudognamptodon*, *Gnaptogaster* and *Liparophleps*, although he expressed some doubt as to whether the last of these was actually related to the others. Subsequently, *Liparophleps* has been removed to synonymy (with *Semirhytus*) within the Doryctinae (Wharton 1997), confirming van Achterberg's suspicion. Another genus, *Neognamptodon*, has been described for a Madagascan species (Belokobylskij 1999), although it seems rather likely that all these other genera render *Gnamptodon* paraphyletic. The subfamily is almost cosmopolitan (Chen et al. 2002).

The aberrant North American exodont genus *Exodontiella*, which was originally considered to be an opiine, has subsequently been shown from DNA sequence data also to belong to the Gnamptodontinae (Wharton et al. 2006). Given that the exodont mandibles of the Alysiinae are used to aid the wasp in cutting its way out of the host Diptera puparium, it seems likely that in the case of *Exodontiella* they are probably involved in cutting the emerging wasp out of a similar situation. However, these small insects (<2 mm) are still known from only a very few specimens (fewer than 10 and these representing more than one species), with no biological data available.

Telengaiinae

A monotypic subfamily based on a single genus and species from Tajikistan was described by Tobias (1962). They are really very similar to gnamptodontines, although they lack the demarked, transverse raised area at the base of the second metasomal tergite. Like many braconines and gnamptodontines, this species has a distinct pair of diagonal grooves near the anterior corners of the third metasomal tergite demarking the antero-lateral areas. The biology of *Telengaia* is unknown, although they are somewhat larger than gnamptodontines so presumably they are not a parasitoid of nepticulids. The genus *Tamdaona* from Vietnam, currently classified in the Exothecinae, has similar metasomal groove patterns and might possibly actually belong to Telengaiinae.

THE NON-CYCLOSTOMES

Several features seem to unite the remaining subfamilies as a monophyletic group. All are koinobiont endoparasitoids of other insects, they have highly modified sperm usually with greatly reduced tails and sinuate nucleus (except Agathidinae), (see Figs 10 and 11 in Quicke et al. 1992b), and the testes are separate and each closely associated with their accessory glands. In addition, the venom apparatus is type 2 of Edson and Vinson (1979), not strongly muscularised and without spiral sculpture on the venom reservoir intima or primary venom duct[10]; however, the polarity of this character is questionable as both the Xoridinae, a putatively very basal subfamily of Ichneumonidae and the Trachypetinae, putatively basal braconid, possess the highly muscularised type 1 venom reservoir. There are generally 4–30 ovarioles per ovary (Fig. 4.1b) but in the microgastroids there are only two (Iwata 1959).

Sigalphoid Complex

This group comprises two obviously closely related subfamilies, the Agathidinae and Sigalphinae (including the Pselaphaninae), possibly together with the highly aberrant Masoninae, although I am by no means confident that the latter belongs with them on the basis of adult (and, in the case of Agathidinae and Sigalphinae, also larval) morphology and DNA sequence data, so I am at present leaving them among the unplaced subfamilies as per Sharanowski et al. (2011). Members of the Agathidinae and Sigalphinae share the presence of hind wing vein CUb and also subpronopes (a pair of pits on either side of the anterior margin of the pronotum), and both have polypodiform first instar larvae, all of which are tentatively synapomorphies.

Evidence that the morphologically highly derived Masoninae belong in this group is very weak. They are not only morphologically aberrant, but are also extremely strange in terms of their 28S rDNA sequence. In molecular analyses they have tended to come together with the agathidoids (Belshaw & Quicke 2002) this could be due to long branch attraction. Unfortunately they were not included in the multi-gene study of Sharanowski et al. (2011).

Agathidinae

The Agathidinae[11] is a rather common and cosmopolitan group that are exclusively koinobiont endoparasitoids of Lepidoptera larvae, most species attacking semi-concealed hosts such as leaf-rollers or tiers and correspondingly possess ovipositors ranging in length from approximately metasoma length up to approximately 1.5 times the length of the body (Fig. 12.38a–c), with one group specialising in exposed hosts and having correspondingly short, downcurved ovipositors (Fig. 12.38d–f). Most species are solitary, but a few species of the New World genus *Zelomorpha* are gregarious (Sarmiento et al. 2004, as *Coccygidium*).

They are nearly all very easily recognised, being small-medium to large wasps with a characteristically long and very narrow fore wing marginal cell bordered by tubular veins (Fig. 12.39a–c). Four tribes were long recognised, the Agathidini, Cremnoptini, Earinini and Disophrini. Sharkey (1992) presented a tribe-level phylogenetic analysis of the group based on adult morphology, but the results were inconclusive. From analysis of the length-variable D2–D3 expansion region of the 28S rDNA gene (Sharkey et al. 2006), a further group, the Eumicrodini, was recognised and a sixth group that would include a group of taxa with reduced venation was also well separated and if recognised at tribe level would be referred to as the Mesocoelini (see below).

A few species have been used successfully in biological control. For example, *Agathis gibbosa* against potato tuberworm (*Phthorimaea operculella*, Gelechiidae), *Alabagrus stigma* (= *Agathis stigmaterus* Cresson) against the sugar cane borer *Diatraea saccharallis* (Pyralidae) and also *Eoreuma* stalk-borers. The last species has been successfully introduced into the southern US states and into other parts of the world as a biological control agent (Jepson 1954). Others species play a variable but an important role in the suppression of forestry pests in their native lands (e.g. Georgiev & Delkov 2003).

Collectively, agathidines appear to be able to attack all the larval instars of their hosts (as do members of the Cremnoptini), although some may show a distinct preference for mid-grown ones (Disophrini; Simmonds 1947a) or younger ones (Agathidini, Earinini, Eumicrodini; Janzen et al. 1998). Some species are highly polyphagous, as in the case of *Bassus brooksi* (Janzen et al. 1998), but most are mono- or oligophagous.

Descriptions of juvenile stages were provided by Quednau (1970) for *Agathis pumila*. Of the few species for which detailed observations have been made, most oviposit within a ganglion of the host caterpillar (Odebiyi & Oatman 1972, 1977), or the egg is attached to a ganglion or the brain (Dondale 1954, Quednau 1970), and in a number of detailed studies where egg placement is not mentioned (e.g. Balduff 1966) it seems fairly likely that the egg may well have been concealed in a ganglion and that this had been, unsurprisingly, overlooked (M.R. Shaw & Quicke 2000). However, at least one species of *Cremnops* (as *Bracon*) deposits its egg free in the haemocoel (Simmonds 1947a). At least some species induce a temporary (or longer term if the host inhabits a semi-concealed situation) paralysis of the host caterpillar during the oviposition bout, although the parasitoid only inserts its ovipositor once (R.V. Flanders & Oatman 1987, Rodriguez-del-Bosque & Smith 1989).

Despite being inserted into a ganglion, the egg is hydropic and increases in volume approximately 30-fold over a 4 day period (Odebiyi & Oatman 1977). Agathidines appear to have three larval instars (Odebiye & Oatman 1972, 1977) and reports of other numbers are likely misinterpretations (see Shaw & Huddleston 1991 for a summary). First instar agathidine larvae are described as polypodiform because they bear paired ventral hook-like protuberances from most of their abdominal segments and are also caudate. They bear a very close resemblance to those of the closely related Sigalphinae (see Fig. 12.41d) (M.R. Shaw & Quicke 2000). After eclosing, they float freely in the haemocoel, and the first instars probably seek out to fight and kill any competitors with their relatively large mandibles and strongly sclerotised head capsule and then remain in that stage without growing while the host develops to the appropriate stage to be killed. The final instar larva emerges from the host caterpillar

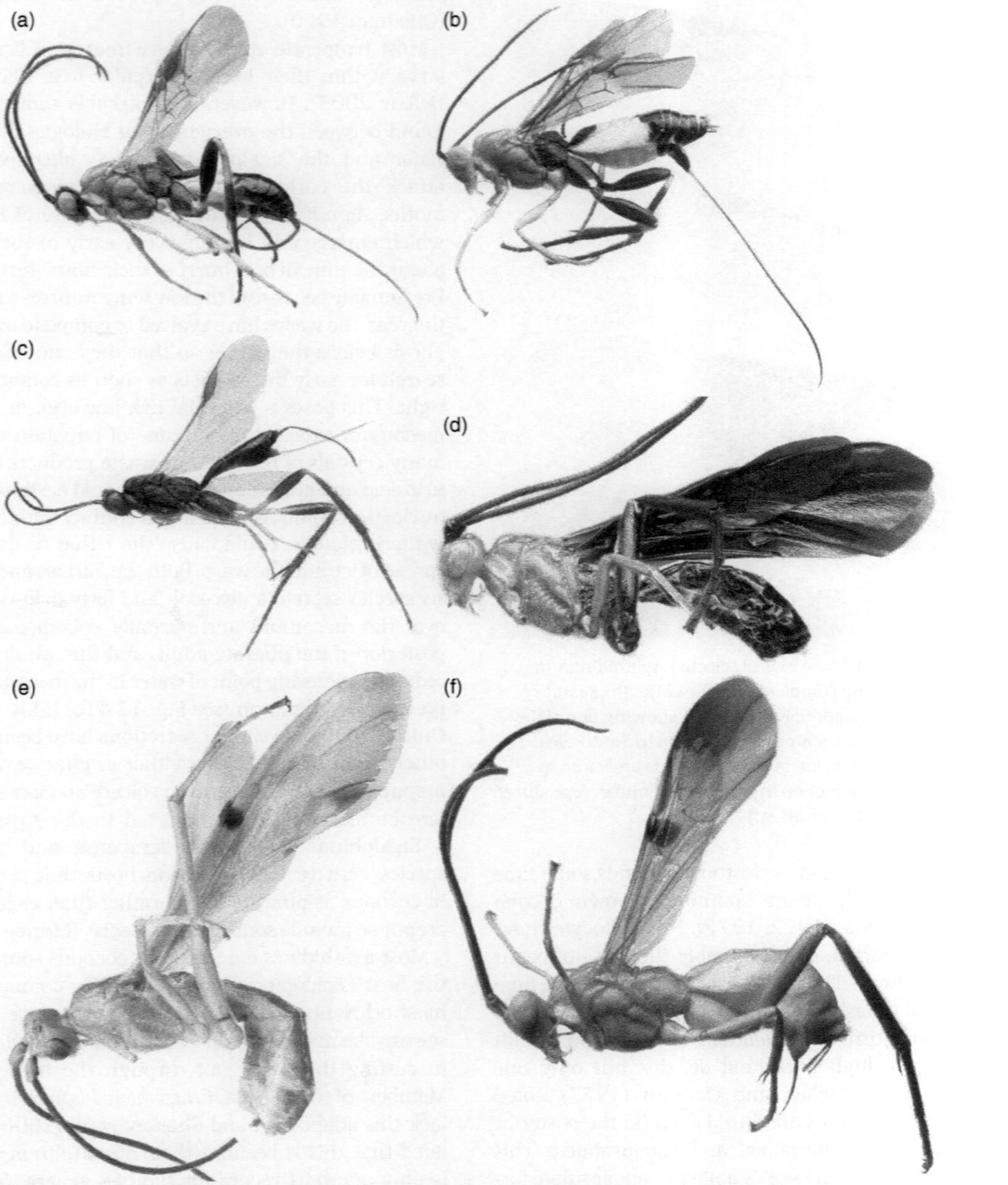

Fig. 12.38 Habitus photographs of female Agathidinae. (a) *Therophilus* sp.; (b) *Bassus* sp.; (c) *Aneurobracon* sp.; (d) *Disophrys* sp.; (e) *Troticus* sp.; (f) *Euagathis* sp. (Source: reproduced by permission of Mike Sharkey.)

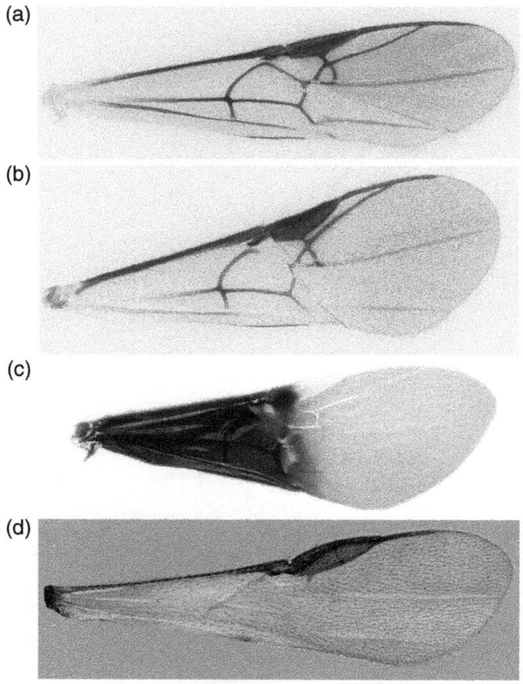

Fig. 12.39 Right fore wings of selected Agathidinae. (a) *Earinus* sp., showing complete vein (Rs+M)a; (b) *Bassus* sp., showing typical venation; (c) *Biroia* sp., showing one of the more unusual wing colour patterns found in South-East Asian and neotropical forest species; (d) *Aneurobracon* sp., showing extreme reduction in venation. (Source: reproduced by permission of Mike Sharkey.)

and, as with most non-cyclostomes, spends some time feeding externally before spinning its own cocoon (Odebiyi & Oatman 1972, 1977). No teratocytes have ever been reported, so presumably they do not occur in this subfamily. The meconium is voided by the final instar larva prior to pupation.

Superparasitism in *A. unicolor* is not common but can occur at high parasitoid density, but only one larva survives. Odebiyi and Oatman (1977) noted that supernumeraries are found dead (in the posterior end of the host caterpillar) and encapsulated. This demonstrates that the host's ability to encapsulate foreign bodies is not affected by being parasitised, and it seems likely that the dead larvae may have been physically attacked, thus exposing their soft tissue to the host immune system. Lack of encapsulation of the surviving larva implies some surface disguise. Realised fecundity for *A. pumila* was reported to be approximately 140, although females carried approximately 300 eggs (Quednau 1970).

Most temperate species overwinter as a first instar larva within their hosts caterpillar (e.g. Georgiev & Delkov 2003). However, a remarkable similarity was found between the overwintering biologies of *Earinus elator* and the sigalphine *Acampsis alternipes*. Both attack the early stages of north temperate spring moths, *Agrochola* (Noctuidae) in the case of *E. elator*, which emerge and lay eggs very early in the year at about the time of bud-burst of their hosts' host plants. Presumably because of the low temperatures so early in the year, the wasps have evolved to complete metamorphosis before the winter so that they can egress and search for early instar hosts as soon as conditions are right. This poses a potential risk however, in that the meconium expelled at the time of pupation contains many crystals of nitrogenous waste products (such as uric acid and guanine) and these could easily act as ice nucleation points, and, being in contact with the overwintering adult, could cause the lethal or damaging spread of ice into the wasp. Both the *Earinus* and *Acampsis* species secrete a glycerol- and fatty acid-rich 'goo' over the meconium and partially encompassing the posterior of the pharate adult, and this would greatly reduce the freezing point of water in the meconium and prevent ice formation (see Fig. 12.41e) (M.R. Shaw & Quicke 2000). No similar secretions have been seen in other taxa that overwinter either as pharate adults or prepupae with the meconium voided, and it may therefore be an innovation restricted to the Agathidinae + Sigalphinae clade. Other temperate and northern species with very early season hosts that overwinter in cocoons as pharate adults rather than as larvae or prepupae include some *Ophion* species (Morley 1915).

Most agathidines emerge from cocoons spun within the host Lepidopteran cocoon and, in common with most other braconids with this biology, have twisted, scissor-like mandibles that are probably an adaptation to cutting their way out through the host cocoon. Members of two genera, *Crassomicrodus* and *Agathirsia*, lack this adaptation and Sharkey et al. (2006) speculated that this is because their hosts (although there is only one host record for the two genera, from the noctuid *Acontia cretata*) do not spin cocoons.

Adults males have very distinctive exocrine glands opening near the anterior margin of the fifth and sixth and seventh metasomal tergites, where there is a specialised transverse setal brush that probably acts as a vaporative surface (Fig. 7.8d) (Buckingham

& Sharkey 1988). Normally these are hidden by the posterior margin of the preceding tergites and are probably exposed during courtship, but there seem to have been no sufficiently detailed descriptions or surveys to confirm this and nothing is known of the chemistry of their secretions. Most tropical species are also brightly coloured and belong to mimicry complexes involving many species (Leathers & Sharkey 2003), so it would be interesting to know whether these glands play a defensive role similar that of the presumably non-homologous ones in Opiinae (see Chapter 7, *Sex-related scent glands*). Goh and Morse (2010) and Ashley Adams and Morse (2014) carried out some interesting experiments on the mating behaviour of *Alabagrus texanus* (see Chapter 7, section *Mate location*) and showed that males that arrived early at a site where a female had recently eclosed had the highest chance of mating but that very young males and old males were always unsuccessful. Also, males of this species were shown to have a short-term memory for sites where females had recently emerged.

Several genera show very reduced wing venation, most particularly of the fore wing, which may be reduced to just veins 1-1A, (M+CU)b, 1cu-a and 1-M. Three of these genera, *Aneurobracon* (Figs 12.38c & 12.39d), *Mesocoelus* and *Plesiocoelus*, have been placed together in their own subtribe, the Mesocoelina within the Agathidini (van Achterberg 1990b). The taxonomic history of the Mesocoelina is complex, with *Aneurobracon* having been placed in its own subfamily, the Aneurobraconinae, by Fahringer (1936), but this was not accepted by van Achterberg (1976b), who synonymised it with the Agathidinae. However, *Mesocoelus* was originally placed in the Agathidinae, but careful consideration of morphological characters led van Achterberg (1984) to remove it, together with the closely related *Aneurobracon*, to the Orgilinae, in which they were treated as comprising their own tribe, the tribe Mesocoelini[12], but subsequently he transferred the Mesocoelini back to the Agathidinae and included within that group his new genus *Plesiocoelus*, also with reduced wing venation (van Achterberg 1990b), and subsequently Braet et al. (2000) again removed it from the Agathidinae and formally synonymised the Mesocoelini with the Orgilinae (Mimagathidini). However, placement of all of these within the Agathidinae was confirmed by molecular studies (Sharkey et al. 2006). However, molecular work also shows that the group as constituted by van Achterberg is artificial and certainly *Plesiocoelus* is not closely related to the other two (Sharkey et al. 2006, Sharkey & Stoelb 2013). In addition, several other species, currently placed in *Bassus* (a dustbin genus within the Agathidinae), also show remarkably reduced fore wing venation. Sharkey et al. proposed a tentative, revised tribal structure with *Aneurobracon*, *Mesocoelus* and the group of '*Bassus*' species with reduced venation as potentially warranting separate tribal status, i.e. as a redefined Mesocoelini.

Interestingly, whereas most Agathidines have fore wing vein (Rs+M)a lost, it is present, complete and tubular in many *Earinus* (Fig. 12.39a), *Amputoearinus* and *Sesioctonus* species, but mapping the character on to the independent molecular trees suggests that this represents a reversal (i.e. a re-acquisition of the vein). While this must remain a possibility, Sharkey et al. suggested that multiple independent losses of the vein in other agathidines was probably a more likely explanation and, if so then the multiple losses of the vein could possibly be functionally associated with the extreme narrowing of the fore wing marginal cell.

Agathidinae appear particularly abundant in regions of the world where fluids are hard to obtain. Many taxa have concealed nectar extraction adaptations (CNEAs; Jervis 1998), which enable them to obtain nectar from flowers, perhaps particularly from Asteraceae. The main CNEAs exhibited in the subfamily are elongate malar regions (Fig. 10.4a), elongate galea (Fig. 10.4a), elongate glossa (Fig. 10.4b) or elongate maxillary palps forming a tube (Fig. 10.4c) (Sharkey & Mason 1986, Pucci & Sharkey 2004, Tucker et al. 2012). Within the subfamily, elongate mouthparts have evolved on six independent occasions (Tucker et al. 2012).

One species of *Agathis* has been observed mating several times on the flowers of a plant in which no suitable hosts occur (Belokobylskij & Jervis 1998), an example of the SFF strategy of Godfray (1994), which is hardly known in the Ichneumonoidea, although it may also occur in some euphorines.

Keys to genera were presented by Sharkey (1997a) for the New World, Sharkey et al. (2009b) and Stevens et al. (2010) for Australia and Sharkey and Clutts (2011) for South-East Asian and oriental genera.

Sigalphinae

These are generally uncommon, medium- to large-sized parasitoids of Lepidoptera larvae, and most species have carapace-like metasomas. The systematics of the subfamily were treated by van Achterberg & Austin

Fig. 12.40 Sigalphinae. (a, b) *Sigalphus irrorator*: a, metasoma, dorsal view; b, wings. (c–e) *Pselaphanus trogoides*: c, habitus, dorsal view; d, detail of wings, showing large second submarginal cell; e, detail of propodeum and base of metasoma. (Source: reproduced by permission, © C. van Achterberg, Naturalis Biodiversity Centre).

(1992), who recognised several new genera, although most of these are known from just a few specimens of species, and biological data are only available for *Sigalphus* (Fig. 12.40a,b) and *Acampsis* (Fig. 12.41). Even then, the limits of the group have been wrongly interpreted and the genus *Afrocampsis*, included by van Achterberg and Quicke (1990) in the Sigalphinae, has more recently shown to belong to an unrelated but convergently similar group of genera based on both morphology and molecular evidence (Quicke et al. 2002). Phylogenetic relationships among the genera of Sigalphinae were explored by Quicke et al. (2008) based on 28S rDNA sequence data, who recovered *Pselaphanus* (Fig. 12.40c–e) as basal and then the endemic Madagascan genus *Malasigalphus* as sister group to the remainder. The genus *Minanga*,

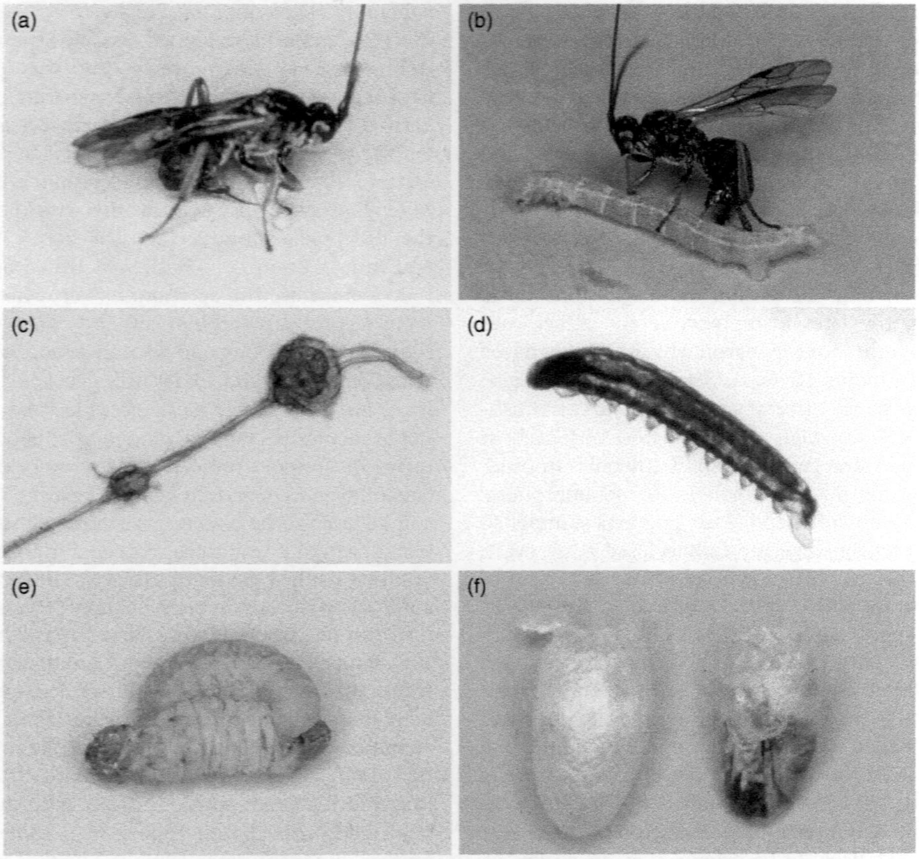

Fig. 12.41 Developmental stages and biology of *Acampsis alternipes* (Sigalphinae). (a, b) Female ovipositing into first and third instar larvae of host *Alsophila aescularia* (Geometridae); (c) part of nerve cord of *Alsophila* larva with *Acampsis* egg and embryo in one ganglion; (d) first instar larva; (e) final external feeding phase; (f) post-eclosion cocoon, with inner part removed and showing glycerol- and glyceride-rich expulsion surrounding meconium. (Source: from M.R. Shaw & Quicke 2000, reproduced by permission of Taylor & Francis Ltd.)

which had been placed in its own tribe, Minangini (De Saeger 1948, van Achterberg & Austin 1992), was not recovered in any remarkable position and therefore its recognition as a separate tribe was considered unwarranted, i.e. the Minangini should be regarded as a synonym of Sigalphini.

Until recently, little was known about sigalphine biology except that several species had been reared from geometrid moth larvae. and in the Neotropics, *Sigalphus romeroi* attacks two species of agaristine erebids (Sharkey & Janzen 1995). In fact, detailed biology is only known in some detail for three species, the Nearctic *S. bicolor* (Cushman 1913), the Costa Rican *S. romerii* (Sharkey & Janzen 1995) and the Euroasian *Acampsis alternipes*, whose larval stages and unusual cocooned-stage biology were described by M.R. Shaw and Quicke (2000).

Egg placement and the early instar larva are only known for the West Palaearctic species *A. alternipes*, and are placed within the ganglia of the ventral nerve cord of the host (Fig. 12.41c). As the embryo develops, it grows and causes the ganglion to swell and become asymmetric (obviously absorbing liquid and nutrient from the host, although the mechanism involved is unknown). The fully developed first instar larva ecloses from the ganglion into the haemocoel, is very active, polypodiform (Fig. 12.41d) and has a tail (although no anal vesicle). Although superparasitism was common, only one adult ever emerged and presumably these active first instars combat one another, leaving only one survivor. Similarly to *S. bicolor* and *S. romeroi*, the final instar larva chews its way out of the host and continues feeding externally (Fig. 12.41e) until only the host's head capsule and skin remain unconsumed.

Acampsis alternipes attacks its hosts, first to third instar larvae of *Alsophila* species (Lepidoptera: Geometridae) (Fig. 12.41a,b), although some hosts attacked in the later instar survive the attack and emerge as moths, but the reason why in unknown (see also the agathidine *Earinus*). *Sigalphus bicolor* is also known to oviposit in very young host larvae (Cushman 1913). *A. alternipes* attacks its host very early in the year when it is generally still fairly cold. In order to be able to do this, it completes its metamorphosis inside its cocoon at the end of the previous summer so the overwintering stage is a fully eclosed adult (M.R. Shaw & Quicke 2000). Such behaviour is typical of several groups with similar requirements, such as the Agriotypinae and some Tersilochinae (Jourdheuil 1961). Dissections of *Acampsis* cocoons have revealed that the eclosed adults overwinter with their anal end stuck inside a sticky, viscous brown liquid, chemical analysis of which showed it to be virtually a liquid soap with high concentrations of glycerol and fatty acids (Fig. 12.41e) (M.R. Shaw & Quicke 2000). This liquid undoubtedly had a very low freezing point, so its sole purpose seems likely to be primarily as an antifreeze and it is speculated that it might function to prevent ice crystal nucleation around the meconium.

The cocoons of both *A. alternipes* and *S. rogeroi* are white and ovoid. Unlike *A. alternipes*, 'overwintering' of *S. rogeroi*, from June until May, as a prepupal larva and emergence from the cocoon is via a neat round hole.

At least one species, *Sigalphus bicolor* from North America, is gregarious and, because of the sex ratios of broods, it has been suggested that it is likely to be polyembryonic, but this requires confirmation. If it is indeed polyembrionic, then it is also likely that its biology is very different from that of the *Acampsis* described above or of *S. romeroi*, whose solitary final instar larva is very large after consuming nearly all of its host, which would preclude more than one individual successfully developing in a host.

The rare, monotypic, Neotropical genus *Pselaphanus* (12.40. 108c–e) was initially treated in the dustbin subfamily Helconinae by Szépligeti (1904) when he described it, and it remained classified there until van Achterberg (1985a) moved it into the Agathidinae as a separated and putatively basal tribe lacking the derived wing venation of other members of that subfamily. As with both Agathidinae and Sigalphinae (and Meteorideinae and Acampsohelconinae), it possesses hind wing vein CUb, a feature otherwise known within the Braconidae only in the cyclostome Apozyginae and various Cretaceous fossil taxa. Unlike agathidines, it does not have the apomorphic very shortened fore wing second submarginal cell and consequently has the marginal cell long and not narrowed, and it has fore wing veins 1RS and CU1b well developed. However, these are all probably plesiomorphic states and so do not in themselves say anything positive about its placement. *Pselaphanus* was later treated as belonging to a subfamily in its own right (van Achterberg 1985a), but still believed to be closely associated with the Agathidinae, where it was often recovered in morphological analyses (Quicke & van Achterberg 1990a). However, contrary to van Achterberg, Sharkey (1997c) included it within the Sigalphinae based on overall morphological similarity, although without any formal analysis. Unlike all other known Sigalphinae, *Pselaphanus* lacks a full metasomal carapace (Fig. 12.40c), although this condition is undoubtedly plesiomorphic. Analysis of 28S rDNA sequence data (Quicke et al. 2008) showed that it forms a monophyletic group with the Sigalphinae under virtually all alignment parameter combinations and is the sister group to all other sigalphines, which is in agreement with its less modified metasoma. Nothing is known about the biology of *Pselaphanus*.

For species-level identification, see Sharkey and Janzen (1995), who revised the world species of *Sigalphus*, Sharkey and Braet (2012), who revised the world species of *Minanga*, and van Achterberg and Austin (1992), who revised the world genera.

Helconoid Complex

Based on molecular data, this is unlikely to be a monophyletic group, but rather a collection of taxa located at the base of the non-cyclostome part of the braconid tree. Within the 'helconoid complex' we can recognise a few almost certainly monophyletic groups, including one referred to as the macrocentroid sub complex comprising the Macrocentrinae, Charmontiinae, Xiphozelinae,

Amicrocentrinae, Homolobinae, Microtypinae and Orgilinae (Sharanowski et al. 2011).

Helconinae

There are virtually no grounds to recognise the Helconinae *sensu auctorum* as monophyletic based only on adult morphology apart from their general *Gestalt*, as the characters of the group are almost all either plesiomorphic or associated with egress from pupation sites in hard, concealed situations. Van Achterberg (1990a) considered the presence of hind wing vein a (as 2-A in his terminology) as an apomorphy rather than a symplesiomorphy, but the status and identity of this vein remain questionable. Three tribes have long been recognised (Helconini, Brachistini and Diospilini) based largely on differences in wing venation that could easily be homoplasic. Molecular phylogenetic analysis clearly demonstrates their monophyly of a group comprising the Helconinae, but also including the Blacinae but with Diospilini, Blacini and the Brulleini intermixed (Sharanowski et al. 2011). Interestingly, Čapek (1983), based on final larval head capsule features, suggested that Blacinae were probably related to the helconine tribe Brachistini, although this was likely influenced by the fact that both lack fore wing vein r-m, whereas DNA sequence data place them among a largely unresolved Diospilini and Brulleini. The typological method is also exposed as misleading since the New World genus *Topaldios*, which lacks hind wing vein a and was therefore previously placed in the Diospilini, is clearly shown to belong to the Helconini.

Helconini

This group comprises relatively large-bodied wasps with body lengths often around 1 cm. Species for which biology is certainly known are all koinobiont endoparasitoids of wood-boring beetles, most records being from Cerambycidae. Their relatively long ovipositors usually have a pre-apical dorsal notch, which would be indicative of oviposition into a larval host, in contrast to the Brachistini, although no notch is present in some genera such as *Austrohelcon*. *Eumacrocentrus* has a particularly long ovipositor (Fig. 12.42a) and *Ussurohelcon*, originally described in the Cenocoeliinae (van Achterberg 1994a), has the metasoma inserted quite high on the propodeum. The long ovipositors fairly clearly indicate that most must oviposit into larval hosts rather than eggs.

Van Achterberg (1997b) added another subtribe, the Vervoortihelconina, for a new genus from Chile with metasomal tergites 2–5 immovably fused, but unlike most other ichneumonoids with this condition[13] the metasoma is elongate and not carapaciform, similar to that of numerous Braconinae, Pimplinae (Ephialtini) and Rhyssalinae, for example. This feature, nevertheless, seems likely to be important for steadying/strengthening the metasoma during oviposition.

Brachistini

The genera traditionally placed in the Brachistini includes a number of genera that are egg–larval parasitoids of concealed beetle larvae such as seed predators and a few subcortical species principally of Curculionidae such as members of the genus *Pissodes* (Haeselbarth 1961, Parnell 1964, Alauzet 1987). A few records also exist of them (e.g. *Nealiolus*) also having been reared from Tephritidae (Diptera) (Hernández-Ortiz et al. 1994, Ovruski et al. 2000), so it is possible that a few species might have host shifted on to them, but no definitive isolated rearings have been made. Some authors have restarted treating it as a separate subfamily. They are largely distinguishable from other Helconinae by the absence of fore wing vein r-m.

As far as is known, they oviposit into the host egg stage (Alauzet 1987). There are few detailed biological studies and there are some differences in the observations that may reflect true interspecific variations. Beirne (1946) illustrated the developmental stages of *Triaspis pallidipes*, which he reported attacks early larval host beech weevils (*Rhynchaenus fagi*), but the details given are rather vague (also see comments on *Colastes*). The eggs that he illustrates are long, thin and slightly curved and tapering, and were dissected from the female ovary rather than a host stage. Rodriguez-Leyva (2006) illustrated the pedunculate egg and some larval stages of *T. eugenii*, an egg–prepupal parasitoid of the pepper weevil (*Anthonomus eugenii*). The first instar larva is mandibulate and hatches while still in the host egg where, apparently if the host egg is superparasitised, the first larva will bite and kill the other eggs. Haeselbarth (1961) clearly shows that the first instar larva of *Eubazus tuberculator* (as *Brachistes atricornis*) is vesiculate. The final instar emerges, except for its rear end, from the fully

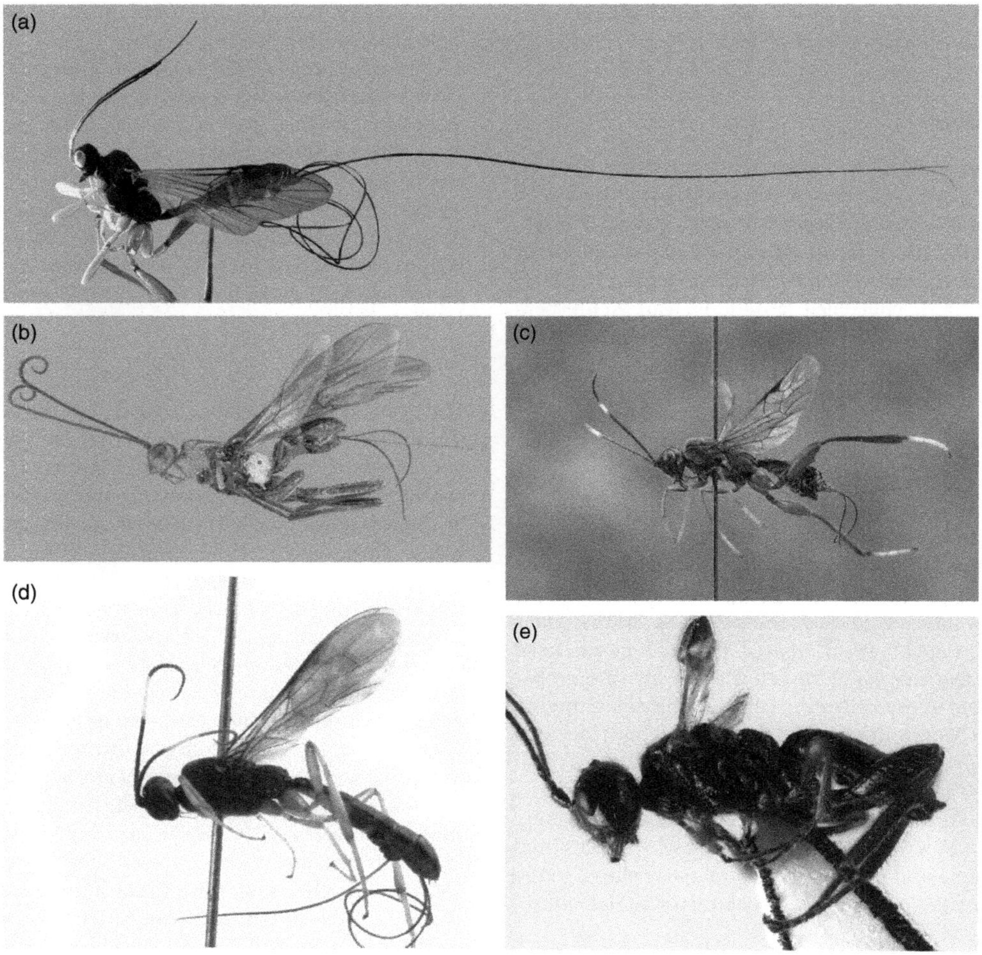

Fig. 12.42 Habitus photographs of representatives of Helconinae *s.l.* (a) *Eumacrocentrus* sp.; (b) *Trichiohelcon* sp.; (c) *Wroughtonia* sp.; (d) *Brulleia oberiae*; (e) *Cuniculobracon verdui*, holotype. (Sources: a, b, reproduced by permission of Barbara Sharanowski; c, reproduced by permission of Hélène Dumas; d, e, reproduced by permission, © C. van Achterberg, Naturalis Biodiversity Centre.)

grown host after it has constructed its pupal cell and feeds externally. Other biological studies of *Triaspis* reported similar features, including Obrtel (1960), who described the biology of *T. caudatus* on *Apion trifolii*, and Parnell (1964), who described life history stages of *T.* sp. nr *obscurellus* on *Bruchidius ater*. As with *T. pallipes*, the first instar is caudate–mandibulate and the 'intermediate' instars are vesiculate. Parnell described how the female wasp locates a host egg with her antennae, arches her metasoma and inserts her ovipositor under the egg and oviposits into it from beneath. All stages of host egg development were attacked and no instances of superparasitism noted. *T.* sp. nr *obscurellus* kills its host in the latter's third instar and exits from its posterior end, leaving only the unconsumed cuticle and head capsule; no external feeding phase has been reported.

Several species are of potential importance in biological control, especially of seed-beetles (e.g. Charlet 2002, Rodriguez-Leyva 2006). *Triaspis aequoris*, studied by Charlet (2002), is capable of limiting populations of its host, the sunflower seed weevil, *Smicronyx fulvus*

(Curculionidae); over 600 reared individuals were all females, strongly suggesting that this species is thelytokous.

Because of its lack of wings, the precise placement of the brachypterous *Cuniculibracon* (Fig. 12.42e) is far from certain, although it is tentatively placed in the Brachistini.

Diospilini

This group, together with the Brulleini, came out intermingled with the Blacinae in the molecular phylogeny of Sharanowski et al. (2011). Members of Diospilini have long been included within the Helconinae, but the morphological features uniting them are either plesiomorphic or of uncertain polarity, such as possession of fore wing vein r-m and usually also 1a and 2a.

Diospilini are cosmopolitan, though poorly studied in the tropics, and are nearly always associated with hosts in woody substrates. Hosts of Diospilini appear to be relatively weakly concealed larvae of xylophagous Anobiidae, Cerambycidae, Curculionidae, Melandryidae and Nitidulidae. No species appear to have been involved in biological control programmes. The biology and early stages of *Diospilus capito*, a parasitoid of nitidulid beetle (*Meligethes* spp.) larvae, were described by Osborne (1960). The hydropic egg has a short pedicel, which was sometimes, but not always, observed to be embedded in the host's fore or hind gut walls. The first instar is mandibulate with a heavily sclerotised head capsule and develops a small anal vesicle as it develops. From what has been written, it appears that some growth occurs before the host enters the soil to pupate, but moulting to the final instar larva happens soon after and the parasitoid larva emerges partially to finish feeding externally, but with its posterior end still partly inside the host. After consuming all the host except for its cuticle, the *Diospilus* spins its cocoon within the pupation chamber in the soil made by its host, the cocoon clearly being composed of two layers. Interestingly, Osborne noted how both the larvae and cocoons of the *Diospilus* have probably been misidentified by other earlier workers as those of a tersilochine and of the beetle itself, and emphasised the need for great care in associating eggs, larvae and adults in situations where multiple parasitoids attack the same host. It also seems that, like brachistines, at least some *Diospilus* may be egg–larval parasitoids (Parrott & Glasgow 1916).

Brulleini

van Achterberg (1983b) erected a new tribe, Brulleiini, for two Old World genera, *Brulleia* (Fig. 12.42d) and *Parabrulleia*, that did not fit well in his concept of either the Helconini or Diospilini because of a number of wing venation features and the absence of a mid-longitudinal carina on the frons. In his hand-drawn phylogeny of the Helconinae they were placed as the sister group of the Brachistini. While this result is not supported by molecular analysis (Sharanowski et al. 2011), the molecular phylogenetic signal was not strong. Biological data are only available for the genus *Parabrulleia*, which has been reared from cerambycid beetle larvae in Japan. Subsequently, van Achterberg (1990) erected a new subtribe in the Brulleiini, the Pseudohelconina, to receive three new genera of nocturnal Afrotropical Helconinae. Nothing is known of their biology.

Helconoid Group *Incertae Sedis*

Genus Eadya

The genus *Eadya* (Fig. 2.2) is known only from Australia. It was originally described as a member of the Euphorinae (Huddleston & Short 1978) and this placement was followed by Shaw (1985), who considered them to be the most basal member of that subfamily exclusive of the Meteorini. Shaw's recognition that they did not fit well with the rest of his euphorines was appropriate, but he misread the paper describing *Eadya* as indicating that they attack adult beetles (Chrysomelidae), and this persuaded him that they were indeed euphorines. As van Achterberg et al. (2000) pointed out, *Eadya* is in fact a larval parasitoid and although the host group is unusual, the biology does not support that they are euphorines either. Many morphological features of the genus give the clue as to their relationships; they have a carinate frons and a bilobed clypeal margin, both features more typically associated with helconines *s.l.*, and subsequently, DNA sequence data from an apparently related but undescribed Australian genus have confirmed that they are more closely related to the *Diospilus* group of genera (Belshaw & Quicke 2002), although it should soon be possible to make a better assessment of its relationships since it turns out that *E. paropsidis* is a potential biological control agent against the *Eucalyptus* defoliator *Paropsis atomaria* (Chrysomelidae)

(Tanton & Epila 1984, Withers et al. 2012). Initial observations showed that the early instar larvae are strongly polypodiform although in a way that differs morphologically from that found in agathidines and sigalphines. Whether this is related to motility and inter-larval combat is unknown, but seems likely.

Genus Korecnomios

This Korean genus was described in the Ecnomiinae, which is now regarded as a tribe in the Euphorinae (Sharanowski et al. 2011), by Park and van Achterberg (1994) principally because it had a rather large hind wing anal lobe. However, unlike *Ecnomios*, the hind wing also has vein a, the fore wing a traces of vein 1a and 2a and a well-developed 2cu-a, features found most often in the Helconini and never in the Euphorinae. Not surprisingly, DNA sequence data support that it is not a Euphorine and place it among the helconoids (D.L.J. Quicke, unpublished work). The biology is unknown, but the moderately long ovipositor suggests parasitism of a concealed host such as a wood-borer.

Genus Plesiotypus

This genus from India, Nepal, China, Hawai'i and Madagascar was described originally in the Microtypinae by van Achterberg (1992c), but is in fact a rather basal representative of the non-cyclostome stem group, the 'Helconinae'. Indeed, van Achterberg noted that *Plesiotypus* 'closely resembles the aberrant genus *Taphaeus* ... which has probably to be transferred from the Helconinae to the Blacinae'. All of the features cited by van Achterberg to separate these genera are highly homoplastic within the Braconidae. DNA sequence data suggest it is either a basal member of the Helconinae grade (unpublished work) or a sister group to the superficially similar agathidoid clade (Quicke et al. 2008). Nothing is known of its biology.

Blacinae

This is a group of almost entirely rather small, dull-looking wasps (Fig. 12.43) that are very common, especially in the north temperate region, and about whose biology remarkably little is known (Shaw & Huddleston 1991). For a long time, *Blacus* was considered to belong to the Euphorinae and to be closely related to the Centistini, but Čapek (1983) placed them in the same group as the Helconinae, which is where DNA sequence data also put them. *Blacus* species are probably koinobiont endoparasitoids, probably all also solitary, of weakly concealed beetle larvae and frequently ones found in damp, even rotting, situations, and probably reliable host records include Nitidulidae, Cryptophagidae, Scolytidae, Sphindidae, Anobiidae, Curculionidae and Cerambycidae (Čapek 1983, Haeselbarth 1973). A single published record from a staphylinid (Lipkow 1965) seems likely to be erroneous, but as so few workers rear these beetles it is hard to draw definite conclusions. Records of Blacinae being reared from Diptera are also very probably incorrect and result from co-occurrence of fly and beetle larvae in the same complex substrate (Shaw & Huddleston 1991), although the possibility that some use Diptera hosts cannot be totally excluded; Čapek (1983) discussed several other likely misidentifications that have confused the matter.

Two genera placed in the subfamily have unusual biologies. The Dyscoletini, comprising *Dyscoletes* (Fig. 12.43a) and *Hellenius*, are generally regarded as belonging to the Blacinae (W.R.M. Mason 1976b, van Achterberg 1988c) and are koinobiont parasitoids of scorpionfly larvae of the genus *Boreus* (Mecoptera), although there are only two published host records and one description of biology, both referring to *Dyscoletes* itself (Aubrook 1939, W.R.M. Mason 1976b). *Xyeloblacus* (Fig. 12.43d) is a parasitoid of *Xyela* sawfly larvae (Xyelidae) (van Achterberg & Altehofer 1997).

Males of some species are known to form swarms (R.B. Benson 1944, Haeselbarth 1973, Syrjämäki 1976, van Achterberg 1977), which have been proposed to be leks. That is the swarm of competing males that attracts females, partly because of its enhanced stimulatory impact. It is not known whether they release pheromones but this seems likely. Mixed-sex swarms have been reported for one species (Čapek 1983). Also, despite generally being dark coloured, individuals of *Blacus* are frequent at light traps and Čapek supposed this to indicate that they may be nocturnal, but it is possible that the light is just attracting individuals that were resting nearby and were disturbed. If they are searching for hosts primarily using chemical and vibratory cues then perhaps some are truly nocturnal.

Almost nothing is known of their larvae, except for final instar head capsule features (Čapek 1983). At least some Palaearctic species of *Blacus* are plurivoltine.

Fig. 12.43 Habitus photographs of representatives of four genera assigned to the Blacinae. (a) *Dyscoletes lancifer*; (b) *Blacometeorus* sp.; (c) *Grypokeros* sp.; (d) *Xyeloblacus leucobasis*. (Source: reproduced by permission, © C. van Achterberg, Naturalis Biodiversity Centre.)

McHugh (1993) noted that remains of the sphindid *Eurysphindus comatulus*, after the *B. koenigi* larva had egressed '... was draped over the top of a white, silky cocoon with the legs wrapped around the cocoon as although holding it'. *Xyeloblacus* larvae emerged from host *Xyela* cocoons up to 3 years after the host larvae entered their pupation substrate, very much like the hosts themselves (van Achterberg & Altehofer 1997). No attempts to use them in biological control appear to have been made, although in Europe *B. nigricornis* is sometimes a common parasitoid of the oilseed rape pollen beetle, *Meligethes aeneus* (Nitidulidae).

Monophyly of the Blacinae is far from certain and, as indicated by Sharkey (1997b), is based on rather shaky foundations with two of the three synapomorphies used by van Achterberg (1988c) being trends rather than fixed conditions – one of these is the presence of only fore wing two submarginal cells (i.e. vein 1r-m absent), but this is highly homoplastic within Braconidae as a whole and even variable even within Blacinae. The molecular study of Sharanowski et al. (2011) included only two species of *Blacus*, two of *Grypokeros* and one *Apoblacus*. While both the first two genera were recovered as monophyletic, there was no support for monophyly of the subfamily and *Dyscoletes*, in particular, has yet to be sequenced.

The world species were revised by van Achterberg (1976a, 1988c) and fossil taxa, some wrongly attributed to the subfamily, were also dealt with by van Achterberg (1982a).

Acampsohelconinae

Few people will have come across this name, which is based on a genus known only from Baltic amber fossils,

Acampsohelcon (Tobias 1987). However, the three extant taxa, *Urosigalphus* and *Canalicephalus*, both with completely fused carapaces (Fig. 12.44a), and *Afrocampsis*, with a carapace but with the first metasomal tergite movably connected to the second (Fig. 12.44b), all possessing at least a trace of hind wing vein CUb, a metasomal carapace and a depressed frons and have a large hind tarsal claw, appear to be related to *Acampsohelcon* and not to the Blacinae, Helconinae or Sigalphinae in which *Urosigalphus*, a common New World genus, was previously placed (see Gibson 1970 for a brief review).

Afrocampsis was originally described in the Sigalphinae by van Achterberg and Quicke (1990) (hence its name) because of its metasomal carapace (Fig. 12.44b) and the presence of a trace of hind wing vein CUb, but that it was not correctly placed there was recognised by M. J. Sharkey (personal communication) because of its very large tarsal claws and Quicke et al. (2002) on the basis of both morphology and DNA sequence data, although its exact affinities were not clear at that time.

Monophyly of the subfamily is questionable. Only one molecular phylogenetic study to date has included data from all three genera (Quicke et al. 2002). Based only on the D2+D3 region of the 28S rDNA gene, it seemed to indicate a close relationship between the three extant genera, and although they did not always come out together as monophyletic in analyses, they are usually grouped together as a grade. In contrast, Quicke et al. (2008), using the same gene fragment, which included only *Afrocampsis* and *Canalicephalus*, recovered them fairly widely separated. However, both studies made the assumption that the Helconinae made an appropriate outgroup for rooting the tree and specifically used a *Helcon* sequences for the outgroups. Sharanowski et al. (2011) only included *Urosigalphus* and recovered them basal to the helconoid complex, with her trees being sequentially rooted with ichneumonids and cyclostomes. All of this suggests, irrespective of whether they form a strictly monophyletic group, that they are derived near the base of one of the major splits among the non-cyclostomes.

Biological information is available only for *Urosigalphus*, which is a koinobiont endoparasitoid of bruchine chrysomelid and curculionid beetle larvae in Fabaceae seeds or, in the case of one subgenus, of gall-inhabiting weevils (Gibson 1970). There appears to be no definite information about the stage attacked or the morphologies of earlier instar larvae. The very slender, tapering and apically simple ovipositor of *Urosigalphus* (Fig. 12.44a) would be consistent with it parasitising an early stage such as the egg or a very early instar larva. Since one of its hosts is associated with cotton (Cushman 1926a), it would be good if more detailed studies could be carried out. One fossil species of *Urosigalphus* has been described from the Florrisant beds (ca 37.2–33.9 Mya) of Colorado (Brues 1910), but the identity of this fossil needs confirmation.

Van Achterberg (2002) recognised three tribes, the Acampsohelconini for the fossil taxon, the Urosigalphini for *Urosigalphus* and *Canalicephalus* and the Afrocampsini for *Afrocampsis*. The metasomal carapace of *Afrocampsis* has a moveable joint between the first and second tergites and has a suture present between tergites 2 and 3, but lacks a postpectal carina. The genera are keyed by van Achterberg.

Macrocentrine Subcomplex

This group is composed mostly of medium-sized to large species, the great majority of which are nocturnal and display an ophionoid facies (Gauld & Huddleston 1976). Those Macrocentrinae and Charmontiinae with long ovipositors share a distinctive ovipositor cross-sectional profile, with two pairs of longitudinal ridges on both the upper and lower valves.

Macrocentrinae

These are a moderately large, common group dominated by nocturnal taxa that are parasitic principally on semi-concealed Lepidoptera larvae. Hosts belong to a variety of families including, among others, Noctuidae, Tortricidae, Pyralidae, Oecophoridae, Sesiidae and Gelechidae (Shaw & Huddleston 1991). Most species belong to the single genus *Macrocentrus*, but a few other genera are known, almost entirely from the tropics (probably several undescribed ones still need description). Many species are crepuscular or nocturnal, come readily to light traps and have an ophionoid facies (Fig. 10.8a). *Macrocentrus* and *Hymenochaona* have long ovipositors and attack weakly concealed hosts whereas *Austrozele* and *Dolichozele* attack exophytic caterpillars. Primitively these were probably parasitoids of moderately concealed Lepidoptera larvae because long ovipositors are found in the basal taxa, and within the Macrocentrinae there have probably been two transitions to very short ovipositors (Belshaw & Quicke 2002).

Fig. 12.44 Acampsohelconinae. (a) *Urosigalphus* sp, female from Peru, showing completely fused carapace; (b) *Afrocampsis griseisetosus*, showing first and second metasomal terga with moveable articulation. (Source: reproduced by permission, © C. van Achterberg, Naturalis Biodiversity Centre.)

Some species play a biological control role. *Macrocentrus cingulum* [often referred to as *grandii*, *cingulatum* (*sic*) or *gifuensis*] is an important parasitoid of the corn-borer genus *Ostrinia* (Crambidae), and populations from Europe show different host preferences from those in Asia and America, the American populations probably having originated from an introduction from Asia, although introductions were originally made from France and Korea (W.A. Baker et al. 1949). Genetic analysis, however, does not indicate a cryptic species pair, although it does indicate significant geographic population structuring (Pélissié et al. 2012). *Macrocentrus ancylivorus* is an important parasitoid of the strawberry leaf roller (*Ancylis comptana*: Tortricidae) (Fink 1926), oriental fruit moth (*Grapholtha molesta*: Tortricidae) and also of the potato tuber moth (*Phthorimaea operculella*: Gelechiidae) (Finney et al. 1947). In 1945, 23 million *M. ancylivorus*, originally from the east coast of the USA, were released in California in an inundative release programme against *G. molesta*, and it established and now appears to exhibit an economically meaningful level of control (see Russell 1987), but many releases of it in other parts of the world have failed. This may be because of the need for an alternative host for overwintering in, such as the strawberry leaf roller in its area of origin. *Hymenochaona delicata*, is an important parasitoid of several pest lepidopterans in the neotropics (Milano et al. 2008).

The subfamily is widely stated as being entirely polyembryonic and this mode of development is fairly well documented for a few species (H.L. Parker 1931, Paillot 1937, Daniel 1932, Finney et al. 1947, Dittrick & Chiang 1982, Krugner et al. 2005). Numerous species, especially the smaller-bodied ones, are gregarious, others are solitary. From the few detailed studies available, it appears that at least some of the solitary species are also polyembryonic, with only one of the two resulting morulae proceeding to develop as an embryo. With solitary taxa, what seems to happen is that the egg initially divides into two or more, but this could easily be overlooked, which might explain why it has not been reported in some studies. Fink (1926) presented an account of *M. ancylivorus* illustrated with photographs of the developing egg, which is hydropic and changes after deposition from an elongated spindle shape to nearly spherical with a 1000-fold increase in volume. The first instar larva of the solitary *M. ancylivora* is mandibulate with large, heavily sclerotised head capsule and large mandibles, and fighting superparasitising larvae occurs. In the gregarious *M. iridescens*, mandibles become apparent only a few days after larvae emerge, suggesting that those of the early first instars are neither strongly sclerotised nor substantial (Krugner et al. 2005).

There appear to be four larval instars, the final one egressing from the host to continue feeding externally (Krugner et al. 2005). Fink (1926) reported that egress from the host was accomplished in an unusual fashion, posterior end first and the head end remaining embedded within the host to continue feeding; however, although this seems entirely possible, it definitely needs confirmation and no other workers have reported it. Cocoon construction in *M. ancylivora* takes a considerable time (24–36 h) and the anterior end is especially thickened (*cf. Meteorus*). This temperate species overwinters as a first instar larva within the host.

For *M. ancylivorus*, all host larval stages seem to be acceptable although survivorship is lowest when very young hosts are attacked (Fink 1926). Further, embryonic development was slower in earlier instar hosts, suggesting that the host is being allowed to develop to a larger size before the parasitoid larva egresses. Protection of the eggs of the polyembryonic *M. cingulum* from host immune attack depends on the fibrous outer chorionic membrane, and protection of the embryos post-hatch from encapsulation has been shown to depend on there being an intact extraembryonic membrane (Lu et al. 2007).

Superparasitism appears to be common with females unable to discriminate between parasitised and unparasitised hosts (Fink 1926). In the gregarious *M. iridescens*, broods that emerged from hosts were either all female, all male or approximately a 50% mix, but in total being approximately twice as large as the mean of an all-male or all-female one. Single or (presumed] doubly oviposited male broods were significantly larger than either singe-sex female broods or mixed broods, respectively. Whether unfertilised eggs show greater polyembryony than female ones or males suffer less juvenile mortality is not known. Shaw and Huddleston (1991), considering only the British fauna, noted that the solitary species form a distinct group and on the whole are larger bodied; however, several very small tropical species are solitary parasitoids of hosts such as Gelechiidae.

Macrocentrines can be fairly easily recognised by the presence of a row or cluster of small pegs on the hind trochantelli and occasionally the basal part of the femur proper; the trochantellus is just a demarked basal part of the femur. However, these pegs are absent in a few African taxa. Similar trochanteral pegs are found also in the orgiline genus *Eleonoria* and the braconine genus *Quickeos* and are no doubt adaptive to some purpose such as egress from their cocoon. Macrocentrines also have a moderately characteristic slender habitus (Fig. 10.8a) and second submarginal cell with a rather distinctive shape.

Cocoon silk, apparently of a fibroin type based on gross amino acid composition, was investigated using X-ray diffraction in *Macrocentrus thoracicus* (Rudall & Kenchington 1971), who found that its beta-pleated sheets had a novel three-dimensional folding structure, but no other X-ray studies have been carried out on parasitoid silks since then.

Physiological work on the subfamily is rather patchy. Some work has been done on heatshock proteins and J. Hu et al. (2003) started to investigate the physiological interactions involved in overcoming host immunity in *M. cingulum*. They found no differences in host caterpillar haemocyte count between parasitised and unparasitised individuals and Sephadex beads injected into parasitised hosts elicited a strong encapsulation response whereas the embryos of the wasp were unaffected. Mature eggs dissected from a wasp and injected into a host larva were similarly passively protected against encapsulation whereas both immature eggs and mature ones treated with Driselase[14] were encapsulated. The protection against encapsulation appears to involve an outer 'fibrous' layer which was damaged by the Driselase treatment. Subsequently, venom or other secretory components from the wasp reduce the host's production of phenoloxidase, via multiple mechanisms (Feng et al. 2004, 2011). Y. Li et al. (2007a) then compared the effects of venom and of ovarian protein extract on host encapsulation of Sephadex beads and found that ovarian proteins had an anti-encapsulation effect but venom had no such effect, although neither venom nor ovary extract seemed to affect the viability or spreading capacity of host haemocytes.

The complex of antennal sensilla of *M. cingulum* (as *cingulatum*) has been described in a detailed SEM and TEM study by Ahmed et al. (2013a) and ovipositor morphology was investigated in detail by Y. Li et al. (2007b). This species has been shown to show a general attraction to host plants and host–host plant complexes with slightly different mixes of non-polar and slightly polar volatile compounds being important (Ding et al. 1989, Udayagiri & Jones 1992).

The Palaearctic fauna was treated in considerable detail by van Achterberg (1993b), who provided a brief introduction to the group's taxonomy.

Charmontiinae

This is a subfamily erected by Quicke and van Achterberg (1990) to accommodate the common Holarctic genus *Charmon* (Fig. 12.45) and a Chilean genus, *Charmontia*, although the possibility that these are not very closely related must be considered. Unfortunately, *Charmontia* is thus far known only from its holotype. *Charmon* is known almost entirely from a few Holarctic species, previously erroneously called *Eubadizon*, some of which appear to have exceptionally broad host ranges for koinobiont endoparasitoids, *C. extensor* having been recorded as attacking members of 12 families of Lepidoptera (Shenefelt 1969, Shaw & Huddleston 1991). Even if this list includes a number of misidentifications and/or miss-associations, host range must still be pretty broad, as indeed it is in some of the relatively closely Homolobinae. Study of the mechanism that allows these to avoid encapsulation by such a large range of hosts would be very interesting. The possibility that *Charmon cruentatus* includes several cryptic species can not be excluded and there is some evidence that this might be the case from cytogenetic studies (Gokhman 2006, Gokhman & Mikhailenko 2012).

In some situations, *Charmon* is an important parasitoid of pest species such as the oriental fruit moth, *Grapholtha molesta* (Tortricidae) (Haeussler 1940). The wide host ranges of these is likely to mean that obtaining permission for release as a biological control agent is unlikely to be possible.

Recently, Belokobylskij et al. (2010) described a fossil genus, *Palaeocharmon*, from the lowermost Eocene Paris Basin amber which they placed in a new tribe, the Palaeocharmontinae. The exact subfamily placement of this taxon is far from certain, however, and it could belong to another subfamily within the helconiformes. What is remarkable about this specimen, however, is that it has a very distinct hind wing vein m-cu, a feature otherwise completely absent from the non-cyclostome Braconidae.

Amicrocentrinae

This monogeneric group restricted to the Afrotropical region (including Madagascar) includes a few

Fig. 12.45 *Charmon extensor* (Charmontinae) from Japan. (Source: reproduced by permission of Kaoru Maetô.)

large to very large (1–3 cm body length) nocturnal braconids that are parasitoids of stem-boring Lepidoptera larvae and which exhibit a typical ophionoid facies (Fig. 12.46a) (van Achterberg 1979a). The only genus, *Amicrocentrum*, was originally placed in the Macrocentrinae, which they superficially resemble, but van Achterberg (1979a) could find no morphological synapomorphy to unite it with the macrocentrines and consequently erected a new subfamily for it. Morphologically they display many peculiarities, including highly reduced labial and maxillary palps, a very large plical lobe on the hind wing with erect setosity, distally expanded hind wing basal cell (Fig. 12.46b) and unusual setosity pattern of metasomal tergites, among others (van Achterberg 1979a). Recent molecular phylogenies concur that they are fairly closely related to the Macrocentrinae (Sharanowski et al. 2011).

Their ovipositors are long and possess the same type of steering mechanism as in Macrocentrinae, Charmontiinae and several Helconinae, which involves a pre-apical swelling of the rhachis of olistheter mechanism (Fig. 3.10a) (Quicke et al. 1995; see Chapter 3, section *Ovipositor steering mechanisms*). Two of the five known species of *Amicrocentrum* have host records, *A. excilis* from the cocoa stem-borer *Eulophonotus myrmeleon* (Cossidae), which bores in the woody stems of cocoa, and *A. curvinervis* from the maize stalk-borer, *Busseola fusca* (Noctuidae). The long ovipositor might indicate that both of these hosts are attacked at a late stage of development when they have bored deep into their host plant.

Xiphozelinae

The Xiphozelinae was erected by van Achterberg (1979b) to accommodate several species that had previously been placed in the Macrocentrinae. The subfamily has only two recognised genera, *Xiphozele* and *Distilirella*, both entirely from the Palaeotropics, including southern China, and with *Xiphozele* also known from New Guinea. These taxa form another principally nocturnal group of parasitoids of Lepidoptera larvae, have a typical ophionoid facies (Fig. 12.47) and are attracted to light at night. *Xiphozele* has been reared from the catocaline noctuid *Dysgonia simillima* (van Achterberg 1979b). These are fairly large wasps, with body lengths up to 20 mm. Both *Xiphozele* and

Fig. 12.46 *Amicrocentrum curvinervis* (Amicrocentrinae). (a) Female habitus; (b) detail of hind wing venation showing distally expanded and glabrous basal cell. (Source: reproduced by permission, © C. van Achterberg, Naturalis Biodiversity Centre.)

Fig. 12.47 *Xiphozele compressiventris* (Xiphozelinae) female pinned specimen. (Source: reproduced courtesy of Kyohei Watanabe.)

Distilirella have highly inclivous hind wing vein cu-a (or CU1 in *Distilirella*; see below) that elsewhere in the braconidae is only found in the Brulleini. *Distilirella* is interesting also in that some species possesses a trace of hind wing vein CUb (which is why the clearly homologous transverse vein in *Xiphozele* is called cu-a whereas that in *Distilirella* is CUa), but the genus seems correctly placed in the Xiphozelinae and not closely related to any of the other non-cyclostomes with that vein (viz. sigalphoid complex members, acampsohelconines and meteorideines).

A key to the 13 known species of *Xiphozele* was provided by van Achterberg (2008).

Homolobinae

This is a relatively common but biologically surprisingly poorly known group. Earlier papers on these wasps often incorrectly referred to them as *Zele*, which are superficially similar, but this was a result of Authors[15] misinterpreting what the original published name referred to. Some clarification came with van Achterberg (1979c), who discovered that the name *Zele* actually applied to a different braconid (see Euphorinae: Meteorini) from what authors had been using it for. With increasing understanding of relationships it became apparent that *Homolobus* (i.e. *Zele* of earlier

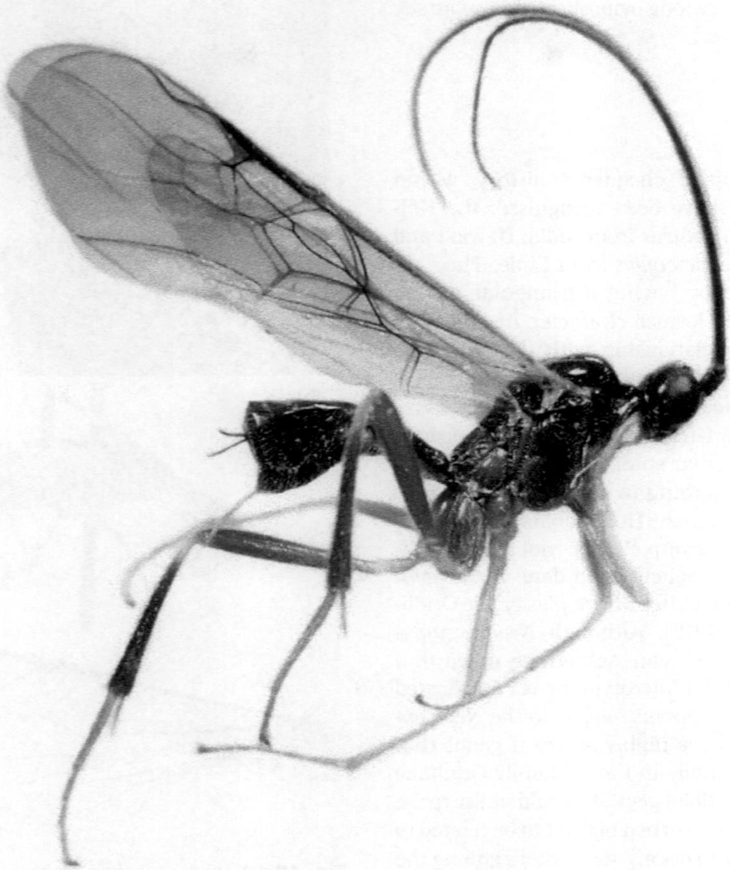

Fig. 12.48 *Homolobus rufiventris* (Homolobinae), habitus. (Source: reproduced courtesy of Kyohei Watanabe.)

Authors) was isolated from the Helconinae where it had been classified, and van Achterberg (1979c) erected the subfamily Homolobinae for it and another genus, *Charmon*, now placed also in a separate subfamily, the Charmontiinae. Shaw and Huddleston (1991) summarised the relevant changes and care should be taken when considering some of the earlier literature on biology published under the name *Zele* and, appropriately, noted that except for rather superficial features, such as generally having an ophionoid facies, there was 'little to suggest a close relationship'.

All homolobines are koinobiont endoparasitoids of Lepidoptera larvae, mainly belonging to the Noctuidae and Geometridae, but with records from other families including Erebidae (Lymantriinae)and Lasiocampidae. Most have short ovipositors and attack exposed hosts (Shaw & Huddleston 1991). There is an obligate external feeding phase following host cocoon formation (A.A. Allen 1977). Most homolobines are crepuscular or nocturnal, have an ophionoid facies and not infrequently come to light traps, although a few are diurnal and some are more darkly coloured (Fig. 12.48).

Most species belong to *Homolobus* and van Achterberg presented an informal phylogenetic tree for these its five subgenera. *Westwoodiella*, known from a single Australian species, differs from the remainder of species included in the subfamily by a number of features, and these led van Achterberg (1992b) to place it in a separate tribe, the Westwoodiellini, but unfortunately nothing is known about its biology. Unlike *Homolobus*,

Westwoodiella has a very long ovipositor, so must attack a deeply concealed host.

Microtypinae

This is a tiny group of chequered history, within which three genera have been recognised: the Holarctic *Microtypus*, *Plesiotypus* from India, Hawai'i and Madagascar and *Neomicrotypus* from Chile. They are largely characterised by having a triangular second submarginal cell, the former character having been taken to indicate a relationship with the Orgilinae. However, the shape of the second submarginal cell is not likely to be a particularly reliable character because the migration of vein r-m towards 2RS has happened so many times. Molecular studies have clearly shown that *Microtypus* has nothing to do with the Orgilinae and is probably related to the Homolobinae and thence to the Macrocentrinae group. Perhaps not surprisingly, the only other genus sequenced to date, *Plesiotypus*, has nothing to do with either *Microtypus* or the Orgilinae (Quicke et al. 2008). Although *Neomicrotypus* has yet to be sequenced, van Achterberg noted that '[t]he relationship of the Microtypinae is complicated by the similarity of ... *Neomicrotypus* to the Neotropical genus *Antestrix* ... a highly aberrant genus that was included provisionally in the subfamily Orgilinae because it has minute tibial pegs'. It would not surprise me if *Neomicrotypus* also turned out not to be related to *Microtypus* but rather to belong somewhere among the basal helconines along with *Plesiotypus*.

Getting back to *Microtypus* itself, the only genus definitely belonging to the subfamily, it is widely distributed in the Holarctic with about half a dozen described species. All species for which biology is known are solitary koinobionts, hosts including a range of weakly concealed lepidopteran larvae including Gelechiidae, Pyralidae, Tortricidae and Yponomeutidae (Čapek & van Achterberg 1992). The cocoon is rather flimsy with an equatorial stripe, and this together with final larval head capsule features suggested to Čapek (1970) a relationship to the Orgilinae.

Orgilinae

This is a cosmopolitan subfamily of generally medium-sized wasps with 13 genera divided among three tribes and just over 300 described species. Most species belong to just two genera, *Orgilus* and *Stantonia*, and virtually all detailed biological observation is based on a few north temperate species in the former. The temperate fauna is dominated by rather small (4–7 mm), dark-coloured species or the genus *Orgilus*, while *Stantonia* (= *Bentonia*) (Fig. 12.49a) and some of the other tropical genera are more conspicuous (Fig. 12.49b). *Stantonia* is a principally tropical genus of usually somewhat larger, principally tropical and subtropical and more brightly coloured species that frequent sunny places.

A few species of *Orgilus* and one of *Stantonia* have been used in classical biological control programmes (van Achterberg 1997d). *O. obscurator* was an early

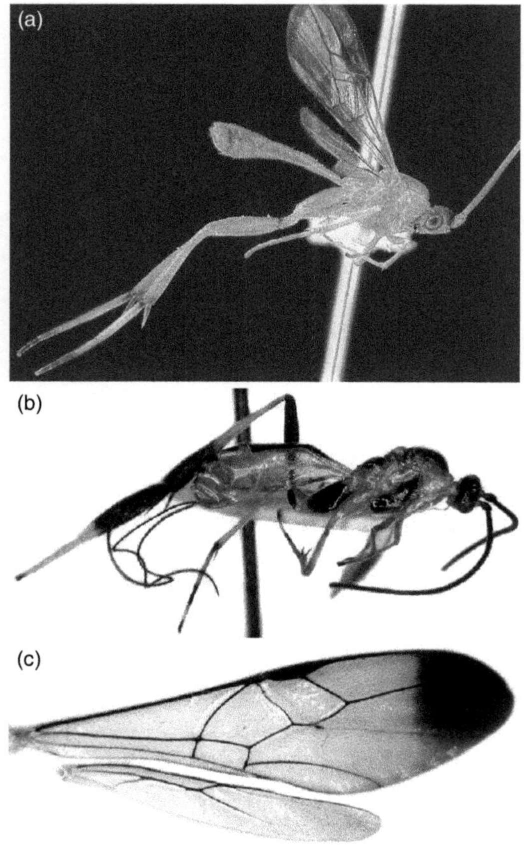

Fig. 12.49 Orgilinae habitus and wing photographs. (a) Male *Stantonia* sp. from Papua New Guinea; (b, c) *Podorgilus nemorensis*. (Sources: a, reproduced by permission Jan Hrcek; b, c, reproduced by permission of Yves Braet.)

biological control agent introduced from Europe into North America in the 1930s to control the European pine shoot moth, *Rhyacionia buoliana* (Tortricidae), and has subsequently been introduced and established in several other parts of the world (e.g. Chile). *O. lepidus* was introduced in the 1960s from South America into Australia, India and the USA and has been a moderately successful control agent against the potato tuber worm, *Phthorimaea operculella* (Gelechiidae). *O. elasmopalpi* is a native parasitoid of the lesser cornstalk borer, *Elasmopalpus lignosellus* (Pyralidae), in the southern USA where it generally causes a low level of mortality (Johnson & Smith 1980). Kula et al. (2010b) reported the less desirable parasitism by *S. pallida*, of a moth, *Neomusotima conspurcatalis*, that is used for the biological control of an invasive climbing fern in Florida, although at present. parasitisation levels only range between 2 and 10%.

The subfamily had generally been considered as comprising two tribes, the Mimagathidini for *Eleonoria*, *Orgilonia* and *Stantonia* (= *Bentonia*) and the Orgilini for the rest. Van Achterberg (1987) added the Antestrigini to include the aberrant Chilean genus *Antestrix*, which has fore wing vein 1RS distinct and the second submarginal cell rhombic. *Antestrix* has not been included in any molecular study and I would not be surprised if it turned out not to be related to the other orgilines. The Mimagathidini, or more strictly, *Stantonia*, has been variously treated as a member of the Agathidinae or as part of the Microtypinae in a lot of earlier literature, almost entirely because of their similarly-shaped, fore wing, triangular second submarginal cells. Some genera (e.g. *Orgilus* and *Podorgilus*) have fore wing vein r-m absent (Fig. 12.49c).

All orgilines appear to be endoparasitoids of Lepidoptera larvae, although there appear to have been numerous transitions between attacking exposed and weakly concealed hosts (such as leaf miners, stem-borers or leaf-rollers): even within the large genus *Stantonia* there may have been up to seven separate transitions from attacking concealed to exposed hosts based upon variation in ovipositor length (Braet & Quicke 2003). Shaw and Huddleston (1991) noted that of the British fauna (i.e. *Orgilus* species), most species attack hosts associated with low vegetation and suggested that a lot of host searching might be carried out by walking over the relevant vegetation, and this must be true of the brachypterous species. The hosts of *Orgilus* are predominantly members of the Coleophoridae, Gelechiidae, Gracillariidae, Lyonetiidae, Oecophoridae, Psychidae, Pyralidae and Tortricidae, while *Stantonia* species are most frequently reared from crambids (Braet & Quicke 2003). Depending on species, hosts are attacked as early, mid- or late instars (Shaw & Huddleston 1991).

The developmental stages have been described in various amounts of detail for *Orgilus obscurator* (Juillet 1959), *O. parcus* (Broodryk 1969b), *O. lepidus* (Oatman et al. 1969) and *O. jenniae* (R.V. Flanders & Oatman 1982); there are no equivalent studies on any of the other genera. Orgilines lay their hydropic eggs directly into the host haemocoel and they appear to become sticky and adhere to host internal organs (see A.R. Kraaijeveld 1994 for similar behaviour of alysiine eggs). There appear to be three larval instars (Oatman et al. 1969, R.V. Flanders & Oatman 1982). The first instar is initially caudate and mandibulate but an anal vesicle then develops, which is apparently anatomically different from that of microgastroids (Thorpe 1932) and although when it first hatches it may be surrounded by serosal cells, these disperse and do not form teratocytes. The cuticle of the first instar is described as being evenly spiculate and lacks distinct rows of strong spines. The wasp remains as a first instar until the host has nearly finished growing and the second instar, which is also vesiculate, is also mandibulate, although its mandibles are considerably smaller than those of the first instars and it feeds on the host fat body. The final instar chews its way out of its host using its serrate mandibles and completes feeding externally.

Females of *O. lepidus* mate only once (Oatman et al. 1969) and Greany and Oatman (1972) showed that females host and trail mark using a substance that originates in the metasoma, and this acts as a powerful deterrent to conspecific females, presumably leading to a reduction in superparasitism.

Species of *Clotildea* and of *Declotila* are rather larger than most of the other species, but very few specimens are known for these rare Afrotropical genera. Braet (1997) noted that males of *Clotildea* have a pair of dorsal protuberances on the sixth metasomal tergite and these seem likely to be associated with exocrine glands, given the preponderance of such glands on the posterior tergites in other subfamilies. The apex of the ovipositor usually has a dorsal notch typical of many larval endoparasitoids (Boring et al. 2009); however, *Declotila* has a long, rather robust and apically rather blunt ovipositor, without any trace of a notch (see plate 58 in van Achterberg 1993a), so it is hard to understand how it might function or penetrate a host without leaving a large hole in it.

The ventral face of the apex of the hind tibia has a cluster of spines or denticles, close to but slightly less medial to where many taxa just have a dense patch of slightly elongate and slightly more erect setae that are used in grooming. The role of these structures is unknown though they seem unsuited for grooming.

The relationships of orgilines to other subfamilies has not been satisfactorily resolved, with molecular evidence variously placing them at the base of the microgastroid complex (Belshaw & Quicke 2002) or associated with the macrocentroid subcomplex among the helconoids (Sharanowski et al. 2011) or somewhere between the two (Belshaw et al. 2000). The latter agrees more with the past, when various workers have placed them with the helconines or microtypines (Braet & Quicke 2003). Čapek (1970), based on larval features, associated them with the Agathidinae and both have a similar general habitus shape, and also these two subfamilies share the same type of ovipositor steering mechanism involving a pair of opposing bosses on the apico-lateral parts of the upper and lower valves (Fig. 3.10b) (Quicke et al. 1995).

Two morphological phylogenetic studies of generic relationships were presented in quick succession (Braet & van Achterberg 2003, Braet & Quicke 2003), the former based on relatively few characters and few exemplar taxa, all but one genus being represented by a single species, the latter with far more extensive species representation. The results of Braet and Quicke varied according to whether or not colour characters were included in the matrix. The trees from these studies show few similarities, although both agree in that *Eleonoria*, *Orgilonia* and *Stantonia* form a clade and that *Antestrix* is basal[16]. Neither recovered the Orgilini as monophyletic, but in both it was paraphyletic with respect to the Mimagathidini.

A key to the genera was provided by Braet and van Achterberg (2003).

Euphoroid Complex

Euphorinae

This subfamily is one of only two in the Ichneumonoidea that include species developing entirely on the adult stages of other insects (the other being the Aphidiinae) or attacking sub-adults of hemimetabolous orders. Indeed, S.R. Shaw (2004) stated, '... the euphorine Braconidae provides a special case for understanding the diverse pathways of the evolution of parasitic strategies in the order Hymenoptera'.

The Euphorinae has been divided into a number of tribes on the basis of morphology and phylogenetic analysis, although that of S.R. Shaw (1985) did not include any outgroups in the analysis and so was unable to test the monophyly of the subfamily. In that paper, he recognised nine tribes excluding Meteorini, which he considers a separate subfamily (largely because of its perceived difference in biology). The latest classification (D.S. Yu 2012) includes 14 with the inclusion of both Meteorini and the Neoneurini.

A relationship between Meteorinae and Euphorinae has long been suspected based on morphology, and several braconidologists who had informally considered the phylogenetics of the group treated meteorines as belonging to the Euphorinae (e.g. Tobias 1966, Čapek 1970, van Achterberg 1976c). However, S.R. Shaw (1985, 1988) followed by Shaw and Huddleston (1991) and Sharkey (1993) re-elevated the Meteorinae to subfamily status, probably largely on the grounds that they display what appeared to be a more ancestral (typical biology); Sharanowski et al. (2011) also proposed retaining Meteorinae as a separate subfamily. The first molecular analysis dealing with the Euphorinae in particular was that of F.F. Li et al. (2003), which was based on mostly previously published data and only one part of the nuclear 28S rDNA gene, and the results were not very informative, although their failure to recover several morphologically-based tribes it likely correct. More recently, Stigenberg & Boring (in preparation) have carried out an extensive multi-gene study and most of the classification aspects below are based on their results.

Recently, molecular studies have not only confirmed that meteorines and euphorines form a monophyletic group but also several suggest that the meteorines, in spite of their biology, could be derived from within the Euphorinae *s.s.*, although this seems unlikely and needs further investigation. Nevertheless, the traditional picture, that the 'Meteorinae' are the sister group of the Euphorinae and the very elegant scenarios proposed to explain the evolution of parasitism of adult stages of holometabolous insects and nymphal stages of hemimetabolous (strictly paurometabolous) ones from ancestral parasitism of Lepidoptera larvae as shown by *Meteorus* (S.R. Shaw 1985), may need revising. Tobias (1965, 1966) first proposed what has now become a widely accepted scenario which involves a switch from attacking exposed Lepidoptera larvae to similarly exposed beetle larvae, and in particular chrysomelids, and subsequently from these to attacking the beetle's early then late pupal stages and finally beetle adults. This whole series is facilitated by the fact that many chrysomelids have their larval, pupal and adult stages occupying the same physical place. Indeed, at least one genus of the normally Lepidoptera-parasitising genus *Meteorus* has been reared from a chrysomelid larva (M.R. Shaw 1988). DNA data also support the inclusion of the Neoneurinae in the subfamily, in agreement with Tobias (1966), while S.R. Shaw in his studies of the group excluded them. However, Tobias's inclusion of the Blacinae with the Euphorinae[17] is firmly rejected (e.g. Sharanowski et al. 2011).

Several other morphologically aberrant tribes and subfamilies are nested within the Euphorinae. Van Achterberg (1995) described Planitorini (comprising the single genus *Planitorus* from Australia) in the cyclostome subfamily Betylobraconinae and the Mannokeraiini (comprising the single genus *Mannokeraia*) within his new subfamily Masoninae and also treated as putative cyclostomes. However, neither of these are morphologically cyclostome (although that condition has been lost on several occasions within the cyclostome lineage), and recent molecular evidence clearly places both *Planitorus* and *Mannokeraia* as sister groups within the Euphorinae (Sharanowski et al. 2011).

Below I deal with each of the accepted tribes of Euphorinae separately, but necessarily there is a great deal more to say about some than others, and the exact compositions of some are questionable and might well need revising in the light of new molecular evidence. While broadly agreeing with S.R. Shaw's (1985, 1988) morphological phylogeny, the system employed below largely follows the multi-gene molecular phylogeny of Stigenberg and Boring (in preparation). As noted by them, molecular analysis revealed a number of strongly supported clades that mapped approximately on to Shaw's tribes, although with several genera having to be transferred to other groups. One genus, *Asiacentistes*, that had been placed in the Centistini actually appears to be extremely basal and possibly the sister group to the Euphorinae+Cenocoeliinae, in which case it may at some point be desirable to place it within its own subfamily.

Meteorini

The Meteorini includes only two genera, the very large, diverse and cosmopolitan *Meteorus* with more than 300 described species and the far less speciose *Zele*. The systematics of this group were partially clarified by van Achterberg (1979c), who showed that *Zele* belonged with *Meteorus*. It now seems that *Zele* might represent just a species group within *Meteorus* (Maetô 1990, S.R. Shaw 1997a, Stigenberg & Ronquist 2011). Biologically, it has often been stated that Meteorini are exclusively larval parasitoids of Lepidoptera and Coleoptera (the latter mostly involving moderately concealed species, but also a few exposed ones, e.g. M.R. Shaw 1988), whereas the Euphorinae are almost exclusively parasitoids of nymphs or adult insects. The practice of recognising the Meteorinae, either as a separate subfamily or as a tribe within Euphorinae, has persisted despite the fact that separation of the two is not at all straightforward. Chen and van Achterberg (1997) addressed this issue further by noting that members of the genus *Meteorus* could not reliably be separated from the euphorine genus *Perilitus* on the basis of morphological characters alone because the main difference is the presence or absence of fore wing vein rs-m and this character even varies intraspecifically in some '*Meteorus*' species.

Several *Meteorus* species are important in biological control. *Meteorus laphygmae* has been widely released around the world as a generalist biological control agent against *Spodoptera frugiperda* and *M. pulchricornis* is similarly a highly polyphagous parasitoid used in biocontrol but accidentally introduced to New Zealand, where it can parasitise 21 Lepidoptera species collectively representing eight families (Berry & Walker 2004). In Hawai'i, *M. laphygmae* is incredibly abundant and constituted 28% of all ichneumonoid specimens collected along a large altitudinal transect by R.W. Peck et al. (2008). The North American indigenous *M. autographae* is an important control agent of both the soybean looper (*Chrysodeixis includens*) and the cabbage looper (*Trichoplusia ni*) in south-east USA and also attacks many other noctuid pests throughout its range (Grant & Shepard 1984, 1986). This species has a typical lifetime fecundity of over 200 and at 26 °C it can complete development in approximately 2 weeks. Another useful feature is that parasitism by it reduces host food consumption and therefore plant damage by as much as 78%. Indeed, just recently, *M. ictericus* has been found to be the dominant parasitoid of the light-brown apple moth (*Epiphyas postvittana*) (Tortricidae) pest in California, whereas previously it had only been known as a minor parasitoid of various European, predominantly non-pestilential, tortricid moths (Bürgi & Mills 2013).

Nearly all temperate Meteorini are solitary, koinobiont endoparasitoids of Lepidoptera larvae, although a number of gregarious species occur in warmer regions (Maetô 1990, Zitani & Shaw 2002), and Maetô considered the Old World gregarious species to form a monophyletic lineage, the *M. rubens* group. G.Z. Jones & Shaw (2012) noted for their rearings of *Meteorus* from caterpillars in Napo Province, Ecuador, that gregarious development was exclusively associated with exposted hosts comprising 41% of species records (n = 29), whereas all of the 11 species reared from concealed hosts were solitary. Although this gives a significant χ^2 p-value, and neither host size nor wasp phylogeny were controlled for, it does suggest that there is some sort of ecological distinction between gregarious and solitary species. Only one species is known to be brachypterous and attacks caterpillars of grass-feeding crambid moths in alpine New Zealand (Aguirre et al. 2014).

A few species, e.g. *M. corax*, have longer ovipositors and attack wood-boring beetle larvae (Huddleston 1980), and one species has been reared from the exposed larvae of a chrysomelid beetle (M.R. Shaw 1988). Two species of *Meteorus* are know to attack snakeflies (Rhaphidioptera) (Aspöck 2002). Some species that attack particularly large and often ground-pupating hosts such as Sphingidae, particularly in the tropics, are gregarious (Shaw & Huddleston 1991, Zitani & Shaw 2002, Sobczak et al. 2012). The majority of species are associated with hosts that are not well defended chemically, but a few have recently been found to attack well-defended hosts such as papillionid butterfly larvae feeding on *Aristolochia* (Zitani et al. 1997), and S.R. Shaw & Jones (2009) discovered a species parasitising ithomine butterfly caterpillars protected by sequestered compounds from their Solanaceae host plants.

Detailed descriptions of biology have been provided for various species of *Meteorus* by Madel (1963), Askari et al. (1977), Thireau et al. (1990), Bell et al. (2000) and Suzuki and Tanaka (2007). The biology of *Zele* was described and illustrated by Kolaib (1987). Eggs are hydropic, increasing in volume within the host many times, with Strickland (1923) reporting that those of *M. dimidiatus* undergo an astonishing approximately 10,000-fold increase in volume. Three larval instars are generally reported (Ibrahim et al. 1987, Kolaib 1987, Thireau et al. 1990), although in some the moult to the third instar takes place as the larva egresses through the host cuticle (Strickland 1923, Madel 1963), and this may allow some host caterpillars to continue roaming after the parasitoid larva has emerged to pupate. Thus there is considerable variation in which instar carries out most of the feeding (Shaw & Huddleston 1991). Oviposition is often but not always into later instar host larvae (e.g. Bürgi & Mills 2013), but some species are capable of attacking a range of instars and indeed may have more than one generation on a single host generation (Shaw & Huddleston 1991). Emergence from the host is often, in solitary species, close to an abdominal spiracle, and it is tempting to speculate that this has something to do with Rao et al.'s (2009) observations of the cardiochiline *Toxoneuron nigriceps*, whose larvae are typically located near abdominal host tracheal tufts, thus presumably increasing their access to oxygen.

First instar larvae break out from the egg chorion surrounded by their serosal membrane (Fig. 12.50) apparently using a combination of their mandibles and caudal appendage. The serosal cells soon dissociate and swell to form teratocytes. In *M. gyrator*, an initial 1000 or so teratocytes decline during the course of parasitisation to approximately 400 (Bell et al. 2004), and a similar pattern is followed by *M. pulchricornis* with initially some 1200 teratocytes, the number falling dramatically around day six (Suzuki & Tanaka 2007). In both species, the teratocytes grow considerably after they dissociate from the serosal membrane. Those of *M. gyrator* inhibit host phenoloxidase activity and so presumably help protect the developing parasitoid larva from encapsulation. In the case of *M. pulchricornis*, many teratocytes adhere to the fat body of the host, *Mythimna* (as *Pseudaletia*) *separata* (Noctuidae), and are responsible for disrupting it and making it easier for the parasitoid larva to consume.

The first instar is caudate and mandibulate (Fig. 12.50b), with a large head relative to the body, which shows little sign of segmentation and has no apparent tracheal system. The second and third instars

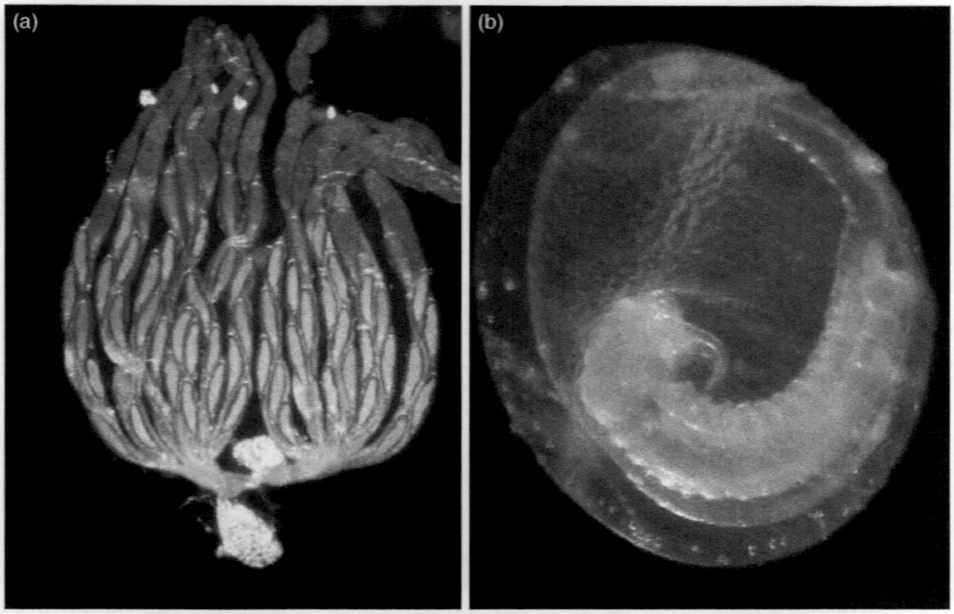

Fig. 12.50 *Meteorus ictericus* (Euphorinae), a parasitoid of the light-brown apple moth, *Epiphyas postvittana (Tortricidae)*. (a) Ovaries slightly squashed to spread ovarioles and show egg number and development; (b) first instar larva within serosal membrane still within chorion, showing the long tail and very large sickle-shaped mandibles. (Source: reproduced by permission of Linda P. Buergi.)

are vesiculate. Only the first and third instars have strongly sclerotised cephalic structures (Thireau et al. 1990) and presumably the second instar feeds only on haemolymph, perhaps enhanced by teratocytes or their products. In *Z. chlorophthalma*, the second instar larvae are found attached to the mid- or hind gut of the host. Most of the host's internal organs are consumed by the internal feeding of the third instar parasitoid, which after emerging through the host cuticle continues feeding externally, although most species have completely lost the external feeding phase and correspondingly the final instar larvae of meteorines are not serrated (Shaw & Huddleston 1991).

Cocoon formation can either be in close association with the host remains or nearby, or the parasitoid larva can wander off or drop to the ground and pupate far away from the host. The emergence hole from the cocoon is irregular. There is a considerable amount of variation in the cocoons of various Meteorini. Those of *Zele* and many *Meteorus* are sessile, as typified by some of the parasitoids of wood-borers such as *M. corax* (Huddleston 1980, Maetô 1986). However, those of quite a few *Meteorus* are pendulous (meteor-like and hence the name of the genus). The pendulous cocoons made by *Meteorus* (as also some campoplegine ichneumonids and the analogous suspended mummies of the *buzurae* group of the rogadine genus *Aleiodes*) are generally considered to be an adaptation against predation by predators such as ants and possibly birds and is supported by experimental evidence for *Meteorus pulchricornis* (Shirai & Maetô 2009), although it has been shown not to be effective against hyperparasitoids (Fig. 10.9) (Proper 1934, Lejeune & Silver 1961, Harvey et al. 2011, Sobczak et al. 2012). In some gregarious *Meteorus* species, the larvae having emerged from the host remains spin their cocoons in a loose, irregular heap (Huddleston 1980), whereas in some others they cooperate to make either masses or chains of cocoons (Zitani & Shaw 2002, S.R. Shaw & Nishida 2005, Zitani et al. 1997). The cocoons of *Meteorus* have a loosely woven pad (cap) internally at the anterior end against which the head of the pupa rests and (Askari et al. 1977, Zitani & Shaw 2002), which possibly provides physical protection (Fig. 12.51) (Barrantes et al. 2011). The same structure, composed of more than 80 fine silk layers, also occurs in

Fig. 12.51 Gregarious cocoons and cocoon features of *Meteorus* spp. (Euphorinae). (a–d) Chain of cocoons made by *M. restionis*, showing multiple vertical threads, pad of silk within the apex of the cocoon, and head of wasp pupa in the normal position near to the silk pad; (e) globular suspended cocoon mass, approximately 12 mm diameter, of an unidentified species from Okinawa. (Sources: a–d, reproduced by permission of Gilbert Barrantes; e, reproduced by permission of Kaoru Maetô.)

cocoons of *Zele* (*Z. deceptor* and *Z. albiditarsus*, personal observation).

At least one species of *Meteorus* shows facultative seasonal host alternation (Maltais et al. 1989), attacking caterpillars of two different species of the tortricid genus *Choristoneura* at different times of year and overwintering as a larva within one of them. Askari and Coppel (1978) described the courtship and mating behaviour of *M. pulchricornis* and successful mating apparently depends on both visual and chemical cues from the female and wing vibrations produced by the male. Before mounting, the male repeatedly touches the female metasoma with his head but it is not clear whether this is to stimulate her, and if so by what modality, or for him to conclude that she is an appropriate mate, or both. Copulation is relatively protracted, lasting for between 80 and 120 s.

Oviposition in at least some species also involves temporary host paralysis (Chau & Maetô 2009) and it has been suggested that this helps the female wasp avoid self-superparasitism through making it easy to recognise recently attacked host individuals, and this was supported by Yamamoto et al. (2009), who showed that oviposition behaviour in *M. pulchricornis* was initiated by host movement.

Several species have been reported to harbour virus-like particles in the female reproductive system (Edson et al. 1981). Those in the venom of *Meteorus pulchricornis* have been shown to induce host haemocyte apoptosis (Suzuki & Tanaka 2006).

Maetô (1990) discussed intrageneric groupings within *Meteorus* and noted that it should probably be divided into a number of separate genera, but that the group needed to be studied better before any such action were taken. Stigenberg and Ronquist (2011), concentrating on the West Palaearctic fauna, revealed very high species diversity, including numerous cryptic species in their analyses. They also recovered the

Table 12.5 Summary of known host associations of Euphorinae genera.

Genus	Host group	Host stage attacked	Ref.
Allurus	Curculionidae: *Sitona*	Adults	Loan 1964
Aridelus	Pentatomidae, Acanthosomatidae, Plataspidae	Nymphs, adults	Čapek & Davidová-Vilímová 1978, S.R. Shaw et al. 2001, Maetô & Kudô 1992
Centistes	Coleoptera: Curculionidae, Chrysomelidae, Carabidae, Staphylinidae	Adults	Lipkow 1965, Loan 1972
Bracteodes	Apidae	Adults	
Centistina	? Coleoptera		
Chrysopophorus	Chrysopidae	Adults	Principi 1948
Cosmophorus	Scolytidae	Adults	
Cryptoxilos	Scolytidae	Adults	Čapek & Čapeki
Dinocampus	Coccinelidae	Adults	
Euphoriella	Psocoptera	Nymphal or nymph–adult koinobiont	Sommerman 1956
Holdawayella	Tingidae	Nymphs and adults	Loan et al. 1971
Leiophron	Miridae	Nymphs	Waloff 1967
Marshiella	Anthicidae		Shaw & Marsh 2000
Microctonus	Various Coleoptera	Adults	
Myiocephalus	Apparently associated with *Formica* ant nests		
Perilitus	Curculionidae, Chrysomelidae, Melandryidae; one sp. on Orthoptera	Adults	Waloff 1961
Peristenus	Miridae; records from Nitidulidae are dubious	Nymphal or nymph–adult koinobiont	
Plynops	Associated with ants		
Pygostolus	Curculionidae; ?Lepidoptera	Adults	Loan & Holdaway 1961
Ropalophorus	Curculionidae: Scolytinae		
Streblocera	Chrysomelidae	Adults	J.H. He 1984, Maetô & Nagai 1985
Syntretomorpha	Apidae and possibly Thynninae	Adults	Walker et al. 1990b
Syntretoriana	Thynninae	Adults	
Syntretus	Apidae (*Bombus, Psithyrus, Trigona*) Ichneumonoidea	Adults	Cole 1958, Alford 1968
Towneselitus	Chrysomelidae		
Wesmaelia	Nabidae	Nymphs and adults	

genus *Zele*, which comprises generally rather larger species, nested within *Meteorus*, but refrained from synonymizing the two. Previous studies had typically only included one representative of each or lacked *Zele* altogether.

General Comments on Non-Meteorine Euphorines

It is a fairly large group with some 500 described species in 54 genera. Its species are broadly equally divided between those attacking Coleoptera and those attacking Hemiptera, with a small number of genera specialising on other host orders (Table 12.5). Most of those that attack beetles attack the adult stage, although some are known, in addition, to attack pupal or larval stages.

S.R. Shaw coined the term 'imagobiosis' to emphasise that the strategy of parasitising adult hosts differs in several important ways from the standard definitions of idiobiont and koinobiont, in that the host is not immediately immobilised but cannot, since it is an adult already, continue development. The evolution of parasitoidism of adult hosts has been explored on several occasions (Tobias 1966, S.R. Shaw 1985, 1988, 2004). Tobias first proposed that parasitism of Chrysomelidae was probably key, as these beetles usually have exophytic larvae, that pupate on the host plant. It would be an easy step for a koinobiont endoparasitoid attacking a late instar larva also to oviposit within pupae in the same niche (possibly 'protected' by the same plant secondary compounds), thence to attacking late pupae with adults about to emerge and then evolving adaptations to deposit eggs within the adult which might be presumed to be physiologically similar. S.R. Shaw (2004) noted that the four tribes he recognised that attack chrysomelids (viz. Centistini, Microctonini, Perilitini and Towneselitini) all occupy basal or medial positions on his (1988) morphological phylogeny, which could suggest that this is indeed an ancestral host group.

Most euphorines, excluding *Meteorus*, are solitary parasitoids, but gregarious development is known in *Syntretus splendidus* parasitising adult bumblebees (*Bombus* species: Apidae), where from six to 77 individual juvenile stages were found depending partly on the size of the host (worker, male, queen) (Alford 1968). Gregarious development is also known in some *Microctonus* species that attack adult beetles. Luff (1976) reported that *Microctonus caudatus* lays up to 92 eggs per host *Harpalus rufipes* (mean 20.8) but as no more than approximately 60 adult wasps emerge, hypothesised that there must be some mortality in the host due to inter-larval competition. As adult ground-dwelling carabids are often available all year round, the parasitoid in this case is bivoltine. In another study, Sasakawa et al. (2011) reported an only partially identified species (probably *Microctonus*) attacking adults of the carabid beetle, *Poecilus versicolor*, with individual hosts having between 9 and 33 parasitoid larvae in them, although parasitism rates were very low (0–3.2%). One of the three species of *Microctonus* that attack weevil pests in pastures, *M. zealandicus*, typically produces broods of 2–6 individuals in its host weevil, *Irenimus aequalis* (McNeill et al. 1993), whereas the other species are all solitary (Fig. 12.52b), and if superparasitism occurs, only one of the larvae completes development and the supernumeraries remain atrophied and die.

Detailed descriptions of biology and egg and larval stages are available for only a few species (e.g. O.J. Smith 1952), possibly because entomologists who collect adult insects often kill them rather keeping them alive to see what might emerge. Oviposition into adult insects with their generally hard cuticle necessitates various adaptations (M.R. Shaw & Huddleston 1991) and H.D. Smith (1932) noted that on warm, sunny days in particular, the host beetle's agility is often enough to thwart the oviposition attempts of *Microctonus vittatae*. The ovipositors are nearly always moderately to strongly compressed either laterally (e.g. *Pygostolus*, *Centistes*, *Centistoides*) or dorso-ventrally (e.g. *Spathicopsis*) and those of some groups are very knife-like. Oviposition in many species is between posterior abdominal sternites, although some oviposit into the host's mouth (*Microctonus*: Wylie 1985) or through the soft membranes of the antennal socket or neck [*Perilitus*: Waloff 1961; *Allurus* (as *Centistes*): Loan 1964]. As with *Meteorus*, eggs are highly hydropic (Loan & Holdaway 1961) and, in the cases of some *Perilitus* and *Microctonus* species, increase in volume once laid by a factor of 1000 or more (D.L. Jackson 1928, O.J. Smith 1952, Waloff 1961). Once laid, the egg tends to come to rest in the host's abdomen or the first instar larva rapidly makes its way there. Teratocytes have been described in some detail for *M. aethiops* (as *aethiopoides*) (Fig. 12.52a) (Barratt & Sutherland 2001, Shelby et al. 2014) and for *Dinocampus coccinellae* (Sluss & Leutenegger 1968) and work on the secretions of the latter is under way (Kadono-Okuda et al. 1998, Gopalapillai et al. 2005). They are also known in nearly all other genera investigated, including Allurus (Loan 1964), *Microctonus* (Barratt & Sutherland 2001) and *Pygostolus* (Loan & Holdaway 1961), although Drea et al. (1972) reported their absence in *M. stelleri*.

Microctonus colesi and *M. larvicida* are unusual for euphorines in that they oviposit consistently in, and complete development in, the larva of their beetle hosts (Drea 1968, van Achterberg et al. 2000), although oviposition in larvae has occasionally been observed in some other species, for example in *Dinocampus coccinellae* and *Perilitus dubius* (Waloff 1961). In the case of *M. brevicollis*, the first generation is apparently completed within the larva of its host, the chrysomelid *Haltica ampelophaga*, and the second generation in the adult (Kunkel D'Herculais & Langlois 1891). In temperate regions, many species are plurivoltine, and this may be facilitated by the potentially risky strategy of causing their host to exit winter diapause earlier than unparasitised individuals (Wylie 1982).

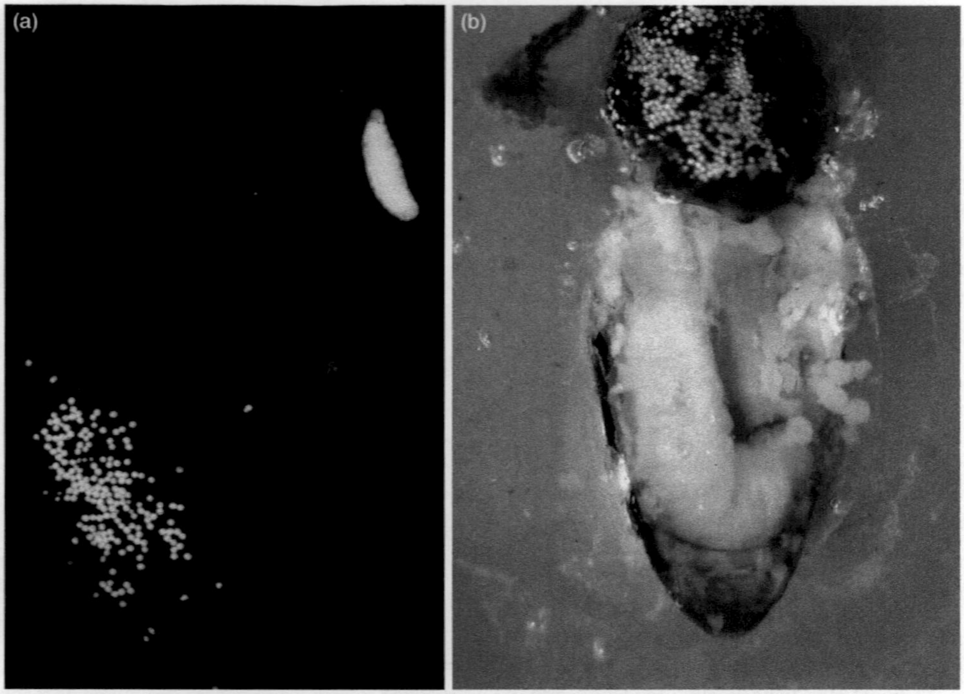

Fig. 12.52 *Microctonus* (Euphorinae) development. (a) Young larva of *M. aethiops* and teratocytes extracted from single *Hypera postica*, a target host for biological control; (b) Near full-grown larva of *M. aethiops* in abdomen of adult *Nicaeana cervina*, a non-target weevil host. (Sources: a, reproduced by permission of Kent Shelby; b, reproduced by permission of Barbara Barratt.)

Little is known about courtship and mating, but some taxa produce pheromones. Females of the genus *Tainiterma* have a large setose pit in the middle of the eighth metasomal tergite (van Achterberg & Shaw 2001), which almost certainly has an associated pheromone gland.

Major rearrangements of taxa between tribes on the basis of molecular evidence means that identification is best done directly to genus. Keys to the genera have been provided for Palaearctic (Belokobylskij & Tobias 2000), China (Chen & van Achterberg 1997) and the New World (S.R. Shaw 1997). The earlier paper by S.R. Shaw (1985) may also be useful, although several new genera and synonymies have been recognised since then.

Centistini

The Centistini comprising *Allurus*, *Centistes* and *Centistoides* are primarily parasitoids of larval and adult weevils (Coleoptera: Curculionidae) but are also known from Chrysomelidae, Coccinellidae and Carabidae (see S.R. Shaw 1985, also van Achterberg 1985b, Schroder & Athanas 2002). *Pygostolus* and *Spathicopsis*, which were included in this tribe by S.R. Shaw are removed to the Pygostolini and Perilitini, respectively, by Stigenberg and Boring (in preparation). Biological studies include those of Loan (1964) on *Allurus* (as *Centistes*), Zijp and Blommers (2002b) on *C. delusorius*, all of which are parasitoids of adult Curculionidae, and Schroder & Athanas (2002) on *C. gasseni*, a parasitoid of Chrysomelidae. *C. delusorius* overwinters inside its adult beetle host as a first instar larva ('young' in the paper). The cocoon is described as thin and delicate and as affording little protection against desiccation or drowning in a well-managed and weedless orchard. Futher, Zijp and Blommers suggested that the absence of *C. delusorius* in several orchards probably reflected a combination of insecticide use and a lack of suitable pupation sites. Problems can occur in rearing in captivity if the parasitoid larvae are disturbed during their external feeding and cocoon construction stages (Schroder & Athanas 2002). *Centistes lucidator* has been reared from adults of four species of the staphylinid beetle genus *Tachyporus* (Lipkow 1965), which are exceedingly uncommon hosts for ichneumonoids. *Centistes* (as *Syrrhizus*) (Loan 1972) and *Allurus* both use their fore legs to grasp their hosts while ovipositing. *Allurus* species also have posteriorly-directed pairs of denticles on some of their metasomal sternites, which seem likely to be involved in host manipulation, but detailed observations are lacking. The type genus *Centistes* is actually very similar in appearance to members of the Brachistini and differs principally in having fore wing vein M+CU largely unsclerotised. All species appear to be solitary, but superparasitism occurs and supernumeraries are eliminated soon after eclosion (Loan & Holdaway 1961, Schroder & Athanas 2002).

Cosmophorini

Van Achterberg & Quicke (2000) recognised two genera, *Cosmophorus* and *Sinuatophorus*; however, the composition of this tribe is expanded as a result of molecular evidence. Belshaw and Quicke (2002) recovered the type genus as sister group to *Rhopalophorus* and Stigenberg and Boring (in preparation) in addition to *Sinuatophorus* (Fig. 12.53) also include, *Tuberidelus*, *Plynops* and *Cryptoxilos* here. *Cosmophorus*, *Cryptoxilos* and *Rhopalophorus* are all apparently specialist parasitoids of adult scolytine beetles (S.R. Shaw & Berry 2005), but the biology of both *Plynops* and *Tuberidelus* is unknown. Perhaps the rest of the group may utilise these hosts also, which is in contrast to S.R. Shaw's (1988) statement that 'Parasitism of bark beetles (Scolytidae) has evolved independently in … *Cosmophorus*, *Cryptoxilos* and *Rhopalophorus*'. Tobias (1986b) had based his Cryptoxilonini on *Cryptoxilos*, which has an unusual distribution stretching from the Nearctic through the Oriental Region to New Zealand and Fiji. As these changes indicate, the tribe is morphologically heterogeneous.

Based on a very few biological observations, *Cosmophorus* females grasp their hosts with their enlarged mandibles, probably aided by the protuberances below the antennal sockets (Fig. 12.53) during oviposition. They also apparently inject a venom that causes some temporary paralysis of the host. The biology of *Cosmophorus cembrae* was described by Seitner and Nötzl

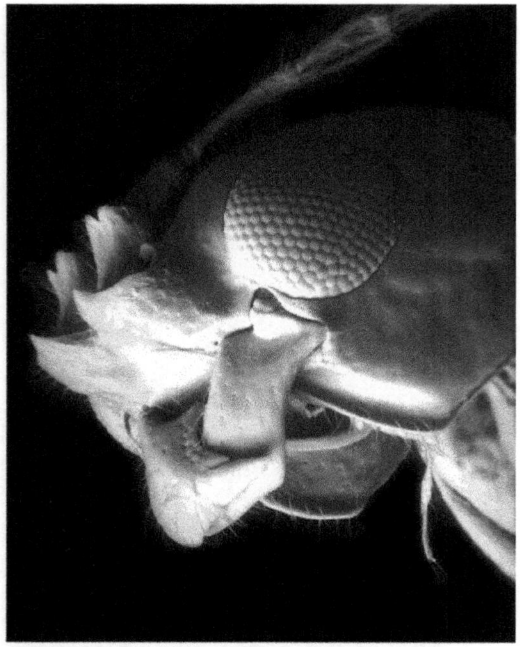

Fig. 12.53 Head of *Sinuatophorus longiceps* (Euphorinae) showing enlarged, overlapping, multitdentate mandibles.

(1925) and that of the Nearctic *C. capeki* by Loan and Matthews (1973). As with *Histeromerus* (Rhyssalinae), they spend much of their time searching tree trunks and not flying so they are therefore rather uncommon in Malaise trap samples. The head of *Plynops* is modified in a way somewhat reminiscent of that of *Cosmophorus*, i.e. concave with large mandibles and, in many species, anteriorly protruding antennal sockets, so perhaps members of this genus also grasp their hosts with their jaws before parasitising them.

The placement of *Cosmophorus* in the Euphorinae has even been questioned because the wasps are virtually cyclostome with large mandibles, have a venom apparatus virtually identical with that of the Rhyssalinae (Jiménez et al. 1997) with a very small reservoir and a pair of unbranched gland filaments that are connected directly to the annularly sculptured primary venom duct, quite unlike the apparatus of other studied euphorines. Muesebeck had tentatively suggested that they might belong to the Pambolinae. They even lack the floor to the placoid sensilla of the antennae (unpublished observations) as also occurs in the Aphidiinae and Mesostoinae group of subfamilies. However, molecular data confirm that they are indeed euphorines.

The brachypterous males of *C. cembrae* and of the Corsican species, *C. laricio*, apparently release a pheromone to attract females and adopt a 'calling' posture sitting on a twig with metasoma raised high in the air at an angle of approximately 60°, and M.R. Shaw (2009a) suggested that the coarse sculpture of the propodeum and first metasomal tergite could form some sort of locking/stabilising mechanism, allowing the males easily to maintain their calling stance. The mating position assumed by males is unusual in that it is tail-to-tail with the male inverted (Seitner & Nötzl 1925).

S.R. Shaw (1985) had recovered the Cosmophorini as the sister group to Syntretini in his morphological analysis, but van Acherberg and Quicke (2000) suggested that the characters concerned were liable to homoplasy and postulated that the Centistini, with their highly laterally compressed ovipositors, are a more likely candidate. Members of *Sinuatophorus* have highly apically sinuous ovipositors (very like those of the rhyssine genus *Myllenyxis*), which are probably capable of being steered, although exactly how this might work given their highly laterally compressed shape is not clear.

In a recent revision of the tribe, based on searching many tropical Old World Malaise samples, a remarkable number (22) of new species were found, most being known from only one specimen (van Achterberg & Quicke 2000). Application of the Chao 1 diversity estimator to these suggested that there might be as many as 80 still to be discovered. *Cosmophorus* appears to be largely Holarctic and Palaeotropical, but I have seen one undescribed species from South America (Colombia). Nothing is known about the biology of any of the tropical species, although their similar modified mandibles and rarity in Malaise samples would suggest that they have the same sort of biology as the temperate species.

Dinocampini

According to Stigenberg and Boring (in preparation), five genera should be placed in this tribe: *Betelgeuse, Centistina, Dinocampus, Ecclitura* and *Napo*. Virtually nothing is known about the biology of any of these apart from the very abundant and cosmopolitan *Dinocampus coccinellae* (often referred to as *Perilitus coccinellae*), which, as its name suggests, is a specialist parasitoid of ladybird beetles (Coccinellidae). Its biology has been very well studied compared with most euphorines (e.g. Balduf 1926, Bryden & Bishop 1945, M.F. Walker 1962, Sluss & Leutenegger 1968, Obrycki et al. 1985). It has a very broad host range among the coccinelids, although some host species are more preferred than others (Richerson & DeLoach 1972, Majerus 1997). In years when it is common it has sometimes been reported in the press as posing a major threat to ladybird populations and hence to the control of aphids in gardens and on crops; it is also sometimes found as a contaminant of coccinellids commercially supplied for biocontrol, and can sometimes cause economically significant depression of coccinellid pest predators (e.g. Silva et al. 2012). This species is predominantly parthenogenetic, although males have very occasionally been found. (M.R. Shaw et al. 1999). Before this, S.R. Shaw (1988) noted that all known members of Dinocampini (in his sense, see below), were only known from females and therefore that the whole tribe might be thelytokous; in the absence of sex, it would be hard to see how such divergent taxa could be generated[18], although the group, if it were monophyletic, might be predisposed towards the development of thelytokous strains which could be successful at least in the short term.

The biology of *D. coccinellae* was described in detail by Balduff (1926). Oviposition is normally into the adult but occasionally, as with some other euphorine parasitoids of holometabolous insects whose larvae live in the same microhabitat as their adults, as do most aphidophagous ladybirds, juvenile hosts may also be attacked and although *D. coccinellae* will oviposit into all host larval instars, maximum survival occurs with adult parasitism (Obrycki et al. 1985). Balduff (1926) reported that this species has four larval instars. *Dinocampus* preferentially oviposits into female ladybirds both in the field and laboratory, and this has been shown not just to be because female hosts tend to be larger as the preference still exists when size is controlled for (Davis et al. 2006). The reason is likely to be because female beetles also actively consume more aphids and thus provide a better food reward for the developing parasitoid larva. Having completed development, the *Dinocampus* larva egresses from the still living coccinelid through its dorsal abdominal sternites and moves to beneath the beetle, where it spins a distinctive cocoon. Maure et al. (2011) demonstrated two very interesting things about this behaviour. First, the host usurpation serves to protect the euphorine from

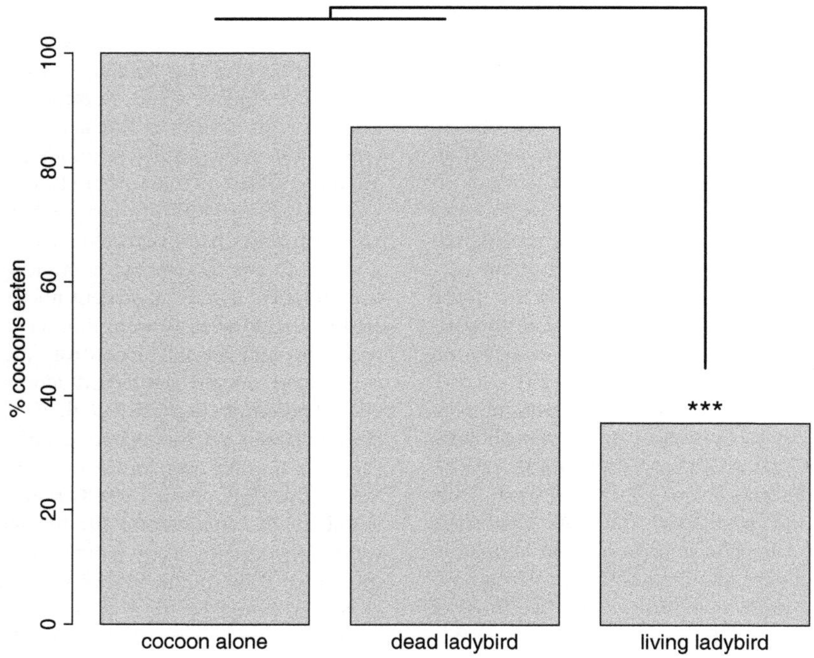

Fig. 12.54 Barplots illustrating the survival benefit to *Dinocampus coccinellae* (Euphorinae) of having a living (though moribund) usurped host sitting above its cocoon when predatory larvae of the lacewing, *Chrysoperla carnea*, were present, whereas there was no significant difference between naked cocoons and ones with dead ladybirds on top (***$p < 0.0001$, Fisher exact test; $n = 20$ for each treatment). (Source: data from Maure et al. 2011.)

predation, and although having a dead ladybird on top of its cocoon did not afford any significant protection (against the lacewing, *Chrysoperla carnea*), having a living one did (Fig. 12.54). Second, when the survival of the ladybird body above the *Dinocampus* cocoon was analysed in relation to the number of mature eggs in the wasps at emergence (allowing for both wasp size and pupal development time), a significant negative relationship was found (Fig. 12.55), indicating that there was a trade-off between allowing the host to survive longer and hence afford protection for longer and the potential fecundity of the wasp if it survived.

Interestingly, in Europe, the *Formica* ant associated species *Coccinella magnifica*, which is superficially very similar to the seven-spot ladybird, *C. septempunctata*, a common host of *D. coccinellae*, is hardly parasitised by it in the field and this had been assumed to be at least partly due to the protection afforded by the ants. However, Sloggett et al. (2004) showed that even in the absence of ants, *C. magnifica* is not a very suitable host for the wasp. Oviposition behaviour in *D. coccinellae* is stimulated by host movement (Richerson & De Loach 1972).

S.R. Shaw (1985) defined this tribe on the basis of three characters and, in addition to *D. coccinellae*, included *Centistina* and *Rhopalophorus*. S.R. Shaw (1988) added another genus to the tribe, *Betelgeuse*, from Mexico, with serrate antennae, while *Rhopalophorus* was removed to the Cosmophorini (Belshaw & Quicke 2002, Stigenberg & Boring in preparation). The latter authors also include *Ecclitura* in the Dinocampini. Unfortunately, *Centistina* has not been sequenced and, along with *Betelgeuse* and *Ecclitura*, its biology is unknown.

Ecnomiini

Ecnomios is an aberrant-looking genus (Fig. 12.56b) first discovered in Papua New Guinea (W.R.M. Mason 1979), although it is now know also to occur in Australia, the East Palaearctic and the Afrotropical region, including Madagascar (Austin & Wharton 1992, Chen

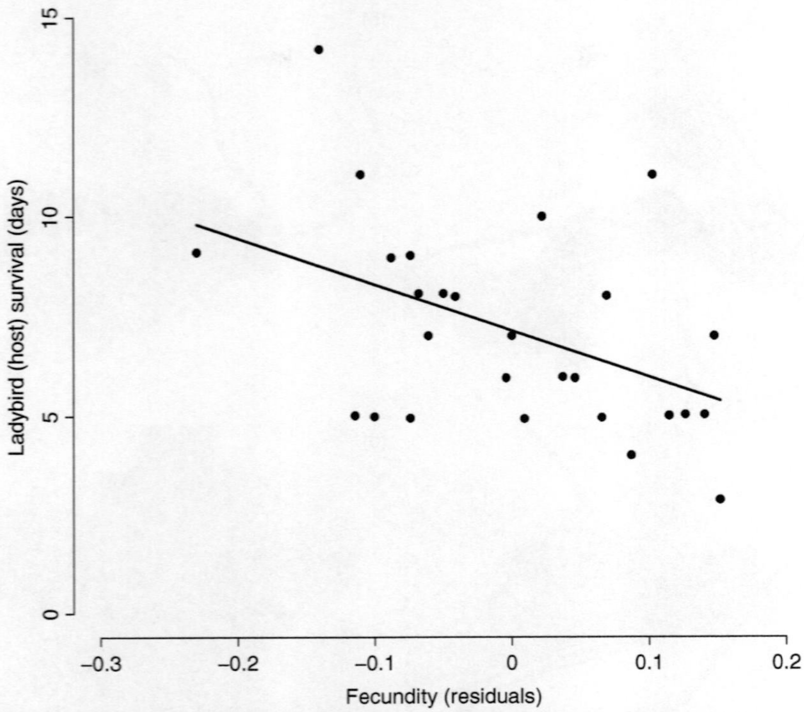

Fig. 12.55 Plot showing trade-off between the potential fecundity of *Dinocampus coccinellae* (Euphorinae) at emergence and the survival duration of its moribund ladybird host, *Coleomegilla maculata*, sitting on top of the parasitoid's cocoon ($r^2 = 0.219$, $p = 0.137$). (Source: data from Maure et al. 2011.)

& Whitfield 2003, Quicke unpublished observations). It was originally placed in the Orgilinae (although Mason also noted that it had some similarities with the Microgastrinae and he was cautious about its exact relationships. Van Achterberg (1985a) noted a number of character states that appeared to exclude *Ecnomios* from the Orgilinae and instead erected the Ecnomiinae to accommodate it, where he considered it to be placed 'half way between the Cheloninae and Neoneurinae'. However, Austin and Wharton (1992) were not convinced that *Ecnomios* was related to the microgastroids and, drawing particular attention to the scutellum, pointed out that it could just as easily be considered as belonging somewhere near the Blacinae or Euphorinae. While not considering its placement in detail, Whitfield and Mason (1994) still placed it among the microgastroids. Morphological analyses, both informal and more formal (Quicke & van Achterberg 1990, Braet & Quicke 2003), failed satisfactorily to resolve its placement and it was recovered in various places either rather basally among microgastroids or towards the base of a large group of non-cyclostomes. Given its chequered history, it did not come as too much of a shock when DNA data placed *Ecnomios* in the Euphorinae (Sharanowski et al. 2011, Quicke unpublished observations). Nothing is known of their biology.

Euphorini

The Euphorini, including *Leiophron* (=*Euphorus*; including *Euphoriella*), *Peristenus*, *Holdawayella* and *Mama*, are mostly parasitoids of nymphal and adult Hemiptera (Heteroptera) and Psocoptera. *Cryptoxilos* and *Stenothremma*, which were included in this group by S.R. Shaw, are moved on the basis of molecular data to the Cosmophorini and Perilitini, respectively. Several species of *Leiophron* are important in the biological control of mirid plant bugs (Loan 1974, 1980) and

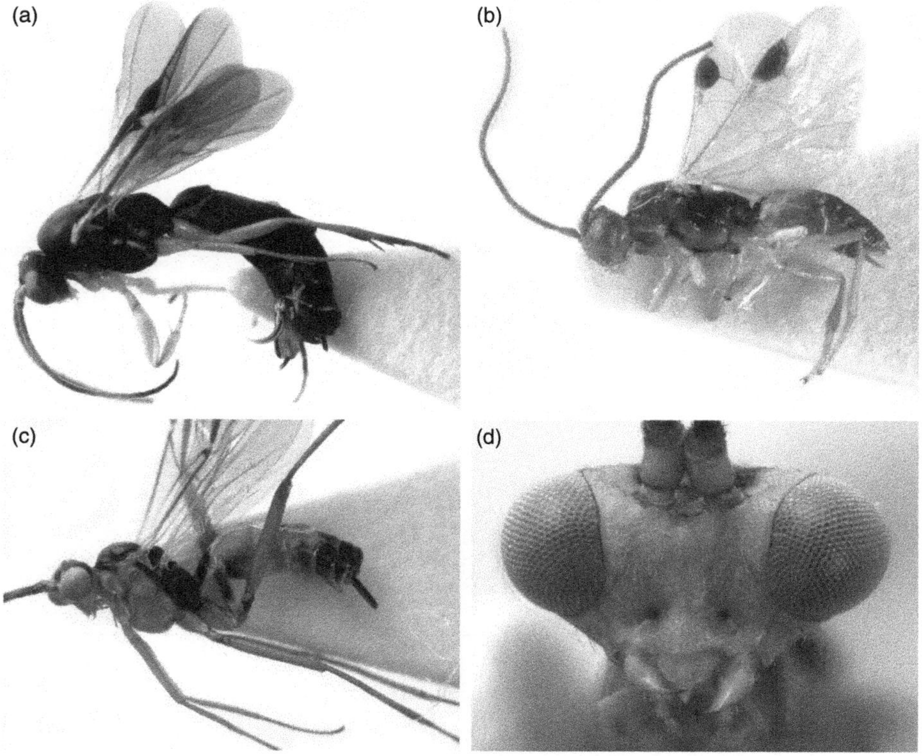

Fig. 12.56 Four diverse members of the Euphorinae. (a) *Neoneurus auctus* (Neoneurini); (b) *Ecnomios* sp. (Ecnomiini); (c, d) *Loxocephalus boops* (Syntretini), habitus and face. (Source: reproduced by permission, © C. van Achterberg, Naturalis Biodiversity Centre).

Peristenus digoneutis is sometimes an important parasitoid of the tarnished plant bug, *Lygus rugulipennis* (Lygaeidae), a pest of several agriculturally important crops. The latter, originally from Europe, was released into the USA as a biological control agent in 1984 and after several attempts became established and is gradually spreading, although it does not do very well at more southern latitudes. Other species of *Peristenus*, such as *P. relictus*, have been released to control various other *Lygus* pests (Pickett et al. 2007, 2009) and, probably due to local climatic variables, may do better than *P. digoneutis* in some places.

Detailed descriptions of biology have been presented for *Peristenus* (sometimes treated as a subgenus of *Leiphron*) *pseudopallipes* (Lim & Stewart 1976), *P. digoneutis* (Carignan et al. 1995) and *P. pallipes* (Loan 1965), all of which are associated with mirid pests of alfalfa. There appear to be four larval instars. The first larval instar is caudate and mandibulate, the second is similar although with slightly less curved mandibles and shorter tail, and the last two are hymenopteriform (Čapek & Davidová-Vilímová 1978). The larvae of *Leiophron* and *Peristenus* leave their hosts and usually form cocoons 2–4 cm under the ground. In temperate regions, *Leiophron* and *Peristenus* overwinter as diapausing adults within their cocoons (Loan 1980, Pansa et al. 2012) and eclose relatively early in the year. Some care needs to be exercised here, though, as Day (2005) has provided biological evidence that *P. pallipes* is in fact a complex of two (or possibly three) species with different diapause durations and hosts.

Helorimorphini

The name of this tribe is based on the generic name *Helorimorpha*, which is a junior synonym of *Aridelus*, and in addition to that genus also includes *Chrysopophorus*, *Ussuraridelus* and *Wesmaelia*. S.R. Shaw had regarded both *Chrysopophorus* and *Wesmaelia* as members of the Euphorini, as did van Achterberg (1994b), although the latter author only recognised four tribes in the subfamily rather than the current significantly larger number. Both tribes attack hemimetabolous nymphs, but they differ markedly morphologically, with the Helorimorphini having petiolate metasomas and longer ovipositors.

Aridelus attacks the soya bean pest *Nezara viridis* (Pentatomidae) and in various parts of Italy was found to attack from 4 to 21% of sampled host nymphs (S.R. Shaw et al. 2001). *Chrysopophthorus* species are parasitic on adult green lacewings (Chrysopidae: *Chrysoperla* and *Mallada* spp.), although apparently many attempted parasitisations fail because the hosts defend themselves forcefully (Principi et al. 1979, van Achterberg 1994b), and this is probably good from a biological control standpoint. The parasitoid larva moults to the final instar upon emerging from the host's abdomen, leaving its skin partially blocking the emergence hole. It then crawls away to find a suitable place for cocoon construction and pupation.

Detailed descriptions of biology have been presented for *Aridelus* by Čapek and Davidová-Vilímová (1978) and S.R. Shaw et al. (2001). Four larval instars are reported, the first being mandibulate, and occupies the host's thorax, apparently without causing damage to internal organs (Čapek & Davidová-Vilímová 1978). The second instar migrates to the abdomen and is orientated with its head towards the host's posterior, and again does not appear to feed on host organs directly. The third and fourth instars occupy most of the host's abdomen and also part of the thorax. Considerable atrophy of the host's gonads was noted. Cocoons are constructed just under the ground (0.5 cm) or sometimes on the ground surface. Maetô and Kudô (1992) described how despite vigorous attempts by the adult female of the subsocial host bug, *Elasmucha putoni* (Acanthosomatidae), the female *Aridelus* is very persistent, constantly circling the brood of nymphs, and finally successfully attacks a high percentage of them.

Neoneurini

The neoneurines have long posed problems as regards their relationships. Tobias (1966) placed them with the Euphorinae largely because they are parasitoids of adult insects, but van Achterberg (1984), in a hand-drawn cladogram, placed them instead with the 'microgastroid' group of subfamilies, citing several putative synapomorphies and shared trends. In the more formal phylogenetic analysis of morphological data by Quicke and van Achterberg (1990), they also associated with the microgastroids. DNA sequence data now clearly show that they are just a derived group within the Euphorinae (Dowton et al. 1998, Belshaw & Quicke 2002, Sharanowski et al. 2011), which is absolutely consistent with their being parasitoids of adult insects and with their reduced and/or aberrant wing venation. These are tiny wasps (Fig. 12.56a) that are most commonly observed in association with formicine ants, hovering around nests or ant trails. The exact biology was not known until recently. S.R. Shaw (1993) presented observations on *Neoneurus mantis* in North America that convincingly showed that they attack adult ants by swooping on them from a perch when they are exposed over the top of their nest. The ants were noted as being very aggressive towards the wasp and to become particularly agitated if *Neoneurus* was present at a mound. The first detailed oviposition observation was by Panis (2007) on *Elasmosoma berolinense* on the ant *Camponotus vagus*, who reported that female wasps oviposited into the worker ant through the intersegmental membrane at the base of the metasoma. Most recently, Gómes Durán and van Achterberg (2011) described and presented video observations (see http://www.pensoft.net/journals/zookeys/article/1754/abstract/oviposition-behaviour-of-four-ant-parasitoids-hymenoptera-braconidae-euphorinae-neoneurini-and-ichneumonidae-hybrizontin and https://www.youtube.com/playlist?list=PLD265E5613AE736AE) of members of three genera. Oviposition by *Kollasmosoma* was exceedingly brief (mean 0.05 s) and that of *Neoneurus* the longest (mean 1.5 s), with *Elasmosoma* (Fig. 12.56a) intermediate. Members of all three genera primarily grasp the adult ant using their fore legs and, in the *Neoneurus* studied, the fore leg is modified and rather raptorial as it is in *N. mantis* and held in a mantid-like posture. Oviposition site varies between the species and, quoting the authors: '*Neoneurus vesculus* lays the egg in the postero-ventral part of the ant's mesosoma, bending its metasoma between the hind leg of the ant and the metasoma, certainly the longer and most complex of the ovipositions observed. *Elasmosoma luxemburgense* lays an egg into the posterior area of the last metasomal segment, probably through the anus. *Kollasmosoma sentum* lays the egg in any location of the ant's metasomal surface, probably through an intersegmental membrane'. The two separate observations on *Elasomosoma* oviposition differ and might indicate adaptations to particular ant host species.

Poinar (2004) described some aspects of the biology of an unidentified *Elasmosoma* species and reported that it has three larval instars. Early development occurred within a trophamnion. The first instar was caudate and had a large head capsule with falcate mandibles that would seem well suited to fighting supernumeraries, although it is hard to imagine that superparasitism occurs very often in the field. However, during ant dissections, it was immediately apparent whether a first instar *Elasmosoma* was going to be present because the ant's fat body was shredded rather than being compact, indicating that the falcate first instar mandible were probably used to macerate it. No mention was made of teratocytes. The second instar is totally different, with a sectorial mouth and no visible mandibles, but with enormous convoluted salivary glands. This form of larva was only rarely encountered in dissections, indicating that it probably is of short duration. The final instar is hymenopteriform and is the first stage to possess spiracles. They subsequently emerge from worker ants usually through the host's anus but occasionally from between abdominal segments, and then go fairly quickly to create a small pupation chamber just below the surface of the soil or other substrate (Lachaud & Perez-Lachaud 2012). Poinar (2004) observed that the larva attached grains of sand to its cocoon in a complicated and specialised behavioural sequence.

All known host associations were listed by Lachaud and Perez-Lachaud (2012). Poinar and Miller (2002) described an interaction between a neoneurine and an ant preserved in amber dating back approximately 40 Mya. A key to the genera was provided by van Achterberg & Argaman (1993).

Oncometeorini

This was erected by Tobias (1987) for a fossil from Baltic amber. Euphorinae are indeed abundant in Baltic amber, but there has been almost no taxonomic work on them.

Perilitini

S.R. Shaw (1985) included only the genus *Perilitus* in this tribe but noted that it is probably paraphyletic as it is defined traditionally only by symplesiomorphies. However, Belokobylskij and Tobias (2000) included many more genera in this tribe but only a few of them appear to be related on the basis of Stigenberg and Boring's (in preparation) molecular study, and the latter authors propose including some other genera in this group, viz. the well-studied *Microctonus*, along with *Heia*, *Litostolus*, *Orionis*, *Rilipertus*, *Spathicopsis* and *Stenothremma*. Most species, including the type species, are parasitoids of adult beetles, principally of the Chrysomelidae, Curculionidae, Melandryidae and Nitidulidae. *M. eathiops* and *M. hyperodae* are important parasitoids of various pest weevils and the latter has been successfully introduced into New Zealand for the control of Argentine stem weevil (*Listronotus bonariensis*). *Perilitus morabinarum* from Australia is very unusual in that it is a parasitoid of nymphal and adult grasshoppers (Orthoptera: Eumastacidae) (Blackith 1967). O.J. Smith (1952) and Loan (1967) described some aspects of development in *Microctonus* species and reported that they have five larval instars. Parasitism of the strawberry sap beetle, *Stelidota geminata* (Nitidulidae), by *M. nitidulidis* results in a significant reduction in host female fecundity but not in the fertility of males (Weiss & Williams 1980). Thelytoky seems to have arisen several times in this group (see Table 7.1).

Several *Perilitus* species are known regularly to superparasitise hosts, both in the same initial oviposition and in multiple oviposition events, although only one individual normally survives to adulthood (rarely two) (Waloff 1961). Other species are clearly gregarious (e.g. *P. eliodis*; D.L. Jackson 1928). Oviposition by *P. dubius* into its chrysomelid host may involve a deceptive mimicry. Females face their beetle host head-to-head and 'palpates its head violently'. In response, the beetle may then vibrate its own antennae rapidly and drive the wasp away, or may also extend its head forwards, exposing the membrane between head and thorax into which the euphorine oviposits (usually rather briefly). Waloff noted that the beetle's antennation and head extension behaviour is the same as happens when two beetles are courting and so suggested that the wasp is exploiting part of its host's premating behaviour. As the *P. dubius* larvae develop, they consume various non-vital host organs and ,in particular, female host ovaries were affected.

Venom gland histology has been studied for *Microctonus hyperodae* (A.M. Crawford et al. 2006) and *M. aethiops* (as *aethiopoides*) has been found to release virus-like particles into its host (Barratt et al. 1999), but their role if any is unknown, and since they are absent from biotypes collected in Europe and North Africa (Barratt et al. 2006) they cannot be essential for overcoming host immunity, although it has been suggested that they may influence host specificity.

Planitorini

These were originally described in the Betylobraconinae, but shown by Sharanowski et al. (2011) to be Euphorinae and comprising the genera *Planitorius* and *Mannokeraia*, both from Australia. Their biology is unknown, but their robust legs produced, often quite bulbous, faces and brachyptery suggest that they may seek hosts in the soil.

Pygostolini

The biology of *Pygostolus falcatus*[19], a parasitoid of adult curculionids, was studied by Loan & Holdaway (1961). It is reported to go through five larval instars and causes host castration, curtailing and then completely preventing host egg production while the parasitoid is still at either the egg or first instar larval stage. Unlike most other euphorines, which typically construct their cocoons in leaf litter or just below the soil surface, *Pygostolus* attach their cocoons to twigs. *P. falcatus* overwinter inside its adult beetle host as a first instar larva ('young' in the paper). *P. falcatus* is largely thelytokous.

Syntretini

The concept of this tribe is expanded by Stigenberg & Boring (in preparation) also to include *Myiocephalus* (= *Loxocephalus*[20]) which was placed previously in its own tribe (Myiocephalini) because of its highly distinctive morphology (Fig. 12.56c,d). In addition, the Syntretini includes *Bactreodes*, *Falcosyntretus*, *Syntretus*, *Syntretomorpha*, *Syntretoriana* and *Syntretellus*. All species with known biology attack various adult Hymenoptera, including bumblebees, stingless bees (e.g. *Trigona*) and ichneumonoids (Cole 1959a); *Myiocephalus* is associated with ants' nests, but its biology is as yet unknown. The developmental stages of the bumblebee (primarily *Bombus* but occasionally *Psithyrus*) parasitoid *Syntretus splendidus* were described by Alford (1968), who reported five larval instars, and Cole (1959a) provided a description of larvae and biology of *S. lyctaea*, a parasite of the ichneumonine *Phaeogenes invisor*.

Tainitermini

Little is known about this monogeneric tribe from South-East Asia (van Achterberg & Shaw 2001) and no sequence data are available at present. It has a long, strongly dorso-ventrally compressed ovipositor sheath and ovipositor and the eighth metasomal tergite has a circular membranous area that seems likely to be glandular. From the published SEM images, the sheaths appear to be fused along their entire length, a condition otherwise known only in one species of the doryctine genus *Leptorhaconotus*. It may be related to the Centistini or Cosmophorini, but that is just speculation.

Towneslitini

Towneslitini of S.R. Shaw originally comprised just *Towneslitus* and later he described *Marshiella*. Subsequently, Belokobilskij (2000) considered *Towneselitus* as a subgenus of *Perilitus*, but DNA data unambiguously show that the two are not closely related and therefore Stigeberg and Boring (in preparation) reinstate the tribe. Molecular data indicate that this group also includes *Streblocera* (Fig. 12.57), which was previously placed in the Perilitini and *Proclithrophorus*, which was placed in its own tribe (Proclithrophorini) by Tobias (1986b). *Towneselitus* and *Streblocera* are parasitoids of adult chrysomelid beetles and *Marshiella* has been reared from an anthicid beetle. Despite a lot of taxonomic attention having been paid to *Streblocera* because of their charismatic and easily recognisable shape (Fig. 12.57), little is known about their biology. Their antennae, with elongate and often toothed scapus and modified, also often toothed, basal set of flagellomeres are clearly raptorial and it has always been assumed that they are used to restrain hosts during oviposition, similar to the use of mandibles in *Cosmophorus*, but it has never been observed. The basal flagellomeres of *Marshiella* are also modified and probably have a similar role.

Excluded from Euphorinae

Two genera included in the Euphorinae or Ecnomiinae, now treated as a euphorine tribe, are clearly not euphorines but rather members of the Helconinae *s.l.* and they are briefly commented on here.

The Australian genus *Eadya*, described in the Euphorinae by Huddleston and Short (1978), was included in its own 'section' by S.R. Shaw (1985),

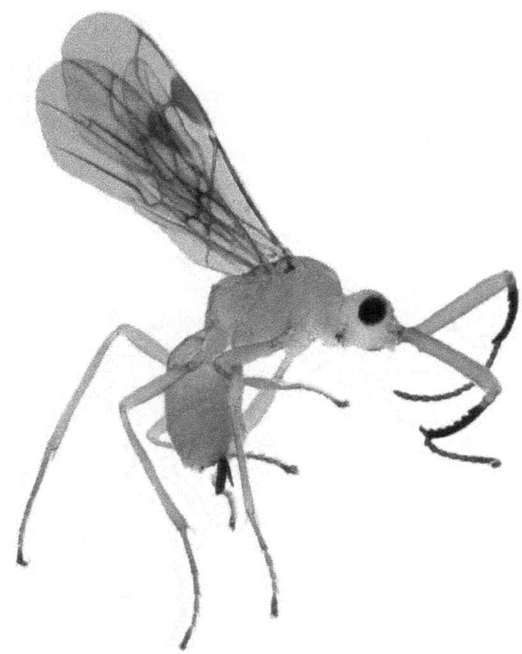

Fig. 12.57 *Streblocera* (Euphorinae) species from West Africa, showing raptorial antennae.

which appeared in his analysis to form a sister group to the rest of the Euphorinae excluding *Meteorus*, although this was partly because he misinterpreted it as being a parasitoid of adult beetles whereas the original description clearly states that it is a larval parasitoid (Huddleston & Short 1978, Withers et al. 2012) *Eadya* is dealt with here under the Helconinae as molecular data from an apparently related genus (*Eadya* itself has not yet been sequenced) suggest that it is not close to the Euphorinae but is better considered a member of the Helconinae *s.l.* (Belshaw & Quicke 2002).

The genus *Korecnomios* was erected by Park and van Achterberg (1994) for a Korean wasp that at the time was known only from the male. It was included in the Ecnomiinae because of the enlarged plical cell of the hind wing, although it differed from *Ecnomios* in a number of other seemingly important respects such as the presence of fore wing vein a, a well-developed vein 2cu-a and also hind wing vein a (called 2A in the original description). It has now been shown that *Korecnomios* is not at all closely related to *Ecnomios* and instead belongs to the *Diospilus* group of Helconinae *s.l.* (D.L.J. Quicke, unpublished work).

Cenocoeliinae

This small, principally tropical subfamily of very distinctive parasitic wasps, with a very high insertion of the metasoma on the propodeum (Fig. 12.58) very reminiscent of the Evaniodea and the only equivalently high insertion in any other iichneumonoid, is in the Neotropical doryctine genus *Evaniodes* (Fig. 12.16e), which is immediately distinguishable through being cyclostome. The subfamily has, however, had a varied taxonomic history and although its distinctiveness was recognised since the beginning of the 20th century when it was afforded subfamily status, it later became more standard to include it as a tribe (Cenocoeliini) within the Helconinae (e.g. Čapek 1970). It was resurrected as a subfamily by van Achterberg (1984). It is a cosmopolitan group of small–medium to large wasps, especially abundant and diverse in the neotropics and recently discovered for the first time in Africa, where it had long been assumed to be absent (Pitz & Sharkey 2005). Within the New World, they are far more species rich in the tropica than the subtropics, with Costa Rica alone having five times more species than the USA and Canada combined (van der Ent & Shaw 1998), and in Costa Rica they only occur below 1600 m, whereas potential hosts are diverse to considerably higher altitudes.

The biologies of very few species are known in detail. The west Palaearctic species have been reared from a variety of beetle larvae in twigs, scolytine curculionids under bark and less commonly Buprestidae (M.R. Shaw & Huddleston 1991, M.R. Shaw 1999b). Some Central and South American species has been reared from seed-feeding curculionids and cerambycids in pods or fruit. The habit of attacking hosts in thinnish twigs, and perhaps in rounded pods or fruit, may explain the highly inserted metasoma. When standing on a thin twig, the wasp's legs will almost inevitably be placed rather lower around the side of the twig, so reducing the amount of space available for manoeuvring the ovipositor. Thus, by having a more highly inserted metasoma, the wasp can compensate for this. Many species possess a groove on the inner face of the hind coxa, which is believed to help support the ovipositor during substrate penetration as in Labeninae. The orientation of this groove probably reflects the angle at which the ovipositor is held during substrate penetration (Pitz & Sharkey 2007).

Pitz & Sharkey (2007) described three unusually small new species of *Capitonius* from Central and South

Fig. 12.58 Four neotropical cenocoeliines showing characteristic habitus with high insertion of petiolate metasoma on the propodeum, and showing some common neotropical colour patterns. (Source: reproduced by permission of Yves Braet.)

America that are distinctly dorso-ventrally compressed (one having been collected from under bark) and have short ovipositors (<3 mm) that would not be long enough to reach hosts if they had to oviposit through tree bark, suggesting that this species group is adapted to searching for hosts in the narrow space between slightly loosened bark and the tree wood proper.

One of the other key characters of the group to distinguish them from some helconines with relatively highly inserted metasomas is the presence of a postpectal carina. However, this carina is almost certainly functionally associated with having to manipulate a highly inserted metasoma and therefore might not be as reliable a character as previously suspected.

Van Achterberg (1994a) included the East Palaearctic and South-East Asian genus *Ussurohelcon* in the Cenocoeliinae, in its own tribe Ussurohelconini, describing it as a 'missing link' between Cenocoeliinae and Helconinae. As with the Cenocoeliinae, *Ussurohelcon* has a relatively highly inserted metasoma, although it is not so highly inserted as in other cenocoeliines, a deep frontal depression and the presence of a trace of a postpectal carina (although again far less developed than in other cenocoeliines). In addition, it lacks the groove on the hind coxa of cenocoeliines proper, hence the morphological evidence for an association is at best weak and, as might therefore be anticipated, DNA sequence data place it among the basal 'helconine' genera (Quicke et al. 2002) or unresolved at the base of the Helconini (Sharanowski et al. 2011), in either case far removed from cenocoeliines.

A key to the world genera was provided by van Achterberg (1994a).

The Microgastroids

The name of this group is based upon the Microgastrinae, which fittingly often have a small metasoma relative to the wings and the rest of their body. For a long time, the Microgastrinae was taken to include not only the members of the subfamily as currently defined but also three tribes, Cardiochilinae, Adeliini and Miracini (Nixon 1965), which have subsequently been treated as subfamilies in their own right, although the adeliines have now been relegated to synonymy within the related Cheloninae. The initial, 'gut feel' taxonomy now appears to have been only approximately correct.

That the Microgastrinae group of subfamilies (i.e. Microgastrinae *sensu stricto*, Miracinae and Cardiochilinae) was probably closely related to the Cheloninae slowly became apparent through both morphological and molecular work and numerous studies indicate also that the Ichneutinae are derived from the base of the same lineage. Thus, Whitfield and Mason (1994), following morphological analyses by Quicke and van Achterberg (1990), additionally included the Ichneutinae, Ecnomiinae and Neoneurinae in their concept and thus in their phylogenetic analyses. However, based on molecular data, it is now clear that the last two of these are not at all close to the microgastroids and are, instead, derived members of the Euphorinae. Thus the microgastroid lineage is now fairly firmly established to comprise Ichneutinae (which may or may not be monophyletic), Dirrhopinae, Cheloninae (including Adeliinae/Adeliini), Cardiochilinae, Khoikhoiinae, Mendesellinae, Microgastrinae and Miracinae (Sharanowski et al. 2011).

Slowly, it has become apparent that the 'microgastroid' clade also includes the Dirrhopiinae, almost certainly the Ichneutinae. Several studies, i.e. by Belshaw et al. (1998, 2000), Dowton et al. (2002) and Belshaw and Quicke (2002), placed the Orgilinae at the base of this clade, but the latest study based on multiple genes (Sharanowski et al. 2011) places them among the helconoids.

The higher microgastroids, viz. Cheloninae, Miracinae, Cardiochilinae and Microgastrinae, all possess polydnaviruses that produced in the female wasps calyx gland (Fig. 4.3a,b) and released into the host where they have strong physiological effects that enable and facilitate parasitism (see Chapter 9, section *Polydnaviruses*). Most likely, given their molecular phylogenetic placements, polydnaviruses will also be discovered in the Khoikhoiinae and Mendesellinae, but to date no one has had living material of these to study and it may also be that they occur more basally in ichneutines and dirrhopiines, but I do not believe that anyone has looked in the former for them and the latter are exceedingly rare.

The relationships between various combinations of the currently recognised microgastroid subfamilies have been investigated in a number of morphological, molecular and combined studies, with little consensus. The trees obtained from a number these are summarised in Fig. 12.59. Whitfield (2002) used molecular clocks calibrated with the fossil record of braconids to place an age of 73.7 ± 10 Mya on the origin of the microgastroid/polydnavirus system and this also is a fairly good approximation to the estimated date of origin of the Ditrysia[21]. The resolution achieved by Whitfield (2002) was not very great, so an effort was made to obtain a better supported tree, and Murphy et al. (2008) presented the results of an eight gene study which partially succeeded, but still left a number of relationships without firm support. The latter investigation also included more data on fossils used for calibration and pushed the estimated age of the lineage back by nearly another 50%, to approximately 100 Mya. However, such dates must be treated with caution not only because the braconid fossil record is not well studied (and not rich at the supposed date of origin), but also because it is not yet known for certain that polydnaviruses do not occur in groups such as the Ichneutinae and Dirrhopinae.

Cardiochilinae

All cardiochilines for which host data are available are koinobiont endoparasitoids of caterpillars and attack early instars, emerging only later when they are fairly well grown and often have entered the soil ready to pupate and sometimes even after they have become prepupal (Danks et al. 1979), but emergence may also take place while the host is still on the food plant (Singh & Baldev Parshad 1970). No gregarious species are known. Hosts belong predominantly to the Noctuidae and Pyralidae with exposed caterpillars, but at least some attack leaf-mining microlepidoptera (Whitfield & Wagner 1991, Whitfield & Dangerfield 1997). It is virtually cosmopolitan in distribution but absent in the far north of the Holarctic, many species being associated with semi-arid habitats, and some are very brightly coloured and probably involved in mimicry complexes (e.g. Fig. 12.60).

The limits of several genera have been ill-defined and many physiologists have tended to confuse the New World genus *Toxoneuron* with the Old World *Cardiochiles*. A recent morphological phylogeny of the subfamily by Dangerfield et al. (1999) clarified many issues and led to the recognition of a number of new taxa. However, it is likely that when molecular

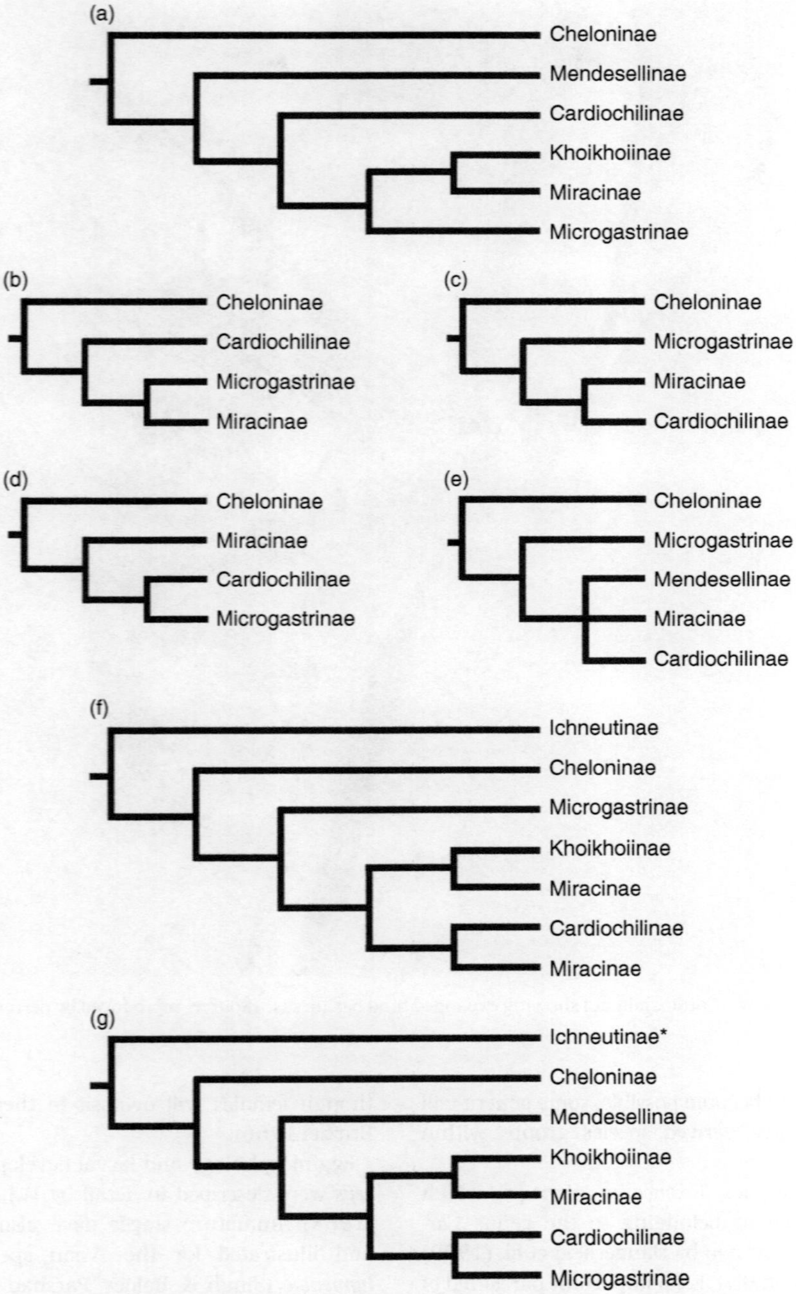

Fig. 12.59 Phylogenetic relationships of the microgastroid subfamilies as recovered by various studies. (a) Redrawn from Whitfield and Mason (1994); (b) as recovered by Whitfield (1997) based on combined molecular and morphological data, also morphological and combined morphological and molecular trees from Mardulyn and Whitfield (1999); (c) from Belshaw et al. (1998); (d) most parsimonious tree of Dowton and Austin (1998), based on 16S and 28S, and Dowton and Austin (2001), 16S, 28S and morphology; (e) strict consensus most parsimonious tree from Banks and Whitfield (2006) based on nine gene fragments of seven genes, the Bayesian majority rule tree resolved differed in having Mendesellinae+[Microgastrinae+(Miracinae+Cardiochilinae)]; (f) from Murphy et al. (2008), maximum likelihood and Bayesian trees based on seven genes but neither with strong backbone support; (g) Bayesian tree from Sharanowski et al. (2011) based on six gene fragments of four genes. * indicates paraphyletic.

Fig. 12.60 *Hartemita* sp. (Cardiochilinae) showing expanded hind basitarsus. (Source: reproduced by permission of Kyohei Watanabe.)

species level studies become possible, some genera will be found to be just derived species groups within others.

The Nearctic species *Toxoneuron nigriceps*, which used to be treated as belonging to the genus *Cardiochiles* until the revision by Dangerfield et al. (1999) (and still is occasionally), is an important parasitoid of the pest noctuid moth, the tobacco budworm, *Heliothis virescens*, and is one of the best studied of all parasitoids (Danks et al. 1979). It is fairly host specific and does not develop on the related corn earworm, *H. zea*, even though females will oviposit in them (W.J. Lewis & Brazzel 1966).

Egg morphology and larval development of *T. nigriceps* were described in detail by W.J. Lewis & Vinson (1968). Immature stages have also been described and illustrated for the Asian species *Cardiochiles hymeniae* (Singh & Baldev Parshad 1970), and the features observed are very similar between the two species. The egg is elongate curved ellipsoid and four larval instars are reported. The egg is hydropic but only increases in volume approximately 12-fold between laying and hatching. The first instar, which remains free in the host haemocoel, feeds on haemolymph and has a large head capsule relative to the body and fairly well developed mandibles. It is caudate and the cuticle is smooth. The second instar additionally develops a fairly large anal vesicle, but both vesicle and tail are largely lost in the third instar, which is the first stage in which a tracheal system is visible but spiracles are still absent. Lewis and Vinson (1968) noted that the anal vesicle of the second instar 'remains attached to the third instar larva for a period after molting', but the exact way in which this might function is not clear. The fourth instar has an open spiracular system and egresses from the host to continue feeding from an external position by reinserting its head into the host and then commensing tissue feeding (Kuriachan et al. 2011), although unlike what these authors assert, a final external feeding phase is more the norm for non-cyclostome braconids rather than it being that unusual. The cocoons are rather rough, longitudinally ridged, with a dark median band and the exit hole chewed by the emerging wasp is very ragged. The tropical dry forest species *Hansonia chavarriai* spends a prolonged period (>6 months) in its extremely hard cocoon among the leaf litter as a larva or pupa (Dangerfield et al. 1996).

Courtship and mating were described by Vinson (1978). Details of the antennal sensilla of *T. nigriceps* were described by Norton and Vinson (1974). Females release pheromones that initiate male courtship behaviour and the chemistry of these has only been partially elucidated (Syvertsen et al. 1995).

There have been many investigations of its polydnaviruses and teratocytes. It is also one of the few koinobiont endoparasitoid species where substantial (and hard-won) progress has been made on *in vitro* culture (Pennacchio et al. 1992a), because of the intimate relationship between its natural host physiology and its own developmental requirements. Having said that, it has still not been reared all the way from egg to adult *in vitro* and media require the incorporation of as yet unidentified host factors and pupation requires a suitable chamber as the larvae cannot spin a cocoon successfully on an open surface (Henderson et al. 2011).

Important work on identification was carried out by Dangerfield et al. (1999), who presented a morphological phylogeny of the group with a substantive set of exemplar species and a key to world genera. Whitfield and Dangerfield (1997) provided a key to the New World genera. The first of these works recovered two large groupings of genera and found that the Neotropical genus *Heteropteron*, which is comprised of rather large wasps, was basal to both of these. Mercado and Wharton (2003) discussed relationships between species in *Toxoneuron* and *Retusigaster*.

Cheloninae (Including Adeliini)

This is a moderately large and taxonomically rather difficult subfamily with some 1300 described species in 15 genera (Kittel & Austin 2014). The concept of this important subfamily became fixed fairly early on when it was realised that not all taxa with a carapace-like metasoma were closely related. Thus sigalphines and various genera of Helconinae *s.l.* that previously had been treated together were excluded, leaving the Cheloninae with three very large cosmopolitan genera [*Chelonus* (including *Microchelonus*), *Ascogaster* and *Phanerotoma*] and a number of smaller ones, sometimes with restricted distributions. In this long-standing sense, the subfamily keys out easily, the included species possessing the almost unique[22] combination of a carapace formed of three fused metasomal tergites and a transverse posterior carina on the mesopleuron (van Achterberg 1976c, 1993a). Several species have been investigated for the biological control of pest species. Sixteen million individuals of *C. blackburni* originating from Hawai'i were bred and released in Texas between 1971 and 1974 in an attempt to control a major cotton pest, the pink bollworm, *Pectinophora gossypiella* (Gelechiidae), after previous failures to get it to establish in the 1930s. It is now established in various parts of the southern USA and releases in India in 1976 to control *Helicoverpa armigera* on cotton led to its establishment there also, where it has effected significant host population control and it has subsequently been found to have colonised *Pectenophora gossypiella*, *Easias* spp. and *Phthorimaea operculella* there.

All chelonines are solitary egg-larval parasitoids of Lepidoptera and have extremely slender, pointed ovipositors. Superparasitism is apparently rare (Ullyett 1949, Kawakami & Kainoh 1985) but does occur (Abdel-Malek 1947, Broodryk 1969a), and can be induced when batches of host eggs are offered to multiple females (Lanzrein et al. 2001) but only one wasp survives to emerge.

Immature stages and development have been described for *Ascogaster reticulatus* (Kawakami 1985), *Chelonus annulipes* (Vance 1932a), *C. blackburni* (C.G. Jackson et al. 1978), *C. inanitus* (Hafez et al.

1980, Grossniklaus-Bürgin et al. 1994), *C. narayani* (Narayanan et al. 1961), *C.* sp. nr. *curvimaculatus* (Hentz et al. 1997), *Phanerotoma toreutae* (Harbo & Kraft 1969) and *P. hendescasiella* (Peter & David 1992b). The eggs are small, seldom more than 0.2 mm long, but only slightly hydropic (Broodryk 1969a). The number of larval instars reported is either three (Vance 1932a, Abdel-Malek 1947, Broodryk 1969a, C.G. Jackson et al. 1978, Peter & David 1992b) or four (Harbo & Kraft 1969, Hentz et al. 1997). Vance's study described larval internal anatomy including musculature in great detail. The first instar larva is mandibulate and vesiculate but very newly emerged ones are caudate, the anal vesicle developing as the instar progresses (Narayanan et al. 1961, Broodryk 1969a). Upon eclosing from the chorion, serosal cells still remain attached for a while. Moulting to the second instar does not take place until the host moults to its final instar (Bradley & Arbuthnot 1938, Broodryk 1969a), and the tracheal system first appears at this stage although with the spiracle closed. The final instar larva emerges from the host head first and continues to feed externally for a while before spinning a thin, papery cocoon, in which the meconium is voided before pupation (Broodryk 1969a, C.G. Jackson et al. 1978). Egress from the host in *C. narayani* is reported to take a long time (10–12 h) and takes place after moult to the third instar, unlike in microgastrines in which the moult occurs as the parasitoid larva is emerging through the host integument (Narayanan et al. 1961; cf. Nakamatsu et al. 2006, 2007). Vance (1932a) showed that if a larva that is just removed from the host is prevented from feeding externally, then it is unable to spin a cocoon. Studied species seem to show very tight developmental synchrony with their hosts and this even extends to hosts that may enter seasonal or multi-annual diapause (Bradley & Arbuthnot 1938, Marais 1955).

Whether or not the age of the host egg is important for the successful development of the chelonine probably varies between species, with some studies indicating that a particular stage either early (e.g. Rosenberg 1934) or late (e.g. Ullyett 1949) is required. Some species have a set of strategies for coping successfully with eggs all the way from freshly laid to just about to emerge. Hawlitzky (1972) studied in detail how the larva of *Phanerotoma* gets into the host larva. If the *Phanerotoma* egg is laid in the host egg's yolk, it hatches fairly quickly and after 3–5 days it moves into the haemocoel of the developing host embryo in the region where the mid-gut will finally form, but at this stage the host's fore and hind guts are not connected. The same wasp will also lay in older host eggs (Hawlitzky 1979), in which case the egg is deposited straight into the host embryo's haemocoel. Kaeslin et al. (2005a) described the same host penetration for *Chelonus inanitus*, so it may be assumed to be a widespread feature of the subfamily.

Phanerotoma leucobasis (as *flavitestacea*), is one of the few parasitoids to have had its precise mode of larval feeding investigated. Hawlitzky and Boulay (1974), using histological methodology, showed that the endoparasitic first instar feeds only on host haemolymph whilst the ectophagous third instar consumes all host organs in no particular order. The consumed host tissue was fully digested by the time of parasitoid pupation. Once in the host haemocoel, chelonine larvae remain as a first instar for a protracted period, although host development is often curtailed and the host may be caused to enter a prepupal state prematurely (Grosniklaus-Bürgin et al. 1994). Moulting to the second instar occurs as a result of host physiological cues associated with the onset of pupation (Harbo & Kraft 1969), and development is completed with killing of the host only after it has entered its pupation retreat or constructed its cocoon.

Since eggs do not normally give off many cues that an ovolarval parasitoid might use to locate them, chelonines seem largely to rely on detecting kairomones left by the female host moth at the oviposition site, including wing scales (Chiri & Legner 1986, DeLury et al. 1999a) and/or female excretory products and abdominal materials (Vinson 1975, Kainoh et al. 1982). In several cases, these compounds have been shown to be heat stable, non-volatile and water soluble, although DeLury et al. identified a number of volatile compounds from scales of the codling moth, *Cydia pomonella*, which attracted, as a blend, *Ascogaster reticulatus*. The combination of volatile and non-volatile kairomones indicates that there is a multi-stage process, with initial attraction caused by the former and then more detailed host egg location probably involving antennal contact and gustatory sensilla. Once eggs have been located, Kainoh et al. (1989) showed that actual oviposition by the parasitoid required chemical cues from within the host egg. Various amino acids were important and it appears that it is a mixture of amino acids that is critical rather than any single one.

Courtship behaviour has been described for *Ascogaster reticulatus* by Kamano et al. (1989), who showed that a female produced a potent hexane-soluble pheromone and that this induced the males, which searched for mates by walking over the substrate, to follow a trail that would lead to a potential mate. Once the male had located a mate, antennal contact was then important for mating to occur. Kainoh and Oishi (1993), by a process of sequential dissection of female wasps and measuring male responsiveness, demonstrated that the female sex pheromone was produced in the hind tibia. As the hind tibia is fairly enlarged and robust in all chelonines, it seems likely that it will be found to be the source of sex pheromones in most or all other species when they are investigated. Subsequently, DeLury et al. (1999b) identified the sex pheromone of *Ascogaster reticulatus* as the unsaturated aldehyde (Z,Z)-9,12-octadecadienal. Gothilf (1969) also briefly described courtship in *P. leucobasis* (as *flavitestacea*) and noted that males mate several times, copulation lasting only about 10 s, but unfortunately no records of numbers of female matings were made, although Broodryk (1969a) noted that females of *C. curvimaculatus* will mate twice in quick succession. S.R. Shaw (1991) reported on lekking behaviour by males of *Chelonus* (*Microchelonus*) *hadrogaster*, where they formed aggregations close to the mounds of the ant *Formica podzolica*. Males of the *Chelonus scrobiculatus* species group have a pair of putative metasomal gland openings at pores postero-laterally on the carapace (Huddleston & Walker 1994) and males of most species of *Chelonus* (*Microchelonus*) have a setose pit at the posterior of the metasoma, which is presumably associated with a gland, but no-one seems to have looked.

Female internal reproductive anatomy has been described by Moreno and Jiménez (1993) for *Phanerotoma ocularis*. As with all microgastroids, there are just two, meroistic ovarioles per ovary and each, at least in recently eclosed adults, has a large germanium anteriorly and a long medial portion where eggs are maturing with clusters of nurse cells attached to them anteriorly. The ovaries have a common lateral oviduct which is very swollen, forming a uterus-like chamber which stores many mature eggs, and is followed immediately by a morphologically distinct calyx gland. Females emerge with a proportion of their eggs fully mature and lifetime fecundity can be very high, Broodryk (1969a) reporting values from 112 to a startling 1537 eggs laid for *C. curvimaculatus*, and C.G. Jackson et al. (1978) reported a mean of 909 eggs per female of the thelytokous *C. blackburni*. A second thelytokous, or perhaps more precisely deterotokous, species is *C. sulcatus* (Abdel-Malek 1947).

Parasitisation by chelonines often induces the host prematurely to enter a prepupal stage, at which point its development is arrested (D. Jones 1987, Soller & Lanzrein 1996, Kaeslin et al. 2005a), and *Chelonus inanitus* causes its *Spodoptera littoralis* host to enter the soil during its fourth instar rather than in the normal sixth instar (Rechav & Orion 1975). Further, at least in the case of *Ascogaster* species, parasitisation leads to host castration (Reed-Larsen & Brown 1990, Brown & Kainoh 1992). All chelonines produce polydnavirus particles in their calyx glands, and those of two species, *Chelonus inanitus* and *Ascogaster reticulatus*, have been the subjects of many investigations (see Chapter 9, section *Polydnaviruses*).

Brown et al. (1993) showed that the larvae of *A. reticulatus* actively secreted ecdysteroids (20-hydroxyecdysone and later ecdysone) when placed in culture medium, suggesting that at least part of the host's altered development might be due to directly influencing host haemolymph hormone levels.

One might therefore have thought that the Cheloninae proper was one of the best characterised subfamilies in the whole of the Braconidae. Few things are so simple, however, because another, previously recognised subfamily, the Adeliinae (Fig. 12.61d,e), with only four genera, now appears, based on DNA sequence data, to be derived from within the Cheloninae and perhaps more specifically from within the Phanerotomini (e.g. fig. 5 in Dowton & Austin 1998, Belshaw et al. 2000, Dowton et al. 2002). The Adeliinae *sensu* Authors share with typical chelonines the fusion between the first two metasomal tergites, but the metasoma is not generally carapace like and the posterior metapleural carina is lacking (Fig. 12.61d). Adeliines have very similar wing venation to that of the Miracinae in that fore wing vein r-rs is missing or very reduced (Fig. 12.61d,e). Adeliines also appear to be exclusively parasitoids of Nepticulidae (Lepidoptera) although, while the possibility that other leaf-miners are attacked cannot be excluded (Whitfield & Wagner 1991), Shaw & Huddleston (1991) noted that larvae of many nepticulids search out pupation sites such as other leaf-mines, and this could easily account for records from non-nepticulid hosts 'at least in Britain'. Shaw & Huddleston (1991), referring to Čapek (personal communication), intimated that oviposition by adeliines must be into the egg stage of

Fig. 12.61 Various members of the Cheloninae. (a) *Ascogaster* sp., showing complete lack of transverse metasomal sutures; (b) *Phanerotomella* sp., metasoma dorsal view; (c) *Phanerotoma* sp.; (d, e) Adeliini spp. (Sources: a, b, d, e, reproduced by permission of Rebecca Kittle; c, reproduced by permission of Kaoru Maetô).

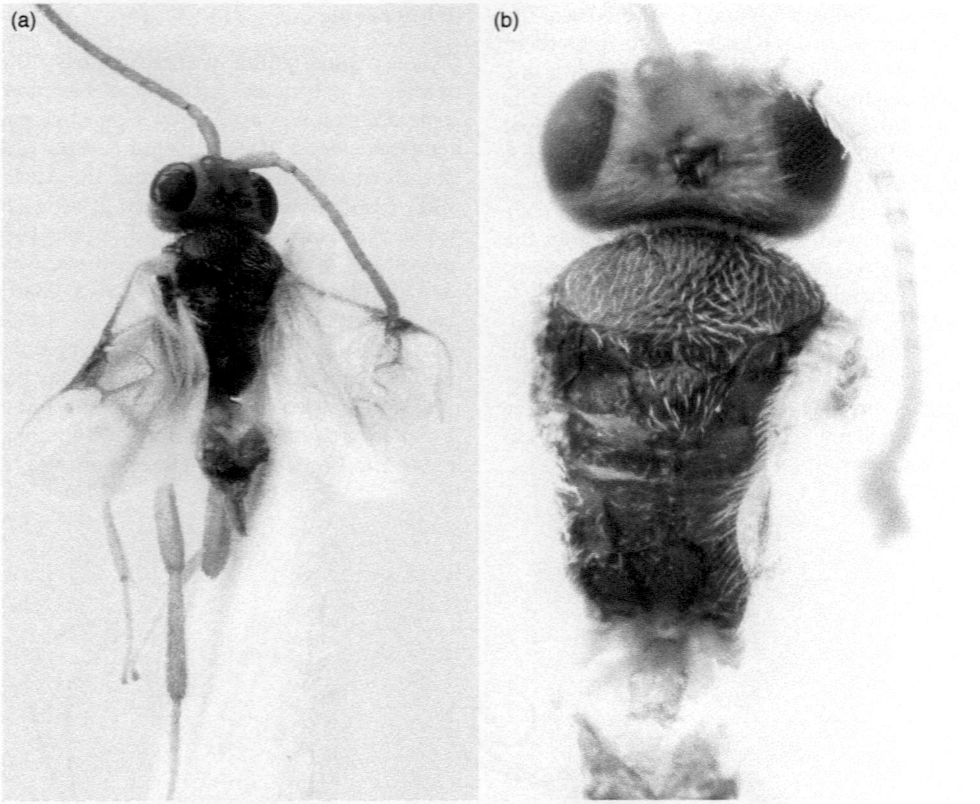

Fig. 12.62 *Dirrhope* sp. (Dirrhopinae) from Panama. (a) Dorsal view of habitus and (b) detail of mesosoma; note the areolate propodeum. (Source: reproduced by permission, © C. van Achterberg, Naturalis Biodiversity Centre.)

the host, which would be completely consistent with them being derived chelonines.

Despite being koinobionts, some species are fairly long-lived as adults and Gothilf (1969) reported that *P. leucobasis* (as *flavitestacea*) males and females, under laboratory rearing conditions and supplied with water and carbohydrate food, have 50% survival lengths of 70 and 97 days, respectively, and that females may live for up to 9 months. Such longevity might be associated with having to locate widely dispersed host eggs.

The genera were revised by Zettel (1990) and the Australian fauna, with the addition of two new genera, by Kittel and Austin (2014).

Dirrhopinae

This is a small subfamily of tiny wasps (Fig. 12.62) that are only very rarely collected. It was only afforded subfamily status relatively recently (van Achterberg 1984a), although its previous inclusion in the Microgastrinae had been questioned by both Nixon (1965) and W.R.M. Mason (1981a). It has also been placed, following break-up of the Microgastrinae *s.l.*, with the Miracinae, largely because of similarities in their wing venation (Fig. 12.62), despite the fact that it has the spiracles of the first metasomal segment in the notum rather than the epipleuron (laterotergite), a character that unites the core microgastroids. Species or specimens are known from the Palaearctic extending into South-East Asia (Vietnam), North and Central America, South Africa and the Australasian Region (Nixon 1965, Belokobylskij et al. 2003), but it may well turn out to be cosmopolitan. Unfortunately, no specimens have been available as yet for molecular study, so it is not included in any of the trees proposed for the microgastroid complex (Fig. 12.59), but on purely morphological evidence it is usually recovered near the base of the microgastroids (Quicke & van Achterberg 1990, Whitfield & Mason 1994). The subfamily is one of only eight that have a fossil record dating back to the Cretaceous (Russia, Taimyr amber, 83.5–85.8 Mya) (Belokobylskij 2012), which again is concordant with it occupying a basal position to the other microgastroids.

Biological information is only available for one North American species which was reared from the leaf mining phleophagan chestnut moth, *Ectoedemia phleophaga* (Nepticulidae) (Shaw & Huddleston 1991, Belokobylskij et al. 2003). No observations on oviposition or juvenile stages have been made but their apparent phylogenetic position near the base of the microgastroids strongly indicates that they will be found to be koinobiont endoparasitoids and possibly to have polydnaviruses.

Ichneutinae

This is a small group of approximately 90 described species that includes a couple of common tropical genera plus a few, apparently somewhat rarer, north temperate ones. The 'Ichneutini' (Fig. 12.63a) are parasitoids of sawfly larvae and the Muesebeckiini (Fig. 12.63b), erected by W.R.M. Mason (1969) for a few very small and principally tropical wasps, are parasitoids of lepidopteran leaf-miners, Nepticulidae and possibly Bucculatricidae (W.R.M. Mason 1969, van Achterberg 1976c). Gauld (1988a) suggested that parasitism of sawfly larvae, a very unusual trait in the Braconidae, probably arose in this group from parasitism of leaf-mining Lepidoptera, but as far as I am aware, no records from leaf-mining sawflies exist. The Proteropini separated from the Ichneutini

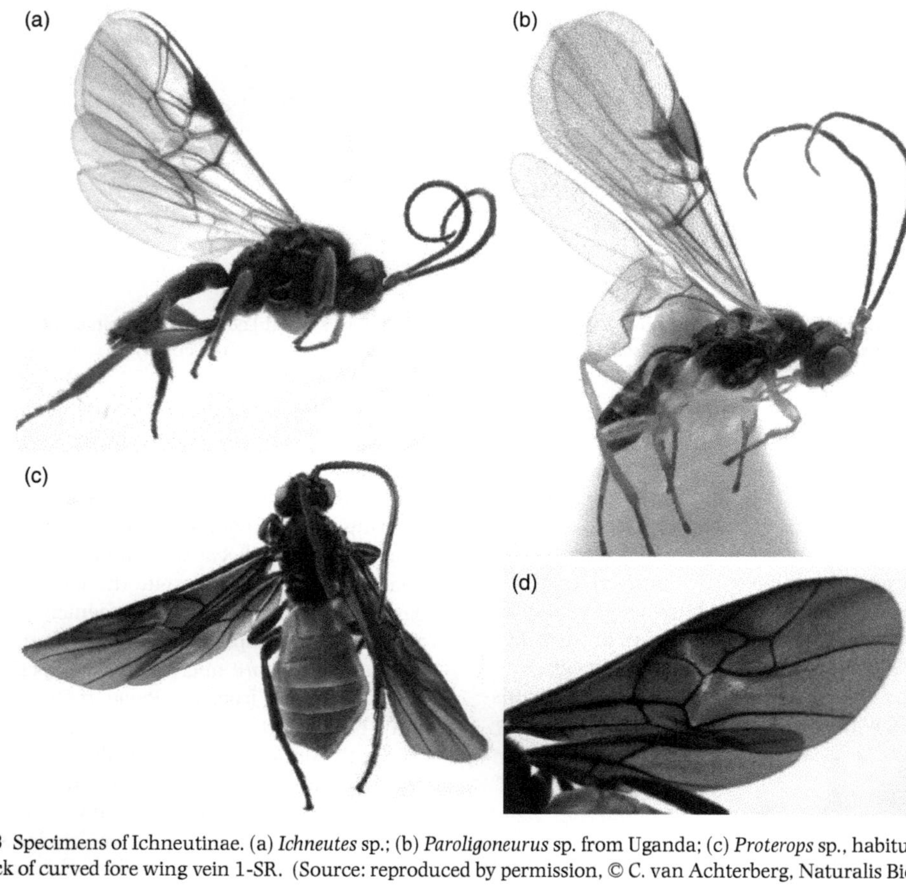

Fig. 12.63 Specimens of Ichneutinae. (a) *Ichneutes* sp.; (b) *Paroligoneurus* sp. from Uganda; (c) *Proterops* sp., habitus, wing detail showing lack of curved fore wing vein 1-SR. (Source: reproduced by permission, © C. van Achterberg, Naturalis Biodiversity Centre.)

by van Achterberg (1976c) are also parasitoids of sawfly larvae. In their traditional sense, the Ichneutini and Muesebeckiini share a strongly curved fore wing vein 1RS (Fig. 12.63a,b) which is lacking in *Proterops* (Fig. 12.63c,d), leaving the subfamily rather difficult to diagnose. Sharkey and Wharton (1994) proposed that the presence of one or more apical spines on the fore tibia was at least a moderately convincing synapomorphy for the group and, in the past, some workers (e.g. Tobias & Belokobylskij 1981) thought that the *Oligoneurus* group (i.e. Muesebeckiini) were related to the Miracinae, but Nixon (1965), in his preliminary splitting of the Microgastrinae, rightly considered their resemblance to Miracines as being due only to the rather weak character of reduced wing venation associated with small body size. Molecular analyses have confirmed that they are at least closely related to the Ichneutini (Sharanowski et al. 2011).

Monophyly of the Ichneutinae is still somewhat questionable (Whitfield & Mason 1994). Neither Belshaw et al. (2000) nor Sharanowski et al. (2011) recovered them as monophyletic in their molecular studies; in both cases they appeared as a grade taxon leading to the microgastroid complex, and similarly they sometimes formed a grade taxon in morphological analyses (Quicke & van Achterberg 1990) although they were recovered as monophyletic by Whitfield and Mason (1994). The comparative rarity of many ichneutine genera means that molecular analyses to date have not included many representatives and therefore monophyly has not been properly tested. It is often the case that expected groups that fail to be recovered as monophyletic when few representative taxa are included, do 'become' monophyletic with better taxonomic representation which helps to break long branches.

The relationships of the Ichneutinae have been the subject of considerable discussion. Several workers and studies have shown similarities between ichneutines and microgastroids; for example, Vinson (1969) noticed the considerable similarity between the ovaries of *Ichneutes* and those of the cardiochiline *Toxoneuron nigriceps*, and Iwata's (1959) illustration of the ovaries of *Proterops* shows a fairly swollen, two-part lateral oviduct which could indicate a fairly well-developed calyx gland. Sharkey and Wharton (1994) paid particular attention to the presence of a pair of submedial pits on either side of the pronotum (the subpronopes) and suggested that these indicated an affinity with the Agathidinae, but this is now disproven.

Further, morphology-based phylogenetic analyses and the discovery of a number of additional Neotropical genera suggest that the division into distinct tribes is not really warranted (Sharkey & Wharton 1994; see also Sharkey 1996). Such studies, however, are potentially flawed because of inadequate character and taxon coverage (all the new genera recognised by Sharkey and Wharton being neotropical, and it is possible that this reflects lower sampling in the Old World tropics) and the use of a hypothetical outgroup, which means that the rooting of the tree obtained may be far from secure. That this might be the case is apparent from Sharanowski's tree, which included *Muesonia*, the second most basal taxon in Sharkey and Wharton's study, but which based on DNA appeared more derived as the sister group of *Proterops* with >95% support. Given this and the absence of host records for all but one of the new genera described by Sharkey and Wharton, it would be unwise to speculate too much about evolutionary transitions at the present time.

The very limited biological data for temperate species were well summarised by Shaw and Huddleston (1991). *Ichneutes* are solitary koinobiont endoparasitoids of Tenthrdinidae. some are ovo-larval parasitoids of sawflies (Nägeli 1936, Zinnert 1969), while others attack only early instar host larvae. They pupate after the host has constructed its cocoon, spinning their own cocoon within that of the host. Overwintering may be as a pupa or as a diapausing first instar larva within the still prepupal host. *Proterops* appears to specialise on Argidae (van Achterberg & Desmier de Chenon 2009) and the Neotropical genus *Masonbeckia* has been reared from a sawfly that might also have been an argiid (Sharkey & Wharton 1994).

Members of several genera have conspicuous male tergal glands opening at one or two pores on metasomal tergite 7 (*Ichneutes*, *Paroligoneurus*) or tergites 6 and 7 (*Oligoneurus*) (W.R.M. Mason 1991, Sharkey & Wharton 1994), These glands differ from other tergal glands apart from those of the rogadine *Aleiodes* (*Hemigyroneuron*) in that they open at the posterior of the tergites rather than basally (Buckingham & Sharkey 1988). This difference could well be reflected in courtship posture or calling.

Two species of *Paroligoneurus* (W.R.M. Mason 1991; described originally as constituting the separate genus as *Anaprixia*) are exodont and, together with the aberrant gnamptodontine *Exodontiella*, are unique among the Ichneumonoidea in having spiracles only present on alternate metasomal tergites[23]. Females of some genera have a large gap between placoid antennal sensilla on their median flagellomeres where there are a cluster of specialised 'setae' (W.R.M. Mason 1991, Sharkey & Wharton 1994), but whether these are involved in host location or acceptance or courtship is unknown.

Khoikhoiinae

This is one of the most recently described subfamilies (Fig. 12.64a,b) (W.R.M. Mason 1983) that so far is known from only a handful of species in two genera, all from the Cape Province of South Africa, and all but one species only from the Western Cape. It seems almost unbelievable that a group of mobile organisms with several species can be restricted to a very small region, but this appears to be the case with the khoikhoines – perhaps it is somehow related to the Cape Floral Province?

Sharkey et al. (2009a) revised the known species and based on morphology (long, thin and sharply pointed ovipositor) and phylogenetic relationships, speculated that at least some species may be egg–larval parasitoids of wood or stem-boring Lepidoptera, although to date there are no known host records.

Mendesellinae

Willim (Bill) Mason spotted a number of Neotropical braconid wasp specimens in the accessions of the Canadian National Insect Collection in Ottawa that did not fit into any of the existing microgastroid subfamilies, and after his death in 1991, this group was formally described as the Mendesellinae with two included genera, *Epsilogaster* (Fig. 12.64c,d) and *Mendesella* (Whitfield & Mason 1994). These are apparently not very common wasps and, despite extensive rearing programmes in Costa Rica, only one host record is available; *Epsilogaster bicolor* having been reared from stem-galling momphid (Lepidoptera) (Whitfield & Mason 1994). The absence of other host records therefore almost certainly reflects the rearing programmes themselves, which are strongly biased towards large exophytic and easily collected and reared hosts. Whereas the ovipositor of *Epsilogaster* is relatively simple, that of *Mendesella* is distinctly wavy at the apex (as in some opiines, cremastines and tersilochines), and this is no doubt part of a mechanism enabling at least the tip to be steered within the host's confines (see Chapter 3, section *Ovipositor steering mechanisms*). Unlike Cardiochilinae, Khoikhoiinae, Microgastrinae and Miracinae, mendeselines have the first metasomal spiracle situated in the tergum rather than in the laterotergite (or membranous area lateral to the tergum) and therefore are likely to be basal to that clade, as is confirmed by purely molecular studies (Fig. 12.59c–g).

Microgastrinae

This is a huge and challenging group and one which has tended to have its own taxonomic specialists. All species for which reliable host records exist (and in the case of this subfamily that still a lot), are koinobiont endoparasitoids of Lepidoptera larvae, principally members of the more 'advanced' Ditrysia, with the single exception of a species, thus far known only from The Netherlands, which is a parasitoid of the larva of a terrestrial Limnophilid caddis fly (Trichoptera) (van Achterberg 2002). This latter case is perhaps not so surprising given that the extant Trichoptera and Lepidoptera are sister groups and the two are frequently united in one order, the Amphismenoptera.

Far more species of microgastrines have been investigated for or used in biological control than those of any other braconid subfamily (Whitfield 1997b). Some of the most important and successful biological control associations include the following, but the list is far from exhaustive: *Apanteles fumiferanae* on spruce budworm, *Choristoneura fumiferanae* (Tortricidae); *Cotesia flavipes* and its close relatives (*chilonis*, *sesamiae* and *nonagriae*) which attack a wide range of pests of sugarcane including *Chilo partellus*, *C. sacchariphagus* and *Diatraea saccharalis*, *flavipes* itself originating from Asia but having been widely introduced in Africa and the Americas; *Cotesia marginiventris* against beet armyworm, *Spodoptera exigua* (Noctuidae) and green cloverworm, *Hypena scabra* (Erebidae); *C. melanoscela* on gypsy moth, *Lymantria dispar* (Lymantriidae); *Microplitis mediator* against *Mamestra brassicae* and *M. croceipes* against the polyphagous *Heliocoverpa zea*, *Heliothis subflexa* and *H. virescens* (all Noctuidae); *Pholetesor circumscriptus* has been widely used against a range of leaf-miners and *P. pedias* on apple leaf miner.

To date, approximately 2000 species have been formally described (D.S. Yu 2012), but various objective estimates suggested that this might only be 50% of the true number (Dolphin & Quicke 2001, O.R. Jones et al. 2009). More recently, Rodriguez et al. (2013), based on a combination of DNA barcodes, revealed biodiversity,

Fig. 12.64 Specimens of Khoikhoinae and Mendesellinae. (a, b) *Sania masneri*, dorsal and lateral habitus, respectively; (c, d) *Epsilogaster* sp., habitus and detail of metasomal tergites, respectively. (Sources: a, b, reproduced by permission of Emily Tucker; c, d, reproduced by permission, © C. van Achterberg, Naturalis Biodiversity Centre.)

notably from the caterpillar rearing programme in Cost Rica, and applying a focal taxon approach suggested that there may be between 16,000 and 40,000 microgastrine species worldwide. This discrepancy in part reflects their generally small size and the fact that many closely related species are extremely hard to distinguish morphologically and may go unsuspected in the absence of either rearing/breeding data or barcodes. Even the Western Palaearctic fauna, which has been extensively and repeatedly revised, based largely on morphology, includes many undescribed species and a lot of these have fairly narrow host ranges (Kankare et al. 2005a,b, M.R. Shaw et al. in preparation). When it comes to understanding host relationships, particularly of the tropical fauna, barcodes are likely to be about the only way forward as few workers are likely to be able to apply careful morphological analysis to such a diverse group, and recently 20,000 barcode sequences were released to facilitate future taxonomic and evolutionary studies (M.A. Smith et al. 2012).

Microgastrines are the most numerous and frequently collected braconids as well as the group most involved in biological control. Taxonomically, the European and to a lesser extent the Nearctic faunas have been comparatively well studied, but the tropical fauna is virtually untouched save for a very few distinctive genera from the Indo-Australian region and Costa Rica and a few species groups of economic importance.

The subfamily has provided somewhat of a case study for resolving genus-level relationships in a group that almost certainly underwent a rapid radiation a long time ago (Banks & Whitfield 2006). Nixon (1965) made the first substantial attempt to understand relationships within the Microgastrinae (then treated as the tribe Microgastrini because the subfamily at that time also incorporated Miracinae and Cardiochilinae). In Nixon's system, the huge cosmopolitan genus '*Apanteles*' was split into a number of species groups. W.R.M. Mason (1981a) took the process further by presenting explicitly cladistic argumentation. First, he restricted the subfamily to its present concept, i.e. excluding Cardiochilinae and Miracinae, and demonstrated that within the Microgastrinae there had been multiple independent migrations of fore wing vein r-m towards 2RS, which results in progressively smaller second submarginal cells and ultimately in loss of that cell. This was of consequence because the genus *Apanteles* had been characterised by the absence of vein r-m (see Fig. 12.71a) and therefore was actually a polyphyletic assemblage, with many of the monophyletic units comprising it corresponding to Nixon's species groups. Mason postulated a tribal system comprising Apantelini, Cotesiini, Forniciini, Microgastrini and Microplitini; he also erected numerous new genera, many to receive groups of *Apanteles sensu* Authors, with the consequence that many familiar species from a biological and biocontrol point of view were placed in other genera and, in particular, in the very species-rich genus *Cotesia*. In total, Mason recognised 50 genera within the subfamily, a very considerable increase from the 20 that Nixon (1965) did. The genus *Apanteles* still exists and is still fairly large, but not many of the species that it includes have been worked on extensively biologically. Whitfield et al. (2002) provided a neat graphical illustration of the changes that the generic classification have undergone, from only one genus in 1802 to 54 in 2002.

Unfortunately, Mason did not carry out a formal analysis of his data and his work was largely restricted to the Nearctic species with which he was most familiar and to which he had ready access. Not surprisingly, some of his conclusions were not liked or approved of by some other workers, which led A.K. Walker et al. (1990a) to codify Mason's characters and conduct a parsimony analysis on them. The resulting trees were not very well supported and only two of Mason's proposed groups were recovered, one largely composed of taxa with the macrolepidoptera suite of characters (see below) and thus potentially a result of morphological convergence. Further, some of the genera included in the analyses may not themselves have been monophyletic, so Walker et al. did not consider their results sufficient to warrant changing the existing tribal classification. Maetô (1996) carried out a detailed study of male genitalia in the subfamily, which broadly supported Mason's recognition of two major groups, but monophylies of his tribes was not supported. Subsequent authors have been divided over how far to follow Mason's generic concepts and some, such as Tobias (1986c), largely rejected them. Van Achterberg (2003) presented a much toned-down view, based on his consideration of the West Palaearctic fauna in recognising only four of the genera proposed by Mason as occurring there. The most recent molecular studies have failed to address the issue fully.

Placement of *Microplitis* is particularly interesting. This genus, along with the closely related *Snellenius*, is partially characterised by having a relatively large, distinctly four-sided fore wing second submarginal cell, which would tend to be regarded as plesiomorphic[24]. However, *Microplitis* species attack relatively derived groups of Lepidoptera, such as Sphingidae, and Shaw & Huddleston (1991) considered them one of the most 'advanced' microgastrines. They display the so-called macrolepidoptera suite of characters (W.R.M. Mason 1981a, A.K. Walker et al. 1990a), which would be consistent with them being a derived group if there was a degree of phylogenetic congruence between the microgastrines and their hosts at genus level. In Walker et al.'s analyses it was recovered as a derived clade, whereas in van Achterberg's it was the sister group to the rest of the subfamily. Indeed, all current evidence points to them being the sister group of the remaining microgastrines.

Whitfield (1997a) in his early molecular analysis of microgastroids did not present trees showing the placement of *Microplitis*. Dowton and Austin (1998) recovered it basal with 28S rDNA but derived with 16S mtDNA data. Mardulyn and Whitfield (1999) extended the earlier studies using three gene fragments but found only a few well-supported clades and many short internal branches likely due to the degree of host specialisation which would probably lead to high speciation rates, although as there's was more or less a study of genus-level relationships the implication must be that the genera themselves mostly also originated during a short period of history. Their separate and combined phylogenetic analyses for their gene fragments and a morphological data set found strong conflict between morphology and DNA very much as with analyses of the subfamilies of Braconidae (Quicke & Belshaw 1999). All three genes placed *Microplitis* as sister group to the rest of the subfamily, but morphology recovered then as derived and they shifted from their most basal position in the combined analysis. There can be little doubt that the morphological data are phylogenetically misleading as a result of convergence of characters associated with parasitism of 'macrolepidoptera'. Whitfield et al. (2002) provided a more detailed comparison of the phylogenetic results from three gene fragments and morphology, using both maximum parsimony and maximum likelihood, but the results were again inconclusive because of conflict. The authors examined how well each of four morphological character partitions fitted on to the best trees obtained from analyses of morphology alone, sequence data alone and combined data. The best performing (i.e. most influential) characters on the morphological tree were from the ovipositor mechanism, whereas these performed worst on both molecular and combined trees. Final instar larval features did about equally well on all trees.

More comprehensive studies by Banks and Whitfield (2006) and Murphy et al. (2008) using 6 and 8 gene fragments respectively, concur on the basal position of *Microplitis* (and *Snellenius* when included).

A completely independent study based on the bulk amino acid compositions of cocoon silks seems to add considerable support to the idea that *Microplitis* is indeed basal within the subfamily (Quicke et al. 2004b). Most non-cyclostome braconds have fibroin-type cocoon silk with high relative abundances of alanine, serine or glycine (of which either alanine or serine was the most abundant) and this is true also of members of the genus *Microplitis*. However, the remaining microgastrines examined have a very unusual, apomorphic silk amino acid composition with aspartate/asparagine comprising 32.4–50.5 mol% and glycine less than 10 mol%, indicating a structurally different fibroin-type protein that they termed xenofibron (Yamada et al. 2003, 2004). Fibroin proteins usually comprise layers of antiparallel beta-pleated sheets (beta sheets) with repeated short sequences, of short side-chain amino acids held together by hydrogen bonds. Given that xenofibron is likely to be a synapomorphy for most genera of microgastrines, the question arises as to why they evolved such a marked chemical shift, and whether it reflects host amino acid availability at all.

Life history and larval stages have been described for numerous species, which reflects largely their role in attacking economically important pests: *Apanteles medicaginis* (W.W. Allen 1952); *Cotesia angaleti* (Narayanan et al. 1956); *C.* (as *Apanteles*) *marginiventris* (Boling & Pitr 1970); *C. flavipes* (Pinheiro et al. 2010); *C. vestalis* (R.-X. Yu et al. 2008); *C. yakutatensis* (Madar & Miller 1983); *Microplitis croceipes* (W.J. Lewis 1970); *M. mediator* (Arthur & Mason 1986); *Microplitis rufiventris* (Altahtawy et al. 1972, Tawfik et al. 1980, Hegazi & Führer 1985).

Most microgastrines oviposit into recently hatched or early instar host larvae, but a few are ovo-larval parasitoids (e.g. *Cotesia hyphantriae*, Tadic 1958) attacking well-developed embryos and some are facultatively so (e.g. *C. marginiventris*: Ruberson & Whitfield 1996), and some species can successfully parasitise older or even the whole range of host larval instars. The eggs are hydropic and swell considerably before hatching (Vance 1931, Fallis 1942, Peter & David 1992a, Arthur & Mason 1986). Extraembryonic membranes of *Cotesia congregata* were described by Beckage and DeBuron (1994). Hosts are exophytic or weakly concealed when attacked, although a few, such as sesiids, may bore deeper into their host plants as they develop. There appear to be three larval instars in most species (Puttler & Thewke 1970, Arthur & Mason 1986, Jugovits 1987, Shaw & Huddleston 1991), although Hegazi and Führer (1985) and Strand et al. (1988) reported four for some *Microplitis* species and K. Porter (1983) reported four for *Cotesia* (as *Apanteles*) *bignellii*,

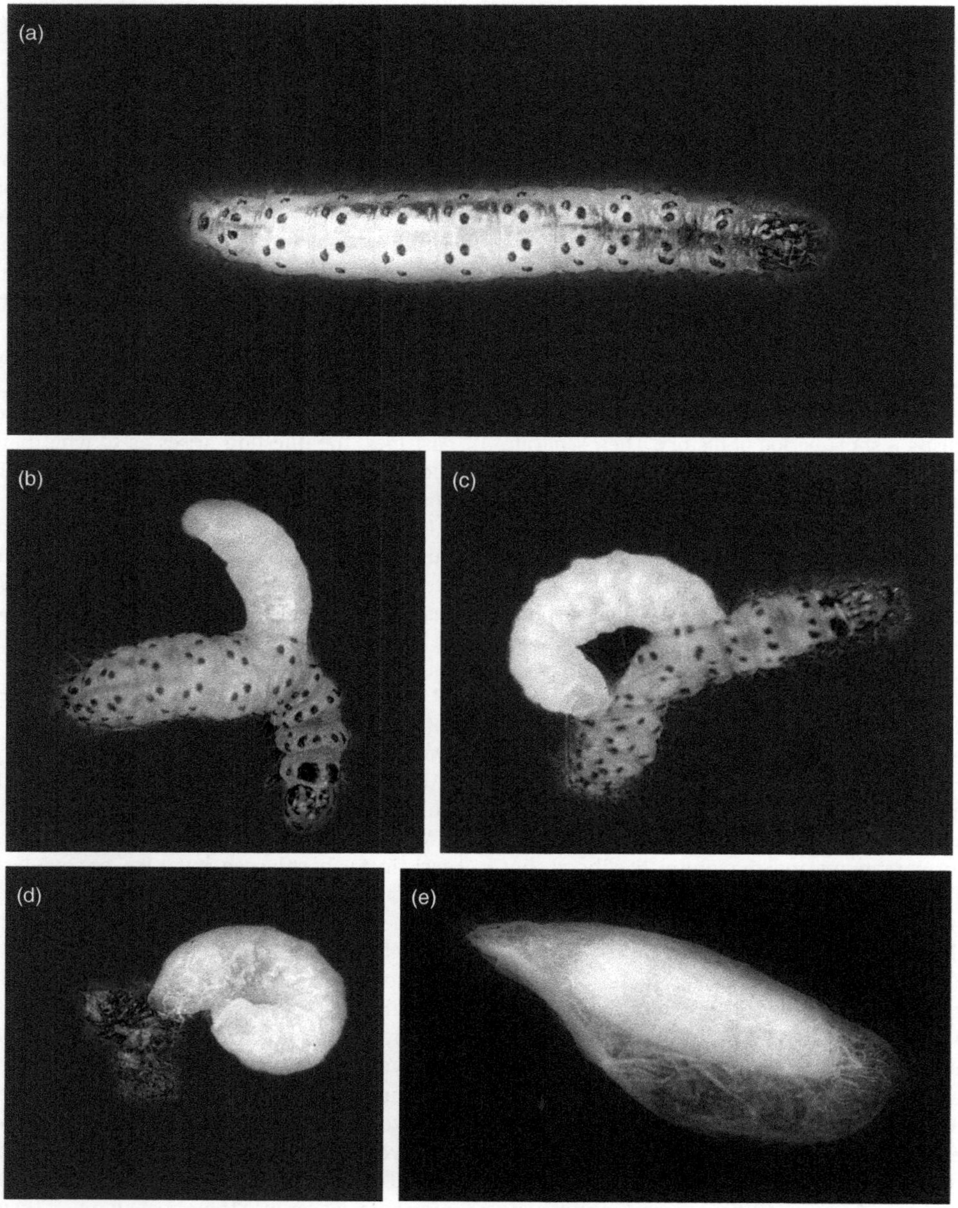

Fig. 12.65 *Microgaster nixalebion* (Microgastrinae). (a–c) Larval egress; (d) external host phase; (e) cocoon. (Source: reproduced by permission of Mark Shaw.)

but these seem to be exceptions. The first instar larva is mandibulate and has a large head, usually much wider than the rest of the body and well-developed mandibles, and has a small tail and a distinct anal vesicle above it (Fig. 5.1), which grows to become a much larger vesicle as the instar progresses (Kaeslin et al. 2006). The posterodorsal surfaces of the posterior thoracic and most abdominal segments are furnished with spines that no doubt aid mobility. Jugovits (1987) noted that in *Microplitis tuberculifera*, the tracheal system is distinctly visible. The second instar is more vesiculate but still has a distinct tail (caudate) (Arthur & Mason 1986), but its mandibles are relatively far smaller than the first and poorly sclerotised, suggesting that they play little part in the feeding process. Hegazi and Führer (1985) reported that the anal vesicle of the final endoparasitic instar of *M. rufiventris* makes a tight association with the host's Malpighian tubules, although nothing is known about what function this might serve. The final moult occurs as the parasitoid exits the host, and the penultimate larval exuvium is left plugging the emergence holes in the caterpillar integument (Fulton 1940). The final instar larva is hymenopteriform with finely serrate and bifurcate mandibles (Čapek 1970). Although most species appear to have lost a final external feeding phase, it is still displayed by a few, e.g. *Microgaster nixalebion* (Fig. 12.65) (Vance 1932b, M.R. Shaw 2004a). *Microgaster* species consequently use hosts which make suitable retreats where external feeding by the parasitoid larva can take place (Shaw & Huddleston 1991). As van Achterberg (2003) put it, to have 'an ectoparasitic phase on an exposed host is biologically speaking suicidal!'. Overwintering in temperate species is typically either as a cocooned prepupa or as a first instar larva within the host (Puttler & Thewke 1970, M.R. Shaw & Huddleston 1991). A few species also overwinter inside their host as a second instar larva and very rarely as a mature larva (Georgiev

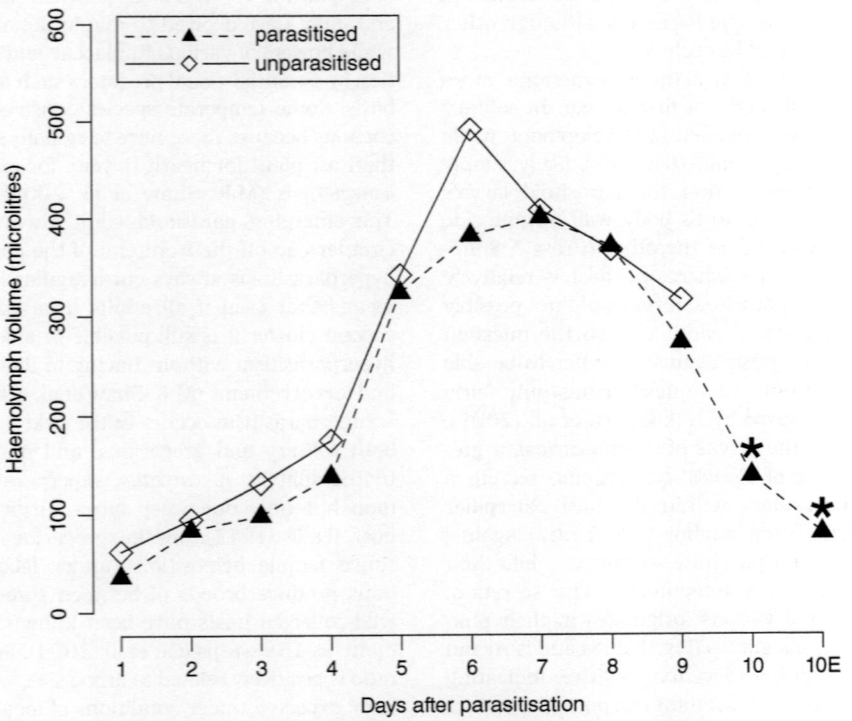

Fig. 12.66 Haemolymph volume of *Pseudaletia separata* (Noctuidae) at various times following parasitisation by *Cotesia kariyai* (Microgastrinae), showing a great reduction at the end of parasitoid development and when they emerge from the host caterpillar. Haemolymph reduction is not associated with weight loss and indicates that it is due to absorption/ingestion by the parasitoid larva. (Source: data from Nakamatsu et al. 2006, fig. 5.)

2001, M.R. Shaw et al. 2009). At least some species show some host alternation (Puttler & Thewke 1970).

In gregarious species, emergence from the host is highly synchronised and the trigger for this is not well understood. It has been noted that the haemolymph volume of the host decreases rapidly shortly before microgastrine larval egress (Fig. 12.66), and as this is not associated with any weight loss, it must be due to the parasitoid larvae ingesting it with a concomitant increase in volume and reduction in space between larvae (Nakamatsu et al. 2006). This increased proximity would help the parasitoid larvae in their struggle to escape the host (see below). However, what these authors did next was very interesting. They artificially bled parasitised hosts shortly before the major decline in host haemolymph volume by cutting one of the larval pro-legs, thus simulating the change in proximity of parasitoid larvae prematurely. The result was startling in that this procedure alone led to premature emergence of parasitoid larvae from the host, suggesting that physical proximity between brood members might be an important cue for egress, although other potential causes cannot be excluded.

Egress from the host is actually sometimes more complicated than it might at first appear. In solitary parasitoids or those completing development while the host is relatively small, they most likely simply orientate themselves so that the parasitoid larva's rear end is against the host's body wall on one side and chew their way out of the other (Cross & Simpson 1972). In the case where the host is relatively huge, however, the parasitoid larva could not possibly stretch across its whole width and so the question arises of what they push against in order to be able to bite their way out. This question has only fairly recently been answered by Nakamatsu et al. (2007), who showed that the larvae of *Cotesia kariyai*, a gregarious parasitoid of *Pseudaletia separata*, secrete a silk-like fibrous protein within the host caterpillar to create a scaffolding anchor (Fig. 12.67) against which they can gain purchase so they can bite their way through the host integument. The secretion, which is sticky and viscous, originates in their pharyngeal/salivary/silk glands (Fig. 12.68) and is Alcian Blue and PAS staining (depicted) positive, indicating mucoglycoproteins, glycoproteins and potentially other carbohydrates.

Cocoons are nearly always formed close to the host, either on top of, next to or underneath the caterpillar (Fig. 12.69b,c,d and see also Fig. 12.71c), although rarely, metamorphosis might be completed without cocoon construction (Fallis 1942). Cocoon construction in *C. marginiventris* was described in some detail by Boling and Pitre (1970). They are normally have a coarse outer silk layer and are ovoid, but those of some genera are longitudinally ridged, and in the rare tropical genus *Distatrix* they may be attached to the substrate via a thick silk stalk approximately half the length of the cocoon proper (Grinter et al. 2009). In gregarious microgastrines, the cocoons may be arranged fairly haphazardly or in really quite precise arrays. In the case of various *Microplitis* species that emerge from large hawkmoth caterpillars, the cocoons are sometimes arranged into small clusters over the host, all aligned perpendicular to the host, where the synchronously emerging parasitoid larvae form spinning groups of about 4–8 individuals (Janzen et al. 2003). In other species, these clusters are much larger. Koptur (1989) fairly convincingly suggested that the cocoons of one gregarious Costa Rican species of *Glyptapanteles*, which are adorned with spurs at the end, may have evolved to mimic the abundant small white flowers of various Rubiaceae and so avoid detection by potential visual predators such as leaf-gleaning birds. Some temperate species construct very robust cocoons because these have to remain stuck *in situ* on the host plant for nearly 1 year, for example, *Cotesia gonepterygis* (M.R. Shaw et al. 2009; Fig. 12.69a). The emerging parasitoid adult always cuts a neat circular cap off the front end of the cocoon, whereas hyperparasitoids always cut irregular holes, and this means that even if all adults have emerged from a cocoon cluster it is still possible to assess the level of hyperparasitism without having to dissect cocoons to find larval remains (M.R. Shaw et al. 2009).

Superparasitism occurs in the field in some species, both solitary and gregarious, and may be adaptive. In the solitary *A. carpatus*, superparasitism is common but only one wasp larva survives to exit the host (Fallis 1942). In the gregarious *C. glomerata*, single female ovipositions under laboratory conditions produce broods of between three and 62, but wild-collected hosts have been known to produce as many as 158 wasps (Gu et al. 2003). Further, the sex ratio is positively related to brood size, which is exactly to be expected under conditions of local mate competition if the host may contain offspring from another female wasp. However, even more interesting is that the dry weights of females emerging from multparasitised hosts were significantly higher (and therefore

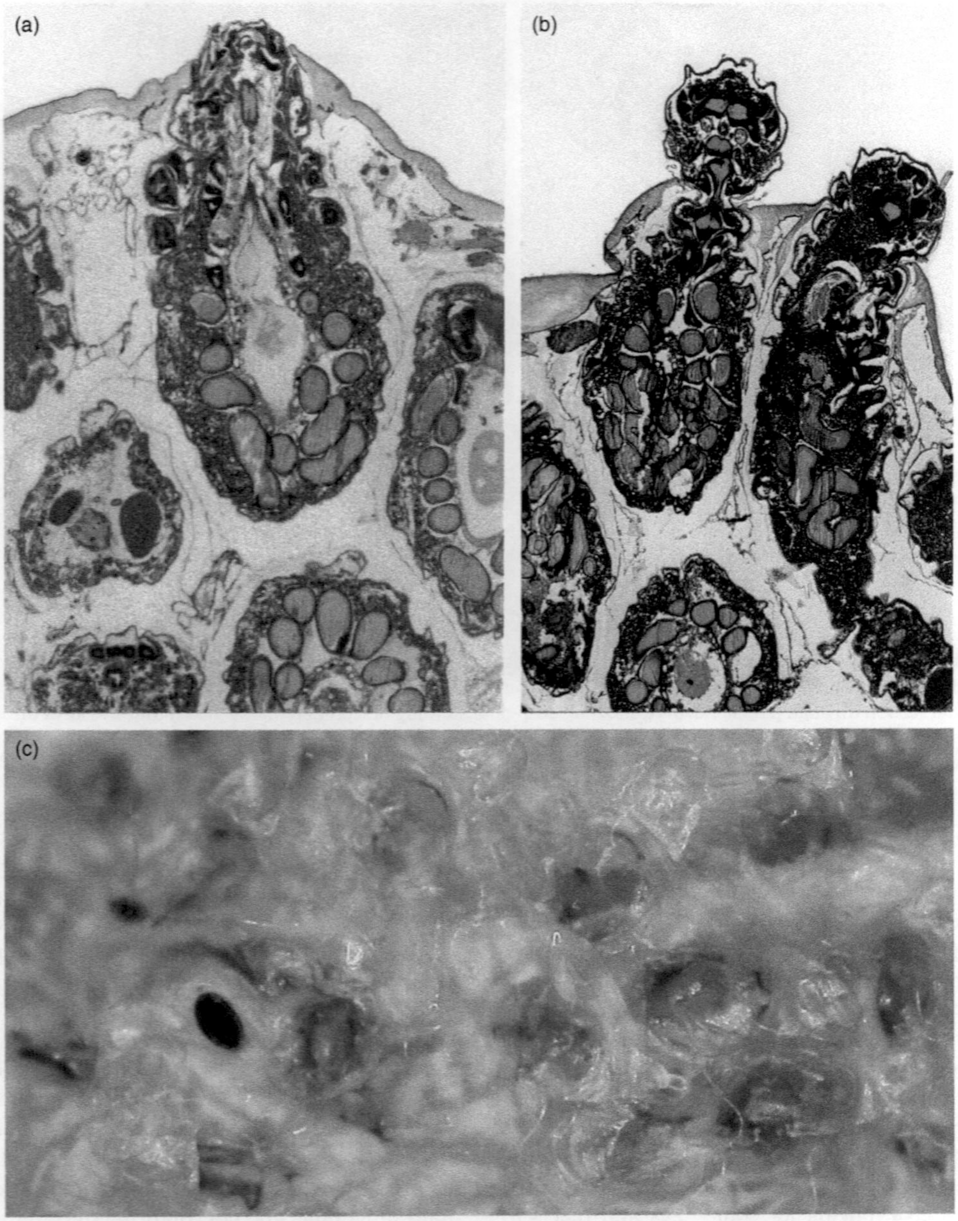

Fig. 12.67 Features of larval egress of *Cotesia kariyai* (Microgastrinae) from its host *Pseudaletia separata* (Noctuidae). (a, b) Haematoxylin/eosin-stained, 8 μm paraffin wax histological sections through *Cotesia* larvae at different stages of exiting through the host cuticle, showing the anchoring material surrounding the parasitoid larval body within the host, and the torn moulted cuticle remains just protruding through the exit holes; (c) photograph of host integument after parasitoid emergence showing penultimate larval exuviae in the exit holes, apparently blocking them. (Source: reproduced by permission of Toshiharu Tanaka.) For a colour version, see Plate 28.

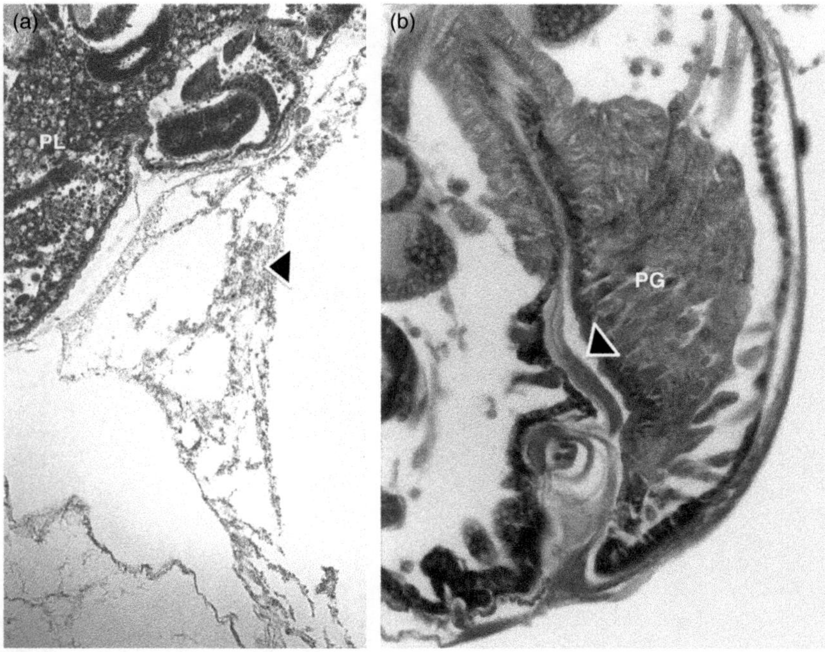

Fig. 12.68 Histological sections of *Cotesia kariyai* (Microgastrinae) within a host. (a) Following PAS staining showing PAS-positive secreted anchor material (arrow head) outside of parasitoid larva (PL); (b) haematoxylin/eosin-stained, longitudinal section through parasitoid larva showing deep-pink staining of secretion (arrow head) that will form the anchoring material, from the pharyngeal gland (PG) spreading out over the anterior surface of the parasitoid larva. (Source: reproduced by permission of Toshiharu Tanaka.) For a colour version, see Plate 29.

presumably more fit) than those from singly parasitised hosts, meaning that superparasitism could be advantageous in this species, especially under circumstances where hosts are rare.

Most species are either solitary or, if gregarious, produce fairly large broods (10 or more), but small broods of two or three are known in some; for example, *Cotesia astrarches* usually has small broods (3–6 individuals emerging) in *Aricia* hosts, but can have larger ones, while *C. subordinaria* on *Rivula sericealis* (Noctuidae) is facultatively gregarious (M.R. Shaw 2012). The host caterpillar is completely moribund by the time the microgastrine larvae complete feeding, but is often capable still of moving for some days and reacts to stimuli such as other passing insects. Some tropical species of geometrid moth caterpillar, that under normal unparasitised conditions spend much of the day in a camouflaging, twig-like pose, when parasitised still do this with the microgastrine cocoons formed just anterior to their hind claspers. The moth caterpillar is still capable of detecting 'threat' and, just as in unparasitised ones, will thrash from side to side to deter other insects such as tachinid flies or, in this case, potential pseudohyperparasitoids of the cocooned wasps (Harvey et al. 2011). This combination of parasitoids pupating next to a still active host has been termed usurpation, since the parasitoids may benefit from the retained defensive reflexes of the host (Fig. 12.70) (Brodeur & Vet 1994, Grosman et al. 2008, Harvey et al. 2008). For this to happen, it is of course a prerequisite that the feeding done by the parasitoid larvae within the host did not destroy any vital tissues, leaving the nerve cord and most musculature intact and at least partially functional, and not rupturing the gut that would release many potential pathogens into the host's body. Maure et al. (2011) demonstrated similar usurped protection in the euphorine *Dinocampus coccinellae* and additionally demonstrated a trade-off between host 'viability' and wasp fecundity; it would be interesting to know to what extent, if any, this might apply

Fig. 12.69 Examples of cocoons of Microgastrinae. (a) Overwintering cocoon of the univoltine microgastrine *Cotesia gonepterygis* showing how securely it is attached to the twig where it will spend the winter months; (b) *Microplitis impressa* (Microgastrinae), univoltine wasp showing hard, ribbed cocoon attached to the host; (c) *Glyptapanteles* sp.; (d) *Parapanteles* sp. (Sources: a, reproduced by permission of Mark R. Shaw; b, reproduced by permission of Marion Friedrich; c, d, reproduced by permission of Lee Dyer.) For a colour version, see Plate 30.

to microgastrines, since they are koinobionts rather than idiobionts and therefore have more potential to control their host's feeding and reserves. That the emergence holes from the host are plugged with the second larval exuvium is important in that it prevents the host rapidly bleeding to death, which would almost certainly be the case otherwise (Fulton 1940, Nakamatsu et al. 2006). Thus, the occurrence of the final larval moult as the parasitoid egresses from the host can be seen as a vital prerequisite for usurpation of the host's defensive behaviours (see below).

The *P. brassicae* host of *C. glomerata* is caused by the parasitoid to perform a more complicated protective role. Once the parasitoid larvae have all emerged from the host caterpillar and spun their yellowish cocoon mass underneath it, the caterpillar then adds an extra layer of silk over the parasitoid cocoons (Fig. 12.71c; *cf.* Fig. 12.71d; see M.R. Shaw et al. 2009), fixing them more firmly to the substrate and no doubt helping to protect them from hyperparasitoids or predators. The caterpillar, although totally moribund, can nevertheless still move and thrash in response to stimuli for several days, indeed long enough for the *Cotesia* to complete their metamorphosis and emerge as adults. To test whether this significantly increases parasitoid survival, as suggested by the so-called 'usurpation hypothesis', Harvey et al. (2008) presented cocoon masses of *C. glomerata* with and without the additional silk layering to the pseudohyperparasitic cryptine *Lysibia nana*. They found that the presence of extra silk was associated with higher survival of the primary parasitoid but that the presence of the still responsive caterpillar had no significant effect, thus providing only partial support for the hypothesis. However, *L. nana* is specialised on the *C. glomerata*/*P. brassicae* system and the behaviour of the *Pieris* larva might be more effective in deterring generalist predators and pseudohyperparasitoids. Such modified host behaviours

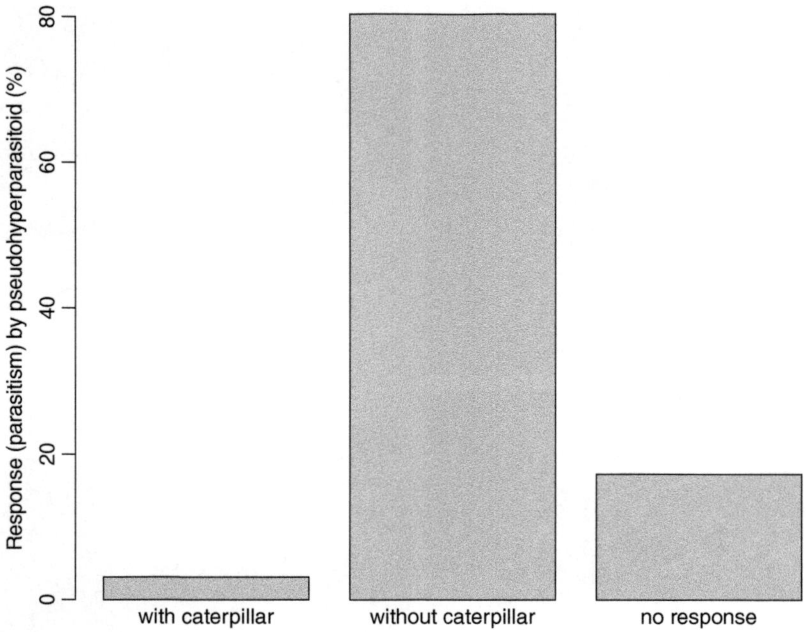

Fig. 12.70 Demonstration of effectiveness of usurpation hypothesis by *Microplitis* sp. (Microgastrinae) attacking *Mythimna separata* caterpillar, with many more host cocoons being attacked successfully by the pseudohyperparasitoid *Gelis agilis* (Cryptinae) when not attached to caterpillar's caudal appendage ($\chi^2 = 72.4$, $p < 0.001$) despite more than 10 times as many attempts to parasitise being made when the cocoon is attached to a caterpillar. (Source: data from Harvey et al. 2011.)

are possibly widespread but seldom observed, and the true nature of what goes on is best understood using time-lapse photography.

Many other species cause hosts to behave abnormally when they are about to egress. Sato et al. (1983) found that *C. kariyai* causes its normally nocturnal host *Leucania separata* caterpillar to emerge from its normal daytime retreat, climb a stalk and remain quiescent there so that the parasitoids egress in full daylight.

Host location has been investigated intensively in *C. glomerata* and *C. rubecula* and these have been shown to differ in important ways in how they learn host-associated cues (Smid et al. 2007: see Chapter 8, section *Associative learning*). As with virtually all microgastrines, *C. glomerata* attacks early instar hosts and can discriminate the cues given off by host/host–plant complex when the *P. brassicae* larvae are in their first instar and thus suitable from those produced when fifth instars are feeding, the latter potentially being risky to attack (Mattiacci & Dicke 1995). Other species that attack rather less formidable hosts, such as *Apanteles carpatus* on tineid clothes moth larvae, can successfully attack all larval instars, including prepupae (Fallis 1942, Plarre et al. 1999, 2000).

Courtship and mating behaviour has been described for several species (Weseloh 1977) and involves male-produced songs (Sivinski & Webb 1989, Field & Keller 1993a, Danci et al. 2010, Bredlau et al. 2013). The male courtship song of *Cotesia marginiventris* is described as very complex, with two distinct forms of pulse trains, compared with that of *Microplitis croceipes*, and Bredlau et al. (2013) showed that the song of *C. congregata* males comprises two parts, starting with amplitude-modulated 'buzzes' when the male was 2–3 cm from the female and changing to a 'boing' sound when he was about 1 cm away from her. Because song vibrations may be transmitted to females through the substrate, the nature of the latter can have a marked influence on a male's mating success (Joyce et al. 2008). In *C. rubecula*, competition for mates between males is very intense and in addition to regular courtship which involves a male singing, other males may adopt either of two alternative mating strategies (Field & Keller 1993b). Some show

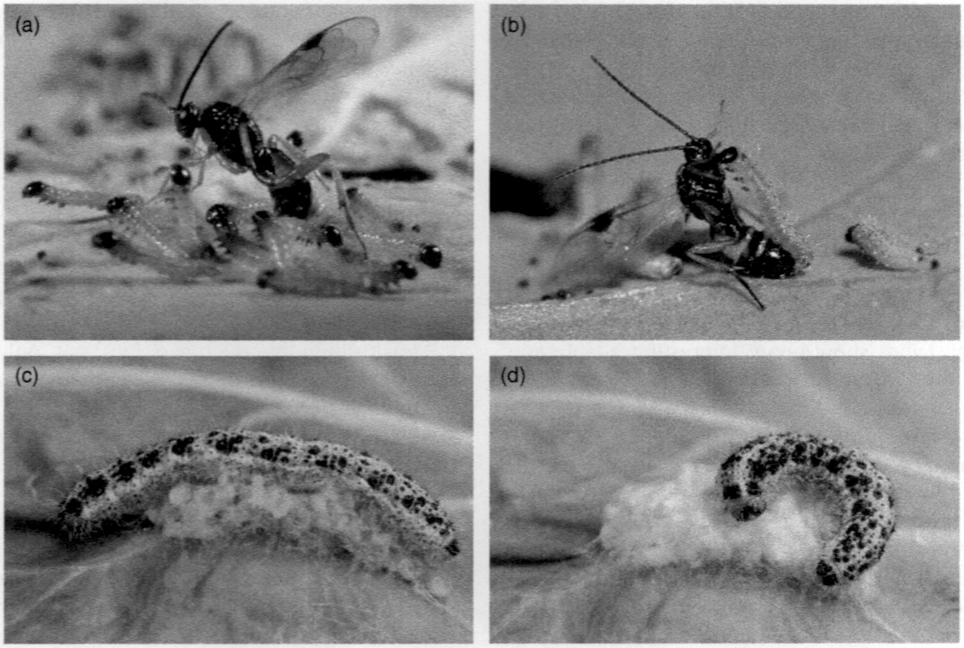

Fig. 12.71 Interactions between *Cotesia glomerata* (Microgastrinae) and its host, *Pieris brassicae* (Pieridae). (a) Female wasp attacking freshly emerged caterpillars; (b) individual caterpillar trying to defend itself; (c) parasitoid larvae emerging synchronously from the host and immediately starting to spin their cocoons in a mass underneath the moribund host; (d) 'usurpation' of the host which, in this species, not only stays on guard but additionally has spun extra silk covering over the parasitoid cocoon mass. (Source: reproduced by permission of Hans Smid/bugsinthepicture.com; http://bugsinthepicture.com.) For a colour version, see Plate 31.

sneaky mating, what the authors term 'short-circuit behaviour', or competing males may distract the attention of a courting male by mimicking a female, which gives the mimic a short opportunity to approach the potentially now receptive female and mate with her. Although most females mate only once, there appears to be a brief window of opportunity for a second male to mate immediately after a female's first copulation. *C. rubecula* males having mated may then adopt a form of mate guarding by themselves mimicking a receptive female and thus distracting the attentions of male competitors away from the real female, thus maximising his probability of obtaining 100% parentage. In their laboratory trials, Field and Keller found that this female mimicry was successful in attracting courtship effort of 55 out of 64 rivals.

Females produce a sex pheromone that induces male courting and singing. Using a variety of dissections and behavioural assays, supported by TEM and SEM, the female sex pheromone glands of several species of *Cotesia* have been localised to the terminal part of the metasoma and probably the genital capsule (Tagawa 1977, Weseloh 1980[25], De Freitas et al. 2004). The sex pheromone gland of *Cotesia rubecula* has been localised to either '... a group of cells located directly underneath surface pores which are adjacent to the mechanosensory hair plate on the genital capsule ... The hair plate was in turn juxtaposed to the second valvifer' of the female genitalia where two pairs of exocrine glands were discovered, although it is not certain what the individual functions of these are (Field & Keller 1994). The cocoon masses of *C. glomerata* apparently give off a sex pheromone which tends to keep a large proportion of the protandrous males waiting by it for females to emerge, and Tagawa & Kitano (1981) estimated that 60% of matings in the field were between siblings, a surprisingly high proportion given that this species has been shown to possess slCSD. It seems likely that the ether-soluble, cocoon-associated sex pheromone might originate from virgin females waiting to eclose, but this is not certain. In the same species, Ruf et al. (2011) found that intersexual kin interactions caused more than one-third of females to disperse and that these females searched for new patches in order to find mates; males showed no such tendency to disperse. These measures of the proportion of females mating at the natal patch are very similar, despite them involving populations in widely different parts of the world, and it might suggest that there is a mixed ESS involved here.

Microgastrines show many interesting adaptations to their hosts. *Microplitis mediator*, a parasitoid of the Bertha armyworm (*Mamestra configurata*), a noctuid pest of cabbage, attacks its host on the leaf, but if the host reacts by dropping down on a silk thread, the wasp will fly down and attack it in mid-air (Arthur & Mason 1986). The Nearctic *Diolcogaster facetosa*, which attacks the green clover worm, *Hypena* (= *Plathypena*) *scabra*, a pest species of noctuid, goes one better. When disturbed, its host usually drops down from its food plant on a silk thread, a common adaptation against parasitism and predation. However, in this instance it does not work because the *Diolcogaster* has evolved a behaviour whereby if it locates the silk thread, it descends down this thread until it is within a couple of millimetres of the host and then thrusts its ovipositor forward into the host and oviposits (Yeargan & Braman 1986). The wasp can also locate threads by themselves, rather than first having to locate a host caterpillar, and then follow them to the ground where the potential host might be resting.

Nearly all microgastrines have a fixed number of antennal segments (16 flagellomeres in both sexes), but recently a new genus from New Zealand has been discovered in which females have 17 but males have 18 flagellomeres (Fernandez-Triana et al. 2011). Detailed studies of antennal sensilla have been presented for a large number of species: *Apanteles cypris* (H. Zhou et al. 2011), *Cotesia plutellae* (G. Yang et al. 2003, Roux et al. 2005), *Microplitis croceipes* (Navasero & Elzen 1991, Ochieng et al. 2000) and *Microplitis pallidipes* (Gao et al. 2007). One consistent finding is that males not only have longer antennae than females but also have a larger number and higher density of chemosensory *sensilla placodea*. Bleeker et al. (2004) compared the sensilla of *C. glomerata* and *C. rubecula*, which differ in their host specificity, but while the latter had more *sensilla placodea*, the sensilla complement of the two species were very similar. In general, these authors reported 7–11 types of sensilla being present, although it is not possible to compare these numbers directly as authors differ in their inclusion of the scape and pedicellus and also in the degree of differentiation of some similar sensilla types. Females of some *Fornicia*, *Alloplitis* and *Diolcogaster* have a ventral patch of basiconic sensilla towards the antennal tip in a region devoid of placode sensilla (A.K. Walker et al. 1990a), which may be important in their host location or acceptance, but no members of these genera have been subjected to detailed investigation.

Parasitisation by microgastrines not only causes behavioural changes but also endocrinological ones in the host (e.g. Beckage & Riddiford 1982, Beckage et al. 1990, 1994). Because of the complexity of their interactions with the hosts, it is very difficult to pinpoint exactly what causes some effects. In normal parasitisation, the female wasp injects venom, calyx gland secretions which include polydnaviruses but also probably other compounds, while the larvae themselves may be secreting substances and the teratocytes are also active secretors that modify host physiology (Dahlman 1991).

All species produce polydnavirus particles in their calyx glands (see Table 9.1) and a very large number of publications have dealt with the genes encoded by these viruses and their actions on the host immune defence system and host development in general. Among the best studied are *Cotesia congregata* bracovirus (CcBV), *Microplitis demolitor* bracovirus (MdBV), *Glyptapanteles indiensis* bracovirus (GiBV) and *Glyptapanteles flavicoxis* bracovirus (GfBV) (Bezier et al. 2009). Whitfield (2000) found very high co-phylogeny (as would be expected) between six *Cotesia* species based on two genes and their associated polydnaviruses based on the glycosylated secreted protein gene (CrV1). Not only were the topologies virtually identical, but also branch lengths were highly concordant.

There is enormous variation in the degree of host specificity displayed by different species. Some such as *C. glomerata*, various species of *Cotesia* attacking melitiine butterflies (Kankare et al. 2005a,b) and many others are very specialised and often utilise only a single host species. Even some plurivoltine species in the north of Europe produce multiple generations on a single generation of their host species, but most temperate plurivoltine species utilise more than one host species at different times of the year (M.R. Shaw et al. 2009). Others such as *C. rubecula* can attack and develop on a small range of hosts, in that case pierid butterfly larvae, while some such as *C. marginiventris* can attack a wide range of host species (mostly but not entirely restricted to the Noctuidae) (Ruberson & Whitfield 1996), and *C. vestalis*, whilst attacking several nymphalid species depending on the time of year, also parasitises the unrelated diamond back moth, *Plutella xylostela* (Plutellidae).

Relationships within the enormous genus *Cotesia* have been investigated using a four-gene data set (Michel-Salzat & Whitfield 2004) although, as the title suggests, the results must be treated as very preliminary as diversity within this enormous genus had barely been scratched. The molecular study revealed four reasonably well-supported species groups, but these did not conform well to groupings that had been recognised previously on the basis of morphology. Mapping biology onto the trees showed that *Cotesia* species were primitively solitary parasitoids and that gregariousness has evolved within the genus on numerous separate occasions. If these gregarious species in large hosts all utilise the same mechanism of egress as found by Nakamatsu et al. (2007) for the gregarious *C. kariyai* are present in all the other independent occurrences of gregariousness, then it would be interesting to know whether primitively solitary species similarly secrete a silk-like scaffold in their hosts.

The evolution of host ranges in microgastrines is only just beginning to be explored in a phylogenetic context. The Neotropical species of *Parapanteles* are not randomly distributed across their host, *Eois*, moths, indicating some degree of either co-speciation or taxonomically constrained host shifts (J.S. Wilson et al. 2012), but Whitfield & Wagner (1988) found that for Nearctic *Pholetesor*, host biology and life history were far more important than their taxonomy in determining which species were attacked.

Genera, in the sense that many users have adopted Mason's hypotheses to a greater or lesser extent, can be identified using the work of W.R.M. Mason (1981a), Austin & Dangerfield (1992: Australia) and Whitfield (1997b: New World).

Miracinae

These are all very small wasps that are koinobiont endoparasitoids of very small, leaf-mining, Lepidoptera larvae and have thus far been reared from Heliozelidae, Gracillariidae, Lyonetiidae, Nepticulidae, Tineidae and Tischeriidae (Shaw & Huddleston 1991, Whitfield and Wagner 1991, Memmott et al. 1994, Maetô 1995, Whitfield 1997b), but there are almost no host records for species from the tropics apart from O.T. Lewis et al. (2002) and one species from Hawai'i (originally described in *Blacus*!; see van Achterberg 1976a), so the list is likely to expand with rearing of tropical 'microlepidoptera'. As far as is known, all species are solitary. Overwintering in temperate species is within the host cocoon, but it is not known whether as a prepupa, pupa or adult (Shaw & Huddleston 1991).

In morphological phylogenetic analyses they tend to come out as the sister group to the Microgastrinae, probably due to features associated with small body size, but in molecular and combined studies their position is highly unstable (see Fig. 12.59). As would be expected from their phylogenetic placement within the microgastroid group of subfamilies, they are now known to produce polydnaviruses (Wharton & Sittertz-Bhatkar 2002). However, these have not been studied any further, probably because of a combination of their lack of importance in any biological control situation and consequent lack of any species in culture.

All miracines have 12 flagellar segments and their wing venation is fairly distinctive (Fig. 12.72). Their taxonomy is very poorly worked and identification is difficult. There are probably many cryptic species and almost nothing can be said meaningfully about the host ranges of species.

Unplaced Subfamilies

Two subfamilies have thus far eluded satisfactory placement even within the broad framework above, the Masoninae and the Meteorideinae.

Masoninae

This is a relatively recently recognised group of seldom-collected, tiny wasps with body lengths of 0.7–1.8 mm, that are largely associated with the ground layer with females of all the four known species being wingless and with very robust legs (Fig. 12.73) (van Achterberg 1995). Females have only been collected in flight intercept traps, into which they presumably walked, and are known from Australia and south-eastern USA down to the Florida Keys. Males are only known for two of the four described species and are winged but with extremely reduced venation, but have been collected in Malaise traps; they are considerably less structurally modified. The females are highly prognathous

Fig. 12.72 *Mirax* sp. (Miracinae). (Source: reproduced by permission of Kyohei Watanabe.)

(Fig. 12.73) and it is impressive that they were recognised as braconids; as mentioned by van Achterberg, Bill Mason, who was going to describe them before his untimely death, intended to place them in their own subfamily based on the large gular sclerite of the prognathous females.

Their small size and tendency to be collected in flight intercept traps (they probably also get caught in yellow pan traps and pitfall traps), which are not so commonly used by ichneumonid wasp workers, is probably why they have gone unnoticed for so long. Additionally, they are now being found in numerous localities and microhabitats where braconidologists have not typically applied themselves. A well-preserved fossil species of *Masona* from Dominican amber (15–20 Mya) has been described (van Achterberg 2001). Nothing is known of their hosts.

The extreme specialisation of the wasps together with their small size makes morphological assessment of their relationships problematic. The only sequence data available (the D2 to D10 region of the 28S rDNA gene) was rather aberrant and did not yield an unambiguous placement (Belshaw & Quicke 2002). Depending on alignment method, they were recovered either with the outgroup or on a long branch at the base of the sigalphoid clade. The Mannokeraiini[26], the other tribe originally included in the Masoninae (van Achterberg 1995), has now been transferred to the Euphorinae, their similarity with *Masona* being due to convergence.

Plate 1 Major regions of the mesosoma illustrated in *Eadya* (Helconinae *s.l.*): (a, b), dorsal view; (c, d), lateral view. Abbreviations: mp, mesopleuron; ms, mesoscutum; mst, mesosternum; mtn, metanotum; mtp, metapleuron; n, notauli; pcs, precoxal sulcus; pn, pronotum; pp, propleuron; s, scutellum; ss, scutellar sulcus; t, tegula.

The Braconid and Ichneumonid Parasitoid Wasps: Biology, Systematics, Evolution and Ecology, First Edition. Donald L. J. Quicke.

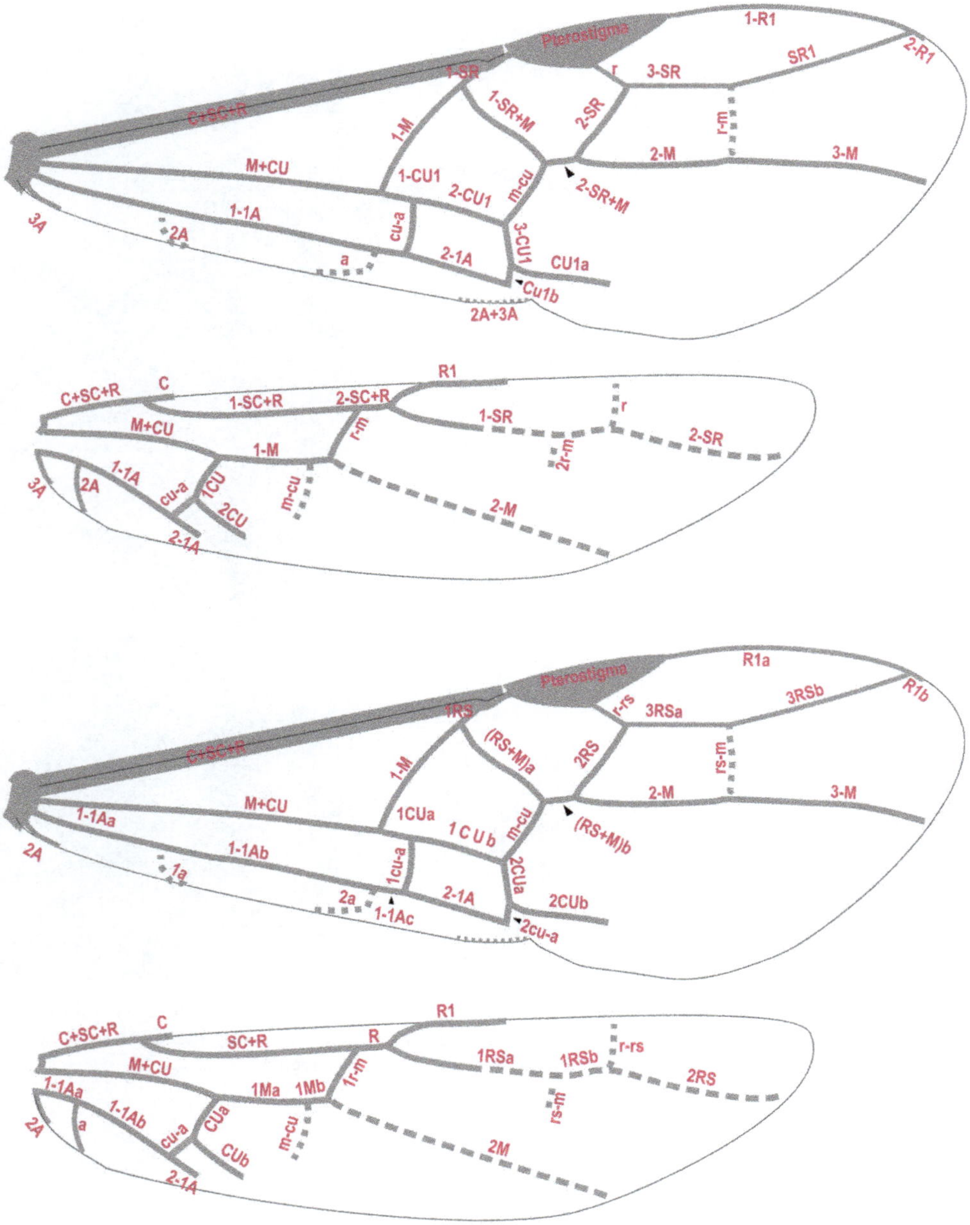

Plate 2 Terminology for venation in hypothetical braconid wasp fore and hind wings: (a) based on Comstock–Needham system as applied by van Achterberg (1979c, 1993a) and (b) by Sharkey & Wharton (1997) with the longitudinal vein elements indicated by capital letters (C, costa; SC, subcostal; R, radius; SR, Sector radialis; M, medius; CU, cubitus; A, anal), cross-veins indicated in lower case, abscissas of vein numbered from base at the beginning, and branches of longitudinal veins indicated by number at end.

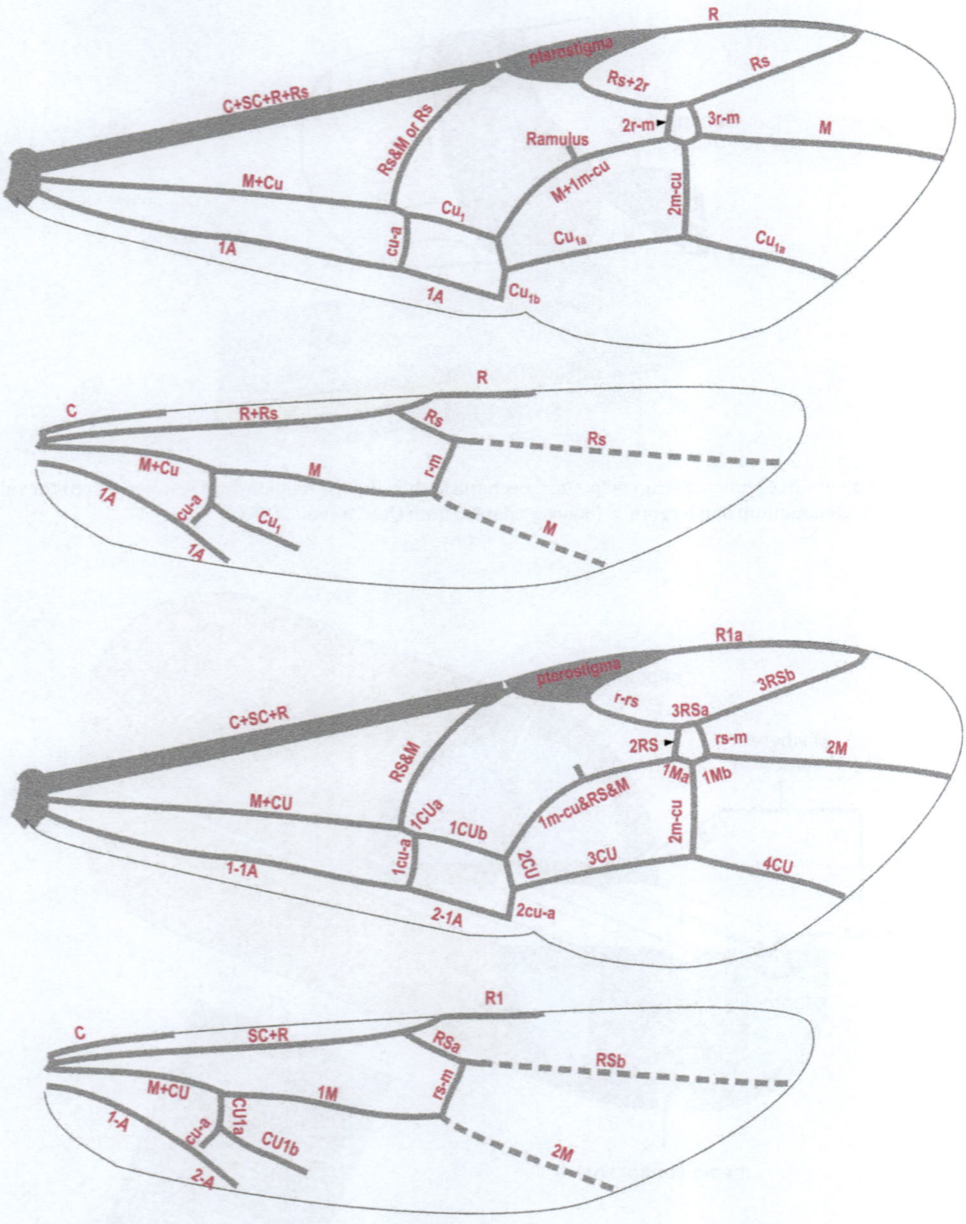

Plate 3 Terminology for venation in hypothetical ichneumonid wasp fore and hind wings: (a) based on Comstock–Needham system and (b) applying the system of Sharkey & Wharton (1997). Longitudinal vein elements indicated by capital letters (C, costa; SC, subcostal; R, radius; SR, Sector radialis; M, medius; CU, cubitus; A, anal), cross-veins indicated in lower case, abscissas of vein numbered from base at the beginning, and branches of longitudinal veins indicated by number at end. See Table 2.2 for the vein names used by Henry Townes and others.

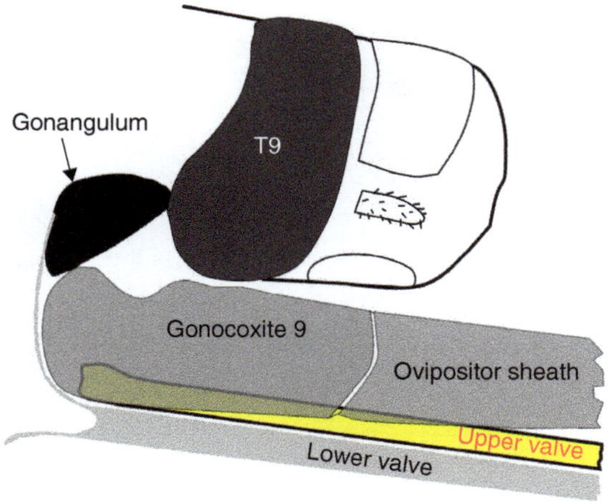

Plate 4 Schematic diagram of hymenopteran ovipositor mechanism showing the relationships between ovipositor valves, sheath, gonocoxite 9, gonangulum and tergum 9. (Source: adapted from Quicke et al. 2000a.)

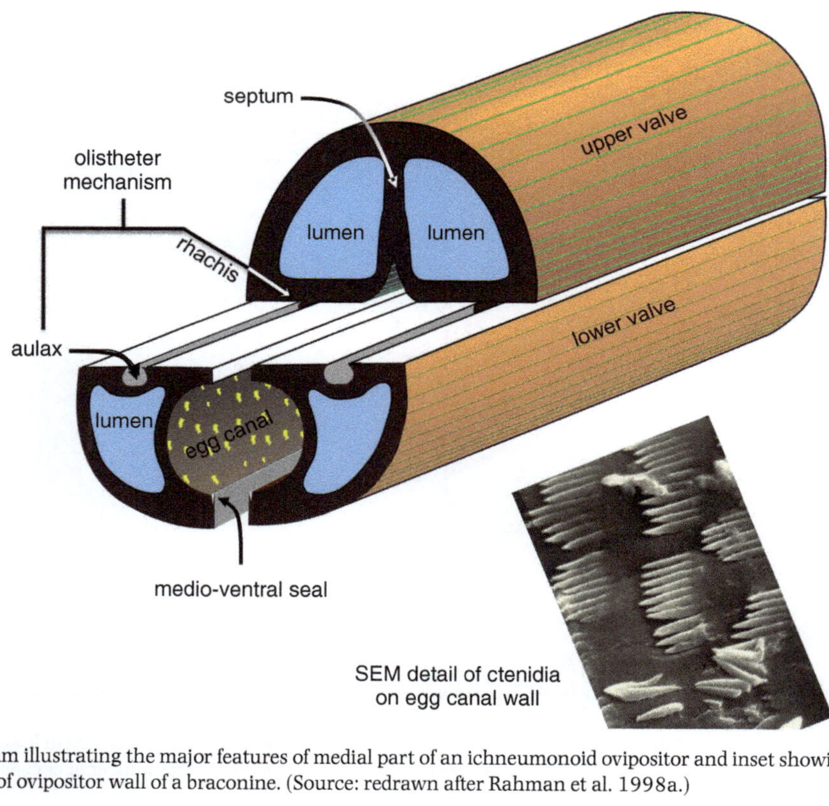

Plate 5 Diagram illustrating the major features of medial part of an ichneumonoid ovipositor and inset showing microsculpture of ovipositor wall of a braconine. (Source: redrawn after Rahman et al. 1998a.)

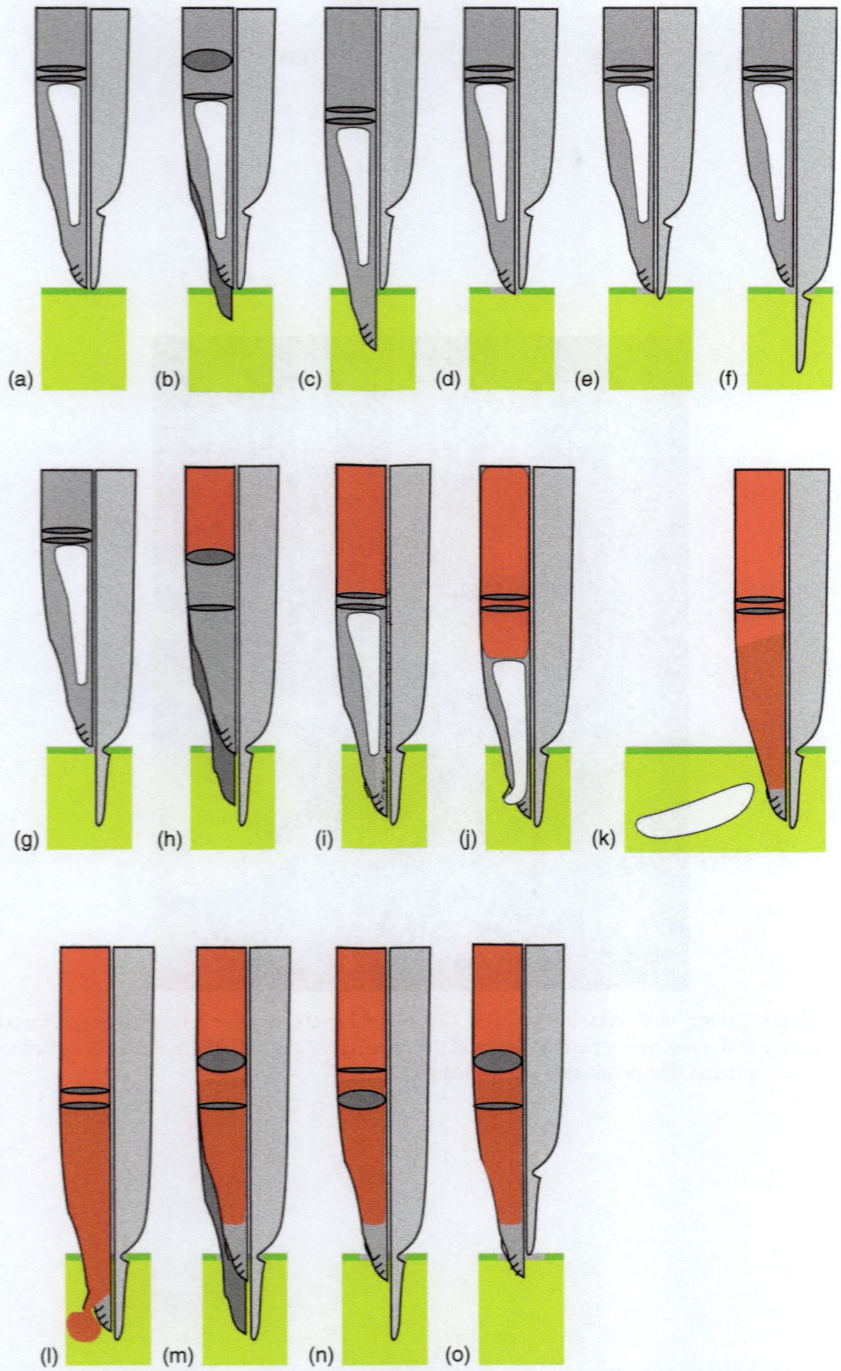

Plate 6 Diagrams of stages in ovipositor use by *Homolobus* (Homolobinae) and other ichneumonoids with a pre-apical dorsal notch on the ovipositor, showing its interaction with the host cuticle and subsequent extrusion of the egg. (Source: adapted from Boring et al. 2009.)

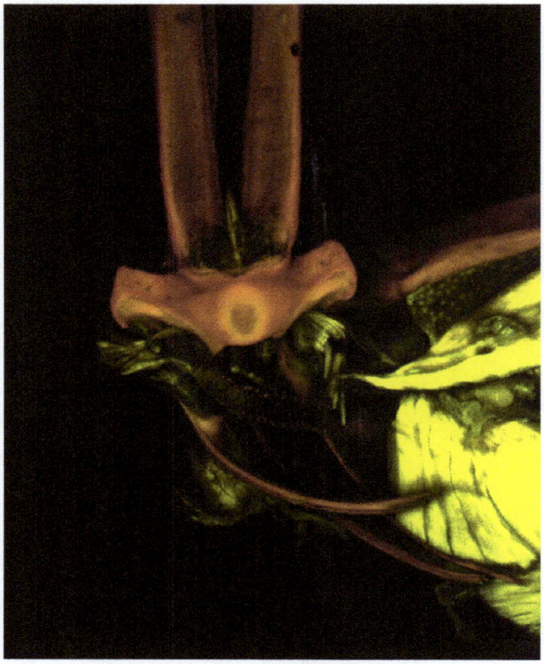

Plate 7 Laser confocal scanning microscope images of the base of the dorsal ovipositor valve of *Glypta* sp. from Korea, showing the transverse basal piece and completely separated halves of the upper valve; the aulaciform rod is visible just to the right of the left-hand valve (Source: reproduced by permission of Istvan Miko.)

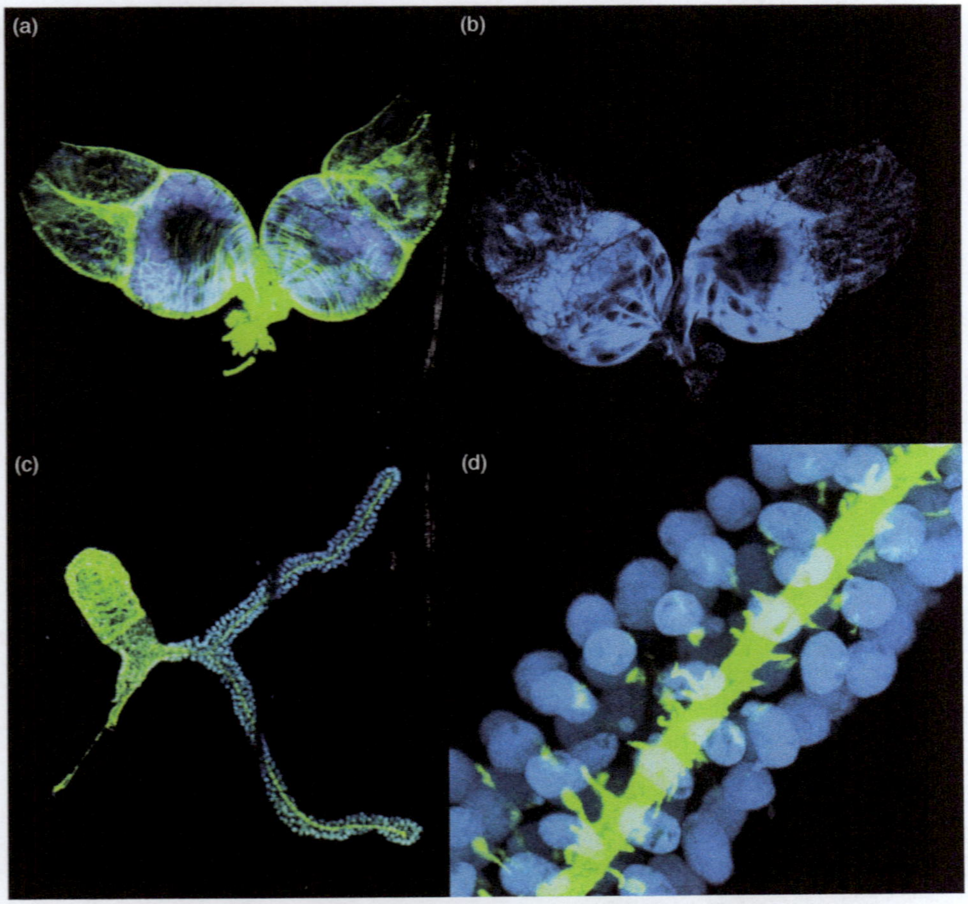

Plate 8 Fluorescence microscopy images of the female internal reproductive system of the microgastrine *Cotesia congregata* showing DNA (both wasp nuclear and polydnavirus) fluorescing blue and actin, stained with phalloidin, fluorescing green. (a, b) Ovaries and calyx gland; note the intense blue fluorescence surrounding eggs in the calyx due to high concentration of polydnavirus particles; (c) entire venom apparatus; (d) detail of venom gland filament. (Source: reproduced by permission of Germain Chevignon).

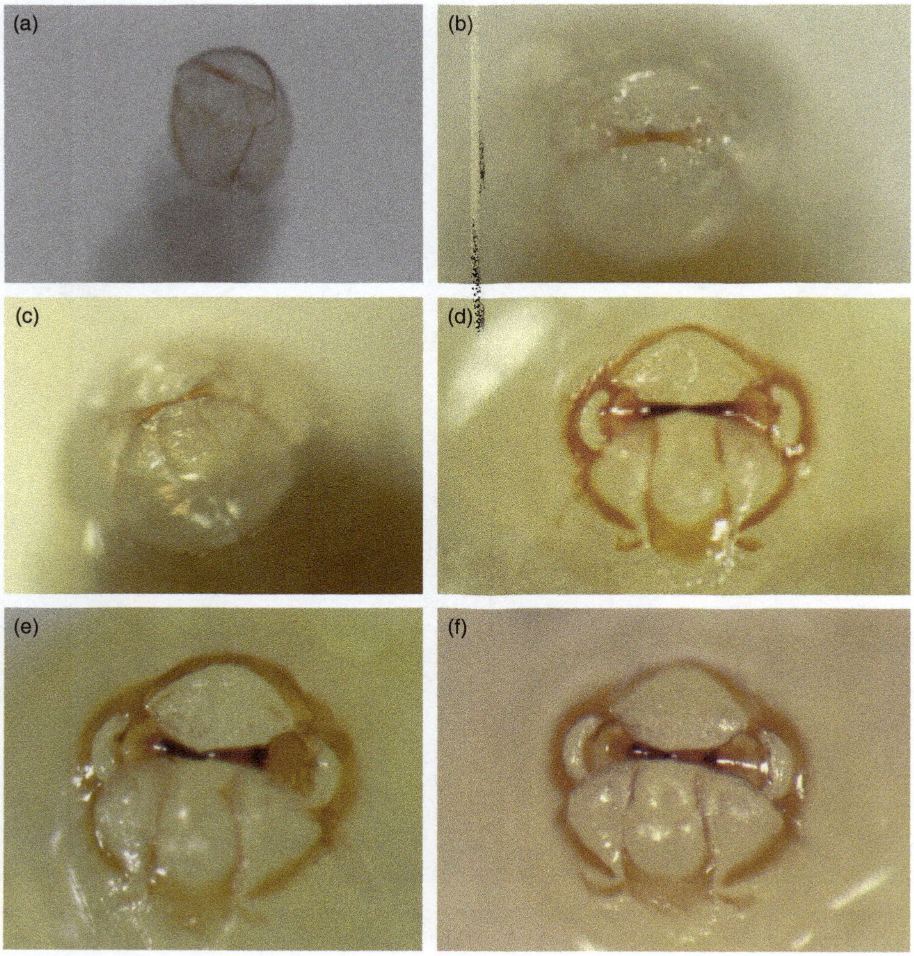

Plate 9 Larval head development in *Itoplectis melanocephala* (Pimplinae). (a, b) two views of first instar head; (c–f) cephalic strictures of living second–fifth instars (not to scale). (Source: reproduced by permission of Ayşegül Özdan.)

Plate 10 Pupae of *Itoplectis melanocephala* (Pimplinae). (a) Male, just moulted; (b–d) female at three stages of darkening; note the ovipositor is curved over the back. (Source: reproduced by permission of Ayşegül Özdan.)

Plate 11 Lethal combat between endoparasitoid larvae. (a) Intraspecific fight between two larvae of *Meteorus pulchricornis* (Euphorinae), the winner biting the loser behind the head, where the damage will expose the immunologically active soft tissue to further attack by host defence mechanisms. (Source: reproduced by permission of Kaoru Maetô.) (b) Interspecific contest between loser, *Hyposoter horticola* (Campopleginae), and winner, the much smaller larva of *Cotesia melitaearum* (Microgastrinae), parasitoids of the Glanville fritillary *Melitaea cinxia* (Nymphalidae). (Source: reproduced by permission of Saskya van Nouhuys.)

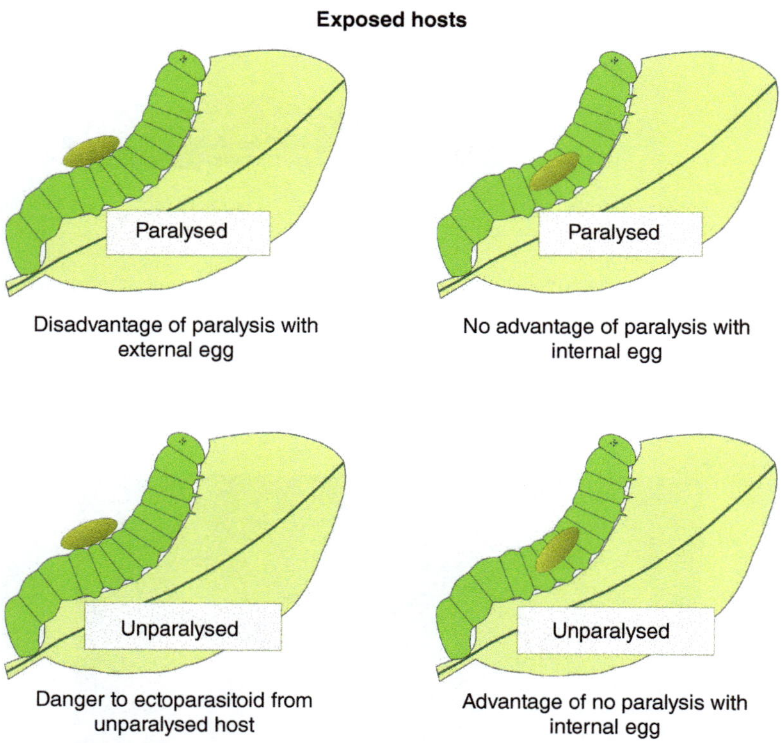

Plate 12 Cartoons illustrating the advantages and disadvantages of four combinations of ecto- and endoparasitism of a paralysed or unparalysed exposed host. (Source: adapted from Quicke et al. 2000a.)

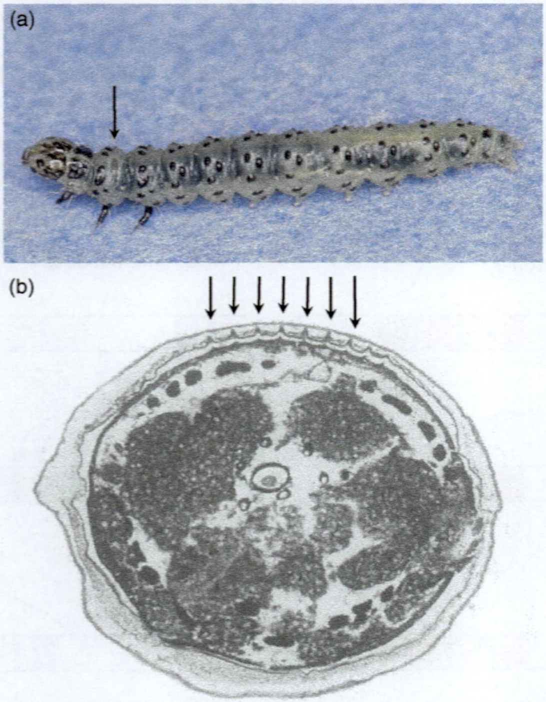

Plate 13 Host of *Clinocentrus cunctator* (Rogadinae). (a) Larva of the choreutid *Anthophila fabriciana* host showing parasitoid egg located transversely just below the cuticle of the second thoracic segment; (b) transverse light microscope section through pre-penultimate instar of host envenomated but not oviposited in (pseudoparasitised), showing the prematurely developed dorsal pupal cuticle with spines that would normally form one instar later. (Source: reproduced by permission of Mark Shaw from M.R. Shaw 1981.)

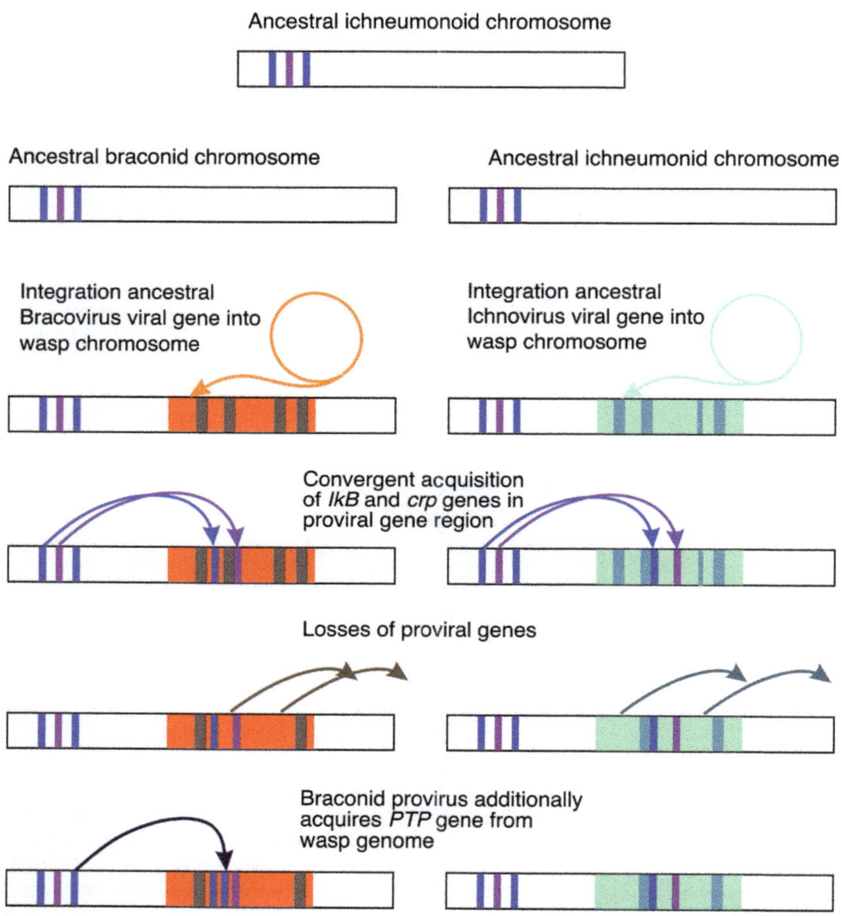

Plate 14 Diagram of early evolutionary events (from top to bottom) leading to bracoviruses (left) and ichnoviruses (right) showing probably convergent evolutionary events in the two separate lineages prior to their subsequent radiation into the forms we know today. (Source: redrawn after part of fig. 3 in Dupuy et al. 2006.)

Plate 15 Features of ichneumonids associated with egress from and ovipositing into a hard wood substrate. (a, b) Transversely ridged mesoscutum of *Megarhyssa* sp. (Rhyssinae) and *Pseudorhyssa* sp. (Poemeniinae), respectively; (c) head of *Podoschistus alpensis* (Poemeniinae) showing dorsal sculpture; (d) ventral surface of anterior of metasoma of *Megarhyssa* sp. (Rhyssinae) showing medial groove and clasps of the ovipositor stabilising mechanism. (Source: reproduced by permission of Rikio Matsumoto.)

Plate 16 Four examples of ichneumonoids displaying the ophionoid facies with brown coloration, large eyes and ocelli and long antennae. (a) Macrocentrinae gen. sp.; (b) three *Netelia* sp. (Tryphoninae), at light sheet; (c) *Cystomastax* sp. (Rogadinae); (d) five *Ophion* sp. (Ophioninae) on underside of leaf, attracted to light. (Source: reproduced by permission of Kenji Nishida.)

Plate 17 Impression fossils of two recently discovered and better preserved species of the extinct family Praeichneumonidae†. (a) *Praeichneumon dzhidensis*†; (b) *P. khamardabanicus*†. (Source: reproduced by permission of Dmitry S. Kopylov.)

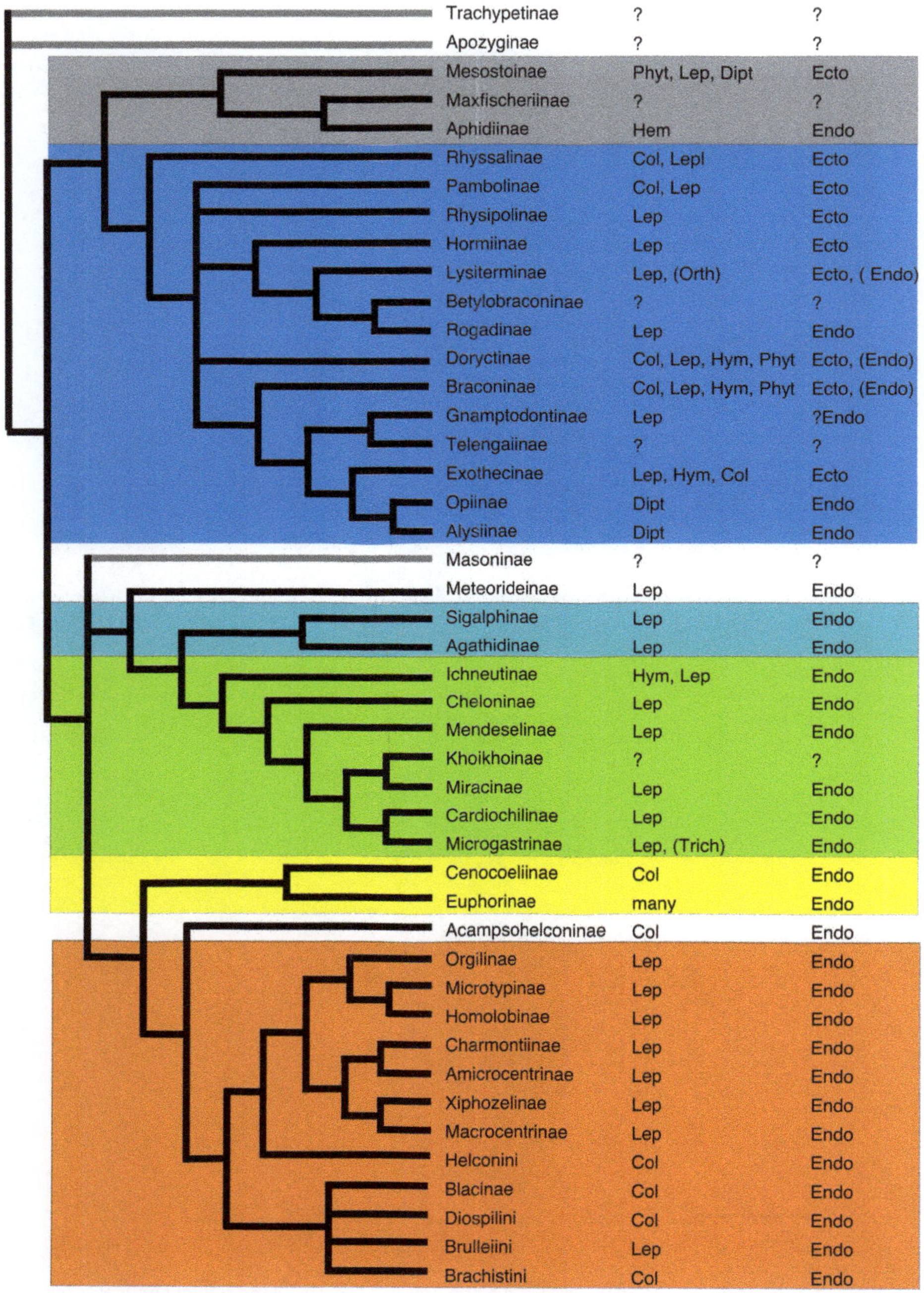

Plate 18 Best-guess phylogenetic tree, major host groups and mode of parasitism for the subfamilies of Braconidae, based largely on the multi-gene analysis of Sharanowski et al. (2011) with some groups added according to Zaldivar-Riverón's (2006) cyclostome study, and Murphy et al.'s (2008) microgastroid study. Host orders: Col, Coleoptera; Hym, Hymenoptera; Dipt, Diptera; Lep, Lepidoptera; Neur, Neuroptera; Orth, Orthoptera; Trich, Trichoptera; Phyt, phytophagous. Mode of parasitism: Ecto, ectoparasitoid; Endo, endoparasitoid. Colour blocks from top to bottom indicate the informal group names following Sharanowski et al.; aphidioids, cyclostomes, sigalphoids, microgastroids, euphoroids and helconoids; non-coloured ones are isolated taxa of uncertain affinity.

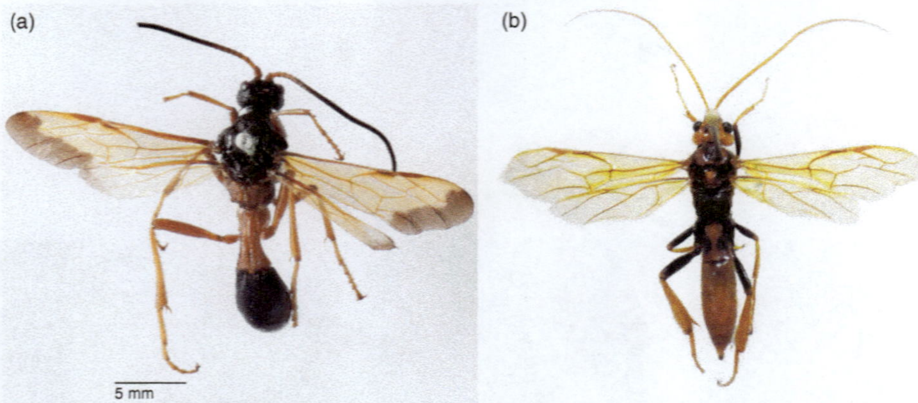

Plate 19 Trachypetinae habitus photographs. (a) *Trachypetus petiolatus*. (Source: reproduced by permission Andrew Austin.) (b) *Megalohelcon* sp. (Source: reproduced by permission, © C. van Achterberg, Naturalis Biodiversity Centre).

Plate 20 Aphidiinae and their interactions. (a) *Pauesia unilachni* attacking *Schizolachnus pineti* on *Pinus sylvestris*; (b) ant, *Lasius niger*, attacking *Trioxys angelicae* among colony of *Aphis fabae* on *Cirsium arvense*. (Source: reproduced by permission of Wolfgang Völkl.)

Plate 21 Features of *Maxfischeria* spp. (Maxfischeriinae). (a) Habitus *of M. folkertsorum*; (b) eggs *in situ* in ovary of dissected female *M. ovumancora*; (c) dissection showing single egg with anchor in the posterior oviduct; (d) detail of single egg; (e) SEM image of ovipositor showing extensive lower valve serrations. (Source: reproduced by permission of Andrew Boring and Barb Sharanowski.)

Plate 22 *Mesostoa kerri* (Mesostoinae). (a, b) Micropterous male; (c, d) macropterous female; (e) close-up dorsal view of head and mesoscutum; (f) front view of face. (Source: reproduced by permission of Andrew Austin.)

Plate 23 Various mounted specimens of Doryctinae showing some of the variation in body form. (a) *Iare* sp.; (b) *Sharkeyella* sp.; (c) *Histeromeroides onkotorebus*; (d) *Leptorhaconotus seyrigi*; (e) *Evaniodes spathiiformis*; (f) *Schlettereriella variegata*. (Source: reproduced by permission of Yves Braet.)

Plate 24 Aspects of the biology of *Histeromerus mystacinus* (Rhyssalinae). (a) Female encountering pupal beetle host; (b) cluster of approximately eight eggs deposited on the side of the host just behind the legs; (c) mature larvae and female wasp still in attendance; (d) holes made in cork stopper of tube from which females escaped by chewing their way out, and in piece of wood substrate. (Source: reproduced by permission Mark R. Shaw.)

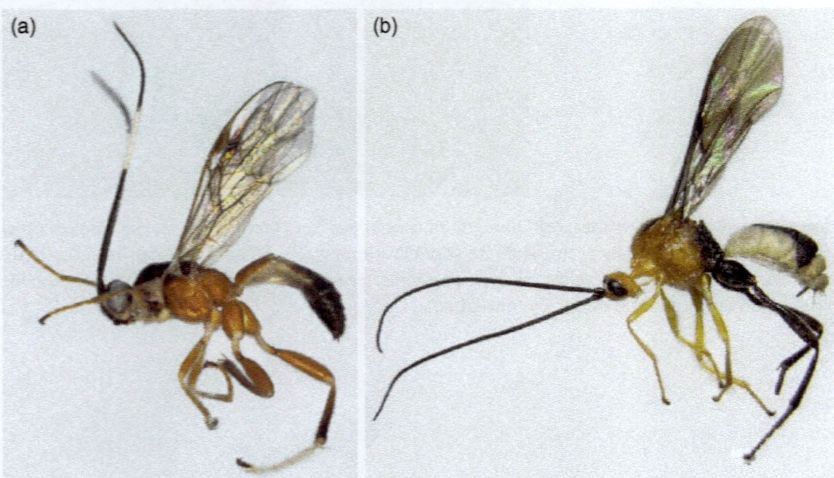

Plate 25 Exemplar Rogadinae. (a) *Aleiodes* (*Arcaleiodes*) sp., from Nepal; (b) *Batothecoides yakushimensis*, from Japan, (Source: part b, reproduced by permission of Kaoru Maetô.)

Plate 26 Three long-ovipositored braconines. (a) *Euurobracon yokahamae* from Japan, body darkened post mortem, in life it is pale orange–yellow. (Source: reproduced by permission of Kyohei Watanabe.) (b) *Monilobracon* sp. from Uganda, searching for host beetle borings on barkless tree trunk. (c) *Glyptomorpha pectoralis*, a reasonably common species from the Mediterranean region. (Source: reproduced by permission of Cor Zonneveld.)

Plate 27 Oviposition and development of the idiobiont, ectoparasitoid *Atanycolus cappaerti* (Braconinae) on larva of the emerald ash borer, *Agrilus planipennis* (Cerambycidae), an invasive pest of trees in North America, accidentally introduced from China. (a) Ovipositing female wasp; (b) egg on paralysed host larva (slightly to right of middle, arrow), note the long thin tail to the right of the egg body; (c, d) medium and nearly fully grown larva feeding on host; (e) isolated fully grown larva; (f) thin, papery parasitoid cocoon in host tunnel. (Source: reproduced by permission of David Cappaert, Michigan State University, Bugwood.org.)

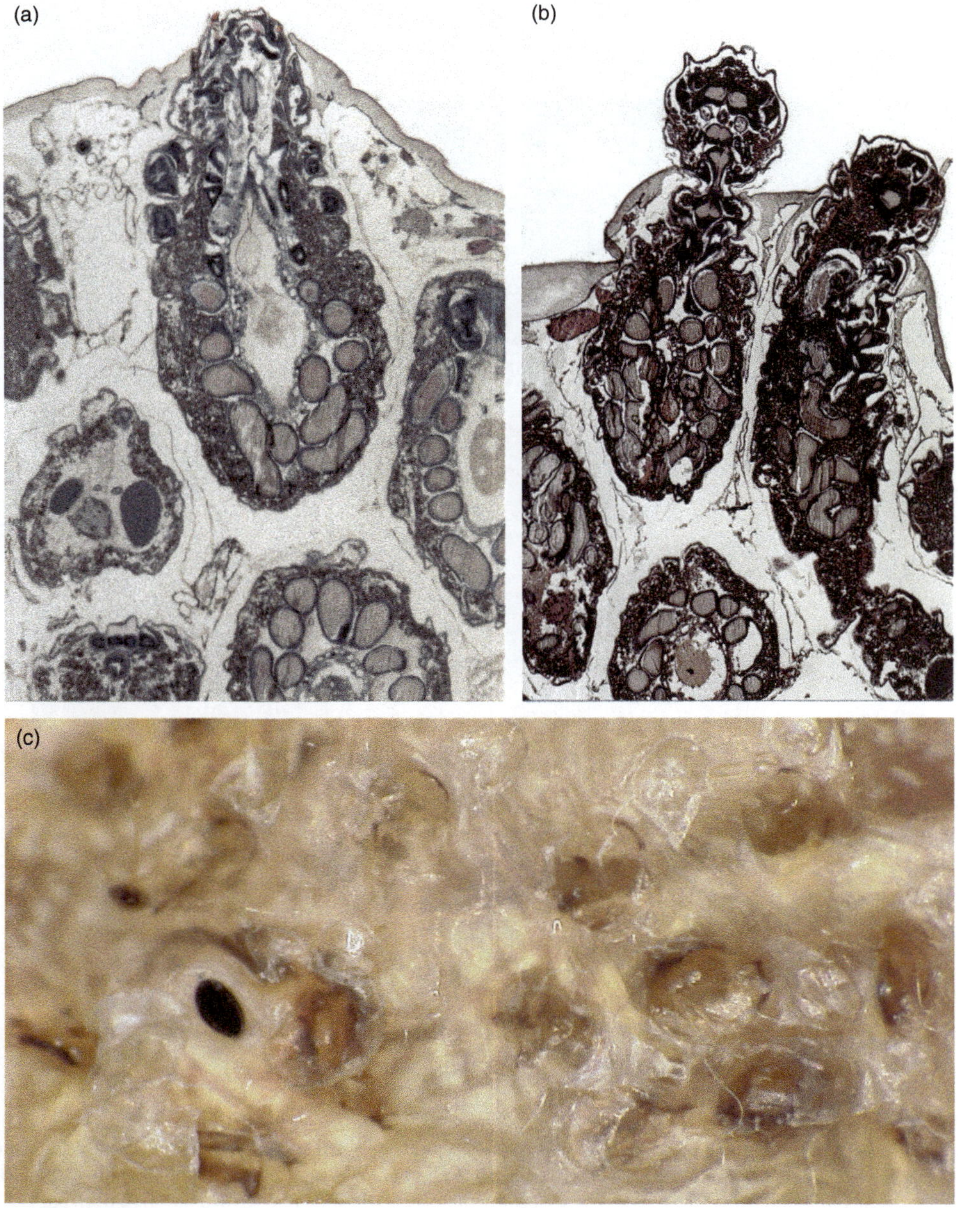

Plate 28 Features of larval egress of *Cotesia kariyai* (Microgastrinae) from its host *Pseudaletia separata* (Noctuidae). (a, b) Haematoxylin/eosin-stained, 8 μm paraffin wax histological sections through *Cotesia* larvae at different stages of exiting through the host cuticle, showing the anchoring material surrounding the parasitoid larval body within the host, and the torn moulted cuticle remains just protruding through the exit holes; (c) photograph of host integument after parasitoid emergence showing penultimate larval exuviae in the exit holes, apparently blocking them. (Source: reproduced by permission of Toshiharu Tanaka.)

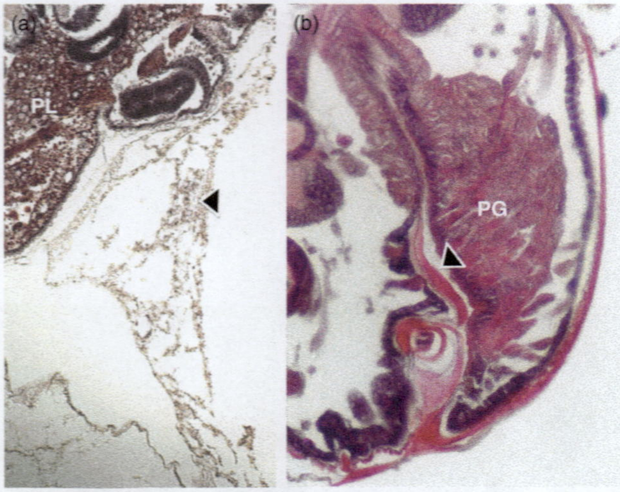

Plate 29 Histological sections of *Cotesia kariyai* (Microgastrinae) within a host. (a) Following periodic acid–Schiff (PAS) staining showing PAS-positive secreted anchor material (arrow head) outside of parasitoid larva (PL); (b) haematoxylin/eosin-stained, longitudinal section through parasitoid larva showing deep-pink staining of secretion (arrow head) that will form the anchoring material, from the pharyngeal gland (PG) spreading out over the anterior surface of the parasitoid larva. (Source: reproduced by permission of Toshiharu Tanaka.)

Plate 30 Examples of cocoons of Microgastrinae. (a) Overwintering cocoon of the univoltine microgastrine *Cotesia gonepterygis* showing how securely it is attached to the twig where it will spend the winter months; (b) *Microplitis impressa* (Microgastrinae), univoltine wasp showing hard, ribbed cocoon attached to the host; (c) *Glyptapanteles* sp.; (d) *Parapanteles* sp. (Sources: a, reproduced by permission of Mark R. Shaw; b, reproduced by permission of Marion Friedrich; c, d, reproduced by permission of Lee Dyer.)

Plate 31 Interactions between *Cotesia glomerata* (Microgastrinae) and its host, *Pieris brassicae* (Pieridae). (a) Female wasp attacking freshly emerged caterpillars; (b) individual caterpillar trying to defend itself; (c) parasitoid larvae emerging synchronously from the host and immediately starting to spin their cocoons in a mass underneath the moribund host; (d) 'usurpation' of the host which, in this species, not only stays on guard but additionally has spun extra silk covering over the parasitoid cocoon mass. (Source: reproduced by permission of Hans Smid/bugsinthepicture.com; http://bugsinthepicture.com.)

Plate 32 Impression fossils of the extinct subfamily Tanychorinae† including two recently described ichneumonids from the Early Cretaceous, Berriasian horizon of Russia, 139.8–145 Mya. (a) *Tanychora petiolata*; (b) *Khasurtella buriatica*; (c) *Amplicella mininae*. (Source: reproduced by permission of Dmitry S. Kopylov.)

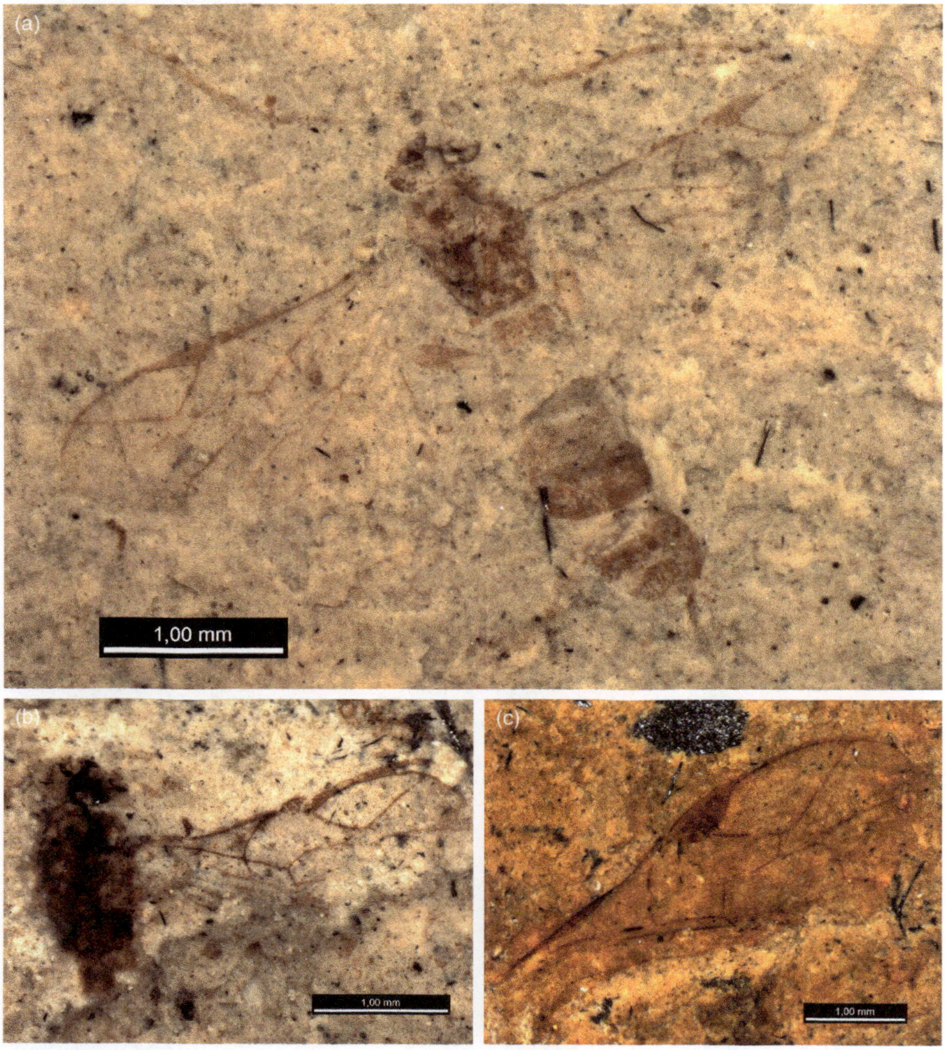

Plate 33 Impression fossils of Palaeoichneumoninae†. (a) *Palaeoichneumon freja*, showing short ovipositor; (b) *P. mirabilis*; (c) fore wing of *Rudimentifera mora* (Palaeoichneumoninae) from middle Lower Cretaceous (Hauterivian/Barremian) Dzun–Bain Formation of Mongolia ~129.4–132.9 Mya. (Source: reproduced by permission of Dmitry S. Kopylov.)

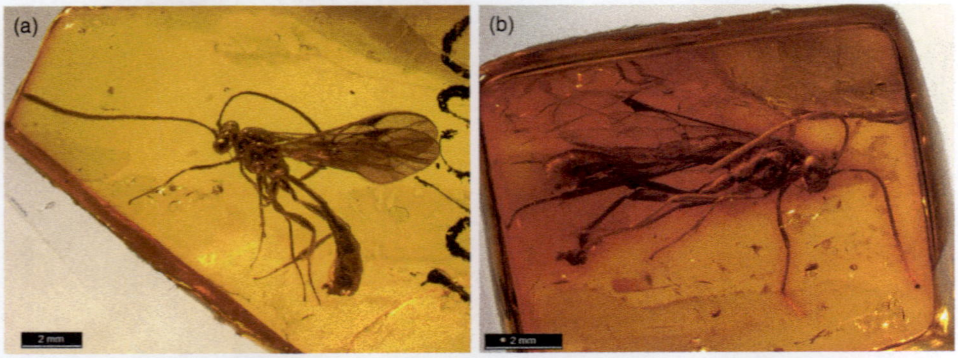

Plate 34 Baltic amber inclusions of *Pherombus antennalis* (extinct subfamily Pherombinae†). (a) Holotype female; (b) paratype male. (Source: reproduced by permission of Dmitry S. Kopylov.)

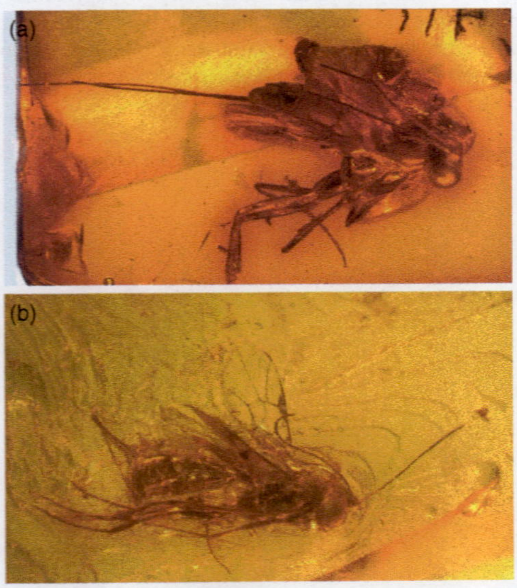

Plate 35 Amber inclusions of extinct ichneumonid subfamily Townesetinae†. (a) Holotype female *Townesites mandibularis*, body length approximately 2.5 mm; (b) specimen of *Marjorietta minor*, body length approximately 1.6 mm. (Source: reproduced by permission of Dmitry S. Kopylov.)

Plate 36 Features of Acaenitinae. (a) *Arotes moiwanus*, showing very large hypopygium typical of all members of subfamily except *Procinetus*; (b) *Coleocentrus excitator*; (c, d), *Procinetus vipioniformes*, showing long ovipositor and detail showing truncate hypopygium. (Sources: a, reproduced by permission of Kyohei Watanabe; b, reproduced by permission of Cor Zonneveld; c, d, reproduced by permission, © C. van Achterberg, Naturalis Biodiversity Centre.)

Plate 37 *Diplazon laetatorius* (Diplazontinae), one of, if not the, commonest ichneumonid wasps in the world. (Source: reproduced by permission of Karol Ox.)

Plate 38 Female of *Stenomacrus brevicornis* (Orthocentrinae). (Source: reproduced by permission of Ilari Sääksjärvi.)

Plate 39 Assorted Pimplinae species. (a) *Pimpla viridescens* (Pimplini); (b) *Xanthopimpla pedator* (Pimplini); (c) *Lamnatibia andina* (Ephialtini, *Polysphincta* group); (d) *Dolichomitus mesocentrus* (Ephialtini). (Sources: a, c, reproduced by permission of Ilari Sääksjärvi; b, reproduced by permission of Chan To; d, reproduced by permission of Cor Zonneveld.)

Plate 40 Larva and behaviour of *Polysphincta* group pimplines on host spiders. (a) *Acrodactyla* sp. on a tetragnathid spider in the UK; (b) *Reclinervellus masumotoi* fully grown larva killing host spider; note dorsal tubercles; (c, d), *Zatypota albicoxa*, infanticide, removing a penultimate instar rival's larva, and removal of a second instar rival larva from host spider, *Parasteatoda tepidariorum*, respectively; (e, f) webs of *Cyclosa argentoalba*, host of *Reclinervellus nielseni*, normal and modified webs, respectively. (Sources: a, reproduced by permission of Roger Key; b–f, reproduced by permission of Keizo Takasuka.)

Plate 41 Host feeding by adult female ichneumonids. (a) *Itoplectis aterrima* (Pimplinae) feeding on its host, *Hyposoter clausus*, through a hole in the host cocoon made by stabbing with ovipositor; (b) *Netelia vinulae* (Tryphoninae) simultaneously ovipositing and host feeding by biting a hole in the host cuticle with mandibles, on a host, *Cerura vinula*, caterpillar. (Source: reproduced by permission of Mark R. Shaw.)

Plate 42 Habitus of *Rodrigama* sp. (Poemeniinae) from Japan. (Source: reproduced by permission of Rikio Matsumoto.)

Plate 43 Sequence of photographs of female *Megarhyssa vagatoria* (Rhyssinae) 'drilling' her ovipositor into a tree trunk female. (Source: reproduced by permission of C. M. Zwakhals.)

Plate 44 Males of the European Rhyssinae showing scramble competition for mates. (a) *Megarhyssa vagatoria*, aggregating around the future point of emergence of (probably) a conspecific female; (b) *Megarhyssa rixator*, showing long slender metasoma; (c, d) *Rhyssa persuasoria*, showing male aggregation and more robust male metasoma, respectively. (Sources: a, reproduced by permission of C. M. Zwakhals; b–d, photographs by Reto Burri, Uppsala, Wildlight.ch, reproduced by permission of Seraina Klopfstein.)

Plate 45 *Adelognathus leucotrochus* (Adelognathinae) development on larva of host *Nematus leucotrochus* (Tenthredinidae). (a, b) Larvae after 3 and 30 h, respectively. (Sauce: reproduced by permission of Mark R. Shaw.)

Plate 46 *Agriotypus gracilis* (Agriotypinae). (a) Parasitised caddis case *in situ* on a boulder in a stream with the silk thread gill protruding; (b) isolated caddis case showing silk ribbon; (c), parasitised caddis case opened to show *Agriotypus* pupa with head at end with ribbon; (d) adult female habitus. (Sources: a–c, reproduced by permission of Toshimasa Mitamura; d, reproduced by permission of Kazuhiko Konishi.)

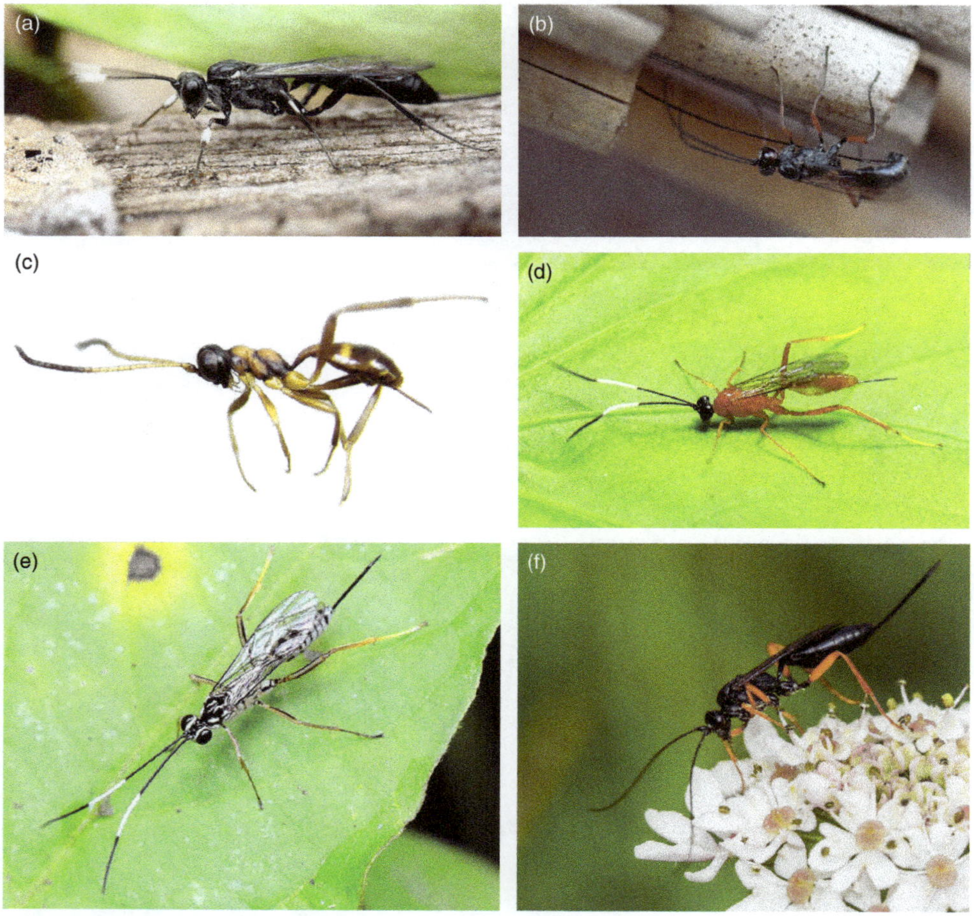

Plate 47 Various Cryptinae. (a) *Echthrus* sp.; (b) *Stenarella domator*, ovipositing in trap nest; (c) apterous *Gelis* sp., showing relatively reduced size of mesothorax; (d, e) unidentified species from Costa Rica; (f) probably *Buathra laborator* (det. G. Broad) feeding at flowers of Apiaceae. (Sources ; a, reproduced by permission of David Cappaert, Michigan State University, Bugwood.org; b, reproduced by permission of Cor Zonneveld; c, reproduced by permission of Kyohei Watanabe; d, e, reproduced by permission of Kenji Nishida; f, reproduced by permission of Paul Lathbury.)

Plate 48 Five female ichneumonines. (a) *Homotherus varipes*; (b, c) *Ichneumon sarcitorius*, female and male, respectively; (d) *Vulgichneumon satuatorius*; (e) gen. sp., showing oxypygous form of metasomal apex indicative of attacking pupal hosts from Australia. (Sources: a, reproduced by permission of Håkon Haraldseide; b, reproduced by permission of Paul Lathbury; c, reproduced by permission of Cor Zonneveld; d, reproduced by permission of Thijs de Graaf; e, reproduced by permission of Malcolm Tattersall.)

Plate 49 *Brachycyrtus nawaii* (Brachycyrtinae) female habitus. (Source: reproduced by permission of Kyohei Watanabe.)

Plate 50 Female habitus of *Orthopelma simile* (Orthopelmatinae). (Source: reproduced by permission of Kyohei Watanabe.)

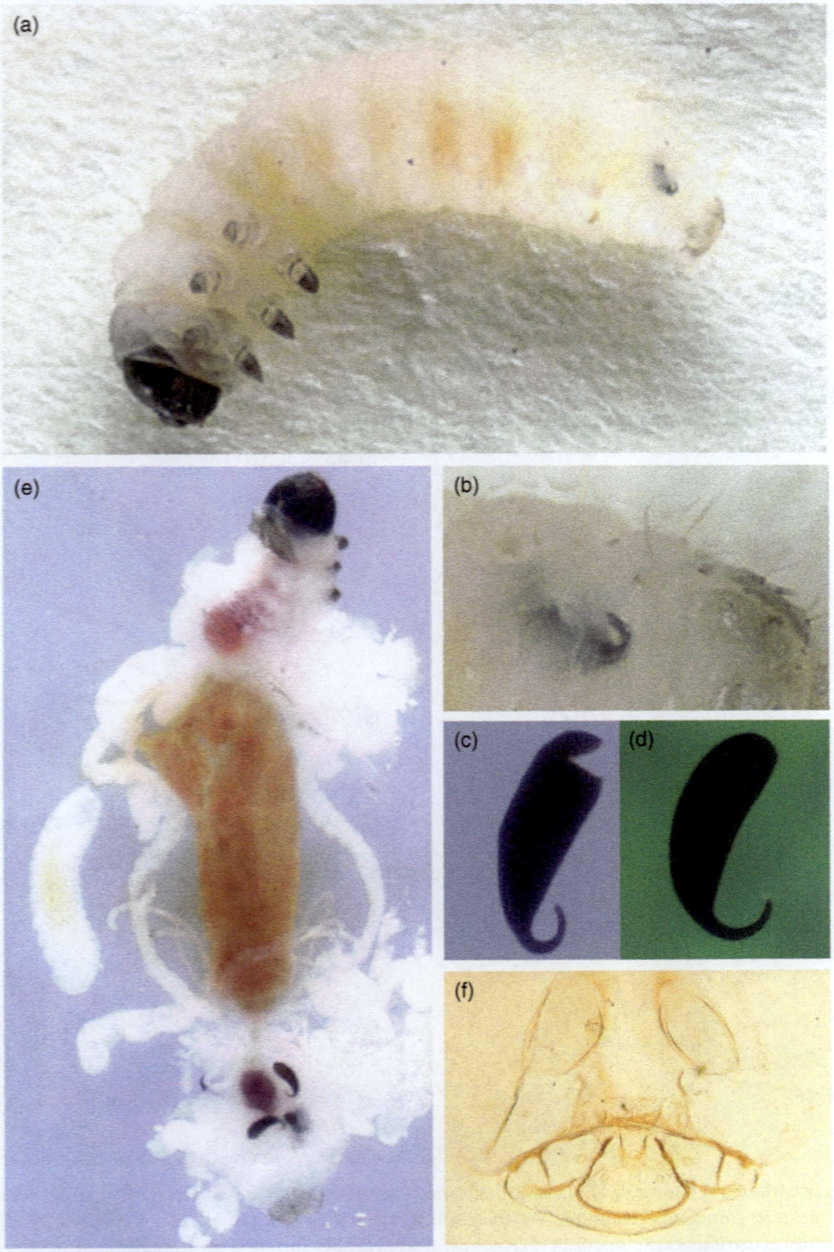

Plate 51 Features of the life history of *Panteles schuetzeanus* (Stilbopinae). (a) Host, *Lampronia fuscatella* (Incurvariidae), larva showing black *Panteles* egg with hooked tail, clearly visible through the cuticle at the posterior end of the body; (b) detail of egg within host; (c, d) closed and uneclosed eggs, respectively, the latter from a case of superparasitism with only one egg having hatched; (e) dissected host caterpillar showing an intermediate instar *Panteles* larva more or less where found, lateral to the mid-gut; (f) final larval instar head capsule. (Source: from Quicke 2005. Reproduced by permission of Taylor & Francis Ltd.)

Plate 52 Adults of various Tryphoninae. (a) *Excavarus apiarus* (Exenterini); (b) *Cladeutes* sp. (Oedemopsini); (c) female specimen of *Phytodietus longicauda* (Tryphoninae: Phytodietini) showing long exserted ovipositor used to attack hosts concealed in leaf-rolls; (d) *Sphinctus nigrithorax* (Sphinctini); (e) *Polyblastus* sp. (Tryphonini) showing multiple mature eggs being retained at the base of the ovipositor with only their anchors inside the ovipositor egg canal; (f) gen. sp. indet. showing single large egg stored at base of ovipositor. (Sources: a, reproduced by permission of Cor Zonneveld; b, reproduced by permission of Håkon Haraldseide; c, d, reproduced by permission of Kyohei Watanabe; e, reproduced by permission of Mark Shaw; f, reproduced by permission of Alan Hadley.)

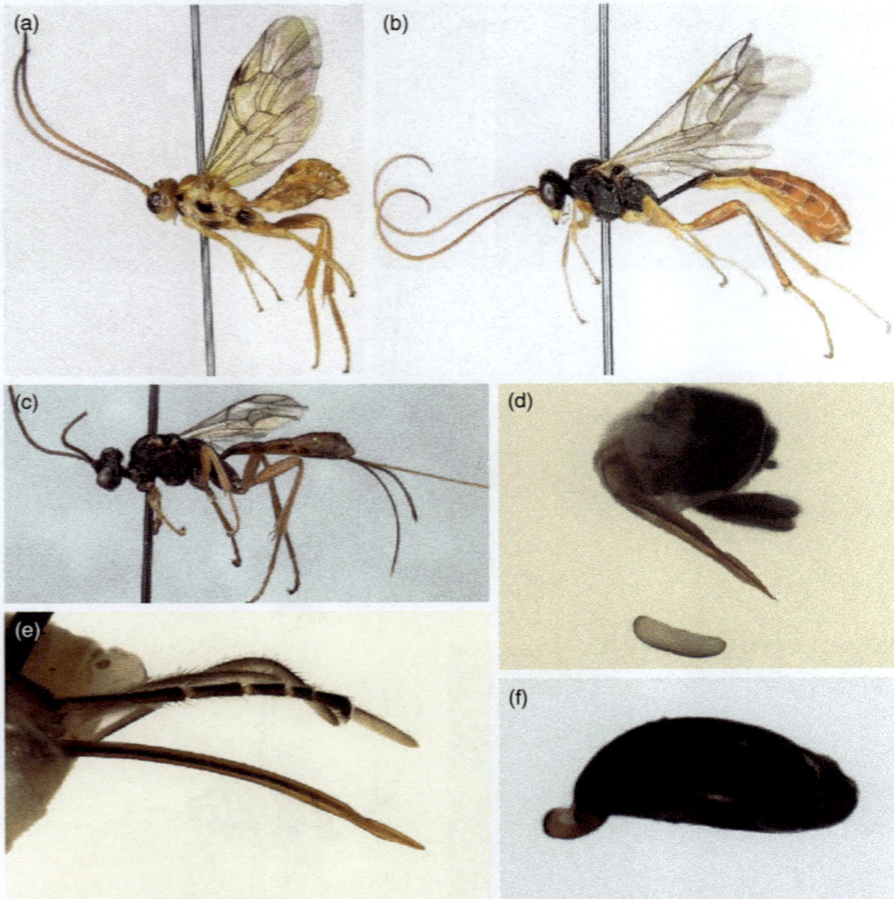

Plate 53 Ctenopelmatinae including features of ovipositor and eggs. (a, b) *Scolobates testaceus* (Scolobatini) and *Hadrodactylus orientalis* (Euryproctini), respectively, showing a short, notched ovipositor typical of most members of subfamily; (c) *Olethrodotis* sp. (Chrionotini), showing a long ovipositor that is largely restricted to this tribe; (d) *Oetophorus pleuralis* (Perilissini) dissected apex of metasoma and isolated egg ; (e) ovipositor of *Tetrambon* (Perilissini); (f) *Westwoodia* egg with putative anchor. (Sources: a, b, reproduced by permission of Kyohei Watanabe; c, e, f, © Robert A. Wharton, reproduced by permission; d, image by Heather Cummins, © Robert A. Wharton, reproduced by permission.)

Plate 54 Various Metopiinae. (a, b) Two views of wasp-mimicking *Metopius* sp.; (c) *Metopius browni*; (d) *Colpotrochia cincta*; (e) *Triclistus dimidiatus*, showing robust legs and protruding face. (Sources: a, b, reproduced by permission of Peter Waikwok Chew; d, reproduced by permission of Karol Ox; c, e, reproduced by permission of Kyohei Watanabe.)

Plate 55 Anomaloninae species. (a) *Anomalon* sp., showing relatively long ovipositor; (b) *Ophionellus* sp. from Peru, showing small wings with reduced venation; (c) *Habronyx insidiator* in the hand. (Sources: a, reproduced by permission of Bernardo Navarette; b, reproduced by permission of Ilari Sääksjärvi; c, reproduced by permission of Kyohei Watanabe.)

Plate 56 Photographs of various campoplegine cocoons. (a) *Scirtetes robusta* (Campopleginae) with host *Cosmia* sp. larval remains, showing typical pale equatorial stripe; (b) suspended cocoon of a campoplegine species; (c) striped, unsuspended cocoon formed underneath host caterpillar remains; (d) cocoon formed within host larval skin, probably by *Hyposoter* sp. (Sources: a, reproduced by permission of Marion Friedrich; b, reproduced by permission of Kenji Nishida; c, d, reproduced by permission of Mark Shaw.)

Plate 57 Hybrizontinae. (a) *Eurypterna* male habitus; (b) *Eurypterna* face; (c) *Gilaromma* habitus; (d) *Gilaromma*, detail of hind wing venation. (Source: reproduced by permission, © C. van Achterberg, Naturalis Biodiversity Centre.)

Plate 58 Features of Nesomesochorinae. (a–d) *Chriodes* sp. ♀: (a) habitus; (b) face showing strongly ventrally convergent eyes; (c) ovipositor (upper valve at top); (d) side view of campoplegine-like first metasomal tergum; (e) *Klutiana* sp. fore wing; (f) *Nonnus* sp. (Sources: a–e, reproduced by permission, © C. van Achterberg, Naturalis Biodiversity Centre; f, reproduced by permission of Andreas Kay.)

Plate 59 *Euceros* adults and juvenile stages (Eucerotinae) habitus photographs. (a, b) *E. sensibus* habitus photographs of female and male, respectively; (c, d) *E. ablitarsus*, stalked eggs and planidial larva on host, *Orthosia gothica* (Noctuidae), respectively. (Sources: a, b, reproduced by permission of Kyohei Watanabe; c, d, reproduced by permission of Mark Shaw.)

Plate 60 *Microleptes splendidulus* (Microleptinae). (Source: reproduced by permission, © C. van Achterberg, Naturalis Biodiversity Centre.)

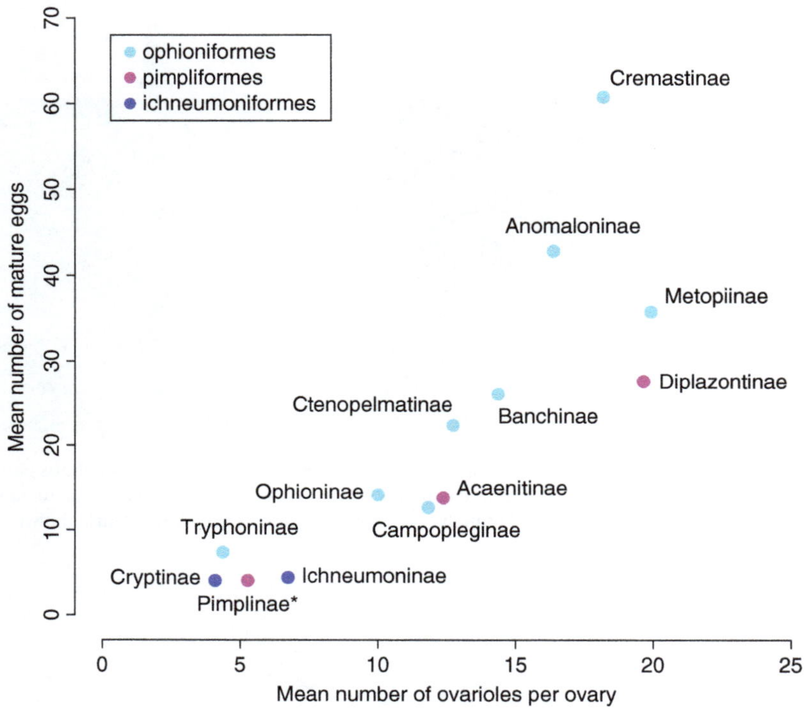

Plate 61 Relationship between number of mature eggs per ovary and number of ovarioles per ovary for 13 subfamilies of Ichneumonidae, coloured to show major subfamily groupings. (Source: data from Price 1973, fig. 3.))

Plate 62 Illustrations of field techniques. (a) Use of pan traps, combination of yellow and blue bowls, along an edge, but often they would be set more widely spaced; (b) collecting ichneumonoids off a light sheet in Taiwan; (c) sweep netting using a Noyes design net fitted with detachable wire mesh guard to prevent entry of large, coarse vegetation fragments enabling sweeping to be carried out for longer and facilitating subsequent sorting; (d) a separating bag modified from van Achterberg's (1981a) design. (Sources: a, d, reproduced by permission of Scott R. Shaw; c, reproduced by permission James B. Woolley.)

Plate 63 Illustrations of field techniques and caterpillar rearing in tropics. (a) Extraction of parasitoids from sweep net using the head in the bag technique and holding the tip of the net towards the light, while pooting (aspirating) interesting specimens; (b) a canopy Malaise trap with four vanes and netting floor, secured by strings from below; (c) Malaise trap set up in moist tropical forest with collecting head end protruding into the trail where it is lightest; (d) rearing bags for field-collected caterpillars in the tropics, housed in a well-ventilated, shady shed. (Sources: a, c, reproduced by permission of Nina M. Laurenne; b, d, reproduced by permission Scott R. Shaw) .

Fig. 12.73 *Masona* sp. (Masoninae). (a) Side view habitus; (b) dorsal view habitus; (c) detail of head. (Source: reproduced by permission, © C. van Achterberg, Naturalis Biodiversity Centre.)

Meteorideinae

This is a widespread, principally tropical group with only two included genera, *Meteoridea* and *Pronkia*, although evidence that these actually form a monophyletic group is scant (see below). *Pronkia* is only known from New Zealand, whereas *Meteoridea* occurs throughout the rest of the world (except for the West Palaearctic). Both are small wasps and both appear to be essentially nocturnal and are correspondingly yellowish in colour. In the Old World tropics and in particular in South-East Asia, *Meoteoridea* are fairly commonly collected at light traps and two species are apparently important in the control of pest species numbers, which make the paucity of our knowledge all the more surprising. The subfamily was erected by Čapek (1970) based largely on the cephalic structure of the final instar larva, *Meteoridea* having a complete and robust epistoma and not fitting well in the Helconinae, where it had previously been placed, largely because of the similarity in shape of the fore wing second submarginal cell.

Biological details, other than simple host records, are only known for *Meteoridea* and, even then, only a few things are certain as there is only one paper on them (Ghosh & Abdurahiman 1984). These wasps are solitary or gregarious, in the latter case emerging as adults in very large numbers (often hundreds of individuals) from the pupae of their lepidopteran hosts. The best known 'species' biologically is *M. 'hutsoni'* which attacks the black-headed caterpillar, *Opsina arenosella*, pests on coconut (Lepidoptera, Xyloricitidae) and *Sylepta* species, including the cotton leaf roller, *Sylepta derogata* (Lepidoptera: Crambidae); the latter species in Africa apparently also being attacked by *M. testacea*. Molecular data (CO1 barcoding gene fragment) indicate that there are potentially very many morphologically extremely similar species of *Meteoridea*, so the possibility exists that what workers have been referring to as *M. hutsoni* might not actually all be the same species. *M. hutsoni* has also been reported to act as a hyperparasitoid of a microgastrine (Ghosh & Abdurahiman 1986).

Ghosh and Abdurahiman (1984) described its biology briefly, although unfortunately did not illustrate the larval stages. Oviposition of the species that they studied is into mid- to late-instar hosts and the parasitoid completes its development after the host has pupated. Although solitary in these hosts, superparasitism apparently occurs but larval competition results in the elimination of all but one of the larvae – the exact stage at which this occurs is not known, but from what has been written it might be concluded that larval fighting does not occur, at least not at the first instar stage, and physiological suppression is a distinct possibility. The final instar larva spins a cocoon within the host pupa and, in the case of the solitary *M. hutsoni*, emergence is through a hole chewed in the anterior end, although in gregarious species there are multiple emergence holes all over the pupa.

Courtship has been observed to commence only when a male is in close proximity to a female and involves wing fanning. Copulation takes place with the male extending his metasoma forward under his body

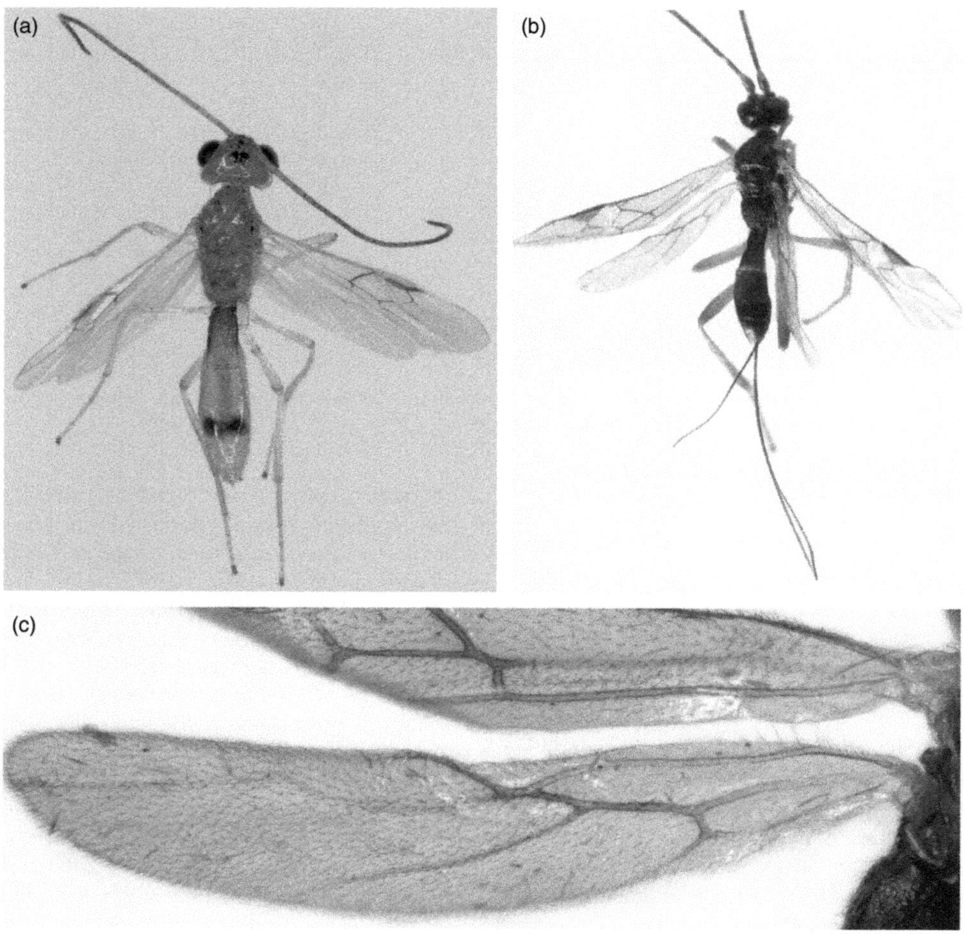

Fig. 12.74 Meteorideinae. (a) *Meteoridea hutsoni*; (b, c) *Pronkia antefurcalis* habitus of holotype and detail of wing venation./ (Sources: a, reproduced by permission of Janakiraman Poorani, National Institute of Nationally Important Insects, Bangalore; b, c, reproduced by permission, © C. van Achterberg, Naturalis Biodiversity Centre.)

before mounting her and lasts between 35 and 45 s. Copulation is accompanied by antennal contact and the female vibrates her antennae in response. Males mate multiply, but females only mate once and do not accept further courtship attempts.

Pronkia is really rather dissimilar to *Meteoridea*, as reflected in the fact that van Achterberg (1990c) placed each in a separate monotypic tribe (Pronkiini). Unlike *Meteoridea* (Fig. 12.74a), *Pronkia* has a long exposed ovipositor (Fig. 12.74b,c), possesses a distinct fore wing vein 1RS which is strongly curved (reminiscent of that of some Ichneutinae), an antefurcal 1cu-a and a desclerotised fore wing vein M+CU, among many other things, and I am far from convinced that the two genera form a monophyletic group, although they may well belong to the same larger clade. Indeed, van Achterberg noted that *Pronkia* superficially resembles members of the diospiline genus *Vadum* and that in many molecular analyses the Diospilinae either forms a clade with the Meteorideinae and Acampsohelconinae or they form a grade, which suggests that van Achterberg's initial feelings may not have been far off the mark.

ENDNOTES

1. Vein 2m-cu is present in approximately 5% of specimens of *Histeromerus* (see Histeromerinae below) and in a tiny percentage of specimens of some Rhyssalinae and Helconinae.
2. Marsh (1977) concluded that this species was probably present in N. America prior to its mass introduction from Europe.
3. Yves Braet is currently preparing a regional online generic key (see http://doryctinaekey.myspecies.info/node/1881)
4. a fossil Mesocentrus species has recently been described from the Palaearctic Eocene, viz. Butcher et al. 2014 Zootaxa 3860: 449–463.
5. In older literature *Rogas* was frequently miss-spelt as *Rhogas*
6. These three are all considered as subgenera of *Coelinius* by some authors.
7. Some taxonomists follow the International Code of Zoological Nomenclature in modifying (when necessary) the gender endings of specific names when species are transferred between genera; some do not.
8. *Habrobracon* is a fairly easily diagnosed genus (Quicke 1987) and I prefer to treat it separately from *Bracon*, although it, along with numerous other genera, may render the latter paraphyletic (Belshaw et al. 2001).
9. Also Microbraconinae based on the belief that correct name for *Bracon* was *Microbracon*.
10. Rhyssalinae, which are undoubtedly members of the cyclostome lineage, do not have the highly muscularised or spirally sculptured venom reservoir but their primary ducts are still strengthened with conspicuous spiral annelli.
11. In some older literature it is referred to as the Braconinae because of misinterpretation of the type species by early authors. Opinion 162 (1945) resolved the issue by suspending the rules of zoological nomenclature and designated *Ichneumon minutator* Fabricius, 1798 as the type of *Bracon*, thus preserving the usage of the name, despite the fact that the true type was *Ichneumon desertor* Linnaeus, 1758, which is the type of the agathidine genus *Cremnops*.
12. The family group name Mesocoelini Viereck, 1918, predates Aneurobraconinae Fahringer, 1936.
13. The doryctine genus *Iare* has the same feature.
14. A mixture containing many polysaccharides and some protein-degrading enzymes, originally of fungal origin.
15. Here and later, the word 'Authors' or the term 'of Authors' does not signify someone called Authors, but instead means in the sense of many authors who wrote about it, usually applying the name incorrectly.
16. In the light of recnt molecular findings, the outgroups employed in these two studies were probably both inappropriate.
17. Blacinae and Euphorinae lack fore wing vein 2cu-a (CU1b) and so have a distally open second subdiscal call, which Tobias considered important *contra* van Achterberg (1977).
18. Although bdelloid rotifers have done similarly over the past 80 or so million years in the absence of sex, albeit with a faster generation time.
19. Van Achterberg (1992d) found that *falcatus* of Authors comprised four species, although the genuine *falcatus* attacks the same host as Loan and Holdaway studied.
20. The generic name *Loxocephalus*, long used in the Braconidae, was found to be pre-occupied by a genus of protist (Foissner & van Achterberg 1997).
21. Approximately 98% of all extant Lepidoptera, and nearly all hosts of Microgastrinae, belong to this group.
22. *Bitomus* and *Coleopus* (Opiinae) and some species of *Minanga* (Sigalphinae) also have a posterior mesopleural carina and carapace-like metasoma; see Chapter 10, section *Carapacisation*.
23. Mason (1991) summarised what is known of reductions in numbers of metasomal spiracles in adult Braconidae.
24. The rare genus *Pelicope* from southern USA has an even more cardiochiline-like second submarginal cell, but it has not yet been included in any published molecular phylogenetic studies.
25. An earlier study by Weseloh (1976) suggested that the Dufour's gland was the source of the sex pheromone, but this was subsequently refuted.
26. As a junior synonym of Planitorini.

Chapter 13

PHYLOGENY AND SYSTEMATICS OF THE ICHNEUMONIDAE

The Braconid and Ichneumonid Parasitoid Wasps: Biology, Systematics, Evolution and Ecology, First Edition. Donald L. J. Quicke.

HISTORY OF ICHNEUMONID CLASSIFICATION

Uncovering relationships within the Ichneumonidae has been no less difficult than for the braconids and, indeed, Gauld and Mound (1982) discussed the high level of homoplasy apparent in the family which necessarily makes the job of the phylogeneticist harder. Historically, the Ichneumonidae was divided into just a small number of subfamilies which were transcribed on the basis of a very few features and have subsequently turned out to be largely unnatural. The first real attempt to arrange the ichneumonids was by Gravenhorst (1815, 1829), who recognised 14 subfamilies (which he referred to as genera), and these were divided into four groupings depending on the combination of whether the metasoma was dorsoventrally or laterally compressed and whether the first tergite was petiolate or sessile (Townes 1969; see also http://www.amentinst.org/GIN/towclas.php). The number of subfamilies recognised gradually increased through the 20th century with attempts to create a more natural classification and now stands at 38, but this number is unlikely to remain stable when a better phylogenetic understanding becomes available for the complex of subfamilies and tribes, particularly those centred around the Ctenopelmatinae.

Gauld (1991, 1997, 2000) provided a broad division of the subfamilies into major groups, with small changes in each case, but significantly left a number of subfamilies as 'unplaced taxa'. A preliminary molecular study (Belshaw et al. 1998) based on very few representative taxa and with many missing subfamilies nevertheless supported many of Gauld's intuitive divisions. In the molecular world things tend to advance quickly, not least the processing speed for DNA samples, so whereas in 1998 Belshaw et al. only had sequences for 101 genera, of which 56 were ichneumonids, the most recent studies (Quicke et al. 2000b, 2009) included 60 and 641 ichneumonid genera respectively. The taxonomic coverage of these newer data sets breaks up many of the longer phylogenetic branches and thus makes substitutions that are globally homoplastic, locally informative (Purvis & Quicke 1997). Of course, there is a trade-off between including large numbers of taxa and the number of genes sequenced, and the larger studies only amplified the D2–D3 expansion region of the nuclear 28S rDNA gene. However, although this data set was not easy to analyse, it has firmed up many aspects of the ichneumonid tree and also shown a number of misplaced taxa. Of course, nothing is that simple: several taxa still remain on long branches, notably the problematic Hybrizontinae. The gene region used has also been criticised and is certainly not straightforward to work with as, despite including many phylogenetically informative sites that are relevant to genus up to subfamily relationship levels, it is length variable. As shown by Laurenne et al. (2006), the homoplastic occurrence of large indels, computerised alignment or homology line estimations fail. Quicke et al. (2009) made an attempt to overcome this in their analyses of their large, 1001 sequence, data set, by eliding 120 separate gene alignments created using a range of gap-opening/gap-extension parameter combinations with the program Clustal X. The most parsimonious trees recovered from analysis of the purely molecular data set were truly dreadful, as appraised by their failure to recover as monophyletic hardly any of the subfamilies whose monophyly seems otherwise well established. It was necessary, therefore, to combine the elided molecular data with the morphological data set of 162 characters, since no particular parameter combination seemed preferable over any other. Since the 120 elided DNA sequences included over 50,000 informative characters, this combination required differential weighting of the data partitions and the one used gave equal weight to DNA and morphology based on the retention index of the best trees from separate analyses. The tree that I am using here as the basis for the classification and discussion of extant taxa is presented in Fig. 13.1. It is largely based on the tree in fig. 21 of Quicke et al. (2009) but with Eucerotinae and Microleptinae excluded and Orthopelmatinae shifted more in accordance with Quicke et al. (2000b).

Gauld's (1991) tryphonoid group (tryphoniformes) has been thoroughly dismantled. The largest of its three subfamilies, the Tryphoninae, are now well established within the ophioniformes and in this respect the placement is supported by internal anatomy with the long lateral oviducts (Quicke et al. 2000b, 2009). The Eucerotinae, which have stalked eggs (see Fig. 13.58c) but of a completely different nature to those of the Tryphoninae, have appeared to be associated with a few other small groups in what we now call the brachycyrtiformes in some analyses, but not in others, but they do not appear to belong to the Tryphoninae. and the Adelognathinae which glue their eggs on to the

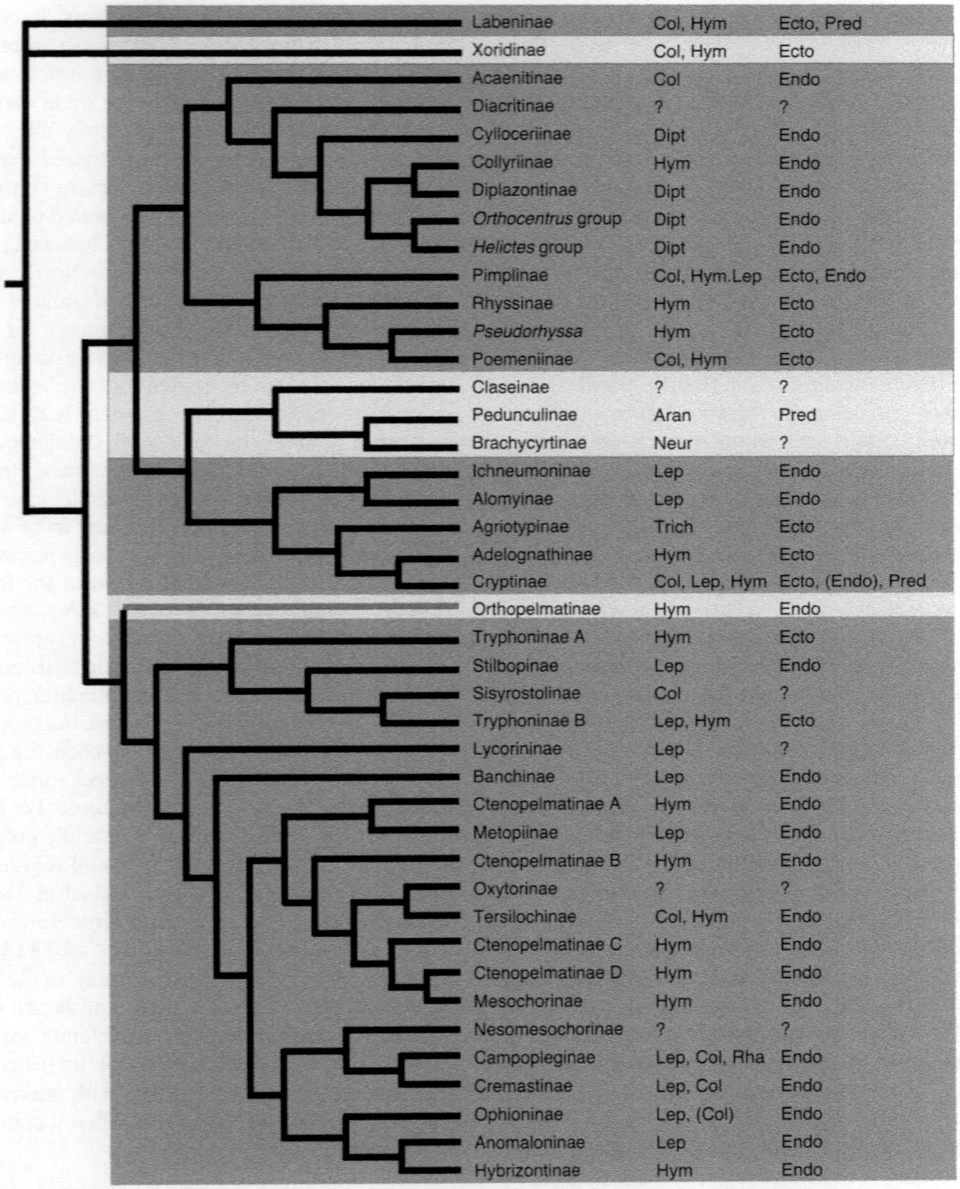

Fig. 13.1 Best-guess phylogenetic tree, major host groups and mode of parasitism for the extant subfamilies of Ichneumonidae, based largely on the elided molecular and morphological analyses of Quicke et al. (2009) but with Orthopelmatinae moved more basal (see text), Eucerotinae, and Microleptinae omitted as their recovered placements are too unreliable. Host orders: Col, Coleoptera; Hym, Hymenoptera; Dipt, Diptera; Lep, Lepidoptera; Neur, Neuroptera; Rha, Rhaphidioptera; Trich, Trichoptera. Mode of parasitism: Ecto, ectoparasitoid; Endo, endoparasitoid; Pred, predatory on egg clusters or brood cells. Shaded blocks from top to bottom indicate the informal group names that have been given in the literature: xoridiformes, labeniformes, pimpliformes, brachycyrtiformes (= Gondwanan clade), ichneumoniformes, orthopelmatiformes and ophioniformes; the old tryphoniformes of Gauld are now completely broken up (Tryphoninae in ophioniformes, Eucerotinae unplaced though possibly in brachycyrtiformes, Adelognathinae in ichneumoniformes).

outsides of their hosts are placed with the Cryptinae and Ichneumoninae group, possibly as a sister group to, or even derived within, the cryptines.

Almost all of Gauld's (1991, 1997) 'unplaced' subfamilies are now satisfactorily associated with major groups. The Metopiinae, Stilbopinae, Neorhacodinae and Phrudinae are all now in the ophioniformes, although with Neorhacodinae along with Phrudinae *sensu stricto*, both synonymised with the Tersilochinae. The situation with the unplaced genera listed by Townes (1969, p. 31) is far from satisfactory and the relationships of *Bremiella*, *Lapton*, *Ischyrocnemis* and *Microleptes*, and also of the Eucerotinae, I consider still highly uncertain. The Hybrizontinae appear most probably to belong to the ophioniformes [despite being recovered as a sister group to the Braconidae by Quicke et al. (1999b), which was probably a result of long-branch attraction]. The Diacritinae reasonably certainly belong to the pimpliformes, although I think this is not absolutely secure.

The 38 currently recognised extant subfamilies of Ichneumonidae are arranged here into seven informal groups (Table 13.1) (Quicke et al. 2009). In addition, five extinct subfamilies are recognised, although some and possibly most of these may be paraphyletic. Gauld (1985) surmised that the ancestral ichneumonid probably had a longish ovipositor without a pre-apical dorsal notch, but with an areolate propodeum, a metasoma that was broad at the base and with a free sternite, simple or lobed claws (i.e. not pectinate) and a mesoscutum with well-impressed notauli. The accepted view has been that they probably attacked xylophagous hosts such as beetle larvae (however, see Chapter 11, section *Ancestral biology of Ichneumonoidea*). Based on these presumptions, Gauld suggested that the most basal of the extant ichneumonids was likely to be found among the Pimplinae, Labeninae and 'perhaps also the Tryphoninae', and these notions have considerably influenced the rooting of morphological phylogenetic studies. Table 13.1 shows the subfamily level classification employed here, along with major informal groupings ('-formes') and synonyms that might be encountered, especially in some of the earlier literature.

HENRY TOWNES (1913–90) AND HIS IDIOSYNCRATIC NOMENCLATURE

One of the most confusing things that workers have to get to grips with is that the names for some subfamilies tribes and genera employed by Henry Townes differ from current usage and, even worse, some of the names used by him for one group are now applied to a completely different taxon so the same name in the literature can refer to two completely different entities. These differences cannot be ignored since Henry Townes made such vast and important contributions to ichneumonid taxonomy over a period of more than 50 years, and his terminology was followed my many other entomologists. Like numerous North American workers, Townes did not approve of some of the rules made by the International Commission for Zoological Nomenclature (ICZN) or many of its quasi-legal decisions. Although he argued his case logically, most subsequent workers have accepted the ICZN as a useful body and followed its Code and Opinions. Townes' nomenclature system differed in two ways. First, he did not accept the ICZN's Opinions when it came to deciding what is the correct generic name to be applied to a group in cases where there was early homonymy or synonymy. Second, he disagreed about the formation of family group[1] names. The ICZN Code states that the correct family group name for a taxon is that relevant family group name that was published earliest (i.e. has priority). Townes and various of his contemporaries in the 1940s thought that that was not the best route to long-term stability and instead applied the principle of strict priority, i.e. that the correct name must be based on the oldest valid generic name. For example, the subfamily Campopleginae is a family group name first published by Förster in 1869 and is based on the genus *Campoplex* which was described in 1829. But, within the Campopleginae is the genus *Porizon*, which was described by Fallén as far back as 1813. That is why Townes referred to the subfamily as the Porizontinae. Table 13.2 gives the dates and hence rationale behind Townes' applications of subfamily names and Table 13.3 gives a quick way to look up his genus and tribe names that are at variance with current usage. Most confusion is likely to occur when it comes to the Pimplinae.

THE EXTINCT SUBFAMILIES

Tanychorinae†

Townes (1973a) described an early fossil ichneumonid, *Tanychora petiolata*, from early Cretaceous amber of Russia, which is still the earliest known

Table 13.1 Classification of Ichneumonidae employed throughout this book with synonyms commonly encountered in post-1900 literature.

Ichneumonidae	Obsolete names, miss-spellings, etc.
Xoridiformes	
Xoridinae	
Labeniformes	
Labeninae	Labiinae, Orthognathelinae, Groteinae
'Gondwanan group'	
Brachycyrtinae	
Clasinae	
Pedunculinae	
Pimpliformes	
Acaenitinae	Coleocentrinae
Colyriinae	
Cylloceriinae	
Diacritinae	
Diplazontinae	Diplazoninae
Orthocentrinae	Helictinae, Plectiscinae
Pimplinae	Ephialtinae
Poemeniinae	
Rhyssinae	
Ichneumoniformes	
Adelognathinae	
Agriotypinae	
Alomyinae	
Cryptinae	Gelinae, Hedycryptinae, Hemigasterinae, Hemitelinae, Latibulinae, Mesoleptinae, Mesosteninae, Osprynchotinae, Pezomachinae, Phygadeuontinae, Stilpninae, Trachyspyrinae
Ichneumoninae	Amblytelinae, Heresiarchinae, Joppinae, Joppocryptinae, Listrodrominae, Phaeogeninae, Troginae
Orthopelmiformes	
Orthopelmatinae	
Ophioniformes	
Anomaloninae	Anomalinae, Gravenhorstiinae, Hymenoptarsalinae, Nototrachinae, Pharsalinae, Theriinae, Therioninae
Banchinae	Exetastinae, Lissonotinae, Megacremastinae, Townesioninae
Campopleginae	Charopsinae, Cryptophioninae, Limnerinae, Porizontinae
Cremastinae	Eiphosominae, Pristomerinae, Xiphosominae
Ctenopelmatinae	Catoglyptinae, Megaceriinae, Mesoleiinae, Scolobatinae, Zemiophorinae
Hybrizontinae	Ghilarovitinae, Pachylommatinae
Mesochorinae	
Metopiinae	Exochinae, Laptoninae, Spudaeinae, Tylocomninae
Nesomesochorinae	Nonninae
Ophioninae	Hellwigiinae, Paniscinae
Oxytorinae	
Stilbopinae	
Sisyrostolinae	Brachyscleromatinae
Tatogastrinae	
Tersilochinae	Neorhacodinae, Phrudinae, Thersilochinae
Tryphoninae	Ctenisсinae, Exenterinae, Idiogrammatinae, Lysiognathinae, Neteliinae, Sphinctinae, Thymaridinae
Unplaced	
Ecerotinae	Eucerinae
Microleptinae	

Source: mainly from Fitton & Gauld (1976).

Table 13.2 Explanation of Henry Townes' idiosyncratic subfamily names Ichneumonidae at variance with current usage.

Correct name, author and date	Type genus and date	Townes' name	Type genus of Townes' name and date	Comments
Campopleginae Förster 1869	*Campoplex* Gravenhorst, 1829	Porizozontinae	*Porizon* Fallén, 1813	The correct identity of the type species of *Porizon* is a campoplegine, but Gravenhorst and subsequent authors misinterpreted it as a tersilochine (Gauld & Wahl 2013)
Ctenopelmatinae Förster, 1869	*Ctenopelma* Holmgren, 1857	Scolobatinae	*Scolobates* Gravenhorst, 1829	Townes was simply applying his principle of absolute priority
Cryptinae	*Cryptus* Fabricius, 1804	Gelinae	*Gelis* Thunberg, 1827	The first use of the name *Cryptus* was by Jurine (1801), validated by Panzer (1804) for a species of sawfly, prior to Fabricius (1804), but ICZN Opinion 1757 suppressed all uses of *Cryptus* prior to Fabricius
Pimplinae Wesmael, 1845	*Pimpla* Fabricius, 1804	Ephialtinae	*Ephialtes* Gravenhorst, 1829	Townes followed by Carson (1969), incorrectly, that Opinion 159 which suppressed *Ephialtes* Schrank 1802 was not valid (Wahl & Mason 1995)

Table 13.3 Correct and generally used interpretations of Townes' non-standard tribe and genus names; the differences in interpretation result from Townes considering Opinion 159 invalid.

Townes' name	Correct name
Ephialtini	Pimplini
Ephialtes	*Apechthis*
Coccygomimus	*Pimpla*
Ichneumonini	Heresiarchini
Ichneumon	*Coelichneumon*
Pimpla	*Ephialtes*
Pimplini	Ephialtini
Joppini	Ichneumonini
Pterocormus	*Ichneumon*

probable member of the Ichneumonidae dating to 121–125 Mya. *Tanychora* shows several plesiomorphic features and has been important in interpreting various characters in modern taxa. In particular, it has a very large second submarginal cell (areolet) compared with most extant ichneumonids, with vein 2RS very oblique and vein 1Ma long (Fig. 13.2a) and fore wing vein (Rs+M)a is also fully developed and several fossils show well-developed and complete sets of propodeal carinae (Fig. 13.2a). Some specimens show a distinct trace of a costal cell apically (e.g. *Khasurtella*), but because they are only known from impression fossils the possibility of some post-mortem separation cannot be excluded. The second and third metasomal tergites appear to show a distinct separation rather than being fused as in most extant braconids and extinct eoichneumonines. Since its original description, two new and putatively related genera have been added to the Tanychorinae† (Fig. 13.2b,c) which share its plesiomorphic venation (Kopylov 2010a, 2011).

Palaeoichneumoninae†

This subfamily was erected by Kopylov (2009) for three genera and 12 species of Lower Cretaceous ichneumonids that are somewhat intermediate in morphology between Tanychorinae† and recent taxa (Fig. 13.3). They have a smaller second submarginal cell (areolet) than tanychorines, but with veins 3RSa and 1Ma (see Fig. 2.4b, but referred to as 2+3M by Kopylov) still rather long, suggesting, as is applied in this book, that the basal-most vein of the ichneumonid areolet is 2RS migrated distally rather than 2rs-m, which would make the distal cross-vein closing the areolet 3rs-m (as in Fig. 2.4a). They also mostly possess a ramulus from 1m-cu&Rs+M, which in these almost

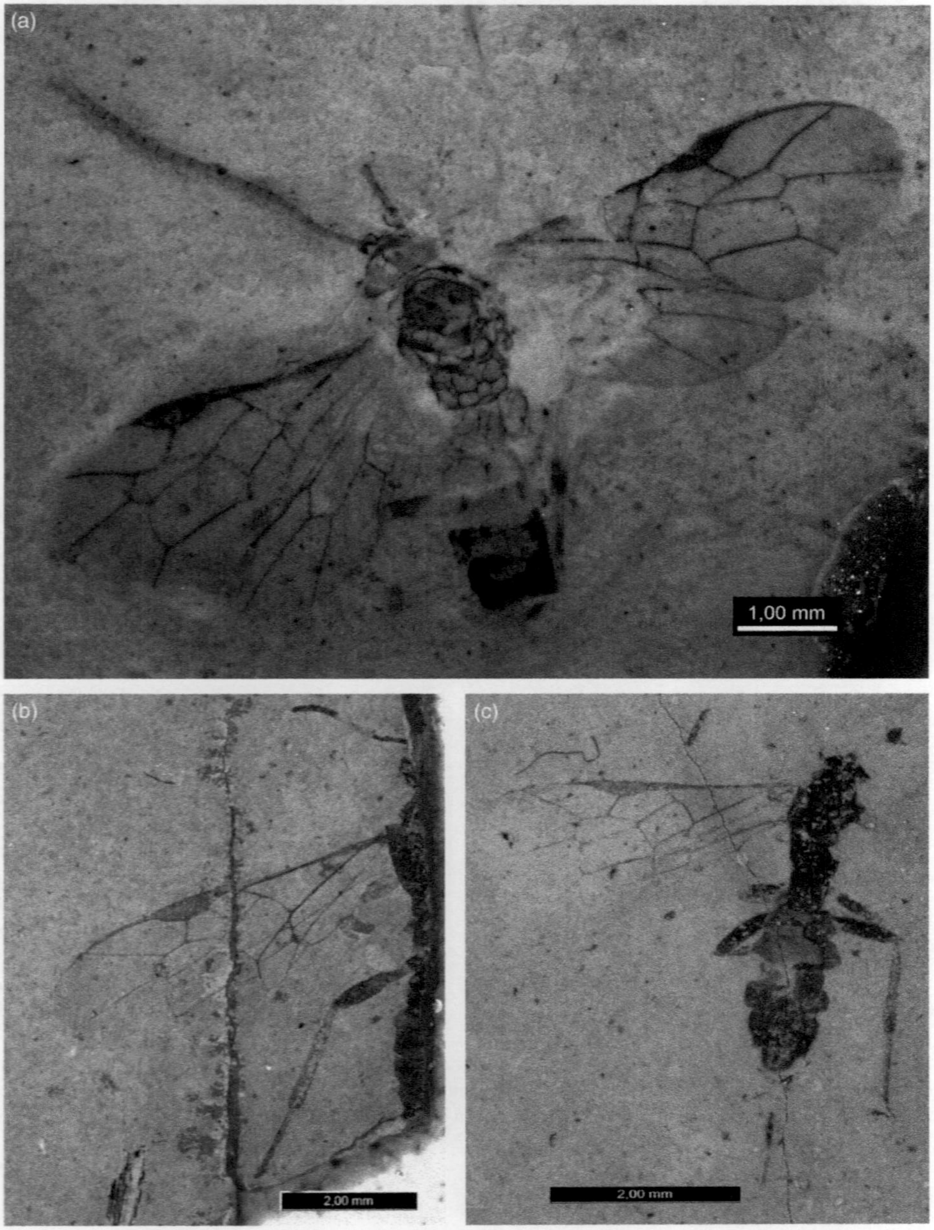

Fig. 13.2 Impression fossils of the extinct subfamily Tanychorinae† including two recently described ichneumonids from the Early Cretaceous, Berriasian horizon of Russia, 139.8–145 Mya. (a) *Tanychora petiolata*; (b) *Khasurtella buriatica*; (c) *Amplicella mininae*. (Source: reproduced by permission of Dmitry S. Kopylov.) For a colour version, see Plate 32.

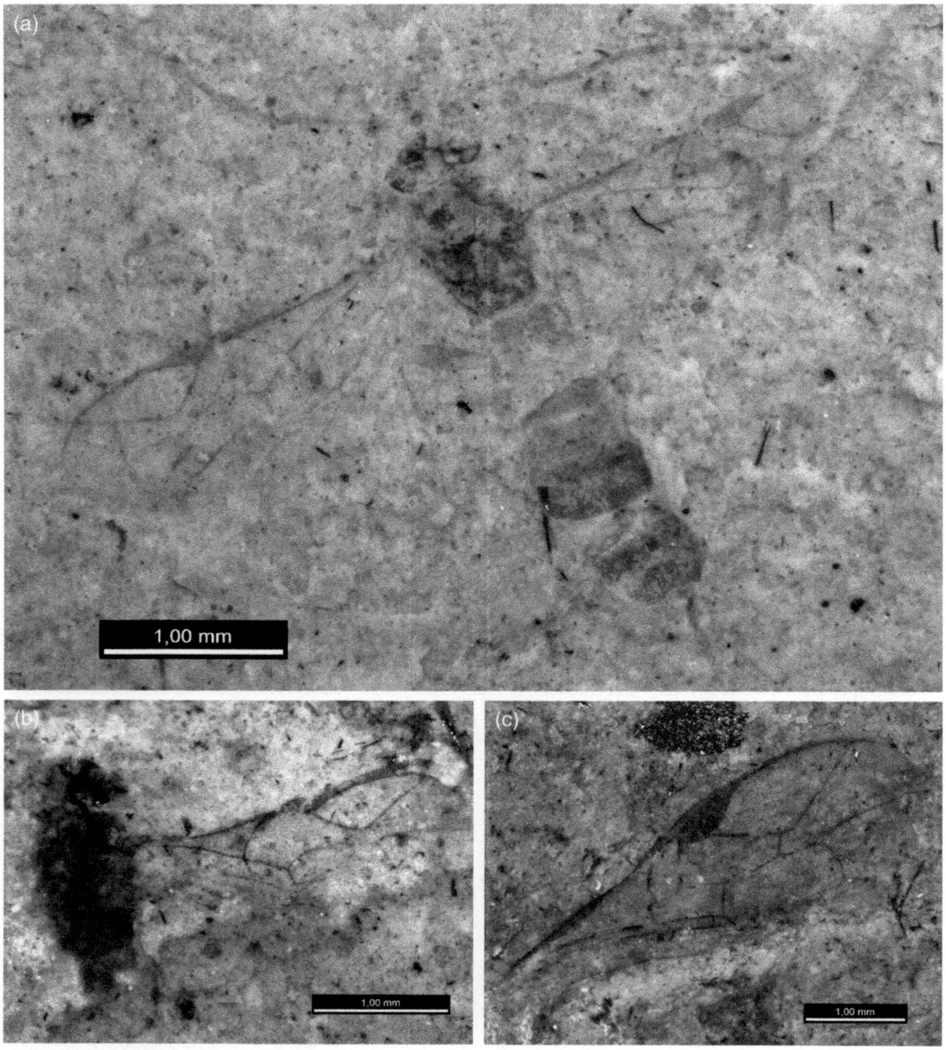

Fig. 13.3 Impression fossils of Palaeoichneumoninae†. (a) *Palaeoichneumon freja*, showing short ovipositor; (b) *P. mirabilis*; (c) fore wing of *Rudimentifera mora* (Palaeoichneumoninae) from middle Lower Cretaceous (Hauterivian/Barremian) Dzun–Bain Formation of Mongolia ~129.4–132.9 Mya. (Source: reproduced by permission of Dmitry S. Kopylov.) For a colour version, see Plate 33.

certainly represents (as Kopylov indicated) a remnant of (Rs+M)a. They also appear to have had rather short ovipositors, although in some of the fossils it is not possible to be certain that the distal parts have not been lost. I suspect, however, given that ovipositors are generally rather tough structures, that they probably did have short ovipositors and thus probably attacked either exposed or very weakly concealed hosts. The group is probably paraphyletic.

Labenopimplinae†

This subfamily was erected by Kopylov (2010b) for five fossil genera (*Labenopimpla*, *Armanopimpla*,

Ramulimonstrum, *Rugopimpla* and *Micropimpla*) from the Upper Cretaceous of the Russian far east. These differ from most other fossil and extant ichneumonids in the shape of the areolet. Two further new Upper Cretaceous genera (*Agapia* and *Agapteron*) and also three fossil genera previously placed in the Tryphoninae by Townes (1973), viz. *Catachora*, *Eubaeus* and *Urotryphon*, also from Taimyr amber were included by Kopylov (2012b). Townes thought that the wasps he described might be related to the tryphonine genera *Idiogramma* and *Grypocentrus*, but this placement was somewhat arbitrary and he also noted a more general resemblance to other small ichneumonids such as *Adelognathus* (Adelognathinae), *Lysibia* and *Gnypetomorpha* (Cryptinae), *Lathrolestes* (Ctenopelmatinae) and *Earobia* and *Pygmaeolus* (Tersilochinae, formerly Phrudinae). Four of the specimens from which these are known are females and none of them has any egg attached to the ovipositor as is common with Tryphoninae, and for this and other reasons, Kopylov was able to exclude them from the Tryphoninae.

These fossil wasps are all small with fore wings usually less than 2.8 mm long, which represents a significantly different range of body sizes compared with the present-day fauna (Kopylov 2012b; see Fig. 15.7). It is possible that other, larger species existed in the Upper Cretaceous and that the amber fauna was biased to only very tiny ichneumonid taxa, but this seems unlikely, also because braconids from the period are also generally small and it is possible that the ichneumonoid fauna of the period had undergone some sort of miniaturisation, due to the availability of different niches (at least in the amber generating forests of the time).

Pherombinae†

Within the Ichneumonidae, Kasparyan (1988) created the monotypic subfamily Pherombinae based on a fossil *Pherombus* (Fig. 13.4). They are known from the Upper Eocene Baltic amber and possessed enlarged eyes and ocelli, strongly suggesting that they or at least some of them, were nocturnal.

Townesitinae†

Kasparyan (1994) erected a new subfamily, the Townesetinae†, for four fossil genera all known only from Baltic amber, which he divided into two tribes. The wasps are rather small (Fig. 13.5) but at least one has an ovipositor rather longer than the body, perhaps indicating a shift towards attacking fairly deeply concealed hosts, at least relative to the wasp's body length.

THE XORIDIFORMES

Xoridinae

Although with four genera, the xoridines are dominated by the single genus, *Xorides*, with a total of just over 200 species currently split into a number of species groups, the previous subgeneric classification having been abandoned (Wahl 1997). The most basal genus based on morphological analysis appears to be *Aplomerus*, a group that was until recently thought to be restricted to North America but which has since

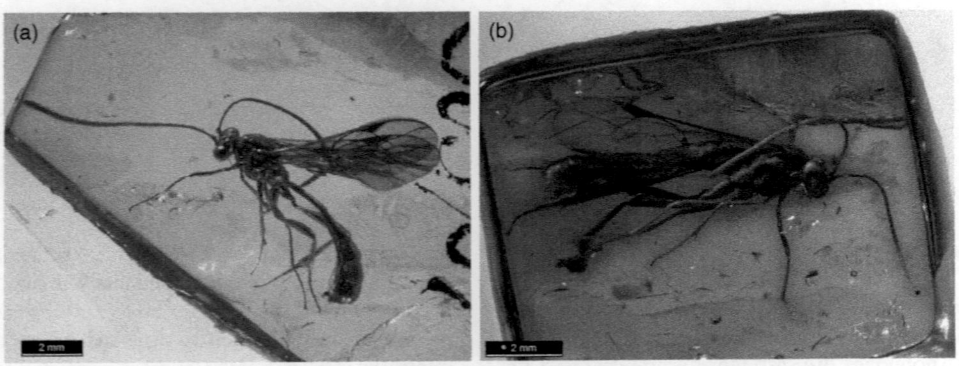

Fig. 13.4 Baltic amber inclusions of *Pherombus antennalis* (extinct subfamily Pherombinae†). (a) Holotype female; (b) paratype male. (Source: reproduced by permission of Dmitry S. Kopylov.) For a colour version, see Plate 34.

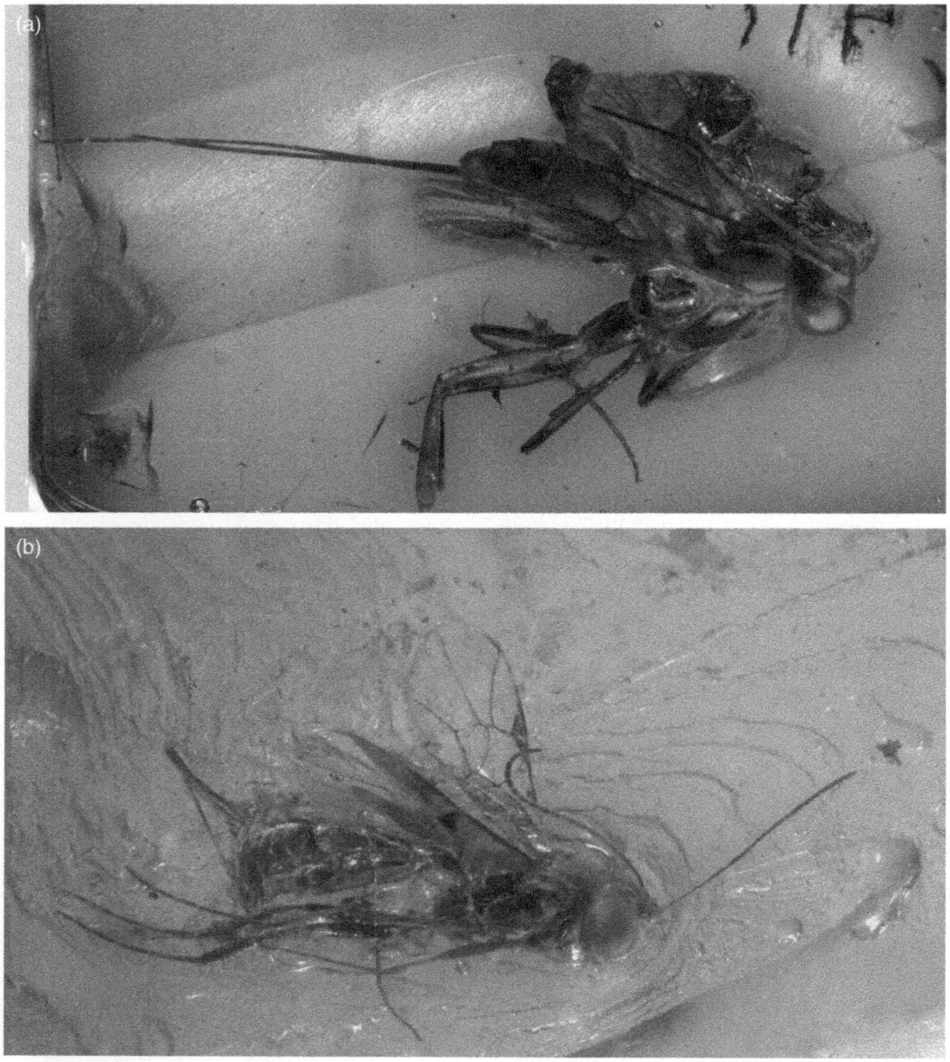

Fig. 13.5 Amber inclusions of extinct ichneumonid subfamily Townesetinae†. (a) Holotype female *Townesites mandibularis*, body length approximately 2.5 mm; (b) specimen of *Marjorietta minor*, body length approximately 1.6 mm. (Source: reproduced by permission of Dmitry S. Kopylov.) For a colour version, see Plate 35.

been discovered in Japan (Watanabe & Matsumoto 2010).

Xoridines are idiobiont ectoparasitoids and all known hosts are xylophagous insects, but these include both beetles [Anobiidae, Buprestidae, Cerambycidae, Curculionidae (including Scolytinae), Elateridae and Melandryidae] and also siricid woodwasps (Chrystal & Skinner 1931, Gauld & Fitton 1981); a record from a scarabaeid is not impossible but requires confirmation. Of these, siricids and cerambycids appear to be the most frequently attacked hosts, although relatively little is known of host relationships of the many tropical *Xorides* species which are unlikely to attack woodwasps. Not only do xoridines parasitise the larvae and prepupae of the host, but also the pupa and even pharate adults, and although the majority are solitary parasitoids (e.g. Chrystal & Skinner 1931, Sharifi & Javadi 1971, Sheng et al. 2012b), gregarious development has been recorded in *Ischnocerus rusticus* (Gauld & Bolton 1988). Although Chrystal and Skinner (1931) never observed it during their study of *Xorides* (as *Xylonomus*) attacking the conifer longhorn, *Tetropium gabrieli*, the paralysed and rather flaccid hosts are probably envenomated by the female wasp and the venom reservoir of xoridines is ellipsoidal and strongly muscularised, suggesting the need to inject venom fairly quickly/forcibly.

Chrystal and Skinner (1931) found that the egg was normally attached to the dorsal part of the host larva (though it could be anywhere) and was once found a few millimetres from the host, and in that case the first instar larva easily located the host and started feeding. The egg has a long tail as is typical of many idiobiont taxa which lay large eggs on to deeply concealed hosts. The egg shell remains affixed to the host larva after the first instar has hatched, which suggests that the female wasp probably exudes a glue along with the egg – as pure speculation, it seems likely that this originates from the wasps large vaginal glands. There are four larval instars, and after the last instar has completed feeding it rests motionless in the host's tunnel for a period of 2–5 days before it again resumes activity and spins its cocoon.

The adult mandibles are rather twisted with the chisel-shaped upper tooth concealing the more posterior lower tooth. Females very commonly have white striped antennae (Fig. 13.6), but those of the males tend to be unicolorous or to have less sharply demarked white bands. In common with many other wood-borer parasitoids (see Eggleton 1989), the mesoscutum is often strongly sculptured (pitted, rugulose), the head is robust with chisel-like mandibles and the fore tibia has numerous denticles that may be used for egress or to enable them to move around under bark or in frass-filled host borings (see below). Females of *Odontocolon* often oviposit under loose bark (Fig. 13.7c). The ovipositor has strong apico-ventral serrations and usually two discrete dorsal nodi, and is typically rather down-curved (Fig. 13.7a). Species spin a buff-coloured, somewhat papery cocoon near the host remain in the wood (Fig. 13.7b).

Dissections of host substrates sometimes reveal xoridines in the host borings and parasitised host remains can be found deep within the dead wood well beyond the reach of an ovipositing female on the outside of the substrate. These observations show that female xoridines may actively seek hosts within the wood, although it is not known whether in such instances the parasitoid has arrived from another tree and entered the substrate or emerged within the same substrate and never left it. *Aplomerus* species have particularly robust legs and are rather dorso-ventrally flattened, which appears to be an adaptation for crawling under loose bark (Watanabe & Matsumoto 2010), although that begs the question of how they might then manipulate their long ovipositors. The hosts of *Aplomerus* are not known with certainty but one

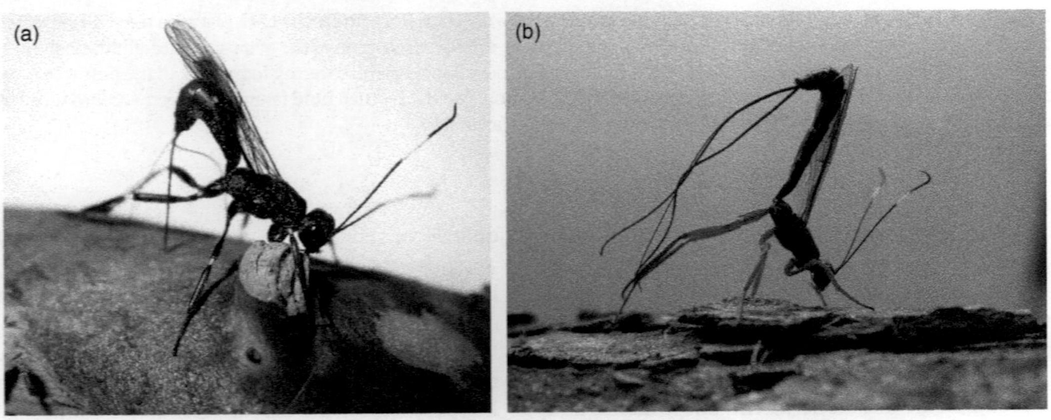

Fig. 13.6 Ovipositing females of two *Xorides* species (Xoridinae): (a) *X. filiformis*; (b) *X. rufipes*. (Source: reproduced by permission of Jacek Hilszczanski.)

Fig. 13.7 The European wasp *Odontocolon dentipes* (Xoridinae). (a) Cocoon from under the bark of a pine tree and extracted pupa with near fully developed adult female enclosed, showing the unique disposition of the ovipositor straight out behind the wasp rather than curved over the back (cf. Fig. 5.4); (b) female ovipositing, with the covering loose bark of the pine tree removed; note the position of the ovipositor sheaths underneath the wasp, and also the toothed hind femur. (Sources: a, reproduced by permission of Mark Shaw; b, reproduced by permission of Tomasz Jaworski.)

species has been associated with buprestid-infested dead wood.

Females of *Xorides* appear to use a form of vibrational sounding to help detect hosts, although unlike other taxa that use this method, it is not the tip of the terminal flagellomere that is modified to form the 'hammers' used to transmit vibrations to the substrate, it is a small series of peg-like (see Fig. 10.3b,c) spines that form a line along a set of flagellomeres near the apical quarter of the antenna (often just beyond the white antennal stripe) (Fig. 13.6). The male antennae lack hammer setae and are sometimes highly setose (Sheng et al. 2012b). Transmission electron microscopy of the antennal pegs shows that they are not sensilla but are solid cuticular structures (Quicke unpublished). In agreement with use of vibrational sounding, the fore tibia of females is strongly swollen and also often has a strong diagonal groove, all associated with their enlarged, vibration detecting, subgenual organ. The whole oviposition process in *Xorides* takes a considerable time (15–40 min, with 30 min being typical) (Chrystal & Skinner 1931). Host location is apparently very precise and in the *Xorides* species investigated by the parasitoid nearly always laid her egg directly on the host larva's body. Despite being idiobiont ectoprasitoids, Chrystal and Skinner indicated some degree of host specificity under experimental conditions, but, given what we now know about the host location mechanism, some of their tests may not have been appropriate and the wasps may not have been able to locate larvae of non-hosts simply placed beneath bark, rather than being in a natural boring.

Chrystal and Skinner observed courtship frequently in captivity. Males approach females from the front with their antennae vibrating, at first stroking her antennae with them and, after mounting, using them to continue stroking her mesosoma. No mention of wing vibration is made. Copulation is a very protracted process in these wasps, lasting typically 20 min, but the authors noted some pairings to last up to 40 min.

THE LABENIFORMES

Labeninae

Comprising just the Labeninae, this might be thought of as a uniform group, but they are not really that tidy, although molecular data suggest that the *Gestalt* of ichneumonidologists has not been too awry. Gauld (1983), who carried out the first formal cladistics analysis of the group, commented that they '… have had a remarkably unsettled taxonomic history'. With the Brachycyrtinae now removed to the 'southern ichneumonid' clade, having previously been placed variously in Labeninae, Campopleginae and Cryptinae, the remaining labenines are now represented by four tribes, the Labenini, Groteini, Poecilocryptini and Xenothyrini (see Wahl 1993a, 1996b), all with different biologies, although the biology of Xenothyrini is not yet known.

Following his phylogenetic analysis, Gauld (1983) emphasised that the Labeninae are both very morphologically and biologically plesiomorphic and the suggestion was that they represent one of the most basal groups of extant ichneumonids that probably radiated in west Gondwanaland prior to the separation of Australia, Antarctica and South America around the end of the Cretaceous. However, while the subfamily is currently restricted to the southern hemisphere, the first Cretaceous fossil labenine has recently been described from Canadian late Cretaceous (Campanian) amber (McKellar et al. 2013), indicating that in the past it was far more widely distributed and possibly cosmopolitan, as with so many 'Gondwanan' groups, (see also Belshaw et al. 2000). Apart from being largely recognised through a combination of putatively plesiomorphic characters, however, Gauld and Holloway (1986) noted that 'It is not clear what the sister group of the labenines is' and that similarities between some taxa and, for example, the Pimplinae, could all be interpreted as either shared ancestral features or derived features associated with attacking wood-boring hosts that are easily subject to convergence (see Eggleton 1989). However, Gauld's argumentation that the similarities between labenines and rhyssines are parallelisms is on very uncertain ground, being based on the two groups possessing two different types of ovipositor stabilisation mechanism – coxal in labenines versus claspers on the metasomal sternites in rhyssines (Fig. 13.8; cf. Fig. 10.5d). Although I see no reason why these taxa should be closely related, it is clear that an ancestral 'intermediate' taxon could easily possess both mechanisms or neither. Gauld (1983) also said that although the Labeninae could not be defined by any single apomorphy, they generally possessed at least two of four characters he listed, although some of the characters can clearly be seen as adaptations to attacking deeply concealed hosts. Gauld's definition at that time also included the Brachycyrtinae. The Labeninae's basal position in the Ichneumonidae is nevertheless indicated by some having a very large areolet, reminiscent of those of some Cretaceous ichneumonids.

The cladistic analysis of relationships within the subfamily provided by Gauld (1983) was subject to re-examination by Wahl (1993a), which although not changing the picture of higher relationships much, importantly led to the conclusion that the Brachycyrtini were wrongly included in the Labeninae, and he accordingly elevated them to subfamily rank, suggesting that they had a close affinity with the Cryptinae+Ichneumoninae. Molecular data provide an alternative picture (see Brachycyrtinae below), but the 28S RNA sequences of remaining labenines are fairly divergent and they do not allow a great deal to be said with confidence either about internal relationships or about the placement of the subfamily. Most recently,

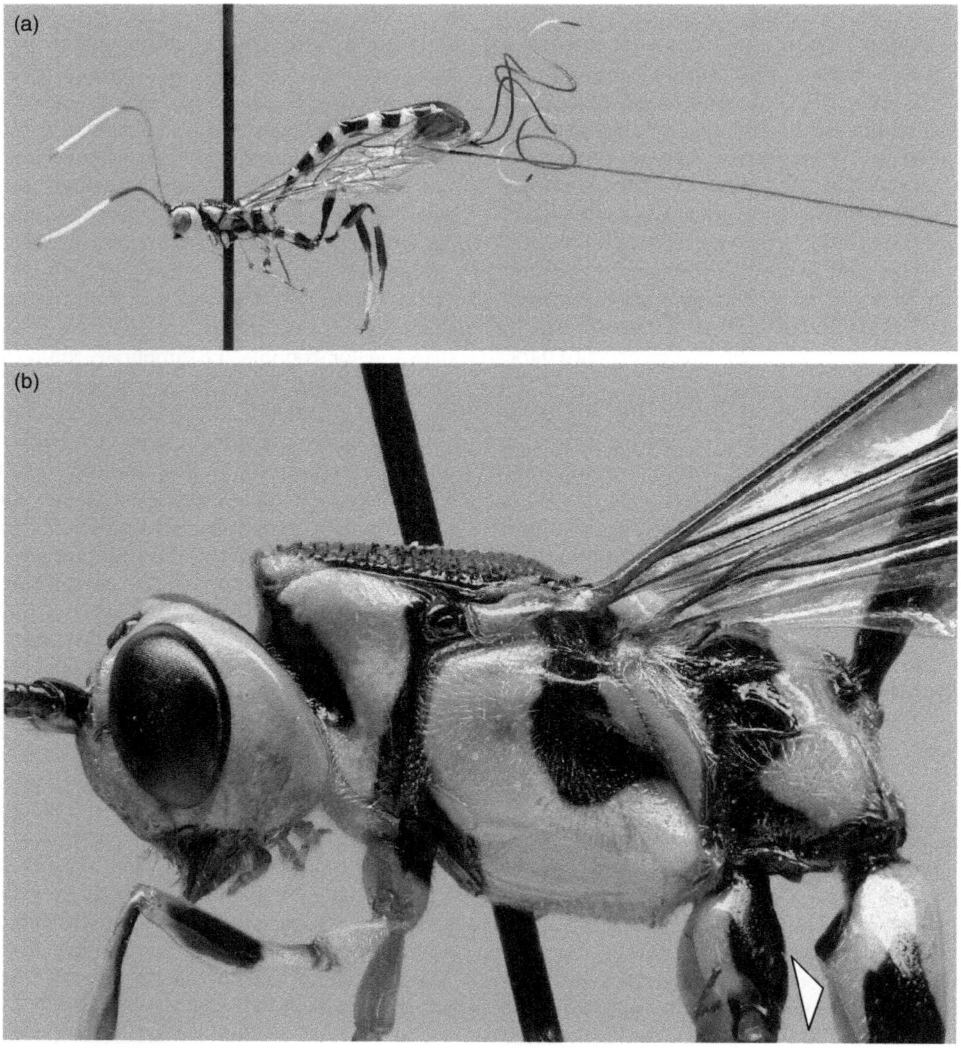

Fig. 13.8 *Apechoneura* spp. (Labeninae). (a) *Apechoneura* sp., habitus; (b) *A, longicauda*, detail of head and mesoscutum showing transverse ridging on the latter, associated with egress, and longitudinal groove on hind coxa (arrow) used for stabilising ovipositor. (Source: reproduced by permission of Ilari Sääksjärvi.)

Gauld and Wahl (2000a) presented a very thorough morphological phylogeny of the group, including many new characters.

The phylogenetic analyses conducted by Gauld and Wahl (2000a) support Gauld's (1983) earlier conclusions that the group originated and diversified in Gondwanaland after the split of Africa, India and Madagascar but before the split between Australia and South America. They showed that those species now found in non-Gondwanan places such as North America are actually derived members of their lineages, and they proposed that these were the result of much more recent colonisation events.

Particularly interesting from Gauld and Wahl's (2000a) study is that the taxa whose females show the most obviously developed specialisations to attacking

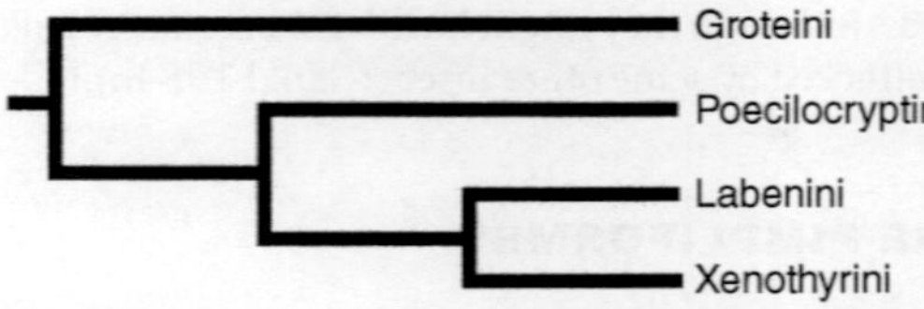

Fig. 13.9 The relationships between the tribes of Labeninae recovered by Gauld and Wahl (2000).

wood-boring hosts are also among the more derived taxa (Fig. 13.9). Hence the possibility that labenines are not ancestrally parasitoids of xylophagous hosts needs to be considered and thus the linked assertion that they are among the most basal of extant ichneumonids may involve erroneous reasoning. Nevertheless, molecular studies confirm their near basal origin in the family, but unfortunately, the major compositional shift in base frequencies in the 28S rDNA gene between ichneumonids and braconids (the only gene fragment investigated to any extent to date) makes rooting the ichneumonid tree problematic.

A key to the tribes and genera was provided by Gauld and Wahl (2000a).

Groteini[2]

This group comprises three genera, *Grotea*, *Labium* and *Ozlabium*. All species appear to be associated with bees and while the early stages are parasitic on the bee larva or simply suck out the bee egg, they go on to complete their feeding using the pollen store (bee bread) that was created to feed the bee larva (Graenicher 1905, Janvier 1933, Slobodchikoff 1967). The larva and pupa of *Grotea californica* were described by Slobodchikoff (1967). *Labium* species seem to have been reared exclusively from nests of Halictidae.

Host location has been described for the neotropical bee parasitoids/predators *Grotea gayi* and *Labena* sp., both of which attack the nests of *Manuelia postica* (Niemeyer & Flores-Prado 2012). Both were observed apparently to use vibrational sounding, in agreement with predictions based on antennal apex morphology and female fore tibia size (Broad & Quicke 2000), and to use this to select large cells in which to oviposit.

The basal genus, *Grotea*, has a rather unspecialised ovipositor which is less sclerotised than in some other labenines and is not supported in any way during penetrating the host substrate, which tends to be relatively weakly lignified plant tissue (Gauld & Wahl 2000a). In more derived taxa, the apex of the ovipositor has multiple dorsal nodi and also numerous ventral serrations similar to the condition seen in the Rhyssinae, but how these function is not fully understood.

Labenini

The Labenini is a small group comprising five southern genera, that are recorded primarily as parasitoids of wood-boring hosts such as buprestid, cerambycid and curculionid beetles, and one species of *Certonotus* has been recorded as attacking the introduced siricid woodwasp pest *Sirex noctilio* in Australia (Hocking 1967b), and it is from this association that we known much about their biology. The record of a *Certonotus* species from Limacodidae (Lepidoptera) (Chadwick & Nikitin 1976) is very dubious given that the latter live in an exposed position on the host plant and *Certonotus* has a long ovipositor. *Labena* has been recorded as a parasite in the nests of the Neotropical bee, *Manuelia postica* (Flores-Prado & Niemeyer 2012).

Gauld & Wahl (2000a) note that species in this tribe exhibit some of the strongest sexual dimorphism in the family. Males of some *Labena* species possess a strong brush of adpressed, thickened and recurved setae on the underside of the fore telotarsus and among the neotropical species, males often have pale antennae with black tips, that also appear to have subterminal antennal glands as there is often a congealed secretion in this area in specimens collected into alcohol. Mating takes place shortly after emergence and there is then a period of a few days before oviposition starts (Hocking 1967b), although it is not known whether this is because of a need fully to mature eggs. They are ectoparasitoids and it is presumed that mobile hosts are paralysed. Hocking noted that one *Certonotus* larva was 'found within the remains of an adult *Sirex* which had been parasitised just prior to emergence', but it is not clear whether it had crawled inside the partly consumed woodwasp (most likely) or had conceivably developed internally (very unlikely). Overwintering takes place as a final instar larva within its cocoon.

As with the Groteini, labenines use vibrational sounding as a method of host location or assessment and have the apex of their antenna modified to form a hammer like structure (see Fig. 10.3d). Associated with this, females have enlarged fore tibiae (Broad & Quicke 2000) and, also probably associated with this, females have the third tarsal segment protruding ventrally past the whole of the fourth segment and a large part of the telotarsus (figs 56 and 57 in Gauld & Wahl 2000a). This latter feature is likely to be associated with passing substrate vibrations more effectively and directly to the tibial subgenual organ.

Most Labenini have a groove on the inside surface of the female hind coxa near the base, which is used for supporting the ovipositor during oviposition (figs 34–48 and 101 in Gauld & Wahl 2000a; see also Fig. 13.8b), although this feature is less developed in those *Apechoneura* species whose females have a very long, slender and flexible ovipositor (especially of the *A. longicauda* group). In addition, the groove is furnished with highly modified, so-called 'grossulariform setae', which was proposed as a synapomorphy for the tribe. The weak coxal groove and absence grossulariform setae make it unlikely that *Apechoneura* 'drill' through hard wood; it was assumed that these most probably reached deeply concealed hosts by threading through pre-existing cracks or holes in wood, and this was recently confirmed by Hanson et al. (2008), who found that *A. longicauda* attacking a larva of a large species of scolytid beetle in Costa Rica use their ovipositors to probe frass-filled burrows. In addition to the hind tibial modifications, many Labenini have particularly extensive intersegmental membranes between the posterior metasomat tergites that allow greater flexibility when manoeuvring their long ovipositors, and in some *Certonotus* the membrane 'balloons out' during the early part of oviposition (Lindsay 1925, cited in Gauld & Wahl 2000a). Some *Labena* species have modified hind third and fifth tarsal segments (fig. 60 in Gauld & Wahl 2000a) that might have a grooming function (see Chapter 10, section *Distitarsal scraper*).

Poecilocryptini

Poecilocryptines were associated with the Brachycyrtini by Townes (1969) but they are not closely related, as shown by Gauld (1983) and later by Wahl (1993a). *Poecilocryptus* species have been reared from a variety of galls, including those made by beetles and by hymenopterans including the braconid wasp *Mesostoa* on *Banksia*, the fly *Fergusonina* on *Melaleuca* (Myrtaceae), the pteromalid *Trichilogaster*, on *Acacia*, and even from Hemiptera galls, those produced by Eriococcidae (Chadwick & Nikitin 1976). Goolsby et al. (2001) found that *P. nigromaculatus* frequently tunnelled between *Fergusonina* galls in the host *Melaleuca* tree, thus acting as a predator. Another Australian genus, *Alaeothyris*, has been reared from *Araucaria* seeds, although they might have been previously galled or attacked by some other insect (Gauld 1984b).

THE PIMPLIFORMES

This group of subfamilies includes the very large Pimplinae (which might not be monophyletic) together with the Collyriinae, Acaenitinae, Diacritinae, Diplazontinae, Orthocentrinae (=Helictinae), Rhyssinae, Poemeniinae and tentatively the extinct Labenopimplinae. Relationships between these subfamilies, have been explored phylogenetically using both morphology alone (Fig. 13.10a; Wahl & Gauld 1998[3]) and in combination with the D2+D3 28S rDNA gene fragment (Fig. 13.10b; Quicke et al. 2009). Wahl (1986) noted that the Acaenitinae and Orthocentrinae[4] were sister groups based on fusion of the final instar larval hypostomal spur and stipital sclerite, to form a large hypostomal plate (Fig. 5.2e), but although they appear to belong to the same clade in formal analyses, the character appears to be homoplastic. The cladograms obtained from both data sets are extremely similar (Fig. 13.10) and both include a clade comprising Cylloceriinae, Diplazontinae and Orthocentrinae, although with differing internal relationships, and as all these are specialist endoparasitoids of Diptera larvae, it suggests a single shift to this group of hosts in the pimpliformes. These phylogenetic analyses suggest that the whole group had had its origins in the north temperate region and various lineages had then diversifying into the tropics.

Along with the previous two extant subfamilies and the ichneumoniformes, the lateral oviducts are short relative to the ovarioles (Fig. 13.11), in marked contrast to the situation in most ophioniformes (see Fig. 13.34).

Acaenitinae

This is a distinctive subfamily of koinobiont endoparasitoids of beetle larvae comprising approximately 27 genera and 260 described species. Many species are largish and often rare, at least in collections. Until recently they were known from almost all parts of the world except South America, but extensive collecting in Peru the past few years has revealed at least one species from the region (Castillo et al. 2011).

The subfamily was traditionally divided into two tribes, the Acaenitini (Fig. 13.12a) with 20 genera and

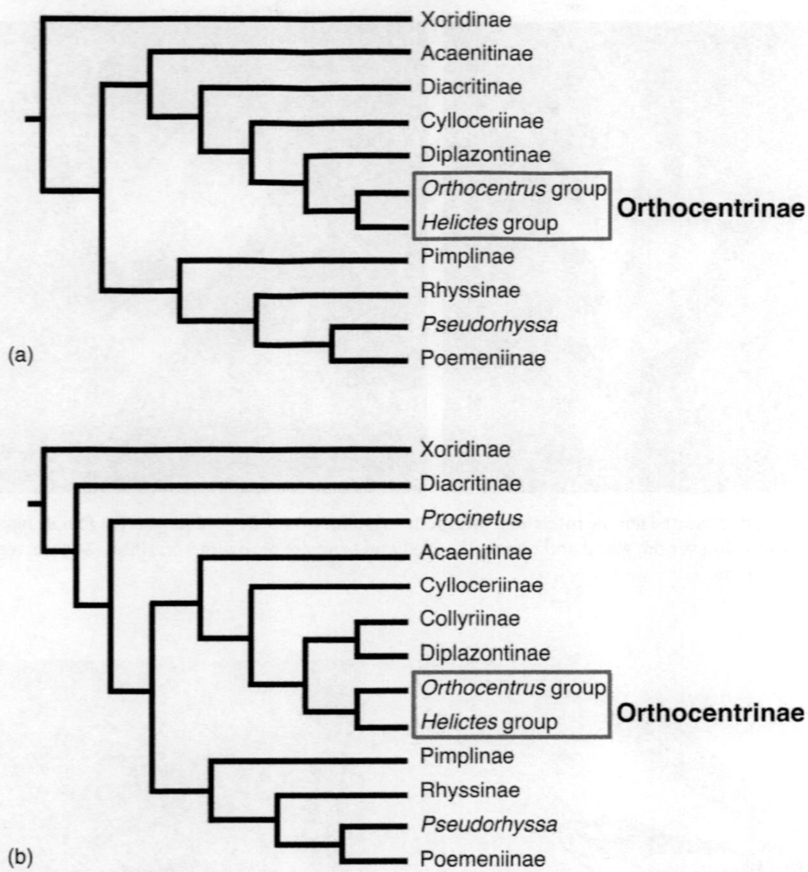

Fig. 13.10 Summarised trees of 'pimpliformes' taxa. (a) Relationships from morphological analysis of Wahl and Gauld (1998) which did not include Collyriinae and did not treat *Procinetus* separately; (b) from the weight balanced, combined concatenated molecular and morphological analyses of Quicke et al. (2009), but which also included the Collyriinae. Note the main difference is the placement of Diacritinae.

the Coleocentrini with seven (Fig. 13.12b). However, Wahl and Gauld (1998) based on morphology, provided evidence that the Acaenitini rendered the latter paraphyletic and therefore the tribes should be abandoned, although the concept of a '*Coleocentrus* grade group' may still be useful.

Despite their large size, little is known about their biology. Species have been reared predominantly from wood-boring beetle larvae (notably Cerambycidae and Melandryidae) whereas others attack stem-borers such as Mordellidae and also cerambycids (Townes & Townes 1960, Gauld 1991, Shaw & Wahl 1989). Most acaenitines have very long ovipositors. Females of most species have a very enlarged hypopygium (Fig. 13.12a,b), which is presumably important in controlling and stabilising the ovipositor. The biology of *Acaenitus dubitator* was described by M.R. Shaw and Wahl (1989) and is an endoparasitoid of a *Cleonus piger* (Curculionidae) larva living in root galls on *Circium*. 'Coleocentrines' are also presumably endoparasitoids, as suggested by their larval features (Wahl 1986), but there are no direct observations. It was not possible to ascertain the number of larval instars, but the first was thought to be caudate although the illustrations show no tail and it was presumed that the tail might have been 'lost in handling' prior to illustration. *A. dubitator*

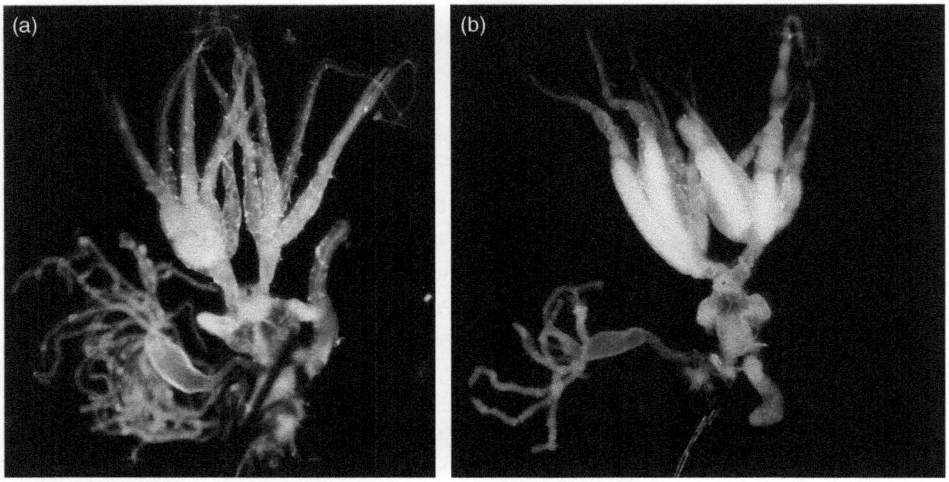

Fig. 13.11 Pimpliformes dissected female internal reproductive system. (a) *Itoplectis naranyae*; (b) *Pimpla nipponica* female reproductive systems showing venom gland and reservoir on left and synovigenic ovaries on right. (Source: reproduced by permission of Takatoshi Ueno.)

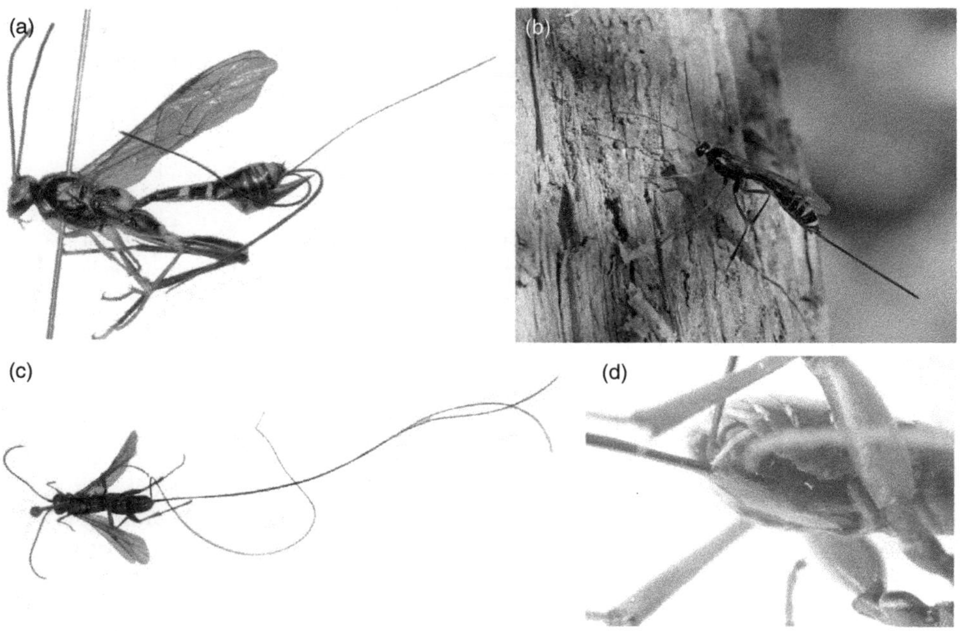

Fig. 13.12 Features of Acaenitinae. (a) *Arotes moiwanus*, showing very large hypopygium typical of all members of subfamily except *Procinetus*; (b) *Coleocentrus excitator*; (c, d), *Procinetus vipioniformes*, showing long ovipositor and detail showing truncate hypopygium. (Sources: a, reproduced by permission of Kyohei Watanabe; b, reproduced by permission of Cor Zonneveld; c, d, reproduced by permission, © C. van Achterberg, Naturalis Biodiversity Centre.) For a colour version, see Plate 36.

overwinters in a cocoon either as a fully developed larva (i.e. still containing the meconium in its rectum) or as a prepupa (i.e. with the meconium voided). Study of the larvae of *Acaenitus dubitator* allowed reappraisal of some of the larval features previously described based on exuviae recovered from cocoons (Wahl 1986) and the labral sclerite appears to be absent, as was stated by Short (1978). Loss of the labral sclerite may therefore be a synapomorphy for the Acaenitini, since it is present in *Coleocentrus*.

Coleocentrines include some of the largest ichneumonids but are generally not common; they seem to be primarily associated with old trees in cool, moist, temperate forests (Townes & Townes 1960, M.R. Shaw 1986) and are believed to be principally parasitoids of cerambycid beetle larvae and maybe also of siricid wood wasps (Townes 1971), but until more detailed rearings are available, such statements should be treated as only provisional. The genus *Procinetus* is probably the most basal of the subfamily in that they lack the highly developed hypopygium of the female (Fig. 12.12c,d; cf. Fig. 12.12a,b), although this might also represent a secondary reduction perhaps associated with the very long, flexible ovipositor of some species which probably threads its way to the host rather than having to be forced through a hard substrate and hence might not need the same support.

At least some species have a row of pores on the inner side of their highly curved hind claw (see fig. 132 in Gauld 1991) and it is tempting to speculate that these may be the source of a 'venom' in a similar way to the poison claws of *Theronia* and other Pimplinae and also in the poemeniine *Eugalta* (q.v.).

The British species can be identified using Fitton (1981) (but see Shaw 1986) and those of North America by using Townes and Townes (1960).

Collyriinae

This is a very small subfamily, for a long while comprising the single genus *Collyria* and only nine described species, all from the Palaearctic region. It is one of very few ovo-larval parasitoids in the Ichneumonidae, the others being members of the ctenopelmatine tribe Pionini, the genus *Stilbops* (Stilbopinae) and some Diplazontinae and a very few Campopleginae.

Nearly all host records of *Collyria* are from cephine (Cephidae) sawflies, whose larvae bore in the stems of grasses. Detailed biology, however, is known for only one species, *C. coxator*, but is likely at least to apply to all *Collyria* species. *C. coxator* is a common and potentially important parasitoid of the wheat-stem sawfly, *Cephus pygmaeus*. Salt (1931) described its biology in depth (as *C. calcitrator*), including how the wasp searches down the stems of the grass to locate the cephid's oviposition scars and then, still facing downwards, inserts her ovipositor and oviposits into the host eggs. The first and second instars are both mandibulate and presumably they eliminate supernumeraries resulting from superparasitism. Moulting to the second instar takes place in a specialised evagination of the host integument. Overwintering is as a second instar and the host is finally killed within its cocoon in the spring. Salt (1931) suggested that there are probably five instars but with the caveat that this is difficult to ascertain. Interestingly, unlike in most ichneumonoids, the elimination of supernumeraries is often carried out by second instar larvae, which also have the more developed mandibles.

Collyria is unique within the Ichneumonidae, so far as is known, in that although it possesses a well-developed venom gland, it lacks any venom reservoir (see fig. 6.4b in Quicke 1997a). This is probably because as an egg parasitoid it cannot inject a very large volume of venom into the host egg and as *Collyria* is not a small wasp, it does not complete development until after its sawfly host has spun its cocoon, and the lumens of its venom gland alone can probably hold sufficient venom for its needs.

Recently, Sheng et al. (2012a) described a highly unusual new genus and species of wasp from China, *Bicurta sinica*, whose subfamilial placement was ambiguous. Including it in a morphological phylogenetic analysis of other pimpliformes recovered it as a sister to *Collyria* and therefore the subfamily diagnosis was modified to accommodate the new taxon and another recently described genus from the Near East. The biology of *Bicurta* is unknown, but its sharply pointed, slender and needle-like ovipositor would be consistent with it also being an ovo-larval parasitoid.

The ovipositor of *Collyria* is unusual in that while tapering gradually to a sharp point, the lower valves have a series of protuberances on them, sometimes referred to as teeth, although they do not have the same scarp shape as true teeth. Their function is unknown but they may be sensory. Nor is it known whether the wasp inserts her ovipositor into the slit in the host food plant made by the sawfly ovipositor, but this seems likely.

Because of its unusual morphology and biology, placement of the Collyriinae among the other ichneumonid subfamilies has been uncertain for a long time. Gauld (1997) and Wahl and Gauld (1998) excluded it from any of their major groups. Bennett (2001) included it within his morphological phylogenetic analysis of the placement of the Agriotypinae among various pimpliformes and ichneumoniformes and stated that in the preferred cladogram it came out basal, but as the cladogram was not rooted with an outgroup, this was simply because for his coded characters, the Collyriinae had more zero states. Molecular analyses clearly place Collyrinae among the pimpliformes group (Quicke et al. 2009).

In the United Kingdom, *C. coxator* is a major parasitoid of *Cephus pygmaeus*, causing up to 60% mortality (Salt 1931), and as this sawfly has become a serious pest of wheat in North America, it is surprising how little work appears to have been conducted on its biology since Salt. Several attempts to introduce it from Europe in the 1930s apparently failed, but it now appears to be established and causing a useful degree of control across a wide area of the east coast (Streams & Coles 1965, Filipy et al. 1985, Shanower & Hoelmer 2004), where it frequently attacks 40–50% of hosts. It is hard to know whether this current level of control stems from individuals released 80 years or so ago, but remaining undetected, perhaps until natural selection led to them being more adapted to local conditions or some later, accidental introductions. An alternative explanation was put forward by Ivie (2001), which suggested that *Cephus cinctus* in North America might not have originated from Europe at all, but rather from East Asia, having been introduced first to the west coast of North America and spreading eastwards. Therefore, parasitoids collected in Europe might well not have been suitably adapted to the strain in North America.

Cylloceriinae

Cylloceriines are relatively unknown, but sometimes locally common insects associated with damp habitats. *Cylloceria* used to be included in the Oxytorinae *s.l.* (Gauld 1984a), but Wahl (1986, 1990) explained that *Oxytorus* was not related to the other 'oxytorines', thus restricting the use of that name. Based on his cladogram of the groups he regarded as comprising the pimpliformes, subfamily Cylloceriinae was erected by to receive two genera, *Cylloceria* and *Allomacrus*. Subsequently a third genus, *Sweaterella*, was described (Wahl & Gauld 1998). Most species come from various parts of the Holarctic and/or Oriental Regions, although *Cylloceria* also extends to Central America and the neotropics (Gauld 1995).

The flagellomeres of females are very elongate (Fig. 13.13a), the medial ones typically being more than four times longer than wide. Male *Cylloceria* are distinctive because their antennae have a specialised concavity formed by the third and fourth flagellomeres (Fig. 13.13a), which has been referred to as a pair of tyloids but does not resemble the tyloids of other ichneumonids in form. This region is apparently glandular as it lacks sensilla but possesses pores (Gokhman & Krutov 1996, Isidoro et al. 1996) and is presumably involved in courtship. The structure is almost certainly used by males to clasp the female's antenna and spread a contact pheromone over a receptive part of hers in a way similar to that known for the antenna coiling Diplazontinae, but there are no observations of courtship in any cylloceriines.

The only available host records are for the genus *Cylloceria* all from tipulid fly larvae (Humala 2002). *Hyperacmus* (Fig. 13.13c), a genus formerly placed in either the Orthocentrinae or Microleptinae, is regarded here as also belonging to the Cylloceriinae, based on results of DNA analysis and because of the venom reservoir which is paired (see fig. 10A in Quicke et al. 2009), a condition unknown elsewhere in the Hymenoptera. Although the venom reservoir is not completely paired in *Cylloceria*, its anterior end is widely divided. *Hyperacmus* has a superficially very different habitus from *Cylloceria*, but virtually all of the differences can be seen as part of its short, robust body facies. *H. crassicornis* has been reared from the mycetophilid fly, *Sciophila varia*, living in *Hydnum repandum* (Sedivy & Sevsik 2003).

A key to the world genera was provided by Wahl and Gauld (1998).

Diacritinae

A small and very poorly known subfamily from the Holarctic region comprising only three genera, *Daschyana* (=*Cressonia*), *Diacritus* (Fig. 13.14) and *Ortholaba*. Nothing is known about the biology of the group. Townes (1944) had included it within the Poemeniini but later transferred it (Townes 1957) to the Orthocentrinae (as his Plectiscinae), essentially because it did not fit well in the former and later as a tribe

Fig. 13.13 Cylloceriinae. (a) *Cylloceria* sp., female habitus; (b) detail of male antennal tyloid structure presumably used to grasp female antenna during courtship; (c) *Hyperacmus* sp., showing protruding face and robust legs. (Sources: a, b, reproduced by permission of Kyohei Watanabe; c, reproduced by permission, © C. van Achterberg, Naturalis Biodiversity Centre.)

Pimplinae (as his Ephiatinae), although many other workers considered them more closely related to the Orthocentrinae, but Eggleton (1989) recognised that they warranted placement in a separate subfamily and publication of that came with Gauld (1991). Combined morphological and DNA sequence data associate them firmly with the pimpliformes, but as a sister group to all the other taxa, whereas purely morphological analysis placed them basal to a combined ichneumoniformes, Gondwanan group and ophioniformes, whilst Gauld (1991) included them amongst his unplaced subfamilies, all emphasising the lack of clear character support for any particular close relative.

Diplazontinae

This is a relatively small and fairly readily recognised subfamily which are specialised koinobiont endoparasitoids of hoverfly larvae (Diptera: Syrphidae) and

Fig. 13.14 Diacritinae specimens. (a) *Diacritus aciculatus japonicas*; (b) *D. muliabris*, habitus and detail of mesosoma. (Sources: a, reproduced by permission of Kyohei Watanabe; b, c, reproduced by permission, © C. van Achterberg, Naturalis Biodiversity Centre.)

nearly exclusively of aphidophagous ones, although at least one genus, *Bioblapsis*, attacks a fungus-feeding syrphid (Rotheray 1990). Probably for this reason, i.e. aphids are largely a north-temperate group of insects, they are a predominantly Holarctic taxon, with only a few genera and very few species occurring within the tropics as far as is known. They have a sessile first metasomal segment and resemble small pimplines to an extent (Fig. 13.15), but they also have tridentate mandibles (see below).

Various diplazontines have different preferences for host developmental stages to attack and Rotheray (1981b) compared stages attacked by six Western European species representing four genera (Fig. 13.16). Most species will oviposit in a fairly broad range of host stages and facultative ovo-larval (ovo-puparial) parasitism occurs in a few species, although none of those studied would attack eggs that were less than 48 h old.

F. Schneider (1950) described the biology of, and carried out some extremely interesting experiments with, *Syrphoctonus* (as *Diplazon*) *fissorius* which parasitises *Epistrophe eligans* (which he called by its synonym *E. bifasciata*). The parasitoid lays directly into the haemocoel of first and second instar hosts and the first instar parasitoid larva floats freely in the haemolymph. Development commences after the host has formed its puparium and proceeds rapidly, but the duration as a first instar in the host may be protracted, depending

Fig. 13.15 *Diplazon laetatorius* (Diplazontinae), one of, if not the, commonest ichneumonid wasps in the world. (Source: reproduced by permission of Karol Ox.) For a colour version, see Plate 37.

on host voltinism and diapause. The first instar has a conical tail but is not caudate; it has no spiracles but does have a well-developed tracheal system. Schneider noted that the early first instar does not appear to feed for some while and the fore gut remains empty. Once the host puparium has formed, the *S. fissorius* first instar larva moves quickly to the host's brain, which it partly destroys along with some imaginal discs, bringing a halt to further host metamorphosis. The final instar, and probably other instars, of most diplazontines have highly reduced mandibles or even have no mandibles at all. Reduced mandibles are also a feature of some other pimpliformes, including Acaenitinae and Orthocentrinae. J.R.T. Short & Nakanishi (1981) noted that the third instar larvae of *D. laetatorius* move their heads up and down, which is probably involved in physically breaking down host tissues and thus serves the same function as mandibles in most other taxa (Wahl 1990). There appear to be a total of five instars, at least in *S. fissorius*, although the first seems, because of its feeding pattern, to have multiple forms, but with the head capsule remaining identical. *D. pectoratorius* shows a somewhat different developmental profile, with the first instar larva commencing active feeding and movement well before the host commences pupari-sation. Emergence from the host puparium in at least some species involves the progressive cutting of a series of thin ribbons in the host cuticle until a sufficiently large exit hole has been made (Rotheray 1981a); perhaps this is related to the tridentate mandibles that characterise the subfamily and it would be interesting to know how general this behaviour was.

Rotheray (1981b) described the oviposition behaviour of seven species, representing four genera, and found that most can attack host larvae of a range of instars and generally oviposit into the mid-region of the host, depositing the egg in haemocoel. The behaviour of *Enizemum ornatum*, which specialises in later instar hosts, was more complicated. The female wasp mounts the back of the hoverfly larva, stabs it with its ovipositor and bites its back, causing it to rear up its front end and defensively regurgitate. At this point, the wasp quickly stabs the host just behind its mouthparts (i.e. in the 'throat') with her ovipositor and oviposits. In a neat experiment, Rotheray then restrained the head end of the host, *Metasyrphus lunifer*, third instar larvae from beneath through a hole in the Petri dish in which they were presented to the *Enizemum*. This allowed control of the host's head movement in response to the biting by the parasitoid. If the host was made to move its head up in the normal response fashion the parasitoid oviposited in it, but if the head of the hoverfly larva was prevented from rearing up, the parasitoid did not oviposit and soon moved away, indicating that host movement or posture was an important releaser for the actual parasitoid oviposition event.

Wahl (1990) surveyed the final larval head capsule morphology of a range of diplazontines and related taxa, supplementing the previous descriptions of J.R.T. Short (1978), and incidentally noted the extremely elastic properties of the larval cuticle, which meant that making good microscopic preparations 'is achieved only by practice'. The larval cephalic structures are greatly reduced (antennae and hypostoma lost, mandibles lost or reduced) and also have a set of clear synapomorphies. The information available at that time, however, precluded detailed internal phylogenetic consideration or production of a key.

The Diplazontines are one of a few groups of ichneumonidae that display tyloids on their antennae (Fig. 13.17), which are release structures for glands within a few, species-specific, flagellomeres. The flagellomeres that they are on in a given species correspond to the points of contact during courtship in which the male wraps his antenna around those of the female during courtship (Klopfstein et al. 2010b, S. Steiner

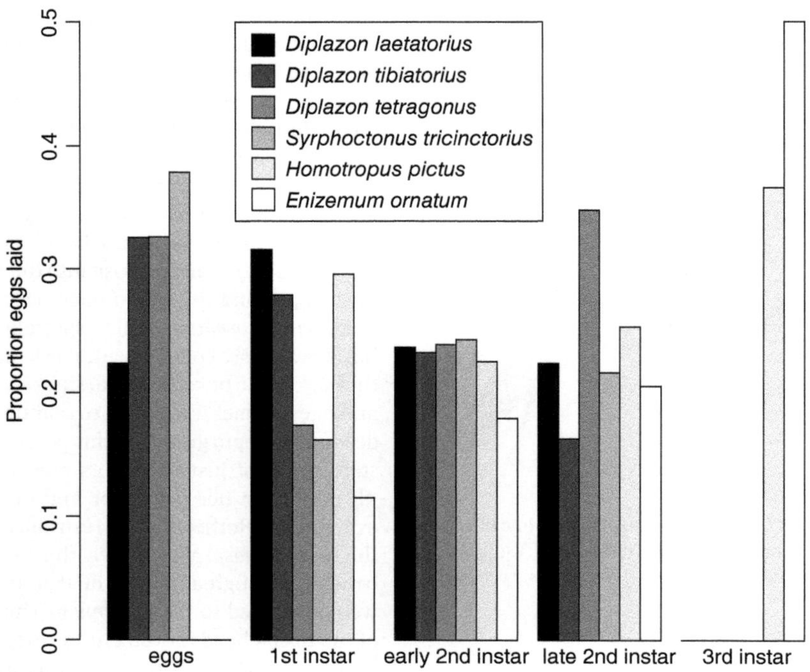

Fig. 13.16 Relative preferences of six diplazontine species for hosts of different developmental stages, expressed a proportion of eggs deposited into each category. (Source: data from Rotheray 1981b.)

et al. 2010). In different species, the males wrap their antennae around those of the female either once or twice. As insect antennae lack internal musculature, the bending is achieved by the interaction of different dimensions of intersegmental membranes and haemolymph pressure. Being an entirely physical phenomenon, S. Steiner et al. (2010) showed that it could be replicated in dead specimens, even sometimes ones of considerable antiquity, by transferring excised antennae from alcohol to fresh water. In this way, it was possible to determine the mode of antennal coiling in taxa whose courtship and mating had never been observed in life.

Destructive host feeding is carried out by numerous species, but unlike most instances of this behaviour, the diplazontines can act as true predators. *D. laetatorius*, *D. tibiatorius*, *Homotropus pictus* and *E. ornatum* all eat host eggs, the first two also consuming first instar larvae and the last species, especially when old, will kill and consume all larval instars of its host (Scott 1939, Rotheray 1981b). It is likely that this behaviour provides females with additional protein and lipid resources for continued egg production (F. Schneider 1950). *D. laetatorius* (Fig. 13.15) is undoubtedly one of the commonest ichneumonids in the world, being spread virtually everywhere in association with syrphids feeding on crop aphids. It can sometimes be extremely numerous and must be considered a harmful species from the pest management point of view (Gauld 1984a) – one of the very few such ichneumonoids, although in the same system, the euphorine ladybird parasitoid *Dinocampus coccinellae* can also cause substantial reduction in aphid control.

The metasomas of females of some species are elongate and strongly laterally compressed (e.g. *Phthorima compressa*) or strongly dorso-ventrally depressed (e.g. *Campocraspedon arcanus*), adaptations to attacking host hoverfly larvae associated with gall-forming aphids while not possessing a long ovipositor (Fitton & Rotheray 1982).

Relationships between the genera have been investigated recently by a molecular phylogenetic analysis using two gene fragments, COI and 28S rDNA (Klopfstein et al. 2010a) and subsequently by combined

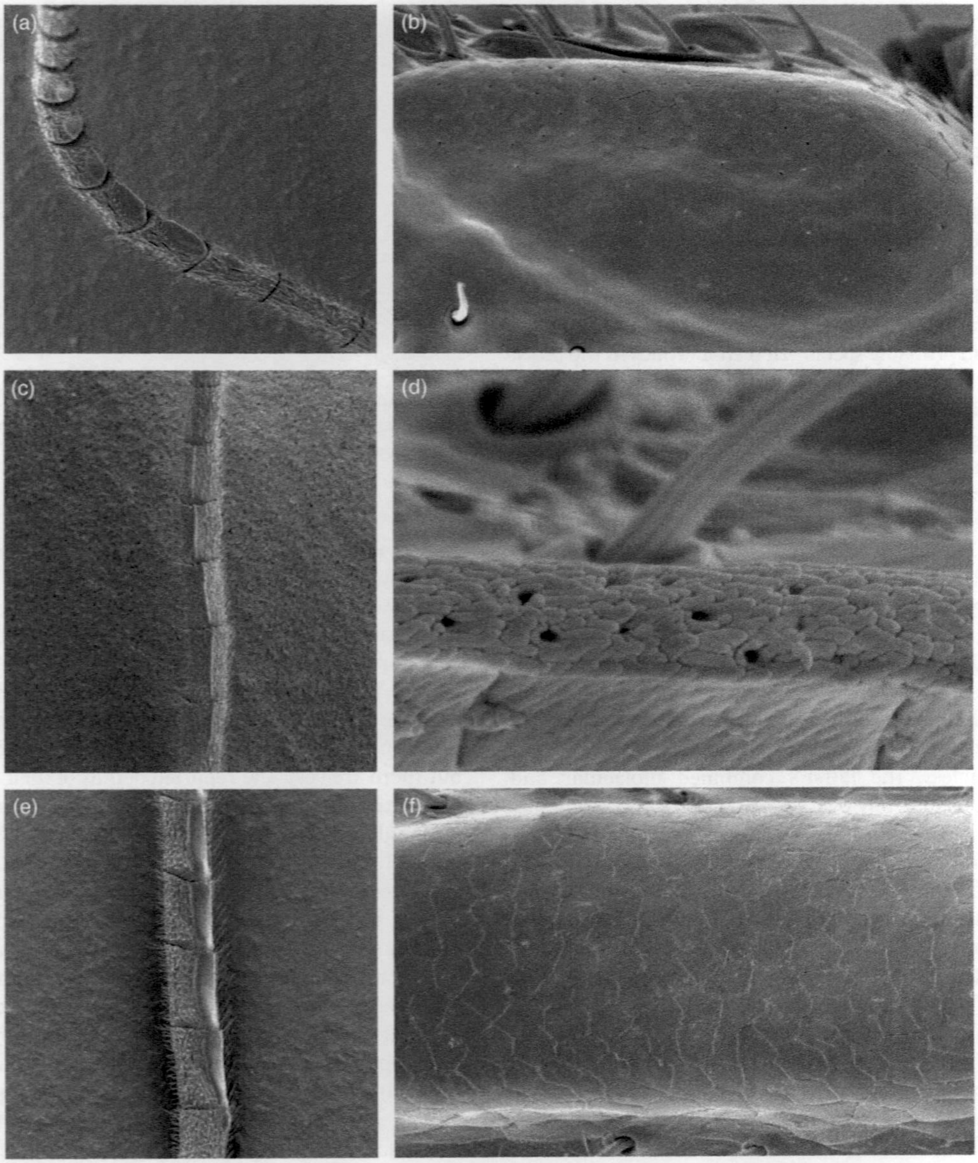

Fig. 13.17 SEM images of tyloids of Diplazontinae, low magnification (left) and high magnification (right). (a, b) *Sussaba pulchella*; (c, d), *Syrphoctonus tarsatorius*; (e, f) *Enizemum ornatum*. (Source: reproduced by permission of Seraina Klopfstein.)

molecular and morphological phylogeny (Klopfstein et al. 2011). Although there was some conflict between the data partitions, the resultant phylogeny generally reflected the DNA data indicating morphological homoplasy. Three well-supported clades were recovered and these corresponded to groups that displayed either morphological or behavioural synapomorphies. All genera with multiple representatives were recovered as monophyletic except for *Syrphoctonus*. Fitton and Rotheray (1982) provided a key to the European genera, although some generic definitions were modified and a new genus was described by Klopfstein et al. (2011).[5]

Orthocentrinae (= Helictinae)

These small wasps, some of the smallest in the Ichneumonidae, are cosmopolitan and, because of their small size (see Fig. 15.7), relatively poorly known. A recent study of ichneumonids caught in Malaise traps in South America showed that these are one of the most abundant and probably species-rich subfamilies in the moist tropics; they are also considerably under-researched in the temperate region (see below). Superficially, many rather resemble small metopiines, particularly because of the rather roundly protruding face.

The taxonomic history of the group is complex and could be confusing for the beginner, and Townes (1971) referred to them (as Microleptinae) as a 'waste-basket' group, reflecting the morphological diversity among the genera he included and indeed among those still included. The name has been associated with Microleptinae and Oxytorinae, which are now not considered to be closely related, and also with the Helictinae, which is regarded as a synonym in agreement with Wahl (1990) and Quicke et al. (2009). Nevertheless, the latter's combined molecular and morphological phylogenies show that the subfamily falls into two groups, the '*Orthocentrus* group' of genera and another, comprising Wahl's *Eusterinx* group and clusters A and B. This arrangement is considerably different from that in Wahl's cladogram, but the separation in that of the cluster B+*Eusterinx* group was only supported by two not very convincing apomorphies, where the first sternite ends, and the loss of glymmae. If the monophyly of the group remains supported, it is as much a matter of preference whether they are treated two separate subfamilies (as by Gauld 1991) or as a single one. Biologically, from what little is known, they are similar.

The '*Orthocentrus* group' of genera are relatively easy to recognise; they have a protruding face and a large scapus (Fig. 13.18). As stated by Broad (2010), their classification has been rather neglected since Townes (1971) and the group displays rather more morphological variation than has generally been appreciated. Broad presented morphological phylogenetic analyses of relationships between exemplar species which support the monophyly of the included genera except for *Stenomacrus*, which appears to be multiply polyphyletic.

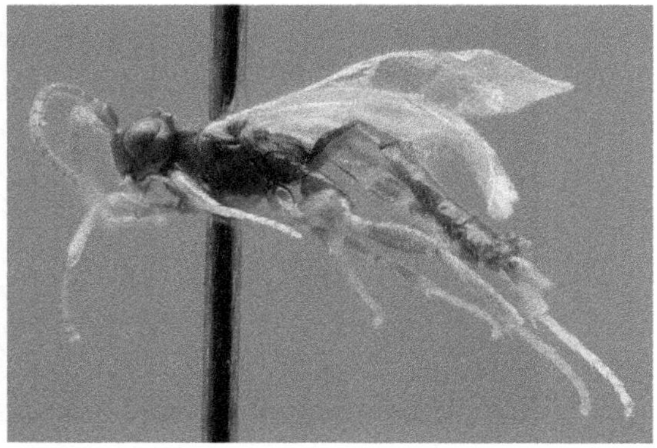

Fig. 13.18 Female of *Stenomacrus brevicornis* (Orthocentrinae). (Source: reproduced by permission of Ilari Sääksjärvi.) For a colour version, see Plate 38.

To date, all rearing records (although it is not a very large number) show that these are primary koinobiont endoparasitoids of nematoceran, fungivorous Diptera larvae, predominantly Mycetophilidae (Wahl 1990, Shaw & Askew 2010) but also Sciaridae (Waterston 1929b); records from other families seem less likely but there are very few reported details of their biology. One of the great difficulties with studying them is that although they can readily be reared from mycetophilid infested fungal fruiting bodies, as with nearly all substrate rearing, a considerable number of potential host species are often reared from a single fruiting body making study of actual host relationships very difficult. The separation of potential host larvae combined with DNA sequencing of both host and parasitoid seems to be the obvious way forward. Interestingly, some mycetophilids make a flimsy silk cocoon and, as Gauld (1988a) noted, host silk seems to be an important link in host transitions within the Ichneumonidae, so perhaps the group colonised mycetophilids by this route. Nothing is known about their early developmental stages and almost nothing about the host stages attacked. *Neurateles papyraceus*, which attacks sciarids under bark, has a markedly telescopic and laterally compressed metasoma with only a short ovipositor, which it presumably uses to reach hosts through cracks or narrow gaps.

The group as an whole is relatively very understudied, especially so in the tropics. Shapiro and

Pickering (2000) found that they are significantly more abundant in Malaise trap samples from wet rather than moist MesoAmerican forests, and it now appears that they may have very high, and almost entirely undescribed, species richness in tropical moist/wet forests (Veijalainen et al. 2012a,b), which is perhaps not surprising given the very high species richness of, for example, soil- and detritus-dwelling flies there. Some species at least appear to be highly host specific (Askew & Shaw 1979). Roman (1923) reported that *Proclitus edwardsi* probably only attacks a single species of mycetophilid, *Brachypeza radiata*, but furthermore, only when it is feeding on a single species of fungus. A great deal of more carefully obtained information will be required to determine whether narrow host ranges are the norm or the exception for the subfamily.

Venom apparatus varies considerably between genera and that of *Proclitus* especially so, such that Quicke et al. (2009) scored it separately for morphology to try to test whether it was monophyletic with the other orthocentrines. It was, however, recovered nested within the subfamily and its position as the sister group to the *Helictes* group of genera probably reflects this separate morphological scoring.

Pimplinae

This large group is heterogeneous biologically and is currently divided into three tribes, the Delomeristini, Ephialtini and Pimplini; Wahl and Gauld (1998) also recognised a fourth tribe, the Perithoini, but this was synonymised with the Delomeristini by Gauld et al. (2002b). Many earlier workers included the Rhyssinae and Poemeniinae under the name Pimplinae (e.g. Finlayson 1967, J.R.T. Short 1978, Gauld 1984a,b). Eggleton (1989) removed *Diacritus* to its own subfamily, the Diacritinae, and this was published effectively by Gauld (1991). Gauld (1991) also formally removed the Rhyssinae and the Poemeniinae into their own separate subfamilies thus completing the splitting of the traditional Pimplinae. What is left is now generally considered to be a monophyletic group (but see below).

The Pimplinae includes idiobiont ectoparasitoids and endoparasitoids, koinobiont ectoparasitoids and egg predators. The egg predators and koinobiont ectoparasitoids belong to the Ephialtinae and for a long while the latter, which attack spiders, were recognised as the separate tribe Polysphinctini (Fig. 13.19c). However, phylogenetic analyses place them firmly as a derived group within the Ephialtini and so now they are referred to as the '*Polysphincta* group', although their biology is so different that I cover them in a separate section from the other ephialtines. The great majority of pimpline species belong to the Ephialtini.

A summary of the various classificatory changes was presented by Gauld et al. (2002b) so only a few key points are discussed here and, as often is the case when dealing with subfamily and tribe level names in the Ichneumonidae, the idiosyncratic terminology employed by Henry Townes inevitably has to be mentioned. For reasons discussed above (see the section *Henry Townes (1913–90) and His Idiosyncratic Nomenclature*, above), Townes referred to this subfamily as the Ephialtinae, the Pimplini as the Ephialtini and the Ephialtini as the Pimplini (Table 13.3).

There is conflict between the results of molecular phylogenetic analyses based on the D2 region of 28S and morphological analysis (Gauld et al. 2002b) as to whether the Pimplinae are recovered as monophyletic. Gauld et al. identified three characters as being probable synapomorphies for the subfamily, of which the most convincing is the presence in males (and females of some groups) of a poison gland indicated by a membranous vesicle on the inside of the hind claw – the degree of development of the gland was surveyed in detail by Gauld et al. (2002b; character 87)[6]. In the case of some of the larger *Xanthopimpla* (Fig. 13.19b) and *Theronia*, these can deliver a moderately painful sting to the human hand. In some taxa, the poison pore is shielded while not in use by a large spatulate setum (Fig. 13.20). This gland is not, however, present in all Pimplinae, being absent in *Pseudopimpla* and scattered members of the *Pimpla* genus group. This absence was interpreted by Gauld et al. as probably representing secondary losses associated, in the case of the Pimplini, with the evolution of a large basal lobe to the claws which would make the claw less effective as a weapon of defence.

One criticism that can be levelled here (and in many other studies for that matter) is that the choice of, and extent of, outgroups used might not have fully tested the monophyly of the group. The available independent molecular studies are also limited because they are essentially based on a single gene fragment, the 28S D2–D3 rDNA gene, but these have all failed to recover the Pimplini as monophyletic with the Ephialtini (Belshaw et al. 1998, Belshaw & Quicke 2002), although they invariably do in combined analyses with

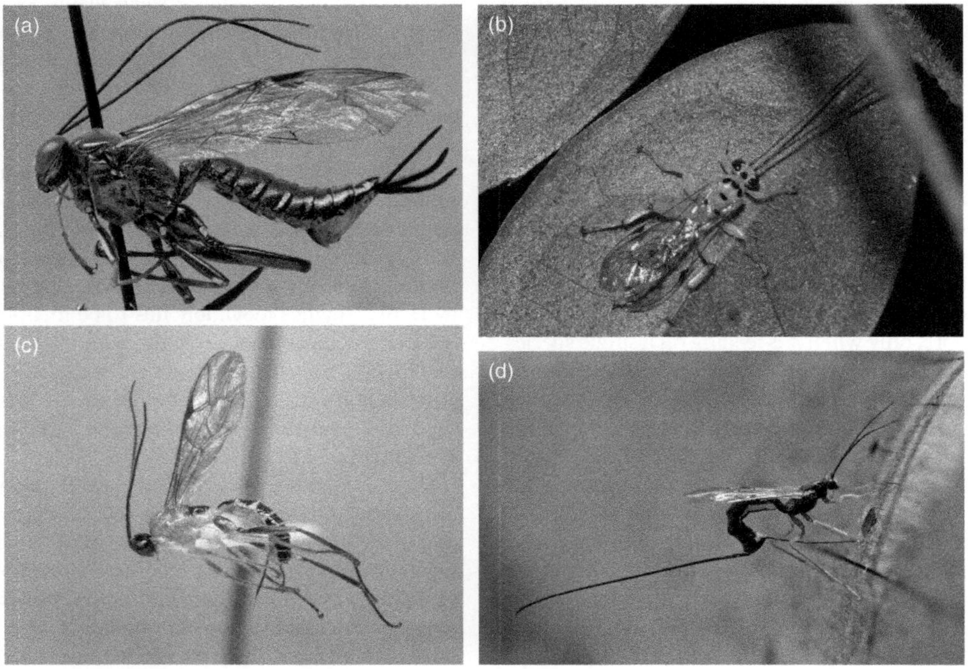

Fig. 13.19 Assorted Pimplinae species. (a) *Pimpla viridescens* (Pimplini); (b) *Xanthopimpla pedator* (Pimplini); (c) *Lamnatibia andina* (Ephialtini, *Polysphincta* group); (d) *Dolichomitus mesocentrus* (Ephialtini). (Sources: a, c, reproduced by permission of Ilari Sääksjärvi; b, reproduced by permission of Chan To; d, reproduced by permission of Cor Zonneveld.) For a colour version, see Plate 39.

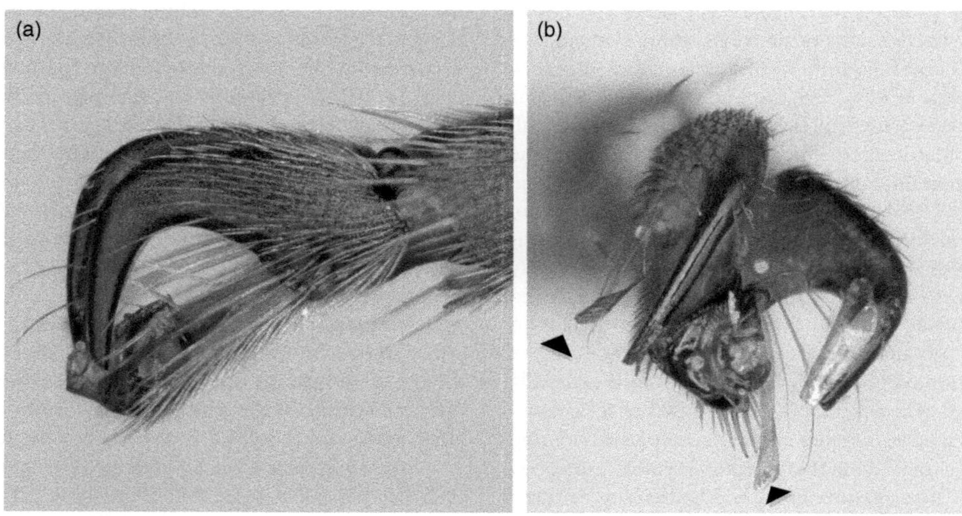

Fig. 13.20 Two views of claw of *Xanthopimpla konowi* (Pimplinae) showing the spatulate seta (arrowhead in b) that protect the end of the claw proper with its venom pore. (Source: reproduced by permission of Rikio Matsumoto.)

morphology included (Quicke et al. 2009). Having said this, the molecular data are not especially robust because the rate of evolution of at least the 28S gene and possibly numerous others within the pimpliformes has not been very great and thus does not provide much signal.

The phylogeny of the group has been considered in detail in two papers, first by Wahl and Gauld (1998) and culminating in a very thorough morphological, genus-level treatment (Gauld et al. 2002b). The latter showed convincingly that the spider parasitoids, formerly placed in the tribe Polysphinctini, are a derived group within the Ephialtini, hence these are now referred to as the *Polysphincta* genus group. In addition, it also indicated that a number of fairly well-studied genera such as *Dolichomitus* (Fig. 13.19d), *Itoplectis* and *Polysphincta* are probably not monophyletic and that was found to be the case in the more detailed morpho-phylogenetic study of Gauld and Dubois (2006). From the studies of Wahl and Gauld (1998) and Gauld et al. (2002), it seems that the Pimpliformes had their origin in the Northern Hemisphere where most of their diversification occurred. In fact, the group is not particularly well represented in either Australia, New Zealand or Chile, and of the taxa found in the latter, there are no obvious links with the other 'Gondwanan' genera and instead they appear to be more associated with North and Central American elements.

Little is known about their courtship and mating. Males of a few genera possess antennal tyloids (four genera distributed among Delomeristini, Ephialtini and Pimplini; Kolarov & Gürbüz 2007), thus probably representing independent origins, but how they are used is not known.

The fossil record for the subfamily includes numerous specimens of Ephialtini (including the *Polysphincta* group) and Pimplini from the Florissant Oligocene of Colorado, ca 37–33.7 Mya, but no definite members from earlier (Kasparyan & Rasnitsyn 1992).

Delomeristini

This small group comprises some medium to large wasps. As currently constituted, it also includes *Perithous*, which was previously placed in its own tribe, Perithoini, by Wahl and Gauld (1998). *Perithous* species are probably exclusively parasitoids of stem, wood and gall nesting aculeate Hymenoptera and records from other hosts probably reflect failure to recognise that their borings in these localities had subsequently become utilised by their true hosts (Fitton et al. 1998). *Delomerista* also appear to be Hymenoptera parasitoids and appear to attack cocooned sawflies and ichneumonids, but taxonomic confusion over the use of the name means that some earlier publications have to be treated with caution (Fitton et al. 1998). *Perithous* is unusual in that once the final instar larva has sucked the remains of its host dry, it proceeds to eat the remaining host cuticle (Danks 1971).

Ephialtini (= Pimplini of Townes)

This is the largest and most diverse group of Pimplinae, with 60 recognised genera. They are all either idiobiont ectoparasitoids or egg predators (such as *Tromatobia* in spider egg cocoons) or, in the case of the *Polysphincta* group of genera (see below), koinobiont ectoparasitoids on spiders; because of the highly specialised biology of these, it is convenient to discuss them separately. Evenhuis and Vlug (1983) commented that in *Scambus brevicornis* 'the first instar parasite larva seems to live endoparasitically; the larval moult may be seen protruding from the caterpillar skin'. This seems unlikely, but it would be very interesting to know precisely what they observed.

Most probably attack concealed Lepidoptera and less commonly Coleoptera larvae or concealed sawfly larvae and a few genera are associated with aculeates such as *Townesia* on *Megachile* bees and probably *Ephialtes manifestator* on sphecids such as *Trypoxylon*. Some species, especially of *Dolichomitus* (Fig. 13.19d), have very long ovipositors (Townes 1975, Zwakhals 2010), that of the Peruvian *D. hypermeces* being up to 18 cm long or eight times the length of the body. In addition to host larvae, some *Dolichomitus* consume prepupae and even sometimes fresh, still soft pupae. *Endromopoda* are associated with galls in graminaceous plants and at lease some attack Diptera hosts (Fitton et al. 1988). Occasionally, species have made the host switch from attacking beetle or moth larvae in wood to parasitising aculeate hymenopterans that have secondarily utilised an empty boring as a nesting hole (Jussila & Kapyla 1975). Some have extremely broad host ranges that are limited by niche rather than host taxonomy. For example, the Eurasian *Scambus annulatus* is fairly reliably recorded from nine families of weakly concealed Lepidoptera, also as a pseudohyperparasitoid of both braconids and ichneumonids attacking hosts in the same habitat, and also from leaf-mining beetles and rarely weakly concealed hosts in other situations (Fitton et al. 1988). *S. brevicornis* has a similarly wide range of lepidopteran, hymenopteran (as pseudohyperparasitoid) and dipteran hosts associated with flowers and heads of Asteraceae. Despite being idiobiont ectoparasitoids, some species may still have narrow host ranges, probably due to behavioural specialisations associated with host location. Zwakhals (1989), for example, reported observations by H. Pschorn-Walcher that *Alophosternum albofascialis* 'attacks only *Heterarthrus* [Tenthredinidae] prepupae in their cocoons in the withered and fallen leaves of their host plants'. Most species are solitary, but members of some genera such as *Exeristes*, *Fredegundia*, *Gregopimpla* and *Iseropus* that have large hosts are facultatively gregarious and can produce small gregarious broods (sometimes only two or three individuals per host; Fitton et al. 1988).

Pseudopimpla attack fully grown cephid sawfly larvae (Matsumoto & Saigusa 1999; although they might also attack Lepidoptera hosts in the same location) and *Hemipimpla* has been found to attack the social polybiine vespid *Ropalidia formosa* (R.W. Brooks & Wahl 1987). Probably nearly all those species attacking Lepidoptera or Coleoptera hosts are at least potentially polyphagous, but no doubt some of those associated with aculeates are more host specific. *Scambus pomorum* is, as far as is known at present, unique within the Ichneumonoidea and possibly parasitic wasps as a whole, in that the females are predatory for a large part of their adult life span on a variety of leafminer larvae (in captivity feeding on lepidopterous, dipterous and sawfly leaf-miners on a variety of tree species) (Zijp & Blommers 2002a). During this predatory phase which lasts all the summer after mating in the spring, the ovaries remain undeveloped and egg maturation only occurs in the following winter during hibernation. The hosts of *S. pomorum* are the larvae of an *Anthonomus* weevil which reach a suitable size in the early spring. Leius (1962, 1963, 1967) presented a series of studies on the effects of different adult dietary components (host haemolymph, pollen) of the polyphagous *Scambus buolianae* (Pimplinae). Host feeding was found to increase female longevity [in contrast to Ueno's (1999a) finding for *Pimpla nipponica*].

Phylogenetic relationships among the genera of Ephialtini are far from certain; however, Wahl & Gauld (1988), based on their morphological cladograms, suggested that parasitisation of Hymenoptera hosts was probably a basal feature of the group with subsequent radiations on to Lepidoptera and then spiders. The latter may have come about through the use of host silk in providing cues for host location, although exactly what the silk-related stimulus might be, if there really is one, is uncertain and it cannot really be the silk proteins themselves as they are both highly insoluble and vary considerably in chemistry.

Exeristes species have been the subject of a considerable number of investigations of the diet of parasitic wasp larvae and their nutritional requirements (see Chapter 5; S.N. Thompson 1974, 1975, 1976a,b, 1979, 1982a, S.N. Thompson & Adams 1976, S.N. Thompson & Barlow 1970, 1972, 1973). *E. roborator* larvae have been cultured successfully on artificial media comprising just L-amino acids, glucose, cholesterol, triolein (a triglyceride of oleic acid), inorganic salts and vitamins (S.N. Thompson 1975, 1976a,b). When sufficient amino acid was supplied, the wasp larvae could synthesise their own carbohydrates from them alone and the triglyceride seemed of little importance for larval development.

Despite their diversity, there are few detailed descriptions of their juvenile stages. Juillet (1959) described the development of *Exeristes* (as *Ephialtes*) *ruficollis*, a parasitoid of the tortricid pine shoot moth, *Rhyacionia buoliana*, an introduced pest in North America. The first instar larva has a relatively large, sclerotised head and tapering body, but the subsequent four instars are more or less hymenopteriform, the last two having finely serrate mandibles. Overwintering is as a prepupal larva. Five larval instars were also reported for *Scambus* (as *Ephialtes*) *brevicornis* by Stuart (1957). *E. ruficollis* apparently detects infested plant shoots by volatiles and then searches using her antennae to find the opening into the host's boring. Once located, she inserts her ovipositor and, having touched a host larva, 'suddenly probes rapidly with her ovipositor … until the latter is motionless' (Juillet 1959). Individual females lay on average 23 eggs during their lifetime, which appears to be fairly typical of the tribe, although dissections indicated that there were many immature eggs remaining to be oviposited. What factors might lead to higher realised fecundities are unknown, as 'females killed and fed on many host larvae just before and throughout the oviposition period', so availability of protein/lipid was not the limiting factor. In addition, the wasps apparently fed on resin exuded from new buds as a result of their piercing them with their ovipositors. Being polyphagous and therefore there being no particularly guaranteed host cue for recently emerged learning host associated cues is important

Fig. 13.21 Larva and behaviour of *Polysphincta* group pimplines on host spiders. (a) *Acrodactyla* sp. on a tetragnathid spider in the UK; (b) *Reclinervellus masumotoi* fully grown larva killing host spider; note dorsal tubercles; (c, d), *Zatypota albicoxa*, infanticide, removing a penultimate instar rival's larva, and removal of a second instar rival larva from host spider, *Parasteatoda tepidariorum*, respectively; (e, f) webs of *Cyclosa argentoalba*, host of *Reclinervellus nielseni*, normal and modified webs, respectively. (Sources: a, reproduced by permission of Roger Key; b–f, reproduced by permission of Keizo Takasuka.) For a colour version, see Plate 40.

for this species and has been investigated by Wardle and Borden (1985, 1990). Female *E. ruficollis* have a protracted preoviposition period that lasts from 10 to 29 days post-mating, but it is not clear whether this is due to the need to mature eggs or the lack of really attractive, i.e. sufficiently mature, hosts (Juillet 1959).

Polysphincta Group

These wasps have a very interesting biology being koinobiont ectoparasitoids exclusively of spiders (Fig. 13.21). Most attack immature hosts, although some *Schizopyga* species are known to attack adults

(Fitton et al. 1987). Other instances of koinobiont ectoparasitism are in the ichneumonid Tryphoninae and Adelognathinae and the braconid *Rhysipolis*. Despite what might be construed from older literature and catalogues, polysphinctines are generally very host specific, no doubt because spiders are dangerous 'prey' for small wasps and a high degree of coevolutionary arms race has no doubt been involved in the history of each species pair. Following very extensive collecting and rearing and checking older records where possible, M.R. Shaw (1994) showed that many of the previously published host data on European species were simply wrong. For example, prior literature records indicated that *Polysphincta boops* attacked the common garden orb-weaver *Aranaeus diadematus* and *Therdion* spp., whereas it attacks neither of these hosts and is instead a specialist on *Lariniodes cornutus*.

As with a very few other ichneumonoids (e.g. *Histeromerus*, M.R. Shaw 1995; *?Adelognathus*), the egg of some *Polysphincta* group genera does not travel down the ovipositor and exit near the tip, but instead exits from the swollen basal portion of the ovipositor (Eberhard 2000a).

The biologies of several *Polysphincta* group species, along with the behavioural changes they induce in their spider hosts, have been described in considerable detail (Eberhard 2000a, Matsumoto & Konishi 2007, Weng & Barrantes 2007, Matsumoto 2008, Gonzaga et al. 2010) and some detailed accounts of their biology and development have also been published (e.g. Fincke et al. 1990, Weng & Barrantes 2007), but one of the most detailed descriptions is that of Eberhard (2000a) for the moderately large Neotropical *Hymenoepimecis argyraphaga*, which attacks the tetragnathid spider *Plesiometa*. Females are synovigenic and emerge with no mature eggs available. Maturation of eggs in captive wasps took 3–4 days and was accomplished by using resources stored in the wasp's fat body (whether this is true of all species is not yet known). Females glue their eggs on to the spider host, although the source of the glue is not yet known. The first instar larva hatches but keeps its posterior end inside the firmly attached egg chorion and starts to feed on host haemolymph, making numerous small feeding wounds (see also Weng & Barrantes 2007). Subsequently, the growing larvae remain attached to the host via a saddle which is probably at least partly congealed spider blood and the wasp larva has a pair of hooked projections (taps) on the eighth and ninth segments that firmly attach into the saddle. The relatively brief final instar larva of several species (Fig. 13.21a,b) has a series of paired retractable tubercles that are furnished with small hooks, and these are used to snag the spider's silk and the parasitoid larva continues to feed on the spider's haemolymph after the spider had died (as indicated by its heart having stopped beating). Eberhard called the final instar of *H. argyraphaga* the 'third', indicating some degree of uncertainty about the total number of instars, and I am rather doubtful that such a relatively large wasp could develop externally through so few skin changes.

Polysphinctines are always solitary parasitoids, although undoubtedly the host-searching females must fairly frequently encounter previously parasitised spiders, something hard to observe in the wild, but easy to engineer under laboratory conditions. When this is done, as Takasuka and Matsumoto (2011b) did with *Zatypota albicoxa*, the female wasps always displayed infanticidal behaviour, removing any other *Zatypota* egg or larva from the spider's back (Fig. 13.21c,d), using her ovipositor to pry off the egg or larval attachment saddle from the host, a process which could take a considerable time if she was removing a late instar rival. That they did this readily suggests that this is a specific evolved behaviour.

The females of some species such as *Zatypota petronae* attack their host spiders in their retreats (Weng & Barrentes 2007), whereas others have different attack strategies. *Z. albicoxa*, a parasitoid of the Japanese theridiid house spider, *Parasteatoda* (=*Achaearanea*) *tepidariorum* (Theridiidae), has two attack strategies: one is a direct attack wherein the wasp climbs up the spider's web to its retreat and attacks it there, the other is an ambush strategy with the wasp hanging from the spider's web and waiting until the spider hauls it up as a potential prey item (Fig. 13.22) (Takasuka et al. 2009). The same species of wasp can also display a different oviposition strategy with the wasp lying on her dorsum at the bottom of one of the vertical web trap-lines which she tugs with her legs, so mimicking the action of a truly captured prey (Takasuka & Matsumoto 2011a). The spider is then attacked when it climbs down the silk line to reach the 'trapped' wasp.

Many species modify the host spider's behaviour and often cause it to make abnormal webs or shelters close to its demise (Fig. 13.21e,f) and in some cases the benefit to the parasitoid is clear. Eberhard (2000a) experimentally removed penultimate and final instar larvae of *H. argyraphaga* from the spider host to determine whether the parasitoid larva's presence was

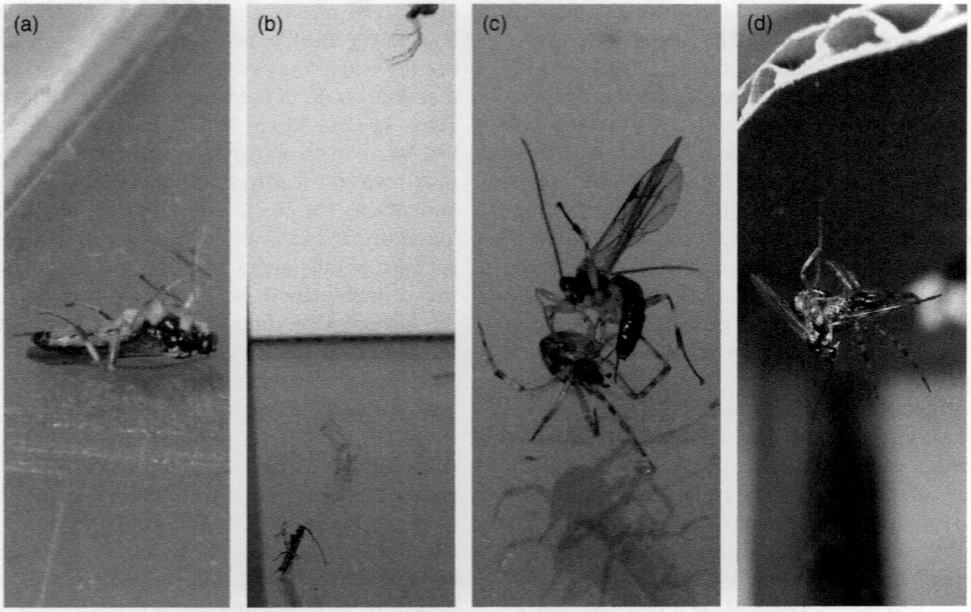

Fig. 13.22 Series of photographs of *Zatypota albixoxa* (Pimplinae, Ephialtini, '*Polysphincta* group') ambushing its host spider in her web. (Source: reproduced by permission of Keizo Takasuka.)

necessary to induce the spider, in its last day of life, to construct a special web, called a cocoon web, in which the parasitoid larva will pupate. Spiders relieved of penultimate instar larvae continued to construct normal webs, but when final instar larvae were removed, the spider produced the modified cocoon web, indicating that the parasitoid had somehow, presumably chemically, caused a persistent change to the spider's web-constructing programme. The modified, either dense or cupola-like webs that the theridiid spiders attacked by *Zatypota percontatoria* construct before they are finally consumed by their parasitoid in the summer actually very closely resemble the webs these spiders make prior to overwintering and therefore the polysphinctine is causing the spider to carry out a behaviour that it already has in its repertoire but at a different time of year (Korenko & Pekár 2011).

Earlier treatments of the 'Polysphinctini' varied in what taxa they included, with some authors considering those specialist spider egg sack predators (*Tromatobia* and allied genera) as also belonging to it. Based on morphological analysis, the *Polysphincta* group does appear to belong to a clade of ephialtines whose members are also associated with spiders, but as egg predators, viz. *Tromatobia*, *Zaglyptus* and *Clistopyga*. However, whilst molecular data strongly support monophyly of the *Polysphincta* group, *Tromatobia* is recovered nested elsewhere within the Ephialtini (Quicke et al. 2009). Whether this is due to morphological convergence to do with biology or misleading molecular signal is unclear at present.

The world genera of the *Polysphincta* group of genera were partially revised and keyed by Gauld and Dubois (2006).

Pimplini

The Pimplini is a relatively small tribe comprised mostly of idiobiont endoparasitoids, nearly always of Lepidoptera pupae, although Gauld (1988a) noted that some members of the paraphyletic basal and almost certainly polyphyletic genus, *Theronia* (see Gauld et al. 2002b) are ectoparasitic on pupae within thick robust cocoons[7]. Given that other species of *Theronia* show typical pupal endoparasitism of both concealed and exposed Lepidoptera pupae, it would be very interesting to have a more detailed and ideally independent molecular phylogenetic picture of the group so that perhaps more understanding of the transition between ecto- and endoparasitism might be gained.

Under laboratory conditions at least, *Pimpla disparsis* can occasionally develop successfully within a prepupal host (Fuester et al. 1989) and Cole (1967) noted that very rarely *Itoplectis* and *Apechthis* are reared from wild-collected prepupae, but that the female wasps more regularly using prepupal hosts for host feeding; perhaps those parasitised were at a rather late stage of development. Pupal hosts have relatively little physiological defence capability, probably largely due to the major body reorganisation that goes on through much of this stage, and pupae therefore offer a nutrient-rich food source that is relatively easy to adapt to. Nevertheless, the eggs of Pimplini become coated in a protective mucopolysaccharide layer originating in the vaginal gland, which prevents encapsulation by host haemocytes and suppresses haemocyte spreading (Führer 1973b, Osman 1978). Osman's study interestingly showed that while the vaginal gland secretion inhibited pseudopodial formation by plasmatocytes and some other haemocyte types in each of the four lepidopteran species tested, it had no effect on one sawfly, although with N=1 it is impossible to know how generalisable that might be. Not surprisingly many species have broad potential host ranges, although some are more specialist either due to behavioural or physiological adaptations to particular hosts.

Some *Itoplectis* species are common endoparasitoids of ichneumonid pupae in cocoons, thus acting as pseudohyperparasitoids (M.R. Shaw 2009b), and the large and common, principally tropical and warm-temperate genus *Theronia* species appear normally to be pseudohyperparasitoids attacking cocooned ichneumonids and some may be kleptoparasitoids (Fitton et al. 1988, Sime 2000). Further, some *Itoplectis* in the wild attack the 'cocoons' made by some beetles such as those of the weevil genus *Cionus* and oviposit within either full grown beetle larvae or into the pupa (Iwata 1966b); interestingly, *Cionus* cocoons are rather smooth and it is easy to see how they could be mistaken for an exposed pupa, and *I. maculator* has been reared from tachinid fly puparia (Zwölfer & Kraus 1957).

The development of Pimplini has been studied in *Pimpla* (Rojas-Rousse & Benoit 1977). The first instar larvae are mandibulate (see Fig. 5.3) and actively seek out to fight and try to kill competitors (Ueno 1997). Many superparasitised eggs were found to have melanised wounds, indicating that physical damage was a major cause of mortality at this stage also. Those of *Pimpla* species also actively make their way to the host pupa's brain where they appear to destroy endocrine tissue and thus prevent the further metamorphosis of the host (Führer & Kilincer 1972). There are five larval instars in total, those after the first being hymenopteriform and with roughly similar head capsule features, all them having functional mandibles (Fig. 5.3). Rojas-Rousse and Benoit also found that it was possible to sex the final instar larvae because of differences in their gonad imaginal discs. Pupal development of *I. melanocephala* progresses in the normal ichneumonoid way, with pigmentation of the eyes even being visible in the prepupa (Fig. 5.4). Pimplines usually spin a cocoon within the pupa of the host, although some do not (e.g. *Itoplectis alternans*: Iwata 1961).

Development is usually more successful in younger pupae (Hailemichael et al. 1994). Because the pupal stage of holometabolous insects has a rather less effective immune capability than larval stages, even endoparasitic pupal parasitoids are less restricted in host range by host immune reactions, so most species of Pimplini have at least potentially large host ranges and in some species a very great number of hosts are used in the wild. Further, since they are idiobionts and do not have to rely on host hormonal cues, they are relatively easy to culture *in vitro* on chemically defined artificial diets. Such methods have been developed for *Itoplectis conquisitor* (Yazgan & House 1970, Yazgan 1972), and, following limited success with *Pimpla turionellae* with a very basic meat-derived diet by Bronskill & House (1957), this species has now been reared much more successfully on the diet proposed by Yazgan (1972) and Emre (1988). Hedgekar and Arthur (1973) found that a mixture of hexoses and amino acids were the components of host haemolymph that triggered oviposition by *I. conquisitor*, and Arthur et al. (1972) developed a medium containing high levels of the amino acids serine, arginine and leucine plus some $MgCl_2$ that acted as an oviposition stimulant for females of *I. conquisitor*, which was actually more effective than host haemolymph. Hedgekar and Arthur were able to stimulate this wasp to oviposit using haemolymph from Coleoptera, Lepidoptera, Hymenoptera and Trichoptera, but not from Hemiptera, Orthoptera, Neuroptera, Plecoptera or Crustacea. House (1978) managed to take this step further by encapsulating the artificial medium in Parafilm so as to create an artificial host, something achieved far more readily for chalcidoid egg parasitoids. Using such culture conditions makes it possible to explore parasitoid nutrient requirements such as those for vitamins, lipids, amino acids, etc. (e.g. Coskun et al. 2005).

The ovipositors of Pimplini are robust, presumably to aid penetration through tough pupal cuticle or thick tough cocoons. *Apechthis* is distinctive in having a very robust, crowbar-shaped ovipositor which has been shown to be used to prise its way into the host pupa between the articulated abdominal segments (Cole 1959b). Although *Apechthis* were able to oviposit in tortricid pupae of all ages, they were, as with *Itoplectis*, far less successful with older, fully hardened ones and frequently gave up before managing to penetrate them. Interestingly, Cole found that when *Apechthis* tried to attack the exposed chrysalis of the nymphalid butterfly *Aglais urticae*, the host would thrash so violently that it often threw the parasitoid off and violent movements by concealed pupae were also often sufficient to protect them.

One species in particular, *P. rufipes* (usually as *hypochondriaca*), has a remarkably broad host range, attacking Lepidoptera pupae that are made above ground (Fitton et al. 1988). However, while most Pimplini are probably generalists to some extent or other (see, for example, Cole 1959b, M.R. Shaw et al. 2009b), a few are quite specialist. Zwakhals (2005) showed that within what several workers had treated as *P. rufipes* there was a specialist parasitoid of the oak processionary moth *Thaumetopoea processionea* (Thaumatopoeidae), indicating the importance of careful taxonomy when it comes to biological control agents.

Pimpla is one of the better studied ichneumonids from a physiological point of view and it is even being used in biological control, which is rather uncommon for large-bodied idiobiont species. It is also one of the few ichneumonids to have been investigated from an endocrinological point of view (Özlük 1991, Özlük et al. 2011). *P. rufipes* and, to a lesser extent, *P. turionellae* have been investigated extensively in terms of their venoms and the actions of their larval secretions. As the normal hosts are more or less immobile, there is unlikely to have been any selection for Pimplini venoms to contain paralysing components and M.R. Shaw (2009) noted that when *Itoplectis* spp. act as hyperparasites of campoplegine prepupae, some of which are quite mobile and can cause their cocoons to jump, the host remains perfectly active for a few days following oviposition. Nevertheless, the venom of *Pimpla rufipes* does contain a paralytic factor, called pimplin, when tested against housefly larvae (Parkinson et al. 2002b), so perhaps pimplines immobilise hosts to facilitate oviposition. *Pimpla rufipes* (as *hypochondriaca*) venom includes components with antibacterial activity (Dani et al. 2003), which makes a lot of sense in that the immunocompetance of Lepidoptera pupae seems to be low and as the parasitoid makes a hole through its cuticle, it might inadvertently open the way for bacterial entry, which could be detrimental to the parasitoid's development. Indeed, Dani et al. (2004) showed that the wasp venom also increased the susceptibility of the host to infection by *Bacillus* and *Beauveria* pathogens. It may be that antibacterial activity is a common feature of parasitic wasp venoms, perhaps particularly of idiobionts, but there are no data available. *P. turionellae* frequently harbours a microsporidian commensal endosymbiont, which, despite being extremely abundant in the ovaries, Dufour's gland, muscle and other tissues, appears to have little if any detrimental effect on the wasp in terms of longevity (Middeldorf & Ruthmann 1984, who considered them to be yeast-like organisms), although there must presumably be some cost unless there are, as yet unknown, benefits to offset them (Pavenstädt-Grupp & Ruthmann 1989).

Willers et al. (1982) discovered that larvae of *P. turionellae* secrete up to 8 μl of an antibacterial agent from their anus and this secretion was particularly effective against *Enterobacter cloacae*, a virulent bacterium found in the gut of the host *Pieris brassicae* pupae. Whereas injection of host pupae with *E. cloacae* resulted in 100% mortality, simultaneous injection with wasp larval anal secretion allowed some hosts to survive. The secretion was also mildly active against a range of other bacteria and some fungi, but its action clearly appears to be targeted at the organism most likely to be harmful to it. Führer and Willers (1986) carried out further extensive work on the anal secretions of this wasp. It was shown to inhibit the DOPA–tyrosinase reaction, which is important for host encapsulation, and the active secretion components were apparently synthesised in the hypertrophied Malpighian tubules. In addition, they showed that the bactericidal activity of the anal secretion, as measured using a *Beauveria* spore germination assay, was caused by secretory components originating from the hind gut.

The host location behaviour of the Pimplini has taken on special interest with early observations by Henaut and Guerdoux (1982) and Henaut (1990) being followed up by detailed behavioural and morphological studies, mainly on *Pimpla* (=*Coccygomimus*) species (Otten et al. 2000, 2002). Genera in this group have the apex of the terminal flagellomere covered in a series of knob-like processes, erroneously thought at various times to be sensory or expandable vesicles. In fact, they are solid chitinous hammers (Fig. 10.1a,b) used by the female to tap potential host substrates so creating vibrations. Enlarged subgenual organs in the female fore tibia then receive back echoes which are interpreted to reveal the locations of potential hosts, in this case Lepidoptera pupae, within voids in the substrate (Wäckers et al. 1998).

Xanthopimpla stemmator has been introduced into both North America and East Africa for the biological control of cereal borers and several other Pimplini pupal parasitoids have variously been used in control programmes, predominantly against leaf-rollers such as tortricids. In the case of *X. stemmator*, which is used to control the crambid stem-borer *Chilo partellus* on maize, its use of vibrational sounding in host location means that it is limited to locating hosts inside the maize stems and has a negligible effect on those within the maize ears, and it may therefore be desirable to use additional control agents that use alternative host location modalities for those (Muli et al. 2006).

In at least some Pimplini, it seems that the sex of an oviposited egg can be ascertained by observing the precise movements made by the female abdomen during oviposition (Cole 1981, Ueno 1995), although how this relates to the functioning of the spermatheca has not been investigated in this subfamily.

Itoplectis has been one of the few ichneumonoids investigated as regards lipid usage (see also *Exeristes* in the section Ephialtini). Parasitic wasps cannot synthesise their own lipids, so must obtain them from their hosts. *Itoplectis* is highly polyphagous and it can develop successfully given a range of different source lipid compositions. No one lipid was found to be essential, although the wasp larva did require a mixture to develop (Aktümsek 1996). D. Jones et al. (1982) found that neither *Itoplectis* nor *Exeristes* were capable of any significant *de novo* glyceride synthesis, although they could acylate mono- and diacylglycerides to produce triglycerides. Nevertheless, S.N. Thompson and Barlow (1970, 1973) found that the fatty acid profiles of *Itoplectis* closely matched those of whatever host it developed on, suggesting that energetically expensive modification of glycerides was only carried out when necessary.

Important morphological studies include the investigation by Borden et al. (1978a) of the sensilla placodea of *Itoplectis conquisitor*.

Mate finding has not been very well studied except for work by Hrabar et al. (2012), who, having noted that males of *P. disparis* aggregate at gypsy moth, *Lymantria dispar*, cocoons when a conspecific is about to emerge, followed this up with a series of behavioural assays that showed that oral secretions of the egressing wasp were the source of the chemical cues, very much as in the distantly related Rhyssinae. Males seeking mates appear not to be able to distinguish males from females that are about to emerge. How widespread the use of cues arising from pre-emerged potential mates is, is not known, nor is it clear what the oral secretions' main role is – it is tempting to speculate that some components may help soften the cocoon, facilitating egress. Robacker et al. (1976) showed that males of *Itoplectis conquisitor* will court newly merged males as well as females, and Robacker and Hendry (1977) identified neral and geranial (two isomers of citral) as major components of this species' sex pheromone, although it was clear that other as yet unidentified minor components were also resent as authentic citral did not elicit exactly the same male response as female sex pheromone. Males of a few Pimplini have some flagellar segments modified with structures called tyloids (see Kolarov & Gürbüz 2007). The shape, number and location of these have for a long time been used in taxonomic studies and a role in courtship suspected, but only more recently has their exact nature been investigated. Bin et al. (1999b), using SEM and TEM, showed that they are release sites for a male contact 'pheromone' involved in courtship with associated internal glands, rather than the 'definite sensory area' postulated by Townes (1969). Males stroke their tyloid area over the female flagellum (Bin et al. 1999a). Not all Pimplini, however, display antennal stroking during courtship, e.g. *Xanthopimpla stemmator* and the males correspondingly lack tyloids (Romani et al., cited in Bin et al. 1999b). Males of one Australian species, *Lissopimpla excelsa*, are duped by *Cryptostylis* tongue orchids, which closely resemble female *L. excelsa* both morphologically and scent wise (Gaskett 2012). This wasp is the sole pollinator of these orchid species.

Fecundity is usually around 50 eggs per female (Meyer 1925). Destructive host feeding is fairly common among *Itoplectis* and may cause considerable host mortality (Campbell 1963, Jervis & Kidd 1986). In *Pimpla nipponica*, host feeding by females on already parasitised hosts leads to far fewer successful emergences, and the wasps that do emerge may be smaller (Ueno 1997). *Itoplectis* uses its ovipositor to mutilate its

Fig. 13.23 Host feeding by adult female ichneumonids. (a) *Itoplectis aterrima* (Pimplinae) feeding on its host, *Hyposoter clausus*, through a hole in the host cocoon made by stabbing with ovipositor; (b) *Netelia vinulae* (Tryphoninae) simultaneously ovipositing and host feeding by biting a hole in the host cuticle with mandibles, on a host, *Cerura vinula*, caterpillar. (Source: reproduced by permission of Mark R. Shaw.) For a colour version, see Plate 41.

host within the primary host's cocoon and host inside it and then drinks the liquids that exude (Fig. 13.23a) (M.R. Shaw 2009b).

The relationships between the genera of Pimplini generated from morphological analysis (Gauld et al. 2002b) and molecular analysis (Quicke et al. 2009) are in good agreement with a paraphyletic genus *Theronia* at the base and *Xanthopimpla*+*Lissopimpla* as sister group to the remaining taxa including *Pimpla* and *Itoplectis*. Since *Xanthopimpla* uses vibrational sounding for host location, it would be interesting to know whether some of the basal *Theronia* do likewise, in which case this might be a synapomorphy for the tribe.

Poemeniinae (= Neoxoridinae)

Superficially, poemeniines are rather like rhyssines, with which they used to be united and both were previously placed in the Pimplinae. It is currently divided into three tribes, the Poemeniini, Pseudorhyssini and Rodrigamini (Wahl & Gauld 1998).

A key to the world genera was provided by Wahl and Gauld (1998).

Poemeniini

Members of the Poemeniini are idiobiont ectoparasitoids of both concealed Coloeoptera larvae and solitary aculeate Hymenoptera (e.g. Sphecidae *s.l.*), mostly in dead, standing wood, although, as pointed out by Fitton et al. (1988), determining the precise host association can be clouded by the fact that many aculeate Hymenoptera may utilise old beetle borings as nesting sites and these holes can also become inhabited by other insects and spiders.

Interestingly, *Eugalta* females have a spatulate seta that protects the end of the hind claw (see fig. 53 in Wahl & Gauld 1998; see also Fig. 13.20), which presumably protects the opening of a gland as the same structures do in many Pimplinae. Whether this represents a homoplastic separate origin or a retention of a character that was part of the pimpliformes groundplan remains to be discovered. Members of the genera *Cnastis* and *Podoschistus* have a strong, ridged sculpture on the temples which are believed to aid in bracing the head as the newly emerged wasp chews its way out of its pupation site (Fig. 10.51c); *Deuteroxorides* species have a rough but markedly finer sculpture.

Pseudorhyssini

The biology of *Pseudorhyssa maculicoxis* (as *sternata*) was described in detail by Spradberry (1969, 1970a) and filmed by Skinner and Thompson (1960). This genus is unusual because they are kleptoparasites of *Rhyssa* and *Rhyssella* (Rhyssinae), but show the same transverse mesoscutal sculpture associated with egress (Fig. 10.5b). Female *Pseudorhyssa alpestris* associate with females of *Rhyssella approximator* ovipositing on to their *Xiphydria* wood wasp hosts in dying alder trees, and when the *Rhyssella* has completed egg-laying on a host and withdrawn its ovipositor, the *Pseudorhyssa* inserts her narrower ovipositor down the *Rhyssella*'s boring and deposits her own egg, which is considerably larger than the *Rhyssella*'s on the same paralysed host. The *Pseudorhyssa* egg hatches before that of the rhyssine and the mandibulate first instar larva searches over the *Xiphydria* until finding and then biting the *Rhyssella* egg, killing it. After that, it develops as an idiobiont ectoparasitoid on the stolen host. In the laboratory, *Pseudorhyssa* females could be stimulated to probe through holes in paper made by *Rhyssa* ovipositors and showed considerable probing on filter papers treated with *Rhyssa* vaginal gland extract.

Wahl and Gauld's (1998) analysis also recovered *Pseudorhyssa* as an isolated branch, forming the sister group to the remaining Poemeniini, and therefore they also gave it tribal status. In the morphological analysis of Quicke et al. (2009), *Pseudorhyssa* formed a clade with the Pimplinae. The placement of *Pseudorhyssa* was equivocal in individual combined analyses of 28S rDNA and morphological data and it did not consistently clade with the rest of the Poemeniinae, instead sometimes associating with either the Pimplinae or Rhyssinae depending on alignment parameters. However, in the overall combined analysis with elided 28S alignments it was placed as in Wahl and Gauld (1998) as sister to the Rodrigamini+Poemeniini and therefore it is best treated as a tribe of Poemeniinae, despite its aberrant biology.

Rodrigamini

Rodrigama was described by Gauld based on a series of large and rather spectacular wasps with a long ovipositor collected at a single dead tree in Costa Rica, but the tree no longer remains and no other specimens of that species have been seen. However, recently Matsumoto and Broad (2011) described two new species (Fig. 13.24), from Japan and Taiwan, and a third Old World species has been seen from Israel. They superficially resemble rhyssines and were placed, with considerable uncertainty, in the Poemeniinae. Wahl and Gauld (1998)'s morphological analysis recovered them as the sister group to the rest of the Poemeniinae (excluding Pseudorhyssini) and consequently they afforded it tribal status.

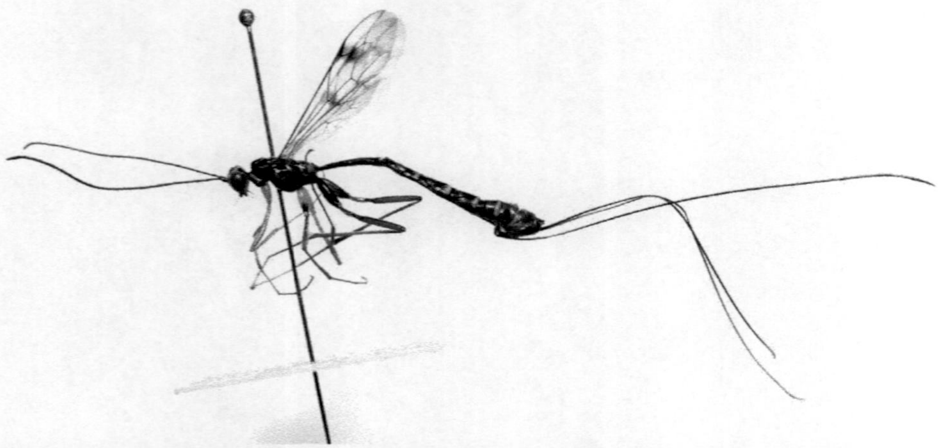

Fig. 13.24 Habitus of *Rodrigama* sp. (Poemeniinae) from Japan. (Source: reproduced by permission of Rikio Matsumoto.) For a colour version, see Plate 42.

Rhyssinae

This is perhaps one of the best studied ichneumonid subfamilies in some respects, because they include some of the largest ichneumonid species with body lengths up to 50 mm and with very impressive ovipositors that can exceed 100 mm. Monophyly of the subfamily is not doubted and it is strongly supported by numerous apomorphies (Wahl & Gauld 1998). Although their ovipositors are not the longest in the superfamily in either absolute or relative terms (see *Dolichomitus*, Townes 1975), their mode of use and associated adaptations are almost unique. For a long time, the rhyssines have been included within the Pimplinae and sometimes specifically within the Ephialtini, but that they are a well-defined and separate group was recognised by Gauld (1991).

They are cosmopolitan although with most species apparently occurring in South East Asia. Within the Holarctic, rhyssines are idiobiont ectoparasitoids of siricid (*Tremex*, *Sirex*, *Urocerus*) and xiphydriid wood wasps, although they may also attack other hosts in the same habitat, and *Rhyssa persuasoria* has been recorded to act as a facultative hyperparasitoid of *Ibalia* (Cynipoidea), a primary parasite of siricids (Hanson 1939). *Rhyssella approximator* has also been reported to kleptoparasitise *Rhyssa* in addition to being, more normally, a primary parasitoid of *Xiphydria* (Coutourier 1949).

Partly because of their economic importance, but also because of their large size and conspicuousness (Fig. 13.25) and unusual aspects of their behaviour, they have attracted considerable attention. The biology of *Rhyssella* (as *Thalessa*) has been described by Chrystal and Skinner (1932), and the behaviour and ecology of *Megarhyssa* spp. were described by Heatwole et al. (1962) and Heatwole & Davis (1965). Biological studies are virtually restricted to temperate species of *Rhyssa*, *Megarhyssa* and *Rhyssella*. Host larvae are consumed quickly, as is typical of idiobionts, and the parasitoid overwinters as a quiescent larva (prepupa), pupating in the spring. Several rhyssine species are important in the regulation of siricid wood wasp pests of conifers in the southern hemisphere and several releases have been made in both Australia and South Africa (Nuttall 1973; see below).

The egg has an ovoid anterior end and a very long posterior part which is the first to enter the ovipositor. When the tail starts to emerge from the tip of the ovipositor, the contents of the anterior portion rapidly transfer to the emerging part. There are apparently four larval instars in *Megarhyssa atrata* and Nénon (1995) suggested that in Quebec its development might take 2 years, but this seems unlikely, because as an idiobiont, its host would deteriorate considerably after such a period. The possibility that some individuals remain in a diapause for an extra year cannot be ruled out. Temperate species show typical

Fig. 13.25 Sequence of photographs of female *Megarhyssa vagatoria* (Rhyssinae) 'drilling' her ovipositor into a tree trunk female. (Source: reproduced by permission of C. M. Zwakhals.) For a colour version, see Plate 43.

idiobiont rapid development on the paralysed host and then overwinter as a prepupa. Females of both *Rhyssa* and *Megarhyssa* locate host-infected trees by detecting volatiles released by the *Amylosterium* fungi that are symbionts of the host which render the wood that the siricid is consuming more nutritious (Madden 1968) and prefer the odours released by fungal colonies that are several months old, which would correspond to the time when the host wood wasp has reached a suitable stage of development. *M. nortoni* and *R. persuasoria* respond to different threshold levels of *Amylosterium* culture odour, which may reduce competition, with the generally smaller bodied *Rhyssa* being attracted to somewhat less mature hosts. In addition to *Amylosterium*, the fairly fresh siricid frass contains yeasts (mainly *Saccharomyces*) and the presence of yeasts enhances the attractiveness of the fungus/wood/frass complex odours to the rhyssines, and perhaps this also increases their accuracy of host location.

Nénon et al. (1995) presented a series of SEM images that suggest that there are pores along the length of the ovipositor of *Megarhyssa* and that these are the sources of a lubricant that lubricates the boring. It is hard to know what to make of these. The images look convincing but there is no evidence that the ovipositor contains glandular tissue, either in this or any other parasitic wasp, and it seems very unlikely that such pores leak haemolymph, which in any case would be expected to coagulate. This is certainly an area requiring further investigation.

Given their size and conspicuousness, it is not surprising that these wasps have attracted considerable attention and there have been quite a few, mostly purely observational, behavioural studies of their host searching and mating. Several of these have been carried out in regions such as Australia and New Zealand where *M. nortoni* has been introduced as a biological control agent against the introduced woodwasp, *Sirex noctilio*, in the mid 1960s (Heatwole et al. 1962, Heatwole & Davis 1965, Nuttall 1973, Crankshaw & Matthews 1981). Along with the deliberately introduced *R. persuasoria* and the accidentally introduced *R. lineolata* and an ibaliid, *Ibalia leucospoides*, these parasitoids generally cause approximately 70% mortality of their pest hosts in the antipodes (Nuttall 1973). Some of the work has used unnatural host associations, and as Spradberry (1968) showed, they can even be reared on honey bee larvae.

In North America, three species of *Megarhyssa* co-occur in *M. atrata*, *M. greenei* and *M. macrura*. Frequently, all three species occur together on the same log and attack the same host, *Tremex columba* (Siricidae). The co-occurrence of these species with ovipositors of different lengths led Gibbons (1979) to propose that the speciation events leading to this could have occurred sympatrically. A very analogous situation occurs in the Afrotropical braconine genus *Monilobracon* (Quicke et al. 2005b).

Emergence and pairing in *Rhyssa* has been described by B.R. Baker (1992) and in *Megarhyssa* by many workers (e.g. Abbott 1934, Nuttall 1973, Matthews et al. 1979) and summarised by Eggleton (1991). In *Megarhyssa* and *Rhyssa*, as in many insects, males emerge before females (protandry) and can detect where females are about to emerge from, where they aggregate and wait for the female[8] to chew through the bark (Fig. 13.26a,c). Nuttall (1973) described how as soon as a female has chewed through the wood to start making her exit hole, males of *Megarhyssa* can insert their very flexible and slender metasomas (Fig. 13.26b) and mate with them (as evidenced by the females producing female offspring subsequently), scrambling for a mating opportunity. Females will mate with several males both whilst still confined in her emergence tunnel and for a while after fully egressing. Interestingly, Nuttall noted that no aggressive behaviour was observed between the males within an aggregation that were trying to mate with a given female. Often mixed-species aggregations (>60%) of up to 28 males can be found over sites on dead wood from where females are going to emerge (Crankshaw & Matthews 1981), but only single species aggregations occur over the precise future emergence hole of a conspecific female; nevertheless, Matthews et al. (1979) reported that in approximately 20% of the aggregations they studied, the emerging female was not conspecific with the associated males. Unlike in *Megarhyssa*, males of *Rhyssa* (Fig. 13.26d) do not have such a flexible metasoma and are therefore unable to mate with a female until after she has egressed fully.

N.W. Davies and Madden (1985) subsequently showed that local aggregations of males are due to secretions from their mandibular glands and that these differ between *Rhyssa persuasoria* and *Megarhyssa nortoni*. Secretions from both species contained 6-methylhept-5-en-2-one, but *Rhyssa* also contained 3-hydroxy-3-methylbutan-2-one while *Megarhyssa* contained a series of spiroacetals, a group of compounds involved in aggregation behaviour in other Hymenoptera. The spiroacetals of *M. nortoni* have subsequently been investigated further by Schwartz et al. (2008).

Matthews et al. (1979) described a male marking behaviour they termed 'tergal stroking' in which the male's metasoma is extended forwards between the legs and the dorsal tip (but not the genitalia) strokes the substrate. The tip of the metasoma in males of all three *Megarhyssa* species has a V-shaped brush structure close to the anus, referred to as an 'anal gland', which is clearly used for spreading a marking substance. Treatment of KOH-macerated male metasomas failed to reveal any associated typical gland and they postulated that active components might originate from cells of the hind gut. The fluid droplets elicited by handling males had no discernible odour and it was not clear that they were of the same composition as what males spread. The role of this brushing is unclear, but it was postulated that it may be associated with inducing the females to copulate as the males extend their metasomas into the burrow of the emerging females and the brush organ must presumably make considerable contact with the female's body during the copulation attempts.

The only other genus for which observations of male behaviour are available is *Lytarmes* from South East Asia, and in this wasp there is a very different behaviour pattern (Eggleton 2008). A large male of *L. maculipennis*, which has a short and robust metasoma (unlike the thin, elongate ones of *Megarhyssa* males), was observed to defend a territory around the site where a female was due to emerge. Other smaller males tried on numerous occasions to attack the larger individual, biting at his wings/legs, but he was able to repulse them – indeed, Eggleton noted that the large male had a chunk of one of its hind wings missing, which could reflect similar previous encounters. When the female finally emerged, the large male pushed his metasoma into the hole but only for about 2 s, and it is not clear whether this represented a successful mating attempt. As with *Rhyssa* and *Megarhyssa*, Eggleton showed that *Lytarmes* also has a peri-anal scent gland and this may have been important in making the emerging female receptive.

Rhyssines display many features associated with parasitising deeply concealed wood-boring hosts. The head is robust, containing relatively large muscles for operating the mandibles, which are themselves chisel-shaped and robust for chewing the wasp's way out of the host substrate. The mesoscutum is furnished with strong transverse ridges that help the egressing wasp brace itself against the wall of its confines (Fig. 10.5a) while chewing. The ovipositor is relatively rigid and is furnished with multiple, strong apico-ventral serrations and usually multiple small nodi on the upper valve. During substrate 'drilling', the ovipositor is usually constrained at approximately half its free

Fig. 13.26 Males of the European Rhyssinae showing scramble competition for mates. (a) *Megarhyssa vagatoria*, aggregating around the future point of emergence of (probably) a conspecific female; (b) *Megarhyssa rixator*, showing long slender metasoma; (c, d) *Rhyssa persuasoria*, showing male aggregation and more robust male metasoma, respectively. (Sources: a, reproduced by permission of C. M. Zwakhals; b–d, photographs by Reto Burri, Uppsala, Wildlight.ch, reproduced by permission of Seraina Klopfstein.) For a colour version, see Plate 44.

length by tubercles on the sub-anterior metasomal sternites (Figs 10.5d and 13.25), an adaptation to increase the force that can be applied before it were to buckle[9] (Fig. 3.6) (see Quicke et al. 2000a). In the oriental and Pacific genus *Myllenyxis*, the ovipositor apex is sinuate in lateral aspect and also rather strongly laterally compressed, which suggests that it may have some steering capability, although the mechanics are not understood and presumably, due to the extreme lateral compression, it cannot function in the same way as those of the braconine *Zaglyptogastra* or the cremastine *Pristomerus* (Quicke 1991a). When set specimens are viewed from above, the two lower valves are often protruding and curving fairly strongly across to the other side, a feature similar to that of several non-pollinating figwasps, and suggests that the ovipositor might be guided left or right depending on which of the lower valves is pushed more into the substrate. A very similar morphology is exhibited by the euphorine *Sinuophorus* and acaenitines also generally have rather laterally compressed ovipositors with lower valves having a tendency to curve contralaterally if they are protruding beyond the apex of the upper valve.

Drilling into a host substrate can take a considerable time and renders the insects vulnerable to predation and also must incur some considerable amount of wear on the ovipositor teeth. Females of *Megarhyssa* are believed to lay approximately 20 eggs during their normal adult lifespan and spends approximately 1 h 'drilling' in order to lay each one (Heatwole et al. 1962), thus typically a female may be actively 'drilling' through wood for about 20 h. The ovipositor teeth of rhyssines have high concentrations of manganese which (in the form of a metal–protein complex) give them considerable hardness, which might explain why the ovipositors of field-caught specimens show relatively sign of little abrasion (Quicke et al. 1998). One can often see broken-off ovipositors sticking out from logs that are infested with rhyssine hosts, the results of predation on ovipositing females by birds such as blue jays and maybe also by squirrels.

Rhyssa persuasoria females, although they are not often observed feeding in the wild, clearly must do so. Under laboratory conditions they live for only about 9 days in the absence of any carbohydrate source, but up to 55 days when provided with leaves coated with dried honey solution to simulate natural honeydew deposition (Hocking 1967a).

THE ICHNEUMONIFORMES

This group comprises the Ichneumoninae, Cryptinae, Adelognathinae and Alomyinae, and also most likely the Agriotypinae, although our understanding of the relationships between them is going through a period of reassessment due largely to molecular analyses. It corresponds poorly to what Gauld (1991) referred to as the phygadeuontoid group of subfamilies in that he included also the Labeninae, Orthopelmatinae and Xoridinae and excluded the Adelognathinae). The possibility that it also includes the Brachycyrtinae cannot be completely excluded, but here they are treated among the 'Gondwanan' clade (Quicke et al. 2009). Gokhman (1995) presented a formal morphological cladistics analysis of the group, including presenting trees with the interesting modification of downweighting characters in proportion to their correlation with biology to try to eliminate the effect of directed homoplasy. However, his results were not very convincing, with his included members of the Labeninae and Xoridinae not coming out monophyletic and depending on weighting scheme, Alomyinae being recovered either derived within Ichneumoninae or basal to them.

Adelognathinae

There is a single, almost entirely north temperate genus comprising this subfamily with approximately 35 described species (Fitton et al. 1982). They are unremarkable in appearance except for their antennae, which have a very reduced number of segments (12 or 13 flagellomeres). All the species for which biology is definitely known are ectoparasitoids of well-grown sawfly larvae, including those of gall-forming species. Probably related to this diversity, they collectively display considerable variation in ovipositor morphology and some have the lower valves rather membranous, similar to the condition seen in some oedemopsine tryphonines. The majority of reliable host records are from Pamphiliidae and Tenthredinidae. They are particularly interesting for a number of reasons, including that some species are ectoparasitic koinobionts (as with tryphonines and the braconid genus *Rhysipolis*; see Fitton et al. 1982), although others appear to be strictly ectoparasitic idiobionts and, more surprisingly, sometimes on a more or less fully exposed host. Most species are gregarious but at least a few are solitary and the gregarious ones have small broods (2–7 larvae per host).

Biology has been described by Kopelke (1987) for *Adelognathus cubiceps*, which attacks gall-forming tenthredinids on *Salix*, and by Rahoo & Luff (1987) for *A. granulatus*, a parasitoid of the small gooseberry sawfly, *Pristiphora pallipes*. The ovoid egg is relatively large, more typical of idiobionts rather than koinobionts, and is firmly glued on to the upper surface of the host larva, and in gregarious species they form a tight cluster. *Adelognathus* species have very large vaginal glands (personal observation) and it is tempting to speculate that these are at least one of the sources of the glue that affixes their eggs to the host. Both species paralyse their hosts, the former gaining access to the host in its gall by chewing its way into the gall and ovipositing while standing on it (Kopelke 1988). Not surprisingly, given their apparent vulnerability, the eggs hatch quickly after being laid (within a few hours) and the subsequent larval development is equally rapid, approximately 4 days (Rahoo & Luff 1987), again far more typical of idiobionts than koinobionts (Fig. 13.27). Fitton et al. (1982) suggested that the very rapid larval development may be an adaptation to overcoming quickly the host's potential ability to dislodge them, and this seems reasonable since the quality of the host, which often continues feeding after parasitisation, is unlikely

Fig. 13.27 *Adelognathus leucotrochus* (Adelognathinae) development on larva of host *Nematus leucotrochus* (Tenthredinidae). (a, b) Larvae after 3 and 30 h, respectively. (Sauce: reproduced by permission of Mark R. Shaw.) For a colour version, see Plate 45.

to decrease that quickly. Unlike many tryphonines with similar biology, the developing *Adelognathus* larva does not show any particular morphological specialisation for staying attached to the host sawfly larva, but instead keeps its posterior end stuck within its egg, which is firmly glued to the host's back, which seems to ensure that the parasitoid larva is not easily dislodged. Gauld (1988a) commented that 'this is a particularly unspecialised form of koinobiosis as the parasitoid fails to exploit the abilities of its host to construct a pupation retreat' and hypothesised that this might be due to the lack of adaptations (such as an anchor) to prevent the parasitoid egg/larva being sloughed off at the next ecdysis. Thus *Adelognathus* larvae attacking exposed hosts, when fully fed, leave the host to spin their own cocoons among leaf litter on the ground. Among the koinobiont species, the need to complete feeding on a single instar of the host in this situation not only necessitates very rapid development, but also begs the question as to whether female wasps specifically attack hosts that have just moulted and therefore provide a sufficient time window for her egg/larvae to complete development.

Agriotypinae

With a single genus now recognised (Bennett 2001), this is a small group of very specialised ichneumonids which all appear to attack caddisfly (*Silo* and *Goera* spp., Trichoptera) larvae under water. They are certainly much better known to freshwater biologists than they are to hymenopterists, because the former often counter parasitised caddis cases, which can be recognised by an extraordinary silk 'ribbon' that is produced from one end of the caddis case and which appears to act as a lung enabling the parasitoid to obtain enough oxygen even though it is not producing a continuous flow of water through the case as would be the case brought about by a living host.

Unlike other ichneumonoids, the sternites of the metasoma are well sclerotised, approximately to the same degree as the tergites (Fig. 13.28d). This feature, along with the structure of the larval head capsule, led W.R.M. Mason (1971) to retain them as a separate family (Agriotypidae), which he associated with the Proctotrupoidea rather than the Ichneumonoidea – illustrating the problem of using a typological approach in systematics. Most workers, however, have correctly treated them as ichneumonids based on recognition that their features are likely derived characters associated with their unusual biology (Sharkey & Wahl 1992), which is also well supported by DNA data (Quicke et al. 2009).

The biology of *Agriotypus armatus* has been described in detail by Clausen (1931), K. Fischer (1932), Grenier (1970) and Elliot (1982, 1983), and a number of workers have carried out experiments on the role of the silk ribbon. Aoyagi and Ishii (1991) described in detail how the Japanese species *A. gracilis* enters the water by walking down protruding rocks with its antenna held back and playing no apparent role in host location. Once a host caddis case is located, the female wasp walks to-and-fro along its length, what they called 'turning', and this appears to be how they assess the host's size and ovipositor probing appeared to be how the wasp assessed the host stage, preferring to oviposit on prepupal and pupal hosts rather than active larvae. A similar behaviour pattern was described for *A. armatus* by Elliot, who reported that initial examination of potential hosts is carried out using the tarsi and that this is enough for them to distinguish their *Silo* hosts from cases of other species which are fairly quickly rejected. Similarly, if caddis cases start to move, indicating that they contain an active larvae, these are also quickly rejected. Handling time per host is approximately 20 min (Elliot 1983), but females do not distinguish cases containing suitable prepupal or pupal hosts from unsuitable larvae (unless they moved) or pharate adults. Also, they did not appear to avoid superparasitism even although this usually led to death.

Mating takes place soon after emergence and is brief (10–20 s), and is easy to achieve in vials (Clausen 1931). In more natural environmental conditions, it is reported to occur both on the ground (on damp moss) (Elliott 1983) or in the air (Grenier 1970). From what Elliott (1983) wrote, it appears that females and males may mate multiply. This is followed by a variable length of time (1–5 days) before females commence host searching, which may indicate a need to finish maturing eggs, possibly under warmer conditions than under water. Having commenced searching, the female initially spends up to 5 h submerged, and then briefly comes up for air before resubmerging. They do this a number of times (at least under laboratory conditions with artificial streams), but appear to exhaust themselves after 15–20 h and then cease foraging, and from then onwards are moribund. Thus lifetlime realised fecundity is limited by the number of suitable

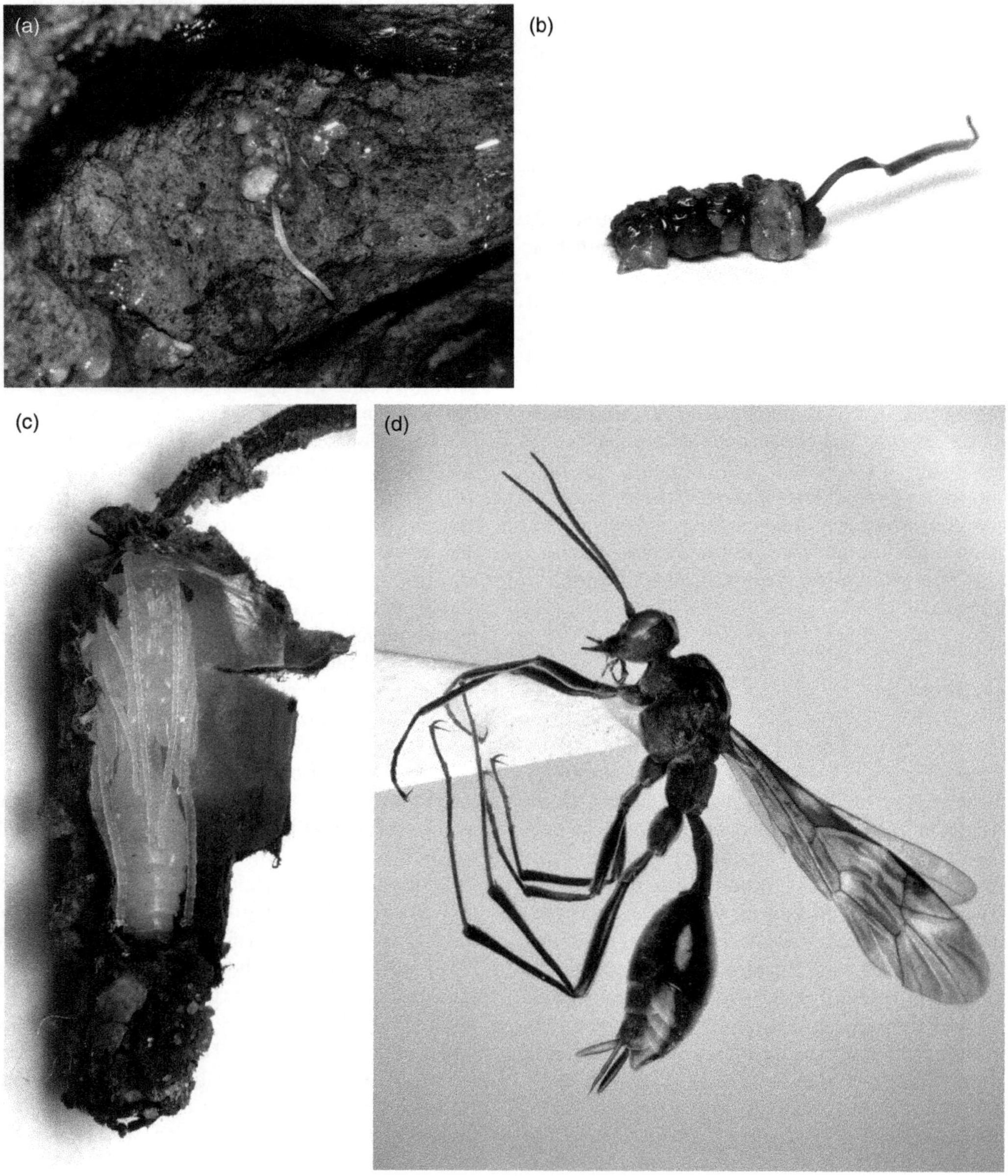

Fig. 13.28 *Agriotypus gracilis* (Agriotypinae). (a) Parasitised caddis case *in situ* on a boulder in a stream with the silk thread gill protruding; (b) isolated caddis case showing silk ribbon; (c), parasitised caddis case opened to show *Agriotypus* pupa with head at end with ribbon; (d) adult female habitus. (Sources: a–c, reproduced by permission of Toshimasa Mitamura; d, reproduced by permission of Kazuhiko Konishi.) For a colour version, see Plate 46.

hosts that a female can encounter during this rather brief period. Clausen (1931) noted that the body of the female under water is covered by a bubble of air that adheres tightly to the body and must act as her oxygen reservoir. The air bubble traps the antennae backwards over the wasp's body, thus preventing them from being used in host location or discrimination.

A female will never oviposit in caddis cases containing active larvae, and always departs from the case if her initial ovipositor probe caused a larva to stick its head out of the case (Clausen 1931); thus the wasp does not use venom to paralyse its host as do most other idiobiont ectoparasitoids, and therefore leaves the prepupal or pupal host still capable of respiratory movements, which have been proposed to serve to provide adequate oxygen for parasitoid egg and larval survival. The egg is attached to the cuticle of the host by a thin stalk. The first and second instar larvae, which are surrounded by water, have paired projections from the tenth abdominal segment, which may be important in maintaining their position in the caddis case. After completing feeding on the host, the *Agriotypus* larva spins a cocoon closely attached to the caddis case, partitioning the host cuticular remains at one end of the case, and when nearly complete it weaves a silk thread a little over 1 mm wide from the end of its cocoon, and extrudes it several centimetres into the water (see Fig. 13.28a). This unique structure is believed to act as a gill, greatly increasing exchange of carbon dioxide produced by the parasitoid with oxygen from the surrounding stream water (Messner & Tashenberger 1981).

Adults are very easily recognised by their possession of a more or less sharp spine from the scutellum (Fig. 13.28d) (more so than in *Banchus* in which this state also occurs) and multiple longitudinal carinae on the propodeum and first metasomal tergite and the heavily sclerotided metasomal sternites. In addition, the claws are especially long and slender (Fig. 13.28d), no doubt an adaptation to enable them to hold on to stones under fast-flowing water. The world fauna comprises only 13 described species, which have been revised be Bennett (2001), who also carried out a morphological phylogeny which led to synonymisation of the *Atopotypus* with *Agriotypus*.

Bennett (2001) presented several putative morphological synapomorphies suggesting that the sister group of the Agriotypinae was the Labeninae. However, in more wide-reaching formal morphological analyses and in combined molecular and morphological phylogenetic trees, it never appears so basal, i.e. with the Labeninae, but instead was recovered among the ichneumoniformes, sometimes as a sister group to the Cryptinae (Quicke et al. 2009).

Alomyinae

This is a small subfamily comprising just two genera distributed in the Holarctic, members of which closely resemble some Ichneumoninae and with which it is frequently lumped as a senior synonym of the ichneumonine tribe Phaeogenini (Wahl 1993, Wahl & Mason 1995). However, DNA data do not support this (see below), and therefore I retain them as a separate subfamily and consider their morphological similarity to some Phaeogenini as being due either to convergence or possibly to symplesiomorphy.

Hinz and Short (1983)[10] showed that *Alomya semiflava* (Fig. 13.29) has a different biology than that of ichneumonines in that it mummifies its host Lepidoptera larvae, which are swift moths (Hepialidae) that feed on roots of grasses and other plants. Experiments by these authors suggested that it may be host specific on *Korscheltellus* (as *Hepialus*) *lupulinus*. Two other genera have been included in the subfamily, *Pseudalomya* and *Megalomya*, both from the eastern Palaearctic. All have a rather strongly modified head, which is one of the main features in their association, but in the case of the former, it appears to be due to convergence. Host mummification is an unusual strategy exemplified by aphidiine and rogadine braconids and some Encyrtidae (Chalcidoidea). *Alomya* has a short ovipositor with very well-developed serrations on both upper and lower valves, a feature shared with several ichneumonines, but whose function is unknown, although in the latter it may be associated with penetration of a hard pupa.

Wahl and Mason (1995), recognising the great similarity between both *Alomya* and *Pseudalomya* and members of the ichneumonine tribe Phaeogenini, synonymised the Alomyinae with the former. However, consideration of the 28S D2–D3 sequence data shows that whereas *Pseudalomya* is indeed a derived 'phaeogenine' (Laurenne et al. 2003, 2006), *Alomya* itself shares no molecular synapomorphies with either the Phaeogenini or with the Ichneumoninae as a whole. Figure 7 in Laurenne et al. (2006) illustrates

Fig. 13.29 Mounted specimen of *Alomya debellator* (Alomyinae). (Source: reproduced by permission of Ilari Sääksjärvi.)

a number of synapomorphic base substitutions found in the Ichneumoninae exclusive of *Alomya* and additional sites supporting monophyly of Phaeogenini. Unfortunately, no sequence data are as yet available for *Megalomya*. However, *Pseudalomya* was recovered within the Phaeogenini group, as was *Colpognathus*, a genus that looks a little like both of them and which in common with *Alomya* mummifies its host. Interestingly, Gokhman (1995), who presented a limited formal phylogenetic analysis to try to resolve the origin of the Ichneumoninae, recovered *Alomya* as the sister group to the Ichneumoninae within a paraphyletic Cryptinae and not with the Phaeogenini, but in his analysis *Pseudalomya* was also recovered outside the Ichneumoninae. In various morphological, molecular and combined analyses, *Alomya* has been recovered as sister group to either the Ichneumoninae [Gokhman 1992 (morphology), Quicke et al. 2000b (morphological, combined and molecular), Quicke et al. 2009 (combined)] or the Cryptinae [Laurenne et al. 2006 (molecular)].

Cryptinae

This is one of the huge subfamilies, in fact the largest within the Ichneumonidae with more than 395 recognised genera (Townes 1970a) and some 4500 described species (D.S. Yu & Horstmann 1997). It was one of the five early major divisions of the Ichneumonidae and its limits have been stable for a long time, but nomenclaturally the group can be expected to give newcomers to ichneumonid studies a headache. For a great deal of its past it was referred to as the Cryptinae, but Henry Townes refused to follow the standard rules of nomenclature and preferred to name all higher taxa after the oldest available generic name in the group, and in this case, the oldest name was not *Cryptus*, but *Gelis*, and hence he referred to the subfamily as the Gelinae (Townes 1970a). However, although the name Cryptinae is now correct and more or less universally accepted, it was realised that the name *Cryptus* did not actually refer to what people thought it did, being a junior homonym (see Table 13.2 for details). Thus, Fitton & Gauld (1978), following the rules of the ICZN, determined that the subfamily should correctly be referred to as the Phygadeuontinae and as such it was in quite a few publications between about 1978 and 1994, i.e., until the ICZN, inappropriately in many people's view, conserved the names *Cryptus* and Cryptinae for the ichneumonid genus and subfamily as treated here.

The Cryptinae has been divided into three tribes and nearly 30 subtribes. The currently recognised

Table 13.4 Application of names for tribes of Cryptinae.

Used here	Townes (1970) and Authors
Aptesini	Echthrini auct. Hemigasterini auct.
Cryptini	Mesostenini
Phygadeuonitini	Gelini

tribes, following the results of Laurenne et al. (2003, 2006), are the Aptesini, Cryptini and Phygadeuontini, but see the discussion under Aptesini and see Gauld (1995) and Sawoniewicz (2008) for why this name is used rather than Hemigasterini or Echthrini. Other names used in the literature, including those used by Henry Townes, are given in Table 13.4. The limits of the three tribes as now defined are fairly well defined morphologically and by DNA; a few taxa[11] in the molecular studies of Laurenne et al. (2006) that seemed clearly misplaced could be due to some artefact such as extreme 28S rDNA sequence length variation or possible contamination.

The character combinations used by Townes (1970a) to recognise most subtribes are often weak and unreliable and he admitted that his classification was 'partly artificial' and, referring to the Cryptini in particular, 'that they are the best that could be contrived with present information'. Therefore, it is not surprising that molecular data support virtually none of them (albeit with far from complete taxon sampling) and therefore only a few that seem relatively well supported are referred to here. Laurenne et al. (2006), for example, discussed the Osprynchotina (referred to as Nemapodiina by Townes 1970a) as an example. These are large wasps that attack aculeate nests and share a long, slender mandible with an elongate upper tooth, a feature Gauld (1984a) considered a strong indication of monophyly. Molecular data, however, clearly show the group to be polyphyletic and the mandible character probably a convergent feature associated with biology. Townes (1970a) similarly commented that his subtribe Baryceratina, all of which parasitise cocooned Limacodidae (called Eucleidae by Townes) were probably not monophyletic, their similar adaptations probably being convergent due to utilising these hard host structures. The problem of assigning some of these taxa to the right group on the basis of morphology alone is exemplified by the predominantly East Asian genus *Nipponaetes*. When a specimen from Thailand was discovered, Gavin Broad noted the resemblance of it to a genus that Ian Gauld had recently described from Costa Rica, *Zurquilla*, but in the Tryphoninae and not in the Cryptinae. Subsequent morphological examination and molecular analysis showed *Nipponaetes* to be placed correctly in the Cryptinae but that *Zurquilla* was, in fact, just a synonym. Hence the lack of clear diagnostic characters had led to its placement not just in the wrong tribe but the wrong, and quite distantly related, subfamily.

Given its size, it is not surprising that the cryptines include taxa with a large range of biologies. In addition to idiobiont ectoparasitoids of weakly to strongly concealed hosts, a few are endoparasitoids [e.g. members of the *Stilpnus* group within dipteran hosts; some attack naked hosts, i.e. butterfly pupae: *Blapsidotes vicinus* (Phygadeuontini), and some *Cryptus*, *Glabridorsum* and *Goryphus*: all Cryptini] (Townes 1970a, M.R. Shaw et al. 2009); some are strictly predators within spider egg cocoons (e.g. some *Gelis*, *Hemiteles*, *Aclastus*, *Nipponaetes*, *Polyaulon* and *Rhacodopteron*: all Phygadeuontini, and *Hidryta*, *Idiolispa* and *Trychosis*: Cryptini) (Fitton et al. 1987), pseudoscorpion egg masses (*Obisiphaga*: Phygadeuontini) or among clusters of other insect eggs (e.g. *Sozites kerichoensis* (Phygadeuontini) on egg masses of *Entypotrachelus* weevils (Kerrich 1968, Benjamin & Demba 1969), cocoons of Neuroptera (Kerrich 1940), or are pseudohyperparasitoids, sometimes facultatively, of cocooned parasitic wasps (e.g. some *Bathythrix*, *Gelis*, *Mastrus* and *Lysibia* species: Horn 1976, Evenhuis & Vlug 1983, Wieber et al. 1995, Harvey et al. 2009).

Aptery and brachyptery have evolved on several occasions within the subfamily, predominantly within the Aptesini, but also occur to various extents in members of other groups, particularly the Phygadeuontini. Apterous *Gelis* (Fig. 13.30c), *Pleolophus* and *Theroscopus* frequently resemble small ants and no doubt gain some protection from the similarity. Whilst being found everywhere, they are particularly prevalent among island faunas.

Many species are important parasitoids of pests (Gauld 1991, 1995), although few have actually been introduced in classical biological control programmes, probably because many are rather generalist in their host utilisation. *Sphecophaga vesparum vesparum* has been successfully introduced into South Island, New Zealand where it exerts a small degree of control on the introduced and pestilentious *Vespula vulgaris* and *V. germanica* populations. The gregarious *Mastrus ridibundus* is an important prepupal ectoparasitoid of the codling

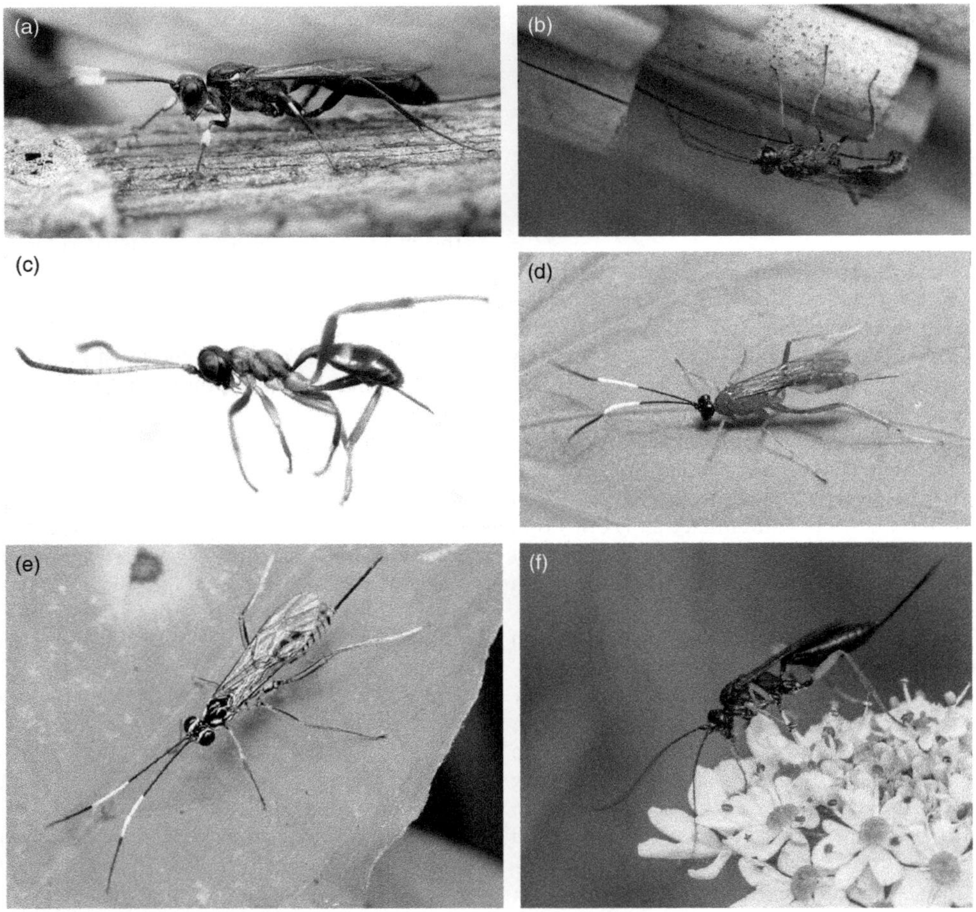

Fig. 13.30 Various Cryptinae. (a) *Echthrus* sp.; (b) *Stenarella domator*, ovipositing in trap nest; (c) apterous *Gelis* sp., showing relatively reduced size of mesothorax; (d, e) unidentified species from Costa Rica; (f) probably *Buathra laborator* (det. G.R. Broad) feeding at flowers of Apiaceae. (Sources ; a, reproduced by permission of David Cappaert, Michigan State University, Bugwood.org; b, reproduced by permission of Cor Zonneveld; c, reproduced by permission of Kyohei Watanabe; d, e, reproduced by permission of Kenji Nishida; f, reproduced by permission of Paul Lathbury.) For a colour version, see Plate 47.

moth fruit pest (*Cydia pomonella*: Tortricidae) and is fairly host specific (Devotto et al. 2010). It has been shown to be attracted to a host aggregation pheromone mixture that is released from its host cocoon soon after it has been spun (Jumean et al. 2005). Behavioural tests showed that at least 11 compounds were involved.

A few members of each of the tribes are semi-aquatic. *Apsilops sericata* (Cryptini) is parasitic on the aquatic nymphalid *Occidentalia comptulatalis* (Frohne 1939), *Medophron dytiscivorus* (Phygadeuontini) attacks dytiscid water beetle pupae in their cocoons (W.R.M. Mason 1968) and others of *Medophron* are associated with various cyclorrhaphous diptera puparia, and Sinu et al. (2007) described the behaviour of a semi-aquatic species of Aptesiini related to the genus *Litochila* that 'dives' into water in search of submerged rice case worm *Parapoynx stagnalis* (Crambidae) pupae, the female wasps remaining submerged for periods of up to 90 s at a time.

Adults of the species studied so far are synovigenic. *Aptesis* females carry few voluminous mature eggs at any one time and egg maturation is slow at fewer than one new egg reaching maturity per day (Babendreier & Hoffmeister 2002). *Mastrus ridibundus* females have a lifetime fecundity of about 20 offspring (Devotto et al. 2010).

The sheer size of the subfamily and high levels of homoplasy mean that identification to genus level requires patience and practice, more so than with most other subfamilies. The best key to the whole subfamily is still that of Townes (1970a). For the Palaearctic, Horstmann (1969) provided an updated key to a group of Phygadeuonitini called the Mastrini and Horstmann (1978) a key to the flightless Phygadeuontini genera.

Aptesini

In a great deal of the literature, the Aptesini are referred to as Echthrini (e.g. Townes 1970a) or Hemigasterini (e.g. Fitton & Gauld 1978). There are two problems here. The first is that the oriental genus *Hemigaster* shows a marked resemblance to another Asian genus, *Rothneyia*, which is placed in the Phygadeuontini. While *Rothneyia* has long been included in the Phygadeuontini, Gauld (1995) suggested that the *Hemigaster* group of genera 'are probably best placed in the Cryptini'. Laurenne et al. (2006) only included *Rothneyia* in their molecular analyses but the trees confirmed that on the basis of DNA alone, it was a member of the Phygadeuontini and well removed from the Aptesini (i.e. *Aptesis* and related genera), but the question remained as to whether *Hemigaster* was related to it. When this genus was included in the analyses of Quicke et al. (2009), it was likewise found to be a member of the Phygaeuontini (and not Cryptini), although it was not recovered close to *Rothneyia*, suggesting that perhaps their similar metasomas with enlarged syntergite is homoplastic. The second is that *Echthrus* (Fig. 13.30a) is, also on the basis of DNA, a derived member of the Cryptini and similarly not close to *Aptesis*. In common with many Cryptini, *Echthrus* also has antennal hammers and locates hosts by vibrational sounding (Fig. 10.2).

Most aptesines are ectoparasitoids of late instar larvae or prepupae of phytophagous insects, although a number occasionally act as pseudohyperparasitoids of braconids, ichneumonids and tachinids. Available host data for European species have been detailed by Sawoniewicz (2008). *Aptesis* species appear to be specialised ectoparasitoids of prepupae and pupae of tenthredinoid sawflies within the host's cocoon (Babendreier 2000, J. Li et al. 2013), although *A. nigrocincta* has been shown occasionally to act as a facultative hyperparasitoid of *Lathrolestes ensator*, another biological control agent against the apple sawfly in North America (Babendreier & Hoffmeister 2003).

Cryptini

This tribe has been divided into 16 subtribes, but many of these were found to lack molecular support by Laurenne et al. (2006) and it was recommended that they be abandoned and instead refer to genus groups. Schwartz & Shaw (1998) reported the gregarious parasitism of a scarabaeid by the cryptine ichneumonid *Cryptus* sp., scarabaeids being very unusual hosts for ichneumonoids, although the absence of records might be due in part to a lack of research effort on the appropriate stages or situations. Parasitism of Apidae[12], Vespidae and Sphecidae nests occurs in a number of genera, including *Acrorictus, Arthula, Buathra* (Fig. 13.30d), *Latibulus, Messatoporus, Nematopodius, Osprynchotus, Pachysomoides, Sphecoctonus, Sphecophaga, Stenarella, Toechorychus* and *Xylophrurus* and possibly some *Cryptus* and *Meringops* species (Casirhagi et al. 2001). The genus *Nipporicnus* and some *Messatoporus* are parasitoids of Pompilidae nests; *Messatoporus* on several genera and *Nipporicnus* on *Pseudagenia* (Iwata 1961). These genera are fairly widely scattered in the phylogeny, indicating multiple transitions to this way of life. Unfortunately, detailed observations of their mode of parasitism are sparse. Daly (1983) noted that *Aritranis graeffei* and *A. heliophilus* consume the contents of five or more cells of their *Ceratitis* bee hosts. Also, interestingly, some studies have shown that the cryptines are at least partially kleptoparasitic within the nests of their hosts. Polidori et al. (2011) showed that *Acrorictus seductor* does not attack the older, larger larvae of its host *Sceliphron caementarium* and that its larvae probably largely feed on the host's spider prey. Casiraghi et al. (2001) found a similar situation with *Buathra tarsoleuca*, which parasitises nests of the sphecids *Ammophila* and *Podalonia*, the cryptine larva feeding ectoparasitically (kleptoparasitically) on the caterpillar prey of the host sphecid with no sign remaining of the sphecid, suggesting that its egg or early larva had been fully consumed prior to consuming the food intended for it.

The *Barycnemis* genus group (Barycerotina) species appear to be rather specialised parasitoids of cocooned Limacodidae pupae or prepupae (Gauld 1987b) and their ovipositors have multiple dorsal teeth that appear to be an adaptation to penetrating these particularly hard structures (see also *Wahlamia* in the Banchinae). Similar ovipositors are also found in *Priotomis* but its hosts and host substrate are not known. *Priotomis* was placed by Townes (1969) in the (as Mesostenini) subtribe Lymeonina, and Porter (1972) considered it more likely also to be a member of the Barycerotina. Molecular data failed to give a clear resolution of this, although it was consistently recovered in a small clade including at least some Lymeonina along with some other genera (Quicke et al. 2009).

Thrybius attacks a stem-feeding eurytomid chalcidoid of the genus *Tetramesa* (Matsumoto & Saigusa 2001). The hosts live gregariously in a cavity in the *Phragmites* reed stem and the *Thrybius* larva is predatory rather than a parasitoid in a similar way to spider egg sack predating *Hidryta*, *Idiolispa*, *Trychosis* and *Gelis* spp. (Fitton et al. 1987). *Thrybius* is particularly interesting, not only because its prey/host is a chalcidoid, but also it is effectively koinobiont in the sense that at the time that its larva hatches, the clutch of host eurytomid larvae are not sufficiently well grown to provide enough food for it. Matsumoto and Saigusa (2001) supposed that it may limit its feeding on the young *Tetramesa* larvae so that some continue to grow and feed such that they will be large enough to allow it to complete development later. *Thrybius* appears to have four instars based on examination of its cast skins (Matsumoto & Saigusa 2001).

Kugler and Wollberg (1967) described and illustrated the development of *Agrothereutes tunetanusi* on the lymantriid *Orgyia dubia* and Ueno (2002) studied the development of *A. lanceolatus* on the unnatural host *Galleria mellonella*. In the latter case, when superparasitism occurred, lethal combat between larvae nearly always resulted in only one surviving to adulthood, although occasionally two parasitoids could develop successfully; it could also develop facultatively as a pseudoparasitoid on pupae of *Pimpla nipponica*. *A. tunetanusi* stings the final instar (prepupal) host caterpillar within its cocoon and lays a small number of eggs on or near to the paralysed host. The life history of *Xoridescopus annulicornis* was described by Ayyar (1943) and four larval instars were reported.

The life histories of the aculeate nest parasitoid/predator *Sphecophaga vesparum* is particularly interesting; it has been described in detail by Semichon (1908) and Donovan (1991) and the marked seasonal dimorphism of the North American *S. vesparum burra* cocoons by Schmeider (1939). These wasps predominantly attack *Vespula*, *Dolicovespula* and *Polistes* species. The European subspecies *S. vesparum vesparum* similarly shows cocoon polymorphism and also adult seasonal polymorphism with overwintering males being produced late in the season and emerging from tough yellow cocoons that are formed at the base of the host's nest cell; macropterous overwintering females and rapidly developing stockier and often brachypterous females emerge from flimsy cocoons attached to the side of the host cell. Early in the season, the flimsy white cocoons produce short-winged, thelytokous females that hatch within 2 weeks (Donovan 1991). These females start ovipositing on other brood cells within the same nest without mating. The next generation may contain some males and emerges from intermediate pale-brown cocoons and their offspring produce more durable darker cocoons that overwinter and yield fully winged males and females – sometimes a prolonged diapause of up to 4 years may follow. All three morphs can occur in a single brood and the summer females are thelytokous and can be maintained in culture for many generations without mating. Thus these wasps combine a mix of cyclical parthenogenesis and normal sexual reproduction in parallel. Despite Godfray's (1994) plea, we still do not have a proper understanding of how *Sphecophaga*'s sex and polymorphism are determined.

Latibulus argiolatus and possibly other members of the genus, which is parasitic on polistine wasp larvae in their nests, also produce two types of cocoon (Frilli 1965, Makino 1983). In the summer the cocoon remains in the host's larval (pedotrophic) cell, but the winter one falls out and, through contortions of the *Latibulus* larva within it, it is capable of jumping until a favourable resting site is reached (see also *Bathyplectes anurus*).

Numerous Cryptini use vibrational sounding (echolocation) to locate their often immobile hosts mostly concealed in wood (Fig. 10.2) and is particularly prevalent in groups such as the traditional Gabuniina. Just as in several members of the Pimplini, they use modified terminal flagellomeres with hammer-like structures (Fig. 10.1c,d) to tap the substrate and have very enlarged fore femora in the females (Fig. 10.2) which house a large, vibration-detecting subgenual organ. Laurenne et al. (2009) mapped the distribution of antennal hammer modifications on a molecular phylogeny of the group and found very strong evidence that antennal hammers, and therefore presumably echolocation of hosts, had been gained and lost on multiple occasions in the subfamily. In an SEM survey, Laurenne and Quicke (2009) presented images indicating that the hammer structures in cryptines derive from sensilla. Presumably, when the ancestral sensilla at the tip of the antenna were exploring the substrate, they generated sufficient vibration for the wasp to detect.

Unlike some of the cryptine parasitoids of aculeates that rely vibrational sounding to determine a suitable point for 'drilling', *Buathra tarsoleuci* appears to 'drill' first and then insert its antenna through the hole thus made to determine where in the penetrated cell to probe with its ovipositor (Casiraghi et al. 2001).

Two species, both ectoparasitic on the fall army worm, *Spodoptera frugiperda* (Noctuidae), have been reared successfully on artificial diets, viz. *Cryptus albitarsus* and *Diapetimorpha introita* (Carpenter & Greany 1998). Hussain and Mathur (1924) noted that *Goryphus* (as *Melcha*) *nursei* lays its egg externally on its host *Earias insulana* (Nolidae) pupa, but the larva may enter it to feed (see also the phygadeuontine *Blapsidotes*). Russell (1987) found the same for *Glabridorsum stokesii*, a parasitoid of the oriental fruit moth, *Grapholitha molesta* (Tortricidae). The female wasps attacks either the late caterpillar, prepupal or pupal stages of its host, which it oviposits on externally through the host cocoon. If a pupa is attacked, the parasitoid larva commences feeding underneath the wing cases and finally bores into the host, completing development internally and spinning its own flimsy cocoon within the pupal shell – very occasionally it does not form a cocoon at all.

The genus *Myrmeleonostenus* and one Chilean species of *Itamuton* are parasitoids of antlions (Myrmeleontidae). *Itamuton* is a (presumably external) parasitoid of the pupa (Porter 1989).

Phygadeuontini

Phygadeuontini are largely recovered as monophyletic in molecular analyses (Laurenne et al. 2006, Quicke et al. 2009[13]). It is a large group with more than 120 recognised genera. The Bathytrichina included in molecular studies (viz. *Bathythrix*, *Chrysocryptus*, *Surculus*) generally formed the sister group to the remainder of the tribe or were distributed more basally within the subfamily. Biologically, phygadeuontines are strongly associated with weakly cocooned hosts, although a number of exceptions occur. *Bathythrix* (Bathytrichina) species are parasitoids of Gyrinidae pupae (F.G. Butcher 1933), as are some other members of the tribe. They are typically entirely ectoparasitic, but a few that attack pupal stages complete development internally, e.g. *Gelis* sp. studied by Russell (1987) (cf. *Glabridorsum* above) and some may oviposit into the pupa (e.g. *Blapsidotes*). The *Stilpnus* genus group (or Stilpnina) are entirely endoparasitic (see below). *Helcostizus* and *Amydraulax*, are parasitoids of beetle larvae in small branches, *Lochetica westoni* is recorded as a parasitoid in the nest of the sphecid *Passaloecus insignis*. All these three were placed in the Mastrina by Townes (1970a) and although this group was not recovered as monophyletic in the molecular studies mentioned above, they were generally found to be fairly closely related.

As noted in the genera discussion of the subfamily, this group includes many species that are egg predators. The larvae of *Gelis apteris*, which feed in the egg sack of *Zodarion* spiders, are true carnivores, feeding on the spiderlings rather than the eggs (Korenko et al. 2013). Many species of *Gelis*, *Polyaulon* and *Thaumatogelis* are apterous (Table 7.2, Fig. 13.30c) and search for hosts predominantly on tree trunks or among leaf litter. Some attack psychids or coleophorids (bagworms) that construct larval/pupal silk bags decorated with substrate. Whilst these apterous species are not as vagile as flying wasps, trees are long-lived and might be expected to support many generations of wasp, making dispersal less of an issue. They typically have broad host ranges and long adult life-spans, which may facilitate the co-occurrence of multiple species on a single host tree with considerable host overlap (Wieber et al. 1995). The European *G. acororum* is unusual among insects as an whole in that it typically has a male-biased sex ratio, which appears to be associated with having patchily distributed hosts and is not a consequence of lack of mates or of inbreeding (B. Visser et al. 2014).

Pseudohyperparasitism is practised by many species of *Gelis* (Horn 1976), *Lysibia* (Harvey et al. 2004, 2008, 2011), *Acrolyta* and *Bathythrix*. Poelman et al. (2012) found that *L. nana*, a relative specialist on *Cotesia* species attacking Brassicaceae-feeding pierids, was attracted not to volatiles emitted from the primary parasitoid cocoons, but to the volatiles released by the cabbage plants being fed on by the *Pieris* host caterpillars. In this species, actual location of host cocoons then seemed to rely largely on bumping into them.

The group of genera comprising Townes' subtribe Stilpnina, viz. *Atractodes*, *Mesoleptus* and *Stilpnus*, appears to be monophyletic and they have a very different biology compared with most of the other Cryptinae in that they are koinobiont larval endoparasitoids of caterpillars and also of Diptera. Beaver (1972) provided a detailed account of a species of *Mesoleptus* that attacks the maggot of *Sarcophaga nigriventris* (Sarcophagidae) which develops within *Cepaea* snails, apparently detecting hosts almost entirely using ovipositor sensilla. Unlike most ectoparasitic cryptines, this *Mesoleptus* is apparently very host specific and does not parasitise other species of sarcophagid even within the same snail.

A few other species have host associations of interest. Two species of *Tropistes* are known to attack snakeflies (Rhaphidioptera) (Aspöck 2002); *Rhembobius* species appear to be specialist puparium endoparasitoids of milesiine syrphids (Diptera) with larvae that feed as saprophages (M.R. Shaw & Askew 2010) and some *Phygadeuon* species feed ectoparasitically on Diptera pupae within their puparia. *P. trichops* is a pupal ectoparasitoid within the puparium of various Diptera, and Monteith (1956) presented a detailed description of its development, internal anatomy and general biology, and various other members of the genus can exert substantial levels of control against nuisance and pestilential farm muscid flies associated with pigs, cattle, etc.

Ichneumoninae

This is an enormous but morphologically relatively uniform group that are nearly all larvae–pupal or pupal endoparasitoids of Lepidoptera. Ichneumonines include some of the largest and most colourful of the ichneumonids and sexual dimorphism of colour pattern is common in the larger species. They are abundant in all faunas and many species are frequently observed hunting for hosts among low vegetation in both the temperate and tropical regions (Gauld 1984a).

Several species have small roles in the biological or integrated control of pest species, such as the *Dentichasmias busseolae*, which was initially described based on material reared from the maize and sorghum stalk borer, *Busseola fusca* (Noctuidae), and which in East Africa has formed a new association with the spotted stalk borer, *Chilo partellus* (Crambidae), an important pest of maize and sorghum, *Diadromus collaris*, a pupal endoparasitoid diamondback moth, *Plutella xylostella* and *D. pulchellus*, of the leek moth (*Acrolepiopsis assectella*). *Tycherus nigridens*, referred to as *Phaeogenes* in most of the literature, was an important parasitoid of the European corn borer, *Ostrinia nubilalis* (Crambidae), in the USA (H.D. Smith 1932), although in Europe it is generally only responsible for a low level of host mortality. *Tycherus osculator* is an important native parasitoid of the pest tortricid *Zeiraphera canadensis* (R.J. West et al. 1999) in parts of Canada.

Basic biological studies on these are, however, rather sparse (see, e.g., S. Liu et al. 2001). Hinz (1983) presented details of the biology of various *Ichneumon* species. There are surprisingly few descriptions of developmental stages for the subfamily. Fairly detailed anatomical descriptions of the larvae of *Tycherus nigridens* were given by H.D. Smith (1932), of *T.* (as *Phaeogenes*) *osculator* by Gerig (1960) and of *Diadromus collaris* by Zhao et al. (2014). Despite a very large number of host dissections, Smith was unable to be certain about the exact number of instars that it went through, but Gerig reported five and likewise Cushman (1926b) cited Chewyreuv that there are five larval instars in *Eutanyacra picta* (as *Amblyteles vadatorius*). However. Carpenter et al. (1994) and Zhao et al. (2014) reported only four for *Ichneumon promissorius* and *D. collaris*. Despite being endoparasitoids, the eggs appear to be anhydropic and are fairly large. Probably most species lay their egg into the host haemocoel (e.g. H.D. Smith 1932), but at least one species, *Poecilostictus cothurnatus*, implants it into the wall of the hind gut (van Veen 1981), presumably as part of a mechanism to avoid encapsulation. The first instar is hymenopteriform, without a conspicuous tail. Chewyreuv reported that spiracles are absent from the first instar, and although the second has the full number these are small (and presumably non-functional). According to Smith, all larval instars appear to feed on host fat body tissue using their mandibles. Unless there is a diapause, the development of species attacking the pupal stage is rapid, for example, *I. promissorius* develop from egg to pupa in just 8 days (Carpenter et al. 1994).

Cocoons are often flimsy on account of pupation occurring within the usually tough pupal case of the host and may be completely absent when the pupa affords sufficient protection on its own (e.g. *Tricyphus respinozae*: Ward & Gauld 1987). The meconium is not voided until after the adult emerges from within the host pupa and Smith noted that whereas males will not mate before voiding it, females will. Various species host feed and drink from the wound inflicted by the ovipositor during oviposition (e.g. Lloyd 1940, Tan

& Takasu 2000, Cole 1967), but its exact prevalence is uncertain and some species do not, despite having the opportunity (H.D. Smith 1932). The ovaries contain relatively few mature eggs (Iwata 1960) and the lifetime fecundity is generally not very high, Clausen (1940) reporting a total of 50 eggs laid for *Phaeogenes nigridus*.

Although most species attack fully grown host caterpillars as they seek pupation sites, others attack the host pupa directly. Hinz (1983) noted that those species that attack mature larvae or prepupae have very short ovipositor and broad hypopygium, termed amblypgous, and those that oviposit directly into pupae have a more pronounced and slightly longer ovipositor and pointed hypopygeum and are termed oxypygous (Fig. 13.31e). Those species that attack pupal hosts (e.g. Cole 1967) generally only accept relatively young ones. *Limerodops elongata* is very unusual in that it attacks early instar host larvae (the web-building hadenid *Hadena gemina*) which it reaches using its elongate metasoma (Heinrich 1961), and a similar ability to attack all host instars is known in *Trogus mactator* (Omata 1984), but the parasitoid only moults into a second instar after the host has pupated. *Tricyphus respinozae* oviposits in its host sphingid caterpillar while the latter is still feeding on its host plant (Ward & Gauld 1987). The subfamily therefore includes both idiobionts and some true koinobionts. At least one genus (*Colpognathus*, see below) mummifies its host still in the caterpillar stage in the same way as *Alomya* of the Alomyinae does (viz. *Colpognathus*: M.R. Shaw & Bennet 2001). Host location in many species appears to involve a lot of walking. Hosts of *D. pulchellus* are located from a distance via volatiles emitted by their frass (Auger et al. 1989).

Gokhman (1988, 1992) proposed that ichneumonines may have been so successful because by evolving to be endoparasitic initially on prepupae and pupae of small Lepidoptera this transition to endoparasitism enabled them to extend their host range to attacking wandering caterpillars, thus releasing them from competition from cryptines that can only attack prepupal hosts in exposed cocoons or cases. He further suggested that ichneumonines probably evolved from phygadeuontines (which he referred to as Gelini) that were ectoparasitoids of prepupae of small moths in exposed cocoons or with exposed pupae. Part of Gokhman's proposal is based on the idea that the Phaeogenini are basal within the Ichneumoninae, thus permitting a direct link to Cryptinae: Phygadeuontini. This relationship was supported by the molecular analyses of Quicke et al. (2009) in which the Phaeogenini were (largely) recovered as the sister group to the rest of the subfamily, but Laurenne et al. (2006) with a rather small set of exemplars recovered them as derived. However, since analyses generally recover both Ichneumoninae and Cryptinae as monophyletic (to the exclusion and Alomyinae), Gokhman's suggestion cannot be accepted exactly as proposed, although the biological events could easily have been as he envisaged.

By far the best studied species is *Diadromus pulchellus*, which has been investigated extensively from a physiological point of view. Overcoming host immunity in this species is dependent on a symbiotic Reovirus and Ascovirus (Renault et al. 2004), although it has also been found to harbour a Cypovirus, although not in all individuals (Renault et al. 2003). *D. collaris* has also attracted considerable attention because it is often important in the control of diamond back moth and in most places is thelytokous, although in South Africa it is predominantly sexual, leading Kfir (1997) to suggest that the diamond back moth itself might have originated in South Africa.

A few species of ichneumonine are specialist parasitoids of some cuckoo *Maculinea* blue butterfly (Lycaenidae) species whose larvae spend much of their lives in brood chambers of ant colonines where they are fed by worker ants. For example, *Ichneumon eumerus* is a specialist parasitoid of the rare blue butterfly *Maculinea rebeli*, which dwells in *Myrmica schencki* nests, and some other myrmecophilous blue butterflies are attacked by other *Ichneumon* spp. Another ichneumonine genus, *Neotypus*, has also recently been shown to parasitise lycaenid larvae in ant nests (Tartally 2005). *I. eumerus* has been investigated to discover how it manages to survive from ant predation. Thomas and Elmes (1993) showed that the wasp first locates ant nests by their odour and can differentiate between nests that contain or do not contain *Maculinea* larvae, only entering the former. Upon emergence from the host butterfly pupa, the wasp is attacked viciously by the ants but, through a combination of physical toughness and the release of an allomone that causes confused behaviour among the ants, it generally survives to exit the ant colony. Compared with other *Maculinea* species, whose larvae also are obligate residents within ant nests, *M. rebeli* is a specialist and its parasitoid's behaviour and selection of larger host larvae (i.e. ones that are most likely to survive to pupation) are postulated as special adaptations

Fig. 13.31 Five female ichneumonines. (a) *Homotherus varipes*; (b, c) *Ichneumon sarcitorius*, female and male, respectively; (d) *Vulgichneumon satuatorius*; (e) gen. sp. from Australia, showing oxypygous form of metasomal apex indicative of attacking pupal hosts. (Sources: a, reproduced by permission of Håkon Haraldseide; b, reproduced by permission of Paul Lathbury; c, reproduced by permission of Cor Zonneveld; d, reproduced by permission of Thijs de Graaf; e, reproduced by permission of Malcolm Tattersall.) For a colour version, see Plate 48.

compared with ichneumonine parasitoids of other *Maculineas* that are less host-ant specific. The high conservation status of some of the ant-associated blue butterflies, several of which are threatened and often live in small isolated colonies, also means that their specialist parasitoids are a conservation issue (e.g. Hochberg et al. 1998).

Colpognathus superficially resembles *Alomya*, especially in the form of its head, which on the basis of molecular evidence does not appear to belong to the Ichneumoninae (Quicke et al. 2009). Interestingly, both genera mummify their host caterpillars (M.R. Shaw & Bennet 2001) and it seems likely that the overall morphological similarity is explained by convergent evolution to this particular mode of life.

The host ranges of ichneumonines are variable. Some such as *Phaeogenes invisor* appear to be restricted to either a single host species or perhaps a few closely related species; Cole (1967) could find no evidence that this species attacks other tortricids living in the same habitat.

Sime (2002) described the host searching behaviour of *Trogus pennator*, a large species that enabled her to follow individuals as they searched for swallowtail butterfly (*Eurytides marcellus*) caterpillars on their host plant (*Asimina*). Interestingly, the wasps appeared to locate host plants primarily visually, although they did approach plants with host feeding damage at a significantly higher rate than undamaged plants, given the relative abundances of the two categories.

In the temperate region, fertilised females of numerous species hibernate (see Chapter 14, section *Cold hardiness, hibernation and overwintering*), typically under loose bark, often in multispecies aggregations of tens of individuals (Rasnitsyn 1964, Dasch 1971). In such aggregations one can frequently encounter individuals of species that otherwise appear fairly rare and therefore presumably spend much of their lives searching for hosts in places where collecting is seldom carried out, perhaps in the canopy.

Relationships within the subfamily have received very little formal consideration. There are many tribes described in the literature, but as only a handful of workers have ever got to grips with the group, it is not surprising that there is almost nothing by way of synthesis. Sime and Wahl (2002) presented a detailed cladistic analysis of the large, often colourful *Callojoppa* genus group and Tschopp et al. (2013) presented a molecular phylogeny of number of European species of *Ichneumon* with known biologies and showed that evolutionary shifts between host groups have been numerous and, likewise, shifts between above- and below-ground host pupae.

The large size of the subfamily (approximately 420 genera) makes identification difficult and, in some parts of the world, even generic level determination may be very problematic. This is the one subfamily for which Henry Townes did not publish a generic treatment. Also, inevitably, there are some nomenclatural issues resulting from Townes' application of names, thus Townes' Ichneumonini are properly called the Heresiarchini, his Joppini are what we call the Ichneumonini and his Notosemini are the Oedicephalini. Heinrich also used some non-standard tribal names in his works and what Gauld (1984a) treats as Protichneumonini are the Heresiarchini, which also includes his Trogini. There is considerable variation between authors as regards its tribal classification. Heinrich (1967) presented a classification for the African genera. The tribes recognised here and by Quicke et al. (2009) are Ceratojoppini, Clypeodromini, Compsophorini, Ctenocalini, Eurylabini, Goedartiini, Heresiarchini, Ichneumonini, Ischnojoppini, Joppocryptini, Listrodromini, Oedicephalini, Phaeogenini, Platylabini and Zimmeriini, although as Gauld (1984a) noted, the 'limits of the tribes are vague in the extreme'. Following earlier workers and supported by molecular data (Laurenne et al. 2006, Quicke et al. 2009), the Alomyinae are regarded as a separate subfamily rather than as a tribe. The reader may encounter the terms 'Ichneumoninae stenopneusticae' and 'Ichneumoninae cyclopneusticae' in the literature and refer to those taxa with the propodeal spiracle slit-shaped or round, respectively. Thus they effectively differentiate the generally larger-bodied species from the small-bodied Phaeogenini and Heterischnini. Because the small-bodied phaeogenines are more taxonomically challenging, with Gauld (1984a) calling them 'a notoriously difficult group', they are often excluded from revisions which consequently state that they deal with the 'Ichneumoninae stenopneusticae'. Most of the 'cyclopneusticae' attack 'microlepidoptera', are oxypygous and thus oviposit into host pupae.

Quicke et al. (2009) included a total of 93 species representing 13 of the 15 generally recognised tribes, the only missing ones being the Clypeodromini and Goedartiini. The phylogenetic relationships recovered from analysis of the elided 28S rDNA data (all ichneumonines were scored identically for morphology) did not support monophyly of any of the larger tribes, although, with the exceptions of *Diadromus* and *Lusius* (Fig. 4.1) or with the inclusion of *Dagoo* and *Macrojoppa* (Fig. 4.2), the Phaeogenini were monophyletic and the Platylabini+Zimmeriini also, except that the Ichneumonini genus *Probolus* was also included. The placements of *Diadromus*, *Dagoo*, *Lusius* and *Macrojoppa* in these trees are certainly wrong and highly likely to be a consequence of their sequence lengths, with long insertions (see fig. 4 in Laurenne et al. 2006).

The British fauna was treated by Perkins (1959, 1960) and Gerd Heinrich provided several important faunistic treatments in a long career: Heinrich (1934) for the Celebes, Heinrich (1967) for Sub-Saharan Africa, Heinrich (1974) for Burma and Heinrich (1961, 1962, 1977) for North America. Gauld (1984a) provided a key to the genera occurring in Australia. Rousse et al. (2013) have revised the Afrotropical Phaeogenini. Tereshkin (2009) provided a key to the tribes (including Alomyinae as a separate tribe).

THE BRACHYCYRTIFORMES

This was an almost completely unsuspected clade until molecular data became available. It includes four small subfamilies, the cosmopolitan Brachycyrtinae, together with two small groups occurring in Chile (Clasinae and Pedunculinae) and Australia (Pedunculinae), and was named the brachycyrtiformes by Quicke et al. (2009). It is also possible that the Eucerotinae belong in this group and it was recovered there in the purely molecular tree of Quicke et al. (2000b). Despite now being predominantly northern in distribution, their origins appear to be Gondwanan; their relationships are currently still far from certain, however.

Brachycyrtinae

This is a small, subfamily whose relationships have been rather controversial with just two recognised genera, the cosmopolitan *Brachycyrtus* (Fig. 13.32) and *Monganella* from Australia, neither of which is very common. Originally, *Brachycyrtus* was described in the Cryptinae, in particular in the tribe Hemitelini, although its odd nature was recognised and some workers considered it better placed near *Sphinctus* in the Tryphoninae, near the Cremastinae or in the Diplazontinae. Townes was initially uncertain about its placement but in Townes and Townes (1951, p. 204) placed it in the Pimplinae. See Walkley (1956) for a more thorough summary of its chequered history. Walkley himself considered it to show affinities with the genera *Poecilocryptus* and *Labium*, which at the time were considered as cryptines but are now in the Labeninae.

Townes' (1969) placed the Brachycyrtini as a tribe in the Labeninae, but also included with it the genera *Adelphion*, *Pedunculus* and *Poecilocryptus*. Gauld (1983) recognised that *Brachycyrtus* was not closely related to *Poecilocryptus* and placed the latter in its own tribe within the Labeninae. Subsequently, Wahl (1993b) elevated the Brachycyrtini (then comprising *Brachycyrtus*, *Adelphion*, *Pedunculus* and *Monganella*) to its own subfamily. The picture was still confused, however, because Porter (1998) removed the Chilean genus *Pedunculus* to its own subfamily, Pedunculinae, but failed formally to move any of the other genera to it even although *Adelphion* and *Pedunculus* are very similar.

Very little is known of their biology. Walkley (1956) reported them as definitely having been reared from *Chrysopa* lacewing cocoons and rightly cast doubt on records from other hosts, Kabissa et al. (1996) reared a species from a *Chrysoperla* larva and Gauld & Ward (2000) also only found confirmed reared specimens from *Chrysopa*. They have generally been believed to be ectoparasitic idiobionts because of both their apparently rather basal position in the family and the features of the final instar larval cephalic structures including the mandible with small teeth on its dorsal edge and a labral sclerite present (J.R.T. Short 1978, Wahl 1993a). However, Kabissa et al. stated that their specimen was reared from the pupa of a field-collected larva, so more detailed observation is clearly required.

Despite being rather uncommon, they are easily recognised by the combination of venation (Fig. 13.32) and fore wing vein 1cu-a very postfurcal, by more than 0.8 its length, and the areolet is pentagonal even although vein rs-m is absent, with vein 1Ma long, and in addition, many species have the antennae quite swollen beyond the middle (clavate).

Claseinae (Clasinae)

The genus *Clasis* was described in the Labeninae (H. Townes in Townes & Townes 1966) for a single

Fig. 13.32 *Brachycyrtus nawaii* (Brachycyrtinae) female habitus. (Source: reproduced by permission of Kyohei Watanabe.) For a colour version, see Plate 49.

species from Chile and the Juan Fernandez Islands. Gauld (1983) transferred it to the Cryptinae, following cladistics analysis and because they do not display any of the four characters that he proposed defined the Labeninae, forming the tribe Claseini along with *Ecphysis*. Without making clear his reasons for doing so, Porter (1998) created a new subfamily, the Clasinae, to accommodate this genus, but Gauld (2000) retained it in Cryptinae. Molecular analysis has, however, shown Porter's judgement to have been sound despite the lack of formal argumentation (Laurenne et al. 2006, Quicke et al. 2009), since *Clasis* was never recovered inside the Cryptinae except in the case of very extreme gap parameters. However, it has been consistently recovered in a clade together with the Pedunculinae and Brachycyrtinae.

Claseines do indeed superficially resemble cryptines, being medium-sized ichneumonids with a petiolate first metasomal tergite and an exserted ovipositor approximately as long as the metasoma with a pre-apical dorsal nodus. Almost nothing is known of the biology of either *Clasis* or *Ecphysis*; however, Broad and Quicke (2000) concluded that the latter probably uses vibrational sounding to locate hosts as the apex of the antenna bears hammer-like structures and the female fore bibiae are especially swollen. *Clasis* was reported to have been reared from pupae of the defoliator *Cerospastus volupis* (Pergidae) on *Nothofagus* (Gara et al. 1980, cited in Bauerle et al. 1997).

Pedunculinae

With only two included genera, *Pedunculus* from Chile and *Adelphion* from Australia and New Guinea, this is a small, rather uniform group that does not seem particularly remarkable at first appearance. Porter (1998) created the subfamily for *Pedunculus* which had previously been placed in the Labeninae, but he did not provide any clear justification for his action, although it has turned out to have been a good decision. Porter was not explicit about whether his action was also intended to transfer *Adelphion*, at that time included in Brachycyrtini (within Labeninae), with *Pedunculus*, but Gauld and Ward (2000) provided a clear summary of the morphological features that define Pedunculinae inclusive of *Adelphion* and also clearly differentiated it from Brachycyrtinae. Recent molecular analyses based on a single gene fragment, include both genera in a group along with the Clasini, Brachycyrtinae and Eucerotinae, but *Pedunculus* clusters with *Clasis* rather than *Adelphion*, so monophyly of the group is not certain.

Adelphion has been reared from the egg sac of a spider [Chadwick & Nikitin 1976, as 'Brachycyrtinae (? genus)'].

THE ORTHOPELMATIFORMES

From analysis of a broad but relatively small sample of taxa sequenced for the D2+D3 region of the nuclear 28S rDNA region, Quicke et al. (2000b) treated the small subfamily Orthopelmatinae as the sole member of its own larger group, the orthopelmatiformes, because in their analyses it always came out in a basal position within the family. Although its exact position varied, in many trees it came out as the sister group to the ophioniformes. In the more recent morphological analysis of Quicke et al. (2009), they similarly appear in a clade with *Microleptes*, *Oxytorus* and Tersilochinae (but excluding Phrudinae *s.s.*) towards the base of the ophioniformes, but several taxa in that tree are clearly misplaced. In the combined molecular and morphological analyses they were recovered in a rather unexpected position next to *Microleptes* and Tersilochinae as one of the clades rendering the Ctenopelmatinae multiply paraphyletic (see Fig. 13.36). Problematic with these placements within the ophioniformes is that they have relatively short lateral oviducts (approximately two-thirds of the length of the ovarioles), whereas most ophioniformes taxa have considerably longer ones. I suspect that rather than being a reversal, as its placement in the trees of Quicke et al. (2009) would indicate, the relatively short lateral oviducts may be a plesiomorphy.

Orthopelmatinae

This is a small and inconspicuous, entirely northern group whose placement has been elusive to morphologists. The sole genus, *Orthopelma* (Fig. 13.33), is distributed throughout the Holarctic and they are reared easily from the galls of the cynipid genus *Diplolepis* which form galls on Rosaceae (*Rosa* and *Rubus*), such as *O. mediator* on the bedeguar gall made by *D. rosae*, where they are koinobiont endoparasitoids

Fig. 13.33 Female habitus of *Orthopelma simile* (Orthopelmatinae). (Source: reproduced by permission of Kyohei Watanabe.) For a colour version, see Plate 50.

of the *Diplolepis* larvae (Blair 1945, Shorthouse & Leggo 2002). The latter authors noted that when they collected maturing larvae of *D. triforma* in their galls and sectioned them, larvae of the endoparasitoid that were almost certainly an *Orthopelma* were found to be feeding on the fat body of the host and the development of the surrounding gall was normal for hosts at that stage of development. My dissections indicate that the lateral oviducts are quite wide and have a distinct whitish ring near their anterior ends and it would be worth investigating whether this represents a calyx gland and indeed whether there are any associated virus-like entities.

THE OPHIONIFORMES

A close relationship between many of these subfamilies was suspected back in the 19th century, based largely on the laterally compressed metasomas with the spiracle of the first metasomal tergite being situated well behind the middle as in Ophioninae, Anomaloninae, Campopleginae, Cremastinae and Tersilochinae. Early on, these were termed the Ophiones and in works such as those of Pampel (1914) many of these were included in his Ophioninae – often referred to as the Ophioninae *s.l.* (see Wahl 1991). When Pampel's observations were actively taken up again, some 70 years later, it had a tremendous influence, because it was not only

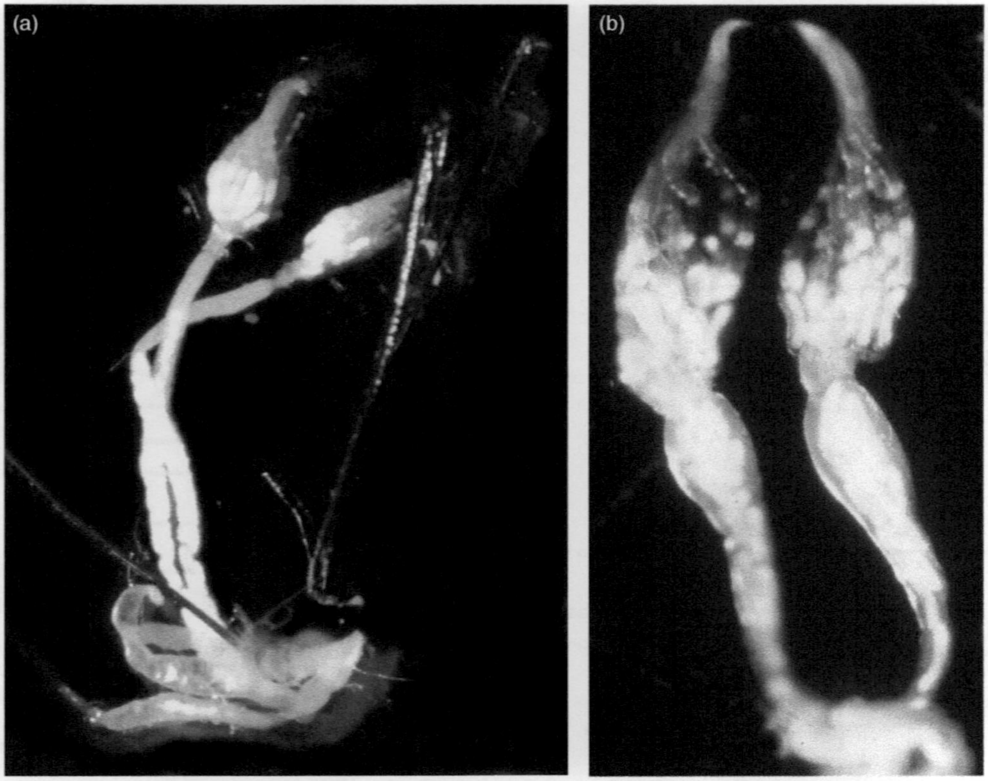

Fig. 13.34 Ophioniformes dissected female internal reproductive system. (a) *Apophua* (Banchinae); (b) *Tranosema rostrale* (Campopleginae) showing long lateral oviducts relative to ovaries, and in *Tranosema* the swollen calyx gland at proximal part of lateral oviduct. (Source: reproduced by permission of Michel Cusson.)

a very early and beautifully illustrated use of internal anatomy, it provided a very clear synapomorphy for the grouping that had earlier been suspected, i.e. the lateral oviducts are very long, often markedly longer than the ovaries themselves (Fig. 13.34).

Gauld (1985) recognised as a natural group the Ophioninae, Banchinae, Campopleginae, Cremastinae, Ctenopelmatinae and Tersilochinae, noting that banchines and ctenopelatines have the metasoma more broadly connected to the propodeum than the others, with more central spiracles and separate sternite, suggesting that these might be more basal in the group. Along with Townes (1969), Gauld did not at that time consider the Anomaloninae to be closely related to the Ophioninae, instead thinking they were closer to the Metopiinae, largely based on larval morphology and the fact that both emerge from the pupa of the host, which might of course be connected. The Metopiinae were not considered to be related to ophioniformes largely because they have a dorso-ventrally compressed metasoma and not a laterally compressed one as in the other taxa. As found by Pampel (1914) and confirmed by Wahl (1991), anomalonines do have highly elongate lateral oviducts, i.e. the main synapomorphy for the group. In addition, those taxa with laterally compressed metasomas have the spiracle of the first tergite situated well beyond the midpoint, an nearly all have a pre-apical notch on the lower ovipositor valve, all of which are features found in the Anomaloninae. With the hindsight provided by molecular data, it is hard to see how Townes and Gauld did not investigate the relationships of the Anomaloninae in more depth – and Gauld was certainly aware of Pampel's findings.

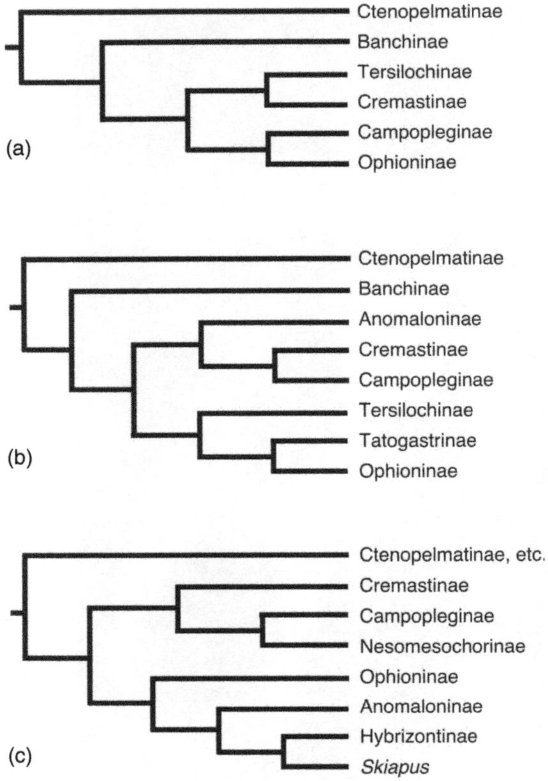

Fig. 13.35 Hypothesised relationships between the subfamilies thought to comprise the 'higher ophioniformes'. (a) from Gauld (1985) in which he was unaware that several additional subfamilies also belonged to the grouping; (b) from Wahl (1991); (c) from combined elided molecular and morphological analyses with 28S rDNA gaps treated as uninformative of Quicke et al. (2009, fig. 22; if indels age treated as uninformative the Neosomesochinae shift to being basal to Campopleginae+Cremastinae).

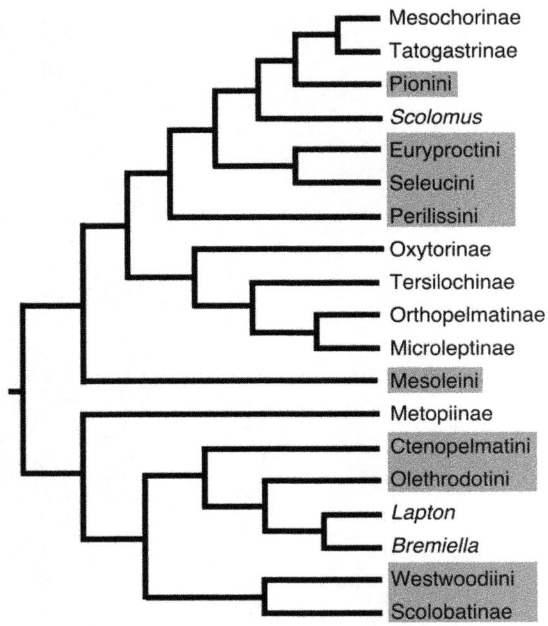

Fig. 13.36 Paraphyletic relationships of Ctenopelmatinae as recovered from the weight-balanced, combined elided molecular and morphological analyses with 28S rDNA gaps treated as uninformative of Quicke et al. (2009, fig. 21). Note that although differing in detail, the results of the comparable analysis with gaps treated as informative (Quicke et al. 2009, fig. 22) were largely similar.

The composition of this grouping, now usually referred to as the ophioniformes, has grown considerably as workers started to investigate internal anatomy more. As currently comprised, this group also includes taxa such as the metopiines, tryphonines, ctenopelmatines and banchines, which have sessile metasomas; and possibly hybrizontines which are just bizarre.

There have been several hypotheses about phylogenetic relationships among a group of subfamilies that are often thought of as the 'higher ophioniformes', but these are not always directly comparable because workers differed in their view of what constituted this grouping. Figure 13.35 summarises three of these, and Fig. 13.36 summarises the findings of Quicke et al. (2009) for what might be thought of as the 'middle ophioniformes', i.e. a group of subfamilies and genera of uncertain placement nested in among various tribes of a polyphyletic Ctenopelmatinae. Basal to all these in Quicke et al.'s trees were the Banchinae, Lycorininae, Tryphoninae (polyphyletic), Sisyrostolinae and Stilbopinae (polyphyletic), which for ease I regard as the 'lower ophioniformes'. These three groups are dealt with in batches below.

Lower ophioniformes

Banchinae

This is a large, cosmopolitan and commonly collected group of medium-sized to large wasps. Banchines are all koinobiont endoparasitoids of Lepidoptera, most being solitary. Gregarious development is known in one European *Lissonota* species (M R Shaw 1999a) and one Costa Rican *Diradops* sp., and a second *Diradops* is apparently facultatively gregarious (Gauld et al. 2002a). In the case of *L. mutator*, only single-sex broods have been reared, which led M.R. Shaw to postulate that it might develop through polyembryony.

The subfamily has long been considered a natural subfamily and divided into three tribes, Banchini, Glyptini and Atrophini (=Lissonotini), with stable boundaries. Here I consider a fourth tribe, Townesionini, based on its phylogenetic position in the combined molecular and morphological analyses of Quicke et al. (2009), although Gauld and Wahl (2000b) considered that *Townesion* was actually a derived member of the Glyptini.

Gauld et al. (2002a) noted that while the banchines belong to the Ophioniformes, they lack several of the putative synapomorphies associated with a number of these, instead displaying the probable ground plan state for the group including an essentially broad attachment of the metasoma to the propodeum and having a broad, non-petiolate first metasomal tergite. Thus they concluded that they are probably one of the more basal of the Ophioniform subfamilies, and this was supported by Quicke et al. (2009).

Atrophini and Glyptini

The Atrophini (Fig. 13.37a,b) and Glyptini (Fig. 13.37c) appear, on the basis of 28S rDNA sequence data, to form a monophyletic group, with each independently being monophyletic (Quicke et al. 2009). They are biologically rather similar. Wahl (1988) commented that final instar larval head capsules of the Glyptini appeared to show no apomorphies and therefore probably represent the groundplan state, whereas the Atrophini possess the derived state of having a reduced hypostoma.

Hosts are generally weakly concealed Lepidoptera larva, principally Torticidae (Gauld et al. 2002a), while Atrophini attack a considerably larger range of hosts including members of the Pyralidae, Cossidae, Sesiidae, Gelechiidae and others, including some in quite concealed situations. Several species have been investigated from a biological control point of view and a few have been introduced into different parts of the world, primarily to control forestry pests (Gauld et al. 2002a). These include *Glypta haesitator* and *Cephaloglypta murinanae* which were introduced into North America from Europe and *G. rufiscutellaris* from North America into Australia. *Australoglypta latrobei* is a useful parasitoid of the light-brown apple moth, *Epiphyas postvittana* (Tortricidae), which is a serious pest of several plants including grape vines; both are indigenous to Australia and New Guinea (Gauld 1977b). In North America, *G. fumiferanae* is an important parasitoid of spruce budworms, *Choristoneura* spp. (Tortricidae). In addition, *Syzeuctus* (Fig. 13.37a) includes species that are important endemic parasitoids of various graminaceous borers, particularly in Africa, and *Leptobaptopsis* includes several species that attack pyralid pests in the Old World tropics.

Some Atrophini genera have short ovipositors [e.g. *Amphirhachis* (Fig. 13.37b), *Arenetra*, *Cryptopimpla*, *Diradops*, *Hadrostethus* and *Hylesicida*] and attack exposed hosts including Erebidae (Arctiinae), Lasiocampidae and Saturniidae (Gauld et al. 2002a). Those that are parasitic on exposed hosts tend to attack early instar larvae, but complete development after the host has constructed its cocoon.

Development has been described for *Glypta rufiscutellaris* (A.W. Crawford 1933). Iwata (1960) illustrated the ovarian egg of a *Glypta* sp. (as *Conoblasta*[14]), which suggests that it has a rather complex morphology, but did not comment on it further. At least under laboratory conditions, some species will host feed and also predate on early instar host larvae (Cossentine et al. 2004). Cocoons of Atrophini have a rather amorphous middle layer in which woven threads and/or brown–red-coloured material were interspersed except in a paler, equatorial band, whereas those of most Glyptini are typically translucent. Those species that pupate in the soil often form very dark cocoons.

Courtship and mating have been described for *Lissonota holocerica* by Zong et al. (2012). In this species, virgin females, at least in captivity, chase after males until mated, and afterwards rebut additional mating attempts by males. Apparently, despite a male's attempts to mate further times, these individuals are reported to be rejected, perhaps indicating that they have somehow been marked by a female.

The ovipositors of some neotropical atrophine genera (*Quillonota*, *Wahlamia*) are quite atypical for members of the ophioniformes, having two to eight dorsal serrations posterior to the first one, which is clearly the anterior rim of the normal pre-apical notch (Gauld et al. 2002a). Both of these have been reared from stem galls so these modifications are no doubt associated

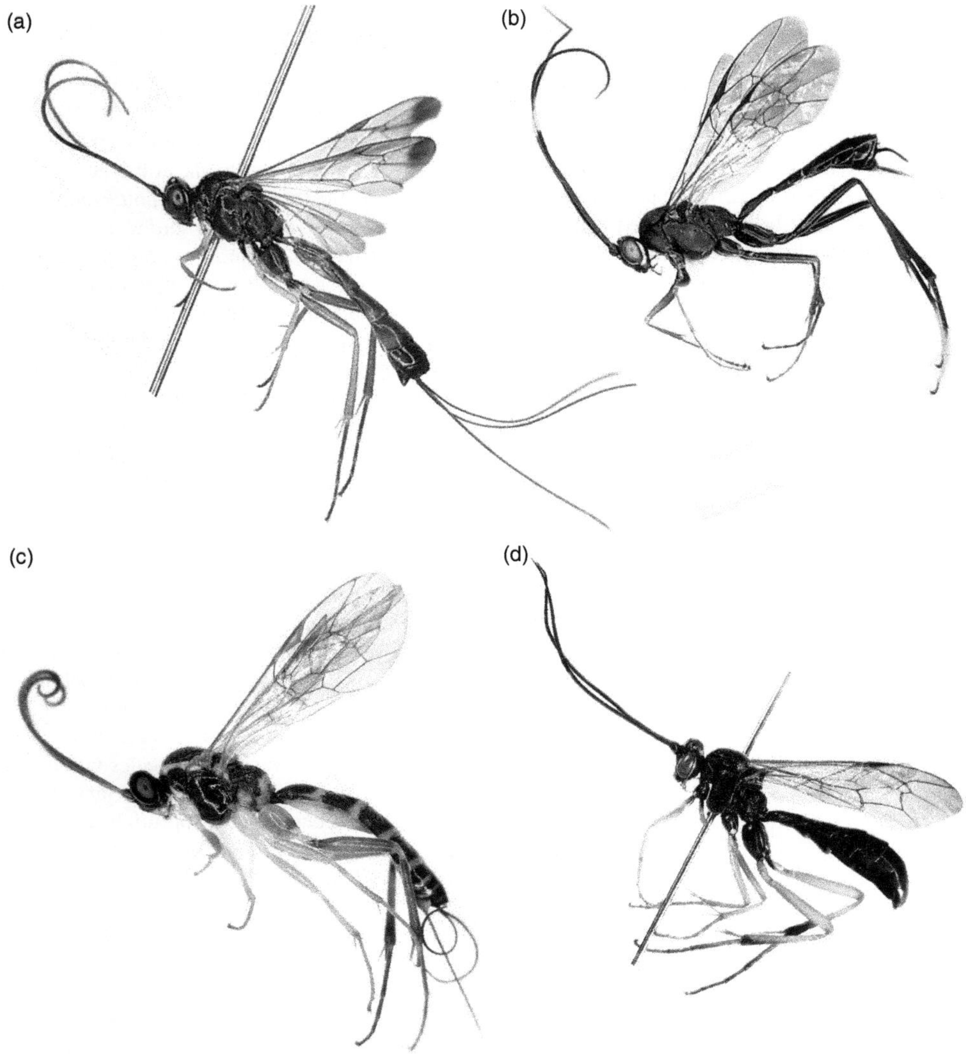

Fig. 13.37 Banchinae. (a) *Syzeuctus apicifer* (Atrophini); (b) *Amphirhachis nigra* (Atrophini); (c) *Glypta uchidai* (Glyptini); (d) *Banchus* sp. (Banchini). (Sources: a–c, reproduced by permission of Kyohei Watanabe; d, reproduced by permission of Kazuhiko Konishi.)

with penetration of hard plant tissue and it is possible that the first dorsal serration acts like the simple notches found in other banchines.

Three species of Glyptini, *Glypta fumiferanae*, *Glypta* sp and *Apophua simplicipes*, and one Atrophini, *Lissonota* sp. (Stoltz & Whitfield 1992, Lapointe et al. 2005, Djoumad et al. 2013), are now known to possess polydnaviruses and these differ considerably in their morphology from those present in the Campopleginae (Fig. 9.2 cf. 9.4), the only other ichneumonid subfamily known to have them. However, *Glypta* and *Apophua* are very closely related and it is not yet known whether these Banchovirus occur outside of the Glyptini and Atrophini.

Banchini

Members of the Banchini have short ovipositors (Fig. 13.37d) and attack exposed caterpillars, reliable records being all from Noctuidae (Fitton 1985, 1987), whereas members of the two other tribes typically have ovipositors about as long as or longer than the metasoma and attack semi-concealed hosts such as leaf rollers. Although some Banchini species have been recorded to attack Lepidoptera of a number of other families (Nymphalidae, Thiatiridae, Geometridae, Sphingidae, Notodontidae, Erebidae (Lymantriinae) and Zygaenidae), Fitton suggested that these should be treated with caution. Even more so, a record from a diprionid sawfly and an ophionine ichneumonid are almost certainly erroneous.

Wahl (1988) considered the phylogeny of the Banchini based on available larval head capsule characters. The tribe appeared to be defined by the synapomorphy of having eight or more sensilla on the larval prelabium, although as noted by Wahl, a large number of sensilla are also found in all known members of the Ophioninae. The Banchini comprise the *Banchus* group of genera, which have tridentate mandibles, and *Exetastes*, which differs from them in a number of features. *Exetastes* has in the past been placed in the Atrophini and also afforded a tribe of its own, but Wahl (1988) clearly demonstrated that it belongs to the Banchini, albeit rather basally derived, and this has been supported by the combined morphological and molecular analysis of Quicke et al. (2009), in which it was recovered as the sister group to the six other banchine genera included.

The Banchini appear to be primarily creatures of open shrubby country and some at least give off a pungent odour when handled, a feature which is probably related to their often, brightly banded, yellow and black coloration, which is broadly vespid like (Townes & Townes 1978), although most European species are rather more sedately coloured. Both *Rhynchobanchus* and another banchine genus, *Agathilla*, have their mouthparts modified for accessing concealed nectar sources, in the latter case with very elongated glossae (Jervis 1998). Fitton (1985) indicated that because they are usually only flying for a fairly short period each spring or early summer most species are probably univoltine.

The biology of one species, *B. femoralis*, has been described in considerable detail by van Veen (1982). Developmental stages have been described for *Exetastes* (Slovák 1983, 1984). The first instar larva of banchines is vesiculate (Bledowski & Krainska 1926). The number of larval instars does not appear to be known. The final instar larva of *Banchus* may wander off after emerging from the host to find a suitable crevice for cocoon formation (Arthur & Mason 1985). The host of *Banchus* is killed after it has entered the soil to pupate and the cocoons of both *Banchus* and *Exetastes* are hard and nearly black. Despite having short ovipositors, some *Banchus* species have strongly elongate and compressed metasomas with which they can access semi-concealed hosts such as leaf-rollers. Courtship and mating have been described for *B. femoralis* (van Veen 1982) and *Exetastes cinctipes* (Slovak 1986).

An unusual form of host physiological defence, called cuticular encystment, by the cabbage looper caterpillar, *Trichoplusia ni*, against the North American *B. flavescens*, was described by Arthur & Ewen (1975) and subsequently shown to be a result of venom ('acid') gland components (Ewen & Arthur 1976). This type of protective response appears to be a feature of plusiine noctuids (Namba et al. 2004, 2008) and is also shown by *Autographa* as a response to parasitism by *Campoletis*.

Several differences between *Exetastes* larval features and those of other Banchini led Wahl (1988) to speculate that '... some sort of exception to the usual biology of Banchini occurs'. Wahl (1988) proposed a morphological phylogeny for the genera of Banchini based on 'hand analysis'. Despite the fact that this did not resolve whether *Agathilla* or *Exetastes* were the sister group to the remaining included genera, the tree he obtained was identical with that resulting from molecular analysis (Quicke et al. 2009).

Townesionini

Kasparyan (1993) erected a new subfamily, the Townesioninae, for a new and highly derived genus *Townesion* from the East Palaearctic and *Sachtlebenia* from China. *Townesion* is almost unique within the Ichneumonidae in that it has a metasomal carapace comprising four segments, the last of which is furnished with a pair of submedial points (Fig. 13.38). Gauld & Wahl (2000b) concluded that *Townesion* was just a derived banchine, but while a close relationship with the banchines is undisputed, there are no clear and convincing morphological synapomorphies. In agreement with Gauld & Wahl, Quicke et al. (2009) recovered *Townesion* in a monophyletic group with the Banchinae, although the latter analyses recovered it basally rather than as

Fig. 13.38 *Townesion japonicus* (Banchinae: Townesionini) male. (Source: reproduced by permission of Kazuhiko Konishi.)

a member of the Glyptini as treated by Gauld & Wahl and therefore it seems best to retain it as a distinct tribe rather than as a synonym of Glyptini, probably being the sister group to the remaining Banchinae.

A few taxa traditionally included in the Banchinae were removed to other subfamilies [Townes & Townes 1978 placed *Stilbops* and *Notostilbops* in a newly recognised Stilbopinae and subsequently Wahl (1988) transferred *Panteles* also to the Stilbopinae[.

Lycorininae

This enigmatic subfamily is now considered to comprise just the single, virtually cosmopolitan genus *Lycorina* (=*Toxophoroides*). Although it is seldom encountered in numbers in regular collecting, it can be fairly regular on some hosts (Lill 1999). Views on their placement have shifted radically. Traditionally they were included, no doubt largely because of their metasomal sculpture and sessile metasoma and general *Gestalt*, in the Pimplinae, but subsequently they were suspected of being related to the Banchinae (Townes & Townes 1951) until, following information on final instar larval morphology, Townes (1970b) placed them in their own subfamily and Gauld (1984b) noted that they appeared morphologically isolated. Because they lack a dorsal notch on the ovipositor and their final instar larvae have long denticulate mandibles that suggested to him that they might be ectoparasitoids, Gauld considered that they did not belong to the ophioniformes. However, combined molecular and morphological data and many molecular analyses alone (Quicke et al. 2009) place them with the ophioniformes and, to me most convincingly, they possess a completely divided upper ovipositor valve with interlocking aulaciform rod (Fig. 3.4h), which is found in many ophioniformes with long ovipositors.

Details of their biology are scant but most host records are from leaf-tiers and leaf-rollers of various microlepidopteran families including Crambidae, Ethmiidae, Gelechiidae, Oecophoridae, Torticidae and Yponomeutidae, but there is some uncertainty involved as there have been rather few rearings and some literature references might just be repeats of previous, and possibly erroneous, statements (M.R. Shaw 2004b). Oviposition has only been observed once, for the European species *L. triangulifera* (M.R. Shaw 2004b), although it is likely that some variation may occur. In the case of *L. triangulifera*, the female wasp attacks leaf spinning tortricid caterpillars. Although the robust exserted ovipositor is approximately 3–4 mm long and used vigorously to probe the host's retreat, apparently without trying to locate the position of the host in the leaf roll, two of the observed ovipositions were into a larva that had been driven to exit its leaf roll 'before becoming subdued', whereupon the *Lycorina* grabbed the caterpillar and inserted its ovipositor into the host's anus. The eggs of lycorines were described by Iwata (1958) as 'leech-like', but more recent dissections of neotropical species by Coronado-Rivera et al. (2004) and of a European species by M.R. Shaw (2004b) have shown that these have a very interesting anchor-like, bifurcate extension at the posterior end (i.e. that which emerges first from the ovipositor), whereas the ovarian egg of the European species is simple, without anchor. The long egg anchor might therefore be used to attach the egg through the wall of the host caterpillar's rectum with the body of the egg remaining external to the host but hidden within the rectum, and perhaps with the parasitoid larva developing as an ectoparasitoid at least for a while. Interestingly, the lower ovipositor valves of some species are somewhat expanded pre-apically, and this part may function to house the body of the egg with the narrower apical part fixing the anchor. This is purely conjecture, however. That *Lycorina* may be fully ectoparasitic is supported by M.R. Shaw's observation of its externally feeding larva 8 days following oviposition after the host had started to spin a cocoon, but no other observations of an externally feeding parasitoid were made.

As noted by Gauld (1984b), the toothed final instar mandibles (Finlayson 1976; as *Toxophoroides*) are consistent with ectoparasitism, but really say nothing about what the earlier stages might be, and similar features are present in several braconids with a final ectoparasitic feeding phase. On the other hand, the flattened, disc-like larval antennae are more consistent with taxa that are endoparasitic in at least their early stages (H.F. Chao 1980). This leaves the possibility that the *Lycorina* hatches within the host rectum and continues feeding and developing there as a concealed ectoparasitoid or that it burrows through the host's rectum and feeds as a typical endoparasitoid for a while before finally exiting the host for a final ectoparasitic feeding phase like many other ichneumonoids.

Sisyrostolinae

This small, principally tropical subfamily was resurrected by Quicke et al. (2009) under the name Brachyscleromatinae[15], to include five genera previously regarded as an aberrant group of Phrudinae, among which the Madagascan *Melanodolius* includes some impressively large species with fore wing lengths up to 26 mm (Townes 1971). *Brachyscleroma* is Oriental, *Lygurus* is Palaearctic, *Icaromimus* is entitely Madagascan, *Melanodolius* occurs in Madagascar and continental Africa, *Erythrodolius* (Fig. 13.39) occurs in Madagascar and Meso-America (Gauld 1997, Bennett et al. 2013), and Sheng and Sun (2011) recently described another genus in the subfamily from China and the East Palaearctic, *Laxiareola*. Bennett et al. (2013) also noted that it is possible that some genera currently still classified within the Phrudini (within Tersilochinae), e.g. *Notophrudus* and *Peucobius*, might, when material becomes available for molecular study, need to be transferred to the Sisyrstolinae.

There are almost no biological data available for the group. *Brachyscleroma* is known to be a parasite of beetles, in its case the curculionid *Apoderus quadripunctatus* (Cushman 1940). The ovipositors of some *Erythrodolius* and *Laxiareola* have a weak dorsal nodus and six weak ventral serrations and so more closely resemble those of ectoparasitoids than endoparasitoids, while the ovipositors of *Melanodolius* are more or less simple. Several species have a distinct fore tibial spine similar to those of ctenopelmatines.

Stilbopinae

The Stilbopinae, based on the European and East Palaearctic genus *Stilbops* (Fig. 13.40) and comprising only three genera, is not well defined and is probably

Fig. 13.39 Female of *Erythrodolius luteus* (Sisyrostolinae). (Source: reproduced by permission of Ilari Sääksjärvi.)

not monophyletic, and it has certainly had a chequered history. Kloet and Hinckes (1945) placed *Stilbops* in the Acaenitinae and this genus was subsequently treated as comprising a tribe in the Tryphoninae by Townes and Townes (1951), and later treated as a tribe within the Banchinae (now with *Panteles* accompanying it). J.R.T. Short (1978) noted major differences in the final larval head capsule of *Stilbops* from those of banchines and stated that *Stilbops* 'had best be placed in a subfamily of its own', but did not formally do so. Townes and Townes (1978) raised the Stilbopini to subfamily status and transferred the monotypic Chilean genus *Notostilbops* to it but removed *Panteles* back to the Banchinae (Atrophini). Wahl's (1988) study of the larval head capsule and other features led him to transfer *Panteles* also to the Stilbopinae, but he had no final instar larval head capsule to study. Nevertheless, Wahl considered the Stilbopinae as probably being monophyletic based on their large triangular female hypygium, with little indication of the apical membranous notch found in banchines. Overall, he concluded that they were 'unrelated to the Banchinae (or Ctenopelmatinae or Ophioninae *s.l.*)', although he did not suggest to what they might be most closely related. Wahl also commented that knowledge of the final larval instar head capsules of *Notostilbops* and *Panteles* might help solve the issue, but that of *Panteles* described by Quicke (2005) (Fig. 13.41f) is extremely different from that of *Stilbops* (J.R.T. Short 1957) so that it does not really say anything about their relationships.

Stilbops is by far the commonest, being distributed through the Holarctic, although its range also extends into South East Asia. Despite the features detailed by Wahl, monophyly of the subfamily is not well supported, and although members of only two of the three genera (*Stilbops* and *Panteles*) have been sequenced, analyses never place them as a monophyletic group, although they often appear as a grade leading to a much larger clade of ophioniformes. Both *Stilbops* and *Panteles* are parasitoids of Incurvariidae, a rather basal group of Lepidoptera, thus adding some support to the notion that they might be fairly closely related.

Fig. 13.40 *Stilbops japonicus* (Stilbopinae). (Source: reproduced by permission of Kyohei Watanabe) .

The biology is known for two species in some detail. *S. vetula* is an ovo-larval parasitoid of *Adela* and *Nemophora* (Hinz 1981; see also M.R. Shaw 1989) and the final instar kills the host after it has spun a cocoon and then emerges from the fully grown moth caterpillar before spinning its own cocoon. The host moths of *S. vetula* lay their eggs underneath the epidermis of the host plant and the female wasp, after examining the host oviposition site, brings her metasoma forward between her legs and oviposits into the host eggs in a fashion very like that of many aphidiine braconids attacking aphids.

Unlike *Stilbops*, *Panteles* is a larval endoparasitoid (Quicke 2005). Its eggs are dark-brown to blackish and can be seen clearly through the host integument (Fig. 13.41a). The egg has an hooked tail-like posterior end which appear to be attached to internal organs (Fig. 13.41b). Hatching takes place with the larva forcing off a cap at the anterior end of the egg (Fig. 13.41c,d); whether its small mandibles are involved is not clear. Superparasitism is common, although only one egg hatches while the unhatched ones show no signs of physical attack, suggesting that there may be some physiological suppression. The first instar larva has a very large sclerotised head capsule but only very small mandibles. Intermediate instar larvae always lie in the host haemocoel parallel to the host gut (Fig. 13.41e). The mature larva exits from the consumed host and pupates within a thin silk cocoon; it is not known whether or not there is any external feeding phase.

Not surprisingly, the apex of the ovipositor of *Panteles* is very different from that of *Stilbops* in that it has well-developed teeth on the lower valve and several dorsal serrations behind what is apparently the dorsal notch, features that are typically associated with penetrating hard substrates (and also normally of being ectoparasitic). The host substrate for *Panteles*,

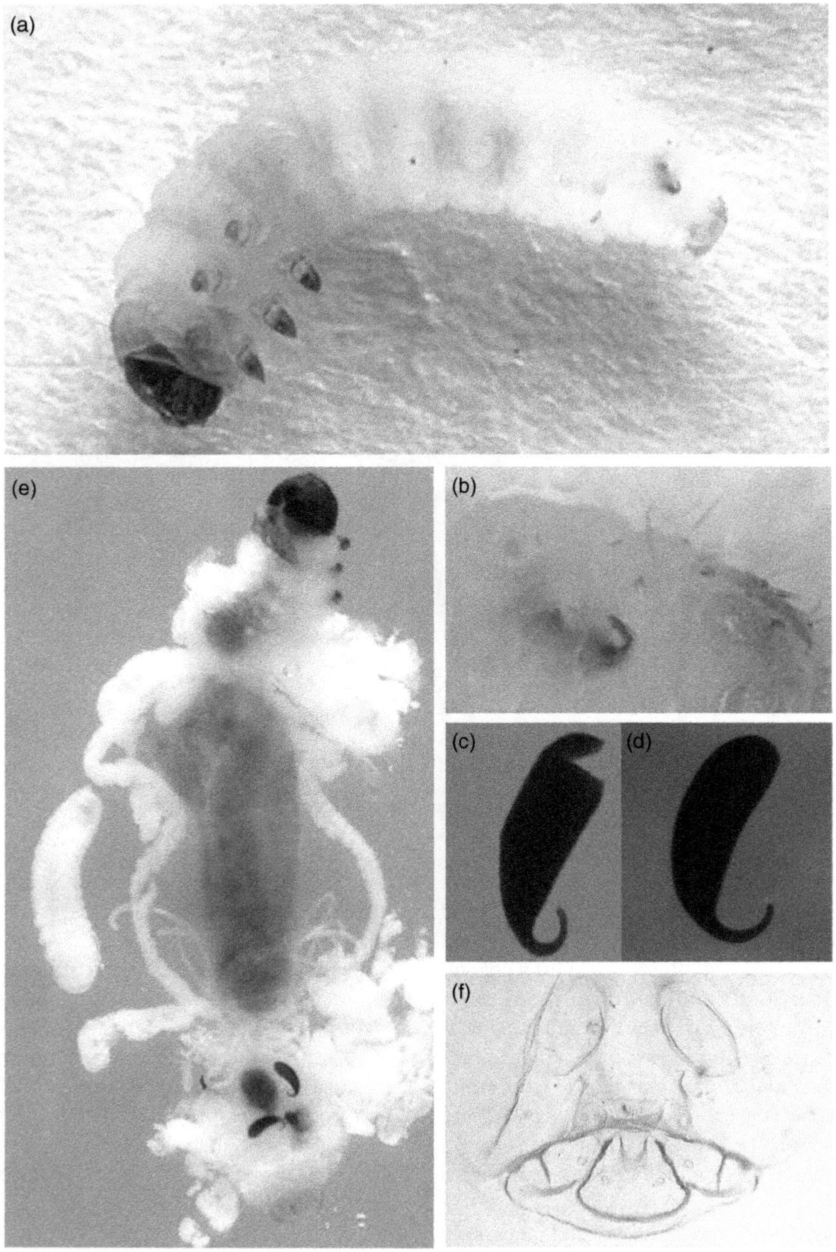

Fig. 13.41 Features of the life history of *Panteles schuetzeanus* (Stilbopinae). (a) Host, *Lampronia fuscatella* (Incurvariidae), larva showing black *Panteles* egg with hooked tail, clearly visible through the cuticle at the posterior end of the body; (b) detail of egg within host; (c, d) closed and uneclosed eggs, respectively, the latter from a case of superparasitism with only one egg having hatched; (e) dissected host caterpillar showing an intermediate *Panteles* larva more or less where found, lateral to the mid-gut; (f) final larval instar head capsule. (Source: from Quicke 2005. Reproduced by permission of Taylor & Francis Ltd.) For a colour version, see Plate 51.

galls formed by the larva of the incurvariid *Lampronia fuscatella* in birch tree (*Betula*) twigs, are fairly hard but it is not known whether the wasp 'drills' through the hard woody tissue or follows an easier access route. Very high levels of parasitism appear to be the norm.

Placement of *Stilbops* in the greater ophioniformes recognised by Quicke et al. (2009) is well supported by DNA and the basal modification of the upper ovipositor valve, which has a separate transverse basal plate articulated with the completely (except at the apex) separated lateral halves of the valve connected by a medial aulaciform rod (Fig. 3.4h,i). However, they do not possess long lateral oviducts.

Nothing is known of the host or biology of *Notostilbops*. Its ovipositor is quite fine though it has a distinct pre-apical dorsal notch and is otherwise rather smooth, features that would be most consistent with it being a larval endoparasitoid, although this form might also be found in egg–larval parasitoids.

Specimens of *Stilbops* are commonly encountered that have orchid pollinia attached to them, perhaps indicating that they are attracted by some mimetic pheromone orchid, but they are not likely to be the principal pollinators.

The British species can be recognised using Fitton (1984) and the European ones using Hinz (1981).

Tryphoninae

These wasps have often been cited as displaying many primitive features and so have often been proposed as being relatively basal in the family and, in some cases, through circular argument, are then used to polarise characters for other groups. Circular arguments should be avoided, however, and molecular data place them in a less basal position than some supposed. The subfamily is currently divided into eight tribes (Ankylophonini, Eclytini[16], Exenterini, Idiogrammatinae, Oedemopsini, Phytodietini, Sphinctini and Tryphonini), all of which appear to be supported by molecular data, with the exception of the first, which is known from only one Australian species for which no molecular data are as yet available (Quicke et al. 2009). Monophyly of the subfamily is questionable and, in the purely morphological and combined analyses of Quicke et al. (2009), the Idiogrammatinae+Phytodietinae were always recovered separated from the rest at least by Stilbopinae and Sisyrostolinae.

A few species have been involved in biological control programmes *Grypocentrus albipes* was introduced from Europe into North America (1994–96) to control the leaf-mining sawfly, *Fennusa pusilla*, but it established at only one site and failed to spread (Langor et al. 2000).

Tryphonine biology is very unusual; first, they are koinobiont ectoparasitoids, and second, they have anchored eggs which allow them to be attached firmly to the larval host (usually a sawfly larva but sometimes Lepidoptera caterpillars) through its cuticle, but the body of the egg remains external. Females of many species carry an egg externally at the base of the ovipositor in readiness (Fig. 13.42f). The egg anchor in tryphonines is of chorionic origin, i.e. it is part of the chorionic membrane that is secreted around the egg. In most cases, only the anchor of the egg passes down the lumen of the ovipositor, the body being external. although in members of the Oedemopsini there is the possibility that the egg body passes along the ovipositor internally until after the rather membranous pre-apical swelling of the lower valves (Fitton & Ficken 1990). When it hatches, the parasitoid larva remains firmly attached, with a specially modified rear end that has anteriorly directed spines that secure it within the base of the egg case (Cushman 1926b, Gerig 1960).

Tryphonine eggs are typically placed just behind the head of the host sawfly larva or caterpillar where they cannot be chewed off by the host (Kasparyan 1981), but a few place them elsewhere, for example, immediately behind the host's prolegs, but again in a place where they cannot easily be dislodged. *Erromenus calcator* is unusual in that the female oviposits into the anus of its host and attaches the egg within the rectum with the anchor (Zinnert 1969), although the larva is strictly still ectoparasitic. Eggs and larval stages have been described in detail by Clausen (1932) and Kasparyan (1981). A few published observations, such as the suggestion that in *Eclytus ornatus* the anchor forms (Pschorn-Walcher & Zinnert 1971) or the apparent lack of anchor in some (Iwata 1958, 1960), seem unlikely and certainly require confirmation (see Quicke et al. 2009). The egg anchor in most tryphonines is broadly T-shaped, but that of the Phytodietini (*Netelia* and *Phytodietus*) is narrow and rather corkscrew-shaped.

W.R.M. Mason (1967), in his study of *Exenterus*, commented that when the anchor passes down the inside of the ovipositor it is 'embedded in a thick, gelatinous mass that fills the internal cavity of the ovipositor', although the origin of this substance is unknown. Kasparyan (1981) suggested that pressure

Fig. 13.42 Adults of various Tryphoninae. (a) *Excavarus apiarus* (Exenterini); (b) *Cladeutes* sp. (Oedemopsini); (c) female specimen of *Phytodietus longicauda* (Tryphoninae: Phytodietini) showing long exserted ovipositor used to attack hosts concealed in leaf-rolls; (d) *Sphinctus nigrithorax* (Sphinctini); (e) *Polyblastus* sp. (Tryphonini) showing multiple mature eggs being retained at the base of the ovipositor with only their anchors inside the ovipositor egg canal; (f) gen. sp. indet. showing single large egg stored at base of ovipositor. (Sources: a, reproduced by permission of Cor Zonneveld; b, reproduced by permission of Håkon Haraldseide; c, d, reproduced by permission of Kyohei Watanabe; e, reproduced by permission of Mark Shaw; f, reproduced by permission of Alan Hadley.) For a colour version, see Plate 52.

of the secretion helps to push the anchor along the egg canal. Mason reasoned that it serves to seal the gap between the lower ovipositor valves where the egg stalk separates them and so helps the wasp use body fluid pressure to help move the egg anchor to the ovipositor, and it may also prevent leakage of venom; the details of the way in which tryphonines oviposit are not well understood.

Parasitoid development sometimes requires particular external conditions. Egg hatching in many tryphonines is delayed until the host has constructed its cocoon, and Clausen (1932) considered that the cue might be an increase in humidity of the surrounding air, although what would happen on a rainy day before this is uncertain. In the case of some species of *Netelia* in which the eggs would normally hatch once the host has constructed its cocoon, the eggs can be caused to hatch prematurely by keeping the host in a sealed container where the humidity becomes fairly high. In such cases, the host may not be sufficiently developed for the successful development of the parasitoid (M.R. Shaw personal communication).

First instar larvae are usually densely setose or spiny (e.g. *Eridolius*: Pschorn-Walcher & Zinnert 1971) and some have lateral tufts of long bristles (e.g. *Exenterus abruptorius*) (Clausen 1932, Kasparyan 1981). These setae aid the larva in moving on the host if necessary. Morris (1937) noted that in *Exenterus tricolor*, the first instar stays almost still on the host, imbibing some haemolymph from the site where it is attached by its mandibles, but in the second instar it is far more active and larvae search to try to kill competitors.

Townes (1973b) described three genera from 80 to 90 Mya old (Cretaceous: Coniacian-Santonian) amber from the Taimyr Peninsula, Siberia, that he believed belonged to the Tryphoninae, but these (*Catachora*, *Urotryphon* and *Eubaeus*) have now been removed to the extinct Labenopimplinae (Kopylov 2012b). Fossils placed reasonably in the subfamily are, however, known from the Florrisant of Colorado (37.2–33.9 Mya), Green River shales (50.3–46.2 Mya) and Baltic amber (54.8–33.7 Mya).

Ankylophonini

Very little can be said about this monotypic tribe from Australia, which is known from very few specimens from the states of New South Wales and Victoria. Gauld (1984a) stated that 'it is quite unlike any other tryphonine in the combination of characters it exhibits'. The first metasomal tergite is petioloate, which resembles only *Sphinctus* within the subfamily. The hind tibia has a 'shelf-like' platform at the end as in some cryptines and the Australian *Cercobarcon* (Braconidae: Trachypetinae), but its function is unknown, although it would seem well suited to transmitting or receiving vibrations. The ovipositor is robust and strongly up-curved through 90° with strongly serrated lower valves. The egg is unknown. No specimens have been available for sequencing (all specimens, as far as I am aware, date from before 1960).

Eclytini

This monotypic tribe was erected by Gupta (1988) following cladistics analysis of the Oedemopsini, which showed it to be an isolated genus and not an oedemopsine as it had hitherto been considered. Parasitoids of Argidae and Tenthredinidae, and can be reared on Lepidoptera in the laboratory (Kasparyan 1981).

Exenterini

This tribe is composed of 14 genera that are all parasitoids of sawfly larvae, mostly tenthredinids but some also on diprionids. *Exenterus abruptorius* (introduced), *E. amictorius* (introduced), *E. nigrifrons* (native) and *E. affinis* (native) are important parasitoids of *Neodiprion* species, including *N. sertifer*, in North America. *E. abruptorius* is a superior competitor to the endoparasitic ctenopelmatines *Lophyroplectus* and *Lamachus* that attack *N. serifer*, but is less efficient at locating hosts (Pschorn-Walcher 1987). During sawfly outbreaks, *E. abruptorius* and *Lophyroplectus luteator* build up to high population densities and, where the pest populations are highest, out-compete *Lamachus eques*, but the latter species does best in areas where *N. sertifer* is less dense, hence collectively these more or less complement one another, although due to the time lag are unable to prevent sawfly outbreaks.

W.R.M. Mason (1967) described and illustrated remarkable eggs of several *Exenterus* species which attack diprionid larvae. In the more specialised *Exenterus* species, the anchor is shaped like an umbrella and is attached to the egg by two stalks, one near either end, such that when the anchor has been inserted into through host's cuticle, it forms a depression in which the body of the egg sits, with the outer face of the egg hardly protruding.

Idiogrammatini

Idiogramma (the only genus of the Idiogrammatini) is unusual in that it is a parasitoid of the basal-most clade of extant Hymenoptera, the xyelid sawflies (*Xyela* spp.), and because it has long, thin mandibles that are exodont in a few species, i.e. the adult mandible can be splayed out, but unlike in alysiines, they can cross in the middle. In the Alysiinae, the mandibles are used principally to aid egress from the host fly puparium, but Burdick (1961) proposed that in *Idiogramma* they are used to separate the surrounding pine cone bracts to enable them to gain access to the host larvae. Cushman (1937) noted that they (as *Lysiognatha*) temporarily paralyse the host.

Oedemopsini

This small, easily recognised group comprises ectoparasitoids of weakly concealed, leaf-rolling and leaf-tying Lepidoptera larvae, notably Tortricidae, although several other families have also been recorded. Their exact mode of oviposition is uncertain; they nearly all have the lower valve of the ovipositor rather membranous along its mid-length (Gauld 1997). The eggs of oedemopsines are also considerably smaller than those of other tryphonines with a very small anchor (Iwata 1960). These two features led Fitton and Ficken (1990) to speculate that the egg body may pass down the lumen of the ovipositor as far as the membranous lower valve zone, rather than externally as in other members of the subfamily, but observations appear to be lacking. The development of *Hercus* (as *Eclytus*) *frontinalis* was described by Gerig (1960), who also found that this species could occasionally develop a brood of two on its tortricid (*Zeiraphera*) host and that there were five larval instars. The egg, once oviposited, naturally curves down so that the first instar larva can feed readily. Many species have the lower half of the body pale or white (Fig. 13.42b).

Gupta (1988) presented a morphological phylogeny of Oedemopsini genera and Bennett (2004) provided a key to world genera.

Phytodietini

This tribe[17] comprises just two genera in modern usage, *Phytodietus* with a rather long ovipositor (Fig. 13.42c) and the far more common, widespread and species-rich *Netelia* (=*Paniscus* of Authors[18]) with short or relatively short ovipositors (Fig. 10.8b). *Netelia* itself is currently divided into eight subgenera, and the monophyly of most of these has been supported by a morphological phylogenetic analysis (Coronado-Rivera 2009). Both genera only attack Lepidoptera, the hosts of *Phytodietus* being weakly concealed and those of *Netelia* exophytic. *Netelia* are very commonly collected at light traps where they superficially resemble *Ophion* or *Enicospilus*, usually having an ophionoid facies and often being fairly large (Fig. 10.8b). Females can give a nasty, although generally short-lived, sting and some of the larger, principally tropical and subtropical, species warrant being handled with care.

Simmonds (1947b) described the biology of *Phytodietus obscurus* (as *pulcherrimus*) in considerable detail. As with most tryphonines, the eggs are deposited at the anterior of the host caterpillar, predominantly between the legs. The egg anchor (as also in *Netelia*) is a long, irregular spiral that is inserted deeply into the host and, after oviposition, the internal part may be melanised and harden, thus securing it firmly, although the external part remains flexibly attached. Chewyreuv (cited by Cushman 1926b) reported that in *Netelia* (as *Paniscus*) the first instar has only one pair of spiracles, and the following four instars each have progressively more pairs. The larvae of both genera remain attached to the host by keeping the posterior of their body firmly inside the egg case, where anteriorly directed setae help to prevent it from being dislodged when the host moults (fig. 46 in Cushman 1926b, Anisimova 1931, cited in Kasparyan 1981).

M.R. Shaw (2001) investigated the oviposition behaviour in five of species of *Netelia* belonging to three of the recognised subgenera. Two species of the subgenus *Bessobates* did sting the host prior to oviposition and did not induce any host paralysis, whereas the *N.* (*Netelia*) and *N.* (*Paropheltes*) species investigated stung their host caterpillars, inducing brief temporary paralysis, before laying their eggs. In another species, *N. producta*, Johns and Whitehouse (2004) found that multiple stings are sometimes needed to induce sufficient temporary paralysis. In all cases the host recovered fully from the sting and there were no long-term effects of the venom on host development. All the species also displayed concurrent, non-destructive host feeding and occasionally would feed from potential hosts without ovipositing on them. In the species investigated by M.R. Shaw, the wasps used their mandibles to host feed (Fig. 13.23b), but *N. producta* only feeds from punctures made with the ovipositor in its initial stinging attempts.

The early larvae of *Netelia* hatch only partly from the egg (Fig. 13.43a) and do not feed, at least substantially, until the host has either made a pupation chamber or constructed a cocoon, whence they then feed rapidly and, after consuming the host, construct their own black cocoons within that of the host (Fig. 3.431b). Various species attacking larger hosts are gregarious with rather small broods (e.g. Johns & Whitehouse 2004).

Sphinctini

This small group of rather large-bodied wasps comprises the single genus *Sphinctus* (Fig. 13.42d) with 14 valid species that are widely distributed except in the

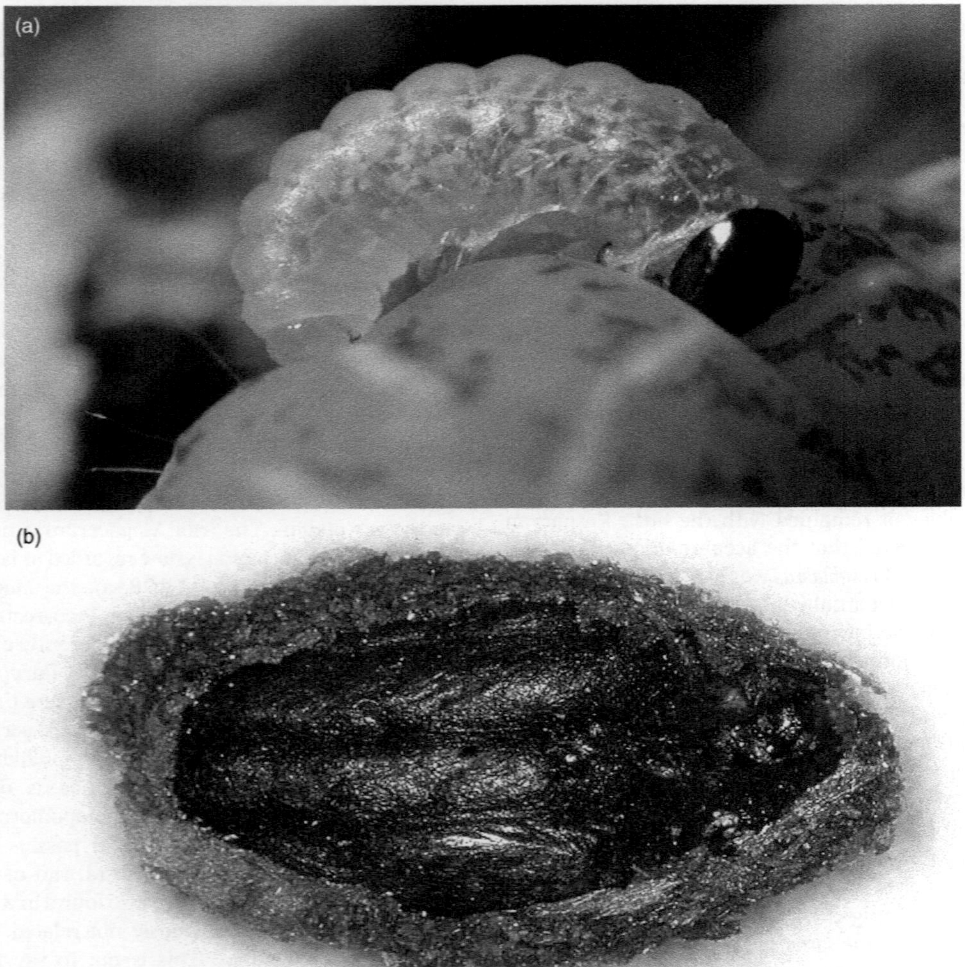

Fig. 13.43 *Netelia* spp. (Tryphoninae). (a) First instar larva of Costa Rican species, with posterior end still firmly attached within black anchored egg, probably carrying out limited feeding on the host caterpillar until the latter builds its pupation retreat; b, cocoons of *N. vinulae* within cocoon of host *Cerura vinula* (Notodontidae). (Sources: a, reproduced by permission of Kenji Nishida; b, reproduced by permission of Mark Shaw.)

Afrotropical region and Australia, and are currently unknown in North and Central America, although it seems likely that it will be found to be there given its known distribution. *Sphinctus* is a specialist parasitoid of limacodid moth larva, the European species *S. serotinus* attacking *Apoda* (Hinz 1976, M.R. Shaw personal communication and in preparation). A record from a tortricid seems highly improbable.

Tryphonini

A very good summary of the biology of the tribe was provided by Kasparyan (1981). *Polyblastus* females are characterised by their typically carrying a number (5–15) of eggs on their ovipositor (Fig. 13.42e). Larval development in these (and indeed in *Netelia*) can clearly be seen even before the eggs are laid and several tryphonines, if deprived of hosts, will dump some of their mature eggs and these may, if kept humid, split, revealing a living larva. The likely reason for this very early development is that the parasitoid larva has to be able to secure itself on the host using its strongly hooked, somewhat down-curved mandibles. Kerrich (1936) showed that if a tenthredinid host, *Dolerus*, larva was parasitised just prior to ecdysis, the anchored egg was sloughed off along with the larval skin, but if the *Polyblastus* larva has hatched and attached itself using its mandibles, it remained with the host. Kasparyan (1981) considered that the accumulation of eggs on the ovipositor of *Polyblastus* was likely an adaptation to overcome the potentially fatal risk of an egg hatching internally, while having them ready to hatch as soon as a host was found.

The biology of *Grypocentrus* was described by Eichhorn and Pschorn-Walcher (1973) and by Guévremont and Quednau (1977). *G. albipes* host-feeds on its leaf-mining host by chewing through the leaf tissue with its mandibles and 'gnawing into it'. Maximum fecundity observed in a laboratory setting was 96, with a mean just three eggs being laid per day. Interestingly, two species of *Grypocentrus* have switched to attacking the leaf-mining caterpillars of an *Eriocrania* moth (Eriocranidae) in the same situation (Jordan 1988a).

Middle ophioniformes

Ctenopelmatinae

Also referred to as the Scolobatinae in the literature following Henry Townes' terminology, this is a largely northern group, commonest where its main hosts, sawfly larvae, are most abundant (and probably most speciose), although the Westwoodiini are a largely Australian group where they attack pergid sawflies. However, the moist tropics are home to numerous species of sawflies: the Pergidae, being principally southern, and the tenthredinid subfamily Selandrinae, whose larvae feed on ferns and are the hosts of many of the tropical and southern ctenopelmatines.

Probably the best known species from a biological perspective is *Mesoleius tenthredinis*[19], which is important in the control of the sawfly *Pristiphora erichsonii* (Tenthedinidae), an important pest of larch (*Larix*) trees, and was introduced from Europe into Canada around 1910 as a control agent against it. Soon after its introduction, it achieved a very high level of control, but 40 years later it was found to be having little effect (Muldrew 1950). Interestingly, in its home range there is considerable geographic variation in the ability of the host to prevent development the *Mesoleius* eggs, with populations from the Alps being effective encapsulators, but populations from other parts of Europe hardly ever showing an encapsulation response (Pschorn-Walcher & Zinnert 1971).

The subfamily is divided into nine tribes: Chrionotini (=Olethroditini), Ctenopelmatini, Euryproctini, Mesoleiini, Perilissini, Pionini, Scolobatinae, Seleucini and Westwoodiini. The tribe Megaceriini is no longer recognised and *Megaceria* is now regarded as belonging to the Euryproctini (Gauld 1984a), the Megaceriini *sensu* Townes having been based on incorrectly identified specimens now placed in the apparently unrelated endemic Australian genus *Pergaphaga*. Euryproctines also lack the tyloid on the first flagellomere that characterises westwoodiines, scolobatines and perilissines. Ctenopelmatinae are parasitoids of Pamphiliidae.

Monophyly of the Ctenopelmatinae is questionable and there are no obvious synapomorphies for the group as a whole. Many taxa possess a small tooth-like process on the outer distal rim of the fore tibia, but similar structures are also found in a number of other taxa that are obviously not related, such as in some campoplegines. This is not to say that the species included in the Ctenopelmatinae are not closely related, they almost certainly are, but the group as a whole may be paraphyletic with respect to a number of other groups, and if the available preliminary molecular results are anything to go by, these other groups could include a substantial proportion of the 'ophioniformes'. When more genetic data become available, it is almost certain that to comply with a fully phylogenetic system, it will be split into a number of separate subfamilies.

Most ctenopelmatines have very short, not or hardly exserted, ovipositors as would be expected for taxa attacking exposed sawfly larvae (Fig. 13.44a,b,d). A few subtropical and tropical species, notably the chrionotines (*Chrionota, Olethrodotis* and *Peakelestes*) (Fig. 13.44c) and the perilissine genus *Tetrambon* (Fig. 184e), have a rather longer one that is approximately as long as the metasoma (see fig. 10.6 in Townes 1969). Species generally seem to show a high degree of host specificity (Gauld 1984a, Heitland & Pschorn-Walcher 1992).

All ctenopelmatines appear to be synovigenic (e.g. Quednau & Guévremont 1975), which is unusual given their koinobiont biology, although in a survey of Malaise trap-collected individuals from Texas a great deal of variation was observed in both egg size and egg load, although in many the chorion is moderately to

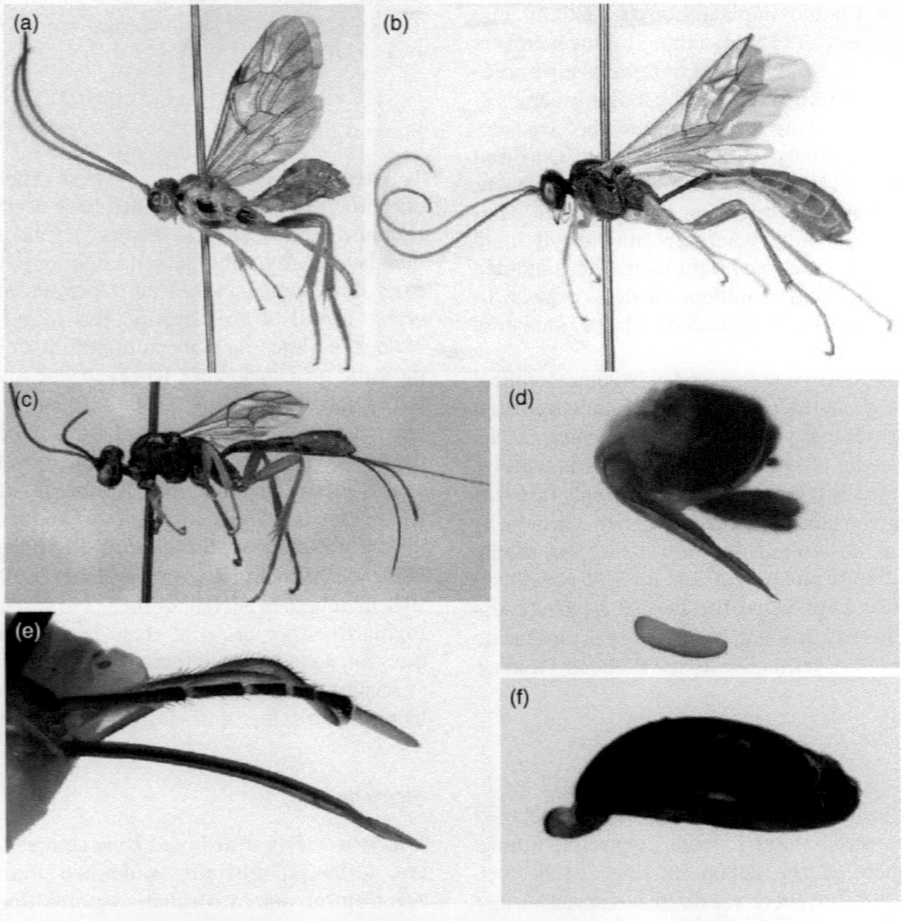

Fig. 13.44 Ctenopelmatinae including features of ovipositor and eggs. (a, b) *Scolobates testaceus* (Scolobatini) and *Hadrodactylus orientalis* (Euryproctini), respectively, showing a short, notched ovipositor typical of most members of subfamily; (c) *Olethrodotis* sp. (Chrionotini), showing a long ovipositor that is largely restricted to this tribe; (d) *Oetophorus pleuralis* (Perilissini) dissected apex of metasoma and isolated egg ; (e) ovipositor of *Tetrambon* (Perilissini); (f) *Westwoodia* egg with putative anchor. (Sources: a, b, reproduced by permission of Kyohei Watanabe; c, e, f, © Robert A. Wharton, reproduced by permission; d, image by Heather Cummins, © Robert A. Wharton, reproduced by permission.) For a colour version, see Plate 53.

strongly sclerotised (Pschorn-Walcher 1967, Cummins et al. 2011). *Lathrolestes* (Perilissini) has been observed to host-feed should liquids ooze out through the host leafminer after it has stabbed a host (Quednau & Guévremont 1975), but little is known about whether other species do. The eggs of *Euryproctus* (at least one species) and *Westwoodia* (Fig. 13.44b) are stalked and thus similar to some tryphonines (Cummins et al. 2011), and it seems likely that this is an adaptation to anchoring it in place in some internal host tissue as with the stilbopine *Panteles* and various Anomaloninae. First instar larvae are caudate but generally not mandibulate, although Heitland and Pschorn-Walcher (1992) noted that in some members of the Euryproctini that show high rates of superparasitism there is a heavily sclerotised head capsule.

Courtship and mating have been described for *Lathrolestes* (as *Priopoda*) *nigricollis* (Perilissini) by Quednau and Guévremont (1975). From a rearing point of view, their observations are interesting. Wasps were completely unwilling to mate when kept in relatively small containers and also needed bright light. They obtained the best success in large, outdoor Plexiglas cages with plenty of ventilation and containing a large number of males and females.

A few species have been involved in biological control programmes. *L. nigricollis* was successfully introduced from Europe to North America to control *Fenusa pusilla*, although because of synchrony and diapause issues it may not be equally efficient everywhere. *Lophyroplectus luteator* was introduced from Europe to Canada in the 1960s in an attempt to control the pest sawfly *Neodiprion sertifer*, but only a low level of parasitism was achieved except when the host is relatively less common, otherwise it is out-competed by ectoparasitic tryphonines of the genus *Exenterus* (Pschorn-Walcher 1987).

Chrionotini

Very little is known about these. Several of the included genera have rather long exerted ovipositors unlike other members of the subfamily (Fig. 13.44c; cf. Fig. 13.44a,b,d), but there are no host records for any of the included genera; they presumably attack hosts within leaf-rolls or similar weakly concealed situations as their ovipositors show no adaptations for 'drilling' into hard substrates. Townes (1970b) referred to this tribe as the Olethrodotini, having overlooked that a senior family group name already existed.

Ctenopelmatini

This is a small tribe with only six genera. All reliable host records (e.g. for *Ctenopelma*, *Homaspis*, *Notopygus*, *Xenoschesis*) for these wasps are from the megalodontoid family Pamphiliidae whose larvae live in webs or leaf rolls. It is absent from the Afrotropical region. *Homaspis* oviposits into a host ganglion (Eichhorn 1988). Members of some species are known to form swarms (Fig. 7.6).

Euryproctini

This tribe comprises 19 recognised genera. Hosts of euryproctines are mostly members of the Argidae. Heitland and Pschorn-Walcher (1992) found that members of this tribe have a trophamnion during its early development. and that it persists for a considerable period of time around the young larva. The biology of *Megaceria* is aberrant in that it appears to be parasitic on Lepidoptera (Geometridae and Notodontidae) (Gauld 1984a), despite some scepticism of lepidopterous hosts expressed by Townes (1970b). Other euryproctines are parasitoids of tenthredinoid sawfly larvae, for example, Tenthredinidae: Dolerini (*Hadrodactylus*, Fig. 13.44b), Diprionidae (*Synomelix*) and Tenthredinidae (*Euryproctus*, *Hypamblys*, *Mesoleptidea*). Heitland and Pschorn-Walcher (1992) reported that members of several of these euryproctine genera regularly superparasitise their hosts and asked the question as to whether there is a relationship between superparasitism and the existence of the protracted trophamnion stage, but with no firm conclusion.

Mesoleiini

This is a widely distributed tribe comprising 25 genera, although currently unknown from both the Afrotropical and Australian regions. Species attack various Tenthredinoidea, for example, Tenthredinidae (*Campodorus*, *Mesoleius*, *Saotis*, some *Lamachus*), Cimbicidae (*Himerta*) and Diprionidae (*Lagarotis*, most

Lamachus). *Lamachus* can be moderately successful in controlling the pest *Neodiprion sertifer* when its population levels are fairly low and it has been employed in biological control programmes against it (Pschorn-Walcher 1987). R.J. West et al. (1999) reported a *Lamachus* species as causing a low level (ca 1%) of parasitism in the pine pest tortricid moth *Zeiraphera canadensis*, but this would certainly be a very unusual host record.

Perilissini

This is a widely distributed tribe comprising 24 genera, although currently unknown from Australia. Most species attack Argidae (*Oetophorus*, Fig. 13.44d), Cimbicidae (*Opheltes*) Diprionidae (*Lophyroplectus*) and Tenthredinidae (*Lathrolestes*, *Perilissus*, *Priopoda*), but the widespread genus *Lathrolestes* contains some members, which, in addition to attacking leaf-mining sawflies, also attack leaf-mining Eriocraniidae moth caterpillars in similar situations (Carlson 1979, Barron 1994) – a switch also made by the tryphonine *Grypocentrus*. *Lophyroplectus* will oviposit into all the feeding stages of its *Neodiprion* host (K. J. Griffiths 1975), but the egg does not hatch until after the host has spun its cocoon, and after completing its own development and exiting the host remains it has an obligate diapause as a prepupa. It is tempting to speculate that the ability to spend a long period at the egg stage is connected both to the relatively large egg size and quite often the sclerotised chorion.

Pionini

Comprising 19 genera, this is a moderately common group found over most of the world except the Afrotropical and Australian regions. Pionines generally attack either early instar host larvae (e.g. *Rhorus* spp.; Pschorn-Walcher & Zinnert 1971), but most others including *Pion* are believed to be are egg–larval parasitoids (M.D. Cameron & Wharton 2011). *Rhorus* has been observed to oviposit through one of its host's dorsal ocelli into its head (Pschorn-Walcher & Zinnert 1971), and it would be interesting to know whether the egg was being deposited in the brain to evade encapsulation. Gauld (1984a) noted that this correlates with a larger number of mature ovarian but small eggs in pionines compared with other ctenopelmatines, probably because they lay relatively smaller eggs and maybe because their younger hosts tend to be more aggregated and hence found in larger numbers at any one site. Development of the ovo-larval parasitoid *Glyptorhaestus tomostethae* has been described by McConnell (1938). Eggs are deposited into host egg yolk cells and thus become located within the host embryo's mesenteron. When the first instar host is about to hatch, the parasitoid larva migrates to a position in the thoracic region between the gut and fat body and remains there until the host is fully grown without moulting but it does increase greatly in size. The first instar also apparently has an anal vesicle, a feature known in only a very few other ichneumonids. This species completes feeding after the host has become a prepupa in its earthen cell in the ground and spins its own cocoon within the cell where, after a period of quiescence, it completes development and overwinters as an adult within its cocoon, a biology typically associated with species that have to locate hosts early in the year in temperate regions (see, for example, *Acampsis*). There are apparently five larval instars in this species.

The definition of the tribe has relied considerably on the morphology of the ovipositor. *Hodostates*, which possesses a distinct pre-apical notch on the ovipositor, a feature typical of larval parasitoids, therefore does not fit the 'definition' of the Pionini. This was discussed at length by M.D. Cameron & Wharton (2011), who stated that '[they were] confident in rejecting *Hodostates* from Pionini as the tribe is traditionally defined'. However, this ignored the fact that in the molecular phylogeny of Quicke et al. (2009) *Hodostates* was recovered nested deep within the Pionini. Inclusion of the only Australian genus, *Austropion*, in the tribe is also not strongly supported because it lacks the apomorphic 'needle-like' ovipositor of most of the other taxa but no molecular data are available to confirm or reject its current placement.

Scolobatini

For the Scolobatini as now constituted (see Zhaurova & Wharton 2009), Gauld (1984a) considered the group also to include the Westwoodiini and considered them to be a monophyleteic group because

Fig. 13.45 Habitus of female *Seleucus cuneiformis* (Ctenopelmatinae). (Source: reproduced by permission, © C. van Achterberg, Naturalis Biodiversity Centre.)

all members possess two rather convincing synapomorphies, a tyloid on the first flagellar segment (see fig. 193 in Gauld 1997) and, in males, an enlarged proctodeal membrane. However, whilst the tyloid is found in both Scolobatini and Westwoodiini it is also present in some members of the Perilissini. Two other putative synapomorphies proposed by Gauld are less convincing. *Scolobates* (Fig. 13.44a), however, is widely distributed in the north temperate region, three genera occur in the neotropics and one in Australia. Host data are available for very few species. *Physotarsus* has been reared once from an argid and *Scolobates* has been reared from Argidae and Cimbicidae.

Seleucini

This tribe is monotypic, based on the rather strange-looking Palaearctic genus *Seleucus* with a very elongate metasoma (Fig. 13.45). *Seleucus* was placed in the Phrudinae but was moved to the Ctenopelmatinae by Kolarov (1987) in Mesoleiini and then transferred to its own tribe by Vikberg and Koponen (2000). It was subsequently recovered as the sister group to the Euryproctini by Quicke et al. (2009), well removed from the mesoleiines. The tooth on the apex of the fore tibia differs slightly from those found in most other ctenopelmatines but is probably homologous.

Newly available biological information on the group (van Achterberg & Altenhofer 2013) is in agreement with it being a ctenopelmatine in that *S. cuneiformis* has been reared from the small and rather rare tenthredinoid sawfly family Blasticotomidae, whose larvae make a stem gall in ferns, and it seems likely that the elongate metasoma is an adaptation to reach the host within the gall through holes into the gall.

Westwoodiini

Gauld (1984a) summarised some of the taxonomic issues with this group and clarified the placement of *Megaceria*, which had been included in the same tribe as *Westwoodia*[20]. Gauld treated the Westwoodiini as a junior synonym of the Scolobatini, but Zhaurova and Wharton (2009) reseparated them and transferred *Tasmabates* from the Westwoodiini to the Scolobatini. Whether the recognition of two tribes is warranted is a matter of opinion, and in the combined molecular and morphological analyses of Quicke et al. (2009) the two form a sister group.

Westwoodiines are restricted to Australia and New Guinea and all host records are from members of the pergid sawfly tribe Pergini. The genera can be recognised by the key in Zhaurova and Wharton (2009). Gauld (1984a) noted that the generic limits of the genera endemic to Australia had been vague and he greatly clarified their limits.

Mesochorinae

This subfamily is dominated by the huge genus *Mesochorus* (including *Plectochorus*, which was treated as a synonym by Wahl 1993b). Knowledge of the biology of the group is based almost entirely on *Mesochorus* species. Almost all mesochorines attack parasitoids of Lepidoptera, tenthredinoid sawflies, adult Coleoptera and nymphs of Hemiptera (Heteroptera), for example, some species attack euphorine braconids that are endoparasitic within a mirid plant bug primary host (Waloff 1967, Day 2002). There is also some indication that a few may attack parasitic tachinid (Diptera) larvae (Schwenke 1999). A few have also been recorded (very rarely) as being endoparasitoids of ectoparasitic tryphonine ichneumonoids (*Phytodietus*, *Oedemopsis*) on weakly concealed lepidopteran hosts (M.R. Shaw 1993a). How such host shifts are achieved is a mystery as it is unlikely that the secondary hosts would be giving off consistent cues about them containing an hymenopteran parasitoid. Hyperparasitism appears to be obligatory within the Mesochorinae, with one possible exception, and the few published records suggesting that mesochorines might be primary parasitoids of phytophagous hosts should be treated with extreme caution (M.R. Shaw 1993a). The exception involves *M.* (as *Plectochorus*) *iwatensis*, which was reared from larvae of the oriental fruit moth, *Grapholitha molesta* (Tortricidae), a pest of various fruit trees including pear, peach and apple, in Japan (Haeussler 1940). Despite numerous rearings from larvae in infested twigs at several localities, no evidence was found in the host cocoons that the wasp was anything other than a primary parasitoid of the moth, but this needs further investigation. Interestingly, *M. nigripes* is reported to lay a small number of eggs (2.5%) in unparasitised hosts (Coseglia et al. 1977), which suggests that were physiology and any other relative factors to permit, primary parasitism might sometimes occur in other species, albeit very rarely.

Mesochorines are one of the easiest groups of ichneumonids to recognise, with the exception of a few species. The areolet is usually relatively large and more or less a symmetrical diamond shape with its posterior veins (1Ma and 1Mb) forming nearly the same angle as 2RS (in the sense shown in Fig. 2.4b) and rs-m. The ovipositor is shorter than the metasoma, usually about one third of its length, and lacks a dorsal pre-apical notch. Males have the parameres long, narrow and 'needle-like' (Fig. 2.7d). The nocturnal genus *Cidaphus*, members of which display a typical ophionoid facies, includes some fairly large species but is relatively uncommon, especially in the north temperate region.

Internally, the long lateral oviducts of mesochorines (Wahl 1993b) indicate membership of the ophioniformes and their external morphology would probably place them within the Ctenopelmatinae or near Metopiinae. Gauld (1988a) suggested that they probably evolved from ichneumonids such as ctenopelmatines that are endoparasitoids of other Hymenoptera (specifically sawfly larvae), and the associated physiological adaptations pre-adapted them to becoming parasitic on ichneumonoids, perhaps initially within a sawfly host. Further, in the combined morphological and molecular analyses of Quicke et al. (2009), they are recovered as the sister group to the Pionini, which, being ovo-larval parasitoids, have very narrow ovipositor tips. If this relationship is true, this latter feature might well have been a pre-adaptation to ovipositing within the small larvae of endoparasitoid ichneumonids. Having thus become specialist hyperparasitoids on Hymenoptera hosts, it is likely that hyperasitism itself became the key feature and additional non-hymenopteran endoparasitic hosts became incorporated into the group's host range.

The fine, needle-like ovipositor of mesochorines enables them to probe within the haemocoel of the Lepidoptera or beetle larva to locate the larva of the ichneumonoid or tachinid parasite that will be its primary host (Blunck 1944). In Europe, Lepidoptera, sawflies and Coleoptera account for approximately 60, 30 and 10% of rearings, respectively. The life history of *M. discitergus* was described by Yeargan and Braman (1989), including its tactics for overcoming the defensive behaviour of its secondary host, the green clover worm, which, when threatened, drops from its host plant on a silk thread. The *Mesochorus* female can reel in the thread and, if the caterpillar is found to have been parasitised by its primary host, the microgastrine *Cotesia marginiventris*, it will then oviposit on it. Why this wasp behaviour and that of the microgastine *Diolocogaster facetosa* have both evolved ways of utilising the green clover worm (*Plathypena scabra*, Noctuidae) silk thread and yet similar behaviours have not been reported from other species, is a mystery. Maybe such behaviours are not uncommon but are only noticed when an association is studied intensively.

The host associations of a given mesochorine seem to be governed by the secondary host, and M.R. Shaw (1993a) noted that when a particular secondary host is attacked in its early stages by several ichneumonid or braconid primary parasitoids, nearly all of these are usually attacked by the same species of mesochorine (i.e. *Mesochorus*). As hyperparasitoids, mesochorines have generally been regarded as enemies of biocontrol programmes, both through accidental introductions and through native ones moving on to introduced parasitoids (Wahl 1993b).

'*Mesochorus nigripes*' is the best-studied species of the subfamily, which occurs in both Europe and North America, where it attacks members of the campoplegine genus *Bathyplectes*, attacking the alfalfa weevil, *Hypera postica*. Its life history and development have been described by Coseglia et al. (1977). There are four larval instars, the first of which has a large head capsule and mandibles. Although *M. curvulus* is best studied in North America, its host preference for (probably) introduced grass-feeding Miridae suggested to Day (2002) that it also might have originated in Europe, from where these hosts appear to have been introduced to North America.

Baur and Yeargan (1994a) showed that *M. discitergus* could discriminate between hosts containing the primary parasitoid *Cotesia marginiventris* and unparasitised ones within one hour of the caterpillar having been attacked, indicating that they did not rely on chemical marking scents left by the ovipositing microgastrine and instead used a response of the host to the parasitism. Crude injections into unparasitised hosts of either *Cotesia* calyx gland extract or of combined venom gland and reservoir extract showed that it was a factor in the latter that led to the detection of parasitised hosts. This ability to detect the presence of a *Cotesia* in the host so soon following parasitisation is somewhat detrimental to the *Mesochorus* since host eggs *per se* are not suitable hosts and it takes a day or so for the host egg to hatch, at which point it becomes suitable. Consequently, handling times are greater for freshly parasitised host caterpillars than for ones containing suitable larval stages.

M. nigripes occurs in both thelytokous and arrhenotokous races, North American wasps being thelytokous whereas at least the majority of European ones are arrhenotokous (Day & Hedlund 1988), and the different races show preferences for different *Bathyplectes* species. Their genetics was studied by Hung et al. (1986), who discovered, using allozyme electrophoretic methods, a surprisingly high level of heterozygosity in a Swedish sample, indeed higher than anything published for any other hymenopteran at the time. Subsequently, Hung et al. (1988) compared allelic diversity between the European and North American sexual and thelytokous populations and although slightly higher diversity was found in the sexual sample, it was not statistically higher than that of the thelytokous ones (0.187 compared with 0.103).

The first cladistic analysis of their relationships was presented by Wahl (1993b), who also described a number of new genera and proposed a number of new genus-level synonyms. *Cidaphus* together with the Chilean genus *Chineator* were recovered as the sister group to the remaining six genera. The 28S rDNA sequences of 22 species of *Mesochorus* included in the analyses of Quicke et al. (2009) showed a very large amount of sequence divergence. Whereas *Astiphromma* was recovered as a sister group to the remaining taxa, *Cidaphus* was recovered as derived within *Mesochorus*.

A key to world genera was provided by Wahl (1993b). At species level, the subfamily is taxonomically challenging and a revision of the European (Schwenke 1999) fauna revealed 300 species in seven genera, of which 231 were new species. Given how well the Palaearctic fauna is generally known, this suggests that if the species of tropical regions are comprehensively revised, thousands of new species might be expected.

Metopiinae

This is a medium-sized, cosmopolitan group with a few common and taxonomically difficult genera and one of the most frequently encountered groups. They are also, for the most part, one of the most easily recognised groups, most species having a distinctive habitus, a completely confluent face and clypeus which forms a typically rather protruding surface (Fig. 13.46e) that is sometimes produced into a point between the antennal sockets. Some Orthocentrinae can have a similar appearance also.

Many metopiines are common and some of them large and spectacular, with *Metopius* species reaching a little over 2 cm, and frequently being boldly black and yellow striped (Fig. 13.46a–c), are probably involved in mimicry complexes with aculeate Hymenoptera such

Fig. 13.46 Various Metopiinae. (a, b) Two views of wasp-mimicking *Metopius* sp.; (c) *Metopius browni*; (d) *Colpotrochia cincta*; (e) *Triclistus dimidiatus*, showing robust legs and protruding face. (Sources: a, b, reproduced by permission of Peter Waikwok Chew; d, reproduced by permission of Karol Ox; c, e, reproduced by permission of Kyohei Watanabe.) For a colour version, see Plate 54.

as eumenines, and buzz aggressively when captured, probably as a sort of '*aide memoire*' mimicry). Some of the smaller taxa, such as *Exochus* species, are noted to produce a pungent odour when captured/handled, and this presumably serves a defensive role (Townes & Townes 1959).

Little is known about their biology, most information coming from the studies of Aeschlimann (1974a,b). Iwata (1960) commented that the ovarian eggs of *Colpotrochia* are exceedingly minute and it would therefore be very interesting to know where they oviposit. Oviposition is into the host larva, often into a precise tissue, but emergence is from the host pupa, with both fully exposed and weakly concealed hosts such as leaf rollers being attacked. The ovipositors are short and barely exserted, so when concealed hosts are attacked the female parasiotoid enters the caterpillar's retreat and their usually short robust legs and protruding smooth face are likely to be adaptations to pushing their way in to the retreat to gain access (Fig. 13.46e). Most hosts of the smaller species are torticids or pyralids, but given the level of identification/association errors in many earlier publications, the 16 families have been recorded as hosts in the literature (D.S. Yu 2012).

The first instar larva is caudate and there appear to be five larval instars. Not surprisingly, given that they pupate within the pupal case of the host, they make only rather flimsy cocoons and adults emerge by cutting off the anterior end of the pupal case. The European *Triclistus species* oviposit behind the head of their host tortricid, *Zeiraphera*, caterpillar hosts, placing its egg either in or adjacent to the suboesophageal ganglion (Gerig 1960, Aeschlimann 1974a, 1975), so presumably at least sometimes evading a host haemocytic immune response. The first instar is not caudate and that of *Triclistus* develops a small anal vesicle (Gerig 1960), but this is far smaller proportionately that that seen in taxa such as the microgastrines or indeed *Banchus*. This species is also unusual in that whereas most koinobiont endoparasitoids of caterpillars hatch rapidly and then remain as first instar larvae until the host has reached a suitable developmental stage, *T. podagricus* remains as an unhatched egg until its host is about to pupate and then completes development in the pupa. Having reached full size, the wasp pupates within the host pupa and emerges quickly and spends the winter as an adult still enclosed by the host pupa. Gerig (1960) reported that *Triclistus* species go through four larval instars. *Chorinaeus funebris*, another parasitoid of *Z. diniana*, oviposits into mature larvae through the host's anus and the parasitoid egg hatches within the host gut, and the first instar then burrows through the gut wall into the haemocoel where it completes its development. If the *Chorinaeus* larva has not managed to chew through to the host's haemocoel by the time that the host moults to the pupal stage, it dies, possibly as a simple result of having been shed along with the larval exuvium since the hind gut is ectodermal in origin and chitin lined. When an *Yponomeuta* host is multiparasitised by both a *Trieces tricarinatus* and a *Triclistus yponomeutae*, Dijkerman and Koenders (1988) found that the former survives only if the host has already pupated before the latter has hatched.

Host feeding, following parasitisation, has been observed in *T. podagricus*, *C. funebris* and in a second *Triclistus* parasitoid of the same host (Aeschlimann 1974a, 1975). As with some *Netelia*, host feeding in *Chorinaeus* involves the female biting an incision in the host integument rather than using her needle-like ovipositor, but *Triclistus* females drink from the wound made by the ovipositor.

A few genera with facial shields resembling those of other metopiines have traditionally been included but with some reservations (Townes 1971). These are *Apolophus*, *Bremiella*, *Ischyrocnemus* and *Lapton*. Sequence data for the other Palaearctic genera, however, confirm Townes' suspicions in that none associate with the Metopiinae (Quicke et al. 2009), and in the future it may be deemed desirable to place several of them in their own monotypic subfamilies.

Gauld and Sithole (2002) concluded that *Apolophus*, which occurs in the New World, is correctly placed in the subfamily, and subsequently Gauld and Wahl (2006) synonymised *Apolophus* with the genus *Scolomus*, which had originally been described as a tryphonine but subsequently transferred to the Ctenopelmatinae, and similarly came to the conclusion that it was a metopiine. *Scolomus* is the senior of the two names and therefore the correct name for the genus. The latter authors also considered it possible, given the difficulty of ascertaining which subfamily *Scolomus* belonged to, that the Metopiinae may have arisen within the Ctenopelmatinae. Broad and Shaw (2005), showed that it (under the name *Apolophus*) is a parasitoid of the lepidopteran family Shreckensteiniidae which is also more in keeping with it being a metopiine. However, in the combined molecular and morphological analyses of Quicke et al. (2009, fig. 21; unfortunately, it was omitted from fig. 22) *Scolomus* was recovered away from the Metopiinae, both rendering the Ctenopelmatinae paraphyletic (Fig. 13.36).

Menier et al. (2004) described a new genus from the lowermost Eocene, approximately 50 Mya old. Many genera have wide geographic distributions, often disjunct ones, and therefore it is possible that future molecular studies may result in considerable reclassification. Identification to genus can be made using the keys of Townes (1971) and Gauld and Sithole (2002).

Oxytorinae

The interpretation of this subfamily has changed greatly because of the recent realisation that the genus *Oxytorus* (Fig. 13.47) is not at all closely related to the other genera that were placed in it. *Oxytorus* was for a long time considered a member of

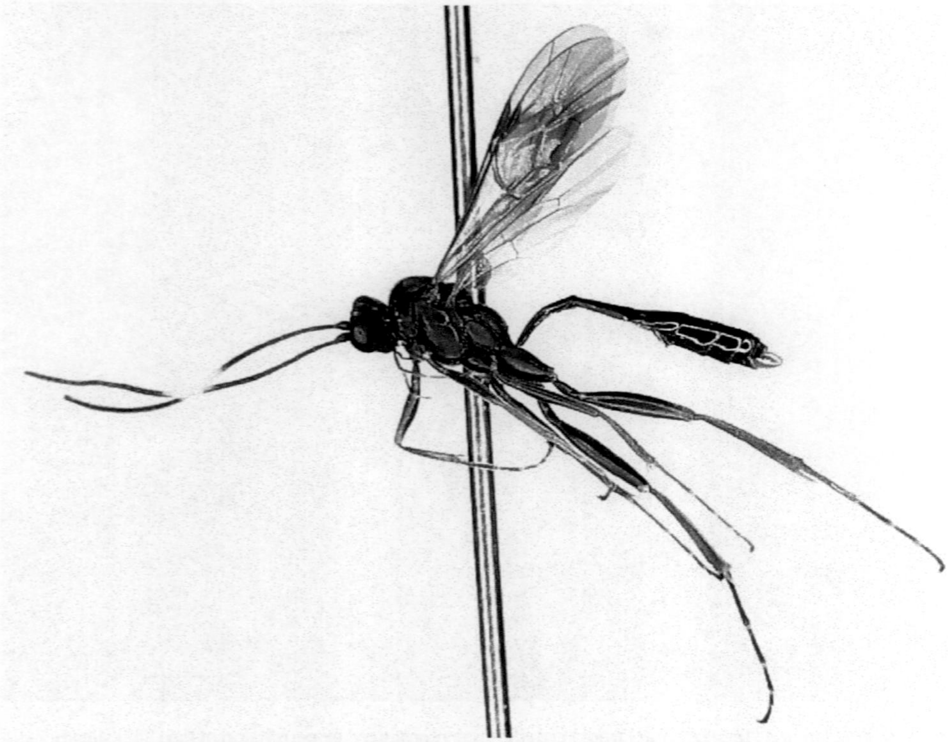

Fig. 13.47 *Oxytorus corniger* (Oxytorinae), part of paraphyletic 'ctenopelmatines'. (Source: reproduced by permission of Kyohei Watanabe.)

the Ctenopelmatinae, but Henry Townes (cited as a personal communication in Gupta 1987) considered that it probably was a Cryptine (Gelinae in Townes' terminology), whilst Wahl (1990) noted that it shared no known synapomorphy with the Cryptinae or with the Cryptinae+Ichneumoninae group. However, he could not find any characters 'indicating relationship of *Oxytorus* to other ichneumonid subfamilies' and thus restricted the Oxytorinae to only the type genus, *Oxytorus*. Other taxa included in the old sense of the subfamily (see, e.g., Wahl 1986) are now distributed among Orthocentrinae, Microleptinae, Cylloceriinae and Tatogastrinae (see Wahl 1990).

It now seems that the more traditional placement of *Oxytorus* was more or less correct, because molecular sequence data show that unlike the other taxa that used to be included with it, it is not even a member of the Pimpliformes but instead is derived somewhere within the Ctenopelmatinae assemblage (Fig. 13.36), where indeed it had originally been placed (Kerrich 1939). Its biology is unknown, although most species have a well-developed pre-apical dorsal notch on the ovipositor, strongly suggesting that they are endoparasitoids.

The single genus *Oxytorus* comprises 18 recognised species and is Holarctic, extending into the neotropics.

Tatogastrinae

This subfamily comprises the single monotypic genus *Tatogaster* which is known from only a small part of Chile and Argentina, which was initially placed by Townes (1971) in his concept of the Microleptinae that included taxa now spread among the Microleptinae, Helictinae and Orthocentrinae. Wahl (1990) moved it to its own subfamily because it shared no obvious synapomorphies with any other former 'microleptines'. The posterior distal margin of the fore wing is distinctly infuscate and sclerotised, a feature Wahl, wrongly in

Fig. 13.48 *Stethantyx loretosa* (Tersilochinae) from Peru. (Source: reproduced by permission of Ilari Sääksjärvi).

my view, thought synapomorphic with the spurious vein that runs along that part of the wing, but which is separate from the wing margin in the Ophioninae. Wahl therefore considered it most likely to be the sister group of the Ophioninae. However, DNA sequence data, albeit based upon a single gene fragment, show that *Tatogaster* is in fact no more than a derived member of the ctenopelmatine group (Quicke et al. 2009), and is perhaps close to *Cacomisthus*, another Chilean genus, and similarly to other ctenopelmatines it possesses a distinct tooth at the apex of the fore tibia. Not surprisingly, nothing is known of its biology.

Tersilochinae (Including Neorhacodinae and Phrudinae *s.s.*)

As a result of molecular and combined molecular and morphological studies (Quicke et al. 2009), the concept of the Tersilochinae has recently undergone an expansion to include two other groups, the small, predominantly north temperate genera of Phrudinae Authors (i.e. the group often referred to as 'microphrudines'), together with the Neorhacodinae. Thus the group now sensibly includes three tribes, Tersilochini (Fig. 13.48) for all of the originally included taxa, Phrudini for the 'microphrudines' and Neorhacodini for the two genera formerly included in Neorhacodinae (Fig. 13.49). The biologies of the first two are similar, being koinobiont endoparasitoids of beetles, whereas Neorhacodini attack stem-nesting aculeate hymenopterans.

Neorhacodini

These are small (ca 3 mm long) Palaearctic and New World wasps that are very poorly known. They are sometimes confused with braconids because fore wing vein 2m-cu may be largely or totally absent (Fig. 13.49c). Also, for a long time they were treated as belonging to the Banchinae, but molecular analyses have consistently recovered both *Neorhacodes* and the *Phrudus* group of the Phrudinae nested within Tersilochinae, and consequently these three subfamilies were synonymised (Quicke et al. 2009). The British

Fig. 13.49 Neorhacodini (Tersilochinae). (a) *Neorhacodes*, habitus; (b, c) habitus and fore wing of *Eremura perepetshaenkoi*. (Source: reproduced by permission, © C. van Achterberg, Naturalis Biodiversity Centre.)

species, *N. enslini*, has regularly been reared from the sphecid wasp genus *Spilomena*, which are nests in stems such as those of *Phragmites* reeds (Horstmann 1968, Danks 1971). Based on his study of the final instar larval head capsule, J.R.T. Short (1978) thought that neorhacodines would turn out to be endoparasitoids, although they have papilliform antennae and toothless mandibles, and the spiracular closing apparatus is long and is far removed from the atrium, both of which would suggest ectoparasitism. Unfortunately, no direct observations of its biology have been made and the egg is unknown. The latter is unfortunate, because molecular data suggest that Tersilochinae may be quite closely related to the tryphonine tribe Phytodietini and, if that is the case, we might expect it to have an anchored egg. The cocoon is hard, elongate and ovoid and the adult emerges at roughly the same time as its hosts do. The group was reviewed by Notton and Shaw (1998).

Only *Neorhacodes* has been included in molecular investigations to date, and studies have varied in where it was recovered. Belshaw and Quicke (2002) recovered it with a range of ophioniformes but lack of monophyly of some expected taxa makes its exact position doubtful. Quicke et al. (2009) recovered it as sister to Tersilochini+'microphrudinae' and together these clading with Lycorininae.

Phrudini

The Phrudinae as constituted previously was a very heterogeneous subfamily including a number of very small, inconspicuous, mainly northern temperate species (the *Phrudus* genus group) and also a few very large tropical ones that are among the larger ichneumonids, specifically the Afrotropical genus *Melanodolius*, members of which can reach 25 mm or more in body length. Together these were included among Gauld's (1997) 'unplaced subfamilies'. However, phrudine monophyly has long been considered far from certain, as voiced by Townes (1971), Wahl (1993c) and Gupta (1994). Wahl commented that 'Although some are superficially similar to Tersilochinae, phrudines ... are almost certainly not related to that subfamily'. A little later, Gauld (1997) identified a character that may constitute a phrudine synapomorphy, a particularly narrow proboscidial fossa, although as he noted, it is not that obvious in *Brachyscleroma*, thereby supporting the idea that they might be a natural group after all, or at least the majority of them. Apart from the question of whether the genera of large species are monophyletic with the small bodied ones including the type genus, *Phrudus* (often referred to 'microphrudines'), two genera included in the Phrudinae by Townes (1971) seemed particularly aberrant: *Brachyscleroma* from the Eastern Palaearctic and Oriental Region and *Seleucus* a rarer, monotypic genus with a very distinctive habitus (Fig. 13.45) from the central and north Palaearctic. Molecular data indicate that neither belong in the Phrudinae in its old sense and combined molecular and morphological analyses supported a grouping of the large phrudines with *Brachyscleroma*, and this led Quicke et al. (2009) to place them in a separate subfamily, the Brachyscleromatinae, which is now properly called the Sisyrostolinae (q.v.), while *Seleucus* appears to better placed among the Ctenopelmatinae.

In common with the traditional tersilochines, members of the *Phrudus* group (i.e. traditional 'microphrudines') attack beetles as larval endoparasitoids: *Earobia* is known to attack third or fourth instars of the derodontid beetle *Laricobius erichsonii* which are temporarily paralysed (Franz 1958). The egg is deposited in the host's fat body and hatches fairly quickly, but then the first instar parasitoid remains inactive until the host has entered the soil to pupate, whereupon the *Earobia* larva completely eats out its host's soft tissues, exits through its cuticle and spins its own, soil-studded flimsy cocoon, emerging as an adult the following year. Some species of *Phrudus* have extremely pectinate claws, suggesting that they may use their feet for holding on to the host (see Chapter 10, section *Pectinate claws and claws with angular basal lobes*).

A basal position within the ophioniformes is consistent with them attacking beetles, a putatively basal host group. The group (as Phrudinae) was first recorded only from South America (Chile) by Porter (1993), although the systematic placement of the genus he described, *Notophrudus*, is somewhat unclear and no molecular sequence data are available for it as yet.

The type genus *Phrudus* is widely distributed, occurring from Europe to Australia (Gauld 1984a), but Gauld noted that even with intensive surveying using Malaise traps in various parts of the world, Phrudini (as Phrudinae, although therefore also including Sisyrostolinae) are seldom common insects. Most of the genera of 'microphrudinae' were keyed by Khalaim and Yurtcan (2011). Townes (1971) provided a key to the genera; however, this also includes taxa now included in the Sisyrostolinae and *Seleucus* from the Ctenopelmatinae.

Tersilochini

Tersilochini have long been considered taxonomically difficult. The majority of species are fairly easily recognised by having the fore wing areolet open, tiny and composed of thickened veins. However, the genus *Stethantyx* (Fig. 13.48) has a much more normal arrangement. Most species are rather small, 3–6 mm, with the largest species being *Megalochus grandis* from southern Central to southern South America, and reach a body length of 15 mm.

Tersilochines in the traditional sense are nearly all koinobiont endoparasitoids of weakly concealed beetle larvae, notably members of the Chrysomelidae, Curculionidae (including Attelabinae and Scolytinae), Nitidulidae, Cerambydae and Bupresitidae, especially the first two (see Aubert & Jourdheuil 1959), although there are also records of the genus *Diaparsis* attacking gall-forming *Pontania* sawfly larvae (Kopelke 1994, Al-Saffar & Aldrich 1998) and others attacking exposed chrysomelid larvae such as *D. jacunda* on the chrysomelid lily pest beetle, *Lilioceris lilii* (Haye & Kenis 2004). *Gelanes* attacks xyelid sawfly larvae in male pine cones (Carlson 1979, van Achterberg & Altenhofer 1997), and one North American and two European species have been reared from an eriocraniid moth caterpillars (Carlson 1979, Jordan 1998b), both of the latter and probably the unidentified tersilochine in the former reference belonging to *Tersilochus*.

Several species of *Tersilochus* are considered important for the biological and integrated control of *Psylliodes* (Chrysomelidae) and *Ceutorhynchus* (Curculionidae), pests of oilseed rape (*Brassica napus*) (Barari et al. 2005 and references therein). A few species have been employed in classical biological control introductions, such as *Stethantyx* (Kerrich 1961). All species are solitary, although superparasitism occurs regularly in some species with supernumeraries apparently being killed by the surviving larva through physical attack (H.L. Parker et al. 1950), although first instars no not have particularly large mandibles.

Many species are united by the behaviours of their hosts forming pupation chambers in the soil in which the wasp completes its development (Cushman 1916), and some species, possibly most, of *Barycnemis* use their robust ovipositors to attack hosts in the soil. The ovipositors of other tersilochines are variable and within the genus *Phradis* alone there are species with typical pre-apical dorsal notch (e.g. *P. minutus* and *P. nigritulus*), some with the apex strongly narrowed and needle-like (e.g. *P. terebrator*) and some with multiple dorsal nodi (e.g. *P. punctus*) indicating a considerable degree of adaptation to attacking hosts in various situations. Several species (e.g., *Sathropterus*, *Stethantyx sinuata* and some *Probles* species) have ovipositors formed into one or two distal arches that are presumed to function as a steering mechanism, although I can find no published observations of their movement capabilities. The needle-like ovipositors of some of these indicate oviposition into very early host stages and the multiple nodi of punctus is very reminiscent of some *Theronia* spp. (Pimplinae) that are pupal parasitoids of Lepidoptera, so it would be very interesting to obtain more detailed information on the host stages attacked by these and on the workings of their ovipositors.

The biologies of various species have been described by Cushman (1916), H.L. Parker et al. (1950) and Osborne (1960). Tersilochine eggs are quite variable. Some are more or less colourless, but those of a number of species are very dark coloured. Eggs may be deposited into the host's haemocoele but Dysart et al. (1973) showed that those of two species of *Diaparsis* have a 'lateral' protuberance that is used to anchor them to tissues within the host thorax and other species place their egg within body wall tissue, which may be an adaptation to avoiding encapsulation. Similarly, Jordan (1998b) illustrated the egg of the *Tersilochus* sp. n. being anchored with a dorsal protuberance through the *Eriocrania* caterpillar cuticle, the anchor remaining external and the body of the egg internal and split with the head of the early first instar larva just protruding. However, the egg of *Tersilochus curvatus*, also illustrated by Jordan and also attacking *Eriocrania*, in addition to three other species illustrated by Osborne (1960; two as *Isurgus*) are simple, with no anchor. Such evolutionary plasticity in the form of the chorion is interesting. It is also tempting to speculate that if the external anchored eggs of the Tryphoninae represent an intermediate stage in the evolution of endoparasitism in the ophioniformes branch of the Ichneumonidae, the condition present in the *Tersilochus* sp. n. of Jordan could be the next stage. However, as there is intrageneric variation within *Tersilochus*, the egg anchor, as well as the anchor in *Diaparsis* sp., of one species cannot be truly plesiomorphic, but could reflect that the group generally maintains the genetic mechanism for creating chorionic anchors from an ancestral group in which all members did.

Developmental stages for several species of *Tersilochus* (=*Isurgus*) attacking *Brassicogethes* (=*Meligethes*) (Nitidulidae) were described by Osborne (1960). In *T. heterocerus*, the tough black egg does not hatch until the host is fully grown and then splits at its anterior, blunter end and the first instar larva protrudes and grows while keeping its posterior end within the egg shell. The eggs of a second species (*T.* sp. A) are not pigmented and swell somewhat before the first instar larva exits the chorion completely when the host is fully grown. There appears to be no trophamnion. Both of these first instars are caudate with mandibles, but the mandibles are not particularly large, suggesting that they are used for feeding rather than aggression (but see above). Slightly different larval features were described by Montgomery and DeWitt (1975) for a *Diaparsis* sp. attacking the cereal leaf beetle, *Oublema melanopus* (Chrysomelidae), and they mentioned that the early instars have two pairs of abdominal pseudopods. The final instar larva exits the largely consumed host and finishes feeding externally (H.L. Parker et al. 1950, Osborne 1960) and then constructing subspherical, rather thin-walled cocoons. At least the majority of temperate species are univoltine, overwintering either as an adult within its cocoon (e.g. *Phradis* and *Tersilochus* species; Jourdheuil 1961) or as a diapausing larva (e.g. *Diaparsis*), although Jourdheuil wrote that some *Aneuclis* species might be bivoltine and host-alternating.

This has long been considered a taxonomically rather difficult group. Townes (1971) keyed the world genera based in part on Klaus Horstmann's revisionary work on the European species (Horstmann 1971, 1981). The Australian genera were treated by Gauld (1984a). The tropical fauna is rather poorly known. More recently, Khalaim and Yurtcan (2011) provided an updated key to the European genera.

Higher ophioniformes

Below I treat the *Belisica* group under the Cremastinae where they are currently classified, and similarly, *Skiapus* is treated as an ophionine to where it was transferred by Quicke et al. (2005a), although I suspect they may both warrant separate subfamily status on further study.

Anomaloninae

Anomalonines are cosmopolitan and moderately frequently encountered medium-sized to large wasps and are well supported as monophyletic based on morphology (Gauld & Bradshaw 1997), although DNA data sometimes suggest paraphyly (Quicke et al. 2009). They are currently divided into two tribes following Gauld's (1976a) amalgamation of the four originally recognised by Townes (1971). These are the Anomalonini, comprising just the genus *Anomalon* (Fig. 13.50a), although with *Neogreeneia* sometimes considered separate, and the Gravenhorstiini (Therionini) (Fig. 13.50b,c), which includes the remaining 42 genera (Gauld 1976a, D.S. Yu 2012). Anomalonini have long exserted ovipositors and are now considered to comprise the single genus *Anomalon* with various synonyms[21]. These wasps attack Coleoptera larvae, primarily those of elaterids and tenebrionids, two groups utilised as hosts by very few other ichneumonoids or, indeed, parasitic wasps in general. Dasch (1984) additionally recorded one North American species as attacking a noctuid moth.

Gravenhorstiines attack a wide range of Lepidoptera, larger species often utilising Noctuidae, Geometridae, Lasiocampidae and Notodontidae, but others attack a range of 'microleps', including Gelechiidae, Oecophoridae, Pyralidae, Tortricidae and Yponomeutidae, and those species with slightly longer ovipositors (a little less than the hind tibia) attack semi-concealed hosts such as pyralids and immids in leaf rolls. A few species attack well-known crop pests such as *Helicoverpa* species (Noctuidae) and may be responsible for a significant amount of host mortality. Attempts to mass rear *Heteropelma scaposum* in Australia for release against *Helicoverpa* ran into the slight problem that in culture the sex ratio increased fairly quickly, probably indicating that these wasps have slCSD (Johns & Whitehouse 2004). In Europe, *H. megathrum* (=*calcator*) is an important parasitoid of the bordered white, also called the pine looper moth[22], *Bupalus piniaria* (Geometridae), an important pest of pine trees (Plotnikov 1914).

The tiny eggs of anomalonines possess anchors or holdfasts very similar to those of tryphonines (Iwata 1958, Gauld 1976a,b), but the whole egg is placed internally within the host. The anchor is used attach the egg to an internal organ (e.g. gut) or tissue (muscle) of the host, although its body remains surrounded by haemolymph (Tothill 1922, Gauld 1976a). Oviposition by gravenhorstiines involves manoeuvring the metasoma forward between the wasp's legs, just as in aphidiines, and 'loading' an egg into the ovipositor 'into the swelling immediately behind the dorsal notch' (Gauld 1976a, 1997), i.e. to a position near the valvillus (see Chapter 3, section *Number, position and possible functions of ovipositor valvilli*). The main source of information on their development comes from Tothill (1922), according to whom the first instar larva upon eclosing remains partly surrounded by a trophamnion, something that is very rare in other subfamilies of Ichneumonidae, although common among several groups of Braconidae. Gauld and Bradshaw (1997) suggested that it may be involved in circumventing the host encapsulation response, but without presenting any evidence. The first instar is caudate but the size of the mandibles appears to differ between genera. Slobodchikoff (1974a) reported large ones in *Therion circumflexum* whereas Plotnikov (1914) noted that the first instar of *Heteropelma megathrum* has only minute mandibles. Ecdysis to the second instar does not occur until after the host has pupated and development is then rapid, but this means that there is a very protracted first instar that, in temperate species, can last for about 6–9 months (Tothill 1922, Porcelli 1988). The number of larval instars is uncertain. The final instar spins a flimsy cocoon within the host pupa. Townes noted the similarity in final instar larval head capsule morphology between Anomaloninae and Metopiinae, but, rightly suggested that this might be convergent due to both groups making only flimsy cocoons within the pupa of the host. Nothing is known about their venom chemistry or effects on the host immune system.

Adults are usually fairly easily recognised by a combination of a fairly distinctive *Gestalt* and a reticulate, often posteriorly produced, propodeum. Although the habitus is a little like that of some ophionines and campoplegines with a long, laterally compressed metasoma, the wings often appear rather short and in flight many, especially males, fly with their hind legs conspicuously splayed outwards and backwards and their antennae outstretched forwards (Slobodchikoff 1973), enhanced by the often pale hind tarsi (Fig. 13.50c), which can be rather expanded and with the metasoma held rather high (Gauld & Mitchell 1977, Gauld & Bradshaw 1997). Most species are medium sized but some are rather larger and can occasionally deliver a fairly mild sting, although they are nor normally too aggressive, unlike some ophionines. I have never confused ophionines with anomalonines in flight. *Ophionellus* species, from the New World and especially the Neotropics, are unusual in that they have rather small wings for its body length (Fig. 13.50b), although they still fly well and the distal venation is highly reduced and they lack fore wing vein 2m-cu, the distal wing membrane having numerous flexion line-like folds.

For a long time, the anomalonines were thought to be closely allied to the Ophioninae because both tend to be slender insects with laterally compressed metasomas and they both share the apparent loss of fore wing vein 2RS. However, there are many differences between them and their similarity to ophionines may be homoplastic (but see Fig. 13.35c).

The first instar larva of Gravenhorstiini is caudate with minute mandibles (Plotnikov 1914), it is similar to that of *Panteles*, suggesting that physical fighting is not used in cases of superparasitism.

The general flight and mating behaviour has been briefly summarised by Gauld (1997). Mating often

Fig. 13.50 Anomaloninae species. (a) *Anomalon* sp., showing relatively long ovipositor; (b) *Ophionellus* sp. from Peru, showing small wings with reduced venation; (c) *Habronyx insidiator* in the hand. (Sources: a, reproduced by permission of Bernardo Navarette; b, reproduced by permission of Ilari Sääksjärvi; c, reproduced by permission of Kyohei Watanabe.) For a colour version, see Plate 55.

takes place in flight although the coupling pair may soon settle and some species will readily fly *in copulo*, e.g. *Trichomma cnaphalocrosis* (Camarao & Morallo-Rejesus 2003). There is no pre-mating period. Males of *Heteropelma* have a depressed area on the second tarsal segment that is probably an exocrine gland or gustatory area.

Campopleginae

This is a large subfamily that has been one of the most widely investigated ichneumonids from a biocontrol point of view, when it comes to both introduced control agents and indigenous ones (Gauld 1995). Because of this, there is a considerable body of biological information available for a range of species and genera. Important biological control examples include such well-known taxa as *Bathyplectes curculionis* and *B. anurus* on the alfalfa weevil, *Hypera postica*, *B. tristis* on the clover weevil, *H. punctata*, *Campoletis sonorensis*, a parasitoid of the *Heliothis* spp., *C. chlorideae* on oriental army worm, *Mythimna separata* (Noctuidae), *Diadegma semiclausum* and relatives, parasitoid of the diamond back moth, *Plutella xylostella* (Plutellidae) (Aziz et al. 2000), *Eriborus terebrans* on the European corn borer, *Ostrinia nubilalis* (Crambidae), *Olesicampe benefactor* against larch sawfly, *Pristiphora erichsonii* (e.g. Drooz et al. (1985), *O. geniculatae* against mountain ash sawfly, *Pristiphora geniculata* (e.g. Quednau 1990) and *Tranosema restrale* against the eastern spruce budworm, *Choristoneura fumiferana* (Tortricidae) (Cusson et al 1998). In the last of these cases, the availability of alternative tortricid hosts may be very important for its effective control of the target species. Other endemic species may also cause high levels of mortality of pest species, for example, *Eriborus argenteopilosus* against cotton bollworm, *Helicoverpa armigera*.

Venturia canescens (often mistakenly referred to as *Nemeritis canescens* or *Devorgilla canescens* in older literature because of a simple misidentification of the latter genera) has been a virtual laboratory rat for much parasitoid research and is important in the control of pyralid stored product pests, notably *Ephestia kuehniella* and *Plodia interpunctella*. Much of what we know and surmise about campoplegines comes from these relatively few species; however, based on the available phylogenic studies including a range of members of the subfamily, they cover it rather well.

Campoplegines are koinobiont endoparasitoids, most of them attacking exposed or weakly concealed Lepidoptera larvae, although a few, e.g. *Bathyplectes*[23], *Nepiesta*, *Pyracmon* and *Rhimphoctona*, which belong to the *Bathyplectes* group (see below), attack beetle larvae, while *Olesicampe* and some *Dolophron*, *Lathrostizus* and *Sinophorus* species attack tenthredinid, cimbicid and pamphiliid sawfly larvae; *Dusona* has also been recorded from a tenthredinid (Lejeune & Hildahl 1954), but this requires confirmation and, in any case, is not a common association. *Pyracmon* has an unusual biology in that it attacks soil-dwelling beetle larvae (Elateridae and Artematopidae) (Barron & Walley 1983). Most species of *Nemeritis* in the correct sense[25] attack Rhaphidioptera (Neuroptera) larvae (Aspöck 2002), but some species also attack beetles. Apart from members of the *Bathyplectes* group, few other genera specialise on Coleoptera, for example, *Lemophagus* is a specialist on chrysomelid larvae and *L. curtus* has been released in the USA to control *Oulema melanopus*, and *L. errebundus* has been suggested as a suitable candidate for biological control introduction against *Lilioceris*, also in North America. Other reliable records from Coleoptera include some *Nemeritis* species and *Lathroplex anthreni* attacks larvae of the museum beetle, *Anthrenus museorum* (Dermestidae) (Vikberg 1999). *Enytus* mostly attack Lepidoptera larvae but are recorded also from Coleoptera and Hymenoptera, although these records seem doubtful and need confirmation. One species of campoplegine has been reared from the larva of the terrestrial trichopteran (see van Achterberg 2002).

Most species are solitary, but one European species of *Olesicampe* is gregarious with broods of 20–48 individuals emerging from a single host (M.R. Shaw 1999a). Most species appear to be fairly host specific. They mostly attack early instar hosts and kill them often before they reach maturity (see Fig. 13.51a), although some members of a few genera have recently been found to be ovo-larval parasitoids. *Hyposoter horticola*, a parasitoid of the Glanville fritillary butterfly, *Melitaea cinxia*, that has been much studied on the islands of the Aland archipelago in the Baltic Sea, is one of the few exceptions in that it is an obligate egg–larval parasitoid but oviposits into the near fully developed first instar host caterpillar just prior to its emergence from the egg (Castelo et al. 2010), and this may indicate that the eggs produce a more reliable host location signal. Another instance of ovo-larval parasitism by a campoplegine was provided by Shimazaki et al. (2011), a member of *Melalophacharops* attacking a lycaenid.

The embryology of *Lathrostizus lugens* (as *Angitia vestigialis*) was described by Ivanova-Kasas (1956). Larval development and morphology have been described for *Bathyplectes curculionis* (Bartell & Pass 1978), *B. anurus* (Bartell & Pass 1980), *B. tristis* (Pottler & Coles 1962), *Campoletis sonorensis* (D.D. Wilson & Ridgway 1975), *Campoplex haywardi* (Leong & Oatman 1968), *Diadegma* (as *Horogenes*) *chrysostictos* (Fisher 1959), *Enytus* sp. (Mazanec 1990), *Lemophagus curtus* (Montgomery & DeWitt 1975), *Diadegma semiclausum* (F. Huang et al. 2009), *D. argentopilosa* (Sathe 1990), *Hyposoter didymator* (Bahena et al. 1999; a very nice SEM study), *H. exiguae* (Puttler 1961) and *Sinophorus* (as *Eulimneria*) *crassifemur* (W.R. Thompson & Parker 1930). There appear to be three larval instars (e.g. *Bathyplectes*, *Campoplex*, *Enytus*, *Hyposoter*), but larger numbers are reported for some, e.g. four larval instars in *Diadegma* (F. Huang et al. 2009) and *Campoletis flavicincta* (Patel & Habib 1987) and five for *D. argentopilosa* (Sathe 1990), *D. chrysostictos* (Fisher 1959) and *B. anurus* (Bartell & Pass 1980), but care should be exercised as the number of instars can easily be overestimated due to changes in apparent larval morphology without actual moulting and growth.

Eggs are laid in the host's haemocoele where they float freely. First instar campoplegine larvae have a well-developed caudal appendage which, in most species, gradually shrinks relative to the rest of the body and only a tiny remnant usually remains in the final instar. Some taxa have highly mandibulate first instars, e.g. *Olesicampe* (Quednau & Lim 1983), and these probably eliminate supernumeraries by physical combat, but others have rather small mandibles; for example, in *Hyposoter*, there is a large head capsule in the first instar but it has a suctorial mouth with the mandibles are not or hardly visible (Bahena et al. 1999). Others such as *Enytus* are rather intermediate, but still capable of eliminating supernumeraries by combat. W.R. Thompson & Parker (1930) reported that *Sinophorus* (as *Eulimneria*) *crassifemur* first instars may eliminate competitors through release of a 'cytolytic' enzyme that the first larva to hatch secretes, and Fisher (1959) found that whereas first instar larvae of *D. chrysostictos* fight one another using mandibles, later instars eliminate supernumerary first instars through humoral inhibition. The second and third instars are usually more or less hymenopteriform and spiracles sometimes do not develop until the parasitoid reaches its final instar (e.g. Mazanec 1990). The caudal appendage is retained as a substantial structure to the final instar of several campoplegines. In *Phobocampe*, rapid movements of it enable them to jump their cocoons until they reach a more secluded place (Gillespie & Finlayson 1979), and Day (1970) showed that *Bathyplectes anurus* cocoons jumped far more when exposed to either high temperatures or very bright light. Day also showed that the jumping cocoons of *B. anurus* were five times less susceptible to parasitism by the pteromalid hyperparasitoid *Dibrachys cavus* than were those of its non-jumping congener *B. curculionis*, and M.R. Shaw (2009b) also postulated that the mobility of some campoplegine prepupae affords them a degree of protection from pseudohyperparasitoids, although at least one species of pimpline, *Itoplectis clavicornis*, appears to be specialised in attacking them.

Final instar larval cephalic structures have been surveyed extensively, notably by Finlayson (1975) and supplemented by Barron & Bisdee (1977), J.R.T. Short (1978) and Wahl (1991), and a detailed SEM study of *Hyposoter didymator* larval development was presented by Bahena et al. (1999).

The tendency for campoplegines of some genera (notably *Campoletis*, *Casinaria*, *Charops*, *Hyposoter* and *Scenocharops*: Gauld & Janzen 2004) to attack early instar hosts whose development they curtail means that the final instar larva exits its host often at the place where it was feeding rather than in the concealed places that most exposed Lepidoptera caterpillars seek out for pupation. This means that the parasitoid has to spin its own cocoon often in a rather exposed and vulnerable place, and campoplegines have evolved a number of modified cocoons that maximise their survival chances. Many campoplegine cocoons are banded black and white (Fig. 13.51); some are made suspended from vegetation by a silk thread, hanging freely in a manner like that of many Meteorini, probably reducing vulnerability to bird and ant predation. Several species spin their cocoon within the remains of their caterpillar host; some create false cocoons which may distract a visual predator's attention from the real one (*Hyposoter parorgyae*: Finlayson 1966). A few species that attack later instar hosts do not complete their development until after the host has spun its own cocoon (rather more typical of ichneumonids) and the egressed campoplegine larva may then spin its own cocoon within that of the host (e.g. some *Campoplex* species: Leong & Oatman 1968); a few other species complete development after the host has pupated and may construct their cocoons within the remains of the host pupa (*Campoplex* sp.; Carlson 1979). The cocoon of *Scenocharops* is nearly spherical, translucent and very smooth and Gauld & Janzen (1994) postulated that its texture and shape might make it hard for pseudohyperparasitoids to hold on to in order to penetrate it with their ovipositors.

Cocoon construction in *Bathyplectes curculionis* was described by Cross and Simpson (1972). Depending on whether a diapausing or non-diapausing cocoon is being constructed, it takes 24–30 and 32–42 h, respectively. The cocoons are typically striped, with a medial equatorial pale region (Fig. 13.51a) and often with flecks of darker colour within this (Fig. 13.51b) and it has been suggested that this breaks up their outline and, in some cases, may give them a resemblance to bird droppings and so afford them some protection from visual predators.

Courtship and mating have been described for *Campoletis sonorensis* (Vinson 1972b), *Bathyplectes anurus* (Gordh & Hendrickson 1976), *B. curculionis* (Dowell & Horn 1978) and *V. canescens* (van Santen & Schneider 2002), the last authors presenting a detailed ethogram and noting the highly stereotyped behaviour involved. Vinson found that 'female tainted air' to lower the threshold for males to exhibit courtship behaviour, although the full response seemed to require actually seeing the female. Barrows (1976) found that copulation in *Hyposoter fugitivus* lasts for approximately 17 min, which is very long for ichneumonoids, and that the pair also assume an unusual posture with the

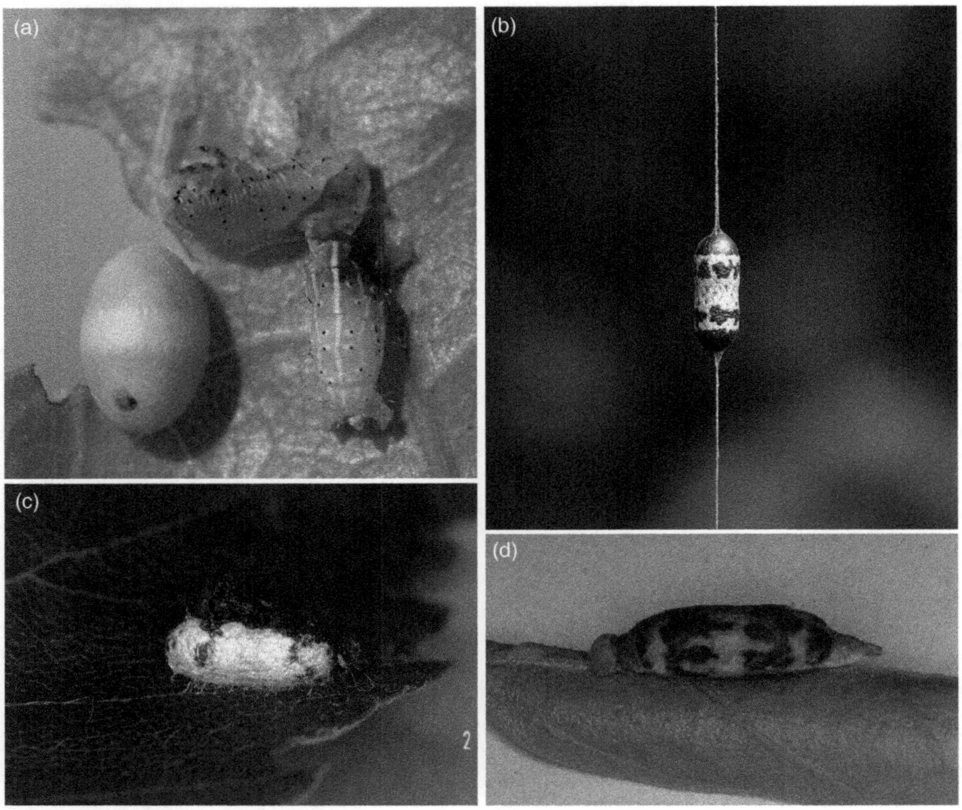

Fig. 13.51 Photographs of various campoplegine cocoons. (a) *Scirtetes robusta* (Campopleginae) with host *Cosmia* sp. larval remains, showing typical pale equatorial stripe (identified by M.R. Shaw); (b) suspended cocoon of a campoplegine species; (c) striped, unsuspended cocoon formed underneath host caterpillar remains; (d) cocoon formed within host larval skin, probably by *Hyposoter* sp. (Sources: a, reproduced by permission of Marion Friedrich; b, reproduced by permission of Kenji Nishida; c, d, reproduced by permission of Mark Shaw.) For a colour version, see Plate 56.

male and female facing forwards, side by side – the male on the right. *B. anurus* also has an unusual mating posture with the male 'standing' on his hind tarsi, the tips of his wings and the dorsum of his metasoma, which is arched forward underneath his body. In this species copulation lasted approximately 4 min. Gordh and Hendrickson also reported that females dissected very soon after copulation had a capsule in their genital tract which they presumed to be a spermatophore.

Females of many species probably emerge with only a proportion of their eggs mature. Khatri et al. (2009) showed that in *D. semiclausum*, there is a sharp increase in the number of mature eggs between 1 and 2 days post-eclosion (Fig. 4.2). However, the preoviposition period of the majority is less than 1 day (e.g. Leong & Oatman 1968).

Despite the importance of some species in biological control and interest in their polydnaviruses, there have been few studies of their mating or egg maturation. Khatri et al. (2009) showed that female *Diadegma semiclausum* do not have any mature eggs at the time of emergence but both sexes were able to mate within 12 h post-emergence, although females did so more readily than males, perhaps because they could store mature sperm until their eggs were mature whereas it would be of no benefit for a male to mate if it did not already have viable sperm available.

The ultrastructure of the venom apparatus has been studied in *Diadegma* (Pan & Chen 2003), ovipositor sensilla have been surveyed in *Venturia canescens* by Ganesalingam (1972) and Shah (2012) and possible ovipositor mechanics were discussed by Shah et al. (2012).

Two species of *Lemophagus* that attack lily beetle, *Lilioceris lilii* (Chrysomelidae), larvae have been shown to differ in their responsiveness to host semiochemicals (Schaffner & Müller 2001), with one species responding positively to extracts from larvae and their faecal shield, whereas the other only shows a strong response to shield-related volatiles.

Interspecific hybridisation has been demonstrated in the laboratory between *D. semiclausum* ♀ and *D. fenestrale* ♂ using sequencing of mitochondrial (COI) and nuclear (ITS2) genes, but this has not been demonstrated in the field situation where both species broadly overlap geographically and utilise the same host (A.P. Davies et al. 2009).

All campoplegines studied to date make use of either polydnaviruses or, in the sole known case of *Venturia canescens*, virus-like particles that lack DNA but which are almost certainly derived from polydnaviruses, to provide protection from host immune defences (see also microgastroid group of Braconidae and ichneumonid Banchinae). Wahl (1987) considered *Venturia* to be a derived member of what, at that time, he called the *Campoplex* group and subsequent molecular and morphological phylogenies place *Venturia* as a relatively derived campoplegine (Fig. 13.52), and therefore its particles are presumably derived from proper polydnaviruses through loss of the genes encoding the circular DNA molecules found in all other polydnaviruses. True polydnaviruses have now been studied in a number of campoplegines, e.g. *Campoletis, Diadegma, Eriborus* (Fig. 13.53), *Hyposoter* and *Tranosema*, and they are known also to occur in many other genera and species (Table 9.1) (Dib-Hajj et al. 1993). *Hyposoter didymator* is interesting in that whereas it larva requires polydnavirus to have been injected into the host to avoid encapsulation, its egg does not (Dorémus et al. 2013a), indicating that its surface must not be recognised as foreign by the host immune system. How widespread this camouflage is, is not yet known.

The classification of and relationships within the subfamily have long been considered difficult and generic boundaries are often ill-defined. There have been few attempts, therefore, at creating phylogenies. Four tribes were recognised by Townes (1969), Campoplegini, Hellwigiini, Nesomesochorini (as Nonnini) and Limneriini (as Porizontini), and Gupta (1987) erected the Cymodusini for five genera previously included in the Limneriini.

Wahl (1991) combined evidence from larvae and adults and, rather than referring to the tribes, recognised five genus groups and rejected the idea of any formal tribe-level classification. The *Bathyplectes* group, comprising five genera characterised within the Campopleginae by possessing a tooth at the apex of the fore tibia, a character otherwise more associated with the Ctenopelmatinae, was placed basally in his analysis, suggesting that perhaps the common ancestor of the subfamily may have been a parasitoid of concealed coleopteran hosts. Wahl also recognised a *Gonotypus* group, a *Nemeritis* group and a *Menaka* group, based on only a few taxa, and placed all other of the included genera in Campopleginae by Townes (1969: as Porizontinae) in a very large *Dusona* group, which by default, therefore, was left including four very aberrant genera, viz. *Hellwigia, Skiapus, Nonnus* and *Chriodes*

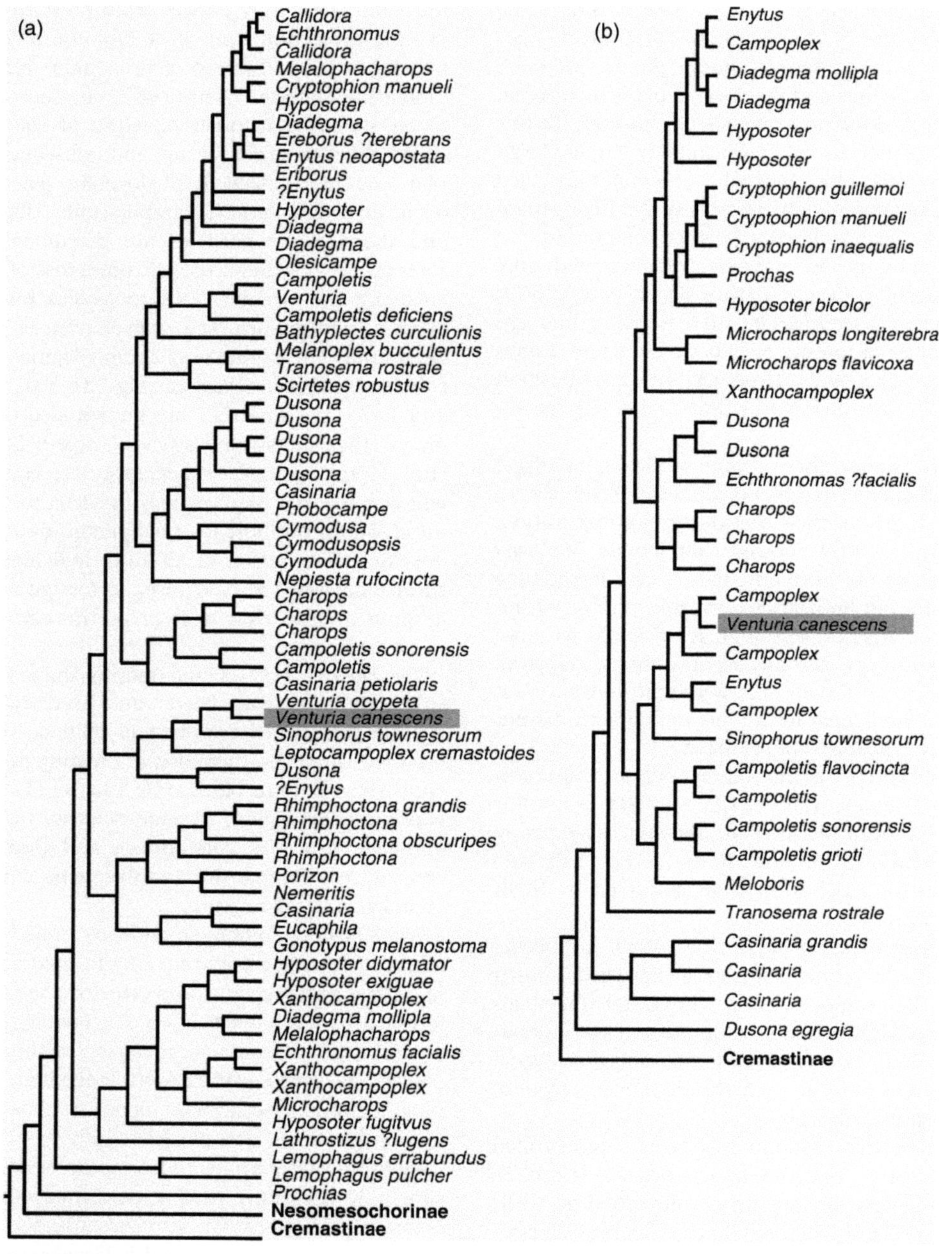

Fig. 13.52 Recovered phylogenetic relationships of genera of Campopleginae showing position of *Venturia canescens*. (a) Maximum parsimony tree from combined elided molecular and morphological analyses with 28S rDNA gaps treated as uninformative of Quicke et al. (2009, fig. 22); (b) maximum likelihood tree based on barcoding region of CO1 gene. Note that in neither does *Venturia* appear as a basally derived group.

Fig. 13.53 Female of *Eriborus vulgaris*. (Source: reproduced by permission of Kyohei Watanabe.)

(including *Klutiana*). *Hellwigia*, and hence the Hellwigiini, were, however, shown fairly conclusively by Quicke et al. (2005a, 2009) actually to be a member of the Ophioninae, and in the same work, Nesomesochorinae were shown to warrant separate subfamily status. *Skiapus* is also currently classified within the Ophioninae.

A morphological phylogeny was generated by Miah and Bhuiya (2001) but there was little support for most groupings. Although the focus of Quicke et al.'s (2005a) combined morphological and molecular analysis was not on the generic relationships of the bulk of the subfamily, it did include 22 representative genera.

Gauld and Janzen (2004) analysed relationships between Costa Rican species of *Cryptophion*, which are specialist parasitoids of Sphingidae and Saturniidae, and concluded that members of this genus originally parasitised macroglossine sphingids. Relationships between the subset of *Diadegma* species that attack diamondback moth, *P. xylostella*, have been investigated using combined CO1 and ITS-2 sequence data (Wagener et al. 2006). *D. semiclausum* has also had its entire mitochondrial genome sequenced (Wei et al. 2009). Its mitochondrial genome was especially large (18,728 bp) and at the time the second largest known for any hexapod, and displays four gene rearrangements compared with the ancestral insect gene sequence [see Dowton (1999) for indormation on mitochondrial gene rearrangements in the basal cyclostome braconids]. It also has a large AT-rich region (1515 bp) separating cytochrome oxidase genes I and II, which probably evolved by a series of tandem duplications of a region at the 3′-end of the cytochrome oxidase I gene. This gives it the highest mitochondrial AT richness overall (87.4%) with the AT-rich region being 99.1% AT.

Identification to genus level is not always easy and Townes' (1970b) key to world genera will not always work properly, although it is the best available. The situation is exacerbated since several genera may be polyphyletic as currently construed.

Cremastinae

In Europe this is a small group mostly associated with drier habitats and it is most speciose in the tropics. Most species are parasitoids of weakly concealed Lepidoptera, but at least some attack Coleoptera larvae in similar situations (Townes 1965), e.g. *Cremastus nordi*, a parasitoid of the poplar branch borer, *Oberea schaumii* (Cerambycidae). They are koinobiont endoparasitoids but the final instar parasitoid emerges from the host and completes feeding externally, often consuming everything except the head capsule. The biologies of some *Cremastus* and *Pristomerus* (Fig. 13.54) species that are associated with European pests such as the codling moth (*Cydia pomonella*: Tortricidae) were fairly well documented about 80 years ago (Bradley & Burgess 1934, Rosenberg 1934) and Oatman and Platner (1974) described the biology of *Temelucha* sp. and *T. platensis*, which is a parasitoid of the potato tuberworm, *Phthorimaea operculella*. Hosts of *Pristomerus* are mostly endophytic (e.g. leaf rollers/tiers and stem-borers), at least in the latter parts of their development, and include tortricids, crambids, sesiids and oecophorids. The common European *P. vulnerator*, in addition to attacking codling moth, has also been reared from six other lepidopteran families, although these are all associated with garden and hedgerow trees (Gauld 1984a), perhaps indicating a preference for a sunnier microclimate. Hosts are usually attacked as an early instar and sometimes before they enter their final substrate (Okada & Oike 1940). More recently, *Eiphosoma vitticole*, a parasitoid of the fall armyworm, *Spodoptera frugiperda* (Noctuidae), has been investigated in considerable detail (Giraldo-Vanegas & Garcia 1992, 1994a,b, 1995). The egg is laid into the host haemocoel and is weakly hydropic, doubling in volume during the 4 days it takes to hatch. Either three (*Cremastus*: Bradley & Burgess 1934; *Temelucha*: Oatman & Platner 1974; *Trathala*: Solleza & Javier 2009) or four (*Eiphosoma*: Giraldo-Vanegas & Garcia 1992) larval instars have been reported. The egg appears to be laid free in the host caterpillar's body cavity. The first instar has a

Fig. 13.54 Female *Pristomerus* sp. (Cremastinae), showing apically sinuate ovipositor and hind femoral tooth. (Source: reproduced by permission of Rikio Matsumoto.)

heavily sclerotised head capsule with large mandibles which it uses to break through the egg chorion, and has a well-developed tail. The relative size of the tail decreases through the third instar and the final instar is hymenopteriform. Giraldo-Vanegas and Garcia (1994b) reported that feeding during the first instar is through cuticular absorption and only reported the presence of a gut from the second instar onwards. Cremastines kill their hosts after they have constructed their cocoons, emerge from the host caterpillar and spin their own cocoon within that of their host.

Host caterpillars of *E. vitticola* following parasitisation reach only 62% of the final weight of unparasitised caterpillars and are killed in the fifth rather than the final sixth instar (Ashley 1983), so crop protection may be aided by this as well as the ultimate death of the moth. Mating in captivity can be difficult to achieve but is facilitated by the presence of a breeze, according to Bradley & Burgess (1934), who reported that copulation in *Cremastus flavoorbitalis* lasts from 1 to 5 min. In at least some species mating is strophandrous, i.e. tail-to-tail with wasps facing in opposite directions. Adult female fecundity is fairly high and Oatman and Greeny reported fed females of *Temeleuca* sp. laying a mean of 556 eggs during an approximately 3 week long adult life span. Females of this species needed less than 12 h before starting ovipositing. One species of *Cremastus* is aquatic, parasitising the larvae of the aquatic moth *Nymphula* (Berg 1949).

The ovipositors of some cremastines (notably *Pristomerus* spp.: Fig. 13.54) are formed into a series of pre-apical thickened regions which are almost certainly part of a specialised ovipositor steering mechanism (Quicke 1991a). The only other ichneumonids known with a similar modification are some tersilochines, although in the latter case the ovipositor is strongly laterally compressed, which probably means that the detailed functioning of the system is different. In the Braconidae, similar ovipositor modifications to those of *Pristomerus* are found in members of the braconine genera *Zaglyptogastra*, *Undabracon* and *Serrundabracon*, and a weaker version in *Diachasmimorpha* (Opiinae) and *Terebratus* (Alysiinae: Dacnusini). The Australian genus *Gahus* is atypical for the subfamily since it only has a very short, upcurved ovipositor, but unfortunately its biology is unknown, although Gauld (1984a) speculated that it may use its long, slender metasoma to probe into cracks and crevices to reach its host.

The atypical Afrotropical genus *Belesica* was originally placed in the Ctenopelmatinae, but J. F. Perkins convinced Henry Townes that it really was better placed in the Cremastinae (Townes 1971) based on its hind tibial spurs being separated from the basitarsal insertion by a cuticular bridge, a synapomorphy for the subfamily. *Belesica* has a short ovipositor and attacks larvae of the chrysomelid beetle, *Cladocera nigroornata*, that local hunters use for making their poison arrows, and its biology was described by Waterston (1929a). Fresh *Belesica* were not available for inclusion in Quicke et al.'s (2009) combined molecular and morphological analysis, but the putatively related *Eurygenys*, a parasitoid of Limacodidae (as Eucleidae), a species of which was also originally described as a ctenopelmatine, was included and was scored separately for morphology. It did not, however, clade with the other cremastines and instead was recovered as a sister group to the higher ophioniformes irrespective of whether sequence gaps were treated as informative or not.

Gauld's (1984a) suggestion that cremastines probably form a monophyletic group with the Tersilochinae has not been upheld by either formal morphological, molecular or combined analyses (Quicke et al. 2000b, 2009); instead, they are consistently recovered in a clade with the Campopleginae and Nesomesochorinae and often as the sister group of the former. Interestingly, both subfamilies include species that attack Lepidoptera and Coleoptera, although unlike campoplegines, cremastines are not known to possess polydnaviruses.

Hybrizontinae

This has long been a problematic group, which, because of the lack of fore wing vein rs-m combined with aberrant looking venation, position of (or identity of) hind wing vein r-m, misinterpretation of metasomal structure and to some extent their general highly derived appearance (Fig. 13.55), has variously been treated as comprising a family in its own right (e.g. Tobias 1968, Kasparyan 1988), a subfamily within the Braconidae (e.g. Shenefelt 1969, Čapek 1970, Matthews 1974, van Achterberg 1976a,b) or a subfamily within Ichneumonidae (e.g. Wahl 1993c). To make things worse for beginners, there has been equal see-sawing of the name thought to be correct for the group, with both Paxylommatidae/-inae (or Pachylommatinae) and Hybryzontidae/-inae being used at various times. For a number of years, the

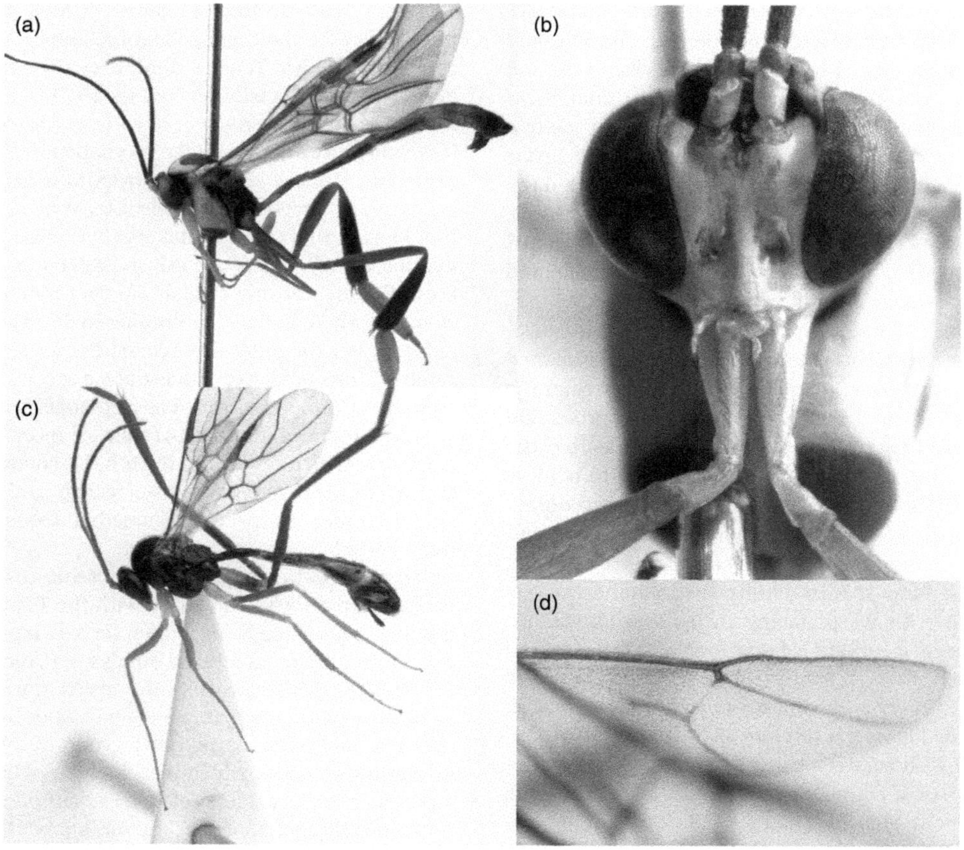

Fig. 13.55 Hybrizontinae. (a) *Eurypterna* male habitus; (b) *Eurypterna* face; (c) *Gilaromma* habitus; (d) *Gilaromma*, detail of hind wing venation. (Source: reproduced by permission, © C. van Achterberg, Naturalis Biodiversity Centre.) For a colour version, see Plate 57.

name Paxylommatinae was thought to be the valid family group name for this group (W.R.M. Mason 1981b), but the situation has now reverted back to Hybrizontinae.

Morphologically particularly confusing is the apparent second metasomal segment, where an apparent single tergum has two pairs of associated sternites. Van Achterberg (1976b), considering the spiracle position and a potentially vestigial one, concluded that the tergite of this segment really was a composite of two, thus showing the same completely fused condition as found in the Braconidae, and therefore placed them there as well as including them as a subfamily of Braconidae in his 1976a preliminary subfamily key. Van Achterberg (1976b) also reported that they have a 'remnant' of a dorsope, a pit at the base of the first metasomal tergite unique to Braconidae, which added to his argument.

The nature of the second tergite needs further study (counting the total number of metasomal segments, examination of ganglia and tracheal system, for example); however, there is now no doubt that the Hybrizontinae are in fact members of the Ichneumonidae based both on molecular data and on the discovery of intermediate fossils that show the loss of fore wing vein 2m-cu.

Molecular data so far have not been much help in that the 28S D2–D3 rDNA gene sequences are highly aberrant, and although analyses often place *Hybrizon* close to the base of the Ichneumonidae (Quicke et al. 1999b), there can be little doubt that this is

Fig. 13.56 Stills captured from video of *Hybrizon* attacking the host ant larva. (Source: reproduced by permission of J.-M. Gómes Durán.)

a case of long branch attraction and of inadequate taxon sampling (Quicke et al. 2000b), and an equally unlikely position for it was recovered by Belshaw and Quicke (2002), where it came out with the Ichneumoniformes. The more comprehensive combined molecular and morphological phylogeny of Quicke et al. (2009) consistently places it within the 'higher' ophioniformes, associating variously with the Anomaloninae, Nesomesochinae, Nonninae and the aberrant genus *Skiapus*, which is currently associated, although with little support, with the Ophioninae (Quicke et al. 2005a).

Of the three recognised extant genera, *Hybrizon* has long been known to be associated with formicine ants. Individuals are apparently attracted to the smell of formic acid and can be seen flying or hovering over ant columns (Fig. 13.56) (Gómes Durán & van Achterberg 2011). *Hybrizon* pupae have been found in ant nests and, interestingly, they are naked, i.e. they are not protected by a cocoon. As long ago as 1852, the large hybrizontine (fore wing up to 11 mm) *Eurypterna cremieri* was observed in Germany to be associated with *Lasius fuliginosus* nests and the wasps are certainly far larger than the ants and therefore it seems likely that they must consume more than one individual to complete development, although the possibility that they might be fed by the ants cannot be excluded. Komatsu and Konishi (2010) observed a Japanese *Eurypterna* and another new genus. Swarming has been noted in *Hybrizon* (Hybrizontinae), but in this case represents an accumulation of females hovering over a colony of host ants rather than males lekking. Nothing is known about courtship and mating in the subfamily.

The ovipositor is not very robust and is unsuitable for penetrating an adult ant. Thus, unlike neoneurines the eggs are laid in host ant larvae as they are transported by workers outside then nest (Fig. 13.56c) (Komatsu & Konishi 2010, Gómes Durán & van Achterberg 2011). All known species are from the Holarctic (which is interesting because all known neoneurine euphorine braconids, which are also associated with formicines, are also restricted to the north temperate region), perhaps suggesting that only temperate formicine ants are accessible for colonisation by parasitoids.

Although rather species-poor today, the group appears likely to have been richer in the past and three species in three genera have been described from Baltic amber (Kasparyan 1988). At that time, Kasparyan considered the group warranting family-level status (as Paxylommatidae), so the four groups that he recognised as subfamilies (two extant, two extinct) are now regarded as tribes. The fossil species are morphologically very similar to the extant hybrizontines, but *Tobiasites* has a distinct fore wing vein 2m-cu present, thus retaining the plesiomorphic state. Among the ichneumonids, loss of 2m-cu is also known in the cryptines *Rhacodopteron* and *Victorovia*, as well as being highly reduced in *Neorhacodes* (Fig. 13.49a,c).

Nesomesochorinae

This small, largely tropical subfamily includes three genera, *Chriodes* (Fig. 13.57a–d), *Klutiana* (Fig. 13.57e) and *Nonnus* (Fig. 13.57f), which until relatively recently were included in the Campopleginae, but they are morphologically aberrant (Gauld 1991). From morphological phylogenetic analysis (Miah & Bhuiya 2001) and subsequently molecular and combined analyses (Quicke et al. 2005a) they were found to clade (usually) with the Anomaloninae. Miah and Bhyuiya suggested that they should be afforded separate subfamily status (which would be Nesomesochorinae); Quicke et al. formally resurrected the Nesomesochorinae for the first two of the above genera which are from the Old World subtropics and tropics but, because of inconsistent placement in their trees, used the Nonninae for the New World tropical genus *Nonnus*. However, they are now uniformly being treated as belonging to a single subfamily (e.g. Veijalainen et al. 2012a).

They characteristically have large eyes which are moderately to strongly convergent ventrally (Fig. 13.57b). Although widely stated that nothing is known of their biology, Conlong (1994) recorded a *Chriodes* sp. as attacking the pyralid stem-borer pest, *Eldana saccharina*, in Africa. The wasps have variously modified (curved, compressed) ovipositors but all have a well-developed pre-apical dorsal notch, suggesting that they attack the larval stage (Fig. 13.57c). *Nonnus* are commonly collected in Malaise traps.

Ophioninae

These are perhaps one of the few ichneumonids that commonly catch the attention of the general public, because some of their larger members are commonly attracted to lights (e.g. *Ophion* and *Enicospilus* spp.) and gave their name to the so-called 'ophionoid facies'[24] (Gauld & Huddleston 1976). Females of most of these can deliver at least a mild sting (stings of some of the large tropical species such as members of the Afrotropical genus *Dicamptus* and some New World *Thyreodon* can be really painful; personal observation), but as the males display prominent pseudostinging behaviour when handled, probing with their highly mobile metasomas and genitalia greatly resembling female stinging behaviour, it is wise to take some care until the sex is evident.

A few species play a significant role in pest control. The North American *O. flavidus* is a parasitoid of the fall armyworm, *Spodoptera frugiperda* (Noctuidae), and under experimental conditions can parasitise up to 79% of hosts (Rohlfs & Mack 1984). Several *Enicospilus* species are regular parasitoids of a variety of agricultural pests, especially in disturbed habitats (Gauld & Mitchell 1981, Gauld 1995), although most species appear to have broad host ranges more defined by host niche than host taxonomy (Gauld 1984a), and therefore they are not particularly suitable for biocontrol release programmes.

The Ophioninae are also one of the most morphologically distinctive subfamilies, nearly all members having a characteristic wing venation with apparent cross-vein closing the discosubmarginal cell far distal to 2m-cu, a distinct adventitious vein running parallel to but separate from the wing margin in the fore wing second subdiscal cell (Perkins 1959, Gauld 1985) and a usually very well-developed and sinuous pectinal comb on the claws (Gauld 1985). Even the final larval head capsule is distinctive, there being more than

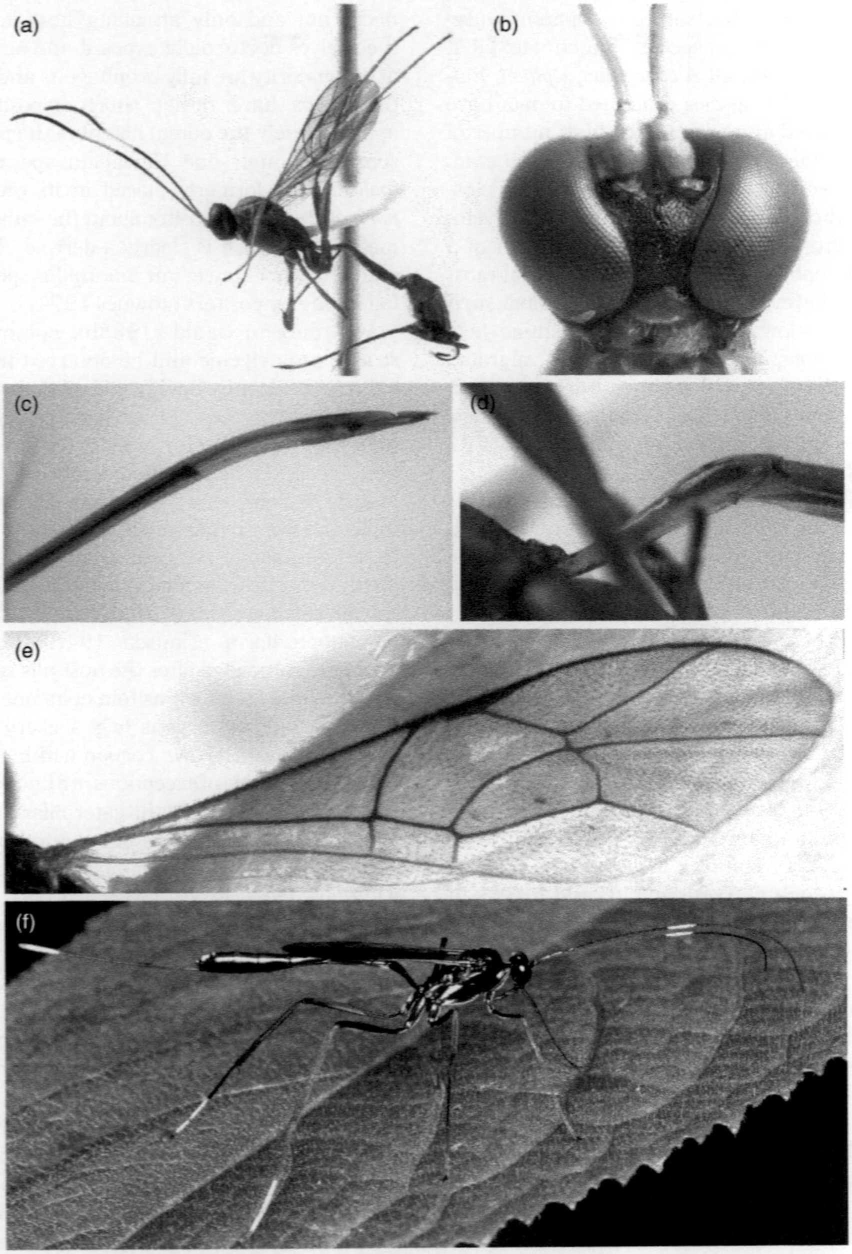

Fig. 13.57 Features of Nesomesochorinae. (a–d) *Chriodes* sp. ♀: (a) habitus; (b) face showing strongly ventrally convergent eyes; (c) ovipositor (upper valve at top); (d) side view of campoplegine-like first metasomal tergum; (e) *Klutiana* sp. fore wing; (f) *Nonnus* sp. (Sources: a–e, reproduced by permission, © C. van Achterberg, Naturalis Biodiversity Centre; f, reproduced by permission of Andreas Kay.) For a colour version, see Plate 58.

eight sensilla on the prelabium (J.R.T. Short 1978, Wahl 1988), although this feature is apparently also displayed by some *Exetastes* species (Banchinae: J.R.T. Short 1978). Internally, all *Alophophion*, *Ophion*, *Enicospilus* and *Thyreodon* species examined to date have the rectum enlarged and with a very high number of small, round rectal pads (Pampel 1914, Quicke et al. 1999a). Some workers have considered the vein closing the discosubmarginal cell to be an r-m cross-vein, implying that from the plesiomorphic condition of a closed areolet, ophionines have lost the basal-most vein, i.e. 2RS. However, given that various taxa such as *Skiapus* (see below) and some Cremastinae (e.g. *Sustenus*, *Pimplomorpha*) clearly have 2RS migrated in line with 2m-cu, as well as having lost the rs-m cross-vein, it seems most likely that in ophionines, the transverse vein is actually homologous to 2RS especially as it lacks a bulla.

In terms of numbers of species, the subfamily is dominated by *Enicospilus* and *Ophion*, currently with 312 and 171 described species, respectively. Both are cosmopolitan but, as Gauld & Mitchell (1981) noted, the two genera 'tend to be mutually exclusive' with *Ophion* being commoner in the Holarctic and Neotropical regions and *Enicospilus* elsewhere. Some tropical ophionines are diurnal and these are usually more or less black, bluish black or dark brown in coloration, sometimes with fairly conspicuous aposematic counter markings; some New World *Thyreodon* and *Rhynchophion* species are very aposematic in being black with yellow head and antennal tips and appear to be in the same mimicry complex as some of the larger pompilid wasps such as *Pepsis* spp. (Gauld p. 395 in Hanson & Gauld 1995, Janzen & Gauld 2004). Given their ability to sting, although not as badly from a human point of view as the pompilids, this is likely to be quasi-Müllerian rather than Batesian.

Virtually all ophionines are solitary koinobiont endoparasitoids of large Lepidoptera larvae principally belonging to the Erebidae (Arctiinae, Lymantriinae), Geometridae, Lasiocampidae, Noctuidae, Saturniidae and Sphingidae, but one North American species apparently attacks soil-dwelling beetle larvae (Townes 1971) and it is possible that one European species occasionally attacks sawfly larvae (W.R. Thompson 1957). Despite their conspicuousness and abundance, very little detail is known about their biology, apart from a considerable number of host records. This may be partly attributed to the majority of species being nocturnal and only attacking host larvae that are themselves nocturnally exposed and active. The hosts of the majority are fully exophagous and consequently the wasps have rather short ovipositors, typically approximately the dorsal height of the posterior metasoma, although one Hawai'ian species, *Enicospilus molokaiensis*, formerly placed in its own genus *Pycnophion*, has an ovipositor about the same length as the metasoma, which is clearly a derived character state, and two other Hawai'ian *Enicospilus* species also have fairly long ovipositors (Townes 1971).

According to Gauld (1988b), ophionines may be strictly pro-ovigenic and he observed that some individuals, apparently having laid all or nearly all of their eggs, show no signs of further egg maturation and they have never been observed to host-feed. Oviposition appears to be into the haemocoel (Vickery 1929, Moutia & Courtois 1952) of late instar hosts and is rapid and not accompanied by any host paralysis. The hosts are usually attacked relatively late in development, typically from about the middle instar onwards. The first instar larva is caudate and subsequent ones hymenopteriform (Clausen 1940). Development is normally completed after the host has spun its cocoon and the final instar parasitoid consumes almost all of its host's internal organs (e.g. Vickery 1929) before egressing to spin its own cocoon within that of its host, although a couple of exceptions are known. In one case, pupation occurs within the caterpillar's cuticle, and in a few other cases, the host metamorphoses to form its pupa within which the ophionine spins and pupates (Quezada 1967, Brock 1982). Cocoons of ophionines are rather hard, blackish and often have a conspicuous white medial band. Those of the New World genus *Thyreodon* also have a medial (equatorial) thickening (see Janzen & Gauld 2004). Host food consumption, at least in the case of *O. flavidus*, is significantly reduced in comparison with that of unparasitised controls (Rohlfs & Mack 1983).

Gauld (1977a, 1979, 1985, 1988b) and Gauld and Mitchell (1978, 1981) revised most of the none European species and in the 1985 paper proposed a phylogeny of the genera which involved abolishing the tribal classification into an Enicospilini and Ophionini that had previously been based on relatively weak characters and without cladistic argumentation by Townes (1971), and instead proposed an informal subdivision into five, putatively monophyletic genus groups based on *Enicospilus*, *Eremotylus*, *Ophion*, *Sicophion* and *Thyreodon*. Gauld and Janzen (2004) presented a morphological phylogeny of the principally diurnal *Thyreodon* genus group. Molecular phylogenetic studies largely support these groupings (Quicke et al. 2009).

In a morphological study by Miah and Bhuiya (2001), two genera that had for a long time been included in the Campopleginae, *Hellwigia* (taken here also to include *Protohellwigia*) and *Skiapus*, appeared as potentially being related to the Ophioninae instead. *Hellwigia* had also been considered as warranting placement in its own subfamily by Horstmann (1969) on account of its highly derived morphology and wing venation. More recently, these taxa have been sequenced for the 28S D2–D3 rDNA gene region (Quicke et al. 2009) and these data have partly confirmed Miah and Bhuiya's findings with *Hellwigia* clearly belonging to the Ophioninae. However, the placement of *Skiapus* is still far from certain. Interestingly, the fore wing venation of both of these is less typically ophionine like with cross-vein 2RS (referred to by some as r-m or rs-m) more or less in line with 2m-cu in *Skiapus* and more typically ophionine like (i.e. postfurcal) in *Hellwigia*. However, the venation of *Hellwigia* is very curvy, and this apomorphic feature could explain its non-ophionine arrangement. Interestingly, while *Hellwigia* is currently restricted to the Palaearctic 'Mediterranean zone', in the lower Oligocene it was also present in North America [the Florissant deposits of Colorado (37–35 Mya): Townes 1966].

The highly aberrant structure of *Skiapus* makes any morphological assessment of its relationships almost impossible. It lacks the adventitious fore wing vein and the far postfurcal 2RS vein of ophionines, but its large claws have a very well-developed and sinuous comb, so there is at least one supporting feature, but it lacks the large number of rectal pads characteristic of Ophioninae and instead has just four, which is a derived state found mostly in Campopleginae (Quicke et al. 1999a). Unfortunately, *Skiapus* specimens are rare in collections but I collected one flying over weedy vegetation on a rubbish heap in Tanzania, and they turn up in Malaise trap samples from disturbed areas with some frequency. In the combined molecular and morphological elided analysis of Quicke et al. (2009), it was recovered with the Hybrizontinae and Anomaloninae, although it was recovered with the Ophioninae with some alignment parameter combinations.

Gauld (1988b) noted it as being one of the few groups of koinobiont ichneumonids that are more species rich in the tropical than the temperate region, but this could be an artefact of sampling and taxonomic attention, although Gauld (1987a) argued that their nocturality is an important factor because if they had to search for scarce hosts by flight, in the day time they would experience too high a predation pressure.

UNPLACED SUBFAMILIES

Eucerotinae

This is one of the most unusual groups of ichneumonids biologically and is generally fairly easy to recognise (Fig. 13.58a,b). Until recently it was known from only a single, almost cosmopolitan genus, *Euceros*, upon which all our biological knowledge is based, comprising nearly 50 recognised species. *Euceros* has a very strange hyperparasitoid biology which has been described in some detail but still with a lot of unknowns, for four species, the Nearctic species, *E. frigidus* (Tripp 1961) and the West Palaearctic, *E. unifasciatus* (Varley 1965) and *E. albitarsus* and *E. pruinosa* (M.R. Shaw 2014).

Females lay many hundreds or thousands of tiny eggs (Fig. 13.58c, see also figs 4 and 5 in Tripp 1961) each, of which is glued near to the edge of a leaf by a long, thin stalk, supposedly a secretory product of the female reproductive tract (along the posterior half of the ovariole) rather than being part of the egg proper. Females are synovigenic and have approximately 100 delicate ovarioles. These soon hatch to yield a tiny triungulin-like[26] larva called a planidium, which is heavily sclerotised, that sits on the empty egg shell and waits until a suitable intermediate host (Lepidoptera or sawfly larva) passes by, and then the *Euceros* larva attaches to it and remains there (Fig. 13.58d), usually hidden under an integumentary fold, until the intermediate makes its cocoon, managing to stay attached through host moults. Through its attachment to the sawfly larva or caterpillar or prepupa, it may absorb nutrient through small bite holes; it does not develop any further unless the caterpillar is parasitised by an endoparasitic ichneumonid belonging to the Anomalonine, Banchinae, Campopleginae, Ctenopelmatinae

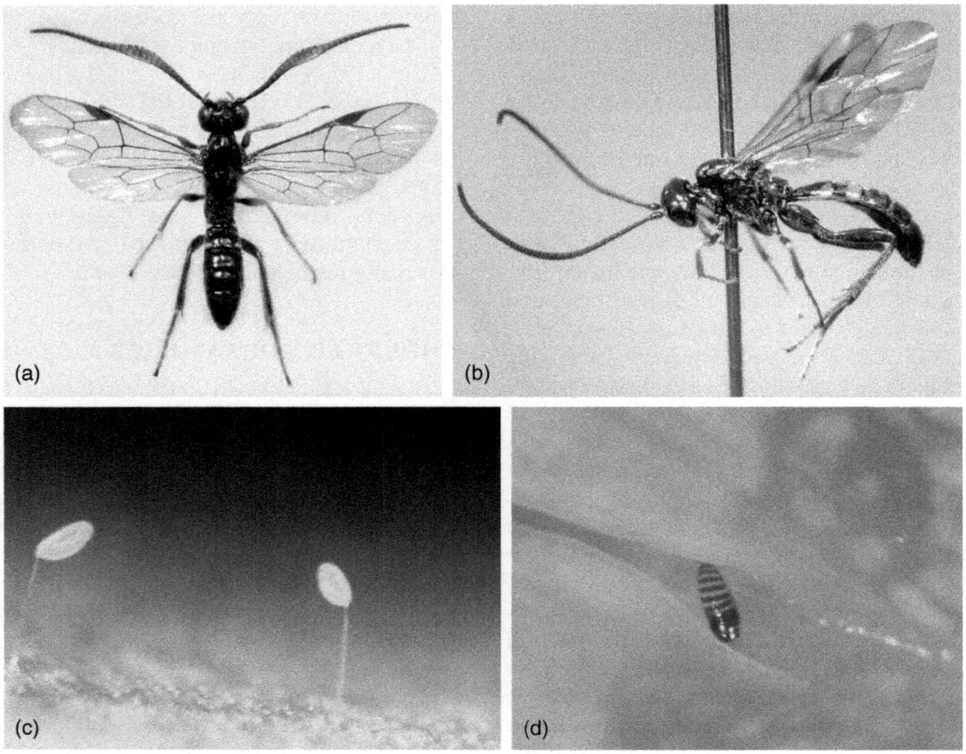

Fig. 13.58 *Euceros* adults and juvenile stages (Eucerotinae) habitus photographs. (a, b) *E. sensibus* habitus photographs of female and male, respectively; (c, d) *E. ablitarsus*, stalked eggs and planidial larva on host, *Orthosia gothica* (Noctuidae), respectively. (Sources: a, b, reproduced by permission of Kyohei Watanabe; c, d, reproduced by permission of Mark Shaw.) For a colour version, see Plate 59.

and Ophioninae, although there is also one record from a braconid (*Meteorus*: Euphorinae). Other species also attack ectoparasitic Cryptinae and Tryphoninae (*Netelia*). The *Euceros* planidium either enters the primary host after it has egressed from the sawfly remains in the sawfly cocoon if it is an endoparasitoid or it may transfer to and enter an ectoparasitic primary host such as a cryptine (Gauld 1984a). Tripp reported also that *E. frigidus* had seven or eight larval instars, the planidium, two or three post-planidial internally feeding ones, one that exits the primary host, and then those (three) that feed externally on the primary, parasitic wasp, host. However, M.D. Shaw (2014) suspected that that might be an overestimate and that there might be only a single post-planidial instar that feeds internally, since his observations showed that the first externally feeding stage was still very small. M.R. Shaw also recorded a 'planidium' within the host ichneumonid head capsule which might suggest that it interacts physiologically with the host's brain or neuroendocrine system as some other taxa do (see Diplazontinae and Pimplini). Hence there is still vast amount waiting to be discovered about exactly when and how the *Euceros* transfers to and utilises its particular hosts. In those instances where an externally feeding parasitoid is attacked, the primary host is allowed to complete development to its cocoon constructing phase, but it is not known whether the planidium simply sits on its outside or not. Then there is the question of how they develop in anomalonines which complete all their development within the secondary host, pupating within its pupal case. Necessarily, the *Euceros* planidium must enter the secondary host to attack the anomalonine primary.

Interestingly, *E. semiothesae*, assuming it is a single species, has been reared from endoparasitoid hosts that exit to pupate externally, from entirely ectoparasitic tryphonines and from entirely endoparasitic anomalonines (Gauld & Wahl 2002), and at least two other species have been recorded attacking hosts in both the first two of these categories, all indicating a great deal of behavioural and developmental plasticity.

This very aberrant biology, which no doubt involves an extremely high level of juvenile mortality, is very similar to that shown by the completely unrelated Trigonalyidae and the chalcidoid families Perilampidae and Eucharitidae. How such hit-and-miss host location strategies can evolve is a mystery, although, of course, with the eggs being so small, the actual cost of failure of individual offspring to the mother is slight. Knowing the relationships of the subfamily and details of the biology of its close relatives might one day help us to reconstruct the transitions that might have occurred.

The antennal flagellum is medially expanded and compressed, especially in males (Fig. 13.58b; cf. Fig. 13.58a) and the placoid sensilla on these have different arrangements in different species (Fitton 1984). It seems likely that these modifications are associated with courtship and probably the corresponding female antennae segments will be found to possess glands.

A more recently discovered and apparently much less derived eucerotine genus from Chile, *Barronia*, prompted a phylogenetic analysis of the subfamily (Gauld & Wahl 2002), which indicated that although this is a principally northern group with few tropical representatives, its origins might have been in Gondwanaland during the lower Cretaceous, and that it spread into the North temperate region when the Indian land mass impacted Eurasia (similar to analyses of the aphidiine braconids: see Belshaw et al. 2000). Gauld and Wahl therefore argued that its current distribution does not reflect an initial association with sawflies as hosts as was previously suspected, because tenthredinoid sawflies (Diprionidae and Tenthredinidae) are a predominantly northern group, but rather that they probably originated as parasitoids of ophioniformes attacking Lepidopteran hosts. All known host records were tabulated by Gauld & Wahl (2002).

The relationships of the Eucerotinae to other ichneumonid subfamilies are very far from certain. As discussed by Gauld and Wahl (2002), various morphological, molecular and combined analyses have variously associated them with the Lycorininae, Metopiinae, Microleptinae, Sphinctini of the Tryphoninae or Brachycyrtinae+Claseinae+Pedunculinae, none of them very convincingly. In Gauld and Wahl's morphological analysis with a few exemplar taxa, they were recovered as a sister group to the Brachycyrtinae+Labeninae. A relationship with the Brachycyrtinae was found by Quicke et al. (1999b) based on just 28S rDNA data and a small subset of taxa and also the molecular and combined trees of Quicke et al. (2000b)[27], although Gauld and Wahl displayed little enthusiasm for molecular data and stated 'We see no reason to give particular credence to the group Brachycyrtinae+Eucerotinae'. Further, they rooted their cladogram with a pimpline, which, out of the taxa represented in their matrix, now appears less suitable than using the labenine or including a xoridine. The positions Eucerotinae were recovered in the combined elided trees of Quicke et al. (2009), i.e. in a clade with *Sphinctus* (Tryphoninae) and the aberrant genus *Ischyrocnemus*, or with Microleptinae in many individual combined analyses, or with Lycorininae in the purely morphological tree all seem improbable, but that begs the question as to what would seem like a sensible placement. Perhaps they are a very basal group as the tree of Gauld and Wahl would suggest if it were rooted with the labenine rather than the pimpline.

Taxonomic revisions are available for the Nearctic (Barron 1976), world (Barron 1978) and United Kingdom (Fitton 1984). Gauld and Wahl (2002) summarised known host associations and presented a morphological cladistics analysis of a representative subset of species.

Microleptinae

Henry Townes, and many others, used the name Microleptinae to include all the species now placed in Orthocentrinae plus *Oxytorus* (Oxytorinae). The subfamily was then restricted by Wahl (1986) to include the single Holarctic genus *Microleptes* (Fig. 13.59) that had previously been placed in the Orthocentrinae. Wahl (1991) seemed uncertain about its affinities and chose it as one of the outgroups for his phylogenetic analysis of the ophioniform subfamilies. However, in his 1986 paper, Wahl noted that it possesses a number of characters of the final instar larva that were only known in Banchinae, Ctenopelmatinae and the subfamilies he referred to as Ophioninae *s.l.*, which roughly corresponds to the 'higher ophioniformes' as

Fig. 13.59 *Microleptes splendidulus* (Microleptinae). (Source: reproduced by permission, © C. van Achterberg, Naturalis Biodiversity Centre.) For a colour version, see Plate 60.

treated here. Therefore, that recent combined molecular and morphological analyses do suggest that it might belong there (Quicke et al. 2009) is not too surprising, but in the trees presented it associated with the morphologically very dissimilar Eucerotinae, so I think its relationships within the Ichneumonidae are still extremely uncertain. Morphologically they show a mixture of traits. The lateral oviducts are approximately 0.5–0.7 the length of the ovarioles, the latter (12 pairs) being packed full of apparently mature or nearly mature eggs, rather like in ophioniformes, but the common oviduct has large paired vaginal glands, which is more a feature of pimpliformes; a single gland is present in some banchines.

Little is known of their biology other than that they are endoparasitoids of Diptera larvae, and all host records to date are from Stratiomyidae, e.g. *Chloromyia* species (Wahl 1986, Schwartz 1991, Fitton & Huddleston personal communication), and M.R. Shaw & Askew (2010) suggested that it is probably a larval–puparial endoparasitoid. Endoparasitism of Diptera within the family is otherwise restricted to members of the pimpliformes. Therefore, if they really do belong with the ophioniformes, then they would represent another independent colonisation of Diptera as hosts within the Ichneumonidae.

ENDNOTES

1. Family group names are, in increasing inclusivity, subtribe, tribe, subfamily, family and superfamily.
2. As explained by Gauld (1983), Henry Townes used the name Labiini contrary to the rules on family group name formation and that name is pre-occupied by a family of Dermaptera (Labiidae); Carlson (1979) recommended using the next available family group name, i.e. Groteini, and that has been adopted by all subsequent workers.
3. Although they excluded Collyriinae from their concept of the pimpliformes.
4. Referring to these as the Oxytorinae.
5. Western Palaearctic species are treated by Klopfstein S (2014) Revision of the Western Palaearctic Diplazontinae (Hymenoptera, Ichneumonidae). Zootaxa 3801, 1–143.
6. An identical-looking claw and spatulate bristle are present in the poemeniine *Eugalta* (Wahl & Gauld 1998) and therefore it might have a more basal origin within the pimpliformes; see also the earlier section Acaenitinae.
7. Indeed, *Theronia* was associated by Gauld (1984a) with the Delomeristini rather than Pimplini.
8. Sometimes, it seems, the males make a mistake and aggregate where a male is going to emerge.
9. Buckling of a structure under compression means that it can no longer exert any addition force.
10. An earlier description of its life history (E. Cameron 1950) actually was the result of a misidentification and the wasp involved was actually *Ichneumon suspiciosus* (Ichneumoninae)
11. Clearly misplaced were *Aclastus*, *Bathyzonus*, *Idiolispa*, *Toechorychus* and *Trafana*.
12. Apidae in the old sense are just a specialised group of Sphedidae in the old sense.
13. Quicke et al.'s trees are based on combined molecular and morphological data, but all cryptines were coded the same for morphology so the trees (apart from rooting) can be considered purely molecular.
14. Though I should point out that he lists it in his table 5 between Metopiinae and Tryphoninae, so there may be some question about its true identity.
15. Bennett et al. (2013) pointed out that Quicke et al. (2009) had missed a pre-existing available name for this group, viz. the Sisyrostolinae, based on a junior synonym of one of the included genera, *Melanodolius*.
16. Until recently included in the Oedemopsini, but appear to be more basal.
17. The tribe takes its name from the first family group name proposed for a member of it.
18. *Netelia* were referred to by workers for nearly 100 years as *Paniscus*, through a misinterpretation of that name based

on rather incomplete original description; the type species of *Paniscus* is actually an ophionine.

19. Frequently referred to as *M. aulicus* due to misidentification, or misspelt, *tenthredinidis*.

20. Megaceriini would actually have taken nomenclatural priority over Westwoodiini, but Townes' (1970b) interpretation of *Megaceria* was due to him having been loaned a specimen that actually belonged to the westwoodiine genus *Pergaphaga* instead.

21. *Neogreeneia* was for a long time regarded as a separate genus and the name will often be encountered in the literature.

22. Pine looper is also used to refer to other species of Geometridae in North America.

23. *Biolysia* and *Bathypiesta* are now regarded as synonyms of *Bathyplectes*.

24. In much of the older literature the name is used incorrectly for what is now called *Venturia*.

25. The ophionoid facies can be summarised as yellowish body colour, large ocelli and eyes and long antennae.

26. Triungulins are the first instar larvae of meloid beetles.

27. Labeninae in the purely molecular analysis were placed far away in an improbable place.

Part III

Ecology and Diversity

Chapter 14

ECOLOGY

The Braconid and Ichneumonid Parasitoid Wasps: Biology, Systematics, Evolution and Ecology, First Edition. Donald L. J. Quicke.

This chapter, and to some extent the following one, deals with the ways in which ichneumoids interact with or are affected by both biotic and non-biotic environmental factors. Ichneumonoids collectively occupy many sub-compartments of the broad parasitoid niche, attacking a number of different host groups which in turn occupy numerous different niches. Yet there are some generalisations that can be made. They are generally larger bodied than the Chalcidoidea, although of roughly similar size to the far less speciose Stephanioidea and Megalyroidea, and unlike the Chalcidoidea, they do not include any egg parasitoids or any parasitoids of non-heteropteran Hemiptera with the exception of the braconid subfamily Aphidiinae, which out of that group only attack Aphidoidea. The chalcidoids and ichneumonoids also show quite different patterns in host utilisation in other ways. For example, although a few ichneumonoids attack leaf miners, predominantly Exothecinae, some Hormiinae and a few Braconinae in the Braconidae and a few ctenopematine ichneumonids, leaf miners are subject to far greater attack by chalcidoids (Fig. 14.1). From Hawkins' (1994) database, the same would appear to be true of gall-forming hosts. However, externally feeding caterpillars and beetle larvae are generally attacked by many more ichneumonoids and similarly leaf-rollers. Although the data in Hawkins' database do not show it, xylophagous hosts, especially larger ones, are far more prone to attack by ichneumonoids than chalcidoids.

ADULT DIET

Nearly all adult hymenopteran parasitoids have a limited diet because of the narrow waist between propodeum and metasoma which restricts what can be eaten to liquids and near-liquids without significant large solid chunks. Further, those that do not host-feed, if they feed at all, usually have a diet restricted largely to sugar resources such as floral or extrafloral nectar or honeydew. Water access is also very important for the longevity of species even in the absence of sugar sources, and Townes (1958) pointed out that moisture, often in the form of early morning dew, often appeared to be a limiting factor for ichneumonids, although with no experimental data.

In addition, a number of idiobionts whose hosts are exposed, or sometimes weakly concealed, will feed on host haemolymph that exudes from wounds made either by the wasp's ovipositor or, less commonly, mandibles, a behaviour called host-feeding, which has been much studied. As Jervis (1998) noted, one type of feeding adaptation might prevent another, and suggested that while the concealed nectar feeding apparatus (CNEA) types 1 and 2 might facilitate host-feeding, types 6 and 7 might prevent it (see Table 10.1). Indeed, almost nothing is known about whether those taxa with CNEA modifications host-feed or not.

Host-Feeding

Although host feeding is best known and perhaps best developed among chalcidoids, some of which construct special feeding tubes to access haemolymph of concealed hosts, host feeding is also shown by several ichneumonoids. In addition to providing all the food for the larval parasitoid wasp, adults of some species regularly imbibe haemolymph that exudes from puncture wounds made in the host by the ovipositor or, less commonly, from wounds made by the mandibles [e.g. *Chorinaeus* (Metopiinae: Aeschliman 1974a, 1975); some *Netelia* (Tryphoninae: Vance 1927, M.R. Shaw 2001; see Fig. 13.23b)]. In the case of *Netelia*, use of the mandibles may be something to do with the fact that her ovipositor and egg are especially adapted to laying an anchored egg through the host cuticle, which in some species might preclude their use for making a suitable puncture for feeding from.

Host-feeding enables the wasp to obtain protein and also other potential nutrients and is almost exclusively associated with idiobiont taxa, which, because they tend to produce larger eggs and are nearly always synovigenic, require this rich food to continue maturing eggs (Jervis & Kidd 1986). It is therefore not surprising that records of host-feeding in the superfamily are strongly clustered in a small number of clades corresponding to the major predominantly idiobiont taxa. One major exception is the braconid subfamily Rogadinae, which, whilst being koinobiont endoparasitoids of Lepidoptera larvae, are synovigenic and produce relatively large, yolky eggs (Fig. 4.1a), and several species of *Aleiodes* in this group do host feed (e.g. Dowden 1938, Ahmad 1944, M.R. Shaw personal communication). Quednau and Guévremont (1975) described host-feeding in another koinobiont, *Priopoda*, which belongs to the Ctenopelmatinae, but as with *Aleiodes*, some ctenopelmatines have relatively large

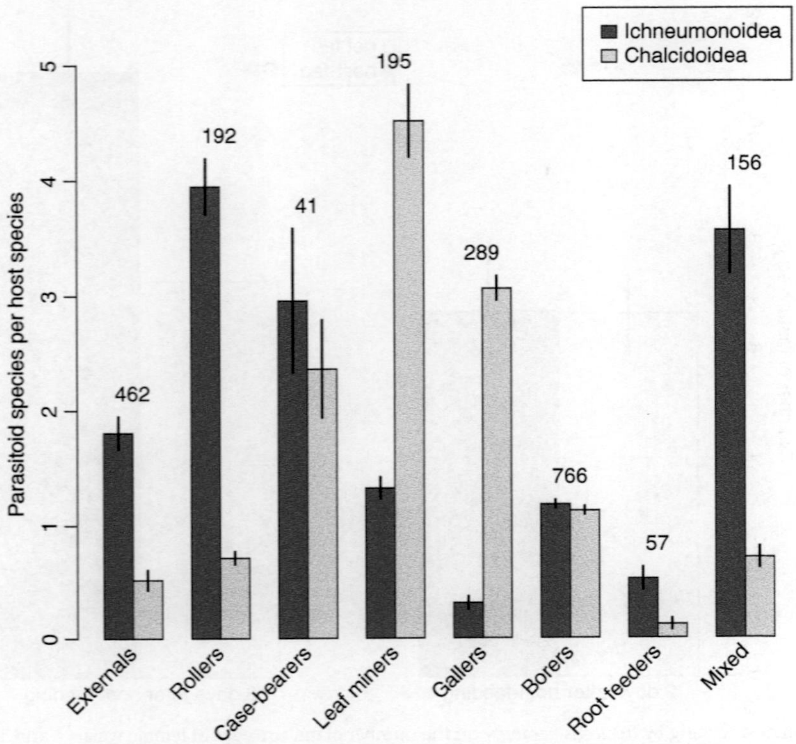

Fig. 14.1 Differences between Ichneumonoidea and Chalcidoidea in their utilisation of hosts occupying different feeding niches expressed as mean numbers of parasitoid species attacking each host species, with standard errors of means indicated by vertical lines. (Source: data from Hawkins 1994, fig. 4.1a,b.)

eggs for koinobionts and are phylogenetically near the base of the Ophioniformes (see Cummins *et al.* 2011).

Within the Ichneumonidae, host-feeding is best known among the Pimplinae and Diplazontinae (Pimpliformes group), the Tryphoninae and the Cryptinae and Ichneumoninae (Ichneumoniformes), but has also been recorded among Adelognathinae (which are closely allied to the Cryptinae and Ichneumoninae), Ctenopelmatinae, Metopiinae, and Tersilochinae [as Phrudinae although not in Tersilochinae *s.s.*, viz. present in *Earobia paradoxa* (Franz 1958)]. Within the Tryphoninae, the behaviour is largely restricted to those Phytodietini (*Netelia* and *Phytodietus* species) that induce temporary host paralysis (Kasparyan 1988), but it has also been noted in the leaf-miner parasitoid *Grypocentrus* (Guévremont & Quednau 1977). In the Braconidae, it is essentially restricted to the cyclostome subfamilies (Braconinae, Exothecinae, Rhysipolinae, Rhyssalinae and Rogadinae).

Host-feeding may be concurrent or destructive (Jervis & Kidd 1986). In concurrent host-feeding, the female imbibes some haemolymph from an individual host that she either has laid or is going to lay an egg on/in. In destructive host-feeding, the individual host attacked is killed as a result of it and the female normally does not attempt to oviposit in it. Destructive host-feeding adds to the host mortality caused by a parasitic wasp and in some cases may exceed that due to oviposition and hence has practical implications for pest management/biocontrol. In destructive host-feeding, the individuals that are fed on are often ones that are unsuitable or less suitable as potential hosts (Leius 1961). At least two species of *Itoplectis* (Ichneumonidae: Pimplinae), which are pupal endoparasitoids of Lepidoptera, selectively feed on hosts that have previously been parasitised either by a conspecific or allospecific wasp (Ueno 1998, Moser et al. 2008).

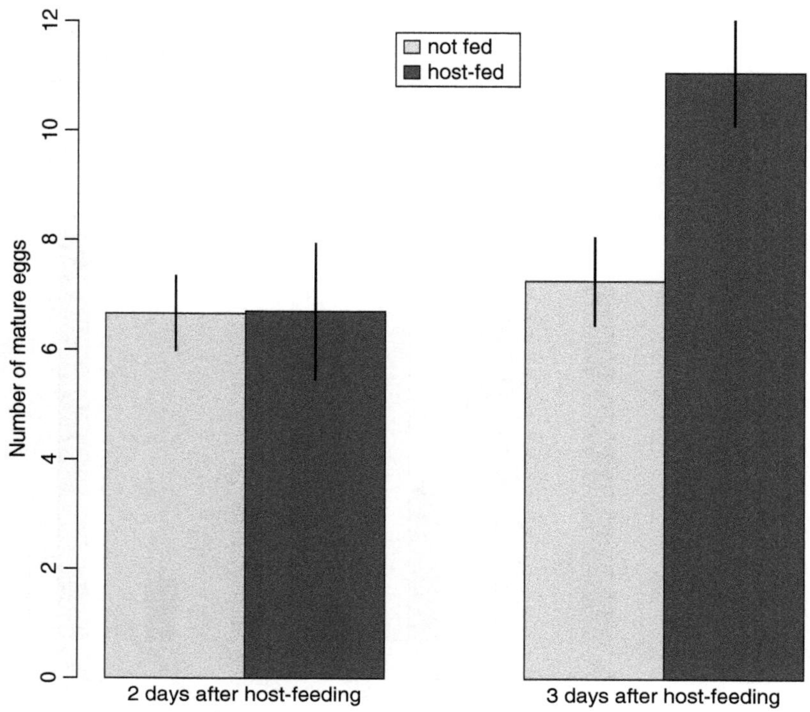

Fig. 14.2 Effects of host-feeding by *Itoplectis naranyae* on the number of mature eggs in female wasps 2 and 3 days after host-feeding on the host (±SE), *Galleria mellonella* pupae, versus controls (ANOVA, $p < 0.05$). (Source: data from Ueno & Ueno 2007.)

Several studies have shown that host-feeding is very important in allowing female idiobionts to continue maturing eggs through their adult life. Thus, for example, in Ueno's (1999a) study of the pimpline ichneumonid *Pimpla nipponica*, a female wasp usually lays only about 20 eggs in her lifetime if prevented from host-feeding (she carries only 4–6 mature eggs at any one time), but that number is greatly increased when host-feeding is permitted. Ueno and Ueno (2007) found that in another pimpline, *Itoplectis naranyae*, it took almost 3 days for nutrients obtained from host-feeding to become incorporated into additional eggs (Fig. 14.2). Interestingly, host-feeding in this case has no significant impact on adult female longevity, so the nutrients acquired from this behaviour must be directed primarily towards egg maturation rather than general body maintenance functions. Many pimplines display both destructive and concurrent host-feeding, probably partly depending upon the size and quality of the host. In addition to providing female parasitoids with proteins and lipids, host-feeding might also have an endocrine aspect, as the haemolymph imbibed will also contain hormones. Bracken and Nair (1967) showed that when the sugar water provided to recently emerged females of the pimpline *Exeristes comstocki* was doped with the juvenile hormone mimic farnesyl methyl ether, yolk deposition was increased. Thus access to hosts might, via host-feeding, lead to direct stimulation of egg production.

Cole (1967) described how members of two other pimpline genera that attack leaf-roller prepupae and pupae, *Itoplectis* and *Apechthis*, stab the prey with the ovipositor and then '... work the ovipositor round and round inside it with vigorous rotatory movements of the abdomen'. The result is that the leaf-roller's haemolymph oozes out, sticking it to the leaf, and after several minutes the wasp withdraws its ovipositor and the prey's body fluids then flow out through the hole made in the leaf and are imbibed. Figure 13.23a shows a similar result for a pseudohyperparasitic *Itoplectis* feeding from a 'mashed' host (M.R. Shaw 2009b). These and other pimplines are quite generalist parasitoids and their host-feeding activities can equally involve multiple species, some of which cannot be actual hosts because they pupate in the soil rather than within their leaf-roll and therefore their appropriate developmental stage would not be available. One species of the pimpline genus *Scambus* takes 'host-feeding' a stage further in preying on a large range of non-host species for a large part of the year when its particular host is not available (Zijp & Blommers 2002a) and therefore acting as a true predator. Similarly, the cosmopolitan diplazontine *Diplazon laetatorius* (Fig. 13.15), an egg–larval parasitoid of aphidophagous syrphid Diptera, will consume eggs of the host if they are not sufficiently far developed to permit oviposition (Rotheray 1981b).

Water, Sugar and Pollen Feeding

Apart from host-feeding, which is practised by a relatively small proportion of taxa, many parasitic wasps seek out both water and also sugar sources as adults to provide energy and possibly some other nutrients. Indeed, probably nearly all adult ichneumonoids will drink water and the great majority will avail themselves of sugar sources such as floral and extrafloral nectar and honeydew. Other, non-standard sugary resources are occasionally used by ichneumonoids, for example, the alysiine braconids *Asobara tabida* and *A. rufescens*, which attack *Drosophila* larvae in decaying/fermenting substrates, both feed directly on the substrate, i.e. fermenting fruit or decaying plants, and this behaviour significantly increases their fat reserves (Eijs et al. 1998). As countless laboratory studies and a huge number of publications in the biological control literature have shown, adult parasitoids will survive far longer and have greater lifetime fecundity if provided with a sugar resource as opposed to water alone (e.g. S.J. Johnson & Smith 1980).

Various ichneumonoid species have also been observed feeding at extrafloral nectaries (Jervis et al. 1992). Those observed in Hawai'i on castor oil plants (*Ricinus communis*) by Nishida (1958) included *Enicospilus*, *Venturia*, *Opius* and '*Apanteles*' species, while Gentry (2003) found that a number of ichneumonids were frequent visitors to the extrafloral nectaries of a *Solanum* species. Gentry suggested that this was part of an adaptive syndrome which led to this plant's lepidopteran herbivores suffering a higher rate of parasitism than two congeners that lacked extrafloral nectaries. Extrafloral nectar may also be important for parasitoids involved in biological control. The microgastrine *Microplitis croceipes*, an important parasitoid of *Heliothis* and *Helicoverpa* (Noctuidae) species that are serious cotton pests, could use extrafloral nectar and fared just as well as wasps fed honey and water (Röse et al. 2006). Importantly, in the field, extrafloral nectar is available early in the growing season before the cotton or many other plants are in flower and also increases the retention time of the wasps in the patch (Staple et al. 1997), and in this species has a more beneficial effect on reproductive performance than honeydew. Extrafloral nectar was also found to be a good food source for *M. mediator*, enhancing their parasitisation rate on the cabbage moth, *Mamestra brassicae*, by a very large amount (Geneau et al. 2012).

Experimentation to establish exactly what sugars are best for adult parasitoid longevity has been carried out on several species. Hogervorst et al. (2007) found that the aphidiine *Aphidius ervi* did just as well on the disaccharide sucrose as on an equivalent, equimolar solution of its monosaccharide constituents, fructose and glucose. Females of both the campoplegine *Bathyplectes curculionis* and the microgastrine *Cotesia glomerata* did better when fed glucose or fructose (or sucrose) than they did when fed trehalose or melezitose, the last two sugars being fairly common components of honeydew (Wäckers 2001, Jacob & Evans 2004). Makatiani et al. (2014) investigated the threshold concentrations of four sugars that initiate a feeding response in the microgastrine *Microplitis mediator* and found that this was affected by prior experience. Wasps preferred, i.e. accepted at a lower concentration and fed for longer, sucrose, glucose and fructose compared with maltose. If the wasp was given experience of feeding on sucrose or glucose, both of which increased female longevity more than fructose or maltose, the threshold concentration of the latter for feeding was significantly increased. This suggests that in the wild, female wasps may assess what sugar sources are available and if the more beneficial sucrose or glucose is present, selectively feed on these, whereas if they were absent, the wasp would make do with the less beneficial available alternatives.

Honeydew is an important source of dietary sugar for many adult parasitoids, especially in the temperate region aphids, and honeydew is far less abundant in the tropics although there are other possible sources of it. There is evidence that honeydew is the preferred source of dietary sugar for two Australian anomalonines, *Habronyx pammi* and *Heteropelma scaposum*, that do not seem to visit flowers in the wild or when provided with them in the laboratory (M.W. Short & Steinbauer 2004). Numerous anomalonines are associated with arid habitats (Gauld 1984a), so it would be of considerable interest to know how general this is among tropical anomalonines and what alternative water/sugar resources they might in use in nature.

Faria et al. (2008) showed that under laboratory conditions, parasitoids fed on sucrose solution did better than those provided with honeydew, and both did better than those only provided with water alone. Wyckhuys et al. (2008) showed the same for the aphid parasitoid *Binodoxys communis* and J.C. Lee et al. (2004) for the campoplegine *Diadegma insulare*. The chemical composition of honeydew differs from that of nectar. Nectar typically contains approximately equal amounts of fructose and glucose (rather than the disaccharide sucrose) whereas honeydew typically contains other plant-derived sugars which might have intrinsically lower nutritive value. Or indeed honeydew might include compounds that have a negative effect on parasitoid survival. Honeydew tends to dry out and become viscous and difficult to consume whereas plants have evolved to maintain their nectar drinkable so as to be attractive to pollinators. By synthesising artificial honeydew with the same sugar complement, Faria et al. (2008) showed that there were no issues with the sugars themselves, but that the viscosity of the honeydew meant that parasitoids had a hard job consuming it. To date no one seems to have done the experiment of dissolving the same dry weight of honeydew and nectar, rehydrating and then comparing their effects on parasitoid longevity.

Despite many studies on the importance of providing sugar sources for maintaining parasitoids in culture, very few of these studies, however, have sought to determine optimal sugar concentration or composition. For culture purposes on a small scale, M.R. Shaw (1997) recommended the use of diluted honey (1 part honey to 4 parts water) on the basis that honey is made by hymenopterans to feed hymenopterans. However, this needs frequent replenishing as it tends to dry-up and so become less available. Azzouz et al. (2004) found with the aphidiine *Aphidius ervi* that the stronger the sugar solution, up to 70% w/v, the greater was the improvement in longevity. They also found that frequency of feeding was important, with wasps faring better if fed daily rather than every other day.

A few other studies have examined the effects of different pollens as supplements to the honey water, as no doubt few flower-visiting parasitoids obtain nectar without some pollen contamination, and pollen is rich in amino acids/proteins whereas nectar generally is not. As ichneumonoids cannot eat solid food, the basic concept is that nectar or other liquid food sources might get 'contaminated' by pollen and as pollen is relatively rich in protein, this might be advantageous to the wasp. Leius (1963) reported that wild carrot and tulip pollens increased both fecundity and longevity in females of the pimpline ichneumonid *Scambus buolianae*, whereas various other pollens added to diet increased either fecundity or longevity, but not both, and suggested that introductions of potential control agents would be more likely to succeed in areas with an abundance and diversity of plants as sources of both nectar and pollen. Eijs et al. (1998) found that alder (*Alnus*) pollen had a significant, positive effect on wasp fat reserves compared with no food alternative for the alysiine *Asobara tabida*, although *Plantago* pollen did not yield any increase, but it is not clear whether this was due to lack of consumption or digestion, and Giraldo-Vanegas & Garcia (1995) found that addition of pollen to sugar water led to an increase in both longevity and realised fecundity in the synovigenic cremastine *Eiphosoma vitticole*.

Balancing the requirements for adult food and hosts poses an interesting dilemma for a female parasitoid, as searching for food resources away from host substrate takes time and costs energy (Eijs et al. 1998). The trade-off between egg-production and life span was examined in detail for the alysiine *Asobara tabida* by Ellers et al. (2000) by manipulating egg production and measuring longevity – the results showed a linear trade-off, with wasps that produced more eggs living significantly less long. This species also displays geographic variation in egg load at emergence, with southern European strains having more mature eggs and lower fat reserves that northern ones, and in addition, egg production increased with the availability of food and also with the availability of hosts (Ellers & van Alphen 1997), the northern strain showing greater plasticity. When females of the small, gregarious, ectoparasitic braconine *Habrobracon gelechiae* are given access to either no food, water alone, water and honey, or water and honey and hosts from which to host-feed, provision of honey greatly increased longevity (Fig. 14.3), but host-feeding opportunity does not lead to any greater longevity, again showing

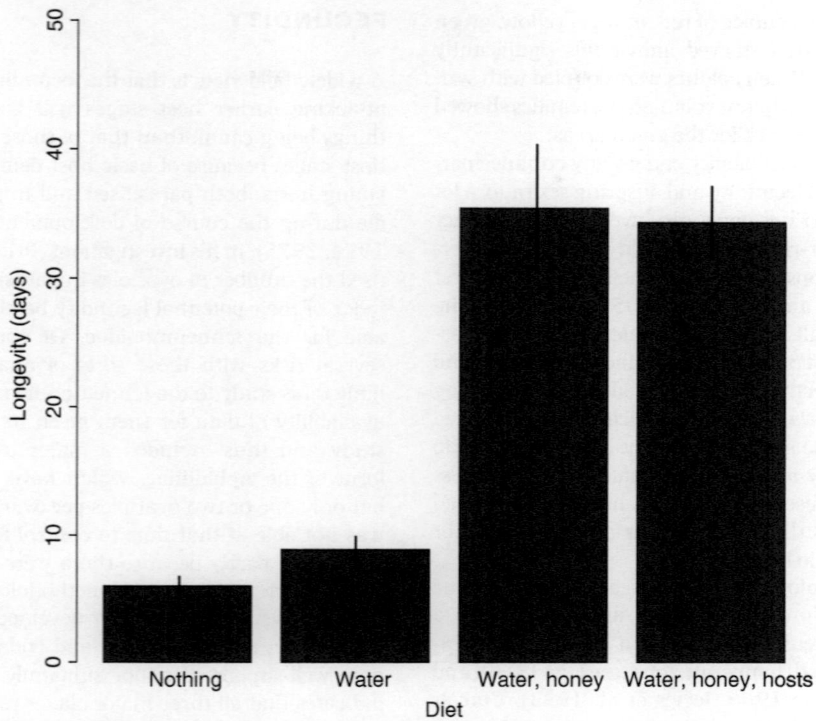

Fig. 14.3 Effects of four different adult 'diets' on adult female longevity (±SE) in the idiobiont ectoparasitoid *Habrobracon gelechiae* (Braconinae). (Source: data from Daane et al. 2013.)

that what resources are derived from host-feeding are important only for egg maturation and not for female survival (Daane et al. 2013).

Several ichneumonoid species have been demonstrated to show a broad attraction to odours from flowering plants. Wäckers (2004) assessed the attractiveness of odours from 11 flowering plant species to two braconids, *Cotesia glomerata* (Microgastrinae) and *Heterospilus prosopidis* (Doryctinae), and one ichneumonid, *Pimpla turionella*, and found a high degree of consistency among the parasitoids in terms of which plants were most to least attractive or repellent and found that attractiveness did not necessarily correlate with nectar availability. Females of the microgastrine *Microplitis mediator* showed strong innate attractiveness towards a range of wild flowers that might be expected in agro-ecosystems, with the wasps showing a clear preference for two of them (Belz et al. 2013). From an applied point of view, these studies indicate that if nectar source plants are going to be provided for the benefit of beneficial parasitoids, then both their attractiveness and nectar availability need to be taken into account, and this could explain some of the variability observed in trials with wild flower strips found in various applied studies (Belz et al. 2013).

In addition to showing attraction towards volatiles from flowering plants, some ichneumonoids show an innate preference for the colour yellow, something that is taken advantage of with yellow pan traps. In the wild colour might be expected to be strongly associated with flowers (Lucchetta et al. 2008), and interestingly this cannot be modified by associative learning when paired with food reward, at least in the species studied. So unlike in bees, it seems that ichneumonids do may not employ learning of a particular colour as indicative of the presence of a nectar-type food resource. Colour preference has been shown to be modulated by the wasp's hunger status in the microgastrine *Cotesia vestalis*, a parasitoid of the diamondback moth, *Plutella xylostella*. Uefune et al. (2013) found in a choice experiment using an arena with differently coloured light zones that satiated females showed no colour preference given a choice of red, orange, yellow, green and white, but that starved individuals significantly preferred yellow. When colours were coupled with synthetic host/host foodplant volatiles, the females showed a significant preference for the green areas.

Because nectar availability can greatly enhance parasitoid longevity, fecundity and offspring sex ratio, a lot of research effort has gone into investigating whether providing flower-rich strips at field margins or interspersed with crops can have a beneficial effect on pest control. Berndt and Wratten (2005) showed that the longevity of adult males and females of the microgastrine *Dolichogenidea tasmanica* was increased three- and sevenfold, respectively, when provided with flowering alyssum (*Lobularia maritima*, Brassicaceae), and this led to greater female lifetime fecundity. Offspring sex ratio was also greatly improved as females without access to the nectar resource produced highly male-biased broods whereas those with nectar produced more or less even sex ratio broods.

Ichneumonidologists are well aware that some umbelliferous flowers (Apiaceae) are attractive to a number of ichneumonids and, to a lesser extent, braconids in the north temperate zone (see Figs 13.30f and 13.31b,c) (Hirose 1966, Jervis et al. 1993). Umbels are the plants of choice for most temperate nectivorous ichneumonoids because they have exposed nectar, although various other plant groups also provide relatively easy access. In more arid areas where water is less available and umbellifers and other plants with exposed nectaries far less common or absent, a number of parasitic wasps have evolved elongated proboscises of various forms, referred to by Jervis (1998) as 'concealed nectar extraction apparatus' (CNEA), that enable them to reach nectar in plants such as various Asteraceae (Sharkey & Mason 1986, Jervis 1998).

Jervis et al. (1992) presented a brief set of guidelines for assessing adult diet, noting both direct and indirect sources of evidence. Technology has moved on somewhat and it is now far easier, given the correct equipment, to determine rather more about diet using either polymerase chain reaction (PCR) to test for host DNA in a parasitoid's gut that might indicate host-feeding or HPLC to obtain detailed quantitative analyses of small amounts of sugars that would reflect feeding on nectar or honeydew. For small parasitoids, a qualitative cold anthrone test can be used to determine whether individuals have been feeding on sugar sources at all (Heimpel et al. 2004, Heimpel & Jervis 2005).

FECUNDITY

A widely held view is that the fecundity of parasitoids attacking earlier host stages will be greater, other things being equal, than that of those attacking later host stages because of basic host demography; many young hosts, both parasitised and unparasitised, will die during the course of development (Price 1973a, 1974, 1975). In his investigations, Price (1974, 1975) used the number of ovarioles in parasitic wasps as an index of their potential fecundity based on data available for the Ichneumonidae. Of course, there are several risks with these sorts of analyses. First, he limited his study to the Ichneumonidae because of the availability of data for them given in Iwata's (1960) study and thus excluded a major exception in the form of the Aphidiinae, which have high egg loads but only one or two ovarioles per ovary. Second, Price was not able at that time to control for the effects of phylogeny, partly because there were no meaningful phylogenetic estimates and methodologies of comparative analysis had not been developed. Figure 14.4 shows Price's data redrawn and coded based on the now well-supported major subfamily groupings and indicates that all three major clades represented show the same trend. Alongside this, Price (1973a) showed a strong negative correlation between the number of ovarioles and the ranked developmental stage of the host attacked, with ichneumonids attacking early instar hosts having larger numbers of ovarioles. This is, of course, not independent of egg size since wasps cannot simultaneously have very large numbers of ovarioles containing very large eggs (Price 1973b). Jervis et al. (2012) re-analysed the data that had led to the paradigm that parasitoids that attack hosts at earlier stages in their life history are more fecund in a phylogenetically independent way and found that the conclusion was still supported, but cautioned that the apparent relationship could well be a sampling artefact and that it should not necessarily be assumed to apply to all natural host–parasitoid systems.

Lifetime fecundity varies enormously between species, in terms of both potential fecundity (i.e. the number of eggs that a female parasitoid could conceivably make and lay) and realised fecundity (how many offspring she actually produces). Some small idiobionts such as *Habrobracon gelechiae* have lifetime fecundities (in the laboratory) of approximately 200 eggs laid (Daane et al. 2013). In idiobiont parasitoids, especially those

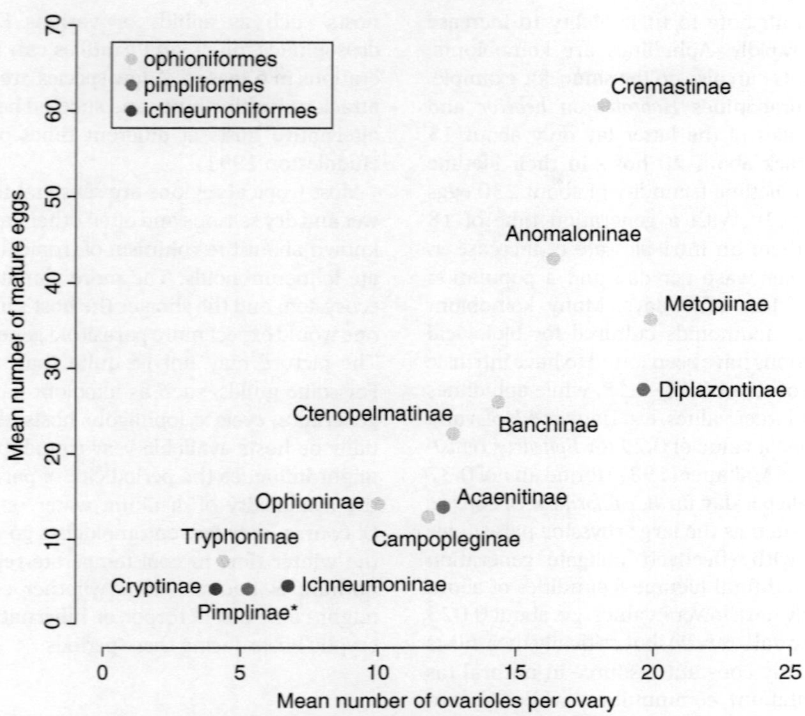

Fig. 14.4 Relationship between number of mature eggs per ovary and number of ovarioles per ovary for 13 subfamilies of Ichneumonidae, coloured to show major subfamily groupings. (Source: data from Price 1973, fig. 3.) For a colour version, see Plate 61.

attacking wood-borers, the numbers of mature eggs at any one time as well as lifetime fecundity is generally much lower. The doryctine braconid *Syngaster lepidus*, used in the biological control of eucalypt stem-borer beetles (*Phoracantha* spp.), even under good conditions, lay only about 20 eggs in their lifetime and the same value has been estimated for the rhyssine *Megarhyssa* (Heatwole et al. 1962).

Some workers have effectively taken potential realised fecundity to equate to whether the parasitoid is an *r*- or a *K*-strategist, those with high fecundities being at the *r*-selection end of the spectrum. However, as Godfray (1994) pointed out, this really serves no purpose and is not necessarily in concordance with the original definitions of the terms *r*- and *K*-selection, which instead relate to whether a species is generally being selected under conditions when its population is either far away from or nearly at *K*, the carrying capacity. Godfray provided a very nice hypothetical example in which two parasitoids that are at equilibrium with the same host simply attack it at different times in its development. The species attacking the earlier host stage will suffer higher juvenile mortality as a consequence of the higher mortality among younger hosts, and therefore it will be selected for higher fecundity; however, both species are nevertheless *K*-selected because their populations are at equilibrium. Of course, there will be a broad correlation between selection mode and fecundity, but the real point is that *r*-selected species will have unstable populations and when opportunity arises, will be able quickly to build up to high population levels before some future crash. Typical of these might be aphidiines, since their hosts typically increase in abundance rapidly as soon as spring arrives and the aphidiines having high fecundities can become very abundant through the host season, but suffer a crash at the end of the year. Associated with their high fecundity is also their relatively short generation time, both of which contribute to their ability to increase population size rapidly. Aphidiines are koinobionts, but some idiobionts can also do the same, for example, the gregarious braconines *Habrobracon hebetor* and *H. gelechiae*. Females of the latter lay only about 15 eggs per day, attack about 20 hosts in their lifetime and have a mean lifetime fecundity of about 230 eggs (Daane et al. 2013). With a generation time of 18 days, this gives them an intrinsic rate of increase, *r*, of 0.24 females per wasp per day and a population doubling time of just 2.88 days. Many koinobiont braconids and ichneumonids cultured for biological control investigations have been found to have intrinsic rates of increase of about 0.14–0.18, while aphidiines often have rather larger values, e.g. Hågvar & Hofsvang (1990) determined a value of 0.29 for *Ephedrus cerasicola* and Cohen and Makauer (1987) found an *r* of 0.37 females per female per day for *E. californica*. Of course, many idiobionts such as the large rhyssine parasitoids of wood-borers with effectively obligate generation times of 1 year and total lifetime fecundities of about 20 will necessarily have lower *r* values, i.e. about 0.025 given a roughly equal sex ratio, but critically their hosts are probably a fairly constant resource in natural (as opposed to plantation) communities, while Akdura et al. (2012) found an intermediate value of 0.098 for the ectoparasitic, idiobiont cryptine *Agrothereutes hospes*.

Female fecundity in numerous species is positively related to body size (M.E. Visser 1994), which is, of course, a great deal more variable in idiobionts than koinobionts. With the idiobiont pupal parasitoid *Itoplectis naranyae*, which host-feeds to continue maturing eggs after its initial load has been used, H. Liu & Ueno (2012) found that body size was also positively correlated with longevity. Thus larger wasps may intrinsically be able to produce more eggs, but also may live longer and so have a greater opportunity to find and oviposit in hosts.

VOLTINISM AND SEASONALITY

In temperate regions, parasitic wasps can usually be categorised easily as being univoltine, bivoltine or, in some cases, plurivoltine. Most temperate univoltine ichneumonoids have emergence patterns well synchronised with those of their hosts, although there are many exceptions. Some smaller bodied taxa with relatively rapid development and attacking plurivoltine hosts such as aphids or various Diptera such as drosophilids, muck and fruitflies can have many generations in a season. A few species are plurivoltine but attack univoltine hosts and succeed because they have alternative hosts at different times of year (Shaw & Huddleston 1991).

Most tropical regions are seasonal too, with distinct wet and dry seasons and often other variants. Far less is known about the voltinism of tropical than of temperate ichneumonoids. The more climatically stable the ecosystem and the shorter the host's life cycle, perhaps one would expect more parasitoid generations per year. The picture may not be quite that simple, however. For some guilds, such as idiobiont parasitoids of long generation cycle xylophagous hosts, there may potentially be hosts available year round, but other factors might influence the periodicity of parasitoids, such as the availability of drinking water or nectar sources. Of course, very few entomologists go out collecting in the winter time in cool temperate regions, so almost nothing is known about whether some parasitoids might come out of torpor or hibernation to forage for cryptic hosts during these periods.

DAILY ACTIVITY PATTERNS

The activity of parasitic wasp adults is likely to be determined by their need to find hosts and in many cases to obtain water and/or carbohydrate food sources, and to some extent these behaviours may be temporally separated. Water in the form of dew is often only present early in the day and Townes postulated that this was particularly important for tropical ichneumonids. As most field entomologists will know, a lot of insects become active when the daytime temperature warms up, but also a lot become less active during the warmest hours of the day. Such daily behaviour patterns are even more pronounced in more arid and warmer regions and activity can be strongly bimodal (González-Moreno et al. 2012), contrasting with the typical unimodality of cooler temperate regions (e.g. Owen et al. 1981). These authors' data suggest a slightly less bimodal pattern for Braconidae than Ichneumonidae but the sample size was smaller and therefore it is not possible to know whether it represents a genuine difference; they did, however, find statistically significant differences between the families, with ichneumonids being more active at intermediate humidities and wind speeds, whereas braconids were more active at lower humidities.

Although most ichneumonoids are diurnal, a substantial number distributed across many subfamilies have evolved to seek hosts at night. In the section *The 'ophionoid facies'* in Chapter 10 some of the anatomical adaptations that are associated with nocturnal behaviour are discussed, but the reason to discuss it here is that it is strongly associated with koinobionts. Nearly all the nocturnally active ichneumonoids are parasitoids of caterpillars or sawfly larvae and probably mostly ones that hide away during the day as part of a strategy to minimise their exposure to predation by birds and other diurnal predators. Hence to access these, the wasps too need to be night-active. However, as discussed above, exposed hosts are almost entirely unsuitable for idiobiont parasitoids – paralysing such a host when it was exposed on a leaf would effectively be suicidal.

More recently, greater effort has been put into documenting all catches of ichneumonoids at light traps and this has revealed potential nocturnal activity in a number of genera that are not normally considered nocturnal and which do not display the ophionoid facies. These include numerous Alysiinae and Blacinae, among others (Lozan 2002). The problem here is that just because insects come to a light trap it does not mean that they are normally nocturnally active – I suspect that many such species are just adventitious, having been resting at night within the range of illumination from the trap and then disturbed and flying towards the light. For many of these groups, particularly within the cooler temperate regions, there is no obvious advantage that can be associated with nocturnal activity such as host accessibility. In hot tropical regions, there may be an advantage to nocturnal activity, i.e. avoiding unnecessary heating or evaporation, but this can only be an efficient strategy if hosts can be located and accessed by non-visual senses.

DIAPAUSE

Many insects undergo periods of diapause, a period of arrested development and low metabolic rate, usually as larvae or sometimes as pupae, as an adaptation to synchronise with climatic or biotic conditions. Diapause is a hormonally controlled condition which can be triggered by both biotic and abiotic environmental cues that indicate that unfavourable conditions will follow soon. Diapause may be obligate, such as during overwintering in a temperate univoltine species or facultative in plurivoltine species, in which it can be triggered by a variety of abiotic or biotic factors. Some parasitoid species have a proportion of their larvae entering diapause at each generation and prolonged diapause has been noted in many ichneumonoids (e.g. Simmonds 1947a, Parrish & Davis 1978, Donovan 1991, van Achterberg & Altehofer 1997). Zijp and Blommers (2002c) discussed the possible evolutionary advantage of variable prolonged diapause in the ctenopelmatine ichneumonid *Lathrolestes ensator*, a parasitoid of the apple sawfly, *Hoplocampa testudinea* (Hym., Tenthredinidae), and suggested that because this species does not avoid superparasitism, this may reduce the risk of local extinction, although this argument seems to be based on group selection.

In many taxa, diapause is facultative, but the cues that wasp larvae, prepupae or pupae use for determining whether they enter diapause are largely not understood and only a few cases have been investigated in detail. For several species, daylength is important (e.g. Omata 1989, Coop & Croft 1990). For example, Claret (1973) showed that in the generalist pupal endoparasitoid *Pimpla rufipes* (as *instigator*), entry into diapause was determined by the photoperiod experienced by the final (fifth) instar parasitoid larva. However, it never reached 100% of individuals, males showed a lower percentage diapause response and it could be broken relatively quickly by longer daylength (12 days at 16 hours). Further, the effect of daylength was found to be moderated by temperature and at 30 °C no larvae would enter diapause (Claret 1978), and a similar thermal moderation has been found for the microgastrine *Cotesia plutellae* (Alvi & Momoi 1994) and the alysiines, *Alysia manducator* and *Aphaereta minuta* (Vinogradova & Zinovjeva 1972a,b, Zinovjeva 1974, 1976, 1978). With the alysiines, entry into diapause depended upon whether they were reared at constant temperature or under fluctuating and hence far more natural conditions. Alvi and Momoi also found geographic differences in the susceptibility of populations to enter diapause. Larvae of the ichneumonine *Trogus mactator* make the decision as to whether or not to enter diapause when they reach their third instar and this is independent of the host papilionid, which commits to diapause immediately after pupation (Omata 1989). The situation with the campoplegine *Bathyplectes curculionis* is more complicated (Parrish & Davis 1978) and although most individuals spend 10 months of the year as a diapausing prepupa within their cocoons, some individuals that emerge early produce non-diapausing larvae. Although there was inter-individual variation, the important factors seemed to be synchronisation of temperature change with photoperiod. Development of the north, temperate braconid *Meteorus trachynotus* (Euphorinae) seems to depend largely or solely on its host caterpillar's physiological state to determine when it enters and emerges from diapause, although parasitised hosts generally spent less time in diapause than unparasitised hosts and diapause duration variation was also reduced (Thireau & Régnière 1995).

In some other temperate species diapause is obligatory and the wasps may require a given period of chilling before emergence [e.g. *Lophyroplectus luteator* (Ctenopelmatinae), K.J. Griffiths 1975; *Banchus flavescens* (Banchinae), Arthur & Mason 1985]. Obligate diapause might be associated with climates that always involve a protracted harsh period and when premature exit from diapause following a brief inclement period would be fatal. Sometimes, only a proportion of parasitoids in any one generation enter diapause, the rest emerging in synchrony with the next available hosts) e.g., *Coeloides dendroctoni* (Braconinae), de Leon 1935].

Emergence from diapause can be signalled by photoperiod, temperature, humidity or host factors. For the North American tryphonine *Phytodietus vulgaris*, Coop and Croft (1990) found that its facultative diapause could be broken by 8 days of exposure to long photoperiod, irrespective of temperature, whereas in *C. plutellae*, high temperature was important and daylength had no effect (Alvi & Momoi 1994). Although the cues for diapause exit have only been investigated in a relatively small number of species, the data suggest that in those of tropical origin, where there is little annual variation in daylength, humidity might be the main cue. This was demonstrated by Seymour and Jones (2000), who investigated diapause escape in the microgastrine *Microplitis demolitor* and found that temperature and daylength had little effect, but once triggered by an increase in humidity, adults eclosed 38 days later more or less independently of temperature (Fig. 14.5).

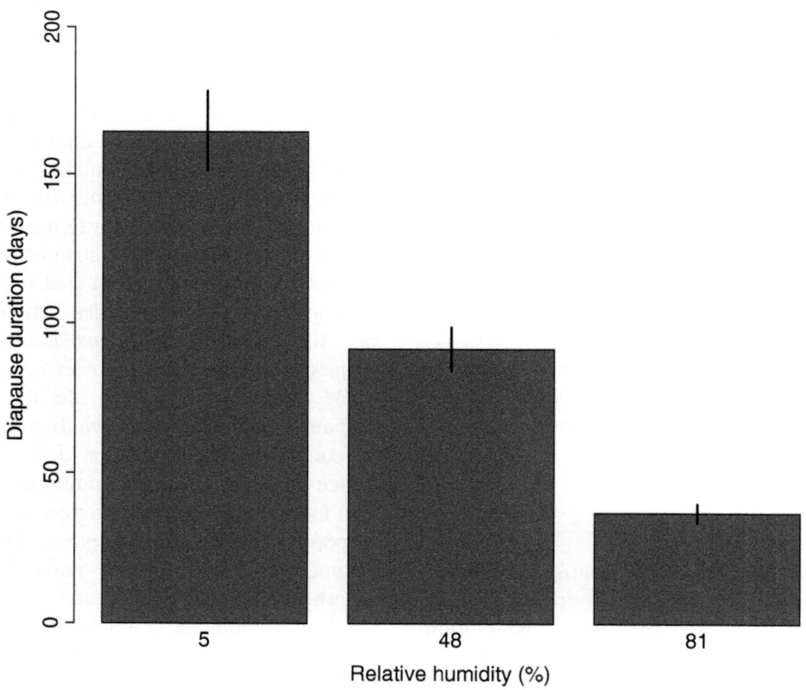

Fig. 14.5 Effect of humidity on diapause duration (±SE) in the tropical microgastrine *Microplitis demolitor*. (Source: data from Seymour & Jones 2000.)

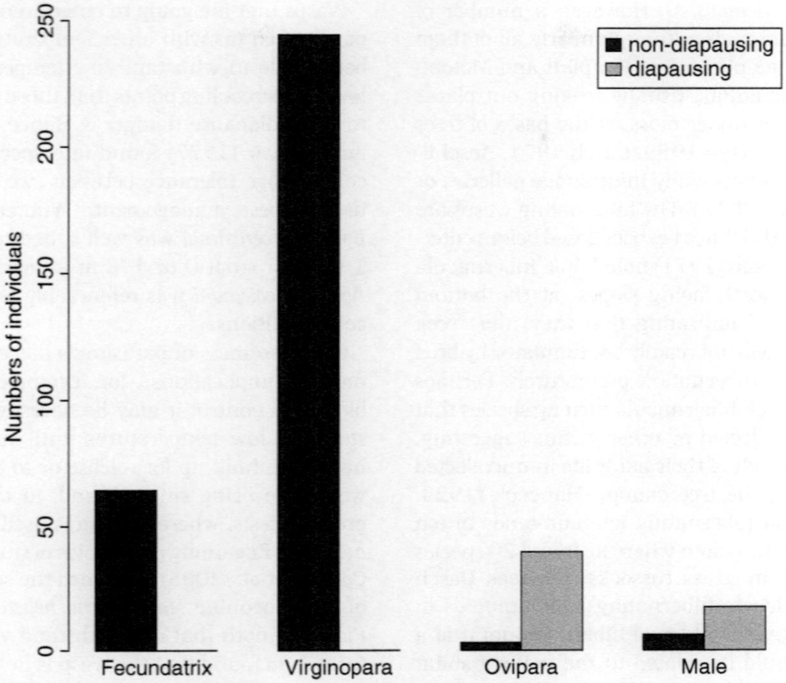

Fig. 14.6 Effect of host aphid, *Schizolachnus pineti*, type on entry of the larva of the aphidiine braconid *Pauesia unilachni* into diapause. (Source: data from Polgár *et al.* 1991.)

Omata (1989) indicated that there may be a difference between those koinobiont endoparasitoids that that are in their first instars when the host and they enter diapause and those that do so at a later stage of parasitoid development as in the case of the *Trogus* she studied. In the former case, the host's physiological state generally appears to be critical to both initiating and terminating diapause, whereas in the latter, the parasitoid may respond partly or wholly independently of the host's condition. Similarly, Brown et al. (1990) found that diapausing first instar larvae of the chelonine *Ascogaster quadridentata* responded to an increase in host 20-hydroxyecdysone which was triggered by a host cephalic factor soon after transfer to long day and warm, diapause-breaking conditions. Ligation of the host's head prevented the hormonal increase post-ligation and the *Ascogaster* larva did not respond to the new environmental conditions of its own accord. Interestingly, the nymphal and adult parasitising, plurivoltine Euphorinae are not known to show any larval diapause unless the host is also in diapause (S.R. Shaw 2004), whereas in the aphidiine *Pauesia unilachi*, a major factor determining whether they enter diapause is the morph/sex of the aphid host attacked (Fig. 14.6) (Polgár et al. 1991, Polgár & Hardie 2000).

In the bark beetle parasitoid *Celoides brunneri* (Braconinae), R.B. Ryan (1965) found that the probability that a wasp larva would enter diapause depended considerably on the conditions experienced by the mother and that the photoperiod that logs containing parasitoid larvae were exposed to had no significant effect on the probability of them entering diapause or not.

COLD HARDINESS, HIBERNATION AND OVERWINTERING

In the temperate region, the parasitoid's overwintering stage is most usually either as an egg or young larva within its hibernating host, as a prepupal larva within a cocoon, sometimes as a pharate adult within the cocoon in species that have to become active very early in the season [see Chapter 12, Agathidinae (*Earinus*) and Sigalphinae (*Acampsis*)]. However, a number of temperate ichneumonids, although nearly all of them are Ichneumoninae plus a few Pimplini and Metopiinae, hibernate as adults, usually seeking out places under loose bark or under moss, at the bases of trees (Aubert 1959, Rasnitsyn 1959, Dasch 1971, Aeschlimann 1974b) or occasionally in drainage galleries or caves (Penado et al. 2013). The hibernating wasps are always females and all have been fertilised before entering hibernation. Dasch (1971) noted that 'hibernacula were located on north-facing slopes, at the bottom of steep ravines, ... ' indicating that they must seek places where they will not readily be stimulated by brief warm days to exit hibernation prematurely. Perhaps not surprisingly, such hibernacula turn up species that are very seldom collected by other means suggesting, that they spend much of their adult life in uncollected places, perhaps in the tree canopy. Hancock (1923, 1925) reported on hibernating ichneumonids in fen land in the United Kingdom where he found 20 species hibernating deep in grass tussocks, whereas Dasch (1971) failed to locate hibernating ichneumonids in this microhabitat, so the type of hibernaculum that a species utilises could be related to the major habitat type that it inhabits.

Salt (1959) investigated the cold hardiness of the wheat stem sawfly parasitoid *Bracon cephi*, which overwinters as a larva, and found that its haemolymph contained up to 5 M glycerol, which depressed its melting point to approximately −17.5 °C and depressed the supercooling temperature even lower. Salt noted that the glycerol therefore protects the braconid larva in two ways, by reducing the likelihood of it freezing and by protecting its tissues if it did freeze. The Arctic and large areas of continental Holarctic can experience very cold temperatures. Humble (2006) investigated freezing tolerance in various Arctic sawflies and their parasitoids, including the ectoparasitoids *Bracon* sp., *Adelognathus* sp. and *Scambus vesicarius* (Pimplinae) and the endoparasitoids *Ichneutes* (Ichneutinae), *Lathrostizus* (Campopleginae) and *Syndipnus* (Ctenopelmatinae). After acclimation, most ectoparasitoids could be supercooled to between −30 and −35 °C, although *Adelognathus* only to −25 °C. *Adelognathus*, however, differed from the other ectoparasitoids in that it showed a degree of freezing tolerance. Freezing of the endoparasitoids, occurred when the host larva froze, as a result of ice crystal nucleation originating at the host's hind gut, but all endoparasitoids were found to be freezing tolerant (Humble & Ring 1985).

Wasps that are going to enter an overwintering diapause often (as with other temperate insects) become better able to withstand low temperatures and have lower supercooling points than those that are not going to enter diapause (Langer & Hance 2000). Hofsvang and Hågvar (1977) found interspecific differences in cold storage tolerance between two aphidiine species use in pest management. Whereas the aphidiine *Ephedrus cerasicola* was well suited to cold storage for 1 week or so at 0 or 1 °C at an early mummy stage, *Aphidius colemani* was remarkably less tolerant of the cold conditions.

Cold hardiness of parasitoids has considerable commercial implications, for example, in inundative biological control it may be necessary to keep wasps stored at low temperatures until sufficient numbers have been built up for release or to prolong life when wasps are being shipped, and, in the case of stored product pests, where cooling is used as a pest control measure. Regarding parasitoids of stored product pests, Carrillo et al. (2005) examined the supercooling point of the braconine *Habrobracon hebetor*, which attacks *Plodia*, a moth that can withstand very low temperatures, and found that the wasp is not freezing tolerant and that supercooling points of eggs and pupae were significantly lower than those of larvae and adults. Mortality at −5 and −12 °C both increased with duration of exposure to the low temperature in a strongly sigmoid way, such that most individuals could withstand brief exposures of a few hours, as might be experienced overnight, but with all individuals being killed by prolonged exposure (e.g. 7–10 days) even at −5 °C. Younger adults of the campoplegine *Venturia canescens* are better able to withstand very low temperatures than older ones and fed ones better still (Andreadis et al. 2014), although the hosts of this species tended to have lower supercooling points than the wasp.

Duffield and Nordin (1970) compared the glycerol contents, after 3 weeks of cold-shock, of females of seven North American species of Ichneumoninae that regularly hibernated with those of five, similarly treated, non-hibernating species belonging to different subfamilies, although unfortunately non of non-hibernating ichneumonines. All of the ichneumonines contained substantial concentrations of glycerol after the treatment whereas there was no detectable glycerol in the other species (Fig. 14.7). Two of the tested ichneumonine species were further tested after being held for a further 1 month at −1 °C, by which point they had very substantially

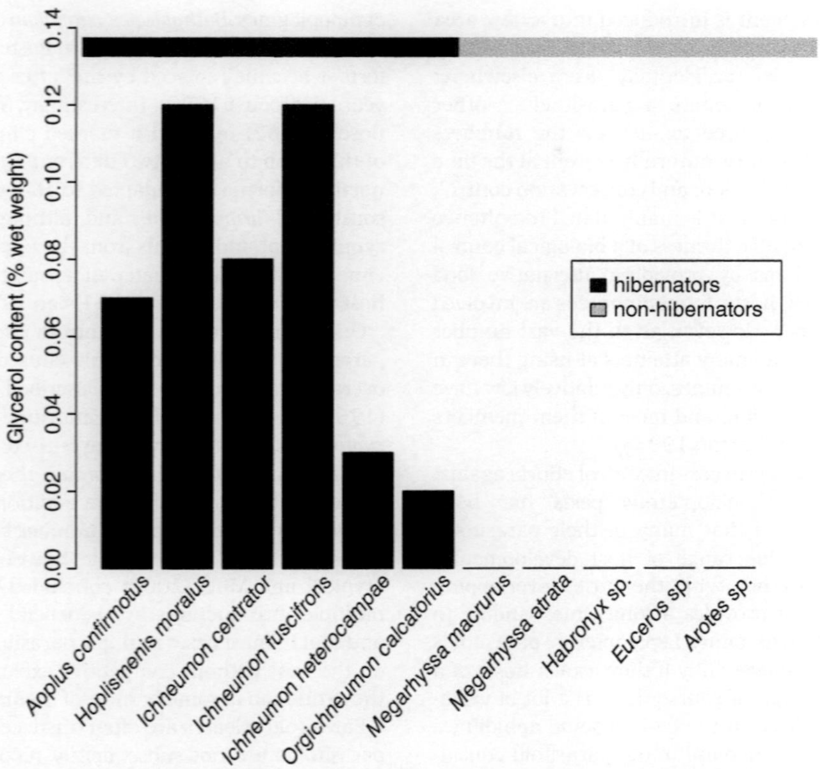

Fig. 14.7 Antifreeze (glycerol) content of adults of six hibernating ichneumonines and five other non-hibernating ichneumonids collectively representing four other subfamilies, following 3 weeks of cold stress at 4 °C. (Source: data from Duffield & Nordin 1970.)

increased glycerol levels, up to 0.46% of body weight. Unfortunately, they did not measure the concentration of sorbitol, another compound used by many hibernating insects to lower the freezing point.

COLORATION AND THERMOREGULATION

There has been surprisingly little work on the role of coloration on the body temperature of parasitic wasps, although it is widely known to be important in many groups of insects. Ichneumonids occurring at high latitudes or high altitudes are predominantly dark bodied and this is generally assumed to because their bodies will warm up more quickly given the available sunlight and thus enable them to forage for longer. Experimental demonstration of the importance of dark bodies in cold conditions has recently been provided by Abe et al. (2013), who investigated the effects of rearing temperature on body colour in various strains of the thelytokous euphorine *Meteorus pulchricornis*. Rearing at lower temperatures led to increasing melanisation and differences in the pattern of darkening were apparent between different thelytokous lines. It was demonstrated that the melanisation allowed darker forms to reach body temperatures 2 °C above ambient in sunshine and that this gave them a considerable advantage in terms of potential flight activity and hence foraging time when ambient conditions were cold.

BIOLOGICAL CONTROL

Biological control programmes are usually one of three types: classical biological control, where an alien species of control agent is introduced into a new area in the hope of controlling a pest which has usually also been introduced (accidentally) from elsewhere; augmentative control, where a parasitoid or other natural enemy is released to increase the numbers present in the field if those naturally present at the time are insufficient to do the job; and conservation control, in which the environment is manipulated to enhance the survivorship or effectiveness of a biological control agent there, perhaps by providing alternative food sources or non-pest hosts. Ichneumonoids are involved in all of these types. However, given the vast number of species and despite many attempts at using them in biological control programmes, only relatively few have proved highly successful, and most of them members of the Braconidae (Wharton 1993a).

The greater general success in control efforts against homopteran than Lepidopterous pests has been attributed to the fact that many of their parasitoids can parasitise a wide range of host developmental stages which, together with the host's overlapping generations, which provides a long time window to effect attack whereas many Lepidopteran parasitoids can only develop successfully if they attack hosts of a given stage, although of course there is a lot of variation. Further, at least in the case of some aphidiines, the intrinsic rate of increase of the parasitoid considerably exceeds that of its host. The small, gregarious, ectoparasitic idiobiont *Habrobracon hebetor* has been used for the biological control of stored product pests for many years and one of the reasons for its success, especially in small-scale operations, is that it attacks the wandering final instar larval stage of hosts such as the almond moth (*Cadra cautella*: Pyralidae) (Hagstrum & Smittle 1977), thus reducing the spread of the pest to nearby uncontaminated stores (Cline & Press 1990).

When pest species are attacked by multiple parasitoids, there has long been the question as to whether it is better to introduce one or multiple parasitoid species in biological control efforts. Will one of the parasitoids outcompete the other(s) with no net benefit? In some cases, one biocontrol agent may out-compete and replace another, although from a control perspective this is not necessarily bad, as the winner is likely to eliminate more of the pest. One example involves the alfalfa weevil, *Hypera postica*, which is a major pest species attacking various herbaceous Fabaceae in a number of countries where it is attacked/controlled by a number of parasitoids and fungi, some of them introduced as biocontrol agents. Two ichneumonids, the campoplegines *Bathyplectes curculionis* and *B. anurus*, attack it in North America and in Eastern Ontario. The former became replaced by the latter over a period of years (Harcourt 1990). Interestingly, Salt and van den Bosch (1967) reported a marked cline in the ability of this wasp to attack two different *Hypera* species. In north California it is adapted to *H. postica* and in the south to *H. brunneipennis* and, although not perfectly symmetrical, individuals from the opposite end of the cline experience far greater encapsulation in the other host weevil species (Maund & Hsiao 1991).

One example where the introduction of a second parasitoid has almost certainly caused a reduction in overall parasitism levels was described by Arthur et al. (1964). Here females of the cremastine *Temelucha interruptor* preferentially multiparasitised hosts attacked by the braconid *Orgilus obscurator*, the first parasitoid. Although they also kill some additional hosts themselves, they must reduce the number killed by the first parasitoid. This latter example, however, seems rather atypical and Mills (2007) concluded that in general multiple introductions are beneficial in pest control and that in most cases multiple parasitoids will co-exist on the host without competitive exclusion, leading to the extinction of some or most of them.

Parasitoid releases are often unsuccessful in that the parasitoids are not subsequently recovered and only about 25% are successful (e.g. Hirose 2006). Hawkins (1994) listed the results of 74 classical biological control attempts against insect pests and these data were utilised by Hawkins et al. (1993) to demonstrate that refuge theory could be applied to decisions about whether to release new parasitoids in classical biological control. The data showed a significant relationship between the maximum recorded level of parasitism achieved and the success (or partial success) of the programme, but there is considerable overlap. In those relationships where the parasitoid never achieves a high level of parasitism it is possible that the host has a refuge, i.e. some proportion of the population is not vulnerable to parasitoid attack. Of these 74 cases, surveyed 17 involved only a braconid and nine only ichneumonids, the remainder being made up predominantly by chalcidoids and tachinids and a few cases in which multiple parasitoid taxa were released together. The results of analysing the ichneumonoid data separately are essentially the same as for the whole data set, although with somewhat lower statistical significance (Fig. 14.8). Twelve of the attempts using Braconidae were deemed as either fully or partially successful.

Some attempts fail because, despite fairly high parasitism levels sometimes being achieved, climatic or seasonal synchrony factors may lead to extinction of the parasitoid.

Sometimes parasitoids cannot be recovered following release for a number of years, but nevertheless still survived at low population densities, later probably adapting better to the local hosts or conditions and becoming detected. This is in some ways to be expected since, even when very large numbers of parasitoid individuals are released, they are likely, especially if they have been subject to a mass-rearing effort, to represent only a part of the natural genetic variation in their home range. Shouba et al. (2005) reported the example of the released *B. anurus* in Japan, where *H. postica* is a major pest of Chinese milk vetch, an important nectar source for honey production. The wasps were introduced from the USA into Japan 1988 and 1989 but were not recovered until 1996, when they were found to be causing rather low host mortality (ca 5%) (Shouba et al. 2005). In the following years, however, the percentage *Hypera* attacked increased to around 30%, which was also associated with decreased loss of milk vetch.

The success of biological control attempts using introduced, non-native parasitoids will always depend on appropriate synchronisation of parasitoid and host life cycles (e.g. Lyons 1999), and, along with the availability of alternative hosts, is probably one of the main reasons underlying failures. As few ichneumonoids are monophagous, alternative hosts are probably of considerable importance to parasitoid persistence in local populations when their main hosts suffer major population crashes, yet, for obvious reasons, the ability to utilise alternative hosts is something that biological introduction legislation frowns upon. Parasitoids may also show local adaptations to climatic conditions. Attempts in the 1950s to control the accidentally introduced walnut aphid, *Chromaphis juglandicola*, in California using the aphidiine *Trioxys pallidus* largely failed despite numerous mass releases, because the strain being released originated in France whereas the bulk of commercial walnut production was in the central valley where conditions are hot and dry. Another release programme was initiated in 1968 using *T. pallidus* obtained from Iran, where the climate is much more similar, and this time it was quickly successful (van den Bosch et al. 1982). Another such example involves the control of the *Brassica* pest *Pieris rapae* in the north-east USA. Initially, the gregarious Eurasian microgastrine *Cotesia glomerata* was important, whereas attempts to establish a European strain of the solitary parasitoid, *C. rubecula* were unsuccessful. However, these failures are believed to have been due to diapause cues that were inappropriate for the New England climate (Nealis 1985). When the same species of parasitoid was sourced from the climatically similar part of China (near Beijing), it rapidly became a very successful control agent (Van Driesche 2008) and supplanted *C. glomerata* as the primary control agent. Dowell and Horn (1977) postulated that the slow increase in parasitisation of the alfalfa weevil, *Hypera postica*, by the campoplegine *Bathyplectes anurus*, and the concomitant decline in parasitisation by *B. curculionis* and the eulophid *Tetrastichus incertus* in Ohio, was due to the better synchronisation of the first of these with its host's phenology.

When recently in Kenya the diamondback moth, *Plutella xylostella*, which is normally an oligophagous herbivore on various *Brassica* spp., showed a host plant switch to two bean crops with economically significant results, it was found that its campoplegine parasitoid, *Diadegma mollipla*, had successfully followed it. When parasitoids from one host plant group were made to attack caterpillars on the other, they were more successful on the novel host, indicating that an independent arms race between host and parasitoid was going on independently in the two groups of host plants, with the hosts having evolved specific defence mechanisms against the population of parasitoids attacking them, which parasitoids from caterpillars on the other host plant were able to avoid (Rossbach et al. 2006).

There are, of course, numerous cases where attempts at biological control have not only failed, but have actually led to undesirable results, with the effect that in a lot of places now, and especially environmentally more sensitive, island situations, obtaining permission for releasing parasitoids has become a political issue, with a strong popular movement against such interventions. Rather unfortunate host shifts by opiine species introduced into Hawai'i to control fruit flies, on to endemic gall-forming tephritids, are a classic example of biological control gone wrong (Duan & Messing 1997). Little is understood about why such problems occur. It is not unlikely that those individuals of an introduced parasitoid that make a host shift have some genetic predisposition to be able successfully to colonise other host species, but perhaps if preferred hosts are sufficiently abundant and attractive and there

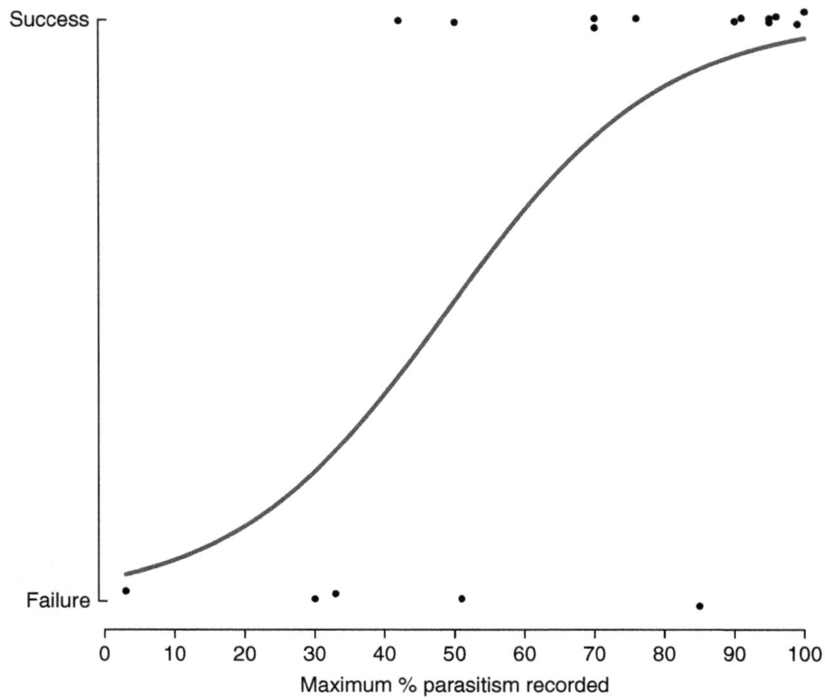

Fig. 14.8 Plot and logistic regression line for success or failure of Braconidae in attempts at classical biological control in relation to the maximum recorded levels of parasitisation recorded (points 'jittered' for clarity); two instances deemed 'partially successful' which had 90% and 7% maximum parasitisation rates were scored as successes (generalised linear model: $p < 0.05$, $z = 2.12$, residual deviance 12.738 on 15 degrees of freedom). (Source: data from Hawkins 1994, table 9.1.)

is a small trade-off between being able to attack a novel host as opposed to the one the parasitoid was released against, then there might not be strong selection in favour of attacking the novel host. The situation on islands is inherently likely to be atypical of mainland ones because population sizes, both of intended and non-target hosts, and population contiguity are likely to be lower. This might be why the less desirable effects of parasitoid introductions appear to be felt more strongly in such situations.

Several introduced braconids used in biocontrol have, in addition to attacking their target species, moved on to attack alternative native species, which can potentially pose a threat both to natural ecosystem function and, in some cases, rare or vulnerable species might be made more vulnerable. In New Zealand, *Microctonus aethiopioides* was introduced to attack the afalfa pest weevil, *Sitona discoideus*, but has moved on to various native weevils and in some localities parasitised more than 20% of *Nicaeana fraudator* (Barratt et al. 2007). Concerns about such issues are particularly high for islands such as those of the Hawai'ian archipelago, where many parasitoids have been introduced to combat introduced pests. In Hawai'i, Hennemann and Memmot (2001) reared over 2000 caterpillars from native vegetation and constructed a quantitative food web of the parasitoids reared. Of these, endemic parasitoid species constituted only 3%, the remainder being species introduced as control agents a long time ago against a range of pests, such as *Meteorus laphygmae*, which was introduced in 1942 to control the African armyworm, *Spodoptera exempta* (Noctuidae). Indeed, *M. laphygmae* has had little effect on the population levels of the pest itself. However, the endemic Hawai'ian moth *Scotorythra paludicola* (Geometriidae), which frequently causes devastating, defoliating outbreaks on its host *Acacia koa* trees, is only parasitised by introduced species. In the longer term, it will be interesting to see whether they will evolve to being able to effect control on the moth outbreaks. A beneficial non-target effect was described by Wellings et al. (1994) in which the polyphagous aphidiine *Aphidius colemani* was introduced on the island of Tonga in an attempt to control banana aphid, Pentalonia nigronervosa, and although the parasitoid did not establish on this host, it did on another introduced pest aphid, *Aphis gossypii*.

De Nardo and Hopper (2004) proposed using a literature survey to identify likely non-target species in a region that an introduced parasitoid might be likely to attack and used the well-known braconid *Macrocentrus cingulum* (often called *grandii*), a parasitoid of the European corn borer, *Ostrinia nubilalis* (Crambidae), as an example. Although their results for that well known species closely matched observations, most parasitoids are far less well known and often the literature will harbour a lot of erroneous information, which I think renders such a tactic rather meaningless and potentially dangerous, unless very extensive and quality-controlled data are available. Nowadays, before parasitoids are permitted to be released for the control of some pest or other, there is usually an expensive and extensive programme of laboratory trials, to assess both its potential success and also any potential harmful effects the parasitoid might have on endemic non-target hosts. Of course, if a parasitoid successfully attacks non-target insects in the laboratory (potential hosts), it does not necessarily follow that it will cause a high level of mortality on the non-target in the field (Haye et al. 2005). The realised host range is often far less, not only due to the preference of the parasitoid given that a field situation is analogous to a choice experiment whereas potential hosts presented in the laboratory are often a no-choice experiment, but also the temporal availability of the non-target hosts might be different and asynchronous with the parasitoid in the field. Even as recently as the 1970s some authorised parasitoid introductions involved species towards the extreme end of the specialist/generalist spectrum. For example, the polyphagous pupal parasitoid *Pimpla disparsis* was introduced into North America to control gypsy moth (part of its known host range of more than 70 species in many families of Lepidoptera) (Schaeffer et al. 1989). However, with far greater awareness about possible non-target effects, I think it is unlikely that any such applications would be successful nowadays, especially in environmentally sensitive situations.

Many idiobionts, and a few koinobionts, destructively host-feed, i.e. they use a subset of hosts for host-feeding, and as a result these individuals are often killed (Jervis & Kidd 1986). Although this has little effect on the population dynamics of the system, it has generally been viewed by biological control workers as a bonus – extra host mortality, even although the individual hosts killed do not directly lead to the production of a new parasitoid in the next generation. This may generally be true, but Ueno (1998) presented data from the pupal parasitoid *Itoplectis naranyae* (Pimplinae) that should add a note of caution. In this species, females were less likely to attempt oviposition in hosts that had already been parasitised 40 h earlier (Fig. 14.9a), but showed a significantly increased probability of destructively host-feeding on them (Fig. 14.9b), thus undoing the benefit of the parasitism by what ever female had oviposited into the host pupa previously.

EFFECT ON HOST FOOD CONSUMPTION

Koinobiont hymenopteran parasitoids generally cause hosts to develop more slowly, to consume less food and to reach lower final body weights than unparasitised ones (Fig. 14.10) (e.g. Vinson & Barrass 1970, Vinson 1972a, Duodu & Davis 1974, Brewer & King 1981, Rohlfs & Mack 1983, Duodu & Antoh 1984, Grant & Shepard 1984, Hawlitzky & Boulay 1986, Kumar & Ballal 1992, Grossniklaus-Bürgin et al. 1994, Bell et al. 2000, Elzinga et al. 2003, Morales et al. 2007) and includes when hosts are carnivores rather than herbivores (Hazell et al. 2005). This is true even although most endoparasitic koinobionts do little by way of feeding until their host is either fully grown or sometimes (particularly Euphorinae: Meteorini, Microgastrinae and Campopleginae) has reached a suitable size for the parasitoid larvae rapidly to complete their feeding. Nevertheless, there are some exceptions (see Next page).

As most studies on the effect of parasitisation on host growth involve species of biological control potential, many studies have focused on the effects of parasitisation on total food consumption by the host, because if a host consumes less it is most likely going to cause less economic damage even before the parasitoid kills it. In many cases the host's growth is retarded and stunted and the total food consumed is markedly less than that by unparasitised individuals. For example, P.G. Mason et al. (2001) compared host food consumption between two parasitoids of the Bertha armyworm (*Mamestra configurata*: Noctuidae), an important noctuid pest of

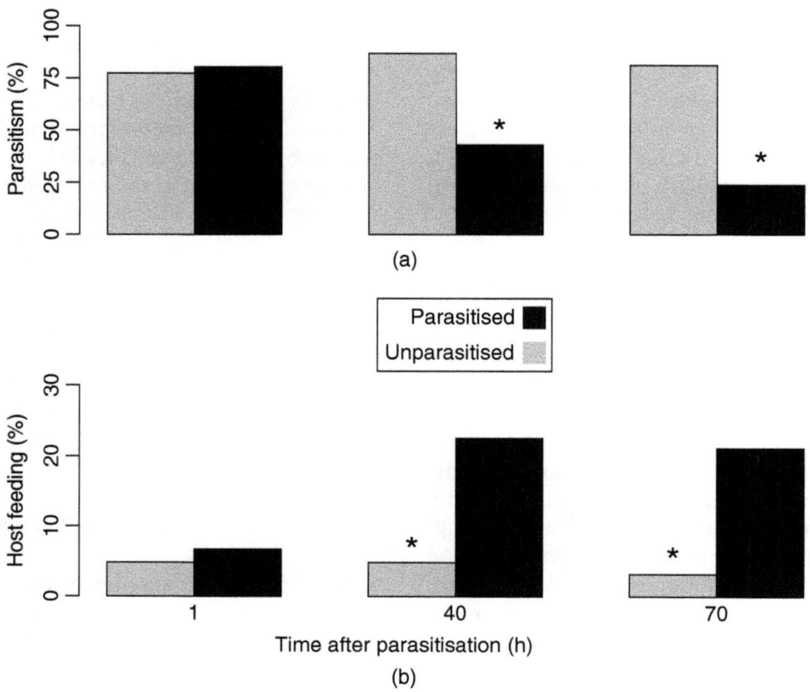

Fig. 14.9 Showing decreased likelihood of superparasitising, and increased likelihood of destructively host-feeding on hosts parasitised 40 h or more earlier, by the pimpline ichneumonid, *Itoplectis naranyae*; asterisks indicate significant differences at $p < 0.05$ level. (Source: data from Ueno 1998.)

Brassicas and flax. Parasitism by both the ichneumonid *Banchus flavescens* and the microgastrine *Microplitis mediator* significantly reduced total host caterpillar food consumption and the latter also reduced the time that the host would be on the crop, thus making it a more preferable biological control candidate than the *Banchus*, which did not affect the duration of the host's feeding. Parasitisation of black cutworm caterpillars (*Agrotis ipsilon*: Noctuidae) by *Meteorus leviventris* led to a 24% reduction in maize foliage consumption and a 36% reduction in cutting of seedlings compared with damage done by unparasitised larvae (Schoenbohm & Turpin 1977); in this case the beneficial effects were attributed to the fact that the *Meteorus* kill their hosts before they are fully grown and so cease feeding shortly before the parasitoid larva emerges. Also, the amount of food consumed by hosts of koinobiont that can attack a range of host instars such as Cardiochilinae may depend on the stage attacked, the parasitoids completing development earlier in hosts attacked at an earlier stage (e.g. H.A. Smith et al. 1994). When trying to rear parasitoids from wild-collected host caterpillars, such retarded growth can give a good clue as to which individuals might be parasitised, i.e. the smaller, less advanced ones (M.R. Shaw 1997).

However, reduced food consumption by a parasitised host is not always the case, and in caterpillars parasitised by some microgastrines that increase the duration of the host larval stage there may even be an increase (e.g. Rehman 1970, Sato et al. 1986). Large white butterfly caterpillars, *Pieris brassicae*, parasitised by the gregarious *Cotesia glomerata* (F.D. Parker & Pinnell 1973) have a greater lifetime food consumption compared with unparasitised hosts and Slansky (1978) showed that small white butterflies, *Pieris rapae*, parasitised by the same wasp, consume and convert into biomass more energy and nitrogen than unparasitised ones, despite having a lower overall metabolic rate. In the case of highly polyphagous *Meteorus pulchricornis*, Harvey et al. (2010) showed that both were possible in the same species, depending on the size of the host species attacked. *M. pulchricornis*

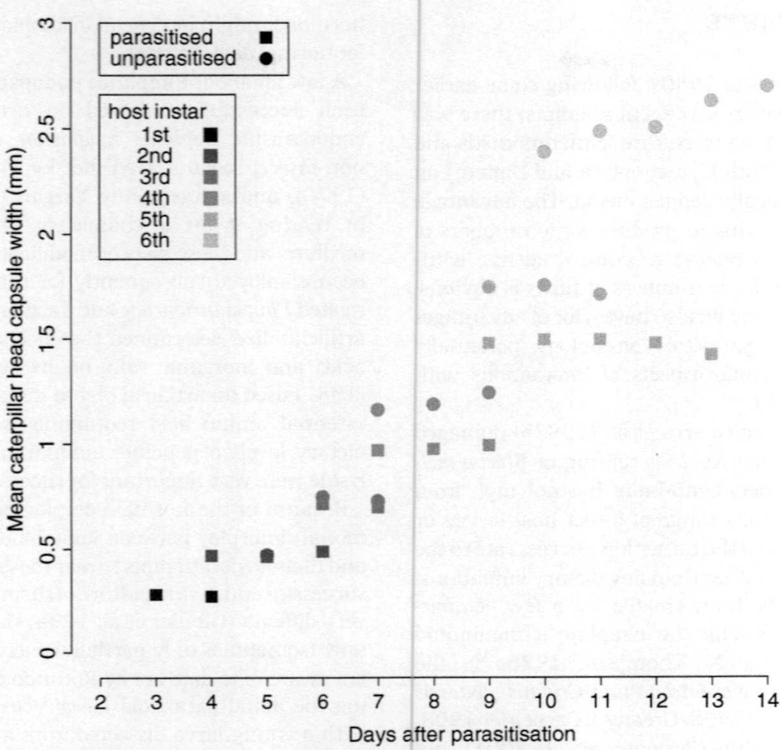

Fig. 14.10 Effect of parasitisation by *Chelonus inanitus* on growth of its host (*Spodoptera littoralis*; Noctuidae), showing that it retards growth from the third host larval instar onwards, and causes premature metamorphosis, preventing the host from reaching the normal final sixth instar. (Source: data from Grossniklaus-Bürgin *et al.* 1994; error bars not shown here, but not overlapping after host reaches fourth instar.)

oviposited into second instar larvae of the large host *Mythimna separata* or similarly sized third instars of the far smaller *Plutella xylostella*. In the first of these, growth of parasitised larvae was about 95% reduced, whereas in the smaller host it was increased by nearly 30%, thus achieving approximately similar final sizes for the parasitoid to complete its development on. How widespread such plasticity in the regulation of host food intake may be is still unknown.

When considering the effect of parasitisation on host food consumption from the parasitoid's point of view there are actually two separate factors to consider. The amount of food consumed by the host in order to grow a certain amount can be defined as the efficiency conversion of ingested food (ECI = 100 × biomass gained/biomass ingested) and the amount digested by the host gives the efficiency of conversion of digested food [ECD = 100 × biomass gained/(biomass ingested – frass)]. The approximate digestibility (AD) of the host's food is defined as the percentage of ingested food that is assimilated, i.e. AD = 100 × (biomass ingested – frass)/biomass ingested. In some systems the parasitism does not affect any of these and the parasitoid may obtain its food by influencing the total food consumed by the host (e.g. Schopf & Steinberger 1996). In other cases parasitism results in a lowered ECD, although AD may remain unchanged or may even be increased (e.g. S.N. Thompson 1982b). Whilst total host food consumption and mass are often affected by parasitism, so too is their body composition. Bischof and Ortel (1996) found that parasitisation of gypsy moth, *Lymantria dispar* (Lymantriidae), larvae by the microgastrine *Glyptapanteles liparidis* caused an increase in host glycogen and reductions in host haemolymph sugars, although total lipid content was not affected.

ARTIFICIAL DIETS

During the 1970s and 1980s, following some earlier unsuccessful or barely successful attempts, there was a drive towards trying to culture ichneumonoids and other parasitoids (both Hymenoptera and Diptera) on synthetic or chemically defined media. The advantage of being able to do this to produce large numbers of individuals for field release as control agents, without having to rear large numbers of hosts is obvious. Chemically defined media also have a lot of advantages for understanding parasitoid physiology, potentially removing the dynamic aspects of interactions with hosts (Grenier 2009).

Guerra (1992) and Guerra et al. (1993b) managed to achieve approximately 25% rearing of *Bracon mellitor* on artificial diets containing haemolymph from a taxonomically broad range of insect host larvae or pupae, and attributed the rather low success rate to the rearing conditions rather than any dietary limitations. Artificial diets have been created for a few ectoparasitoid species including the pimpline ichneumonid *Exeristes roborator* (S.N. Thompson 1976a,b), the cryptines *Diapetimorpha introita* and *Cryptus albitarsis* (Carpenter & Greany 1998, Greany & Carpenter 1998, Ferkovich et al. 1999, Carpenter et al. 2001) and the braconine braconid *Habrobracon* (Xie et al. 1989, Yazlovetsky et al. 1992, Magro & Parra 2004, Magro et al. 2006). Larvae of the polyphagous pimpline corn-borer parasitoid *Exeristes roborator* can survive on a very simple diet and, given a sufficiently high amino acid concentration, do not even requite the presence of a carbohydrate source such as glucose. S.N. Thompson (1976a) cultured it on a holidic diet and showed that they had an absolute requirement for 10 amino acids: arginine, histidine, isoleucine, leucine, lysine, methionine, phenylalanine, threonine, tryptophan and valine. Although larvae could survive on diets lacking in any of the other non-essential amino acids, there were significant differences in development time between some of them and, indeed, the wasps could not survive on the essential amino acids alone. Several species reared on artificial diets take far longer to develop than those reared on natural hosts, suggesting that dietetic or hormonal factors might be involved. Gelman et al. (2000) compared the ecdysteroid levels of *D. introita* reared on natural hosts and artificial diet and identified two main peaks, but these were higher in those fed insect hosts, suggesting that intake of host hormones might be crucial in idiobiont ectoparasitoids for normal development.

A few idiobiont Pimplinae endoparasitoids have also been successfully cultured on artificial diets. The endoparasitic *Itoplectis conquisitor* was reared on a non-insect, meat-derived diet by Yazgan and House (1970), and subsequently Yazgan (1972) succeeded in rearing it on a completely chemically defined medium, and these diets or modifications of them have been employed subsequently for this species and the related *Pimpla turionellae* and Yazgan (1981), using an artificial diet, determined the effects of different fatty acids and inorganic salts on its development. Çelik (1987) used an artificial diet to investigate this wasps' essential amino acid requirements and found that dietary levels of leucine, methionine, isoleucine and lysine were very important for successful development.

Because of the normal complex physiological (hormonal) interplay between koinobiont endoparasitoids and their hosts, attempts to rear these have been far less successful and *in vitro* culture of them is likely to remain very difficult (Grenier et al. 1994, Grenier 2009). The only (sometimes only partially) successful cases that I am aware of to date are by Rotundo et al. (1988) rearing the aphid parasitoid *Lysiphlebus fabarum* starting with a young larva dissected from a host and rearing of the campoplegine *Venturia canescens* by Nakahara et al. (1997) from pre-germband egg stage through to pharate adult stage, development being dependent on a protein component(s) of the host haemolymph. Rotundo et al.'s medium contained aphidiine teratocytes but no host cells. Although adults were obtained, they were unable to fly because they were contaminated with residue of the culture medium. J.S. Hu & Vinson (1997) also managed to obtain development of *Campoletis sonorensis* to third instar, and this was followed by rearing the prepupal stage (J.S. Hu & Vinson 1998), although physical factors appeared to be important in preventing further development. A great deal of effort has gone into trying to rear the cardiochiline *Toxoneuron nigriceps*, a major parasitoid of *Heliothis virescens*, and the microgastrine gypsy moth parasitoid *Microplitis croceipes in vitro*, but this has met with only limited success and no adults have yet been obtained. Since these wasps have small eggs that are not greatly distorted during oviposition, stimulation of embryonic development might be triggered by host factors. Many koinobiont endoparasitoids do not develop beyond the first instar until the host has reached a suitable developmental stage, i.e. often when it starts seeking a pupation site or has started to construct its cocoon (Ferkovitch et al. 1991). Thus triggering ecdysis to the second instar is a second crucial stage (Pennacchio et al. 1992a). One of the other important issues that needs to be overcome with many endoparasitoids is the transition to an external feeding phase and in the microgastroid braconids which includes *Toxoneuron*, the normal egress from the host larvae is associated with the moult to the final instar with the penultimate instar's cuticle being left behind in the exit hole through the host's integument. In the absence of this physical struggle through the host cuticle (Nakamatsu et al. 2006), it is likely that moulting will be largely unsuccessful. Following egress, the parasitoid larva has also to construct a cocoon in which to pupate, and this too may require special conditions (Henderson et al. 2011).

Interestingly, in Pennacchio et al.'s study, although they managed to get *Toxoneuron* (as *Cardiochiles*) explanted eggs to develop to second instar, their teratocytes died early on in the medium, which was entirely free of insect-derived components, showing that they were not crucial to the early stages of parasitoid larval development. Another important factor, however, was that they were unable to get freshly laid eggs to undergo embryogenesis *in vitro* and had to use ones that had been laid naturally in the host and remained there for at least 7 h. A similar limitation was found by Digilio (1999) in her attempts to rear another aphidiine, *Aphidius ervi*, in a defined medium. Ferkovitch et al. were able to get *Microplitis* eggs to undergo embryogenesis, but the process was only completed in media that had been preconditioned with cell lines.

The fact that those ichneumonoids that are easiest to culture tend to be generalists, either idiobiont ectoparasitoids of paralysed larval hosts or endoparasitoids attacking pupae which have low immune capability, is unfortunate from the point of view of rearing wasps for biological control, since generally it is desirable to utilise specialists. Perhaps some of the above difficulties associated with attempts at *in vitro* rearing of endoparasitic koinobionts might be less prominent in species that can attack a very wide range of host taxa and ages and that also can kill the host at a variety of stages of its development depending only on their having received sufficient nutrition. Taxa that spring to mind are *Charmon*, some *Meteorus* and perhaps some other Campopleginae, or perhaps even the euphorine *Dinocampus coccinellae*. In general, it seems that the mass *in vitro* rearing of koinobiont ichneumonoids for commercial purposes is a long way off and the requirement of several investigated systems for factors derived either from host haemolymph or host cell cultures would probably make it prohibitively expensive until the critical compounds have been identified and can then be synthesised.

ARTIFICIAL HOSTS

In terms of rearing parasitoids on artificial hosts to make mass culturing easier, there has been very little progress made with the Ichneumonoidea. There are three stages involved in the successful use of artificial hosts – the female wasp must be interested enough in them to penetrate them with her ovipositor, the internal medium needs to provide the right feedback for her to lay an egg, preferably female, and the offspring has to develop and survive. A few ichneumonoids have been found to accept (i.e. oviposit in) artificial hosts, for example the aphidiine *Aphidius ervi* will even oviposit into Parafilm bubbles containing host haemolymph that have been surface sprayed with aphid extract (Larocca et al. 2005), and Hegdekar and Arthur (1973) found that a fraction of lepidopteran pupal haemolymph containing a mixture of hexoses and 19 common amino acids were necessary and sufficient to induce oviposition in the pimpline *Itoplectis conquisitor* and has been induced to oviposit into an artificial Parafilm host containing synthetic or meridic culture medium (House 1978), although only one male was successfully reared. In the first case the need for host extract and haemolymph rather defeats the purpose and a lot more development would be needed to get the *Itoplectis* system working.

USE OF ALTERNATIVE HOSTS

Fairly often it is either difficult or less practicable to rear a parasitoid in culture on the target host species than on another host species. For example, maybe the parasitoid can develop satisfactorily on a different species that is already in culture. The efficiency of such parasitoids reared on non-target hosts in release programmes then needs to be considered. Will they be as good at attacking the target pest as they were the one they have been cultured upon? There are essentially two issues here: first, whether the parasitoid has practical and meaningful host races or host strains in the wild, such that an inappropriate one may have been bought into culture, and second, whether the culture conditions and, in particular, the alternative host used, impose selection on the parasitoid that will have a negative effect on its utility as a pest control agent. Powell and Wright (1988) found evidence for both of these effects in two species of aphidiine. Differences in ability to physically parasitise (host handling) affected the ability of *Aphidius ervi* successfully to parasitise *Acyrthosiphon pisum* and *Microlophium carnosum* if they had been reared on the former, whereas those reared on the latter were good at handling and parasitising both species. Crosses showed that if females reared on *A. pisum* were mated with males from lines cultured on *M. carnosum*, the offspring's ability to utilise *M. carnosum* was markedly improved, suggesting a strong genetic basis to host handling behaviour. Females of a second species in culture, *Aphidius rhopalosiphi*, when compared with wild-collected stock showed they had evolved a differential ability to attack two wheat-feeding host aphids (*Metopolophium dirhodeum* and *Sitobium avenae*) that was not apparent in wild-collected stock, suggesting that some aspect of the culture conditions had created an inadvertent selection pressure. Genetic studies have in fact suggested that *A. ervi* exists as different host-related biotypes in the field with little gene transfer between those attacking different hosts. This means that under agricultural conditions, field margin reservoirs of what might at first sight seem to be a suitable alternative host, i.e. *M. carnosum* on stinging nettle, *Urtica*, for maintaining a level of *A. ervi*, might not work in practice because different biotypes are involved. In another study involving aphidiine biotypes, Baer et al. (2004) used the analysis of rapidly evolving sites, single-strand conformation polymorphisms, to show that the biotypes of *Diaeretiella rapae* are not completely genetically isolated and that gene flow between them is currently sufficient to prevent speciation occurring.

Given the ability of laboratory culture to exert unwanted selection on a parasitoid, this also opens up the possibility of carrying out selective breeding to improve the ability of a given parasitoid to attack a target species. Indeed, this ought to be possible but there are two important practical limitations: first, that the culturing conditions do not impose unwanted and unexpected selection effects because culturing conditions are seldom very similar to field situations, and second, that the numbers of individuals/lines investigated is adequate. With large field populations and enough original genetic variation, if an introduction establishes, natural selection on to the available hosts will almost automatically ensue. One problem is that sometimes the requirements of biocontrol in distant parts of the world, where it is not wanted that the released agent should attack other endemic hosts, means that those species selected for release should have a low chance of shifting on to native species, whereas in agricultural pest management, sometimes the opposite might be desirable.

HYPERPARASITISM AND KLEPTOPARASITISM

Many ichneumonoids are themselves the targets for hyperparasitoids, both by other ichneumonids and by various Chalcidoidea, and in the case of aphidiines also by hyperparasitic Megaspilidae (notably *Dendrocerus carpenteri*) and Cynipidae (*Alloxysta* spp.). Ichneumonoid larvae parasitising caterpillars are generally vulnerable to attack by mesochorines, which are endoparasitoids of the primary parasitoid within its host. Exposed cocooned ichneumonoids are vulnerable to attack by ectophagous pseudohyperparasitoids, notably Cryptinae, although occasionally ephialtine pimplines. In all these situations, the level of hyperparasitism can be very high and a real nuisance if one is trying to study the biology of a species reared from wild-collected parasitised hosts. Other hyperparasitoids include the generally rare Eucerotinae, a few specialist pseudohyperparasitic, *Itoplectis* (Pimplinae) species (see M.R. Shaw 2009b).

In biological control, hyperparasitoids and pseudohyperparasitoids are generally considered detrimental if they are attacking a primary control agent and there is evidence that they have caused increased damage to crops. For example, Simpson et al. (1979) reported that hyperparasitism of the alfalfa weevil, *Hypera postica*, primary campoplegine parasitoid, *Bathyplectes curculionis*, could reduce the population of the latter by 50%, and Kfir (1997) noted that some samples of diamond back moth (*Plutella xylostella*) being surveyed for parasitoids yielded only hyperparasitoids. Nevertheless, while high local levels of hyperparasitoids must necessarily reduce numbers of primaries significantly, if the ones that survive are reasonably good dispersers and have high fecundities, the effect of hyperparasitoids on crop production might not be that bad. Also, in situations where a hyperparasitoid does not attack all the primary parasitoid species on a pest, the ones that are immune might be released from competition with the more vulnerable one and so take over. On the other hand, in weed control, high levels of hyperparasitism of parasitoids of herbivores may be a good thing.

Mackauer and Völkl (2005) found that the aphidiine *Lysiphlebus hirticornis* experienced approximately 60% mortality from the encyrtid *Syrphophagus aphidivorus*. Further, as this hyperparasitoid attacks the mummy stage of its host, the female hyperparasitoid was found preferentially to target larger mummies, which are predominantly those of female *Lysiphlebus* larvae/pupae, so their parasitism shifted the sex ratio of the surviving aphidiines from female biased to even – certainly not a desirable effect in a biological control situation.

Two instances are known in which a chalcidoid species (Eurytomidae) kleptoparasitises an ichneumonoid. Roques (1976) showed that *Eurytoma waatchi* (Eurytomidae) only attacks the larvae of its host *Pissodes* weevil after they have been paralysed and oviposited on by another of their parasitoids, species of the pimpline genus *Scambus*, and Hougardy and Grégoire (2003) showed that kleptoparasitism of the braconine bark beetle parasitoid *Coeloides bostrichorum* by the pteromalid *Rhopalicus tutela* greatly increased the latter's host-finding ability.

PREDATION

Adult parasitoids may, of course, be vulnerable to predation by various common insectivores such as birds, although probably the practical and taxonomic difficulties of identifying parasitoid wasp remains in bird faeces or collar samples means that they are greatly under-reported as prey items, and this is not helped by their often high local species diversity. Despite their numerical abundance, there have been almost no studies on the role of predation in regulating numbers of parasitic wasps. Sparks and Valdez (2003) reported that ichneumonoids constituted approximately 19% of the volume of guano from a particular species of bat and, given the large size and numerical abundance of nocturnal ichneumonids such as ophionines and tryphonines at light traps, this does not seem particularly surprising, although how the bats cope with the stings, which are painful to humans, is not known. Ichneumonoids that attack deeply concealed hosts such as rhyssines attacking woodwasps (Siricidae) or other species attacking deeply concealed wood-boring beetle larvae, where oviposition ('drilling') may take hours, are well known to suffer considerable losses to bird predation (Heatwole et al. 1962, Spradberry 1970b, Yamazaki & Matsumoto 2009), but there appear to have been no quantitative studies. Examination of host infected logs that are being attacked by parasitoids such as *Rhyssa* and *Megarhyssa* (Rhyssinae) often reveals protruding remains of parasitoid (and sometimes host) ovipositors where a predator has eaten the insect during oviposition and the ovipositor has broken off. In Uganda, I caught an adult female *Zaglyptogastra* in flight with only the basal few millimetres of her ovipositor remaining, clearly having escaped a predator attack, but equally no longer capable of parasitising any more hosts.

Harvey and Strand (2002) have provided some evidence that the larval development time of parasitoids is not just influenced by the idiobiont/koinobiont spectrum which considers primarily the resource provided by the host, but also, within koinobionts, by the risk of predation that the host faces. Thus, parasitoids of hosts that might be more vulnerable to predation by a third party, such as a bird, will do better if they can complete their larval development more quickly. Tropical ichneumonoids and their exposed hosts seem to develop very quickly, meaning that the windows for both predation and observation are narrow.

PATHOGENS

Like everything else, ichneumoids get exposed to various pathogens and some have adverse effects upon them. Microsporidians, mostly *Nosema* spp., a group of fungus-like organisms belonging to the Microspora, are fairly common pathogens of ichneumonoids and can be especially important and harmful under culture conditions where the densities of hosts and parasitoids are typically far greater and less variable than in the wild. Nevertheless, there is great variability in the level of pathogenicity caused, ranging from severe to undetectable. In many cases, heavily infected wasps can be recognised by conspicuous white patches on the underside of the metasoma and often a swollen or deformed appearance (H.W. Allen & Brunson 1945).

Nosema bordati, which affects *Chilo partellus* and its parasitoid, *Cotesia flavipes*, is highly pathogenic to the parasitoid and development is only successful in lightly infected hosts (Bordat et al. 1994). W.M. Brooks and Cranford (1972) found that the campoplegine

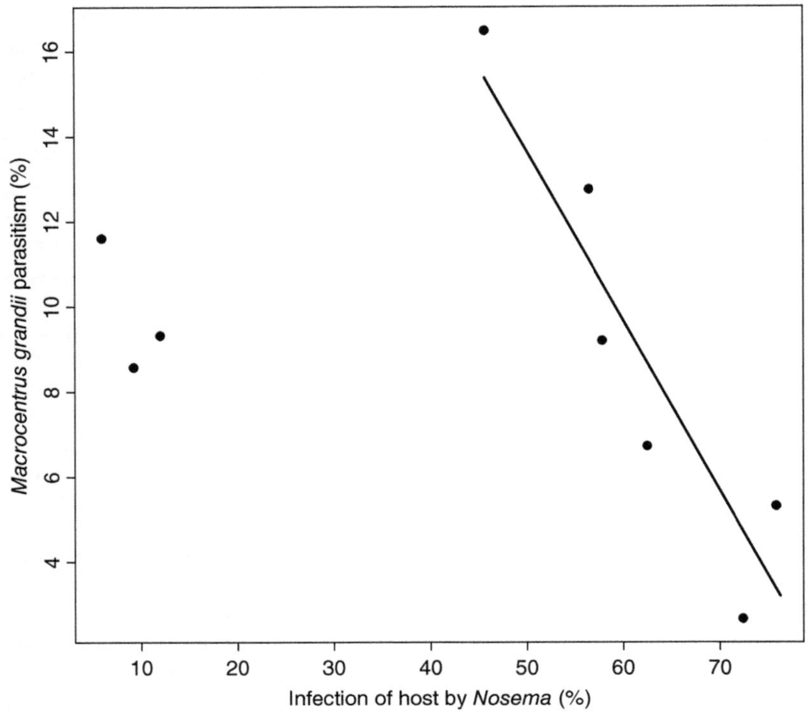

Fig. 14.11 Field measures of parasitism of the European corn borer, *Ostrinia nubilalis*, suffering from varying levels of infection by the microsporidian *Nosema pyrausti*, on the parasitism rate by the braconid wasp *Macrocentrus cingulum* (as *grandii*). The regression line is fitted only to those points corresponding to infection rates by *Nosema* greater than 45% ($p < 0.01$, $r^2 = 0.88$). (Source: data from Andreadis 1982.)

Campoletis sonorensis became infected in virtually all tissues when reared on *Heliothis zea* and adult wasps transmitted the *Nosema heliothidis* to their offspring transovarially, although it does cause some mildly harmful effects. However, this system was found to be far more complicated, because field populations of *C. sonorensis* are routinely infected with another transovarially transmitted microsporidian, *N. campoletidis*, which appears to have no harmful effect on its host and could not infect *Heliothis*. Another common and important parasitoid of *Heliothis* is the cardiochiline *Toxoneuron nigriceps*, and this species has also been also found to harbour its own specific Microsporidian, *N. cardiochilis*, which was unable to infect the lepidopteran host.

Nosema pyrausta is an important natural control agent of the European corn-borer, *Ostrinia nubilalis* (Crambidae), whose caterpillars it infects and is transmitted from the host to the emerging biocontrol parasitoid, *Macrocentrus cingulum* (Macrocentrinae), which as a consequence have lower survival and live less long as adults, consequently attacking fewer hosts (Andreadis 1980). In the field, the infection of *M. cingulum* is highest where the host is most dense and correspondingly parasitism levels declined significantly with increasing *Nosema* infection levels (Fig. 14.11)[1]. *Nosema* appear likely to be responsible for a decline in the role of this parasitoid as a control agent in the USA (Andreadis 1982). However, that the infection is positively density dependent should mean that the parasitoid will remain on the scene in areas where there are lower host densities and thus a smaller likelihood of picking up *Nosema* infections. Laboratory cultures of *Drosophila* also frequently become infected by another microsporidian, *Tubulinosema kingi* (Franzen et al. 2006). This species is transmitted to the alysiine *Asobara tabida* and causes increased mortality in the wasp in the host puparium with reduced adult emergence (Futerman et al. 2006). Field survey in this case failed to reveal any natural infections of *Drosophila* populations, at least in southern UK. Infection of the microgastrine *Cotesia flavipes* by a *Nosema* species, again under mass rearing/laboratory conditions, caused a number of changes, all unfavourable from a mass-rearing perspective (Simões et al. 2012). Parasitoids failed to develop in hosts that had the highest levels of *Nosema* infection and those that developed in less heavily infected hosts took longer to develop, were smaller, had reduced adult longevity and had lower lifetime fecundity. Further, whereas uninfected *Cotesia* were able to discriminate odours from host–host plant complexes with uninfected hosts and infected hosts, preferring the uninfected ones, wasps that were infected with *Nosema* were no longer able to do so. Another species, *T. hippodamiae*, has a 100% transmission from infected host convergent ladybird beetles, *Hippodamia convergens*, to the euphorine imagobiont parasitoid *Dinocampus coccinellae*, and has been shown histologically to infect all of the emerging parasitoid's major organs and tissues, although not its ovary (Saito & Bjørnson 2013). In this case, although the microsporidian infection reduced the fecundity of the wasp, it had only a small, non-significant effect on their host attack rate; the number of attacked hosts that contained mature parasitoid larvae when surveyed was significantly lower than for uninfected wasps, although the precise mechanism behind this is unclear.

In most of these systems, infection is a two-way process, with wasps picking up the microsporidian from the host during development and also transmitting it to new hosts during parasitisation (Bordat et al. 1994, Simões et al. 2012), but this was not the case with the *Tubulinosema kingi*/*Asobara tabida* system, nor was *Asobara* able to transmit the fungus vertically to its offspring (Futerman et al. 2006).

Pimpla turionellae harbour and infect hosts with another yeast-like organism that is probably a microsporidian (Middeldorf & Ruthmann 1984). This fungus is more a commensal in the wasp than a parasite and Pavenstädt-Grupp and Ruthmann (1989) were unable to detect any harmful effects. The fungus cells particularly infected adult wasp ovocytes and ovarian cells, fat body, Malpighian tubules, gut muscles, nervous tissue and Dufour's gland[2].

In addition to all these microsporidian infections, some species are attacked by neogregarine protists (Apicomplexa), especially of the genus *Mattesia*, some of which are used in biological pest management. The best known example involves *Bracon mellitor*, a parasitoid of the boll weevil, *Anthonomus grandis*. The host is frequently attacked by a specific neogregarine protist, *Mattesia grandis*, which can also infect the ectoparasitic *Bracon*, through consumption of its spores. No transmission from wasp to host occurs, but infected wasps show reduced fecundity and host parasitisation behaviour (McLaughlin & Adams 1966).

No surveys seem to have been conducted of the many species of ichneumonoid that might occur in any given natural habitat and it is possible that the species studied, i.e. parasitoids of pest species that have regularly high population densities, might be atypical. One might also expect that the occurrence of host-specific microsporideans in ichneumonoids would be found to be related to the stability and density of their host populations.

Surprisingly little is known about the defence mechanisms of parasitoids against bacterial and fungal pathogens, although they are presumably much the same as in many other insects. *Pimpla* larvae have been shown to release anti-fungal substances from the anus (Führer & Willers 1986). Führer and El-Sufty (1979) found that teratocytes of the microgastrine *Cotesia glomerata* released a fungicidal compound, which might be related to the fact that parasitised hosts are rendered more susceptible to the fungal pathogen *Beauveria* and Z-Z Wang et al. (2013) have just started to investigate the putative defensin genes of *C. vestalis* and found that the three candidate genes were expressed from the second larval instar onwards.

TRANSMISSION OF HOST PATHOGENS

As parasitic wasps penetrate their hosts with their ovipositors but sometimes, for whatever reason, do not kill the host, they have the potential to transmit small host pathogens from one individual to another, including viruses, bacteria and fungi. Young and Yearian (1990) showed that the microgastrine braconid *Microplitis croceipes* could transmit nuclear polyhedrosis virus between larvae of *Heliothis*, and subsequently Sait et al. (1996) showed inter-host transmission of baculovirus-infected *Plodia interpunctella* by *Venturia canescens*, Lopez et al. (2002) reported transmission of an iridescent virus between individual fall armyworm, *Spodoptera frugiperda*, caterpillars by the cremastine *Eiphosoma vitticolle* and Caballero et al. (1991) demonstrated that females of two braconids and one ichneumonid parasites of *Agrotis segetum* caterpillars were all capable of transmitting granulosis virus from infected to uninfected hosts, and in the case of the two braconids [*Cotesia* (as *Apanteles*) *telengai* and *Aleiodes gastritus*], transmission of the pathogen did not requite oviposition, merely host penetration by the ovipositor, indicating that the virus particles were probably just contaminants on the ovipositor surfaces, although possibly internal ones as well as external. Hence the phenomenon is widespread and possibly universal. Interestingly, Caballero et al. (1991) also showed that all three parasitoids preferentially oviposited in uninfected hosts,

DISPERSAL

Despite the importance of the rate of dispersal of ichneumonids and parasitoids for biological control programmes, very little is known about it in the field or the reasons why some species spread far more quickly than others and landscape features may play an important role in addition to general vagrancy. Juillet (1959) reported that females of the pimpline ichneumonid *Exeristes* (as *Ephialtes*) *ruficollis* display two sorts of flight posture, one associated with host searching in which the body is held at a steep angle and the other associated with dispersal when it is held at an angle of only 20–30° from the horizontal. Depending somewhat on the wind[3], this species was recorded as being able to disperse 85 m on average during a 48 h period. Following field release in Montana, USA, the braconid *Agathis pumila* spread at a rate of approximately 100 m/year, while Barlow et al. (1998) found a dispersal rate of 1.3 km/year for *Sphecophaga vesparum* in New Zealand after its introduction into sites to control the introduced invasive pests *Vespula vulgaris* (L.) and *V. germanica* (F.). Lei and Camara (1999) used mark-release-recapture to investigate dispersal by the specialist microgastrine *Cotesia melitaearum*, a parasitoid of the Glanville fritillary on islands off the coast of Finland, and detected no movement between habitat patches. Although such studies are notoriously bad at detecting rare events that may nevertheless be important evolutionarily or for the long-term survival of a species, such results demonstrate that specialist parasitoids attacking highly fragmented host populations might be particularly vulnerable to local extinction.

Although, in general, dispersal of ichneumonoids must rely on the adult wasps flying, in the interesting case of some Aphidiinae that attack late instar of mature alate aphid hosts, the hosts themselves may contribute to the dispersal of the parasitoid. In this group of wasps, it is potentially particularly important in that many temperate aphids are heteroecious, i.e. they show obligate seasonal host–plant alternation and, in particular, those associated with many crops migrate to tree or shrub hosts for overwintering. Although this probably does occur in some species, Asplen et al. (2011) found that the imported soybean aphid parasitoid *Binodoxys communis* did not develop well in late alatoid nymphs, and although they could develop if they attacked a migratory morph nymph early in its development, they then mummified it before it could fly. Being able to develop in a flying adult could also be a significant factor in dispersal of some Euphorinae with similar biology, through attacking adult Coleoptera and some other orders and also various non-aphid Hemiptera; however, no work seems to have been done on this.

COLORATION AND MIMETIC RINGS

Many ichneumonoids, especially the larger ones, are brightly coloured, usually with contrasting patterns of black, white, red, orange and/or yellow (see, e.g., Figs 12.32, 12.38 and 12.58) (Leathers & Sharkey 2003), and in a small number of cases the black may have an iridescent bluish, purplish or greenish sheen. Coloration is often stated as being unreliable as a taxonomic character although it is often employed in keys. We do know for certain that some taxa show considerable intraspecific variation in colour, but it does also seem to provide some good specific characters – everything depends upon the particular case. However, what is much more interesting is the fact that the same colour patterns or same features keep reoccurring in different groups and also that the patterns found in particular parts of the world are often similar in ichneumonids and braconids but different from other regions. This strongly suggests that some mimicry (or at the very least adaptive resemblance) is involved.

In general, ichneumonoids from the temperate regions are much more conservative in their coloration than tropical ones, although there are still a good proportion of large and brightly coloured ones (see, e.g., Figs 13.25 and 13.31b–d). Temperate taxa, especially north temperate ones, seldom have their wings strongly coloured, although some infuscation is present. In the tropics, brightly coloured and/or patterned wings, or very dark wings, are very common, especially within the Braconidae. Strong wing coloration is particularly seen in large species, small ones usually having hyaline or pale smoky wings. W.R.M. Mason (1964) noted that coloured wings are almost absent in the Holarctic (especially for ichneumonids) and relatively uncommon in the Oriental region but are abundant in Africa. He also noted that the Neotropical region contained the greatest variety of colour patterns, although there are many with largely transparent wings. A particularly interesting observation was that the colour patterns displayed by Chilean taxa bore little resemblance to those of the rest of South America, thus confirming the idea that the Chilean fauna has evolved in considerable isolation from elsewhere – neither the wasps nor their predators seem to intermix frequently with those of neighbouring countries. My own observations are broadly in agreement, although it seems likely that the apparently greater diversity of patterns that Mason was aware of in Central and South America might have been a sampling artefact since he was based in Ottawa and their collection is relatively depauperate in Afrotropical material. There is also a serious problem of recognising what constitutes a distinct pattern.

The homeochromatic complexes that many larger ichneumonoids are involved in, especially tropical species, also include numerous and diverse other insects including members of several orders such as flies, beetles, moths and bugs (Linsley 1959, Slobodchikoff 1974b, Quicke 1986, Leathers & Sharkey 2003, Namyatova & Cassis 2013, Pereira et al. 2013). That the ichneumonoids are the models, at least in some cases, is strongly supported by a couple of morphological and behavioural adaptations of mimics. One of the most bizarre involves another braconid, a very rare Neotropical genus of braconine, *Pheloura*, which has three long protuberances from its anus that resemble greatly the ovipositor and its sheaths of some of the large ichneumonoids such as *Dolichomitus* or *Compsobracon* spp. (van Achterberg 1989). Several ctenuchine (Arctiinae) moths also have pseudo-ovipositors sticking out their rear ends that, combined with their ichneumonoid-like colour patterns, indicate that they are not only mimicking ichneumonoids but ones with fairly long ovipositors. Several tropical flies (especially Micropezidae and Stratiomyidae) look and behave remarkably similarly to cryptine ichneumonids on dead logs or among the leaf litter – these are typically black but with prominent white stripes on their legs. They wave their fore legs so they closely resemble the wasp's antennae and their locomotion is very wasp-like, but the most interesting thing is that they also hold one of their white-tipped hind legs out directly behind the end of their abdomens, so again obviously mimicking an exserted ovipositor.

Nearly all braconid females attempt to sting when handled/attacked (see, e.g. Buckingham 1975, Buckingham & Sharkey 1988) and males show the same sorts of metasomal movements referred to as pseudostinging, which is a form of behavioural mimicry, which is probably most effective against vertebrate predators. Sometimes, aposematic parasitoids and their hosts can strongly resemble one another, indicating that they belong to the same Batesian or Müllerian mimicry complexes, and when this occurs, the parasitoid is nearly always acting as either a Batesian mimic or as a Müllerian homeochrome of its host (Quicke et al. 1992d), although in some cases it seems clear that the host is mimicking a wasp pattern (Harris 1978; see also Hespenheide 2010). The possibility that the linked population dynamics of the parasitoid and its host in this direction favours the evolution of such mimetic resemblances still needs to be investigated.

Several workers have commented on the ant-like appearance of some apterous or brachypterous ichneumonoids (e.g. Aguirre et al. 2014) and have sometimes considered this to be ant mimicry. However, although a resemblance to ants in these cannot be denied, the taxa involved do not seem to display any particular resemblances other than wing reduction and associated changes to thoracic morphology (Reid 1941; see also Chapter 10, section *Brachyptery and Aptery*) and thus both ants and brachypterous ichnemonoids are small wingless hymenopterans. Hence any protection gained through their resemblance to ants may be incidental.

PALATABILITY AND ODOURS

It has long been known that some ichneumonoids give off a strong odour when disturbed (Townes 1939). Within the Ichneumonidae, various species are reported to produce a pungent odour, e.g., the ctenopelmatines *Alexeter* and *Mesoleius*, the banchine *Banchus* (Townes & Townes 1978), the metopiine *Exochus*, numerous Pimplini, especially *Apechthis*, *Pimpla* and *Theronia* species, the rhyssine *Megarhyssa* and some *Hemiteles* (Cryptinae) (Townes 1970a) and *Phaeogenes*

(Ichneumoninae). Eberhard (2000a) noted that even newly eclosed adults of the ephialtine spider parasitoid *Hymenoepimecis argyraphaga* (Pimplini) give off a pungent, presumably protective, odour when handled, but this did not seem to protect them from attack by orb-weaving *Gasteracantha* spiders. *Megarhyssa* species have a distinctive odour that appears to emanate from their heads and is probably due to the secretions of their mandibular glands (N.W. Davies & Madden 1985). Whether this is protective as was at least first assumed is not clear, however, and the compounds produced also have other functions to do with aggregation (see Chapter 13, section *Rhyssinae*), although there is no reason that they could not fulfil both roles.

The only pungent and sharp odour known within the Braconidae is from a small North American adeliine (see Chapter 12, section *Cheloninae*), *Paradelius* (Whitfield 1988), which reportedly released a formic acid-like irritant. Protective odours/secretions against ants and spiders have been demonstrated for the Hagen's gland secretions of opiines and an alysiine by Buckingham (1975), but these do not smell unpleasant to humans. Both males and females of virtually all Braconinae have two or three pairs of eversible pouch-like glands between the tergum and sternum of the first metasomal tergite which appears to be the source of an oily volatile that they release when disturbed or threatened (both by humans and, in the case of the Afrotropical wood-borer parasitoid genus *Monilobracon*, sometimes other members of the same genus; Quicke et al. 2005b). Another braconid, the physically rather small euphorine *Leiophron uniformis*, has been shown to produce a significant quantity of (*S*)-4-methyl-3-heptanone in its head and releases this into the air when biting and struggling to escape attack (or simulated attack by grabbing the wasp with forceps) (Byers & Zada 2010). (*S*)-4-Methyl-3-heptanone is an alarm pheromone in some insects (ants), but is a defensive volatile in harvestmen (Opiliones: Blum & Edgar 1971) and therefore likely to be so in *Leiophron*.

COMPETITION

Most host species act as hosts to multiple species of parasitoid, often multiple species of ichneumonoid, and this begs the question of what enables their coexistence. There are several ways in which we can see niche-space to be divided; for example, Mills (1992) considered the multiple stages that a host insect might be attacked and the developmental spans of parasitoids. In addition, the precise host ranges of two parasitoids may differ slightly such that even if one is outcompeted on one host it may still survive on another.

Bauer (1985) considered the coexistence of two, apparently biologically very similar, ichneumonids, a species of *Campoplex* (Campopleginae) and *Glypta similis* (Banchinae) attacking the rose tortricid *Pardia tripunctata*. The former parasitoid was superior in terms of competitive ability when hosts were multiparasitised, but this is a generalist species and it was less efficient at finding the tortricid hosts that the specialist *Glypta*. Paine et al. (2000) examined the way in which four doryctine parasitoids of the Australian eucalyptus pest cerambycid *Phoracantha* spp. partition this resource. The largest of the four is *Syngaster lepidus*, which is solitary, whereas the three *Jarra* species are all gregarious. Somewhat counter-intuitively, *Syngaster* preferentially selects smaller hosts. The *Jarra* species differ in ovipositor length and the two with the shortest ovipositors (can) only attack host larvae that are under thin bark (Fig. 12.17). Similarly, Aluja et al. (2013) considered the case of two competing opiine parasitoids of tephritid fruit flies. *Doryctobracon areolatus* and *Utetes anastrephae* compete for *Anastrepha* hosts over a broad range. The slightly longer ovipositor of the former means that it can access hosts that are slightly deeper and in larger fruit whereas the second can persist on hosts in smaller fruit. However, *D. areolatus* can also attack maggots in small fruit but it cannot out-compete the *Utetes* there as the latter is a superior larval competitor irrespective of which species has oviposited first if the interval between ovipositions was less than 1 day.

APPARENT COMPETITION

Apparent competition, which is when the population dynamics of two species suggest that they are competing but the effect is mediated by a third party such as a predator or parasitoid, was a popular research area a few years ago (van Veen et al. 2006). It is particularly likely in insect host parasitoid situations because parasitoids often maintain host population levels, even if the hosts might be feeding on the same resource, at levels where direct competition between hosts is negligible. Holt and Lawton (1993) explored the likely outcomes of apparent competition in such situations in a purely theoretical study and concluded that the most likely outcome was that the parasitoid would bring all but one of its hosts to extinction. The winning host species in these models was the one that supported the highest density of the polyphagous parasitoid. This conclusion has been investigated in a range of laboratory and field situations. Bonsall and Hassell (1998) used a laboratory set-up involving two grain moths, *Plodia interpunctella* and *Ephestia kuehniella*, and their shared campoplegine parasitoid *Venturia cansecens*. In this case, the *E. kuehniella* population always went extinct although the parasitoid and *Plodia* showed long-term persistence as a two-species system because of delayed density dependence, thus supporting Holt and Lawton's prediction. The problem here is that the world supports many cases where hosts with shared parasitoids coexist, so understanding what mechanisms allow that to happen is important. One likely explanation alluded to in the title of Holt and Lawton's paper is that the more vulnerable hosts in terms of the model may have enemy-free refuges, perhaps because of incomplete overlap with the shared parasitoid in space or seasonality or perhaps effects due to different parasitoid success depending on host plants.

Apparent competition in parasitoids is not limited to primary parasitoids but can also be mediated via shared hyper- or pseudohyperparasitoids. Van Nouhuys and Hanski (2000) described such a situation involving the specialist microgastrine *Cotesia melitaearum* and the generalist pseudohyperparasitic cryptine *Gelis agilis* on the Åland islands of south-west Finland where the primary parasitoid and its host exist in many small metapopulations. These authors provided an alternative host to the system, *C. glomerata*, which is a main host of the *Gelis* in many places, and this resulted in a large increase in *Gelis* numbers and a concomitant decline in the populations of *C. melitaearum*, which of course might be seen as advantageous to its Glanville fritillary butterfly host.

HOST RANGES OF PARASITOIDS

A distinction needs to be made between possible host range and that which is normally attacked in the field (realised host range, 'usual' host range). These two may differ widely and probably not enough distinction is made between the two. As M.R. Shaw (1983) pointed out, very few temperate parasitoids are strictly monophagous and while most 'specialists' have a preferred host, they may attack other species perhaps less avidly and/or with less success. When strict monophagy occurs it is not always clear why, for example, the pimpline *Pimpla processioneae* is monophagous on the oak processionary moth (*Thaumetopoea processionea*: Thaumatopoeidae) but most other species are highly polyphagous (Zwakhals 2005). Shaw formulated a working definition of usual host range as 'including only the species of potential hosts that the parasitoid is usually able to attack successfully, following a pattern of searching behaviour enabling it to encounter them regularly'. This definition excludes those potential hosts that are attacked only under unusual conditions, perhaps by old females who cannot locate any of the usual host and dump there eggs in anything vaguely suitable, because although the successful development of an offspring in such a host may be unlikely, it may nevertheless have a small chance of survival.

There are numerous problems with determining the host range of a parasitoid (M.R. Shaw 1994). One, which is often not appreciated, is that host ranges may vary geographically. The most abundant data almost exclusively involve hosts that are pest species and it is always possible that there might be something different about pest species than the majority of other hosts, even controlling for obvious factors such as host-feeding niche, let alone the fact that pests are typically only pests outside of their native ranges. However, very rarely does sampling extend to the full range of other potential hosts in a given region. The sorts of data that are most useful are those such as Shaw has collected over years of rearing parasitoids from a huge range of caterpillars and spiders, predominantly in the UK and the much more recent mass rearing programmes of parasitoids in a few tropical places (see Chapter 15).

M.R. Shaw (2002b) recognised that host ranges could be taxonomically narrow, sometimes with only one host species attacked, to very broad. Broad host ranges are shaped by two types of factor, host phylogeny and host ecology such as location, feeding niche, seasonality and also morphology and behaviour. Broad host ranges occupy a spectrum from continuous, where the hosts attacked are all fairly closely related to one another, and what he termed disjunct (or discontinuous), i.e. comprising hosts that belong to two or more taxonomically unrelated groups but often sharing some other ecological or behavioural attribute. Disjunct ranges are typical of plurivoltine, host alternating species such as various temperate microgastrines and rogadines (*Aleiodes*) in which two (or sometimes a few more) taxonomically unrelated hosts are used at different times of year (M.R. Shaw 2002b). Another interesting example from Shaw's *Aleiodes* research shows how host-feeding place can be important. Two closely related species, *A. pulchripes* and *A. rugulosus*, are both largely specialised on noctuid larvae of the subfamily Acronyctinae, but the former exclusively attacks hosts that feed on trees and shrubs and the latter only hosts that feed in the herb layer. Although it is not known, it seems likely that specific differences in flight height are crucial in limiting their host ranges.

The large-scale caterpillar rearing programmes that have been under way during the past few years in Costa Rica, Papua New Guinea and Ecuador are starting to yield very interesting data on parasitoid host ranges in these tropical communities (Janzen & Gauld 1997, Hrcek et al. 2011, G.Z. Jones & Shaw 2012). Hrcek et al.'s (2013) analysis of a diverse community of 266 Lepidoptera species and their parasitoids in lowland forest in Papua New Guinea found that there were both far more species of Braconidae involved than Ichneumonidae and also that the number of host individuals yielding braconids was far higher. These differences were true for both exposed and semi-concealed hosts but far more marked with the latter. Further, far more species and a higher proportion of species of Braconidae were host species specific, although the total number of reared ichneumonid species was fairly low, which would be consistent with that family showing only rather low species richness on exophytic tropical hosts. There are really no exactly equivalent temperate studies, but Sheehan (1991) analysed rearing of macrolepidopteran parasitoids from caterpillars across nine north-eastern US states that were carried out between 1915 and 1933 in an attempt to discover alternative hosts for gypsy moth (*Lymantria dispar*) and found the same pattern.

PARASITOID GUILDS AND FOOD WEBS

Far more straightforward in many respects it to try to determine the range of parasitoids that attack a particular host species, since in practice, very large samples of the host might be collected throughout its development, a necessary factor to try to make the resulting parasitoid rearings more quantitative, although this involves an enormous amount of fieldwork and subsequent rearing effort even to produce only semi-quantitative data. Several things need to be considered. If potential hosts are only collected at one time, perhaps when they are nearly fully grown, then numerous parasitoids might be missed. Apart from egg parasitoids, which do not involve any ichneumonoids, some campoplegines, for example, might have completed their larval development beforehand and already killed the host (Fig. 13.51a), several groups cause hosts prematurely to become prepupal and construct cocoons early (see Chapter 9, section *Effects on host moulting pattern*) and, of course, pupal parasitoids will be missed entirely. Also, the behaviours of parasitised hosts are sometimes different from those of unparasitised hosts, so sampling could involve biases. Many caterpillars hide during the daytime and come out to feed at night when mostly they are collected by bulk methods such as sweep-netting, so if parasitisation affects this behaviour, quantifying results might be virtually impossible. In fact, it would be extremely difficult to build up a full picture of the parasitoids attacking a given exophagous host, which is why so much experimental work to understand host–parasitoid population dynamics utilised hosts that stay put throughout all their development, such as leaf miners. In addition to the problems of collecting quantifiable samples of hosts for rearing parasitoids, there will almost always be some host mortality before a parasitoid has emerged or mortality of the parasitoid between egress from the host and emergence of an identifiable adult, although molecular techniques are now available to obtain data even from these if the bodies are preserved appropriately.

One of the puzzles of parasitoid biology is how so many different parasitoid species can persist on a single host, without one outcompeting the rest (Godfray 1994). Probably the majority of host species experience attack by more than one parasitoid species, and this is particularly true for hosts in particular feeding site categories (Fig. 14.1) (Hawkins 1994). In many of these cases, competition on a given host may be alleviated by the parasitoids involved having wide host ranges (e.g. Wieber et al. 1995). Another factor that emerged from analysis of the rearing programme for alternative gypsy moth hosts (Sheehan 1991) is that the number of parasitoid wasps recorded per host species correlates significantly with host abundance (Sheehan 1994), which means that it would be unwise to draw conclusions about parasitoid species richness on a given host without data on the latter's abundance and number of rearings attempted.

However, multiple parasitoid species often compete for the same host species even when they are specialists rather than generalists. If the parasitoids have too similar niches, perhaps both are specialists attacking the same host stage, then coexistence could still be achieved if they had different responses to other, fluctuating conditions, perhaps one doing better in warmer conditions, the other under cooler ones. Such a scenario would also suggest that the geographic ranges of the two parasitoids might differ. Probably in most situations where multiple parasitoids attack the same host, a whole range of factors, such as differences in stage attacked, induced host behaviour changes, climatic responses, host foodplant and host immune variations, alternative hosts and hyperparasitoids, must play a role in their long-term coexistence.

Dissecting what goes on is difficult and can take a lot of detailed study. For example, two species of *Mesochorus* are hyperparasitoids of the campoplegine *Bathyplectes curculionis* in North America. Simpson et al. (1979) found that *M. agilis* primarily emerged from non-diapausing *Bathyplectes* cocoons, whereas *M. nigripes* were mostly associated with diapausing ones. Does this indicate differences in host selection, perhaps mediated by some feature of the primary host or some influence of the parasitoid on whether its host enters diapause? And if the latter scenario is the case, what would be the advantage to the parasitoid? Perhaps reducing niche overlap?

Van Nouhuys and Punju (2010) reported on an in-depth study of two specialist parasitoids of the Glanville fritillary butterfly, *Melitaea cinxia*, on the Finnish Åland islands over many years. The microgastrine *Cotesia melitaearum* is gregarious and multivoltine with two or three generations on its univoltine host, whereas *Hyposoter horticola* is larger, solitary and univoltine. Earlier expectations that they coexist because the *H. horticola* is a better disperser but *C. melitaerarum* is a better local competitor have not been borne out.

Mills (1994a,b) presented a graphical way of expressing the way in which different parasitoids attack a given host species with a circle representing the host's integument and positions around the circle representing its developmental stages. The way that it is utilised by each of its parasitoid species can then be superimposed as either an arrowed line touching the circle tangentially for an ectoparasitoid or passing in and out of it for an endoparasitoid.

Entire host–parasitoid systems have traditionally been studied by rearing hosts and parasitoids, but barcoding of host remains or dissecting parasitoid larvae from either living host caterpillars offers an alternative method or at least a back-up for identifying wasps and/or hosts that had died during rearing (Hrcek et al. 2011). The fact that one can obtain usable DNA from larvae that die in captivity as well as their parasitoids helps to overcome one of the great biases that could pertain to rearing programmes where potentially parasitised hosts died before egress of either adult host or parasitoid, which might well not be a random occurrence. Dissecting hosts to locate and extract juvenile parasitoids is also likely to include biases, as some parasitoid eggs/embryos might be easy to overlook – for example, some species deposit their eggs within the host's brain or in a ganglion. As with everything else, practice and knowing what to look for will lead to the dissector having greater success. Even in highly diverse tropical systems, such as the Papua New Guinea caterpillar system studied by Hrcek et al., existing databases of sequences allowed putative species-level determination of 87 and 36% of hosts and parasitoids, respectively. Although the barcoding gene region is fast evolving, it has also contributed to a number of phylogenetic studies, although is obviously unsuitable normally for working out relationships on its own. However, Quicke et al. (2012a) found that when sufficiently large numbers of identified barcode sequences are available for a group, running unknown ichneumonoid sequences through a phylogenetic analysis with all the others would likely give an accurate identification to subfamily. As of that date, barcodes were available for 4174 ichneumonoids, although exceedingly unevenly distributed among subfamilies. Using both maximum likelihood and parsimony nevertheless recovered as monophyletic those subfamilies for which monophyly could confidently be postulated that included a large number of representatives with good taxonomic coverage. Thus, it should soon be possible to barcode ichneumonoid larvae and determine with some certainty what subfamily or perhaps even tribe or genus it belongs. It is important to emphasise, however, that for the barcoding approach to work properly, it is necessary to have a complete sequence from the barcoding CO1 region and when only 60–80% of the sequence is known, the results are frequently inaccurate even to family.

The possibility of resolving host relationships and possibly of discovering what hosts are for species that have not been reared became a possibility through the initially accidental discovery that host (Lepidoptera) sequences were sometimes obtained when attempting to CO1 barcode a parasitoid. This was explored further by Rougerie et al. (2011), who used a combination of Lepidoptera-specific primers for the barcoding gene region and high-fidelity *Taq* (DNA polymerase) in the PCR to show that at least some host DNA might remain detectable in the emerging adult parasitoid. Probably the extent to which this occurs will depend on the duration of the parasitoid pupal stage, but this has yet to be investigated. Finding potential host DNA in the parasitoid adult might also reflect host-feeding, but could therefore be misleading in some cases where parasitoids feed on non-host species as, for example, Zijp and Blommers (2002a) found for an essentially predatory *Scambus* (Pimplinae) species. This sort of forensic host detection has recently been applied to an Arctic host–parasitoid community by Wirta et al. (2014), who found that a surprisingly large 24% of parasitoids contained detectable host DNA. This procedure has been dubbed 'molecular analysis of parasitoid linkages' (MAPL) and can make a very substantial difference to our understanding of food webs. Of course, that only 24% of wasps provided sequencable host DNA means that there is a potentially substantial gap between even the most thoroughly constructed food webs made by rearing combined with MAPL and what is probably really going on in the field.

Hawkins (1990) and Hawkins et al. (1990), using the lists of hosts and parasitoids available, analysed the factors determining the numbers of parasitoid species that attack hosts of different types, in different situations and at different latitudes. A very consistent pattern that he found was that weakly concealed hosts, such as leaf-rollers and case-bearers, supported significantly more parasitoid species than either exposed or deeply concealed ones (Fig. 14.1). I suspect that there are two reasons for this. First, attacking fully exposed hosts means that the wasp is highly likely to be endoparasitic, and this entails physiological adaptation to overcome the host's immune defence systems and parasitoids and hosts are effectively always in a physiological arms race. Thus, parasitoids will tend to be more host specific, which limits the overlap between species on any one given host. Second, attacking deeply concealed hosts requires specialisations of host location and oviposition mechanisms, again making it harder to be a generalist and so reducing overlap. In relation to global diversity patterns, Gauld (1988a) suggested that the high tropical species and generic richness of ectoparasitic idiobiont parasitoids of concealed hosts are potentially able to search for hosts throughout the whole year, whereas in temperate regions they are constrained by the winter period.

What would it mean if one community were dominated by larval parasitoids whereas another had predominantly pupal ones? The former would probably include more specialists than the latter, but we are a long way from understanding what might cause such differences. One might reasonably expect that communities rich in alternative hosts will have more parasitoid species records from a given host, especially if the sample size is large, but what this would mean in terms of parasitoid host range (see M.R. Shaw 1994) is not clear. Price (1972) found that larval and pupal (cocoon) parasitoids of the jack pine sawfly, *Neodiprion swainei*, had different responses to host density. Pupal parasitoids, especially the cryptines *Pleolophus basizonus* and *P. indistinctus*, 'gained dominance' when hosts were abundant and locating them was easy, whereas larval parasitoids (predominantly tryphonines and ctenopelmatines) did better at lower to intermediate host densities, and it would be particularly interesting to know how general this was and, if it was a general effect, how it might be related to latitude.

EVOLUTION OF HOST RANGES AND SPECIATION

Host shifts might be largely fortuitous and result largely, over evolutionary time, in happy accidents when a parasitoid makes what would normally be a bad decision to accept and oviposit in a non-normal host. In this way, parasitoids routinely foraging in a given host environment, are likely to encounter a number of other non-host species that live in the same environment on a regular basis. Occasionally they may attack these non-hosts and it is well known that older parasitoids are more likely to make such attack and oviposition decisions than recently emerged ones (e.g. D.C. Griffiths 1960). M.R. Shaw (2002b) hypothesised that the occasional success in these non-hosts, probably through fortuitous coincidence of particularly appropriate parasitoid genotype and perhaps rather immunologically favourable hosts, probably represents the mechanism by which parasitoid species accrue new hosts into their host range. For endoparasitoids, penetration of a potential host by the ovipositor may be part of the host recognition–acceptance–learning process and particularly for those wasps that oviposit on the first penetration, there would be little to be lost by depositing an egg even into a novel host, although the individual may revise its external cue acceptance criteria subsequently.

Following the above scenario through, we can easily envisage how an originally monophagous parasitoid, over evolutionary time, accrues more and more host species into its normal host range (see M.R. Shaw et al. 2009), the majority probably being fairly closely related because of physiological adaptations, but also perhaps including the occasional phylogenetically distant host that regularly co-occurs. A nice example is provided by the European microgastrine *Microgaster nixalebion*, which parasitises red admiral and occasionally small tortoiseshell butterflies, *Vanessa atalanta* and *Aglais urticae*, respectively (both Nymphalidae), but also the choreutid moth *Anthophila fabriciana*, all of which have, at least as early instars in the case of the nymphalids, gregarious caterpillars living under silk retreats that feed on stinging nettle, *Urtica dioica* (M.R. Shaw 2004a). *M. nixalebion* also attacks another choreutid that feeds on the lamiacian plant *Scutellaria* and the leaf-rolling, mother of pearl moth, *Pleuroptya ruralis* (Crambidae), also on *Urtica*. Significantly, the very closely related *M. alebion*, with which it had previously been confused or just considered a form of, is a parasitoid of a different microlepidopteran, the pterophorid *Platyptilia gonodactyla* feeding in *Tussilago* (Asteraceae) stems. Thus, *M. nixalebion*, through feeding on choreutids including one feeding on *Urtica*, almost certainly initially made oviposition 'mistakes' into other semi-concealed caterpillars in the same *Urtica* micro-habitat and ultimately accrued, to its host range, the crambid and nymphalid hosts, since most *Microgaster* species are parasitoids of microleps. This is a particularly interesting example from the point of view of the interaction of polydnaviruses which play such an important role in successfully overcoming host immunity in microgastrines, as it seems likely that the great phylogenetic distance between nymphalids and choreutids must be reflected in significant differences in their immune systems. The rogadine genus *Aleiodes*, also a specialist parasitoid of Lepidoptera caterpillars, appears to have done things in the opposite direction, with macrolepidopterans forming the main host group but with occasional transitions to microleps scattered within the phylogeny, each again through close proximity to macrolepidopteran hosts used by close relatives (see Zaldivar- Riverón et al. 2008).

Such colonisation of new hosts might be especially likely at the edge of the normal hosts' range, which could be either geographic range or perhaps height in vegetation (see M.R. Shaw 2002b), where the normal hosts are hard to find, making occasional 'egg dumping' into non-hosts more likely. The abundance of the novel host is also likely to play a role if the normal host is absent (e.g. Hill & Smith 1931). If these lead to the accrual of additional hosts into the host's range, we might envisage a parasitoid having a patchwork of hosts across its range with peripheral populations progressively evolving to utilise other hosts and perhaps into host space where the original hosts do not occur. Similar things must occur commonly in nature, but in ichneumonoids, with their widespread use of associative learning to maximise/optimise host location, gene flow might be rather lower than in many other taxa. Envisage a marginal parasitoid population that for a few generations has largely been attacking a locally abundant new host and spreading further into that host's range and further away from the original host. Individuals that developed on the new host may learn the associated odours and natural selection will be leading to improved adaptations for overcoming the new host's immune defence system. Simultaneously, natural selection is likely to be leading to a loss of some of those adaptations that were previously important for overcoming the defence systems of the host/hosts in the main part of the parasitoid species' range, as the ability to overcome host immunity has its own costs (Kraaijeveld et al. 2001). These additional factors, associative learning, some spatial (or even temporal) degree of population separation and the costs of being able to overcome the defence mechanisms of a host that the local/peripheral population rarely encounters must necessarily increase the rate at which true new species emerge – at least in a peripatric situation. Whether they might also permit sympatric speciation is open to debate and must, also necessarily, depend on the relative strengths of the factors in any particular situation.

Henry et al. (2008) used the aphidiine *Aphidius ervi* to consider how host range might evolve, and in particular considered the possibility that parasitisation of different host species might result in disruptive selection because 'no single genotype is uniformly superior in all environments'. By rearing and selecting wasps on ancestral (pea aphid *Acyrthosiphon pisum*) and novel (foxglove aphid, *Aulacorthum solani*) hosts, they showed that those reared on the former make malapted oviposition decisions when reared on the latter, choosing older and less suitable instars, probably because pea aphids of suitable instars are larger than the same instar in foxglove aphids. Whether host selection traits will be strongly genetically controlled, controlled by the experience of the developing parasitoid or moderated by experience of the adult parasitoid will reflect the long- and medium-term reliability of the cues used, i.e. environmental stability, and also affect the potential of a species to colonise new hosts. Strong genetic preferences for particular sets of cues will require strong selection pressure to overcome, thus limiting the opportunities for colonising new host species because the female wasp will be unlikely to 'make mistakes'.

ENDNOTES

1. Of course, if the six right-hand points were not analaysed separately there would be no significant trend in the whole data set.
2. Information was not provided on whether the venom apparatus was also infected.
3. Wasps do not fly much when it is windy, in this case wind speeds of 8 km/h or more.

Chapter 15

LOCAL AND GLOBAL PATTERNS IN DIVERSITY

The Braconid and Ichneumonid Parasitoid Wasps: Biology, Systematics, Evolution and Ecology, First Edition. Donald L. J. Quicke.

Thanks to the mammoth cataloguing works of Shenefelt, M. Fischer, Mackauer and Marsh, we have a very accurate estimate of the total number of described species of Braconidae up to the mid-1970s for all the subfamilies (Mackauer 1968, Shenefelt 1969, 1970a,b, 1972, 1973a,b, 1974, 1975, 1978, Shenefelt & Marsh 1976), and the more recent catalogue of D.S. Yu (2012) serves the same purpose for the ichneumonids. For the Ichneumonidae, a complete listing was provided by D.S. Yu and Horstmann (1997) and D.S. Yu (2012). The big problem is that the number of described species is going to be a major underestimate in such poorly known groups as the parasitic wasps and therefore more accurate estimates of the world fauna require other factors to be taken into account. Various attempts to do this and to understand the biases are described below. Estimating global diversity is complicated, however, because the Ichneumonidae in particular may not show the same pattern of temperate to tropical species richness that most other groups of organisms do.

FIELD RESEARCH IN THE TROPICS AND ANOMALOUS DIVERSITY

Most biodiversity is located in the moist tropics and it is there that the majority of undescribed biodiversity is likely to be, so various studies at particular tropical localities have played a considerable role in trying to estimate possible global species totals. The ichneumonoids have been the source of one of the most intriguing hypotheses for explaining tropical species richness or, in their case, the possible absence of it. With the advent of mass insect collecting techniques and particularly the invention and popularisation of Malaise trapping (see Chapter 16, section *Malaise traps*), came an apparently easy way of assessing relative species diversity, or more typically species richness, between different places. When Jennifer Owen set up a Malaise trap in a garden in Leicestershire in the English Midlands in 1972 and 1973, in collaboration with the Henry Townes who sorted the collected Ichneumonidae to morphospecies, a remarkable 455 species were collected, of which 141 were represented by singletons (J. Owen et al. 1981). The sample included a large proportion of taxa that are associated with aphidophages, spiders, 'microlepidoptera' and flies associated with decaying matter and soil. A comparative study in which Malaise traps were also in similar habitats in Sweden, Freetown, Sierra Leone and Kampala, Uganda surprisingly yielded fewer species in the tropical samples (D.F. Owen & Owen 1974). This is perhaps more surprising given that Sierra Leone is a continental country whereas Britain is an island and typically has a rather reduced fauna compared with adjacent continental Europe, indeed often having fewer than half as many species. Further comparisons made using other techniques such as sweep net sampling (Janzen & Pond 1975) supported the trend discovered by Owen's work and even suggested that they were even less diverse in the tropics. This phenomenon was termed 'anomalous diversity' because it bucked the well-known trend of increasing species richness in the tropics. Thus the Ichneumonidae joined a short list of taxa mentioned in ecological textbooks that were apparently more species rich at higher latitudes, other well known examples being penguins, freshwater zooplankton and bumblebees (Noyes 1989b, Stevens 1989, Hillebrand 2004, Kindlmann et al. 2007). Whether ichneumonids also defy Rapoport's rule, that geographic or latitudinal range is positively correlated with absolute latitude, is as yet untested and will require many more comparable and accurate species lists for many localities before it can be answered.

Actually demonstrating that ichneumonids do display anomalous diversity is far harder. All studies show that samples have many singletons and Morrison et al. (1979) pointed out that valid comparisons of faunas at different latitudes requires that an adequate proportion of the local fauna has been sampled. Despite the fact that small chalcidoids do not show anomalous diversity, the consistent pattern found in ichneumonids led Sime and Brower (1998) to state that '... the unexpected decline in ichneumonid diversity in the lowland tropics has been found repeatedly worldwide and is accepted as a real trend and not a sampling artifact'. This begs the question, of course, of why the Ichneumonidae and not the Braconidae also? After all, they each largely occupy the same spectrum of biologies. True, two large groups of ichneumonids (most tryphonines and virtually all ctenopelmatines) are largely specialised parasitoids of sawflies, a group which has its peak abundance, and probably also species richness, at higher latitudes, whereas only some of the relatively uncommon and not very species-rich Ichneutinae occupy this niche in the Braconidae. However, this alone accounts for only part of the apparent deficit of ichneumonid species in tropical samples. Conversely, the only ichneumonoids that attack aphids, another

Table 15.1 Taxonomic effect of intensive study of a previously poorly known tropical ichneumonid fauna*.

Subfamily	No. of valid species recognised before Ichneumonidae of Costa Rica	New species described in Ichneumonidae of Costa Rica	No. of new synonyms	Change
Acaenitinae	0	1	0	∞
Anomaloninae	8	60	3	8.12
Banchinae	15	222	4	15.53
Brachycyrtinae	6	6	2	1.33
Cremastinae	27	122	3	5.41
Ctenopelmatinae	1	41	0	41
Cylloceriinae	0	4	0	∞
Diplazontinae	3	18	0	7
Labeninae	14	33	2	3.21
Lycorinae	1	7	0	8
Metopiinae	9	121	1	14.33
Oxytorinae	0	4	0	∞
Pimplinae	39	106	4	3.03
Poemeniinae	0	2	0	∞
Rhyssinae	6	7	1	2.0
Tryphoninae	7	32	0	5.57
Xoridinae	0	9	0	∞

*Includes a small percentage of taxa given as species A, B, C, etc.
Source: data from Gauld 1988b, 1991, 1997, 2002.

temperate host group, are the aphidiine braconids. So both families have sizeable subsets of species that are going to be largely restricted to the temperate zone.

Ian Gauld in collaboration with INBio[1] and Dan Janzen have carried out a long-term collecting programme with Malaise traps set all over Cost Rica, a country of approximately 51,000 km^2, i.e. approximately one-fifth of the size of the UK. The Ichneumonidae in these samples have been very thoroughly researched, many groups having been revised in the process, and the data show that Costa Rica has approximately 1.14 times more species overall than the UK based on those subfamilies that have already been revised for Costa Rica (see Table 15.1), which is of course a far smaller multiplier than would be found for most other groups, for example, six for 'macrolepidoptera' (Rodriguez et al. 2013). Of course, this study is not comparable to that of the Owens because of its far greater scale in both space and time. The Costa Rican study reached more than 100 Malaise trap years and in combination with associated taxonomic revisions has made a very substantial difference to the number of known species (Table 15.1). However, it must always be borne in mind that this does not give a degree of coverage comparable to that which has led to the checklist of British Ichneumonidae, which has involved hundreds of collectors, in many more localities, using many more techniques and over a real period of roughly 200 years. Hence the ratios of numbers of species known in the two faunas (Table 15.2) have to be treated with caution. Some differences are easy to explain, for example, the relatively fewer Ctenopelmatinae in Costa Rica reflects the fact that their sawfly hosts are a predominantly north temperate group, and similarly for the Diplazontinae, which are almost entirely restricted to aphidophagous syrphids whose aphid prey are also poorly represented and generally less abundant in the tropics.

Compared with other Hymenoptera, the Ichneumonoidea as a whole appear to comprise a smaller proportion of species in Costa Rica than in the UK (Gauld & Gaston 1995), but this conclusion is likely heavily biased by the Ichneumonidae and there has been no comparable taxonomic effort on Costa Rican braconids. In contrast, Rodriguez et al. (2013), working on microgastrines, revealed a startling diversity in Costa Rica. Based upon more than 20,000 DNA barcodes for microgastrines from 75 countries, with 12,274

Table 15.2 Comparison of number of recorded species of ichneumonidae in Britain and Costa Rica by subfamily.

Ichneumonid subfamily (and biology)	No. of British species	No. of Costa Rican species	Ratio
Acaenitinae	7	1	0.14
Adelognathinae	~15	0	0
Agriotypinae	1	0	0
Anomaloninae	~40	65	1.6
Banchinae	~110	233	2.1
Campopleginae	>250	Not known	–
Collyriinae	2	0	0
Cremastinae	~13	146	11
Ctenopelmatinae	>200	42	0.2
Cylloceriinae	~5	4	0.8
Diacritinae	1	0	0
Diplazontinae	~50	21	0.42
Eucerotinae	3	0	0
Hybrizontinae	2	0	0
Labeninae	0	45	∞
Lycorininae	1	7	7
Mesochorinae	>50	Not known	–
Metopiinae	>60	129	2.1
Ophioninae	~23	28	1.3
Orthocentrinae	~50	Not known	–
Orthopelmatinae	2	0	0
Pimplinae	~100	141	1.4
Poemeniinae	5	2	0.4
Rhyssinae	2	12	6
Stilbopinae	4	0	0
Tersilochinae	>50	Not known	–
Tryphoninae*	~147	39	0.26
Xoridinae	~14	9	0.6

*Excluding *Netelia*.
Sources: data for Britain from Fitton in Barnard 1999; data for Costa Rica from Gauld 1988b, 1991, 1997, 2002.

from Cost Rica, M.A. Smith et al. (2012) obtained 'molecular species' accumulation curves very similar qualitatively to that shown in Fig. 15.1. For Costa Rica, there are now over 800 MOTUs recognised and approximately 900 when additional ecological/host data are taken into account, such that some sequences that differ by less than 2% are taken to represent more than one species given the ancillary evidence.

ESTIMATION OF GLOBAL ICHNEUMONOID SPECIES RICHNESS

Several methods have been used to try to assess the undescribed numbers of species, almost entirely for the Braconidae. A common form of estimation uses a focal taxon which is taxonomically taken to be nearly completely known globally and data from more limited regions where the group of interest is considered sufficiently well known. Through lack of alternatives, birds, butterflies, mammals and plants have tended to be used for these purposes. All of these groups have their own particular biases, but unless ichneumonoids differ markedly then the focal taxon approach could be expected to give at least a reasonable estimate of species richness, all other things being equal. For the Braconidae, Dolphin and Quicke (2001) first applied the focal taxon approach based on the comparatively well-studied Palaearctic fauna and using mammals and butterflies as examples of groups for which both

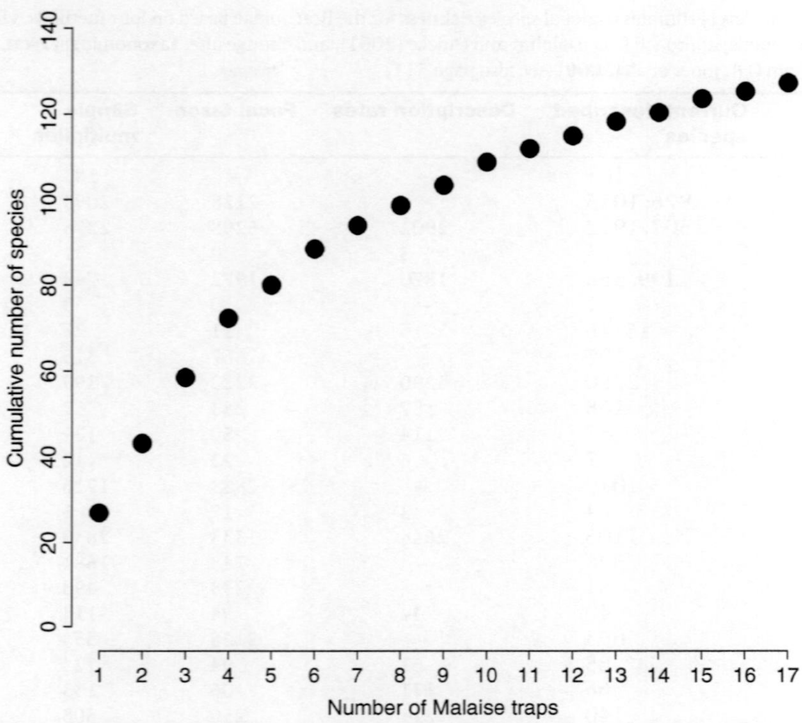

Fig. 15.1 Accumulation of species of Pimplinae *s.l.* in Malaise traps (smoothed by randomisation) at a total of 17 sites in Costa Rica, and representing 100 Malaise trap years. (Source: data from Gaston & Gauld 1993.)

Palaearctic and world species numbers are known with reasonable accuracy. The mammal data suggested that the global number of braconid species would be double the currently described total, whilst the data for butterflies suggested that it would be approximately four times greater, i.e. between 25 and 50% of species had already been described.

Dolphin and Quicke (2001) also used the rates of taxonomic description, corrected for taxonomic effort, as a way of predicting species richness in groups that were taxonomically incompletely known, again using braconid subfamilies as test cases. The rationale was that the rate of taxonomic description in terms of new species described for a given amount of taxonomic effort declines, non-linearly, with number of species (of mammal). After excluding a lag phase which plausibly results from the initial process of taxonomic organisation, which is required before serious and comprehensive taxonomic description can really get under way, they found that most subfamilies were likely to be increased by a factor of between two and four when all species are known. The estimates of actual numbers of species obtained are shown in Table 15.3. This independent approach thus lent support for the focal taxon results, i.e., that approximately 50% of all braconid species had already been described, which seems rather remarkable and indeed unbelievable.

A different approach for estimating total numbers of species, considering what happens to numbers of valid species in genera when a new taxonomic revision is carried out and allowing for the geographic region(s) concerned, date of the revision and body size of the wasps, was used by O.R. Jones et al. (2009). Tropical faunas are well known to be less thoroughly investigated than temperate ones and therefore revisions based of taxa in the tropics are more likely to discover a larger number of undescribed species (Fig. 15.2). Physically larger species are also more likely to have come to the attention both of collectors and of taxonomists who have preferentially studied larger bodied taxa. Thus, irrespective of origin, a large wasp is far more likely already to have been described than a small one.

Table 15.3 Subfamily level estimates of global species richness for the Braconidae based on four methods, viz. description rates and focal taxon (mammals) approach from Dolphin and Quicke (2001), and change after taxonomic revision, raw and using Chao 1 estimator from O.R. Jones et al. (2009) (see also page 511).

Subfamily	Current described species*	Description rates	Focal taxon	Simple multiplier	Chao 1-derived multiplier
Acampsohelconinae	101	–	–	183	242
Agathidinae	976, 1015	–	2128	2094	2810
Alysiinae	1503, 1912	2902	5298	2378	3177
Amicrocentrinae	5	5	0	31	41
Aphidiinae	499, 558	1870	1972	749	1004
Apozyginae	1	–	0	3	4
Betylobraconinae	15, 16	15	11	28	38
Blacinae	197	–	667	315	423
Braconinae	2760	3290	3222	7399	9858
Cardiochilinae	168	187	283	359	476
Cenocoeliinae	67	114	50	131	175
Charmontiinae	7	6	22	12	16
Cheloninae	1010	–	2883	1716	2282
Dirrhopinae	4	3	17	6	8
Doryctinae	1105	2644	1333	2839	3787
Euphorinae	598	–	745	1681	2256
Exothecinae†	81	–	233	493	815
Gnamptodontinae	40	44	94	113	151
Helconinae	693	–	2128	353	471
Homolobinae	65	61	94	124	167
Ichneutinae	66	131	106	133	178
Macrocentrinae	140	211	250	308	411
Masoninae	5	5	0	11	14
Mendesellinae	9	–	0	40	53
Mesostoinae	9	–	–	15	121
Meteorideinae	11	8	11	29	38
Microgastrinae	1856	4178	3617	4025	5370
Microtypinae	13	–	17	13	17
Miracinae	29	–	78	56	75
Opiinae	1734	3242	2422	3469	4640
Orgilinae	310	909	656	539	716
Pambolinae	85	–	278	141	188
Rhysipolinae	17	–	67	94	126
Rogadinae	580	1188	844	1256	1673
Sigalphinae	33	50	72	76	102
Trachypetinae	6	10	14	3	5
Xiphozelinae	15	–	–	12	16

*Where two figures are given, they refer to numbers described at the time of Dolphin and Quicke (2001) and O.R. Jones et al. (2009), respectively.
†Including also Hormiinae, Lysiterminae and Rhyssalinae.

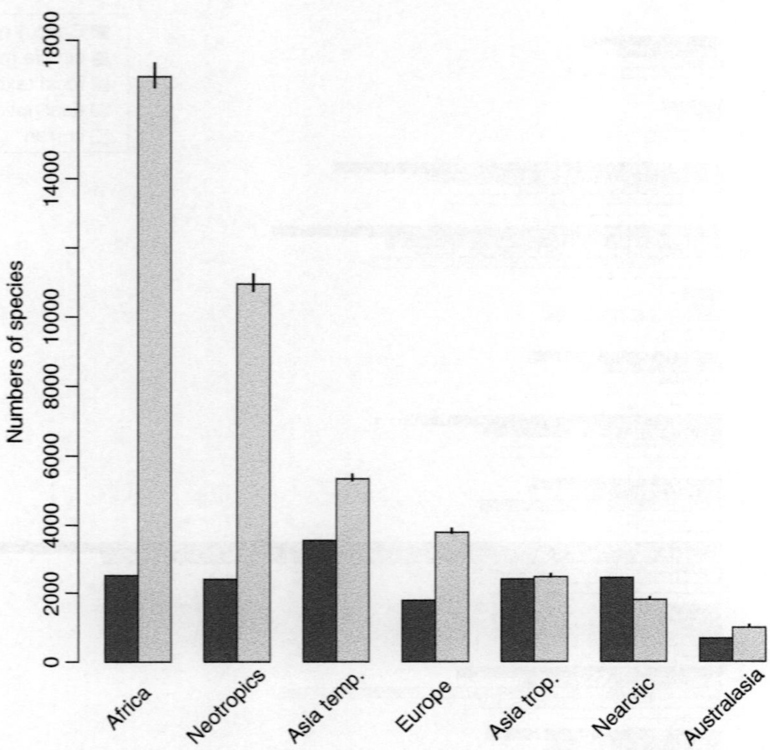

Fig. 15.2 Numbers of described braconid species (left-hand bars) and estimated numbers (right-hand bars) for seven biogeographic regions (±SE). (Source: data from O.R. Jones et al. 2009.)

The fitted best model again predicted that if all genera were revised, the total number of species would be expected to increase by approximately 90%.

In addition, O.R. Jones et al. (2009) made use of the number of individuals upon which each species in the revision was based and used the Chao 1 non-parametric estimator to adjust the multiplier of species recognised after and before revision for those that are likely to have been missed by the revision (A. Chao 1984; see also *Estimating local species diversity*, below). Thus, for genera where all species are known from multiple individuals, it is unlikely that there are many more species still to be discovered, whereas, if many species are known only from a single individual, it is fairly likely that many species were not sampled at all. The Chao 1 estimator is non-parametric because it assumes that all species in the population from which the sample is drawn are equally abundant, which is obviously not the case for biological communities, which often approximate log-normal, log-series or neutral species abundance distributions, which are characterised by having a few common species and many rarer ones. The point here is that in such situations, especially where there are many singletons relative to total sample size, Chao 1 is likely still to be a considerable underestimate. Incorporating the Chao 1 estimator into the model suggested that total known species would increase about 2.5-fold if all genera were revised, although the predictions varied markedly between different subfamilies (Fig. 15.3), with the Braconinae seemingly likely to increase by the greatest proportion.

In very marked contrast to the four objective estimates described above, Rodriguez et al. (2013), concentrating just on the Microgastrinae and incorporating new and unpublished data based on ongoing DNA barcoding studies in various parts of the world and using a range of focal taxa, came up with greatly increased estimates of global species richness ranging from 17,000 to >46,000 compared with the 2067

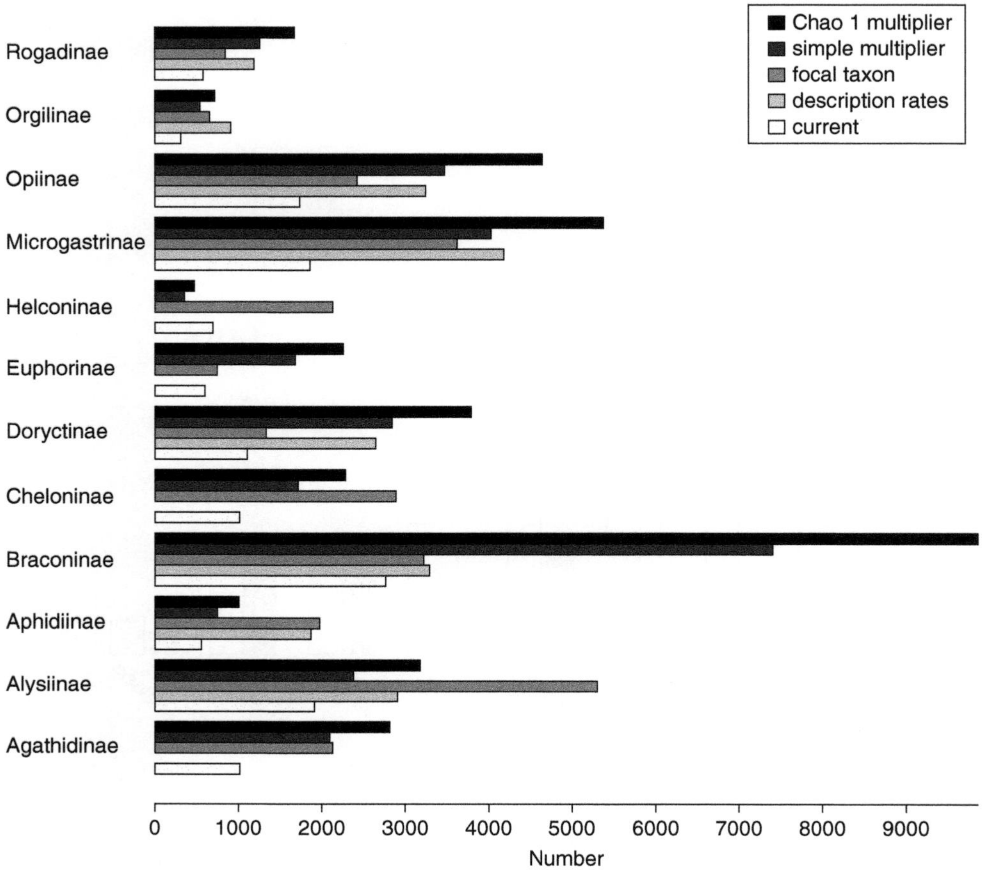

Fig. 15.3 Current and estimated global species numbers for 12 large braconid subfamilies comparing results from focal taxon comparison and description rates (source: data from Dolphin & Quicke 2001) and from multipliers based on changes in total numbers of recognised species following taxonomic revision, and a correction to that based on the numbers of species known from only singletons and doubletons. (source: data from O.R. Jones et al. 2009.)

species described to date. They recently pointed out that the estimates of O.R. Jones et al. (2009) had various underlying assumptions, principally that comprehensive morphology-based taxonomic revisions do indeed capture the (near) total diversity of a group and that morphospecies are an adequate surrogate for real biological species. I concur on both points. The major limitation of taxonomic revisions is that they are based on what material is in already assembled museum collections, sometimes augmented by a few fresh samples that may have prompted the revision in the first place. Museums do hold wonderful collections and often have material from collecting localities that no-longer exist, but they are neither comprehensive nor structured samples and owe a great deal to serendipity. The recent barcoding studies on groups such as the microgastrines have also revealed many morphospecies that are in fact complexes of hard to separate but distinct species. What we do not yet know, however, is how generalisable findings such as these are – are they perhaps functions of latitude or systematic group or body size or biology?

DISTRIBUTION RELATED TO CLIMATE AND LATITUDE

Idiobiont ichneumonoids dominate in saproxylic communities in boreal forests (Stenbacka et al. 2010) and ichneumonids appear to be more speciose in arctic communities than braconids (Várkonyi & Roslina 2013).

A surprisingly high species richness of parasitoids in general and ichneumonoids in particular has also been demonstrated using DNA barcoding of samples collected at very high latitudes around Churchill in Manitoba, Canada (Stahlhut et al. 2013). In agreement with perhaps general views on the diversity of the Ichneumonidae and Braconidae in relation to latitude, they found nearly four times more molecular taxonomic units in their Ichneumonidae sample than they did in the Braconidae, but the total MOTUs for the superfamily, comprising well over 50% of the total for Hymenoptera, is rather startling.

Whether this latter observation has any more general implications relating to the comparative latitudinal distributions of the two families is not clear – I cannot think of any necessary causal connection when it comes to very high latitudes. Wharton (1993a) commented on the widely held view that ichneumonids have a less tropicocentric distribution than braconids, many groups of the latter being spectacularly more diverse in the tropics, and noted that there is always the possibility that the pattern observed in the Ichneumonidae is a sampling artefact. Almost everyone who has collected ichneumonoids in the tropics will have noticed that arid habitats have very different faunas to moist habitats, and Townes (1958) suggested that the presence of ichneumonids was largely limited by available water for adults to drink and that tropical coastal lowlands, where there is seldom any dew, could therefore support far fewer ichneumonids. Shapiro and Pickering (2000) compared the ichneumonid faunas of a moist and a wet forest area in Central America using Malaise trapping and found that pupal parasitoids were more abundant in the former whereas lepidopteran larval parasitoids were relatively more common in the wet forests. Overall, wet forest traps caught 2.3 times more ichneumonids than moist forest traps and old-growth sites yielded nearly 1.9 times more individuals than secondary growth sites. Of the subfamilies with large enough sample sizes, Tersilochinae (Tersilochini) and Orthocentrinae, both dominated by small-bodied species (see Fig. 15.7), were both significantly more abundant in the lowland wet forest of La Selva, Costa Rica, than in the moist forest at Barro Colorado Island, Panama. As the former are parasitoids of Diptera associated with fungi and damp situations (i.e. Myceophilidae and Sciaridae), this is not surprising. Having said that, there is a problem of statistical non-independence with their data as they lacked multiple, and geographically intermixed, examples of their forest types, so it is impossible to know if this is a general trend.

An interesting and rather surprising pattern has been found in North America in that the greatest diversity of Ichneumonidae appears to occur at intermediate latitudes (Janzen 1981). Originally based on catalogued data, this pattern was repeated by Skillen et al. (2000) for two subfamilies, the Campopleginae and Ichneumoninae, based on a Malaise trap transect in old-growth forests ranging from 30 to 49°N. This pattern was largely replicated for Braconidae by Quicke and Kruft (1995) and a higher species richness away from the equator was also shown for most groups of Australian ichneumonids (Gauld 1986b) although, as generally seems the case, ophionines bucked the trend. However, these data do not really support the anomalous diversity idea because the area covered is all far from the equator and the pattern almost certainly reflects the increasing aridity of a good number of the more southern states.

To summarise, the focal taxon approach for braconids gave broadly similar estimates of total global species richness, as did two other independent methods, thus seemingly validating the focal taxon approach for this family. In other words, the different estimation methods did not suggest that braconids showed markedly lower increase in species richness towards the tropics than other well-known groups. Going back to the fieldwork comparing Ichneumonidae species richness between tropical and temperate samples (D.F. Owen & Owen 1974, Janzen & Pond 1975, Gauld & Gaston 1994), there would appear to be a considerable difference between the two families. Taking that to be the case, for the time being, I discuss below the theories proposed to explain why ichneumonids might be less tropicocentric.

THE NASTY HOST HYPOTHESIS

In order to explain the apparent anomalous diversity of ichneumonids, various hypotheses have been put forward. Janzen and Pond (1975) proposed a resource fragmentation hypothesis and Rathke and Price (1976) a predation hypothesis, and Gauld et al. (1992) and Gauld & Gaston (1994) developed the 'nasty host hypothesis' (NHH). The predation hypothesis works on the assumption that predation pressure is higher in the tropics and so long-term persistence of species is more difficult there. The resource fragmentation hypothesis is essentially that the number of parasitoids in a community increases non-linearly with the numbers of host species because some hosts become too rare to support viable populations of parasitoids. The NHH, broadly stated, says that individuals in many, largely phytophagous and maybe exophytic, host groups are considerably more toxic in the tropics than are members of equivalent temperate niches. These are not in any way mutually exclusive and the NHH has received most attention and gains marginal support from the available evidence (Sime & Brower 1998). It is also one of the most biologically interesting if it is true.

Nearly all work on the NHH has been secondary, that is, based upon host associations, and these involve caterpillars that feed on food plants that are known to contain high levels of toxins. Levin (1976) showed that the proportion of alkaloid-containing plants is higher in tropical than temperate floras [see also Moody (1978) for consideration of the anomalously high percentage found in India and Pakistan]. Indeed, feeding on toxic plants and sequestering their toxins are a major defensive strategy of many Lepidoptera species and there is no doubt that in many cases this can effectively put them in enemy-free space (Greeney et al. 2012), i.e. they themselves become so toxic that they are avoided as potential food by both invertebrate predators such as ants and parasitoids.

Relatively little work has been carried out on the effects of sequestered plant toxins in caterpillars on their parasitoids. There are two extreme ways in which they may act in principle, and probably both are true of different host plant –parasitoid combinations. The NHH says that the caterpillar's acquired toxins will render it immune from attack by parasitoids as they may be unable to cope with ingesting these toxins. On the other hand, if a parasitoid can overcome the toxin's effects, then a caterpillar may form a safe haven (Greeney et al. 2012) because it is less likely that its host will be predated by ants, birds, etc. Certainly some highly toxic caterpillars have specialist parasitoids. Lampert et al. (2011) found that larvae of the microgastrine *Cotesia congregata*, dissected from within caterpillars of the catalpa sphinx moth, *Ceratomia catalpae* (Sphingidae), contained the host plant iridoid glycoside catalpol, which is sequestered by the caterpillar, but at lower concentration than in the host tissue, probably indicating reduced absorbance. The wasp was able to develop with a high level of success in hosts containing high levels of catalpol and only a slight negative association was found between host catalpol concentration and *Cotesia* survival (Lampert et al. 2010), so in this case, rather contrary to the NHH idea, the catalpa sphinx is actually providing a safe haven for this wasp, since the catalpol protects its caterpillars from many other vertebrate and invertebrate predators. Of course, this actually says very little about the NHH, since the NHH predicts that caterpillars with high levels of sequestered toxin may be defended against parasitoids in general, but we know that some parasitoids overcome these defences. More telling would be whether the parasitoid guild size of these protected caterpillars was the same as, or smaller than, that of a range of equivalent unprotected ones or whether, if the sphinx could be reared on a diet low in or lacking catalpol, it would become an accessible host for other parasitoids that it is currently protected against. In a different system, Nieminen et al. (2003) found that parasitisation levels of Glanville fritillary, *Meliaea cinxia* (Nymphalidae), larvae against the specialist microgastrine parasitoid *Cotesia mellitearum* were lower in individuals that had higher dietary concentrations of catalpol and another iridoid glycoside, aucubin, which would be consistent with the parasitoid perhaps avoiding ovipositing in those hosts that contained more glycoside. However, Laurentz et al. (2012) subsequently found that the caterpillars with more glycosides were better at encapsulating their parasitoid, perhaps suggesting that the glycosides rendered the parasitoid more vulnerable to encapsulation, perhaps through directly enhancing the host's immune capability.

Sime (2000) reported on the occasional rearings of two generalist pimpline pupal endoparasitoids, an *Apechthis* and a *Theronia* species, from the toxic swallowtail, *Battus philenor*, in North America which feeds on Aristolochiaceae and sequesters aristolochic acids. *B. philenor* is almost never attacked by larval parasitoids; it seems to be in a virtually enemy-free space. She suggested that these generalists might be more capable of overcoming a range of toxic plant allelochemicals in the host pupae than larval and larval–pupal parasitoids. If this is indeed a widespread phenomenon, it might have an influence on the relative abundances and diversities of wasps with these life history strategies in tropical compared with temperate zones.

Smilanich et al. (2011) have shown that caterpillars of the polyphagous arctiine (formerly Arctiidae) moth

Grammia incorrupta, which feeds on a range of plants with different types of toxins, change their feeding preference when parasitised by the monophagous microgastrine *Cotesia* nr. *phobetri*[2]. Caterpillars parasitised by this wasp fed more on *Malva* leaves, which contain high levels of antioxidant flavanoids and phenolic compounds. The interpretation of this effect on host behaviour is complex, but it is suggested that it may be beneficial in mopping up free radicals produced by the melanisation aspect of the hosts encapsulation process and also perhaps because the antimicrobial properties of the *Malva* leaves may reduce the chance of secondary infection through the parasitoid's oviposition puncture wound. Whether this would be significant is hard to say, as most hosts parasitised do not seem to die from secondary infection at the wound site.

One of the supporting pieces of information for the NHH is that members of what appear to be a closely related species group of ophionine, the *Enicospilus americanus* group, which ranges from southern USA into South Central America and has become widely quoted. In North America there are two species, each with a broad host range attacking saturniids, whereas in Costa Rica there are approximately 19 species in the complex, each, as far as is known, being monophagous (Gauld 1988b), and some do so only when the host is feeding on one of its host plants. Another, older potential example involves the highly polyphagous *Habrobracon hebetor* (as *brevicornis*), which, when attacking the pest noctuid *Heliothis armigera* in South Africa, only does so when it is feeding on *Antirrhinum* (Taylor 1932, cited in Askew 1971).

More recent data on the host ranges of tropical parasitoids of exophagous hosts, such as from the caterpillar-rearing programmes of Dan Janzen and Winnie Harbachs in the Area de Conservación Guanacaste in Costa Rica and Vojtech Novotny and co-workers in Papua New Guinea are starting to provide a consistent picture and indicate that at least some groups, such as microgastrines, display very high levels of host specificity in the tropics, with many species being truly monophagous (M.A. Smith et al. 2008). It will be very interesting to see whether other groups of braconids and ichneumonids with similar biologies follow suit. Microgastrines could potentially be somewhat unusual in that high levels of host specificity have been found in species groups attacking temperate hosts such as Melitaeni butterflies (Kankare et al. 2004, 2005a,b). This group of butterflies, however, often feed on plants well known to contain toxins, such

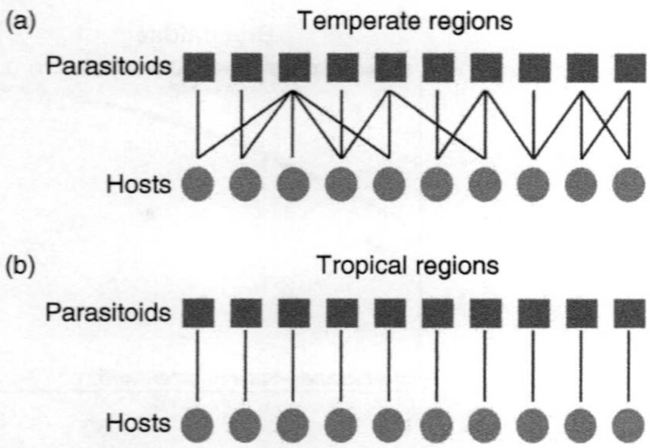

Fig. 15.4 Cartoon depicting relative host specificities of parasitoids in temperate and tropical regions as predicted by the nasty host hypothesis, (Source: adapted from Sime & Brower 1998.)

as *Plantago* (Plantaginaceae), which produce iridoid glycosides that the butterfly caterpillar sequesters. Such toxins are likely to render hosts unsuitable to many parasitoids, but the situation can be much more complicated and, when the plant is eaten by a generalist herbivore, the latter may have its immune defence mechanism compromised as a result, perhaps making it more vulnerable to attack by some parasitoids (Reudler et al. 2011).

Sime and Brower (1998), having accepted the validity of the NHH, showed that the relative numbers of potential butterfly host species belonging to groups of known toxicity (various groups of butterflies including Troidini, which sequester aristolochic acid) increase dramatically in species richness towards the tropics whereas their specialist ichneunomine parasitoids of the Trogini, especially *Trogus*, do not. Their data therefore supported the NHH, whereas they are either ambiguous or contradictory for other hypotheses. Figure 15.4 depicts the sort of food-web extremes that the NHH suggests, with temperate parasitoids tending to be polyphagous and with hosts having more than one equivalent parasitoid attacking them, and tropical parasitoids being predominantly monophagous and few hosts having multiple parasitoids. Further, they noted that some abundant troidines appear so well protected by chemicals that even intensive rearing efforts have failed to recover any parasitoids, so they appear to have achieved the enemy-free space that is central to the NHH. The situation described above for *Enicospilus americanus* is what one would expect from the NHH with parasitoids having evolved to cope with

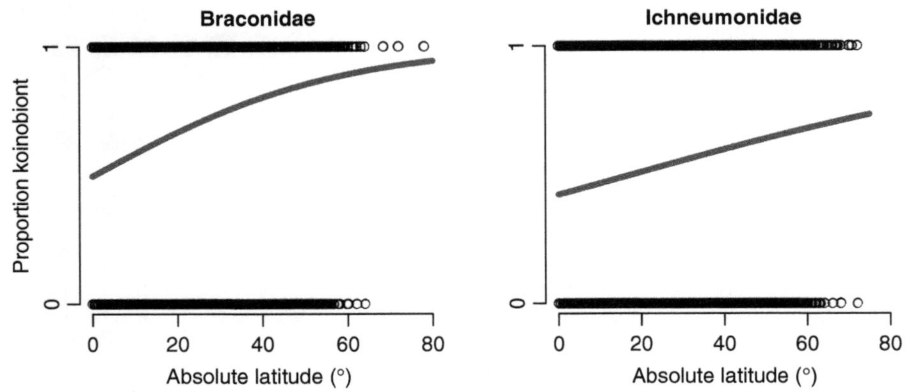

Fig. 15.5 Logistic regressions of koinobiont to idiobiont ratio versus absolute mean latitude for braconids and ichneumonids. (Source: data from D.S. Yu et al. 2005, reanalyzed and redrawn as per Quicke 2012.)

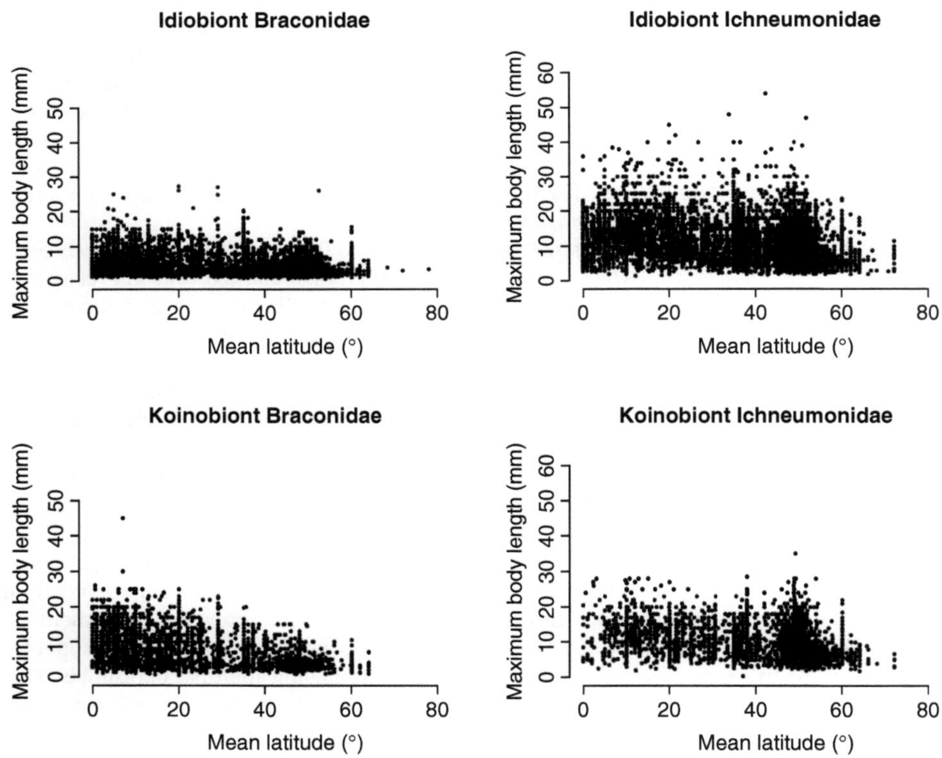

Fig. 15.6 Braconid and ichneumonid species body size versus mean latitude plotted separately for described idiobionts and koinobionts and showing the probably artificial peaks and clusters in both categories of ichneumonids at approximately latitude 50° representing taxonomic effort. (Source: data from D.S. Yu et al. 2004, latitudes determined as in Quicke 2012, euphorines excluded.)

specialised particular host species and their associated plant-derived toxins and being unable to utilise hosts with different sets of toxins, albeit they are closely related. The increase in host specificity of these *Enicospilus* at lower latitudes (or, more precisely, in Costa Rica as data from other places are either sparse or have not been analysed) was mirrored by Sime and Brower's (1998) pattern for the *Trogus* species, although in their case based on a personal communication from D. Janzen.

One of the predictions of the NHH is that the proportion of parasitoid species attacking folivorous hosts, i.e. the ones most likely to have high concentrations of toxic plant secondary compounds, relative to ones parasitising less chemically protected wood-borers, for example, should decline towards the tropics. Since most taxonomic groups of ichneumonoids are fairly internally consistent in the groups they attack (e.g. Wharton 1993), it is possible to infer, with a high level of reliability, the host group or other life-history feature that a species of a given subfamily or tribe has.

Apart from carrying out what are hoped to be comparable samplings in different geographic regions to test this, another approach might be to use taxonomic literature as a crude estimate of numbers of species at given latitudes. This approach has several flaws, of course. Description effort is concentrated in the temperate region and so too is taxonomic revision, and revisions mean that not only are new species likely to be described, but also others may be placed into synonymy. O.R. Jones et al. (2012) found that the proportion of described species of Ichneumonidae that are found to be synonyms is significantly affected by both region and latitude. Fortunately, the potential biases that this might introduce are not large enough to invalidate other major trends recovered from analysis of species list data. The equivalent analysis has yet to be done for the Braconidae.

Using the idiobiont–koinobiont distinction as a surrogate for whether hosts are likely to be exophytic and hence potentially chemically protected, Quicke (2012) showed that the proportion of idiobiont ichneumonids increased significantly towards the equator, but so also did that for braconids (Fig. 15.5). However, Santos & Quicke (2011) showed that body size increased significantly towards lower latitudes for both ichneumonids and braconids and the same broad pattern is observed when data for idiobionts and koinobionts are examined separately (Fig. 15.6), i.e. more large-bodied species occur at lower latitudes. It is also apparent that braconids show a somewhat more linear trend than do ichneumonids, the latter being perhaps best described has having fewer large bodied species at extreme high latitudes. This is because ichneumonids are generally larger bodied than braconids, but both families have roughly the same upper size limit which is imposed by their hosts. Further, braconids have more gregarious koinobiont species (e.g. in Microgastrinae, Euphorinae (Meteorini) and Macrocentrinae), which are generally smaller than solitary species attacking the same host.

Santos and Quicke (2011) explored these data further by including subfamily as a factor in the model. Because different subfamilies are characterised to some extent by having different body size distributions (Fig. 15.7), it could be that overall family trends just reflected latitudinal patterns in the distributions of described species in different subfamilies. With both families, the direction of the body size trend with absolute latitude changed from negative to positive, but only the ichneumonids now showed a significant relationship (F = 173.4, 74,16498 d.f., adjusted $r^2 = 0.435$, $p < 0.01$), i.e. described ichneumonids become smaller nearer to the equator once subfamily is taken into account.

The numbers of species in braconid and ichneumonid genera are plotted against the mean absolute latitude of their included species in Fig. 15.8 (Quicke 2012). Both show a similar pattern, but what it indicates is open to question. Perhaps it simply reflects the taxonomic effort applied to braconids and ichneumonids from different latitudes or perhaps it indicates real differences in speciation/extinction rates with respect to absolute latitude, since genera are proxies for clades. The Braconidae appear to have more large genera centred closer to the equator, whereas the genera of ichneumonids show a distinctly bimodal pattern with a far more pronounced absence of genera, with species spread evenly across a wide range of latitudes/or large genera with centres around 30° away from the equator. The dip around 30° is almost certainly explained by large, arid zones of north central America and North Africa. Both families do have an approximately similar pattern of genus sizes for tropico-centric genera, indeed ichneumonids have slightly more very large tropicocentric genera than braconids do, but they have many more large genera that have the majority of their species at absolute latitudes of 40–50°. If this is because these genera are mostly ones that specialise on sawflies, then another way of expressing anomalous diversity would be to say that unlike braconids, the ichneumonids have adapted more successfully on to predominantly temperate host groups.

Several pieces of information, however, draw into question the validity of current estimates of the relative

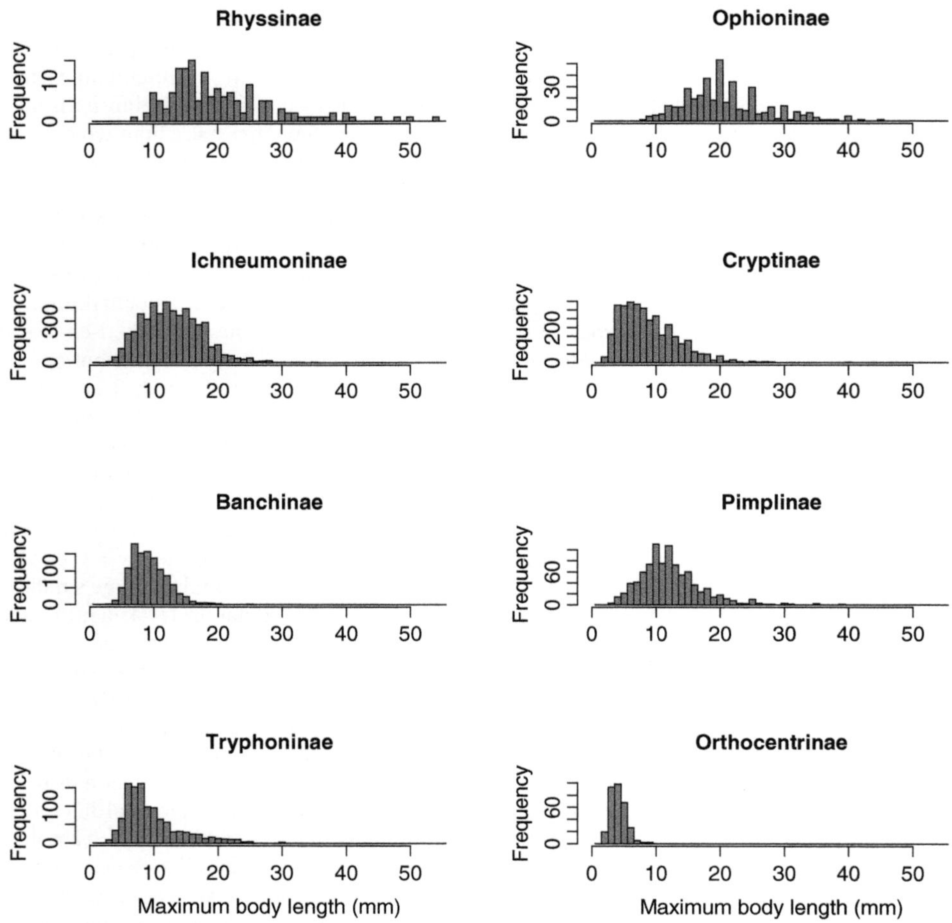

Fig. 15.7 Body size distributions of described species of selected subfamilies of Ichneumonidae. (Source: data from D.S. Yu et al. 1997.)

species richness of various ichneumonoid subfamilies in the tropics. Gaston (1993) drew attention to the fact that, for the three largest insect orders, the majority of species currently being described were from the temperate region, which goes completely against the expected spatial distributions, and certainly no-one has suggested that Coleoptera, Lepidoptera or Hymenoptera as a whole are anything other than much more species rich in the tropics. This simply reflects the enormous bias in the number of taxonomic entomologists in the temperate region, many of whom are amateurs and are particularly interested in their regional fauna. A second source of bias is that larger bodied species tend to get described earlier than smaller bodied species (Blackburn & Gaston 1994, Gaston et al. 2005), so, given that a greater proportion of the temperate fauna has been described than tropical fauna, the ichneumonoids and other groups are likely to include significantly more small-bodied taxa awaiting description in the tropics. This has even had an effect on the groups that have been targeted for studies on temperate/tropical diversity, which are almost entirely ichneumonids[3] and largely limited until recently to the large-bodied members of the pimpliformes and Ophioninae as opposed to the far more poorly known small-bodied subfamilies such as Orthocentrinae, Tersilochinae and Tryphoninae (see Fig. 15.7). Another aspect is that groups that

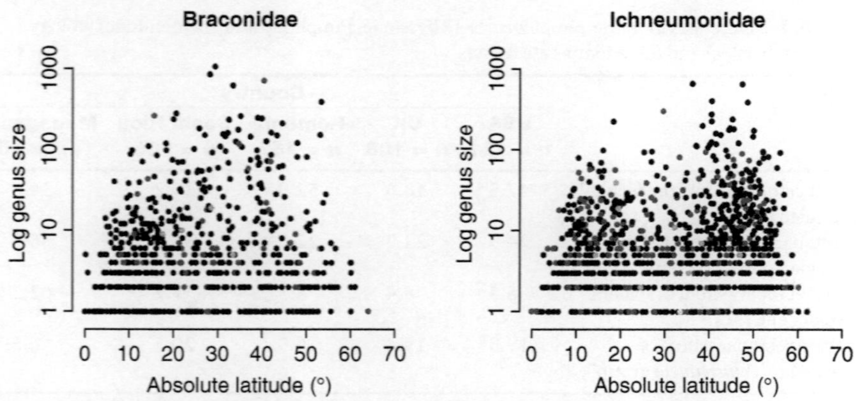

Fig. 15.8 Numbers of described species in genera of Braconidae and Ichneumonidae plotted against mean absolute latitude of the genus showing that in the former the majority of large genera have a mean distribution at low latitudes, but for ichneumonids, a more bimodal pattern is apparent. (Source: data from D.S. Yu et al. 2005, latitudes determined as in Quicke 2012, euphorines excluded.)

are very species rich require a great deal of effort to study, especially morphologically, as they probably include many morphologically hard to differentiate species. For such groups, full taxonomic revisions are only rarely undertaken and taxonomists often choose to work on taxa that are tractable – even more so when it comes to taxonomists working in institutions that want to see regular and frequent publications from their staff. Porter (1980), for example, noted that the New World ichneumonine genus *Carinodes* is represented in the catalogue by Townes (1966) by just 31 species, but Porter estimated, based on his neotropical fieldwork and specimens in museum collections, that there could well be as many as 500 species. *Carinodes* species are not small bodied, but the Ichneumoninae is a large and taxonomically difficult group to work on and they have been largely ignored in all the papers dealing with estimates of tropical biodiversity. The recent revisions of several genera of fairly small-bodied braconids from the tropics, each based more or less on all available material, have each yielded large numbers of new species (van Achterberg & Quicke 2000, Areekul & Quicke 2006, Butcher et al. 2012), and each includes many singleton species. Of these, probably only *Aleiodes* would have had a prior expectation of having many tropical species. I cannot imagine that anyone in the near future will attempt to tackle the formal taxonomy of all the world species of the large number of widespread genera of small-bodied ichneumonoids such as *Bracon* (Braconinae), various Microgastrinae and Orthocentrinae or even the slightly larger Phaeogenini (Ichneumoninae), and it seem that estimates for these will be based largely on barcoding as in Rodriguez et al. (2013).

Thus, whether any of the above-described patterns are biologically meaningful rather than biases in description remains to be seen (Quicke 2012).

BIOGEOGRAPHY

There are some remarkable similarities between the phylogenies of the Ichneumonidae and Braconidae that are emerging from molecular studies. Both groups include two very large monophyletic clades, one of which is composed entirely of koinobiont endoparasitoids and the other dominated by idiobiont ectoparasitoids but with several, perhaps even numerous, independent origins of endoparasitism within it. This would tend to suggest that the evolution of endoparasitism from ectoparasitism is relatively easy but that the reverse is very much less so (see Chapter 6, section *Evolution of life history strategies*). In both families, there is evidence for a small clade that forms the sister group to the ectoparasitoid clade and which includes a diverse array of principally Gondwanan genera. Evidence for this is strongest within the Braconidae (Sharanowski et al. 2011) with the assemblage of Aphidiinae+Mesostoinae+Maxfischeriinae *s.l.*, which includes koinobiont aphid parasitoids, gall

Table 15.4 Differences in 'lower pimpliformes' (Rhyssinae, Pimplinae and Poemeniinae) biology between two tropical and three temperate faunas.

Biology	Country*				
	USA $n = 181$	UK $n = 108$	Romania $n = 157$	Costa Rica $n = 156$	Madagascar $n = 46$
Idiobiont ectoparasitoid on deeply concealed host	47.5	48.3	52.9	37.2	34.8
Idiobiont parasitoid on weakly concealed pupae/cocoons	24.9	23.9	22.3	34.6	56.5
Idiobiont ectoparasitoid of spiders or spider egg sacs	8.3	8.4	8.3	7.7	2.2
Koinobiont ectoparasitoid of spiders, i.e. '*Polysphincta* group'	19.3	19.3	16.5	20.5	6.5

*Values given are percentage species in each fauna with given biology; n = total number of species in fauna.
Source: data from Gauld (1991).

formers on *Banksia* plants (*Mesostoa* spp.), parasitoids of gall-forming cecidomyiid flies (Hydrangeocolini), gregarious koinobiont parasitoids of weakly concealed Lepidoptera (*Andesipolis*) and several other groups whose biologies are unknown. Maximum likelihood and parsimony estimates based on two ribosomal gene fragments provide fairly strong support for a southern hemisphere origin of this clade, although with more recent northern divergences within three large tribes of Aphidiinae (Belshaw et al. 2000). The Ichneumonidae have been lagging far behind the Braconidae in terms of molecular studies and conclusions have to be based upon the limited data available, which might of course be misleading.

The Palaearctic and Nearctic faunas are often remarkably similar (see Table 15.4). Although no thorough study has been conducted (and to do so before phylogeny is much better known might be misleading), it is clear that in many groups the two regions share a large proportion of genera. Townes (1963) proposed that the Bering land bridge probably acted as an important migration route for the colonisation of the Nearctic by members of the Pimplinae and Gauld et al. (2002b) noted that most of those pimpline genera that are found in either the Nearctic or Palaearctic but not both simply represent incursions of Neotropical or Afrotropical elements, respectively. The same is true for nearly all other subfamilies (e.g. Agathidinae: Sharkey et al. 2011).

A similar major separation of the neotropical fauna from that of the rest of the world, apart from some representatives of clades found in the Holarctic, is apparent in those few truly cosmopolitan genera, e.g. *Stantonia* (Braet & Quicke 2003), *Yelicones* (Areekul & Quicke 2006) and *Aleiodes* (Quicke et al. in preparation), and it would be very good to have global phylogenies with dense species sampling for other large cosmopolitan genera (e.g. the braconids *Blacus*, *Bracon* and *Meteorus* and ichneumonids *Pimpla*, *Enicospilus* and *Netelia*) to test how general this is. Of course, genera are just convenient clades of species with similar morphology and usually biology, so if they are of comparable age or affected by the same ecological features as subfamilies, then there is no good reason to expect them to behave differently.

Several ichneumonid and a few braconid groups have what are often referred to as Gondwanan distributions, i.e. being largely or entirely restricted to some combination of Australia, New Zealand, southernmost Africa and south America, and in the last case, particularly to Chile. The ichneumonid fauna of Chile can be divided into three quite distinct areas, the one of particular interest being that associated with *Nothofagus* forests termed the neantarctic realm (Porter 1991). This fauna differs strongly from that of the rest of South America and includes an high proportion (32%) of endemic genera. In addition to being home to the enigmatic Apozyginae, it also includes a number of probably basal genera (e.g. Blacinae: van Achterberg 1988c; Aphidiinae: Belshaw et al. 2000; Eucerotinae: Gauld & Wahl 2002). The Hydrangeocolini (Mesostoinae) and the Pedunculinae have a particularly interesting distributions with some predominantly Chilean and some Australian genera. Such distributions might reflect origins in the early Tertiary while South America, Antarctica and Australia were still close together but after the African+Indian plate had separated (see Gauld 1984a).

The ichneumonoid fauna of Australia (and New Zealand and adjacent islands) is also quite distinct, with many endemic genera, some putatively very basal. Notable is the Trachypetinae, which might warrant placement in a separate family. Many of its endemic elements probably reflect a long period of isolation. The close proximity of Cape York to Papua New Guinea means that numerous Australian genera or genus groups extend to that island, and also the Cape York fauna includes a number of genera of Oriental/Old World origin.

ISLANDS AND THEIR PARASITOID FAUNAS

Islands, especially oceanic ones, make excellent natural experiments. If they are hard to colonise by plants and herbivores, it ought to be especially the case with parasitoids such as ichneumonoids, since it would seem that not only does the wasp first have to reach the island but it must, once there, also encounter a suitable host that is already living there, unless of course they rafted in on some piece of substrate, such as a log, along with some of their hosts. This would lead one to hypothesise that, all other things being equal, island faunas ought to be dominated by generalists (i.e. idiobionts) because even if their regular host has not already colonised the island, an ecologically similar one that might be accepted by a tired female wasp might be there, and given enough such opportunities, a successful parasitoid colonisation event should occur. Hence, using idiobiont and koinobiont as surrogates for generalist and specialist, respectively, it should be easy to test the above hypothesis by comparing island ichneumonoid faunal lists with those of adjacent mainland regions. A few ichneumonid genera are particularly widespread on islands, notably *Enicospilus* (Ophioninae) and *Echthromorpha* (Pimplinae), both of which have reached many remote oceanic islands. These are large-bodied wasps and strong fliers (e.g. Gauld & Mitchell 1981). Whilst *Echthromorpha*, especially the widespread *E. agrestoria*, are highly polyphagous idiobiont endoparasitoids of Lepidoptera pupae and would be expected to be able to colonise the islands fairly easily, ophionines being koinobionts might be thought to face more of a problem. However, ophionines seem to be quite polyphagous at species level with host ranges more specified by niche, and therefore they too presumably have a better than average chance of finding suitable hosts once they have made landfall.

Maetô and Thornton (1993) were probably the first to obtain evidence that could test this by examining the parasitoid fauna of the Krakatau Islands in Indonesia, which had been defaunated by the volcanic explosion of 1883. Their results showed a preponderance of koinobiont endoparasitoids of Lepidoptera on the youngest island, Anak Krakatau, which had been devastated by another explosive eruption in 1952. The three older islands additionally had several parasitoids of Diptera, mainly associated with grasses. These sampled parasitoids were all heavily biased towards koinobionts. However, it could be argued that the time scale of 100 years since the first volcanic explosion was not long enough to provide a test of the hypothesis, since there might be something peculiar about very early island colonists, perhaps associated with a particular type of early colonising plant community. Another test was carried out by Santos et al. (2011), who, instead of finding that island ichneumonoid faunas dominated by idiobionts, found that their faunas were most significantly determined by those of the nearest mainland. Such analyses are not without potential sampling problems and it cannot be ruled out that taxonomists working in a given region including both mainland and islands might have taxonomic biases towards either idiobiont or koinobiont groups. Nevertheless, the results at the minimum suggest that there is no strong bias towards generalists on islands, and this leads to the question, 'why not?'.

It really cannot be as likely for a generalist and a specialist to find a suitable host on an island if the range of hosts on islands and their immune abilities roughly resemble mainland host communities. One explanation could be that there is a counter-balancing bias towards koinobionts/specialists actually making it to islands in the first place, perhaps owing to smaller body size or some similar factor. No data are available on this, so it therefore seems sensible to seek alternative explanations. One such might be that the potential host insect pool on islands is itself either biased or modified compared with mainland source communities. Perhaps the taxa that successfully colonise islands end up with a lesser degree of physiological defence capability against parasitoids, just like island birds' susceptibility to human hunting was often high, e.g. the dodo;

hosts may have evolved in a way that leads to reduced immune defensive capabilities, since being capable of having an immune response is evolutionarily costly in the absence of selection to maintain it (A.R. Kraaijeveld et al. 2002).

Island faunas also tend to be of special conservation interest as they often include endemic species and radiations that have small populations and indeed have a high extinction probability. Probably the fauna of the Hawai'ian islands has been most thoroughly studied and perturbed. R.W. Peck et al. (2012) found that introduced species dominated the parasitoid fauna, and indeed more than 100 alien ichneumonoid species have been introduced either deliberately or accidentally to these islands. Hawai'i is also where it seems likely that at least one endemic species and genus has gone extinct. Despite extensive collecting effort at its last known sites, no-one has managed to find any more specimens of the endemic ophionine genus *Banchogastra*, and it was suggested that ophionines might be especially prone to extinction because they attack late instar hosts and although they did not state as much explicitly, this life history strategy is likely associated with lower fecundity, all other things being equal (see Chapter 14, section *Fecundity*).

SPECIES ACCUMULATION CURVES

Species accumulation curves for ichneumonoids seldom seem to plateau, even after very extensive sampling, and Fraser et al. (2008) suggested that simply extending the sampling period will not provide a solution since samples will still contain singletons. Effectively, it is impossible to know all the species in any moderately sized piece of habitat. Gaston and Gauld (1993) plotted the cumulative number of morphospecies of Pimplinae *s.l.* collected in Costa Rica at 17 Malaise trap sites between 1986 and 1990, representing a total of 100 Malaise trap years (Fig. 15.1), and although new species are certainly being encountered at a progressively lower rate as sampling effort continues, the curve equally has not reached an asymptote. A similar result was found for Pimplinae and Rhyssinae in lowland rainforest in Peru by Sääksjärvi et al. (2004), who operated 27 Malaise traps for a total of more than six Malaise trap years. It is difficult to be sure exactly what this means. If the sampling is only at one place or a limited number of places and not representing all the local habitats or is not close to all the potential host host plants, then species that frequent these other places will be relatively rare vagrants at the trap sites and will be caught only serendipitously as they venture further afield, either accidentally or in search of new patches. Given the heterogeneity of tropical forests, often with high overall plant species richness but local clustering, it is not at all surprising that any given trap will continue to collect species that may live fairly close nearby but whose host's host plants are somewhat removed. It is also worth noting that some guilds of parasitoids might not do much flying (see, for example, Rhyssalinae: Histeromerini and Euphorinae: Cosmophorini), and these will inevitably occur only rarely in Malaise trap samples.

Short-term sampling results will also be biased by a number of factors; for example, Fig. 15.9 shows how, in broad terms, season will affect not just the actual species collected in a tropical environment, but also their overall biology. Thus it would be very misleading to try to estimate and compare say the species richnesses of idiobionts and koinobionts at different places unless each place was sampled throughout the year. Inter-year variation is also important, as any field entomologist can tell you. Marked differences may be apparent between adjacent years, with a species being abundant one year and perhaps apparently absent the next. This can be seen for parasitoids, including ichneumonoids, even in rearing programmes (e.g. Lill 1999). A very nice anecdotal example of this was provided by Aldrich (1912) and concerns the North American pimpline, *Theronia fulvescens*, a pupal parasitoid of the pine white butterfly, *Neophasia menapia* (Pieridae). For years the butterfly was considered rare, but in 1896–98 it became abundant, causing much damage to forestry. To quote Aldrich, 'The parasite reached its maximum in 1898, at which time it swarmed in the woods in late summer in incredible numbers. In places the air was full of them and they made a very perceptible humming sound like a swarm of bees'. The wasp 'exterminated' the butterfly all over the north-east USA and in the following 10 years Aldrich saw only one live specimen.

Beta diversity, i.e. spatial turnover in species, has been little studied in the Ichneumonoidea and more work is desperately needed. Gaston and Gauld's (1993) study of Costa Rican pimplines with Malaise traps operating at 17 sites all over the country found that most species were widespread, with only a slight indication of habitat specialisation, with a group of species associated with high altitudes and one restricted to mangrove swamps. Indeed, as they noted, many of those species

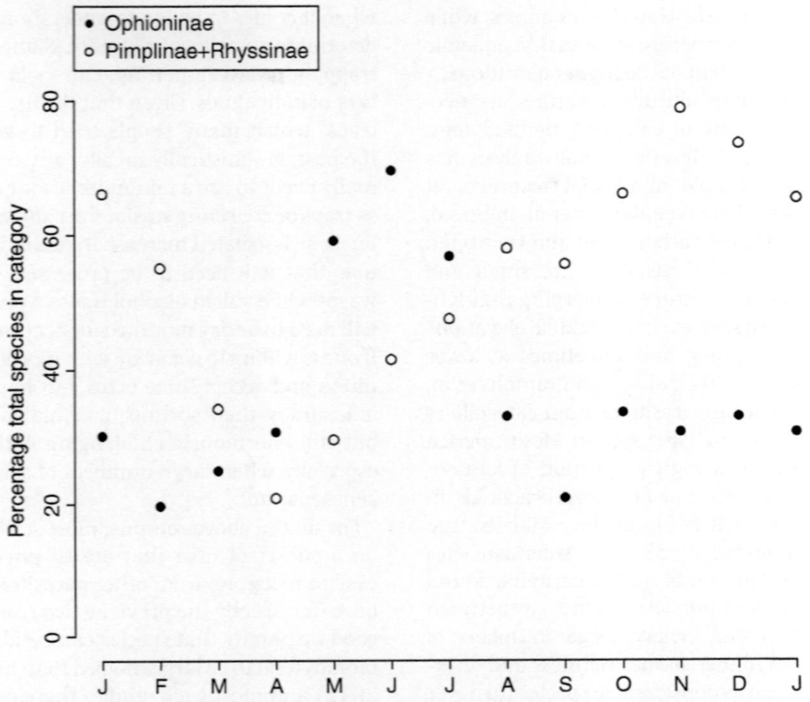

Fig. 15.9 Monthly species richness of selected idiobiont (Pimplinae and Rhyssinae) and koinobiont (Ophioninae) collected in Pacific dry forest at Santa Rosa (Guanacaste National Park, Costa Rica), presented as percentage out of totals of 38 and 61 species respectively. (Source: data from Gauld 1991, fig. 28.)

are possibly very widespread through the neotropics and the principally tropical ophioninae also seems to include many widespread species. Across a larger distance, Bartlett et al. (1999) compared the species composition of beetles and of Pimplinae between Barro Colorado Island, Panama, and Costa Rica and between Las Cuevas, Belize, and Costa Rica, and in both cases there was a very high degree of overlap (96% and 84%, respectively), which contrasted dramatically with the 2.6% overlap of beetles between a site in Peru and one in Brazil. However, pimplines are often generalists and are rather strong fliers, and at least one appears to show seasonal migration. As noted above, *Echthromorpha agrestoria* is widespread on tropical and subtropical islands, apparently having reached many on their own accord. Ophionines are towards the larger end of the ichneumonid size spectrum and they too might be expected to be strong fliers. Thus the taxa on which these observations of low beta diversity are based might not be representative of ichneumonoids as a whole. There is a very high need for work on other groups, both braconids and parasitoids of foliage-feeding hosts, to understand how generalisable, for example, Bartlett et al.'s results are.

ALTITUDINAL GRADIENTS

There has been slowly growing interest in the effect of altitude on ichneumonoid faunas since the climate at altitude in the tropics can be far more like in the temperate region at lower altitudes. In the tropics, a higher proportion of koinobiont ichneumonids are nocturnal at lower altitudes (Gauld, p. 396 in Hanson & Gauld 1995) and this may have something to do either with night-time temperatures being too cool at high elevation or with daytime temperatures being too hot for much active host searching lower down. R.W. Peck et al. (2008) ran Malaise traps along a 1700 m altitudinal transect on the island of Hawai'i and found that both abundance and species richness of ichneumonoids increased with elevation, especially species richness. Further, the Hawai'ian samples, while being dominated by alien species, showed that endemic species were more prevalent at the higher elevations.

The few surveys across altitudinal ranges are necessarily uneven in terms of sampling methodology, altitudes sampled, etc. Only one formal analysis has been based upon multiple altitudinal transects at widely separated locations (Veijalainen et al. in press). Often the sample sizes are rather small and very often the sampling periods and durations are small and unequal. Nevertheless, a picture is emerging that Ichneumonidae are more species rich at middle elevations (Veijalainen et al. in press) and sometimes at lower elevations (van Noort 2004), although Connahs et al. (2009), based on rearing parasitoids from caterpillars of geometrid *Eois* spp. on *Piper* spp. in MesoAmerica and Ecuador, obtained a high proportion of ichneumonids at high altitudes and far more braconids in the lowlands. Veijalainen et al., in their Malaise trap study of Ichneumonidae at 25 MesoAmerican sites across Guatemala, Honduras and Nicaragua, found that in general mid-elevation sites (defined as between 1033 and 1709 m) had yielded larger numbers of individuals and had higher species richness and diversities. Further, there was considerable species turnover with altitude. Out of a total of more than 250 species collected at the high and low altitude sites, only five were common to both and there was also a fairly small overlap between both low and high with the mid-elevation sites.

In Mediterranean-type climates, ichneumonids in particular are most abundant and diverse in mountainous areas, and this is generally attributed to the fact that, especially in the summer, the nights are still cool allowing the formation of dew, since, as widely quoted, Townes noted that most ichneumonids need to drink at least once per day. In contrast, lowlands in the Mediterranean region are very hot and dry in the summer, with accessible water often scarce, so in these regions ichneumonids tend to be most prevalent in the spring and again in the autumn, when rain occurs (Rodríguez-Berrío et al. 2008). Unfortunately, the pattern displayed by braconids is not so well known.

ESTIMATING LOCAL SPECIES DIVERSITY

The high species richness and the taxonomic challenge posed by ichneumonoids are actually prohibitive when it comes to trying to understand the factors that determine their local diversity. Sampling, by Malaise trapping or sweep-netting, can yield very large numbers of individuals. Given that the use of single Malaise traps, which many people tried to get away with in the past, is statistically invalid, any comparative study really needs to use a minimum of four and ideally 10 or so traps or collecting sessions per site with an obviously linearly associated increase in the numbers of individuals that will need to be processed. Identification of wasps while still in alcohol is also a problem and many will need to be dry mounted for accurate identification. If one is willing to get away with morpho-species recognition and accept some extra, but hopefully unbiased, inaccuracy then sorting in liquid might be feasible, but still enormously challenging with large samples, especially when large numbers of samples need to be kept separate.

For all the above reasons, most studies have focused on a subset of taxa that are (i) physically large, (ii) easy to recognise from other parasitoids and (iii) often have (for exactly the previous two reasons) sufficiently good taxonomy that species can be identified. Ichneumonids tend to be larger bodied than braconids (Quicke 2012) and most such studies therefore deal with ichneumonids. Among the ichneumonids, pimplines and other larger pimpliformes subfamilies such as poemeniines and rhyssines are relatively easy, and with Gauld (1977a, 1988b) having revised the ophionines of much of the tropics, it means they are also a tractable group in many places. Hence, with few exceptions, conclusions about such important questions as the effects of plant species richness or of logging or oil palm plantations, etc., or trends with altitude or season, on the parasitic Hymenoptera tend to boil down to the effect of these factors on the Pimplinae and a few other groups. There is no doubt that for some of these questions, the results obtained based, for example, on pimplines are going to be completely contrary to the real trend across all ichneumonoids.

Within the UK, Fraser et al. (2007) showed that the species richness of Pimplinae was highest in woodland with high broad-leaf tree richness and poorest in coniferous woodlands, but for the other subfamily investigated, the Diplazontinae, no effect of woodland type was found and they instead displayed considerable local within-woodland spatial heterogeneity. It is good to see such results, but the findings are more or less what would be expected as there is generally a higher species richness of insects within broad-leaf woodland. An issue here is that broad-leaf woodlands generally tend to have higher plant species richness than coniferous woodlands at adjacent sites.

There are other major considerations, the most important of which is the length of time that the sampling is carried out, and in comparative studies between localities these need to be equivalent. Differences in season affect not only the total numbers of individuals and species that are likely to be sampled, but also the balance of biologies. Gauld (1991) showed how in Costa Rica the relative abundances of idiobiont and koinobiont ichneumonids (although again represented by only the usual suspect subfamilies) swapped over at different times of year (Fig. 15.9) and Gaasch et al. (1998) showed that the abundance of adult ichneumonids attacking similar orders of hosts were broadly correlated. To obtain a representative sample of the taxa present in a place, it is therefore necessary to sample through at least 1 year.

More insidious, however, and potentially of great importance if the question being asked concerns conservation and true species richness, is whether the number of species present in a small sample may truly indicate whether one habitat is more species rich than another. Indeed, I would argue that it can be quite the opposite. This is because of the effect of the species abundance distribution and species evenness. Several studies have purported to show, for example, that species richness of ichneumonids is fairly high in disturbed or modified habitats as opposed to pristine forest (Idris et al. 2001a, 2003). Species abundance distributions always show that there are a small number of very abundant species and a larger number of rarer ones. Small samples will pick up most or all of the commoner species, whereas only a few of the rare ones may be obtained, nearly all being represented by only one individual. To illustrate the problem of small sample size and the possibly dangerous interpretation that can be made, the study by Idris et al. (2003) of the pimpline genus *Xanthopimpla* in recently and long-ago logged forests in Selangor, Malaysia, is a good example. Based on a total sample size of just 48 individuals across two sites, they found a larger number of species and greater diversity in the forest that was logged only 5 years ago than in that logged 26 years ago, which presumably must be recovering well. With such small sample sizes, of course, it is impossible to know what the total numbers of species might be in each area and with only two areas studied, there is no way of estimating sampling variance – perhaps it was simply physically easier to collect these conspicuous wasps in the recently logged areas. However, I suspect that the results might easily be representative of what is actually present, and that the more recently disturbed areas could have relatively higher numbers of individuals of more generalist parasitoids, perhaps associated with 'trash species' hosts. Semi-arid areas, such as savannahs, are often rich in Braconinae and these are also fairly common in trashy habitats such as road-sides in the tropics which are dry to their high insolation. These habitats tend to have different common subfamilies than inside forests, with Cardiochilinae and Agathidinae also often being very abundant there, but their species and generic diversity is low.

Meaningful comparisons can only be made with very extensive, long-term sampling and I find it remarkable that there seem to be no large published studies, despite there being so much current interest in landscape ecology. Such work is undoubtedly hampered by both the numbers of specimens that would need to be processed and identified (by people with reasonable expertise) and the sheer logistics of sampling, over a long period, from 10 or more comparable sites, even if paired statistics could be used.

In order to overcome some of the issues with small and incomplete samples, it might nevertheless be possible, if the samples are not too small, to make use of some of the methods that have been developed to try to estimate how many species have not been sampled (Colwell & Coddington 1994, Magurran 2004). For example, Chao (1984) presented an estimator, referred to as Chao 1, of the total number of species in a community (S_{est}) based on the total number observed in a sample (S_{obs}) and the numbers represented by just one individual and those represented by precisely two individuals, termed *singletons* and *doubletons*, respectively, and is formulated as follows:

$$S_{est} = S_{obs} + \frac{singletons^2}{2 \times doubletons}$$

Chao 1 is based on the assumption that all the entities being sampled are equally abundant and detectable, which is why it is non-parametric. However, this is far from the case with natural communities whose members differ in both true abundance and catchability, abundance often approximating either a log-normal or log-series distribution. In these cases, Chao 1 underestimates the total number, especially with small samples, so even applying such corrections, which is seldom done, may not solve the problem. In addition to Chao 1, there are several other non-parametric estimators such as ones based on jack-knifing and bootstrapping (see Magurran 2004). I draw particular attention here to Chao 1 since it has been applied not just to field-collected samples, but also to samples of numbers of individuals of described species (O.R. Jones et al. 2009), although there are issues about the nature of the samples.

ICHNEUMONOIDEA AS BIODIVERSITY INDICATORS

Several workers have tried to assess the potential of parasitic wasps, and Ichneumonoidea in particular, as biodiversity indicators. There is no doubt that their high trophic level and the collective range of their host taxa ought to make them suitable for this purpose and, as Mayhew et al. (2009) pointed out, 'rare parasitic wasps may suffer a double-jeopardy from extinction risks' and therefore be potentially very good indicator species. The key word here is probably 'rare'. Rare taxa, especially small ones that might be difficult to spot and collect, are intrinsically going to be poor indicators, and with truly rare species, the need to kill individuals in order to identify them, which is normally going to be necessary for ichneumonoids, makes it even worse. In addition, the general difficulty with identifying many to species level (an obvious requirement) probably means that they are not really a practical option and there are also other important issues, for example, hosts are not known for many species and certainly detailed knowledge of host ranges is not available for the great majority, even in the faunistically better studied parts of the world. Thus, interpreting the significance of particular species will be limited. For most areas there are few reliable baseline studies. Further, although this applies to most studies, sampling over short time periods is likely to miss many species.

Sääksjärvi et al. (2005) provided some evidence that the species richness of ichneumonids is related to that of selective plant groups within a tropical forest mosaic, but the pattern was only evident when sampling effort was accounted for (not surprising) and only found for those idiobiont parasitoids of weakly concealed hosts. However, as with many other studies, this one only considered the Pimplinae *s.l.* and, as these tend to be generalists, perhaps a lack of strong correlation might be expected and I would strongly suspect that more specialist, koinobiont parasitoids of ectophytic hosts would show a stronger correlation, but no-one has as yet tested that assumption.

Mayhew et al. (2009), after several years of extensive Malaise trapping in a number of habitats in Yorkshire, UK, not a mega-diverse country, generated numerous new county records and found species new to the British Isles, suggesting that previous recording was far from complete. Their study, even across a limited set of habitats, showed no sign that all species had been collected. Some of this is undoubtedly because insect populations can fluctuate considerably from year to year (see *Species accumulation curves*, above). Such fluctuations appear to be no less marked for tropical systems than they are for temperate ones (Wolda 1978a,b). The implication is that no inventory can be reliable if based on a single or even a few seasons' sampling.

CONSERVATION

Insect parasitoids are one of the least understood groups of animals when it comes to conservation issues, and those that do occasionally get thought of are usually specialist parasitoids of other, more charismatic insects that are also endangered, such as the myrmecophilous large blue butterflies (Lycaenidae) (Hochberg et al. 1998). This is largely due to the relatively small number of people who are interested in them and also have the expertise to identify them. Recently, a little more attention has been paid to this issue, but it is hard to see what easy progress can be made as so little is known about the real abundances or even the presence and absence of particular species. Even ignoring very rare species (see above), regional species lists are usually very incomplete and knowledge of biology or how insects in general, and parasitic wasps in particular, relate to other habitat classifications is scant. Mayhew et al. (2009) pointed out that parasitoids provide ecosystem services, many probably maintaining their hosts at low population levels and therefore even important ones are likely to exist themselves at low population densities much of the time. Their high trophic level indeed renders them more vulnerable to both local and global extinction (MR Shaw & Hochberg 2001). If there is any truth to the resource fragmentation hypothesis (see above), then hosts may be so scattered under some circumstances that they are too scarce and unreliably present to support a viable population of a specialist parasitoid, and specialist parasitoids might be more prevalent in the tropics because of the NHH (see Previous page).

EFFECT OF HABITAT DEGRADATION ON ICHNEUMONOID COMPOSITION

One possible role for ichneumonoids, as members of a relatively high trophic level but with diverse biologies, is as indicator species of the 'health' of a habitat. Obviously I have no argument with that in principle, but the practice is far more difficult and, with insufficient knowledge, might even be dangerous. Parasitoids are very often likely to be less numerous than their hosts, just as in Africa there are fewer lions than antelopes, and they tend to be harder to spot. At least with lions we are pretty sure that we know what species we are looking at and have a pretty good idea of its diet. None of these things are true of most parasitoid wasps. Even in very well-studied countries we have no good numerical data on host utilisation by the great majority of ichneumonoids. Certainly it is likely to be the case that the more members of specialist/koinobiont species are sampled, the more species of their host groups there are in a given habitat – it is almost a truism. However, ichneumonoids and other parasitoids are generally far less well known taxonomically than their hosts and require a great deal of effort to identify. But even if all of the ichneumonoid taxa in a sample have correctly been identified, we still will not know many of their hosts and certainly not understand their host ranges fully.

As one might expect, Horstmann et al. (2005) compared ichneumonid numbers and species richness between primary and secondary forests in Malaysia and found nearly twice as many species in the canopy fogging samples from the primary forest. But there is a very real danger if sampling is not performed thoroughly and over a long period of time because the evenness of the commoner species in a disturbed habitat can mean that numerous 'trash' species may be collected with relatively little effort, but in the disturbed/degraded habitat continued sampling will soon start to yield new species at a very low rate whereas primary/pristine systems may continue to reveal new species at a fairly constant rate for a long period. The higher the organisms being sampled are up the food chain, the lower their population densities relative to those of their hosts are likely to be and the harder they will be to sample to the same extent.

There has been a growing amount of entomological conservation concern, especially in Europe, about saproxylic beetles, as many forests are 'managed', which often means that dead timber is cleared away and old trees are felled. Many of the beetles associated with old-growth forests support parasitoids of varying degrees of specialisation, and these rightly also ought to be of conservation concern. Hilszczański et al. (2005) showed that no one arboriculture scenario is good for all ichneumonoid parasitoids of saproxylic hosts in central European spruce (*Picea abies*) forests. As so little is known about what determines the host relationships of tropical idiobiont parasitoids of wood-boring beetles, it is impossible to know what effects impoverishing forest tree species richness will have in situations where selective logging occurs, but it must surely be the case that they will suffer exceedingly when monocultures of oil palm or other trees are planted. It might be that selective logging is not so bad for them if, as appears to be the case, many xylophagous beetles are quite polyphagous, but there again, the parasitoid wasps might be limited by things such as wood texture. Anecdotally, the large, brightly coloured Afrotropical braconine genus *Rhamnura* is fairly well represented in European museum collections from the 1950s to the 1970s, but I have only seen one recent specimen which was caught next to a dead log in Uganda by Jim Mallett. These wasps may well have fared far better when intensive tropical logging was going on due to all the dead wood habitat that it created.

SIGNIFICANCE OF CRYPTIC SPECIES

All studies of diversity are based on their relevant species concepts and, for the greatest part, these have been purely morphological/typological. If that species concept were to be fundamentally flawed, then estimated total global numbers based on them will be equally flawed. It is therefore worth asking what the available evidence tells us about the general validity of the traditional morphotaxonomic species concept, particularly with respect to parasitoids. Not surprisingly, there have been few studies and certainly no unified approach. Sometimes cryptic species are revealed through their host interactions, when what was thought to be a single parasitoid species attacking two or more host species actually comprises separate species that will not develop on the other's host.

Such life history studies are, of course, time consuming and more or less require having the species in culture. Day (2005) provided a nice example. The euphorine *Peristenus pallipes* was initially thought to attack a range of mirid bug hosts, but Loan (1965) concluded that there were in fact two very similar species and named the new one *P. pseudopallipes*. Day investigated the duration of diapause (defined as the delay in emergence of the adult wasp after its cocoon had been removed from cold storage) and found that it varied according to the host that the nominal *P. pallipes* had been reared from, and concluded that the rump *pallipes* still comprises at least two separate species and possibly three.

Information on the existence of cryptic species can come from many different kinds of source. Gokhman investigated the chromosomes in several species of ichneumonine and discovered some individuals with atypical karyotypes and some at least of these probably indicated the existence of previously unrecognised species. Jordan (1998a) found that what had been thought to be a single species of the tryphonine genus *Grypocentrus* attacking leaf-mining moths produced eggs with eight different morphologies and, coupled with host specificity data, concluded that eight biologically distinct species were involved. However, nowadays, more and more, it is molecular data that are revealing that what appear at first sight to be single species are in fact complexes of more than one. Of course, sometimes it works in reverse and M.R. Shaw et al. (2011) found that what had appeared to be three separate species of the pimpline genus *Scambus* were actually just teneral and fully hardened forms of the spring generation of *S. calobatus*, and the late summer morph of this species.

Final resolution of whether or not cryptic species suggested by DNA barcoding are truly different or might represent some mitochondrial polymorphism that has not yet gone to fixation will usually need a combination of approaches (e.g. M.A. Smith et al. 2008). In the case of Smith et al.'s study of microgastrines reared from known hosts in Costa Rica, barcoding (backed up in difficult cases by additional sequence markers) increased the number of putative species based on morphology from 171 to 313. These species largely mapped well on to host data, the great majority being very host specific, many monophagous. The story of one morphospecies in particular, *Apanteles leucostigmus*, was remarkable, in that what was thought to be a polyphagous wasp with 32 different hosts turned out to be an aggregate of 36 provisional species with very narrow host ranges.

ENDNOTES

1. Instituto Nacional de Biodiversidad, Costa Rica.
2. An invalid species name.
3. Ichneumonids are generally larger bodied than braconids and almost completely lack the very small-bodied taxa.

Chapter 16

COLLECTING AND REARING ICHNEUMONOIDEA

The Braconid and Ichneumonid Parasitoid Wasps: Biology, Systematics, Evolution and Ecology, First Edition. Donald L. J. Quicke.

This section, which is aimed at providing a few guidelines about the important business of collecting, rearing and culturing ichneumonoids, is included so that the most suitable specimens may be obtained and the most reliable possible biological data can be associated with specimens. In the latter respect, it is worth re-emphasising a point that has been mentioned repeatedly in preceding chapters, namely that a great number of the published host records for parasitic wasps, probably the majority, are erroneous (Noyes 1994, M.R. Shaw 1997), and this is startlingly illustrated by Shaw's rearings of a relatively common British braconid, *Aleiodes alternator* (Rogadinae) (Table 16.1). Not only are only four of the published host records supported, but 10 of its hosts, some of them quite frequently attacked, were not previously known. Experiments with two reasonably close relatives of the two definite hosts, but ones that feed on the trees rather than herbs, conducted with 23 either previously reported, actual or close relatives of actual hosts, revealed only one potential host, *Spilosoma luteum*, that the parasitoid would accept and develop in successfully, that it had not been reared from. In summary, *A. alternator* appears to be restricted to Erebidae (Arctiinae, Lymantriinae) and Lasiocampidae and entirely species that feed in the herb layer. It should be noted that lasiocampids are not closely related to erebids, indeed they are in different superfamilies, so the host range of this wasp is determined in part by host morphology. Its host range can be broadly classified according to Shaw as low-feeding hairy caterpillars in the above-mentioned groups. The situation might turn out, based on unpublished DNA sequence data, to be more complicated, because *A. alternator* might be a complex of cryptic species, each with more restricted host ranges, but nevertheless, the point about the unreliability of literature records is clear. In the same paper, M.R. Shaw also illustrates the marked discrepancies between published host records of British polysphinctines (Pimplinae) and records from carefully controlled rearings (see page 374).

Records from economically important hosts are perhaps especially liable to include wrong associations, as such rearings are often done in bulk and might involve much associated collected vegetation upon which non-associated parasitoid cocoons or other parasitised hosts might be lurking (Čapek 1983, M.R. Shaw 1994).

There are three major reasons for these erroneous records, and they are not mutually exclusive. First, the identification of the parasitoid is often wrong, sometimes badly so. Second, the identification of the host species can be wrong. Third, and probably more misleadingly, the published association between a parasitic wasp and a host can be wrong, even if both species have been identified correctly. The association errors that plague the literature are generally because the rearing vessel contains the host that the worker is interested in and other parasitised hosts that the worker is unaware of. Perhaps a twig of leaves with a caterpillar of interest has not been searched thoroughly for other possible sources of parasitoids, maybe parasitised aphids or mummies (yes, aphidiines have been recorded wrongly from lepidopteran hosts) or parasitoid cocoons (perhaps a microgastine or campoplegine had previously killed another caterpillar on that bit of host plant and its cocoon gone unnoticed or any of several other possibilities). For these reasons, it is necessary to keep a record of all the target species' individuals, throughout the rearing period, and to account for all emerging parasitoids in terms of one fewer host. Even whether a parasitoid is ecto- or endoparasitic has sometimes been mistaken, although the authors have clearly observed parasitised hosts – parasitoid eggs are often small and parasitoid larvae can 'hide' under the host, or perhaps within a cuticular fold in some cases, until they are quite far grown.

Improving taxonomy and the availability of better identification keys is helping to reduce instances where the taxonomic identification of parasitoids and hosts is wrong, but will never eliminate this problem completely, especially as there is increasing evidence that many systems involve complexes of morphologically indistinguishable, but biologically well-separated, species. Numerous such examples have come to light only because of rearing programmes, either in the field or as part of biological control efforts. Correctly associating parasitoids and hosts is, however, far from simple and may take considerable effort. The protocols described below, largely based on M.R. Shaw's incredibly helpful publication, provide a good baseline for studies and, although his work is based almost entirely on rearing parasitoids from temperate region lepidopteran hosts, the issues involved can easily be adapted to other situations.

Rearing from multiple hosts in a substrate can be even more problematic. Although perhaps extreme, imagine a situation where a piece of wood is placed in a rearing container (for larger pieces, pillow cases are good for this as they allow air flow, prevent mould and

Table 16.1 Host range of the koinobiont rogadine braconid *Aleiodes alternator*, comparing prior literature references and accurate rearing records.

Potential and actual hosts	Host feeding site	Literature host records*	Host records from accurate rearings*	Experimetal rearings*
Tortricidae				
Spilonota ocellana	Trees	√		
Epiblema cynosbatella	Herbs/trees	√		
Hedya sp.	Trees	√		
Archips rosana	Trees	√		
Lasiocampidae				
Trichiura crataegi	Trees			×
Malacosoma neustria	Trees			×
Lasiocampa quercus	Herbs		79	√
Macrothylacia rubi	Herbs		1	√
Philudoria potatoria	Herbs	√	71	
Thaumetopoeidae				
Thaumatopoea pityocampa	Trees	√		
Thaumatopoea processionea	Trees	√		
Erebidae (Lymantriinae)				
Orgyia antiqua	Herbs/trees	√	16	√
Orgyia aurolimbata	Herbs	√		
Orgyia dubia	Herbs		4	
Dasychira fascelina	Herbs		4	√
Dasychira pudibunda	Trees			×
Dasychira selenitica	Herbs	√		
Euproctis chrysorrhoea	Trees	√		×
Euproctis similis	Trees	√		×
Leucoma salicis	Trees			×
Lymantria dispar	Trees			×
Erebidae (Arctiinae)				
Coscinia cribraria	Herbs	√		
Arctia caja	Herbs	√	9	√
Arctia villica	Herbs	√	5	√
Arctia flava	Herbs		1	
Diacrysia scannio	Herbs		2	√
Spilosoma lubricipeda	Herbs		2	√
Spilosoma luteum	Herbs			√
Spilosoma urticae	Herbs		2	
Diaphora mendica	Herbs		3	√
Ocnogyna baetica	Herbs	√		
Phragmatobia fuliginosa	Herbs		17	√
Callimorpha dominula	Herbs			×
Tyria jacobaeae	Herbs			×
Noctuidae				
Noctua pronuba	Herbs	√		
Noctua fimbriata	Herbs	√		
Acronicta populi	Trees	√		
Acronicta psi	Trees	√		×
Acronycta menyanthidis	Herbs			×
Acronicta rumicis	Herbs			×
Simyra nervosa	Herbs	√		

*Ticks indicate that the host permitted a successful parasitoid rearing and crosses indicate that no parasitoid rearing could be obtained.
Source: data simplified from M.R. Shaw 1994, table 9.4.

are fine-woven enough to prevent insects escaping), and after some time the researcher finds a number of parasitic wasps and a number of beetles, one species of each. Well, one possibility certainly is that the wasp is a parasitoid of the beetle species recovered. But another possibility is that the wasp had parasitised all of its suitable hosts and the beetles that emerged actually belong to a species that the wasp cannot attack. Even worse, and more likely, is that several species of potential host and several species of parasitoid are found in the rearing container. But which is the host to which? Relative size can only help narrow the options in one direction, a small host cannot give rise to a far larger parasitoid, but a large host can certainly have smaller, often gregarious, parasitoids feeding on or in it.

Molecular techniques such as DNA barcoding can be used to confirm both host and parasitoid identities. The former can be of great value when incomplete host remains that are effectively impossible to identify with certainty morphologically are involved (e.g. Quicke et al. 2012c, Hrcek et al. 2013), and the latter is particularly useful when only parasitoid larvae are available and the chance of successfully rearing them nearly negligible (e.g. Laurenne et al. 2000).

FIELD COLLECTING ADULTS

Ichneumonoids can be collected using a wide range of passive and active collecting techniques just as with most other insects, and each method has its own advantages and disadvantages and all have biases (Noyes 1989, Idris et al. 2001b). Noyes compared the efficiencies of five methods for sampling parasitic Hymenoptera, viz. sweep-netting, Malaise trapping, yellow pan (Möricke) trapping, flight intercept trapping and canopy fogging. Individual hand netting, by which many of the larger, more spectacular specimens in museums were obtained, is not really quantifiable on the same axes and indeed all the other methods have their own particular advantages and disadvantages. The numbers obtained by Noyes are worth noting. In the forest habitat, his samplings with the first three methods were equivalent to approximately 10 Malaise trap days, 100 yellow pan trap days and 1 h of sweep netting, and these methods yielded 1666, 470 and 533 ichneumonoids, respectively. The first two methods trapped approximately equal numbers of braconids and ichneumonids, whereas sweep netting produced more than four times as many braconids – whether this was due to the nature of the vegetation being swept and their communities or because ichneumonids tended to avoid being collected is not known.

Although yellow pan trapping has long been advocated for collecting small chalcidoids but generally rather disparaged regarding ichneumonoids, some studies have shown it to be fairly effective. Indeed, some ichneumonoids show an innate preference for the colour yellow (Luchetta et al. 2008, T. Li et al. 2012), and in at least one species the exact preference was affected by the wasp's hunger status and the presence or absence of host odour cues (Uefune et al. 2013).

The main considerations when it comes to collecting are the time taken in trap setting and the frequency with which the traps need to be serviced. T. Li et al. (2012) investigated the effect of flight intercept trap[1] netting colour on ichneumonid catches at three sites in China and showed that colour had a profound effect on catch, with yellow and green ones consistently catching considerably more individuals that the traditional black or white ones at most sites, but there was a strong location effect also on the relative success of black netting traps. They also showed significant colour effects on the subfamily composition of the catch (Fig. 16.1). Interestingly, they noted that three subfamilies (Rhyssinae, Metopiinae and Poemeniinae) were common at all three sites but seldom collected in the flight intercept traps, which could indicate a difference in their flight activity or in colour preference or both. Certainly the large catches generally obtained using yellow or green material ought to be investigated further, as it could potentially greatly improve trap efficiency.

In terms of using either Malaise of Möricke traps alone to try to inventory a site, a major issue is that they show very strong taxonomic biases, even at genus level. In an extensive comparison in Brazilian Atlantic and Amazon forest settings, involving the equivalent of 4.8 Malaise trap years and nearly 90 Möricke trap years, Aguiar and Santos (2010) found that Möricke traps caught disproportionately more Cryptinae than did Malaise traps, but, far more worryingly when it comes to comparing different studies, they found that nearly all genera (not just species) were collected by only one type of trap or the other. Mazón and Bordera (2008) obtained a similar result in various scrubland and woodland habitats in Spain, yellow pan traps

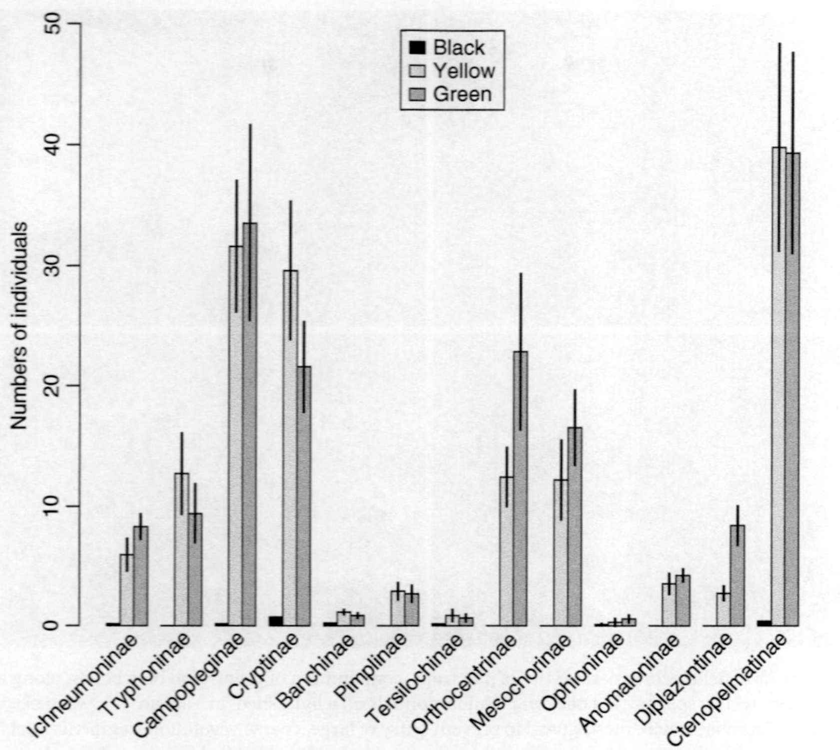

Fig. 16.1 Relative numbers of individuals of different ichneumonid subfamilies collected in black, yellow and green flight intercept traps (modified Malaise traps), suspended between trees slightly above the ground, at Liupanshan (Shaanxi Province, China). Bars represent standard errors. Traps at two other sites produced broadly similar results, but black traps were relatively more successful at Daxinggou. (Source: data from T. Li et al. 2012.)

being more effective in open habitats where they would be most visible and also more effective at catching Orthocentrinae, a group that parasitise dipteran larvae and therefore probably are particularly associated with the ground layer – they are not particularly often caught visiting flowers. Hence for maximising taxonomic coverage, it is important to use multiple trap types simultaneously. No doubt there were genera that neither of these techniques were particularly good at collecting, such as those that spend most of their time searching on trees and seldom fly or seldom fly near to the ground. Part of the interpretation of why Gaston and Gauld's (1993) species accumulation plot (Fig. 16.2) does not reach a plateau despite 100 Malaise trap years of sampling effort might be that they were doing a poor job of catching a subset of the species present that simply do not often fly up into their collecting heads, as well as them only slowly accumulating species from a long tail of rare ones.

Malaise traps typically take approximately 20 min each to set up well, but normally only require a relatively small amount (say 200–400 ml of alcohol per trap), and can usually be left for at least 1 week between collecting samples and refreshing the sample bottle. Yellow pan traps may take only a few seconds each to place and fill with collecting fluid, but as water can evaporate quickly or traps can be washed out by heavy rain, they typically need frequent visits to remove catch and to refill them.

Two other trap types deserve some mention. Flight intercept traps, consisting of a transparent (nowadays usually clear plastic sheet) mounted vertically above

Fig. 16.2 Illustrations of field techniques. (a) Use of pan traps, combination of yellow and blue bowls, along an edge, but often they would be set more widely spaced; (b) collecting ichneumonoids off a light sheet in Taiwan; (c) sweep netting using a Noyes design net fitted with detachable wire mesh guard to prevent entry of large, coarse vegetation fragments enabling sweeping to be carried out for longer and facilitating subsequent sorting; (d) a separating bag modified from van Achterberg's (1981a) design. (Sources: a, d, reproduced by permission of Scott R. Shaw; c, reproduced by permission James B. Woolley.) For a colour version, see Plate 62.

a row of pan traps or a preservative-filled gutter, are popular with coleopterists because many beetles whose flight path is abruptly interrupted, drop downwards and so are caught in the pans. However, numerous Hymenopterans and members of other insect orders are usually collected also. These traps have been the main sources for some rarely collected and poorly known groups that appear habitually to live near to the ground, such as Masoninae. No doubt pitfall traps would also yield these but the catch rate per trap makes them unacceptably inefficient unless the hugely greater volume of by-catch is also of interest.

A trap specially designed for collecting insects visiting standing trees (dead or alive) was devised by A.M. Tereshkin (1990, as a widely circulated privately published note) and consists of a conical cloth bag analogous to a Malaise trap roof that is wrapped tightly around a tree trunk and has on one side, to which it slopes slightly upwards, a Malaise trap-type collecting head. Insects visiting the trunk walk or fly upards under the cloth canopy and enter the collecting jar. Large numbers of larger wood-borer parasitoids such as xoridines, ephialtines and rhyssines can be collected using this sort of arrangement. It would be most interesting to know what it yielded in tropical forests where xylophilous braconids appear to be more abundant, and it might well provide a relatively easy way of assessing the importance of tree species and decay stage in beetle and parasitoid community assembly.

Pan Traps

Micro-hymenopterists have found that yellow pans containing liquid (Fig. 16.2a) are very attractive to many chalcidoids and proctotrupoids and routinely employ them on collecting trips, often in large numbers. An individual pan trap seldom collects many ichneumonoids, but they do work and collect numerous species that tend not to be captured so often by other techniques. Some workers combine pan traps with Malaise traps, setting pans at the bottom of the central curtain. The collecting tray of flight intercept traps can also be coloured, and yellow and green are most productive in terms of numbers of individuals collected. Some workers use a mix of different coloured pan traps, and while yellow catches most parasitic Hymenoptera under most circumstances, it is possible that blue might be particularly attractive for some forest-dwelling species. Of the numerous studies carried out on the effects of different colours on insect catches, almost none have been properly scientifically controlled to take into account either UV reflectance or simple radiance or purity of colour, and indeed few report what actual pigments were used. Pucci (2008) compared parasitic wasp catches between yellow pan traps placed on the ground and those suspended from tree limbs and elevated above the ground by 6.7–11.9 m (mean 9 m). Not surprisingly, given that parasitoids forage for hosts at different heights and in/on different substrates, he found marked differences in the catches between the two categories of trap. In his survey, far more Ichneumonidae were collected in the elevated traps, although as this study is effectively a single data point and the results strongly dominated by a few common species, it is impossible to generalise, even as to what might be found in other temperate forests, let alone the tropics.

Finally, I urge all yellow pan trap users never to fill them with ethylene glycol as this is deadly to mammals and birds and those who use them are often greeted by an array of dead and dying animals around the traps. If yellow pan traps are to be left for a few days where evaporation is high, strong salt (NaCl) solution with a small amount of glycerol will help prevent unnecessary death and still maintain some decent specimens, although subsequently they will need extra washing to remove glycerol traces as it is a highly hygroscopic compound. If evaporation is not a major issue and traps are to be left for a few days, then preventing specimens from rotting is an issue. A small amount of sodium benzoate (often used as an antimicrobial, non-toxic food preservative in soft drinks as E211) can be included. In any case, addition of a small quantity of unperfumed washing-up liquid (<1 drip per pan) will greatly increase catch by facilitating the submersion and drowning of the specimens.

Sweep Netting

In certain habitats, and obviously limited to the vegetation within reach, sweep netting (Fig. 16.2c) is an extremely efficient method for collecting parasitic wasps and warrants some consideration. It is most effective in grassy, meadow and low scrub situations or on low trees that are not particularly thorny. Despite the fact that sweep net material is fairly tough, operating nets in areas with lots of thorny vegetation such as *Rubus*, *Acacia* and *Pandanus* can be extremely difficult and frustrating.

The mode of operating a sweep net might appear easy, but what is obtained at the end of a sequence of sweeps depends hugely on what is done next, and a lot depends on the aim of the collecting. If the insects of interest are reasonably large and visible to the naked eye, one of the best methods is to hold the end of the net up towards to lightest part of the sky, and insert your head and pooter (aspirator) into it, largely blocking the lower end with your upper torso (Fig. 16.3a). In the period between stopping sweeping and inserting your head, most of the large stinging aculeates (bees, vespids) will already have flown out as they are very quick, and you might lose a small number of equally active larger ichneumonoids at this stage. Although a few big ichneumonoids may also be lost, most smaller 'parasitica' will remain in the net along with plant debris, and over the course of 2–3 min will emerge to fly up to the top of the net towards the light, from where they may be pooted quite easily. It should be borne in mind that in some habitats, ticks can be a serious issue and, where serious tick-borne diseases are prevalent, appropriate preventative medical measures should be taken. Most other creatures that might be swept, such as ants and, in the moist tropics, terrestrial leeches, are really just nuisances and not likely to be dangerous – nevertheless, I do tend to be rather more cautious if I have been sweeping in a place that has many vicious ants or large numbers of leeches.

For those less inclined to insert their head into the sweep bag, a very good alternative is quickly to transfer all the plant debris and insects into a separation bag. Van Achterberg (1981a) presented a simple design for a light, portable sweep sample sorter (see Fig. 16.2d) and with a bit of imagination it is fairly easy to construct something functional from simple resources, such as a shoe box and jam jar. The principle by which they operate is that most Hymenoptera (and some

Fig. 16.3 Illustrations of field techniques and caterpillar rearing in tropics. (a) Extraction of parasitoids from sweep net using the head in the bag technique and holding the tip of the net towards the light, while pooting (aspirating) interesting specimens; (b) a canopy Malaise trap with four vanes and netting floor, secured by strings from below; (c) Malaise trap set up in moist tropical forest with collecting head end protruding into the trail where it is lightest; (d) rearing bags for field-collected caterpillars in the tropics, housed in a well-ventilated, shady shed. (Sources: a, c, reproduced by permission of Nina M. Laurenne; b, d, reproduced by permission Scott R. Shaw.) For a colour version, see Plate 63.

other insects) will fly up and towards the light. In general, you should not rush to dispose of the leafy debris as it can take several minutes for many of the smaller parasitoids to make it to the top, after which they generally fly towards the light.

It is virtually impossible to do sweep netting in a quantifiable way (see, for example, Hespenheide 1979). Some studies tried to standardise sampling by having collectors carry out a prescribed number of sweeps and walking at a more or less constant pace. There are nevertheless potentially large sources of bias in the samples obtained, as some individuals may be stronger and/or carry out more forceful sweeps, some may sweep higher up the vegetation and some, in spite of not wanting to, might choose to include some interesting flowering plant that might otherwise have not been included in their transect. Having said that, in perhaps some agro-ecosystem studies, where there are only a few target taxa, the results of standardised sweeps might have some comparative value.

Light Trapping

Bright lights, especially ones containing a moderate amount of soft UV and to a lesser extent, UV light alone, are good for attracting a subset of insects, notably many moths, some beetles and several, principally nocturnal, groups of parasitic Hymenoptera. For collecting the major nocturnal groups of ichneumonoid, it is generally preferable to use a white sheet (Fig. 16.23b) with the light source near the top centre and to man this from dusk to as late as possible. Several groups, such as Rogadinae and Cheloninae, tend to settle on the sheet well away from the bulb, often on the reverse side, and annoyingly also often on nearby vegetation. For these reasons, standard turn-on and return in the morning traps, such as Heath or Robinson traps, probably miss large numbers of individuals. Patrolling the sheet regularly with collecting bottles yields the best results, and if you are particularly interested in large ophionines or *Netelia* (Tryphoninae), then a pair of long-handled forceps might be an aid, as picking them up by hand, especially in the tropics, can result in a painful sting. Some species are notably more aggressive than others.

Where the light trap is set is important. It should be far away from other competing light sources and with a good, large visibility range. Thus, a trap set in a large forest clearing or on the edge of woodland or on the side of a hill is likely to yield larger catches. For setting vertical sheets, the side of a buildings may seem convenient, but a wall hides the light from 50% of the surroundings. It is far better to string the sheet between two poles or tree trunks and to look for parasitoids on both sides of the sheet.

Despite much informal knowledge among lepidopterists that the best catches (of moths) are generally on warm nights, aided by cloud cover and when the moon is far from full and with very little wind, there have been almost no systematic studies of the effects of these on attracting nocturnal Ichneumonoidea. M.W. Short et al. (2006) presented a chart of catch numbers against date and showing moon phase, for large nocturnal Australian ichneumonoids, but it only covers the six summer months of one year. The catch was strongly cyclical, with larger catches closer to the new moon than full moon, as expected. Ichneumonoids arrived at the trap throughout the night (Fig. 16.4), which means that if you stop collecting around midnight, which tends to be fairly general practice, you will probably miss a number of species that tend to fly in the early hours. Surprisingly little is known about the temporal habits of nocturnal ichneumonoids and it would be well worth investigation. Certainly many species are active around dusk and are hardly caught later on, but whether these might also be active around dawn is unknown.

Canopy Fogging

This technique is widely used for obtaining samples of many hard to ge,t largely immobile insects from the canopy, especially in tropical forests where this habitat is rather inaccessible for other forms of study and collecting. It involves hauling into the canopy, by means of ropes, a two-stroke engine attached to a long exhaust pipe modified so that pesticide (usually a biodegradable pyrethroid) enters the exhalent fumes and fogs the surrounding branches and foliage. Insects affected by pyrthroids show the so-called DDT jitters (pyrethroids and DDT act in the same way, holding open voltage-activated sodium channels in the nerves), and as a consequence they lose their grip and fall to the ground. Fogging is always carried out when there is little or no breeze, and in temperate woodland this usually means very early in the morning. Having said that, it is not particularly successful for catching larger ichneumonoids as these insects tend to be rather mobile and during the time that the insecticide is taking effect, often fly out of the catching zone. Very small wasps also have a larger surface area to mass ratio than a similarly sized beetle and so are more prone to drifting out of the catch zone before reaching the ground.

Malaise Traps

The best, passive collecting method and certainly the most widely used is Malaise trapping (Mazón & Bordera 2008), which works 24/7 and even in the tropics can usually be left for 1–2 weeks between collecting the samples. The commonest design in current use comprises three vertical netting curtains forming a letter H

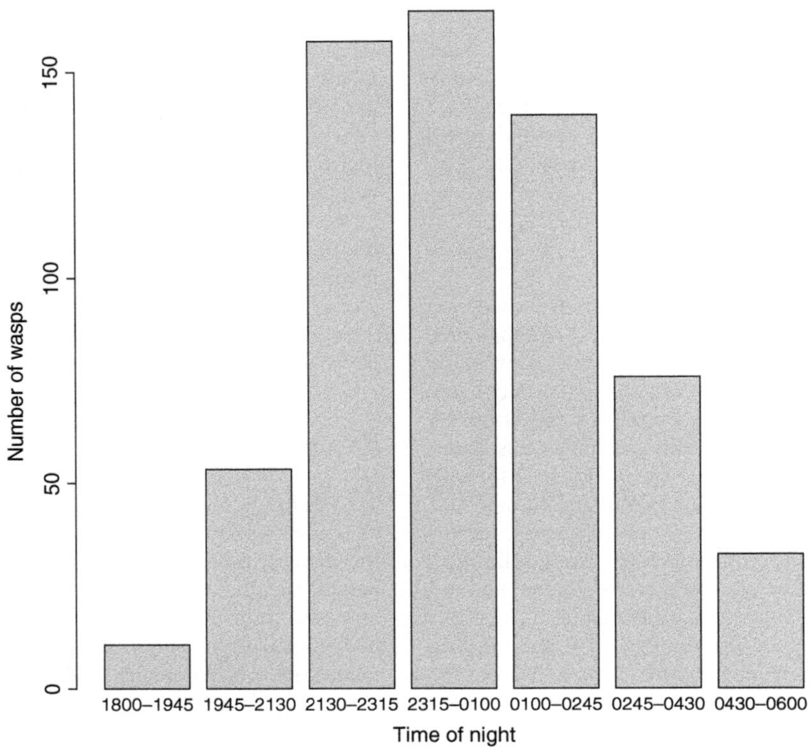

Fig. 16.4 Numbers of large nocturnal ichneumonoids caught in a UV light trap at different times of night in a *Eucalyptus* plantation in the Australian Capital Territory from December to May inclusive. (Souce: data from M.W. Short et al. 2006.)

in dorsal view and a pitched roof that slopes up at one end where there is a collecting bottle filled with preservative, normally strong alcohol (Fig. 16.3c). Although the original design required several supporting poles, nowadays traps need just one pole, at the collecting head end; and the rest of the erected structure (and indeed the main pole) is stabilised by guy strings rather than having to be fixed firmly in the ground. The main pole can usually be improvised from material near the trap site such as a tree sapling or length of bamboo. Alternatively, domestic plastic water pipe lengths can be bought nowadays in almost any town, are robust and light to carry; slight flexibility is useful if it is windy where the trap will be set, as it may reduce the possibility of the tent pegs coming loose.

Insects that blunder into the central, vertical curtain of a Malaise trap may fly directly upwards towards the collecting head, but others may drop down, so ensuring that the central curtain is as flush as possible with the ground along its length can be important. The sloping roof of the trap should be made as wrinkle free as possible by adjusting the positions of the various tie-strings, but sometimes it simply is not time-effective to try to eliminate every wrinkle, so the next best thing is to make sure that any creases in the roof netting lead up towards the collecting head rather than away from it, thus maximising the chance that wasps will end up in the collecting bottle. Trap mesh size is not particularly important and most ichneumonoids are collected with slightly coarser mesh than usual, although fine meshes are preferable and make little difference to apparency, once the wasp is under the Malaise trap canopy, especially when standard black mesh is used.

If Malaise traps are to be left set up for prolonged periods and serviced once every week or so, it is advisable to carry a repair kit when visiting them to collect samples. Insectivorous birds often notice the insects buzzing around at the top near the collecting head and will perch on the trap and peck through the netting to eat them. A needle and cotton or some duct-tape may be easy ways of repairing such holes. Mammals such as rabbits will frequently chew through the cords holding the trap upright and so spare strings are a good idea. Termites might chew through supporting sticks or improvised 'tent pegs', so some means of making fresh ones would be handy – better still, proper metal or bright plastic ones should be used. The double-ended, threaded, fitting employed to collect the collecting jar to the collecting head often breaks apart when attaching/detaching the sample bottle, so one or two spares of these are also a good idea. In the tropics in particular, small lizards, ants and spider webs may also have made their home near the entrance to the collecting head and will need to be removed. Versions of Malaise traps that can be hauled up into the forest canopy zone are becoming fairly widely used (Fig. 16.34b) and may be expected to yield different sets of species.

REARINGS FROM WILD-COLLECTED HOSTS

If only people working on different groups would interact more! There can be little doubt that many parasitic wasps that still lack biological data have indeed already been reared, probably by a lepidopterist of coleopterist, but these workers, horrified that their nurtured larva did not yield a beautiful moth, butterfly or beetle, discarded, squashed or similarly annihilated the parasitoids they had reared. Luckily, attitudes are slowly changing and this has been helped both by increased scientific publicity and also by the increasing availability of good identification guides to hymenopterans, which are slowly making this difficult order more accessible.

Mark R. Shaw has probably done the best work ever on obtaining accurate host records, principally from exposed Lepidoptera larvae, and he has described the sorts of precautions that need to be taken to prevent contaminated rearings. General principles include hygiene and mimicking, as far as possible, natural conditions (M.R. Shaw 1997). In addition, Shaw (1990) published a useful short note on rearing parasitoids, particularly Ephialtini, from spider egg sacs and nests and also *Polysphincta* group pimplines from living spiders.

Keeping the insects alive is also a skillful task, requiring attention to hygiene, since viral, bacterial and fungal pathogens can easily infect hosts, and not only wipe out those in one rearing container, but can also be very easily spread to others if equipment and hands are not sterilised. Moisture is a major problem as condensation on plants and on the insides of rearing containers encourages fungal and bacterial infections. Therefore, rearing containers should always be kept in situations where sudden temperature changes, leading to condensation, cannot occur – placing containers on metal tables is bad because metal conducts heat quickly and enhances rates of temperature change. Containers should never be left in direct sunlight, and this is also important whilst collecting in the field as inside temperatures may soon become lethal and even if not they will stress the hosts and their parasitoids (perhaps even differentially killing the latter), and will increase food plant transpiration and thus the formation of condensation later.

When collecting herbivorous hosts on plants that do not occur locally near your rearing facility, it is wise to collect additional food plant, as this can often be stored for a week or two in a cool refrigerator and used to feed the potential host either when all the earlier lot has been consumed or it has become too dry or fouled with droppings or fungus.

Containers should be relatively large compared with the combined hosts and substrate as that will help keep humidity from rising too high too quickly, so reducing problems with condensation, helping prevent mould and disease, including maintaining the food plant in a healthy condition. To minimise condensation further, each tube or box should contain a thick layer (i.e. 8–10 sheets) of absorbant tissue such as toilet paper or kitchen towel roll. This will help to buffer the moisture inside the container and reduce condensation. It is also easy to replace when it becomes soiled, and for many parasitoids and hosts that would normally leave the host/food plant to pupate, it provides a medium in which they can construct their cocoons.

Following from the above, it is important to keep track of the numbers of hosts in a rearing container. M.R. Shaw (1997) recommended keeping a piece of paper with collection data and numbers of individuals counted on it within each rearing box. When it comes to replenishing food plant or other things, you should account for every individual. If you cannot find the right number of caterpillars, for example, it could mean that one or more have entered the tissue paper to pupate or gone there prior to a parasitoid larva egressing from them, so the paper substrate and also plant remains need careful searching before being discarded. If a caterpillar has been parasitised by a rogadine braconid, it may have formed a rather inconspicuous mummy on a food plant twig or leaf (see Fig. 12.27). Any hosts that look sick should be removed to a separate container in case they are diseased, and remaining ones transferred to fresh plant in a fresh container – it could also be that a parasitoid larva, perhaps of a campoplegine, has egressed from them, in which case an exit wound will probably be visible and the container and substrate can be searched to locate the parasitoid cocoon (see Fig. 13.51a).

Effective sterilants depend on the nature of the material that needs to be cleansed. Clear plastic boxes will be adversely affected by phenolic disinfectants (such as Jaye's fluid), which will not only dissolve the plastic and make it go cloudy, but also leave residues on any absorbant material, which will later evaporate and can harm or kill caterpillars in the boxes, etc. Sodium bisulfite ($NaHSO_3$), the active ingredient of Campden tablets, is an efficient non-volatile agent that can be safely used to sterilise corks, plastic and glassware.

Maintaining stock under near-natural conditions has several important benefits. It is less likely to cause them stress and so rearing may be more successful, it provides approximate information on seasonality or natural diapause and it also means that if you wish to use reared hosts and parasitoids for experimentation, they will be roughly in synchrony. In tropical caterpillar-rearing programmes, hosts are normally reared in plastic bags, ideally with very fine perforations to allow some gas exchange, suspended from the roof of an open shed, in the shade (Fig. 16.3d).

Rearing Leaf Rollers and Tiers

In principle, such hosts can yield relatively accurate information on levels of parasitoidism, but care has to be taken. First, before collected leaves with hosts are put into their final rearing containers, they should each be carefully opened to check that they do indeed contain an active host caterpillar – some may already have been predated or a parasitoid may have emerged to spin its cocoon, or there may be predators such as spiders or ants, which, if more than one is put into a rearing vessel, can proceed to consume other hosts. If a caterpillar is not moving vigorously, the likelihood is that it has an ectoparasitoid egg or larva attached to it and it should definitely be put into a separate container and if possible closely searched to check for parasitoids.

Substrate Rearings

Substrate rearings in which a piece of host substrate is put into a container and potential hosts and parasitoids reared are particularly unreliable. For example, if a piece of wood is put into a container and a parasitoid and potential host reared, the assumption that the one attacks the other is often ill-founded; the hosts of that parasitoid may have all been successfully attacked, so what emerges are of an non-host species. Sometimes size will give the lie, but often not.

An additional issue that often arises with xylophagous hosts is that they take a long time to develop. Hence if they have been attacked by a koinobiont endoparasitoid, it may be necessary to keep them in the rearing container for weeks, months or even years, and this additional time increases the risk that both hosts and parasitoids may die. A particular issue is that the substrate needs to be kept in an airy environment in as close to natural conditions of temperature and humidity as possible, because changes in these parameters are likely to cause abnormal development of microorganisms in the wood that might render the host environment too anoxic or otherwise unsuitable. Although typical insect rearing cages may be suitable, an alternative is to use containers such as pillow cases (see above), although they have the additional problem that one can not see into them to find if anything has hatched, without opening them, which therefore has to be done frequently.

Unless you are rearing very large hosts, it is best to work with relatively small pieces of substrate, and if you are only interested in ectoparasitoids it may be advisable to open the wood to locate hosts with their parasitoid eggs/larvae on them (the hosts will invariably be paralysed) and then reassemble the relevant pieces, making note of what was found, until the parasitoid emerges. Then the host remains can be kept for confirmation of identification.

CULTURING

In order to do any experimental work on parasitoids, it is nearly always necessary to get them into culture. Working with temperate species that are obligately univoltine and on hosts that are univoltine requires considerable patience, but the rewards can be considerable and, indeed, getting them into culture might be the only way to elucidate many aspects of their biology. There are so many idiosyncrasies that it is hard to give any generalisations except to say that the closer you can get to natural conditions, probably the better.

M.R. Shaw (1997), working in the UK, advocated the use of a pared-down garden shed, in a shady position, with doors and windows replaced by fine chicken wire to allow free flow of air and prevent entry of larger animals such as cats or squirrels, which could wreak havoc with rearing materials and livestock. In addition, the roof of the shed should be painted white to reflect as much heat as possible. Heat kills, and it is vital not to let livestock get too hot. This simple set-up means that hosts and parasitoids will essentially experience the same conditions as they would in the wild, which not only means that dates will have some biological meaning, such as emergence dates, etc, but also that if you should need to replenish cultures with new material from the wild, it is likely that wild hosts would be at a similar, and therefore usable, stage of development.

Having a near-natural regimen of temperature fluctuations may in fact be more important to rearing success than just keeping synchrony between hosts and parasitoids. Iştvan et al. (2011) have recently shown that the in the braconine *Habrobracon hebetor*, maintaining them under a fluctuating temperature regime accumulated more proteins than from wasps reared at either of the three test temperatures.

Mating in Captivity

It is not always easy to get parasitoids to mate in captivity (e.g. Bradley & Burgess 1934, Quednau & Guévremont 1975, Schroder & Athanas 2002). When dealing with small numbers of individuals, it is always sensible to introduce male and female to a new container since a high build-up of female sex pheromone in a small container may confuse any male introduced and he may not be able to orientate properly. In general, better results are likely to be obtained by having individuals in a large, airy container such as a muslin cage, and again, under as natural conditions as possible (e.g. H.L. Parker 1931). Obtaining matings in many species is not a large problem, but workers have found that some species are very reluctant to do so and probably only experimentation with conditions and observing what they do in the field will solve these cases.

MASS REARING

Mass culturing of parasitoids for biological control, and in particular for inundative field release, has concentrated on small species, notably members of the Chalcidoidea, and relatively few ichneumonoids have been involved in inundative releases to control pests, although there is growing interest in utilising larger species in this way. Finney et al. (1947) described mass rearing of *Macrocentrus ancylivorus* for control of potato tuberworm (*Phthorimaea operculella*), which was one of the earliest uses of augmentative release of a braconid for pest management. Many other species have now been successfully reared for bulk release, including the fruitfly parasitoid *Fopius arisanus* (Bautista et al. 1999), stemborer parasitoids, xylophagous beetle parasitoids, etc.

Culturing parasitoids of deeply concealed hosts such as wood-borers presents an entirely new set of problems. Spradbery (1968) developed a system for culturing woodwasp parasitoids. More recently, several groups have been trying to optimise rearing conditions for parasitoids of bark beetles and wood-borers. Millar et al. (2002) described a method for rearing two doryctine parasitoids of *Phoracantha* beetle larvae which feed on the cambium and then, prior to pupation, the phloem of *Eucalyptus* trunks. Parasitoids of concealed hosts will often only accept them if they are encountered in a concealed situation into which the wasp can probe her ovipositor and will never attack an exposed host no matter how long they are left together (de Leon 1935). Culturing of *Digonogastra kimballi*, a braconine parasitoid of the boll weevil, *Anthonomus grandis*, were successfully duped in culture to lay on host larvae in largely closed Perspex wells with small holes drilled through the tops. A need to be able to probe through a hole or substrate is probably widespread among parasitoids of concealed hosts and it is rather obvious that species with very long ovipositors are unlikely to be able to deal with exposed hosts even if they wanted to.

During culturing, many ichneumonoids show an increased sex ratio presumably due to complementary sex determination and the loss of sex alleles due to inbreeding, and this can be a major practical bottleneck and it has certainly prevented mass releases of several species. Maintaining multiple separate lines with occasional mixing is one solution (Cook 1993). This also has the added advantage of helping preserve other genetic variations that might be beneficial in facilitating the adaptation of released parasitoids to a new environment. Unruh et al. (1983) found that the effective population size of cultures of *Aphidius ervi* was only about half the number of individuals present due to unequal reproductive contributions of the various females, and this should clearly be taken into account when planning mass rearing programmes. Obviously, the larger the population used to start a culture and the larger the maintained population size, the slower will allelic variation be lost.

MOUNTING SPECIMENS FOR TAXONOMIC STUDY

These next few sections are aimed at giving workers who do not have much experience in handling specimens that may be required as vouchers or to be sent to experts for identification or who need to build a collection some basic guidelines on how to mount and handle specimens and to package them for safe posting. More detailed advice can be found in Carter and Walker (1999), in some general entomology text books such as Johnson and Triplehorn (2005) and in particular in M.R. Shaw's (1997) Amateur Entomologists' Society handbook.

There is usually too little time these days to mount individual ichneumonoids in the beautiful wing-spread way that most Lepidoptera are. Instead, mounting techniques and styles should be used that allow all taxonomically important parts to be clearly visible and that afford the specimens a good degree of protection from physical damage from handling or transportation. Even packaging for the latter, although not different from that used for other insect groups, needs careful consideration and it is something that many ecologists and physiologists are perhaps not as aware of as they ought to be.

How specimens should be mounted depends on their size. Appropriate techniques are direct pinning, gluing to side of pin and card-mounting for specimens of decreasing size. Micro-pinning into a stage takes extra time and adds another 'movable part' that can increase the risk of damage in addition to limiting finger room for manipulation. Some workers prefer card-pointing to mounting on a flat card, although it also increases the risk of damage. It is essentially a trade-off between visibility of parts and protection afforded.

Even the size (length and diameter) of the pins used is important, in addition to what they are made from. Effort should be made to obtain modern stainless-steel pins with nylon heads, ideally so-called Continental pins, which are 38 mm long. Headless pins as recommended in some older, mainly lepidopterological guides should never be used as they can lead to injury to the user and make handling more difficult and increase the risk of damage to the specimen. Pins usually available are graded as sizes 000 to number 7, from thinnest (0.25 mm) to most robust (0.71 mm); they are 38 mm long. Ichneumonoid workers should restrict them to sizes 3 (0.5 mm) and 5 (0.6 mm). Thinner pins are prone to bending, while size 3s are suitable for direct pinning medium- to large-bodied ichneumonoids and size 5s when card or other secondary mounting stages are used.

Preparing Specimens from Alcohol Storage

The increasing use of mass collecting techniques such as Malaise trapping means that very large numbers of individuals are frequently obtained in a single lot (often many lots), so storage for some time in alcohol is inevitable. Fortunately, ichneumonoids are tough individuals and even small ones are sufficiently robust not to collapse when dried from liquid preservation, so the use of critical point driers or organic solvent intermediates is not normally necessary as it is with, for example, small dipterans and many chalcidoids.

Specimens are normally mounted from alcohol in which they have been collected (as in Malaise trapping) or hand netting and subsequently stored. For all mounting styles, the wasps need to be nearly but not completely dried. First, it is a good idea, having sorted the specimens for mounting, to transfer them to fresh, clean alcohol, for a sort of wash, before mounting; Malaise samples and others with many insects stored in a small volume of alcohol lead to the alcohol becoming contaminated with dissolved lipid compounds that will dry on the specimen's surface, rendering them dirty-looking and sometimes making it difficult to see some fine details or judge subtle differences in sculpturation.

Drying is best achieved by having the specimens in a Petri dish of alcohol and placing a stiff piece of paper or filter-paper underneath the wasp, and removing it with the wasp on top out of the alcohol and placing to one side, possibly on another piece of paper/filter-paper to speed removal of the excess fluid. With ichneumonids in particular, the fore wings may be folded at the pterostigma/parastigma junction, which will make the measurement of veins and general appreciation of venation difficult. While there is still a film of alcohol over the specimen and paper, it is usually possible, with the aid of a pin head or fine forceps, to persuade at least the bottom wing to spread out flat. Once it has nearly dried it should, with luck, retain the fully spread position. Aguiar (2012) has recently proposed a different technique that I have not yet personally tried, which at first sight looks promising. It involves first micro-pinning and staging while the wasp is still soaked in alcohol and then pinning the mounted, but wet, specimen firmly and blowing air head-on over it from a small pump – several specimens can be prepared simultaneously in this way but they must be retained wet in alcohol until all are ready to be batch-processed with the air drier.

It is traditional to mount specimens all facing the same way, with the head facing left or such that the left side of the body is visible. Although this may at times be virtually impossible if an insect has died in a very awkward posture, it facilitates the comparison of specimens. It is also good to try to ensure that card rectangles or other staging are at a uniform height up the pin so that when series of specimens need to be compared there is less need to change the microscope's focus from one specimen to another.

Direct Pinning

Large specimens, meaning ones that are approximately 15 mm or more in length (excluding ovipositor and antennae), are usually best direct pinned using at least a No. 3 entomological pin. Although this method is well suited for specimens that have already been dried and it works best with recently collected specimens that have not been in alcohol, it can also be used for those that have been in alcohol, except that the effect of the alcohol on their internal proteins means that they will not adhere to the pin as securely as will a fresh specimen.

The mesosoma of the specimen should be held firmly between two fingers and the point of the pin inserted through the mesoscutum, a little in front of the scutellar sulcus and off-centre, traditionally slightly to the right of the midline (see, e.g., Figs 6.8f, 12.31b, 12.47 and 13.39) because most people are right-handed and therefore when the pin is picked up the observer will be able to see a nearly complete left side and middle without the pin shaft being in the way. The pin should then be pushed down through the mesosoma in a slightly posterior direction and vertically as possible in frontal aspect, so that it emerges at the anterior of the mesosternum (see Fig. 13.24) and therefre does not push off the propleuron and front legs; this is usually made easier by twisting the pin as it is being pushed through.

Side Gluing

A technique popularised by Henry Townes and widely adopted that is suitable for preparing large batches of medium-sized to fairly large specimens rather quickly is to use a small amount of glue to affix specimens to the side of an entomological pin. The glue should be one that remains slightly flexible when dry, as very brittle glues are likely to shatter if the pin with specimen is accidentally 'twanged'. The Natural History Museum in London recommends white shellac for this purpose (Gauld & Bolton 1988). The set of specimens, for example ones extracted from Malaise trap samples and quickly washed in fresh alcohol to minimise surface residues, are laid out on filter-paper facing right and allowed almost to dry – there should not be any free liquid on the mesosoma surface, but nor should they be completely dry if possible. If the wasp is too wet then the glue will mix with surface liquid and run over the specimen, obscuring vital details. A ring of the glue is made around the pin shaft approximately 0.6–0.7 of the way up and then the pin is placed such that the glue attaches to the side of the mesosoma, and after a few seconds the specimen will be firmly attached (see Figs 12.60, 13.18, 13.47 and 13.50b). The height up the pin and, to an extent, the angle at which the wasp is glued, need to be judged so that when the pin head is held, the wings and antennae are sufficiently out of the way but also the legs are not so low down that they will be broken when labels are attached to the pin. The glue used should be one that is soluble and does not leave residue (e.g. shellac is soluble in 70% ethanol), so that the specimen can be removed cleanly if it is required for other examination.

Card Rectangles and Card Points

Smaller specimens are often best mounted on a card rectangle. Rectangles and points, which can purchased from entomological suppliers, come in a range of sizes and are made from archival quality and sturdy card. I prefer rectangles on which nearly all the specimen can be mounted so that little protrudes that can be accidentally broken. For example, it is easy to imagine how the specimens in Fig. 12.24a or 13.55a could have accidents. Specimens need to be glued to the card using a small drop of soluble glue – a liquid fish glue such as Seccotine is ideal – placed centrally, then the specimen is lowered carefully on to it so that the glue sticks just to the lower right-hand side of the mesosoma and ideally the basal part of both right-hand wings. This will enable those two wings to be spread so that venation is clearly visible, and the wings are very firmly attached to the insect so they help to increase the security of the mount. When the glue is nearly dry, it is usually possible to use a pin to manipulate the other wings upwards slightly and the legs, antennae, etc., so that all parts are clearly visible. Occasionally specimens die in a posture that prohibits perfect mounting and so it may be necessary to glue them ventrally on to card (e.g. Fig. 12.45) or on to the tip of the card point.

Secondary Staging

In this technique, a small, headless, stainless-steel pin called a minutin is used to pin a small wasp on to a piece of supporting 'staging', which is then pinned into the drawer or store box using a regular entomological pin. The staging most often used nowadays is a fine Plastazote (expanded, cross-linked polyethylene), but older specimens were often mounted on square-section bars of polyporous fungus. Expanded polystyrene should never be used for this purpose as it quickly deforms permanently and loosens on the main pin shaft so it will rotate, and it and other specimens will inevitably get damaged; the other materials retain their springiness and do not come loose. Usually the pin is pushed upwards through the stage and then into the underside of the mesopleuron of the wasp (e.g. Fig. 13.59). I do not personally recommend this technique except under special circumstances when perhaps particular photographs need to be taken, as it is fiddly and the specimens are easily broken.

Labelling

It is always important to attach data labels to specimens, ideally at the time they are mounted. Labels should be written in insoluble ink (such as pigma pens) or printed, on reasonable thickness, acid-free paper or thin card. Printers now make it very easy to create large batches of labels quickly. Workers should never use codes with records kept in log-books, because books easily go astray and one cannot see the data immediately when examining a specimen. As M.R. Shaw (1997) noted, 'you are not writing the label just for yourself, but rather you are providing a scientific document for posterity'. The amount of information put on a label obviously will depend somewhat on the role of the specimen, but country, unambiguous location and collection date should be regarded as the minimum. I say 'unambiguous location' because some place names are used widely, even within one country. For example, there are many places called Santa Rosa, at least four in the USA alone. Again, nowadays it is usually fairly easy to add latitude and longitude data. However, where there is extra biological information it is always worth adding this, using either larger labels (preferable) or several labels on the same pin. A single, larger label is preferable for three reasons: it makes all the information clearly available, it affords the specimen extra protection and it minimises the extra handling and inevitable loosening of multiple labels if they need to be separated to enable them to be read. If the specimen has been reared from a wild-collected host, it is important to note extra information such as the identity of the host (see below), when the host was collected and when the wasp emerged. Hosts are sometimes known for certainty, which is ideal. However, sometimes there are various degrees of uncertainty and these should be captured in what is written on the label. For example, if the host genus is known for certain, but the host species is unsure, it is sensible to write something like *Aus ?bus;* if it might be something else, then *?Aus bus*.

If there are host remains associated with the specimen, these should be added to the same pin – a gelatin capsule containing these can usually be used and placed below the specimen. It is increasingly possible to obtain DNA sequence data from host remains, so even if the remains cannot be identified with certainty morphologically, it may be possible with the new tools (e.g. Quicke et al. 2012b,c).

PRESERVING SPECIMENS FOR DNA ANALYSIS

Techniques for obtaining at least short DNA fragments from old specimens are improving all the time;

Table 16.2 Sequencing success rate of Ichneumonid specimens in relation to age and mode of preservation.

Years since collection	Preserved in ethanol		Preserved dry	
	No. of specimens	Sequencing success (%)	No. of specimens	Sequencing success (%)
0–5	64	81	5	60
5–10	7	42	23	17.4
10–15	7	14	10	40
15–20	6	50	24	0
20–25	2	0	12	0.8
25–35	2	100	7	0
>35	–	–	10	0.9

Source: data from Quicke et al. (1999c).

nevertheless, if new or recently collected specimens are to be of most use for future molecular studies, it is best to preserve them in ways that minimise DNA degradation.

Quicke et al. (1999c) presented the results of their successes and failures at amplifying the 28S rDNA D2 expansion region of a range of ichneumonids that had either been stored in alcohol (usually 70%) or dry pinned (Table 16.2). From these very few data, it is nevertheless apparent that DNA degrades fairly rapidly either when in liquid or when specimens are kept dry. However, it may be that the 'dry' collection material has not always been perfectly dry and that old museum material can frequently be seen to have traces of fungi upon it, and, if brass pins were used, frequently verdigris resulting from chemical reaction between lipids in the specimen and copper in the pin. I have certainly noticed that old pinned material from some museums has far higher sequencing success rates than that from similarly aged specimens in other collections. In a slightly more controlled study, Andersen and Mills (2012) obtained very similar result for dry museum specimens of four species of braconid and showed that the sequencing success of old specimens was gene sensitive, with amplification of CO1 being far less successful than amplification of 28S rDNA (Fig. 16.5). Keeping specimens cool or even in a freezer is likely to prolong the half-life of DNA considerably even when in dilute alcohol because of simple reaction kinetics, with degradatory DNA hydrolysis being slowed at lower temperatures, so samples kept in refrigerators and freezers are likely to have better DNA.

What is probably generally of equal or greater importance might be how the insect was killed, because although a number of chemical changes can occur to DNA with storage (e.g. acid hydrolysis) these are slow chemical processes; when cells in an animal or plant die, they autolyse, which means that enzymes, including nucleases, packed in small sub-cellular compartments called lysosomes, are released into the surrounding milieu. Methods of killing wasps that denature these enzymes are probably the single most important factor in preserving DNA as, in the short term, these enzymes can cut up a great deal of DNA very quickly. In experiments carried out in Belize, I and several colleagues ran Malaise traps for nearly 2 weeks, filled with either 70 or 96% ethanol, and with trapped insects removed for subsequent storage in 96% ethanol after various durations. Although not involving ichneumonoids, the insects (in this case beetles) were randomly selected from catches and subjected to PCR and sequencing for cytochrome oxidase I (Pat/Jerry primers). Figure 16.6 shows the effect of the initial killing medium and subsequent field storage conditions. A statistically significantly greater proportion of specimens that were collected directly into 96% ethanol yielded sequences than those collected into 70% ethanol and subsequently transferred to 96% ethanol, with no significant effect of length of subsequent field storage in 70% alcohol, suggesting that the early denaturing of enzymes is critical for good subsequent DNA availability. Fitton et al. (1988), speaking at the time from an entirely morphological perspective, suggested that isopropanol (propan-2-ol) could be substituted for ethanol with the advantage that in many countries it could be purchased without the licenses that drinkable ethanol require. However, it is very inferior when it comes to

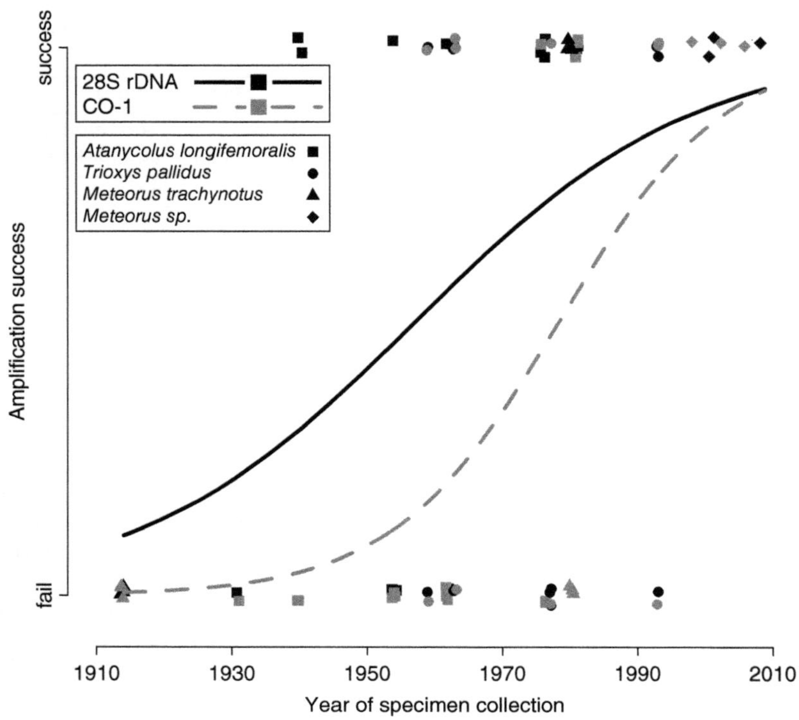

Fig. 16.5 Logistic regression plots of proportions of museum specimens of braconids that amplified successfully for two gene fragments (often only short stretches but sufficient for species identification) of the multi-copy nuclear 28S rDNA gene and the mitochondrial cytochrome oxidase 1 (CO-1) gene; points jittered slightly for clarity. (Source: based on data from Andersen & Mills 2012.)

preserving DNA, perhaps because, as a slightly larger molecule, it penetrates insect cuticle more slowly.

From the above, it can easily be imagined that the 'traditional' field protocol for hand-netted wasps, that involves putting them either in a killing jar containing an organic solvent such as ethyl acetate or in a cyanide killing jar, as used to be very common (but are almost impossible to make in many countries because of health and safety issues, despite, I believe, no one ever having been harmed, let alone killed, as a result of an entomological killing jar or even as a result of its manufacture), are really bad for DNA preservation because the associated procedure normally involves the dead insects being kept after death in the killing vessel at ambient temperature for at least hours and possibly overnight, until they are subsequently processed. This prolonged post-mortem storage with no neutralisation of DNA-degrading enzymes, sometimes in the warm conditions of the field entomologist's trouser pocket, is probably virtually the worst thing one could do to preserve easily sequenceable DNA. Even the old Victorian method widely used for killing beetles, of putting them in boiling water, is likely to have been far more DNA friendly.

PACKAGING AND POSTING SPECIMENS TO OTHER WORKERS

It is really important to know how to send specimens safely to taxonomic experts and other workers. Pin-mounted, dry specimens are intrinsically fragile and also they can often come loose from their mount if they are subjected to shock. There are several essential aspects to packing specimens for postage. (1) they should be securely fastened to their mount, (2) if mounted on a card, the card should not be able to rotate on its pin (use sufficiently robust card), (3) the

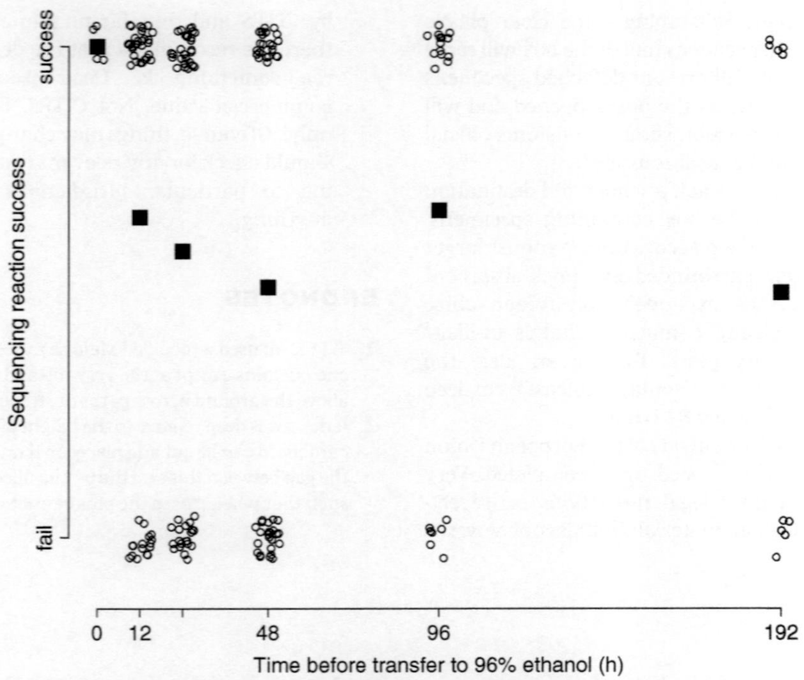

Fig. 16.6 Sequencing success for field collected beetles showing the effect of initial ethanol concentration used for killing and non-significant subsequent effect of field storage in 70% ethanol for up to nearly 2 weeks in a Malaise trap. (Filled squares indicate means; open circles are 'jittered' raw data.)

pin(s) should be securely stuck into the box so they can not come free[2], (4) the box should be sealed such that pests cannot enter, (5) the box that the specimens are in should be surrounded by a large thickness of shock-absorbing material and included in a robust larger box.

1. In general, if the advice on mounting given above has been followed, specimens should not easily become detached from their pin or mount. For side-glued specimens, with use of white shellac, which retains a tiny amount of flexibility, this is less likely to be a problem than if a glue that goes brittle is used.
2. Use of proper archival quality card rectangles or card points, and also thick paper for labels, is important, because it will grip the pin securely, whereas many types of card lack elasticity and very soon work loose on a pin and rotate freely, which means that not only is the specimen attached to them unsafe, so too may be others nearby in the box. A procedure called cross-pinning is often used whereby two additional entomological pins are used, one on either side of the specimen card, to prevent it rotating if it should get jolted. Cross-pinning is also often used to stabilise direct-pinned large wasps.
3. Although easy to obtain, the worst pinning substrate is expanded polystyrene because it deforms permanently so every small movement of a pin inexorably loosens it and unless the lid is tight fitting and more or less flush with the pin head, the pins are very likely to fall out and certainly rotate, possibly causing damage to specimens. The best substrate for pinning is Plastazote or alternatively the more traditional cork. In addition, it is good practice to pin securely in one corner of the specimen box a wad of teased cotton-wool such that should a specimen or part of a specimen become detached from its pin, it will soon be caught among the cotton-wool strands and so restrained from moving around in the box and causing further damage either to itself or to other

specimens. Tight Sellotaping some clear plastic over the lower (specimen) half of the box will serve two purposes: it will prevent detached specimens from falling out when the box is opened and will provide some protection should a customs official open the box to see what is inside it.

4. It is a good idea to attach a sender and destination address label to the box containing specimens. Then it needs to be placed within a robust larger box completely surrounded by shock-absorbent material such as expanded polystyrene chips or similar – ideally, a material that is unlikely to be harbouring pests. To give an idea, the shock-absorbent layer should be at least 6 cm deep all around, and ideally 8–10 cm.
5. If posting abroad or outside of the European Union a customs form will need to be completed. Very few countries have legal restrictions on receiving entomological material that is not covered by CITES and thus far no ichneumonoids are. I therefore recommend that the declaration should read something like, 'Dead insect specimens. No commercial value. Not CITES. For scientific use only'. Of course, things may change and the reader should check for any relevant restrictions pertaining to particular jurisdictions before posting anything.

ENDNOTES

1. T. Li et al. used a modified Malaise trap design but without end curtains except at the very top and suspended slightly above the ground across paths or streams.
2. If the box is deep relative to the height of the pins, a sheet of cardboard can be cut and placed in it on top of the pins and the gap between this and the box lid filled with cotton-wool such that when closed the pins cannot work loose.

Chapter 17

EPILOGUE

The Braconid and Ichneumonid Parasitoid Wasps: Biology, Systematics, Evolution and Ecology, First Edition. Donald L. J. Quicke.

Throughout this book, I have been writing phrases such as, 'unfortunately nothing is known about ... '. Clearly, there is an enormous amount that we still need to discover and I thought that this would be a good opportunity to present some of my 'top 10 wants' lists. I have divided this into four sections, 10 'wants' in each. The first concerns obtaining phylogenetic estimates with significant support for various taxa, which will be valuable for determining the transitions in important characters and also stabilising nomenclature. The second is a list of taxa for which basic biological data such as host group and/or mode of parasitism are not yet known. The third concerns questions about physiology and the fourth about ecology.

PHYLOGENETIC QUESTIONS

1. Are the Apozyginae the sister group to the remaining extant Braconidae? That is, are they truly better considered as eoichneumonines?
2. What are the relationships of the Trachypetinae – DNA sequence data for additional genes are desperately needed in addition to knowledge about its overall base composition.
3. Where do the Masoninae belong? These very small wasps inhabit the ground layer and the available 28S rDNA sequence is on a long branch, so additional gene sequencing of these will be needed to place them with confidence.
4. Are the large ichneumonid subfamilies Ctenopelmatinae, Tryphoninae and Pimplinae monophyletic? If not, there is a need to redefine their taxonomic groups, but should this be towards greater lumping or greater splitting? If the Tryphonines are really not monophyletic, then that could well indicate multiple shifts from endo- to ectoparasitism, assuming that their placement in the best current phylogenetic trees quite far from the base of the ophioniformes+ orthopelmatiformes is correct.
5. Is the Stilbopinae monophyletic – and is *Notostilbops* related to either or both of the other two genera?
6. Where do the Hybrizoninae belong? I have chosen to include them among the ophioniformes, which is at least reasonably consistent with their endoparasitism, and they show some resemblance to Anomaloninae, but their 28S rDNA sequences are on very long branches and so their placement is still far from certain.
7. Are the Ophioninae+*Skiapus* monophyletic or should *Skiapus* be removed to its own subfamily?
8. Where do *Euceros*, *Microleptes*, *Ischyrocnemus*, *Lapton* and *Bremiella* belong? Although all were included in the analyses of Quicke et al. (2009), their placements in the best trees are either not consistently supported or seem extremely unlikely based on morphology.
9. What are the relationships between the various Helconinae and Blacinae? Sharanowski et al.'s (2011) phylogeny suggests that this grouping will need major rearrangement. It will be important to include as many genera as possible, especially of the doubtfully monophyletic Blacininae, and also additional genes to resolve this.
10. There are many genera, predominantly southern and small bodied, that are currently placed in the mess of subfamilies including the Hormiinae *s.l.*, Exothecinae, Lysiterminae, Rhysipolinae and Rhyssalinae, that are probably misplaced. There is a great need to obtain and sequence a far better representation of these, as well as using additional gene markers, to obtain a better understanding of the basal phylogeny of the cyclostome lineage which is currently pretty shaky.

HOST AND PARASITISM QUESTIONS

1. Are Lycorininae ectoparasitic but living in the host rectum or endoparasitic? These wasps appear to occupy a fairly basal position within the ophioniformes, but not completely basal, so if they are ectoparasitic, but within a moist environment, perhaps this represents an unusual biological reversal.
2. What is the biology of the Trachypetinae? These large Australian wasps are reasonable candidates for being the sister group to the Braconidae. If they are, and if predictions based on morphology that they are endoparasitoids are correct, then it slightly increases the probability that the ancestral braconid was also an endoparasitoid.

3. What are the biologies of the termite-associated doryctines including the Ypsistocerini? – their ovipositors do not suggest they are ectoparasitoids.
4. What are the hosts and mode of parasitism of the Maxfischeriinae? The very strange eggs of this group that seem most likely to belong to the same group as the Aphidiinae and Mesostoinae, as a combined sister group to the cyclostome braconids, infer that they have an unusual biology.
5. What is the biology of the Helconinae *s.s.*? Apart from being predominantly endoparasitoids of deeply concealed beetles, the host stages attacked and details of larval development are unknown.
6. What is the biology of the Gnamptodontinae? Much circumstantial evidence points to them being endoparasitoids, yet some features suggest ectoparasitism. Their apparent phylogenetic position among the 'alysioid' subcomplex of subfamilies suggests that were they to be shown to be endoparasitic, then they would represent yet another separate transition to that way of life within the group.
7. What are the biologies of the members of the brachycyrtiformes? *Brachycyrtus* attacks Chrysopidae and Claseinae apparently pergid sawflies, but in neither case is it known whether they are ecto- or endoparasitoids. If one or more of them are endoparasitoids then this would most likely represent yet another independent origin of endoparasitism.
8. What is the biology of the extremely small-bodied Masoninae whose females live close to the ground.
9. What is the biology of the Betylobraconinae *s.s.* (i.e. *Mesocentrus* and *Betylobracon*)? They may be the sister group of the Rogadinae, so it would be very interesting to know what host group they attack, whether they are endoparasitoids and whether they similarly mummify the host. If so, then undoubtedly the two subfamilies will need to be synonymised.
10. The biology of the ichneumonid subfamily Diacritinae is completely unknown and if its placement in current trees at the base of the pimpliformes is confirmed, then knowing its biology would greatly assist reconstruction of biology in that group.

PHYSIOLOGICAL QUESTIONS

1. Determine lipid synthesis capabilities of taxa other than Pimplinae – thus far we have incredibly sporadic evidence that among ichneumonids there are some quite different degrees of biochemical capability for synthesising and transforming fatty acids, but the data are extremely phylogenetically biased.
2. Determine the precise limits of bracoviruses in the Braconidae, most importantly, do they occur in any Ichneutinae or Dirrhopinae? Some molecular analyses have also suggested that the Orgilinae might belong to this clade, yet there seem to be no TEM studies of the calyx regions of these wasps either.
3. Determine the precise limits of Ichnoviruses in the Ichneumonidae. While the Banchinae appear monophyletic, some other related taxa have yet to be investigated. Subfamilies such as the Stilbopinae and Sisyrostolinae would be well worth investigating.
4. What is the mechanism (or mechanisms) by which highly polyphagous koinobiont endoparasitoids such as *Charmon extensor* overcome host physiological defences? To date, studies of defences against host immunity have involved only species with moderately broad and sometimes misleadingly dichotomous host ranges, whilst *C. extensor* can successfully parasitise members of more than 12 host Lepidoptera families?
5. Campoplegines collectively attack Lepidoptera, Coleoptera and Rhaphidioptera, yet the roles of their polydnaviruses have been investigated largely in the former host order. How do they function in the hosts belonging to other orders? Are the same genes involved and do they have the same effects across host orders?
6. What sensory roles do ovipositor sheaths play and what, if any, generalisations might be made? It seems extraordinary that almost no work, even simple ablation/excision experiments on these organs, has been carried out. Totally a speculation, but it would seem entirely plausible that they might be involved in detecting previous female's oviposition marking compounds.
7. What function does the secretory product of the spermathecal glands have? Does it continue to

enter the spermatheca after fertilisation? Does it serve a role in stimulating sperm swimming?

8. When (and where) in relation to the release of venom-, egg- and ovary-related compounds do ichneumonoids release Dufour's gland components? The latter tend to be at least moderately hydrophobic hydrocarbons and potentially well suited to acting as signals to conspecifics if they were released on to the surface of an host rather than into its haemolymph., but no-one to date has specifically localised them *in situ*.
9. Far more information is needed on the stimuli that trigger embryogenesis in different groups of ichneumonoids. In many groups it appears that the trigger is the egg distortion that occurs during oviposition, but in other groups, such as in many tryphonines, it can initiate within the female oviducts. The latter is a potentially risky business as it can result in matricide. In most of those wasps in which oviposition is the cue, the egg may actually be laid while still in the middle of meiosis. One has to wonder, therefore, whether it is the completion of meiosis which is essential or post-meiotic factors.
10. Why is the protection against encapsulation of many endoparasitoid eggs provided by products from female oviduct glands, such as VLPs, polydnaviruses and mucopolysaccharides, that adhere to the surface of the egg rather than the follicle (chorion) cells secreting a protecting outer layer.

ECOLOGICAL QUESTIONS

1. If ichneumonid koinobiont endoparasitoids of exophytic Lepidoptera larvae do indeed show anomalous diversity, whereas braconids in the equivalent niche do not, what differences in the toxin tolerance physiologies of the two families would explain this?
2. Do all subfamilies of ichneumonoid show the same low beta diversity that appears to be the case with Pimplinae and Ophioninae, the groups that to date have been most widely studied (e.g. Bartlett et al. 1999).
3. How does beta diversity of parasitoids of exophytic hosts relate to beta diversity of hosts and of their host plants?
4. How widespread in Nature is the selective multiparasitism of hosts that have already been attacked by species that cause a general reduction in host immune capability?
5. Island parasitoid faunas appear to reflect those of the nearby mainland sources, whereas logic would suggest that generalists should be more capable of successfully colonising them. It seems likely that host species on islands, in the absence of maintained selection pressure by their normal parasitoids, might evolve generally reduced immune asininities because of the associated costs (cf. dodos). Are potential island hosts more physiologically vulnerable and therefore more easily attacked by parasitoids that are otherwise obliged to be specialists because of host responses?
6. Why are there so many generalist, idiobiont parasitoids attacking what seems like a small range of resources? Taking tropical idiobiont ectoparaditoids of wood-boring hosts as an example, one might reasonably divide hosts into a number of depths in wood and size categories, each combination of which might constitute a niche – say, 4–6 mm deep, 1.0–1.5 cm, long. Perhaps even broader categories apply to some taxa. Yet in Africa alone there are 100 plus species of braconine that could probably successfully develop on such a host, not to mention doryctines, pimplines, stephanids, etc.
7. Do fungi and wood-fungal associations play an important role in the attraction of parasitoids to xylophagous hosts? Wood, once attached to or detached from a tree, undergoes a semi-predictable succession of fungal associates, which may produce volatile organic compounds that signal parasitoids of likely sources of hosts.
8. What are the palatability spectra of large and clearly aposematic tropical ichneumonids and braconids? Not only are many species (see, e.g., Figs 12.16e,f and 12.37) conspicuously brightly coloured, they sometimes seem more likely to be the models in mimicry systems than the dupes.
9. Are there any generalisations that can be made about species host ranges for endoparasitoid taxa that attack members of more than one major host group, e.g. Lepidoptera and Coleoptera, Lepidoptera and Hymenoptera, Lepidoptera and Diptera, and do these reflect more physiology or host niche? So little is known about the hosts of the Diptera parasitizng ichneumonoids that this question will require a great deal more rearing,

probably backed up by molecular identification tools.

10. What are the actual mechanisms behind the apparent use of host silk as a general set of host-finding/acceptance cues in ichneumonids, given that the silks of the many different host groups concerned (Lepidoptera, Hymenoptera, spiders) are massively chemically diverse?

Serendipity will no doubt play some large part in answering these questions, but hard work can be very rewarding. If one or more host records are known (or at least in the literature), then intensive targeted fieldwork in appropriate localities has a good chance of locating material that could be bought into culture or otherwise studied.

GLOSSARY

For anatomical terms the reader is also recommended to consult the structure section (pp. 13–59) in Goulet and Huber (1993), the detailed morphological descriptions of exemplar opiines by Karlsson and Ronquist (2012) and the *Hymenoptera Anatomy Ontogeny* portal (http://bioportal.bioontology.org/ontologies/HAO).

abscissa a part of a longitudinal wing vein between adjacent cross-veins or between one of its ends and the nearest cross-vein.

accessory nuclei structures originating from the oöcyte nucleus possibly involved in gradient formation.

alecithal eggs having (virtually) no yolk.

alitrunk another word for mesosoma.

allomone a chemical substance produced by species X that affects the behaviour of a member of species Y such that X, but not Y, benefits.

amblypygous with the female hypopygium long such that almost all of the ovipositor and sheaths are concealed (cf. oxypygous). A term used almost entirely in connection with Ichneumoninae.

anal vesicle a posterior eversion of the rectum in early larval instars of several groups of Braconidae, which usually places the aorta internally in close proximity to the outside medium and is involved variously in host regulation, respiration, nutrition and excretion.

antefurcal when the anterior junction of a posterior cross vein is basal to the posterior junction of the crossvein that it is routinely nearly opposite to.

antennomere any segment of the whole antenna, i.e. including the scapus and pedicellus.

Apocrita the monophyletic group of Hymenoptera that have (or ancestrally had) a narrow waste, and comprising all the paraphyletic 'Parasitica' plus the monophyletic Aculeata.

apomorphy a derived character state (*see* autapomorphy, synapomorphy).

apophyses tooth-like tubercles on the mediolateral part of the propodeum, formed variously on the lateral carina, on the transverse posterior carina or in the second lateral area.

apoptosis programmed cell death.

aposymbiotic devoid of symbiotic organisms, especially bacteria.

apterous without wings or with them extremely reduced.

areolet a name frequently given to the small second submarginal cell in Ichneumonidae and sometimes to braconids with an equivalently small cell (e.g. Microgastrinae, Orgilinae). There are different views about the homologies of the basal and apical veins bordering this cell in the two families.

arrhenotoky parthenogenetic production of males, i.e. from unfertilised eggs.

associative learning learning more reliable cues connected to finding hosts or mates as a result of positive experiences.

atavistic, atavism pertaining to features once lost in evolution, but that reappear rarely in some individuals.

The Braconid and Ichneumonid Parasitoid Wasps: Biology, Systematics, Evolution and Ecology, First Edition. Donald L. J. Quicke.

aulax the longitudinal groove on each lower ovipositor valve that interlocks with the rhachis at the ventro-lateral part of the upper valve.

autapomorphy an apomorphic (derived) character state present only in a single taxon and therefore not providing evidence of its relationships.

Banchovirus genus of Polydnaviridae occurring in the banchine ichneumonid subfamily Banchinae.

basitarsus the basal-most segment of the tarsus adjacent to the tibia.

β-conditioning *see* associative learning.

biotype otherwise known as 'host race', a genetic line of a parasitoid (or other organism) that displays features indicating strong selection to perform well on a particular host, and sometimes showing partial practical denetic isolation from biotypes adpted to different hosts.

brachypterous having wings of reduced size, usually non-functional.

Bracovirus genus of polydnaviruses found in members of the microgastroid lineage of Braconidae.

bulla a weakened, unpigmented, part of a wing vein where a flexion line passes through it.

caudal appendage the elongate, pointed tail of in early larval instars of several groups of Ichneumonoidea.

caudal vesicle *see* anal vesicle.

cerci paired, small, sensory structures derived from posterior margin of the tenth metasomal tergum but sometimes arising from the ninth if the tenth is reduced.

chorion the outer covering of the egg.

clypeus part of the front of the head below the face.

CNEA concealed nectar extraction apparatus, any of several modifications of the head and mouthparts that enable extraction of concealed nectar.

coxa the basal-most part of the leg of an insect.

ctenidia comb-like scales or rows of teeth on the ovipositor egg canal wall that are involved in egg movement during oviposition.

cyclostome having a cavity above the mandibles formed by a recession of the lower part of the clypeus and exposure of the often concave labrum; usually dorsally rounded.

cystatin a family of cyteine protease inhibitors.

deuterotoky parthenogenetic reproduction that leads to production of both male and female offspring from unfertilised eggs.

DOPA–tyrosinase reaction initial stage in the production of melanin in which tyrosinase catalyses the reactions: tyrosine → DOPA → dopaquinone.

dorsope a moderately to very deep pit located at the anterior end of the first metasomal tergite (predominantly in Braconidae) just behind the junction of the dorsal and dorso-lateral carinae.

dsDNA double-stranded DNA.

Dufour's gland an unpaired, more or less tubular, gland in female parasitic wasps and other Hymenoptera that opens into the base of the ovipositor and produces fatty volatile compounds that are predominantly involved in host-marking.

elision, elided combining in a single sequence for analysis multiple different alignments of the same sets of bases in the hope that alignment parameter-specific anomalies will cancel one another out.

endodont of adult wasp mandibles having the normal configuration, at rest with the tips of the teeth overlapping.

eonymph another name for prepupa.

evapatoria cuticular modifications with large surface area that enhance evaporation of pheromones.

exodont variously defined as having the adult mandibles either not overlapping when closed or protruding from the head laterally; normally these have more than two teeth.

exuvium a shed larval skin.

factitious host not a natural host of a parasitoid but one on which it can be reared readily either for experimentation or release in a pest control programme.

flagellomere any segment of the flagellum.

flagellum that part of the antenna beyond the scapus and pedicellus (i.e. from segment 3 onwards).

frons part of top of head between antennal sockets and ocelli.

functional response the number of hosts attacked (or prey consumed) as a function of host (prey) density.

galea distal part of the maxilla.

gaster term often used to mean that part of the abdomen behind the propodeum, or in some groups, behind the first metasomal tergite.

gastrocoelus a depression anterolaterally on either side of the second metasomal tergite in ichneumonines, including an area of modified texture, the thyridium.

germarium the part of an ovariole where germline cells are dividing.

germ band the length of blastoderm that thickens to form the embryo proper.

glabrous lacking setae.

glossa distal part of the labium.

glymma an elongate groove on either side of the base of the first metasomal tergite in some Ichneumonidae, usually more posterior in position than the laterope (q.v.) in Braconidae.

gonangulum sclerotised plate to which lower ovipositor valve is attached.

gonocoxite (in female) basal plate-like part of ovipositor valve, to which muscles attach, homologous to the coxa of the primitive abdominal limbs from which the ovipositor is derived.

gonoforceps parameres.

gonoplac same as ovipositor sheath.

gonostylus same as paramere.

gregarious of species that habitually oviposits more than one egg on/in a host and more than one of these usually develops successfully to adulthood.

gynogenesis reproduction involving stimulation of the egg to develop but without the genome of the sperm playing any role; gamete nuclear fusion does not take place.

Hallez's law that the head end of a developing insect embryo forms at the end of the egg opposite the end with the pedicel, i.e. the anterior end.

hamuli hook-like setae on hind wing vein R1 (and on C in many other Hymenoptera) that interlock with the curved posterior ridge of the fore wing, making the wings functionally dipterous in flight; also used to refer to the enlarged setae on vein C or C+SC+R that restrain the basal part of the fore wing anal cell in flight.

hemimetabolous often used to mean having juvenile instars that become progressively more adult-like in form, i.e. paurometabolous, but sometimes restricted to just those taxa with gilled aquatic nymphal stages.

holidic diet a completely chemically defined diet; *cf.* meridic.

homonym in zoological nomenclature, when the same name has been used independently for two (usually) different genera, or for two different species within a given genus; only the oldest (most senior) name is potentially a valid name that can be used.

homoplasy the result of parallel or convergent evolution leading to highly similar character states, or reversals such that a derived character state is lost. Homoplasy obscures the discovery of phylogenetic relationships.

hyperparasitoid a parasitoid of another parasitoid while the latter is in its parasitic phase; *see also* pseudohyperparasitoid.

hypoclypeus the lower part of the clypeus in cyclostome braconids which is angled backwards and forms the 'roof' of the hypoclypeal depression.

hypopygium the postero-most visible metasomal sternite; in females this is the sixth metasomal sternite, in males the seventh.

ichnovirus member of the Ichnoviridae which really means polydnaviruses associated with the Campopleginae (but see also Banchovirus).

imagine, imago the adult, reproductive stadium of an insect.

imagobiont term coined to indicate the strategy of ovipositing in and completing development in the imaginal stage of the host; imagobionts show a mix of idiobiont and koinobiont features.

interstitial when the anterior junction of a posterior cross vein is more or less directly in line with the posterior junction of the cross vein that it is routinely nearly opposite to.

kairomone a semiochemical produced by species X that benefits an individual of species Y, but is detrimental to the emitter (X).

labium part of mouthparts usually largely concealed, posterior to the mouth opening, it can be thought of as the lower lip complex. The distal part of it bears the labial palps.

labrum part of mouthparts at the front of the mouth opening connected to the clypeus dorsally; usually concealed, but exposed in cyclostome braconids and weakly exposed notably in some ichneumonids.

lancet term sometimes applied to the lower ovipositor valve.

laterope a pit at the base of the first metasomal lergite located at the confluence of the dorso-lateral and lateral carinae.

laterotergite the lateral parts of a metasomal tergum that are typically thin and pendulous, and often separated from the generally much more heavily sclerotised medial tergite by a crease; but which may be reduced to a narrow strip or be completely absent. In Ichneumonidae absent from the first metasomal tergite, but frequently present in braconids. In higher microgastroids those of the first metasomal tergite are membranous and identifiable because the spiracle opens in them.

lekking behaviour in which males aggregate or swarm collectively to attract mates.

LMC *see* local mate competition.

local mate competition situation in which siblings are likely to mate with one another.

macropterous wings fully developed.

malar space the region between the lower part of the eye and the base of the mandible.

Malpighian tubules tubular excretory glands that join to the anterior of the hind gut.

meconium the first liquid/solid waste products/faeces expelled from the anus by the parasitoid larva (usually) or adult (a few instances) just prior to pupation or immediately after eclosion, respectively.

meridic diet a synthetic diet with some but not all of its components chemically defined.

meroistic of ovaries/ovarioles that have oöcytes attached to a cluster of nutritive (nurse) cells as they pass to the uterus.

metacarp vein running anterior margin of fore wing distal to apex of pterostigma; here referred to as 1-R1.

micropyle area at or near the end of an egg where narrow pores allow sperm to reach the oöcyte plasma memrane and penetrate it.

micropterous having very reduced, non-functional wings.

multiparasitism when two species of parasitoid attack the same host individual.

nervellus fore wing cross-vein 1cu-a.

nervulus hind wing cross-vein cu-a.

non-cyclostome of braconids not having the cyclostome (q.v.) condition, or members of the lineage of braconids that are the sister group to the extant cyclostomes.

notauli pair of submedial lines on the mesoscutum, sometimes or wholely indicated by a groove, that demarks the medial border of the dorsal insertion of the main mesopleuro–mesonotal muscle.

nurse cell in meroistic ovaries, one of cluster of interconnected cells that as a cluster are also connected to the oocyte by intercellular bridges and are responsible for delivering yolk precursors and other material to the developing egg cell.

occiput posterior part of head.

olistheter the combination of the rhachis and aulax on touching parts of the upper and lower ovipositor valves, respectively, that form the longitudinal running 'dovetail'-type joint that holds them together while permitting the valves to slide independently with respect to one another.

Opinions in zoological taxonomy, rulings published by the International Commission for Zoological Nomenclature that specify, in cases where there are conflicting views, the correct name to be applied to a taxon.

ostiole the narrow, distal opening into a fig syconium.

ovipositor sheeth normally protective coverings for the ovipositor proper that originate dorsally from the gonangulum.

ovisorption a process shown by some parasitoids (principally Chalcidoidea) in which mature eggs that have not been laid are digested in the oviduct and their nutrient components recycled.

oxypygous with the female hypopygium short such that most of the ovipositor and sheaths are exposed (cf. amblypygous). A term used almost entirely in connection with Ichneumoninae.

pabula, pabulum things an animal will eat.

parameres the outer, usually protruding, parts of the male genitalia.

paurometabolous gradual development through a series of nymphal instars progressively becoming more adult like as in Hemiptera.

pedicellus the second segment of the antenna, located between the scapus and the most basal flagellomere.

pensile hanging or attached by a thread.

petiolate 'shaped like a petal', i.e. narrow basally and widening fairly abruptly at about the mid-length.

phenoloxidase enzyme, also called tyrosinase, responsible for initiating melanin synthesis.

pheromone a chemical compound released by one individual of a species and received by another of the same species, whose behaviour and/or physiology they affect.

plesiomorphy the ancestral state of a character rather than a derived one (*see* symplesiomorphy; cf. apomorphy).

plurivoltine having more than one generation per year.

polar bodies the three nuclei formed by meiosis which do not fuse with the sperm nucleus to form the zygote.

polydnavirus a group of virus-like entities whose segmented genomes are completely encoded within the wasp genome, but which are replicated and packaged into virions within cells of the calyx region of the female reproductive tract and injected into the host, where they infect cells and kill or alter their development.

polyembryonic of taxa in which the egg within the host divides one or more time to give rise to cells that will each give rise to separate embryos.

postfurcal when the anterior junction of a posterior cross vein is distal to the posterior junction of the cross vein that it is routinely nearly opposite to.

post-pectal carina a carina running transversely across the posterior part of the mesosternum just in front of the mid coxae.

prepupa the last larval instar of a holometabolous insect after it starts to change shape and becomes immobile before actually moulting into a pupa.

priority in zoological taxonomy, the principle that when more than one names have been applied to a taxon, the oldest one (i.e. the name published at the earliest date), if not otherwise prohibited from being used, will be the correct name for the taxon.

pronymph another name for prepupa (q.v.).

pro-ovigenic technically, emerging as an adult female with all her life's complement of eggs fully mature; in reality, being close to the end of the spectrum of degree of egg maturation at eclosion; *see* ovigeny index.

protandry the relatively earlier adult emergence of males compared with females.

pseudohyperparasitoid a parasitoid of another parasitoid after the latter has exited from the body of the primary host.

pseudoparasitism when a host is attacked by a parasitoid with injection of venom and/or other secretions, but without either deposition of an egg, or with deposition of a non-viable egg (e.g. after X-ray or gamma-ray irradiation of a female wasp).

pseudovirginity the result of female wasps mating with sterile males (such as diploid males in some species).

pterostigma an approximately triangular sclerotised area just beyond the middle of the fore wing involving the base of vein 1-R1, and from which r-rs arises.

puparium the hardened final larval instar skin in higher Diptera that forms before pupation and then encloses the pupa after moulting.

pygidium name given to the last tergum of the abdomen in insects and other arthropods.

pygostyles *see* cerci.

quasigregarious wasps are not gregarious parasitoids, but multiple individual hosts in a small patch are attacked such that at parasitoid emergence, many siblings are likely to be in close proximity, e.g. is the case for many aphid-parasitising Aphidiinae.

rectal pads structures on the wall of the rectum of an insect that may be involved with water balance and inionic regulation; also called rectal papillae.

recurrent vein either 1m-cu (or 1m-cu&RS+M) (first recurrent vein) or 2m-cu (second recurrent vein).

rhachis the longitudinal T-section ridge on the ventro-lateral part of the upper ovipositor valve that interlocks with the aulax groove on the lower valve.

rhinaria occasionally used alternative name for *sensilla placodea*.

scapus the first segment of the antenna.

scrobe a depression on the head into which the scapus of the antenna can be partially concealed/protected.

secondary hyperparasite/-oid another name for pseudohyperparasitoid (q.v.).

secondary parasite/-oid another name for hyperparasite (q.v.).

solitary of species that only have one individual developing in and emerging from a host.

spermatodesmata clusters of sperm with their anterior ends united within an extracellular matrix, and which swim as group in a coordinated fashion.

spurious vein a wing vein of *de novo* origin and not having continuous ancestry with veins in a similar position as were primitively present.

stenopneusticae essentially the generally larger bodied tribes of Ichneumoninae, i.e. excluding the Phaeogenini and Heterischnini.

subgenital plate another name for hypopygium (q.v.).

superparasitism laying more eggs in a host that has already been oviposited in by a wasp of the same species; this may be by the same individual, in which case it is called *self-superparasitism*.

superparasitoid sometimes used to refer to pseudohyperparasitoid (q.v.).

symplesiomorphy the shaing of character state because it was present in a common ancestor.

synapomorphy an apomorphic (derived) character state shared by two or more taxa suggesting that the transition to the apomorphic state occurred in a common ancestor.

synomone a semiochemical that benefits both the receiving individual of another species as well as the emitter.

synonymy in zoological nomenclature, when two or more different species names have been applied to what is believed to be the same species, or genus names to what is believed to be the same genus; the oldest (most senior) name, if valid, will be the correct name.

synovigeny more or less continuous maturation of eggs through at least the early stages of adult life.

tarsus the multisegmented part of the leg distal to the tibia, and bearing claws on the last segment (telotarsus).

telotarsus the distal-most, claw-bearing segment of the tarsus.

teratocyte an enlarged (giant) cell that originates from the break-up of the serosal membrane after the parasitoid larva egresses from the egg, and survives and grows within the host separate from the parasitoid larva but benefits the latter as a direct food source and through regulation of host physiology.

teratological abnormal morphology resulting from the features of the organisms genome or its expression during development.

terebra another name for ovipositor used in older literature.

terga *see* tergum

tergite medial subpart of a tergum; cf. laterotergite.

tergum dorsal body (metasomal) sclerite (*pl.* terga).

thelytoky parthenogenetic reproduction in which unfertilised females produce diploid female offspring; sometimes induced by the endosymbiotic bacterium *Wolbachia*.

torulus another name for antennal socket.

tribe a formal classificatory group between genus and subfamily; one tribe name is always based on the same genus as the subfamily name.

trochantellus a demarcated basal part of the femur immediately distal to the trochanter but immovably joined to the rest of the femur (not differentiated from femur in *Alomya*).

trochanter second segment of an insect leg between coxa and femur.

trophamnion a cellular membrane formed around the developing embryo in some endoparasitoids that serves to transport food molecules from the host to the embryo. In some taxa formed from cells derived from the polar bodies.

trophocyte essentially another name for nurse cell (q.v.).

type specimen the specimen upon which the concept of a species is based.

type species the species upon which the concept of a genus is based.

uniparental *see* thelytoky.

univoltine having only a single generation per year.

uterine gland *see* vaginal gland.

vaginal gland a sac-like, single, bilobed or paired, exocrine gland (and its reservoir) located at the anterior part of the common oviduct.

valve the upper or either of the two lower parts of the ovipositor.

valvillus a hinged, flap-like protuberance from the egg canal wall of a lower ovipositor valve that may be involved in controlling egg and/or venom movement into the host.

valvula any of the ovipositor valves or the ovipositor sheath (*see* Table 3.1).

vannal cell/vein an obsolete name for the anal cell/vein, hence some authors use abbreviations v and cu-v.

vertex postero-dorsal part of head, behind and to some extent lateral to the ocellar triangle and bordered posteriorly by the occiput, from which it may be separated by an occipital carina.

vesiculate of a parasitoid larva, usually an early instar one, having an anal vesicle (q.v.)

REFERENCES

Abbott C (1934) How *Megarhyssa* deposits her eggs. *Journal of the New York Entomological Society*, **42**, 127–33 .

Abe Y, Nishimura T, Maetô K (2013) Causes of polymorphic melanism and its thermoregulatory function in a parasitoid wasp *Meteorus pulchricornis* (Hymenoptera: Braconidae). *European Journal of Entomology*, **110**, 627–32.

Abdel-Malek A (1947) A study of the biology of *Chelonella* Sulcata Nees (Hymenoptera: Braconidae). *Ohio Journal of Science*, **47**, 206–16.

Ables JR, Vinson SB (1981) Regulation of host larval development by the egg-larval endoparasitoid *Chelonus insularis* [Hym.: Braconidae]. *Entomophaga*, **26**, 453–8.

Achtelig M, Krause G (1971) Experimente am ungefurchten Ei von *Pimpla turionellae* L. (Hymenoptera) zur Funktionsanalyse des Oosombereichs. *Wilhelm Roux Archiv fur Entwicklungsmechanik der Organismen*, **167**, 164–82.

Adams CH, Cross WH, Mitchell HC (1969) Biology of *Bracon mellitor*, a parasite of the boll weevil. *Journal of Economic Entomology*, **62**, 889–96.

Aeschlimann J-P (1974a) Biologie et comportement de *Chorinaeus funebris* Gravenhorst (Hymenoptera: Ichneumonidae). *Annales de Zoologie Ecologie Animale*, **6**, 529–38.

Aeschlimann J-P (1974b) Hibernation chez trois espèces de métopiines: Hymenoptera, Ichneumonidae. *Entomologia Experimentalis et Applicata*, **17**, 487–92.

Aeschlimann J-P (1975) Biologie, comportement et lacher experimental de *Triclistus pygmaeus* Cresson (Hym., Ichn.). *Mittleilungen der Schweizerischen Anstalt für das forstliche Versuchswesen*, **48**, 165–71.

Aeschlimann J-P (1990) Simultaneous occurrence of thelytoky and bisexuality in hymenopteran species, and its importance for the biological control of pests. *Entomophaga*, **35**, 3–5.

Aguiar AP (2012) A technique to dry mount Hymenoptera (Hexapoda) from alcohol in a few seconds, and its application to other insect orders. *Zootaxa*, **3412**, 53–61.

Aguiar AP, Santos BF (2010) Discovery of potent, unsuspected sampling disparities for Malaise and Möricke traps, as shown for Neotropical Cryptini (Hymenoptera, Ichneumonidae). *Journal of Insect Conservation*, **14**, 199–206.

Aguirre H, Shaw SR, Berry JA, de Sassi C (2014) Description and natural history of the first micropterous *Meteorus* species: *M. orocrambivorus* sp. n. (Hymenoptera, Braconidae, Euphorinae), endemic to New Zealand. *Journal of Hymenoptera Research*, in press.

Ahmad M (1944) The morphology and biology of *Rhogas testaceus* (Spin.) a braconid internal parasite of spotted bollworms of cotton. *Indian Journal of Entomology*, **5**, 189–205.

Ahmed T, Zhang T-T, Wang Z-Y, He K-L, Bai S-X (2013a) Morphology and ultrastructure of antennal sensilla of *Macrocentrus cingulum* Brischke (Hymenoptera: Braconidae) and their probable functions. *Micron*, **50**, 35–43.

Ahmed T, Zhang T-T, He K-L, Bai S-X, Wang Z-Y (2013b) Sense organs on the ovipositor of *Macrocentrus cingulum* Brischke (Hymenoptera: Braconidae): their probable role in stinging, oviposition and host selection process. *Journal of Asia-Pacific Entomology*, **16**, 343–8.

Akdura N, Aksoylar Y, Karaca I, Özgökçe MS (2012) Fecundity and life table parameters of *Agrothereutes hospes* Tschek (Hymenoptera: Ichneumonidae) on *Galleria mellonella* (L.) (Lepidoptera, Pyralidae). *Türkiye Biyolojik Mücadele Derneği*, **3**, 3–12.

Aktümsek A (1996) Effect of host and sex on fatty acid composition in the parasitoid *Itoplectis maculator* F. (Hymenoptera: Ichneumonidae). *Turkish Journal of Zoology, Supplement*, **20**, 7–10.

Alam SM (1952) A contribution to the biology of *Stenobracon deesae* Cameron (Braconidae, Hymenoptera) and the anatomy of its pre-imaginal stages. *Zeitschrift für Parasitenkunde*, **15**, 159–82.

Alam SM (1954) A preliminary survey of the alimentary canal in *Stenobracon deesae* Cameron (Braconidae, Hymenoptera). *Indian Journal of Entomology*, **16**, 55–66.

The Braconid and Ichneumonid Parasitoid Wasps: Biology, Systematics, Evolution and Ecology, First Edition. Donald L. J. Quicke.

Alam SM (1955) The skeleto-muscular mechanism of *Stenobracon deesae* Cameron (Braconidae, Hymenoptera): abdomen and internal anatomy. *Aligarh Muslim University Publications, Zoology Series*, 1–75.

Alauzet C (1987) Bioecologie de *Eubazus semirugosus*, *Coeloides abdominalis* et *C. sordidator* (Hym.: Braconidae) parasites de *Pissodes notatus* (Col.: Curculionidae) dans le sud de la France. *Entomophaga*, **32**, 39–47.

Albrecht U, Wyler T, Pfister-Wilhelm R, Gruber A, Stettler P, Heiniger P, Kurt E, Schümperli D, Lanzrein B (1994) Polydnavirus of the parasitic wasp *Chelonus inanitus* (Braconidae): characterization, genome organization and time point of replication. *Journal of General Virology*, **75**, 3353–63.

Aldrich JM (1912) Notes on *Theronia fulvescens*. *Journal of Economic Entomology*, **5**, 87–8.

Alford DV (1968) The biology and immature stages of *Syntretus splendidus* (Marshall) (Hymenoptera: Braconidae, Euphorinae), a parasite of adult bumblebees. *Transactions of the Royal Entomological Society of London*, **120**, 375–93.

Allen AA (1977) Exhibits. *Proceedings of the British Entomological and Natural History Society*, 111–12.

Allen HW, Brunson MH (1945) A microsporidian in *Macrocentrus ancylivorus*. *Journal of Economic Entomology*, **38**, 393.

Allen WW (1952) The biology of *Apanteles medicaginis* Muesebeck (Hymenoptera: Braconidae) and factors influencing its efficiency. PhD thesis, University of California, 210 pp.

Almohamad R, Hance T (2014) Encounters with aphid predators or their residues impede searching and oviposition by the aphid parasitoid *Aphidius ervi* (Hymenoptera: Aphidiinae). *Insect Science*, **21**, 181–8.

Al-Saffar ZY, Aldrich JC (1998) *Pontania proxima* (Tenthredinidae: Hymenoptera): natural enemies and defensive behavior against *Pnigalio nemati* (Eulophidae: Hymenoptera). *Annals of the Entomological Society of America*, **91**, 858–62.

Altahtawy MM, El-Sawaf SM, Shalaby FF (1972) Studies on morphology and development of the immature stages of *Microplitis rufiventris* Kokujev (Hym., Bracon.). *Zeitschrift für Angewandte Entomologie*, **71**, 134–40.

Althoff DM (2003) Does parasitoid attack strategy influence host specificity? A test with New World braconids. *Ecological Entomology*, **28**, 500–2.

Altson AM (1920) The life-history and habits of two parasites of blowflies. *Proceedings of the Zoological Society of London*, **17**, 195–243.

Aluja M, Ovruski SM, Sivinski J, Córdova-García G, Schliserman P, Nuñez-Campero SR, Ordano M (2013) Inter-specific competition and competition-free space in the tephritid parasitoids *Utetes anastrephae* and *Doryctobracon areolatus* (Hymenoptera: Braconidae: Opiinae). *Ecological Entomology*, **38**, 485–96.

Alvi SM, Momoi S (1994) Environmental regulation and geographical adaptation of diapause in *Cotesia plutellae* (Hymenoptera: Braconidae), a parasitoid of the diamondback moth larvae. *Applied Entomology and Zoology*, **29**, 89–95.

Amos WB, Salt G (1974) An atlas of the development of eggs of an ichneumon wasp. *Journal of Entomology* (B), **43**, 11–18.

Amy RL (1961) The embryology of *Habrobracon juglandis* (Ashmead). *Journal of Morphology*, **109**, 199–217.

Andersen JC, Mills NJ (2012) DNA extraction from museum specimens of parasitic Hymenoptera. *PLoS ONE*, **7** (10), e45549.

Andreadis SS, Spanoudis CG, Athanassiou CG, Savopoulou-Soultani M (2014) Factors influencing supercooling capacity of the koinobiont endoparasitoid *Venturia canescens* (Hymenoptera: Ichneumonidae). *Pest Management Science*, **70**, 814–18.

Andreadis TG (1980) *Nosema pyrausta* infection in *Macrocentrus grandii*, a braconid parasite of the European corn borer, *Ostrinia nubilalis*. *Journal of Invertebrate Pathology*, **35**, 229–233.

Andreadis TG (1982) Impact of *Nosema pyrausta* on field populations of *Macrocentrus grandii*, an introduced parasite of the European corn borer, *Ostrinia nubilalis*. *Journal of Invertebrate Pathology*, **39**, 298–302.

Angalet GW, Fuester R (1977) The aphidius parasites of the pea aphid *Acyrthosiphon pisum* in the eastern half of the USA. *Annals of the Entomological Society of America*, **70**, 87–96.

Anisimova M (1931) Biology of *Phytodietus segmentator* Gr., parasite of the prairie moth. In *Lugovoi Motylek v, 1929–1930 gg*, Kiev, pp. 161–70.

Antolin MF, Strand MR (1992) Mating system of *Bracon hebetor* (Hymenoptera: Braconidae). *Ecological Entomology*, **17**, 1–7.

Antolin MF, Ode PJ, Strand MR (1995) Variable sex ratios and ovicide in an outbreeding parasitic wasp. *Animal Behaviour*, **49**, 589–600.

Antolin MF, Bosio CF, Cotton J, Sweeney W, Strand MR, Black WC IV, (1996) Intensive linkage mapping in a wasp (*Bracon hebetor*) and a mosquito (*Aedes aegypti*) with single-strand conformation polymorphism analysis of random amplified polymorphic DNA markers. *Genetics*, **143**, 1727–38.

Antropov AV, Belokobylskij SA, Compton SG, Dlussky GM, Khalaim AI, Kolyada VA, Kozlov MA, Perfilieva KS, Rasnitsyn AP (2014) The wasps, bees and ants (Insecta: Vespida = Hymenoptera) from the Insect Limestone (Late Eocene) of the Isle of Wight, UK. *Earth and Environmental Science Transactions of the Royal Society of Edinburgh*, **104**, 335–446.

Aoyagi M, Ishii M (1991) Host acceptance behavior of the Japanese aquatic wasp *Agriotypus gracilis* (Hymenoptera: Ichneumonidae) toward the caddisfly host *Geora japonica* (Trichoptera: Limnephilidae). *Journal of Ethology*, **9**, 113–19.

Arnaud PH (1952) Swarming of *Blacus* sp. in California (Hymenoptera: Braconidae). *Entomological News*, **63**, 149–50.

Ardila-Garcia AM, Umphrey GJ, Gregory TR (2010) An expansion of the genome size dataset for the insect order Hymenoptera, with a first test of parasitism and eusociality as possible constraints. *Insect Molecular Biology*, **19**, 337–46.

Areekul Butcher B, Quicke DLJ (2011) Revision of *Aleiodes* (*Hemigyroneuron*) parasitic wasps (Hymenoptera: Braconidae: Rogadinae) with reappraisal of subgeneric limits, descriptions of new species and phylogenetic analysis. *Journal of Natural History*, **45**, 1403–76.

Areekul B, Quicke DLJ (2006) Systematics of the parasitic wasp genus *Yelicones* Cameron (Hymenoptera: Braconidae: Rogadinae) and revision of the genus from North, Central and South America. *Systematics & Biodiversity*, **4**, 255–376.

Aron S, de Menten L, Van Bockstaele DR, Blank SM, Roisin Y (2005) When hymenopteran males reinvented diploidy. *Current Biology*, **15**, 824–7.

Arthur AP (1965) Absence of preimaginal conditioning in *Itoplectis conquisitor* (Say) (Hymenoptera: Ichneumonidae). *The Canadian Entomologist*, **97**, 1000–1.

Arthur AP (1971) Associative learning by *Nemeritis canescens* (Hymenoptera: Ichneumonidae). *The Canadian Entomologist*, **103**, 1137–41.

Arthur AP, Ewen AB (1975) Cuticular encystment: a unique and effective defense reaction by cabbage looper larvae against parasitism by *Banchus flavescens* (Hymenoptera: Ichneumonidae). *Annals of the Entomological Society of America*, **68**, 1091–4.

Arthur AP, Mason PG (1985) The life history and immature stages of *Banchus flavescens* (Hymenoptera: Ichneumonidae), a parasitoid of the Bertha armyworm *Mamestra configurata* (Lepidoptera: Noctuidae) in western Canada. *The Canadian Entomologist*, **117**, 1249–55.

Arthur AP, Mason PG (1986) Life history and immature stages of the parasitoid *Microplitis mediator* (Hymenoptera: Braconidae), reared on the Bertha armyworm *Mamestra configurata* (Lepidoptera: Noctuidae). *The Canadian Entomologist*, **118**, 487–91.

Arthur AP, Stainer JER, Turnbull AL (1964) The interaction between *Orgilus obscurator* (Nees) (Hymenoptera: Braconidae) and *Temelucha interruptor* (Grav.) (Hymenoptera: Ichneumonidae), parasites of the pine shoot moth, *Rhyacionia buoliana* (Schiff.) (Lepidoptera: Olethreutidae). *The Canadian Entomologist*, **96**, 1030–4.

Arthur AP, Hegdekar BM, Batsch WW (1972) A chemically defined, synthetic medium that induces oviposition in the parasite *Itoplectis conquisitor* (Hymenoptera: Ichneumonidae). *The Canadian Entomologist*, **104**, 1251–8.

Asgari S (2006) Venom proteins from polydnavirus-producing endoparasitoids: their role in host–parasite interactions. *Archives of Insect Biochemistry and Physiology*, **61**, 146–56.

Asgari S, Rivers DB (2011) Venom proteins from endoparasitoid wasps and their role in host–parasite interactions. *Annual Review of Entomology*, **56**, 313–35.

Asgari S, Schmidt O (1994) Passive protection of eggs from the parasitoid, *Cotesia rubecula*, in the host, *Pieris rapae*. *Journal of Insect Physiology*, **40**, 789–95.

Asgari S, Zareie S, Zhang G, Schmidt O (2003a) Isolation and characterization of a novel venom protein from an endoparasitoid, *Cotesia rubecula* (Hymenoptera: Braconidae). *Archives of Insect Biochemistry and Physiology*, **53**, 92–100.

Asgari S, Zhang G, Zareie R, Schmidt O (2003b) A serine proteinase homolog venom protein from an endoparasitoid wasp inhibits melanization of the host hemolymph. *Insect Biochemistry and Molecular Biology*, **33**, 1017–24.

Ashley TR (1983) Growth pattern alterations in fall armyworm, *Spodoptera frugiperda*, larvae after parasitization by *Apanteles marginiventris*, *Campoletis grioti*, *Chelonus insularis*, and *Eiphosoma vitticole*. *Florida Entomologist*, **66**, 260–6.

Ashley Adams S, Morse DH (2014) Condition-dependent mate choice of a parasitoid wasp in the field. *Animal Behaviour*, **88**, 225–32.

Askari A, Coppel HC (1978) Observations on courtship and mating behavior of *Meteorus pulchricornis*, a gypsy moth parasitoid. *Annals of the Entomological Society of America*, **71**, 364–6.

Askari A, Mertins JW, Coppel HC (1977) Developmental biology and immature stages of *Meteorus pulchricornis* in the laboratory. *Annals of the Entomological Society of America*, **70**, 655–9.

Askew RR (1971) *Parasitic Insects*, Heinemann Educational Books, London, 316 pp.

Askew RR, Shaw MR (1986) Parasitoid communities: their size, structure and development. In J Waage and D Greathead (eds), *Insect Parasitoids*, Academic Press, London, pp. 225–64.

Asplen MK, Whitfield JB, de Boer JG, Heimpel GE (2009) Ancestral state reconstruction analysis of hymenopteran sex determination mechanisms. *Journal of Evolutionary Biology*, **22**, 1762–9.

Asplen MK, Wyckhuys KAG, Heimpel GE (2011) Parasitism of autumnal morphs of the soybean aphid (Hemiptera: Aphididae) by *Binodoxys communis* (Hymenoptera: Braconidae) on buckthorn. *Annals of the Entomological Society of America*, **104**, 935–44.

Aspöck H (2002) The biology of Rhaphidioptera: a review of present knowledge. *Acta Zoologica Academiae Scientiarum Hungaricae*, **48** (Suppl. 2), 35–50.

Aubert JF (1959) Biologie de quelques Ichneumonidae Pimplinae et examen critique de la théorie de Dzierzon. *Entomophaga*, **4**, 75–188.

Aubert JF (1982) Sex determination in ichneumonid flies Pimplinae. *Annales des Sciences Naturelles. Zoologie et Biologie Animale*, **4**, 33–40.

Aubert JF, Jourdheuil P (1959) Nouvelle description et biologie de quelques Ichneumonides appartenant aux genres *Aneuclis* Först., *Isurgus* Först. et *Thersilochus* Holm. *Revue de Pathologie Vègètale et de Entomologie Agricole de France*, **37**, 175–93.

Aubrook EW (1939) A contribution to the biology and distribution in Great Britain of *Boreus hyemalis* L. (Mecopt., Boreidae). *Journal of the Society for British Entomology*, **2**, 13–21.

Auger J, Lecomte C, Paris J, Thibout E (1989) Identification of leek-moth and diamondback-moth frass volatiles that stimulate parasitoid, *Diadromus pulchellus*. *Journal of Chemical Ecology*, **15**, 1391–8.

Austin AD (1989) Revision of the genus *Buluka* de Saeger (Hymenoptera: Braconidae: Microgastrinae). *Systematic Entomology*, **14**, 149–63.

Austin AD, Browning TO (1981) A mechanism for the movement of eggs along insect ovipositors. *International Journal of Insect Morphology and Embryology*, **10**, 93–108.

Austin AD, Dangerfield PC (1992) Synopsis of Australasian Microgastrinae (Hymenoptera: Braconidae), with a key to genera and description of new taxa. *Invertebrate Taxonomy*, **6**, 1–76.

Austin AD, Dangerfield PC (1998) Biology of the *Mesostoa kerri* Austin and Wharton (Insecta: Hymenoptera: Braconidae: Mesostoinae), an endemic Australian wasp that causes stem galls on *Banksia marginata* Cav. *Australian Journal of Botany*, **46**, 559–69.

Austin AD, Faulds W (1989) Two new Australian species of *Bracon* F. (Hymenoptera: Braconidae) parasitic on *Phylacteophaga* spp. (Hymenoptera: Pergidae). *Journal of Australian Entomological Society*, **28**, 207–13.

Austin AD, Wharton RA (1992) New records of subfamilies, tribes and genera of Braconidae (Insecta: Hymenoptera) from Australia, with descriptions of seven new species. *Transactions of the Royal Society of South Australia*, **16**, 41–63.

Austin AD, Wharton RA, Dangerfield PC (1993a) Revision of the endemic Australian subfamily Trachypetinae Schulz s.l. (including Cercobarconinae Tobias) (Hymenoptera: Braconidae). *Systematic Entomology*, **18**, 97–119.

Austin AD, Dangerfield PC, Adams M (1993b) Differentiation of the Australian species in the genus *Microplitis* Foerster (Hymenoptera: Braconidae) and the *Microplitis demolitor* complex, using comparative morphology and electrophoretic data. In SA Corey, DJ Dall and WM Milne (eds), *Pest Control and Sustainable Agriculture*, CSIRO, Melbourne, pp. 219–22.

Austin AD, Quicke DLJ, Marsh PM (1994) Systematics of the hymenopteran parasitoids of the eucalypt longicorns, *Phoracantha* spp. (Coleoptera: Cerambycidae) in Australia, with a description of a new doryctine genus (Hymenoptera: Braconidae). *Bulletin of Entomological Research*, **84**, 145–74.

Avilla J, Albajes R (1984) The influence of female age and host size on the sex ratio of the parasitoid *Opius concolor*. *Entomologia Exeperimentalis et Applicata*, **35**, 43–7.

Ayyar PNK (1943) Biology of a new ichneumonid parasite of the *Amaranthus* stem weevil of south India. *Journal of the Straits Branch of the Royal Asiatic Society*, **48**, 27–36.

Ayyar PNK, Narayanaswami PS (1940) On the biology of *Spathius vulnificus* Wilk. a possible effective parasite of *Pempheres affinis* in South India. *Indian Journal of Entomology*, **2**, 79–86.

Aziz A, Fitton MG, Quicke DLJ (2000) Key to the *Diadegma* species (Hymenoptera: Ichneumonidae) parasitising diamondback moth, *Plutella xylostella*, with the description of a new species. *Bulletin of Entomological Research*, **90**, 375–89.

Azzouz H, Giordanengo P, Wäckers FL, Kaiser L (2004) Effects of feeding frequency and sugar concentration on behavior and longevity of the adult aphid parasitoid: *Aphidius ervi* (Haliday) (Hymenoptera: Braconidae). *Biological Control*, **31**, 445–52.

Babendreier D (2000) Life history of *Aptesis nigrocincta* (Hymenoptera: Ichneumonidae) a cocoon parasitoid of the apple sawfly, *Hoplocampa testudinea* (Hymenoptera: Tenthredinidae). *Bulletin of Entomological Research*, **90**, 291–7.

Babendreier D, Hoffmeister TS (2002) Superparasitism in the solitary ectoparasitoid *Aptesis nigrocincta*: the influence of egg load and host encounter rate. *Entomologia Experimentalis et Applicata*, **105**, 63–9.

Babendreier D, Hoffmeister TS (2003) Facultative hyperparasitism by the potential biological control agent *Aptesis nigrocincta* (Hymenoptera: Ichneumonidae). *European Journal of Entomology*, **100**, 205–7.

Baccetti B (1958) Ghiandole labiali e fabbricazione del bozzolo negli Imenotteri. Ricerche comparative su Icneumonidi e Braconidi. *Redia*, **43**, 215–94.

Baer CF, Tripp DW, Bjorksten A, Antolin MF (2004) Phylogeography of a parasitoid wasp (*Diaeretiella rapae*): no evidence of host-associated lineages. *Molecular Ecology*, **13**, 1859–69.

Bahena F, Budia F, Adán A, del Estal P, Viñuela E (1999) Scanning electron microscopy of *Hyposoter didymator* (Hymenoptera: Ichneumonidae) in host *Mythimna umbrigera* (Lepidoptera: Noctuidae) larvae. *Annals of the Entomological Society of America*, **92**, 144–52.

Bai B (1991) Conspecific superparasitism in two parasitoid wasps, *Aphidius ervi* Haliday and *Aphelinus asychis* Walker: reproductive strategies influence host discrimination. *The Canadian Entomologist*, **123**, 1229–37.

Baker BR (1992) Emergence and pairing of *Rhyssa persuasoria* L. (Hymenoptera: Ichneumonidae). *British Journal of Entomology and Natural History*, **5**, 184–5.

Baker JE, Fabrick JA (2000) Host hemolymph proteins and protein digestion in larval *Habrobracon hebetor* (Hymenoptera: Braconidae). *Insect Biochemistry and Molecular Biology*, **30**, 937–46.

Baker JE, Fabrick JA (2002) Unusual responses of Indian meal moth larvae (Lepidoptera: Pyralidae) to envenomation and parasitization by a braconid ectoparasitoid. *Journal of Entomological Science*, **37**, 370–4.

Baker JE, Howard R, Morrill W, Meers S, Weaver D (2005) Acetate esters of saturated and unsaturated alcohols (C12–C20) are major components in Dufour glands of *Bracon cephi* and *Bracon lissogaster* (Hymenoptera: Braconidae), parasitoids of the wheat stem sawfly, *Cephus cinctus* (Hymenoptera: Cephidae). *Biochemical Systematics and Ecology*, **33**, 757–69.

Baker WA, Bradley WG, Clark CA (1949) Biological control of the European corn borer in the United States. *USDA Technology Bulletin*, **983**, 185 pp.

Balduf WV (1926) The bionomics of *Dinocampus coccinellae* Schrank. *Annals of the Entomological Society of America*, **19**, 465–98.

Balduf WV (1966) Life of *Acrobasis rubrifasciella* (Lepidoptera: Phycitidae), its main parasite, *Agathis calcarata* (Hymenoptera: Braconidae), and three hyperparasites (Chalcidoidea). *Annals of the Entomological Society of America*, **59**, 1038–49.

Banks JC Whitfield JB (2006) Dissecting the ancient rapid radiation of microgastrine wasp genera using additional nuclear genes. *Molecular Phylogenetics and Evolution*, **41**, 690–703.

Barari H, Ferguson AW, Piper RW, Smith E, Quicke DLJ, Williams IH (2005) The separation of two hymenopteran parasitoids, *Tersilochus obscurator* and *Tersilochus microgaster* (Ichneumonidae), of stem-mining pests of winter oilseed rape using DNA, morphometric and ecological data. *Bulletin of Entomological Research*, **95**, 299–307.

Barbalho SM, Penteado-Dias AM (1997) Análise morfológica do aparelho de veneno nos Braconidae cyclostome (Hymenoptera). *Revista Brasileira de Zoologia*, **14**, 65–71.

Barbarossa IT, Muroni P, Dardani M, Casula P, Angioy AM (1998) New insight into the antennal chemosensory function of *Opius concolor* (Hymenoptera, Braconidae). *Italian Journal of Zoology*, **65**, 367–70.

Barlin MR, Vinson SB (1981) The multiporous plate sensillum and its potential use in braconid systematics (Hymenoptera: Braconidae). *The Canadian Entomologist*, **113**, 931–8.

Barlow ND, Beggs JR, Moller H (1998) Spread of the wasp parasitoid *Sphecophaga vesparum vesparum* following its release in New Zealand. *New Zealand Journal of Ecology*, **22**, 205–8.

Barnard PC (1999) *Identifying British Insects and Arachnids: an Annotated Catalogue of Key Works*. Natural History Museum, London.

Barrantes G, Triana E, Shaw SR, Jones GZ (2011) Characteristics of the cocoon and natural history of the gregarious *Meteorus restionis* sp. n. (Hymenoptera, Braconidae, Meteorinae) from Costa Rica. *Journal of Hymenoptera Research*, **20**, 9–21.

Barratt BIP, Sutherland M (2001) Development of teratocytes associated with *Microctonus aethiopoides* Loan (Hymenoptera: Braconidae) in natural and novel host species. *Journal of Insect Physiology*, **47**, 257–62.

Barratt BIP, Evans AA, Stoltz DB, Vinson SB, Easingwood R (1999) Virus-like particles in the ovaries of *Microctonus aethiopoides* Loan (Hymenoptera: Braconidae), a parasitoid of adult weevils (Coleoptera: Curculionidae). *Journal of Invertebrate Pathology*, **73**, 182–18.

Barratt BIP, Murney R, Easingwood R, Ward VK (2006) Virus-like particles in the ovaries of *Microctonus aethiopoides* Loan (Hymenoptera: Braconidae): comparison of biotypes from Morocco and Europe. *Journal of Invertebrate Pathology*, **91**, 13–18.

Barratt BIP, Ferguson CM, Bixley AS, Crook KE, Barton DM, Johnstone PD (2007) Field parasitism of nontarget weevil species (Coleoptera: Curculionidae) by the introduced biological control agent *Microctonus aethiopoides* Loan (Hymenoptera: Braconidae) over an altitude gradient. *Environmental Entomology*, **36**, 826–39.

Barron JR (1976) Systematics of Nearctic *Euceros*. *Naturaliste Canadien*, **103**, 285–375.

Barron JR (1978) Systematics of world Eucerotinae. Part II: Non-Nearctic species. *Naturaliste Canadien*, **105**, 327–74.

Barron JR (1981) The Nearctic species of *Ctenopelma* (Hymenoptera, Ctenopelmatinae). *Naturaliste Canadien*, **108**, 17–56.

Barron JR (1994) The Nearctic species of *Lathrolestes* (Hymenoptera, Ichneumonidae, Ctenopelmatinae). *Contributions of the American Entomological Institute*, **28**, 1–135.

Barron JR, Bisdee HE (1977) The final instars if *Leptocampoplex cremastoides* and *Macrus parvulus* (Hymenoptera: Ichneumonidae: Campopleginae), and notes on the classification of the genera. *Annals of the Entomological Society of America*, **70**, 290–2.

Barron JR, Walley GS (1983) Revision of the Holarctic genus *Pyracmon* (Hymenoptera: Ichneumonidae). *The Canadian Entomologist*, **115**, 227–41.

Barrows EM (1976) Sexual behavior In *Hyposoter fugitivus* (Hymenoptera Ichneumonidae). *Entomological News*, **87**, 101–2.

Bartell DP, Pass BC (1978) Morphology, development, and behavior of the immature stages of the parasite *Bathyplectes curculionis*. *Annals of the Entomological Society of America*, **71**, 23–30.

Bartell DP, Pass BC (1980) Morphology, development, and behavior of the immature stages of the parasite *Bathyplectes anurus* (Hymenoptera: Ichneumonidae). *The Canadian Entomologist*, **122**, 481–7.

Bartlett R, Pickering J, Gauld I, Windsor D (1999) Estimating global biodiversity: tropical beetles and wasps send different signals. *Ecological Entomology*, **24**, 118–21.

Basibuyuk HH, Quicke DLJ (1994) Evolution of antennal cleaner structure in the Hymenoptera (Insecta). *Norwegian Journal of Agricultural Science, Supplement*, **16**, 199–206.

Basibuyuk HH, Quicke DLJ (1995) Morphology of the antenna cleaner in the Hymenoptera with particular reference to non-aculeate families (Insecta). *Zoologica Scripta*, **24**, 157–77.

Basibuyuk HH, Quicke DLJ (1997) Hamuli in the Hymenoptera (Insecta) and their phylogenetic implications. *Journal of Natural History*, **31**, 1563–85.

Basibuyuk HH, Quicke DLJ (1999a) Gross morphology of multiporous plate sensilla in the Hymenoptera (Insecta). *Zoologica Scripta*, **28**, 51–67.

Basibuyuk HH, Quicke DLJ (1999b) Grooming behaviours in the Hymenoptera (Insecta): potential phylogenetic significance. *Zoological Journal of the Linnean Society*, **125**, 349–82.

Basibuyuk HH, Rasnitsyn AP, van Achterberg C, Fitton MG, Quicke DLJ (1999) A new Cretaceous fossil braconid and a reinterpretation of hind wing venation in the family (Hymenoptera: Braconidae). *Zoologica Scripta*, **28**, 211–14.

Basinger AJ (1938) The orange tortrix, *Argyrotaenia citrana*. *Hilgardia*, **11**, 635–69.

Basio NA, Kim Y ((2004)) (2005) A short review of teratocytes and their characters in *Cotesia plutellae* (Braconidae: Hymenoptera). *Journal of Asia-Pacific Entomology*, **8**, 211–17.

Battaglia D, Pennacchio F, Romano A, Tranfaglia A (1995) The role of physical cues in the regulation of host recognition and acceptance behavior of *Aphidius ervi* Haliday (Hymenoptera, Braconidae). *Journal of Insect Behavior*, **8**, 739–50.

Battaglia D, Isidoro N, Romani R, Bin F, Pennacchio F (2002) Mating behaviour of *Aphidius ervi* (Hymenoptera: Braconidae): the role of antennae. *European Journal of Entomology*, **99**, 451–6.

Bauer G (1985) Strategies in parasitoid associations utilizing the same developmental stages of 2 rose tortricids. *Entomologia Experimentalis et Applicata*, **37**, 275–81.

Bauerle P, Rutherford P, Lanfranco D (1997) Defoliadores de roble (*Nothofagus obliqua*), raulí (*N. alpina*), coigüe (*N. dombeyi*) y lenga (*N. pumilio*). *Bosque*, **18**, 97–107.

Baur ME, Yeargan KV (1994a) Developmental stages and kairomones from the primary parasitoid *Cotesia marginiventris* (Hymenoptera: Braconidae) affect the response of the hyperparasitoid *Mesochorus discitergus* (Hymenoptera: Ichneumonidae) to parasitised caterpillars. *Annals of the Entomological Society of America*, **87**, 954–61.

Bautista RC, Mochizuki N, Spencer JR, Harris EJ, Ichimura DM (1999) Mass-rearing of the tephritid fruit fly parasitoid *Fopius arisanus* (Hymenoptera: Braconidae). *Biological Control*, **15**, 137–44.

Beard RL (1952) The toxicity of *Habrobracon* venom: a study of a natural insecticide. *Connecticut Experimental Station Bulletin*, **562**, 3–27.

Beaver RA (1972) Ecological studies on Diptera breeding in dead snails. 1. Biology of the species found in *Cepaea nemoralis* (L.). *Entomologist*, **105**, 41–52.

Beck M, Siekmann G, Theopold U, Schmidt O, Li DM (1999) A maternal gene mutation correlates with an ovary phenotype in a parthenogenetic wasp population. *Insect Biochemistry and Molecular Biology*, **29**, 453–60.

Beck M, Theopold U, Schmidt O (2000a) Evidence for serine protease inhibitor activity in the ovarian calyx fluid of the endoparasitoid *Venturia canescens*. *Journal of Insect Physiology*, **46**, 1275–83.

Beck M, Theopold U, Schmidt O, Austin AD, Dowton M (2000b) Two genetically distinct *Venturia canescens* strains display different reproductive strategies. In AD. Austin and M Dowton (eds), *Hymenoptera: Evolution, Biodiversity and Biological Control*, CSIRO, Canberra, pp. 38–45.

Beck M, Reineke A, Lorenz H, Theopold U, Schmidt O (2001) Two distinct reproductive strategies are correlated with an ovarian phenotype in co-existing parthenogenetic strains of a parasitic wasp. *Journal of Insect Physiology*, **47**, 1189–95.

Beckage NE, de Buron I (1994) Extraembryonic membranes of the endoparasitic wasp *Cotesia congregata*: presence of a separate amnion and serosa. *Journal of Parasitology*, **80**, 389–96.

Beckage NE, Drezen J-M (eds) (2012) *Parasitoid Viruses. Symbionts and Pathogens*, Academic Press, London.

Beckage NE, Riddiford LM (1982) Effects of parasitism by *Apanteles congregatus* on the endocrine physiology of the tobacco hornworm *Manduca sexta*. *General and Comparative Endocrinology*, **47**, 308–22.

Beckage NE, Metcalf JS, Nesbit DJ, Schleifer KW, Zetlan SR, de Buron I (1990) Host hemolymph monophenoloxidase activity in parasitised *Manduca sexta* larvae and evidence for inhibition by wasp polydnavirus. *Insect Biochemistry*, **20**, 285–94.

Beckage NE, Tan FF, Schleifer KW, Lane RD, Cherubin LL (1994) Characterization and biological effects of *Cotesia congregata* polydnavirus on host larvae of the tobacco hornworm, *Manduca sexta*. *Archives of Insect Biochemistry and Physiology*, **26**, 165–95.

Beeson CFC, Chatterjee SN (1935) On the biology of the Braconidae (Hymenoptera). *Indian Forest Records*, **1**, 105–38.

Beirne BP (1941) A consideration of the cephalic structures and spiracles of the final instar larvae of the Ichneumonidae. *Transactions of the Society for British Entomology*, **7**, 123–90.

Beirne BP (1942) Obervations on the life-history of *Praon volucre* Haliday (Hym.: Braconidae), a parasite of the mealy plum aphis (*Hyalopterus arundinis* Fab.). *Proceedings of the Royal Entomological Society of London (B)*, **17**, 42–47.

Beirne BP (1946). Notes on the biology of some hymenopterous parasites of the beech weevil (*Rhynchaenus fagi* L.) (Col.). *Proceedings of the Royal Entomological Society of London (A)*, **21**, 7–11.

Bekkaoui A, Thibout E (1993) Role of the cocoon of Acro*lepiopsis assectella* (Lep., Hyponomeutoidae) in host recognition by the parasitoid *Diadromus pulchellus* (Hym., Ichneumonidae). *Entomophaga*, **38**, 101–13.

Béliveau C, Laforge M, Cusson M, Bellemare G (2000) Expression of a *Tranosema rostrale* polydnavirus gene in the spruce budworm, *Choristoneura fumiferana*. *Journal of General Virology*, **81**, 1871–80.

Bell HA, Marris GC, Bell J, Edwards JP (2000) The biology of *Meteorus gyrator* (Hymenoptera: Braconidae), a solitary endoparasitoid of the tomato moth, *Lacanobia oleracea* (Lepidoptera: Noctuidae). *Bulletin of Entomological Research*, **90**, 299–308.

Bell HA, Kirkbride-Smith AE, Marris GC, Edwards JP (2004) Teratocytes of the solitary endoparasitoid *Meteorus gyrator* (Hymenoptera: Braconidae): morphology, numbers and possible functions. *Physiological Entomology*, **29**, 335–43.

Belokobylskij SA (1984) O razdelenii triby Exothecini s.l. (Hymenoptera, Braconidae) na dve s opisaniem novogo roda i podroda. *Zoologicheski Zhurnal*, **63**, 1019–26.

Belokobylskij SA (1993) On the classification and phylogeny of the braconid wasps subfamilies Doryctinae and Exothecinae (Hymenoptera, Braconidae). Part I. On the classification, 2. *Entomologicheskogo Obozrenie*, **71**, 900–98 (in Russian).

Belokobylskij SA (1998) Subfamily Doryctinae. In PA Lehr (ed.) *Key to Insects of the Russian Far East. IV. Neuropteroidea, Mecoptera, Hymenoptera. Part 4*, Dal'nauka, Vladivostok, pp. 50–109.

Belokobylskij SA (1999) New genera of the subfamilies Rhyssalinae, Exothecinae and Gnamptodontinae from the Old World (Hymenoptera: Braconidae). *Zoosystematica Rossica*, **8**, 155–69.

Belokobylskij SA (2002) Two new Oriental genera of Doryctinae (Hymenoptera, Braconidae) from termite nests. *Journal of Natural History*, **36**, 953–62.

Belokobylskij SA (2004) Taxonomic reclassification of the East Asian species of the genus *Oncophanes* Förster (Hymenoptera: Braconidae, Rhyssalinae). *Proceedings of the Russian Entomological Society, St. Petersburg*, **75**, 106–17.

Belokobylskij SA (2009) A new enigmatic apterous cyclostome genus (Hymenoptera: Braconidae) from the Greek Islands. *Annales de la Société Entomologique de France (Nouvelle Série)*, **45**, 77–82.

Belokobylskij SA (2012) Cretaceous braconid wasps from the Magadan Province of Russia. *Acta Palaeontologica Polonica*, **57**, 351–61.

Belokobylskij SA, Austin AD (2013) New species of flightless doryctine parasitoid wasps (Hymenoptera: Braconidae: Doryctinae) from Australia and New Zealand. *Australian Journal of Entomology*, **52**, 338–55.

Belokobylskij SA, Jervis MA (1998) Descriptions of two new species of the genus *Agathis* Latreille (Hymenoptera, Braconidae, Agathidinae) from Spain, with a record of mating by one species on flowers. *Journal of Natural History*, **32**, 1217–25.

Belokobylskij SA, Kula RR (2012) Review of the brachypterous, micropterous, and apterous Braconidae of the cyclostome lineage (Hymenoptera: Ichneumonoidea) from the Palearctic Region. *Zootaxa*, **3240**, 1–66.

Belokobylskij SA, Long KD (2005) The braconid wasps of the subfamily Betylobraconinae (Hymenoptera, Braconidae) in Vietnam. *Deutsche Entomologische Zeitschrift*, **52**, 155–64.

Belokobylskij SA, Maetô K (2006) A new species of the genus *Parachremylus* Granger (Hymenoptera: Braconidae), a parasitoid of *Conopomorpha* Lychee pests (Lepidoptera: Gracillariidae) in Thailand. *Journal of Hymenoptera Research*, **15**, 181–6.

Belokobylskij SA, Maetô K (2009) Doryctinae (Hymenoptera: Braconidae) of Japan. *Fauna Mundi*, **1**, 1–806.

Belokobylskij SA, Quicke DLJ (1999) A new genus and two new species of brachypterous Lysiterminae (Hymenoptera: Braconidae). *Journal of Hymenoptera Research*, **8**, 120–5.

Belokobylskij SA, Tobias VI (2000) Fam. Braconidae. In PA Lehr (ed.) *Key to the Insects of the Russian Far East. IV. Neuropteroidea, Mecoptera, Hymenoptera. Part 4*, Dal'nauka, Vladivostok, pp. 8–571 (in Russian).

Belokobylskij SA, Iqbal M, Austin AD (2003) First record of the subfamily Dirrhopinae (Hymenoptera: Braconidae) from the Australian region, with a discussion of relationships and biology. *Australian Journal of Entomology*, **42**, 260–5.

Belokobylskij SA, Zaldivar-Riverón A, Quicke DLJ (2004) Phylogeny of the parasitic wasp genera of the subfamily Doryctinae (Hymenoptera: Braconidae) from morphological evidence. *Zoological Journal of the Linnean Society*, **142**, 369–404.

Belokobylskij SA, Zaldivar-Riverón A, León-Regagnon V, Quicke DLJ (2007) A new genus of Lysitermini (Hymenoptera: Braconidae: Lysiterminae) from Madagascar and its taxonomic placement based on 28S rDNA sequence data. *Zootaxa*, **1461**, 25–37.

Belokobylskij SA, Zaldivar-Riverón A, Maetô K, Sáez AG (2008) Asian Betylobraconinae (Hymenoptera, Braconidae), with description of a new genus and phylogenetic affinities of the tribe Facitorini. *Insect Systematics and Evolution*, **39**, 133–54.

Belokobylskij SA, Nel A, Waller A, De Plöeg G (2010) New fossil noncyclostome braconid wasps from the lowermost Eocene amber of Paris Basin. *Acta Palaeontologica Polonica*, **55**, 519–27.

Belshaw R, Quicke DLJ (1997) A molecular phylogeny of the Aphidiinae (Hymenoptera: Braconidae). *Molecular Phylogenetics and Evolution*, **7**, 281–93.

Belshaw R, Quicke DLJ (2002) Robustness of ancestral state estimates: evolution of life history strategy in ichneumonoid parasitoids. *Systematic Biology*, **51**, 450–77.

Belshaw R, Quicke DLJ (2003) The cytogenetics of thelytoky in a predominantly asexual parasitoids wasp with covert sex. *Genome*, **46**, 170–3.

Belshaw R, Fitton MG, Herniou E, Gimeno C, Quicke DLJ (1998) A phylogenetic reconstruction of the Ichneumonoidea (Hymenoptera) based on the D2 variable region of 28S ribosomal DNA. *Systematic Entomology*, **23**, 109–23.

Belshaw R, Quicke DLJ, Völkl W, Godfray HCJ (1999) Molecular markers indicate rare sex in a predominantly asexual parasitoid wasp. *Evolution*, **53**, 1189–99.

Belshaw R, Dowton M, Quicke DLJ, Austin AD (2000). Estimating ancestral geographic distributions: a Gondwanan origin for the aphid parasitoids. *Proceedings of the Royal Society B: Biological Sciences*, **267**, 491–96.

Belshaw R, Lopez-Vaamonde C, Degerli N, Quicke DLJ (2001) Paraphyletic taxa and taxonomic chaining: evaluating the classification of braconine wasps (Hymenoptera: Braconidae) using 28S D2–3 rDNA sequences and morphological characters. *Biological Journal of the Linnean Society*, **73**, 411–24.

Belshaw R, Grafen A, Quicke DLJ (2003) Inferring life history from ovipositor morphology in parasitoid wasps using phylogenetic regression and discriminant analysis. *Zoological Journal of the Linnean Society*, **139**, 213–38.

Belz E, Kölliker M, Balmer O (2013) Olfactory attractiveness of flowering plants to the parasitoid *Microplitis mediator*: potential implications for biological control. *BioControl*, **58**, 163–73.

Bender JC (1943) Anatomy and histology of the female reproductive organs of *Habrobracon juglandis* (Ashmead) (Hym., Braconidae). *Annals of the Entomological Society of America*, **36**, 537–45.

Bénédet F, Leroy T, Gauthier N, Thibaudeau C, Thibout E, Renault S (2002) Gustatory sensilla sensitive to protein kairomones trigger host acceptance by an endoparasitoid. *Proceedings of the Royal Society B: Biological Sciences*, **269**, 1879–86.

Benelli G, Canale A (2013a) Do tephritid-induced fruit volatiles attract males of the fruit flies parasitoid *Psyttalia concolor* (Szépligeti) (Hymenoptera: Braconidae)? *Chemoecology*, **23**, 191–9.

Benelli G, Canale A (2013b) Male–male sexual behavior in the parasitic wasp *Psyttalia concolor*. *Journal of Insect Science*, **13**, 25.

Benelli G, Carpita A, Simoncini S, Raspi A, Canale A (2014) For sex and more: attraction of the tephritid parasitoid *Psyttalia concolor* (Hymenoptera: Braconidae) to male sex pheromone of the olive fruit fly, *Bactrocera oleae*. *Journal of Pest Science*, in press.

Benham GS Jr, (1972) Digestive and reproductive systems of *Eriborus* molestae* Uchida (Hymenoptera: Ichneumonidae). *International Journal of Insect Morphology & Embryology*, **1**, 153–61.

Benjamin DM, Demba WM (1969) Notes on the biology of three parasites of the eggs of *Entypotrachelus meyeri* Kolke and *E. micans* Hurst. (Col., Curculionidae). *Bulletin of Entomological Research*, **59**, 473–8.

Bennett AMR (2001) Phylogeny of Agriotypinae (Hymenoptera: Ichneumonidae), with comments on the subfamily relationships of the basal Ichneumonidae. *Systematic Entomology*, **26**, 329–56.

Bennett AMR (2004) A new genus of Oedemopsini (Hymenoptera: Ichneumonidae: Tryphoninae) with a key to the genera of the tribe. *The Canadian Entomologist*, **138**, 464–72.

Bennett AMR, Sääksjärvi IE, Broad GR (2013) Revision of the New World species of *Erythrodolius* (Hymenoptera: Ichneumonidae: Sisyrostolinae), with a key to the world species. *Zootaxa*, **3702**, 5.

Benson JF (1973) Intraspecific competition in the population dynamics of *Bracon hebetor* Say (Hymenoptera: Braconidae). *Journal of Animal Ecology*, **42**, 105–24.

Benson RB (1944) Swarming flight of *Blacus tripudians* Haliday. *Entomologist's Monthly Magazine*, **80**, 21.

Berberet RC (1986) Relationship of temperature to embryogenesis and encapsulation of eggs of *Bathyplectes curculionis* (Hymenoptera: Ichneumonidae) in larvae of *Hypera postica* (Coleoptera: Curculionidae). *Annals of the Entomological Society of America*, **79**, 985–8.

Berberet RC, Willson LJ, Odejar M (1987) Probabilities for encapsulation of eggs of *Bathyplectes curculionis* (Hymenoptera: Ichneumonidae) by larvae of *Hypera postica* (Coleoptera: Curculionidae) and resulting reduction in effective parasitism. *Annals of the Entomological Society of America*, **80**, 483–5.

Berg CO (1949) Limnological relations of insects to plants of the genus *Potamogeton*. *Transactions of the American Microscopical Society*, **68**, 279–91.

Bernal JS, Gillogly PO, Griset J (2001) Family planning in a stemborer parasitoid: sex ratio, brood size and size-fitness relationships in *Parallorhogas pyralophagus* (Hymenoptera: Braconidae), and implications for biological control. *Bulletin of Entomological Research*, **91**, 255–64.

Berndt LA, Wratten SD (2005) Effects of alyssum flowers on the longevity, fecundity, and sex ratio of the leafroller parasitoid *Dolichogenidea tasmanica*. *Biological Control*, **32**, 65–9.

Berry JA, Walker GP (2004) *Meteorus pulchricornis* (Wesmael) (Hymenoptera: Braconidae: Euphorinae): an exotic polyphagous parasitoid in New Zealand. *New Zealand Journal of Zoology*, **31**, 33–44.

Beukeboom LW (2001) Single-locus complementary sex determination in the ichneumonid *Venturia canescens* (Gravenhorst) (Hymenoptera). *Netherlands Journal of Zoology*, **51**, 1–15.

Beukeboom LW, Pijnacker LP (2000) Automictic parthenogenesis in the parasitoid *Venturia canescens* (Hymenoptera: Ichneumonidae) revisited. *Genome*, **43**, 939–44.

Beukeboom LW, Ellers J, Van Alphen JJM (2000) Absence of single-locus complementary sex determination in the braconid wasps *Asobara tabida* and *Alysia manducator*. *Heredity*, **84**, 29–36.

Bézier A, Herbinière J, Serbielle C, Lesobre J, Wincker P, Huguet E, Drezen J-M (2008) Bracovirus gene products are highly divergent from insect proteins. *Archives of Insect Biochemistry and Physiology*, **67**, 172–87.

Bézier A, Annaheim M, Herbinière J, Wetterwald C, Gyapay G, Bernard-Samain S, Wincker P, Roditi I, Heller M, Belghazy M, Pfister-Wilhelm R, Periquet G, Dupuy C, Huguet

E, Volkoff AN, Lanzrein B, Drezen J-M, (2009). Polydnaviruses of braconid wasps derive from an ancestral nudivirus. *Science*, **323**, 926–30.

Bigot Y, Rabouille A, Doury G, Sizaret P-Y, Belbost F, Hamelin M-H, Periquet G (1997) Biological and molecular features of the relationship between *Diadromus pulchellus* ascovirus, a parasitoid hymenopteran wasp (*Diadromus pulchellus*) and its lepidopteran host, *Acrolepiopsis assectella*. *Journal of General Virology*, **78**, 1149–63.

Bigot Y, Samain S, Augé-Gouillou C, Federici BA (2008) Molecular evidence for the evolution of ichnoviruses from ascoviruses by symbiogenesis. *BMC Evolutionary Biology*, **8**, 253.

Bilinski S (1991a) Are accessory nuclei involved in the establishment of developmental gradients in hymenopteran oocytes? Ultrastructural studies. *Wilhelm Roux Archives of Developmental Biology*, **199**, 423–6.

Bilinski S (1991b) Morphological markers of anteroposterior and dorsoventral polarity in developing oocytes of the hymenopteran, *Cosmoconus meridionator* (Ichneumonidae). *Wilhelm Roux Archives of Developmental Biology*, **200**, 330–5.

Bilinski S, Klag J, Kubrakiewicz J (1993) Morphogenesis of accessory nuclei during final stages of oögenesis in *Cosmoconus meridionator* (Hymenoptera, Ichneumonidae). *Wilhelm Roux Archives of Developmental Biology*, **203**, 100–3.

Billany DJ, Winter TG, Gauld ID (1985) *Olesicampe monticola* (Hedwig) (Hymenoptera: Ichneumonidae) redescribed together with notes on its biology as a parasite of *Cephalcia lariciphila* (Wachtl) (Hymenoptera: Pamphiliidae). *Bulletin of Entomological Research*, **75**, 267–74.

Bin F, Isidoro N, Romani R (1999a) Antennal structures of Hymenoptera: sensilla or glands? *Atti dell'Accamedemia Nazionale Italiana di Entomologia Rendiconti*, **47**, 251–63.

Bin F, Wackers F, Romani R, Isidoro N (1999b) Tyloids in *Pimpla turionellae* (L.) are release structures of male antennal glands involved in courtship behaviour (Hymenoptera: Ichneumonoidea). *International Journal of Insect Morphology and Embryology*, **28**, 61–8.

Biondi A, Desneux N, Amiens-Desneux E, Siscaro G, Zappalà L (2013) Biology and developmental strategies of the Palaearctic parasitoid *Bracon nigricans* (Hymenoptera: Braconidae) on the neotropical moth *Tuta absoluta* (Lepidoptera: Gelechiidae). *Journal of Economic Entomology*, **106**, 1638–47.

Bischof C, Ortel J (1996) The effects of parasitism by *Glyptapanteles liparidis* (Braconidae: Hymenoptera) on the hemolymph and total body composition of gypsy moth larvae (*Lymantria dispar*, Lymantriidae: Lepidoptera). *Parasitology Research*, **82**, 687–92.

Blackburn TM (1991) A comparative examination of lifespan and fecundity in the parasitoid Hymenoptera. *Journal of Animal Ecology*, **60**, 151–64.

Blackburn TM, Gaston KJ (1994) Animal body size distributions change as more species are described. *Proceedings of the Royal Society B: Biological Sciences*, **257**, 293–97.

Blackith RE (1967) A hymenopterous primary parasite of morabine grasshopper. *Australian Journal of Zoology*, **15**, 93–102.

Blackwell J, Weih MA (1980) Structure of chitin protein complexes: ovipositor of the ichneumon fly *Megarhyssa*. *Journal of Molecular Biology*, **137**, 49–60.

Blair KG (1945) Notes on the economy of the rose-galls formed by *Rhodites* (Hymenoptera, Cynipidae). *Proceedings of the Royal Entomological Society of London (A)*, **20**, 26–31.

Blass S, Ruthmann A (1989) Fine structure of the accessory glands of the female genital tract of the ichneumonid *Pimpla turionellae* (Insecta: Hymenoptera). *Zoomorphology (Berlin)*, **108**, 367–77.

Bledowski R, Krainska MK (1926) Die Entwicklung von *Banchus femoralis* Thoms. (Hymenoptera, Ichneumonidae). *Varsaviae*, **16**, 1–50.

Bleeker MAK, Smid HM, Van Aelst AC, Van Loon JJA, Vet LEM (2004) Antennal sensilla of two parasitoid wasps: a comparative scanning electron microscopy study. *Microscopy Research and Technique*, **63**, 266–73.

Blum MS, Edgar AL (1971) 4-Methyl–3-heptanone: identification and role in opilionid exocrine secretions. *Insect Biochemistry*, **1**, 181–8.

Blumberg D, Ferkovich SM (1994) Development and encapsulation of the endoparasitoid, *Microplitis croceipes* (Hym.: Braconidae), in six candidate host species (Lep.). *Entomophaga*, **39**, 293–302.

Blunck H (1944) Zur Kenntnis der Hyperparasiten von *Pierris brassicae L. Zeitschrift für Angewandte Entomologie*, **30**, 418–91.

Boivin G, Jacob S, Damiens D (2005) Spermatogeny as a life-history index in parasitoid wasps. *Oecologia*, **143**, 198–202.

Boling JC, Pitre HN (1970) Life history of *Apanteles marginiventris* with descriptions of immature stages. *Journal of the Kansas Entomological Society*, **43**, 465–70.

Bonsall MB, Hassell MP (1998) Population dynamics of apparent competition in a host-parasitoid system. *Journal of Animal Ecology*, **67**, 918–29.

Bordat D, Goudegnon AE, Bouix G (1994) Relationships between *Apanteles flavipes* (Hym.: Braconidae) and *Nosema bordati* (Microspora, Nosematidae) parasites of *Chilo partellus* (Lep.: Pyralidae). *Entomophaga*, **39**, 21–32.

Borden JH, Chong L, Rose A (1978a) Morphology of the elongate placoid sensillum on the antennae of *Itoplectis conquistor*. *Annals of the Entomological Society of America*, **71**, 223–7.

Borden JH, Rose A, Chorney RJ (1978b) Morphology of the elongate sensilla placodeum on the antennae of *Aphidius smithi* (Hymenoptera: Aphidiidae). *Canadian Journal of Zoology*, **56**, 519–25.

Bordera S, Hernández- Rodríguez E (2003) Description of two new species of *Enclisis* (Hymenoptera: Ichneumonidae) and support for the secretory role of tyloids in ichneumonoid males. *European Journal of Entomology*, **100**, 401–9.

Boring AC, Sharkey MJ, Nychka JA (2009) Structure and functional morphology of the ovipositor of *Homolobus truncator* (Hymenoptera: Ichneumonoidea: Braconidae). *Journal of Hymenoptera Research*, **18**, 1–24.

Boring CA, Sharanowski BJ, Sharkey MJ (2011) Maxfischeriinae: a new braconid subfamily (Hymenoptera) with highly specialised egg morphology. *Systematic Entomology*, **36**, 529–48.

Bourdais D, Vernon P, Krespi L, Le lannic J, van Baaren J (2006) Antennal structure of male and female *Aphidius rhopalosiphi* DeStefani-Peres (Hymenoptera: Braconidae): description and morphological alterations after cold storage or heat exposure. *Microscopy Research and Technique*, **69**, 1005–13.

Boush GM, Baerwald RJ (1967) Courtship behavior and evidence for a sex pheromone in the apple maggot parasite, *Opius alloeus*. *Annals of the Entomological Society of America*, **60**, 865–6.

Bracken GK, Barlow JS (1967) Fatty acid composition of *Exeristes comstockii* (Cress) reared on different hosts. *Canadian Journal of Zoology*, **45**, 57–61.

Bracken GK, Nair KK (1967) Stimulation of yolk deposition in an ichneumonid parasitoid by feeding synthetic juvenile hormone. *Nature*, **216**, 483–4.

Bradley WG, Arbuthnot KD (1938) The relation of host physiology to development of the braconid parasite, *Chelonus annulipes* Wesmael. *Annals of the Entomological Society of America*, **31**, 359–65.

Bradley WG, Burgess ED (1934) The biology of *Cremastus flavoorbitalis* (Cameron). *United States Department of Agriculture, Technical Bulletin*, **441**, 1–15.

Braet Y (1997) The Orgilinae (Hymenoptera: Braconidae) in the Ethiopian and Malagasy aera [*sic*]. *Bulletin et Annales de la Société royale belge d'Entomologie*, **133**, 457–67.

Braet Y, Quicke DLJ (2003) A phylogenetic analysis of the Mimagathidini with revisionary notes on the genus *Stantonia* Ashmead, (1904) (Hymenoptera: Braconidae: Orgilinae). *Journal of Natural History*, **38**, 1489–589.

Braet Y, van Achterberg C (2003) *Doryctorgilus* gen. nov. and other new taxa, with a study of the internal microsculpture of the ovipositor in the subfamily Orgilinae Ashmead (Hymenoptera: braconidae). *Zoologische Mededelingen, Leiden*, **77**, 127–52.

Braet Y, van Achterberg C, Chen X (2000) Notes on the tribe Mimagathidini Enderlein, with the description of a new genus (Hymenoptera: Braconidae: Orgilinae). *Zoologische Mededelingen*, **73**, 12–33.

Braga SMP, Barbalho SM, Penteado-Dias AM (2002) *Glaucia* gen. nov. (Hymenoptera: Braconidae: Doryctinae) from Brazil. *Zoologische Mededelingen, Leiden*, **76**, 79–82.

Bragg D (1974) Ecological and behavioral studies on *Phaeogenes cynarae*: ecology; host specificity; search and oviposition; and avoidance of superparasitism. *Annals of the Entomological Society of America*, **67**, 931–6.

Brajković M, Stanisavljević L, Nikolić Z, Ćurčić SB, Živić I, Stojanović D (2010a) Male genitalia of the species of the subfamily Agathidinae (Hymenoptera: Braconidae) and their importance in taxonomy. *Arch. Biol. Sci., Belgrade*, **62**, 455–67.

Brajković M, Nikolić Z, Ćurčić SB, Živić I, Stojanović D (2010b) Morphological changes of the ovipositor in species of Cheloninae (Hymenoptera: Braconidae) in the course of adaptation to egg-larval parasitism. *Arch. Biol. Sci., Belgrade*, **62**, 469–77.

Bredlau JP, Mohajer YJ, Cameron TM, Kester KM, Fine ML (2013) Characterization and generation of male courtship song in *Cotesia congregata* (Hymenoptera: Braconidae). *PLoS ONE*, **8** (4), e6(2051).

Brewer FD, King EG (1981) Food consumption and utilization by sugarcane borers parasitised by *Apanteles flavipes*. *Journal of Georgia Entomological Society*, **16**, 181–5.

Broad GR (2010) Status of *Batakomacrus* Kolarov (Hymenoptera: Ichneumonidae: Orthocentrinae), with new generic combinations and description of a new species. *Zootaxa*, **2394**, 51–68.

Broad GR, Quicke DLJ (2000) The adaptive significance of host location by vibrational sounding in parasitoid wasps. *Proceedings of the Royal Society B: Biological Sciences*, **267**, 2403–9.

Broad GR, Shaw MR (2005) The species of four genera of Metopiinae (Hymenoptera: Ichneumonidae) in Britain, with new host records and descriptions of four new species. *Journal of Natural History*, **39**, 2389–407.

Broad GR, Laurenne NM, Quicke DLJ (2004) The genus *Nipponaetes* Uchida (Hymenoptera: Ichneumonidae: Cryptinae) in Costa Rica, with a reassessment of the generic limits. *European Journal of Entomology*, **101**, 651–5.

Brock JP (1982) A systematic study of the genus *Ophion* in Britain (Hymenoptera: Ichneumonidae). *Tijdschrift voor Entomologie*, **125**, 51–91.

Brodeur J, McNeil JN (1990) Overwintering microhabitat selection by an endoparasitoid (Hymenoptera: Aphidiidae): induced phototactic and thigmokinetic responses in dying hosts. *Journal of Insect Behavior*, **3**, 751–63.

Brodeur J, Vet LEM (1994) Usurpation of host behaviour by a parasitic wasp. *Animal Behaviour*, **48**, 187–92.

Bronskill JF (1959) Embryology of *Pimpla turionellae* (L.) (Hymenoptera: Ichneumonidae). *Canadian Journal of Zoology*, **37**, 655–88.

Bronskill JF (1964) Embryogenesis of *Mesoleius tenthredinis* Morl. (Hymenoptera: Ichneumonidae). *Canadian Journal of Zoology*, **42**, 439–53.

Bronskill JF, House HL (1957) Notes on rearing a pupal endoparasite, *Pimpla turionellae* (L.) (Hymenoptera: Ichneumonidae), on unnatural food. *The Canadian Entomologist*, **89**, 483.

Broodryk SW (1969a) The biology of *Chelonus* (*Microchelonus*) *curvimaculatus* Cameron (Hymenoptera: Braconidae).

Journal of the Entomological Society of Southern Africa, **32**, 169–89.

Broodryk SW (1969b) The biology of *Orgilus parcus* Turner (Hymenoptera: Braconidae). *Journal of the Entomological Society of Southern Africa*, **32**, 243–57.

Brooks RW, Wahl DB (1987) Biology and mature larva of *Hemipimpla pulchripennis* (Saussure), a parasite of *Ropalidia* (Hymenoptera: Ichneumonidae, Vespidae). *Journal of the New York Entomological Society*, **95**, 547–52.

Brooks WM, Cranford JD (1972) Microsporidioses of the hymenopterous parasites, *Campoletis sonorensis* and *Cardiochiles nigriceps*, larval parasites of *Heliothis* species. *Journal of Invertebrate Pathology*, **20**, 77–94.

Brown JJ, Kainoh Y (1992) Host castration by *Ascogaster* spp. (Hymenoptera: Braconidae). *Annals of the Entomological Society of America*, **85**, 67–71.

Brown JJ, Reed-Larsen D, Ahl J (1990) Physiological relationship between a diapausing endoparasitoid (*Ascogaster quadridentata*) and its dormant host (*Cydia pomonella*). *Archives of Insect Biochemistry and Physiology*, **13**, 229–38.

Brown JJ, Kiuchi M, Kainoh Y, Takeda S (1993) *In vitro* release of ecdysteroids by an endoparasitoid, *Ascogaster reticulatus* Watanabe. *Journal of Insect Physiology*, **39**, 229–234.

Brues CT (1910) The parasitic Hymenoptera of the Florissant, Colo. *Bulletin of the Museum of Comparative Zoology, Harvard College*, **54**, 1–125.

Brues CT (1923) *Termitobracon*, a termitophilous braconid from British Guiana. *Zoologica*, **3**, 426–32.

Bryden JW, Bishop MHW (1945) *Perilitus coccinellae* (Schrank) (Hymenoptera: Braconidae) in Cambridgeshire. *Entomologist's Monthly Magazine*, **81**, 51–2.

Buckingham GR (1968) Pygidial glands in male *Opius* (Hymenoptera: Braconidae). *Annals of the Entomological Society of America*, **61**, 233–4.

Buckingham GR (1975) The parasites of walnut husk flies (Diptera: Tephritidae: *Rhagoletis*) including comparative studies on the biology of *Biosteres juglandis* Mues. and on the male tergal glands of Braconidae (Hymenoptera). Unpublished PhD thesis, University of California, Berkeley.

Buckingham GR, Sharkey MJ (1988) Abdominal exocrine glands in Braconidae (Hymenoptera). In VK Gupta (ed.), *Advances in Parasitic Hymenoptera Research*, E.J. Brill, Leiden, pp. 199–242.

Burdick DJ (1961) A taxonomic and biological study of the genus *Xyela* Dalman in North America. *University of California Publications in Entomology*, **17**, 285–356.

Bürgi LP, Mills NJ (2013) Developmental strategy and life history traits of Meteorus ictericus, a successful resident parasitoid of the exotic Light Brown Apple Moth in California. *Biological Control*, **66**, 173–82.

Butcher FG (1933) Hymenopterous parasites of Gyrinidae with descriptions of new species of *Hemiteles*. *Annals of the Entomological Society of America*, **26**, 76–85.

Butcher BA, Smith MA, Sharkey MJ, Quicke DLJ (2012) A turbo-taxonomic study of Thai *Aleiodes* (*Aleiodes*) and *Aleiodes* (*Arcaleiodes*) (Hymenoptera: Braconidae: Rogadinae) based largely on COI barcoded specimens, with rapid descriptions of 179 new species. *Zootaxa*, **3457**, 1–232.

Butcher BA, Zaldivar-Riverón, van de Kamp T, Dos Santos Rolo T, Baumbach T, Quicke DLJ (2014) Extension of historical range of Betylobraconinae (Hymenoptera: Braconidae) into Palaearctic Region based on a Baltic amber fossil, and description of a new species of *Mesocentrus* Szépligeti from Papua New Guinea. *Zootaxa*, **3860**, 449–63.

Butcher RDJ, Whitfield WGF, Hubbard SF (2000a) Complementary sex determination in the genus *Diadegma* (Hymenoptera: Ichneumonidae). *Journal of Evolutionary Biology*, **13**, 593–606.

Butcher RDJ, Whitfield WGF, Hubbard SF (2000b) Single-locus complementary sex determination in *Diadegma chrysostictos* (Gmelin) (Hymenoptera: Ichneumonidae). *Journal of Heredity*, **91**, 104–11.

Byers JA, Zada A (2010) Individual variation of (*S*)-4-methyl-3-heptanone in heads of braconid wasp, *Leiophron uniformis*, and *Pogonomyrmex* ants indicates costs of semiochemical production. *Chemoecology*, **21**, 35–44.

Caccia S, Grimaldi A, Casartelli M, Falabella P, de Eguileor M, Pennacchio F, Giordana B (2012) Functional analysis of a fatty acid binding protein produced by *Aphidius ervi* teratocytes. *Journal of Insect Physiology*, **58**, 621–627.

Caballero P, Vargas-Osuna E, Santiago-Alvarez C (1991) Parasitization of granulosis-virus infected and noninfected *Agrotis segetum* larvae and the virus transmission by three hymenopteran parasitoids. *Entomologia Experimentalis et Applicata*, **58**, 55–60.

Cai D-Z, Bai S-F, Li X, Chen X-X (2006) Effects of parasitization by three endoparasitoids on spermatogenesis in their hosts, *Plutella xylostella* larvae. *Acta Entomologica Sinica*, **49**, 908–917 [in Chinese with English summary].

Calvert DJ, van den Bosch R (1972) Behaviour and biology of *Monoctonus paulensis* (Hymenoptera: Braconidae), a parasite of dactynotine aphids. *Annals of the Entomological Society of America*, **65**, 773–9.

Camarao GC, Morallo-Rejesus B (2003) Parasitoids of the Asian corn borer, *Ostrinia furnacalis* (Guenee), and their biological attributes. *Philippine Agricultural Scientist*, **86**, 17–26.

Cameron E (1941) The biology and post-embryonic development of *Opius ilicis* n. sp., a parasite of the holly leaf-miner (*Phytomyza ilicis* Curt.). *Parasitology*, **33**, 8–39.

Cameron E (1950) The biology and economic importance of *Alomya debellator* (F.), a remarkable parasite of the swift moth, *Hepialus lupulinus* (L.). *Bulletin of Entomological Research*, **41**, 429–38.

Cameron MD, Wharton RA (2011) Revision of *Hodostates* (Hymenoptera: Ichneumonidae: Ctenopelmatinae), with a discussion of tribal placement. *The Canadian Entomologist*, **143**, 136–56.

Campbell RW (1963) Some ichneumonid-sarcophagid interactions in the gypsy moth *Porthetria dispar* (L.) (Lepidoptera: Lymantriidae). *The Canadian Entomologist*, **95**, 337–45.

Canale A (2003) *Psyttalia concolor* (Hymenoptera Braconidae): role of host movement and host substrate in ovipositor-probing behaviour. *Bulletin of Insectology*, **56**, 211–13.

Canale A, Raspi A (2000) Host location and oviposition behaviour in *Opius concolor* (Hymenoptera: Braconidae). *Entomological Problems*, **31**, 25–32.

Canale A, Benelli G, Lanzo F, Giannotti P, Mazzoni V, Lucchi A (2013) The courtship song of fanning males in the fruit fly parasitoid *Psyttalia concolor* (Szépligeti) (Hymenoptera: Braconidae). *Bulletin of Entomological Research*, **103**, 303–9.

Čapek M (1970) A new classification of the Braconidae (Hymenoptera) based on the cephalic structures of the final instar larva and biological evidence. *Canadian Entomologist*, **102**, 846–75.

Čapek M (1973) Key to the final instar larvae of the Braconidae. *Acta Instituti Forestalis Zvolenensis*, **4**, 259–68.

Čapek M (1983) Relationships and taxonomic placement of the genus *Blacus* (Hymenoptera: Braconidae). *Contributions of the American Entomological Institute*, **20**, 119–24.

Čapek M, Čapecki Z (1979) A new genus and new species of Euphorinae (Braconidae, Hymenoptera) from southern Poland. *Polski Pismo Entomologiczne*, **49**, 215–21.

Čapek M, Davidová-Vilímová J (1978) On the life history of *Aridelus egregius* (Hymenoptera, Braconidae) a parasite of European *Coptosoma* (Heteroptera, Plataspidae). *Acta Entomologica Bohemoslovaca*, **75**, 243–54.

Čapek M, van Achterberg C (1992) A revision of the genus *Microtypus* Ratzeburg (Hymenoptera: Braconidae). *Zoologische Mededelingen, Leiden*, **66**, 323–38.

Carabajal-Paladino LZ, Papeschi AG, Cladera JL (2010) Immature stages of development in the parasitoid wasp, *Diachasmimorpha longicaudata*. *Journal of Insect Science*, **10**, 56.

Carignan S, Boivin G, Stewart RK (1995) Developmental biology and morphology of *Peristenus digoneutis* Loan (Braconidae, Euphorinae). *Biological Control*, **5**, 553–60.

Carlson RW (1979) Family Ichneumonidae. In KV Krombein, PD Hurd Jr, DR Smith and BD Burks (eds), *Catalog of Hymenoptera in America North of Mexico*, vol. 1, Smithsonian Institution, Washington, DC, pp. 315–740.

Carpenter JE, Greany PD (1998) Comparative development and performance of artificially reared versus host-reared *Diapetimorpha introita* (Cresson) (Hymenoptera: Ichneumonidae) wasps. *Biological Control*, **11**, 203–8.

Carpenter JE, Pair SD, Fitt GP (1994) *Ichneumon promissorius* (Hymenoptera: Ichneumonidae): development on North American hosts. *Journal of Economic Entomology*, **87**, 929–32.

Carpenter JE, Ferkovich SM, Greany PD (2001) Fecundity and longevity of *Diapetimorpha introita* (Cresson) (Hymenoptera: Ichneumonidae) reared on artificial diets: effects of a lipid extract from host pupae and culture media conditioned with an insect cell line. *The Florida Entomologist*, **84**, 43–9.

Carrillo MA, Heimpel GE, Moon RD, Cannon CA, Hutchison WD (2005) Cold hardiness of *Habrobracon hebetor* (Say) (Hymenoptera: Braconidae), a parasitoid of pyralid moths. *Journal of Insect Physiology*, **51**, 759–68.

Carter D, Walker AK (eds) (1999) *Care and Conservation of Natural History Collections*, Butterworth Heinemann, Oxford.

Casiraghi M, Andrietti F, Bonasoro F, Martinoli A (2001) A note on host detection by *Buathra tarsoleuca* (Schrank) (Hymenoptera: Ichneumonidae), a parasite of *Ammophila sabulosa* (L.) and *Podalonia affinis* (Kirby) (Hymenoptera: Sphecidae). *Journal of Insect Behavior*, **14**, 299–312.

Cassidy JD, King RC (1972) Ovarian development in *Habrobracon juglandis* (Ashmead) (Hymenoptera: Braconidae). I. The origin and differentiation of the oocyte–nurse cell complex. *Biological Bulletin*, **143**, 483–505.

Castelo MK, van Nouhuys S, Corley JC (2010) Olfactory attraction of the larval parasitoid, *Hyposoter horticola*, to plants infested with eggs of the host butterfly, *Melitaea cinxia*. *Journal of Insect Science*, **10**, 53.

Castillo C, Sääksjärvi IE, Bennett AMR, Broad GR (2011) First record of Acaenitinae (Hymenoptera, Ichneumonidae) from South America with description of a new species and a key to the world species of *Arotes* Gravenhorst. *ZooKeys*, **137**, 77–88.

Caudri LWD (1941) The braconid *Alysia manducator* Panzer and its relation to the blowfly *Calliphora erythrocephala* Meigen. *Archives Néerlandaises de Zoologie*, **5**, 413–97.

Çelik S (1987) Quantitative essential amino acid requirements of *Pimpla turionellae* L (Hymenoptera Ichneumonidae). *DOGA Turk Kimya Dergisi*, **11**, 89–99 (in Turkish).

Chanda S, Panda RN, Chakravorty S (2002) Hormonal implication in *Bracon* venom-induced paralysation of the host larva of *Corcyra cephalonica* (Lepidoptera: Pyralidae). *Indian Journal of Experimental Biology*, **40**, 918–23.

Chadwick CE, Nikitin MI (1976) Records of parasitism in the families Ichneumonidae, Braconidae and Aulacidae. *Journal of the Entomological Society of Australia (New South Wales)*, **9**, 28–38.

Chao A (1984) Nonparametric estimation of the number of classes in a population. *Scandinavian Journal of Statistics*, **11**, 265–70.

Chao HF (1980) Description of *Lycorina spilonotae* Chao, sp. nov., with notes on its final instar larva (Hymenoptera: Ichneumonidae: Lycorinae). *Entomotaxonomia*, **2**, 165–8 (in Chinese with English summary).

Charlet LD (2002) Parasitization of the red sunflower seed weevil (Coleoptera: Curculionidae) by its larval parasitoid *Triaspis aequoris* (Hymenoptera: Braconidae) in cultivated sunflower. *Environmental Entomology*, **31**, 844–51.

Chau NNB, Maetô K (2009) Temporary host paralysis and avoidance of self-superparasitism in the solitary endoparasitoid *Meteorus pulchricornis*. *Entomologia Experimentalis et Applicata*, **132**, 250–5.

Chaubet B, Derocles SAP, Hullé M, LeRalec A, Outreman Y, Simon J-C, Tomanović Z (2013) Two new species of aphid parasitoids (Hymenoptera, Braconidae, Aphidiinae) from the high arctic (Spitsbergen, Svalbard). *Zoologischer Anzeiger*, **252**, 34–40.

Chauvin G, El Agoze M, Hamon C, Huignard J (1988) Ultrastructure des spermatozoïdes des mâles haploïdes et diploïdes de *Diadromus pulchellus* Wesmeal (Hymenoptera: Ichneumonidae). *International Journal of Insect Morphology and Embryology*, **17**, 359–366.

Chavarría L, Hanson P, Marsh PM, Shaw SR (2009) A phytophagous braconid, *Allorhogas conostegia* sp. nov. (Hymenoptera: Braconidae), in the fruits of *Conostegia xalapensis* (Bonpl.) D. Don (Melastomataceae). *Journal of Natural History*, **43**, 2677–2689.

Chen X, He J (1997) Revision of the subfamily Rogadinae (Hymenoptera: Braconidae) from China. *Zoologische Verhandelingen, Leiden*, **308**, 1–187.

Chen X, van Achterberg C (1997) Revision of the subfamily Euphorinae (excluding the tribe Meteorini Cresson) (Hymenoptera: Braconidae) from China. *Zoologische Verhandelingen, Leiden*, **313**, 1–217.

Chen X, Whitfield JB (2002) A new species of the rare genus *Katytermus* (Hymenoptera: Braconidae) in China. *Entomological News*, **113**, 239–42.

Chen X, Whitfield JB (2003) The discovery of the genus *Ecnomios* Mason (Hymenoptera: Braconidae) in China, with description of a new species. *Proceedings of the Entomological Society of Washington*, **105**, 348–51.

Chen X, Whitfield JB, He J (2002) The discovery of the genus *Gnamptodon* Haliday (Hymenoptera: Braconidae) in China, with description of one new species. *Pan-Pacific Entomologist*, **78**, 184–7.

Chien C-C, Chang S-C (2012) Morphology and life history of *Opius caricivorae* (Hymenoptera: Braconidae). *Journal of Taiwan Agriculture Research*, **61**, 144–57.

Chiri AA, Legner EF (1986) Response of three *Chelonus* (Hymenoptera: Braconidae) species to kairomones in scales of six Lepidoptera. *Canadian Entomologist*, **118**, 329–33.

Chiu S-C (1984) *Keys to the Genera of Ichneumonidae (Hymenoptera) of Taiwan*. Special Publication No. 17, TARI, Taiwan, 58 pp.

Chorney RJ, Mackauer M (1979) The larval instars of *Aphidius smithi* (Hymenoptera: Aphidiidae). *Canadian Entomologist*, **111**, 631–4.

Chou L-Y (1984) The phylogeny of Aphidiidae (Hymenoptera) of Taiwan. *Journal of Agricultural Research, China*, **33**, 437–46 (in Chinese with English summary).

Chow FJ, Mackauer M (1984) Inter and intraspecific larval competition in *Aphidius smithi* and *Praon pequodorum* (Hymenoptera: Aphidiidae). *The Canadian Entomologist*, **116**, 1097–107.

Chow FJ, Sullivan DJ (1984) Developmental stages of *Praon pequodorum* Viereck (Hymenoptera: Aphidiidae), a pea aphid parasitoid. *Annals of the Entomological Society of America*, **77**, 319–22.

Christiansen-Weniger P, Hardie J (1998) Wing development in parasitised male and female *Sitobium fragariae*. *Physiological Entomology*, **23**, 208–13.

Chrystal RN, Skinner ER (1931) Studies in the biology of *Xylonomus brachylabris* Kr. and *X. irrigator* F., parasites of the larch longhorn beetle, *Tetropium gabrieli* Weise. *Forestry*, **5**, 21–33.

Chrystal RN, Skinner ER (1932) Studies on the biology of the wood wasp Xiphydria *prolongata* Geoffr. and its parasite *Thalessa curvipes* Grav. *Scottish Forestry Journal*, **46**, 36–51.

Chu J (1935) Notes on the biology of *Cedria paradoxa* Wilkinson, a hymenopterous parasite of the mulberry pyralid (Mararonia pyloalis Walker). *Chekiang Province Bureau of Entomology Yearbook*, **1933**, 193–202.

Chudek JA, Crook AM, Hubbard SF, Hunter G (1996) Nuclear magnetic resonance microscopy of the development of the parasitoid wasp *Venturia canescens* within its host moth *Plodia interpunctella*. *Magnetic Resonance Imaging*, **14**, 679–86.

Chumakova BM (1968) Regulative mechanisms of fertilization and sexual determination in progenies of parasitic Hymenopters. *Proceedings of the All-Union Scientific Research Institute of Plant Protection, Leningrad*, **31**, 121–63 (in Russian).

Claret J (1973) La diapause facultative de *Pimpla instigator* (Hym.: Ichneumonidae) I. Rôle de la photopériode. *Entomophaga*, **18**, 409–18.

Claret J (1978) La diapause facultative de *Pimpla instigator* (Hym.: Ichneumonidae) II. Rôle de la tempeérature. *Entomophaga*, **23**, 411–15.

Clancy DW, Marucci PE, Dresner E (1952) Importation of natural enemies to control the oriental fruit fly in Hawaii. *Journal of Economic Entomology*, **45**, 85–90.

Clark AM, Rubin MA (1961) The modification by X-radiation of the life span of haploids and diploids of the wasp, *Habrobracon* sp. *Radiation Research*, **15**, 244–53.

Clausen CP (1931) Biological observations in *Agriotypus* (Hymenoptera). *Proceedings of the Entomological Society of Washington*, **33**, 29–37.

Clausen CP (1932) The early stages of some tryphonine Hymenoptera parasitic on sawfly larvae. *Proceedings of the Entomological Society of Washington*, **34**, 49–60.

Clausen CP (1940) *Entomophagous Insects*, McGraw-Hill, New York, 688 pp.

Clausen CP (1956) The egg–larval host relationship among parasitic Hymenoptera. *Bolletino del Laboratorio di Zoologia Generale e Agraria Portici*, **33**, 119–33.

Cline LD, Press JW (1990) Reduction in almond moth (Lepidoptera: Pyralidae) infestation using commercial packaging of foods in combination with the parasitic wasp, *Bracon hebetor* (Hymenoptera: Braconidae). *Journal of Economic Entomology*, **83**, 1110–13.

Cloutier C, Douglas AE (2003) Impact of a parasitoid on the bacterial symbiosis of its aphid host. *Entomologia Experimentalis et Applicata*, **109**, 13–19.

Cohen MB, Mackauer M (1987) Intrinsic rate of increase and temperature coefficients of the aphid parasite *Ephedrus californicus* Baker (Hymenoptera: Aphidiidae). *The Canadian Entomologist*, **119**, 231–7.

Cole LR (1959a) On a new species of *Syntretus* Foerster (Hym., Braconidae) parasitic on an adult ichneumonid, with a description of the larva and notes on its life history and that of its host, *Phaeogenes invisor* (Thunberg). *Entomologist's Monthly Magazine*, **95**, 18–21.

Cole LR (1959b) On the defenses of lepidopterous pupæ in relation to the oviposition behaviour of certain Ichneumonidae. *Journal of the Lepidopterists' Society*, **13**, 1–10.

Cole LR (1967) A study of the life-cycles and hosts of some Ichneumonidae attacking pupae of the green oak-leaf roller moth, *Tortrix viridana* (L.) (Lepidoptera: Tortricidae) in England. *Transactions of the Royal Entomological Society of London*, **119**, 267–81.

Cole LR (1970) Observations on the finding of mates by *Phaeogenes invisor* and *Apanteles medicaginis* (Hymenoptera: Ichneumonidae). *Animal Behaviour*, **18**, 184–92.

Cole LR (1981) A visible sign of a fertilization act during oviposition by an ichneumonid wasp, *Itoplectis maculator*. *Animal Behavior*, **29**, 299–300.

Colinet D, Mathé-Hubert H, Allemand R, Gatti J-L, Poirié M (2013) Variability of venom components in immune suppressive parasitoid wasps: from a phylogenetic to a population approach. *Journal of Insect Physiology*, **59**, 205–12.

Colwell RK, Coddington JA (1994) Estimating terrestrial biodiversity through extrapolation. *Philosophical Transactions of the Royal Society B*, **345**, 101–18.

Comstock JH, Needham DG (1898) The wings of Insects IX. The venation of the wings of Hymenoptera. *The American Naturalist*, **32**, 413–24.

Conlong DE (1994) A review and perspectives for the biological control of the African sugarcane stalkborer *Eldana saccharina* Walker (Lepidoptera: Pyralidae). *Agriculture, Ecosystems & Environment*, **48**, 9–17.

Connahs H, Rodríguez-Castañeda G, Walters T, Walla T, Dyer L (2009) Geographic variation in host-specificity and parasitoid pressure of an herbivore (Geometridae) associated with the tropical genus *Piper* (Piperaceae). *Journal of Insect Science*, **9**, 28.

Cônsoli FL, Kitajima EW (2006) Symbiofauna associated with the reproductive system of *Cotesia flavipes* and *Doryctobracon areolatus* (Hymenoptera, Braconidae). *Braz. Journal of Morphological Science*, **23**, 463–70.

Cônsoli FL, Vinson SB. (2004) Host regulation and the embryonic development of the endoparasitoid *Toxoneuron nigriceps* (Hymenoptera: Braconidae). *Comparative Biochemistry & Physiology B*, **137**, 463–73.

Cônsoli FL, Brandt SL, Coudron TA, Vinson SB (2005) Host regulation and release of parasitism-specific proteins in the system *Toxoneuron nigriceps–Heliothis virescens*. *Comparative Biochemistry & Physiology B*, **142**, 181–91.

Cook JM (1993) Inbred lines as reservoirs of sex alleles in parasitoid rearing programs. *Environmental Entomology*, **22**, 1213–15.

Cook JM, Butcher RDJ (1999) The transmission and effects of *Wolbachia* bacteria in parasitoids. *Research in Population Ecology*, **41**, 15–28.

Cook JM, Rivero Lynch AP, Godfray HCJ (1994) Sex ratio and foundress number in the parasitoid wasp *Bracon hebetor*. *Animal Behaviour*, **47**, 687–96.

Coop LB, Croft BA (1990) Diapause and life history attributes of *Phytodietus vulgaris* (Hymenoptera: Ichneumonidae), a parasitoid of *Argyrotaenia citrana* (Lepidoptera: Tortricidae). *Annals of the Entomological Society of America*, **83**, 1148–51.

Corbet SA (1968) The influence of *Ephestia kuehniella* on the development of its parasite *Nemetris canescens*. *Journal of Experimental Biology*, **48**, 291–304.

Coronado-Rivera J (2009) Filogenia de avispas del género *Netelia* (Hymenoptera: Ichneumonidae) con énfasis en las especies costarricenses. *Revista Biologia Tropicana*, **57** (Suppl. 1), 213–38.

Coronado-Rivera J, Gonzalez-Herrera A, Gauld ID, Hanson P (2004) The enigmatic biology of the ichneumonid subfamily Lycorininae. *Journal of Hymenoptera Research*, **13**, 223–7.

Coseglia AF, Simpson RG, Eklund LR (1977) Biology of *Mesochorus nigripes*: a hyperparasite of *Bathyplectes* spp. *Annals of the Entomological Society of America*, **70**, 695–8.

Coskun M, Özalp P, Emre I (2005) Effects of vitamin E concentrations on sex ratio of *Pimpla turionellae* (Hymenoptera: Ichneumonidae) adults. *Annals of the Entomological Society of America*, **98**, 336–9.

Cossé AA, Petroski RJ, Zilkowski BW, Vermillion K, Lelito JP, Cooperband MF, Gould JR (2012) Male-produced pheromone of *Spathius agrili*, a parasitoid introduced for the biological control of the invasive emerald ash borer, *Agrilus planipennis*. *Journal of Chemical Ecology*, **38**, 389–99.

Cossentine JE, Deglow EK, Jensen LBM, Bennett AMR (2004) A biological assessment of *Apophua simplicipes* (Hymenoptera: Ichneumonidae) as a parasitoid of the oblique banded leafroller, *Choristoneura rosaceana* (Lepidoptera: Tortricidae). *Biocontrol Science and Technology*, **14**, 691–9.

Couchman JR, King PE (1977) Morphology of the larval stages of *Diaeretiella rapae* (M'intosh) (Hymenoptera: Aphidiidae). *International Journal of Insect Morphology and Embryology*, **6**, 127–36.

Couturier A (1949) Observations sur *Rhyssa approximator* F. cleptoparasite de *Rh. persuasoria* L. (Hym. Ichneumonidae). *Bulletin de la Société Entomologique de France*, **54**, 62–3.

Crankshaw OS, Matthews RW (1981) Sexual behaviour among parasitic *Megarhyssa* wasps (Hymenoptera, Ichneumonidae). *Behavioural Ecology and Sociobiology*, **9**, 1–7.

Crawford AM, Still L-A, Smith P (2006) A histological examination of the venom apparatus of *Microctonus hyperodae* Loan (Hymenoptera: Braconidae). *New Zealand Entomologist*, **29**, 103–6.

Crawford AM, Brauning R, Smolenski G, Ferguson C, Barton D, Wheeler TT, McCulloch A (2008) The constituents of *Microctonus* sp. parasitoid venoms. *Insect Molecular Biology*, **17**, 313–24.

Crawford AW (1933) *Glypta rufiscutellaris* Cresson, an ichneumonid larval parasite of the oriental fruit moth. *New York State Agricultural Experimental Station Technical Bulletin*, **217**, 1–29.

Croft P, Copland JW (1994) Larval morphology and development of the parasitoid *Dacnusa sibirica* (Hym., Braconidae) in the leafminer host *Chromatomyia syngenesiae*. *Entomophaga*, **39**, 85–93.

Cross JE, Simpson RG (1972) Cocoon construction by *Bathyplectes curculionis* and attack by secondary parasites. *Environmental Entomology*, **1**, 631–3.

Cummins HM, Wharton RA, Colvin AM (2011) Eggs and egg loads of field-collected Ctenoplematinae (Hymenoptera: Ichneumonidae): evidence for phylogenetic constraints and life-history trade-offs. *Annals of the Entomological Society of America*, **104**, 465–75.

Cushman RA (1913) Biological notes on a few rare or little known parasitic Hymenoptera. *Proceedings of the Entomological Society of Washington*, **15**, 153–60.

Cushman RA (1916) *Thersilochus conotrachelli* a parasite of the plum *Curculio*. *Journal of Agricultural Research*, **6**, 847–55.

Cushman RA (1926a) The new *Urosigalphus* parasitic on *Eulechriops gossypii* Barber Hymenoptera: Braconidae). *Proceedings of the Entomological Society of Washington*, **28**, 63.

Cushman RA (1926b) Address of retiring president. Some types of parasitism among the Ichneumonidae. *Proceedings of the Entomological Society of Washington*, **28**, 51.

Cushman RA (1937) The genus *Lysiognatha* Ashmead. *Journal of the Washington Academy of Science*, **27**, 10, 438–44.

Cushman RA (1940) New genera and species of Ichneumon-flies, with taxonomic notes. *Proceedings of the United States National Museum*, **88**, 355–72.

Cusson M, Barron JR, Goulet H, Régnière J, Doucet D (1998) Biology and status of *Tranosema rostrale* (Hymenoptera: Ichneumonidae), a parasitoid of the eastern spruce budworm (Lepidoptera: Tortricidae). *Annals of the Entomological Society of America*, **91**, 87–93.

Cusson M, Stoltz D, Lepointe R, Béliveau C, Nisole A, Volkoff A-N, Drezen J-M, Maaroufi H, Levesque RC (2012) Genomics of banchine ichnoviruses: insights into their relationship to bracoviruses and campoplegine ichnoviruses. In NE Beckage and J-M Drezen (eds), *Parasitoid Viruses: Symbionts and Pathogens*, Elsevier Academic Press, Amsterdam, pp. 79–87.

Daane KM, Wang X, Duerr SS, Kuhn EJ, Son Y, Yokota GY (2013) Biology of *Habrobracon gelechiae* (Hymenoptera: Braconidae), as a parasitoid of the obliquebanded leafroller (Lepidoptera: Tortricidae). *Environmental Entomology*, **42**, 107–15.

Dahlman DL (1990) Evaluation of teratocyte functions: an overview. *Archives of Insect Biochemistry and Physiology*, **13**, 159–66.

Dahlman DL (1991) Teratocytes and host/parasitoid interactions. *Biological Control*, **1**, 118–26.

Dahlman DL, Rana RL, Schepers EJ, Schepers T, DiLuna FA, Webb BA (2003) A teratocyte gene from a parasitic wasp that is associated with inhibition of insect growth and development inhibits host protein synthesis. *Insect Molecular Biology*, **12**, 527–34.

Dahms EC (1984) An interpretation of the structure and function of the antennal sense organs of *Melittobia australica* (Hymenoptera: Eulophidae) with the discovery of a large dermal gland in the male scape. *Memoirs of the Queensland Museum*, **21**, 361–85.

Daly H (1983) Taxonomy and ecology of Ceratinini of north Africa and the Iberian Peninsula (Hym., Apoidea). *Systematic Entomology*, **8**, 29–62.

Dangerfield PC, Austin AD, Whitfield JB (1999) Systematics of the world genera of Cardiochilinae (Hymenoptera: Braconidae). *Invertebrate Taxonomy*, **13**, 917–76.

Danci AT, Gries R, Schaefer PW, Gries G (2006) Evidence for four-component close-range sex pheromone in the parasitic wasp *Glyptapanteles flavicoxis*. *Journal of Chemical Ecology*, **32**, 1539–54.

Danci AT, Takács S, Schaefer PW, Gries G (2010) Evidence for acoustic communication in the parasitoid wasp *Glyptapanteles flavicoxis*. *Entomologia Experimentalis et Applicata*, **136**, 142–50.

Danci A, Hrabar M, Ikoma S, Schaefer PW, Gries G (2013) Learning provides mating opportunities for males of a parasitoid wasp. *Entomologia Experimentalis et Applicata*, **149**, 229–40.

Dangerfield PC, Whitfield JB, Sharkey MJ, Janzen DH, Mercado I (1996) *Hansonia*, a new genus of cardiochiline Braconidae (Hymenoptera) from Costa Rica, with notes on its biology. *Proceedings of the Entomological Society of Washington*, **98**, 592–6.

Dangerfield PC, Austin AD, Whitfield JB (1999) Systematics of the world genera of Cardiochilinae (Hymenoptera: Braconidae). *Invertebrate Taxonomy*, **13**, 917–76.

Dani MP, Richards EH (2009) Cloning and expression of the gene for an insect haemocyte anti-aggregation protein (VPr3), from the venom of the endoparasitic wasp, *Pimpla hypochondriaca*. *Archives of Insect Biochemistry and Physiology*, **71**, 191–204.

Dani MP, Richards EH, Isaac RE, Edwards JP (2003) Antibacterial and proteolytic activity in venom from the endoparasitic wasp *Pimpla hypochondriaca* (Hymenoptera: Ichneumonidae). *Journal of Insect Physiology*, **49**, 945–54.

Dani MP, Richards EH, Edwards JP (2004) Venom from the pupal endoparasitoid, *Pimpla hypochondriaca*, increases the susceptibility of larval *Lacanobia oleracea* to the entomopathogens *Bacillus cereus* and *Beauveria bassiana*. *Journal of Invertebrate Pathology*, **86**, 19–25.

Dani MP, Edwards JP, Richards EH (2005) Hydrolase activity in the venom of the pupal endoparasitic wasp, *Pimpla hypochondriaca*. *Comparative Biochemistry and Physiology Part B: Biochemistry and Molecular Biology*, **141**, 373–81.

Daniel DM (1932) *Macrocentrus ancylivorus* Rohwer, a polyembryonic braconid parasite of the oriental fruit moth. *Technical Bulletin of the New York State Agriculture Station*, **187**, 1–101.

Danks HV (1971) Biology of some stem-nesting aculeate Hymenoptera. *Transactions of the Royal Entomological Society, London*, **122**, 323–99.

Danks HV, Rabb RL, Southern PS (1979) Biology of insect parasites of *Heliothis* larvae in North Carolina. *Journal of the Georgia Entomological Society*, **14**, 36–64.

Darwish E, El-Shazly M, El-Sherif H (2003) The choice of probing sites by *Bracon hebetor* Say (Hymenoptera: Braconidae) foraging for *Ephestia kuehniella* Zeller (Lepidoptera: Pyralidae). *Journal of Stored Products Research*, **39**, 265–76.

Das P, Fadamiro HY (2013) Species and sexual differences in antennal lobe architecture and glomerular organization in two parasitoids with different degree of host specificity, *Microplitis croceipes* and *Cotesia marginiventris*. *Cell and Tissue Research*, **352**, 227–35.

Das P, Chen L, Sharma KR, Fadamiro HY (2011) Abundance of antennal chemosensilla in two parasitoid wasps with different degree of host specificity may explain sexual and species differences in their response to host-related volatiles. *Microscopy Research and Technique*, **74**, 900–9.

Dasch CE (1971) Hibernating Ichneumonidae of Ohio (Hymenoptera). *Ohio Journal of Science*, **71**, 270–83.

Dasch CE (1984) Ichneumon flies of America north of Mexico: 9. Subfamilies Theriinae and Anomaloninae. *Memoirs of the American Entomological Institute*, **36**, 1–610.

Davies AP, Takashino K, Watanabe M, Miura K (2009) Parental genetic traits in offspring from inter-specific crosses between introduced and indigenous *Diadegma* Foerster (Hymenoptera: Ichneumonidae): possible effects on conservation genetics. *Applied Entomology and Zoology*, **44**, 535–41.

Davies DH, Burghardt RL, Vinson SB (1986) Oogenesis of *Cardiochiles nigriceps* viereck (Hymenoptera: Braconidae): histochemistry and development of the chorion with special reference to the fibrous layer. *International Journal of Insect Morphology and Embryology*, **15**, 363–74.

Davies NW, Madden JL (1985) Mandibular gland secretions of two parasitoid wasps (Hymenoptera: Ichneumonidae). *Journal of Chemical Ecology*, **11**, 1115–27.

Davis DS, Stewart SL, Manica A, Majerus MEN (2006) Adaptive preferential selection of female coccinellid hosts by the parasitoid wasp *Dinocampus coccinellae* (Hymenoptera: Braconidae). *European Journal of Entomology*, **103**, 41–5.

Day WH (1970) The survival value of its jumping cocoons to *Bathyplectes anurus*, a parasite of the alfalfa wevil. *Journal of Economic Entomology*, **63**, 586–9.

Day WH (2002) Biology, host preferences, and abundance of *Mesochorus curvulus* (Hymenoptera: Ichneumonidae), a hyperparasite of *Peristenus* spp. (Hymenoptera: Braconidae) parasitizing plant bugs (Miridae: Hemiptera) in alfalfa-grass forage crops. *Annals of the Entomological Society of America*, **95**, 218–22.

Day WH (2005) Diapause duration as a synchronizer of parasite (*Peristenus* spp.: Hymenoptera: Braconidae) and host (Hemiptera: Miridae) life cycles, and its use in separating morphologically-similar parasite species. *Transactions of the American Entomological Society*, **131**, 87–99.

Day WH, Hedlund RC (1988) Biological comparisons between arrhenotokous and thelytokous biotypes of *Mesochorus nigriceps*. *Entomophaga*, **33**, 201–10.

De Boer JG, Ode PJ, Vet LEM, Whitfield JB, Heimpel GE (2007a) Complementary sex determination in the parasitoid wasp *Cotesia vestalis* (*C. plutellae*). *Journal of Evolutionary Biology*, **20**, 340–8.

De Boer JG, Ode PJ, Vet LEM, Whitfield JB, Heimpel GE (2007b) Diploid males sire triploid daughters and sons in the parasitoid wasp *Cotesia vestalis*. *Heredity*, **3**, 288–94.

De Boer JG, Ode PJ, Rendahl AK, Vet LEM, Whitfield JB, Heimpel GE (2008) Experimental support for multiple-locus complementary sex determination in the parasitoid *Cotesia vestalis*. *Genetics*, **180**, 1525–35.

De Boer JG, Kuijper B, Heimpel GE, Beukeboom LW (2012) Sex determination meltdown upon biological control introduction of the parasitoid *Cotesia rubecula*. *Evolutionary Applications*, **5**, 444–54.

de Buron I, Beckage NE (1992) Characterization of a polydnavirus (PDV) and virus-like filamentous particle (VLFP) in the braconid wasp *Cotesia congregata* (Hymenoptera: Braconidae). *Journal of Invertebrate Pathology*, **59**, 315–27.

Decker UM, Powell W, Clark SJ (1993) Sex pheromones in the cereal aphid parasitoids Praon volucre and *Aphidius rhopalosiphi*. *Entomologia Experimentalis et Applicata*, **69**, 33–9.

Dedeine F, Vavre F, Shoemaker DD, Boulétreau M (2004) Intra-individual coexistence of a *Wolbachia* strain required for host oogenesis with two strains inducing cytoplasmic incompatibility in the wasp *Asobara tabida*. *Evolution*, **58**, 2167–74.

Dedeine F, Boulétreau M, Vavre F (2005) *Wolbachia* requirement for oogenesis: occurrence within the genus *Asobara* (Hymenoptera, Braconidae) and evidence for intraspecific variation in *A. tabida*. *Heredity*, **95**, 394–400.

De Eguileor M, Grimaldi A, Tettamanti G, Valvasson R, Leonardi MG, Giordana B, Tremblay E, Digilio MC, Pennacchio F (2001) Larval anatomy and structure of absorbing epithelia in the aphid parasitoid *Aphidius ervi* Haliday (Hymenoptera, Braconidae). *Arthropod Structure and Development*, **30**, 27–37.

De Freitas MdLT, De Lima Mendonza A, Rufino do Nascimento R, Sant'Ana AEG (2004) Behavioural evidence for a female sex pheromone in *Cotesia flavipes* (Hymenoptera: Braconidae). *Physiological Entomology*, **29**, 183–7.

De Leon D (1935) The biology of *Coeloides dendroctoni* Cushman (Hymenoptera: Braconidae), an important parasite of the mountain pine beetle (*Dendroctonus monticolae* Hopk.). *Annals of the Entomological Society of America*, **28**, 411–24.

DeLury NC, Gries R, Gries G, Judd GJR, Khaskin G (1999a) Moth scale-derived kairomones used by egg-larval parasitoid *Ascogaster quadridentata* to locate eggs of its host, *Cydia pomonella*. *Journal of Chemical Ecology*, **25**, 2419–31.

DeLury NC, Greis G, Greis R, Judd GJR, Brown JJ (1999b) Sex pheromone of *Ascogaster quadridentata*, a parasitoid of *Cydia pomonella*. *Journal of Chemical Ecology*, **25**, 2229–45.

de Macêdo MV, Monteiro RF (1989) Seed predation by a braconid wasp, *Allorhogas* sp. (Hymenoptera). *Journal of the New York Entomological Socety*, **97**, 358–62.

de Macêdo MV, Pimentel MCP, Vieira RC (1998) Response of *Pithecellobium tortum* Martius (Leguminosae) seeds to the attack of the phytophagous braconid *Allorhogas dyspistus* Marsh (Hymenoptera: Braconidae). *Journal of Hymenoptera Research*, **7**, 274–9.

De Marzo L (2000) Note anatomiche sui genitali interni femminili nelle Aphidiinae (Hymenoptera Braconidae). *Redia*, **83**, 133–9.

Demmon AS, Nelson AJ, Ryan PJ, Ives AR, Snyder WE (2004) *Aphidius ervi* (Hymenoptera: Braconidae) increases its adult size by disrupting host wing development. *Environmental Entomology*, **33**, 1523–7.

De Nardo EAB, Hopper KR (2004) Using the literature to evaluate parasitoid host ranges: a case study of *Macrocentrus grandii* (Hymenoptera: Braconidae) introduced into North America to control *Ostrinia nubilalis* (Lepidoptera: Crambidae). *Biological Control*, **31**, 280–95.

De Saeger H (1948) *Exploration du Parc National Albert. Mission G. F. de Witte (1933–1935)*, Institut des Parcs Nationaux du Congo Belge, Brussels.

Desneux N, Barta RJ, Delebecque CJ, Heimpel GE (2009) Transient host paralysis as a means of reducing self-superparasitism in koinobiont endoparasitoids. *Journal of Insect Physiology*, **55**, 321–7.

Devotto L, Del Valle C, Ceballos R, Gerding M (2010) Biology of *Mastrus ridibundus* (Gravenhorst), a potential biological control agent for area-wide management of *Cydia pomonella* (Linneaus) (Lepidoptera: Tortricidae). *Journal of Applied Entomology*, **134**, 243–50.

Dhillon MK, Sharma HC (2011) Effect of mating and parasitism regimes on progeny production and sex-ratio of *Campoletis chlorideae* Uchida. *Indian Journal of Experimental Biology*, **49**, 786–90.

Dib-Hajj SD, Webb BA, Summers MD (1993) Structure and evolutionary implications of a "cysteine-rich" *Campoletis sonorensis* polydnavirus gene family. *Proceedings of the National Academy of Sciience of the United States of America*, **90**, 3765–9.

Dieckhoff C, Heimpel GE (2010) Role of cuticular hydrocarbons of aphid parasitoids in their relationship to aphid-attending ants. *Entomologia Experimentalis et Applicata*, **136**, 254–61.

Digilio MC (1999) *In vitro* rearing of the aphid parasitoid *Aphidius ervi* Haliday. *Bollettino del Laboratorio di Entomologia Agraria, Filippo Silvestri*, **55**, 151–61.

Digilio MC, Isidoro N, Tremblay E, Pennacchio F (2000) Host castration by *Aphidius ervi* venom proteins. *Journal of Insect Physiology*, **46**, 1041–50.

Dijkerman HJ, Koenders JTH (1988) Competition between *Trieces tricarinatus* and *Triclistus yponomeutae* in multiparasitised hosts. *Entomologia Experimentalis et Applicata*, **47**, 289–95.

Ding D, Swedenborg PD, Jones RL (1989) Chemical stimuli in host-seeking behavior of *Macrocentrus grandis* (Hymenoptera: Braconidae). *Annals of the Entomological Society of America*, **82**, 232–6.

Dion E, Zélé F, Simon J-C, Outreman Y (2011) Rapid evolution of parasitoids when faced with the symbiont-mediated resistance of their hosts. *Journal of Evolutionary Biology*, **24**, 741–50.

Dittrick LE, Chiang HC (1982) Developmental characteristics of *Macrocentrus grandii* as influenced by temperature and instar of its host, the European corn borer. *Journal of Insect Physiology*, **28**, 47–52.

Djoumad A, Stoltz D, Béliveau C, Boyle B, Kuhn L, Cusson M (2013) Ultrastructural and genomic characterization of a second banchine polydnavirus confirms the existence of shared features within this ichnovirus lineage. *Journal of General Virology*, **94**, 1888–95.

Dmoch J, Lewis WJ, Martin PB, Nordlund DA (1985) Role of host-produced stimuli and learning in host selection behavior of *Cotesia* (=*Apanteles*) *marginiventris* (Cresson) (Hymenoptera, Braconidae). *Journal of Chemical Ecology*, **11**, 453–63.

Dolphin KP, Quicke DLJ (2001) Estimating the global species-richness of incompletely described taxa: an example using parasitoid wasps (Hymenoptera: Braconidae). *Biological Journal of the Linnean Society*, **73**, 279–86.

Dondale CD (1954) Biology of *Agathis laticinctus* (Cress.) (Hymenoptera: Braconidae), a parasite of the eye-spotted bud moth, in Nova Scotia. *The Canadian Entomologist*, **86**, 40–4.

Dong K, Zhang D, Dahlman DL (1996) Down-regulation of juvenile hormone esterase and arylphorin production in *Heliothis virescens* larvae parasitised by *Microplitis croceipes*. *Archives of Insect Biochemistry and Physiology*, **32**, 237–48.

Donisthorpe HStJ, Wilkinson DS (1930) Notes on the genus *Paxylomma* (Hym. Brac.), with the description of a new species taken in Britain. *Transactions of the Entomological Society of London*, **78**, 87–93.

Donovan BJ (1991) Life cycle of *Sphecophaga vesparum* (Curtis) (Hymenoptera: Ichneumonidae), a parasitoid of some vespid wasps. *New Zealand Journal of Zoology*, **18**, 181–92.

Dorémus T, Jouan V, Urbach S, Cousserans F, Wincker P, Ravallec M, Wajnberg E, Volkoff A-N (2013a) *Hyposoter didymator* uses a combination of passive and active strategies to escape from the *Spodoptera frugiperda* cellular immune response. *Journal of Insect Physiology*, **59**, 500–8.

Dorémus T, Urbach S, Jouan V, Cousserans F, Ravallec M, Demettre E, Wajnberg E, Poulain J, Azéma-Dossat C, Darboux I, Escoubas JM, Colinet D, Gatti JL, Poirié M, Volkoff A-N (2013b) Venom gland extract is not required for successful parasitism in the polydnavirus-associated endoparasitoid *Hyposoter didymator* (Hym. Ichneumonidae) despite the presence of numerous novel and conserved venom proteins. *Insect Biochemistry and Molecular Biology*, **43**, 292–307.

Dowden PB (1938) *Rogas unicolor* (Wesm.), a braconid parasite of the satin moth. *Journal of Agricultural Research*, **56**, 523–36.

Dowden PB (1941) Parasites of the birch leaf-mining sawfly (*Phyllotoma nemorata*). *Technical Bulletin, United States Department of Agriculture*, **757**, 1–55.

Dowell RV (1978) Ovary structure and reproductive biologies of larval parasitoids of the alfalfa weevil (Coleoptera: Curculionidae). *The Canadian Entomologist*, **110**, 507–12.

Dowell RV, Horn DJ (1977) Adaptive strategies of larval parasitoids of the alfalfa weevil. *The Canadian Entomologist*, **109**, 641–8.

Dowell RV, Horn DJ (1978) Mating behavior of *Bathyplectes curculionis* (Hym.: Ichneumonidae) a parasitoid of the alfalfa weevil, *Hypera postica* (Col.: Curculionidae). *Entomophaga*, **23**, 271–3.

Dowton M (1999) Relationships among the cyclostome braconid (Hymenoptera: Braconidae) subfamilies inferred from a mitochondrial tRNA gene rearrangement. *Molecular Phylogenetics and Evolution*, **11**, 283–7.

Dowton M, Austin AD (1997) The evolution of strand-specific compositional bias. A case study in the hymenopteran mitochondrial 16S rDNA gene. *Molecular Biology and Evolution*, **14**, 109–12.

Dowton M, Austin AD (1998) Phylogenetic relationships among the microgastroid wasps (Hymenoptera: Braconidae): combined analysis of 16S and 28S rDNA genes. *Molecular Phylogenetics and Evolution*, **10**, 354–66.

Dowton M, Austin AD (2001) Simultaneous analysis of 16S, 28S, COI and morphology in the Hymenoptera: Apocrita – evolutionary transitions among parasitic wasps. *Biological Journal of the Linnean Society*, **74**, 87–111.

Dowton M, Austin AD, Antolin MF (1998) Evolutionary relationships among the Braconidae (Hymenoptera: Ichneumonoidea) inferred from partial 16S rDNA gene sequences. *Insect Molecular Biology*, **7**, 129–50.

Dowton M, Belshaw R, Austin AD, Quicke DLJ (2002) Simultaneous molecular and morphological analysis of braconid relationships (Insecta: Hymenoptera: Braconidae) indicates independent mt-tRNA gene inversions within a single wasp family. *Journal of Molecular Evolution*, **54**, 210–26.

Drea JJ (1968) Castration of male alfalfa weevils by *Microctonus* spp. *Journal of Economic Entomology*, **61**, 1291–5.

Drea JJ, Dysart RJ, Coles LW, Loan CC (1972) *Microctonus stelleri* (Hymemoptera: Braconidae, Euphorinae), a new parasite of the alfalfa weevil introduced into the United States. *The Canadian Entomologist*, **104**, 1445–56.

Drezen JM, Bézier A, Lesobre J, Huguet E, Cattolico L, Periquet G, Dupuy C (2006) The few virus-like genes of *Cotesia congregata* bracovirus. *Archives of Insect Biochemistry and Physiology*, **61**, 110–22.

Drooz AT, Quimby JW, Thompson LC, Kulman HM (1985) Introduction and establishment of *Olesicampe benefactor* Hinz (Hymenoptera: Ichneumonidae), a parasite of the larch sawfly, *Pristiphora erichsonii* (Hartig) (Hymenoptera: Tenthredinidae), in Pennsylvania. *Environmental Entomology*, **14**, 420–3.

Du Y-J, Poppy GM, Powell W (1996) Relative importance of semiochemicals from first and second trophic levels in host foraging behaviour of *Aphidius ervi*. *Journal of Chemical Ecology*, **22**, 1591–605.

Duan JJ, Messing RH (1997) Effect of two opiine parasitoids (Hymenoptera: Braconidae) introduced for fruit fly control on a native Hawaiian tephritid *Trupanea dubautiae* (Diptera: Tephritidae). *Biological Control*, **8**, 177–84.

Dudarenko GP (1974) Formation of the abdominal carapace in braconids (Hymenoptera: Braconidae) and some aspects of the classification of the family. *Entomological Review*, **53**, 80–90.

Duffield RM, Nordin JH (1970) Hibernation and the production of glycerol in the Ichneumonidae. *Nature*, **228**, 381.

Duodu YA, Antoh FF (1984) Effects of parasitism by *Apanteles sagax* (Hym.: Braconidae) on growth, food consumption and food utilization in *Sylepta derogata* larvae (Lep.: Pyralidae). *Entomophaga*, **29**, 63–71.

Duodu YA, Davis DW (1974) A comparison of growth, food consumption, and food utilization between unparasitised alfalfa weevil larvae and those parasitised by *Bathyplectes curculionis* (Thomson). *Environmental Entomology*, **3**, 705–10.

Dupuy C, Huguet E, Drezen J-M (2006) Unfolding the evolutionary story of polydnaviruses. *Virus Research*, **117**, 81–9.

Dweck HKM, Gadallah NS (2007) Description of the antennal sensilla of *Habrobracon hebetor*. *BioControl*, **53**, 841–56.

Dweck HK, Gadallah NS, Darwish E (2008) Structure and sensory equipment of the ovipositor of *Habrobracon hebetor* (Say) (Hymenoptera: Braconidae). *Micron*, **39**, 1255–1261.

Dysart RJ, Maltby HL, Brunson MH (1973) Larval parasites of *Oulema melanopus* in Europe and their colonization of the United States. *Entomophaga*, **18**, 133–67.

Eady RD (1974) The present state of nomenclature of wing venation in the Braconidae (Hymenoptera); its origins and comparison with related groups. *Journal of Entomology Series B, Taxonomy*, **43**, 63–72.

Eberhard WG (2000a) The natural history and behaviour of *Hymenoepimecis argyraphaga* (Hymenoptera: Ichneumonidae) a parasitoid of *Plesiometa argyra* (Araneae: Tetragnathidae). *Journal of Hymenoptera Research*, **9**, 220–40.

Eberhard WG (2000b) Spider manipulation by a wasp larva. *Nature*, **406**, 255–6.

Eberhard WG (2001) Under the influence: webs and building behavior of *Plesiometa argyra* (Araneae, Tetragnathidae) when parasitised by *Hymenoepimecis argyraphaga* (Hymenoptera, Ichneumonidae). *Journal of Arachnology*, **29**, 354–66.

Eberhard WG (2007) Swarming behaviour of two parasitic wasps, *Leurus caeruliventris* (Hymenoptera: Ichneumonidae) and *Conura* sp. (Hymenoptera: Chalcididae), at the same site. *Journal of the Kansas Entomological Society*, **80**, 255–8.

Eberhard WG (2010a) New types of behavioral manipulation of host spiders by a parasitoid wasp. *Psyche*, **2010**, 950614.

Eberhard WG (2010b) Recovery of spiders from the effects of parasitic wasps: implications for fine-tuned mechanisms of manipulation. *Animal Behaviour*, **79**, 375–383.

Edgar WD (1971) Aspects of the ecology and energetics of the egg sac parasites of the wolf spider *Pardosa lugubris* (Walckenaer). *Oecologia*, **7**, 155–63.

Edson KM, Vinson SB (1976) The function of the anal vesicle in respiration and excretion in the braconid wasp, *Microplitis croceipes*. *Journal of Insect Physiology*, **22**, 1037–43.

Edson KM, Vinson SB (1977) Nutrient absorption by the anal vesicle of the braconid wasp, *Microplitis croceipes*. *Journal of Insect Physiology*, **23**, 5–8.

Edson KM, Vinson SB (1979) A comparative morphology of the venom apparatus of female braconids (Hymenoptera, Braconidae). *The Canadian Entomologist*, **111**, 1013–24.

Edson KM, Norton WN, Vinson SB (1977) The ultrastructure of the anal vesicle of the parasitoid, *Microplitis croceipes* (Hymenoptera: Braconidae). *Journal of Morphology*, **154**, 243–58.

Edson KM, Vinson SB, Stoltz DB, Summers MD (1981) Virus in a parasitoid wasp: suppression of the cellular immune response in the parasitoid's host. *Science*, **211**, 582–3.

Edson KM, Barlin MR, Vinson SB (1982) Venom apparatus of braconid wasps: Comparative ultrastructure of reservoirs and gland filaments. *Toxicon*, **20**, 553–62.

Eggleton P (1989) The phylogeny and evolutionary biology of the Pimplinae (Hymenoptera: Ichneumonidae). Unpublished Ph.D. thesis, University of London.

Eggleton P (1991) Patterns in male mating strategies of the Rhyssini: a holophyletic group of parasitoid wasps (Hymenoptera: Ichneumonidae). *Animal Behaviour*, **41**, 829–38.

Eggleton P (2008) Male reproductive behaviour of the parasitoid wasp *Lytarmes maculipennis* (Hymenoptera: Ichneumonidae). *Ecological Entomology*, **15**, 357–60.

Eichhorn O (1988) Untersuchungen über die Fichtengespinsblattwespen *Cephalcia* spp. Panz. (Hym., Pamphiliidae). *Journal of Applied Entomology*, **105**, 105–40.

Eichhorn O, Pschorn-Walcher H (1973) The parasites of the birch leaf-mining sawfly (*Fenusa pusilla* [Lep.], Hym.: Tenthredinidae) in central Europe. *Technical Bulletin of the Commonwealth Institute of Biological Control*, **16**, 79–104.

Eijs EA, Ellers J, van Duinen G-J (1998) Feeding strategies in drosophilid parasitoids: the impact of natural food resources and energy reserves in females. *Ecological Entomology*, **23**, 133–8.

El Agoze M, Drezen J-M, Renault S, Periquet G (1994) Analysis of the reproductive potential of diploid males in the wasp *Diadromus pulchellus* (Hymenoptera: Ichneumonidae). *Bulletin of Entomological Research*, **84**, 213–18.

El Agoze M, Poirié M, Periquet G (1995) Precedence of the 1st male sperm in successive matings in the Hymenoptera *Diadromus pulchellus*. *Entomologia Experimentalis et Applicata*, **75**, 251–5.

Eller FJ, Bartelt RJ, Jones RL, Kulman HM (1984) Ethyl (Z)-9-hexadecenoate a sex-pheromone of *Syndipnus rubiginosus*, Hymenoptera, Ichneumonidae a sawfly parasitoid. *Journal of Chemical Ecology*, **10**, 291–300.

Ellers J (1996) Fat and eggs: an alternative method to measure the trade-off between survival and reproduction in insect parasitoids. *Netherlands Journal of Zoology*, **46**, 227–35.

Ellers J, Jervis MA (2004) Why are so few parasitoid species pro-ovigenic? *Evolutionary Ecology Research*, **6**, 993–1002.

Ellers J, van Alphen JJM (1997) Life history evolution in *Asobara tabida*: plasticity in allocation of fat reserves to survival and reproduction. *Journal of Evolutionary Biology*, **10**, 771–85.

Ellers J, van Alphen JJM, Sevenster JG (1998) A field study of size–fitness relationships in the parasitoid *Asobara tabida*. *Journal of Animal Ecology*, **67**, 318–24.

Ellers J, Driessen G, Sevenster JG (2000) The shape of the trade-off function between egg production and life span in the parasitoid *Asobara tabida*. *Netherlands Journal of Zoology*, **50**, 29–36.

Elliott JM (1982) The life cycle and spatial distribution of the aquatic parasitoid *Agriotypus armatus* (Hymenoptera: Agriotypidae) and its caddis host, *Silo pallipes* (Trichoptera: Goeridae). *Journal of Animal Ecology*, **51**, 923–41.

Elliott JM (1983) The responses of the aquatic parasitoid *Agriotypus armatus* (Hymenoptera: Agriotypidae) to the spatial distribution and density of its caddis host *Silo pallipes* (Trichoptera: Goeridae). *Journal of Animal Ecology*, **52**, 315–30.

El-Sawaf BM, Zohdy MZM (1976) Host–parasite relationship. 2 – Cholinesterase activity of the larvae of the rice moth *Corcyra cephalonica* [Lep.: Pyralidae] parasitized by *Bracon hebetor* [Hym.: Braconidae]. *Entomophaga*, **21**, 99–101.

El-Sufty R, Führer E (1981) Wechselbeziehungen zwischen *Pieris brassicae* L. (Lep., Pieridae), *Apanteles glomeratus* L. (Hym., Braconidae) und dem Pilz *Bedauveria bassiana* (Bals.) Vuill. *Zeitschrift fur Angewandte Entomologie*, **92**, 321–9.

Elzinga JA, Harvey JA, Biere A (2003) The effects of host weight at parasitism on fitness correlates of the gregarious koinobiont parasitoid *Microplitis tristis* and consequences for food consumption by its host, *Hadena bicruris*. *Entomologia Experimentalis et Applicata*, **108**, 95–106.

Emre I (1988) Effects of meridic diet on fecundity of adult females of *Pimpla turionellae* L. (Hymenoptera: Ichneumonidae). *Doga Turkish Journal of Biology*, **12**, 101–5.

Er A, Uçkan F, Rivers DB, Ergin E, Sak O (2010) Effects of parasitization and envenomation by the endoparasitic wasp *Pimpla turionellae* (Hymenoptera: Ichneumonidae) on hemocyte numbers, morphology, and viability of its host *Galleria mellonella* (Lepidoptera: Pyralidae). *Annals of the Entomological Society of America*, **103**, 273–82.

Eslin P, Prévost G (1996) Variation in *Drosophila* concentration of haemocytes associated with different ability to encapsulate *Asobara tabida* larval parasitoid. *Journal of Insect Physiology*, **42**, 549–55.

Eslin P, Giordanengo P, Fourdrain Y, Prévost G (1996) Avoidance of encapsulation in the absence of VLP by a braconid parasitoid of *Drosophila* larvae: an ultrastructural study. *Canadian Journal of Zoology*, **74**, 2193–8.

Espelie KE, Brown JJ (1990) Cuticular hydrocarbons of species which interact on four trophic levels: apple, *Malus pumila* Mill.; codling moth, *Cydia pomonella* L.; a hymenopteran parasitoid, *Ascogaster quadridentata* Wesmael; and a hyperparasite, *Perilampus fulvicornis* Ashmead. *Comparative Biochemistry and Physiology*, **95B**, 131–6.

Evans AC (1933) Comparative observations on the morphology and biology of some hymenopterous parasites of carrion-infesting Diptera. *Bulletin of Entomological Research*, **24**, 385–405.

Evenhuis HH, Vlug HJ (1983) The hymenopterous parasites of leaf-feeding apple tortricids (Lepidoptera, Tortricidae) in The Netherlands. *Tijdschrift voor Entomologie*, **126**, 109–35.

Ewen AB, Arthur AP (1976) Cuticular encystment in three noctuid species (Lepidoptera): induction by acid gland secretion from an ichneumonid parasite (*Banchus flavescens*). *Annals of the Entomological Society of America*, **69**, 1087–90.

Fahringer J (1936) Über einige merkwürdige und seltene Hymenopteren-Gattungen aus Afrika. *Festschrift 60. Geburtst. E. Strand, Riga*, **1**, 568–90.

Falabella P, Tremblay E, Pennacchio F (2000) Host regulation by the aphid parasitoid *Aphidius ervi*: the role of teratocytes. *Entomologia Experimentalis et Applicata*, **97**, 1–9.

Falabella P, Perugino G, Caccialupi P, Riviello L, Varricchio P, Tranfaglia A, Rossi M, Malva C, Graziani F, Moracci M, Pennacchio F (2005) A novel fatty acid binding protein produced by teratocytes of the aphid parasitoid *Aphidius ervi*. *Insect Molecular Biology*, **14**, 195–205.

Falabella P, Riviello L, Caccialupi P, Rossodivita T, Teresa Valente M, Luisa De Stradis M, Tranfaglia A, Varricchio P, Gigliotti S, Graziani F, Malva C, Pennacchio F (2007) A γ-glutamyl transpeptidase of *Aphidius ervi* venom induces apoptosis in the ovaries of host aphids. *Insect Biochemistry and Molecular Biology*, **37**, 453–65.

Fallis AM (1942) The life cycle of *Apanteles carpatus* (Say) (Hymenoptera: Braconidae)., a parasite of the webbing clothes moth, *Tineola bisselliella* Hum. *The Canadian Journal of Research (D)*, **20**, 13–19.

Faria CA, Wäckers FL, Turlings TCJ (2008) The nutritional value of aphid honeydew for non-aphid parasitoids. *Basic and Applied Ecology*, **9**, 286–97.

Farris SM, Schulmeister S (2011) Parasitoidism, not sociality, is associated with the evolution of elaborate mushroom bodies in the brains of hymenopteran insects. *Proceedings of the Royal Society B: Biological Sciences*, **278**, 940–51.

Feddersen I, Sander K, Schmidt O (1986) Virus-like particles with host protein-like antigenic determinants protect an insect parasitoid from encapsulation. *Experientia*, **42**, 1278–81.

Federici BA, Bigot Y (2003) Origin and evolution of polydnaviruses by symbiogenesis of insect DNA viruses in endoparasitic wasps. *Journal of Insect Physiology*, **49**, 419–32.

Federici BA, Bigot Y (2009) Evolution of immunosuppressive organelles from DNA viruses in insects. In P Pontarotti (ed.), *Evolutionary Biology: Concepts, Molecular and Morphological Evolution. Proceedings of the 13th Evolutionary Biology Meeting, Marseille*, pp. 229–48.

Fellowes MDE, Kraaijeveld AR, Godfray HCJ (1999) Cross resistance following artificial selection for increased host defence against parasitoids in *Drosophila melanogaster*. *Evolution*, **53**, 966–72.

Feng CJ, Qiu HG, Qiu ZL, Fu WJ (2004) Effects of parasitism by *Macrocentrus cingulum* Brischke (Hymenoptera: Braconidae) on the activity of phenoloxidase in larvae of the Asian corn borer *Ostrinia furnacalis* Guénée (Lepidoptera: Pyralidae). *Acta Entomologica Sinica*, **47**, 298–304 (in Chinese).

Feng CJ, Huang JH, Song QS, Stanley D, Lu WJ, Zhang Y, Huang YP (2011) Parasitism by *Macrocentrus cingulum* (Hymenoptera: Braconidae) influences expression of prophenoloxidase in Asian corn borer *Ostrinia furnacalis*. *Archives of Insect Biochemistry and Physiology*, **77**, 99–117.

Ferkovich SM, Dillard CR (1986) A study of uptake of radiolabeled host proteins and synthesis during development of eggs of the endoparasitoid, *Microplitis croceipes* (Cresson) (Braconidae). *Insect Biochemistry*, **16**, 337–45.

Ferkovich SM, Dillard CR, Oberlander H (1991) Stimulation of embryonic development in *Microplitis croceipes* (Braconidae) in cell culture media preconditioned with a fat body cell line derived from a nonpermissive host, gypsy moth, *Lymantria dispar*. *Archives of Insect Biochemistry and Physiology*, **18**, 169–75.

Ferkovich SM, Morales-Ramos JA, Rojas MG, Oberlander H, Carpenter JE, Greany P (1999) Rearing of ectoparasitoid *Diapetimorpha introita* on an artificial diet: supplementation with insect cell line derived factors. *BioControl*, **44**, 29–45.

Fernandez-Triana JL, Ward DF, Whitfield JB (2011) *Kiwigaster* gen. nov (Hymenoptera: Braconidae) from New Zealand: the first Microgastrinae with sexual dimorphism in number of antennal segments. *Zootaxa*, **2932**, 24–32.

Ferrari J, Müller CB, Kraaijeveld AR, Godfray HCJ (2001) Clonal variation and covariation in aphid resistance to parasitoids and a pathogen. *Evolution*, **55**, 1805–14.

Field SA, Keller MA (1993a) Courtship and intersexual signaling in the parasitic wasp *Cotesia rubecula* (Hymenoptera, Braconidae). *Journal of Insect Behavior*, **6**, 737–50.

Field SA, Keller MA (1993b) Alternative mating tactics and female mimicry as post-copulatory mate-guarding behaviour in the parasitic wasp *Cotesia rubecula*. *Animal Behaviour*, **46**, 1183–9.

Field SA, Keller MA (1994) Localization of the females sex pheromone gland in *Cotesia rubecula* Marshall (Hymenoptera: Braconidae). *Journal of Hymenoptera Research*, **3**, 151–6.

Filipy FL, Burbutis PB, Fuester RW (1985) Biological control of the European wheat stem sawfly in Delaware (Hymenoptera: Cephidae). *Environmental Entomology*, **14**, 665–8.

Fincke OM, Higgins L, Rojas E (1990) Parasitism of *Nephila clavipes* (Araneae: Tetragnathidae) parasitised by an ichneumonid wasp (Hymenoptera: Polysphinctini) in Panama. *Journal of Arachnology*, **18**, 321–9.

Fink DE (1926) The biology of *Macrocentrus ancylivora* Rohwer, an important parasite of the strawberry leaf roller (*Ancylus comptana* Froehl.). *Journal of Agricultural Research*, **32**, 1121–34.

Finlayson T (1964) The caudal appendage of final-instar larvae of some Porizontinae (Hymenoptera: Ichneumonidae). *The Canadian Entomologist*, **96**, 1155–8.

Finlayson T (1966) The false cocoon of *Hyposoter parorgyiae* (Vier.) (Hymenoptera: Ichneumonidae). *The Canadian Entomologist*, **98**, 139.

Finlayson T (1967) A classification of the subfamily Pimplinae (Hymenoptera, Ichneumonidae) based on final instar larval characteristics. *The Canadian Entomologist*, **99**, 1–8.

Finlayson T (1975) The cephalic structures and spiracles of final instar larvae of the subfamily Campopleginae (Hymenoptera: Ichneumonidae), tribe Campoplegini. *Memoirs of the Entomological Society of Canada*, **107**, 9–137.

Finlayson T (1976) Cephalic structures and spiracles of final instar larvae of the genus *Toxophoroides* (Hymenoptera, Ichneumonidae, Lycorininae). *Canadian Entomologist*, **108**, 981–4.

Finlayson T (1990) The systematics and taxonomy of final-instar larvae of the family Aphidiidae (Hymenoptera). *Memoirs of the Entomological Society of Canada*, **152**, 1–74.

Finney GL, Flanders SE, Smith HS (1947) Mass culture of *Macrocentrus ancylivorus* and its host, the potato tuber moth. *Hilgardia*, **17**, 437–83.

Fischer K (1932) *Agriotypus armatus* (Walk.) (Hymenoptera) and its relations with its hosts. *Proceedings of the Zoolical Society*, 451–61.

Fischer M (1961) Ein neues Opiinen-Genus aus Afrika (Hymenoptera: Braconidae). *Annales Historico-Naturales Musei Nationalis Hungarici*, **53**, 439–41.

Fischer M (1972) Hymenoptera Braconidae (Opiinae I). *Das Tierreich*, **91**, 1–620.

Fischer M, Tormos J, Pardo X, Asís JD (2008) Description of adults, preimaginal phases, and the venom apparatus of a new species of *Aspilota* Förster from Spain (Hymenoptera: Braconidae). *Zoological Studies*, **47**, 247–57.

Fischer S, Samietz J, Wäckers F L, Dorn S (2004) Perception of chromatic cues during host location by the pupal parasitoid *Pimpla turionellae* (L.) (Hymenoptera: Ichneumonidae). *Environmental Entomology*, **33**, 81–7.

Fisher RC (1959) Life history and ecology of *Horogenes chrysostictos* Gmelin (Hymenoptera, Ichneumonidae), a parasite of *Ephestia sericarium* (Lepidoptera, Phycitidae). *Canadian Journal of Zoology*, **37**, 429–46.

Fisher RC (1963) Oxygen requirements and physiological suppression of supernumerary insect parasitoids. *Journal of Experimental Biology*, **40**, 531–40.

Fitton MG (1981) The British Acaenitinae (Hymenoptera: Ichneumonidae). *Entomologist's Gazette*, **32**, 185–92.

Fitton MG (1984) A review of the British Collyriinae, Eucerotinae, Stilbopinae and Neorhacodinae (Hymenoptera: Ichneumonidae). *Entomologist's Gazette*, **35**, 185–95.

Fitton MG (1985) The ichneumon-fly genus *Banchus* (Hymenoptera) in the Old World. *Bulletin of the British Museum (Natural History)*, **51**, 1–60.

Fitton MG (1987) A review of the *Banchus* group of ichneumon-flies, with a revision of the Australian genus *Philogalleria* (Hymenoptera, Ichneumonidae). *Systematic Entomology*, **12**, 33–45.

Fitton MG, Ficken L (1990) British ichneumon flies of the tribe Oedemopsini (Hym., Ichneumonidae). *The Entomologist*, **109**, 200–14.

Fitton MG, Gauld ID (1976) The family-group names of the Ichneumonidae (excluding Ichneumoninae) (Hymenoptera). *Systematic Entomology*, **1**, 247–58.

Fitton MG, Gauld ID (1978) Further notes on family group names in Ichneumonidae. *Systematic Entomology*, **3**, 245–7.

Fitton MG, Rotheray GE (1982) A key to the European genera of diplazontine ichneumon-flies with notes on the British Fauna. *Systematic Entomology*, **7**, 311–20.

Fitton MG, Gauld ID, Shaw MR (1982) The taxonomy and biology of the British Adelognathinae (Hymenoptera: Ichneumonidae). *Journal of Natural History*, **16**, 275–83.

Fitton MG, Shaw MR, Austin AD (1987) The Hymenoptera associated with spiders in Europe. *Zoological Journal of the Linnean Society*, **90**, 65–93.

Fitton MG, Shaw MR, Gauld ID (1988) Pimpline ichneumon-flies. Hymenoptera, Ichneumonidae (Pimplinae). *Handbooks for the Identification of British Insects*, **7**, 1–110.

Flanders RV, Oatman ER (1982) Laboratory studies on the biology of *Orgilus jenniae* (Hymenoptera: Braconidae), a parasitoid of the potato tuberworm, *Phthorimaea operculella* (Lepidoptera: Gelechiidae). *Hilgardia*, **50**, 1–33.

Flanders RV, Oatman ER (1987) Competitive interactions among endophagous parasitoids of potato tuberworm larvae in southern California. *Hilgardia*, **55**, 1–34.

Flanders SE (1945) The role of the spermatophore in the mass propagation of *Macrocentrus ancylivorus*. *Journal of Economic Entomology*, **38**, 323–7.

Flanders SE (1950) Regulation of ovulation and egg disposal in the parasitic Hymenoptera. *The Canadian Entomologist*, **82**, 134–40.

Fleming JG, Summers MD (1986) *Campoletis sonorensis* endoparasitic wasps contain forms of *C. sonorensis* virus DNA suggestive of integrated and extrachromosomal polydnavirus DNAs. *Journal of Virology*, **57**, 552–62.

Fleming JG, Summers MD (1991) Polydnavirus DNA is integrated in the DNA of its parasitoid wasp host. *Proceedings of the National Academy of Sciences of the United States of America*, **88**, 9770–4.

Fleming JA, Blissard GW, Summers MD, Vinson SB (1983) Expression of *Campoletis sonorensis* virus in the parasitized host, *Heliothis virescens*. *Journal of Virology*, **48**, 74–8.

Flores S, Nassar JM, Quicke DLJ (2005) Reproductive phenology and pre-dispersal seed predation in *Protium tovarense* (Burseraceae), with description of the first known phytophagous '*Bracon*' species (Hymenoptera: Braconidae: Braconinae). *Journal of Natural History*, **39**, 3663–85.

Flores-Prado L, Niemeyer HM (2012) Host location by ichneumonid parasitoids is associated with nest dimensions of the host bee species. *Neotropical Entomology*, **41**, 283–7.

Foerster A (1862) Synopsis der Familien und Gattungen der Braconen. *Verhandlungen des naturhistorischen Vereins der preussischen Rheinlande, Westfalens u. des reg.-bez Osnabruch*, **19**, 224–88.

Foissner W, van Achterberg C (1997) The valid name for the genus *Loxocephalus* Foerster, 1862 (Insecta, Hymenoptera: Braconidae), preoccupied by *Loxocephalus* Eberhard, 1862 (Protozoa: Ciliophora). *Zoologische Mededelingen, Leiden*, **71**, 31–2.

Foray V, Henri H, Martinez S, Gibert P, Desouhante E (2013) Occurrence of arrhenotoky and thelytoky in a parasitic wasp *Venturia canescens* (Hymenoptera: Ichneumonidae): effect of endosymbionts or existence of two distinct reproductive modes? *European Journal of Entomology*, **110**, 103–7.

Fortier JC, Nishida K (2004) A new species and host association biology of neotropical *Compsobraconoides* Quicke (Hymenoptera: Braconidae). *Journal of Hymenoptera Research*, **13**, 228–33.

Fortier JC, Shaw SR (1999) Cladistics of the *Aleiodes* lineage of the subfamily Rogadinae (Hymenoptera: Braconidae). *Journal of Hymenoptera Research*, **8**, 204–37.

Franz JM (1958) Studies on *Laricobius erichsonii* Rosenh. (Coleoptera: Derodontidae) a predator on dermesids. Part I, distribution, life-history and ecology. *Entomophaga*, **3**, 109–64.

Franzen C, Futerman PH, Schroeder J, Salzberger B, Kraaijeveld AR (2006) An ultrastructural and molecular study of *Tubulinosema kingi* Kramer (Microsporidia: Tubulinosematidae) from *Drosophila melanogaster* (Diptera: Drosophilidae) and its parasitoid *Asobara tabida* (Hymenoptera: Braconidae). *Journal of Invertebrate Pathology*, **91**, 158–67.

Fraser SEM, Dytham C, Mayhew PJ (2007) Determinants of parasitoids abundance and diversity in woodland habitats. *Journal of Applied Ecology*, **44**, 352–61.

Fraser SEM, Dytham C, Mayhew PJ (2008) The effectiveness and optimal use of Malaise traps for monitoring parasitoid wasps. *Insect Conservation and Diversity*, **1**, 22–31.

Frasson L, Ferroni F, Ko SY, Dogangil G, Rodriguez y Baena F (2012) Experimental evaluation of a novel steerable probe with a programmable bevel tip inspired by nature. *Journal of Robotic Surgery*, **6**, 189–97.

Frilli F (1965) Studi sugli imenotteri ichneumonidi. III. Il *Latibulus argiolus* (Rossi) parassita delle larve di *Polistes* spp. *Entomologica*, **2**, 21–53.

Fritzsche-Hoballah MEF, Turlings TCJ (2001) Experimental evidence that plants under caterpillar attack may benefit from attracting parasitoids. *Evolutionary Ecology Research*, **3**, 553–65.

Frohne WC (1939) Semiaquatic Hymenoptera in north Michigan lakes. *Transactions of the American Microscopical Society*, **58**, 228–40.

Fuester RW, Taylor PB, Eisenberg A (1989) Suitability of prepupae and pupae of the gypsy moth (Lepidoptera: Lymantriidae) as hosts of *Coccygomimus disparis* (Hymenoptera: Ichneumonidae). *Environmental Entomology*, **18**, 338–41.

Führer E (1973a) Der Einfluss zweier Endoparasiten auf die Gonadenentwicklung in den Larven von *Epiblema* (=*Eucosma*) *tedella* Cl. (Lep., Tortricidae). *Zeitschrift für Parasitenkunde*, **41**, 187–206.

Führer E (1973b) Sekretion von Mucopolysacchariden im weiblichen Geschlechtsapparat von *Pimpla turionellae* L. (Hym., Ichneumonidae). *Zeitschrift für Parasitenkunde*, **41**, 207–13.

Führer E, El-Sufty R (1979) Produktion fungistatischer metabolite durch Teratocyten von *Apanteles glomeratus* L. (Hym., Braconidae). *Zeitschrift für Parasitenkunde*, **59**, 21–5.

Führer E, Kilincer N (1972) Die motorische Aktivität der endoparasitischen Larven von *Pimpla turionellae* L. und *Pimpla aquiloniae* Cress. (Hym., Ichneumonidae) in der Wirtspuppe. *Entomophaga*, **17**, 149–63.

Führer E, Willers D (1986) The anal secretion of the endoparasitic larva *Pimpla turionellae*: sites of production and effects. *Journal of Insect Physiology*, **32**, 361–5, 367.

Fukushima J, Kuwahara Y, Yamada A, Suzuki T (1990) New noncyclic homo-diterpene from the sting glands of *Bracon hebetor* Say (Hymenoptera: Braconidae). *Agricultural and Biological Chemistry*, **54**, 809–10.

Fulton BB (1940) The hornworm parasite *Apanteles congregates* Say and the hyperparasite, *Hypopteromalus tabacum* (Fiych). *Annals of the Entomological Society of America*, **33**, 231–44.

Furihata SX, Matsumoto H, Kimura MT, Hayakawa Y (2013) Venom components of *Asobara japonica* impair cellular immune responses of host *Drosophila melanogaster*. *Archives of Insect Biochemistry and Physiology*, **83**, 86–100.

Futerman PH, Layen SJ, Kotzen ML, Franzen C, Kraaijeveld AR, Godfray HCJ (2006) Fitness effects and transmission routes of a microsporidian parasite infecting *Drosophila* and its parasitoids. *Parasitology*, **132**, 479–92.

Gaasch CM, Pickering J, Moore CT (1998) Flight phenology of parasitic wasps (Hymenoptera: Ichneumonidae) in Georgia's piedmont. *Environmental Entomology*, **27**, 606–14.

Gandolfi M, Mattiacci L, Dorn S (2003) Preimaginal learning determines adult response to chemical stimuli in a parasitic wasp. *Proceedings of the Royal Society B: Biological Sciences*, **270**, 2623–9.

Ganesalingam VK (1972) Anatomy and histology of the sense organs of the ovipositor of the ichneumonid wasp, *Devorgilla canescens*. *Journal of Insect Physiology*, **18**, 1857–67.

Gangrade GA (1964) The biology of *Campoletis perdistinctus* (Hymenoptera: Ichneumonidae) in Madhya Pradesh, India. *Annals of the Entomological Society of America*, **57**, 570–4.

Gao Y, Luo LZ, Hammond A (2007). Antennal morphology, structure and sensilla distribution in *Microplitis pallidipes* (Hymenoptera: Braconidae). *Micron*, **38**, 684–93.

Gärdenfors U (1986) Taxonomic and biological revision of Palearctic *Ephedrus* Haliday (Hymenoptera: Braconidae, Aphidiinae). *Entomologica Scandinavica, Supplement*, **27**, 1–95.

Gärdenfors U (1990) *Trioxys apterus* sp. n. from Ecuador, a new wingless species of Aphidiinae (Hymenoptera: Braconidae). *Insect Systematics & Evolution*, **21**, 67–9.

Gaskett AC (2012) Floral shape mimicry and variation in sexually deceptive orchids with a shared pollinator. *Biological Journal of the Linnean Society*, **106**, 469–81.

Gaston KJ (1993) Spatial patterns of species description: how is our knowledge of the global insect fauna growing? *Biological Conservation*, **67**, 37–40.

Gaston KJ, Gauld ID (1993) How many species of pimplines (Hymenoptera: Ichneumonidae) are there in Costa Rica? *Journal of Tropical Ecology*, **9**, 491–9.

Gaston KJ, Blackburn TM, Loder N (2005) Which species are described first?: the case of North American butterflies. *Biodiversity and Conservation*, **4**, 119–27.

Gauld ID (1976a) The classification of the Anomaloninae (Hymenoptera: Ichneumonidae). *Bulletin of the British Museum (Natural History): Entomology*, **33**, 1–135.

Gauld ID (1976b) The taxonomy of the genus *Heteropelma* Wesmael (Hymenoptera: Ichneumonidae). *Bulletin of the British Museum (Natural History): Entomology*, **34**, 155–219.

Gauld ID (1977a) A revision of the Ophioninae (Hymenoptera: Ichneumonidae) of Australia. *Australian Journal of Zoology (Supplementary Series)*, **49**, 1–112.

Gauld ID (1977b) *Australoglypta* gen. n. In: Danthanarayana W, Farrugia D, Gauld ID (1977) Studies on the biology and systematic position of a new species of ichneumonid parasitizing the light brown apple moth, *Epiphyas postvittana* (Walker) (Lepidoptera: Tortricidae), in Australia. *Bulletin of Entomological Research*, **67**, 607–17.

Gauld ID (1979) An analysis of the classification of the *Ophion* genus-group (Ichneumonidae). *Systematic Entomology*, **5**, 59–82.

Gauld ID (1983) The classification, evolution and distribution of the Labeninae, an ancient southern group of Ichneumonidae (Hymenoptera). *Systematic Entomology*, **8**, 167–78.

Gauld ID (1984a) *An Introduction to the Ichneumonidae of Australia. with a Contribution on the Metopiinae by M. G. Fitton*. British Museum (Natural History), London, 413 pp.

Gauld ID (1984b) The Pimplinae, Xoridinae, Acaenitinae and Lycorininae of Australia. *Bulletin of the British Museum (Natural History): Entomology*, **49**, 235–339.

Gauld ID (1985) The phylogeny, classification and evolution of parasitic wasps of the subfamily Ophioninae (Ichneumonidae). *Bulletin of the British Museum (Natural History): Entomology*, **51**, 1–185.

Gauld ID (1986a) Taxonomy, its limitations and its role in understanding parasitoid biology. In J Waage and D Greathead (eds), *Insect Parasitoids*, Academic Press, London, pp. 1–21.

Gauld ID (1986b) Latitudinal gradients in ichneumonid species-richness in Australia. *Ecological Entomology*, **11**, 155–61.

Gauld ID (1987a) Some factors affecting the composition of tropical ichneumonid faunas. *Biological Journal of the Linnean Society*, **30**, 299–312.

Gauld ID (1987b) The ichneumonid parasitoids associated with South-east Asian Limacodidae. In MJW Cock, HCJ Godfray and JD Holloway (eds), *Slug and Nettle Caterpillars*, CAB International, Wallingford, pp. 123–38.

Gauld ID (1988a) Evolutionary patterns of host utilization by ichneumonid parasitoids (Hymenoptera: Ichneumonidae and Braconidae). *Biological Journal of the Linnean Society*, **35**, 351–77.

Gauld ID (1988b) A survey of the Ophioninae (Hymenoptera: Ichneumonidae) of tropical Mesoamerica with special reference to the fauna of Costa Rica. *Bulletin of the British Museum (Natural History): Entomology*, **57**, 1–309.

Gauld ID (1991) The Ichneumonidae of Costa Rica, 1. *Memoirs of the American Entomological Institute*, **47**, 1–589.

Gauld ID (1995) Ichneumonidae. In PE Hanson, and ID Gauld (eds), *The Hymenoptera of Costa Rica*, Oxford University Press, New York and The Natural History Museum, London, pp. 390–431.

Gauld ID (1997) The Ichneumonidae of Costa Rica, 2. *Memoirs of the American Entomological Institute*, **57**, 1–485.

Gauld ID (2000) The Ichneumonidae of Costa Rica, 3. *Memoirs of the American Entomological Institute*, **63**, 1–453.

Gauld ID (2002) The Ichneumonidae of Costa Rica, 4. Introduction and keys to species of subfamilies: Metopiinae by Ian Gauld & Rudo Sithole; Banchinae (Atrophini, Banchini) by Jesus Ugalde Gomez and Ian Gauld; and Banchinae (Glyptini) by Carolina Godoy and Ian Gauld. *Memoirs of the American Entomological Institute*, **66**, 1–768.

Gauld ID, Bolton B (eds) (1988) *The Hymenoptera*, Oxford University Press, Oxford and The Natural History Museum, London, 332 pp.

Gauld ID, Bradshaw K (1997) The subfamily Anomaloninae. In ID Gauld (1997) The Ichneumonidae of Costa Rica, 2. *Memoirs of the American Entomological Institute*, **57**, 1–485.

Gauld ID, Dubois J (2006) Phylogeny of the *Polysphincta* group of genera (Hymenoptera: Ichneumonidae; Pimplinae): a taxonomic revision of spider ectoparasitoids. *Systematic Entomology*, **31**, 529–64.

Gauld ID, Fitton MG (1981) Keys to the British xoridine parasitoids of the wood-boring beetles (Hymenoptera: Ichneumonidae). *Entomologist's Gazette*, **32**, 259–67.

Gauld ID, Fitton MG (1987) Sexual dimorphism in the Ichneumonidae: a response to Hurlbutt. *Biological Joural of the Linnaean Society*, **31**, 291–300.

Gauld ID, Gaston KJ (1994) The taste of enemy-free space: parasitoids and nasty hosts. In BA Hawkins and W Sheehan (eds), *Parasitoid Community Ecology*, Oxford University Press, Oxford, pp. 279–99.

Gauld ID, Gaston KJ (1995) The Costa Rican Hymenoptera Fauna. In P Hanson and ID Gauld (eds), *The Hymenoptera of Costa Rica*, Oxford University Press, Oxford, pp. 13–19.

Gauld ID, Hanson PE (1995) Carnivory in the larval Hymenoptera. In P Hanson and ID Gauld (eds), *The Hymenoptera of Costa Rica*, Oxford University Press, Oxford, pp. 40–5.

Gauld ID, Holloway GA (1986) Australian ichneumonids of the tribes Labenini and Poecilocryptini. *Bulletin of the British Museum (Natural History): Entomology*, **53**, 107–49.

Gauld ID, Huddleston T (1976) The nocturnal Ichneumonoidea of the British Isles, including a key to the genera. *Entomologist's Gazette*, **27**, 35–49.

Gauld ID, Janzen DH (1994) The classification, evolution and biology of the Costa Rican species of *Cryptophion* (Hymenoptera, Ichneumonidae). *Zoological Journal of the Linnean Society*, **110**, 297–324.

Gauld ID, Janzen DH (2004) The systematics and biology of the Costa Rican species of parasitic wasps in the *Thyreodon* genus-group (Hymenoptera: Ichneumonidae). *Zoological Journal of the Linnean Society*, **141**, 297–351.

Gauld ID, Mitchell PA (1977) Ichneumonidae. Orthopelmatinae & Anomaloninae. *Handbooks for the Identification of the British Insects*, **7** (2b), 1–32.

Gauld ID, Mitchell PA (1978) *The Taxonomy, Distribution and Host Preferences of African Parasitic Wasps of the Subfamily Ophioninae (Hymenoptera: Ichneumonidae)*, Commonwealth Institute of Entomology, Slough, 287 pp.

Gauld ID, Mitchell PA (1981) *The Taxonomy, Distribution and Host Preferences of Indo-Papuan Parasitic Wasps of the Subfamily Ophioninae*, Commonwealth Institute of Entomology, Slough, 611 pp.

Gauld ID, Mound LA (1982) Homoplasy and the delineation of holophyletic genera in some insect groups. *Systematic Entomology*, **7**, 73–86.

Gauld ID, Sithole R (2002) Subfamily Metopiinae. In ID Gauld (ed.), The Ichneumonidae of Costa Rica, 4. *Memoirs of the American Entomological Institute*, **66**, 11–262.

Gauld ID, Wahl DB (2000a) The Labeninae (Hymenoptera: Ichneumonidae): a study in phylogenetic reconstruction and evolutionary biology. *Zoological Journal of the Linnean Society*, **129**, 271–374.

Gauld ID, Wahl DB (2000b) The Townesioninae: a distinct subfamily of Ichneumonidae (Hymenoptera) or a clade of the Banchinae? *Transactions of the American Entomological Society*, **126**, 279–92.

Gauld ID, Wahl DB (2002) The Eucerotinae: a Gondwanan origin for a cosmopolitan group of Ichneumonidae? *Journal of Natural History*, **36**, 2229–48.

Gauld ID, Wahl DB (2006) The relationship and taxonomic position of the genera *Apolophus* and *Scolomus* (Hymenoptera: Ichneumonidae). *Zootaxa*, **1130**, 35–41.

Gauld ID, Wahl DB (2013) *Genera Ichneumonorum Nearcticae*, American Entomological Institute, http://www.amentinst.org/GIN/ (accessed 18 December 2013).

Gauld ID, Ward S (2000) Subfamily Brachycyrtinae. In ID Gauld (ed.), *Memoirs of the American Entomological Institute*, **63**, 1–453.

Gauld ID, Gaston KJ, Janzen DH (1992) Plant allochemicals, tritrophic interactions and the anomalous diversity of tropical parasitoids: The "nasty" host hypothesis. *Oikos*, **65**, 353–7.

Gauld ID, Ugalde Gomez J, Godoy C (2002a) Subfamily Banchinae. in ID Gauld (ed.), The Ichneumonidae of Costa Rica, 4. *Memoirs of the American Entomological Institute*, **66**, 263–743.

Gauld ID, Wahl DB, Broad GR (2002b) The suprageneric groups of the Pimplinae (Hymenoptera: Ichneumonidae): a cladistic re-evaluation and evolutionary biological study. *Zoological Journal of the Linnean Society*, **136**, 421–85.

Gauthier N, Mandon N, Renault S, Bénédet F (2004) The *Acrolepiopsis assectella* silk cocoon: kairomonal function and chemical characterisation. *Journal of Insect Physiology*, **50**, 1065–74.

Geervliet JBF, Vet LEM, Dicke M (1994) Volatiles from damaged plants as major cues in long-range host-searching by the specialist parasitoid *Cotesia rubecula*. *Entomologia Experimentalis et Applicata*, **73**, 289–97.

Gelman DB, Carpenter JE, Greany PD (2000) Ecdysteroid levels/profiles of the parasitoid wasp, *Diapetimorpha introita*, reared on its host, *Spodoptera frugiperda* and on an artificial diet. *Journal of Insect Physiology*, **46**, 457–65.

Genduso P (1969) Morfologia dell'addome e dell'apparato riproduttore femminile dell'*Opius concolor siculus* Mon. *Bollettino dell'Istituto di Entomologia Agraria e dell'Osservatorio di Fitopatologia di Palermo*, **7**, 235–55.

Geneau CE, Wäckers FL, Luka H, Daniel C, Balmer O (2012) Selective flowers to enhance biological control of cabbage pests by parasitoids. *Basic and Applied Ecology*, **13**, 85–93.

Genieys P (1925) *Habrobracon brevicornis* Wesm; the effects of the environment and the variation it produces. *Annals of the Entomological Society of America*, **18**, 143–202.

Gentry G (2003). Multiple parasitoid visitors to the extrafloral nectaries of *Solanum adherens*. Is *S. adherens* an insectary plant? *Basic and Applied Ecology*, **4**, 405–11.

Georgiev G (2001) Notes on the biology and ecology of the parasitoids of the poplar clearwing moth, *Paranthrene tabaniformis* (Rott.) (Lep., Sesiidae) in Bulgaria. I. *Apanteles evonymellae* (Bouché, 1834) (Hym., Braconidae). *Journal of Applied Entomology*, **125**, 141–5.

Georgiev G, Delkov A (2003) Bioecological characteristics of *Bassus tumidulus* (Nees) (Hym., Braconidae), a parasitoid of the poplar twig borer, *Gypsonoma aceriana* (Dup.) (Lep., Tortricidae) in Bulgaria. *Journal of Applied Entomology*, **127**, 99–102.

Gerig L (1960) Zur Morphologie der Larvenstadien einiger parasitischen Hymenopteren des Grauen Larchwicklers (*Zeiraphera griseana* Hübner). *Zeitschrift für Angewandte Entomologie*, **46**, 121–77.

Gerling D, Rotary N (1974) Structure and function of the seminal vesicles and the spermatheca in *Bracon hebetor* Say (Hymenoptera: Braconidae). *International Journal of Insect Morphology and Embryology*, **3**, 159–62.

Ghosh SM, Abdurahiman UC (1984) Bioethology of *Meteoridea hutsoni* (Nixon) (Hymenoptera: Braconidae), a parasite of *Opisina arenosella* Walker, the black headed caterpillar pest of coconut. *Entomon*, **9**, 31–4.

Ghosh SM, Abdurahiman UC (1986) Two new records of hyperparasites on *Opisina arenosella* Walker, the caterpillar pest of coconut. *Entomon*, **11**, 75–6.

Gibbons JRH (1979) A model for sympatric speciation in *Megarhyssa* (Hymenoptera: Ichneumonidae): competitive speciation. *The American Naturalist*, **114**, 719–41.

Gibson LP (1970) Revision of the genus *Urosigalphus* of the United States and Canada (Hymenoptera: Braconidae). *Miscellaneous Publications of the Entomological Society of America*, **8**, 85–157.

Gillespie OR, Finlayson T (1979) The function of the caudal appendage cocoon jumping of *Phobocampe* sp. (Hymenoptera: Ichneumonidae: Campopleginae). *Journal of the Entomological Society of British Columbia*, **76**, 39–42.

Gillespie OR, Finlayson T (1983) Classification of final-instar larvae of the Ichneumoninae (Hymenoptera: Ichneumonidae). *Entomological Society of Canada. Memoirs*, No. 124.

Gilstrap FE, Hart WG (1987) *Biological Control of Mediterranean Fruit Fly in the United States and Central America*, ARS 56, United States Department of Agriculture, Washinton, DC, pp. 1–64.

Gimeno C, Belshaw R, Quicke DLJ (1997) Phylogenetic relationships of the Opiinae/Alysiinae (Hymenoptera: Braconidae) and the utility of cytochrome b, 16S and 28S D2 rDNA. *Insect Molecular Biology*, **6**, 273–84.

Giordana B, Milani A, Grimaldi A, Farneti R, Casartelli M, Ambrosecchio MR, Digilio MC, Leonardi MG, de Eguileor M, Pennacchio F (2003) Absorption of sugars and amino acids by the epidermis of *Aphidius ervi* larvae. *Journal of Insect Physiology*, **49**, 1115–24.

Giraldo-Vanegas H, Garcia JL (1992) Determinación del número de instares de *Eiphosoma vitticolle* Cresson (Hymenoptera: Ichneumonidae). *Boletin de Entomologia Venezolana*, **7**, 133–7.

Giraldo-Vanegas H, Garcia RJL (1994a) Behaviour of adults of *Eiphosoma vitticolle* Cresson (Hymenoptera: Ichneumonidae), parasitoid of *Spodoptera frugiperda* (Smith) (Lepidoptera: Noctuidae). *Boletin de Entomologia Venezolana*, **9**, 15–20.

Giraldo-Vanegas H, Garcia RJL (1994b) Comportiamento, descripcion y tiempo de desarrollo de los estados inmadurosde *Eiphosoma vitticolle* Cresson (Hymenoptera: Ichneumonidae), parasite de *Spodoptera frugiperda* (J.E. Smith) (Lepidoptera: Noctuidae). *Agronomia Tropical, Maracay*, **44**, 645–65.

Giraldo-Vanegas H, Garcia RJL (1995) Influence of feeding on the capacity for reproduction of *Eiphosoma vitticolle* Cresson (Hymenoptera: Ichneumonidae), parasitoid of *Spodoptera frugiperda* (J.E. Smith). *Agronomia Tropical Maracay*, **45**, 159–70 (in Spanish).

Giraud-Guille M-M, Chanzy H, Vuong R (1990) Chitin crystals in arthropod cuticles revealed by diffraction contrast transmission electron microscopy. *Journal of Structural Biology*, **103**, 232–40.

Given BB (1944) Notes of the anatomy of *Diadromus* (*Thyraeella*) *collaris* Grav. (Ichneumonidae, Hymenoptera). *Transactions of the Royal Society of New Zealand*, **74**, 154–64.

Glas PCG, Vet LEM (1983) Host-habitat location and host location by *Diachasma alloeum* Muesebeck (Hym.; Braconidae), a parasitoid of *Rhagoletis pomonella* Walsh (Dipt.; Tephritidae). *Netherlands Journal of Zoology*, **33**, 41–54.

Glinwood RT, Du YJ, Powell W (1999) Responses to aphid sex pheromones by the pea aphid parasitoids *Aphidius ervi* and *Aphidius eadyi*. *Entomologia Experimentalis et Applicata*, **92**, 227–32.

Glover PM (1934) The developmental stages of *Bracon tachardiae*, Cam. (Hym.). *Bulletin of Entomological Research*, **25**, 521–39.

Glover PM (1939) Notes on the biology and larval growth of *Aphrastobracon flavipennis* Ashmead) (Braconidae). *Indian Journal of Entomology*, **1**, 1–14.

Godfray HCJ (1994) *Parasitoids: Behavioral and Evolutionary Ecology*, Princeton University Press, Princeton, NJ.

Goh MZ, Morse DH (2010) Male mate search for female emergence sites by a parasitic wasp. *Animal Behaviour*, **80**, 391–8.

Gokhman VE (1988) Ecological and morphological aspects of formation and evolution of the ichneumonid flies of the subfamily Ichneumoninae (Hymenoptera, Ichneumonidae). *Entomologicheskoe Obozrenie*, **67**, 821–5 (in Russian with English summary).

Gokhman VE (1992) On the origin of endoparasitism in the subfamily Ichneumoninae (Hym., Ichneumonidae). *Zhurnal Obshchei Biologii*, **53**, 600–8.

Gokhman VE (1995) Trends of biological evolution in the subfamily Ichneumoninae and related groups (Hymenoptera Ichneumonidae): an attempt of phylogenetic reconstruction. *Russian Entomological Journal*, **4**, 91–103.

Gokhman VE (2001) Chromosomes of the family Ichneumonidae (Hymenoptera). *Zoologichesky Zhurnal*, **80**, 968–75 (in Russian with English summary).

Gokhman VE (2002) Chromosomal analysis of the superfamilies Ichneumonoidea and Chalcidoidea (Hymenoptera). In G Melika and C Thuróczy (eds), *Parasitic Wasps: Evolution, Systematics, Biodiversity and Biological Control*, Agroinform Kiadó, Budapest, pp. 243–8.

Gokhman VE (2004) Chromosomes of the family Braconidae (Hymenoptera). *Trudy Russkogo Entomologicheskogo Obshchestva*, **75**, 96–101 (in Russian with English summary).

Gokhman VE (2006) Karyotypes of parasitic Hymenoptera: diversity, evolution and taxonomic significance. *Insect Science*, **13**, 237–41.

Gokhman VE (2009) *Karyotypes of Parasitic Hymenoptera*, Springer, Dordrecht, 183 pp.

Gokhman VE, Krutov VV (1996) On external structure of male antennae in the subfamily Ichneumoninae (Hymenoptera Ichneumonidae) and related groups. *Zoologichesky Zhurnal*, **75**, 1182–94.

Gokhman VE, Mikhailenko AP (2008) Chromosomes of Ichneumonidae (Hymenoptera): new results and conclusions. *Entomological Review*, **88**, 391–5.

Gokhman VE, Mikhailenko AP (2012) Karyotype of *Homolobus infumator* (Lyle, 1914) (Hymenoptera: Braconidae). *Russian Entomological Journal*, **21**, 325–6.

Gokhman VE, Quicke DLJ (1995) The last twenty years of parasitic Hymenoptera karyology: an update and phylogenetic implications. *Journal of Hymenoptera Research*, **4**, 41–63.

Gokhman VE, Westendorff M (2003) Chromosomes of *Aphidius ervi* Haliday, 1834 (Hymenoptera, Braconidae). *Beiträge zur Entomologie*, **53**, 161–5.

Gómes Durán J-M, van Achterberg C (2011) Oviposition behaviour of four ant parasitoids (Hymenoptera, Braconidae, Euphorinae, Neoneurini and Ichneumonidae, Hybrizontinae), with the description of three new European species. *ZooKeys*, **125**, 59–106.

Gonzaga MO, Sobczak JF, Penteado-Dias AM, Eberhard WG (2010) Modification of *Nephila clavipes* (Araneae: Nephilidae) webs by the parasitoids *Hymenoepimecis bicolor* and *H. robertsae* (Hymenoptera: Ichneumonidae). *Ethology Ecology and Evolution*, **22**, 151–65.

González PI, Montoya P, Perez-Lachaud G, Cancino J, Liedo P (2007) Superparasitism in mass reared *Diachasmimorpha longicaudata* (Ashmead) (Hymenoptera: Braconidae), a parasitoid of fruit flies (Diptera: Tephritidae) *Biological Control*, **40**, 320–6.

González-Moreno A, Bordera S, Leirana-Alcocer J, Delfín-González H (2012) Diurnal flight behavior of Ichneumonoidea (Insecta: Hymenoptera) related to environmental factors in a tropical dry forest. *Environmental Entomology*, **41**, 587–593.

Goolsby JA, Burwell CJ, Makinson J, Driver F (2001) Investigation of the biology of Hymenoptera associated with *Fergusonina* sp. (Diptera: Fergusoninidae), a gall fly of *Melaleuca quinquenervia*, integrating molecular techniques. *Journal of Hymenoptera Research*, **10**, 163–80.

Gopalapillai R, Kadono-Okuda K, Okuda T (2005) Molecular cloning and analysis of a novel teratocyte-specific carboxylesterase from the parasitic wasp, *Dinocampus coccinellae*. *Insect Biochemistry and Molecular Biology*, **35**, 1171–80.

Gordh G, Hendrickson R Jr, (1976) Courtship behavior in *Bathyplectes anurus* (Thomson) (Hymenoptera: Ichneumonidae). *Entomological News*, **87**, 271–4.

Gorgiev G (2006) *Fenusella hortulana* (Hymenoptera: Tenthredinidae) and *Shawiana catenator* (Hymenoptera: Braconidae) – new species to the fauna of Bulgaria. *Acta Zoologica Bulgaarica*, **58**, 275–8.

Gothilf S (1969) A contribution to the biology of *Phanerotoma flavitestacea* Fi. a parasite of *Ectomyelois ceratoniae* Zell. *Israel Journal of Entomology*, **1**, 55–71.

Goulet H, Huber JT (1993) *Hymenoptera of the World: an Identification Guide to Families*, Centre for Land and Biological Resources Research, Ottawa.

Graenicher S (1905) On the habits of two ichneumonid parasites of the bee *Ceratina dupla* Say. *Entomological News*, **16**, 43–9.

Grant JF, Shepard M (1984) Laboratory biology of *Meteorus autographae* (Hymenoptera: Braconidae), an indigenous parasitoid of soybean looper (Lepidoptera: Noctuidae) larvae. *Environmental Entomology*, **13**, 838–42.

Grant JF, Shepard M (1986) Seasonal incidence of *Meteorus autographae* on soybean looper larvae on soybean in South Carolina, and the influence of the host density on parasitization. *Journal of Entomological Science*, **21**, 338–45.

Gravenhorst JLC (1815) *Monographia Ichneumonum Pedestrium, Praemisso Prooemio de Transituét Mutabilitate Specierum et Varietatum*, G.I. Göschen, Lipsiae.

Gravenhorst JLC (1829) *Ichneumonologia Europaea*, Sumtibus Auctoris, Vratislaviae, 10.5962/bhl.title.11531 (accessed 28 June 2014).

Grbic M, Strand MR (1998) Shifts in the life history of parasitic wasps correlate with pronounced alterations in early development. *Proceedings of the National Academy of Sciences of the United States of America*, **95**, 1097–101.

Greany PD, Carpenter JE (1998) Culture medium for parasitic and predaceous insects. US Patent 5799607.

Greany PD, Oatman ER (1972) Analysis of host discrimination in the parasite *Orgilus lepidus*. *Annals of the Entomological Society of America*, **65**, 377–83.

Greany PD, Hawke SD, Carlysle TC, Anthony DW (1977) Sense organs in the ovipositor of *Biosteres* (*Opius*) *longicaudatus*, a parasite of the Caribbean fruit fly *Anastrepha suspensa*. *Annals of the Entomological Society of America*, **70**, 319–21.

Greeney HF, Dyer LA, Smilanich AM (2012). Feeding by lepidopteran larvae is dangerous: a review of caterpillars' chemical, physiological, morphological, and behavioral defenses against natural enemies. *Invertebrate Survival Journal*, **9**, 7–34.

Grenier S (1970) Biologie dí *Agriotypus armatus* Curtis (Hymenoptera: Agriotypidae), parasite de nymphes de Trichoptéres. *Annales de Limnologie*, **6**, 317–61.

Grenier S (2009) *In vitro* rearing of entomophagous insects - Past and future trends: a minireview. *Bulletin of Insectology*, **62**, 1–6.

Grenier S, Greany PD, Cohen AC (1994) Potential for mass release of insect parasitoids and predators through development of artificial culture cechniques. In D Rosen, FD Bennett and JL Capinera (eds), *Pest Management in the Subtropics Biological Control – a Florida Perspective*, Intercept, Andover, Hampshire, pp. 181–205.

Griffiths DC (1960) The behaviour and specificity of *Monoctonus paludum* Marshall (Hym., Braconidae), a parasite of *Nasonovia ribis-nigri* (Mosley) on lettuce. *Bulletin of Entomological Research*, **51**, 303–3-19 plus plates X–XI.

Griffiths DC (1961) The development of *Monoctonus paludum* Marshall (Hym., Braconidae), a parasite of *Nasonovia ribis-nigri* (Mosley) on lettuce, and immunity reactions in other lettuce aphids. *Bulletin of Entomological Research*, **52**, 147–63.

Griffiths GCD (1964) The Alysiinae (Hym. Braconidae) parasites of Agromyzidae (Diptera). I. General questions of taxonomy, neology and evolution. *Beiträge zur Entomologie*, **14**, 823–914.

Griffiths KJ (1971) Discrimination between parasitised and unparasitised hosts by *Pleolophus basizonus* (Hymen. Ichn.). *Proceediongs of the Entomological Society of Ontario*, **102**, 83–91.

Griffiths KJ (1975) The biology of *Lophyroplectus luteator* (Hym.: Ichneumonidae) on *Neodiprion sertifer* (Hym.: Diprionidae in Ontario. *Entomophaga*, **20**, 97–103.

Griffiths KJ (1976) Interractions between *Pleolophus basizonus* and *Lophyroplectus luteator* (Hym.: Ichneumonidae), two parasites of *Neodiprion sertifer* (Hym.: Diprionidae). *Entomophaga*, **21**, 13–17.

Grinter CC, Whitfield JB, Connahs H, Dyer LA, Hallwachs W, Janzen DH (2009) A key to New World *Distatrix* Mason (Hymenoptera: Braconidae), with descriptions of six new reared Neotropical species. *Journal of Insect Science*, **9**, 29.

Grosch DS (1947) The importance of antennae in mating of male *Habrobracon*. *Journal of Comparative Physiology and Psychology*, **40**, 23–9.

Grosch DS, Kratsas RG, Petters RM (1977) Variation in *Habrobracon juglandis* ovariole number. I. Ovariole number increase induced by extended cold shock of fourth-instar larvae. *Journal of Embryology and Experimental Morphology*, **40**, 245–51.

Grosman AH, Janssen A, de Brito EF, Cordeiro EG, Colares F, Oliveira Fonsesca J, Lima ER, Pallini A, Sabelis MW (2008) Parasitoid increases survival of its pupae by inducing hosts to fight predators. *PLoS ONE*, **3** (6), e2276.

Grossniklaus-Bürgin C, Wyler T, Pfister-Wilhelm R, Lanzrein B (1994) Biology and morphology of the parasitoid *Chelonus inanitus* (Braconidae, Hymenoptera) and effects on the development of its host *Spodoptera littoralis* (Noctuidae, Lepidoptera). *Invertebrate Reproduction & Development*, **25**, 143–58.

Grossniklaus-Bürgin C, Pfister-Wilhelm R, Meyer V, Treiblmayr K, Lanzrein B (1998) Physiological and endocrine changes associated with polydnavirus/venom in the parasitoid-host system *Chelonus inanitus–Spodoptera littoralis*. *Journal of Insect Physiology*, **44**, 305–21.

Gu H, Dorn S (2000) Genetic variation in behavioural response to herbivore-infested plants in the parasitic wasp, *Cotesia glomerata* (L. Hymenoptera: Braconidae). *Journal of Insect Behavior*, **13**, 141–56.

Gu H, Dorn S (2003) Mating system and sex allocation in the gregarious parasitoid *Cotesia glomerata*. *Animal Behavior*, **66**, 259–64.

Gu H, Wang Q, Dorn S (2003) Superparasitism in *Cotesia glomerata*: response of host and consequences for parasitoids. *Ecological Entomology*, **28**, 422–31.

Guerra AA (1992) *In vitro* rearing of *Bracon mellitor* and *Catolaccus grandis* with different insect hemolymph-based diets. *Southwestern Entomologist*, **17**, 123–6.

Guerra AA, Robacker KM, Martinez S (1993a) Free amino acid and protein titers in *Anthonomus grandis* larvae venomized by *Bracon mellitor*. *Entomophaga*, **38**, 519–25.

Guerra AA, Robacker KM, Martinez S (1993b) *In vitro* rearing of *Bracon mellitor* and *Catolaccus grandis* with artificial diets devoid of insect components. *Entomologia Experimentalis et Applicata*, **68**, 303–7.

Guerrieri E, Poppy GM, Powell W, Tremblay E, Pennacchio F (1999) Induction and systemic release of herbivore-induced plant volatiles mediating in-flight orientation of *Aphidius ervi*. *Journal of Chemical Ecology*, **25**, 1247–61.

Guertin DS, Ode PJ, Strand MR, Antolin MF (1996) Host-searching and mating in an outbreeding parasitoid wasp. *Ecological Entomology*, **21**, 27–33.

Guévremont HC, Quednau FW (1977) Morphologie et biologie de *Grypocentrus albipes* (Hymenoptera: Ichneumonidae), parasie de la petite mineuse du bouleau, *Fenusa pusilla* (Hymenoptera: Tenthredinidae). *The Canadian Entomologist*, **109**, 1417–24.

Guillot FS, Joiner RL, Vinson SB (1974) Host discrimination: Isolation of hydrocarbon from Dufour's gland of a braconid parasitoid. *Annals of the Entomological Society of America*, **67**, 720–1.

Guinnee MA, Bernal JS, Bezemer TM, Fidgen JG, Hardy ICW, Mayhew PJ, Mills NJ, West SA (2005) Testing predictions of small brood models using parasitoid wasps. *Evolution and Ecology Research*, **7**, 779–94.

Gundersen-Rindal D, Dupuy C, Huguet E, Drezen J-M (2013) Parasitoid polydnaviruses: evolution, pathology and applications. *Biocontrol Science and Technology*, **23**, 1–61.

Guppy JC, Meloche F (1987) Life history and description of the immature stages of *Dacnusa dryas* (Nixon) (Hymenoptera: Braconidae), a European parasite of the alfalfa blotch leafminer (Diptera: Agromyzidae) in eastern Canada. *The Canadian Entomologist*, **119**, 281–5.

Gupta V (1987) The Ichneumonidae of the Indo-Australian area (Hymenoptera). *Memoirs of the American Entomological Institute*, **41**, 1–1210.

Gupta V (1988) Relationships of the genera of the tryphonine tribe Oedemopsini and a revision of *Acaenitellus* Morley (Hymenoptera: Ichneumonidae: Tryphoninae). In VK Gupta (ed.), *Advances in Parasitic Hymenoptera Research*, E.J. Brill, Leiden, pp. 243–58.

Gupta V (1994) A review of the genus *Brachyscleroma* with descriptions of new species from Africa and the Orient (Hymenoptera: Ichneumonidae: Phrudinae). *Oriental Insects*, **28**, 353–82.

Guryanova TM, Molchanov MI, Kochetova NI (1977) Changes in the haemolymph protein composition under infestation of the fox-coloured sawfly larvae with ectoparasites. *Zoologicheskii Zhurnal*, **56**, 648–50 (in Russian).

Gutierrez-Ibanez C, Villagra CA, Niemeyer H (2007) Pre-pupation behaviour of the aphid parasitoid *Aphidius ervi* (Haliday) and its consequences for pre-imaginal learning. *Naturwissenschaften*, **94**, 595–600.

Güven T, Yel M (1990) The ultrastructure of the venom gland in *Pimpla turionellae* (Hymenoptera: Ichneumonidae). *Doğa, Türk Biyoloji Dergisi*, **14**, 17–31.

Guzo D, Stoltz DB (1985) Obligatory multiparasitism in the tussock moth, *Orgyia leucostigma*. *Parasitology*, **90**, 1–10.

Ha D-M, Choi S-H, Shim J-K, Jung D-O, Song K-S, Nho S-K, Lee K-Y (2006) Behavioral attraction of two parasitoids, *Venturia canescens* and *Bracon hebetor*, to silk extracts of a host *Plodia interpunctella*. *Journal of Asia-Pacific Entomology*, **9**, 287–92.

Haeselbarth E (1961) Zur Biologie, Entwicklungsgeschichte und Ökologie von *Brachistes atricornis* Ratz. (Hym., Brac.) als eines Parasiten von *Pissodes piceae* (Ill.) (Col., Curc.). *Zeitschrift für Angewandte Entomologie*, **49**, 233–89.

Haeselbarth E (1973) Die *Blacus*-Arten Europas und Zentral-Asiens (Hymenoptera, Braconidae). *Veroffentlichungen der zoologischen Staatssammlung Muenchen*, **16**, 69–170.

Haeselbarth E (1979) Zur Parasitierung der Puppen von Forleule (*Panolis flammea* [Schiff.]), Kiefernspanner (*Bupalus piniarius* [L.]) und Heidelbeerspanner (*Boarmia bistortana* [Goeze]) in bayerischen Keifernwäldern. *Zeitschrift für Angewandte Entomologie*, **87**, 186–202.

Haeussler GJ (1940) Parasites of the oriental fruit moth in Japan and Chosen and their introduction into the United States. *Technical Bulletin of the United States Department of Agriculture*, **728**, 1–62.

Hafez M, Tawfik MFS, Ibrahim AA (1980) The immature stages of *Chelonus inanitus* (L.), a parasite of the cotton leafworm, *Spodoptera littoralis* (BOISD.), in Egypt (Hym., Braconidae). *Deutsche Entomologische Zeitschrift*, **27**, 29–38.

Hagen KS (1953) A premating period in certain species of genus *Opius* (Hymenoptera: Braconidae). *Proceedings of the Hawaiian Entomological Society*, **15**, 115–16.

Hagstrum DW, Smittle BJ (1977) Host-finding ability of *Bracon hebetor* and its influence upon adult parasite survival and fecundity. *Environmental Entomology*, **6**, 437–9.

Hågvar EB (1988) Multiparasitism of the green peach aphid, *Myzus persicae*: competition in the egg stage between *Aphidius matricariae* and *Ephedrus cerasicola*. *Entomologia Experimentalis et Applicata*, **47**, 275–82.

Hågvar EB, Hofsvang T (1990) Fecundity and intrinsic rate of increase of the aphid parasitoid *Ephedrus cerasicola* Starý (Hym., Aphidiidae). *Journal of Applied Entomology*, **109**, 261–7.

Hågvar EB, Hofsvang T (1991) Aphid parasitoids (Hymenoptera, Aphidiidae): biology, host selection and use in biological control. *Biocontrol News and Information*, **12**, 13–42.

Hailemichael Y, Smith JW, Wiedenmann RN (1994) Host-finding behavior, host acceptance, and host suitability of the parasite *Xanthopimpla stemmator*. *Entomologia Experimentalis et Applicata*, **71**, 155–66.

Halfhill JE, Featherston PE, Dickie AG (1972) History of the *Praon* and *Aphidius* parasites of the pea aphid in the Pacific Northwest. *Environmental Entomology*, **1**, 402–5.

Hallez P (1886) Loi de l'orientation de l'embryon chez les insectes. *Compte Rendu de l'Academie des Sciences, Paris*, **103**, 606–8.

Hancock GLR (1923) On some hibernating Ichneumonidae from the Cambridgeshire fens. *Entomologistís Monthly Magazine*, **59**, 152–8.

Hancock GLR (1925) Notes on the hibernation of Ichneumonidae and on some parasites of *Tortrix viridana* L. *Entomologistís Monthly Magazine*, **61**, 23–8.

Hanson HS (1939) Ecological notes on *Sirex* wood wasps and their parasites. *Bulletin of Entomological Research*, **30**, 27–65.

Hanson PE, Gauld ID (1995) *The Hymenoptera of Costa Rica*, Oxford University Press, Oxford. 893 pp.

Hanson PE, Cambra T. RA, Santos MA (2008) *Apechoneura longicauda* Kriechbaumer (Hymenoptera: Ichneumonidae: Labeninae), parasitoid of *Phloeoborus punctatorugosus* Chapuis (Coleoptera: Curculionidae: Scolytinae). *Tecnociencia*, **10**, 119–24.

Haramoto FH (1957) Observations on the biology of *Opius oophilus* Fullaway (Hymenoptera: Braconidae). *Proceedings of the 8th Pacific Science Congress*, **3(A)**, 1275–89.

Harbo JR, Kraft KJ (1969) A study of *Phanerotoma toreutae*, a parasite of the pine cone moth *Laspeyresia toreuta*. *Annals of the Entomological Society of America*, **62**, 214–20.

Harcourt DG (1990) Displacement of *Bathyplectes curculionis* (Thoms.) (Hymenoptera: Ichneumonidae) by *B. anurus* (Thomson) in eastern Ontario populations of the alfalfa weevil, *Hypera postica* (Gyll.) (Coleoptera: Curculionidae). *The Canadian Entomologist*, **122**, 641–5.

Hardie J, Hick AJ, Höller C, Mann J, Merritt L, Nottingham SF, Powell W, Wadhams LJ, Witthinrich J, Wright AF (1994) The response of *Praon* spp. parasitoids to aphid sex pheromone components in the field. *Entomologia Experimentalis et Applicata*, **71**, 95–9.

Hardy ICW (1994) Sex ratio and mating structure in the parasitoid Hymenoptera. *Oikos*, **69**, 3–20.

Harris AC (1978) Mimicry by a longhorn beetle, *Neocalliprason elegans* (Coleoptera: Cerambycidae), of its parasitoid, *Xanthocryptus novozealandicus* (Hymenoptera: Ichneumonidae). *New Zealand Entomologist*, **6**, 406–8.

Harrison EG, Fisher RC, Ross KM (1985) The temporal effects of Dufour's gland secretion in host discrimination by *Nemeritis canescens*. *Entomologia Experimentalis et Applicata*, **38**, 215–20.

Hartzer KL, Zhu KY, Baker JE (2005) Phenoloxidase in larvae of *Plodia interpunctella* (Lepidoptera: Pyralidae): molecular cloning of the proenzyme cDNA and enzyme activity in larvae paralyzed and parasitised by *Habrobracon hebetor* (Hymenoptera: Braconidae). *Archives of Insect Biochemistry and Physiology*, **59**, 67–79.

Harvey JA, Strand MR (2002) The developmental strategies of endoparasitoid wasps vary with host feeding ecology. *Ecology*, **83**, 2439–51.

Harvey JA, Strand MR (2003) Sexual size and development dimorphism in a parasitoid wasp: an exception to the rule? *European Journal of Entomology*, **100**, 485–92.

Harvey JA, Witjes LMA, Wagenaar R (2004) Development of hyperparasitoid wasp *Lysibia nana* (Hymenoptera: Ichneumonidae) in a multitrophic framework. *Environmental Entomology*, **33**, 1488–96.

Harvey JA, Vet LEM, Witjes LMA, Bezemer TM (2006) Remarkable similarity in body mass of a secondary hyperparasitoid *Lysibia nana* and its primary parasitoid host *Cotesia glomerata* emerging from cocoons of comparable size. *Archives of Insect Biochemistry & Physiology*, **61**, 170–83.

Harvey JA, Kos M, Nakamatsu Y, Tanaka T, Dicke M, Vet LEM, Brodeur J, Bezemer TM (2008) Do parasitised caterpillars protect their parasitoids from hyperparasitoids? A test of the 'usurpation hypothesis'. *Animal Behaviour*, **76**, 701–8.

Harvey JA, Wagenaar R, Bezemer TM (2009) Interactions to the fifth trophic level: secondary and tertiary parasitoid wasps show extraordinary efficiency in utilizing host resources. *Journal of Animal Ecology*, **78**, 686–92.

Harvey JA, Sano T, Tanaka T (2010) Differential host growth regulation by the solitary endoparasitoid, *Meteorus pulchricornis* in two hosts of greatly differing mass. *Journal of Insect Physiology*, **56**, 1178–83.

Harvey JA, Gols R, Tanaka T (2011) Differing success of defense strategies in two parasitoid wasps in protecting their pupae against a secondary hyperparasitoid. *Annals of the Entomological Society of America*, **104**, 1005–11.

Harvey JA, Poelman EH, Tanaka T (2013) Intrinsic inter- and intraspecific competition in parasitoid wasps. *Annual Review of Entomology*, **58**, 333–51.

Hashimoto Y, Lawrence PO (2005) Comparative analysis of selected genes from *Diachasmimorpha longicaudata* entomopoxvirus and other poxviruses. *Journal of Insect Physiology*, **51**, 207–20.

Haviland MD (1922) On the larval development of *Dacnusa areolaris* Nees (Braconidae), a parasite of Phytomyzinae (Diptera), with a note on certain chalcid parasites of phytomyzids. *Parasitology*, **14**, 167–73.

Hawke SD, Farley RD, Greany PD (1973) Fine structure of the sense organs in the ovipositor of the parasitic wasp, *Orgilus lepidus* Muesabeck. *Tissue and Cell*, **5**, 171–84.

Hawkins BA (1990) Global patterns of parasitoid assemblage size. *Journal of Animal Ecology*, **59**, 57–72.

Hawkins BA (1994) *Pattern and Process in Host-Parasitoid Interactions*, Cambridge University Press, Cambridge.

Hawkins BA, Askew RR, Shaw MR (1990) Influences of host feeding-niche and food plant type on generalist and specialist parasitoids. *Ecological Entomology*, **l5**, 275–80.

Hawkins BA, Thomas MB, Hochberg ME (1993) Refuge theory and biological control. *Science*, **262**, 1429–32.

Hawlitzky N (1972) Mode de pénétration d'un parasite ovo-larvaire *Phanerotoma flavitestacea* Fish. (Hym.: Braconidae) dans son hote embryonnaire, *Anagasta kuehniella* Zell. (Lep.: Pyralidae). *Entomophaga*, **17**, 375–89.

Hawlitzky N (1979) Devenir de l'oeuf et comportement de la larve de *Phanerotoma flavitestacea* (Hym.: Braconidae) lorsque la femelle pond dans les oeufs d'*Anagasta kuehniella* (Lep.: Pyralidae) ayant atteint des stades de develeoppement varies. *Entomophaga*, **24**, 237–45.

Hawlitzky N, Boulay C (1974) Identification des régimes alimentaires de la larve d'un insecte entomophage *Phanerotoma flavitestacea* [Hym.: Braconidae]. *Entomophaga*, **19**, 395–408.

Hawlitzky N, Boulay C (1986) Effects of the egg-larval parasite, *Phanerotoma flavitestacea* Fisch. (Hymenoptera, Braconidae), on the dry weight and chemical composition of its host *Anagasta kuehniella* Zell. (Lepidoptera, Pyralidae). *Journal of Insect Physiology*, **12**, 269–74.

Hayakawa Y, Yazaki K (1997) Envelope protein of parasitic wasp symbiont virus, polydnavirus, protects the wasp eggs from cellular immune reactions by the host insect. *European Journal of Biochemistry*, **246**, 820–6.

Haye T, Kenis M (2004) Biology of *Lilioceris* spp. (Coleoptera: Chrysomelidae) and their parasitoids in Europe. *Biological Control*, **29**, 399–408.

Haye T, Goulet H, Mason PG, Kuhlmann U (2005) Does fundamental host range match ecological host range? A retrospective case study of a *Lygus* plant bug parasitoid. *Biological Control*, **35**, 55–67.

Hazell SP, Wenlock C, Bachel S, Fellowes MDE (2005) The costs and consequences of parasitoid attack for the predatory hoverfly, *Episyrphus balteatus*. *Evolutionary Ecology Research*, **7**, 669–79.

He JH (1984) Six new records of braconidflies from China (Hymenoptera, Braconidae). *Acta Agriculturae Zhejiangensis*, **10**, 199–205.

He X-J, Wang Q (2008) Reproductive strategies of *Aphidius ervi* Haliday (Hymenoptera: Aphidiidae). *Biological Control*, **45**, 281–7.

Heatwole H, Davis DM (1965) Ecology of three sympatric species of parasitic insects of the genus *Megarhyssa* (Hymenoptera: Ichneumonidae). *Ecology*, **46**, 140–50.

Heatwole H, Davis DM, Wenner AM (1962) The behavior of *Megarhyssa*, a genus of parasitic hymenopterans (Ichneumonidae: Ephialtinae). *Zeitschrift für Tierpsychologie*, **19**, 652–4.

Hedgekar BM, Arthur AP (1973) Host haemolymph chemicals that induce oviposition in the parasite *Itoplectis conquisitor* (Hymenoptera: Ichneumonidae). *The Canadian Entomologist*, **105**, 787–93.

Hedqvist K-J (1963) Notes on Hormiinae with description of new genera and species (Hymenoptera, Braconidae). *Entomologisk Tidskrift*, **84**, 30–61.

Hedqvist K-J (1976) *Mithotynia aptera* n. gen. n. sp., a wingless ichneumonid from South Sweden (Hymenoptera: Ichneumonidae). *Entomologia Scandinavica*, **7**, 314–15.

Hegazi EM, Führer E (1985) Instars of *Microplitis rufiventris* [Hym., Braconidae] and their relative developmental speed under different photoperiods. *Entomophaga*, **30**, 231–43.

Hegazi EM, Khafagi WE (2005) Gregarious development of the solitary endo-parasitoid, *Microplitis rufiventris* in its habitual host, *Spodoptera littoralis*. *Journal of Applied Entomology*, **129**, 134–41.

Hegazi E, Khafagi W, Schlyter F (2013) Egg maturation dynamics of the parasitoid *Microplitis rufiventris*: starvation speeds maturation in early life. *Physiological Entomology*, **38**, 233–40.

Heimpel GE, Jervis MA (2005) Does floral nectar improve biological control by parasitoids? In F Wäckers, P van Rijn and J Bruin, (eds), *Plant-Provided Food and Plant–Carnivore Mutualism*, Cambridge University Press, Cambridge, pp. 267–304.

Heimpel GE, Antolin MF, Franqui RA, Strand MR (1997) Reproductive isolation and genetic variation between two "strains" of *Bracon hebetor* (Hymenoptera: Braconidae). *Biological Control*, **9**, 149–56.

Heimpel GE, Antolin MF, Strand MR (1999) Diversity of sex-determining alleles in *Bracon hebetor*. *Heredity*, **82**, 282–91.

Heimpel GE, Lee JC, Wu Z, Weiser L, Wäckers F, Jervis MA (2004). Gut sugar analysis in field-caught parasitoids: adapting methods originally developed for biting flies. *International Journal of Pest Management*, **50**, 193–8.

Heinrich GH (1934) Die Ichneumoninae von Celebes. *Mitteilungen aus der zoologischen Museum in Berlin*, **20**, 1–263.

Heinrich GH (1961) Synopsis of Nearctic Ichneumoninae Stenopneusticae with particular reference to the northeastern region. *Canadian Entomologist, Supplement*, **18**, 89–205 and **23**, 369–505.

Heinrich GH (1962) Synopsis of Nearctic Ichneumonidae Stenopneusticae with particular reference to the Northeastern Region. Pt. VI. Synopsis of the Ichneumonini (Genus Plagiotrypes), Acanthojoppini, Listrodromini and Platylabini. *Canadian Entomologist, Supplement*, **27**, 679–802.

Heinrich GH (1967) *Synopsis and Reclassification of the Ichneumoninae Stenopneusticae of Africa south of Sahara (Hym.)*, vol. I, Farmington State College Press, Altötting, Germany.

Heinrich GH (1974) Burmesische Ichneumoninae IX. *Annales Zoologici*, **32**, 10.

Heinrich GH (1977) Ichneumoninae of Florida and neighboring states (Hymenoptera: Ichneumonidae, subfamily Ichneumoninae). *Arthropods of Florida and Neighboring Land Areas*, **9**, 1–350.

Heitland W, Pschorn-Walcher H (1992) A brief note on unusually high superparasitism in some sawfly parasitoids (Hym., Ichneumonidae: Euryproctini). *Mitteillungen der Schweizerischen Entomologischen Gesellschaft*, **65**, 75–9.

Henaut A (1990) Study on the sound produced by *Pimpla instigator* (Hymenoptera: Ichneumonidae) during host selection. *Entomophaga*, **35**, 127–39.

Henaut A, Guerdoux J (1982) Location of a lure by the drumming insect *Pimpla instigator* (Hymenoptera, Ichneumonidae). *Experientia*, **38**, 346–347.

Henderson RE, Kuriachan I, Bradleigh Vinson SB (2011) Suitability of various artificial pupation chambers for use by koinobiont parasitoid *Toxoneuron nigriceps* (Hymenoptera:

Braconidae). *Annals of the Entomological Society of America*, **104**, 1355–9.

Hendry LB, Greany PD, Gill RJ (1973) Kairomone mediated host-finding behavior in the parasitic wasp *Orgilus lepidus*. *Entomologia Experimentalis et Applicata*, **16**, 471–7.

Henneman ML, Memmott J (2001) Infiltration of a Hawaiian community by introduced biological control agents. *Science*, **293**, 1314–16.

Henry LM, Roitberg BD, Gillespie DR (2008) Host-range evolution in *Aphidius* parasitoids: fidelity, virulence and fitness trade-offs on an ancestral host. *Evolution*, **62**, 689–99.

Henschen W (1928) Über die Entwicklung der Geschlechtsdrüsen von *Habrobracon juglandis* Ash. *Zeitschrift für Morphologie und Ökologie der Tiere*, **13**, 144–78.

Henter HJ, Via S (1995) The potential for coevolution in a host–parasitoid system. I. Genetic variation within an aphid population in susceptibility to a parasitic wasp. *Evolution*, **49**, 427–38.

Hentz M, Ellsworth, P, Naranjo S (1997) Biology and morphology of *Chelonus* sp. nr. *curvimaculatus* (Hymenoptera: Braconidae) as a parasitoid of *Pectinophora gossypiella* (Lepidoptera: Gelechiidae). *Annals of the Entomological Society of America*, **90**, 631–9.

Hernández-Ortiz V, Pérez-Alonso R, Wharton RA (1994) Native parasitoids associated with the genus *Anastrepha* (Dipt.: Tephritidae) in Los Tuxtlas, Veracruz, Mexico. *Entomophaga*, **39**, 171–8.

Herniou EA, Huguet E, Thézé J, Bézier A, Periquet G, Drezen J-M (2013) When parasitic wasps hijacked viruses: genomic and functional evolution of polydnaviruses. *Philosophical Transactions of the Royal Society, B, Biological Sciences*, **368**, 20130051.

Heron RJ (1960) The relative effects of cocoon submergence on the mortality of the larch sawfly, *Pristiphora erichsonii* (Hymenoptera: Tenthredinidae) and its parasite *Bessa harveyi* (Diptera: Tachinidae). *Annals of the Entomological Society of America*, **53**, 476–81.

Hespenheide HA (1979) Are there fewer parasitoids in the tropics? *The American Naturalist*, **113**, 766–9.

Hespenheide HA (2010) New *Agrilus* Curtis (Coleoptera: Buprestidae) from México and Costa Rica mimicking parasitic wasps. *Zootaxa*, **2545**, 39–46.

Hilszczański J, Gibb H, Hjältén J, Atlegrim O, Johansson T, Pettersson RB, Ball JP, Danell K (2005) Parasitoids (Hymenoptera, Ichneumonoidea) of saproxylic beetles are affected by forest successional stage and dead wood characteristics in boreal spruce forest. *Biological Conservation*, **126**, 456–64.

Hill CC, Smith HD (1931) *Heterospilus cephi* Rohwer, a parasite of the European wheat sawfly, *Cephus pygmaeus* (L.). *Journal of Agricultural Research*, **43**, 597–609.

Hillebrand H (2004) On the generality of the latitudinal diversity gradient. *The American Naturalist*, **163**,192–211.

Hilpert H (1992) Zur Systematik der Gattung *Ichneumon* Linnaeus, 1758 in der Westpalaearktis (Hymenoptera, Ichneumonidae, Ichneumoninae). *Entomofauna. Supplement*, **6**, 1–389.

Hinz R (1976) Zur Systematik und Ökologie der Ichneumoniden V. *Deutsche Entomologische Zeitschrift (N.F.)*, **23**, 99–105.

Hinz R (1981) Die europäischen Arten der Gattung *Stilbops* Förster (Hym., Ichneumonidae). *Nachrichtenblatt der Bayerischen Entomologen*, **30**, 62–4.

Hinz R (1983) The biology of the European species of the genus *Ichneumon* and related species (Hym., Ichneumonidae). *Contributions of the American Entomological Institute*, **20**, 151–2.

Hinz R, Short J (1983) Life-history and systematic position of the European *Alomya* species (Hymenoptera: Ichneumonidae). *Entomologica Scandinavica*, **14**, 462–6.

Hirose Y (1966) Parasitic Hymenoptera visiting flowers of carrot planted in the truck crop field. *Science Bulletin of the Faculty of Agriculture, Kyushu University*, **22**, 217–23.

Hirose Y (2006) Biological control of aphids and coccids: a comparative analysis. *Population Ecology*, **48**, 307–15.

Hirose Y, Vinson SB, Hirosie Y (1988) Protandry in the parasitoid *Cardiochiles nigriceps*, as related to its mating system. *Ecological Research*, **3**, 217–26.

Hochberg ME, Elmes GW, Thomas JA, Clark RT (1998) Effects of habitat reduction on the persistence of *Ichneumon eumerus* (Hymenoptera: Ichneumonidae), the specialist parasitoid of *Maculinea rebeli* (Lepidoptera: Lycaenidae). *Journal of Insect Conservation*, **2**, 59–66.

Hochuli A, Pfister-Wilhelm R, Lanzrein B (1999) Analysis of endoparasitoid-released proteins and their effects on host development in the system *Chelonus inanitus* (Braconidae) - *Spodoptera littoralis* (Noctuidae). *Journal of Insect Physiology*, **45**, 823–33.

Hocking H (1967a) The influence of food on longevity and oviposition in *Rhyssa persuasoria* (L.) (Hymenoptera: Ichneumonidae). *Australian Journal of Entomology*, **6**, 83–8.

Hocking H (1967b) A native ichneumonid *Certonotus tasmanicus* Turner, parasitizing *Sirex noctilio* (F.) (Siricidae). *Journal of the Australian Entomological Society*, **6**, 57–70.

Hoek GV (1971) The larval instars of *Aphidius pulcher* Baker (Hymenoptera: Aphidiidae). *Proceedings of the Entomological Society of Ontario*, **101**, 55–9.

Hofsvang T, Hågvar EB (1977) Cold storage tolerance and supercooling points of mummies of *Ephedrus cerasicola* Starý and *Aphidius colemani* Viereck (Hym: Aphidiidae). *Norwegian Journal of Entomology*, **24**, 1–6.

Hofsvang T, Hågvar EB (1978) Larval morphology and development of *Aphidius colemani* Viereck and *Ephedrus cerasicola* Starý (Hym., Aphidiidae). *Norwegian Journal of Entomology*, **25**, 1–8.

Hogervorst PAM, Wäckers FL, Romeis J (2007) Effects of honeydew sugar composition on the longevity of *Aphidius ervi*. *Entomologia Experimentalis et Applicata*, **122**, 223–232.

Holloway AK, Heimpel GE, Strand MR, Antolin MF (1999) Survival of diploid males in *Bracon* sp. near *hebetor* (Hymenoptera: Braconidae). *Annals of the Entomological Society of America*, **92**, 110–16.

Holloway AK, Strand MR, Black WC IV, Antolin MF (2000) Linkage analysis of sex determination in *Bracon* sp. near *hebetor* (Hymenoptera: Braconidae). *Genetics*, **154**, 205–12.

Holt RD, Lawton JH (1993) Apparent competition and enemy-free space in insect host-parasitoid communities. *The American Naturalist*, **142**, 623–45.

Hong YC (1988) Early Cretaceous Orthopera, Neuroptera, Hymenoptera (Insecta) of Kezuo in west Liaoning Province. *Entomotaxonomia*, **10**, 119–30 (in Chinese, with English abstract).

Horn DJ (1976) Behavior of *Gelis* species parasitizing *Bathyplectes*, parasitoids of the alfalfa weevil. *Ohio Journal of Science*, **76**, 279–80.

Horstmann K (1968) Zur Systematik und Biologie von *Neorhacodes enslini* (Ruschka) (Hym., Ichneumonidae). *Entomologische Nachrichten*, **12**, 33–6.

Horstmann K (1969) Revision der Hellwigiinae Förster (Hymenoptera, Ichneumonidae). *Mittelungen der Deutschen Entomologischen Gesellschaft*, **28**, 30–2.

Horstmann K (1971) Revision der europäischen Tersilochinen 1 (Hymenoptera: Ichneumonidae). *Veröffentlichungen der Zoologischen Staatsammlung (München)*, **15**, 47–138.

Horstmann K (1978) Revision der Gattungen der Mastrina Townes (Hym., Ich.). Zeitschrift der Arbeitsgemeinschaft Österreich. *Entomologen*, **30**, 65–70.

Horstmann, K. (1981) Revision der europäischen Tersilochinen II (Hymenoptera, Ichneumonidae). *Spixiana, Supplement* **4** (1980), 1–76.

Horstmann K (1993) Revision der brachypteren Weibchen der westpaläarktischen Cryptinae (Hymenoptera, Ichneumonidae). *Entomofauna*, **14**, 85–148.

Horstmann K, Floren A, Linsenmair E (2005) Ichneumonidae (Hymenoptera) from the canopy of tropical forests in Sabah, Malaysia: a comparison between primary and secondary forests. *Ecotropica*, **11**, 42–52.

Hotta M, Okuda T, Tanaka T (2001) *Cotesia kariyai* teratocytes: growth and development. *Journal of Insect Physiology*, **47**, 31–41.

Hougardy E, Grégoire J-C (2003) Cleptoparasitism increases the host finding ability of a polyphagous parasitoid species, *Rhopalicus tutela* (Hymenoptera: Pteromalidae). *Behavioral Ecology and Sociobiology*, **55**, 184–9.

House HL (1978) An artificial host: encapsulated synthetic medium for *in vitro* oviposition and rearing the endoparasitoid *Itoplectis conquisitor* (Hymenoptera: Ichneumonidae). *The Canadian Entomologist*, **110**, 331–3.

Howard RW, Baker JE (2003) Cuticular hydrocarbons and wax esters of the ectoparasitoid *Habrobracon hebetor*: ontogenetic, reproductive, and nutritional effects. *Archives of Insect Biochemistry and Physiology*, **53**, 1–18.

Howard RW, Baker JE, Morgan ED (2003) Novel diterpenoids and hydrocarbons in the Dufour gland of the ectoparasitoid *Habrobrbacon hebetor* (Say) (Hymenoptera: Braconidae). *Archives of Insect Biochemistry and Physiology*, **54**, 95–109.

Howell J, Fisher RC (1977) Food conversion efficiency of a parasitic wasp, *Nemeritis canescens*. *Ecological Entomology*, **2**, 143–51.

Hoy HL, Dahlman DL (2002) Extended *in vitro* culture of *Microplitis croceipes* teratocytes and secretion of TSP14 protein. *Journal of Insect Physiology*, **48**, 401–9.

Hrabar M, Danci AT, Schaefer PW, Gries G (2012) In the nick of time: males of the parasitoid wasp *Pimpla disparis* respond to semiochemicals from emerging mates. *Journal of Chemical Ecology*, **38**, 253–61.

Hrcek J, Miller SE, Quicke DLJ, Smith MA (2011) Molecular detection of trophic links in a complex insect host-parasitoid food web. *Molecular Ecology Resources*, **11**, 786–94.

Hrcek J, Miller SE, Whitfield JB, Shima H, Novotny V (2013) Parasitism rate, parasitoid community composition and host specificity on exposed and semi-concealed caterpillars from a tropical rainforest. *Oecologia*, **173**, 521–32.

Hu J, Zhu X-X, Fu W-J (2003) Passive evasion of encapsulation in *Macrocentrus cingulum* Brischke (Hymenoptera: Braconidae), a polyembryonic parasitoid of *Ostrinia furnacalis* Guenee (Lepidoptera: Pyralidae). *Journal of Insect Physiology*, **49**, 367–75.

Hu JS, Vinson SB (1997) *In vitro* rearing of *Campoletis sonorensis*, a larval endoparasitoid of *Heliothis virescens* from egg to third instar in an artificial medium devoid of insect sources. *Entomologia Experimentalis et Applicata*, **85**, 263–73.

Hu JS, Vinson SB (1998) The *in vitro* development from egg to prepupa of *Campoletis sonorensis* (Hymenoptera, Ichneumonidae) in an artificial medium: importance of physical factors. *Journal of Insect Physiology*, **44**, 455–61.

Huang F, Yu R-X, Chen X-X (2007) Ultrastructure of the male reproductive system of *Cotesia vestalis* (Hymenoptera: Braconidae) with preliminary characterization of the secretions. *Microscopy Research and Technique*, **70**, 563–71.

Huang F, Shi M, Chen Y-F, Cao T-T, Chen X-X (2008) Oogenesis of *Diadegma semiclausum* (Hymenoptera: Ichneumonidae) and its associated polydnavirus. *Microscopy Research and Technique*, **71**, 676–83.

Huang F, Shi M, Chen XX, Ye GY, He JH (2009) External morphology and development of immature stages of *Diadegma semiclausum* (Hymenoptera: Ichneumonidae), an important endoparasitoid of *Plutella xylostella* (Lepidoptera: Plutellidae). *Annals of the Entomological Society of America*, **102**, 532–8.

Huang XF (1986) Use of *Habrobracon hebetor* Say in granary pest control. *Chinese Journal of Biological Control*, **2**, 78–80.

Huddleston T (1980) A revision of the Western Palaearctic species of the genus *Meteorus* (Hymenoptera: Braconidae). *Bulletin of the British Museum (Natural History)*, **41**, 1–58.

Huddleston T, Short JRT (1978) A new genus of the Euphorinae (Hymenoptera: Braconidae) from Australia with a description of the final instar larva of one species. *Journal of the Australian Entomological Society*, **17**, 317–21.

Huddleston T, Walker AK (1994) A revision of the *Chelonus scrobiculatus* species-group of Cheloninae (Insecta: Hymenoptera: Braconidae). *Annalen des Naturhistorischen Museums in Wien*, **96B**, 153–68.

Humala AE (2002) A review of parasitic wasps of the genera *Cylloceria* Schiodte, 1838, and *Allomacrus* Förster, 1868 (Hymenoptera, Ichneumonidae) of the fauna of Russia. *Entomological Review*, **82**, 301–13.

Humble LM (2006) Overwintering adaptations in Arctic sawflies (Hymenoptera: Tenthredinidae) and their parasitoids: cold tolerance. *The Canadian Entomologist*, **138**, 59–71.

Humble LM, Ring RA (1985) Inoculative freezing of a larval parasitoid within its host. *CryoLetters*, **6**, 59–66.

Hung AF, Hedlund RC, Day WH (1986) High level of genetic heterozygosity in the hyperparasitie wasp, *Mesochorus nigripes*. *Experientia*, **42**, 1050–1.

Hung AF, Day WH, Hedlund RC (1988) Genetic variability in arrhenotokous and thelytokous forms of *Mesochorus nigripes* [Hym.: lchneumonidae]. *Entomophaga*, **33**, 7–15.

Hurlbutt B (1987) Sexual size dimorphism in parasitoid wasps. *Biological Journal of the Linnean Society*, **30**, 63–89.

Hussain MA, Mathur CB (1924) Some parasites of the cotton bollworm (*Earias insulana* and *E. fabia*) in the Punjab. *Pusa, 5th Entomological Meeting Report and Proceedings*, 34–52.

Ibrahim AA, Fayad YH, Kares EA, Shalaby FF (1987) Biological and ecological studies on the parasite *Zele nigricornis* Walk. (Hymenoptera, Braconidae). *Bulletin de la Société Entomologique d'Égypte*, **66**, 13–25.

Idar M (1981) Revision of the European species of the genus *Hadrodactylus* Förster (Hymenoptera: Ichneumonidae). Part 2. *Insect Systematics & Evolution*, **12**, 231–9.

Idris AB, Gonzaga AD, Nor Zaneedarwaty N, Hasnah BT, Natasha BY (2001a) Does habitat disturbance has adverse effect on the diversity of parasitoid community? *Online Journal of Biological Sciences*, **1**, 1040–2.

Idris AB, Nor Zaneedarwaty N, Gonzaga AD, Zaidi MI, Azman S, Salmah Y (2001b) A study on four methods of sampling Ichneumonidae and Braconidae at two different habitats of Fraser's Hill, Malaysia. *Pakistan Journal of Biological Sciences*, **4**, 1515–17.

Idris AB, Hanidah J, Gonzaga AD, Nur Azura A (2003) Diversity, abundance, species composition and similarity of genus *Xanthopimpla* (Ichneumonidae: Pimplinae) in logged and fragmented forests of the Langat Basin in Selangor, Malaysia. *Journal of Asia-Pacific Entomology*, **6**, 55–62.

Inaba F (1940) Impaternate females of the parasitic wasp *Habrobracon* produced by colchicine treatment. *Proceedings of the Imperial Academy, Tokyo*, **16**, 411–13.

Infante F, Hanson P, Wharton R (1995) Phytophagy in the genus *Monitoriella* (Hymenoptera: Braconidae) with description of new species. *Annals of the Entomological Society of America*, **88**, 406–15.

Iqbal M, Austin AD, Belokobylskij SA (2003) Diversity and biogeography of reduced-winged doryctine wasps (Hymenoptera: Braconidae) in Australasia. *Records of the South Australian Museum, Monograph Series*, **7**, 243–7.

Isidoro NF, Bin F, Colazza S, Vinson SB (1996) Morphology of antennal gustatory sensilla and glands in some parasitoids Hymenoptera with hypothesis on their role in sex and host recognition. *Journal of Hymenoptera Research*, **5**, 206–39.

Iştvan ÖV, Gündüz NEA, Gülel A (2011) Protein and lipid amounts of the parasitoid *Bracon hebetor* Say (Hymenoptera: Braconidae) at constant and alternating temperatures. *Turkish Journal of Zoology*, **35**, 747–53.

Ivanova-Kasas OM (1956) Comparastive study of embryonal development in aphidiids (*Aphidius* and *Ephedrus*). *Entomologicheskoe Obozrenie*, **35**, 245–61 (in Russian).

Ivanova-Kasas OM (1960) Embryologische Entwicklung von *Angitia vestigialis* Ratz. (Hymenoptera, Ichneumonidae) – des Endoparasiten von *Pontania capreae* L. (Hymemoptera, Tenthredinidae). *Entomologicheskoe Obozrenie*, **39**, 284–95.

Ivie MA (2001) On the geographic origin of the wheat stem sawfly (Hymenoptera: Cephidae): a new hypothesis of introduction from northeastern Asia. *American Entomologist*, **47**, 84–97.

Iwata K (1958) Ovarian eggs of 233 species of the Japanese Ichneumonidae (Hym.). *Acta Hymenopterologica*, **1**, 63–74.

Iwata K (1959) The comparative anatomy of the ovary in Hymenoptera. Part III. Braconidae (including Aphidiidae) with descriptions of ovarian eggs. *Kontyû, Tokyo*, **27**, 231–8.

Iwata K (1960) The comparative anatomy of the ovary in Hymenoptera. Part V. Ichneumonidae. *Acta Hymenopterologica*, **1**, 115–69.

Iwata K (1961) Biological observations on Ichneumonidae Part I (Hymenoptera). *Acta Hymenopterologia*, **1**, 315–25.

Iwata K (1966a) The comparative anatomy of the ovary in Hymenoptera. Supplement on Ichneumonidae. *Acta Hymenopterologica*, **2**, 133–5.

Iwata K (1966b) Biological observations on Ichneumonidae. Part II (Hymenoptera). *Acta Hymenopterologia*, **2**, 137–43.

Jacas J, Budia F, Rodríguez-Cerezo E, Viñuela E (1997) Virus-like particles in the poison gland of the parasitic wasp *Opius concolor*. *Annals of Applied Biology*, **130**, 587–92.

Jackson CG, Butler GD Jr, (1984) Development time of three species of *Bracon* (Hymenoptera: Braconidae) on the pink bollworm (Lepidoptera: Gelechiidae) in relation to temperature. *Annals of the Entomological Society of America*, **77**, 539–42.

Jackson CG, Delph JS, Neemann EG (1978) Development, longevity and fecundity of *Chelonus blackburni* (Hym.: Braconidae) as a parasite of *Pectinophara gossypiella* (Lep.: Gelechiidae). *Biocontrol*, **23**, 35–42.

Jackson DL (1928) The biology of *Dinocampus* (*Perilitus*) *rutilus* (Nees), a braconid parasite of *Sitona lineata* L. *Proceedings of the Zoological Society of London*, (1928), 597–630.

Jacob HS, Evans EW (2004) Influence of different sugars on the longevity of *Bathyplectes curculionis* (Hym., Ichneumonidae). *Journal of Applied Entomology*, **128**, 316–20.

Janvier H (1933) Etude biologique de quelques Hymenopteres du Chili. *Annales des Sciences Naturelles*, 10e Série, 209–356.

Janzen DH (1981) The peak in North American ichneumonid species richness lies between 38 and 42°N. *Ecology*, **62**, 532–7.

Janzen DH, Gauld ID (1997) Patterns of use of large moth caterpillars (Lepidoptera: Saturniidae and Sphingidae) by ichneumonid parasitoids (Hymenoptera) in Costa Rican dry forest. In AD Watt, NE Stork and MD Hunter (eds), *Forests and Insects*, Chapman & Hall, London, pp. 251–71.

Janzen DH, Gauld ID (2004) The systematics and biology of the Costa Rican species of parasitic wasps in the *Thyreodon* genus-group (Hymenoptera: Ichneumonidae). *Zoological Journal of the Linnean Society*, **141**, 297–351.

Janzen DH, Pond CM (1975) A comparison by sweep-net sampling of the arthropod fauna of secondary vegetation in Michigan, England and Costa Rica. *Transactions of the Royal Entomological Society, London*, **127**, 33–50.

Janzen DH, Sharkey MJ, Burns JM (1998) Parasitization biology of a new species of Braconidae (Hymenoptera) feeding on larvae of Costa Rican dry forest skippers (Lepidoptera: Hesperiidae: Pyrginae). *Tropical Lepidoptera*, **9**, 33–41.

Janzen DH, Walker AK, Whitfield JB, Delvare G, Gauld ID (2003) Host-specificity and hyperparasitoids of three new Costa Rican species of *Microplitis* Foerster (Hymenoptera: Braconidae: Microgastrinae), parasitoids of sphingid caterpillars. *Journal of Hymenoptera Research*, **12**, 42–76.

Jell PA, Duncan PM (1986) Invertebrates, mainly insects, from the freshwater Lower Cretaceous, Koonwarra fossil bed (Korumburra group), South Gippsland, Victoria. In PA Jell and J Roberts, *Plants and Invertebrates from the Lower Cretaceous Koonwarra Fossil Bed*, south gippsland, Victoria. *Memoirs of the Association of Australasian Palaeontologists*, **3**, 111–205.

Jepson WF (1954) *A Critical Review of the World Literature on the Lepidopterous Stalk-Borers of Tropical Graminaceous Crops*. Commonwealth Institute of Entomology, London.

Jerman EJ, Gauld ID (1988) *Casinaria*, a paraphyletic ichneumonid genus (Hymenoptera) and a revision of the Australian species. *Journal of Natural History*, **22**, 589–609.

Jervis MA (1998) Functional and evolutionary aspects of mouthpart structure in parasitoid wasps. *Biological Journal of the Linnean Society*, **6**, 461–93.

Jervis MA, Kidd NAC (1986) Host-feeding strategies in hymenopteran parasitoids. *Biological Reviews*, **61**, 395–434.

Jervis MA, Kidd NAC, Walton M (1992) A review of methods for determining dietary range in adult parasitoids. *Entomophaga*, **37**, 565–74.

Jervis MA, Kidd NAC, Fitton MG, Huddleston T, Dawah HA (1993) Flower-visiting by hymenopteran parasitoids. *Journal of Natural History*, **27**, 67–105.

Jervis MA, Heimpel GE, Ferns PN, Harvey JA, Kidd NAC (2001) Lifehistory strategies in parasitoid wasps: a comparative analysis of 'ovigeny'. *Journal of Animal Ecology*, **70**, 442–58.

Jervis MA, Moe A, Heimpel GE (2012) The evolution of parasitoid fecundity: a paradigm under scrutiny. *Ecology Letters*, **15**, 357–64.

Jewett DK, Carpenter JE (1998) Chemical communication between males and females of a pupal parasitoid *Diapetimorpha introita* (Hymenoptera: Ichneumonidae). *Annals of the Entomological Society of America*, **91**, 748–53.

Jiménez-Pérez A, Villa-Ayala P, López-Martínez V (2007) First record of *Diachasmimorpha longicaudata* (Ashmead) (Hymenoptera: Braconidae) attacking *Toxotrypana curvicauda* Gerstaecker (Diptera: Tephritidae). *Neotropical Entomology*, **36**, 482–3.

Jiménez Peydró R, Falco Garí JVC, Gimeno Martos C, Moreno Marí J (1996) Comparative morphology of the venom apparatus of some subfamilies of braconid wasps (Hymenoptera: Braconidae). *Boletin de la Asociacion Espanola de Entomologia*, **20**, 149–59.

Jiménez R, Falco JV, Moreno J (1997) Descripción del aparato del venemo de *Cosmophorus cembrae* Ruschka (Hymenoptera, Braconidae). *Bulletin de la Société Entomologique de France*, **102**, 385–9.

Johner A, Lanzrein B (2002) Characterization of two genes of the polydnavirus of *Chelonus inanitus* and their stage-specific expression in the host *Spodoptera littoralis*. *Journal of General Virology*, **83**, 1075–85.

Johns CV, Whitehouse MEA (2004) Mass rearing of two larval parasitoids of *Helicoverpa* spp. (Lepidoptera: Noctuidae): *Netelia producta* (Brullé) and *Heteropelma scaposum* (Morley) (Hymenoptera: Ichneumonidae) for field release. *Australian Journal of Entomology*, **43**, 83–7.

Johnson JB (1985) *Aleiodes* sp. (Hymenoptera: Braconidae) reared from an anomalous host, an adult of the *Eremochrysa punctinervis* species group (Neuroptera: Chrysopidae). *Pan-Pacific Entomologist*, **61**, 24–5.

Johnson NF, Triplehorn CA (2005) *Borror and DeLong's Introduction to the Study of Insects*, 7th edn, Thomson Brooks/Cole, Belmont, CA.

Johnson SJ, Smith JW Jr, (1980) Biology of *Orgilus elasmopalpi* (Hym.: Bracondiae) with *Elasmopalpus lignosellus* (Lep.; Pyralidae) as host. *Annals of the Entomological Society of America*, **73**, 572–5.

Jones D, (1987) Material from adult female *Chelonus* sp. directs expression of altered developmental programme of host lepidoptera. *Journal of Insect Physiology*, **33**, 129–34.

Jones D, Jones G, Hammock BD (1981) Developmental and behavioural responses of larval *Trichoplusia ni* to parasitization by an imported braconid parasite *Chelonus* sp. *Physiological Entomology*, **6**, 387–94.

Jones D, Barlow JS, Thompson SN (1982) *Exeristes, Itoplectis, Aphaereta, Brachymeria,* and *Hyposoter* species: *in vitro* glyceride synthesis and regulation of fatty acid composition. *Experimental Parasitology*, **54**, 340–51.

Jones D, Jones G, Rudnicka M, Click A, Reck-Malleczewen V, Iwaya M (1986) Pseudoparasitism of host *Trichoplusia* ni by *Chelonus* spp. as a new model system for parasite regulation of host physiology. *Journal of Physiological Entomology*, **32**, 315–28.

Jones GZ, Shaw SR (2012) Ten new species of *Meteorus* (Hymenoptera: Braconidae) from Ecuador reared at the Yanayacu Biological Center for Creative Studies. *Zootaxa*, **3547**, 1–23.

Jones OR, Purvis A, Baumgart E, Quicke DLJ (2009) Using taxonomic revision data to estimate the geographic and taxonomic distribution of undescribed species richness in the Braconidae (Hymenoptera: Ichneumonoidea). *Insect Conservation and Diversity*, **2**, 204–12.

Jones OR, Purvis A, Baumgart E, Quicke DLJ (2012) Latitudinal gradients in taxonomic overdescription rate affect macroecological inferences using species list data. *Ecography*, **35**, 333–40.

Jordan T (1998a) Egg structure and life history of two species groups of *Grypocentrus* (Hymenoptera: Ichneumonidae), monophagous parasitoids of spring leaf-mining *Eriocrania*-moths on birch (Lepidoptera: Eriocraniidae). *Entomologia Generalis*, **22**, 223–31.

Jordan T (1998b) *Tersilochus curvator* Horstmann und *Tersilochus* sp. n. (Ichneumonidae, Tersilochinae), neue Parasitoidenfer an Birken minierenden Trugmotten (Lepidoptera, Eriocraniidae). *Bonner zoologische Beiträge*, **47**, 411–19.

Jørgensen OF (1975) Competition among larvae of *Pimplopterus dubius* Hgn. (Hymenoptera: Ichneumonidae), a parasitoid of *Epinotia tedella* Cl. (Lepidoptera: Tortricidae). *Zeitschrift für Angewandte Entomologie*, **79**, 301–9.

Jourdheuil P (1961) Influence de quelques facteurs écologiques sur les fluctuations de population d'une biocénose parasitaire: étude relative à quelques Hyménoptères (Ophioninae, Diospilinae, Euphorinae) parasites de divers Coléoptères inféodés aux Crucifères. *Annales Épiphyties, Series C*, **11**, 445–658.

Joyce AL, Millar JG, Paine TD, Hanks LM (2002) The effect of host size on the sex ratio of *Syngaster lepidus*, a parasitoid of *Eucalyptus* longhorned borers (*Phoracantha* spp.). *Biological Control*, **24**, 207–13.

Joyce AL, Hunt RE, Bernal JS, Vinson SB (2008) Substrate influences mating success and transmission of courtship vibrations for the parasitoid *Cotesia marginiventris*. *Entomologia Experimentalis et Applicata*, **127**, 39–47.

Joyce AL, Millar JG, Gill JS, Singh M, Tanner D, Paine TD (2011) Do acoustic cues mediate host finding by *Syngaster lepidus* (Hymenoptera: Braconidae)? *BioControl*, **56**, 145–53.

Jugovits T (1987) Entwicklung von *Microplitis tuberculifera* (Wesmael) (Hymenoptera: Braconidae), ein Parasitoid von *Mamestra brassicae* (Lepidoptera: Noctuidae). *Pflanzenschutzberichte*, **48**, 24–32.

Juillet JA (1959) The morphology of immature stages, life history, and behaviour of three hymenopterous parasites of the European pine shoot moth, *Rhyacionia buoliana* (Schiff.) (Lepidoptera: Olethreutidae). *The Canadian Entomologist*, **91**, 709–19.

Jumean Z, Gries R, Gries G, Unruh T (2005) *Mastrus ridibundus* parasitoids eavesdrop on cocoon-spinning codling moth, *Cydia pomonella*, larvae. *Naturwissenschaften*, **92**, 20–5.

Junnikkala E (1985) Testis development in *Pieris brassicae* parasitised by *Apanteles glomeratus*. *Entomologia Experimentalis et Applicata*, **37**, 283–8.

Jussila R, Kapyla M (1975) Observations on *Townesia tenuiventris* (Hlmgr.) (Hym., Ichneumonidae) and its hosts *Chelostoma maxillosum* (L.) (Hym, Megachilidae) and *Trypoxylon figulus* (L.) (Hym., Sphecidae). *Annales Entomologici Fennici*, **41**, 81–6.

Jussila R, Säärksjärvi I, Bordera S (2010) Revision of the western Palaearctic *Mesoleptus* (Hymenoptera: Ichneumonidae). *Annales de la Société Entomologique de France (Nouvelle Série)*, **46**, 499–518.

Kadono-Okuda K, Weyda F, Okuda T (1998) *Dinocampus* (=*Perilitus*) *coccinellae* teratocyte-specific polypeptide: its accumulative property, localization and characterization. *Journal of Insect Physiology*, **44**, 1073–80.

Kabissa JCB, Kayumbo HY, Yarro JG (1996) Seasonal abundance of chrysopids (Neuroptera: Chrysopidae) preying on *Helicoverpa armigera* (Hubner) (Lepidoptera: Noctuidae) and *Aphis gossypii* (Glover) (Homoptera: Aphididae) on cotton in eastern Tanzania. *Crop Protection*, **15**, 5–8.

Kaeslin M, Pfister-Wilhelm R, Lanzrein B (2005a) Influence of the parasitoid *Chelonus inanitus* and its polydnavirus on host nutritional physiology and implications for parasitoid development. *Journal of Insect Physiology*, **51**, 1330–9.

Kaeslin M, Wehrle I, Grossniklaus-Bürgin C, Wyler T, Guggisberg U, Schittny JC, Lanzrein B (2005b) Stage-dependent strategies of host invasion in the egg-larval parasitoid *Chelonus inanitus*. *Journal of Insect Physiology*, **51**, 287–96.

Kaeslin M, Wyler T, Grossniklaus-Bürgin C, Lanzrein B (2006) Development of the anal vesicle, salivary glands and gut in the egg-larval parasitoid *Chelonus inanitus*: tools to take up nutrients and to manipulate the host? *Journal of Insect Physiology*, **52**, 269–81.

Kainoh Y, Oishi Y (1993) Source of sex pheromone of the egg-larval parasitoid, *Ascogaster reticulatus* Watanabe (Hymenoptera: Braconidae). *Journal of Chemical Ecology*, **19**, 963–9.

Kainoh Y, Hiyori T, Tamaki Y (1982) Kairomone of the egg-larval parasitoid, *Ascogaster reticulatus* Watanabe (Hymenoptera: Braconidae). *Applied Entomology and Zoology*, **17**, 102–10.

Kainoh Y, Tatsuki S, Sugie H, Tamaki Y (1989) Host egg kairomones essential for egg-larval parasitoid, *Ascogaster reticulatus* Watanabe (Hymenoptera: Braconidae). *Journal of Chemical Ecology*, **15**, 1219–29.

Kainoh Y, Nemoto T, Shimuzy K, Tatsuki S, Kusano T, Kuwahara Y (1991) Mating behavior of *Ascogaster reticulatus* Watanabe (Hymenoptera: Braconidae), an egg-larval parasitoid of the smaller tea tortrix, *Adoxophyes* sp. (Lepidoptera: Tortricidae). III. Identification of a sex pheromone. *Applied Entomology and Zoology*, **26**, 543–9.

Kamano Y, Shimizu K, Kainoh Y (1989) Mating behavior of *Ascogaster reticulatus* Watanabe (Hymenoptera: Braconidae), an egg-larval parasitoid of the smaller tea tortrix, *Adoxophyes* sp. (Lepidoptera: Tortricidae): II. Behavioral sequence and a role of sex pheromone. *Applied Entomology and Zoology*, **24**, 372–8.

Kalmes R, Hedderwick M-P, Rojas-Rousse D, Chessel D (1983) Interactions entre deux parasitoïdes *Diadromus collaris* et *D. pulchellus* (Ichneumonides): survie des parasitoïdes mâles dans des chrysalides d'*Acrolepiopsis assectella* soumis ou non à une compétition intra- et interspécifique. *Canadian Journal of Zoology*, **61**, 627–32.

Kambhampati S, Völkl W, Mackauer M (2000) Phylogenetic relationships among genera of Aphidiinae (Hymenoptera: Braconidae) based on DNA sequence of the mitochondrial 16S rRNA gene. *Systematic Entomology*, **25**, 437–45.

Kankare M, Shaw MR (2004) Molecular phylogeny of *Cotesia* (Hymenoptera: Braconidae: Microgastrinae) parasitoids associated with Melitaeini butterflies (Lepidoptera: Nymphalidae: Melitaeini). *Molecular Phylogenetics and Evolution*, **32**, 207–20.

Kankare M, van Nouhuys S, Hanski I (2005a) Genetic divergence among host specific cryptic species in *Cotesia melitaerum* aggregate (Hymenoptera: Braconidae), parasitoids of checkerspot butterflies. *Annals of the Entomological Society of America*, **98**, 382–94.

Kankare M, Stefanescu C, van Nouhuys S, Shaw MR (2005b) Host specialization by *Cotesia* wasps (Hymenoptera: Braconidae) parasitizing species- rich Melitaeini (Lepidoptera: Nymphalidae) communities in north-eastern Spain. *Biological Journal of the Linnean Society*, **86**, 45–65.

Karlsson D, Ronquist F (2012) Skeletal morphology of *Opius dissitus* and *Biosteres carbonarius* (Hymenoptera: Braconidae), with a discussion of terminology. *PLoS ONE*, **7** (4), e32573.

Kasparyan DR (1981) *Fauna of the USSR: Ichneumonidae (Subfamily Tryphoninae) Tribe Tryphonini. Hymenoptera*, vol. 3, vssue 1, E.J. Brill, Leiden (translation of 1973 work in Russion published originally by Nauka, Leningrad).

Kasparyan DR (1988) New taxa of ichneumonids of the family Paxylommatidae (Hymenoptera, Ichneumonidae) from Baltic amber. *Trudy Vsesoyuznogo Entomologicheskogo Obshchestva*, **70**, 125–31 (in Russian).

Kasparyan DR (1993) Townesioninae, a new ichneumonid subfamily from the Eastern Palaearctic (Hymenoptera: Ichneumonidae). *Zoosystematica Rossica*, **2**, 155–9.

Kasparyan DR (1994) Review of Ichneumon flies of the Townesitinae subfam. nov. (Hymenoptera, Ichneumonidae) from the Baltic ambers. *Paleontologicheskii Zhurnal*, **4**, 86–96 (in Russian).

Kasparyan DR, Rasnitsyn AP (1992) On *the systematic position* of the Oligocene *Lithoserix williamsi* Brown (Hymenoptera: Ichneumonidae) from Colorado (USA). *Paleontologicheskiy Zhurnal*, 113–14 (in Russian, with English translation in *Paleontological Journal*, **26**, 134–5).

Kawakami T (1985) Development of the immature stages of *Ascogaster reticulatus* Watanabe (Hymenoptera: Braconidae), an egg-larval parasitoid of the smaller teat tortrix moth, *Adoxophyes* sp. (Lepidoptera: Tortricidae). *Applied Entomology and Zoology*, **20**, 380–6.

Kawakami T, Kainoh Y (1985) Host discrimination and competition in the egg-larval parasitoid *Ascogaster reticulatus* Watanabe (Hymenoptera: Braconidae). *Applied Entomology and Zoology*, **20**, 362–4.

Kenis M (1997) Biology of *Coeloides sordidator* (Hymenoptera: Braconidae), a possible candidate for introduction against *Pissodes strobi* (Coleoptera: Curculionidae) in North America. *Biocontrol Science and Technology*, **7**, 153–64.

Kerrich GJ (1936) Notes on larviposition in *Polyblastus*. *Proceedings of the Entomological Society of London (A)*, **11**, 108–10.

Kerrich GJ (1939) Systematic notes on the Oxytorina (Hym., Ichneumonidae, Mesoleptini Auctt.). *Opuscula Entomologica*, **4**, 126–8.

Kerrich GJ (1940) Notes on some ichneumonid and figitid (Hym.) parasites of Neuroptera. *Entomologist's Monthly Magazine*, **76**, 15–17.

Kerrich GJ (1961) A study of the tersilochine parasites of vegetable weevils of the genus *Listroderes* (Hym., Ichneumonidae). *Eos*, **37**, 497–503.

Kerrich GJ (1968) Description of an ichneumonid (Hym.) that preys on egg-masses of weevils harmful to tea culture in Kenya. *Bulletin of Entomological Research*, **59**, 469–72.

Ketseoglou I, Bouwer G (2008) A simple technique for the preliminary screening of parasitic wasps for polydnavirus-like particles. *African Entomology*, **16**, 296–300.

Kfir R (1997) The diamondback moth with special reference to its parasitoids in South Africa. In A Sivapragasam, WH Loke, AK Hussan and GS Lim (eds), *Proceedings: The Management of Diamondback Moth and Other Crucifer Pests*, Malaysian Agricultural Research and Development Institute (MARDI), Kuala Lumpur. pp. 54–60.

Khafagi WE, Hegazi EM (2008) Does superparasitism improve host suitability for parasitoid development? A case study in the *Microplitis rufiventris–Spodoptera littoralis* system. *BioControl*, **53**, 427–38.

Khalaim AI, Yurtcan M (2011) A survey on Tersilochinae (Hymenoptera: Ichneumonidae) species of Turkey, with a key to European genera. *Turkish Journal of Zoology*, **35**, 381–94.

Khatri D, He XZ, Wang Q (2009) Mating behavior and egg maturation in *Diadegma semiclausum* Hellen (Hymenoptera: Ichneumonidae). *New Zealand Plant Protection*, **62**, 174–8.

Khoo CCK, Lawrence PO (2002) Hagen's glands of the parasitic wasp *Diachasmimorpha longicaudata* (Hymenoptera: Braconidae): ultrastructure and the detection of entomopoxvirus and parasitism-specific proteins. *Arthropod Structure & Development*, **31**, 121–30.

Kindlmann P, Schödelbauerová I, Dixon AFG (2007) Inverse latitudinal gradients in species diversity. In D Storch, PA Marquet and JH Brown, (eds), *Scaling Biodiversity*, Cambridge University Press, Cambridge, pp. 246–57.

King PE, Fordy MR (1970) The formation of "accessory nuclei" in the developing oocytes of parasitoid hymenopterans *Ophion luteus* (L.) and *Apanteles glomeratus* (L.). *Zeitschrift für Zellforschung und mikroskopische Anatomie*, **109**, 158–70.

King RC, Cassidy JD (1973) Ovarian development in *Habrobracon juglandis* (Ashmead) (Hymenoptera: Braconidae) – II. Observations on growth and differentiation of component cells of egg chamber and their bearing upon interpretation of radiosensitivity data from *Habrobracon* and *Drosophila*. *International Journal of Insect Morphology and Embryology*, **2**, 117–36.

King RC, Ratcliffe NA, Fordy MR (1971) Oogenesis in a braconid, *Apanteles glomeratus* (L.) possessing an hydropic type of egg. *Zeitschrift für Zellforschung und mikroskopische Anatomie*, **119**, 43–57.

Kistner DH, Jacobson HR, Elliott DE (2000) Some interesting Braconidae that live with termites and their behavior (Hymenoptera; Isoptera: Termitidae, Nasutitermitinae). *Sociobiology*, **36**, 79–82 and 257–64.

Kittel RN, Austin AD (2014) Synopsis of Australian cheloninae wasps (Hymenoptera: Braconidae: Cheloninae) with description of two new genera. *Austral Entomology*, **53**, 183–202.

Kloet GS, Hincks WD (1945) *Check List of British Insects*. Kloet and Hncks, Stockport.

Klopfstein S (2014) Revision of the Western Palaearctic Diplazontinae (Hymenoptera, Ichneumonidae). *Zootaxa*, **3801**, 1–143.

Klopfstein S, Kropf C, Quicke DLJ (2010a) An evaluation of phylogenetic informativeness profiles and the molecular phylogeny of Diplazontinae (Hymenoptera, Ichneumonidae). *Systematic Biology*, **59**, 226–41.

Klopfstein S, Quicke DLJ, Kropf C (2010b) The evolution of antennal courtship in diplazontine parasitoid wasps (Hymenoptera, Ichneumonidae, Diplazontinae). *BMC Evolutionary Biology*, **10**, 218.

Klopfstein S, Quicke DLJ, Kropf C, Frick H (2011) Molecular and morphological phylogeny of Diplazontinae (Hymenoptera, Ichneumonidae). *Zoologica Scripta*, **40**, 379–402.

Klopfstein S, Vilhelmsen L, Heraty JM, Sharkey M, Ronquist F (2013) The hymenopteran tree of life: evidence from protein-coding genes and objectively aligned ribosomal data. *PLoS ONE*, **8** (8), e69344.

Ko SY, Frasson L, Rodriguez y Baena F (2011) Closed-loop planar motion control of a steerable probe with a "programmable bevel" inspired by nature. *IEEE Transactions in Robotics*, **27**, 970–83.

Kolaib MO (1987) Biology of *Zele chlorophthalma* (Nees). *Indian Journal of Agricultural Sciences*, **57**, 491–3.

Kolarov JA (1987) A new Ctenopelmatinae genus and species from Bulgaria (Hymenoptera, Ichneumonidae). *Entomofauna*, **8**, 69–76.

Kolarov J, Gürbüz MF (2007) New data on the structure of the flagellum in males of *Perithous devinator* (Rossi) (Hymenoptera, Ichneumonidae, Pimplinae). *Acta Entomologica Serbica*, **12**, 59–66.

Komatsu T, Konishi K (2010) Parasitic behaviors of two ant parasitoid wasps (Ichneumonidae: Hybrizontinae). *Sociobiology*, **56**, 575–84.

Kopelke JP (1987) *Adelognathus cubiceps* Roman (1924) (Ichneumonidae: Adelognathinae) – ein ungewöhnlicher Parasitoid der gallenbidenden *Pontania*-Arten (Tenthredinidae: Nematinae). *Senckenbergiana Biologica*, **67**, 253–9.

Kopelke JP (1988) Zur Biologie und Ökologie der Arten des Brutparasiten–Parasitoiden Komplexes von gallenbeldenden Blattwespen der Gattung *Pontania* (Hymenoptera: Tenthredinidae: Nematinae). *Mitteilungen der Deutschen Gesellschaft für Allgemeine und Angewandte Entomologie*, **6**, 150–5.

Kopelke JP (1994) Der Schmarotzerkomplex (Brutparasiten und Parasitoide) der gallenbildenden *Pontania*-Arten (Hymenoptera: Tenthredinidae). *Senckenbergiana Biologica*, **73**, 83–133.

Koptur S (1989) Mimicry of flowers by parasitic wasp pupae. *Biotropica*, **21**, 93–5.

Kopylov DS (2009) A new subfamily of ichneumonids from the Lower Cretaceous of Transbaikalia and Mongolia (Insecta: Hymenoptera: Ichneumonidae). *Paleontological Journal*, **43**, 83–93.

Kopylov DS (2010a) Ichneumonids of the Subfamily Tanychorinae (Insecta, Hymenoptera, Ichneumonidae) from the Lower Cretaceous of Transbaikalia and Mongolia. *Paleontological Journal*, **44**, 179–86.

Kopylov DS (2010b) A new subfamily of ichneumon wasps (Insecta: Hymenoptera: Ichneumonidae) from the Upper Cretaceous of the Russian Far East. *Paleontological Journal*, **44**, 422–33.

Kopylov DS (2011) Ichneumon wasps of the Khasurty locality in Transbaikalia (Insecta, Hemenoptera, ichneumonidae). *Paleontological Journal*, **45**, 406–13.

Kopylov DS (2012a) New species of Praeichneumonidae (Hymenoptera, Ichneumonoidea) from the Lower Cretaceous of Transbaikalia. *Paleontological Journal*, **46**, 66–72.

Kopylov DS (2012b) New Ichneumonidae (Hymenoptera) from the Upper Cretaceous ambers of the Taimyr Peninsula. *Paleontological Journal*, **46**, 383–91.

Korenko S, Pekár S (2011) A parasitoid wasp induces overwintering behaviour in its spider host. *PLoS ONE*, **6** (9), e24628.

Korenko S, Schmidt S, Schwarz M, Gibson GAP, Pekár S (2013) Hymenopteran parasitoids of the ant-eating spider *Zodarion styliferum* (Simon) (Araneae, Zodariidae). *ZooKeys*, **262**, 1–15.

Kraaijeveld AR (1994) Local adaptations in a parasitoid-host system. A coevolutionary arms race. Unpublished doctoral thesis, University of Leiden.

Kraaijeveld AR (1999) Kleptoparasitism as an explanation for paradoxical oviposition decisions of the parasitoid *Asobara tabida*. *Journal of Evolutionary Biology*, **12**, 129–33.

Kraaijeveld AR, Godfray HCJ (1997) Trade-off between parasitoid resistance and larval competitive ability in *Drosophila melanogaster*. *Nature*, **389**, 278–80.

Kraaijeveld AR, van Alphen JJM (1994) Geographical variation in resistance of the parasitoid *Asobara tabida* against encapsulation by *Drosophila melanogaster* larvae: the mechanism explored. *Physiological Entomology*, **19**, 9–14.

Kraaijeveld AR, van der Wel NN (1994) Geographic variation in reproductive success of the parasitoid *Asobara tabida* in larvae of several *Drosophila* species. *Ecological Entomology*, **19**, 221–9.

Kraaijeveld AR, Adriaanse ICT, van den Bergh B (1999) Parasitoid size as a function of host sex: potential for different sex allocation strategies. *Entomologia Experimentalis et Applicata*, **92**, 289–94.

Kraaijeveld AR, Hutcheson KA, Limentani EC, Godfray HCJ (2001) Costs of counter defenses to host resistance in a parasitoid of *Drosophila*. *Evolution*, **55**, 1815–21.

Kraaijeveld AR, Ferrari J, Godfray HCJ (2002) Costs of resistance in insect-parasite and insect-parasitoid interactions. *Parasitology*, **125**, S71–S82.

Kraaijeveld K, Reumer BM, Mouton L, Kremer N, Vavre F, van Alphen JJM (2011) Does a parthenogenesis-inducing *Wolbachia* induce vestigial cytoplasmic incompatibility? *Naturwissenschaften*, **98**, 175–80.

Krell PJ, Stoltz DB (1979) Unusual baculovirus of the parasitoid wasp *Apanteles melanoscelus*: isolation and preliminary characterization. *Journal of Virology*, **29**, 1118–30.

Kremer N, Charif D, Henri H, Bataille M, Prévost G, Kraaijeveld K, Vavre F (2009) A new case of *Wolbachia* dependence in the genus *Asobara*: evidence for parthenogenesis induction in *Asobara japonica*. *Heredity*, **103**, 248–56.

Krishnan A, Nair PN, Jones D (1994) Isolation, cloning, and characterization of new chitinase stored in active form in chitin-lined venom reservoir. *Journal of Biological Chemistry*, **269**, 20971–6.

Kröder S, Samietz J, Dorn S (2006) Effect of ambient temperature on mechanosensory host location in two parasitic wasps of different climatic origin. *Physiological Entomology*, **31**, 299–305.

Kröder S, Samietz J, Dorn S (2007a) Temperature affects interaction of visual and vibrational cues in parasitoid host location. *Journal of Comparative Physiology A, Sensory Neural and Behavioral Physiology*, **193**, 223–31.

Kröder S, Samietz J, Schneider D, Dorn S (2007b) Adjustment of vibratory signals to ambient temperature in a host-searching parasitoid. *Physiological Entomology*, **32**, 105–12.

Kroemer JA, Webb BA (2004) Polydnavirus genes and genomes: emerging gene families and new insights into polydnavirus replication. *Annual Review of Entomology*, **49**, 431–56.

Kugler J, Wollberg Z (1967) Biology of *Agrothereutes tunetanus* Haber. [Hym. Ichneumonidae] an ectoparasite of *Orgyia dubia* Tausch. [Lep. Lymantriidae]. *Entomophaga*, **12**, 363–79.

Krugner R, Daane KM, Lawson AB, Yokota GY (2005) Biology of *Macrocentrus iridescens* (Hymenoptera: Braconidae): a parasitoid of the obliquebanded leafroller (Lepidoptera: Tortricidae). *Environmental Entomology*, **34**, 336–43.

Kruidhof HM, Pashalidou FG, Fatouros NE, Figueroa IA, Vet LEM, Schmid HM, Huigens ME (2012) Reward value determines memory consolidation in parasitic wasps. *PLoS ONE*, **7**, e39615.

Kryukov VY, Kryukova NA, Glupov VV (2013) Susceptibility of *Galleria mellonella* larvae to anamorphic entomopathogenic *Ascomycetes* under envenomation and parasitization by *Habrobracon hebetor*. *Russian Journal of Ecology*, **44**, 89–92.

Kryukova NA, Dubovskiy IM, Chertkova EA, Vorontsova YL, Slepneva IA, Glupov VV (2011) The effect of *Habrobracon hebetor* venom on the activity of the prophenoloxidase system, the generation of reactive oxygen species, and encapsulation in the haemolymph of *Galleria mellonella* larvae. *Journal of Insect Physiology*, **57**, 796–800.

Kula RR (2011) Two new brachypterous species of *Heterospilus* Haliday (Hymenoptera, Braconidae, Doryctinae) from the Nearctic Region. *Journal of Hymenoptera Research*, **21**, 53–64.

Kula RR, Lill JT, Murphy SM, Stoepler T (2010a) First host records for the Nearctic species *Triraphis discoideus* (Hymenoptera: Braconidae: Rogadinae). *Entomological News*, **120**, 380–6.

Kula RR, Boughton AJ, Pemberton RW (2010b) *Stantonia pallida* (Ashmead) (Hymenoptera: Braconidae) reared from *Neomusotima conspurcatalis* Warren (Lepidoptera: Crambidae), a classical biological control agent of *Lygodium microphyllum* (Cav.) R. Br. (Polypodiales: Lygodiaceae).

Proceedings of the Entomological Society of Washington, **112**, 61–8.

Kumar P, Ballal CR (1992) The effect of parasitism by *Hyposoter didymator* (Hym.: Ichneumonidae) on food consumption and utilization by *Spodoptera litura* (Lep.: Noctuidae). *Entomophaga*, **37**, 197–203.

Künkel d'Herculais J, Langlois C (1891) Moeurs et metamorphoses de *Perilitus brevicollis* Haliday (Hymenoptere, Braconide) parasite de l'altise de la vigne en Algerie. *Annales de la Société Entomologique de France*, **60**, 457–66.

Kuriachan I, Henderson R, Laca R, Vinson SB (2011) Post-egression host tissue feeding is another strategy of host regulation by the koinobiont wasp, *Toxoneuron nigriceps*. *Journal of Insect Science*, **11**, 3.

Labeyrie V (1960) Contribution à l'étude de la dynamique des populations d'insectes. I. Influence stimulatrice de l'hôte *Acrolepia assectella* Z. sur la multiplication d'un hyménoptère Ichneumonidae (*Diadromus* sp.), Thèse d'Etat, Université de Paris. 193 pp.

Labeyrie V, Rojas-Rousse D (1985) Superparasitism reconsidered: is it an adaptive competition? The example of *Diadromus pulchellus*. *Experientia*, **41**, 15–18.

Lacey LA, Unruh TR, Headrick HL (2003) Interactions of two idiobiont parasitoids (Hymenoptera: Ichneumonidae) of codling moth (Lepidoptera: Tortricidae) with the entomopathogenic nematode *Steinernema carpocapsae* (Rhabditida: Steinernematidae). *Journal of Invertebrate Pathology*, **83**, 230–9.

Lachaud J-P, Perez-Lachaud G (2012) Diversity of species and behavior of Hymenopteran parasitoids of ants: a review. *Psyche*, **2012**, 134746.

Lampert EC, Dyer LA, Bowers MD (2010) Caterpillar chemical defense and parasitoid success: *Cotesia congregata* parasitism of *Ceratomia catalpae*. *Journal of Chemical Ecology*, **36**, 992–8.

Lampert EC, Dyer LA, Bowers MD (2011) Chemical defense across three trophic levels: *Catalpa bignonioides*, the caterpillar *Ceratomia catalpae*, and its endoparasitoid *Cotesia congregata*. *Journal of Chemical Ecology*, **37**, 1063–70.

Langer A, Hance T (2000) Overwintering strategies and cold hardiness of two aphid parasitoid species (Hymenoptera: Braconidae: Aphidiinae). *Journal of Insect Physiology*, **46**, 671–6

Langor DW, Digweed SC, Williams DJM, Spence JR, Saunders C (2000) Establishment and spread of two introduced parasitoids (Ichneumonidae) of the birch leafminer, *Fenusa pusilla* (Lepeletier) (Tenthredinidae). *BioControl*, **45**, 415–23.

Lanzrein B, Pfister-Wilhelm R, von Niederhäusern F (2001) Effect of an egg-larval parasitoid and its polydnavirus on development and the endocrine system of the host. In JP Edwards and RJ Weaver (eds), *Endocrine Interactions of Insect Parasites and Pathogens*, BIOS Scientific Publishers, Oxford, pp. 95–109.

Lapointe R, Tanaka K, Barney WE, Whitfield JB, Banks JC, Béliveau C, Stoltz D, Webb BA, Cusson M (2007) Genomic and morphological features of a banchine polydnavirus: comparison with bracoviruses and ichnoviruses. *Journal of Virology*, **81**, 6491–501.

Larocca A, Fanti P, Romano VA, Marsicovetere E, Pennacchio F, Battaglia D (2005) An 'artificial aphid' for *Aphidius ervi* (Hym., Braconidae). *Journal of Applied Entomology*, **129**, 580–2.

Latham DR, Mills NJ (2010) Life history characteristics of *Aphidius transcaspicus*, a parasitoid of mealy aphids (*Hyalopterus* species). *Biological Control*, **54**, 147–52.

Lathrop FH, Newton RC (1933) The biology of *Opius melleus* Gahan a parasite of the blueberry maggot. *Journal of Agricultural Research*, **46**, 142–60.

Laurenne N, Quicke DLJ (2009) Antennal hammers: echos of sensilla past. In P Pontarotti (ed.), *Evolutionary Biology: Concepts, Molecular and Morphological Evolution. Proceedings of the 13th Evolutionary Biology Meeting, Marseille*, pp 271–82.

Laurenne NM, Belshaw R, Broad GR, Quicke DLJ (2000) Molecular confirmation of host records for ichneumonoid parasitoids of wood-boring beetle larvae. *Journal of Hymenoptera Research*, **9**, 241–5.

Laurenne NM, Broad GR, Quicke DLJ (2003) Preliminary molecular phylogenetic analysis of Cryptinae and related taxa based on 28S D2+D3 rDNA analysed using POY. In G Melika and C Thuróczy (eds), *Parasitic Wasps: Evolution, Systematics, Biodiversity and Biological Control*, Agroinform Kiadó, Budapest, pp. 229–33.

Laurenne NM, Broad GR, Quicke DLJ (2006) Direct optimization and multiple alignment of 28S D2–D3 rDNA sequences: problems with indels on the way to a molecular phylogeny of the cryptine ichneumon wasps (Insecta: Hymenoptera). *Cladistics*, **22**, 442–73.

Laurenne NM, Karatolos N, Quicke DLJ (2009) Hammering homoplasy: multiple gains and losses of vibrational sounding in cryptine wasps (Insecta: Hymenoptera: Ichneumonidae). *Biological Journal of the Linnean Society*, **96**, 82–102.

Laurentz M, Reudler JH, Mappes J, Friman V, Ikonen S, Lindstedt C (2012) Diet quality can play a critical role in defense efficacy against parasitoids and pathogens in the Glanville fritillary (*Melitaea cinxia*). *Journal of Chemical Ecology*, **38**, 116–25.

Lawrence PO (1981a) Host vibration - a cue to host location by the parasitic wasp *Biosteres longicaudatus*. *Oecologia*, **48**, 249–51.

Lawrence PO (1981b) Interference competition and optimum host selection in the parasitic wasp *Biosteres longicaudatus*. *Annals of the Entomological Society of America*, **74**, 540–4.

Lawrence PO (1982) *Biosteres longicaudatus*: developmental dependence on host (*Anastrepha suspensa*) physiology. *Experimental Parasitology*, **5**, 396–405.

Lawrence PO (1986) Host–parasite hormonal interactions: an overview. *Journal of Insect Physiology*, **32**, 295–8.

Lawrence PO (1988) Intraspecific competition among first instars of the parasitic wasp *Biosteres longicaudatus*. *Oecologia*, **74**, 607–11.

Lawrence PO (1990) Serosal cells of *Biosteres longicaudatus* (Hymenoptera: Braconidae): ultrastructure and release of polypeptides. *Archives of Insect Biochemistry and Physiology*, **13**, 199–216.

Lawrence PO (2002) Purification and partial characterization of an *Entomopoxvirus* (DIEPV) from a parasitic wasp of tephritid fruit flies. *Journal of Insect Science*, **2**, 1–12.

Lawrence PO (2005) Morphogenesis and cytopathic effects of the *Diachasmimorpha longicaudata* entomopoxvirus in host haemocytes. *Journal of Insect Physiology*, **51**, 221–33.

Lawrence PO, Akin D (1990) Virus-like particles from the poison glands of the parasitic wasp *Biosteres longicaudatus* (Hymenoptera: Braconidae). *Canadian Journal of Zoology*, **68**, 539–46.

Lawrence PO, Matos LF (2005) Transmission of the *Diachasmimorpha longicaudata* rhabdovirus (DlRhV) to wasp offspring: an ultrastructural analysis. *Journal of Insect Physiology*, **51**, 235–41.

Leach WE (1815) Entomology. In D Brewster (ed.), *Edinburgh Encyclopaedia*, William Blackwood, Edinburgh, vol. **9**, pp. 57–172.

Leathers JW, Sharkey MJ (2003) Taxonomy and life history of Costa Rican *Alabagrus* (Hymenoptera: Braconidae), with a key to world species. *Contributions in Science*, **497**, 1–82.

Lecomte C, Thibout E (1993) Pre- and post-imaginal experience in a specialist parasitoid *Diadromus pulchellus* (Hym.: Ichneumonidae). *Entomophaga*, **38**, 175–84.

Lee JC, Heimpel GE, Leibee GL (2004) Comparing floral nectar and aphid honeydew diets on the longevity and nutrient levels of a parasitoid wasp. *Entomologia Experimentalis et Applicata*, **111**, 189–99.

Lee S, Nalini M, Kim Y (2008) A viral lectin encoded in *Cotesia plutellae* bracovirus and its immunosuppressive effect on host hemocytes. *Comparative Biochemistry and Physiology (A)*, **149**, 351–61.

Lei G-C, Camara MD (1999) Behaviour of a specialist parasitoid, *Cotesia melitaearum*: from individual behaviour to metapopulation processes. *Ecological Eentomology*, **24**, 59–72.

Leibenguth F (1967) Regulation of tryptophan metabolism in the parasitic wasp, *Habrobracon Juglandis*. *Experientia*, **23**, 1069–71.

Leius K (1961) Influence of food on fecundity and longevity of adults of *Itoplectis conquisitor* (Say) (Hymenoptera: Ichneumonidae). *Canadian Entomologist*, **93**, 771–80.

Leius K (1962) Effects of the body fluids of various host larvae on fecundity of females of *Scambus buolianace* (Htg.) (Hym.: Ichneumonidae). *Canadian Entomologist*, **94**, 1078–82.

Leius K (1963) Effects of pollens on fecundity and longevity of adult *Scambus buolianae* (Htg.) (Hymenoptera: Ichneumonidae). *Canadian Entomologist*, **95**, 202–7.

Leius K (1967) Food sources and preferences of adults of a parasite, *Scambus buolianae* (Hym.: Ichneumonidae) and their consequences. *The Canadian Entomologist*, **99**, 865–71.

Lejeune RR, Hildahl V (1954) A survey of parasites of the larch sawfly (*Pristiphora erichsonii* (Hartig)) in Manitoba and Saskatchewan. *The Canadian Entomologist*, **86**, 8, 337–47.

Lejeune RR, Silver GT (1961) Parasites and hyperparasites of the satin moth, Stilpnotia salicis Linnaeus, (Lymantriidae) in British Columbia. *The Canadian Entomologist*, **93**, 456–67.

Le Lannic J, Nénon JP (1999) Functional morphology of the ovipositor in *Megarhyssa atrata* (Hymenoptera, Ichneumonidae) and its penetration into wood. *Zoomorphology*, **119**, 73–9.

Leluk J, Schmidt J, Jones D (1989) Comparative studies on the protein composition of hymenopteran venom reservoirs. *Toxicon*, **27**, 105–14.

Leong JKL, Oatman ER (1968) The biology of *Campoplex haywardi* (Hymenoptera: Ichneumonidae), a primary parasite of the potato tuberworm. *Annals of the Entomological Society of America*, **61**, 26–36.

LeRalec A (1993) Les hyménoptères parasitoïdes: adaptations, de l'appareil reproducteur femelle; morphologie et ultrastructure de l'ovaire de l'œuf et de l'ovipositeur. These doctorat, Universitaire de Rennes I.

LeRalec A (1995) Egg contents in relation to host-feeding in some parasitic Hymenoptera. *Entomophaga*, **40**, 87–93.

LeRalec A, Barbier R, Nénon JP, Rabasse JM (1986) Anatomie et fonctionnement de l'ovaire et de l'ovipositeur d'*Aphidius uzbekistanicus* (Hymenoptera: Aphidiidae), parasitoïde des pucerons des céréales. *Bulletin de la Société Zoologique de France*, **111**, 211–20.

LeRalec A, Rabasse JM, Wajnberg E (1996) Comparative morphology of the ovipositor of some parasitic Hymenoptera in relation to characteristics of their hosts. *Canadian Entomologist*, **128**, 413–33.

Levie A, Legrand M-A, Dogot P, Pels C, Baret PV, Hance T (2005) Mass releases of *Aphidius rhopalosiphi* (Hymenoptera: Aphidiinae), and strip management to control of wheat aphids. *Agriculture, Ecosystems and Environment*, **105**, 17–21.

Levin D (1976) Alkaloid-bearing plants: an ecogeographic perspective. *The American Naturalist*, **110**, 261–84.

Lewis OT, Memmott J, LaSalle J, Lyal CHC, Whitefoord C, Godfray HCJ (2002) Structure of a diverse tropical forest insect-parasitoid community. *Journal of Animal Ecology*, **71**, 855–73.

Lewis WJ (1970) Life history and anatomy of *Microplitis croceipes* (Hymenoptera, Braconidae), a parasite of *Heliothis* spp. (Noctuidae). *Annals of the Entomological Society of America*, **63**, 67–70.

Lewis WJ, Brazzel JR (1966) Incidence of parasitic insects on the bollworm in Mississippi. *Mississippi State College of Agriculture Experimental Station Bulletin*, No. 727, 7 pp.

Lewis WJ, Brazzel JR (1968) A three-year study of parasites of the bollworm and the tobacco budworm in Mississippi. *Journal of Economic Entomology*, **61**, 673–6.

Lewis WJ, Vinson SB (1968) Egg and larval development of *Cardiochiles nigriceps*. *Annals of the Entomological Society of America*, **61**, 561–5.

Lewis WJ, Sonnet PE, Nordlund DA (1988) Responses of braconid parasitoids *Microplitis croceipes* (Cresson) and *M. demolitor* Wilkinson to stereoisomers of kairomone 13-methylhentriacontane. *Journal of Chemical Ecology*, **14**, 883–8.

Li FF, Chen XX, Piao MH, He JH, Ma Y (2003) Phylogenetic relationships of the Euphorinae (Hymenoptera: Braconidae) based on the D2 variable region of 28S ribosomal RNA. *Entomotaxonomia*, **25**, 217–26.

Li J, Wang Z-Y, Bourguet D, He K-L (2013) *Wolbachia* infection in populations of *Ostrinia furnacalis*: diversity, prevalence, phylogeny and evidence for horizontal transmission. *Journal of Integrative Agriculture*, **12**, 283–95.

Li QT (1984) Preliminary observations on thelytoky in *Meteorus cypris*. *Kunchong Zhishi*, **21**, 82–3.

Li S, Falabella P, Giannantonio S, Fanti P, Battaglia D, Digilio MC, Völkl W, Sloggett JJ, Weisser W, Pennacchio F (2002) Pea aphid clonal resistance to the endophagous parasitoid *Aphidius ervi*. *Journal of Insect Physiology*, **48**, 971–80.

Li S, Falabella P, Kuriachan I, Vinson SB, Borst DW, Malva C, Pennacchio F (2003) Juvenile hormone synthesis, metabolism, and resulting hemolymph titre in *Heliothis virescens* larvae parasitized by *Toxoneuron nigriceps*. *Journal of Insect Physiology*, **49**, 1021–30.

Li T, Sheng M-L, Sun S-P, Chen G-F, Guo Z-H (2012) Effect of the trap color on the capture of ichneumonids wasps (Hymenoptera). *Revista Colombiana de Entomología*, **38**, 338–42.

Li WD, Yu RX, Chen XX, He JH (2006a) Venom gland of the ichneumonid *Diadromus collaris*: morphology, ultrastructure and age-related changes. *Insect Science*, **13**, 137–43.

Li WD, Huang F, Chen YF, Chen XX (2006b) The immunosuppression effects of venom of pupal endoparasitoid wasp, *Diadromus collaris* on its host, *Plutella xylostella* pupae. *Acta Entomologica Sinica*, **49**, 206–12.

Li X-Y, van Achterberg C, Tan J-C (2013) Revision of the subfamily Opiinae (Hymenoptera, Braconidae) from Hunan (China), including thirty-six new species and two new genera. *ZooKeys*, **268**, 1–186.

Li Y, Lu J-F, Feng C-J, Ke X, Fu W-J (2007a) Role of venom and ovarian proteins in immune suppression of *Ostrinia furnacalis* (Lepidoptera: Pyralidae) larvae parasitised by *Macrocentrus cingulum* (Hymenoptera: Braconidae), a polyembryonic parasitoid. *Insect Science*, **14**, 93–100.

Li Y, Lu J-F, Ke X, Fu W-J (2007b) Morphology and structure of the ovipositor in *Macrocentrus cingulum* (Hymenoptera: Braconidae), a polyembryonic parasitoid. *Journal of Life Sciences*, **1**, 34–9.

Liepert C, Dettner K (1996) Role of cuticular hydrocarbons of aphid parasitoids in their relationship to aphid-attending ants. *Journal of Chemical Ecology*, **22**, 695–707.

Lill JT (1999) Structure and dynamics of a parasitoid community attacking larvae of *Psilocorsis quercicella* (Lepidoptera: Oecophoridae). *Environmental Entomology*, **28**, 1114–23.

Lim KP, Stewart RK (1976) Laboratory studies on *Peristenus pallipes* and *P. pseudopallipes* (Hymenoptera: Braconidae), parasitoids of the tarnished plant bug, *Lygus lineolaris* (Hemiptera: Miridae). *The Canadian Entomologist*, **108**, 815–21.

Linsley EG (1959) Mimetic form and coloration in the Cerambycidae (Coleoptera). *Annals of the Entomological Society of America*, **52**, 125–31.

Lipkow E (1965) Braconiden als Schmarotzer von Staphyliniden. *Entomophaga*, **10**, 377–81.

Liu H, Ueno T (2012) The importance of food and host on the fecundity and longevity of a host-feeding parasitoid wasp. *Journal of the Faculty of Agriculture, Kyushu University*, **57**, 121–5.

Liu S, Wang X, Shi Z, Gebremeskel ZH (2001) The biology of *Diadromus collaris* (Hymenoptera: Ichneumonidae), a pupal parasitoid of *Plutella xylostella* (Lepidoptera: Plutellidae), and its interactions with *Oomyzus sokolowskii* (Hymenoptera: Eulophidae). *Bulletin of Entomological Research*, **91**, 461–9.

Lloyd DC (1940) Host selection by hymenopterous parasites of the moth *Plutella maculipennis* Curtis. *Proceedings of the Royal Society B: Biological Sciences*, **128**, 451–84.

Loan CC ((1963)) (1964) Observations on the biology of *Centistes excrucians* Haliday (Hymenoptera, Braconidae). *Proceedings of the Entomological Society of Ontario*, **94**, 56–60.

Loan CC (1965) Life cycle and development of *Leiophron pallipes* Curtis (Hymenoptera: Braconidae, Euphorinae) in five mirid hosts in the Belleville district. *Proceedings of the Entomological Society of Ontario*, **95**, 115–21.

Loan CC (1967) Studies on the taxonomy and biology of the Euphorinae (Hymenoptera: Braconidae). II. Host relations of six *Microctonus* species. *Annals of the Entomological Society of America*, **60**, 236–40.

Loan CC (1970) Two new parasites of the tarnished plant bug in Ontario: *Leiophron pseudopallipes* and *Euphoriana lygivora* (Hymenoptera: Braconidae, Euphorinae). *Proceedings of the Entomological Society of Ontario*, **100**, 188–95.

Loan CC (1972) Parasitism of adult *Notoxus anchora* Henz. (Coleoptera: Anthicidae) by *Syrrhizus agilis* (Cress.) (Hymenoptera: Braconidae). *Proceedings of the Entomological Society of Ontario*, **103**, 76.

Loan CC (1974) The European species of *Leiophron* Nees and *Persistenus* Foerster (Hymenoptera, Braconidae, Euphorinae). *Transactions of the Royal Entomological Society of London*, **126**, 207–38.

Loan CC (1980) Plant bug hosts (Heteroptera: Miridae) of some euphorine parasites (Hymenoptera: Braconidae) near Belleville, Ontario, Canada. *Le Naturaliste Canadiene*, **101**, 821–60.

Loan CC, Holdaway FG (1961) *Pygostolus falcatus* (Nees) (Hymenoptera: Braconidae), a parasitoid of *Sitona* species (Coleoptera: Curculionidae). *Bulletin of Entomological Research*, **52**, 473–88.

Loan CC, Matthews R (1973) *Cosmophorus capeki* n. sp., from New York (Hymenoptera: Braconidae: Euphorinae). *Proceedings of the Entomological Society of Washington*, **71**, 205–8.

Loan CC, Gerber GH, Reid DG (1971) Biosystematics of the tinged parasite *Holdawayella* in Ontario (Hymenoptera: Braconidae, Euphorinae). *The Canadian Entomologist*, **103**, 1273–84.

Lopez M, Rojas JC, Vandame R, Williams T (2002) Parasitoid-mediated transmission of an iridescent virus. *Journal of Invertebrate Pathology*, **80**, 160–70.

Lo Pinto M, Cangelosi B, Colazza S (2013) Female-released sex pheromones mediating courtship behavior in Lysiphlebus testaceipes males. *Journal of Insect Science*, **13**, 53.

Lovallo, N, McPheron BA, Cox-Foster DL (2002) Effects of the polydnavirus of *Cotesia congregata* on the immune system and development of non-habitual hosts of the parasitoid. *Journal of Insect Physiology*, **48**, 517–26.

Lozan A (2002) Light traps' braconids (Hymenoptera, Braconidae) from the Mrtvy Luh bog (Central Europe, Bohemian Forest). *Silva Gabreta*, **8**, 247–54.

Lu JF, Feng CJ, Hu J, Fu WJ (2007) Extraembryonic membrane of the polyembryonic parasitoid *Macrocentrus cingulum* Brischke (Hym., Braconidae) is essential for evasion of encapsulation. *Journal of Applied Entomology*, **131**, 472–7.

Lucchetta P, Bernstein C, Théry M, Lazzari C, Desouhant E (2008) Foraging and associative learning of visual signals in a parasitic wasp. *Animal Cognition*, **11**, 525–33.

Luff ML (1976) The biology of *Microctonus caudatus* (Thomson), a braconid parasite of the ground beetle *Harpalus rufipes* (Degeer). *Ecological Entomology*, **1**, 111–16.

Lum PTM, Flaherty BR (1973) Influence of continuous light on oöcyte maturation in *Bracon hebetor*. *Annals of the Entomological Society of America*, **66**, 355–7.

Luo L, Zeng L (2010) A new rod-shaped virus from parasitic *wasp Diachasmimorpha longicaudata* (Hymenoptera: Braconidae). *Journal of Invertebrate Pathology*, **103**, 165–9.

Lyngnes R (1960) Shape and function of the ovipositor in the three hymenopterous species: *Ephialtes extensor* Thom. (Ichneumonidae), *Spathius exarator* L. (Braconidae) and *Plutothrix coelius* Walk. (Chalicidae). *Norsk Entomologisk Tidsskrift*, **11**, 122–34.

Lyons DB (1999) Phenology of the native parasitoid *Sinophorus megalodontis* (Hymenoptera: Ichneumonidae) relative to its introduced host, the pine false webworm (Hymenoptera: Pamphiliidae). *Canadian Entomologist*, **131**, 787–800.

Ma L, Gu S-H, Liu Z-W, Wang S-N, Guo Y-Y, Zhou J-J, Zhang Y-J (2014) Molecular characterization and expression profiles of olfactory receptor genes in the parasitic wasp, *Microplitis mediator* (Hymenoptera: Braconidae). *Journal of Insect Physiology*, **60**, 118–26.

Ma W-J, Kuijper B, de Boer JG, van de Zande L, Beukeboom LW, Wertheim B, Pannebakker BA (2013) Absence of complementary sex determination in the parasitoid wasp genus *Asobara* (Hymenoptera: Braconidae). *PLoS ONE*, **8** (4), e60459.

Mabiala-Moundoungou ADN, Doury G, Eslin P, Cherqui A, Prévost G. (2010). Deadly venom of *Asobara japonica* parasitoid needs ovarian antidote to regulate host physiology. *Journal of Insect Physiology*, **56**, 35–41.

Mackauer M (1961) Die Gattungen der Familie Aphidiidae und ihre verwandtschaftliche Zuordnung. *Beiträge zur Entomologie*, **11**, 791–803.

Mackauer M (1968) Pars 3. Aphidiidae. In C Ferriere and J van der Vecht (eds), *Hymenopterorum Catalogus*, nova editio, Dr W. Junk, The Hague, pp. 1–103.

Mackauer M (1969) Sexual behaviour of and hybridization between three species of *Aphidius* Nees parasitic on the pea aphid. *Proceedings of the Entomological Society of Washington*, **71**, 339–51.

Mackauer M, Chow FJ (2012) Facultative gregarious development in a solitary koinobiont parasitoid, *Ephedrus californicus*: implications of larval ecology and host constraints. *Entomologia Experimentalis et Applicata*, **142**, 203–10.

Mackauer M, Finlayson T (2012) *Choreopraon totarae* (Hymenoptera: Braconidae: Aphidiinae), a new parasitoid of *Neophyllaphis totarae* (Hemiptera: Aphidoidea: Drepanosiphidae) in New Zealand, *New Zealand Journal of Zoology*, **39**, 77–84.

Mackauer M, Henkelman DH (1975) Effect of light–dark cycles on adult emergence in aphid parasite *Aphidius smithi*. *Canadian Journal of Zoology*, **53**, 1201–6.

Mackauer M, Starý P (1967) World Aphidiidae. In V Delluchi and G Remaudière (eds), *Index of Entomophagous Insects*, vol. 2, Le Francois, Paris, pp. 11–195.

Mackauer M, Völkl W (2005) Sex ratio shift caused by hyperparasitism in the solitary parasitoid *Lysiphlebus hirticornis* (Hymenoptera: Braconidae: Aphidiinae). *European Journal of Entomology*, **102**, 475–81.

Madanagopal N, Kim Y (2006) Parasitism by *Cotesia glomerata* induces immunosuppression of *Pieris rapae*: effects of ovarian protein and polydnavirus. *Journal of Asia-Pacific Entomology*, **9**, 339–46.

Madar RJ, Miller JC (1983) Developmental biology of *Apanteles yakutatensis* (Hymenoptera: Braconidae), a primary parasite of *Autographa californica* (Lepidoptera: Noctuidae). *Annals of the Entomological Society of America*, **76**, 683–7.

Madden JL (1968) Behavioral responses of parasites to the symbiotic fungus associates with *Sirex noctilio* F. *Nature*, **218**, 189–90.

Madel G (1963) Beiträge zur morphologie und biologie von *Meteorus fagilis* Wesm. (Hym., Braconidae) – ein endoparasit des mondvogels *Phalera bucephala* L. (Lep. Notodontidae). *Zeitschrift fur Angewandte Entomologie*, **53**, 1–47.

Madel G, Mühlen D, Happe M (1975) *Diadegma semiclausum* Hellen (Hymenoptera, Ichneumonidae): copulation, spermatophore transfer, and offspring. *Zeitschrift für Angewandte Zoologie*, **77**, 347–56.

Maetô K (1986) Systematic studies on the tribe Meteorini from Japan (Hymenoptera: Braconidae). II. The *corax* group of the genus *Meteorus* Haliday. *Kontyû, Tokyo*, **54**, 405–13.

Maetô K (1987) A comparative morphology of the male internal reproductive organs of the family Braconidae (Hymenoptera, Ichneumonoidea). *Kontyû, Tokyo*, **55**, 32–42.

Maetô K (1990) Phylogenetic relationships and host associations of the subfamily Meteorinae Cresson (Hymenoptera, Braconidae). *Japanese Journal of Entomology*, **58**, 383–96.

Maetô K (1995) The genus *Mirax* (Hymenoptera: Braconidae: Miracinae) from Japan. *Japanese Journal of Entomology*, **63**, 649–56.

Maetô K (1996) Intergeneric variation in the external male genitalia of the subfamily Microgastrinae (Hymenoptera, Braconidae), with a reassessment of Mason's tribal system. *Journal of Hymenoptera Research*, **5**, 38–52.

Maetô K, Arakaki N (2005) Gregarious emergence of *Macrostomion sumatranum* (Hymenoptera: Braconidae; Rogadinae) from the mummified, full-grown larvae of *Theretra silhetensis* (Lepidoptera: Sphingidae). *Entomological Science*, **8**, 131–2.

Maetô K, Kudô S (1992) A new euphorine species of *Aridelus* (Hymenoptera, Braconidae) associated with a subsocial bug *Elasmucha putoni* (Heteroptera, Acanthosomatidae). *Japanese Journal of Entomology*, **60**, 77–84.

Maetô K, Nagai K (1985) Notes on braconid parasitoids of *Medythia nigrobilineata* (Motschulsky) (Coleoptera, Chrysomelidae), with description of a new species of *Centistes* Haliday (Hymenoptera, Braconidae). *Kontyû, Tokyo*, **53**, 729–33.

Maetô K, Thornton IWB (1993) A preliminary appraisal of the braconid (Hymenoptera) fauna of the Krakatau Islands, Indonesia, in 1984–1986, with comments on the colonizing abilities of parasitoid modes. *Japanese Journal of Entomology*, **61**, 787–801.

Magdaraog PM, Harvey JA, Tanaka T, Gols R (2012) Intrinsic competition among solitary and gregarious endoparasitoid wasps and the phenomenon of 'resource sharing'. *Ecological Entomology*, **37**, 65–74.

Magro SR, Parra JRP (2004) Comparison of artificial diets for rearing *Bracon hebetor* Say (Hymenoptera: Braconidae). *Biological Control*, **29**, 341–7.

Magro SR, Dias AB, Terra WR, Parra JR (2006) Biological, nutritional, and histochemical basis for improving an artificial diet for *Bracon hebetor* Say (Hymenoptera: Braconidae). *Neotropical Entomology*, **35**, 215–22.

Magurran AE (2004) *Measuring Biological Diversity*, Blackwell, Malden, MA, 256 pp.

Majerus MEN (1997) Parasitization of British ladybirds by *Dinocampus coccinellae* (Schrank) (Hymenoptera: Braconidae). *British Journal of Entomology and Natural History*, **10**, 15–24.

Makatiani JK, Le KH, Olson DM, Wäckers FL, Takasu K (2014) An acquired distaste: sugar discrimination by the larval parasitoid *Microplitis croceipes* (Hymenoptera: Braconidae) is affected by prior sugar exposure. *Journal of Experimental Biology* in press.

Makino S (1983) Biology of *Latibulus argiolus* (Hymenoptera, Ichneumonidae), a parasitoid of the paper wasp *Polistes biglumis* (Hymenoptera, Vespidae). *Kontyu, Tokyo*, **51**, 426–34.

Maltais J, Régnière J, Cloutier C, Hébert C, Perry DF (1989) Seasonal biology of *Meteorus trachynotus* Vier. (Hymenoptera: Braconidae) and of its overwintering host *Chorisoneura rosaceana* (Harr.) (Lepidoptera: Tortricidae). *The Canadian Entomologist*, **121**, 745–56.

Marais SJS (1955) *The Biological Control Project of the Karoo Caterpillar, Loxostege frustralis Zell. in South Africa*. Unpublished report, Department of Agriculture Technical Service, Republic of South Africa.

Mardulyn P, Whitfield JB (1999) Phylogenetic signal in the COI, 16S, and 28S genes for inferring relationships among genera of Microgastrinae (Hymenoptera; Braconidae): evidence of a high diversification rate in this group of parasitoids. *Molecular Phylogenetics and Evolution*, **12**, 282–94.

Marris GC, Casperd J (1996) The relationship between conspecific superparasitism and the outcome of *in vitro* contests staged between different larval instars of the solitary endoparasitoid *Venturia canescens*. *Behavioral Ecology and Sociobiology*, **39**, 61–9.

Marris GC, Hubbard SF, Scrimgeour C (1996) The perception of genetic similarity by the solitary parthenogenetic parasitoid *Venturia canescens*, and its effects on the occurrence of superparasitism. *Entomologia Experimentalis et Applicata*, **78**, 167–74.

Marsh PM (1977) Notes on the taxonomy and nomenclatures of *Aphidius* species (Hymenoptera: Aphidiidae) parasitic on the pea aphid in North America. *Entomophnga*, **22**, 365–72.

Marsh PM (1997) Subfamily Doryctinae. In RA Wharton, PM Marsh and MJ Sharkey (eds), *Identification Manual to the New World Genera of Braconidae*, Special Publication of the International Society of Hymenopterists 1, pp. 207–33.

Marsh PM, Melo GAR (1999) Biology and systematics of New World *Heterospilus* (Hymenoptera; Braconidae) attacking Pemphredoninae (Hymenoptera; Sphecidae). *Journal of Hymenoptera Research*, **8**, 13–22.

Marsh PM, de Macedo MV, Pimentel MCP (2000) Descriptions and biological notes on two new phytophagous species of the genus *Allorhogas* from Brasil (Hymenoptera:

Braconidae: Doryctinae). *Journal of Hymenoptera Research*, **9**, 292–7.

Marti D, Grossniklaus-Bürgin C, Wyder S, Wyler T, Lanzrein B (2003) Ovary development and polydnavirus morphogenesis in the parasitic wasp *Chelonus inanitus*. I. Ovary morphogenesis, amplification of viral DNA and ecdysteroid titres. *Journal of General Virology*, **84**, 1141–50.

Martínez-Martínez L, Leyva-Vázquez JL, Mojica HB (1993) Sperm competition in the female *Diachasmimorpha longicaudata* (Hymenoptera: Braconidae). *Southwestern Entomologist*, **18**, 293–9.

Masler EP, Kovaleva ES (1999) Inhibition of larval growth in the gypsy moth (Lepidoptera: Lymantriidae) by venom from the parasitic wasp *Microbracon hebetor* (Hymenoptera: Braconidae). *Journal of Entomological Science*, **34**, 435–44.

Mason HC (1948) *Chremylus rubiginosus* (Nees), a braconid parasite of the casemaking clothes moth. *Annals of the Entomological Society of America*, **41**, 28–40.

Mason PG, Erlandson MA, Youngs BJ (2001) Effects of parasitism by *Banchus flavescens* (Hymenoptera: Ichneumonidae) and *Microplitis mediator* (Hymenoptera: Braconidae) on the bertha armyworm, *Mamestra configurata* (Lepidoptera: Noctuidae). *Journal of Hymenoptera Research*, **10**, 81–90.

Mason WRM (1964) Regional colour patterns in the parasitic Hymenoptera. *The Canadian Entomologist* **96**, 132–4.

Mason WRM (1967) Specialisation in the egg structure of *Exenterus* (Hymenoptera; Ichneumonidae) in relation to distribution and abundance. *The Canadian Entomologist* **99**, 375–84.

Mason WRM (1968) New Canadian Cryptinae (Ichneumonidae: Cryptinae). *The Canadian Entomologist*, **100**, 17–23.

Mason WRM (1969) Muesebeckiini, a new tribe of Braconidae (Hymenoptera). *Proceedings of the Entomological Society of Washington*, **71**, 263–78.

Mason WRM (1971) An Indian *Agriotypus* (Hymenoptera: Agriotypidae). *The Canadian Entomologist*, **103**, 1521–4.

Mason WRM (1976a) The identity of *Macrocentrus uniformis* Provancher (nec Cresson), description of sibling species and a possible grooming organ (Hym.: Braconidae). *Naturaliste Canadiene*, **103**, 513–15.

Mason WRM (1976b) A revision of *Dyscoletes* Haliday (Hym: Braconidae). *The Canadian Entomologist*, **108**, 855–8.

Mason WRM (1978) A new genus and species and family of Hymenoptera (Ichneumonoidea) from Chile. *Proceedings of the Entomological Society of Washington*, **80**, 606–10.

Mason WRM (1979) A new genus and species of Orgilini (Hymenoptera: Braconidae) from New Guinea. *Proceedings of the Entomological Society of Washington*, **81**, 640–4.

Mason WRM (1981a) The polyphyletic nature of *Apanteles* Foerster (Hymenoptera: Braconidae): a phylogeny and reclassification of Microgastrinae. *Memoirs of the Entomological Society of Canada*, **115**, 1–147.

Mason WRM (1981b) Paxylommatidae – The correct family-group name for *Hybrizon* Fallen (Hymenoptera, Ichneumonoidea), with figures of unusual antennal sensilla. *Canadian Entomologist*, **113**, 427–31.

Mason WRM (1983). A new South African subfamily related to Cardiochilinae (Hymenoptera: Braconidae). *Contributions to the American Entomological Institute*, **20**, 49–62.

Mason WRM (1985) The correct spelling for *Adelius* Haliday (Hymenoptera: Braconidae). *Proceedings of the Entomological Society of Washington*, **87**, 896.

Mason WRM (1987) Discovery of female *Apozyx* (Hymenoptera: Apozygidae) and comments on its taxonomic position. *Proceedings of the Entomological Society of Washington*, **89**, 226–229.

Mason WRM (1991) A new genus of exodont lchneutinae (Hymenoptera: Braconidae). *Insecta Mundi*, **5** (2), paper 413.

Mateo Leach I, van der Zande L, Bolhuis H, Beukeboom LW (2004) No microbial endosymbionts associated with parthenogenesis in *Venturia canescens* (Hymenoptera: Ichneumonidae). *Proceedings of the Netherlands Entomological Society*, **15**, 119–22.

Mateo Leach I, Ferber S, van der Zande L, Beukeboom LW (2012) Genetic variability of arrhenotokous and thelytokous *Venturia canescens* (Hymenoptera). *Genetica*, **140**, 53–63.

Mathur RN (1959) Breeding technique and utilisation of *Cedria paradoxa* Wilk. (Hym., Braconidae) for biological control of teak defoliator. *Entomologist's Monthly Magazine*, **95**, 248–50.

Matsumoto R (2008) "Veils" against predators: modified web structure of a host spider induced by an ichneumonid parasitoid, *Brachyzapus nikkoensis* (Uchida) (Hymenoptera). *Journal of Insect Behavior*, **22**, 39–48.

Matsumoto R, Broad GR (2011) Discovery of *Rodrigama* Gauld in the Old World, with description of two new species (Hymenoptera, Ichneumonidae, Poemeniinae). *Journal of Hymenoptera Research*, **20**, 65–75.

Matsumoto R, Konishi K (2007) Life histories of two ichneumonid parasitoids of *Cyclosa octotuberculata* (Araneae): *Reclinervellus tuberculatus* (Uchida) and its new sympatric congener (Hymenoptera: Ichneumonidae: Pimplinae). *Entomological Science*, **10**, 267–78.

Matsumoto R, Saigusa T (1999) Biology of *Pseudopimpla glabripropodeum* He et Chen (Hymenoptera: Ichneumonidae, Pimplinae) with description of final instar larva and redescription of adults. *Entomological Science*, **2**, 467–71.

Matsumoto R, Saigusa T (2001) The biology and immature stages of *Thrybius togashii* Kusigemati (Hymenoptera: Ichneumonidae: Cryptinae), with description of the male. *Journal of Natural History*, **35**, 1507–16.

Matthews RW (1974) Biology of Braconidae. *Annual Review of Entomology*, **19**, 15–32.

Matthews RW, Matthews JR, Crankshaw O (1979) Aggregation in male parasitic wasps of the genus *Megarhyssa*: I. Sexual discrimination, tergal stroking behavior, and description of associated anal structures. *Florida Entomologist*, **62**, 3–8.

Mattiacci L, Dicke M (1995) The parasitoid *Cotesia glomerata* (Hymenoptera: Braconidae) discriminates between first and fifth larval instars of its host *Pieris brassicae*, on the basis of contact cues from frass, silk, and herbivore-damaged leaf tissue. *Journal of Insect Behavior*, **8**, 485–98.

Maund CM, Hsiao TH (1991) Differential encapsulation of two *Bathyplectes* parasitoids among alfalfa weevil strains, *Hypera postica* (Gyllenhal). *The Canadian Entomologist*, **123**, 197–203.

Maure F, Brodeur J, Ponlet N, Doyon J, Firlej A, Elguero1 E, Thomas F (2011) The cost of a bodyguard. *Biology Letters*, **7**, 843–6.

Mayhew PJ, Blackburn TM (1999) Does development mode organise life-history traits in the parasitoid Hymenoptera? *Journal of Animal Ecology*, **68**, 906–16.

Mayhew PJ, Glaizot O (2001) Integrating theory of clutch size and body size evolution for parasitoids. *Oikos*, **92**, 372–6.

Mayhew PJ, van Alphen JJM (1999) Gregarious development in alysiine parasitoids evolved through a reduction in larval aggression. *Animal Behaviour*, **58**, 131–41.

Mayhew PJ, Dytham C, Shaw MR, Fraser SEM (2009) Collections of ichneumonid wasps (subfamilies Diacritinae, Diplazontinae, Pimplinae and Poemeniinae) from woodlands near York and their implications for conservation planning. *Nauralist*, **134**, 3–24.

Mazanec Z (1990) Immature stages and life history of *Enytus* (Hymenoptera: Ichneumonidae), a parasitoid of *Perthida glyphopa* Common (Lepidoptera: Incurvariidae). *Australian Journal of Entomology*, **9**, 57–66.

Mazón M, Bordera S (2008) Effectiveness of two sampling methods used for collecting Ichneumonidae (Hymenoptera) in the Cabañeros National Park (Spain). *European Journal of Entomology*, **105**, 879–88.

McAllister MK, Roitberg BD (1987) Adaptive suicidal behaviour in pea aphids. *Nature*, **328**, 797–9.

McBrien H, Mackauer M (1990) Heterospecific larval competition and host discrimination in two species of aphid parasitoids: *Aphidius ervi* and *Aphidius smithi*. *Entomologia Experimentalis et Applicata*, **56**, 145–53.

McClure M, Wisthlecraft J, McNeil JN (2007) Courtship behaviour in relation to the female sex pheromone in the parasitoid, *Aphidius ervi* (Hymenoptera: Braconidae). *Journal of Chemical Ecology*, **33**, 1946–59.

McConnell HS (1938) Additional notes on *Oocenteter tomostethae*. *Proceedings of the Entomological Society of Washington*, **40**, 23–34.

McHugh JV (1993) First records of parasitoids for slime mold beetles of the family Sphindidae (Coleoptera: Cucujoidea)! *Entomological News*, **104**, 136–8.

McKellar RC, Kopylov DS, Engel MS (2013) Ichneumonidae (Insecta: Hymenoptera) in Canadian Late Cretaceous amber. *Fossil Record*, **16**, 217–27.

McLaughlin RE, Adams CH (1966) Infection of *Bracon mellitor* (Hymenoptera: Braconidae) by *Mattesia grandis* (Protozoa: Neogregarinida). *Annals of the Entomological Society of America*, **59**, 800–2.

McNeil JN, Brodeur J (1995) Pheromone mediated mating in the aphid parasitoid *Aphidius nigripes* (Hymenoptera: Aphididae). *Journal of Chemical Ecology*, **21**, 959–72.

McNeill MR, Phillips CB, Goldson SL (1993) Diagnostic characteristics and biology of three *Microctonus* spp. (Hymenoptera: Braconidae, Euphorinae) parasitoids of weevils (Coleoptera: Curculionidae) in New Zealand pasture and lucerne. *New Zealand Entomologist*, **16**, 39–44.

Memmott J, Godfray HCJ, Gauld ID (1994) The structure of a tropical host-parasitoid community. *Journal of Animal Ecology*, **63**, 521–40.

Meng C (1968) Strukturwandel und histochemische Befunde insbesondere am Oosom w/ihrend der Oogenese und nach der Ablage des Eies von *Pimpla turionallae* L. (Hymenoptera, Ichneumonidae). *Roux's Archives of Developmental Biology*, **161**, 162–208.

Menier J-J, Nel A, Waller A, de Ploëg G (2004) A new fossil ichneumon wasp from the Lowermost Eocene amber of Paris Basin (France), with a checklist of fossil Ichneumonoidea s.l. (Insecta: Hymenoptera: Ichneumonidae: Metopiinae). *Geologica Acta*, **2**, 83–94.

Mercado I, Wharton RA (2003) Mexican cardiochiline genera (Hymenoptera: Braconidae), including a preliminary assessment of species-groups in *Toxoneuron* Say and *Retusigaster* Dangerfield, Austin and Whitfield. *Journal of Natural History*, **37**, 845–902.

Messner B, Taschenberger D (1981) Zur Funktionsmorphologie des Atembandes von *Agriotypus armatus* Walk. (Hym., Agriotypidae). *Berliner Entomologische Zeitschrift*, **28**, 7–9.

Meyer NF (1925) Zur Biologie und Morphologie von *Pimpla examinator* Fabr. (Hymenoptera, Ichneumonidae). *Journal of Applied Entomology*, **11**, 203–12.

Miah MI, Bhuyia BA (2001) The relationships of the subfamily Campopleginae (Hymenoptera, Ichneumonidae) with its related subfamilies in cladistic assessment. *Proceedings of the Zoological Society (Calcutta)*, **54**, 27–37.

Michaud JP, Mackauer M (1994) The use of visual cues in host evaluation by aphidiid wasps. 1. Comparsison betweenthree *Aphidius* parasitoids of the pea aphid. *Entomologia Experimentalis et Applicata*, **70**, 273–83.

Michaud JP, Mackauer M (1995) The use of visual cues in host evaluation by aphidiid wasps. 2. Comparsison between *Ephedrus californicus*, *Monoctonus paulensis*, and *Praon pequodorum*. *Entomologia Experimentalis et Applicata*, **74**, 267–75.

Michel-Salzat A, Whitfield JB (2004) Preliminary evolutionary relationships within the parasitoid wasp genus

Cotesia (Hymenoptera: Braconidae: Microgastrinae): combined analyses of four genes. *Systematic Entomology*, **29**, 371–82.

Middeldorf J, Ruthmann A (1984) Yeast-like endosymbionts in an ichneumonid wasp. *Zeitschrift für Naturforschung C*, **39**, 322–26.

Milano P, Berti Filho E, Parra JRP, Carvalho SS (2008) Aspectos biológicos e parasitismo de *Hymenochaonia delicata* Cresson (Hymenoptera: Braconidae) sobre *Gynandrosoma aurantianum* Lima (Lepidoptera: Tortricidae) em laboratório. *Revista de Agricultura (Piracicaba)*, **83**, 19–27.

Millar JG, Millar JG, Paine TD, Campbell CD, Hanks LM (2002) Methods for rearing *Syngaster lepidus* and *Jarra phoracantha* (Hymenoptera: Braconidae), larval parasitoids of the phloem-colonizing longhorned beetles *Phoracantha semipunctata* and *P. recurva* (Coleoptera: Cerambycidae). *Bulletin of Entomological Research*, **92**, 141–6.

Mills NJ (1991) Searching strategies and attack rates of parasitoids of the ash bark beetle (*Leperisinus varius*) and its relevance to biological control. *Ecological Entomology*, **16**, 461–70.

Mills NJ (1992) Parasitoid guilds, life-styles, and host ranges in the parasitoid complexes of tortricoid hosts (Lepidoptera: Tortricidae). *Environmental Entomology*, **21**, 320–9.

Mills NJ (1994a) Parasitoid guilds: defining the structure of the parasitoid communities of endopterygote insect hosts. *Environmental Entomology*, **23**, 1066–83.

Mills NJ (1994b) Parasitoid guilds: a comparative analysis of the parasitoid communities of tortricids and weevils. In BA Hawkins and W Sheehan (eds), *Parasitoid Community Ecology*, Oxford University Press, Oxford, pp. 30–46.

Mills NJ (2007) Single versus multiple introductions, and differential success in the biological control of homopteran and lepidopteran pests. *Journal of Insect Science*, **7**, 15–16.

Mills NJ, Krüger K, Schlup J (1991) Short-range host location mechanisms of bark beetle parasitoids. *Journal of Applied Entomology*, **111**, 33–43.

Ming G, Xia XM, Xiang-Xiong Z, Cheng DD (2003) Developmental stages of *Lysiphlebus japonicus* Ashmead (Hymenoptera: Aphidiidae), a bean aphid parasitoid. *Journal of Entomological Science*, **38**, 519–24.

Mochiah MB, Ngi-Song AJ, Overholt WA, Stouthamer R (2002) *Wolbachia* infection in *Cotesia sesamiae* (Hymenoptera: Braconidae) causes cytoplasmic incompatibility: implications for biological control. *Biological Control*, **25**, 74–80.

Mohamed SA, Overholt WA, Wharton RA, Lux SA, Eltoum EM (2003) Host specificity of *Psyttalia cosyrae* (Hymenoptera: Braconidae) and the effect of different host species on parasitoid fitness. *Biological Control*, **28**, 155–63.

Monteith AE (1956) *Phygadeuon trichops* Thoms. (Hymenoptera: Ichneumonidae), an occasional parasite of *Hylemya* spp. (Diptera: Anthomyiidae). *The Canadian Entomologist*, **88**, 69–73.

Montgomery VE, DeWitt PR (1975) Morphological differences among immature stages of three genera of exotic larval parasitoids attacking the cereal leaf beetle in the United States. *Annals of the Entomological Society of America*, **68**, 574–8.

Moody S (1978) Latitude, continental drift, and the percentage of alkaloid-bearing plants in flora. *The American Naturalist*, **112**, 965–8.

Mook JH (1961) Observations on the oviposition behaviour of *Polemon liparae* Gir. (Hym., Braconidae). *Archives Néerlandaises de Zoologie*, **14**, 423–30.

Morales J, Medina P, Viñuela E (2007) The influence of two endoparasitic wasps, *Hyposoter didymator* and *Chelonus inanitus*, on the growth and food consumption of their host larva *Spodoptera littoralis*. *BioControl*, **52**, 145–60.

Moreau SJM, Dingremont A, Doury G, Giordanengo P (2002) Effect of parasitism by *Asobara tabida* (Hymenoptera: Braconidae) on the development, survival and activity of *Drosophila melanogaster* larvae. *Journal of Insect Physiology*, **48**, 337–47.

Moreau SJM, Eslin P, Giordanengo P, Doury G (2003) Comparative study of the strategies evolved by two parasitoids of the genus *Asobara* to avoid the immune response of the host, *Drosophila melanogaster*. *Developmental and Comparative Immunology*, **27**, 273–82.

Moreau SJM, Cherqui A, Doury G, Dubois F, Fourdrain Y, Sabatier L, Bulet P, Saarela J, Prévost G, Giordanengo P (2004) Identification of an aspartylglucosaminidase-like protein in the venom of the parasitic wasp *Asobara tabida* (Hymenoptera: Braconidae). *Insect Biochemistry and Molecular Biology*, **34**, 485–92.

Moreau SJM, Vinchon S, Cherqui A, Prévost G (2009) Components of *Asobara* venoms and their effects on hosts. *Adv. Parasitol.*, **70**, 217–32.

Moreira J, Araújo VA, Báo SN, Lino-Neto J (2010) Structural and ultrastructural characteristics of male reproductive tract and spermatozoa in two Cryptinae species (Hymenoptera: Ichneumonidae). *Micron*, **41**, 187–92.

Moreno J, Jiménez R (1993) Morphology of the female reproductive system of *Phanerotoma* (*Phanerotoma*) *ocularis* Kohl (Hymenoptera: Braconidae). *Research & Reviews in Parasitology*, **53**, 53–6.

Morley C (1915) *Entomologica Britannica, V. The Ichneumons of Great Britain. Ophioninae*. H. & W. Brown, London, 400 pp.

Morris KRS (1937) The prepupal stage in Ichneumonidae, illustrated by the life-history of *Exenterus abruptorius* Thunb. *Bulletin of Entomological Research*, **28**, 525–34.

Morrison G, Auerbach M, McCoy ED (1979) Anomalous diversity of tropical parasitoids: a general phenomenon? *The American Naturalist*, **114**, 303–7.

Moser SE, Alleyne M, Wiedenmann R, Hanks LM (2008) Influence of oviposition experience on multiparasitism by *Pimpla disparis* Viereck and *Itoplectis conquisitor* Say (Hymenoptera: Ichneumonidae). *Environmental Entomology*, **37**, 1307–12.

Moutia LA, Courtois CM (1952) Parasites of the moth borers of sugar-cane in Mauritius. *Bulletin of Entomological Research*, **43**, 325–59.

Mudd A, Fisher RC, Smith MC (1982) Volatile hydrocarbons in the Dufour's gland of the parasite *Nemeritis canescens* (Grav.) (Hymenoptera: Ichneumonidae). *Journal of Chemical Ecology*, **8**, 1035–42.

Muldrew JA (1950) *Mesoleius aulicus*, a parasite of the Larch sawfly. *Bi-monthly Progress Report Forest Insect Investigations, Department of Agriculture, Canada*, **6** (6), 2.

Muli BK, Schulthess F, Maranga RO, Kutima HL, Jiang N (2006) Interspecific competition between *Xanthopimpla stemmator* Thunberg and *Dentichasmias busseolae* Heinrich (Hymenoptera: Ichneumonidae), pupal parasitoids attacking *Chilo partellus* (Lepidoptera: Crambidae) in East Africa. *Biological Control*, **36**, 163–70.

Murata Y, Ideo S, Watada M, Mitsui H, Kimura MT (2009) Genetic and physiological variation among sexual and parthenogenetic populations of *Asobara japonica* (Hymenoptera: Braconidae), a larval parasitoid of drosophilid flies. *European Journal of Entomology*, **106**, 171–8.

Muratori F, Le Lannic J, Nénon JP, Hance T (2004) Larval morphology and development of *Aphidius rhopalosiphi* (Hymenoptera: Braconidae: Aphidiinae). *The Canadian Entomologist*, **136**, 169–80.

Murphy N, Banks JC, Whitfield JB, Austin AD (2008) Phylogeny of the parasitic microgastroid subfamilies (Hymenoptera: Braconidae) based on sequence data from seven genes, with an improved time estimate of the origin of the lineage. *Molecular Phylogenetics and Evolution*, **47**, 378–95.

Myers JG (1927) The Habits of *Alysia manducator* (Hym., Braconidae). *Bulletin of Entomological Research*, **17**, 219–29.

Nägeli W (1936) Die kleine Fichtenblattwespe. (*Lygaeonematus pini* Ratz. = *Nematus abietinus* Christ.). *Mittleilungen der Schweizerischen Anstalt für das forstliche Versuchswesen*, **19**, 213–381.

Nakahara Y, Iwabuchi K, Mitsuhashi J (1997) *In vitro* rearing of the larval endoparasitoid, *Venturia canescens* (Gravenhorst) (Hymenoptera: Ichneumonidae). III. Growth promoting ability of extract of *Galleria mellonella* pupae. *Applied Entomology and Zoology*, **32**, 381–8.

Nakamatsu Y, Fujii S, Tanaka T (2002) Larvae of an endoparasitoid, *Cotesia kariyai* (Hymenoptera: Braconidae), feed on the host fat body directly in the second stadium with the help of teratocytes. *Journal of Insect Physiology*, **48**, 1041–52.

Nakamatsu Y, Harvey JA, Tanaka T (2006) The mechanism of the emergence of *Cotesia kariyai* (Hymenopotera: Braconidae) larvae from the host. *European Journal of Entomology*, **103**, 355–60.

Nakamatsu Y, Tanaka T, Harvey JA (2007) *Cotesia kariyai* larvae need an anchor to emerge from the host *Pseudaletia separata*. *Archives of Insect Biochemistry and Physiology*, **66**, 1–8.

Namba O, Saigusa M, Tanaka T (2004) Cuticular encystment in larvae of plusiinae looper (Lepidoptera: Noctuidae) parasitised by the gregarious endoparasitoid *Cotesia glomerata*. *Annals of the Entomological Society of America*, **97**, 976–81.

Namba O, Nakamatsu Y, Miura K, Tanaka T (2008) *Autographa nigrisigna* looper (Lepidoptera: Noctuidae) excludes parasitoid egg using cuticular encystment induced by parasitoid ovarian fluid. *Applied Entomology and Zoology*, **43**, 359–67.

Namyatova AA, Cassis G (2013) Systematics, phylogeny and host associations of the Australian endemic monaloniine genus *Rayieria* Odhiambo (Insecta: Heteroptera: Miridae: Bryocorinae). *Invertebrate Systematics*, **27**, 689–726.

Narayanan ES, Subba Rao BR, Gangrade GA (1956) The biology and rate of reproduction and the morphology of the immature stages of *Apanteles angaleti* Muesebeck. *Beiträge zur Entomologie*, **6**, 298–320.

Narayanan ES, Venkatraman TV (1952) Technique of mass multiplication of *Stenobracon deesae* (Cam.) (Hymenoptera: Braconidae) for use against sugarcane and maize borers. *Proceedings of the Indian Academy of Sciences – Section A Part 3: Mathematical Sciences*, **36**, 12–18.

Narayanan ES, SubbaRao BR, Thakere KR (1961) The biology and some aspects of morphology of the immature stages of *Chelonus narayani* Subba Rao (Braconidae: Hymenoptera). *Proceedings of the National Institute of Sciences of India*, **27**, 68–82.

Navasero RC, Elzen GW (1991) Sensilla on the antennae, fore tarsi and palpi of *Microplitis croceipes* (Cressen) (Hymenoptera: Braconidae). *Proceedings of the Entomological Society of Washington*, **93**, 737–47.

Nazzi F, Powell W, Wadhams LJ, Woodcock CM (1996) Sex pheromone of aphid parasitoid *Praon volucre* (Hymenoptera, Braconidae). *Journal of Chemical Ecology*, **22**, 1169–75.

Nealis V (1985) Diapause and the seasonal ecology of the introduced parasite, *Cotesia* (*Apanteles*) *rubecula* (Hymenoptera: Braconidae). *The Canadian Entomologist*, **117**, 333–42.

Nees von Esenbeck CG) (1819) Appendix ad J. L. C. Gravenhorst conspectum generum et familiarum ichneumonidum. Genera et familias ichneumonidum adscitorum exhibens. *Isis*, **9**, 299–310.

Nees von Esenbeck CG (1834) *Hymenopterorum Ichneumonibus Affinium Monographiae, Genera Europaea et Species Illustrantes*, vol. 1. Cotta, Stuttgart and Tübingen. 320 pp.

Nénon JP (1995) Bioecology of *Megarhyssa atrata* [Hymenoptera: Ichneumonidae], an ectoparasitoid of *Tremex columba* [Hymenoptera: Siricidae] in Quebec. *Phytoprotection*, **76**, 115–22.

Nénon JP, Le Lannic J, Kacim N, Barbier R, Allo M-R (1995) Micromorphologie de l'ovipositeur des hyménoptères et évolution des symphytes phytophages aux apocrites parasitoïdes. *Comptes Rendu de l'Academie des Sciences*, **318**, 1045–51.

Nénon JP, Kacem N, Le Lannic J (1997) Structure, sensory equipment, and secretions of the ovipositor in a giant species of Hymenoptera: *Megarhyssa atrata* F. (Ichneumonidae, Pimplinae). *Canadian Entomologist*, **129**, 789–99.

Newman TM, Quicke DLJ (1998) Sperm development in the imaginal testes of *Aleiodes coxalis* (Hymenoptera: Braconidae: Rogadinae). *Journal of Hymenoptera Research*, **7**, 25–37.

Nguyen TT, Magnoli I, Cloutier C, Michaud D, Muratori F, Hance T (2013) Early presence of an enolase in the oviposition injecta of the aphid parasitoid *Aphidius ervi* analyzed with chitosan beads as artificial hosts. *Journal of Insect Physiology*, **59**, 11–18.

Nielsen E (1935) A third supplementary note upon the life histories of the Polysphinctas (Hym. Ichneum.). *Entomologiske Meddeleleser*, **19**, 191–215.

Niemeyer HM, Flores-Prado L (2012) Host location by ichneumonid parasitoids is associated with nest dimensions of the host bee species. *Neotropical Entomology*, **41**, 283–87.

Nieminen M, Suomi J, van Nouhyus S, Sauri P, Riekkola M-L (2003) Effect of iridoid glycoside content on oviposition host plant choice and parasitism in a specialist herbivore. *Journal of Chemical Ecology*, **29**, 823–44.

Nishida T (1956) An experimental study of the oviposition behavior of *Opius fletcheri* Silvestri (Hymenoptera: Braconidae), a parasite of the melon fly. *Proceedings of the Hawaiian Entomological Society*, **16**, 126–34.

Nishida T (1958) Extrafloral glandular secretions, a food source for certain insects. *Proceedings of the Hawaiian Entomological Society*, **16**, 379–86.

Nixon GEJ (1965) A reclassification of the tribe Microgasterini (Hymenoptera: Braconidae). *Bulletin of the British Museum (Natural History): Entomology*, **2**, 1–284.

Niyibigira EI, Overholt WA, Stouthamer R (2004a) *Cotesia flavipes* Cameron and *Cotesia sesamiae* (Cameron) (Hymenoptera: Braconidae) do not exhibit complementary sex determination: evidence from field populations. *Applied Entomology and Zoology*, **39**, 705–15.

Niyibigira EI, Overholt WA, Stouthamer R (2004b) *Cotesia flavipes* Cameron (Hymenoptera: Braconidae) does not exhibit complementary sex determination (ii) Evidence from laboratory experiments. *Applied Entomology and Zoology*, **39**, 717–25.

Noirot C, Quennedey A (1974) Fine structure of insect epidermal glands. *Annual Review of Entomology*, **19**, 61–80.

Norberg RÅ (1972) The pterostigma of insect wings an inertial regulator of wing pitch. *Journal of Comparative Physiology*, **81**, 9–22.

Nordlund DA, Lewis WJ (1985) Response of females of the braconid parasitoid *Microplitis demolitor* to frass of larvae of the noctuids, *Heliothis zea* and *Trichoplusia ni* and to 13-methylhentriacontane. *Entomologia Experimentalis et Applicata*, **38**, 109–12.

Norton WN, Vinson SB (1974) A comparative ultrastructural and behavioral study of the antennal sensory sensilla of the parasitoid *Cardiochiles nigriceps* (Hymenoptera: Braconidae). *Journal of Morphology*, **142**, 329–49.

Norton WN, Vinson SB, Thurston EL (1974) An ultrastructural study of the female reproductive tract of *Campoletis sonorensis* (Hymenoptera). *Proceedings of the Electron Microscopy Society of America*, (1975), 140–1.

Norullahoğlu ZÜ, Uçkan F, Sak O, Ergin E (2004) Total lipid and fatty acid composition of *Apanteles galleriae* and its parasitised host. *Annals of the Entomological Society of America*, **97**, 1000–6.

Nose Y, Lee J, Ueno T, Hatekayama M, Oishi K (1997) Cloning of cDNA for vitellogenin of the parasitoid wasp *Pimpla nipponica* (Hymenoptera: Apocrita: Ichneumonidae): vitellogenin primary structure and evolutionary considerations. *Insect Biochemistry and Molecular Biology*, **27**, 1047–56.

Notton DG, Shaw MR (1998) A review of the Palaearctic Neorhacodinae Hymenoptera, Ichneumonidae) with *Eremura* Kasparyan, (1995) new to the west Palaearctic. *Bulletin of the British Museum (Natural History): Entomology*, **67**, 209–18.

Noyes JS (1989a) A study of five methods of sampling Hymenoptera (Insecta) in a tropical rainforest, with special reference to the Parasitica. *Journal of Natural History*, **23**, 299–310.

Noyes JS (1989b) The diversity of Hymenoptera in the tropics with special reference to Parasitica in Sulawesi. *Ecological Entomology*, **14**, 197–207.

Noyes JS (1994) The reliability of published host–parasitoid records: a taxonomist's view. *Norwegian Journal of Agricultural Sciences*, **16**, 59–69.

Nuttall MJ (1973) Pre-emergence fertilisation of *Megarhyssa nortoni nortoni* (Hymenoptera: Ichneumonidae). *New Zealand Entomologist*, **5**, 112–17.

Nyabuga FN, Outreman Y, Simon JC, Heckel DG, Weisser W (2010) Effects of pea aphid secondary endosymbionts on aphid resistance and development of the aphid parasitoid *Aphidius ervi*: a correlative study. *Entomologia Experimentalis et Applicata*, **136**, 243–53.

Oatman ER, Platner GR (1974) The biology of *Temelucha* sp., *platensis* group (Hymenoptera: Ichneumonidae), a primary parasite of the potato tuberworm. *Annals of the Entomological Society of America*, **67**, 275–80.

Oatman ER, Platner GR, Greany PD (1969) The biology of *Orgilus lepidus* (Hymenoptera: Braconidae), a primary parasite of the potato tuberworm. *Annals of the Entomological Society of America*, **62**, 1407–14.

Obara M, Kitano H (1974) Studies on the courtship behavior of *Apanteles glomeratus* L. Experimental studies on releaser of wing-vibrating behaviour of the male. *Kontyû, Tokyo*, **42**, 208–14.

Obonyo M (2011) Sensilla on antennae, ovipositor and tarsi of the larval parasitoids, *Cotesia sesamiae* (Cameron 1906) and *Cotesia flavipes* Cameron 1891 (Hymenoptera: Braconidae): a comparative scanning electron microscopy study. *Annales de la Société Entomologique de France*, **47**, 119.

Obrtel R (1960) Premature stages of *Triaspis caudatus* (Nees) (Hymenoptera: Braconidae). *Folia Zoologica*, **9**, 35–8.

Obrycki JJ, Tauber MJ, Tauber CA (1985) *Perilitus coccinellae* (Hymenoptera: Braconidae): parasitization and development in relation to host-stage attacked. *Annals of the Entomological Society of America*, **78**, 852–4.

Ochieng SA, Park KC, Zhu JW, Baker TC (2000) Functional morphology of antennal chemoreceptors of the parasitoid *Microplitis croceipes* (Hymenoptera: Braconidae). *Arthropod Structure and Development*, **29**, 231–40.

Oda RAM, de Macêdo MV, Quicke DLJ (2001) First biological data for *Aspilodemon* Fischer (Hymenoptera: Braconidae: Hydrangeocolinae): parasitoids of cecidomyiid fly galls on Asteraceae in Brazil. *Journal of Hymenoptera Research*, **10**, 126–30.

Ode PJ, Antolin MF, Strand MR (1995) Brood-mate avoidance in the parasitic wasp *Bracon hebetor* Say. *Animal Behaviour*, **49**, 1239–48.

Ode PJ, Antolin MF, Strand MR (1997) Constrained oviposition and female-biased sex allocation in a parasitic wasp. *Oecologia*, **109**, 547–55.

Odebiyi JA, Oatman ER (1972) Biology of *Agathis gibbosa* (Hymenoptera: Braconidae), a primary parasite of the potato tuberworm. *Annals of the Entomological Society of America*, **65**, 1104–14.

Odebiyi JA, Oatman ER (1977) Biology of *Agathis unicolor* (Schrottky) and *Agathis gibbosa* (Say) (Hymenoptera: Braconidae), primary parasites of the potato tuberworm. *Hilgardia*, **45**, 123–51.

O'Donnell DJ (1987) Larval development and the determination of the number of instars in aphid parasitoids (Hymenoptera: Aphidiidae). *International Journal of Insect Morphology and Embryology*, **16**, 3–15.

O'Donnell DJ (1989) A morphological and taxonomic study of first instar larvae of Aphidiinae (Hymenoptera: Braconidae). *Systematic Entomology*, **14**, 197–219.

Oeser RI (1961) Vergleichend-morphologische Untersuchungen über den Ovipositor der Hymenopteren. *Mitteilungen aus der zoologischen Museum in Berlin*, **37**, 3–119.

Okada I & Oike K (1940) Biological observations on *Pristomerus chinensis* Ashm., a parasite of *Grapholitha glycinivorella* Mats. *Transactions of the Biological Society of Manchuria*, **3**, 87–90.

Oliver KM, Moran NA, Hunter MS (2005) Variation in resistance to parasitism in aphids is due to symbionts not host genotype. *Proceedings of the National Academy of Sciences of the United States of America*, **102**, 12795–800.

Olkowski W, Olkowski H, van den Bosch R, Hom R, Zuparko R, Klitz W (1982) The parasitoid *Trioxys tenuicaudus* Starý (Hymenoptera: Aphidiidae) established on the elm aphid *Tinocallis platani* Kaltenbach (Homoptera: Aphididae) in Berkeley, California. *Pan-Pacific Entomologist*, **58**, 59–63.

Omata K (1984) Influence of the physiological control of the host, *Papilio xuthus* L. (Lepidoptera: Papilionidae), on the development of the parasite, *Trogus mactator* Tosquinet (Hymenoptera: Ichneumonidae). *Applied Entomology and Zoology*, **19**, 430–5.

Omata K (1989) Control mechanisms of larval diapause and quiescence in *Trogus mactator*, an endoparasite of papilionids. *Entomologia Experimentalis et Applicata*, **53**, 31–7.

Opinion 159 (1945) On the status of the names *Ephialtes* Schrank, 1802, *Ichneumon* Linnaeus, 1758, *Pimpla* Fabricius 1804–1805, and *Ephialtes* Gravenhorst, 1829 (Class Insecta, Order Hymenoptera). Opinions of the International Commission on Zoological Nomenclature, London.

Opinion 162 (1945) *Suspension of the rules for Bracon Fabricius [1804–1805] (Class Insecta Order Hymenoptera). Opinions of the International Commission on Zoological* Nomenclature, London.

Opinion 1757 (1994) *Cryptus* Fabricius, 1804, and Cryptinae Kirby, 1837 (Insecta: Hymenoptera): conserved. *Bulletin of Zoological Nomenclature*, **51**, 74–5.

Ortega-Blanco J, Bennett DJ, Declos X, Engel MS (2009) A primitive aphidiine wasp in Albian amber from Spain and a Northern Hemisphere origin for the subfamily (Hymenoptera: Braconidae: Aphidiinae). *Journal of the Kansas Entomological Society*, **82**, 273–82.

Ortega-Blanco J, Declos X, Engel MS (2011) A protorhyssaline wasp in Early Cretaceous amber from Spain (Hymenoptera: Braconidae). *Journal of the Kansas Entomological Society*, **84**, 51–7.

Ortega-Blanco J, Singh H, Engel MS (2012) First amber fossil Rhysipolini (Hymenoptera: Braconidae): a new genus and species in early Eocene Cambay amber. *Acta Entomologica Musei Nationalis Pragae*, **52**, 585–94.

Osborne P (1960) Observations on the natural enemies of *Meligethes aeneus* (F.) and *M. viridescens* (F.) [Coleoptera: Nitidulidae]. *Parasitology*, **50**, 91–110.

Osman SE (1974) Parasitentoleranz von Schmetterlingspuppen Maskierung der Parasiteneier mit Mucopolysacchariden. *Naturwissenschaften*, **61**, 453–4.

Osman SE (1978) Die Wirkung der Sekrete der weiblichen Genitalhangsdrüsen von *Pimpla turionellae* (Hym., Ichneumonidae) auf die Hämocyten und die Einkapselungsreakion von Wirtspuppen. *Zeitschrift fur Parasitenkunde*, **57**, 89–100.

Osman SE, Führer E (1979) Histochemical analysis of accessory genital gland secretions in female *Pimpla turionellae* L. (Hymenoptera: Ichneumonidae). *International Journal of Invertebrate Reproduction*, **1**, 323–32.

Otten H, Wäckers FL, Battini M, Dorn S (2000) Efficiency of vibrational sounding in the parasitoids *Pimpla turionellae* is affected by female size. *Animal Behaviour*, **61**, 671–7.

Otten H, Wäckers FL, Isidoro N, Romani R, Dorn S (2002) The subgenual organ in *Pimpla turionellae* L. (Hymenoptera,

Ichneumonidae): ultrastructure and behavioral evidence for its involvement in vibrational sounding. *Redia*, **85**, 61–76.

Ovruski SM, Aluja M, Sivinski J, Wharton RA (2000) Hymenopteran parasitoids on fruit-infesting Tephritidae (Diptera) in Latin America and the southern United States: diversity, distribution, taxonomic status and their use in fruit fly biological control. *Integrated Pest Management Reviews*, **5**, 81–107.

Owen DF, Owen J (1974) Species diversity in temperate and tropical Ichneumonidae. *Nature*, **249**, 583–4.

Owen J, Townes H, Townes M (1981) Species diversity of Ichneumonidae and Serphidae (Hymenoptera) in an English suburban garden. *Biological Journal of the Linnean Society*, **16**, 315–36.

Özlük A (1991) The cerebral neurosecretory system and the diversity of the neurosecretory cell types in *Pimpla turionellae* L. (Hymenoptera: Ichneumonidae). *University of Ankara, Faculty of Sciences, Communications, Series C*, **9**, 33–47.

Özlük A, Gül N, Muhammed A (2011) Ultrastructural description of the corpora allata of *Pimpla turionellae* L. (Hymenoptera: Ichneumonidae). *African Journal of Biotechnology*, **10**, 3784–9.

Paine TD, Paine EO, Hanks LM, Millar JG (2000) Resource partitioning among parasitoids (Hymenoptera: Braconidae) of *Phoracantha semipunctata* in their native range. *Biological Control*, **19**, 223–31.

Paine TD, Joyce AL, Millar JG, Hanks LM (2004) Effect of variation in host size on sex ratio, size, and survival of *Syngaster lepidus*, a parasitoid of *Eucalyptus* longhorned beetles (*Phoracantha* spp.): II. *Biological Control*, **30**, 374–81.

Paillot A (1937) Sur le développement polyembryonnaire d'*Amicroplus collaris* Spin. parasite des chenilles d'*Euxoa segetum* Schiff. *Compte Rendu Hebdomadaire des Séances de l'Académie des Sciences, Paris*, **204**, 810–12.

Pampel W (1914) Die weiblichen Geschlectsorgane der Ichneumoniden. *Zeitschrift für Wissenschaftliche Zoologie (Leipzig)*, **108**, 290–357.

Pan J, Chen X-X (2003). A comparative study on the ultrastructure of the venom apparatus of two species of parasitic wasps (Hymenoptera) of *Plutella xylostella* (Lepidoptera). *Journal of the Chinese Electron Microscopy Society*, **22**, 298–303.

Panis A (2007) Les braconides parasitoïdes de fourmis et observations biologiques sur *Elasmosoma berolinense* Ruthe (Hymenoptera: Braconidae) et les ouvrières parasitées de *Camponotus vagus* Scopoli (Hymenoptera: Formicidae). *Bulletin Mensuel de la Société Linnéenne de Lyon*, **76**, 57–62.

Pansa MG, Guidone L, Tavella L (2012) Distribution and abundance of nymphal parasitoids of *Lygus rugulipennis* and *Adelphocoris lineolatus* in northwestern Italy. *Bulletin of Insectology*, **65**, 81–7.

Papp J (1974) Study on the systematics of Braconidae (Hymenoptera). *Folia Entomologica Hungarica (Ser Nov)*, **27**, 109–33.

Papp J (1989) A contribution to the braconid fauna of Israel (Hymenoptera), 2. *Israel Journal of Entomology*, **22**, 45–59.

Papp J (1994) *Maxfischeria tricolor* gen. n. et sp. n. from Australia (Insecta: Hymenoptera: Braconidae). *Annalen des Naturhistorischen Museum in Wien*, **96**, 143–7.

Pare F, Cloutier C, Huot L, McNeil JN (1979) Description of egg and larval stages of *Aphidius nigripes*. *Annals of the Entomological Society of America*, **72**, 62066.

Park J-S, van Achterberg C (1994) A new genus of the subfamily Ecnomiinae van Achterberg (Hymenoptera: Braconidae) from Korea. *Zoologische Mededelingen, Leiden*, **68**, 49–54.

Parker FD, Pinnell RE (1973) Effect of food consumption of the imported cabbageworm when parasitized by two species of *Apanteles*. *Environmental Entomology*, **2**, 216–19.

Parker HL (1931) *Macrocentrus gifuensis* Ashmead, a polyembryonic parasite in the European corn borer. *Technical Bulletin. United States Department of Agriculture*, No. 230, 66 pp.

Parker HL, Berry PA, Silveria A (1950) Vegetable weevils and their natural enemies in Argentina and Uruguay. *Technical Bulletin. United States Department of Agriculture*, No. 1016, 28 pp.

Parkinson N, Weaver RJ (1999) Noxious components of venom from the pupa-specific parasitoid *Pimpla hypochondriaca*. *Journal of Invertebrate Pathology*, **73**, 74–83.

Parkinson N, Smith I, Weaver R, Edwards JP (2001) A new form of arthropod phenoloxidase is abundant in venom of the parasitoid wasp *Pimpla hypochondriaca*. *Insect Biochemistry and Molecular Biology*, **31**, 57–63.

Parkinson N, Richards EH, Conyers C, Smith I, Edwards JP (2002a) Analysis of venom constituents from the parasitoid wasp *Pimpla hypochondriaca* and cloning of a cDNA encoding a venom protein. *Insect Biochemistry and Molecular Biology*, **32**, 729–35.

Parkinson N, Smith I, Audsley N, Edwards JP (2002b) Purification of pimplin, a paralytic heterodimeric polypeptide from venom of the parasitoid wasp *Pimpla hypochondriaca*, and cloning of the cDNA encoding one of the subunits. *Insect Biochemistry and Molecular Biology*, **32**, 1769–73.

Parkinson NM, Conyers CM, Keen JN, MacNicoll AD, Smith I., Weaver RJ (2003) cDNAs encoding large venom proteins from the parasitoid wasp *Pimpla hypochondriaca* identified by random sequence analysis. *Comparative Biochemistry and Physiology*, **134C**, 513–20.

Parnell JR (1964) The parasite complex of the two seed beetles *Bruchidius ater* (Marsham) (Coleoptera: Bruchidae) and *Apion fuscirostris* Fabricius (Coleoptera: Curculionidae). *Transactions of the Royal Entomological Society of London*, **116**, 73–88.

Parrish DS, Davis DW (1978) Inhibition of diapause in *Bathyplectes curculionis* a parasite of the alfalfa weevil. *Annals of the Entomological Society of America*, **71**, 103–7.

Parrott PJ, Glasgow H (1916) The leaf-weevil (*Polydrusus impressifrons* Gyll.). *Technical Bulletin of the New York State Agriculture Station*, **56**, 1–22.

Patel PN, Habib MEM (1987) Biological studies on *Campoletis flavicincta* (Ashmead, 1890) (Hym., Ichneumonidae), an endoparasite of the fall armyworm, *Spodoptera frugiperda* (Abbot & Smith, 1797) (Lepid., *Noctuidae*). *Journal of Applied Entomology*, **104**, 28–35.

Pavenstädt-Grupp I, Ruthmann A (1989) Microsporidian infection in *Pimpla turionellae* (Ichneumonidae, Hymenoptera): characteristics and reaction with Calcofluor White. *Parasitology Research*, **76**, 74–9.

Peck O (1937a) The male genitalia of Hymenoptera, especially the Ichneumonidae. I. Comparative morphology. *Canadian Journal of Research*, **15d**, 221–52.

Peck O (1937b) The male genitalia of Hymenoptera, especially the Ichneumonidae. II. Taxonomy of the ichneumonid genitalia. *Canadian Journal of Research*, **15d**, 253–74.

Peck RW, Banko PC, Schwarzfeld M, Euaparadorn M, Brinck KW (2008) Alien dominance of the parasitoid wasp community along an elevation gradient on Hawai'i Island. *Biological Invasions*, **10**, 1441–55.

Pélissié B, Ponsard S, Bourguet D, Kergoat GJ (2012) Phylogenetics and population genetics of the Eurasian parasitoid *Macrocentrus cingulum* based on mitochondrial and nuclear loci. *Entomologia Experimentalis et Applicata*, **145**, 1–14.

Pemberton CE, Willard HF (1918a) Interrelations of fruit-fly parasites of Hawaii. *Journal of Agricultural Research*, **12**, 285–95.

Pemberton CE, Willard HF (1918b) A contribution to the biology of fruit-fly parasites of Hawaii. *Journal of Agricultural Research*, **15**, 419–66.

Penado A, Sampaio M, Madeira M, Selfa J, Rosa GM (2013) Where to spend the winter in Serra da Estrela? The first record of *Diphyus quadripunctorius* (Mueller, 1776) overwintering in a subterranean habitat in Portugal. *Entomological News*, **123**, 81–5.

Pennacchio F, Digilio MC (1990) Morphology and development of larval instars of *Aphidius ervi* Haliday (Hymenoptera, Braconidae, Aphidiinae). *Bolletino del Laboratorio di Entomologia Agrarian 'Filippo Silvestri'*, **46**, 163–74.

Pennacchio F, Strand MR (2006) Evolution of developmental strategies in parasitic Hymenoptera. *Annual Review of Entomology*, **51**, 233–58.

Pennacchio F, Vinson SB, Tremblay E (1992a) Preliminary results on *in vitro* rearing of the endoparasitoid *Cardiochiles nigriceps* from egg to second instar. *Entomologia Experimentalis et Applicata*, **64**, 209–16.

Pennacchio F, Vinson SB, Tremblay E (1992b) Host regulation effects on *Heliothis virescens* (F.) larvae induced by teratocytes of *Cardiochiles nigriceps* Viereck (Lepidoptera: Noctuidae – Hymenoptera: Braconidae). *Archives of Insect Biochemistry and Physiology*, **19**, 177–92.

Pennacchio F, Fanti P, Falabella P, Digilio MC, Bisaccia F, Tremblay E (1999) Development and nutrition of the braconid wasp *Aphidius ervi* in aposymbiotic host aphids. *Archives of Insect Biochemistry and Physiology*, **40**, 53–63.

Penteado-Dias AM (2000) Biological notes and larval morphology of *Donquickeia* (Hymenoptera: Braconidae: Doryctinae). In AD Austin and M Dowton (eds), *Hymenoptera: Evolution, Biodiversity and Biological Control*, CSIRO, Canberra, pp. 296–9.

Pereira AIA, Andrade GS, Zanuncio JC, Penteado-Dias AM, Serrão JE (2013) A brief observation of morphological and behavioral similarities between the Ichneumonidae wasp *Cryptanura* sp. and its presumed mimic, *Holymenia clavigera* (Heteroptera: Coreidae), in Brazil. *Brazilian Journal of Biology*, **73**, 903–9.

Pérez J, Rojas JC, Montoya P, Liedo P, González FJ, Castillo A (2012) Size, shape and hue modulate attraction and landing responses of the braconid parasitoid *Fopius arisanus* to fruit odour-baited visual targets. *BioControl*, **57**, 405–14.

Periquet G, Hedderwick MP, El Agoze M, Poirie M (1993) Sex determination in the hymenopteran *Diadromus pulchellus* (Ichneumonidae): validation of the one-locus multi-allele model. *Heredity*, **70**, 420–7.

Perioto NW, Lara RIR, Ferreira CS, Fernandez DRR, Do Carmo Pedroso E, Volpe HXL, Nais J, Correa LRB, Viel SR (2011) A new phytophagous *Bracon* Fabricius (Hymenoptera, Braconidae) associated with *Protium ovatum* Engl. (Burseraceae) fruits from Brazilian savannah. *Zootaxa*, **3000**, 59–65.

Perkins JF (1959) Hymenoptera, Ichneumonoidea, Ichneumonidae, key to subfamilies and Ichneumoninae – 1. *Handbooks for the Identification of British Insects*, **7** (2ai), 1–116.

Perkins JF (1960) Hymenoptera, Ichneumonoidea, Ichneumonidae. Ichneumoninae – 2. *Handbooks for the Identification of British Insects*, **7** (2aii), 117–213.

Perrichot V, Nel A, Quicke DLJ (2009) New braconid wasps from French Cretaceous amber (Hymenoptera, Braconidae): synonymisation with Eoichneumonidae and implications for the phylogeny of Ichneumonoidea. *Zoologica Scripta*, **38**, 79–88.

Persad AB, Hoy MA, Nguyen R (2007) establishment of *Liplexis oregmae* (Hymenoptera: Aphidiidae) in a classical biological control program directed against the brown citrus aphid (Homoptera: Aphididae) in Florida. *Florida Entomologist*, **90**, 204–13.

Peter C, David BV (1992a) Biology of *Apanteles taragamae* Viereck (Hymenoptera: Braconidae) a parasitoid *of Diaphania indica* (Saunders) (Lepidoptera: Pyralidae). *International Journal of Tropical Insect Science*, **13**, 7–17.

Peter C, David BV (1992b) Biology of *Phanerotoma hendecasisella* (Hym: Braconidae) a parasitoid of *Diaphania indica* (Lep.: Pyralidae). *Biocontrol*, **37**, 3–9.

Petters RM (1977) A morphogenetic fate map constructed from *Habrobracon juglandis* gynandromorphs. *Genetics*, **85**, 279–87.

Petters RM, Mettus RV (1980) Decreased diploid male viability in the parasitic wasp, *Bracon hebetor*. *Journal of Heredity*, **71**, 353–6.

Petters RM, Stefanelli J (1983) Developmental arrest of endoparasitoid wasp larvae (*Nemeritis canescens* Grav.) caused by an ectoparasitoid wasp (*Bracon hebetor* Say). *Journal of Experimental Zoology*, **225**, 459–65.

Petters RM, Kendall ME, Taylor RAJ, Mettus RV (1985) Time required for mating and degree of genetic relatedness in the parasitoid wasp, *Bracon hebetor* Say (Hymenoptera: Braconidae). *Melsheimer Entomological Series*, **35**, 21–7.

Pettersson EM, Birgersson G, Witzgall P (2001) Synthetic attractants for the bark beetle parasitoid *Coeloides bostrichorum* Giraud (Hymenoptera: Braconidae). *Naturwissenschaften*, **88**, 88–91.

Pexton JJ, Mayhew PJ (2004) Competitive interactions between parasitoid larvae and the evolution of gregarious development. *Oecologia*, **141**, 179–90.

Pfister-Wilhelm R, Lanzrein B (1996) Precocious induction of metamorphosis in *Spodoptera littoralis* (Noctuidae) by the parasitic wasp *Chelonus inanitus* (Braconidae): identification of the parasitoid larva as the key regulatory element and the host corpora allata as the main target. *Archives of Insect Biochemistry and Physiology*, **32**, 511–25.

Pickett CH, Rodriguez R, Brown J, Coutinot D, Hoelmer KA, Kuhlmann U, Goulet H, Schwarz MD, Goodell PB (2007) Establishment of *Peristenus digoneutis* and *P. relictus* (Hymenoptera: Braconidae) in California for the control of *Lygus* spp. (Heteroptera: Miridae). *Biocontrol Science and Technology*, **17**, 261–72.

Pickett CH, Swezey SL, Nieto DJ, Bryer JA, Erlandson M, Goulet H, Schwartz MD (2009) Colonization and establishment of *Peristenus relictus* (Hymenoptera: Braconidae) for control of *Lygus* spp. (Heteroptera: Miridae) in strawberries on the California Central Coast. *Biological Control*, **49**, 27–37.

Piek T (1990) Neurotoxins from venoms of the Hymenoptera twenty-five years of research in Amsterdam. *Comparative Biochemistry and Physiology. C, Comparative Pharmacology*, **96**, 223–33.

Piek T, Veenendaal RL, Mantel P (1982) The pharmacology of *Microbracon* venom. *Comparative Biochemistry and Physiology. C, Comparative Pharmacology*, **72**, 303–309.

Pinheiro DO, Rossi GD, Cônsoli FL (2010) External morphology of *Cotesia flavipes* (Hymenoptera: Braconidae) during larval development. *Zoologia (Curitiba)*, **27**, 986–92.

Pitz KM, Sharkey MJ (2005) *Cenocoelius huggerti*, the first record of the subfamily Cenocoeliinae (Hymenoptera: Braconidae) from Africa. *Journal of Hymenoptera Research*, **14**, 92–5.

Pitz KM, Sharkey MJ (2007) Three new species of Cenocoeliinae (Hymenoptera: Braconidae) with novel morphological characteristics and habitat records. *Journal of Hymenoptera Research*, **16**, 167–77.

Plarre R, Lieber K, Burkholder W, Phillips J (1999) Host and host instar preference of *Apanteles carpatus* (Say) (Hymenoptera: Braconidae) a possible parasitoid for biological control of clothes moths (Lepidoptera: Tineidae). *Journal of Stored Products Research*, **35**, 197–213.

Plarre R, Lieber K, Burkholder W, Phillips J (2000) Host instar preference of *Apanteles carpatus* (Say) (Hymenoptera: Braconidae) a parasitoid of *Tineola bisselliella* (Hummel) (Lepidoptera: Tineidae). *Mitteilungen der Deutschen Gesellschaft für Allgemeine und Angewandte Entomologie*, **12**, 499–504.

Plotnikov V (1914) Contribution à la biologie de *Bupalus piniarius* L. (Lep. Geom.) et des quelques uns de ces parasites. *Entomologicheskoe Obozrenie*, **14**, 23–43 (in Russian).

Poelman EH, Bruinsma M, Zhu F, Weldegergis BT, Boursault AE, Jongema Y, van Loon JJA, Vet LEM, Harvey JA, Dicke M (2012) Hyperparasitoids use herbivore-induced plant volatiles to locate their parasitoid host. *PLoS Biology*, **10** (11), e1001435.

Poinar G Jr, (2004) Behaviour and development of *Elasmosoma* species (Braconidae: Hymenoptera), and endoparasite of *Formica* ants (Formicidae: Hymenoptera). *Parasitology*, **128**, 1–11.

Poinar G Jr,, Miller JC (2002) First fossil record of endoparasitism of adult ants (Formicidae: Hymenoptera) by Braconidae (Hymenoptera). *Annals of the Entomological Society of America*, **95**, 41–3.

Polaszek A (1986) The effects of two species of hymenopterous parasitoid on the reproductive system of the pea aphid, *Acyrthosiphon pisum*. *Entomologia Experimentalis et Applicata*, **40**, 285–92.

Polgár LA, Hardie J (2000) Diapause induction in aphid parasitoids. *Entomologia Experimentalis et Applicata*, **97**, 21–7.

Polgár L, Mackauer M, Völkl W (1991) Diapause induction in two species of aphid parasitoids: the influence of aphid morph. *Journal of Insect Physiology*, **37**, 699–702.

Polidori C, Federici M, Mendiola P, Selfa J, Andrietti F (2011) Host detection and rate of parasitism by *Acroricnus seductor* (Hymenoptera: Ichneumonidae), a natural enemy of mud-dauber wasps (Hymenoptera: Sphecidae). *Animal Biology*, **61**, 57–73.

Porcelli F (1988) Morfologia degli stadi larvali di *Habronyx heros* Wesmael (Ichneumonidae – Anomaloninae) con note di biologia. *Entomologica*, **23**, 171–89.

Porter CC (1972) The genus *Priotomis* in Argentina (Hymenoptera, Ichneumonidae). *Psyche*, **79**, 231–5.

Porter CC (1980) Joppini (Hymenoptera, Ichneumonidae) of Tarapaca. *Florida Entomologist*, **63**, 226–42.

Porter CC (1989) New Chilean *Itamuton* (Hymenoptera: Ichneumonidae: Mesostenini) reared from *Elicura litigator* (Neuroptera: Myrmeleontidae). *Florida Entomologist*, **72**, 660–4.

Porter CC (1991) Biogeography of Chilean ichneumonid flies (Hymenoptera: Ichneumonidae). *Acta Entomologica Chilena*, **16**, 37–68.

Porter CC (1993) First record of Phrudinae (Hym., Ichneumonidae) from South America with notice of a new genus and species from Chile. *Journal of the New York Entomological Society*, **101**, 130–4.

Porter CC (1998) Guía de los géneros de Ichneumonidae en la región neantárctica del sur de Sudamérica. *Opera Lilloana*, **42**, 1–234.

Porter K (1983) Multivoltinism in *Apanteles bignellii* and the influence of weather on synchronization with its host *Euphydryas aurinia*. *Entomologia Experimentalis et Applicata*, **34**, 155–62.

Potting RPJ, Otten H, Vet LEM (1997) Absence of odour learning in the stemborer parasitoid *Cotesia flavipes*. *Animal Behaviour*, **53**, 1211–23.

Pottler B, Coles LW (1962) Biology of *Biolysia tristis* (Hymenoptera, Ichneumonidae) and its role as a parasite of the clover leaf weevil (*Hypera punctata*). *Journal of Economic Entomology*, **55**, 831–3.

Powell W, Wright AF (1988) The abilities of the parasitoids *Aphidius ervi* Haliday and *A. rhopalosiphi* De Stefani Perez (Hymenoptera: Braconidae) to transfer between different known host species and the implications for the use of alternative hosts in pest control strategies. *Bulletin of Entomological Research*, 683–93.

Powell W, Kardie J, Hick AJ, Höller C, Mann J, Merritt L, Nottingham SF, Wadhams LJ, Witthinrich J, Wright AF (1993) Responses of the parasitoid *Praon volucre* (Hymenoptera: Braconidae) to aphid sex pheromone lures in cereal fields in autumn: implications for parasitoid manipulation. *European Journal of Entomology*, **90**, 435–8.

Powell W, Pennacchio F, Poppy GM, Tremblay E (1998) Strategies involved in the location of osts by the parasitoid *Aphidius ervi* Haliday (Hymenoptera: Braconidae: Aphidiinae). *Biological Control*, **11**, 104–12.

Pratt HD (1939) Studies on the Ichneumonidae of New England – 2: The male genitalia of the subfamily Ichneumoninae. *Annals of the Entomological Society of America*, **37**, 727–42.

Prévost G, Lewis WJ (1990) Heritable differences in the response of the braconid wasp *Microplitis croceipes* to volatile allelochemicals. *Journal of Insect Behavior*, **3**, 277–87.

Prévost G, Davies DH, Vinson SB (1990) Evasion of encapsulation by parasitoid correlated with the extent of host hemocyte pathology. *Entomologia Experimentalis et Applicata*, **55**, 1–10.

Prévost G, Eslin P, Doury G, Moreau SJM, Guillot S (2005) *Asobara*, braconid parasitoids of *Drosophila* larvae: unusual strategies to avoid encapsulation without VLPs. *Journal of Insect Physiology*, **51**, 171–9.

Prévost G, Eslin P, Cherqui A, Moreau S, Doury G (2012) When parasitoids lack polydnaviruses, can venoms subdue the hosts? The case study of *Asobara* species. In NE Beckage and J-M Drezen (eds), *Parasitoid Viruses. Symbionts and Pathogens*, Academic Press, London, pp. 255–66.

Price PW (1970) Trail odors: recognition by insects parasitic on cocoons. *Science*, **170**, 546–7.

Price PW (1972) Behavior of the parasitoid *Pleolophus basizonus* in response to change in host and parasitoid density. *The Canadian Entomologist*, **104**, 129–40.

Price PW (1973a) Parasitoid strategies and community organization. *Environmental Entomology*, **2**, 623–6.

Price PW (1973b) Reproductive strategies in parasitoid wasps. *The American Naturalist*, **107**, 685–93.

Price PW (1974) Strategies for egg production. *Evolution*, **28**, 76–84.

Price PW (1975) Reproductive strategies of parasitoids. In PW Price (ed.), *Evolutionary Strategies of Parasitic Insects and Mites*, Plenum Press, New York, pp. 87–111.

Prince GJ (1976) Laboratory biology of *Phaenocarpa persimilis* Papp (Braconidae: Alysiinae), a parasitoid of *Drosophila*. *Australian Journal of Zoology*, **24**, 249–64.

Principi MM (1948) Contributi allo studio dei Neurotteri italiana. VII. *Bollettino dell'Istituto di Entomologia della Università di Bologna*, **17**, 93–121.

Principi MM, Memmi M, Sgobba D (1979) Reperti su *Chrysopophthorus chrysopimarginis* Goidanich, parasita solitario delle imaginni di Neurotteri Crisopidi. *Bollettino dell'Istituto di Entomologia della Università di Bologna*, **34**, 247–73.

Prokopy RJ, Webster RP (1978) Oviposition-deterring pheromone of *Rhagoletis pomonella*, a kairomone for its parasitoid *Opius lectus*. *Journal of Chemical Ecology*, **4**, 481–94.

Proper AB (1934) Hyperparasitism in the case of some introduced lepidopterous tree defoliators. *Journal of Agricultural Research*, **48**, 359–76.

Provost B, Varricchio P, Arana E, Espagne E, Falabella P, Huguet E, La Scaleia R, Cattolico L, Poirie M, Malva C, Olszewski JA, Pennacchio F, Drezen JM (2004) Bracoviruses contain a large multigene family coding for protein tyrosine phosphatases. *Journal of Virology*, **78**, 13090–103.

Pschorn-Walcher H (1967) Biology of the ichneumonid parasites of *Neodiprion sertifer* (Geoffrey) (Hym., Diprionidae) in Europe. *Technical Bulletin of the Commonwealth Institute of Biological Control*, **8**, 7–52.

Pschorn-Walcher H (1987) Interspecific competition between the principal larval parasitoids of the pine sawfly, *Neodiprion sertifer* (Geoff.) (Hym.: Diprionidae). *Oecologia*, **73**, 621–5.

Pschorn-Walcher H, Zinnert KD (1971) Investigations on the ecology and natural control of the larch sawfly (*Pristiphora erichsonii* Htg.) (Hym., Tenthredinidae) in central Europe. Part II. Natural enemies: their biology and ecology and their role as mortality factors in *P. erichsonii*. *Technical Bulletin of the Commonwealth Institute of Biological Control*, **14**, 1–50.

Pucci T (2008) A comparison of the parasitic wasps (Hymenoptera) at elevated versus ground yellow pan traps in a beech-maple forest. *Journal of Hymenoptera Research*, **17**, 116–23.

Pucci T, Sharkey MJ (2004) A revision of *Agathirsia* Westwood (Hymenoptera: Braconidae: Agathidinae) with notes on mouthpart morphology. *Journal of Hymenoptera Research*, **13**, 64–107.

Purvis A, Quicke DLJ (1997) Phylogeny reconstruction: Are the big easy? *Trends in Ecology & Evolution*, **12**, 49–50.

Puttler B (1961) Biology of *Hyposoter exiguae* (Hymenoptera: Ichneumonidae), a parasite of lepidopterous larvae. *Annals of the Entomological Society of America*, **54**, 25–30.

Puttler B, Coles LW (1962) Biology of *Biolysia tristis* (Hymenoptera, Ichneumonidae) and its role as a parasite of the clover leaf weevil (*Hypera punctata*). *Journal of Economic Entomology*, **55**, 831–3.

Puttler B, Thewke SE (1970) Biology of *Microplitis feltiae* (Hymenoptera: Braconidae), a parasite of black cutworm *Agrotis ipsilon*. *Annals of the Entomological Society of America*, **63**, 645–8.

Qin QL, Gong H, Ding T (2000) Two collagenases are secreted by teratocytes from *Microplitis mediator* (Hymenoptera: Braconidae) cultured *in vitro*. *Journal of Invertebrate Pathology*, **76**, 79–80.

Quednau FW (1970) Notes on life-history, fecundity, longevity and attack pattern of *Agathis pumila* (Hymenoptera: Braconidae), a parasite of the larch case-bearer. *The Canadian Entomologist*, **102**, 736–45.

Quednau FW (1990) Introduction, permanent establishment, and dispersal in Eastern Canada of *Olesicampe geniculatae* Quednau and Lim (Hymenoptera, Ichneumonidae), an important biological control agent of the mountain ash sawfly, *Pristiphora geniculata* (Hartig) (Hymenoptera: Tenthredinidae). *The Canadian Entomologist*, **122**, 921–34.

Quednau FW, Guèvremont H (1975) Observations on mating and oviposition behaviour of *Priopoda nigricollis* (Hymenoptera: Ichneumonidae), a parasite of the birch leaf-miner, *Fenusa pusilla* (Hymenoptera: Tenthredinidae). *The Canadian Entomologist*, **107**, 1199–204.

Quednau FW, Lim KP (1983) *Olesicampe geniculatae*, a new Palaearctic ichneumonid parasite of *Pristiphora geniculata* (Hymenoptera: Tenthredinidae). *The Canadian Entomologist*, **115**, 109–13.

Quezada JR (1967) Notes on the biology of *Rothschildia ?aroma* (Lepidoptera: Saturniidae), with special reference to its control by pupal parasites in El Salvador. *Annals of the Entomological Society of America*, **60**, 595–9.

Quicke DLJ (1982) Hamuli number in the Braconinae (Hymenoptera: Braconidae): an inter- and intraspecific, size-dependent, taxonomic character. *Oriental Insects*, **15**, 235–40.

Quicke DLJ (1984) Evidence for the role of white-tipped ovipositor sheath in the Braconinae (Hymenoptera: Braconidae). *Proceedings and Transactions of the British Entomology and Natural History Society*, **17**, 71–9.

Quicke DLJ (1986) Preliminary notes on homeochromatic associations within and between the Afrotropical Braconinae (Hym., Braconidae) and Lamiinae (Col., Cerambycidae). *Entomologists' Monthly Magazine*, **122**, 97–109.

Quicke DLJ (1987a) The Old World genera of braconine wasps (Hymenoptera: Braconidae). *Journal of Natural History*, **21**, 43–157.

Quicke DLJ (1987b) A new subfamily of Braconidae, the Vaepellinae, based on a new genus and species from Ghana (Insecta, Hymenoptera). *Zoologica Scripta*, **16**, 73–7.

Quicke DLJ (1987c) First host record for the genus *Hyboteles* van Achterberg (Hymenoptera: Braconidae). *Proceedings and Transactions of the British Entomology and Natural History Society*, **20**, 122.

Quicke DLJ (1988a) The higher classification, zoogeography and biology of the Braconinae. In VK Gupta (ed.), *Advances in Parasitic Hymenoptera Research*, E.J. Brill, Leiden, pp 117–38.

Quicke DLJ (1988b) Inter-generic variation in the male genitalia of the Braconinae (Insecta, Hymenoptera, Braconidae). *Zoologica Scripta*, **17**, 399–409.

Quicke DLJ (1991a) Ovipositor mechanics of the braconine wasp genus *Zaglyptogastra* and the ichneumonid genus *Pristomerus*. *Journal of Natural History*, **25**, 971–7.

Quicke DLJ (1991b) Tergal and inter-tergal glands of male Braconinae. *Zoologica Scripta*, **19**, 413–23.

Quicke DLJ (1992) Nocturnal Braconinae (Hym., Braconidae). *Entomologists' Monthly Magazine*, **128**, 33–7.

Quicke DLJ (1993) The polyphyletic origin of endoparasitism in the cyclostome lineages of Braconidae (Hymenoptera): a reassessment. *Zoologisches Mededelingen, Leiden*, **67**, 159–77.

Quicke DLJ (1997a) *Parasitic Wasps*, Chapman & Hall, London, 470 pp.

Quicke DLJ (1997b) Subfamily Braconinae. In RA Wharton, PM Marsh and MJ Sharkey (eds), *Identification Manual to the New World Genera of Braconidae*, Special Publication of the International Society of Hymenopterists 1, pp. 148–74.

Quicke DLJ (2005) Biology and immature stages of *Panteles schnetzeanus* (Hymenoptera: Ichneumonidae) a parasitoid of *Lampronia fuscatella* (Lepidoptera: Incurvariidae). *Journal of Natural History*, **39**, 431–43.

Quicke DLJ (2012) We know too little about parasitoid wasp distributions to draw any conclusions about latitudinal trends in species richness, body size and biology. *PLoS ONE*, **7** (2), e32101.

Quicke DLJ, Belshaw R (1999) Incongruence between morphological data sets: an example from the evolution of endoparasitism among parasitic wasps (Hymenoptera: Braconidae). *Systematic Biology*, **48**, 436–54.

Quicke DLJ, Falco JV (1998) A putative pheromone gland associated modification of the hind tibia in *Vipio moneilemae* (Hymenoptera: Braconidae: Braconinae). *Journal of Hymenoptera Research*, **7**, 118–21.

Quicke DLJ, Huddleston T (1989) The Australian braconid wasp subfamily Mesostoinae (Hymenoptera: Braconidae) with the description of a new species of *Mesostoa*. *Journal of Natural History*, **23**, 1309–17.

Quicke DLJ, Huddleston T (1991) Redescription of *Gnathobracon* Costa (Hym., Braconidae, Braconinae) with the description of a new species. *Entomologists' Monthly Magazine*, **127**, 191–5.

Quicke DLJ, Kruft RA (1995) Latitudinal gradients in North American braconid wasp species richness and biology. *Journal of Hymenoptera Research*, **4**, 194–203.

Quicke DLJ, Laurenne NM (2005) Notes on host searching by the parasitic wasp *Zaglyptogastra* Ashmead (Hymenoptera: Braconidae: Braconinae) in Kibale Forest, Uganda. *Journal of Hymenoptera Research*, **14**, 177–81.

Quicke DLJ, Marsh PM (1992) Two new species of Neotropical parasitic wasps with highly modified ovipositors (Hymenoptera: Braconidae: Braconinae & Doryctinae). *Proceedings of the Entomological Society of Washington*, **94**, 559–67.

Quicke DLJ, Shaw MR (2004) Cocoon silk chemistry in parasitic wasps (Hymenoptera, Ichneumonoidea) and their hosts. *Biological Journal of the Linnean Society*, **81**, 161–70.

Quicke DLJ, Shaw MR (2005a) First host record for the rogadine genus *Pholichora* van Achterberg (Hymenoptera: Braconidae) with description of a new species and notes on convergent wing venation features. *Journal of Natural History*, **39**, 531–7.

Quicke DLJ, Shaw MR (2005b) First host records for the rogadine genera *Rogasodes* Chen and He and *Canalirogas* van Achterberg & Chen (Hymenoptera: Braconidae) with description of a new species and survey of mummy types within Rogadinae s.str. *Journal of Natural History*, **39**, 3525–42.

Quicke DLJ, Stanton ML (2005) *Trigastrotheca laikipiensis* n. sp. (Hymenoptera: Braconidae): a new species of brood parasitic wasp that attacks foundress queens of three coexisting acacia-ant species in Kenya. *Journal of Hymenoptera Research*, **14**, 182–90.

Quicke DLJ, van Achterberg C (1990) Phylogeny of the subfamilies of the family Braconidae (Hymenoptera, Ichneumonoidea). *Zoologische Verhandelingen*, **258**, 1–95.

Quicke DLJ, Ficken L, Fitton MG (1992a) New diagnostic ovipositor characters for doryctine wasps (Hymenoptera, Braconidae). *Journal of Natural History*, **26**, 1035–46.

Quicke DLJ, Ingram SN, Bailley H, Gaitens PV (1992b) Sperm structure and ultrastructure in the Hymenoptera. *Zoologica Scripta*, **21**, 381–402.

Quicke DLJ, Ingram SN, Fitton MG (1992c) Phylogenetic implications of the structure and distribution of ovipositor valvilli in the Hymenoptera. *Journal of Natural History*, **26**, 587–608.

Quicke DLJ, Ingram SN, Proctor J, Huddleston T (1992d) Batesian and Müllerian mimicry between species with connected life histories with a new example involving braconid wasp parasites of *Phoracantha* beetles. *Journal of Natural History*, **26**, 1013–34.

Quicke DLJ, Tunstead J, Falco JV, Marsh PM (1992e) Venom gland apparatus in cyclostome braconid wasps with special reference to the subfamily Doryctinae (Insecta, Hymenoptera, Braconidae). *Zoologica Scripta*, **21**, 403–16.

Quicke DLJ, Fitton MG, Tunstead J, Ingram SN, Gaitens PV (1994) Ovipositor structure and relationships within the Hymenoptera, with special reference to the Ichneumonoidea. *Journal of Natural History*, **28**, 635–82.

Quicke DLJ, Fitton MG, Harris J (1995) Ovipositor steering mechanisms in braconid wasps. *Journal of Hymenoptera Research* **4**, 110–20.

Quicke DLJ, Wharton RA, Sittertz-Bhatkar H (1996) Recto-tergal fusion in the Braconinae (Hymenoptera, Braconidae): Distribution and structure. *Journal of Hymenoptera Research*, **5**, 73–9.

Quicke DLJ, van Achterberg C, Godfray HCJ (1997a) Comparative morphology of the venom gland and reservoir in opiine and alysiine braconid wasps (Insecta, Hymenoptera, Braconidae). *Zoologica Scripta*, **26**, 23–50.

Quicke DLJ, Wharton RA, Sittertz-Bhatkar H (1997b) Antero-lateral abdominal scent glands of braconine wasps (Hymenoptera, Braconidae). *Journal of Hymenoptera Research*, **6**, 219–30.

Quicke DLJ, Wyeth P, Fawke JD, Basibuyuk HH, Vincent JFV (1998) Manganese and zinc in the ovipositors and mandibles of hymenopterous insects. *Zoological Journal of the Linnean Society*, **124**, 387–96.

Quicke DLJ, Basibuyuk HH, Fitton MG, Rasnitsyn AP (1999a) Morphological, palaeontological and molecular aspects of ichneumonoid phylogeny (Hymenoptera, Insecta). *Zoologica Scripta*, **28**, 175–202.

Quicke DLJ, Lopez-Vaamonde C, Belshaw R (1999b) The basal Ichneumonidae (Insecta: Hymenoptera): 28S rDNA considerations of the Brachycyrtinae, Labeninae, Paxylommatinae and Xoridinae. *Zoologica Scripta*, **28**, 203–10.

Quicke DLJ, Belshaw R, Lopez-Vaamonde C (1999c) Preservation of hymenopteran specimens for subsequent molecular and morphological study. *Zoologica Scripta*, **28**, 261–7.

Quicke DLJ, LeRalec A, Vilhelmsen L (2000a) Ovipositor structure and function in the parasitic Hymenoptera with an exploration of new hypotheses. *Rendiconti*, **47**, 197–239.

Quicke DLJ, Fitton MG, Notton DG, Belshaw R, Broad GR, Dolphin K (2000b) Phylogeny of the Ichneumonidae (Hymenoptera): a simultaneous molecular and morphological analysis. In AD Austin and M Dowton (eds), *Hymenoptera: Evolution, Biodiversity and Biological Control*, CSIRO, Canberra, pp. 74–83.

Quicke DLJ, Brandt AP, Falco JV (2000c) Revision of the Afrotropical species of *Curriea* Ashmead (Hymenoptera: Braconidae: Braconinae): a genus with diverse ovipositor morphology. *African Entomology*, **8**, 109–39.

Quicke DLJ, Manzari S, van Achterberg C (2002) The systematic placement of *Afrocampsis* van Achterberg & Quicke (Hymenoptera: Braconidae): molecular and morphological evidence indicate that it belongs to Helconinae s.l. not Sigalphinae. *Zoologische Mededelingen, Leiden*, **76**, 443–50.

Quicke DLJ, Laurenne NM, Broad GR, Barclay M (2003) Host location behaviour and a new host record for *Gabunia* aff. *togoensis* Krieger (Hymenoptera: Ichneumonidae: Cryptinae) in Kibale Forest National Park, West Uganda. *African Entomology*, **11**, 308–10.

Quicke DLJ, Palmer-Wilson J, Burrough A, Broad GR (2004a) Discovery of calcium enrichment in cutting teeth of parasitic wasp ovipositors (Hymenoptera: Ichneumonoidea). *African Entomology*, **12**, 259–64.

Quicke DLJ, Shaw MR, Takahashi M, Yanechin B (2004b) Cocoon silk chemistry of noncyclostome Braconidae, with remarks on phylogenetic relationships within the Microgastrinae (Hymenoptera: Braconidae). *Journal of Natural History*, **17**, 2167–81.

Quicke DLJ, Fitton MG, Broad GR, Crocker B, Laurenne NM, Miah MI (2005a) The parasitic wasp genera *Skiapus*, *Hellwigia*, *Nonnus*, *Chriodes* and *Klutiana* (Hymenoptera, Ichneumonidae): recognition of the Nesomesochorinae stat. rev. and Nonninae stat. nov. and transfer of *Skiapus* and *Hellwigia* to the Ophioninae. *Journal of Natural History*, **39**, 2559–78.

Quicke DLJ, Laurenne NM, Barclay M (2005b) Host, host location and aggressive behaviour in a tropical wood-borer parasitoid genus (Hymenoptera: Braconidae: *Monilobracon* Quicke), parasitoids of Lymexylidae (Coleoptera) in Kibale Forest National Park, West Uganda. *African Entomology*, **13**, 213–20.

Quicke DLJ, Mori M, Zaldivar-Riverón A, Laurenne NM, Shaw MR (2006) Suspended mummies in *Aleiodes* species (Hymenoptera: Braconidae: Rogadinae) with descriptions of six new species from western Uganda based largely on DNA sequence data. *Journal of Natural History*, **40**, 2663–80.

Quicke DLJ, Sharkey MJ, Laurenne NM, Dowling A (2008) A preliminary molecular phylogeny of the Sigalphinae (Hymenoptera: Braconidae), including *Pselaphanus* Szépligeti, based on 28S rDNA, with descriptions of new Afrotropical and Madagascan *Minanga* and *Malasigalphus* species. *Journal of Natural History*, **42**, 2703–19.

Quicke DLJ, Laurenne NM, Fitton MG, Broad GR (2009) A thousand and one wasps: a 28S rDNA and morphological phylogeny of the Ichneumonidae (Insecta: Hymenoptera) with an investigation into alignment parameter space and elision. *Journal of Natural History*, **43**, 1305–421.

Quicke DLJ, Smith MA, Janzen DH, Hallwachs W, Fernandez-Triana J, Laurenne NM, Zaldívar-Riverón A, Shaw MR, Broad GR, Klopfstein S, Shaw SR, Hrcek J, Hebert PDN, Miller SE, Rodriguez JJ, Whitfield JB, Sharkey MJ, Sharanowski BJ, Jussila R, Gauld ID, Chesters D, Vogler AP (2012a) Utility of the DNA barcoding gene fragment for parasitic wasp phylogeny (Hymenoptera: Ichneumonoidea): data release and new measure of taxonomic congruence. *Molecular Ecology Resources*, **12**, 676–85.

Quicke DLJ, Smith MA, Miller SE, Hrcek J, Butcher BA (2012b) *Colastomion* Baker (Braconidae, Rogadinae): nine new species from Papua New Guinea reared from Crambidae. *Journal of Hymenoptera Research*, **28**, 85–121.

Quicke DLJ, Smith MA, van Achterberg C, Miller SE, Hrcek J (2012c) A new genus and three new species of parasitoid wasp from Papua New Guinea and redescription of *Trigonophatnus* Cameron (Hymenoptera, Braconidae, Rogadinae). *Journal of Natural History*, **46**, 1369–85.

Quimio GM, Walter GH (2000) Swarming, delayed sexual maturation of males and mating behavior of *Fopius arisanus* (Sonan) (Hymenoptera: Braconidae). *Journal of Insect Behavior*, **13**, 797–813.

Quistad GB, Nguyen Q, Bernasconi P, Leisy DJ (1994) Purification and characterization of insecticidal toxins from venom glands of the parasitic wasp, *Bracon hebetor*. *Insect Biochemistry and Molecular Biology*, **24**, 955–61.

R Development Core Team (2009) *R: a Language and Environment for Statistical Computing*, R Foundation for Statistical Computing, Vienna.

Rabouille A, Bigot Y, Drezen JM, Sizaret PY, Hamelin MH, Periquet G (1994) A member of the Reoviridae (DpRV) has ploidy-specific genomic segment in the wasp *Diadromus pulchellus* (Hymenoptera). *Virology*, **205**, 228–37.

Rahbé Y, Digilio MC, Febvay G, Guillaud J, Fanti P, Pennacchio F (2002) Metabolic and symbiotic interactions in amino acid pools of the pea aphid, *Acyrthosiphon pisum*, parasitised by the braconid *Aphidius ervi*. *Journal of Insect Physiology*, **48**, 507–16.

Rahman H, Fitton MG, Quicke DLJ (1998a) Ovipositor internal microsculpture in the Braconidae (Insecta, Hymenoptera). *Zoologica Scripta*, **27**, 319–31.

Rahman H, Fitton MG, Quicke DLJ (1998b) Ovipositor internal microsculpture and other features in doryctine wasps (Insecta, Hymenoptera, Braconidae). *Zoologica Scripta*, **27**, 333–43.

Rahoo GM, Luff ML (1987) The biology of *Adelognathus granulatus* Perkins (Hym., Ichneumonidae) a parasitoid of the small gooseberry sawfly, *Pristiphora pallipes* (Lep.) (Hym., Tenthredinidae). *Zeitschrift fur Angewandte Entomologie*, **104**, 480–4.

Ramadan MM, Wong TTY, A Wong MA (1991) Influence of parasitoid size and age on male mating success of opiinae (Hymenoptera:Braconidae), larval parasitoids of fruit flies (Diptera:Tephritidae). *Biological Control*, **1**, 248–55.

Ramadan MM, Wong TTY, Beardsley JW (1992) Reproductive behavior of *Biosteres arisanus* (Sonan) (Hymenoptera: Braconidae), an egg-larval parasitoid of the Oriental fruit fly. *Biological Control*, **2**, 28–34.

Ramirez BW, Marsh PM (1996) A review of the genus *Psenobolus* (Hymenoptera, Braconidae) from Costa Rica, an inquiline fig wasp with brachypterous males, with descriptions of two new species. *Journal of Hymenoptera Research*, **5**, 64–72.

Rao A, Henderson RE, Vinson SB (2009) The probable significance of tracheal tufts in the 8th abdominal segment of *Heliothis virescens* (F.) on the development of its parasitoid, *Toxoneuron nigriceps* (Viereck). *Journal of Insect Physiology*, **55**, 769–73.

Rappaport N, Page M (1985) Rearing *Glypta fumiferanae* [Hym.: Ichneumonidae] on a multivoltine laboratory colony of the western spruce budworm (*Choristoneura occidentalis*) [Lep.: Tortricidae]. *Entomophaga*, **30**, 347–52.

Rasch EM, Cassidy JD, King RC (1977) Evidence for dosage compensation in parthenogenetic Hymenoptera. *Chromosoma*, **59**, 323–40.

Rasekh A, Michaud JP, Kharazi-Pakdel A, Allahyari H (2010) Ant mimicry by an aphid parasitoid, *Lysiphlebus fabarum*. *Journal of Insect Science*, **10**, 126.

Rasnitsyn AP (1959) Hibernation sites of ichneumon-flies (Hymenoptera, Ichneumonidae). *Entomologicheskoe Obozrenie*, **38**, 546–53 (in Russian, with English translation in *Entomological Review, Washington*, **38**, 491–8).

Rasnitsyn AP (1964) On hibernation of ichneumon-flies (Hymenoptera, Ichneumonidae). *Entomologicheskoe Obozrenie*, **43**, 46–51 (in Russian, with English translation in *Entomological Review, Washington*, **38**, 491–8).

Rasnitsyn AP (1975) Hymenoptera Apocrita of Mesozoic. *Transactions of the Palaeontological Institute of the Academy of Sciences of the USSR*, **147**, 1–134.

Rasnitsyn AP (1983) Ichneumonoidea (Hymenoptera) from the Lower Cretaceous of Mongolia. *Contributions of the American Entomological Institute*, **20**, 259–65.

Rasnitsyn AP (1990) Hymenopterans: Vespida. In AP Rasnitsyn (ed), *Late Mesozoic Insects of Eastern Transbaikalia*, Trudy Paleontologicheskogo Instituta Akademii Nauk SSSR, vol. 239, pp. 177–205.

Rasnitsyn AP (2002) Superorder Vespidea Laicharting, 1781, Order Hymenoptera Linné, 1758 (=Vespida Laicharting, 1781). In AP Rasnitsyn and DLJ Quicke (eds), *History of Insects*, Kluwer, Dordrecht, pp. 242–54.

Rasnitsyn AP, Quicke DLJ (eds) (2002) *History of Insects*, Kluwer, Dordrecht, 517 pp.

Rasnitsyn AP, Sharkey M (1988) New Eoichneumonidae from Early Cretaceous of Siberia and Mongolia (Hymenoptera: Ichneumonoidea). In VK Gupta (ed.), *Advances in Parasitic Hymenoptera Research. Proceedings of the 2nd Conference on the Taxonomy and Biology of Parasitic Hymenoptera held at the University of Florida, Gainesville, Florida, November 19–21, 1987*, E.J. Brill, Leiden, pp. 167–97.

Rasnitsyn AP, Jarzembowski EA, Ross AJ (1998) Wasps (Insecta: Vespida = Hymenoptera) from the Purbeck and Wealden (Lower Cretaceous) of southern England and their biostratigraphical and palaeoenvironmental significance. *Cretaceous Research*, **19**, 329–91.

Rathcke BJ, Price PW (1976) Anomalous diversity of tropical ichneumonoid parasitoids: a predation hypothesis. *The American Naturalist*, **110**, 889–93.

Rathmayer W, Walther C (1976) Mode of action and specificity of *Habrobracon* venom (Hymenoptera, Braconidae). In A Ohsaka, K Hayashi, Y Sawai, R Murata, M Funatsu and N Tamiya (eds), *Animal, Plant, and Microbial Toxins*, Springer, New York, pp. 299–307.

Rechav Y, Orion T (1975) The development of the immature stages of *Chelonus inanitus*. *Annals of the Entomological Society of America*, **68**, 111–18.

Reed DA, Beckage NE (1997) Inhibition of testicular growth and development in *Manduca sexta* larvae parasitised by the braconid wasp *Cotesia congregata*. *Journal of Insect Physiology*, **43**, 29–38.

Reed-Larsen DA, Brown JJ (1990) Embryonic castration of the codling moth, *Cydia pomonella* by an endoparasitoid, *Ascogaster quadridentata*. *Journal of Insect Physiology*, **36**, 111–18.

Rehman M (1970) Effect of parasitism on food consumption of *Pieris rapae* larvae. *Journal of Economic Entomology*, **63**, 820–1.

Reid JA (1941) The thorax of the wingless and short-winged Hymenoptera. *Transactions of the Royal Entomological Society, London*, **91**, 367–446.

Renault S, Bigot S, Lemesle M, Sizaret P-Y, Bigot Y (2003). The cypovirus *Diadromus pulchellus* DpRV-2 is sporadically associated with the endoparasitoid wasp *D. pulchellus* and modulates the defence mechanisms of pupae of the parasitised leek-moth *Acrolepiopsis assectella*. *Journal of General Virology*, **84**, 1799–807.

Renault S, Stasiak K, Federici B, Bigot Y (2004) Commensal and mutualistic relationships of reoviruses with their parasitoid wasp hosts. *Journal of Insect Physiology*, **51**, 137–48.

Reudler JH, Biere A, Harvey JA, van Nouhuys S (2011) Differential performance of a specialist and two generalist herbivores and their parasitoids on *Plantago lanceolata*. *Journal of Chemical Ecology*, **37**, 765–78.

Richards EH, Dani MP (2008) Biochemical isolation of an insect haemocyte anti-aggregation protein from the venom of the endoparasitic wasp, *Pimpla hypochondriaca*, and identification of its gene. *Journal of Insect Physiology*, **54**, 1041–9.

Richards EH, Parkinson NM (2000) Venom from the endoparasitic wasp *Pimpla hypochondriaca* adversely affects the morphology, viability, and immune function of hemocytes from larvae of the tomato moth, *Lacanobia oleracea*. *Journal of Invertebrate Pathology*, **76**, 33–42.

Richards EH, Dani MP, Bradish H (2013) Immunosuppressive properties of a protein (rVPr1) from the venom of the endoparasitic wasp, *Pimpla hypochondriaca*: mechanism of action and potential use for improving biological control strategies. *Journal of Insect Physiology*, **59**, 213–22.

Richerson JV, Borden JH (1972) Host finding by heat perception in *Coeloides brunneri* (Hymenoptera: Braconidae). *The Canadian Entomologist*, **104**, 1877–81.

Richerson JV, DeLoach CJ (1972) Some aspects of host selection by *Perilitus coccinellae*. *Annals of the Entomological Soiciety of America*, **65**, 834–9.

Richerson JV, Borden JH, Hollingdale J (1972) Morphology of a unique sensillum placodeum on the antennae of *Coeloides brunneri* (Hymenoptera: Braconidae). *Canadian Journal of Zoology*, **50**, 909–13.

Ridley M (1993) Clutch size and mating frequency in parasitic Hymenoptera. *The American Naturalist*, **142**, 893–910.

Riegel GT (1948) The wings of Braconidae (Hym.). *Annals of the Entomological Society of America*, **41**, 439–49.

Rivers DB, Dani MP, Richards EH (2009) The mode of action of venom from the endoparasitic wasp *Pimpla hypochondriaca* (Hymenoptera: Ichneumonidae) involves Ca^{+2}-dependent cell death pathways. *Archives of Insect Biochemistry and Physiology*, **71**, 173–90.

Robacker DC, Hendry LB (1977) Neral and geranial: components of the sex pheromone of the parasitic wasp *Itoplectis conquisitor*. *Journal of Chemical Ecology*, **3**, 563–77.

Robacker DC, Weaver KM, Hendry LB (1976) Sexual communication and associative learning in the parasitic wasp *Itoplectis conquisitor* (Say). *Journal of Chemical Ecology*, **2**, 39–48.

Rocha KL, Mangine T, Harris EJ, Lawrence PO (2004) Immature stages of *Fopius arisanus* (Hymenoptera: Braconidae) in *Bactrocera dorsalis* (Diptera: Tephritidae). *Florida Entomologist*, **87**, 164–8.

Rodriguez JJ, Fernández-Triana JL, Smith MA, Janzen DH, Hallwachs W, Erwin TL, Whitfield JB (2013) Extrapolations from field studies and known faunas converge on dramatically increased estimates of global microgastrine parasitoid wasp species richness (Hymenoptera: Braconidae). *Insect Conservation and Diversity*, **6**, 530–6.

Rodríguez-Berrío A, Mazón M, Bordera S (2008) Estudio de la fauna de Ichneumonidae cenobiontes (Hymenoptera) en un ecosistema de Montaña mediterránea. I. Subfamilia Banchinae. *Boletin de la Asociacion Espanola de Entomologia*, **32**, 55–80.

Rodriguez-del-Bosque LA, Smith JW Jr, (1989) Parasitization of *Diatraea lineolata* pupae and diapausing larvae by several exotic parasites. *Florida Entomologist*, **72**, 703–5.

Rodriguez-Leyva E (2006) Life history of *Triaspis eugenii* Wharton and Lopez-Martinez (Hymenoptera: Braconidae) and evaluation of its potential for biological control of pepper weevil *Anthonomus eugenii* (Coleoptera: Curculionidae). Unpublished PhD thesis, University of Florida.

Rodríguez-Pérez MA, Pérez-Rodríguez MA, Mohammed-Ali A, de Jesús de Luna-Santillana E, Guo X-W, Reyes-Villanueva F, Beckage NE (2011) Enhancing the virulence of baculovirus as biopesticides with wasp parasitoid polydnavirus genes to control lepidopteran insect pests. In NM Rosas-Garcïa (ed.), *Biological Control of Insect Pests*, Studium Press, Houston, TX, pp. 259–69.

Rodriguez-Saona CR, Frost CJ (2010) New evidence for a multi-functional role of herbivore-induced plant volatiles in defense against herbivores. *Plant Signaling & Behavior*, **5**, 58–60.

Rogers D (1972) The ichneumon wasp *Venturia canescens*, oviposition and avoidance of superparasitism. *Entomologia Experimentalis et Applicata*, **15**, 190–4.

Rohlfs WM III, Mack TP (1983) Effect of parasitization by *Ophion flavidus* Brullé (Hymenoptera: Ichneumonidae) on consumption and utilization of a pinto bean diet by fall armyworm (Lepidoptera: Noctuidae). *Environmental Entomology*, **12**, 1257–9.

Rohlfs WM III,, Mack TP (1984) Functional response of *Ophion flavidus* Brulle (Hymenoptera: Ichneumonidae) females to various densities of fall armyworm (*Spodoptera frugiperda* J. E. Smith) (Lepidoptera: Noctuidae). *Environmental Entomology*, **13**, 708–10.

Rojas-Rousse D (1972) Description et fonctionnement de l'appareil genital interne de *Diadromus pulchellus* Wesmael (Hymenoptera: Ichneumonidae) – I. Testicules, canaux deferents, glandes annexes, canaux collecteurs, anal ejaculateur du male. *International Journal of Insect Morphology and Embryology*, **1**, 225–32.

Rojas-Rousse D, Benoit M (1977) Morphology and biometry of larval instars of *Pimpla instigator* (F.) (Hymenoptera: Ichneumonidae). *Bulletin of Entomological Research*, **67**, 129–41.

Rojas-Rousse D, Kalmes R (1978) The development of male *Diadromus pulchellus* (Hymenoptera: Ichneumonidae) in the pupae of *Acrolepiopsis assectella* (Lepidoptera: Plutellidae): comparison of assimilation and energy losses under two temperature regimes. *Environmental Entomology*, **7**, 469–81.

Rojas-Rousse D, Palevody C (1981) Structure et fonctionnement de la spermatheque chez l'endoparasite solitaire *Diadromus pulchellus* Wesmael (Hymenoptera: Ichneumonidae). *International Journal of Insect Morphology and Embryology*, **10**, 309–20.

Roman A (1923) Ichneumonids reared from Diptera Nematocera. *Entomologist's Monthly Magazine*, **59**, 71–6.

Roques A (1976) Observations sur la biologie et le comportement cleptoparasite *d'Eurytoma waachtli* (Chalc.: Eurytomidae), parasite de *Pissodes validirostris* (Col.: Curculionidae) dans les cones de pin sylvestre à Fontainebleau. *Entomophaga*, **21**, 289–95.

Röse USR, Lewis J, Tumlinson JH (2006) Extrafloral nectar from cotton (*Gossypium hirsutum*) as a food source for parasitic wasps. *Functional Ecology*, **20**, 67–74.

Rosenberg HT (1934) The biology and distribution in France of the larval parasites of *Cydia pomonella* L. *Bulletin of Entomological Research*, **25**, 201–56.

Ross HH (1936) The ancestry and wing venation of the Hymenoptera. *Annals of the Entomological Society of America*, **29**, 99–111.

Rossbach A, Lohr B, Vidal S (2006) Host shift to peas in the diamondback moth *Plutella xylostella* (Lepidoptera: Plutellidae) and response of its parasitoid *Diadegma mollipla* (Hymenoptera: Ichneumonidae). *Bulletin of Entomological Research*, **96**, 413–19.

Rossi GD, Labate MTV, Labate CA, Vinson SB, Cônsoli FL (2012) Characterization of a *Toxoneuron nigriceps* (Viereck) (Hymenoptera: Braconidae)-derived chitinase and its potential for pest control. *Pesticide Biochemistry and Physiology*, **104**, 96–102.

Rotheram S (1967) The immune surface of eggs of a parasitic wasp. *Nature*, **214**, 700.

Rotheram S (1973a) The surface of the egg of a parasitic insect. I. The surface of the egg and first-instar larva of *Nemeritis*. *Proceedings of the Royal Society B: Biological Sciences*, **183**, 179–94.

Rotheram S (1973b) The surface of the egg of a parasitic insect. II. The ultrastructure of the particulate coat on the egg of *Nemeritis*. *Proceedings of the Royal Society B: Biological Sciences*, **183**, 195–204.

Rotheray GE (1981a) Emergence from the host puparium by *Diplazon pectoratorius* (Gravenhorst) (Hymenoptera: Ichneumonidae), a parasitoid of aphidophagous syrphid larvae. *Entomologistís Gazette*, **32**, 39–41.

Rotheray GE (1981b) Host searching and oviposition behaviour of some parasitoids of aphidophagous Syrphidae. *Ecological Entomology*, **6**, 79–87.

Rotheray GE (1981c) Courtship, male swarms and a sex pheromone of *Diplazon pectoratorius* (Grav.). *Entomologist's Gazette*, **32**, 193–6.

Rotheray GE (1990) A new species of *Bioblapsis* (Hym., Ichneumonidae) from Scotland parasitising a mycophagous hoverfly *Cheilosia longula* (Dipt., Syrphidae). *Entomologica Scandinavica*, **21**, 277–80.

Rotundo G, Cavalloro R, Tremblay E (1988) *In vitro* rearing of *Lysiphlebus fabarum* (Hym.: Braconidae). *Entomophaga*, **33**, 261–7.

Rougerie R, Smith MA, Fernandez-Triana J, Lopez-Vaamonde C, Ratnasingham S, Hebert PDN (2011) Molecular analysis of parasitoid linkages (MAPL): gut contents of adult parasitoid wasps reveal larval host. *Molecular Ecology*, **20**, 179–86.

Rouleux-Bonnin F, Renault S, Rabouille A, Periquet G, Bigot Y (1999) Free serosal cells originating from the embryo of the wasp *Diadromus pulchellus* in the pupal body of parasitised leekmoth *Acrolepiosis assectella*. Are these cells teratocytes-like? *Journal of Insect Physiology*, **45**, 479–84.

Rousse P, Chiroleu F, Domerg C, Quilici S (2007) Naive *Fopius arisanus* females respond mainly to achromatic cues. *Biological Control*, **43**, 41–8.

Rousse P, van Noort S, Diller E (2013) Revision of the Afrotropical Phaeogenini (Ichneumonidae, Ichneumoninae), with description of a new genus and twelve new species. *ZooKeys*, **354**, 1–85.

Roux O, van Baaren J, Gers C, Arvanitakis L, Legal L (2005) Antennal structure and oviposition behavior of the *Plutella xylostella* specialist parasitoid *Cotesia plutellae*. *Microscopy Research and Technique*, **68**, 36–44.

Rowher SA, Gahan AB (1916) Horismology of the hymenopterous wing. *Proceedings of the Entomological Society of Washington*, **18**, 20–76.

Ruberson JR, Whitfield JB (1996) Facultative egg-larval parasitism of the beet armyworm, *Spodoptera exigua* (Lepidoptera: Noctuidae) by *Cotesia marginiventris* (Hymenoptera: Braconidae). *Florida Entomologist*, **79**, 296–302.

Rudall KM, Kenchington W (1971) Arthropod silks: the problem of fibrous proteins in animal tissues. *Annual Review of Entomology*, **16**, 73–96.

Ruf D, Dorn S, Mazzi D (2011) Females leave home for sex: natal dispersal in a parasitoid with complementary sex determination. *Animal Behaviour*, **81**, 1083–9.

Rungrojwanich K, Walter G (2000) The Australian fruit fly parasitoid *Diachasmimorpha kraussii* (Fullaway): mating behavior, modes of sexual communication and crossing tests with *D. longicaudata* (Ashmead) (Hymenoptera: Braconidae: Opiinae). *Pan-Pacific Entomologist*, **76**, 12–23.

Russell DA (1987) Parasitism of the oriental fruit moth *Grapholita molesta* (Lepidoptera: Tortricidae): the New Zealand position in a world perspective. *New Zealand Entomologist*, **10**, 13–26.

Russo G (1938) VI contributo alla conoscenza del coleotteri scolitidi. Fleotribo: *Phloeotribus scarabaeoides* (Bern.) Fauv. Parte secoda. Biografia, simbionti, danni e lotta. *Bollettino del Regio Laboratorio di Entomologia Agraria di Portici*, **2**, 3–420.

Ryan RB (1965). Maternal influence on diapause in a parasitic insect, *Coleoides brunneri* Vier. (Hymenoptera, Braconidae). *Journal of Insect Physiology*, **11**, 1131–6.

Ryan RB (1971) Interaction between two parasites, *Apechthis ontario* and *Itoplectis quadricingulatus*. 1. Survival in singly attacked, super-, and multiparasitised greater wax moth pupae. *Annals of the Entomological Society of America*, **64**, 205–8.

Ryan RB, Rudinsky JA (1962) Biology and habitats of the Douglas fir beetle parasite, *Coeloides brunneri* Viereck (Hymenoptera: Braconidae) in Western Oregon. *Canadian Entomologist*, **94**, 748–63.

Ryan RE, Milligan B (2009) Observations of late instar larva survival of *Cotesia* (Hymenoptera: Braconidae) parasitoid attack of *Euphydryas anicia cloudcrofti* (Lepidoptera: Nymphalidae). *Journal of Research on the Lepidoptera*, **41**, 76–83.

Sääksjärvi IE, Haataja S, Neuvonen S, Gauld ID, Jussila R, Salo J, Burgos AM (2004) High local species richness of parasitic wasps (Hymenoptera: Ichneumonidae; Pimplinae and Rhyssinae) from the lowland rainforests of Peruvian Amazonia. *Ecological Entomology*, **29**, 735–43.

Sääksjärvi IE, Ruokolainen K, Tuomisto H, Haataja S, Fine PVA, Cárdenas G, Mesones I, Vargas V (2005) Comparing composition and diversity of parasitoid wasps and plants in an Amazonian rain-forest mosaic. *Journal of Tropical Ecology*, **22**, 167–76.

Sabri A, Hance T, Leroy PD, Frère I, Haubruge E, Destain J, Compère P, Thinart P (2011) Placenta-like structure of the aphid endoparasitic wasp *Aphidius ervi*: a strategy of optimal resources acquisition. *PLoS ONE*, **6** (4), e18847.

Sait SM, Begon M, Thompson DJ, Harvey JA (1996) Parasitism of baculovirus-infected *Plodia interpunctella* by *Venturia canescens* and subsequent virus transmission. *Functional Ecology*, **10**, 586–91.

Saito T, Bjørnson S (2013) The convergent lady beetle, *Hippodamia convergens* Guérin-Méneville and its endoparasitoid *Dinocampus coccinellae* (Schrank): The effect of a microsporidium on parasitoid development and host preference. *Journal of Invertebrate Pathology*, **113**, 18–25.

Salkeld EH (1959) Notes on anatomy, life-history, and behaviour of *Aphaereta pallipes* (Say) (Hymenoptera: Braconidae), a parasite of the onion maggot, *Hylemya antiqua* (Meig.). *The Canadian Entomologist*, **91**, 93–7.

Salkeld EH (1967) Histochemistry of the excretory system of the endoparasitoid *Aphaereta pallipes* (Say) (Hymenoptera: Braconidae). *Canadian Journal of Zoology*, **45**, 967–73.

Sallam MN, Overholt WA, Kairu E (2002) Intraspecific and interspecific competition between *Cotesia flavipes* and *Cotesia sesamiae* (Hymenoptera: Braconidae), gregarious larval endoparasitoids of lepidopteran stemborers. *Biocontrol Science and Technology*, **12**, 493–506.

Salom SM, Ascoli-Christensen A, Birgersson G, Payne TL, Berisford CW (1992) Electroantennogram responses of the southern pine beetle parasitoid *Coeloides pissodis* (Ashmead) (Hym., Braconidae) to potential semiochemicals. *Journal of Applied Entomology*, **114**, 472–9.

Salt G (1931) Parasites of the wheat-stem sawfly, *Cephus pygmaeus*, in England. *Bulletin of Entomological Research*, **22**, 479–545.

Salt G (1952) Trimorphism in the ichneumonid parasite *Gelis corruptor*. *Quarterly Journal of Microscopical Science*, **93**, 453–74.

Salt G (1959) Role of glycerol in the cold-hardening of *Bracon cephi* (Gahan). *Canadian Journal of Zoology*, **37**, 59–69.

Salt G (1961) Competition among insect parasitoids. Mechanisms in biological competition. *Symposium of the Society for Experimental Biology*, **15**, 96–119.

Salt G (1965) Experimental studies in insect parasitism. XIII. The haemocytic reaction of a caterpillar to eggs of its habitual parasite. *Proceedings of the Royal Society B: Biological Sciences*, **162**, 303–18.

Salt G (1966) Experimental studies in insect parasitism. XIV. The haemocytic reaction of a larva to its habitual parasite. *Proceedings of the Royal Society B: Biological Sciences*, **165**, 155–78.

Salt G (1970) *The Cellular Defense Reactions of Insects*. Cambridge Monographs in Experimental Biology, 16. Cambridge University Press, Cambridge, 118 pp.

Salt G (1973) Experimental studies in insect parasitism. XVI. The mechanism of the resistance of *Nemeritis* to defense reactions. *Proceedings of the Royal Society B: Biological Sciences*, **183**, 337–50.

Salt G (1977) Problems of orientation associated with cocoon-spinning in *Nemeritis*. *Ecological Entomology*, **2**, 171–7.

Salt G, van den Bosch R (1967) The defense reactions of three species of *Hypera* (Coleoptera, Curculionidae) to an ichneumon wasp. *Journal of Invertebrate Pathology*, **9**, 164–77.

Salvador G, Cônsoli FL (2008) Changes in the hemolymph and fat body metabolites of *Diatraea saccharalis* (Fabricius) (Lepidoptera: Crambidae) parasitized by *Cotesia flavipes* (Cameron) (Hymenoptera: Braconidae). *Biological Control*, **45**, 103–10.

Sanchis A, Latorre A, Gonzalez-Candelas F, Michelena JM (2000) An 18S rDNA-based molecular phylogeny of Aphidiinae (Hymenoptera: Braconidae). *Molecular Phylogenetics and Evolution*, **14**, 180–94.

Sandlan K (1979) Sex ratio regulation in *Coccygomimus turionellae* Linnaeus (Hymenoptera: Ichneumonidae) and its ecological implications. *Ecological Entomology*, **4**, 365–78.

Sandrock C, Schirrmeister BE, Vorburger C (2011) Evolution of reproductive mode variation and host associations in a sexual-asexual complex of aphid parasitoids. *BMC Evolutionary Biology*, **11**, 348.

Santos AMC, Quicke DLJ (2011) Large-scale diversity patterns of parasitoid insects. *Entomological Science*, **14**, 371–82.

Santos AMC, Quicke DLJ, Borges PAV, Hortal J (2011) Species pool structure determines the level of generalism of island parasitoid faunas. *Journal of Biogeography*, **38**, 1657–67.

Sarmiento CE, Sharkey MJ, Janzen DH (2004) The first gregarious species of the Agathidinae (Hymenoptera: Braconidae). *Journal of Hymenoptera Research*, **13**, 295–301.

Sasakawa K, Ikeda H, Sutou M, Dobata S, Itô M (2011) Parasitism of adult *Poecilus versicolor* (Coleoptera: Carabidae) by hymenopteran larvae. *The Canadian Entomologist*, **143**, 211–23.

Sathe TV (1990) The biology of *Diadegma argenteopilosa* Cameron (Hymenoptera, Ichneumonidae) an internal larval parasitoid of *Spodoptera litura* (Fab.). *The Entomologist*, **109**, 2–7.

Sato Y, Tanaka T, Imafuku M, Hidaka T (1983) How does diurnal *Apanteles kariyai* parasitize and egress from a nocturnal host larva? *Kontyû, Tokyo*, **51**, 128–39.

Sato Y, Tagawa J, Hidaka T (1986) Effects of the gregarious parasitoids, *Apanteles ruficrus* and *A. kariyai*, on host growth and development. *Journal of Insect Physiology*, **12**, 269–74.

Sawoniewicz J (2008) *Hosts of the World Aptesini (Hymenoptera, Ichneumonidae, Cryptinae)*. Wydawnictwo Mantis, Olsztyn, 150 pp.

Scatolini D, Penteado-Dias AM, van Achterberg C (2002) *Pseudorhysipolis* gen. nov. (Hymenoptera: Braconidae: Rhysipolinae), with nine new species from Brazil, Suriname and Panama. *Zoologische Mededelingen, Leiden*, **76**, 109–31.

Schaefer PW, Fuester RW, Chianese RJ, Rhodes LD, Tichenor RB Jr, (1989) Introduction and North American establishment of *Coccygomimus disparis* (Hymenoptera: Ichneumonidae), a polyphagous pupal parasite of Lepidoptera, including gypsy moth. *Environmental Entomology*, **18**, 1117–25.

Schaffner U, Müller C (2001) Exploitation of the fecal shield of the lily leaf beetle, *Lilioceris lilii* (Coleoptera: Chrysomelidae), by the specialist parasitoid *Lemophagus pulcher* (Hymenoptera: Ichneumonidae). *Journal of Insect Behavior*, **14**, 739–57.

Schepers EJ, Dahlman DL, Zhang D (1998) *Microplitis croceipes* teratocytes: *in vitro* culture and biological activity of teratocyte secreted protein. *Journal of Insect Physiology*, **44**, 767–77.

Schlinger EI (1974) Continental drift, *Notofagus* (Fagaceae) and some ecologically associated insects. *Annual Review of Entomology*, **19**, 323–43.

Schlinger EI, Hall JC (1960) The biology, behavior, and morphology of *Praon palitans* Muesebeck, an internal parasite of the spotted alfalfa aphid, *Therioaphis maculata* (Buckton) (Hymenoptera: Braconidae, Aphidiinae). *Annals of the Entomological Society of America*, **53**, 144–60.

Schlinger EI, Hall JC (1961) The biology, behavior, and morphology of *Trioxys (Trioxys) utilis*, an internal parasite of the spotted alfalfa aphid, *Therioaphis maculata* (Hymenoptera: Braconidae, Aphidiinae). *Annals of the Entomological Society of America*, **54**, 34–45.

Schmeider RG (1939) On the dimorphism of cocoons of *Sphecophaga burra* (Cresson). (Hymenoptera: Ichneumonidae). *Entomological News*, **50**, 91–7.

Schneider F (1950) Die Entwicklung des Syrphidenparasiten *Diplazon fissorius* Grav. (Hymenoptera: Ichneumonidae) in Uni-, Oligo- und Polyvoltinen Wirten und sein Verhalten bei parasitaerer Aktivierung der Diapauselarven durch *Diplazon pectoratorius* Grav. *Mitteilungen der Schweizerischen Entomologischen Gesellschaft*, **23**, 155–94.

Schneider MI, Viñuela E (2007) Improvements in rearing method for *Hyposoter didymator* (Hymenoptera: Ichneumonidae), considering sex allocation and sex determination theories used for Hymenoptera. *Biological Control*, **43**, 271–7.

Schoenbohm RB, Turpin FT (1977) Effect of parasitism by *Meteorus leviventris* on corn foliage consumption and corn seedling cutting by the black cutworm. *Journal of Economic Entomology*, **70**, 457–9.

Schopf A, Steinberger P (1996) Influence of the endoparasitic wasp *Glyptapanteles liparidis* (Hymenoptera: Braconidae) on the growth, food consumption, and food utilization of its host larva, *Lymantria dispar* (Lepidoptera: Lymantriidae). *European Journal of Entomology*, **93**, 555–68.

Schröder D (1974) A study of the interactions between the internal larval parasites of *Rhyacionia buoliana* (Lepidoptera: Olethreutidae). *Entomophaga*, **19**, 145–71.

Schroder RFW, Athanas MM (2002) Biological observations on *Centistes gasseni* Shaw (Hymenoptera: Braconidae), a parasitoid of *Diabrotica* spp. (Coleoptera: Chrysomelidae). *Proceedings of the Entomological Society of Washington*, **104**, 554–62.

Schwartz BD, Moore CJ, Rahm F, Hayes PY, Kitching W, De Voss JJ (2008) Spiroacetal biosynthesis in insects from Diptera to Hymenoptera: the giant ichneumon wasp *Megarhyssa nortoni nortoni* Cresson. *Journal of the American Chemical Society*, **130**, 14853–60.

Schwartz M (1991) Eine neue Art der Gattung *Microleptes* Gravenhorst (Ichneumonidae, Hymenoptera) aus Österreich. *Linzer Biologische Beiträge*, **23**, 399–405.

Schwartz M (2007) Revision der wespalaarktischen Arten der Gattung *Hoplocryptus* Thomson (Hymenoptera, Ichneumonidae). *Linzer Biologische Beiträge*, **39**, 1161–219.

Schwartz M, Shaw MR (1998) Western Palaearctic Cryptinae (Hymenoptera: Ichneumonidae) in the National Museums of Scotland, with nomenclatural changes, taxonomic notes, rearing records and special reference to the British check list. Part 1. Tribe Cryptini. *Entomologist's Gazette*, **49**, 101–27.

Schwarz M, Shaw MR (2000) Western Palaearctic Cryptinae (Hymenoptera: Ichneumonidae) in the National Museums of Scotland, with nomenclatural changes, taxonomic notes, rearing records and special reference to the British check list. Part 3. Tribe Phygadeuontini, subtribes Chiroticina, Acrolytina, Hemitelina and Gelina (excluding *Gelis*), with descriptions of new species. *Entomologist's Gazette*, **51**, 147–86.

Schwenke W (1999) Revision der europäischen Mesochorinae (Hymenoptera, Ichneumonoidea, Ichneumonidae). *Spixiana*, Supplement, 1–124.

Scott EI (1939) An account of the developmental stages of some aphidophagous Syrphidae (Dipt.) and their parasites (Hymenopt.). *Annals of Applied Biology*, **26**, 509–32.

Scudder GGE (1961) The comparative morphology of the insect ovipositor. *Transactions of the Royal Entomological Society of London*, **113**, 25–40.

Sedivy J, Sevsik J (2003) Ichneumonid (Hymenoptera: Ichneumonidae) parasitoids of fungus gnats (Diptera: Mycetophilidae): rearing records from the Czech Republic. *Studia Dipterologica*, **10**, 153–8.

Sedlag U (1957) Über den Bau des weiblichen Geschlechtsapparates der Aphidiiden (Hymenoptera). *Verhandlungen der Deutschen Zoologischen Gesellschaft in Hamburg*, **1956**, 356–61.

Segura DF, Viscarret MM, Ovruski SM, Cladera JL (2012) Response of the fruit fly parasitoid *Diachasmimorpha longicaudata* to host and host-habitat volatile cues. *Entomologia Experimentalis et Applicata*, **143**, 164–76.

Seitner M, Nötzl P (1925) *Pityophthorus henscheli* Seitner und sein Parasit *Cosmophorus henscheli* Ruschka. *Zeitschrift für Angewandte Entomologie*, **11**, 187–96.

Semichon L (1908) Note biologique sur le *Sphecophaga vesparum* Curtis [Hym.]. *Bulletin de la Société Entomologique de France*, 79–81.

Serbielle C, Chowdhury S, Pichon S, Dupas S, Lesobre J, Purisima EO, Drezen JM, Huguet E (2008) Viral cystatin evolution and three-dimensional structure modelling: a case of directional selection acting on a viral protein involved in a host–parasitoid interaction. *BMC Biology*, **6**, 38.

Serbielle C, Dupas S, Perdereau E, Héricourt F, Dupuy C, Huguet E, Drezen JM (2012) Evolutionary mechanisms driving the evolution of a large polydnavirus gene family coding for protein tyrosine phosphatases. *BMC Evolutionary Biology*, **12**, 253.

Seymour JE, Jones RE (2000) Humidity-terminated diapause in the tropical braconid parasitoid *Microplitis demolitor*. *Ecological Entomology*, **25**, 481–5.

Shah ZA (2012) Morphology, ultrastructure, and probable functions of the sense organs on the ovipositor stylets of the hymenoptran parasitoid, *Venturia canescens* (Gravenhorst). *Microscopy Research and Technique*, **75**, 876–83.

Shah ZA, Blackwell A, Hubbard SF (2012) Ultramorphology of the ovipositor of *Venturia canescens* (Gravenhorst) and possible mechanisms for oviposition. *International Journal of Agriculture and Biology*, **14**, 908–14.

Shanower TG, Hoelmer KA (2004) Biological control of wheat stem sawflies: past and future. *Journal of Agricultural and Urban Entomology*, **21**, 197–221.

Shapiro A (1976) Beau geste? *The American Naturalist*, **110**, 900–2.

Shapiro BA, Pickering J (2000) Rainfall and parasitic wasp (Hymenoptera: Ichneumonoidea) activity in successional forest stages at Barro Colorado Nature Monument, Panama, and La Selva Biological Station, Costa Rica. *Agricultural and Forest Entomology*, **2**, 39–47.

Sharanowski BJ, Dowling APG, Sharkey MJ (2011) Molecular phylogenetics of Braconidae (Hymenoptera: Ichneumonoidea), based on multiple nuclear genes, and implications for classification. *Systematic Entomology*, **36**, 549–72.

Sharifi S, Javadi I (1971) Biology of *Xorides corcyrensis* Kriech. (Hymenoptera: Ichneumonidae), a parasite of the Rosaceae branch borer *Osphranteria coerulescens* Redt. (Coleoptera: Cerambycidae). *Zeitschrift für Angewandte Entomologie*, **68**, 25–31.

Sharkey MJ (1992) Cladistics and tribal classification of the Agathidinae (Hymenoptera: Braconidae). *Journal of Natural History*, **26**, 425–47.

Sharkey MJ (1993) Superfamily Braconidae. In H. Goulet and J.T. Huber (eds), *Hymenoptera of the World: an Identification Guide to Families*, Centre for Land and Biological Resources Research, Agriculture Canada, Ottawa, pp. 362–94.

Sharkey MJ (1996) Cladistics of the Ichneutinae revisited. *Journal of Natural History*, **30**, 1577–8.

Sharkey MJ (1997a) Subfamily Agathidinae. In RA Wharton, PM Marsh and MJ Sharkey (eds), *Identification Manual to the New World Genera of Braconidae*, Special Publication of the International Society of Hymenopterists 1, pp. 69–83.

Sharkey MJ (1997b) Subfamily Blacinae. In RA Wharton, PM Marsh and MJ Sharkey (eds), *Identification Manual to the New World Genera of Braconidae*, Special Publication of the International Society of Hymenopterists 1, pp. 141–6.

Sharkey MJ (1997c) Subfamily Sigalphinae. In RA Wharton, PM Marsh and MJ Sharkey (eds), *Identification Manual to the New World Genera of Braconidae*, Special Publication of the International Society of Hymenopterists 1, pp. 414–17.

Sharkey MJ (1997d) Key to New World subfamilies of the family Braconidae. In RA Wharton, PM Marsh and MJ Sharkey (eds), *Identification Manual to the New World Genera of Braconidae*, Special Publication of the International Society of Hymenopterists 1, pp. 39–63.

Sharkey MJ, Braet Y (2012) New species of the rare genera *Dentigaster* Zettel, 1990 (Hymenoptera, Braconidae, Cheloninae) and *Minanga* Cameron, 1906 (Sigalphinae) from French Guiana. *Journal of Hymenoptera Research*, **25**, 93–102.

Sharkey MJ, Clutts SA (2011) A revision of Thai Agathidinae (Hymenoptera, Braconidae), with descriptions of six new species. *Journal of Hymenoptera Research*, **22**, 69–132.

Sharkey MJ, Janzen DH (1995) Review of the world species of *Sigalphus* (Hymenoptera: Braconidae: Sigalphinae) and biology of *Sigalphus romeroi*, new species. *Journal of Hymenoptera Research*, **4**, 99–109.

Sharkey MJ, Mason WRM (1986) The generic validity of *Aenigmostomus* and *Asiacardiochiles* (Hymenoptera: Braconidae). *Proceedings of the Entomological Society of Washington*, **88**, 300–2.

Sharkey MJ, Stoelb SAC (2013) Revision of *Agathacrista* new genus (Hymenoptera, Braconidae, Agathidinae, Agathidini). *Journal of Hymenoptera Research*, **33**, 99–112.

Sharkey MJ, Wahl DB (1992) Cladistics of the Ichneumonoidea. *Journal of Hymenoptera Research*, **1**, 15–24.

Sharkey MJ, Wharton RA (1994) A revision of the genera of the world Ichneutinae (Hymenoptera: Braconidae). *Journal of Natural History*, **28**, 873–912.

Sharkey MJ, Wharton RA (1997) Morphology and terminology. In RA Wharton, PM Marsh and MJ Sharkey (eds), *Identification Manual to the New World Genera of Braconidae*, Special Publication of the International Society of Hymenopterists 1, pp. 19–37.

Sharkey MJ, Laurenne NM, Quicke DLJ, Sharanowshi BJ, Murray D (2006) Revision of the Agathidinae (Hymenoptera: Braconidae) with comparisons of static and dynamic alignments. *Cladistics*, **22**, 546–67.

Sharkey MJ, van Noort S, Whitfield JB (2009a) Revision of Khoikhoiinae (Hymenoptera, Braconidae). *ZooKeys*, **20**, 299–348.

Sharkey MJ, Yu DS, van Noort S, Seltmann K, Penev L (2009b) Revision of the Oriental genera of Agathidinae (Hymenoptera, Braconidae) with an emphasis on Thailand including interactive keys to genera published in three different formats. *ZooKeys*, **21**, 19–54.

Sharkey MJ, Clutts S, Tucker EM, Janzen D, Hallwachs W, Dapkey T, Smith MA (2011) *Lytopylus* Förster (Hymenoptera, Braconidae, Agathidinae) species from Costa Rica, with an emphasis on specimens reared from caterpillars in Area de Conservación Guanacaste. *ZooKeys*, **130**, 379–419.

Sharkey MJ, Carpenter JM, Vilhelmsen L, Heraty J, Liljeblade J, Dowling APG, Schulmeister S, Murray D, Deans AR,

Ronquist F, Krogmann L, Wheeler WC (2012) Phylogenetic relationships among superfamilies of Hymenoptera. *Cladistics*, **27**, 1–33.

Shaumar N (1966) Anatomie du système nerveux et analyse des facteurs externes pouvant intervenir dans le déterminisme du sexe chez les Ichneumonidae Pimplinae. *Annales des Sciences Naturelles. Zoologie et Biologie Animale*, 12e Série, **8**, 391–3.

Shaw MR (1976) Noctuid (Lep.) larvae consuming ichneumonoid (Hym.) larvae. *Entomologist's Monthly Magazine*, **111**, 188.

Shaw MR (1981) Delayed inhibition of host development by the non paralysing venoms of parasitic wasps. *Journal of Invertebrate Pathology*, **37**, 215–21.

Shaw MR (1983) On[e] evolution of endoparasitism: the biology of some genera of Rogadinae (Braconidae). *Contributions of the American Entomological Institute*, **20**, 307–28.

Shaw MR (1986) *Coleocentrus excitator* (Poda) (Hymenoptera: Ichneumonidae) new to Britain. *Entomologist's Gazette*, **37**, 221–4.

Shaw MR (1988) *Meteorus brevicauda* Thomson (Hymenoptera: Braconidae) reared from larvae of *Zeugophora subspinosa* (Fabricius) (Coleoptera: Chrysomelidae). *Entomologist's Gazette*, **39**, 205–6.

Shaw MR (1989) A host record for *Stilbops limneriaeformis* Schmiedeknecht (Hymenoptera: Ichneumonidae, Stilbopinae) in Scotland. *Entomologist's Gazette*, **40**, 5–6.

Shaw MR (1990) Rearing parasitic wasps from spiders and their egg sacs. In *British Arachnological Society – Members' Handbook*, Section 2.8, 2 pp.

Shaw MR (1993a) Species of *Mesochorus* (Hymenoptera: Ichneumonidae) reared as hyperparasitoids of Lepidoptera via koinobiont ectoparasitoid Tryphoninae (Ichneumonidae). *Entomologist's Gazette*, **44**, 181–2.

Shaw MR (1993b) An enigmatic rearing of *Dolopsidea indagator* (Haliday) (Hymenoptera: Braconidae). *Entomologist's Record and Journal of Variation*, **105**, 31–6.

Shaw MR (1994) Parasitoid host ranges. In BA Hawkins and W Sheehan (eds), *Parasitoid Community Ecology*, Oxford University Press, Oxford, pp. 111–44.

Shaw MR (1995) Observations on the adult behaviour and biology of *Histeromerus mystacinus* Wesmael (Hymenoptera: Braconidae). *The Entomologist*, **114**, 1–13.

Shaw MR (1997) *Rearing Parasitic Hymenoptera*, Amateur Entomologist's Society, Orpington, 46 pp.

Shaw MR (1999a) Gregarious development in endoparasitic koinobiont Ichneumonidae (Hymenoptera). *Entomologist's Gazette*, **50**, 55–6.

Shaw MR (1999b) Rearing records of two species of *Cenocoelius* Haliday from Britain (Hymenopera: Braconidae, Cenocoeliinae). *Entomologist's Gazette*, **50**, 283–6.

Shaw MR (2001) Interactions between adults of some species of *Netelia* Gray (Hymenoptera: Ichneumonidae: Tryphoninae) and their caterpillar hosts (Lepidoptera). *Journal of Hymenoptera Research*, **10**, 101–11.

Shaw MR (2002a) A new species of *Macrostomion* Szépligeti (Hymenoptera: Braconidae: Rogadinae) from Papua New Guinea, with notes on the biology of the genus. *Zoologische Mededelingen, Leiden*, **76**, 133–40.

Shaw MR (2002b) Host ranges of *Aleiodes* species and an evolutionary hypothesis. In G Melika and C Thuróczy (eds), *Parasitic Wasps: Evolution, Systematics, Biodiversity and Biological Control*, Agroinform Kiadó, Budapest, pp. 321–27.

Shaw MR (2004a) *Microgaster alebion* and its 'var A': description of a new species and biological notes (Hymenoptera: Braconidae: Microgastrinae). *Entomologist's Gazette*, **55**, 217–24.

Shaw MR (2004b) Notes on the biology of *Lycorina triangulifera* Holmgren (Hymenoptera: Ichneumonidae: Lycorininae). *Journal of Hymenoptera Research*, **13**, 302–8.

Shaw MR (2009a) A new, highly sexually dimorphic species of *Cosmophorus* Ratzeburg (Hymenoptera: Braconidae: Euphorinae) reared from Corsica. *British Journal of Entomology and Natural History*, **22**, 151–7.

Shaw MR (2009b) Notes on the host-feeding and hyperparasitic behaviours of *Itoplectis* species (Hymenoptera: Ichneumonidae, Pimplinae). *Entomologist's Gazette*, **60**, 113–16.

Shaw MR (2012) Larval parasitoids of *Rivula sericealis* (Scopoli) (Lepidoptera: Noctuidae) in Britain, including notes on the biology of *Cotesia subordinaria* (Tobias) (Hymenoptera: Braconidae, Microgastrinae), a solitary-cum-gregarious parasitoid. *Entomologist's Gazette*, **63**, 251–7.

Shaw MR (2014) Illustrated notes on the biology of two European species of *Euceros* Gravenhorst (Hymenoptera: Ichneumonidae: Eucerotinae). *Proceedings of the Russian Entomological Society, St Petersburg*, **85**, 122–32.

Shaw MR, Askew RR (2010) Hymenopterous parasitoids of Diptera. In PJ Chandler (ed.), *A Dipterist's Handbook*, Amateur Entomologist's Society, Orpington, pp. 347–61.

Shaw MR, Bennet FD (2001) Host prepupal mummification by *Colpognathus* Wesmael (Hymenoptera: Ichneumonidae, Ichneumoninae). *Entomologist's Gazette*, **52**, 201–3.

Shaw MR, Hochberg ME (2001) The neglect of parasitic Hymenoptera in insect conservation strategies: the British fauna as a prime example. *Journal of Insect Conservation*, **5**, 253–63.

Shaw MR, Huddleston T (1991) Classification and biology of braconid wasps (Hymenoptera: Braconidae). *Handbooks for the Identification of British Insects*, **7**, 1–126.

Shaw MR, Quicke DLJ (2000) The biology and early stages of *Acampsis alternipes* (Nees) and implications for the relationships of the Sigalphinae (Hymenoptera: Braconidae). *Journal of Natural History*, **34**, 611–28.

Shaw MR, Wahl DB (1989) The biology, egg and larva of *Acaenitus dubitator* (Panzer). *Systematic Entomology*, **14**, 117–25.

Shaw MR, Geoghegan IE, Majerus MEN (1999) Males of *Dinocampus coccinellae* (Schrank) (Hym.: Braconidae: Euphorinae). *Entomologist's Record*, **111**, 195–6.

Shaw MR, Stefanescu C, van Nouhuys S (2009) Parasitoids of European butterflies. In J Settele, TG Shreeve, M Konvicka and H Van Dyck (eds), *Ecology of Butterflies of Europe*, Cambridge University Press, Cambridge, pp. 130–56.

Shaw MR, Jennings MJ, Quicke DLJ (2011) The identity of *Scambus planatus* (Hartig, 1838) and *Scambus ventricosus* (Tschek, 1871) as seasonal forms of *Scambus calobatus* (Gravenhorst, 1829) in Europe (Hymenoptera, Ichneumonidae, Pimplinae, Ephialtini). *Journal of Hymenoptera Research*, **23**, 55–64.

Shaw SR (1985) A phylogenetic study of the subfamilies Meteorinae and Euphorinae (Hymenoptera: Braconidae). *Entomography*, **3**, 277–370.

Shaw SR (1988) Euphorine phylogeny: the evolution of diversity in host-utilization by parasitoid wasps (Hymenoptera: Braconidae). *Ecological Entomology*, **13**, 323–35.

Shaw SR (1991) An unusual manner of aggregation in the braconid *Chelonus (Microchelonus) hadrogaster* McComb (Hymenoptera). *Journal of Insect Behavior*, **4**, 537–42.

Shaw SR (1993) Observations on the ovipositional behavior of *Neoneurus mantis*, an ant-associated parasitoid from Wyoming (Hymenoptera: Braconidae). *Journal of Insect Behavior*, **6**, 649–58.

Shaw SR (1997a) Subfamily Euphorinae. In RA Wharton, PM Marsh and MJ Sharkey (eds), *Manual of the New World Genera of the Family Braconidae (Hymenoptera)*, Special Publication of the International Society of Hymenopterists 1, pp. 235–54.

Shaw SR (1997b) Subfamily Meteorinae. In RA Wharton, PM Marsh and MJ Sharkey (eds), *Manual of the New World Genera of the Family Braconidae (Hymenoptera)*, Special Publication of the International Society of Hymenopterists 1, pp. 326–30.

Shaw SR (1997c) Subfamily Rogadinae s.s. In RA Wharton, PM Marsh and MJ Sharkey (eds), *Manual of the New World Genera of the Family Braconidae (Hymenoptera)*, Special Publication of the International Society of Hymenopterists 1, pp. 404–12.

Shaw SR (2004) Essay on the evolution of adult-parasitism in the subfamily Euphorinae (Hymenoptera, Braconidae). *Proceedings of the Russian Entomological Society*, **75**, 82–95.

Shaw SR (2006) *Aleiodes Wasps of Eastern Forests: a Guide to Parasitoids and Associated Mummified Caterpillars*, US Department of Agriculture Forest Service, Washington, DC. 121 pp.

Shaw SR (2012) A new genus and new species of the tribe Dinocampini from Napo Province in Ecuador (Hymenoptera: Braconidae: Euphorinae). *International Journal of Tropical Insect Science*, **32**, 101–7.

Shaw SR, Berry JA (2005) Two new *Cryptoxilos* species (Hymenoptera: Braconidae: Euphorinae) from New Zealand and Fiji parasitising adult Scolytinae (Coleoptera). *Invertebrate Systematics*, **19**, 371–81.

Shaw SR, Edgerly J (1986) A new braconid genus parasitizing webspinners (Embiidina) in Trinidad. *Psyche*, **92**, 505–11.

Shaw SR, Jones GZ (2009) A new species of solitary *Meteorus* (Hymenoptera: Braconidae) reared from caterpillars of toxic butterflies (Lepidoptera: Nymphalidae) in Ecuador. *Journal of Insect Science*, **9**, 34.

Shaw SR, Marsh PM (2000) Revision of the enigmatic genus *Marshiella* Shaw in the New World with the description of three new species (Hymenoptera: Braconidae: Euphorinae). *Journal of Hymenoptera Research*, **9**, 277–87.

Shaw SR, Nishida K (2005) A new species of gregarious *Meteorus* (Hymenoptera: Braconidae) reared from caterpillars of *Venadicodia caneti* (Lepidoptera: Limacodidae) in Costa Rica. *Zootaxa*, **1028**, 49–60.

Shaw SR, Salerno G, Colazza S, Peri E (2001) First record of *Aridelus rufotestaceus* Tobias (Hymenoptera: Braconidae, Euphorinae) parasitizing *Nezara viridula* nymphs (Heteroptera: Pentatomidae) with observations on its immature stages and development. *Journal of Hymenoptera Research*, **10**, 131–7.

Shaw SR, Marsh PM, Talluto MA (2013) Revision of North American *Aleiodes* (Part 9): the *pallidator* (Thunberg) species-group with description of two new species (Hymenoptera: Braconidae, Rogadinae). *Zootaxa*, **3608**, 204–14.

Sheehan W (1991) Host range patterns of hymenopteran parasitoids of exophytic lepidopteran folivores. In EA Bernays (ed.), *Insect-Plant Interactions*, vol. 3, CRC Press, Boca Raton, FL, pp. 209–48.

Sheehan W (1994) Parasitoid community structure: effects of host abundance, phylogeny, and ecology. In BA Hawkins and W Sheehan (eds), *Parasitoid Community Ecology*, Oxford University Press, Oxford, pp. 90–107.

Shelby KS, Webb BA (1997) Polydnavirus infection inhibits translation of specific growth-associated host proteins. *Insect Biochemistry and Molecular Biology*, **27**, 263–70.

Shelby KS, Habibi J, Puttler B (2014). An ultrastructural and fluorescent study of the teratocytes of *Microctonus aethiopoides* Loan (Hymenoptera: Braconidae) from the hemocoel of host alfalfa weevil, *Hypera postica* (Gyllenhal) (Coleoptera: Curculionidae). *Psyche*, **2014**, 652518.

Shenefelt RD (1969) Pars 4. Braconidae 1, Triaspinae, Hybrizoninae, Euphorinae, Cosmophorinae, Neoneurinae, Macrocentrinae. In Ch Ferriere and J van der Vecht (eds), *Hymenopterorum Catalogus*, nova editio, Dr W. Junk, The Hague, pp. 1–176.

Shenefelt RD (1970a) Pars 5. Braconidae 2, Heliconinae, Calyptinae, Mimagathidinae. In Ch Ferriere and J van der Vecht (eds), *Hymenopterorum Catalogus*, nova editio, Dr W. Junk, The Hague, pp. 177–306.

Shenefelt RD (1970b) Pars 6. Braconidae 3, Agathidinae. In Ch Ferriere and J van der Vecht (eds), *Hymenopterorum Catalogus*, nova editio, Dr W. Junk, The Hague, pp. 307–428.

Shenefelt RD (1972) Pars 8. Braconidae 4, Microgasterinae *Apanteles*. In J van der Vecht and RD Shenefelt (eds), *Hymenopterorum Catalogus*, nova editio, Dr W. Junk, The Hague, pp. 429–668.

Shenefelt RD (1973a) Pars 9. Braconidae 5, Microgasterinae and Ichneutinae. In J van der Vecht and RD Shenefelt (eds), *Hymenopterorum Catalogus*, nova editio, Dr W. Junk, The Hague, pp. 669–812.

Shenefelt RD (1973b) Pars 10. Braconidae 6, Cheloninae. In J van der Vecht and RD Shenefelt (eds), *Hymenopterorum Catalogus*, nova editio, Dr W. Junk, The Hague, pp. 813–936.

Shenefelt RD (1974) Pars 11. Braconidae 7, Alysiinae. In J van der Vecht and RD Shenefelt (eds), *Hymenopterorum Catalogus*, nova editio, Dr W. Junk, The Hague, pp. 937–1113.

Shenefelt RD (1975) Pars 12. Braconidae 8, Exothecinae Rogadinae. In J van der Vecht and RD Shenefelt (eds), Hymenopterorum Catalogus, *nova editio*, Dr W. Junk, The Hague, pp. 1115–262.

Shenefelt RD (1978) Pars 15. Braconidae 10, Braconinae, Gnathobraconinae, Mesostoinae, Pseudodicrogeniinae, Telengaiinae, Ypsistocerinae plus Braconidae in general, major groups, unplaced genera and species. In C van Achterberg and RD Shenefelt (eds), *Hymenopterorum Catalogus*, nova editio, Dr W. Junk, The Hague, pp. 1425–872.

Shenefelt RD, Marsh PM (1976) Pars 13. Braconidae 9, Doryctinae. In J van der Vecht and RD Shenefelt (eds), *Hymenopterorum Catalogus*, nova editio, Dr W. Junk, The Hague, pp. 1263–424.

Sheng M-L, Sun S-P (2011) A new genus and species of Brachyscleromatinae (Hymenoptera: Ichneumonidae) from China, *Laxiareola ochracea*. *Journal of Insect Science*, **11**, 1–6.

Sheng M-L, Broad GR, Sun S-P (2012a) A new genus and species of Collyriinae (Hymenoptera, Ichneumonidae). *Journal of Hymenoptera Research*, **25**, 103–25.

Sheng M-L, Zhou R-X, Sun S-P (2012b) A new species of *Xorides* Latreille (Hymenoptera, Ichneumonidae, Xoridinae) parasitizing *Pterolophia alternata* (Coleoptera, Cerambycidae) in *Robinia pseudoacacia*. *ZooKeys*, **246**, 39–49.

Shevyrev IY (1912) *Parasites and Hyperparasites in the Insect World*, St Petersburg, 216 pp. (in Russian; translation by J. Millman available).

Shi M, Chen XX (2005) Molecular phylogeny of the Aphidiinae (Hymenoptera: Braconidae) based on DNA sequences of 16S rRNA, 18S rDNA and ATPase 6 genes. *European Journal of Entomology*, **102**, 133–8.

Shi M, Chen XX, van Achterberg C (2005) Phylogenetic relationships among the Braconidae (Hymenoptera: Ichneumonoidea) inferred from partial 16S rDNA, 28S rDNA D2, 18S rDNA gene sequences and morphological characters. *Molecular Phylogenetics and Evolution*, **37**, 104–16.

Shimazaki M, Watanabe K, Shimazaki Y (2011) A record of the koinobiont endoparasitoid wasp, *Melalophacharops everese* (Hymenoptera, Ichneumonidae), attacking eggs of a lycaenid butterfly, *Acytolepis puspa*. *Lepidoptera Science*, **62**, 151–5.

Shirai S, Maetô K (2009) Suspending cocoons to evade ant predation in *Meteorus pulchricornis*, a braconid parasitoid of exposed-living lepidopteran larvae. *Entomological Science*, **12**, 107–9.

Short JRT (1952) The morphology of the head of larval Hymenoptera with special reference to the head of Ichneumonoidea, including a classification of the final instar larvae of the Braconidae. *Transactions of the Royal Entomological Society of London*, **103**, 27–84.

Short JRT (1957) On the final instar larva of *Stilbops (Aphanoroptrum) abdominale* (Grav.) (Hymenoptera, Ichneumonidae). *Proceedings of the Royal Entomological Society of London (B)*, **26**, 175–6.

Short JRT (1959) A description and classification of the final instar larvae of the Ichneumonidae (Insecta, Hymenoptera). *Proceedings of the United States National Museum*, **110**, 391–511.

Short JRT (1970) On the classification of the final instar larvae of Ichneumonidae (Hymenoptera). Supplement. *Transactions of the Royal Entomological Society of London*, **122**, 185–210.

Short JRT (1977) A description and classification of some final-instar larvae of the Phytodietini and Eclytini (Hymenoptera, Ichneumonidae, Tryphoninae). *Proceedings of the Linnean Society of New South Wales*, **101**, 114–19.

Short JRT (1978) The final larval instars of the Ichneumonidae. *Memoirs of the American Entomological Institute*, **25**, 1–508.

Short JRT (1982) The final-instar larvae of two Anomaloninae (Hymenoptera: Ichneumonidae) from Australia. *Proceedings of the Linnean Society of New South Wales*, **105**, 237–40.

Short JRT, Nakanishi A (1981) The final-instar larvae of some Diplazontinae (Hymenoptera, Ichneumonidae) from Japan. *Kontyû, Tokyo*, **49**, 668–72.

Short MW, Steibauer MJ (2004) Floral nectar versus honeydew as food for wasp partasitoids: implications for pest management in eucalypt plantations. *Australian Forestry*, **67**, 199–203.

Short MW, Schmidt S, Steibauer MJ (2006) A key to some Australian genera of large nocturnal Ichneumonidae (Hymenoptera), including flight periodicities and influence of moon phase on light trap catches. *Australian Entomologist*, **33**, 49–55.

Shorthouse JD, Leggo JJ (2002) Immature stages of the galler *Diplolepis triforma* (Hymenoptera: Cynipidae) with comments on the role of its prepupa. *The Canadian Entomologist*, **134**, 433–46.

Shu S, Jones RL (1993) Evidence for a multicomponent sex pheromone in *Eriborus terebrans* (Gravenhorst) (Hym.: lchneumonidae), a larval parasitoid of the European corn borer. *Journal of Chemical Ecology*, **19**, 2563–76.

Shuja-Uddin (1977) Observations on normal and diapausing cocoons of the genus *Lipolexis* Foerster (Hymenoptera: Aphidiidae) from India. *Bollettino del Laboratorio Entomologia Agraria "Filippo Silvestri"*, **34**, 51–4.

Sigwalt B (1977) La nomenclature alaire de Comstock-Needham dans l'étude des braconides: distinction des genres *Phanerotoma* et *Phanerotomella* (Hym. Braconidae, Cheloninae). *Annales de la Société Entomologique de France (Nouvelle Série)*, **13**, 525–34.

Silva RB, Cruz I, Figueiredo MLC, Pereira AG, Penteado-Dias AM (2012) Occurrence and biology of *Dinocampus coccinellae* (Schrank, 1802) (Hymenoptera; Braconidae: Euphorinae) parasitising different species of Coccinellidae (Coleoptera) in Neotropical region. *Brazilian Journal of Biology*, **72**, 215–19.

Sime KR (2000) Two new records of pimpline ichneumonids attacking *Battus philenor* (Linnaeus) (Lepidoptera: Papilionidae). *Journal of Hymenoptera Research*, **9**, 210–12.

Sime KR (2002) Experimental studies of the host-finding behavior of *Trogus pennator*, a parasitoid of swallowtail butterflies. *Journal of Chemical Ecology*, **28**, 1377–92.

Sime KR, Brower AVZ (1998) Explaining the latitudinal gradient anomaly in ichneumonid species richness: evidence from butterflies. *Journal of Animal Ecology*, **67**, 387–99.

Sime KR, Wahl DB (2002) The cladistics and biology of the *Callajoppa* genus-group (Hymenoptera: Ichneumonidae, Ichneumoninae). *Zoological Journal of the Linnean Society*, **134**, 1–56.

Sime KR, Daane KM, Andrews JW, Hoelmer K, Pickett CH, Nadel H, Johnson MW, Messing RH (2006) The biology of *Bracon celer* as a parasitoid of the olive fruit fly. *BioControl*, **51**, 553–67.

Simmonds FJ (1947a) The biology of the parasites of *Loxostege sticticalis* L., in North America – *Bracon vulgaris* (Cress.) (Braconidae: Agathinae). *Bulletin of Entomological Research*, **38**, 145–55.

Simmonds FJ (1947b) The biology of *Phytodietus pulcherrimus* (Cress.) (Ichneumonidae, Tryphoninae) parasitic of *Loxostege sticticalis* L. in North America. *Parasitology* **38**, 150–6.

Simões RA, Reis LG, Bento JMS, Solter LF, Delalibera I Jr, (2012) Biological and behavioral parameters of the parasitoid *Cotesia flavipes* (Hymenoptera: Braconidae) are altered by the pathogen *Nosema* sp. (Microsporidia: Nosematidae). *Biological Control*, **63**, 164–71.

Simpson RG, Cross JE, Best RL, Ellsbury MM, Coseglia AF (1979) Effects of hyperparasites on population levels of *Bathyplectes curculionis* in Colorado. *Environmental Entomology*, **8**, 96–100.

Singh RP, Baldev Parshad (1970) Biological notes on *Cardiochiles hymeniae* Max Fischer and Baldev Parshad (Braconidae: Hymenoptera). *Indian Journal of Entomology*, **32**, 127–9.

Sinu PA, Nasser M, Dharmarajan P (2007) Host searching behavior and potential of an aquatic ichneumonid pupal parasitoid of rice caseworm (*Parapoynx stagnalis*) in an upland rice paddy agro-ecosystem of the Western Ghats, India. *Biocontrol Science and Technology*, **17**, 1037–45.

Sivinski J, Webb JC (1989) Acoustic signals produced during courtship in *Diachasmimorpha* (=*Biosteres*) *longicaudata* (Hymenoptera: Braconidae) and other Braconidae. *Annals of the Entomological Society of America*, **82**, 116–20.

Skillen EL, Pickering J, Sharkey MJ (2000). Species richness of the Campopleginae and Ichneumoninae (Hymenoptera: Ichneumonidae) along a latitudinal gradient in eastern North American old-growth forests. *Environmental Entomology*, **29**, 460–6.

Skinner ER, Thompson GH (1960) *The Alder Woodwasp and Its Insect Enemies*, World Educational Film, Oxford.

Sláma K, Lukáš J (2011) Myogenic nature of insect heartbeat and intestinal peristalsis, revealed by neuromuscular paralysis caused by the sting of a braconid wasp. *Journal of Insect Physiology*, **57**, 251–9.

Slansky F (1978) Utilization of energy and nitrogen by larvae of the imported cabbage worm *Pieris rapae*, as affected by parasitism by *Apanteles glomeratus*. *Environmental Entomology*, **7**, 179–85.

Slobodchikoff CN (1967) Bionomics of *Grotea californica* Cresson, with a description of the larva and pupa (Hym., Ichneumonidae). *Pan-Pacific Entomologist*, **43**, 161–8.

Slobodchikoff CN (1973) Behavioral studies of three morphotypes of *Therion circumflexum* (L.) (Hymenoptera: Ichneumonidae). *Pan-Pacific Entomologist*, **49**, 197–206.

Slobodchikoff CN (1974a) Notes on the biology of *Therion circumflexum* (L.) with a description of the immature stages. (Hymenoptera: Ichneumonidae). *Pan-Pacific Entomologist*, **50**, 111–17.

Slobodchikoff CN (1974b) Behavioral and morphological mimicry in a cranefly and an ichneumonid. *Pan-Pacific Entomologist*, **50**, 154–9.

Slobodchikoff CN (1983) Why asexual reproduction?: variation in populations of the parthenogenetic wasp, *Venturia canescens* (Hymenoptera: Ichneumonidae). *Annals of the Entomological Society of America*, **76**, 23–9.

Sloggett JJ, Webberley KM, Majerus MEN (2004) Low parasitoid success on a myrmecophilous host is maintained in the absence of ants. *Ecological Entomology*, **29**, 123–7.

Slovák M (1983) The larval development of the ichneumonid *Exetastes cinctipes*. *Ustavu Experimentálnej Fytopatológie a Entomológie*, **2**, 233–44.

Slovák M (1984) New data about developmental stages of *Exetastes cinctipes* (Hym., Ichneumonidae). *Biologiá (Bratislava)*, **41**, 543–8.

Slovak M (1986) Mating behaviour in laboratory reared *Exetastes cinctipes* Hym. *Ichneumonidae. Biologia (Bratislava)*, **416**, 543–8.

Sluss RR, Leutenegger R (1968) The fine structure of the trophic cells of *Perilitus coccinellae* (Hymenoptera: Braconidae). *Journal of Ultrastructure Research*, **25**, 441–51.

Smart CB, Mayhew PJ (2009) A comparative study of sex ratio and clutch size in gregarious ichneumonoid wasps. *Journal of Insect Behavior*, **22**, 273–88.

Smid HM, Bleeker MAK, Van Loon JJA, Vet LEM (2003) Three-dimensional organization of the glomeruli in the antennal lobe of the parasitoid wasps *Cotesia glomerata* and *C. rubecula*. *Cell and Tissue Research*, **312**, 237–48.

Smid HM, Wang G, Bukovinsky T, Steidle JLM, Bleeker MAK, van Loon JJA, Vet LEM (2007) Species-specific acquisition and consolidation of long-term memory in parasitic wasps. *Proceedings of the Royal Society B: Biological Sciences*, **274**, 1539–46.

Smilanich AM, Mason PA, Sprung L, Chase TR, Singer MS (2011) Complex effects of parasitoids on pharmacophagy and diet choice of a polyphagous caterpillar. *Oecologia*, **165**, 995–1005.

Smith EL (1968) Biosystematics and morphology of Symphyta. I. Stem-galling *Euura* of the California region, and a new female genitalic nomenclature (Hymenoptera: Tenthredinidae). *Annals of the Entomological Society of America*, **61**, 1389–407.

Smith EL (1969) Evolutionary morphology of external insect genitalia. 1. Origin and relationship to other appendages. *Annals of the Entomological Society of America*, **62**, 1051–79.

Smith EL (1970) Evolutionary morphology of external insect genitalia. 2. Hymenoptera. *Annals of the Entomological Society of America*, **63**, 1–27.

Smith HA, Capinera JL, Pena JE, Linboterhaar B (1994) Parasitism of pickleworm and melonworm (Lepidoptera, Pyralidae) by *Cardiochiles diaphaniae* (Hymenoptera, Braconidae). *Environmental Entomology*, **23**, 1283–93.

Smith HD (1932) *Phaeogenes nigridens* Wesmael, an important ichneumonid parasite of the pupa of the European corn borer. *Technical Bulletin, United States Department of Agriculture*, **331**, 1–45.

Smith MA, Rodriguez JJ, Whitfield JB, Deans AR, Janzen DH, Hallwachs W, Hebert PDN (2008) Extreme diversity of tropical parasitoid wasps exposed by iterative integration of natural history, DNA barcoding, morphology, and collections. *Proceedings of the National Academy of Sciences of the United States of America*, **105**, 12359–64.

Smith MA, Bertrand C, Crosby K, Eveleigh ES, Fernandez-Triana J, et al. (2012) *Wolbachia* and DNA barcoding insects: patterns, potential and problems. *PLoS ONE*, **7** (5), e36514.

Smith OJ (1952) Biology and behaviour of *Microctonus vittatae* Muesebeck (Braconidae) with descriptions of its immature stages. *University of California Publications in Entomology*, **9**, 315–44.

Smith OJ, Diboll AG, Rosenberger JH (1955) Laboratory studies of *Pelecystoma harrisinae* (Ashmead), an adventive braconid parasite of the western grape leaf skeletonizer. *Annals of the Entomological Society of America*, **38**, 232–7.

Smith PT, Kambhampati S (1999) Evolutionary transitions in Aphidiinae (Hymenoptera: Braconidae). In AD Austin and M Dowton (eds), *Hymenoptera: Evolution, Biodiversity and Biological Control*, CSIRO, Canberra, pp. 106–13.

Smith PT, Kambhampati S, Völkl W, Mackauer M (1999) A phylogeny of aphid parasitoids (Hymenoptera: Braconidae: Aphidiinae) inferred from mitochondrial NADH 1 dehydrogenase gene sequence. *Molecular Phylogenetics and Evolution*, **11**, 236–45.

Smith Trail DR, (1980) Behavioural interactions between parasites and hosts: host suicide and the evolution of complex life cycles. *The American Naturalist*, **116**, 77–91.

Sobczak JF, Maia DP, Moura JCMS, Costa VAA, Vasconcellos-Neto J (2012) Natural history of interaction between *Meteorus* sp. Haliday, 1835 (Hymenoptera: Braconidae) and its hyperparasitoid *Toxeumella albipes* Girault, 1913 (Hymenoptera: Pteromalidae). *Brazilian Journal of Biology*, **71**, 211–14.

Soldán T, Starý P (1981) Parasitogenic effects of *Aphidius smithi* (Hymenoptera, Aphidiidae) on the reproductive organs of the pea aphid, *Acyrthosiphon pisum* (Homoptera, Aphididae). *Acta Entomologica Bohemoslovaca*, **78**, 243–53.

Soller M, Lanzrein B (1996) Polydnavirus and venom of the egg-larval parasitoid *Chelonus inanitus* (Braconidae) induce developmental arrest in the prepupa of its host *Spodoptera littoralis* (Noctuidae). *Journal of Insect Physiology*, **42**, 471–81.

Solleza MO, Javier PA (2009) Ovipositional preference of *Trathala flavoorbitalis* Cameron (Hymenoptera: Ichneumonidae), a larval parasitoid of eggplant fruit and shoot borer, *Leucinodes orbonalis* Guenee. *Philippine Entomologist*, **23**, 131–48.

Sommerman KM (1956) Parasitization of nymphal and adult psocids (Psocoptera). *Proceedings of the Entomological Society of Washington*, **58**, 149–52.

Southwood TRE (1957) Observations on swarming in Braconidae and Coniopterygidae. *Proceedings of the Royal Entomological Society of London*, **32**, 80–2.

Spanjer W, Grosu L, Piek T (1977) Two different paralyzing preparations obtained from a homogenate of the wasp *Microbracon hebetor* (Say). *Toxicon*, **15**, 413–21.

Sparks DW, Valdez EW (2003) Food habits of *Nyctinomops macrotis* at a maternity roost in New Mexico, as indicated by analysis of guano. *Southwestern Naturalist*, **48**, 132–5.

Speicher BR (1936) Oögenesis, fertilization and early cleavage in *Habrobracon*. *Journal of Morphology*, **59**, 401–17.

Speicher BR, Speicher KG (1940) The occurrence of diploid males in *Habrobracon brevicornis*. *The American Naturalist*, **74**, 379–82.

Speicher KG (1934) Impaternate females in *Habrobracon*. *Biological Bullettin*, **67**, 277–93.

Spradbery JP (1968) A technique for artificially culturing ichneumonid parasites of woodwasps (Hymenoptera: Siricidae). *Entomologia Experimentalis et Applicata*, **11**, 257–60.

Spradbery JP (1969) The biology of *Pseudorhyssa sternata* Merrill (Hym., Ichneumonidae), a cleptoparasite of siricid woodwasps. *Bulletin of Entomological Research*, **59**, 291–7.

Spradbery JP (1970a) The immature stages of European ichneumonid parasites of siricine woodwasps. *Proceedings of the Royal Entomological Society of London (A)*, **45**, 14–28.

Spradbery JP (1970b) Host finding by *Rhyssa persuasoria* (L.), an ichneumonid parasite of siricid woodwasps. *Animal Behaviour*, **18**, 103–14.

Stamp NE (1981) Behavior of parasitized aposematic caterpillars: advantageous to the parasitoid or the host? *The American Naturalist*, **118**, 715–25.

Staple JO, Cortesero AM, de Moraes CM, Tumlinson JH, Lewis JW (1997) Extrafloral nectar, honeydew, and sucrose effects on searching behavior and efficiency of *Microplitis croceipes* (Hymenoptera: Braconidae) in cotton. *Environmental Entomology*, **26**, 617–23.

Stahlhut JK, Fernandez-Triana J, Adamowicz SJ, Buck M, Goulet H, Hebert PDN, Huber JT, Merilo MT, Sheffield CS, Woodcock T, Smith MA (2013) DNA barcoding reveals diversity of Hymenoptera and the dominance of parasitoids in a sub-arctic environment. *BMC Ecology*, **13**, 2.

Starý P (1960) The generic classification of the family Aphidiidae (Hymenoptera). *Acta Societatis Entomologicae Cechosloveniae*, **57**, 238–52.

Starý P (1966) *Aphid Parasites of Czechoslovakia. A Review of the Czechoslovak Aphidiidae (Hymenoptera)*, Dr W. Junk, The Hague, 242 pp.

Starý P (1970) *Biology of Aphid Parasities (Hymenoptera: Aphidiidae) with Respect to Integrated Control*, Dr W. Junk, The Hague.

Starý P (1974) The emergence hole of aphid parasites (Hymenoptera, Aphidiidae): its significance in a natural system. *Acta Entomologica Bohemoslovaca*, **71**, 209–16.

Starý P (1975) *Pseudopraon mindariphagum* gen. n., sp. n. (Hymenoptera, Aphidiidae) – description and life history of a parasite of *Mindarus abietinus* (Homoptera, Mindaridae) in Central Europe. *Acta Entomologica Bohemoslovaca*, **72**, 249–58.

Starý P (1987) Aphid–ant–parasitoid association on the creeping thistle, *Cirsium arvense*, in agroecosystems in Czechoslovavia. *Acta Entomologica Bohemoslovaca*, **84**, 15–21.

Starý P (1988) *Autriquella aptera* gen. n., sp. n. from the Afrotropical region (Hymenoptera, Aphidiidae). *Acta Entomologica Bohemoslovaca*, **85**, 283–6.

Starý P (1999) Biology and distribution of microbe-associated thelytokous populations of aphid parasitoids (Hym., Braconidae, Aphidiinae). *Journal of Applied Entomology*, **123**, 231–5.

Starý P, Rejmanek M (1981) Number of parasitoids per host in different systematic groups of aphids: the implications for introduction strategy in biological control (Homoptera: Aphidoidea; Hymenoptera: Aphidiidae). *Entomologica Scandinavica, Supplement*, **15**, 341–51.

Starý P, Schlinger EI (1967) *Revision of the Far East Asian Aphidiidae (Hymenoptera)*, Dr W. Junk, The Hague.

Stasiak K, Renault S, Federici BA, Bigot Y (2005) Characteristics of pathogenic and mutualistic relationships of ascoviruses in field populations of parasitoid wasps. *Journal of Insect Physiology*, **51**, 103–15.

Steiner S, Kropf C, Graber W, Nentwig W, Klopfstein S (2010) Antennal courtship and functional morphology of tyloids in the parasitoid wasp *Syrphoctonus tarsatorius* (Hymenoptera: Ichneumonidae: Diplazontinae). *Arthropod Structure & Development*, **39**, 33–40.

Steiner WWM, Teig DA (1989) *Microplitis croceipes*: genetic characterization and developing insecticide resistant biotypes. *The Southwestern Entomologist*, **12**, 71–87.

Stelfox AW (1944) The swarming flights of *Blacus* (Hymenoptera: Braconidae). *Entomologist's Monthly Magazine*, **80**, 208.

Stelinski LL, Oakleaf R, Rodriguez-Saona C (2007) Oviposition-deterring pheromone deposited on blueberry fruit by the parasitic wasp, *Diachasma alloeum*. *Behaviour*, **144**, 429–45.

Stelinski LL, Boina DR Meyer WL (2010) Oviposition marking behavior of *Diachasma alloeum*, (Hymenoptera: Braconidae), parasitizing *Rhagoletis pomonella*, (Diptera: Tephritidae). *Journal of Insect Behavior*, **23**, 419–30.

Stenbacka F, Hjältén J, Hilszczański J, Ball JP, Gibb H, Johansson T, Pettersson RB (2010) Saproxylic parasitoid (Hymenoptera, Ichneumonoidea) communities in managed boreal forest landscapes. *Insect Conservation and Diversity*, **3**, 114–23.

Stepper J, Becker C, Schmidt K (1983) Feinbau und Ontogenese der Porenplatten af den Antennen von *Pimpla turionellae* (Hymenoptera, Ichneumonidae). *Zoomorphology*, **102**, 1–32.

Stevens GC (1989) The latitudinal gradient in geographical range: how so many species coexist in the tropics. *The American Naturalist*, **133**, 240–56.

Stevens NB, Austin AD, Jennings JT (2010) Synopsis of Australian agathidine wasps (Hymenoptera: Braconidae: Agathidinae). *Zootaxa*, **2480**, 1–26.

Stigenberg J, Ronquist F (2011) Revision of the Western Palearctic Meteorini (Hymenoptera, Braconidae), with a molecular characterization of hidden Fennoscandian species diversity. *Zootaxa* **3084**, 1–95.

Stoltz DB (1981) A putative baculovirus in the ichneumonid parasitoid, *Mesoleius tenthredinidis*. *Canadian Journal of Microbiology*, **27**, 116–22.

Stoltz DB (1990) Evidence for chromosomal transmission of polydnavirus DNA. *Journal of General Virology*, **71**, 1051–6.

Stoltz DB, Krell P (2012) The origins and early history of polydnavirus research. In NE Beckage and J-M Drezen (eds), *Parasitoid Viruses. Symbionts and Pathogens*, Elsevier Academic Press, Amsterdam, pp. 5–13.

Stoltz DB, Makkay A (2000) Co-replication of a reovirus and a polydnavirus in the ichneumonid parasitoid *Hyposoter exiguae*. *Virology*, **278**, 266–75.

Stoltz DB, Whitfield JB (1992) Viruses and virus-like entities in the parasitic Hymenoptera. *Journal of Hymenoptera Research*, **1**, 125–39.

Stoltz DB, Krell P, Cook D, Mackinnon EA, Lucarotti CJ (1988) An unusual virus from the parasitic wasp *Cotesia melanoscela*. *Virology*, **162**, 311–20.

Stouthamer R (2003) The use of unisexual wasps in biological control. In JC van Lenteren (ed.), *Quality Control and Production of Biological Control Agents*, CABI, Wallingford, pp. 93–112.

Strand MR (1999) The effects of life history on development of the Hymenoptera. In AD Austin and M Dowton (eds), *Hymenoptera: Evolution, Biodiversity and Biological Control*, CSIRO, Canberra, pp. 11–16.

Strand MR, Godfray HCJ (1989) Superparasitism and ovicide in parasitic Hymenoptera: theory and a case study of the ectoparasitoid *Bracon hebetor*. *Behavioural Ecology and Sociobiology*, **24**, 421–32.

Strand MR, Pech LL (1995) Immunological basis for compatibility in parasitoid–host relationships. *Annual Review of Entomology*, **40**, 31–56.

Strand MR, Vinson SB (1982) Behavioural response of the parasitoid, *Cardiochiles nigriceps*, to a kairomone. *Entomologia Experimentalis et Applicata*, **31**, 308–15.

Strand MR, Wong EA (1991) The growth and role of *Microplitis demolitor* teratocytes in parasitism of *Pseudoplusia includens*. *Journal of Insect Physiology*, **37**, 503–15.

Strand MR, Johnson JA, Culin JD (1988) Developmental interactions between the parasitoid *Microplitis demolitor* (Hymenoptera: Braconidae) and its host *Heliothis virescens* (Lepidoptera: Noctuidae). *Annals of the Entomological Society of America*, **81**, 822–30.

Streams FA, Coles LW (1965) Wheat stem sawfly parasites, *Collyria calcitrator* and *Pediobius nigritarsis*, in eastern United States. *Journal of Economic Entomology*, **58**, 303–6.

Strickland EH (1923) Biological notes on parasites of prairie cutworms. *Bulletin of the Department of Agriculture, Canada, Entomological Branch*, **26**, 1–40.

Stuart AM (1957) *Ephialtes brevicornis* (Grav.) as an external parasite of the diamond-back moth, *Plutella maculipennis* (Curt.). *Bulletin of Entomological Research*, **48**, 477–88.

Subba Rao BR, Sharma AK (1962) Studies on the biology of *Trioxys indicus* Subba Rao and Sharma 1958, a parasite of *Aphis gossypii* Glover. *Proceedings of the National Institute of Science, India (B) Biological Sciences*, **28**, 164–82.

Sudheendrakumar VV, Mohamed UVK, Abdurahiman UC, Narendran TC (1978) Mating behavior of *Bracon brevicornis* Wesmael (Hymenoptera: Braconidae), a larval parasite of *Nephantis serinopa* Meyrick, the black headed caterpillar of coconut. *Agricultural Research Journal of Kerala*, **16**, 224–6.

Sudheendrakumar VV, Mohamed UVK, Narendran TC, Abdurahiman UC (1982) Biology and morphology of immature stages of *Bracon brevicornis* (Hymenoptera: Braconidae) an important biological control agent of the black headed caterpillar pest of coconut. *Journal of the Bombay Natural History Society*, **79**, 254–60.

Sugimoto T, Shiinono Y, Hata Y, Nakai A, Yahara M (1988) Foraging for patchily-distributed leaf-miners by the parasitoid, *Dapsilarthra rufiventris* (Hymenoptera: Braconidae) III. Visual and acoustic cues to a close range patch-location. *Applied Entomology and Zoology*, **23**, 113–21.

Sutherland TD, Young JH, Weisman S, Hayashi CY, Merritt DJ (2010) Insect silk: one name, many materials. *Annual Review of Entomology*, **55**, 171–88.

Suzuki M, Tanaka T (2006) Virus-like particles in venom of *Meteorus pulchricornis* induce host hemocyte apoptosis. *Journal of Insect Physiology*, **52**, 602–11.

Suzuki M, Tanaka T (2007) Development of *Meteorus pulchricornis* and regulation of its noctuid host, *Pseudaletia separata*. *Journal of Insect Physiology*, **53**, 1072–8.

Suzuki M, Miura K, Tanaka T (2008) The virus-like particles of a braconid endoparasitoid wasp, *Meteorus pulchricornis*, inhibit hemocyte spreading in its noctuid host, *Pseudaletia separata*. *Journal of Insect Physiology*, **54**, 1015–22.

Swedenborg PD, Jones RL (1992a) Multicomponent sex pheromone in *Macrocentrus grandii* Goidanich (Hymenoptera: Braconidae). *Journal of Chemical Ecology*, **18**, 1901–12.

Swedenborg PD, Jones RL (1992b) (Z)-4-Tridecenal, a pheromonially active air oxidation product from a series of (Z,Z)-9,13-dienes in *Macrocentrus grandii* (Goidanich) (Hymenoptera: Braconidae). *Journal of Chemical Ecology*, **18**, 1913–31.

Swedenborg PD, Jones RL, Liu H-W, Krick TP (1993) (3*R**, 5*S**,6*R**)-3,5-Dimethyl-6-(methylethyl)-3,4,5,6-tetrahydropyran-2-one, a third sex pheromone component for *Macrocentrus grandii* (Goidanich) (Hymenoptera: Braconidae) and evidence for its utility at eclosion. *Journal of Chemical Ecology*, **19**, 485–502.

Swedenborg PD, Jones RL, Zhou HQ, Shin IJ, Liu H-W (1994) (3*R**,5*S**,6*R**)-3,5-Dimethy-6-(methylethyl)-3,4,5,6-tetrahydropyran-2-one and (3*S**,5*R**,6*S**)-3,5-dimethyl-6-(methylethyl)-3,4,5,6-tetrahydropyran-2-one, a pheromone of *Macrocentrus grandii* (Goidanich) (Hymenoptera: Braconidae). *Journal of Chemical Ecology*, **20**, 3373–80.

Syme PD, Green GW (1972) The effect of *Orgilus obscurator* (Hymenoptera: Braconidae) on development of the European pine shoot moth (Lepidoptera: Olethreutidae) *The Canadian Entomologist*, **104**, 523–30.

Syrjämäki J (1976) The mystery of the missing females in connexion with male swarming of *Blacus ruficornis* Nees (Hym., Braconidae). *Annales Entomologica Fennici*, **42**, 66–8.

Syvertsen TC, Jackson LL, Blomquist GJ, Vinson SB (1995) Alkadienes mediating courtship in the parasitoid

Cardiochiles nigriceps (Hymenoptera: Braconidae). *Journal of Chemical Ecology*, **21**, 1971–89.

Szépligeti G (1904) Über *Gnathobracon*, A. Costa. *Annuario del Museo Zoologico della Università Napoli (Nuova Serie)*, **1**, 1–2.

Tadic MD (1958) *Apanteles hyphantrae* Riley, an egg parasite of the fall webworm. *Proceedings of the X Congress of Entomology*, **4**, 859–61.

Tagashira E, Tanaka T (1998) Parasitic castration of *Pseudaletia separata* by *Cotesia kariyai* and its association with polydnavirus gene expression. *Journal of Insect Physiology*, **44**, 733–44.

Tagawa J (1977) Localisation and histology of the sex pheromone-producing gland in the parasitic wasp, *Apanteles glomeratus*. *Journal of Insect Physiology*, **23**, 49–56.

Tagawa J (1983) Female sex pheromone glands in the parasitic wasps, genus *Apanteles*. *Applied Entomology and Zoology*, **18**, 416–27.

Tagawa J (1987) Post-mating changes in the oviposition tactics of the parasitic wasp, *Apanteles glomeratus* L. (Hymenoptera: Braconidae). *Applied Entomology and Zoology*, **22**, 537–42.

Tagawa J (1996) Function of the cocoon of the parasitoid wasp, *Cotesia glomerata* L. (Hymenoptera: Braconidae): protection against desiccation. *Applied Entomology and Zoology*, **31**, 99–103.

Tagawa J, Kitano H (1981) Mating behaviour of the braconid wasp, *Apanteles glomeratus* L. (Hymenoptera: Braconidae) in the field. *Applied Entomology and Zoology*, **16**, 345–50.

Takada H, Hashimoto Y (1985) Association of the root aphid parasitoids *Aclitus sappaphis* and *Paralipsis eikoae* (Hymenoptera, Aphidiidae) with the aphid-attending ants *Pheidole fervida* and *Lasius niger* (Hymenoptera, Formicidae). *Kontyû, Tokyo*, **53**, 150–60.

Takasu K, Lewis WJ (2003) Learning of host searching cues by the larval parasitoid, *Microplitis croceipes*. *Entomologia Experimentalis et Applicata*, **108**, 77–86.

Takasu K, Ode PJ, Antolin MF, Strand M (1997) Environmental and genetic determinants of ovicide in parasitic wasp *Bracon hebetor*. *Behavioural Ecology*, **8**, 647–54.

Takasuka K, Matsumoto R (2011a) Lying on the dorsum: unique host-attacking behaviour of *Zatypota albicoxa* (Hymenoptera, Ichneumonidae). *Journal of Ethology*, **29**, 203–7.

Takasuka K, Matsumoto R (2011b) Infanticide by a solitary koinobiont ichneumonid ectoparasitoid of spiders. *Naturwissenschaften*, **98**, 529–36.

Takasuka K, Matsumoto R, Ohbayashi N (2009) Oviposition behavior of *Zatypota albicoxa* (Hymenoptera, Ichneumonidae), an ectoparasitoid of *Achaearanea tepidariorum* (Araneae, Theridiidae). *Entomological Science*, **12**, 232–7.

Tamò C, Roelfstra L-L, Guillaume S, Turlings TCJ (2006) Odour-mediated long-range avoidance of interspecific competition by a solitary endoparasitoid: a time-saving foraging strategy. *Journal of Animal Ecology*, **75**, 1091–9.

Tan VT, Takasu K (2000) Host age selection by the host-feeding pupal parasitoid *Diadromus subtilicornis* (Gravenhorst) (Hymenoptera: Ichneumonidae). *Applied Entomology and Zoology*, **35**, 549–56.

Tanaka K, Lapointe R, Barney WE, Makkay AM, Stoltz D, Cusson M, Webb BA (2007) Shared and species-specific features among ichnovirus genomes. *Virology*, **363**, 26–35.

Tanaka T (1987) Calyx and venom fluids of *Apanteles kariyai* (Hymenoptera, Braconidae) as factors that prolong larval period of the host, *Pseudaletia separata* (Lepidoptera, Noctuidae). *Annals of the Entomological Society of America*, **80**, 530–3.

Tanaka T, Vinson SB (1991a) Interaction of venoms with the calyx fluids of three parasitoids, *Cardiochiles nigriceps*, *Microplitis croceipes* (Hymenoptera: Braconidae), and *Campoletis sonorensis* (Hymenoptera: Ichneumonidae) in effecting a delay in the pupation of *Heliothis virescens* (Lepidoptera: Noctuidae). *Annals of the Entomological Society of America*, **84**, 87–92.

Tanaka T, Vinson SB (1991b) Depression of thoracic gland activity of *Heliothis virescens* by venom and calyx fluids from the parasitoid, *Cardiochiles nigriceps*. *Journal of Insect Physiology*, **37**, 139–44.

Tanaka T, Wago G (1990) Ultrastructural and functional maturation of teratocytes of *Apanteles kariyai*. *Archives of Insect Biochemistry and Physiology*, **13**, 187–97.

Tanton MT, Epila JSO (1984) Parasitization of larvae of *Paropsis atomaria* Ol. (Coleoptera: Chrysomelidae) in the Australian Capital Territory. *Australian Journal of Zoology* **32**, 251–9.

Tardieux I, Rabasse JM (1988) Induction of a thelytokous reproduction in the *Aphidius colemani* complex (Hym., Aphidiidae). *Journal of Applied Entomology*, **106**, 58–61.

Tartally A (2005) *Neotypus melanocephalus* (Hymenoptera: Ichneumonidae): the first record of a parasitoid wasp attacking *Maculinea teleius* (Lycaenidae). *Nota Lepidopterologica*, **28**, 65–7.

Tawfik MFS, Hafez M, Ibrahim AA (1980) Immature stages of *Microplitis rufiventris* Kok (Hym, Braconidae). *Deutsche Entomologische Zeitschrift (N.F.)*, **27**, 39–50.

Taylor AD (1988) Host effects on functional responses of *Bracon hebetor*. *Journal of Animal Ecology*, **57**, 173–84.

Teder T (2005) Male-biased size dimorphism in ichneumonine wasps (Hymenoptera: Ichneumonidae)-the role of sexual selection for large male size. *Ecological Entomology*, **30**, 342–9.

Tereshkin AM (1990) A trap for the xylobiontes and theirs parasites. http://tereshkin.info/papers/a-trap-for-the-xylobiontes-and-theirs-parasites-in-russian.html (accessed 30 June 2014) (in Russian).

Tereshkin AM (2009) Illustrated key to the tribes of the subfamily Ichneumoninae and the genera of the tribe Platylabini of world fauna (Hymenoptera, Ichneumonidae). *Linzer Biologische Beiträge*, **41**, 1317–608.

Thézé J, Bézier A, Periquet G, Drezen JM, Herniou EA (2011) Paleozoic origin of insect large dsDNA viruses. *Proceedings of the National Academy of Sciences of the United States of America*, **108**, 15931–5.

Thibout E (2005) Role of caterpillar silk thread in location of host pupae by the parasitoid *Diadromus pulchellus*. *Journal of Insect Behavior*, **18**, 817–26.

Thireau JC, Régnière J (1995) Development, reproduction, voltinism and host synchrony of *Meteorus trachynotus* with its hosts *Choristoneura fumiferana* and *C. rosaceana*. *Entomologia Experimentalis et Applicata*, **76**, 67–82.

Thireau JC, Régnière J, Cloutier C (1990) Biology and morphology of immature stages of *Meteorus trachynotus* Vier. (Hymenoptera: Braconidae). *Canadian Journal of Zoology*, **68**, 1000–4.

Thomas JA, Elmes GW (1993) Specialised searching and the hostile use of allomones by a parasitoid whose host, the butterfly *Maculinea rebeli*, inhabits ant nests. *Animal Behaviour*, **45**, 593–602.

Thompson SN (1974) Aspects of amino acid metabolism in the endoparasitic wasp, *Exeristes roborator* Fabricius (Hymenoptera: Ichneumonidae). *Journal of Insect Physiology*, **20**, 1515–28.

Thompson SN (1975) Defined meridic and holidic diets and asceptic feeding procedures for artificially rearing the ectoparasitoid *Exeristes roborator* (Fabricius). *Annals of the Entomological Society of America*, **68**, 220–6.

Thompson SN (1976a) The amino acid requirements for larval development of the hymenopterous parasitoid *Exeristes roborator* Fabricius (Hymenoptera: Ichneumonidae). *Comparative Biochemistry and Physiology A*, **53**, 211–13.

Thompson SN (1976b) Effects of dietary amino acid level and nutritional balance on larval survival and development of the parasite *Exeristes roborator*. *Annals of the Entomological Society of America*, **69**, 835–8.

Thompson SN (1979) Effect of dietary glucose on *in vivo* fatty acid metabolism and *in vitro* synthetase activity in the insect parasite, *Exeristes roborator* (Fabricius). *Insect Biochemistry*, **9**, 645–51.

Thompson SN (1980) Preliminary characterization of phenyloxidase activity in the salivary secretion of the parasite, Exeristes roborator. *Journal of Insect Physiology*, **26**, 505–7, 509.

Thompson SN (1982a) Characterization of the phospholipids in the insect parasite, *Exeristes roborator* (Fabricius) and the effects of dietary glucose on their distribution. *Comparative Biochemistry and Physiology (B)*, **71**, 127–31.

Thompson SN (1982b) Immediate effects of parasitization by the insect parasite, *Hyposoter exiguae*, on the nutritional physiology of its host, *Trichoplusia ni*. *Journal of Parasitology*, **68**, 936–41.

Thompson SN, Adams JD (1976) Characterization of selected lipids of the parasite *Exeristes roborator* (Fabricius). *Comparative Biochemistry and Physiology (B)*, **55**, 591–3.

Thompson SN, Barlow JS (1970) The change of fatty acid pattern of *Itoplectis conquisitor* (Say) reared on different hosts. *Journal of Parasitology*, **56**, 845–6.

Thompson SN, Barlow JS (1972) Synthesis of fatty acids by the parasite *Exeristes comstockii* (Hymenop.) and two hosts, *Galleria mellonella* (Lep.) and *Lucilia sericata* (Dip.). *Canadian Journal of Zoology*, **50**, 1105–10.

Thompson SN, Barlow JS (1973) The inconsistent phospholipid fatty acid composition in an insect parasitoid *Itoplectis conquisitor* (Say) (Lepidoptera: Pyralidae). *Comparative Biochemistry and Physiology (B)*, **44**, 59–64.

Thompson WR (1957) Hosts of the Hymenoptera (Ichneumonidae). *A Catalogue of the Parasites and Predators of Insect Pests*, **2**, 333–561.

Thompson WR, Parker HJ (1930) Morphology and biology *of Eulimneria crassifemur*, an important parasite of the corn-borer. *Journal of Agricultural Research*, **40**, 321–45.

Thorpe WH (1932) Experiments upon respiration in the larvae of certain parasitic Hymenoptera. *Proceedings of the Royal Society B: Biological Sciences*, **109**, 450–71.

Tian S, Zhang J, Yan Y, Wang C (2008) Interspecific competition between the ichneumonid *Campoletis chlorideae* and the braconid *Microplitis mediator* in their host *Helicoverpa armigera*. *Entomologia Experimentalis et Applicata*, **127**, 10–19.

Tillinger NA, Hoch G, Schopf A (2004) Effects of parasitoid associated factors of the endoparasitoid *Glyptapanteles liparidis* (Hymenoptera: Braconidae). *European Journal of Entomology*, **101**, 243–9.

Tillman PG, Powell JE (1992a) Intraspecific discrimination and larval competition among *Microplitis croceipes*, *Microplitis demolitor*, *Cotesia kazak* (Hym, Braconidae), and *Hyposoter didymator* (Hym, Ichneumonidae), parasitoids of *Heliothis virescens* (Lep, Noctuidae). *Entomophaga*, **37**, 229–37.

Tillman PG, Powell JE (1992b) Interspecific discrimination and larval competition among *Microplitis croceipes*, *Microplitis demolitor*, *Cotesia kazak* (Hym, Braconidae), and *Hyposoter didymator* (Hym, Ichneumonidae), parasitoids of *Heliothis virescens* (Lep, Noctuidae). *Entomophaga*, **37**, 439–51.

Timberlake PH (1910) Observations on the early stages of two aphidiine parasites of aphids. *Psyche*, **17**, 125–30.

Tobias VI (1962) A new subfamily of braconids (Hymenoptera, Braconidae) from Central Asia. *Trudy Zoologicheskogo Instituta, Akademia Nauk SSSR*, **30**, 268–70.

Tobias VI (1965) Generic groupings and evolution of parasitic Hymenoptera of the subfamily Euphorinae (Hymenoptera: Braconidae). I. *Entomologicheskoe Obozrenie*, **44**, 841–65 (in Russian, with English translation in *Entomological Review, Washington*, **44**, 494–508).

Tobias VI (1966) Generic groupings and evolution of parasitic Hymenoptera of the sub- family Euphorinae (Hymenoptera: Braconidae). II. *Entomologicheskoe Obozrenie*, **45**, 612–33.

(in Russian, with English translation in *Entomological Review, Washington*, **45**, 348–58).

Tobias VI (1967) Review of the system, phylogenesis and evolution of the family Braconidae (Hymenoptera). *Entomologicheskoe Obozrenie*, **46**, 465–9.

Tobias VI (1968) Voprosy klassifikatsii I filogenii sem. Braconidae (Hymenoptera). [Problems of classification and phylogeny of the family Braconidae (Hymenoptera)]. In *Chteniya Pamyati N.A. Kholodkovskogo*, Nauka, Moscow, pp. 3–43.

Tobias VI (1979) Two new and little known subfamilies of Braconidae (Hymenoptera) from Australia. *Entomologicheskoe Obozrenie*, **58**, 128–42 (in Russian).

Tobias VI (1986a) Subfamily Alysiinae. In GS Medvedev (ed.), *Handbook to the Insects of the European Part of the USSR. 3. Hymenoptera. Part 4*, Akademia Nauk SSSR, Leningrad, pp. 100–231 (in Russian).

Tobias VI (1986b) Subfamily Euphorinae. In GS Medvedev (ed.), *Handbook to the Insects of the European Part of the USSR. 3. Hymenoptera. Part 4*, Akademia Nauk SSSR, Leningrad, pp. 181–250 (in Russian).

Tobias VI (1986c) Subfamily Microgasterinae. In GS Medvedev (ed.), *Handbook to the Insects of the European Part of the USSR. 3. Hymenoptera. Part 4*, Akademia Nauk SSSR, Leningrad, pp. 344–459 (in Russian; English translation available from Amerind Publishing, New Delhi).

Tobias VI (1987) Novyi taksony brakonid (Hymenoptera, Braconidae) izbaltijskogo jantarja. *Entomologicheskoe Obozrenie*, **66**, 845–59 (in Russian, with English translation in *Entomological Review* 67, 18–32).

Tobias VI (1988) A new subfamily of braconids (Hymenoptera, Braconidae) from Australia. *Entomologicheskoe Obozrenie*, **67**, 644–8.

Tobias VI (1989) Application of W. Hennig's method of phylogenetic analysis for the construction of the phylogenetic tree of the family Braconidae (Hymenoptera). *Trudy Zoologicheskogo Instituta, Leningrad*, **202**, 67–83 (in Russian).

Tobias VI (2002) Male swarming in parasitic braconid wasps (Hymenoptera, Braconidae) and its significance. *Entomologicheskoe Obozrenie*, **81**, 3–10, 259 (in Russian).

Tobias VI, Belokobylskij SA (1981) Braconid genera (Hymenoptera, Braconidae) new to science and to the fauna of the USSR from the Maritime territory. *Entomologicheskoye Obozreniye*, **60**, 354–363 (in Russian, with English translation in *Entomological Review*, **60**, 75–86).

Tobias VL, Kyriak IG (1971) The parasite *Areopraon pilosum* Mackauer, 1959, and aspects of the phylogeny and evolution of the family Aphidiidae (Hymenoptera). *Entomological Review*, **50**, 6–8.

Tobias VI, Potapova ES (1982) Morphological peculiarities of the head capsule in braconids (Hymenoptera, Braconidae) and main trends of its evolution. *Entomologicheskoe Obozrenie*, **61**, 30–42 (in Russian).

Togashi I (1963) A comparative morphology of the poison glands in the adults of ichneumon-flies (Hymenoptera: Ichneumonidae) (1). *Kontyû, Tokyo*, **31**, 297–304 (in Japanese, with English summary).

Tomberlin JK, Rains GC, Sanford MR (2008) Development of *Microplitis croceipes* as a biological sensor. *Entomologia Experimentalis et Applicata*, **128**, 249–57

Tormos J, Pardo X, Jiménez R, Asís JD, Gayubo SF (2003) Descriptions of adults, immature stages and venom apparatus of two new species of Dacnusini: *Chorebus pseudoasphodeli* sp. n., parasitic on *Phytomyza chaerophili* Kaltenbach and *C. pseudoasramenes* sp. n., parasitic on *Cerodontha phragmitophila* Hering (Hymenoptera: Braconidae: Alysiinae; Diptera: Agromyzidae). *European Journal of Entomology*, **100**, 393–400.

Tothill JD (1922) The natural control of the fall webworm (*Hyphantria cunea* drury) with an account of its several parasites. *Bulletin of the Department of Agriculture of Canada, Entomology Branch*, **19**, 1–107.

Townes H[1] (1939) Protective odors among the lchneumonidae (Hymenoptera). *Bulletin of the Brooklyn Entomological Society*, **34**, 29–30.

Townes H (1944) A catalogue and reclassification of the Nearctic Ichneumonidae (Hymenoptera). Part 1. *Memoirs of the American Entomological Society*, **11**, 1–477.

Townes H (1957) A revision of the genera of Poemeniini and Xoridini (Hymenoptera, Ichneumonidae). *Proceedings of the Entomological Society of Washington*, **59**, 15–23.

Townes H (1958) Some biological characteristics of the Ichneumonidae (Hymenoptera) in relation to biological control. *Journal of Economic Entomology*, **51**, 650–2.

Townes H (1963) Evidence among the ephialtine Ichneumonidae for a Bering migration route. In JL Gressitt (ed.), *Pacific Basin Biogeography*, Bishop Museum Press, Honolulu.

Townes H (1965) A new *Cremastus* (Ichneurnonidae), parasitic on *Oberea*. *Papers of the Michigan Academy of Science, Arts, and Letters*, **50**, 105–6.

Townes H (1966) Notes on three fossil genera of Ichneumonidae (Hymenoptera). *Proceedings of the Entomological Society of Washington*, **68**, 132–5.

Townes H (1969) The genera of Ichneumonidae, part 1. Ephialtinae to Agriotypinae. *Memoirs of the American Entomological Institute*, **11**, 1–300.

Townes H (1970a) The genera of Ichneumonidae, part 2. Gelinae. *Memoirs of the American Entomological Institute*, **12**, 1–537.

Townes H (1970b) The genera of Ichneumonidae, part 3. Banchinae, Scolobatinae & Porizontinae. *Memoirs of the American Entomological Institute*, **13**, 1–307.

[1]In his publications, Henry Keith Townes variously called himself just by his first name, but sometimes used his middle initial. For simplicity he is referred to throughout here as Townes H.

Townes H (1971) The genera of Ichneumonidae, part 4. Cremastinae to Diplazontinae. *Memoirs of the American Entomological Institute*, **17**, 1–372.

Townes HK (1973a) Two ichneumonids from the early Cretaceous. *Proceedings of the Entomological Society of Washington*, **75**, 216–19.

Townes H (1973b) Three tryphonine ichneumonids from Cretaceous amber (Hymenoptera). *Proceedings of the Entomological Society of Washington* **75**, 282–7.

Townes H (1975) The parasitic Hymenoptera with the longest ovipositors, with descriptions of two new Ichneumonidae. *Entomological News*, **86**, 123–7.

Townes H, Townes M (1951) Ichneumonidae. In CFW Muesebeck, KV Krombein and H Townes (eds.), *Hymenoptera of America North of Mexico – Synoptic Catalog*, United States Department of Agriculture, Agricultural Monograph 2, pp. 184–409.

Townes H, Townes M (1959) Ichneumon-flies of America north of Mexico: 1. Subfamily Metopiinae. *United States National Museum Bulletin*, **216**, 1–318.

Townes H, Townes M (1960) Ichneumon-flies of America north of Mexico: 2. Subfamilies Ephialtinae, Xoridinae, Acaenitinae. *Bulletin of the United States National Museum*, **216**, 1–676.

Townes H, Townes M (1966) A catalogue and reclassification of the Neotropic Ichneumonidae. *Memoirs of the American Entomological Institute*, **8**, 1–367.

Townes H, Townes M (1978) Ichneumon-flies of America North of Mexico: 7. Subfamily Banchinae, tribes Lissonotini and Banchini. *Memoirs of the American Entomological Institute*, **26**, 1–614.

Townsend AC, Shaw SR (2009) A new species of *Andesipolis* Whitfield & Choi from the eastern Andes of Ecuador with notes on biology and classification (Hymenoptera: Braconidae: Rhysipolinae). *Journal of Insect Science*, **9**, 36.

Trang TTK, Dey D (2013) External morphology of antennal sensilla in relation to electroantennogram respomses of *Apanteles angaleti* Muesebeck parasitoids (Hymenoptera: Braconidae) for host searching behaviour. *Omnirice*, **19**, 97–117.

Traynor RE, Mayhew PJ (2005) Host range in solitary versus gregarious parasitoids: a laboratory experiment. *Entomologia Experimentalis et Applicata*, **117**, 41–9.

Tremblay E (1964) Richerche sugli imenotteri parasiti. I. Studio morfo-biologico sul *Lysiphlebus fabarum* (Marshall) (Hymenoptera: Braconidae: Aphidiinae). *Bollettino del Laboratorio Entomologia Agraria "Filippo Silvestri"*, **22**, 1–122.

Tremblay E (1979) Le armature genitali maschili mi generi degli Aphidiinae (Hym., Braconidae). I. *Bollettino del Laboratorio Entomologia Agraria "Filippo Silvestri"*, **36**, 221–33.

Tremblay E (1981) Le armature genitali maschili mi generi degli Aphidiinae (Hym., Braconidae). II. *Bollettino del Laboratorio Entomologia Agraria "Filippo Silvestri"*, **38**, 251–63.

Tremblay E (1983) Le armature genitali maschili mi generi degli Aphidiinae (Hym., Braconidae). II. *Bollettino del Laboratorio Entomologia Agraria "Filippo Silvestri"*, **40**, 147–58.

Tremblay E, Calvert D (1971) Embryosytematics in the aphidiines (Hymenoptera: Braconidae). *Bollettino del Laboratorio Entomologia Agraria "Filippo Silvestri"*,**29**, 223–49.

Tremblay E, Iaccarino FM (1971) Notizie sull'ultrastruttura dei trofociti di Aphidius matricariae Hal. (Hymenoptera: Braconidae). *Bollettino del Laboratorio Entomologia Agraria "Filippo Silvestri"*, **29**, 305–13.

Tremblay E, Pennacchio F, Digilio MC (1998) Host regulation effects of ovary fluid and venom of *Aphidius ervi* (Hymenoptera: Braconidae). *Journal of Insect Physiology*, **44**, 779–84.

Tripp HA (1961) The biology of a hyperparasite *Euceros frigidus* Cress. (Ichneumonidae) and description of the planidial stage. *Canadian Entomologist*, **93**, 40–58.

Tschopp A, Riedel M, Kropf C, Nentwig W, Klopfstein S (2013) The evolution of host associations in the parasitic wasp genus *Ichneumon* (Hymenoptera: Ichneumonidae): convergent adaptations to host pupation sites. *BMC Evolutionary Biology*, **13**, 74.

Tsutsui Y, Maeto K, Hamaguchi K, Isaki Y, Takami Y, Naito T, Miura K (2014) Apomictic parthenogenesis in a parasitoid wasp *Meteorus pulchricornis*, uncommon in the haplodiploid order Hymenoptera. *Bulletin of Entomological Research*, **104**, 307–13.

Tucker E, Sharkey MJ, Stoelb S (2012) A new species-group of *Camptothlipsis* (Braconidae, Agathidinae) from South Africa, with notes on the evolution of long mouthparts. *Journal of Hymenoptera Research*, **24**, 59–74.

Turnbull M, Webb B (2002) Perspectives on polydnavirus origins and evolution. *Advances in Virus Research*, **58**, 203–54.

Turlings TCJ, van Batenburg FDH, van Strien-van Liempt WTFH (1985) Why is there no interspecific host discrimination in the two coexisting larval parasitoids of *Drosophila* species; *Leptopilina heterotoma* (Thomson) and *Asobara tabida* (Nees). *Oecologia (Berlin)*, **67**, 352–9.

Turlings TCJ, Tumlinson JH, Heath RP, Proveaux AT, Doolittle RE (1991) Isolation and identification of allelochemicals that attract the larval parasitoid, *Cotesia marginiventris* (Cresson), to the microhabitat of one of its hosts. *Journal of Chemical Ecology*, **17**, 2235–51.

Uçkan F (1999) The morphology of the venom apparatus and histology of venom gland of *Pimpla turionella* (L.) (Hym; Ichneumonidae) females. *Turkish Journal of Zoology*, **23**, 461–6.

Uçkan F, Sinan S, Savasci S, Ergin E (2004) Determination of venom components from the endoparasitoid wasp *Pimpla turionellae* L. (Hymenoptera: Ichneumonidae). *Annals of the Entomological Society of America*, **97**, 775–80.

Uçkan F, Ergin E, Rivers DB, Gençer N (2006) Age and diet influence the composition of venom from the endoparasitic wasp *Pimpla turionellae* L. (Hymenoptera: Ichneumonidae). *Archives of Insect Biochemistry and Physiology*, **63**, 177–87.

Uefune M, Kugimiya S, Shimoda T, Takabayashi J (2013) Starvation and herbivore-induced plant volatiles affect the color preferences of parasitic wasps. *BioControl*, **58**, 187–93.

Udayagiri S, Jones RL (1992) Role of plant odour in parasitism of European corn borer by braconid specialist parasitoid *Macrocentrus grandii* Goidanich: isolation and characterisation of plant synomones eliciting parasitoid flight response. *Journal of Chemical Ecology*, **18**, 1841–55.

Ueno T (1994) Self-recognition by the parasitic wasp *Itoplectis naranyae* (Hymenoptera: Ichneumonidae). *Oikos*, **70**, 333–9.

Ueno T (1995) Abdominal tip movements during oviposition by two parasitoids (Hymenoptera: Ichneumonidae) as an index predicting the sex of depositing eggs. *Applied Entomology and Zoology*, **30**, 588–90.

Ueno T (1997) Effects of superparasitism, larval competition, and host feeding on offspring fitness in the parasitoid *Pimpla nipponica* (Hymenoptera: Ichneumonidae). *Annals of the Entomological Society of Amererica*, **90**, 682–8.

Ueno T (1998) Selective host-feeding on parasitized hosts by the parasitoid *Itoplectis naranyae* (Hymenoptera: Ichneumonidae) and its implications for biological control. *Bulletin of Entomological Research*, **88**, 461–6.

Ueno T (1999a) Reproduction and host-feeding in the solitary parasitoid wasp *Pimpla nipponica* (Hymenoptera: Ichneumonidae). *Invertebrate Reproduction and Development*, **35**, 231–7.

Ueno T (1999b) Multiparasitism and host feeding by solitary parasitoid wasps (Hymenoptera: Ichneumonidae) based on the pay-off from parasitised hosts. *Annals of the Entomological Society of Amererica*, **92**, 601–8.

Ueno T (2002) Biology of the ectoparasitoid wasp *Agrothereutes lanceolatus* (Hymenoptera: Ichneumonidae): host acceptance and larval development on a laboratory host. *Journal of the Faculty of Agriculture Kyushu University*, **47**, 37–43.

Ueno T, Tanaka T (1994) Comparative biology of six polyphagous solitary pupal endoparasitoids (Hymenoptera: Ichneumonidae): differential host suitability and sex allocation. *Annals of the Entomological Society of America*, **87**, 592–8.

Ueno T, Ueno K (2007) The effects of host-feeding on synovigenic egg development in an endoparasitic wasp, *Itoplectis naranyae*. *Journal of Insect Science*, **7**, 1–13.

Ullyett GC (1944) On the function of the caudal appendage in primary larvae of parasitic Hymenoptera. *Journal of the Entomological Society of South Africa*, **7**, 30–7.

Ullyett GC (1949) Distribution of progeny of *Chelonus texanus* Cress. (Hymenoptera: Braconidae). *The Canadian Entomologist*, **81**, 25–44.

Unruh TR, White W, Gonzalez D, Gordh G, Luck RF (1983) Heterozygosity and effective size in laboratory populations of *Aphidius ervi* (Hym.: Aphidiidae). *BioControl*, **28**, 245–58.

Urano T, Hijii N (1995) Resource utilization and sex allocation in response to host size in 2 ectoparasitoid wasps on subcortical beetles. *Entomologia Experimentalis et Applicata*, **74**, 23–35.

Vagina NP (1982) Morphology of the neuroendocrine system of *Alysia manducator* Panz. (Hymenoptera, Braconidae). *Entomological Review*, **66**, 82–90.

Vagina NP (1987) Role of neurosecretory cells of the brain in regulation of larval development of *Alysia manducator* Panz. (Hymenoptera, Braconidae) larvae. *Entomologicheskogo Obozrenie*, **61**, 245–51 (in Russian).

van Achterberg C (1974) The features of the petiolar segment in some Braconidae. *Entomologische Berichten, Amsterdam*, **34**, 21–3.

van Achterberg C (1975). A new genus, *Mesostoa* gen. nov., from W. Australia, belonging to a new subfamily (Hymenoptera: Braconidae). *Entomologische Berichten, Amsterdam*, **35**, 158–60.

van Achterberg C (1976a) A revision of the tribus Blacini (Hymenoptera, Braconidae, Helconinae). *Tijdschrift voor Entomologie*, **118**, 159–322.

van Achterberg C (1976b) Hybrizontinae or Hybrizontidae? (Hymenoptera, Ichneumonoidea). *Entomologische Berichten, Amsterdam*, **36**, 61–4.

van Achterberg C (1976c) A preliminary key to the subfamilies of the Braconidae (Hym.). *Tijdschrift voor Entomologie*, **119**, 33–78.

van Achterberg C (1977) The function of swarming in *Blacus* species (Hymenoptera, Braconidae, Helconinae). *Entomologische Berichten, Amsterdam*, **37**, 151–2.

van Achterberg C (1979a) A revision of the species of Amicrocentrinae, a new subfamily (Hym., Braconidae), with a description of the final instar of *Amicrocentrum curvinervis* by J. R. T. Short. *Tijdschrift voor Entomologie*, **122**, 1–28.

van Achterberg C (1979b) A revision of the new subfamily Xiphozelinae (Hym., Braconidae). *Tijdschrift voor Entomologie*, **122**, 29–46.

van Achterberg C (1979c) A revision of the subfamily Zelinae auct. (Hymenoptera, Braconidae). *Tijdschrift voor Entomologie*, **122**, 241–479.

van Achterberg C (1981a) A simple separation bag, a device to sort swept insects. *Entomologische Berichten*, **41**, 24–5.

van Achterberg C (1981b) The position of the genus *Rhamnura* Enderlein among Braconidae (Hymenoptera). *Entomologische Berichten, Amsterdam*, **41**, 88–91.

van Achterberg C (1982a) The fossil species of the subfamily Blacinae described by C. T. Brues (Hym.: Braconidae). *Entomologische Berichten, Amsterdam*, **42**, 91–6.

van Achterberg C (1982b) A new genus of the Rogadinae-Lysitermini from Kazakhstan (Hym.: Braconidae). *Entomologische Berichten, Amsterdam*, **42**, 125–8.

van Achterberg C (1983a) Revisionary notes on the Palaearctic genera and species of the tribe Exothecini Foerster (Hymenoptera, Braconidae). *Zoologische Mededelingen*, **57**, 339–55.

van Achterberg C (1983b) A revision of the new tribe Brulleiini (Hymenoptera: Braconidae). *Contributions of the American Entomological Institute*, **20**, 281–306.

van Achterberg C (1983c) Revisionary notes on the subfamily Gnaptodontinae, with description of 11 new species (Hymenoptera, Braconidae). *Tijdschrift voor Entomologie*, **126**, 25–57.

van Achterberg C (1984a) Essay on the phylogeny of Braconidae (Hymenoptera: Ichneumonoidea). *Entomologisk Tidskrift*, **105**, 41–58.

van Achterberg C (1984b) Revision of the genera of Braconini with first and second metasomal tergites immovably joined (Hymenoptera, Braconidae, Braconinae). *Tijdschrift voor Entomotogie*, **127**, 137–64.

van Achterberg C (1985a) Notes on Braconidae V. The systematic position of the genera *Ecnomios* Mason and *Pselaphanus* Szépligeti (Hymenoptera: Braconidae). *Zoologische Mededelingen, Leiden*, **59**, 341–8.

van Achterberg C (1985b) Notes on Braconidae VI. The genera and subgenera of Centistini, with the description of two new taxa from the Nearctic region (Hymenoptera: Braconidae: Euphorinae). *Zoologische Mededelingen, Leiden*, **59**, 348–62.

van Achterberg C (1986) The oviposition behaviour of parasitic Hymenoptera with very long ovipositors (Ichneumonoidea: Braconidae). *Entomologische Berichten, Amsterdam*, **46**, 113–15.

van Achterberg C (1987) Revisionary notes on the subfamily Orgilinae (Hymenoptera: Braconidae). *Zoologische Verhandelingen, Leiden*, **242**, 1–111.

van Achterberg C (1988a) The genera of the *Aspilota*-group and some descriptions of fungicolous Alysiini from the Netherlands (Hymenoptera, Braconidae: Alysiinae). *Zoologisches Verhandelingen, Leiden*, **247**, 1–88.

van Achterberg C (1988b) Parallelisms in the Braconidae (Hymenoptera) with special reference to the biology. In VK Gupta (ed.), *Advances in Parasitic Hymenoptera Research*, E.J. Brill, Leiden, pp. 85–115.

van Achterberg C (1988c) Revision of the subfamily Blacinae Foerster (Hymenoptera, Braconidae). *Zoologisches Verhandelingen, Leiden*, **249**, 1–324.

van Achterberg C (1989) *Pheloura* gen. nov., a neotropical genus with an extremely long pseudo-ovipositor (Hymenoptera: Braconidae). *Entomologische Berichten, Amsterdam*, **49**, 105–8.

van Achterberg C (1990a) Revision of the subtribe Pseudohelconina nov. (Hymenoptera: Braconidae). *Zoologische Mededelingen, Leiden*, **63**, 283–308.

van Achterberg C (1990b) Revision of the subtribe Mesocoelina Viereck (Hymenoptera: Braconidae). *Zoologische Mededelingen, Leiden*, **64**, 31–57.

van Achterberg C (1990c) *Pronkia*, a new genus of Meteorideinae (Hymenoptera: Braconidae) from New Zealand. *Zoologische Mededelingen, Leiden*, **64**, 169–75.

van Achterberg C (1990d) Illustrated key to the subfamilies of the Holarctic Braconidae (Hymenoptera: Ichneumonoidea). *Zoologische Mededelingen, Leiden*, **64**, 1–20,.

van Achterberg C (1991a) Revision of the genera of the Afrotropical and W. Palaeartical Rogadinae Foerster (Hymenoptera: Braconidae). *Zoologische Verhandelingen, Leiden*, **273**, 1–102.

van Achterberg C (1991b) The taxonomic position of the genus *Argamania* Papp (Hymenoptera: Braconidae). *Zoologische Mededelingen, Leiden*, **65**, 203–7.

van Achterberg C (1992a) Revision of the genus *Histeromerus* Wesmael (Hymenoptera: Braconidae). *Zoologische Mededelingen, Leiden*, **66**, 189–96.

van Achterberg C (1992b) Revisionary notes on the subfamily Homolobinae (Hymenoptera: Braconidae). *Zoologische Mededelingen, Leiden*, **66**, 359–68.

van Achterberg C (1992c) Revision of the genera of the subfamily Microtypinae (Hymenoptera: Braconidae). *Zoologische Mededelingen, Leiden*, **66**, 369–80.

van Achterberg C (1992d) Revision of the genus *Pygostolus* Haliday (Hymenoptera: Braconidae: Euphorinae), with a key to the Holarctic species. *Zoologische Mededelingen, Leiden*, **66**, 349–58.

van Achterberg C (1993a) Illustrated key to the subfamilies of the Braconidae (Hymenoptera: Ichneumonoidea). *Zoologisches Verhandelingen, Leiden*, **283**, 1–189.

van Achterberg C (1993b) Revision of the subfamily Macrocentrinae Föerster (Hymenoptera, Braconidae) from the Palaearctic region. *Zoologische Verhandelingen, Leiden*, **286**, 1–110.

van Achterberg C (1994a) Generic revision of the subfamily Cenocoeliinae Szépligeti (Hymenoptera: Braconidae). *Zoologisches Verhandelingen, Leiden*, **292**, 1–52.

van Achterberg C (1994b) The Palaearctic species of the genus *Chrysopophthorus* Goidanich (Hymenoptera: Braconidae: Euphorinae). *Zoologische Mededelingen, Leiden*, **68**, 301–7.

van Achterberg C (1995) Generic revision of the subfamily Bethylobraconinae (Hymenoptera: Braconidae) and other groups with modified fore tarsus. *Zoologisches Verhandelingen, Leiden*, **298**, 1–242.

van Achterberg C (1997a) *Braconidae. An Illustrated Key to all Subfamilies*, ETI World Biodiversity Database CD-ROM Series, ETI Bioinformatics, Amsterdam.

van Achterberg C (1997b) *Vervoortihelcon*, a new genus of the subfamily Helconinae Foerster (Hymenoptera: Braconidae) from Chile. *Zoologisches Verhandelingen, Leiden*, **323**, 401–5.

van Achterberg C (1997c) Subfamily Aphidiinae s.s. In RA Wharton, PM Marsh and MJ Sharkey (eds), *Identification Manual to the New World Genera of Braconidae*, Special Publication of the International Society of Hymenopterists 1, pp. 119–131.

van Achterberg C (1997d) Subfamily Orgilinae s.s. In RA Wharton, PM Marsh and MJ Sharkey (eds), *Identification Manual to the New World Genera of Braconidae*, Special

Publication of the International Society of Hymenopterists 1, pp. 397–400.
van Achterberg C (1998) *Phaenocarpa helophilae* spec. nov. (Hymenoptera: Braconidae), a gregarious parasitoid of *Helophilus* larvae (Diptera: Syrphidae), with notes on related species. *Zoologische Mededelingen, Leiden*, **72**, 15–27.
van Achterberg C (1999) The West Palaearctic species of the subfamily Paxylommatinae (Hymenoptera: Ichneumonidae), with special reference to the genus Hybrizon Fallén. *Zoologische Mededelingen, Leiden*, **73**, 11–26.
van Achterberg C (2000) *Carinichremylus* gen. nov. (Hymenoptera: Braconidae: Pambolinae) from Peru. *Zoologische Mededelingen, Leiden*, **73**, 251–4.
van Achterberg C (2001) The first known fossil Masoninae (Hymenoptera: Braconidae) from Miocene Dominican amber. *Zoologische Mededelingen, Leiden*, **75**, 393–6.
van Achterberg C (2002) *Apanteles* (*Choeras*) *gielisi* spec. nov. (Hymenoptera: Braconidae: Microgastrinae) from The Netherlands and the first report of Trichoptera as host of Braconidae. *Zoologische Mededelingen, Leiden*, **76**, 53–60.
van Achterberg C (2003) Western Palaearctic genera of the subfamily Microgastrinae: a re-appraisal of the generic and tribal division (Hymenoptera: Braconidae). In G Melika and C Thuróczy (eds), *Parasitic Wasps: Evolution, Systematics, Biodiversity and Biological Control*, Agroinform Kiadó, Budapest, pp. 19–35.
van Achterberg C (2008) A new secies of *Xiphozele* Cameron (Hymenoptera: Braconidae) from South Vietnam. *Zoologische Mededelingen, Leiden*, **82**, 1–8.
van Achterberg C, Altenhofer E (1997) *Xyeloblacus* gen. nov. (Hymenoptera: Braconidae: Blacinae) parasitoid of Xyelinae (Xyelidae: Hymenoptera). *Zoologische Mededelingen, Leiden*, **71**, 291–8.
van Achterberg C, Altenhofer E (2013) Notes on the biology of *Seleucus cuneiformis* Holmgren (Hymenoptera, Ichneumonidae, Ctenopelmatinae). *Journal of Hymenoptera Research*, **31**, 97–104.
van Achterberg C, Argaman Q (1993) *Kollasmosoma* gen. nov. and a key to the genera of the subfamily Neoneurinae (Hymenoptera: Braconidae). *Zoologische Mededelingen, Leiden*, **67**, 63–74.
van Achterberg C, Austin AD (1992) Revision of the genera of the subfamily Sigalphinae (Hymenoptera: Braconidae), including a revision of the Australian species. *Zoologische Verhandelingen* **280**, 1–44.
van Achterberg C, Berry J (2004) *Neptihormius* gen. nov. (Hymenoptera: Braconidae: Hormiinae), a parasitoid of Nepticulidae (Lepidoptera) from New Zealand. *Zoologische Mededelingen, Leiden*, **78**, 291–9.
van Achterberg C, Desmier de Chenon R (2009) The first report of the biology of *Proterops borneoensis* Szépligeti (Hymenoptera: Braconidae: Ichneutinae), with the description of a new species from China. *Journal of Natural History*, **43**, 619–33.
van Achterberg C, Haeselbarth E (2003) Revision of the genus *Syntretus* Foerster (Hymenoptera: Braconidae: Euphorinae) from Europe. *Zoologische Mededelingen, Leiden*, **77**, 9–78.
van Achterberg C, Prinsloo GL (2012) Braconidae (Hymenoptera: Opiinae, Alysiinae) reared from aquatic leaf-mining Diptera on *Lagarosiphon major* (Hydrocharitaceae) in South Africa. *African Entomology*, **20**, 124–33.
van Achterberg C, Quicke DLJ (1990) *Afrocampsis*, a new genus belonging to the Sigalphinae (Hymenoptera: Braconidae) from the Afrotropical region. *Zoologische Mededelingen, Leiden*, **64**, 1550–60.
van Achterberg C, Quicke DLJ (1991) A new genus of Braconinae with depressed ovipositor-tip from the Oriental region (Hymenoptera: Braconidae). *Zoologische Mededelingen, Leiden*, **64**, 199–202.
van Achterberg C, Quicke DLJ (1992) Phylogeny of the subfamilies of the family Braconidae. A reassessment assessed. *Cladistics*, **8**, 237–64.
van Achterberg C, Quicke DLJ (2000) The palaeotropical species of the tribe Cosmophorini (Hymenoptera: Braconidae: Euphorinae) with descriptions of twenty-two new species. *Zoologische Mededelingen, Leiden*, **74**, 283–338.
van Achterberg C, Shaw SR (2001) *Tainiterna*, A new genus of the subfamily Euphorinae (Hymenoptera: Braconidae) from Vietnam and China. *Zoologische Mededelingen, Leiden*, **75**, 69–78.
van Achterberg C, Steiner H (1996) A new genus of Tetratermini (Hymenoptera: Braconidae: Lysiterminae) parasitic on grasshoppers (Gryllacrididae). *Zoologische Mededelingen, Leiden*, **70**, 249–59.
van Achterberg C, Weiblen GD (2000) *Ficobracon brusi* gen. nov. & spec. nov. (Hymenoptera: Braconidae), a parasitoid reared from figs in Papua New Guinea. *Zoologische Mededelingen, Leiden*, **74**, 51–5.
van Achterberg C, Kenis M, Sileshi G (2000) *Perilitus* (*Microctonus*) *larvicida* spec. nov. (Hymenoptera: Braconidae: Euphorinae) from Zambia. *Zoologische Mededelingen, Leiden*, **74**, 57–62.
van Alphen JJM, Nell HW (1982) Superparasitism and host discrimination by *Asobara tabida* Nees (Braconidae, Alysiinae), a larval parasitoid of *Drosophila*. *Netherlands Journal of Zoology*, **32**, 232–60.
van Alphen JJM, Visser ME (1990) Superparasitism as an adaptive strategy for insect parasitoids. *Annual Review of Entomology*, **35**, 59–79.
Van Baarlen P, Topping CJ, Sunderland KD (1996) Host location by *Gelis festinans*, an eggsac parasitoid of the linyphiid spider *Erigone atra*. *Entomologia Experimentalis et Applicata*, **81**, 155–63.
Vance AM (1927) On the biology of some ichneumonids of the genus *Paniscus* Schrk. *Annals of the Entomological Society of America*, **20**, 405–16.
Vance AM (1931) *Apanteles thompsoni* Lyle, a braconid parasite of the European corn borer. *Technical Bulletin, United States Department of Agriculture*, No. 233, 28 pp.

Vance AM (1932a) The biology and morphology of the braconid *Chelonus annulipes* Wesm., a parasite of the European corn borer. *Technical Bulletin, United States Department of Agriculture*, No. 294, 48 pp.

Vance AM (1932b) *Microgaster tibialis* Nees as a hymenopterous parasite of *Pyrausta nubilalis* Hubn. [sic] in Europe. *Annals of the Entomological Society of America*, **25**, 121–34.

van den Bosch R, Messenger PS, Gutiérrez AP (1982) *An Introduction to Biological Control*, Plenum Press, New York, 247 pp.

van der Ent L-J, Shaw SR (1998) Species richness of Costa Rican Cenocoeliini (Hymenoptera: Braconidae): a latitudinal and altitudinal search for anomalous diversity. *Journal of Hymenoptera Research*, **7**, 15–24.

Van Driesche RG. (2008). Biological control of *Pieris rapae* in New England: host suppression and displacement of *Cotesia glomerata* by *Cotesia rubecula* (Hymenoptera: Braconidae). *Florida Entomologist*, **91**, 22–5.

van Emden HF, Soreck AP, Douloumpaka S, Eleftherianos I, Poppy GM, Powell W (2008) Plant chemistry and aphid parasitoids (Hymenoptera: Braconidae): imprinting and memory. *European Journal of Entomology*, **105**, 477–83.

van Lenteren JC, Isidoro N, Bin F (1998) Functional anatomy of the ovipositor clip in the parasitoid *Leptopilina heterotoma* (Thompson) (Hymenoptera: Eucoilidae), a structure to grip escaping larvae. *International Journal of Insect Morphology and Embryology*, **27**, 263–8.

van Marle J (1977) Structure and histochemistry of the venom glands of the wasps *Microbracon hebetor* Say and *Philanthus triangulum* F. *Toxicon*, **15**, 529–39.

van Noort S (2004) Ichneumonid (Hymenoptera: Ichneumonoidea) diversity across an elevational gradient on Monts Doudou in southwestern Gabon. *California Academy of Science Memoir*, **28**, 187–216.

van Nouhuys S, Hanski I (2000) Apparent competition between parasitoids mediated by a shared hyperparasitoid. *Ecology Letters*, **3**, 82–4.

van Nouhuys S, Punju E (2010) Coexistence of competing parasitoids: which is the fugitive and where does it hide? *Oikos*, **119**, 61–70.

van Santen MCP, Schneider MV (2002) Courtship description in the solitary parasitoid wasp, *Venturia canescens* (Hymenoptera). *Proceedings of the Section Experimental and Applied Entomology of the Netherlands Entomological Society*, **13**, 87–92.

van Steenis MJ (1994) Intrinsic rate of increase of *Lysiphlebus testaceipes* Cresson (Hymenoptera, Braconidae), a parasitoid of *Aphis gossypii* Glover (Hem, Aphididae) at different temperatures. *Zeitschrift für Angewandte Entomologie*, **118**, 399–406.

van Veen FJF, Morris RJ, Godfray HCJ (2006) Apparent competition, quantitative food webs, and the structure of phytophagous insect communities. *Annual Review of Entomology*, **51**, 187–208.

van Veen JC (1981) The biology of *Poecilostictus cothurnatus* (Hymenoptera, Ichneumonidae) an endoparasite of *Bupalus piniarius* (Lepidoptera, Geometridae). *Annales Entomologici Fennici*, **47**, 77–93.

van Veen JC (1982) Notes on the biology of *Banchus femoralis* Thomson (Hym., Ichneumonidae) an endoparasitoid of *Panolis flammea* (D. & S.) (Lep., Noctuidae). *Zeitschrift für Angewandte Entomologie*, **94**, 300–11.

Várkonyi G, Roslina T (2013) Freezing cold yet diverse: dissecting a high-Arctic parasitoid community associated with Lepidoptera hosts. *The Canadian Entomologist*, **145**, 193–218.

Varley GC (1965) A note on the life-history of the ichneumon-fly *Euceros unifasciatus* Voll. with a description of its planidium larva. *Entomologist's Monthly Magazine*, **100**, 113–16.

Vavre F, Fleury F, Lepetit D, Fouillet P, Boulétreau M (1999) Phylogenetic evidence for horizontal transmission of *Wolbachia* in host-parasitoid associations. *Molecular Biology and Evolution*, **16**, 1711–23.

Veijalainen A, Sääksjärvi IE, Erwin TL, Gómez IC, Longino JT (2012a) Subfamily composition of Ichneumonidae (Hymenoptera) from western Amazonia: insights into diversity of tropical parasitoid wasps. *Insect Conservation and Diversity*, **6**, 28–37.

Veijalainen A, Wahlberg N, Broad GR, Erwin TL, Longino JT, Sääksjärvi IE (2012b) Unprecedented ichneumonid parasitoid wasp diversity in tropical forests. *Proceedings of the Royal Society B: Biological Sciences*, **279**, 4694–8.

Veijalainen A, Saaksjarvi I, Tuomisto H, Broad G, Bordera S, Jussila R (in press) Altitudinal trends in species richness and diversity of Mesoamerican parasitoid wasps (Hymenoptera: Ichneumonidae). *Insect Conservation and Diversity*, doi: 10.1111/icad.1(2073).

Vet LEM (1999) Evolutionary aspects of plant–carnivore interactions. In DJ Chadwick and JA Goode (eds), *Insect–Plant Interactions and Induced Plant Defence. Novartis Foundation Symposium 223*, John Wiley & Sons, Ltd, Chichester, pp. 3–13.

Vet LEM, Meyer M, Bakker K, van Alphen JJM (1984) Intra- and interspecific host discrimination in *Asobara* (Hymenoptera) larval endoparasitoids of Drosophilidae: comparison between closely related and less closely related species. *Animal Behaviour*, **32**, 871–4.

Vet LEM, Wäckers FL, Dicke M (1991) How to hunt for hiding hosts: the reliability- detectability problem in foraging parasitoids. *Netherlands Journal of Zoology*, **41**, 202–13.

Vickery RA (1929) Studies on the fall armyworm in the Gulf Coast district of Texas. *Technical Bulletin of the United States Department of Agriculture*, **138**, 1–63.

Viereck HL (1914) Type species of the genera of ichneumon flies. *Bulletin of the United States National Museum*, **83**, 1–186.

Vikberg V (1999) *Lathroplex anthreni* sp n. (Hymenoptera, Ichneumonidae), a parasitoid on *Anthrenus museorum*

(Linnaeus) in southern Finland. *Entomologica Fennica*, **10**, 239–44.

Vikberg V, Koponen M (2000) On the taxonomy of *Seleucus* Holmgren and the European species of Phrudinae (Hymenoptera: Ichneumonidae). *Entomologica Fennica*, **11**, 195–228.

Vilhelmsen L (1997) Head capsule concavities accommodating the antennal bases in hymenoptera pupating in wood: possible emergence-facilitating adaptations. *International Journal of Insect Morphology*, **2**, 129–38.

Vilhelmsen L (2003) Flexible ovipositor sheats in parasitoid Hymenoptera (Insecta). *Arthropod Structure and Development*, **32**, 277–87.

Vilhelmsen L, Turrisi GF (2011) *Per arborem ad astra*: morphological adaptations to exploiting the woody habitat in the early evolution of Hymenoptera. *Arthropod Structure and Development*, **40**, 2–20.

Vilhelmsen L, Isidoro N, Romani R, Basibuyuk HH, Quicke DLJ (2001) Host location and oviposition in a basal group of parasitic wasps: the subgenual organ, ovipositor apparatus, and associated structures in the Orussidae (Hymenoptera, Insecta). *Zoomorphology*, **121**, 63–84.

Villagra CA, Vásquez RA, Niemeyer HM (2005) Associative odour learning affects mating behaviour in *Aphidius ervi* males. *European Journal of Entomology*, **102**, 557–9.

Villagra CA, Pinto CF, Penna M, Niemeyer HM (2011) Male wing fanning by the aphid parasitoid *Aphidius ervi* (Hymenoptera: Braconidae) produces a courtship song. *Bulletin of Entomological Research*, **101**, 573–9.

Vincent B, Kaeslin M, Roth T, Heller M, Poulain J, Cousserans F, Schaller J, Poirié M, Lanzrein B, Drezen J-M, Moreau SJ (2010) The venom composition of the parasitic wasp *Chelonus inanitus* resolved by combined expressed sequence tags analysis and proteomic approach. *BMC Genomics*, **11**, 693.

Vincent JFV, King MJ (1996) The mechanism of drilling by wood wasp ovipositors. *Biomimetics*, **3**, 187–201.

Vincent JFV, Wegst UGK (2004) Design and mechanical properties of insect cuticle. *Arthropod Structure & Development*, **33**, 187–99.

Vinchon S, Moreau SJ, Drezen JM, Prévost G, Cherqui A (2010) A molecular and biochemical analysis of an aspartylglucosaminidase from the venom of the parasitoid wasp *Asobara tabida* (Hymenoptera: Braconidae). *Insect Biochemistry and Molecular Biology*, **40**, 38–48.

Vinogradova EB, Zinovjeva KB (1972a) The control of seasonal development in parasites of blowflies. IV. Patterns of photoperiodic reactions in *Alysia manducator* Panz. (Hymenoptera, Braconidae). In VA Zaslavskii (ed.), *Host Parasite Relation in Insects*, Nauka, Leningrad, pp. 112–17.

Vinogradova EB, Zinovjeva KB (1972b) Experimental investigation of the seasonal aspect of the relationship between blowflies and their parasites. *Journal of Insect Physiology*, **18**, 1629–38.

Vinson SB (1969) General morphology of the digestive and internal reproductive systems of adult *Cardiochiles nigriceps* (Hymenoptera: Braconidae). *Annals of the Entomological Society of America*, **62**, 1414–19.

Vinson SB (1972a) Effect of the parasitoid, *Campoletis sonorensis*, on the growth of its host, *Heliothis virescens*. *Journal of Insect Physiology*, **18**, 1508–14.

Vinson SB (1972b) Courtship behavior and evidence for a sex pheromone in the parasitoid *Campoletis sonorensis* (Hymenoptera: Ichneumonidae). *Environmental Entomology*, **1**, 409–14.

Vinson SB (1972c) Factors involved in successful attack on *Heliothis virescens* by the parasitoid *Cardiochiles nigriceps*. *Journal of Invertebrate Pathology*, **20**, 118–23.

Vinson SB (1974) The role of the foreign surface and female parasitoid reactions on the immune response of an insect. *Parasitology*, **68**, 27–33.

Vinson SB (1975) Source of material in the tobacco budworm which initiates host-searching by the egg-larval parasitoid, *Chelonus texanus*. *Annals of the Entomological Society of America*, **68**, 381–4.

Vinson SB (1976) Host selection by insect parasitoids. *Annual Review of Entomology*, **211**, 109–33.

Vinson SB (1978) Courtship behavior and source of a sexual pheromone from *Cardiochiles nigriceps*. *Annals of the Entomological Society of America*, **71**, 832–7.

Vinson SB (1998) The general host selection behavior of parasitoid Hymenoptera and a comparison of initial strategies utilised by larvaphagous and oophagous species. *Biological Control*, **11**, 79–96.

Vinson SB (2012) Foreword – The discovery of polydnaviruses and the influence of Dr. George Salt. In NE Beckage and J-M Drezen (eds), *Parasitoid Viruses: Symbionts and Pathogens*, Elsevier, San Diego, CA, pp. xi–xv.

Vinson SB, Barrass DJ (1970) Effects of the parasitoid, *Cardiochiles nigriceps*, on the growth, development, and tissues of *Heliothis virescens*. *Journal of Insect Physiology*, **16**, 1329–38.

Vinson SB, Guillot FS (1971) Host marking: source of a substance that results in host discrimination in insect parasitoids. *Entomophaga*, **17**, 241–5.

Vinson SB, Jang H-S (1987) Activation of *Campoletis sonorensis* (Hymenoptera: Ichneumonidae) eggs by artificial means. *Annals of the Entomological Society of America*, **80**, 486–9.

Vinson SB, Lewis WJ (1973) Teratocytes: growth and numbers in the haemocoel of *Heliothis virescens* attacked by *Microplitis croceipes*. *Environmental Entomology*, **2**, 593–6.

Vinson SB, Mourad AK (2000) The response of *Cardiochiles nigriceps* Viereck (Hymenoptera: Braconidae) larvae to conspecific competitors. In AD Austin and M Dowton (eds), *Hymenoptera Evolution, Biodiversity and Biological Control*, CSIRO, Melbourne, pp. 46–9.

Vinson SB, Scott JR (1975) Particles containing DNA associated with oocyte of an insect parasitoid. *Journal of Invertebrate Pathology*, **25**, 375–8.

Vinson SB, Henson RD, Barfield CS (1976) Ovipositional behavior of *Bracon mellitor* Say, a parasitoid of boll weevil (*Anthonomus grandis* Boh.). I. Isolation and identification of a synthetic releaser of ovipositional probing. *Journal of Chemical Ecology*, **2**, 431–40.

Vinson SB, Barfield CS, Henson RD (1977) Ovipositional behavior of *Bracon mellitor* Say, a parasitoid of boll weevil (*Anthonomus grandis* Boh.). II. Associative learning. *Journal of Physiological Entomology*, **2**, 157–64.

Visser B, Ellers J (2008) Lack of lipogenesis in parasitoids: a review of physiological mechanisms and evolutionary implications. *Journal of Insect Physiology*, **54**, 1315–22.

Visser B, Le Lann C, Snaas H, Hardy ICW, Harvey JA (2014) Consequences of resource competition for sex allocation and discriminative behaviors in a hyperparasitoid wasp. *Behavioral Ecology and Sociobiology*, **68**, 105–13.

Visser ME (1994) The importance of being large: the relationship between size and fitness in females of the parasitoid *Aphaereta minuta* (Hymenoptera: Braconidae). *Journal of Animal Ecology*, **63**, 963–78.

Völkl W (1994) Searching at different spatial scales: the foraging behaviour of the aphid parasitoid *Aphidius rosae* in rose bushes. *Oecologia*, **100**, 177–83.

Völkl W (2001) Parasitoid learning during interactions with ants: how to deal with an aggressive antagonist. *Behavioral Ecology and Sociobiology*, **49**, 135–44.

Völkl W, Mackauer M (1993) Interactions between ants attending *Aphis fabae* spp. *cirsiiacanthoidis* on thistles and foraging parasitoid wasps. *Journal of Insect Behavior*, **6**, 301–12.

Völkl W, Mackauer M (1996) "Sacking" the host: oviposition behaviour of a parasitoid wasp, *Dyscritulus planiceps* (Hymenoptera: Aphidiidae). *Journal of Insect Behavior*, **9**, 975–80.

Völkl W, Mackauer M (2000) Oviposition behaviour of aphidiine wasps (Hymenoptera: Braconidae, Aphidiinae): morphological adaptations and evolutionary trends. *The Canadian Entomologist*, **132**, 197–212.

Völkl W, Novak H (1997) Foraging behaviour and resource utilization of the aphid parasitoid, *Pauesia pini* (Hymenoptera: Aphidiidae) on spruce: influence of host species and ant attendance. *European Journal of Entomology*, **94**, 211–20.

Völkl W, Liepert C, Birnbach R, Hubner G, Dettner K (1996) Chemical and tactile communication between the root aphid parasitoid *Paralipsis enervis* and trophobiotic ants: consequences for parasitoid survival. *Experientia*, **52**, 731–8.

Volkoff AN, Jouan V, Urbach S, Samain S, Bergoin M, Wincker P, Demettre E, Cousserans F, Provost B, Coulibaly F, Legeai F, Béliveau C, Cusson M, Gyapay G, Drezen JM (2010) Analysis of virion structural components reveals vestiges of the ancestral ichnovirus genome. *PLoS Pathogens*, **6** (5), e1000923

von Borstel RC, Smith RH, Grosch DS, et al. (1968). Mutational response of *Habrobracon* in the BIOSATELLITE II experiment. *Bioscience*, **18**, 598–601.

Wäckers FL (2001) A comparison of nectar- and honeydew sugars with respect to their utilization by the Hymenopteran parasitoid *Cotesia glomerata*. *Journal of Insect Physiology*, **47**, 1077–84.

Wäckers FL (2004) Assessing the suitability of flowering herbs as parasitoid food sources: flower attractiveness and nectar accessibility. *Biological Control*, **29**, 307–14.

Wäckers FL, Lewis WJ (1994) Olfactory and visual learning and their combined influence on host site location by the parasitoid *Microplitis croceipes* (Cresson). *Biological Control*, **4**, 105–12.

Wäckers FL, Mitter E, Dorn S (1998) Vibrational sounding by the pupal parasitoid *Pimpla* (*Coccygomimus*) *turionellae*: an additional solution to the reliability-detectability problem. *Biological Control*, **11**, 141–6.

Wagener B, Reineke A, Lohr B, Zebitz PW (2006) Phylogenetic study of *Diadegma* species (Hymenoptera: Ichneumonidae) inferred from analysis of mitochondrial and nuclear DNA sequences. *Biological Control*, **37**, 131–40.

Wago H, Tanaka T (1989) Synergistic effects of calyx fluid and venom of *Apanteles kariyai* Watanabe (Hymenoptera: Braconidae) on the granular cells of *Pseudaletia separata* Walker (Lepidoptera: Noctuidae). *Zoological Science*, **6**, 691–6.

Wahl DB (1986) Larval structures of oxytorines and their significance for the higher classification of some Ichneumonidae (Hymenoptera). *Systematic Entomology*, **11**, 117–27.

Wahl DB (1987) A revision of *Venturia* north of Central America (Hymenoptera: Ichneumonidae). *University of Kansas, Science Bulletin*, **57**, 275–356.

Wahl DB (1988) A review of the mature larvae of the Banchini and their phylogenetic significance, with comments on the Stilbopinae (Hymenoptera: Ichneumonidae). In VK Gupta (ed.), *Advances in Parasitic Hymenoptera Research. Proceedings of the 2nd Conference on the Taxonomy and Biology of Parasitic Hymenoptera held at the University of Florida, Gainesville, Florida, November 19–21, 1987*, E.J. Brill, Leiden, pp. 147–61.

Wahl DB (1990) A review of the mature larvae of Diplazontinae, with notes on larvae of Acaenitinae and Orthocentrinae and proposal of two new subfamilies (Insecta: Hymenoptera, Ichneumonidae). *Journal of Natural History*, **24**, 27–52.

Wahl DB (1991) The status of *Rhimphoctona*, with special reference to the higher categories within Campopleginae and the relationships of the subfamily (Hymenoptera: Ichneumonidae). *Transactions of the American Entomological Society*, **117**, 193–213.

Wahl DB (1993a) Cladistics of the genera of Labeninae (Hymenoptera: Ichneumonidae). *Entomologia Generalis*, **18**, 91–105.

Wahl DB (1993b) Cladistics of the genera of Mesochorinae (Hymenoptera: Ichneumonidae). *Systematic Entomology*, **18**, 371–87.

Wahl D (1993c) Family *Ichneumonidae*. In H Goulet and JT Huber (eds), *Hymenoptera of the World: an Identification Guide to Families*, Centre for Land and Biological Resources Research, Agriculture Canada, Ottawa, pp 395–509.

Wahl DB (1996a) Further notes on the family-group names of the Ichneumoninae (Hymenoptera: Ichneumonidae). *Journal of Hymenoptera Research*, **5**, 273.

Wahl DB (1996b). Xenothyrini, a new tribe of Labeninae, and a key to the tribes and genera of Labeninae (Hymenoptera: Ichneumonidae). *Proceedings of the Entomological Society of Washington*, **98**, 158–61.

Wahl DB (1997) The cladistics of the genera and subgenera of Xoridinae. In ID Gauld, The Ichneumonidae of Costa Rica, 2. *Memoirs of the American Entomological Institute*, **57**, 454–60.

Wahl DB, Gauld ID (1998) The cladistics and higher classification of the Pimpliformes (Hymenoptera: Ichneumonidae). *Systematic Entomology*, **23**, 265–98.

Wahl DB, Mason WRM (1995) The family-group names of the Ichneumoninae (Hymenoptera: Ichneumonidae). *Journal of Hymenoptera Research*, **4**, 285–93.

Wahl DB, Sharkey MJ (1993) Superfamily Ichneumonoidea. In H Goulet and JT Huber (eds), *Hymenoptera of the World: an Identification Guide to Families*, Centre for Land and Biological Resources Research, Agriculture Canada, Ottawa, pp. 358–62.

Wahl DB, Gauld ID, Ward S (2002) Subfamily Brachycyrtinae. http://www.amentinst.org/GIN/Brachycyrtinae/ (accessed 19 January 2014).

Walker AK, Kitching IJ, Austin AD (1990a) A reassessment of the phylogenetic relationships within the Microgastrinae (Hymenoptera: Braconidae). *Cladistics*, **6**, 291–306.

Walker AK, Joshi NK, Verma SK (1990b) The biosystematics of *Syntretomorpha szaboi* Papp (Hymenoptera: Braconidae: Euphorinae) attacking the Oriental honey bee, *Apis cerana* Fabricius (Hymenoptera: Apidae), with a review of braconid parasitoids attacking bees. *Bulletin of Entomological Research*, **80**, 79–83.

Walker MF (1962) Some observations on the biology of the ladybird parasite *Perilitus coccinellae* (Schrank) (Hymenoptera, Braconidae) with special reference to host selection and recognition. *Entomologist's Monthly Magazine*, **97**, 240–4.

Walkley LM (1956) A tribal revision of the Brachycyrtine wasps of the world (Cryptinae- Ichneumonidae). *Proceedings of the United States National Museum*, **106**, 315–29.

Wallner WE, Weseloh RM, Grinberg PS (1982) Intrinsic competition between *Apanteles melanoscelus* Hym, Braconidae and *Rogas lymantriae* Hym. Braconidae reared on *Lymantria dispar* Lep, Lymantriidae. *Entomophaga*, **27**, 99–103.

Waloff N (1961) Observations on the biology of *Perilitus dubius* (Wesmael) (Hymenoptera: Braconidae), a parasite of the chrysomelid beetle *Phytodecta olivacea* (Forster). *Proceedings of the Royal Entomological Society of London, Series A, General Entomology*, **36**, 96–102.

Waloff N (1967) Biology of three species of *Leiophron* (Hymenoptera: Braconidae, Euphorinae) parasitic on Miridae on broom. *Transactions of the Royal Entomological Society of London*, **119**, 187–213.

Waloff N (1968) Studies on the insect fauna on Scotch broom *Sarothamnus scoparius* (L.) Wimmer. *Advances in Ecological Research*, **5**, 87–208.

Walther C, Reinecke M (1983) Block of synaptic vesicle exocytosis without block of Ca^{2+}-influx. An ultrastructural analysis of the paralysing action of *Habrobracon* venom on locust motor nerve terminals. *Neuroscience*, **9**, 213–24.

Wan Z-W, Wang H-Y, Chen X-X (2006) Venom apparatus of the endoparasitoid wasp *Opius caricivorae* Fischer (Hymenoptera: Braconidae): morphology and ultrastructure. *Microscopy Research and Technique*, **69**, 820–5.

Wang Q, Gu H, Dorn S (2003) Selection on olfactory response to semiochemicals from plant–host complex in a parasitic wasp. *Heredity*, **91**, 430–5.

Wang Y-P, Wu Z-Y, Chen X-X, Wu H (2008) SEM Observation of the sensilla of *Aleiodes esenbeckii*. *Chinese Journal of Biological Control*, **24**, 193–8.

Wang X, Daane KM (2014) Cannibalism of parasitoid-attacked conspecifics in a non-carnivorous caterpillar. *Entomologia Experimentalis et Applicata*, **151**, 112–121.

Wang X, Messing R (2003) Intra- and interspecific competition by *Fopius arisanus* and *Diachamimorpha tryoni* (Hymenoptera: Braconidae), parasitoids of tephritid fruit flies. *Biological Control*, **27**, 251–9.

Wang X-Y, Yang Z-Q, Gould JR (2010a) Sensilla on the antennae, legs and ovipositor of *Spathius agrili* Yang (Hymenoptera: Braconidae), a parasitoid of the emerald ash borer *Agrilus planipennis* Fairmaire (Coleoptera: Buprestidae). *Microscopy Research and Technique*, **73**, 560–71.

Wang X-Y, Yang Z-Q, Gould JR Wu H, Ma J-H (2010b) Host-seeking behavior and parasitism by *Spathius agrili* Yang (Hymenoptera: Braconidae), a parasitoid of the emerald ash borer. *Biological Control*, **52**, 24–9.

Wang Z-Z, Shi M, Zhao W, Bian Q-L, Ye G-Y, Chen X-X (2013) Identification and characterization of defensin genes from the endoparasitoid wasp *Cotesia vestalis* (Hymenoptera: Braconidae). *Journal of Insect Physiology*, **59**, 1095–103.

Ward S, Gauld ID (1987) The callajoppine parasitoids of sphingids in central America (Hym., Ichneumonidae). *Systematic Entomology*, **12**, 503–8.

Wardle AR, Borden JH (1985) Age-dependent associative learning by *Exeristes roborator* (F.) (Hymenoptera: Ichneumonidae). *Canadian Entomologist*, **117**, 605–16.

Wardle AR, Borden JH (1990) Learning of host microhabitat form by Exeristes roborator (F.) (Hymenoptera: Ichneumonidae). *Journal of Insect Behavior*, **3**, 251–63.

Washburn JO, Haas-Stapleton EJ, Tan FF, Beckage NE, Volkman LE (2000) Co-infection of *Manduca sexta* larvae with polydnavirus from *Cotesia congregata* increases susceptibility to fatal infection by *Autographa californica* M nucleopolyhedrovirus. *Journal of Insect Physiology*, **46**, 179–90.

Watanabe K, Matsumoto R (2010) Disjunctive distribution of the basal genus *Aplomerus* (Hymenoptera: Ichneumonidae: Xoridinae) in East Asia and North America, with a new species from Japan. *Entomological Science*, **13**, 375–80.

Waterston J (1929a) Additional notes on *Belesica braconoides* Waterst. *Annals and Magazine of Natural History*, **3**, 628–32.

Waterston J (1929b) A new fungus gnat parasite. *Annals and Magazine of Natural History* **3**, 632–6.

Webb BA, Luckhart S (1994) Evidence for an early immunosuppressive role for related *Campoletis sonorensis* venom and ovarian proteins in *Heliothis virescens*. *Archives of Insect Biochemistry and Physiology*, **26**, 147–63.

Webb BA, Strand MR, Dickey SE, Beck MH, Hilgarth RS, Barney WE, Kadash K, Kroemer JA, Lindstrom KG, Rattanadechakul W, Shelby KS, Thoetkiattikul H, Turnbull MW, Witherell RA (2006) Polydnavirus genomes reflect their dual roles as mutualists and pathogens. *Virology*, **347**, 160–74.

Wei S-J, Shi M, He J-H, Sharkey MJ, Chen X-X (2009) The complete mitochondrial genome of *Diadegma semiclausum* (Hymenoptera: Ichneumonidae) indicates extensive independent evolutionary events. *Genome*, **52**, 308–19.

Wei S-J, Shi M, Sharkey MJ, van Achterberg C, Chen X-X (2010) Comparative mitogenomics of Braconidae (Insecta: Hymenoptera) and the phylogenetic utility of mitochondrial genomes with special reference to holometabolous insects. *BMC Genomics*, **11**, 371.

Weiss MJ, Williams RN (1980) Some host-parasite relationships of *Microctonus nitidulidis* and *Stelidota geminata*. *Annals of the Entomological Society of America*, **73**, 323–6.

Wellings PW, Hart PJ, Kami V, Morneau DC (1994) The introduction and establishment of *Aphidius colemani* Viereck (Hym, Aphidiinae) in Tonga. *Zeitschrift für Angewandte Entomologie*, **118**, 419–28.

Weng J-L, Barrantes G (2007) Natural history and larval behavior of the parasitoid *Zatypota petronae* (Hymenoptera: Ichneumonidae). *Journal of Hymenoptera Research*, **16**, 326–35.

Went DF, Krause G (1974) Egg activation in *Pimpla turionellae* (Hym.). *Naturwissenschaften*, **61**, 407–8.

Weseloh RM (1976) Dufour's gland: source of sex pheromone in a hymenopterous parasitoid. *Science*, **193**, 695–7.

Weseloh RM (1977) Mating behavior of the gypsy moth parasite, *Apanteles melanoscelus*. *Annals of the Entomological Society of America*, **70**, 549–54.

Weseloh RM (1980) Sex pheromone gland of the gypsy moth parasitoid, *Apanteles melanoscelus*: revaluation and ultrastructural survey. *Annals of the Entomological Society of America*, **73**, 576–80.

Wesmael C (1835) Monographie des Braconides Belgique. *Nouveaux Mémoires de l'Académie Royale des Sciences et Belles-Lettres de Bruxelles*, **9**, 1–252.

West RJ, Kenis M, Butt GW, Bennett SM (1999) Parasitoid complex of *Zeiraphera canadensis* (Lepidoptera: Tortricidae) and evaluation of *Tycherus osculator* (Hymenoptera: Ichneumonidae) as a biological control agent. *The Canadian Entomologist*, **131**, 465–74.

West SA, Cook JM, Werren JH, Godfray HCJ (1998) *Wolbachia* in two insect host-parasitoid communities. *Molecular Ecology*, **7**, 1457–65.

Wharton RA (1980) Review of the Nearctic Alysiini (Hymenoptera: Braconidae) with discussion of generic relationships within the tribe. *University of California Publications in Entomology*, **88**, 1–112.

Wharton RA (1983) Variation in *Opius hirtus* Fischer and discussion of *Desmiostoma* Förster (Hymenoptera: Braconidae). *Proceedings of the Entomological Society of Washington*, **85**, 327–30.

Wharton RA (1984) Biology of the Alysiini (Hymenoptera: Braconidae), parasitoids of cyclorrhaphous Diptera. *Technical Monographs. Texas Agricultural Experimental Station*, **11**, 1–39.

Wharton RA (1988) Classification of the braconid subfamily Opiinae (Hymenoptera). *The Canadian Entomologist*, **120**, 333–60.

Wharton RA (1989) Classical biological control of fruit infesting Tephritidae. In A Robinson and G Harper (eds), *World Crop Pests. Fruit Flies: Their Biology, Natural Enemies, and Control*, vol. 3b, Elsevier, Amsterdam, pp. 303–13.

Wharton RA (1993a) Bionomics of the Braconidae. *Annual Review of Entomology*, **38**, 121–43.

Wharton RA (1993b) Review of the Hormiini (Hymenoptera: Braconidae) with a description of new taxa. *Journal of Natural History*, **27**, 107–71.

Wharton RA (1997) Generic relationships of opiine Braconidae (Hymenoptera) parasitic on fruit-infesting Tephritidae (Diptera). *Contributions of the American Entomological Institute*, **30**, 1–53.

Wharton RA (2000) Can braconid classification be restructured to facilitate portrayal of relationships? In AD Austin and M Dowton (eds), *Hymenoptera Evolution, Biodiversity and Biological Control*, CSIRO, Melbourne, pp. 143–53.

Wharton RA (2002) Revision of the Australian Alysiini (Hymenoptera: Braconidae). *Invertebrate Systematics*, **16**, 7–105.

Wharton RA (2006) The species of *Sternaulopius* Fischer (Hymenoptera: Braconidae, Opiinae) and the braconid sternaulus. *Journal of Hymenoptera Research*, **15**, 317–47.

Wharton RA, van Achterberg, C (2000) Family group names in Braconidae (Hymenoptera: Ichneumonoidea). *Journal of Hymenoptera Research*, **9**, 254–70.

Wharton RA, Austin AD (1991) Revision of Australian Dacnusini (Hymenoptera: Braconidae: Alysiinae), parasitoids of cyclorrhaphous Diptera. *Journal of the Australian Entomological Society*, **30**, 193–206.

Wharton RA, Hanson PE (2004) Biology and evolution of braconid gall wasps (Hymenoptera). In A Raman, CW Scharfer and TM Withers (eds), *Biology, Ecology, and Evolution of Gall-Inducing Arthropods*, Science Publishers/CRC Press, Enfield, NH, pp. 495–505.

Wharton RA, Sittertz-Bhatkar H (2002) Polydnaviruses in the genus *Mirax* Haliday (Hymenoptera: Braconidae). *Journal of Hymenoptera Research*, **11**, 358–64.

Wharton RA, Smith JW, Quicke DLJ, Browning HW (1989) Two new North American species of *Digonogastra* Viereck (Hym., Braconidae) parasitic on cane boring Lepidoptera. *Bulletin of Entomological Research*, **79**, 401–10.

Wharton RA, Shaw SR, Sharkey MJ, Wahl DB, Woolley JB, Whitfield JB, Marsh PM, Johnson JB (1992) Phylogeny of the subfamilies of the family Braconidae (Hymenoptera, Ichneumonoidea) – a reassessment. *Cladistics*, **8**, 199–235.

Wharton RA, Marsh PM, Sharkey MJ (eds) (1997) *Manual of the New World Genera of the Family Braconidae (Hymenoptera)*. International Society of Hymenopterists, Washington, DC.

Wharton RA, Yoder MJ, Gillespie JJ, Patton JC, Honeycutt RL (2006) Relationships of *Exodontiella*, a non-alysiine, exodont member of the family Braconidae (Insecta, Hymenoptera). *Zoologica Scripta*, **35**, 323–40.

Wheeler EW (1923) Some braconids parasitic on aphids and their life history. *Annals of the Entomological Society of America*, **16**, 1–29.

Whitfield JB (1987) Male swarming by a microgastrine braconid, *Apanteles coniferae* (Haliday) (Hymenoptera). *Proceedings and Transactions of the British Entomological and Natural History Society*, **18**, 133.

Whitfield JB (1988) Two new species of *Paradelius* (Hymenoptera: Braconidae) from North America with biological notes. *Pan-Pacific Entomologist*, **64**, 313–19.

Whitfield JB (1992) The polyphyletic origin of endoparasitism in the cyclostome lineages of Braconidae (Hymenoptera). *Systematic Entomology*, **17**, 273–86.

Whitfield JB (1997a) Molecular and morphological data suggest a single origin of the polydnaviruses among braconid wasps. *Naturwissenschaften*, **84**, 502–7.

Whitfield JB (1997b) Subfamily Microgastrinae. In RA Wharton, PM Marsh and MJ Sharkey (eds), *Identification Manual to the New World Genera of Braconidae*. Special Publication of the International Society of Hymenopterists 1, pp 333–64.

Whitfield JB (1998) Phylogeny and evolution of the host/parasitoid relationship in the Hymenoptera. *Annual Review of Entomology*, **43**, 129–51.

Whitfield JB (2000) Phylogeny of microgastroid braconid wasps, and what it tells us about polydnavirus evolution. in A.D. Austin and M. Dowton (eds), *The Hymenoptera: Evolution, Biodiversity and Biological Control*, CSIRO, Melbourne, pp. 97–105.

Whitfield JB (2002) Estimating the age of the polydnavirus/braconid wasp symbiosis. *Proceedings of the National Academy of Sciences of the United States of America*, **99**, 7508–13.

Whitfield JB (2006) Revision of the Nearctic species of the genus *Pholetesor* Mason (Hymenoptera: Braconidae). *Zootaxa*, **1144**, 1–94.

Whitfield JB, Asgari S (2003) Virus or not? Phylogenetics of polydnaviruses and their wasp carriers. *Journal of Insect Physiology*, **49**, 397–405.

Whitfield JB, Dangerfield P C (1997) Subfamily Cardiochilinae. in RA Wharton, PM Marsh and MJ Sharkey (eds), *Identification Manual to the New World Genera of Braconidae*, Special Publication of the International Society of Hymenopterists 1, pp. 177–83.

Whitfield JB, Mason WRM (1994) Mendesellinae, a new subfamily of braconid wasps (Hymenoptera: Braconidae) with a review of relationships within the microgastroid assemblage. *Systematic Entomology*, **19**, 61–76.

Whitfield JB, Wagner DL (1988) Patterns in host ranges within the nearctic species of the parasitoid genus *Pholetesor* Mason (Hymenoptera: Braconidae). *Environmental Entomology*, **17**, 608–15.

Whitfield JB, Wagner DL (1991) Annotated key to the genera of Braconidae (Hymenoptera) attacking leafmining Lepidoptera in the Holarctic region. *Journal of Natural History*, **25**, 733–54.

Whitfield JB, Wharton RA (1997) Subfamily Hormiinae. In RA Wharton, PM Marsh and MJ Sharkey (eds), *Manual of the New World Genera of the Family Braconidae (Hymenoptera)*, International Society of Hymenopterists, Washington, DC, pp. 285–301.

Whitfield JB, Mardulyn P, Austin AD, Dowton M (2002) Phylogeny of the Microgastinae (Hymenoptera: Braconidae) based on multiple genes and morphology. *Systematic Entomology*, **27**, 337–59.

Whitfield JB, Choi WY, Suh KI (2004) *Andesipolis*, a puzzling new genus of cyclostome Braconidae (Hymenoptera) from the Chilean Andes, with descriptions of three new species. *Zootaxa*, **438**, 1–15.

Whiting AR (1934) Eye colours in the parasitic wasp *Habrobracon* and their behaviour in multiple recessives and in mosaics. *Journal of Genetics*, **29**, 99–107.

Whiting AR (1939) Sex determination and reproductive economy in *Habrobracon*. *Genetics*, **24**, 110–11.

Whiting AR (1943) Multiple alleles in complementary sex determination of *Habrobracon*. *Genetics*, **28**, 365–82.

Whiting AR (1961) Genetics of *Habrobracon*. *Advances in Genetics*, **10**, 295–348.

Whitman DW, Eller FJ (1990) Parasitic wasps orient to green leaf volatiles. *Chemocology*, **1**, 69–75.

Wieber AM, Cook SP, Webb RE, Tatman KM, Reardon RC (1995) Niche partitioning by four *Gelis* spp. (Hymenoptera: Ichneumonidae) hyperparasitoids of the primary gypsy moth parasitoid *Cotesia melanoscela* (Hymenoptera: Braconidae). *Annals of the Entomological Society of America*, **88**, 427–33.

Wild AL, Marsh PM, Whitfield JB (2013) Fast-evolving homoplastic traits are best for species identification in a group of Neotropical wasps. *PLoS ONE*, **8** (9), e74837.

Willers D, Lehmann-Danzinger H, Führer E (1982) Antibacterial and antimycotic effect of a newly discovered secretion from larvae of an endoparasitic insect, *Pimpla turionellae* L. (Hym.). *Archives of Microbiology*, **133**, 225–9.

Williams HJ, Wong M, Wharton RA, Vinson SB (1988) Hagen's gland morphology and chemical content analysis for three species of parasitic wasps (Hymenoptera: Braconidae). *Journal of Chemical Ecology*, **14**, 1727–36.

Williams FX (1928) Studies in tropical wasps – their hosts and associates (with descriptions of new species). *Hawaiian Sugar Planters Association, Entomology Series, Bulletin*, **19**, 19–20.

Wilson DD, Ridgway RL (1975) Morphology, development, and behavior of immature stages of parasitoid, *Campoletis sonorensis* (Hymenoptera-Ichneumonidae). *Annals of the Entomological Society of America*, **68**, 191–6.

Wilson JS, Forister ML, Dyer LA, O'Connor JM, Burls K, Feldman CR, Jaramillo MA, Miller JS, Rodriguez-Castaneda G, Tepe EJ, Whitfield JB, Young B (2012) Host conservatism, host shifts and diversification across three trophic levels in two Neotropical forests. *Journal of Evolutionary Biology*, **25**, 532–46.

Wirta HK, Hebert PDN, Kaartinen R, Prosser SW, Várkonyi G, Roslin T (2014) Complementary molecular information changes our perception of food web structure. *Proceedings of the National Academy of Sciences of the United States of America*, **111**, 1885–90.

Withers TM, Allen GR, Patel VS, Satchell D, Maleny G (2012) Investigating the potential of *Eadya paropsidis* (Braconidae) from Tasmania as a biocontrol agent for *Paropsis charybdis* in New Zealand. Combined Australian Entomological Society – 43rd AGM and Scientific Conference and Australasian Arachnological Society – 2012 Conference, 25–28 November 2012, Hobart, Tasmania, p. 292.

Wolda H (1978a) Fluctuations in abundance of tropical insects. *The American Naturalist*, **112**, 1017–45.

Wolda H (1978b) Seasonal fluctuations in rainfall, food, and abundance of tropical insects. *Journal of Animal Ecology*, **47**, 369–81.

Wolf R, Wolf D (1988) Activation by calcium ionophore injected into unfertilized ovarian eggs explanted from *Pimpla turionellae* (Hymenoptera). *Zoologische Jahrbucher Abteilung fur Anatomie und Ontogenie der Tiere*, **92**, 501–12.

Wu Z, Hopper KR, Ode PJ, Fuester RW, Tuda M, Heimpel GE (2005) Single-locus complementary sex determination absent in *Heterospilus prosopidis* (Hymenoptera: Braconidae). *Heredity*, **95**, 228–34.

Wyatt TD, Foster WA (1989) Parental care in the subsocial intertidal beetle, *Bledius spectabilis*, in relation to parasitism by the ichneumonid wasp, *Barycnemis blediator*. *Behaviour*, **110**, 76–92.

Wyckhuys KAG, Koch RL, Heimpel GE (2007) Physical and ant-mediated refuges from parasitism: Implications for non-target effects in biological control. *Biological Control*, **40**, 306–13.

Wyckhuys KAG, Strange-George JE, Kulhanek CA, Wäckers FL, Heimpel GE (2008) Sugar feeding by the aphid parasitoid *Binodoxys communis*: how does honeydew compare with other sugar sources? *Journal of Insect Physiology*, **54**, 481–91.

Wylie WG (1982) An effect of parasitism by *Microctonus vittatae* (Hymenoptera: Braconidae) on emergence of Phyllotreta cruciferae and *Phyllotreta striolata* (Coleoptera: Chrysomelidae) from overwintering sites. *The Canadian Entomologist*, **114**, 727–32.

Wylie WG (1985) Posterior dispersal of eggs and larvae of *Microctonus vittatae* (Hymenoptera: Braconidae) in crucifer-infesting flea beetles (Coleoptera: Chrysomelidae). *The Canadian Entomologist*, **117**, 541–5.

Wyler S, Lanzrein B (2003) Ovary development and polydnavirus morphogenesis in the parasitic wasp *Chelonus inanitus*. II. Ultrastructural analysis of calyx cell development, virion formation and release. *Journal of General Virology*, **84**, 1151–63.

Xi Y, Yin X, Li X, Zhu C, Zhang Y (2010) Scanning electron microscopy studies of antennal sensilla of *Lysiphlebus fabarum* (Marshall) (Hymenoptera: Braconidae). *Acta Entomologica Sinica*, **53**, 936–42.

Xie Z, Li L, Xie Y (1989) *In vitro* culture of *Habrobracon hebetor* (Say) (Hymenoptera: Braconidae). *Chinese Journal of Biological Control*, **5**, 49–51.

Xu P, Wan Z-W, Chen XX, Liu S, Feng M-G (2007) Immature morphology and development of *Opius caricivorae* (Hymenoptera: Braconidae), an endoparasitoid of the leafminer *Liriomyza sativae* (Diptera: Agromyzidae). *Annals of the Entomological Society of America*, **100**, 425–32.

Yamada H, Igarashi Y, Takasu Y, Tsubouchi K, Kato Y, Akai H (2003) Sericin-like protein from the cocoons of the parasitic wasp *Apanteles glomeratus*. *International Journal of Wild Silkmoths and Silk*, **8**, 19–24.

Yamada H, Shigesada K, Igarashi Y, Takasu Y, Tsubouchi K, Kato Y (2004) A novel asparagine-rich fibrous protein (xenofibron) from the cocoons of the parasitic wasp *Cotesia (=Apanteles) glomerata*. *International Journal of Wild Silkmoths and Silk*, **9**, 61–6.

Yamaguchi H (1987) The role of venom in host-discrimination of *Ascogaster reticulatus*. *Japanese Journal of Applied Entomology and Zoology*, **3**, 80–2 (in Japanese).

Yamamoto M, Chau NNB, Maetô K (2009) Host movement initiates oviposition behavior of *Meteorus pulchricornis*, a braconid parasitoid of free-living lepidopteran larvae. *Applied Entomology and Zoology*, **44**, 53–9.

Yamazaki K, Matsumoto R (2009) Predation on the woodwasp *Tremex longicollis* Konow (Hymenoptera: Siricidae) and its parasitoid *Megarhyssa jezoensis* (Matsumura) (Hymenoptera: Ichneumonidae) adults during oviposition. *Journal of Asia-Pacific Entomology*, **12**, 313–15.

Yang G, Wei H, You MS (2003) Ultrastructure of the antennae of *Cotesia plutellae*. *Journal of the Fujian Agriculture and Forestry University*, **32**, 32–35 (in Chinese).

Yang ZQ, Wang XY, Gould JR, Reardon RC, Zhang YN, Liu GJ, Liu ES (2010) Biology and behavior of *Spathius agrili*, a parasitoid of the emerald ash borer, *Agrilus planipennis*, in China. *Journal of Insect Science*, **10**, 30.

Yazgan S (1972) A chemically defined synthetic diet and larval nutritional requirements of the endoparasitoid *Itoplectis conquisitor* (Hymenoptera). *Journal of Insect Physiology*, **18**, 2123–41.

Yazgan S (1981) A meridic diet and quantitative effects of tween 80, fatty acid mixtures and inorganic salts on devolopment and survival of endoparasitoid *Pimpla turionellae* L. *Zeitschrift für Angewandte Entomologie*, **91**, 433–41.

Yazgan S, House HL (1970) An hymenopterous insect the parasitoid *Itoplectis conquisitor*, reared axenically on a chemically defined synthetic diet. *The Canadian Entomologist*, **102**, 1304–6.

Yazlovetsky IG, Neponnyashaya AM, Ageeva LI (1992) Optimization of the artificial nutritional diet for the larvae of *Bracon hebetor* (Hymenoptera, Braconidae). *Zoologicheskii Zhurnal*, **71**, 134–8.

Yeargan KV, Braman K (1986) Life history of the parasite *Diolcogaster facetosa* (Weed) (Hymenoptera: Braconidae) and its behavioural adaptation to the defensive response of a lepidopteran host. *Annals of the Entomological Society of America*, **79**, 1029–33.

Yeargan KV, Braman K (1989) Life history of the hyperparasitoid *Mesochorus discitergus* (Hymenoptera: Ichneumonidae) and tactics used to overcome the defensive behavior of the green cloverworm (Lepidoptera: Noctuidae). *Annals of the Entomological Society of America*, **82**, 393–8.

Yin L, Zhang C, Qin J, Wang C (2003) Polydnavirus of *Campoletis chlorideae*: characterization and temporal effect on host *Helicoverpa armigera* cellular immune response. *Archives of Insect Biochemistry and Physiology*, **52**, 104–13.

Young SY, Yearian WC (1990) Transmission of nuclear polyhedrosis virus by the parasitoid *Microplitis croceipes* (Hymenoptera: Braconidae) to *Heliothis virescens* (Lepidoptera: Noctuidae) on soybean. *Environmental Entomology*, **19**, 251–6.

Yu DS (1998) *Interactive Catalogue of World Ichneumonidae*. Taxapad: Scientific Names for Information Management, Ottawa, available online at http://www.taxapad.com (accessed 30 June 2014).

Yu DS (2012) *World Ichneumonoidea. Taxonomy, Biology, Morphology and Distribution (Braconidae)*, Taxapad: Scientific Names for Information Management, Ottawa, available online at http://www.taxapad.com (accessed 30 June 2014).

Yu DS, Horstmann K (1997) Catalogue of world Ichneumonidae (Hymenoptera). *Memoirs of the American Entomological Institute*, **58**, 1–1558.

Yu DW, Quicke DLJ (1997) *Compsobraconoides* (Braconidae: Braconinae), the first hymenopteran ectoparasitoid of adult *Azteca* ants (Hymenoptera: Formicidae). *Journal of Hymenoptera Research*, **6**, 419–21.

Yu R-X, Shi M, Huang F, Chen X-X (2008) Immature development of *Cotesia vestalis* (Hymenoptera: Braconidae), an endoparasitoid of *Plutella xylostella* (Lepidoptera: Plutellidae). *Annals of the Entomological Society of America*, **101**, 189–96.

Zaka-ur-Rab M (1978) Morphology of the head of *Bathyaulax alami* Zaka. (Hymenoptera: Braconidae). *Zoologische Jahrbucher Abteilung fur Anatomie und Ontogenie der Tiere*, **99**, 421–36.

Zaldivar-Riverón A, Quicke DLJ (2002) First host record for the braconid wasp genus *Notiopambolus* Achterberg and Quicke (Pambolinae). *Journal of Hymenoptera Research*, **11**, 370–1.

Zaldivar-Riverón A, Areekul B, Shaw MR, Quicke DLJ (2004) Comparative morphology of the venom apparatus in the braconid wasp subfamily Rogadinae (Insecta, Hymenoptera, Braconidae) and related taxa. *Zoologica Scripta*, **33**, 223–38.

Zaldivar-Riverón A, Mori M, Quicke DLJ (2006) Systematics of the cyclostome subfamilies of braconid parasitic wasps (Hymenoptera: Ichnuemonoidea): a simultaneous molecular and morphological Bayesian approach. *Molecular Phylogenetics and Evolution*, **38**, 130–45.

Zaldivar-Riverón A, Belokobylskij SA, León-Regagnon V, Martinez JJ, Briceno R, Quicke DLJ (2007) A single origin of gall association in a group of parasitic wasps with disperate morphologies. *Molecular Phylogenetics and Evolution*, **44**, 981–99.

Zaldivar-Riverón A, Belokobylskij SA, León-Regagnon V, Briceno R, Quicke DLJ (2008) Molecular phylogeny and historical biogeography of the cosmopolitan parasitic wasp subfamily Doryctinae (Hymenoptera: Braconidae). *Invertebrate Systematics*, **22**, 345–63.

Zaldivar-Riverón A, Shaw MR, Sáez AG, Mori M, Belokobylskij SA, Shaw SR, Quicke DLJ (2009) Evolution of the parasitic wasp subfamily Rogadinae (Braconidae): phylogeny and evolution of lepidopteran host ranges and mummy characteristics. *BMC Evolutionary Biology*, **8**, 329.

Zaldivar-Riverón A, Martínez JJ, Belokobylskij SA, Pedraza-Lara C, Shaw SR, Hanson PE, Varela-Hernández F (2014) Systematics and evolution of gall formation in the plant-associated genera of the wasp subfamily Doryctinae (Hymenoptera: Braconidae). *Systematic Entomology*, in press; doi: 10.1111/syen.12078.

Zettel H (1990) Eine Revision der Gattungen der Cheloninae (Hymenoptera, Braconidae) mit Beschreibungen neuer Gattungen und Arten. *Annales Naturhistorisches Museum Wien*, **91**, 147–96.

Zhang D, Dahlman DL, Jarlfors UE, Southgate HH, Wiley SP (1994) ultrastructure of *Microplitis croceipes* (Cresson) (Braconidae: Hymenoptera) teratocytes. *International Journl of Insect Morphology & Embryology*, **23**, 173–87.

Zhang D, Dahlman DL, Jarlfors UE (1997) Effects of *Microplitis croceipes* teratocytes on host haemolymph protein content and fat body proliferation. *Journal of Insect Physiology*, **43**, 577–85.

Zhang JF, (1991) A new species of *Tanychora* Townes (Hymenoptera, Ichneumonidae). *Acta Palaeontologica Sinica*, **30**, 502–4 (in Chinese, with English abstract).

Zhang HC, Rasnitsyn AP (2003) Some ichneumonids (Insecta, Hymenoptera, Ichneumonoidea) from the Upper Mesozoic of China and Mongolia. *Cretaceous Research*, **24**, 193–202.

Zhang J-H, Gu L-Q, Wang C-Z (2010) Superparasitism behavior and host discrimination of *Campoletis chlorideae* (Ichneumonidae: Hymenoptera) toward *Mythimna separata* (Noctuidae: Lepidoptera). *Environmental Entomology*, **39**, 1249–54.

Zhang L, Feng Y-Q, Ren L-L, Luo Y-Q, Wang F, Zong S-X (2014) Sensilla on antenna, ovipositor and leg of Eriborus applicitus (Hymenoptera: Ichneumonidae), a parasitoid wasp of Holcocerus insularis staudinger (Lepidoptera: Cossidae). *Acta Zoologica*, in press; doi: 10.1111/azo.12073.

Zhang Y, Wyckhuys KAG, Asplen MK, Heimpel GE, Wu K-M (2012) Effect of *Binodoxys communis* parasitism on flight behavior of the soybean aphid, *Aphis glycines*. *Biological Control*, **62**, 10–15.

Zhao W, Dong S, Shi M, Chen X-X (2014) Morphology and development of immature stage of *Diadromus collaris* (Hymenoptera: Ichneumonidae), an important endoparasitoid of *Plutella xylostella* (Lepidoptera: Plutellidae). *Annals of the Entomological Society of America*, **107**, 234–41.

Zhaurova K, Wharton RA (2009) Recognition of Scolobatini and Westwoodiini (Hymenoptera, Ctenopelmatinae) and revision of the component genera. *Contributions of the American Entomological Institute*, **35**, 1–77.

Zhou H, Wu Y-J, Zhnag Z-F, Zhnag Y (2011) Antennal sensilla of *Apanteles cypris* Nixon (Hymenoptera: Braconidae), a larval endoparasitoid of *Cnaphalocrocis medinalis* Guenée (Lepidoptera: Pyralidae). *Microscopy Research and Technique*, **74**, 1199–208.

Zhou Y, Gu H, Dorn S (2006) Single-locus sex determination in the parasitoid wasp *Cotesia glomerata* (Hymenoptera: Braconidae). *Heredity*, **96**, 487–92.

Zijp JP, Blommers LHM (2002a) Survival mode between the yearly reproductive periods, and reproductive biology of *Scambus pomorum* (Hymenoptera: Ichneumonidae: Pimplinae), a parasitoid of the apple blossom weevil *Anthonomus pomorum* (Coleoptera: Curculionidae). *Entomologia Generalis*, **26**, 29–46.

Zijp JP, Blommers LHM (2002b) Biology of *Centistes delusorius*, a parasitoid on apple blossom weevil. *Agricultural and Forest Entomology*, **4**, 275–82.

Zijp JP, Blommers LHM (2002c) Apple sawfly *Hoplocampa testudinea* (Hym., Tenthredinidae) and its parasitoid *Lathrolestes ensator* in Dutch apple orchards (Hym., Ichneumonidae, Ctenopelmatinae). *Journal of Applied Entomology*, **126**, 265–74.

Žikić V, van Achterberg C, Stanković SS, Ilić M (2011) The male genitalia in the subfamily Agathidinae (Hymenoptera: Braconidae): morphological information of species on generic level. *Zoologischer Anzeiger*, **250**, 246–57.

Zinnert KD (1969) Vergleichende Untersuchungen zur Morphologie und Biologie der Larvenparasiten (Hymenoptera: Ichneumonidae und Braconidae) mitteleuropäischer Blattwespen aus der Subfamilie Nematinae (Hymenoptera: Tenthredinidae). *Zeitschrift für Angewandte Entomologie*, **64**, 180–217 (Teil I) and 277–306 (Teil II).

Zinovjeva KB (1974) Stimulation of pupation of larvae of *Calliphora vicina* R.-D. (Diptera, Calliphoridae) by the parasite *Aphaereta minuta* Nees (Hymenoptera, Braconidae). *Doklady Akademii Nauk SSSR*, **216**, 702–3.

Zinovjeva KB (1976) Role of photo- and thermorhythms in diapause induction in *Alysia manducator* Panz. (Hymenoptera, Braconidae). *Entomologicheskogo Obozrenie*, **55**, 517–24.

Zinovjeva KB (1978) Relationship of diapause induction to daily rhythms of temperature and photoperiod in Alysia manducator (Panz.) (Hymenoptera, Braconidae). *Trudy Zoologicheskogo Instituta, Leningrad*, **69**, 124–46.

Zitani NM, Shaw SR (2002) From meteors to death stars: variations of a silk thread (Hymenoptera: Braconidae: Meteorinae). *American Entomologist*, **48**, 228–35.

Zitani NM, Shaw SR, Janzen DH (1997) Description and biology of a new species of *Meteorus* Haliday (Hymenoptera: Braconidae: Meteorinae) from Costa Rica, parasitizing

larvae of *Papilio* and *Parides* (Lepidoptera: Papilionidae). *Journal of Hymenoptera Research*, **6**, 178–85.
Zong S-X, Sheng ML, Luo Y-Q, Lu C-K (2012) *Lissonota holcocerica* Sheng sp.n (Hymenoptera: Ichneumonidae) parasitizing Holcocerus hippophaecolus (Lepidoptera: Cossidae) from China. *Journal of Insect Science*, **12**, 112.
Zwakhals CJ (1989) The male of *Alophosternum albofacialis*, with a note on the biology of the species (Hymenoptera: Ichneumonidae, Pimplinae). *Entomologische Berichten*, **49**, 103–4.
Zwakhals CJ (2005) *Pimpla processioneae* and *P. rufipes*: specialist versus generalist (Hymenoptera: Ichneumonidae, Pimplinae). *Entomologische Berichten*, **65**, 14–16.
Zwakhals K (2010) Identification of Western Palearctic *Dolichomitus* species (Hymenoptera: Ichneumonidae: Pimplinae). *Entomologische Berichten, Amsterdam*, **70**, 111–27.
Zwölfer H, Kraus M (1957) Biocoenotic studies on the parasites of two fir- and two oak-tortricids. *Entomophaga*, **2**, 173–96.

AUTHOR INDEX

A

Abbott C, 41, 44, 381
Abdel-Malek A, 315–317
Abdurahiman UC, 91, 120, 263, 337–8
Abe Y, 467
Ables JR, 153
Achtelig M, 74
Adamowicz SJ, 497
Adams CH, 85, 261, 479
Adams JD, 370
Adán A, 433
Adriaanse ICT, 157
Aeschlimann J-P, 117, 423–4, 466
Ageeva LI, 265, 474
Aguiar AP, 516, 527
Aguirre H, 296, 481
Ahl J, 465
Ahmad M, 78, 157, 247–9, 454
Ahmed T, 13, 53–4, 287
Akai H, 86, 325
Akdura N, 462
Akin D, 155
Aksoylar Y, 462
Aktümsek A, 376
Alam SM, 10, 58–9, 263, 266
Alauzet C, 279
Albajes R, 111
Albrecht U, 84, 149
Aldrich JC, 428, 506
Alford DV, 299–300, 309
Allahyari H, 218
Allemand R, 142–3
Allen AA, 291
Allen GR, 282, 310
Allen HW, 477
Allen WW, 325
Alleyne M, 97–8, 455
Allo M-R, 39, 53–54, 379–80
Almohamad R, 221–2
Al-Saffar ZY, 428
Altahtawy MM, 325
Altenhofer E, 420, 428, 463
Althoff DM, 89–90
Altson AM, 253
Aluja M, 97, 257, 275, 482
Alvi SM, 463–4
Ambrosecchio MR, 82, 216
Amiens-Desneux E, 261
Amos WB, 75
Amy RL, 41, 75
Andersen JC, 529–30
Andrade GS, 481
Andreadis SS, 466
Andreadis TG, 478
Andrews JW, 261
Andrietti F, 54, 161, 166, 391, 393
Antropov AV, 193
Angalet GW, 214
Angioy AM, 13, 260
Anisimova M, 414
Annaheim M, 152, 334
Anthony DW, 54, 259
Antoh FF, 471
Antolin MF, 108, 110, 113, 117, 122–3, 204, 264–7, 307, 478
Aoyagi M, 385
Arakaki N, 91, 110, 248
Arana E, 149
Araújo VA, 69
Arbuthnot KD, 316
Ardila-Garcia AM, 125
Areekul B (*see* Butcher BA),
Argaman Q, 308
Arnaud PH, 120
Aron S, 125
Arthur AP, 99, 130, 132, 141, 325, 337, 334, 374, 405, 463–4, 475
Arvanitakis L, 334
Ascoli-Christensen A, 13
Asgari S, 138–9, 144–5, 152, 154
Ashley TR, 439
Ashley Adams S, 118, 275
Asís Béliveau JD, 253
Askari A, 296–8
Askew RR, 88, 251, 257, 366–7, 394, 448, 499
Asplen MK, 109, 214, 480
Aspöck H, 296, 394, 432
Athanas MM, 301–2, 525
Athanassiou CG, 466
Atlegrim O, 511
Aubert JF, 108, 428, 466
Aubrook EW, 282
Audsley N, 144, 375
Auerbach M, 490
Augé-Gouillou C, 152
Auger J, 395
Austin AD, 39, 55, 102–3, 117–18, 153, 170, 172, 204–6, 208, 224, 226, 230–31, 235, 250, 262, 275, 277, 294, 304–5, 312–15, 317, 319–21, 324–5, 334–5, 353, 372, 389, 391, 447, 504
Avilla J, 111
Ayyar PNK, 61, 233, 392
Azéma-Dossat C, 154
Aziz A, 342
Azman S, 516
Azzouz H, 458

B

Babendreier D, 391
Baccetti B, 78
Bachel S, 141, 471
Baer CF, 478
Baerwald RJ, 259
Bahena F, 433

The Braconid and Ichneumonid Parasitoid Wasps: Biology, Systematics, Evolution and Ecology, First Edition. Donald L. J. Quicke.

Bai B, 93, 97
Bai S-F, 157
Bai S-X, 13, 53–4, 287
Bailley H, 69
Baker BR, 381
Baker JE, 64–5, 67, 82, 144, 156, 268
Baker TC, 13, 334
Baker WA, 286
Bakker K, 97
Baldev Parshad, 312, 314
Balduf WV, 272, 303
Ball JP, 496, 511
Ballal CR, 471
Balmer O, 457, 459
Banko PC, 295, 506–7
Banks JB,
Banks JC, 151, 153, 206, 312–13, 324–5, 404
Báo SN, 69
Barari H, 428
Barbalho SM, 64, 236, 266
Barbarossa IT, 13, 260
Barbier R, 39, 53–54, 379–80
Barclay M, 100, 165, 181, 266, 381, 482
Baret PV, 213
Barfield CS, 136
Barlin MR, 12, 52, 63
Barlow JS, 82–3, 370, 376
Barlow ND, 480
Barnard PC, 492
Barney WE, 149, 151, 404
Barrantes G, 86, 297, 372
Barrass DJ, 471
Barratt BIP, 300, 308–9, 470
Barron JR, 182, 419, 432–3, 447
Barrows EM, 434
Barta RJ, 143, 221
Bartell DP, 433
Bartelt RJ, 68
Bartlett R, 507, 536
Barton D (*see* Barton DM),
Barton DM, 142, 470
Basibuyuk HH, 12, 17–18, 25, 27, 38, 59, 164, 190, 193–4, 205, 209–10, 444–5
Basinger AJ, 244
Basio NA, 158
Bataille M, 114
Batsch WW, 374
Battaglia D, 14, 160, 218, 220, 475
Battini M, 375
Bauer G, 482
Bauerle P, 399
Baumbach T, 339
Baumgart E, 322, 493–6, 501, 510
Baur ME, 422
Bautista RC, 525
Beard RL, 263
Beardsley JW, 257
Beaver RA, 394
Beck MH, 60, 117–18, 149
Beckage NE, 149, 155, 325, 334
Becker C, 13
Beeson CFC, 245
Beggs JR, 480
Begon M, 479
Beirne BP, 79, 216, 270, 279
Bekkaoui A, 130
Belbost F, 152
Belghazy M, 152, 334
Béliveau C, 151–2, 154, 404
Bell HA, 296, 471
Bell J, 296, 471
Bellemare G, 154
Belokobylskij SA, 118, 170, 172 178, 186, 193–4, 209, 229, 231, 234–7, 238, 242–3, 245–6, 271, 275, 287, 301, 308, 319–21, 487
Belshaw R, 38–9, 62, 90, 99–100, 114–15, 178, 196, 209, 218, 223–4, 228–9, 238, 250–51, 255, 260, 268, 270, 272, 281, 284, 294, 302, 304, 307, 310, 312–13, 317, 321, 325, 336, 339, 342, 344, 353, 367, 388, 427, 439, 441, 447, 504, 516, 529
Belz E, 459
Bender JC, 265
Bénédet F, 13, 130
Benelli G, 117, 258–9
Benham GS Jr, 58, 68
Benjamin DM, 389
Bennet FD, 395, 397
Bennett AMR, 356, 360, 385, 387, 403, 407, 414, 448
Bennett DJ, 196, 224
Bennett SM, 394, 419
Benoit M, 76, 88, 374
Benson JF, 267
Benson RB, 120, 282
Bento JMS, 479
Berberet RC, 145, 159
Berg CO, 439
Bergoin M, 152
Berisford CW, 13
Bernal JS, 92, 110–111, 254, 332
Bernard-Samain S, 152, 334
Bernasconi P, 267
Berndt LA, 460
Bernstein C, 259
Berry JA, 295–6, 302, 481
Berry PA, 429
Berti Filho E, 286
Bertrand C, 114, 324, 492
Best RL, 476, 485
Beukeboom LW, 109, 114–15, 117, 125–6
Bezemer TM, 82, 92, 161, 330–31
Bézier A, 149, 152, 334
Bhuyia BA, 105, 437, 442, 445
Bian Q-L, 479
Biere A, 110, 471, 499
Bigot S, 395
Bigot Y, 152, 155, 158, 395
Bilinski S, 74, 75
Billany DJ, 121
Bin F, 14, 39, 218, 376
Biondi A, 261
Birgersson G, 13, 109, 129
Birnbach R, 218
Bisaccia F, 220
Bischof C, 473
Bisdee HE, 433
Bishop MHW, 303
Bixley AS, 470
Bjorksten A, 478
Bjørnson S, 479
Black WC 4th, 108, 267
Blackburn TM, 88–9, 502
Blackith RE, 308
Blackwell A, 435
Blackwell J, 43
Blair KG, 400
Blank SM, 125
Blass S, 63
Bledowski R, 405
Bleeker MAK, 13, 58, 131–4, 137, 332, 334
Blissard GW, 146
Blommers LHM, 301, 370, 457, 363, 386
Blomquist GJ, 64, 68, 315
Blum MS, 482
Blumberg D, 145
Blunck H, 421
Boina WL, 93
Boivin G, 70, 306
Bolhuis H, 126
Boling JC, 86, 325, 328
Bolton B, 2, 197–8, 351, 527
Bonasoro F, 166, 391, 393
Bonsall MB, 483
Bordat D, 477, 479
Borden JH, 12, 13, 130, 371, 376
Bordera S, 14, 462, 508, 516, 521
Borges PAV, 505
Bjørnson S, 479

Boring CA, 39, 41, 44–5, 50–52, 64, 73, 224, 293
Borst DW, 160
Bosio CF, 108, 267
Boughton AJ, 293
Bouix G, 477, 479
Boulay C, 313, 316, 471
Boulétreau M, 114
Bourdais D, 222
Bourguet D, 164, 286, 391
Boursault AE, 393
Boush GM, 259
Bouwer G, 148
Bowers MD, 498
Boyle B, 151, 404
Bracken GK, 83, 456
Bradish H, 144
Bradley WG, 286, 316, 438–9, 525
Bradshaw K, 430–31
Braet Y, 275, 278, 293–4, 305, 504
Braga SMP, 236
Bragg D, 93
Brajković M, 32, 39
Braman K, 121, 334, 421
Brandt AP, 177, 179, 265
Brandt SL, 156
Brajković M, 32, 39
Brauning R, 142
Brazzel JR, 314
Bredlau JP, 332
Brewer FD, 471
Briceno R, 178, 186, 231, 234, 245, 487
Brinck KW, 295, 506–7
Broad GR, 18, 30, 39, 100, 164–5, 198, 342, 355–6, 359, 366–7, 369, 377–8, 387–9, 391–3, 395, 397, 399, 407, 424,439, 441, 485, 448, 504, 516
Brock JP, 444
Brodeur J, 161, 217–18, 222, 303–5, 330–31
Bronskill JF, 72, 75, 374
Broodryk SW, 122, 293, 315–7,
Brooks RW, 370
Brooks WM, 477
Brower AVZ, 490, 498–9, 501
Brown J, 306
Brown JJ, 67–8, 83, 157, 160, 317, 465
Browning HW, 68
Browning TO, 39
Brues CT, 209, 236, 284
Bruinsma M, 393
Brunson MH, 429, 477
Bryden JW, 303
Bryer JA, 306
Buck M, 497
Buckingham GR, 119, 123, 126, 249, 259, 271, 274, 321, 481–2
Budia F, 155, 259, 433
Bukovinsky T, 131–4, 137, 332
Bulet P, 144, 255
Burbutis PB, 360
Burdick DJ, 414
Burgess ED, 438–9, 525
Burghardt RL, 72
Bürgi L, 296
Burgos AM, 506
Burkholder W, 332
Burls K, 335
Burns JM, 272
Burrough A, 42
Burwell CJ, 263, 356
Butcher BA, 15, 63–4, 92, 123, 173, 198, 229, 248–50, 339, 503–4, 528
Butcher FG, 393
Butcher RDJ, 109, 115
Butler GD Jr, 88, 266
Butt GW, 157–8, 221, 394, 419
Byers JA, 482

C

Caballero P, 479–80
Caccia S, 157–9, 221
Caccialupi P, 158
Cai D-Z, 157
Calvert DJ, 75, 214,217, 221
Camara MD, 480
Camarao GC, 121, 432
Cambra RA, 181, 356
Cameron E, 258, 448
Cameron MD, 419
Cameron TM, 332
Campbell CD, 231, 525
Campbell RW, 376
Canale A, 117, 258–9
Cancino J, 259
Cangelosi B, 218
Cannon CA, 466
Cao T-T, 59, 148–9, 156
Čapeki Z, 299
Čapek M, 4, 79–80, 183, 202–3, 231, 253, 260, 279, 282, 292, 294, 299, 306–7, 310, 327, 337, 439, 514
Capinera JL, 472
Carabajal-Paladino LZ, 258–9
Cárdenas G, 510
Carignan S, 306
Carlson RW, 419, 428, 433, 448
Carlysle TC, 54, 259
Carpenter JE, 394, 474
Carpenter JM, 51, 192
Carpita A, 258
Carrillo MA, 466
Carter D, 526
Carvalho SS, 286
Casartelli M, 82, 159, 216
Casiraghi M, 166, 391, 393
Casperd J, 93
Cassidy JD, 59, 72, 125
Cassis G, 481
Castelo MK, 182, 432
Castillo A, 258
Castillo C, 356
Casula P, 13, 260
Cattolico L, 149, 152
Caudri LWD, 85, 253
Cavalloro R, 474
Ceballos R, 390–91
Çelik S, 474
Chadwick CE, 263, 355–6, 400
Chakravorty S, 143
Chanda S, 143
Chang S-C, 257
Chanzy H, 43
Chao A, 495
Chao HF, 407
Charif D, 114
Charlet LD, 280
Chase TR, 498
Chatterjee SN, 245
Chau NNB, 93, 298
Chaubet B, 221
Chauvin G, 69
Chavarría L, 234
Chen G-F, 516–17, 532
Chen L, 13
Chen X, 245–6, 249–50, 271, 275, 295, 301, 304–5
Chen X-X, 54, 59, 64, 66, 69, 145, 148–9, 156–7, 195, 205, 223–4, 257–8, 294, 325, 394, 433, 435, 437, 479
Chen Y-F, 59, 66, 145, 148–9, 156
Cheng DD, 215–16, 220
Cherqui A, 64, 143–5, 254
Chertkova EA, 268
Cherubin LL, 334
Chessel D, 97
Chesters D, 485
Chianese RJ, 471
Chiang HC, 286
Chien C-C, 257
Chiri AA, 316
Chiroleu F, 258
Chiu S-C, 199
Choi S-H, 4
Choi WY, 229
Chong L, 13, 376

Chorney RJ, 12, 13, 216
Chou L-Y, 223
Chow FJ, 92, 215–16, 220
Chowdhury S, 150
Christiansen-Weniger P, 158, 214
Chrystal RN, 350–51, 353, 379
Chu J, 245
Chudek JA, 82
Chumakova BM, 61
Cladera JL, 258–9
Clancy DW, 257
Claret J, 463
Clark AM, 109
Clark CA, 286
Clark RT, 397, 510
Clark SJ, 217
Clausen CP, 78, 88, 182, 184, 385, 387, 395, 411, 413, 444
Click A, 152
Cline LD, 468
Cloutier C, 72,78, 215, 221, 296–8, 464
Clutts SA, 275, 504
Coddington JA, 509
Cohen AC, 474
Cohen MB, 214, 462
Colares F, 330
Colazza S, 14, 218, 307, 360
Cole LR, 99, 120, 299, 309, 374–6, 397, 456
Coles LW, 300, 360, 433
Colinet D, 142–3, 154
Colvin AM, 59, 73, 148, 418, 455
Colwell RK, 509
Compère P, 214
Compton SG, 193
Comstock JH, 20–25
Conlong DE, 442
Connahs H, 328, 508
Cônsoli FL, 114, 156, 159, 325
Conyers CM, 144
Cook D, 155
Cook JM, 31, 112, 114, 115, 525
Cook SP, 100, 389, 484
Coop LB, 463–4
Cooperband MF, 124, 236
Copland JW, 253
Coppel HC, 296–8
Corbet SA, 82
Cordeiro EG, 330
Córdova-García G, 97, 257, 482
Corley JC, 182, 432
Coronado-Rivera J, 73, 407, 414
Correa LRB, 263
Cortesero AM, 457
Coseglia AF, 421–2, 476, 485
Coskun M, 375
Cossé AA, 124, 236
Cossentine JE, 403
Costa VAA, 184, 296–7
Cotton J, 108, 267
Couchman JR, 78, 215–16
Coudron TA, 156
Coulibaly F, 152
Courtois CM, 444
Cousserans F, 143, 152, 154, 435
Coutinot D, 306
Coutourier A, 375
Cox-Foster DL, 154
Cranford JD, 477
Crankshaw O, 124, 381
Crankshaw OS, 380–81
Crawford AM, 142, 308
Crawford AW, 403
Crocker B, 430, 437, 441–2
Croft BA, 156
Croft P, 253
Crook AM, 82
Crook KE, 470
Crosby K, 114, 324, 492
Cross JE, 261, 434, 476, 485
Cross WH, 85
Cruz I, 303
Culin JD, 325
Cummins HM, 59, 73, 148, 418, 455
Ćurčić SB, 32, 39
Cushman RA, 76, 277–8, 284, 394, 407, 411, 414, 429
Cusson M, 149, 151–2, 154, 404, 432

D

Daane KM, 261, 286, 459–60, 462
Dahlman DL, 158–9, 334
Dahms EC, 14
Daly H, 391
Damiens D, 70
Danci AT, 67, 118, 332, 376
Danell K, 511
Dangerfield PC, 85, 208–9, 215 226, 312, 314–15, 335
Dani MP, 144, 375
Daniel C, 457
Daniel DM, 184, 286
Danks HV, 312, 314, 369, 427
Dapkey T, 504
Darboux I, 154
Dardani M, 13, 260
Darwish E, 49, 54, 129
Das P, 13, 58
Dasch CE, 397, 430, 466
David BV, 316, 325
Davidová-Vilímová J, 299, 306–7
Davies AP, 435
Davies DH, 72, 138
Davies NW, 12, 381
Davis DM, 379–80, 383, 461, 477
Davis DS, 303
Davis DW, 463, 471
Dawah HA, 460
Day WH, 11–17, 306, 421–2, 433, 512
Deans AR, 51, 192, 499, 512
de Boer JG, 109
de Brito EF, 330
de Buron I, 149, 155, 325, 334
Decker UM, 217
Declos X, 194, 196, 210, 224
Dedeine F, 114
de Eguileor M, 52, 85, 159, 215–16
De Freitas MdLT, 333
Degerli N, 260, 268, 339
Deglow EK, 403
de Jesús de Luna-Santillana E, 155
Delalibera I Jr, 479
Delebecque CJ, 143, 221
De Leon D, 113, 262, 464, 525
del Estal P, 433
Delfín-González H, 462
De Lima Mendonza A, 333
Delkov A, 272, 274
DeLoach CJ, 303–4
Delph JS, 315–7
DeLury NC, 68, 316, 317
Del Valle C, 390–91
Delvare G, 328
de Macêdo MV, 185, 228, 234
De Marzo L, 61–2, 64, 214, 217
Demba WM, 389
de Menten L, 125
Demettre E, 152, 154
de Moraes CM, 457
Demmon AS, 158, 221
De Nardo EAB, 471
De Plöeg G, 287
Derocles SAP, 221
De Saeger H, 277
de Sasst, 296, 481
Desmier de Chenon R, 321
Desneux N, 143, 221, 261
Desouhante E, 114
Destain J, 214
Dettner K, 66, 218
De Voss JJ, 381
Devotto L, 390–91
DeWitt PR, 429, 433
Dey D, 13
Dhillon MH, 122
Dharmarajan P, 390
Dias AB, 265
Dib-Hajj SD, 435

Diboll AG, 247–8
Dicke M, 128–9, 161, 330–32
Dickey SE, 149, 214
Dickie AG, 214
Dieckhoff C, 217–18
Digilio MC, 52, 156–7, 160, 215–16, 220–21, 475
Digweed SC, 411
Dijkerman HJ, 242
Dillard CR, 72, 74
Diller E, 398
DiLuna FA, 158
Ding D, 287
Ding T, 159
Dingremont A, 253
Dion E, 220
Dittrick LE, 286
Dixon AFG, 490
Djoumad A, 151, 404
Dlussky GM, 193
Dmoch J, 128
Dobata S, 157, 300
Do Carmo Pedroso E, 263
Dogangil G, 39
Dogot P, 213
Dolphin KP, 322, 492–4, 496
Domerg C, 258
Dondale CD, 272
Dong K, 158
Dong S, 394
Donisthorpe HStJ, 85
Donovan BJ, 392, 463
Doolittle RE, 129
Dorémus T, 154, 435
Dorn S, 92–3, 95, 109, 128, 132, 164, 166, 328, 334, 375–6
Dos Santos Rolo T, 339
Doucet D, 432
Douglas AE, 221
Douloumpaka S, 133
Doury G, 140, 143–5, 152, 253, 255
Dowden PB, 78, 95, 246–8, 270, 454
Dowell RV, 60, 121–2, 434, 469
Dowling APG, 48, 51, 126, 171, 192, 205–6, 212, 224, 229–30, 237–9, 243–4, 250, 269, 271–3, 276, 278–9, 281–4, 288, 294–5, 305, 307, 309, 311–13, 321, 503, 534
Dowton M, 117–18, 204–5, 224, 229–30, 250, 294, 303, 312–13, 317, 321, 325, 353,437, 447, 504
Doyon J, 303–5, 330
Drea JJ, 157, 300
Dresner E, 257
Drezen J-M, 145, 48–50, 152, 155
Driessen G, 458
Driver F, 263, 356
Drooz AT, 432
Du Y-J, 129, 220
Duan JJ, 257, 469
Dubois F, 144, 255
Dubois J, 369, 373
Dubovskiy IM, 268
Dudarenko GP, 177
Duerr SS, 459–60, 462
Duffield RM, 466–7
Duncan PM, 192, 194, 205
Duodu YA, 471
Dupas S, 150
Dupuy C, 148, 150, 152, 334
Dweck HKM, 13, 15, 49, 54
Dyer LA, 328, 335, 498, 508
Dysart RJ, 300, 429
Dytham C, 506, 508, 510

E

Eady RD, 20
Easingwood R, 308–9
Eberhard WG, 120, 161, 372, 482
Edgar AL, 482
Edgar WD, 82, 482
Edgerly J, 91, 236
Edson KM, 52, 63, 77, 217, 223, 271, 298
Edwards JP, 144–5, 296, 375, 471
Eggleton P, 39, 120, 124, 170, 231, 351, 353, 361, 367, 381
Eichhorn O, 139, 416, 418
Eijs EA, 457–8
Eisenberg A, 374
Eklund LR, 421–2
El Agoze M, 69, 109, 152, 334
El-Sawaf BM, 143
El-Sawaf SM, 325
El-Shazly M, 129
El-Sherif H, 129
El-Sufty R, 158–9, 479
Eleftherianos I, 133
Elguero1 E, 303–5, 330
Eller FJ, 68
Ellers J, 61, 82–3, 108–9, 457–8
Elliott DE, 236
Elliott JM, 385
Ellsbury MM, 476, 485
Ellsworth p, 316
Elmes GW, 395, 397, 510
Eltoum EM, 257
Elzen GW, 15, 18, 334
Elzinga JA, 110, 471
Emre I, 374–5
Engel MS, 194, 196, 210, 224, 237, 353
Epila JSO, 282
Er A, 145
Ergin E, 83, 115
Erlandson MA, 306
Erwin TL, 105, 146, 322, 367, 442, 491
Escoubas JM, 154
Eslin P, 140–41, 143–5, 253
Espagne E, 149
Espelie KE, 67
Euaparadorn M, 295, 506–7
Evans AA, 308
Evans AC, 253
Evans EW, 457
Eveleigh ES, 114, 324, 492
Evenhuis HH, 100, 369, 389
Ewen AB, 141, 405

F

Fabrick JA, 82, 156, 268
Fadamiro HY, 13, 58
Fahringer J, 269, 275, 339
Falabella P, 149, 157–60, 220–21
Falco Garí JVC, 64,177, 179, 236, 238, 241, 265–6, 302
Fallis AM, 325, 328, 332
Fanti P, 156, 160, 220–21, 475
Faria CA, 458
Farley RD, 54–5
Farneti R, 82, 216
Farris SM, 58, 130
Fatouros NE, 131
Faulds W, 362
Fawke JD, 42, 124, 383
Fayad YH, 296
Featherston PE, 214
Febvay G, 156, 221
Feddersen I, 139
Federici BA, 152
Federici M, 54, 155, 161, 391, 395
Feldman CR, 335
Fellowes MDE, 145
Feng C-J, 286–7
Feng M-G, 257–8
Ferber S, 117, 125
Ferguson AW, 428
Ferguson CM, 142, 470
Ferkovich SM, 72, 74, 145, 474
Fernandez DRR, 263
Fernández-Triana JL, 105, 114, 146, 322, 324, 334, 485–6, 491–2, 495, 497, 503
Ferns PN, 460, 462
Ferrari J, 141, 506
Ferreira CS, 263
Ferroni F, 39
Ficken L, 39, 43, 236, 411, 414
Fidgen JG, 92

Field SA, 67, 123, 332–3
Figueiredo MLC, 303
Figueroa IA, 131
Filipi FL, 360
Fincke OM, 372
Fine ML, 332
Fine PVA, 510
Fink DE, 86, 120, 286
Finlayson T, 79–80, 214, 216–17, 223–4, 367, 407, 433
Finney GL, 69, 184, 286, 525
Firlej A, 303–5, 330
Fischer K, 385
Fischer M, 241, 253, 260, 490
Fischer S, 128
Fisher RC, 15, 64–5, 82, 94, 433
Fitt GP, 394
Fitton MG, 25, 27, 31, 36–9, 41, 43, 47–49, 52–3, 59, 69, 76, 100, 102–3, 164, 193–4, 196, 202, 205, 208–10, 236, 238, 250, 271,288, 294, 312–13, 342–3, 350, 359, 364, 366–7, 369, 372, 374–5, 378, 388, 383–4, 388–9, 391, 405, 411, 414, 430, 437, 439, 44–21, 444–5, 447, 460, 492
Flaherty BR, 72
Flanders RV, 113, 120, 272, 293
Flanders SE, 59, 69, 184, 286, 525
Fleming J-AGW (also as 'JA' or 'GW'), 146
Fleury F, 114
Floren A, 511
Flores S, 18, 43, 165, 186, 263–4, 355
Flores-Prado L, 355
Foerster A, 202
Foissner W, 339
Foray V, 114
Fordy MR, 72–4
Forister ML, 335
Fortier JC, 250, 263
Foster WA, 173
Fouillet P, 114
Fourdrain Y, 140, 144, 255
Franqui RA, 265
Franz JM, 428, 455
Franzen C, 478–9
Fraser SEM, 506, 508, 510
Frasson L, 39
Frère I, 214
Frick H, 365–6
Frilli F, 392
Friman V, 498
Fritzsche-Hoballah MEF, 129
Frohne WC, 390
Frost CJ, 130
Fu W-J, 139, 286–7
Fuester RW, 214, 374, 471
Führer E, 62–3, 83, 138, 157–9, 325, 327, 374–5, 479
Fujii S, 159
Fukushima J, 65, 268
Fulton BB, 327, 331
Furihata SX, 143–4
Futerman PH, 478–9

G

Gaasch CM, 509
Gadallah NS, 13, 15, 49, 54
Gahan AB, 19–20
Gaitens PV, 36, 41, 47, 49, 52–3
Gandolfi M, 132
Ganesalingam VK, 54, 435
Gangrade GA, 182, 325
Gao Y, 334
Garcia JL (also as 'Garcia R, JL'), 438–9, 458
Gärdenfors U, 221–2
Gaskett AC, 376
Gaston KJ, 491, 493, 497, 502, 506, 517
Gatti J-L, 142–3, 154
Gauld ID, 2, 4, 14–15, 17–18, 21–2, 30–31, 39, 42–3, 49, 73, 76, 85, 88, 90, 92, 100–104, 120–21, 124, 164, 174, 179–81, 183 ,196–9, 284, 320, 342–6, 350–51, 353–7, 359–61, 367, 369–70, 372–3, 374–5, 377–9, 383, 385, 388–9, 391, 394–5, 397–9, 401–3, 405–7, 413, 414, 416–21, 424, 428–34, 437–9, 442, 444–8, 458, 484–6, 491–3, 497, 499, 504–9, 517, 527, 536
Gauthier N, 13, 130
Gayubo SF, 253
Gebremeskel ZH, 97, 394
Geervliet JBF, 129
Gelman DB, 474
Gençer N, 115
Genduso P, 260
Geneau CE, 457
Genieys P, 113, 122, 265
Gentry G, 457
Geoghegan IE, 303
Georgiev G, 272, 274, 327
Gerber GH, 299
Gerding M, 390–91
Gerig L, 76, 394, 411, 414, 424
Gerling D, 61–2, 69, 123
Gers C, 334
Ghosh SM, 91, 337–8
Giannantonio S, 160
Giannotti P, 259
Gibb H, 496, 511
Gibbons JRH, 380
Gibert P, 114
Gibson GAP, 393
Gibson LP, 284
Gigliotti S, 157, 221
Gilbert P, 114
Gill JS, 128–9
Gillespie DR, 487
Gillespie JJ, 250, 260, 269, 271
Gillespie OR, 433
Gillogly PO, 110–111, 254
Gilstrap FE, 256
Gimeno Martos C, 64, 196, 208, 238, 250–51, 255, 260, 312–13, 342, 367
Giordana B, 82, 85, 159, 215–16
Giordanengo P, 140–41, 253, 458
Giraldo-Vanegas H, 438–9, 458
Giraud-Guille M-M, 43
Given BB, 58
Glaizot O, 92
Glas PCG, 258
Glasgow H, 281
Glinwood RT, 220
Glover PM, 90, 263, 267
Glupov VV, 168, 268
Godfray HCJ, 31, 88, 96,110, 112, 114–15, 117, 132, 134, 140–41, 145, 161, 218, 265, 275, 335, 392, 461, 479, 482, 484, 487, 506
Godoy C, 92, 403
Goh MZ, 118–19, 275
Gokhman VE, 14, 108, 115, 125–6, 287, 360, 383, 388, 395, 512
Goldson SL, 300
Gols R, 97
Gómes Durán J-M, 307, 367, 441–2
Gómez IC, 92, 403
Gong H, 159
Gonzaga AD, 509, 516
Gonzaga MO, 372
Gonzalez D, 526
González FJ, 258
González PI, 259
Gonzalez-Candelas F, 223–4
Gonzalez-Herrera A, 73, 407
González-Moreno A, 462
Goodell PB, 306
Goolsby JA, 263, 356
Gopalapillai R, 300
Gordh G, 70, 121, 434–5, 526
Gorgiev G, 270
Gothilf S, 112, 317, 319
Goudegnon AE, 477, 479
Gould JR, 54, 124, 231, 236

Goulet H, 19–22, 24, 27–8, 198, 306, 432, 471, 497, 539
Graber W, 14, 211, 363–4
Graenicher S, 355
Grafen A, 38–9, 90, 178, 196, 209, 238, 270
Grant JF, 296, 471
Gravenhorst JLC, 342, 346
Graziani F, 157–8, 221
Grbic M, 75
Greany PD, 54–5, 128, 259, 293, 474
Green GW, 157
Greeney HF, 498
Grégoire J-C, 477
Gregory TR, 125
Greis G, 67, 118, 332
Greis R, 67, 332
Grenier S, 385, 474
Gries G, 376, 390
Gries R, 390
Griffiths DC, 105, 139, 141, 219–21, 486
Griffiths GCD, 251, 255
Griffiths KJ, 95, 97, 419, 464
Grimaldi A, 82, 85, 159, 215–16
Grinberg PS, 96
Grinter CC, 328
Griset J, 110–111, 254
Grosch DS, 119, 264, 266–7
Grosman AH, 330
Grossniklaus-Bürgin C, 77–8, 83–4, 149, 160, 316, 327, 471, 473
Grosu L, 267
Gruber A, 84, 149
Gu H, 92–3, 95, 109, 132 328
Gu L-Q, 95
Gu S-H, 13
Guerdoux J, 164, 375
Guerra AA, 156, 265
Guerrieri E, 129, 220
Guertin DS, 113
Guévremont HC, 416–8, 454–5, 525
Guggisberg U, 83
Guidone L, 306
Guillaud J, 156, 221
Guillaume S, 96
Guillot FS, 64
Guillot S, 140
Guinnee MA, 92
Gül N, 375
Gülel A, 83, 525
Gundersen-Rindal D, 152
Gündüz NEA, 83, 525
Guo Y-Y, 13
Guo X-W, 155
Guo Z-H, 516–17, 532
Guppy JC, 253
Gupta V, 413–14, 425, 428, 435
Gürbüz MF, 369
Guryanova TM, 156, 376
Gutiérrez AP, 219, 469
Gutierrez-Ibanez C, 133, 217
Güven T, 64
Guzo D, 99
Gyapay G, 152, 334

H

Ha D-M, 4
Haas-Stapleton EJ, 155
Haataja S, 506, 510
Habib MEM, 433
Habibi J, 300
Haeselbarth E, 88, 99, 279, 282
Haeussler GJ, 287, 421
Hafez M, 315, 325
Hagen KS, 120, 123, 257, 259
Hagstrum DW, 468
Hågvar EB, 213, 215, 221, 462, 466
Hailemichael Y, 374
Halfhill JE, 214
Hall JC, 215–8
Hallez P, 74
Hallwachs W, 105, 146, 322, 328, 485, 491, 499, 504, 512
Hamaguchi K, 115
Hamelin M-H, 152, 155
Hammock BD, 160
Hammond A, 334
Hamon C, 69
Hance T, 72, 213–5, 221–2, 466
Hancock GLR, 466
Hanidah J, 509
Hanks LM, 97–8, 110, 231, 233, 455, 482, 525
Hanski I, 324, 334, 483, 499
Hanson HS, 100, 379
Hanson PE, 73, 88, 181, 185, 198, 234, 407, 444, 507
Happe M, 70
Haramoto FH, 123
Harbo JR, 316
Harcourt DG, 468
Hardie J, 158, 214, 217, 220, 465
Hardy ICW, 92, 393
Harris AC, 481
Harris EJ, 258, 525
Harris J, 47–8, 288, 294
Harrison EG, 64–5
Hart PJ, 471
Hart WG, 256
Hartzer KL, 144, 268
Harvey JA, 75–6, 78, 82, 88, 97, 109–10, 161, 184–5, 316, 327–8, 330–32, 335, 389, 393, 460, 462, 471–2, 475, 477, 479, 499
Hashimoto Y, 67, 218, 259
Hasnah BT, 509
Hassell MP, 483
Hata Y, 129
Hatekayama M, 73
Haubruge E, 214
Haviland MD, 253
Hawke SD, 54–5, 259
Hawkins BA, 454–5, 468, 470, 484, 486
Hawlitzky N, 76, 313, 316, 471
Hayakawa Y, 139, 143–4
Hayashi CY, 78
Haye T, 428, 471
Hayes PY, 381
Hazell SP, 141, 471
He J-H, 66, 145, 246, 249–50, 294, 299, 433
He K-L, 13, 53–4, 164, 287, 391
He X-J, 122
He XZ, 60–61, 120–22, 435
Headrick HL, 98
Heath RP, 129
Heatwole H, 379–80, 383, 461, 477
Hébert C, 298
Hebert PDN, 485–6, 497, 499, 512
Heckel DG, 220
Hedderwick M-P, 97, 109, 152, 334
Hedgekar BM, 374
Hedlund RC, 115–7, 422
Hedqvist K-J, 234, 238, 244–5
Hegazi EM, 61, 95–6, 325, 327
Hegdekar BM, 374
Heimpel GE, 108–9, 143, 214, 217–18, 221, 265, 267, 458, 460, 462, 466, 480
Heiniger P, 84, 149
Heinrich GH, 395, 397–8
Heitland W, 417–18
Heller M, 143, 152, 334
Henaut A, 164, 375
Henderson RE, 83, 296, 315, 475
Hendrickson R, 70, 121, 434–5
Hendry LB, 68, 128, 259, 376
Henkelman DH, 112
Henneman ML, 470
Henri H, 114
Henry LM, 487
Henschen W, 72
Henson RD,
Henter HJ, 219
Hentz M, 316
Heraty JM, 51, 192
Herbinière J, 149, 152, 334
Héricourt F, 150

Hernández -Ortiz V, 279
Hernández -Rodriguez E, 14
Herniou EA, 196, 152, 208, 238, 250, 312–13, 342, 367
Heron RJ, 85
Hespenheide HA, 481, 520
Hick AJ, 220
Hidaka T, 31, 161, 332, 472
Higgins L, 372
Hijii N, 231
Hildahl V, 432
Hilgarth RS, 149
Hill CC, 233
Hillebrand H, 490
Hilpert H, 31
Hilszczański J, 496, 511
Hincks WD, 408
Hinz R, 173, 387, 394–5, 409, 411, 416
Hirose Y, 112, 213, 460, 468
Hirosie Y, 112
Hiyori T, 316
Hjältén J, 496, 511
Hoch G, 152
Hochberg ME, 397, 510
Hochuli A, 160
Hocking H, 355, 383, 468
Hoek GV, 216
Hoelmer KA, 261, 306, 360
Hoffmeister TS, 391
Hofsvang T, 213, 215, 462, 466
Hogervorst PAM, 457
Holdaway FG, 157, 299–300, 303, 309, 339
Höller C, 220
Hollingdale J, 13
Holloway AK, 108, 267
Holloway GA, 180, 353
Holt RD, 482–3
Hom R, 214
Honeycutt RL, 250, 260, 269, 271
Hong YC, 194
Hopper KR, 471
Horn DJ, 121–2, 214, 389, 393, 434, 469
Horstmann K, 322, 391, 427, 429, 445, 490, 511
Hortal J, 505
Hotta M, 158
Hougardy E, 477
House HL, 374, 474–5
Howard RW, 64–5, 67, 268, 286
Howell J, 82
Hoy HL, 159
Hoy MA, 214
Hrabar M, 118, 376
Hrcek J, 92, 198, 248, 484–5, 516, 528
Hsiao TH, 468
Hu J, 139, 286–7
Hu JS, 474
Huang F, 59, 66, 69, 145, 148–9, 156, 325, 433
Huang JH, 287
Huang XF, 260
Huang YP, 287
Hubbard SF, 82, 97, 109, 435
Huber JT, 19–22, 24, 27–8, 198, 497, 539
Hubner G, 218
Huddleston T, 19–20, 80, 81, 167, 179, 182, 197, 233, 238–40, 244, 247–8, 253, 258–9, 269–70, 272, 281–2, 284, 286–7, 291, 293, 296–7, 300, 309–10, 317, 320–21, 324–5, 327, 442, 460, 462, 481
Huguet E, 148–50, 152, 334
Huigens ME, 131
Huignard J, 69
Hullé M, 221
Humala AE, 360
Humble LM, 466
Hung AF, 116–17, 422
Hunt RE, 332
Hunter MS, 220
Hunter G, 82
Huot L, 215
Hurlbutt B, 31
Hussain MA, 393
Hutcheson KA, 487
Hutchison WD, 466

I

Iaccarino FM, 215
Ibrahim AA, 296, 315, 325
Ichimura DM, 525
Idar M, 121
Ideo S, 114, 116
Idris AB, 509, 516
Igarashi Y, 86, 325
Ikeda H, 157, 300
Ikoma S, 118
Ikonen S, 498
Imafuku M, 161, 332
Inaba F, 115, 126
Infante F, 234
Ingram SN, 36, 41, 47, 49, 51–3, 69, 202, 271, 481
Iqbal M, 170, 319–20
Isaac RE, 375
Isaki Y, 115
Ishii M, 385
Isidoro NF, 14, 18, 39, 157, 164, 218, 221, 360, 375–6
Iştvan ÖV, 83, 525
Itô M, 157, 300
Ivanova-Kasas OM, 75, 214, 433
Ives AR, 158, 221
Ivie MA, 360
Iwabuchi K, 474
Iwata K, 41, 59, 72–3, 148, 150, 258, 263, 271, 321, 374, 391, 395, 403, 407, 411, 414, 423, 430, 460
Iwaya M, 152

J

Jacas J, 155, 259
Jackson CG, 88, 266, 315–7
Jackson DL, 116, 122, 158, 300, 308
Jackson LL, 64, 68, 315
Jacob HS, 457
Jacob S, 70
Jacobson HR, 236
Jang H-S, 74
Janssen A, 330
Janvier H, 355
Janzen DH, 85, 105, 146, 272, 277–8, 296–7, 315, 322, 328, 433–4, 437,484–5, 490–91, 497, 499, 504, 512
Jaramillo MA, 335
Jarlfors UE, 158
Jarzembowski EA, 192
Javadi I, 351
Javier PA, 438
Jell PA, 192, 194, 205
Jennings JT, 275
Jennings MJ, 512
Jensen LBM, 403
Jepson WF, 272
Jerman EJ, 88
Jervis MA, 61, 118, 168, 275, 376, 405, 454–5, 457, 460, 462, 471
Jewett DK, 13
Jiang N, 96, 98, 376
Jiménez R, 253
Jiménez-Pérez A, 258
Jiménez Peydró R, 64, 302
Johansson T, 496, 511
Johner A, 153
Johns CV, 95–6, 112, 414–15, 430
Johnson JA, 325
Johnson JB, 105, 246
Johnson MW, 261
Johnson NF, 526
Johnson SJ, 293, 457
Johnstone PD, 470
Joiner RL, 64
Jones D, 82, 141, 143, 152, 160, 317, 376

Jones G, 160
Jones GZ, 86, 296–7, 494
Jones OR, 296, 322, 484, 492–6, 501, 510
Jones RE, 464
Jones RL, 67–8, 287, 297
Jongema Y, 393
Jordan T, 416, 428–9, 512
Jørgensen OF, 95
Joshi NK, 299
Jouan V, 152, 154, 435
Jourdheuil P, 278, 428–9
Joyce AL, 110, 129, 332
Judd GJR, 68, 316, 317
Jugovits T, 325, 327
Juillet JA, 61, 112, 293, 370–71, 480
Jumean Z, 390
Jung D-O, 4
Junnikkala E, 157
Jussila R, 14, 369, 485, 506

K

Kaartinen R, 486
Kabissa JCB, 398
Kacem N, 39, 53–5, 379–80
Kadash K, 149
Kadono-Okuda K, 300
Kaeslin M, 76–7, 83, 143, 160, 316–17, 327
Kainoh Y, 68, 83, 123, 157, 160, 315–17
Kairu E, 95
Kaiser L, 458
Kalmes R, 82, 97
Kamano Y, 316
Kambhampati S, 223–4
Kami V, 471
Kankare M, 324, 334, 499
Kapyla M, 369
Karaca I, 462
Karatolos N, 164–6
Kares EA, 296
Karlsson D, 10
Kasparyan DR, 14, 74, 195, 349, 369, 505, 411, 413–14, 416, 439, 442, 455
Kato Y, 86, 325
Kawakami T, 315
Kayumbo HY, 398
Ke X, 287
Keen JN, 144
Keller MA, 67, 123, 332–3
Kenchington W, 85, 287
Kendall ME, 109
Kenis M, 96, 281, 300, 394, 419, 428
Kergoat GJ, 286
Kerrich GJ, 389, 416, 425, 429
Kester KM, 332
Ketseoglou I, 148
Kfir R, 115, 395, 476
Khafagi WE, 61, 95–6
Khalaim AI, 193, 428–9
Kharazi-Pakdel A, 218
Khaskin G, 316
Khatri D, 60–61, 120–22, 435
Khoo CCK, 259
Kidd NAC, 376, 454–5, 457, 460, 462, 471
Kilincer N, 374
Kim Y, 154, 158
Kimura MT, 114, 116, 143–4
Kindlmann P, 490
King EG, 471
King MJ, 43
King PE, 78, 215–16
King RC, 59, 72–4, 125, 471
Kirkbride-Smith AE, 296
Kistner DH, 236
Kitajima EW, 114
Kitano H, 67, 109–10, 333
Kitching IJ, 55, 324, 334
Kitching W, 381
Kittel RN, 315, 319
Kiuchi M, 83, 160, 317
Klag J, 74
Klitz W, 214
Kloet GS, 408
Klopfstein S, 14, 192, 211, 363–6, 485, 448
Ko SY, 39
Koch RL, 218
Kochetova NI, 156
Koenders JTH, 242
Kolaib MO, 296
Kolarov JA, 192, 369, 376, 420
Kölliker M, 459
Kolyada VA, 193
Komatsu T, 441–2
Konishi K, 372, 441–2
Kopelke JP, 384, 428
Koponen M, 420
Koptur S, 328
Kopylov DS, 192, 194, 196, 346, 348–9, 353, 413
Korenko S, 373, 393
Kos M, 161, 330–31
Kotzen ML, 479
Kovaleva ES, 143
Kozlov MA, 193
Kraaijeveld AR, 140–41, 145, 159, 253, 293, 478–9, 487, 506
Kraaijeveld K, 114
Kraft KJ, 316
Krainska MK, 405
Kratsas RG, 266
Kraus M, 374
Krause G, 74
Krell PJ, 138, 145–6, 149, 152, 155
Kremer N, 114
Krespi L, 222
Krick TP, 68
Krishnan A, 143
Kröder S, 166
Kroemer JA, 149
Krogmann L, 51, 192
Kropf C, 14, 363–6
Kruft RA, 497
Krüger K, 129
Krugner R, 286
Kruidhof HM, 131
Krutov VV, 14, 360
Kryukov VY, 168
Kryukova NA, 168, 268
Kubrakiewicz J, 74
Kudô S, 299, 307
Kugimiya S, 259, 516
Kugler J,
Kuhlmann U, 306
Kuhn EJ, 459–60, 462
Kuhn L, 151, 404
Kuijper B, 109
Kula RR, 170, 172, 235–6, 238, 293
Kulhanek CA, 458
Kulman HM, 68, 432, 471
Kumar P, 471
Kunkel D'Herculais J, 300
Kuriachan I, 160, 315, 475
Kurt E, 84, 149
Kusano T, 68
Kutima HL, 96, 98, 376
Kuwahara Y, 65, 68, 268
Kyriak IG, 217

L

Labate CA, 159
Labate MTV, 159
Labeyrie V, 55, 93, 95
Laca R, 315
Lacey LA, 98
Lachaud J-P, 308
Laforge M, 154
Lampert EC, 498
Lane RD, 334
Lanfranco D, 399
Langer A, 466
Langlois C, 300
Langor DW, 411
Lanzo F, 259

Lanzrein B, 76–8, 83–4, 143, 149, 152–3, 160, 315–7, 327, 334, 471, 478
Lapointe R, 149, 151, 404
Lara RIR, 263
Larocca A, 475
LaSalle J, 335
La Scaleta, 149
Latham DR, 214
Lathrop FH, 257–8, 432
Latorre A, 223–4
Laurenne NM, 39, 46, 100, 164–6, 171, 181,184, 198, 247–9, 258, 265–6, 272–5, 276, 278, 282, 284, 342, 381, 387, 389, 391–3, 395, 397, 430, 437, 441–2, 482, 485, 516
Laurentz M, 498
Lawrence PO, 94, 138, 155, 258–9
Lawson AB, 286
Lawton JH, 482–3
Layen SJ, 479
Lazzari C, 459, 516
Leach WE, 202
Leathers JW, 275, 480–81
Lecomte C, 130, 132, 395
Lee J, 73
Lee JC, 458, 460
Lee K-Y, 4
Lee S, 154
Legal L, 334
Legeai F, 152
Leggo JJ, 400
Legner EF, 316
Legrand M-A, 213
Lehmann-Danzinger H, 375
Lei G-C, 480
Leibee GL, 458
Leibenguth F, 84
Leirana-Alcocer J, 462
Leisy DJ, 267
Leius K, 370, 455, 458
Lejeune RR, 297, 432
Le Lann C, 393
Le Lannic J, 39, 41, 44, 53–5, 215, 222, 379–80
Lelito JP, 124, 236
Leluk J, 141
Lemesle M, 395
León-Regagnon V, 178, 186, 231, 234, 245–6, 487
Leonardi MG, 52, 85, 215–16
Leong JKL, 433, 435
Lepetit D, 114
Lepointe R, 152
LeRalec A, 36, 38, 43, 54–5, 63, 72–3, 101, 102, 104, 214, 217, 221, 383
Leroy PD, 214
Leroy T, 13, 130
Lesobre J, 149–50, 152
Leutenegger R, 158, 300, 303
Levesque RC, 152
Levie A, 213
Levin D, 498
Lewis JW, 457
Lewis OT, 335
Lewis WJ, 128 131, 158, 314–15, 325, 497
Leyva-Vázquez JL, 123
Li DM, 117
Li FF, 294
Li J, 164, 391
Li L, 265, 474
Li S, 160
Li T, 516–17, 532
Li WD, 66, 145
Li X, 222
Li X-Y, 157, 260
Li Y, 287
Lieber K, 332
Liedo P, 258–9
Liepert C, 66, 218
Liljeblade J, 51, 192
Lill JT, 248, 406, 506
Lim KP, 306, 433
Lima ER, 330
Limentani EC, 487
Linboterhaar B, 472
Lindstedt C, 498
Lindstrom KG, 149
Lino-Neto J, 69
Linsenmair E, 411
Linsley EG, 481
Lipkow E, 282, 299, 302
Liu ES, 84, 88, 233
Liu GJ, 84, 88, 233
Liu H-W, 68
Liu H, 462
Liu S, 97, 257–8, 394
Liu Z-W, 13
Lloyd DC, 96, 394
Loan CC, 299–302, 305–6, 308, 339, 512
Loder N, 502
Lohr B, 469
Long KD, 243
Longino JT, 367, 442
Lopez M, 479
López-Martínez V, 258
Lopez-Vaamonde C, 260, 268, 339, 344. 440, 447, 486, 529
Lo Pinto M, 218
Lorenz H, 117
Lovallo N, 154
Lozan A, 180, 463
Lu C-K, 403
Lu J-F, 286–7
Lu WJ, 287
Lucarotti CJ, 155
Lucchetta P, 459, 516
Lucchi A, 259
Luck RF, 526
Luckhart S, 154
Luff ML, 300, 384
Luisa De Stradis M, 157, 221
Luka H, 457
Lukáš J, 263
Lum PTM, 72
Luo L, 259
Luo LZ, 334
Luo Y-Q, 403
Lux SA, 257
Lyal CHC, 335
Lyngnes R, 38, 42
Lyons DB, 468

M

Ma J-H, 231
Ma L, 13
Ma W-J, 109
Ma Y, 294
Maaroufi H, 152
Mabiala-Moundoungou ADN, 143
Mack TP, 442, 444, 471
Mackauer M, 92, 97, 112, 212, 214, 216–218, 220–21, 223–4, 462, 465, 477, 490
Mackinnon EA, 155
MacNicoll AD, 144
Madanagopal N, 154
Madar RJ, 325
Madden JL, 12, 380–81
Madeira M, 466
Madel G, 70, 296
Maetô K, 32, 68, 88, 91, 93, 110, 115, 184, 166, 202, 236–7, 243, 248, 295–9, 307, 324, 335, 467, 505
Magdaraog PM, 97
Magnoli I, 72
Magro SR, 265, 474
Magurran AE, 509–10
Maia DP, 184, 296–7
Majerus MEN, 303–4
Makino S, 392
Makinson J, 263, 356
Makkay AM, 149, 155
Maleny G, 282, 310
Maltais J, 298
Maltby HL, 429

Malva C, 149, 157–8, 160, 221
Mandon N, 130
Mangine T, 258
Manica A, 303
Mann J, 220
Mantel P, 267
Manzari S, 276, 284, 311
Mappes J, 498
Marais SJS, 316
Maranga RO, 96, 98, 376
Mardulyn P, 230, 313, 325
Marris GC, 93, 97, 296, 471
Marsh PM, 46, 64, 85, 89, 181, 185, 214, 231, 233–6, 238, 241, 246, 265, 339, 490
Marsicovetere E, 475
Marti D, 84, 149
Martin PB, 128
Martinez JJ, 186, 234, 245
Martinez S, 114, 156, 265
Martínez-Martínez L, 123
Martinoli A, 166, 391, 393
Marucci PE, 257
Masler EP, 143
Mason HC, 237
Mason PA, 498
Mason PG, 325, 327, 334, 405, 464, 471
Mason WRM, 55, 175, 190, 195–6, 210, 282, 286, 304–5, 312–13, 319–22, 324, 335, 346, 339, 385, 387, 390, 411, 413, 440, 460, 481
Mateo Leach I, 117, 125–6
Mathé-Hubert H, 142–3
Mathur RN, 245, 393
Matos LF, 159
Matsumoto H, 143–4
Matsumoto R, 161, 171, 302, 350–51, 370, 372, 392, 477
Matthews JR, 124, 381
Matthews RW, 119, 122, 124, 380–81, 439
Mattiacci L, 129, 132, 332
Maund CM, 468
Maure F, 303–5, 330
Mayhew PJ, 88, 92, 253–4, 506, 508, 510
Mazanec Z, 433
Mazón M, 508, 516, 521
Mazzi D, 334
Mazzoni V, 259
McAllister MK, 161
McBrien H, 97
McClure M, 218
McConnell HS, 76, 419
McCoy ED, 490
McCulloch A, 142
McHugh JV, 283
McKellar RC, 353
McLaughlin RE, 479
McNeil JN, 215, 217–18, 222
McNeill MR, 300
McPheron BA, 154
Medina P, 471
Meers S, 65, 286
Melo GAR, 85, 233–4
Meloche F, 253
Memmi M, 88, 307
Memmott J, 335, 470
Mendiola P, 54, 161, 391
Meng C, 72, 74
Menier J-J, 194, 424
Mercado I, 315
Merilo MT, 497
Merritt DJ, 78
Merritt L, 220
Mertins JW, 296–7
Mesones I, 510
Messenger PS, 219, 469
Messing RH, 94, 257, 261, 469
Messner B, 387
Metcalf JS, 334
Mettus RV, 108
Meyer M, 97, 160
Meyer NF, 376
Meyer V, 160
Meyer WL, 93
Miah MI, 105,430, 437, 441–2, 445
Michaud D, 218, 220
Michaud JP, 72
Michel-Salzat A, 335
Michelena JM, 223–4
Middeldorf J, 375, 479
Mikhailenko AP, 125–6, 287
Milani A, 82, 216
Milano P, 286
Millar JG, 110, 129, 231,233, 482, 525
Miller JC, 308, 325
Miller JS, 335
Miller SE, 92, 198, 248, 484–5, 516, 528
Milligan B, 88
Mills NJ, 92, 100, 129, 214, 296, 468, 482, 485, 529–30
Ming G, 215–16, 220
Mitchell HC, 85, 261
Mitchell PA, 431, 442, 444, 505
Mitsuhashi J, 474
Mitsui H, 114, 116
Mitter E, 128, 376
Miura K, 115, 141,155, 405, 435
Mochiah MB, 114
Mochizuki N, 525
Moe A, 460
Mohajer YJ, 332
Mohamed SA, 257
Mohamed UVK, 120, 263
Mohammed-Ali A, 155
Mojica HB, 123
Molchanov MI, 156
Moller H, 480
Momoi S, 463–4
Monteiro RF, 185, 234
Monteith AM, 394
Montgomery VE, 429, 433
Montoya P, 258–9
Moody S, 498
Mook JH, 15, 255
Moon RD, 466
Moore CJ, 381
Moore CT, 509
Moracci M, 158
Morales J, 471
Morales-Ramos JA, 474
Morallo-Rejesus B, 121, 432
Moran NA, 220
Moreau S, 64, 141, 144, 253, 255
Moreau SJ, 145
Moreira J, 69
Moreno Marí J, 64, 302
Morgan ED, 64–5, 268
Mori M, 91, 184, 206, 212, 228–9, 237–9, 241–2, 245, 247–50, 260, 268–9
Morley C, 274
Morneau DC, 471
Morrill W, 65, 286
Morris KRS, 413
Morris RJ, 482
Morrison G, 490
Morse DH, 118–19, 275
Moser SE, 97–8, 455
Mound LA, 164, 342
Moura JCMS, 184, 296–7
Mourad AK, 93
Moutia LA, 444
Mouton L, 114, 164, 342
Mudd A, 15, 64–5
Muhammed A, 375
Mühlen D, 70
Muldrew JA, 416
Muli BK, 96, 98, 376
Müller CB, 131, 141, 435
Murata Y, 114, 116
Muratori F, 72, 215
Murney R, 309
Muroni P, 13, 260
Murphy N, 153, 206, 312–13, 325

Murphy SM, 248
Murray D, 51, 192
Myers JG, 255

N
Nadel H, 261
Nagai K, 299
Nägeli W, 321
Nair KK, 456
Nair PN, 143
Nais J, 263
Naito T, 115
Nakahara Y, 474
Nakai A, 129
Nakamatsu Y, 78, 141, 159, 161, 316, 328, 330–31,327–8, 335, 405, 475
Nakanishi A, 79, 82, 363
Nalini M, 154
Namba O, 141, 405
Namyatova AA, 481
Naranjo S, 316
Narayanan ES, 76, 121, 261, 316, 325
Narayanaswami PS, 61, 233
Narendran TC, 120, 263
Nassar JM, 43, 186, 263–4
Nasser M, 390
Natasha BY, 509
Nats J, 263
Navasero RC, 15, 18, 334
Nazzi F, 218
Nealis V, 469
Needham DG, 20–25
Neemann EG, 315–7
Nees von Esenbeck CG, 202
Nel A, 25, 287, 190, 192–4, 205, 209–11
Nell HW, 41, 44, 93
Nelson AJ, 158, 221
Nemoto T, 68
Nénon JP, 39, 41, 44, 53–5, 215, 379–80
Nentwig W, 14, 211, 363–4
Neponnyashaya AM, 265, 474
Nesbit DJ, 334
Neuvonen S, 506
Newman TM, 70
Newton RC, 257–8, 432
Ngi-Song AJ, 114
Nguyen Q, 267
Nguyen R, 214
Nguyen TT, 72
Nho S-K, 4
Nielsen E, 103
Niemeyer HM, 18, 120, 133,165, 217–18, 355
Nieminen M, 498
Nieto DJ, 306
Nikitin MI, 263, 355–6, 400
Nikolić Z, 32, 39
Nishida K, 263, 297
Nishida T, 260, 457
Nishimura T, 467
Nisole A, 152
Nixon GEJ, 312, 319, 321, 324
Niyibigira EI, 109
Noirot C, 64, 67, 78
Nor Zaneedarwaty N, 509, 516
Norberg RÅ, 20
Nordin JH, 466–7
Nordlund DA, 128
Norton WN, 77, 146, 149, 315
Norullahğlu ZÜ, 83
Nose Y, 73
Nottingham SF, 220
Notton DG, 100, 342, 388, 427, 439, 441
Nötzl P, 119, 123, 302–3
Novak H, 218
Novotny V, 484, 516
Noyes JS, 199, 246, 490, 514, 516
Nuñez-Campero SR, 97, 257, 482
Nur Azura A, 509
Nuttall MJ, 172, 379–81
Nyabuga FN, 220
Nychka JA, 39, 41, 44–5, 50–52, 64, 293

O
O'Connor JM, 335
O'Donnell DJ, 76, 182, 216, 220, 247
Oakleaf R, 93
Oatman ER, 112–13, 120, 122, 272, 274, 293, 433, 435, 438
Obara M, 67
Oberlander H, 74, 474
Obonyo M, 18, 54
Obrtel R, 280
Obrycki JJ, 303
Ochieng SA, 13, 334
Oda RAM, 228
Ode PJ, 108–9, 113, 265
Odebiyi JA, 112–13, 272, 274
Odejar M, 159
Oeser RI, 36
Ohbayashi N, 372
Oike K, 438
Oishi K, 73
Oishi Y, 123, 317
Okada I, 438
Okuda T, 158, 300
Oliveira Fonsesca J, 330
Oliver KM, 220
Olkowski H, 214
Olkowski W, 214
Olszewski JA, 149
Omata K, 395, 463, 465
Ordano M, 97, 257, 482
Orion T, 317
Ortega-Blanco J, 194, 196, 210, 224, 237
Ortel J, 473
Osborne P, 281, 429
Osman SE, 62–3, 139, 144, 374
Otten H, 131, 164, 375
Outreman Y, 220–21
Overholt WA, 95, 109, 114, 257
Ovruski SM, 97, 257–8, 279, 482
Owen DF, 490, 497
Owen J, 462, 490, 497
Özalp P, 375
Özgökçe MS, 462
Özlük A, 375

P
Page M, 122
Paillot A, 286
Paine EO, 231, 233, 482
Paine TD, 110, 129, 231, 233, 482, 525
Pair SD, 394
Palevody C, 61
Pallini A, 330
Palevody C, 61
Palmer-Wilson J, 42
Pampel W, 58, 59, 400–401, 444
Pan J, 64, 435
Panda RN, 143
Panis A, 307
Pannebakker BA, 109
Pansa MG, 306
Papeschi AG, 258–9
Papp J, 202, 212, 224, 269,
Pardo X, 253
Pare F, 215
Paris J, 395
Park J-S, 282, 310
Park KC, 282, 310
Parker FD, 472
Parker HJ, 433
Parker HL, 122, 286, 429, 525
Parkinson NM, 144, 145, 375
Parnell JR, 279–80
Parra JRP, 265, 286, 474
Parrish DS, 463
Parrott PJ, 281
Pashalidou FG, 131
Pass BC, 433
Patel PN, 433
Patel VS, 282, 310

Patton JC, 250, 260, 269, 271
Pavenstädt-Grupp I, 375, 479
Payne TL, 13
Pech LL, 139
Peck O, 30, 32, 68, 121
Peck RW, 295, 506–7
Pedraza-Lara C, 234
Pekár S, 373, 393
Pélissié B, 286
Pels C, 213
Pemberton CE, 258
Pemberton RW, 293
Pena JE, 472
Penado A, 466
Penev L, 275
Penna M, 120, 218
Pennacchio F, 14, 52, 85, 129, 143, 149, 156–60, 215–16, 218, 220–201, 273, 303, 315, 475
Penteado-Dias AM, 64, 234, 236, 266, 372, 481
Perdereau E, 150
Pereira AG, 303, 481
Pérez J, 258
Pérez-Alonso R, 279
Perez-Lachaud G, 259, 308
Pérez-Rodríguez MA, 155
Perfilieva KS, 193
Peri E, 307
Perioto NW, 263
Periquet G, 109, 152, 155, 158, 334
Perkins JF, 2, 31, 199, 398, 439, 442
Perrichot V, 25, 190, 192–4, 205, 209–11
Perry DF, 298
Persad AB, 214
Perugino G, 158
Peter C, 316, 325
Petroski RJ, 124, 236
Petters RM, 75, 108–9, 129, 143, 266
Pettersson EM, 129
Pettersson RB, 496, 511
Pexton JJ, 254
Pfister-Wilhelm R, 76, 78, 84, 149, 152–3, 160, 315–7, 334, 471, 473
Phillips CB, 300
Phillips J, 332
Piao MH, 294
Pichon S, 150
Pickering J, 366–7, 497, 507, 509, 536
Pickett CH, 261, 306
Piek T, 267
Pijnacker LP, 114–15
Pimentel MCP, 185, 234
Pinheiro DO, 325
Pinnell RE, 472
Pinto CF, 120, 218
Piper RW, 428
Pitre HN, 86, 325, 328
Pitz KM, 43, 171, 310
Plarre R, 332
Platner GR, 122, 293, 438
Plotnikov V, 430–31
Poelman EH, 76, 393
Poinar G Jr, 158, 308
Poirié M, 109, 142–3, 149, 152, 154, 334,
Polaszek A, 157
Polgár LA, 217, 465
Polidori C, 54, 161, 391
Pond CM, 490, 497
Ponlet N, 303–5, 330
Ponsard S, 286
Poppy GM, 129, 220
Porcelli F, 430
Porter CC, 392–3, 398–9, 428, 503–4
Porter K, 325
Potapova ES, 12
Potting RPJ, 131
Pottler B, 433
Poulain J, 129, 143, 154, 220
Powell JE, 93–4, 98
Powell W, 129, 217–18, 220, 476
Pratt HD, 32, 33
Press JW, 468
Prévost G, 64, 114, 131, 138, 140–41, 143–5, 152, 255
Price PW, 97, 460–61, 486, 497
Prince GJ, 253
Principi MM, 88, 299
Prinsloo GL, 257
Proctor J, 481
Prokopy RJ, 258
Proper AB, 297
Prosser SW, 486
Proveaux AT, 129
Provost B, 149
Pschorn-Walcher H, 41–8, 370, 411, 413, 416, 418–19
Pucci T, 168–9, 275, 519
Punju E, 97, 485
Purisima EO, 150
Purvis A, 322, 342, 493–6, 501, 510
Puttler B, 300, 325, 327–8, 433

Q

Qin J, 154
Qin QL, 159
Qiu HG, 287
Qiu ZL, 287
Quednau FW, 272, 274, 416–8, 432–3, 454–5, 525
Quennedey A, 64, 67, 78
Quezada JR, 444
Quicke DLJ xiii, 12, 17–18, 25, 27, 32–3, 36–9, 41–3, 46–9, 51–3, 62, 55, 59, 62–4, 68–70, 72, 76, 78–81, 85–6, 88, 90–92, 95, 99, 100–105, 114–15, 123–5, 139, 164–7, 170, 173, 177–9, 181, 186, 190, 192–4, 196, 198, 205–6, 209–12, 218, 223–4, 228–9, 231, 234–5, 237–9, 241–2, 245–6, 249–51, 255, 260, 263–4, 268–70, 272, 274, 277–8, 281, 284, 293–4, 302, 304–5, 307, 310, 312–13, 317, 321–2, 336, 339, 342–4, 352–3, 355, 357, 359–60, 366–7, 369,373, 377–8, 381, 383, 385, 387–9, 391–3, 395, 397–400, 402–3, 405–11, 419–28, 430, 436–7, 439–442, 444–5, 447–8, 481–2, 485, 487, 492–7, 501, 503–5, 508, 510, 512, 516, 528, 529, 534
Quilici S, 258
Quimby JW, 432
Quimio GM, 120, 257, 259
Quistad GB, 267

R

Rabasse JM, 54, 115
Rabb RL, 312, 314
Rabouille A, 152, 155, 158
Rahbé Y, 156, 221
Rahm F, 381
Rahman H, 37, 39, 52
Rahoo GM, 384
Rains GC, 134
Ramadan MM, 31, 257
Ramirez BW, 235
Rana RL, 158
Rao A, 83, 296
Rappaport N, 122
Rasch EM, 125
Rasekh A, 218
Rasnitsyn AP, 17, 25, 27, 38, 59, 190, 192–4, 205, 209–10, 369, 397, 444–5, 466
Raspi A, 258
Ratcliffe NA, 72–3
Rathcke BJ, 497
Rathmayer W, 143
Rattanadechakul W, 149
Ratnasingham S, 486
Ravallec M, 154, 435
Reardon RC, 84, 88, 100, 233, 389, 393, 484
Rechav Y, 317

Reck-Malleczewen V, 152
Reed DA, 157
Reed-Larsen DA, 157, 317, 465
Régnière J, 78, 296–8, 432, 464
Rehman M, 472
Reid DG, 299
Reid JA, 170
Reinecke M, 117
Reineke A, 437
Reis LG, 479
Rejmanek M, 218–19
Renault S, 13, 130, 155, 158, 395
Rendahl AK, 109
Reudler JH, 498–9
Reumer BM, 114
Reyes-Villanueva F, 155
Rhodes LD, 471
Richards EH, 144, 375
Richerson JV, 13, 129, 303–4
Riddiford LM, 334
Ridgway RL, 433
Ridley M, 120
Riedel M, 397
Riegel GT, 20, 22
Riekkola M-L, 498
Ring RA, 466
Rivero Lynch AP, 31
Rivers DB, 115, 138, 144, 154
Riviello L, 157–8, 221
Robacker DC, 26, 259
Robacker KM, 156, 265
Rocha KL, 258
Roditi I, 152, 334
Rodriguez JJ, 105, 146, 322, 485, 491, 495, 499, 503, 512
Rodriguez R, 306
Rodríguez-Berrío A, 508
Rodríguez-Castañeda G, 335, 508
Rodríguez-Cerezo E, 155, 259
Rodriguez-del-Bosque LA, 272
Rodriguez-Leyva E, 279–80
Rodríguez-Pérez MA, 155
Rodriguez-Saona CR, 93, 130
Rodriguez y Baena F, 39
Roelfstra L-L, 96
Rogers D, 52
Rohlfs WM III, 442, 444, 471
Roisin Y, 125
Roitberg BD, 161, 487
Rojas E, 372
Rojas JC, 258, 479
Rojas MG, 474
Rojas-Rousse D, 61, 68–9, 76, 82, 88, 93, 95, 97, 374
Roman A, 367, 376
Romani R, 14, 18, 164, 218, 375–6
Romano A, 220
Romano VA, 475
Romeis J, 457
Ronquist F, 51, 294, 298, 192
Roques A, 477
Rosa GM, 466
Rose A, 12, 13, 376
Röse USR, 457
Rosenberg HT, 316, 438
Rosenberger JH, 247–8
Roslina T, 486, 495
Ross AJ, 192
Ross HH 21-Feb,
Ross KM, 64–5
Rossbach A, 469
Rossi GD, 159, 325
Rossi M, 158
Rossodivita T, 157, 221
Rotary N, 61–2, 69, 123
Roth T, 143
Rotheram S, 72, 139, 146
Rotheray GE, 120, 362–4, 366, 457
Rotundo G, 474
Rougerie R, 486
Rouleux-Bonnin F, 158
Rousse P, 258, 398
Roux O, 334
Rowher SA, 19–20
Ruberson JR, 325, 335
Rubin MA, 109
Rudall KM, 85, 287
Rudinsky JA, 128
Rudnicka M, 152
Ruf D, 334
Rufino do Nascimento R, 333
Rungrojwanich K, 259
Ruokolainen K, 510
Russell DA, 286, 393
Russo G, 263
Rutherford P, 399
Ruthmann A, 63, 375, 479
Ryan PJ, 158, 221
Ryan RB, 98, 128, 465
Ryan RE, 88

S

Sääksjärvi IE, 14, 356, 367, 407, 442, 448, 506, 508, 510
Saarela J, 144, 255
Sabatier L, 144, 255
Sabelis MW, 330
Sabri A, 214
Sáez AG, 243, 246–7, 249–50
Saigusa M, 141, 405
Saigusa T, 370, 392
Sait SM, 479
Saito T, 479
Sak O, 83
Salerno G, 307
Salkeld EH, 59, 61, 76, 122, 253
Sallam MN, 95
Salmah Y, 516
Salo J, 506
Salom SM, 13
Salt G, 75, 86, 96, 138–9, 145, 170, 172, 255, 359–60, 466, 468
Salvador G, 156
Salzberger B, 478
Samain S, 152
Samietz J, 128, 166
Sampaio M, 466
Sanchis A, 223–4
Sander K, 139
Sandlan K, 111
Sandrock C, 218
Sanford MR, 134
Sano T, 472
Sant'Ana AEG, 333
Santiago-Alvarez C, 479–80
Santos AMC, 501, 505
Santos BF, 516
Santos MA, 181, 356
Sarmiento CE, 272
Sasakawa K, 157, 300
Satchell D, 282, 310
Sathe TV, 433
Sato Y, 31, 161, 332, 472
Saunders C, 411
Sauri P, 498
Savasci S, 115
Savopoulou-Soultani M, 466
Sawoniewicz J, 389, 391
Scatolini D, 273
Schaefer PW, 67, 118, 332, 376, 471
Schaffner U, 131, 435
Schaller J, 143
Schepers EJ, 158–9
Schepers T, 158
Schirrmeister BE, 218
Schittny JC, 83
Schleifer KW, 334
Schlinger EI, 78, 215–8, 222
Schlup J, 129
Schlyter F, 61
Schmeider RG, 85, 392
Schmid HM, 131
Schmidt J, 141
Schmidt K, 13
Schmidt O, 60, 117–18, 139, 144–5
Schmidt S, 393, 521–2
Schneider D, 166
Schneider F, 95, 362–4

Schneider MI, 109, 112
Schneider MV, 434
Schödelbauerová I, 490
Schoenbohm RB, 472
Schopf A, 152, 473
Schröder D, 99
Schroder RFW, 301–2, 525
Schroeder J, 478
Schulmeister S, 51, 58, 130, 192
Schulthess F, 96, 98, 376
Schümperli D, 84, 149
Schwartz BD, 381
Schwartz M, 39, 100, 391, 393, 448
Schwartz MD, 306
Schwarzfeld M, 295, 506–7
Schwenke W, 421–2
Scliserman P, 97, 257, 482
Scott EI, 364
Scott JR, 146
Scrimgeour C, 97
Scudder GGE, 36
Sedivy J, 360
Sedlag U, 59, 217
Segura DF, 258
Seitner M, 119, 123, 302–3
Selfa J, 54, 161, 391, 466
Seltmann K, 275
Semichon L, 392
Serbielle C, 149–150
Serrão JE, 481
Sevenster JG, 108, 458
Sevsik J, 360
Seymour JE, 464
Sgobba D, 88, 307
Shah ZA, 53–4, 435
Shalaby FF, 296, 325
Shanower TG, 360
Shapiro A, 161
Shapiro BA, 366–7, 497
Sharanowshi BJ, 39, 41, 44–5, 48, 50–52, 64, 73, 126, 205–6, 212, 224, 229–30, 237–9, 243–4, 250, 269, 271–2, 279, 281–4, 293–5, 305, 307, 309, 311–13, 321, 485, 503, 534
Sharifi S, 351
Sharkey MJ, 15, 19–24, 26, 39, 41, 43–5, 48, 50–52, 64, 73, 85, 89, 119, 123, 126, 168–9, 171, 173, 190, 192–3, 195–6, 205–6, 210, 212, 224, 229–30, 237–9, 243–4, 249–50, 259, 269, 271–7, 278–9, 281–4, 286, 288, 293–5, 305, 307, 309–10, 311–13, 315, 321–2, 385, 437, 460, 480–81, 485, 503–4, 534
Sharma AK, 216
Sharma HC, 122
Sharma KR, 13
Shaumar N, 12, 58
Shaw MR, 17–20, 25, 39, 49, 63–4, 73, 78, 85–6, 88, 91–2, 100–103, 105, 109–12, 119, 139, 143, 160, 164, 176–7, 179, 181–2, 184, 192, 197, 199, 205, 229, 237–40, 243–4, 246–51, 253, 257–9, 270, 272, 274, 277–8, 282, 284, 286–7, 291, 293–5, 297, 300, 303, 310, 317, 320–21, 324–5, 327–8, 330–31, 334–5, 357, 359, 366–7, 369–70, 372, 374–5, 377–8, 389, 391, 394–5, 403, 407, 409, 414, 416, 421–2, 424, 427, 432–3, 445, 448, 454, 456, 458, 462, 472, 476, 483–7, 499, 510, 512, 514–15, 523, 525–6, 528
Shaw SR, 86, 91, 119–20, 182, 229, 234, 236–7, 243, 246–7, 249–50, 281, 294–7, 299–305, 307–10, 317, 335, 465, 481, 484, 485
Sheehan W, 484
Sheffield CS, 497
Shelby KS, 149, 156, 300
Shenefelt RD, 260, 265–6, 287, 439, 490
Sheng M-L, 351–2, 359, 403, 407, 516–17, 532
Shepard M, 296, 471
Shevyrev IY, 74
Shi M, 59, 148–9, 156, 205, 223–4, 325, 394, 433, 437, 479
Shi Z, 97, 394
Shigesada K, 86, 325
Shiinono Y, 129
Shim J-K, 4
Shima H, 484, 516
Shimazaki M, 182, 432
Shimazaki Y, 182, 432
Shimizu K, 68, 316
Shimoda T, 259, 516
Shimuzy K, 68
Shin IJ, 68
Shirai S, 88, 184, 297
Shoemaker DD, 114
Short JRT, 79–82, 100, 173, 183, 281, 309–10, 359, 363, 367, 387, 398, 408, 427, 433, 444
Short MW, 458, 521
Shorthouse JD, 400
Shu S, 67
Shuja-Uddin, 217
Siekmann G, 117
Sigwalt B, 20
Sileshi G, 281, 300
Silva RB, 303
Silver GT, 297
Silveria A, 429
Sime KR, 131, 261, 374, 397, 490, 498–9, 501
Simmonds FJ, 61, 165, 272, 414, 463
Simões RA, 479
Simon J-C, 220–21
Simoncini R, 258
Simpson RG, 421–2, 434, 476, 485
Sinan S, 115
Singer MS, 498
Singh H, 237
Singh M, 129
Singh RP, 312, 314
Sinu PA, 390
Siscaro G, 261
Sithole R, 424
Sittertz-Bhatkar H, 146, 265–6, 335
Sivinski J, 97, 257, 259, 279, 332, 482
Sizaret P-Y, 152, 155, 395
Skillen EL, 497
Skinner ER, 350–51, 353, 379
Sláma K, 263
Slansky F, 472
Slepneva IA, 268
Slobodchikoff CN, 81, 185, 355, 430–31, 481
Sloggett JJ, 160, 304
Slovák M, 405
Sluss RR, 158, 300, 303
Smart CB, 92
Smid HM, 13, 58, 13143, 137, 332, 334
Smilanich AM, 498
Smith E, 428
Smith EL, 36
Smith HA, 76, 180, 300, 494–5
Smith HD, 233
Smith HS, 69, 184, 286, 525
Smith I, 144, 145, 375
Smith JW Jr, 68, 272, 293, 374, 457
Smith MA, 15, 92, 105, 114, 146, 173, 198, 248, 322, 324, 484–6, 491–2, 497, 499, 503–4, 512, 528–9
Smith MC, 15, 64–5
Smith OJ, 72, 158, 247–8, 300
Smith P, 308
Smith PT, 223, 224
Smith RH, 267
Smittle BJ, 468
Smith Trail DR, 161
Smolenski G, 142
Snaas H, 393
Snyder WE, 158, 221
Sobczak JF, 184, 296–7, 372

Soldán T, 156, 216
Soller M, 317
Solleza MO, 438
Solter LF, 479
Sommerman KM, 299
Son Y, 459–60, 462
Song K-S, 4
Song QS, 287
Sonnet PE, 128
Soreck AP, 133
Southern PS, 312, 314
Southgate HH, 158
Southwood TRE, 120
Spanjer W, 267
Spanoudis CG, 466
Sparks DW, 477
Speicher BR, 41, 74, 108–9
Speicher KG, 109, 115
Spence JR, 411
Spencer JR, 525
Spradbery JP, 122, 128, 378, 380, 477, 525
Sprung L, 498
Stahlhut JK, 497
Stainer JER, 99, 463
Stamp NE, 161
Stanisavljević L, 32
Stanković SS, 32, 39
Stanley D, 287
Stanton ML, 263
Staple JO, 457
Starý P, 156, 216–9, 221, 224
Stasiak K, 155, 395
Stefanelli J, 143
Stefanescu C, 324, 334, 375, 377, 499
Steibauer MJ, 458
Steidle JLM, 131–4, 137, 332
Steinberger P, 473
Steiner H, 91–2, 245
Steiner S, 14, 211, 363–4
Steiner WWM, 109
Stelfox AW, 120
Stelinski LL, 93
Stenbacka F, 496
Stepper J, 13
Stettler P, 84, 149
Stevens GC, 490
Stevens NB, 275
Stewart RK, 306
Stewart SL, 303
Stigenberg J, 294, 298
Still L-A, 308
Stoelb SAC, 275
Stoepler T, 248
Stojanović D, 32, 39
Stoltz DB, 99, 138, 141, 145–6, 148–52, 155, 298, 308, 404
Stouthamer R, 109, 114–7
Strand MR, 75, 108, 109–10, 113, 117, 122–3, 129–30, 139, 143, 149, 158–9, 199, 264–7, 477
Strange-George JE, 458
Streams FA, 360
Strickland EH, 296
Stuart AM, 370
Subba Rao BR, 76, 121, 216, 316, 325
Sudheendrakumar VV, 120, 263
Sugie H, 316
Sugimoto T, 129
Suh KI, 229
Sullivan DJ, 215, 220
Summers MD, 146, 298, 435
Sun S-P, 407, 351–2, 359, 516–17, 532
Sunderland KD, 93
Suomi J, 498
Sutherland M, 300
Sutherland TD, 78
Sutou M, 157, 300
Suzuki M, 145, 155, 296, 298
Suzuki T, 65, 268
Swedenborg PD, 67–8, 287
Sweeney W, 108, 267
Swezey SL, 306
Syme PD, 157
Syrjämäki J, 120, 282
Syvertsen TC, 64, 68, 315
Szépligeti G, 269, 278

T

Tadic MD, 325
Tagashira E, 157
Tagawa J, 31, 67, 85, 109–10, 113, 123, 333, 472
Takabayashi J, 259, 516
Takács S, 332
Takada H, 67, 218
Takahashi M, 86
Takashino K, 435
Takasu K, 131, 265, 394–5
Takasu Y, 86, 325
Takasuka K, 372
Takeda S, 83, 160, 317
Talluto MA, 246
Tamaki Y, 115, 316
Tamò C, 96
Tan FF, 155, 334
Tan J-C, 260
Tan VT, 394–5
Tanaka K, 149
Tanaka T, 76, 78, 88, 97, 112, 141, 145, 151, 154–5, 157–9, 161, 184–5, 296–8, 330–32, 316, 327–8, 331–2, 335, 404–5, 472, 475
Tanner D, 129
Tanton MT, 282
Tardieux I, 115
Tartally A, 395
Taschenberger D, 387
Tatman KM, 100, 389, 393
Tatsuki S, 68, 316
Tauber CA, 303
Tauber MJ, 303
Tavella L, 306
Tawfik MFS, 315, 325
Taylor AD, 267
Taylor PB, 374
Taylor RAJ, 109
Teder T, 31, 113
Teig DA, 109
Tepe EJ, 335
Teresa Valente M, 157, 221
Tereshkin AM, 398, 518
Terra WR, 265, 474
Tettamanti G, 85, 215–16
Thakere KR, 76, 121, 316
Theopold U, 60, 117–18
Théry M, 459, 516
Thewke SE, 325, 327–8
Thézé J, 152
Thibaudeau C, 13, 130
Thibout E, 13, 130, 132, 395
Thinart P, 214
Thireau JC, 78, 296–7, 464
Thoetkiattikul H, 149
Thomas F, 303–5, 330
Thomas JA, 395, 397, 510
Thomas MB, 468
Thompson DJ, 479
Thompson GH, 41, 52, 378
Thompson LC, 432
Thompson SN, 78, 82–3, 370, 376, 473–4
Thompson WR, 433–4
Thornton IWB, 505
Thorpe WH, 76–7, 293
Thurston EL, 146, 149
Tian S, 97
Tichenor RB Jr, 471
Tillinger NA, 152
Tillman PG, 93–4, 98
Timberlake PH, 94, 158, 215
Tobias VI, 301, 308
Tobias VL, 12, 15, 32, 82, 121, 202–3, 209, 217, 223, 225, 229, 243, 251, 269, 271, 284, 294–5, 300, 307–9, 321 , 324, 339, 439
Togashi I, 63–4, 66
Tomanović Z, 221
Tomberlin JK, 134

Topping CJ, 93
Tormos J, 253
Tothill JD, 88, 430
Townes H[K], 2, 21, 22, 36, 44, 103, 179, 194, 199, 342, 344, 349, 356, 357, 359, 360, 366, 369, 376, 379, 388, 389, 391–393, 398, 401, 405–8, 413, 417, 418, 423, 425, 428–30, 435, 437–9, 445, 448, 449, 454, 462, 481, 490, 497, 503, 504
Townes M, 103, 357, 359, 398, 405, 406, 408, 423, 462, 481, 490
Townsend AC, 229, 237
Tranfaglia A, 157–8, 220–21
Trang TTK, 13
Traynor RE, 253
Treiblmayr K, 160
Tremblay E, 30, 32, 75, 85, 157–9, 129, 214–7, 220–21,224, 315, 457, 474
Triana E, 86, 297
Triplehorn CA, 526
Tripp DW, 478
Tripp HA, 73, 445–6
Tschopp A, 397
Tsubouchi K, 86, 325
Tsutsui Y, 115
Tucker E, 275
Tucker EM, 504
Tuda M, 109
Tumlinson JH, 129, 457
Tunstead J, 64, 236, 238, 241, 265
Tuomisto H, 508, 510
Turlings TCJ, 96, 129, 458
Turnbull AL, 99, 463
Turnbull MW, 149, 152
Turpin FT, 472
Turrisi GF, 196

U

Uçkan F, 64, 83, 115
Udayagiri S, 297
Uefune M, 259, 516
Ueno K, 456
Ueno T, 73, 93, 97–8, 370, 374, 376, 392, 395, 456, 462, 471, 472
Ugalde J, 92, 403
Ullyett GC, 76, 315–16
Umphrey GJ, 125
Unruh TR, 98, 390, 526
Urano T, 231
Urbach S, 152, 154, 435

V

Vagina NP, 81
Valdez EW, 477
Valvasson R, 85, 215–16
van Achterberg C, 15, 17, 19–20, 22–23, 26–8, 30, 32–3, 36, 42, 44, 53, 59, 62–64, 69, 76, 80, 91–2, 99, 120, 164, 169, 173–4, 177, 193–6, 198–9, 202–5, 208–10, 212, 224–5, 228–30, 236–9, 242–3, 246–6, 249–51, 253, 255–7, 260, 263, 265–6, 273, 275–9, 281–4, 287–8, 290–95, 300–33, 305, 307–12, 315, 319–22, 325–5, 327, 335–7, 339, 420, 428, 432, 437, 439–2, 463, 481, 503–4, 516, 519, 528
Van Aelst AC, 13, 334
van Alphen JJM, 92–3, 97, 108–9, 114, 140, 159, 254, 458
Van Baarlen P, 93
van Baaren J, 222, 334
van Batenburg FDH, 96
Van Bockstaele DR, 125
Vance AM, 61, 78, 315–16, 325, 327, 454
Vandame R, 479
van de Kamp T, 339
van den Bergh B, 157
van den Bosch R, 214, 219, 221, 468–9
van der Ent L-J, 210
van der Wel NN, 160
van der Zande L, 109, 117, 125–6
Van Driesche RG., 469
van Duinen G-J, 457–8
van Emden HF, 133
van Lenteren JC, 39
van Loon JJA, 13, 58, 131–4, 137, 332, 334
van Marle J, 64
van Noort S, 275, 322, 398, 508
van Nouhuys S, 97, 182,324, 334, 375, 377, 432, 483, 485, 498–9
van Santen MCP, 434
van Steenis MJ, 214
van Strien-van Liempt WTFH, 96
van Veen FJF, 482
van Veen JC, 39, 52, 120. 394, 405
Varela-Hernández F, 234
Vargas V, 510
Vargas-Osuna E, 479–80
Várkonyi G, 486, 495
Varley GC, 73, 445
Varricchio P, 149, 157–8, 221
Vasconcellos-Neto J, 184, 296–7
Vásquez RA, 218
Vavre F, 114
Veenendaal RL, 267
Veijalainen A, 367, 442, 508
Venkatraman TV, 261
Verma SK, 299
Vermillion K, 124, 236
Vernon P, 222
Vet LEM, 58, 82, 97, 109, 128–9, 131–4, 137, 161, 258, 330–32
Via S, 219
Vickery RA, 444
Vidal S, 469
Vieira RC, 185, 234
Viel SR, 263
Viereck HL, 199, 202, 339
Vikberg V, 420, 432
Vilhelmsen L, 18, 36, 43, 45, 51, 55, 63, 101–2, 104, 164, 170, 192, 196, 383
Villa-Ayala P, 258
Villagra CA, 120, 133, 217–18
Vincent B, 143
Vincent JFV, 42–3, 124, 383
Vinchon S, 64, 253
Vinogradova EB, 463
Vinson SB, 12, 14, 52, 58–9, 63–64, 68, 72,74, 77, 83, 93, 112, 119, 120, 122, 128–30, 140, 145–6, 149, 153–4, 156, 158–9, 217, 223, 271, 296, 298, 308, 314–6, 332, 360, 434, 457, 471, 474–5
Viñuela E, 109, 112, 155, 259, 433, 471
Viscarret MM, 258
Visser B, 82, 393
Visser ME, 31, 93, 108, 110, 254, 462
Vlug HJ, 100, 369, 389
Vogler AP, 485
Völkl W, 114–15, 160, 212, 218–19, 221, 223–4, 465, 477
Volkman LE, 155
Volkoff A-N, 152, 154, 334, 435
Volpe HXL, 263
von Niederhäusern F, 152–3, 315
von Borstel RC, 267
Vorburger C, 218
Vorontsova YL, 268
Vuong R, 43

W

Wäckers FL, 14, 128, 131, 164, 375–6, 457–9, 460
Wadhams LJ, 218, 220
Wagenaar R, 82, 389, 393
Wagener B, 437
Wagner DL, 317
Wago G, 159
Wago H, 154

Wahl DB, 15, 18,28, 42–4, 78–84, 120, 124, 180–81, 183–4, 190, 192–3, 195–6, 198, 210, 346, 349, 353–7, 359–60, 363, 366, 367, 369–70, 378–9, 385, 387, 397–8, 400–403, 405–6, 408, 421–2, 424–8, 433, 435, 439, 444, 447–8, 504
Wahlberg N, 367, 442
Wajnberg E, 54, 154, 435
Walker AK, 55, 299, 317, 324, 328, 334, 526
Walker GP, 295
Walker MF, 303
Walkley LM, 398
Walla T, 508
Waller A, 287
Walley GS, 432
Wallner WE, 96
Waloff N, 158, 299–300, 308, 421
Walter GH, 120, 257, 259
Walters T, 508
Walther C, 143
Walton M, 457, 460
Wan Z-W, 64, 195, 257–8
Wang C, 97, 154
Wang C-Z, 95
Wang G, 131–4, 137, 332
Wang H-Y, 64, 122, 195
Wang Q, 60–61, 92–3, 95, 120–22, 132, 328, 435
Wang S-N, 13
Wang X, 94, 257, 459–60, 462
Wang X, 161
Wang X-Y, 54, 84, 88, 231, 233
Wang Y-P, 54
Wang Z-Y, 13, 53–4, 164, 287, 391
Wang Z-Z, 479
Ward DF, 334
Ward S, 85, 394–5, 399
Ward VK, 309
Wardle AR, 130, 371
Washburn JO, 155
Watada M, 114, 116
Watanabe K, 171, 182, 432, 350–51
Watanabe M, 435
Waterston J, 366, 439
Weaver D, 65, 286
Weaver KM, 259, 376
Weaver RJ, 144–5, 375
Webb BA, 149, 151–2, 154, 156, 158, 404, 435
Webb JC, 259, 332
Webb RE, 100, 389, 484
Webberley KM, 304
Webster RP, 258
Wegst UGK, 43
Wehrle I, 83
Wei H, 324
Wei S-J, 205, 437
Weiblen GD, 263
Weih MA, 43
Weiser L, 460
Weisman S, 78
Weiss MJ, 308
Weisser W, 160, 220
Weldegergis BT, 393
Wellings PW, 471
Weng J-L, 372
Wenlock C, 141, 471
Wenner AM, 379–80, 383, 461, 477
Went DF, 74
Werren JH, 114
Wertheim B, 109
Weseloh RM, 64, 67, 96, 332–3, 339
Wesmael C, 202
West RJ, 394, 419
West SA, 92, 114
Westendorff M, 115, 125
Wetterwald C, 152, 334
Weyda F, 300
Wharton RA, 4, 17, 19–24, 26, 33, 59, 64, 68, 73, 89, 99, 122–3, 148, 170, 177–8, 185, 202, 204–5, 208–9, 221, 228–30, 234, 236, 238, 241, 243–5, 250–51, 253–7, 259–60, 265–6, 269, 271, 279, 304–5, 313, 315, 317, 320–22, 335, 418–21, 455, 468, 484, 497, 501, 516
Wheeler EW, 78
Wheeler TT, 142
Wheeler WC, 51, 192
White W, 526
Whitefoord C, 335
Whitehouse MEA, 95–6, 112, 414–15, 430
Whitfield JB, 85, 90–91, 109, 120, 148, 150–53, 181, 184, 206, 229–30, 236, 245, 271, 304–5, 312–15, 320–22, 324–5, 328, 334–5, 404, 482, 499, 512
Whitfield WGF, 109
Whiting AR, 108–9, 266–7
Whitman DW, 129
Wieber AM, 100, 389, 484
Wiedenmann RN, 97–8, 374, 455
Wild AL, 181
Wiley SP, 158
Wilkinson DS, 85
Willard HF, 258
Willers D, 83, 138, 375, 479
Williams DJM, 411
Williams FX, 263
Williams HJ, 123, 259
Williams IH, 428
Williams RN, 308
Williams T, 479
Willson LJ, 159
Wilson DD, 433
Wilson JS, 335
Wincker P, 149, 152, 334
Windsor D, 507, 536
Winter TG, 121
Wirta HK, 486
Wisthlecraft J, 218
Witherell RA, 149
Withers TM, 282, 310
Witjes LMA, 82, 393
Witthinrich J, 220
Witzgall P, 109, 129
Wolda H, 510
Wolf D, 74
Wolf R, 74
Wollberg Z, 392
Wong EA, 158–9
Wong MA, 31, 123, 259
Wong TTY, 31, 257
Woodcock CM, 218
Woodcock T, 497
Woolley JB, 204
Wratten SD, 460
Wright AF, 220, 476
Wu H, 231
Wu K-M, 214
Wu Y-J, 334
Wu Z-Y, 54
Wu Z, 460
Wyatt TD, 173
Wyckhuys KAG, 214, 218, 458, 480
Wyder S, 84, 149
Wyeth P, 42, 84, 124, 149, 383
Wyler S, 149
Wyler T, 77–8, 83–4, 149, 160, 316, 327, 471, 473
Wylie WG, 300

X

Xi Y, 222
Xia XM, 215–16, 220
Xiang-Xiong Z, 215–16, 220
Xie Y, 265, 474
Xie Z, 265, 474
Xu P, 257–8

Y

Yahara M, 129
Yamada A, 65, 268

Yamada H, 86, 325
Yamaguchi H, 93
Yamamoto M, 298
Yamazaki K, 477
Yan Y, 97
Yanechin B, 86
Yang G, 324
Yang Z-Q, 54, 84, 88, 231, 233
Yarro JG, 398
Yazaki K, 139
Yazgan S, 374, 474
Yazlovetsky IG, 265, 474
Ye GY, 433, 479
Yeargan KV, 121, 334, 421–2
Yearian WC, 479
Yel M, 64
Yin L, 154
Yin X, 222
Yoder MJ, 250, 260, 269, 271
Yokota GY, 286, 459–60, 462
You MS, 324
Young B, 335
Young JH, 78
Young SY, 479
Youngs BJ, 471
Yu DS, 260, 275, 294, 322, 424, 430, 490, 500, 502–3
Yu DW, 263
Yu R-X, 66, 69, 145, 325
Yurtcan M, 428–9

Z

Zada A, 482
Zaidi MI, 516
Zaka-ur-Rab M, 15
Zaldivar-Riverón A, 63–4, 79, 91, 178, 184, 205–6, 212, 228–9, 231, 234, 237–9, 241–3, 245–50, 260, 268–9, 339, 485
Zanuncio JC, 481
Zappalà L, 261
Zareie R, 144–5
Zareie S, 144–5
Zebitz PW, 437
Zélé F, 220
Zeng L, 259
Zetlan SR, 334
Zettel H, 319
Zhang C, 154
Zhang D, 158
Zhang G, 144–5
Zhang HC, 193–4
Zhang J-H, 95
Zhang J, 97
Zhang JF, 194
Zhang T-T, 13, 53–4, 287
Zhang Y, 214, 222, 287
Zhang J-J, 13
Zhang YN, 84, 88, 233
Zhao W, 394, 479
Zhaurova K, 419–21
Zhnag Y, 334
Zhnag Z-F, 334
Zhou H, 334
Zhou HQ, 68
Zhou J-J, 13
Zhou R-X, 351–2
Zhou Y, 109
Zhu C, 222
Zhu F, 393
Zhu JW, 13, 334
Zhu KY, 144, 268
Zhu XX, 139, 287, 351–2
Zijp JP, 301, 370, 457, 363, 386
Žikić V, 32
Zilkowski BW, 124, 236
Zinnert KD, 370, 411, 413, 416, 419
Zinovjeva KB, 463
Zitani NM, 296–7
Živić I, 32, 39
Zohdy MZM, 143
Zong S-X, 403
Zuparko R, 214
Zwakhals CJ, 369–70, 375, 380, 382, 483
Zwölfer H, 374

GENERAL INDEX

28S rDNA, 196, 204, 205, 209, 212, 223, 239, 250, 255, 268, 272, 276, 278, 284, 294, 313, 325, 336, 342, 353, 355, 356, 364, 367, 369, 378, 387, 389, 397, 400, 402, 403, 422, 436, 440, 445, 447, 529, 530, 534

A

acid gland (*see* gland, venom)
acrosome, 69–70, 202
adult diet, 254, 259
aggregations, 117, 264, 317, 381, 397
aggressive behaviour, 381
alimentary canal (*see* gut)
alkaline gland (*see* gland, Dufour's)
allomones, 395
amber
  Baltic, 104, 193, 195, 228, 238, 243, 245, 283,308, 349, 413, 432, 442
  Cretaceous, 209, 211, 344
amino acids, 144, 156, 159, 220, 316,
  diet, 82, 86, 370, 374–5, 458, 474–5
  essential, 474
  silk, 85–6, 287, 325
Amylosterium, 375
anal vesicle, 3, 76–8, 83, 138, 281, 293, 315–16, 327, 419, 424
  absence of, 253, 278
anhydropic (*see* eggs)
antennae
  glands (*see* glands, antennal)
  larval, 78–9, 100, 104, 182–3, 226, 233, 264, 270, 427, 407
  sensilla, 12–14, 130, 222, 259, 287, 315, 322, 334
antennation, 68, 308
anthrone, test for sugars, 460
antibiotics, 114, 220
antifreeze, 278, 467
ants
  and aphids, 218
  as hosts, 307–8, 263, 439–442
  as predators, 213
  trophylaxis, 218
anomalous diversity, 490–92, 497, 501, 536
aphids, 4, 66–67, 88, 97, 141, 143, 159–61, 172, 214–15, 217–24, 303, 362, 462, 476, 488, 490
  honeydew, 457
  secondary symbionts, 160, 220
Apiaceae, 265, 390, 460
apophyses, 173
aptery, 119, 170–72, 390, 393, 481, 539
aquatic ichneumoids, 256 385–7, 390, 439
artificial hosts, 374, 475
artificial selection, 145
associative learning, 58, 127–8, 130–34, 218, 332, 459, 487
Asteraceae, 3, 213, 228, 365, 275, 370, 460, 487

B

Bacillus, 144, 375
barcoding, 114, 212, 250, 322–4, 337, 436, 485, 486, 491, 495–7, 503, 512, 516
Beauveria, 144, 159, 261, 375, 479
behaviour
  aggressive, 258, 266
  courtship, 14, 64, 67, 109, 111, 113, 115, 117, 119–25, 217–18, 249, 253, 259, 263, 266, 275, 298, 301, 315–17, 321–2, 332–3, 338, 353, 360–61, 363–4, 369, 376, 403, 405, 418, 434, 442, 447
  wing fanning, 120, 218, 249, 253, 259, 338,
biocontrol (*see* biological control)
biodiversity
  global, 490–503, 510
  local, 477, 508
biological control*, 467–41
  and host races, 475–6
  mass rearing, 525
  multiple parasitoids, 468
  nematodes, 261
biosensors, 134
body size
  early ichneumonoids, 104, 349
  fecundity, 110, 462
  fitness, 61, 254, 462
  in diversity estimation, 493
  islands, 505
  larval instars and, 182
  lifespan, 89
  morphology, 18, 31, 321, 335
  islands, 505
  sexual dimorphism in, 31
  subfamilies, 501–2
brachyptery, 119, 170–72, 235–6, 244, 281, 293, 296, 303, 309, 335, 389, 392, 481, 540
bracoviruses, 146, 148–150, 152–4, 334, 535
brain, 10, 13, 58, 125, 130, 162, 272, 363, 374, 419, 446, 485
brood size, 92, 96, 111, 245, 248, 328
bullae, 27–9
Buchnera, 159, 220–21

C

calcium, 42, 74, 85
carina
  dorsal, 30, 346

The Braconid and Ichneumonid Parasitoid Wasps: Biology, Systematics, Evolution and Ecology, First Edition. Donald L. J. Quicke.

carina (*continued*)
hypostomal, 12
mesopleural, 339
occipital, 10, 12, 238, 251, 260, 269
castration, 156–7, 221, 309, 317
caudate larvae, 76, 216, 253, 272, 280, 293, 296, 308, 315–16, 327, 357, 418, 424, 429–31, 444
cecidogenesis (*see* galls, formation of)
Chalcidoidea, 110, 114, 119, 173, 184, 192, 387, 454, 455, 476, 525, 542
Chao, 1 509–10, 303
chitin, 10, 33, 43, 52–3, 58, 59, 63–4, 78, 143, 159, 376, 424
chitinase, 143, 159
chorion (*see* egg, chorion)
chromosomes, 108, 115, 125, 157, 512
citral, 68, 376
claws, 17–18, 175–6, 268, 284, 344, 359, 367–8, 378, 387, 428, 442, 445, 448
clypeus, 10, 11, 422, 540, 541
CNEA, 15, 168–9, 265, 277, 454, 540
cocoons
of host, 4, 100, 102–3, 130, 181, 274, 291, 332, 377, 390, 393, 421
of wasp, 84–86, 291, 298, 303, 326, 331, 338, 351, 387, 428, 430, 434, 444
suspended, 74, 184–5, 298, 433–4
colchicine, 115
colour preference, 459, 516
cold shock, 266, 466
cold tolerance, 465–7
coloration
dimorphism, 31, 396
temperature, 467
Comstock-Needham system, 20, 22–5
competition, 380, 395, 482
apparent, 482–3
complementary sex determination (CSD), 108–10, 115, 117, 120, 264, 334
and culturing/breeding, 112, 430, 525
diploid males, 108–10, 112
conservation, 397, 468, 506, 509–10, 511
corpora allata, 160
cotton, 4, 261, 284, 315, 337, 432, 457
courtship (*see* behavior, courtship)
Cretaceous period, 5, 17, 25–6, 49, 190, 192–4, 196–7, 205, 209–11, 344, 278, 320, 346, 349, 353, 413, 447
cryptic species (*see* species, cryptic)
cuticle hardness, 43, 383
cuticular encystment, 141, 405
cuticular hydrocarbons, 67–8, 218
cytoplasmic incompatibility, 113–14

D

deuterotoky (*see* parthenogenesis)
development time, 32, 88–9, 474, 477
diapause, 81, 85, 255, 300, 306, 316, 363, 379, 392, 394, 418–19, 463–6, 469, 485, 512, 524
Dibrachys, 433
digitus, 32
diploid males, 108–10, 112
dispersal, 110, 117, 214, 233, 393, 480
distitarsal scraper, 175, 356
diversity
altitudinal trends, 508
as indicators of, 510
estimation
latitudinal trends, 490–97, 500–3
DNA
base composition, 125, 196, 204–5
preservation, 528–30
dorsope, 30, 440
doseage compensation, 125
dorso-ventral flattening, 171–2, 351

E

ecdysteroid, 83–4, 154, 160, 317, 465
echolocation (*see* vibrational sounding)
ectoparasitism, 3, 90–91, 100–101, 103, 104, 226, 270, 372, 407, 427, 503, 534–5
effective population size, 526
egg
activation, 74
anchors, 72–4, 78, 100, 143, 224–5, 253, 385, 407, 411, 412–18, 427, 429–30, 454
anhydropic, 72–3, 394
distortion, 41, 52–3, 74, 536
chorion, 72–6, 139, 247, 254, 259, 286, 296–7, 316, 372, 411, 417, 419, 429, 439, 536
coating, 139–40, 145
dumping, 74, 416, 483, 487
hydropic, 72–3,
load, 96, 417, 458, 460
resorption, 257
stickiness, 140, 143, 254, 293
yolk, 72, 73, 74, 75, 88, 89, 91, 101, 111, 419, 454, 456
egg predation, 2, 4, 82, 93, 103, 263, 367, 369, 373, 393
egg-larval parasitism, 52, 76, 152–3, 162, 178, 180, 182, 248, 253, 255, 257, 279, 281, 315, 321–2, 325, 359, 362, 409, 419, 421, 432
embryogenesis, 41, 74, 75, 475, 536
embryonic membranes, 75–6
serosal, 75–6, 138, 158, 214, 253, 257, 293, 296–7, 316
trophamnion, 75, 104, 253, 308, 418, 429, 430
encapsulation, 62, 83, 99, 138, 139–41, 144, 145, 152, 154, 156, 159, 219, 247, 254, 268, 274, 286–7, 296, 374–5, 394, 416, 419, 429–30, 435, 468, 498–9, 536
endocrinology
of host, 153, 160, 334
of parasitoid, 375, 456
endoparasitism
transitions from ectoparasitism, 100–104
endosymbiont, 375
enemy free space, 498–9
Enterobacter, 375
Eocene, 5, 193, 349, 424
epistoma, larval, 79–81, 184
Eucoilidae, 39, 99, 105, 145
excretory products, 59, 316, 542
extinction, 2, 463, 468–9, 480, 483, 501, 506, 510
extrafloral nectaries, 457

F

fat body
of host, 82, 140, 157, 159, 253, 293, 296, 308, 394, 400, 419
of wasp, 59, 73, 83, 372, 428
fate map, 75
fatty acids, 83, 159, 274, 278, 376, 474, 535
fecundity, 61, 89, 108, 110, 213–14, 219, 248, 255, 274, 296, 304–5, 308, 317, 330, 376, 385, 391, 395, 416, 439, 457–8, 460–62, 479, 506
feeding
and associative learning, 459
flowers, 118–9, 275, 390, 458–60, 517
of larva, 81–2, 316
feeding tubes, 265, 454
femur, 17, 76, 99, 148, 173–5, 352, 433, 544
fertilisation, 123, 536
figs, 53, 263
food conversion efficiency, 76, 473
food, host consumption, 296, 444, 471–3,
food webs, 470, 484–6, 499
formic acid, 441, 482

fossils, 5, 20, 25, 48, 103–4, 152, 190, 192, 193–6, 344, 346, 348–9, 353, 369, 440, 442
freezing point, 274, 278, 467,
fructose, 457–8
fungi, 98, 128, 138, 158, 179, 261, 375, 380, 468, 479, 497, 528, 536
fungicide, 158, 479

G

galls
  Bedeguar, 400
  formation, 81, 90, 172, 184–7, 225–6, 234–6, 503
  parasitism of, 18, 38, 95, 177–9, 185–6, 197, 217, 228, 234–5, 255, 257, 263, 284, 322, 356–7, 369, 383–4, 400, 403, 411, 420, 428, 454–6, 469, 504
genitalia, 121
  female, 333
  male, 30, 32–3, 68, 121, 210, 224, 238, 268, 324
genome
  AT richness, 125, 196, 204–5
  size, 125
geranial, 68, 376
germarium
  of host, 156, 159, 221
  of parasitoid, 59, 72, 217
germ band, 75
gland
  acessory, of males, 68–70, 202, 209
  acid (*see* venom)
  alkaline (*see* Dufours)
  anal, 381
  antennal, 14–15, 130, 218, 355, 360, 363, 376, 447
  antero-lateral, 124, 266, 482
  calyx, 59, 65, 99, 138–9, 144, 146, 148, 151, 153, 155, 159, 312, 317, 321, 334, 400–401, 422
  claw, 367–8, 378
  collaterial (*see* vaginal)
  Dufour's, 54, 60, 64–6, 138, 268, 334, 375, 479, 536
  exocrine, 123, 259, 293, 333
  female sex pheromone, 333
  Hagen's, 123, 126, 259, 271, 482
  inter-tergal, 123
  labial (*see also* salivary), 78, 216, 138, 259, 328, 330
  mandibular, 12, 381, 482
  oviduct, 52, 62–3, 536
  pheromone, 67–8
  pygidial (*see* Hagen's)
  salivary, 58, 78, 82, 85, 138, 216, 302, 328, 330
  silk, 78, 85
  spermathecal, 61–2, 535
  sternal, 266
  tarsal, 174, 432
  tergal, 119, 124, 249, 266, 274–5, 293, 301, 309, 317, 321
  tibial, 266
  uterine (*see* vaginal)
  vaginal, 54, 62–3, 70, 138–9, 351, 374, 378, 384, 448
  venom, 52, 54, 60, 63–6, 138, 141, 143–4, 153–5, 216–17, 249, 251, 259–60, 266, 268–9, 271, 308, 358–9, 405, 422
glucose, 370, 457–8, 474
glycerol, 82, 274, 278, 466–7, 519
glycerides, 82–83, 376
glymma, 30, 366
gonangulum, 36
gonapophyses, 36
Gondwanan distribution, 212, 222, 224, 343, 354, 361, 398, 503, 504
gonocoxite, 36, 67
gregarious development, 31, 91–2, 95, 97, 103, 109–10, 113, 120, 131, 157, 184, 351, 370, 384, 389, 391–2, 403, 415, 458, 462, 468–9, 472, 482, 485, 487, 501, 504, 516
  facultative, 92, 248, 330, 370, 403
guilds, 462, 498, 506
gut
  adult, 58–9, 68, 155, 210, 266, 268, 381, 460, 479
  larval, 76–78, 81–2, 216, 258–9, 363, 375, 439

H

habitat degradation, 511
haemocytes, host, 62, 72, 83, 138–41, 144–5, 154, 156, 287, 374, 424
Hagen's glands (*see* glands)
Hamiltonella defensa, 220
hamuli, 18
handling time, 385, 422
head capsule (larval), 78–81
head morphology, 10–12
hibernation, 370, 397, 462, 465–7
homoplasy, 16–7, 126, 164, 174, 178, 198, 268, 271, 303, 342, 365, 383, 391
honey, 73, 380, 383, 454, 457–60, 469
honeydew, 383, 454, 457–8, 460
host
  acceptance, 14, 130, 220
  alternation, 298, 328, 429, 493
  alternative, 249, 262, 286, 376, 462, 469, 475–6, 483–6
  defences against parasitisation, 248–9, 334, 421
  discrimination, 13, 65, 97,
  location, 4, 13, 18, 55, 103–4, 127–9, 131, 164, 166, 231, 258, 322, 324, 353, 355, 370, 375–6, 378, 380, 385, 387, 395, 432, 447, 486–7
  marking, 15, 64, 138, 268
  quality, 88, 110, 256
  suicide, 161
  toxin sequestration, 496, 498
  trail following, 93
host-feeding, 61, 89, 454–7, 248, 257, 364, 370, 374, 376–7, 397, 414, 424, 454–7
  and biological control, 471
host range, 471, 482–4, 486
  biotypes, 309, 476
  evolution of, 486–8
  realised range, definition, 483
  generalists and specialists, 89
hyperparasitoids, 3, 4, 67, 85, 88, 99–100, 102, 105, 184–5, 197, 369–70, 374–5, 379, 389, 391, 421–2, 433–4, 445, 476, 477, 483, 485
hypoclypeal depression, 10, 11, 230, 241, 256
hypopygium, 33, 67, 357, 358–9, 395

I

Ibalia, 379–80
idiobiont, definition and traits, 88–89
immunosuppression of host, 145
infanticide (*see also*, ovicide), 371
intrinsic rate of increase, 462, 468
island parasitoid faunas, 470, 505–6

J

juvenile hormone, 153, 160
juvenile hormone esterase, 153

K

kairomones, 13, 117, 128–30, 258, 316,
kleptoparasitism, 99–100, 374, 378–9, 391, 476–7
koinobiont, definition and traits, 88–89

L

labial sclerite, 79, 81–2, 234
labrum, 10, 11, 58

larval competition
combat, 92–5, 220, 235, 247, 254, 272, 282, 286, 308, 338, 374, 451
physiological suppression, 93–5, 97, 249, 338, 409
larval external feeding phase, 82, 160, 277, 280, 291, 297, 302, 315, 327, 409, 475
larval-pupal parasitism, 183, 498
Lasius niger, 66, 213, 218
laterope, 30
leaf miner, roller, tier, hosts, 46, 128–9, 147, 182, 184, 231, 238, 253, 270, 293, 317, 320, 322, 370, 375, 405, 407, 412, 414, 418, 423, 430, 438, 454–7, 484, 486
rearing parasitoids from, 524
learning (*see* associative learning)
legs, 173–4
fossorial/robust, 173–4, 243, 269, 335, 351, 361
Leptopilina, 99, 105, 145
life expectancy, 61, 83
ligation, 465
light traps, 521–2
linkage map, 108, 267
lipids, 73, 82–3, 364, 370, 375–6, 456, 473, 526, 529, 535
local mate competition, 92, 109–11, 113, 239, 248, 328
longevity, 83, 108, 112, 214, 319, 370, 375, 454, 456–8, 460, 462, 479

M

maize, 129, 288, 376, 394, 472
Malaise traps, 417, 428, 442, 445, 490–91, 493, 497, 506–8, 510, 516–520
setting one up, 521–2
Malpighian tubules, 59, 85, 125, 155, 327, 375, 479
mandibles
adult, 10–12, 58, 167–8, 186, 190, 209, 229, 239, 251–2, 256, 268–9, 271, 274, 302–3, 351, 362–3, 377, 381, 405, 413–14, 454
exodont, 202, 251–2, 255, 271, 321, 414
hardening, 42
larval, 76, 78–9, 81–2, 92–95, 182–3, 216, 220, 226, 234, 327, 247, 258, 260, 263–4, 272, 286, 293, 296–7, 306, 308, 315, 327, 359, 363, 370, 374, 389, 394, 398, 406–7, 409, 416, 422, 427, 429, 430–31, 433, 439
mate loction, 13, 31, 117–19, 226, 275
associative learning and, 118–19
mating, 14, 31, 10–25, 5, 364, 369–71, 380, 381, 392, 403, 405, 418, 431–2, 434, 435, 439, 442, 490, 493, 495, 509, 525
in captivity, 525–6
multiple by females, 121–3
meconium, 59, 76, 216, 253, 258, 274, 277–8, 316, 359, 394, 542
meiosis, 41, 115, 536
mesopleuron, 15, 16, 173
mesosoma, 10, 15–17, 48, 84, 170, 173, 190, 353, 362, 527, 528
metanotum, 16–17
petiolate, 177, 307, 311, 342, 399
metasoma, 29–30
Microsporidea, 479
mimicry, 138, 218
aggressive, 208, 372
as dupe of orchid, 376
Batesian/Müllerian, 181, 265, 275, 312, 422–3, 444, 480–81, 536
camouflage/masquerade, 67, 138, 218, 328, 435, 328
chemical, 67, 138, 218
female by male, 333
mitochondrial derivative, 69–70, 202
mounting specimens, 526–9
multiparasitism, 96–9, 143, 145, 221, 254, 424, 536, 542
and biological control, 468, 482
obligate, 99
mummies, 85, 88, 95, 102, 133, 184, 217, 220, 248, 397, 466, 477, 514, 524, 535
mushroom bodies, 58, 130

N

nasty host hypothesis (NHH), 497–9, 501, 511
nematodes, 98, 261
neogregarines, 479
neral, 68, 376
nocturnality, 176–7, 179–80, 196, 246, 249, 281–2, 284, 288, 291, 332, 337, 349, 421, 444, 463, 477, 507, 521–2
Nosema, 477–9
notauli, 15–16, 344, 542
nucleocapsid, 149, 155–6
nurse cells, 59, 72, 74, 317

O

ocelli, 10, 58, 179–80, 349, 449
olistheter, 36–7, 46, 52–3, 288
ommochromes, 84
oosorption, 217
ophionoid facies, 10, 179–80, 246, 284, 288, 291, 414, 421, 442, 449, 463
Opinions (ICZN), 339, 344, 346
Orussidae, 164, 196
outbreeding, 110, 113
ovariole, 59, 60, 72, 89, 217, 248, 255, 266, 271, 297, 317, 356, 400, 445, 448, 460–61,
number, 89, 460–61
ovary, 59–60, 65, 73, 74, 138, 145, 157, 225, 258, 271, 279, 297, 308, 317, 321, 358, 370, 395, 401, 460, 461, 479, 536, 542
germarium, 59, 72, 156, 159
overwintering, 85, 247–8, 278, 286, 298, 321, 327, 331, 335, 359, 370, 373, 392, 429, 463, 465–7
ovicide, 265
oviduct(s), 52, 54, 59, 61–3, 74, 138–9, 143, 217, 225, 342, 400–401, 411, 421, 448, 536
ovigeny index, 60, 61, 89, 543, 544
oviposition, 18, 30, 37, 39, 41, 44, 52, 53, 54, 61, 63, 73–4, 93, 94, 95, 97, 98, 100, 104, 110, 117, 133, 134, 139, 143, 152, 157, 166, 167, 177, 179, 182, 195, 353, 355, 356, 359, 363, 371, 372, 374–6, 394, 407, 409, 414, 429, 435, 455, 457, 471, 474, 475, 477, 480, 482, 486, 488, 536, 540
ovipositor
buckling force, 43, 383
ctenidia, 37, 44
evolutionary transitions, 38
guides, 43, 44, 383
hardening, 42
notch, 37–9, 44, 45, 48–50, 279, 293, 344, 401, 401–4, 406, 408, 409, 411, 417, 419, 421, 425, 429–30, 442
secretory pores, 53
sensilla, 54–5, 260, 394, 435
serrations, 37–9, 42, 49, 224–5, 263, 265, 351, 355, 381, 387, 403, 407, 409
stabilisation, 43, 175, 353, 383
steering, 36–7, 44, 46–50, 288, 294, 322, 383, 429, 439
ovo-larval parasitoids 52, 152–3, 162, 166, 178, 180, 182 248, 253, 255, 257, 279, 281, 315, 321–2, 325, 329, 362, 409, 411, 419, 421, 432, 457

P

pan traps, 336, 459, 516–19
paralysis
  permanent, 89–91, 101, 141, 143, 231
  temporary, 37, 52, 141, 143–4, 161, 221, 237, 247, 258, 272, 298, 302, 414, 455
parameres, 32, 421
parental care, 239–40, 245
parthenogenesis
  arrhenotoky, 114, 422, 158
  deuterotoky, 115, 540
  thelytoky, mechanisms, 113–17, 126
  thelytoky, distribution of, 116
  thelytoky, relative competitiveness, 118
  thelytoky, 218, 246, 281, 303, 308–9, 317, 393, 395, 422, 467
Perilampidae, 447
petiole (*see* metasoma)
phenol oxidase, 78, 142, 144–5, 159, 268, 287, 296
pheromones, 65, 67, 117–21, 123–4, 128, 258–9, 266, 301, 303, 315, 317, 333–4, 339, 376, 390, 411, 482, 525
physiological suppression (*see* laral competition)
phytophagy, 2, 184, 186, 206, 234–5, 263–4
planidium, 445–6
ploidy, 78, 115, 125, 158
polar bodies, 75, 158, 217
polydnaviruses (*see* virus)
polyembryony, 92, 184, 278, 286, 403
polyploidy, 78, 125, 158
Polytoma, 76–7
precoxal sulcus, 16–17
predation, by birds, etc., 161, 184, 186, 248, 299, 304, 383, 433, 463, 477
premating period, 109, 117, 120
premature metamorphosis, of host, 111, 473, 153
pro-ovigenic, 59, 61, 89, 214, 217, 444
propodeum, 10, 15, 17, 30, 33, 173, 192, 344, 387, 401, 403, 430, 454
  areas and carinae, 17, 192, 344, *see also* Fig. 13.2a
  spines (apophyses), 173
pseudohyperparasitism, 3–4, 88, 99, 102, 184, 248, 330–32, 369–70, 374, 389, 391, 393, 433–4, 456, 476, 483
pseudo-ovipositor, 36, 481
pseudoparasitism, 185
pseudostigma, 124, 235–6
pterostigma, 20, 22, 27, 193, 527, 542, 543
pupal parasitoids, 82, 98, 111, 128, 133, 138, 181, 183, 279, 374, 376, 462, 471, 484, 486, 497–8, 506

R

rearing
  in vitro, 374,474–5
  mass, 112, 262, 469, 479, 483, 525–6
  wild collected hosts, 523, 528
recombination, 115, 125
rectal pads/papilla, 59, 444–5
Regiella insecticola, 220
resource fragmentation hypothesis, 497–8, 410
rhachis, 36, 46–8, 403–4
rhinaria (*see* antennal sensilla)
r- and *K*-selection, 462

S

seed predation, 185–6, 234, 260, 263–4
seminal vesicle, 69, 257
sensilla (*see* ovipositor, antenna)
Sephadex, 72, 144, 154
serosal membrane (*see* embryonic membranes)
Serratia symbiotica, 220
sex determination (*see* complementary sex determination CSD)
sex ratio, 525
silk press, 78–9, 81
silks
  and host location, 4
  chemistry, 85–6
size (*see* body size)
speciation, 105, 380, 476, 487, 501
species
  accumulation curves, 492–3, 506–7
  abundance distributions, 495, 509
  cryptic, 126, 245, 265, 286–7, 298, 335, 511–12, 514
  morphospecies, 126, 490, 496, 506, 512
  richness, 254, 367, 484, 490
sperm
  competition and precedence, 121–3
ultrastructure, 69–70
spermathecae, 122, 123
  colour, 62, 179
spermatophores, 49, 70, 121
spiders
  as hosts, 2, 91, 103, 161, 367, 36–73
  as predators, 123, 482
  as prey, 391, 393, 400
  egg predators of, 2, 4, 82, 93, 102, 369, 389, 392
  rearing parasitoids from, 483, 523
spiracles
  adult, 30, 177, 319, 321, 339, 401
  larval, 76, 78, 81, 83, 91, 247, 308, 315, 363, 394, 414, 433
spurs (*see* tibial spurs)
Stephanidae, 190, 192, 196
sternaulus, 17
stigma (*see* pterostigma)
stinging, 145, 265, 442
subgenual organ, 164, 166, 353, 356, 376, 392
subpronope, 271, 321
superparasitism, 93–7, 117, 143, 152, 359, 385, 392, 410, 418, 429, 431, 463
survival after parasitism, 88
sweep netting, 484, 490, 508, 516, 518–21
synovigeny, 59–60, 72, 108, 118, 219, 247, 257, 358, 372, 390, 417, 445, 454, 458

T

teratocytes, 75, 76, 83, 104, 138, 158–60, 214–15, 217, 258, 274, 293, 296–7, 300–31, 315, 334, 474–5, 479
termitophily, 10, 172, 236, 535
thelytoky (*see* parthenogenesis)
tibial spurs, 17, 269
trade-offs, life history, 330, 342, 358, 470
Trichogramma, 261
tritrophic interactions, 128–9
trochanter, 17, 297
trochantellus, 17, 286
trophamnion (*see* embryonic membranes)
trophylaxis, 218–19
Tubulinosema, 478–9
tyloids, 14, 360, 363, 365, 369, 376, 416, 420

U

unisexual (*see* parthenogenesis, thelytoky)
usurpation hypothesis, 303–4, 330–33

V

valvifer (*see* gonocoxite)
valvilli, 38, 44, 50–53, 190, 210, 430
venation, 18–29

venom apparatus
  glands (*see* glands, venom)
  type, 1 52, 60, 63–4, 165, 249, 271
  type, 2 63–4, 271
venoms, 30, 37, 39, 41, 44, 49–54, 60, 63–6, 78, 93, 95, 99, 112, 138, 140–45, 152–6, 159–60, 178, 351, 358–60, 368, 375, 387, 413–14, 422, 430, 435, 488
  from larval anal vesicle, 138
  paralysing (*see* paralysis)
  synergy with polydnavirus, 145, 154
vibrational sounding, 4, 18, 54, 164, 166, 181, 186, 352, 353, 355, 356, 376–7, 391–3, 399
virginity, 112
virus
  ascovirus, 152, 155, 395
  baculovirus, 146, 152, 155, 156, 479
  banchovirus, 146, 404
  bracovirus, 146, 148–50, 152, 153–4, 535
  cypovirus, 155, 395
  entomopox, 155, 259
  ichnovirus, 148–50, 152
  polydnavirus, 138, 145–60, 162, 312, 315, 317, 320, 334–5, 404, 435, 439, 487
  reovirus, 155, 395
  rhabdovirus, 259
virus-like particles (VLPs), 59, 72, 117, 138–9, 145, 155, 259, 308, 400, 435, 536
vitamins, 370, 375

W

water, adult need for, 319, 454–60, 462, 497
wings
  cell names, 25
  flexion lines, 27–9
  venation, 18–25
winglessnessness (*see* aptery, brachyptery)
wing vibration, 120, 298, 353
wood borers, parasitism of, 14, 18, 38–9, 46, 52, 54–55, 90, 100, 103, 110, 120, 128–9, 135, 170, 174, 181, 196, 239, 260, 263, 265, 282, 297, 351, 461–2, 482, 501, 508

X

X-rays
  creating mutants, 267
  diffraction, 43, 287
  sterilising wasps, 152

Y

yellow
  colour preference, 258
  pan trap, 336, 459, 516–7

Z

zinc, 42

HOST INDEX

A
Acanthosomatidae, 299, 307
Achaearanea tepidariorum, *See Parasteatoda tepidariorum*
Achroia grisella (Pyralidae), 83
Acontia cretata (Noctuidae), 274
Acrolepiopsis assectella (Acrolepiidae), 97, 130, 158, 394
Acronycta (Noctuidae), 515
Acronyctinae, 484
Acyrthosiphon pisum (Aphididae), 158, 160, 214, 220, 476, 487
African sugar-cane borer, *See Eldana saccharina*
Aglais urticae (Nymphalidae), 375, 487
Agrilus planipennis (Buprestidae), 231, 261–2
Agrochola (Noctuidae), 274
Agromyzidae, 251, 253, 255–7
alfalfa aphid, *See Macrosiphon creelii*
alfalfa weevil, *See Hypera postica*
almond moth, *See Cadra cautella*
Alsophila (Geometridae), 277–8
Ammophila (Sphecidae), 391
Anagasta kuehniella, *See Ephestia kuehniella*
Anastrepha (Tephritidae), 257, 482
Ancylosis fulvobasella (Pyralidae), 244
Angoumois grain moth, *See Sitotroga cerealella*
Anobiidae, 242, 281–2, 350
Anthicidae, 299, 309
Anthonomus (Curculionidae), 370
  A. eugenii (Curculionidae), 279
  A. grandis (Curculionidae), 261, 470, 525
Anthophila fabriciana (Choreutidae), 112, 160, 487
Anthrenus museorum (Dermestidae), 432
ants, 67, 85, 161, 263, 299, 307–8, 441–2
aphid, 4, 66, 88, 97, 141, 143, 159–61, 172, 212–22
Aphis
  A. fabae (Aphididae), 66, 213, 218
  A. glycines (Aphididae), 213
  A. gossypii (Aphididae), 471
Apoderus quadripunctatus (Curculionidae), 407
Archips argyrospira (Tortricidae), 147
Arctiidae
Argentine sem weevil, *See Listronotus bonariensis*
Argidae, 321, 413, 418–20
Artematopidae, 432
Asian cornborer, *See Ostrinia furnacalis*
Aulacorthum solani (Aphididae), 214, 487
Autographa (Noctuidae), 141, 405
  A. californica (Noctuidae), 111, 155
Azteca (Formicidae), 263

B
Bactrocera oleae (Tephritidae), 257, 261
banana aphid, *See Pentalonia nigronervosa*
Banksia (Proteaceae), 172, 226–7
Battus philenor (Papilionidae), 498
bedeguar gall, *See Diplolepis rosae*
beech weevil, *See Rhynchaenus fagi*
beet armyworm, *See Spodoptera exigua*
Bertha armyworm, *See Mamestra configurata*
black bean aphid, *See Aphis fabae*
black-headed caterpillar, *See Opsina arenosella*
black snout, *See Hypena scabra*
Blasticotomidae, 420
Bledius spectabilis (Staphylinidae), 173
boll weevil, *See Anthonomus grandis*
Bombus (Apidae), 299–300, 309
Boreus (Mecoptera: Boreidae), 282
Brachypeza radiata (Mycetophilidae), 367
Brassicogethes (often as *Meligethes*; Nitidulidae), 281, 283, 429
brown ambrosia aphid, *See Uroleucon ambrosiae*
brown citrus aphid, *See Toxoptera citricola*
Bucculatricidae, 320
budworm, *See Choristoneura*
Bupalus piniaria (Geometridae), 430
Buprestidae, 231, 236, 239, 261, 310, 350, 352, 355
burnet moth, *See Zygaena filipendula*
bushfly, *See Musca vetustissimma*
Busseola fusca (Noctuidae), 147, 288, 394
butterfly
  blue (*see* Lycaenidae)
  catalpa sphinx (*see Ceratomia catalpae*)
  Glanville fritillary (*see Melitaea cinxia*)
  large white (*see Pieris brassicae*)
  pine white (*see Neophasia menapia*)
  pipevine swallowtail (*see Battus philenor*)
  red admiral (*see Vanessa atalanta*)
  small tortoiseshell (*see Aglais urticae*)
  swallowtail (*see* Papilionidae)

C
cabbage looper moth, *See Trichoplusia ni*
cabbage moth, *See Mamestra brassicae*
cabbage white, *See Pieris*
cabbageworm, *See Pieris rapae*
caddis fly, 322, 385–7
Cadra cautella (Pyralidae), 468
Calliphoridae, 254
Camararia (Gracillariidae), 147

The Braconid and Ichneumonid Parasitoid Wasps: Biology, Systematics, Evolution and Ecology, First Edition. Donald L. J. Quicke.

Camponotus vagus (Formicidae), 307
Carabidae, 157, 299–301
case-bearing clothes moth, *See Tinea pellionella*
catalpa sphinx, *See Ceratomia catalpae* (Sphingidae)
Catocala (Noctuidae), 148
Cecidomyidae, 185, 228, 234, 242, 263, 504
Cephidae, 359, 370
Cephus
  C. cinctus (Cephidae), 65, 261, 360
  C. pygmaeus (Cephidae), 233, 359–60
Cerambycidae, 46, 231, 233, 239, 262, 279, 281–2, 310, 350, 355, 357, 359, 438, 482
Ceratitis (Apidae), 391
Ceratomia catalpae (Sphingidae), 498
cereal leaf beetle, *See Oublema melanopus*
Cerospastus volupis (Pergidae), 399
Ceutorhynchus (Curculionidae), 428
Cheiracanthium (Araneae: Miturgidae), 103
chestnut moth, See *Ectoedemia phleophaga*
Chilo
  C. partellus (Crambidae), 322, 376, 394, 477
  C. sacchariphagus (Crambidae), 322
Chloromyia (Stratiomyidae), 448
Chloropidae, 255
Chlosyne harrissii (Nymphalidae), 161
Choreutidae, 112, 160, 487
Choristoneura (Tortricidae), 146, 403
  C. fumiferanae (Tortricidae), 147–8, 154, 322, 432
  C. rosaceana (Tortricidae), 147, 161
Chromaphis juglandicola (Aphididae), 214, 469
Chrysodeixis includens (Noctuidae), 147, 295
Chrysomelidae, 178, 231, 237, 253, 281, 284, 295–6, 299–301, 308–9, 428–9, 432, 435, 439
Chrysopa (Chrysopidae), 398
Chrysoperla (Chrysopidae), 304, 307, 398
Chrysopidae, 299, 304, 307, 398, 535
Cimbicidae, 418–20, 432
Cinara (Aphididae), 218
Cladocera nigroornata (Chrysomelidae), 439
Cleonus piger (Curculionidae), 357
Clothoda urichi (Embioptera), 236
clover [leaf] weevil, *See Hypera punctata*
Coccinella
  C. magnifica (Coccinellidae), 304
  C. septempunctata (Coccinellidae), 304
cocoa stem borer, *See Eulophonotus myrmeleon*)
codling moth, *See Cydia pomonella*
Coleomegilla maculata (Coccinellidae), 305
Coleophoridae, 244, 293, 393
common nettle-tap, *See Anthophila fabriciana*
convergent lady beetle, *See Hippodamia convergens*
corn borer, *See Ostrinia*
corn earworm, *See Heliothis zea*
Cosmia (Noctuidae), 434
Cossidae, 288, 403
cotton boll worm, *See Helicoverpa armigera*
Crambidae, 114, 173, 286, 293, 296, 337, 376, 390, 394, 407, 432, 438, 471, 478, 487
Cryptophagidae, 282
Curculionidae, 46, 129, 157, 261, 279, 281–2, 284, 299, 301, 308–10, 350, 355, 357, 407, 428
Cyclosa argentoalba (Araneae: Aranaeidae), 371
Cydia pomonella (Tortricidae), 87, 98, 147, 157, 258, 316, 390, 438

D

Dendroctonus monticola (Curculionidae: Scolytinae), 262
Dermestidae, 432
Derodontidae, 428
diamondback moth, *See Plutella xylostella*
Diatraea saccharallis (Pyralidae), 147, 322
Diplolepis (Cynipidae), 197, 400
Diprionidae, 97, 405, 413, 418–19, 447
Drosophila (Drosophilidae), 75, 96–7, 99, 114, 125, 140–1, 145, 157, 159, 254, 457, 478–9
Dytiscidae, 390

E

Earias insulana (Nolidae), 393
Ecdytolopha insiticiana (Tortricidae), 147
Ectoedemia phleophaga (Nepticulidae), 320
Ectomyelois ceratoniae (Pyralidae), 147
Egyptian alfalfa weevil, *See Hypera brunneipennis*
Egyptian bollworm, *See Earias insulana*
Elasmopalpus lignosellus (Pyralidae), 293
Elateridae, 105, 239, 350, 430, 432
Eldana saccharina (Pyralidae), 442
elm aphid, *See Tinocallis platani*
Embioptera, 236
emerald ash borer, *See Agrilus planipennis*
Eois (Geometridae), 335, 508
Eoreuma loftini (Crambidae), 111, 231
Ephestia kuehniella (Pyralidae), 130, 139, 147, 260, 266, 432, 483
Ephydridae, 255, 257
Epiblema tedella (Tortricidae), 157
Epiphyas postvittana (Tortricidae), 296–7, 403
Epistrophe eligans (Syrphidae), 362
Episyrphus balteatus (Syrphidae), 95, 141, 221–2
Eriococcidae, 263, 356
Eriocrania (Eriocranidae), 416, 429
Eriocranidae, 419
Ethmiidae, 407
Eublemma (Noctuidae), 197
  E. amabilis (Noctuidae), 267
Eulophonotus myrmeleon (Cossidae), 288
European corn borer, *See Ostrinia nubilalis*
European pine shoot moth, *See Rhyacionia buoliana*
Eurysphindus comatulus (Sphindidae), 283
Eurytides marcellus (Papilionidae), 397
Eurytomidae, 263, 392

F

fall army worm, *See Spodoptera frugiperda*
Fennusa pusilla (Tenthredinidae), 411, 418
Fergusonina (Fergusoninidae), 263, 356
foxglove aphid, See *Aulacorthum solani*
fruitfly, *See Drosophila* OR Tephritidae

G

Galleria mellonella (Pyralidae), 98, 111, 266, 393, 456
Gelechiidae, 113, 229, 244, 260, 272, 292–3, 296, 315, 403, 407, 430
Geometridae, 148, 277–8, 291, 330, 405, 418, 430, 444, 449, 470, 508
glass-house potato aphid *See Aulacorthum solani*
Glossina (Glossinidae), 254
Goera (Trichoptera), 385
gooseberry sawfly, *See Pristiphora pallipes*
Gracillariidae, 293, 335
Grapholitha molesta (Tortricidae), 393, 421
greater was moth, *See Galleria mellonella*
green clover worm, *See Hypena scabra*
green-peach aphid, *See Myzus persicae*
Gryllacrididae, 245
gypsy moth, *See Lymantria dispar*
Gyrinidae, 393

H

Hadena gemina (Noctuidae), 395
Haematobia irritans (Muscidae), 254
Harpalus rufipes (Carabidae), 300
Helicoverpa armigera (Noctuidae), 96, 148, 261, 315, 432, 499
Heliocheilus albipunctella (Noctuidae), 261
Heliothis
  H. virescens (Noctuidae), 83, 147–8, 153–4, 156, 158, 160, 314, 322, 474
  H. zea (Noctuidae), 314, 322, 478
Heliozelidae, 335
Hepialidae, 173, 387
Hepialus lupulinus, See Korscheltellus lupulinus
Hesperiidae, 245
Heterarthrus (Tenthredinidae), 370
Heteroptera, 181, 305, 421, 454
Hippodamia convergens (Coccinellidae), 479
Homoptera, 468
hornfly, *See Haematobia irritans*
Hydrellia (Ephydridae), 255–6
Hypena scabra (Erebidae), 147, 421
Hypera
  H. brunneipennis (Curculionidae), 468
  H. postica (Curculionidae), 148, 157, 301
  H. punctata (Curculionidae), 432

I

Ibalia (Ibaliidae), 379–80
Incurvariidae, 95, 408, 410–11
Indian gypsy moth, *See Lymantria obfuscata*
Indian meal moth, *See Plodia interpunctella*
Ips (Curculionidae: Scolytinae), 261
Irenimus aequalis (Curculionidae)

J

jack pine sawfly, *See Neodiprion swainei*

K

Korscheltellus lupulinus (Hepialidae), 387

L

Lacanobia oleracea (Noctuidae), 144
ladybird, *See* Coccinelidae
Lampronia fuscatella (Incurvariidae), 95, 410–11
larch sawfly, *See Pristiphora erichsonii*
Laricobius erichsonii (Derodontidae), 428
Lasiocampidae, 148, 249, 291, 403, 430, 444, 514–15
Lasius fuliginosus (Formicidae), 441
Lathrolestes (Ichneumonidae), 391
leek moth, *See Acrolepiopsis assectella*
lesser cornstalk borer, *See Elasmopalpus lignosellus*
Leucanthiza (Gracillariidae), 147
Leucoma (=Stilpnotia) salicis (Lymantriidae), 246, 515
light brown apple moth, *See Epiphyas postvittana*
Lilioceris lilii (Chrysomelidae), 131, 428, 435
Limacodidae, 178, 229, 355, 389, 391, 416, 419, 439
Lipara (Chloropidae), 355
Liriomyza (Agromyzidae), 255, 257
Listronotus bonariensis (Curculionidae), 308
locust bean moth, *See Ectomyelois ceratoniae*
locust twig borer, *See Ecdytolopha insiticiana*
Lucanidae, 239
Lycaenidae, 395, 432, 510
Lyctidae, 239
Lygus (Lygaeidae), 306
Lymantria (Erebidae), 146
  L. dispar (Erebidae), 147, 148, 322, 376, 473, 484, 515
  L. obfuscata (Erebidae), 147
Lymantriinae (Erebidae), 246, 291, 392, 405, 444, 514–5
Lymexeliidae, 181
Lyonetiidae, 293, 335

M

Macrosiphon
  M. avenae (see Sitobium avenae)
  M. creelii (Aphididae), 220
Macrosiphum euphorbiae (Aphididae), 222
Maculinea (Lycaenidae), 395, 397
  M. rebeli (Lycaenidae), 395
maize stalk borer, *See Busseola fusca*
Malacosoma (Lasiocampidae), 249, 515
Mamestra
  M. brassicae (Noctuidae), 322, 457
  M. configurata (Noctuidae), 322, 471
Manduca sexta (Sphingidae), 147, 155, 157
Manuelia postica (Apidae), 355
Mecoptera, 282
Mediterranean flour moth, *See Ephestia kuehniella*
Megachile (Apidae), 369
Melandryidae, 281, 299, 308, 350
Meligethes, See Brassicogethes
Melitaea cinxia (Nymphalidae), 94, 432, 485, 498
Metasyrphus lunifer (Syrphidae), 363
Metopolophium dirhodeum (Aphididae), 476
Microlophium carnosum (Aphididae), 476
millet head miner, *See Heliocheilus albipunctella*
Miridae, 299, 305–6, 421–2, 512
mother of pearl moth, *See Pleuroptya ruralis*
mountain ash sawfly, *See Pristiphora geniculata*
mountain pine beetle, *See Dendroctonus monticola*
Musca
  M. domestica (Muscidae), 260
  M. vetustissimma (Muscidae), 254
Muscidae, 253–4, 260
museum beetle, *See Anthrenus museorum*
Mycetophilidae, 85, 360, 366–7
Myrmeleontidae, 393
Mythimna separata (often as *Pseudaletia*: Noctuidae), 34–5, 97, 147, 155, 157, 296, 327–9, 332, 432, 473
Myzus persicae (Aphididae), 214

N

Nabidae, 299
Neodiprion
  N. sertifer (Diprionidae), 97, 413, 418–19
  N. swainei (Diprionidae), 486
Neomusotima conspurcatalis (Crambidae), 293
Neophasia menapia (Pieridae), 506
Nepticulidae, 244, 270–1, 317, 320, 335
Neuroptera, 73, 206, 343, 374, 389, 432
Nezara viridis (Pentatomidae), 307
Nicaeana fraudator (Curculionidae), 470
Nitidulidae, 281–3, 299, 308, 428–9
Noctuidae, 31, 83, 96, 129, 146, 148, 153, 393, 394, 405, 421, 430, 432, 438, 442, 444, 446, 457, 470–3, 515
Nolidae, 245, 393
northern armyworm, *See Mythimna separate*
Notodontidae, 405, 415, 418, 430
Nymphalidae, 94, 161, 335, 375, 390, 405, 487, 498

O

oak processionary moth, *See Thaumetopoea processionea*

Oberea schaumii (Cerambycidae), 438
Occidentalia comptulatalis (Nymphalidae), 390
Oecophoridae, 239, 245, 284, 293, 407, 430, 438
olive fruitfly, *See Bactrocera oleae*
Opsina arenosella (Xylorictidae), 337
Orgyia spp. (Erebidae), 515
O. leucostigma (Erebidae), 99
O. dubia (Erebidae), 392
oriental armyworm, *See Mythimna separate*
oriental fruit moth, *See Grapholtha molesta*
oriental leafworm, *See Spodoptera litura*
Orthoptera, 206, 245, 299, 308, 374
Orthosia (Noctuidae), 249, 446
Ostrinia (Crambidae), 286
O. furnacalis (Crambidae), 114
O. nubilalis (Crambidae), 148, 394, 432, 471, 478
Oublema melanopus (Chrysomelidae), 429, 432

P

Paleacrita (Geometridae), 147
Pamphiliidae, 383, 416, 418, 432
Papilionidae, 131, 463, 498, *See also Battus*
Parapoynx stagnalis (Crambidae), 390
Parasteatoda tepidariorum (Araneae: Theridiidae), 371–2
Pardia tripunctata (Tortricidae), 482
Parnara guttata (Hesperiidae), 245
Paropsis atomaria (Chrysomelidae), 281
Passaloecus insignis (Sphecidae), 393
pea aphid, *See Acyrthosiphon pisum*
pea leafminer, *See Liriomyza huidobrensis*
Pectinophora gossypiella (Gelechiidae), 147, 315
Peleopoda (Oecophoridae), 245
Pentalonia nigronervosa (Aphididae), 471
Pentatomidae, 299, 307
pepper weevil, *See Anthonomus eugenii*
Pergidae, 399, 416, 421, 535
Phaeogenes invisor (Ichneumonidae), 309
Phoracantha (Cerambycidae), 110, 231, 233, 461, 482, 525
Phthorimaea operculella (Gelechiidae), 113, 272, 286, 293, 315, 438, 525
Phyllonorycter (Gracillariidae), 147
Pieridae, 157, 159, 331, 333, 506
Pieris
P. brassicae (Pieridae), 128, 131, 147, 154, 157, 164, 333, 375, 393, 469, 472
P. rapae (Pieridae), 111
pine looper moth, *See Bupalus pinaria*
pine white butterfly, *See Neophasia menapia*
pink bollworm, *See Pectinophora gossypiella*
pink stalkborer, *See Sesamia nonagrioides*
pipevine swallowtail, *See Battus philenor*
Pissodes (Curculionidae), 279, 477
Plataspidae, 299
Plathypena scabra, *See Hypena scabra*
Platyptilia gonodactyla (Pterophoridae), 487
Plesiometa (Tetragnathidae), 372
Pleuroptya ruralis (Crambidae), 487
Plodia interpunctella (Pyralidae), 130, 156, 260, 264, 266, 432, 479, 483
Plutella xylostella (Plutellidae), 97, 147, 157, 335, 394, 432, 437, 459, 469, 473, 476
Podalonia (Sphedidae), 391
pollen beetle, *See Meligethes*
Pompilidae, 391, 444
Pontania (Tenthredinidae), 428
poplar branch borer, *See Oberea schaumii*
potato tuber worm, *See Phthorimaea operculella*
Pristiphora
P. erichsonii (Tenthedinidae), 148, 416, 432
P. geniculata (Tenthedinidae), 148, 432
P. pallipes (Tenthredinidae), 384
Protium (Burseraceae), 263, 264
Pseudaletia separata, *See Mythimna separata*
Pseudoplusia includens, *See Chrysodeixis includens*
Psithyrus (Apidae), 299, 309
Psocoptera, 299, 305
Psychidae, 237, 245, 293, 393
Psylliodes (Chrysomelidae), 428
Pterophoridae, 148, 487
Ptinidae, 239
Pyralidae, 83, 111, 156, 173, 229, 231, 244, 260, 272, 284, 292–3, 312, 403, 424, 430, 432, 442, 468

R

Raphidioptera, 296, 343, 394, 432, 535
red admiral, *See Vanessa atalanta*
Rhagoletis (Tephritidae), 93, 257–8
Rhyacionia buoliana (Tortricidae), 99, 157, 293, 370
Rhynchaenus fagi (Curculionidae), 279
Rivula sericealis (Noctuidae), 330
Ropalidia (Vespidae), 370
rose-grain aphid, *See Metopolophium dirhodeum*

S

Sarcophaga nigriventris (Sarcophagidae), 394
Sarcophagidae, 253–4, 394
Saturniidae, 148, 403, 437, 444, 499
Scarabaeidae, 238, 350, 391
Schizolachnus (Aphidae), 213, 465
Sciophila varia (Mycetophilidae), 360
Scolytidae, *See* Scolytinae
Scolytinae, 46, 129, 167, 261–2, 299, 302, 310, 350, 428
Scotorythra paludicola (Geometriidae), 470
serpentine leafminer, *See Liriomyxa trifolii*
Sesamia nonagrioides (Noctuidae), 148
Sesiidae, 264, 325, 403, 438
Silo (Trichoptera: Sericostomatidae), 385
Sirex noctilio (Siricidae), 355, 380
Siricidae, 379–80
Sitobium avenae (Aphididae), 476
Sitona discoideus (Curculionidae), 470
Sitotroga cerealella (Gelechiidae), 260
small gooseberry sawfly, *See Pristiphora pallipes*
small mottled willow moth, *See Spodoptera exigua*
small tortoiseshell, *See Aglais urticae*
sorghum stalk borer, *See Busseola fusca*
soybean aphid, *See Aphis glycines*
soybean looper, *See Chrysodeixis includens*
Sphecidae, 54, 85, 166, 233–4, 369, 378, 391, 393, 427
Sphindidae, 282–3
Sphingidae, 85, 155, 157, 248, 296, 323, 395, 405, 437, 444, 498
spiders, 2, 4, 39, 82, 91, 93, 102–3, 123, 161, 367, 369–73, 378, 389, 391–3, 482–3, 490, 504, 523–4, 537
Spilomena (Sphecidae), 427
Spodoptera
S. exigua (Noctuidae)
S. frugiperda (Noctuidae), 96, 148, 154, 295, 393, 438, 442, 479
S. litura (Noctuidae), 93, 147
spotted stalk borer, *See Chilo partellus*
spruce budworm, *See Choristoneura fumiferanae*
spruce needle miner, *See Epiblema tedella*
Staphylinidae, 282, 299, 302
Stelidota geminata (Nitidulidae), 308
stinging nettle apid, *See Microlophium carnosum*
Stratiomyidae, 448, 481
straw dot moth, *See Rivula sericealis*

strawberry sap beetle, *See Stelidota geminata*
submerged rice case worm, *See Paraponyx stagnalis*
sugar cane borer, *See Diatraea saccharallis*
Sylepta derogata (Crambidae), 337
Syrphidae, 95, 221–2, 361–2, 364, 394, 457, 491

T

Tachinidae, 99, 374, 391, 421, 468
Tachyporus (Staphylinidae), 302
Tenthredinidae, 75, 270, 370, 383–4, 391, 413, 416, 418–21, 432, 447, 449, 463
Tephritidae, 94, 256–61, 279, 469, 482
Tetragnathidae, 371–2
Tetramesa (Eurytomidae), 392
Tetropium gabrieli (Cerambycidae), 351
Thaumetopoea processionea (Thaumetopoeidae), 375, 483
Thaumetopoeidae, 515
Thiatiridae, 405
Tinea pellionella (Tineidae), 237
Tineidae, 335
Tingidae, 299
Tinocallis platani (Aphididae), 214
Tischeria (Tischeriidae), 147
Tischeriidae, 335
tobacco budworm, *See Heliothis virescens*
tobacco hornworm, *See Manduca sexta*
tomato aphid, *See Macrosiphum euphorbiae*
tomato leafminer, *See Liriomyza bryoniae*
tomato moth, *See Lacanobia oleracea*
Tortricidae, 148, 157, 229, 244, 284, 286–7, 292–3, 296–8, 322, 370, 375–6, 390, 393–4, 397, 403, 407, 414, 416, 419, 421, 424, 430, 432, 438, 482, 515
Toxoptera citricola (Aphididae), 214
Tremex columba (Siricidae), 380
Trichoplusia ni (Noctuidae), 111, 147–8, 152, 295, 405
Trichoptera, 105, 206, 322, 343, 374, 385, 432
Trypoxylon (Sphecidae), 369
tsetse fly, *See Glossina*

U

Uroleucon ambrosiae (Aphidae), 213

V

Vanessa atalanta (Nymphalidae), 487
Vespidae, 100, 171–2, 370, 391
Vespula (Vespidae), 389, 392, 480

W

walnut aphid, *See Chromaphis juglandicola*
wheat aphid, *See Sitobium avenae*
wheat stem sawfly, *See Cephus*

X

Xiphydria (Xiphydriidae), 128, 378–9
Xyela (Xyelidae), 282–3, 414
Xyelidae, 428
Xylorictidae, 245, 337

Y

Yponomeuta (Yponomeutidae), 424
Yponomeutidae, 97, 130, 292, 407, 430

Z

Zeiraphera (Tortricidae), 394, 414, 419, 424
Zodarion (Zodariidae), 393
Zygaena filipendula (Zygaenidae), 181
Zygaenidae, 405

ICHNEUMONOID GENUS, TRIBE AND SUBFAMILY INDEX

A

Acaenitinae (Ich.), 40, 59, 79, 104, 174, 345, 357–9, 363, 408, 448, 461, 492
Acaenitus (Ich.: Acaenitinae), 357, 359
Acampsis (Brac.: Sigalpginae), 139, 274, 276–8, 419, 466
Acampsohelcon (Brac.: Acampsohelconinae), 284
Acampsohelconinae (Brac.), 27, 173, 178, 207, 278, 283–5, 290, 339, 494
Acanthocaudus (Brac.: Aphidiinae), 221
Acanthormius (Brac.: Lysiterminae), 178
Achaiabracon (Brac.: Rhyssalinae), 172, 238
Achaiabraconini (Brac.: Rhyssalinae), 172, 238
Aclastus (Ich.: Cryptinae), 389, 448
Aclitini (Brac.: Aphidiinae), 116, 223
Aclitus (Brac.: Aphidiinae), 26, 214–15, 224
Acoelini (*see* Adeliinae)
Acrisidini (Brac.: Rhyssalinae), 238, 242
Acrisis (Brac.: Rhyssalinae), 11, 172, 239, 242
Acrocerillia (Brac.: Braconinae), 91
Acrodactyla (Ich.: Pimplinae), 371
Acrolyta (Ich.: Cryptinae), 393
Acrorictus (Ich.: Cryptinae), 54, 391
Adeliinae, -ini (*see* Cheloninae)
Adelius (Brac.: Cheloninae), 172, 208
Adelognathinae (Ich.), 342–3, 345, 349, 372, 383–4, 455, 492
Adelognathus (Ich.: Adelognathinae), 12, 101, 349, 372, 384, 466
Adelphion (Ich.: Pedunculinae), 399, 400
Ademon (Brac.: Opiinae), 256, 260
Ademonini (Brac.: Opiinae), 260
Aenigmobracon†, 49, 193–4, 210–11
Afrocampsis (Brac.: Acampsohelconinae), 276, 284–5
Agathidinae (Brac.), 26–7, 32, 78, 91, 113, 116, 123, 168–9, 175–6, 183, 199, 202–3, 207–9, 271–5, 278, 293–4, 321, 465, 496, 504, 509
Agathidini (Brac.: Agathidinae), 203, 272, 275
Agathilla (Ich.: Banchinae), 405
Agathirsia (Brac.: Agathidinae), 168–9, 274
Agathis (Brac.: Agathidinae), 32, 48, 112, 113, 116, 118, 124, 169, 272–3, 275, 480
Agriotypinae (Ich.), 18, 345, 360, 383, 385–7, 492
Agriotypus (Ich.: Agriotypinae), 61, 385–7
Agrothereutes (Ich.: Cryptinae), 392, 462
Alabagrus (Brac.: Agathidinae), 118–19, 272, 275
Aleiodes (Brac.: Rogadinae), 15, 18, 40, 54, 60, 66, 70, 92, 102, 104–5, 116, 123, 157, 172, 173, 177–9, 181, 184, 246–50, 297, 321, 454, 480, 483, 487, 503, 504, 514, 515
Aleiodini (Brac.: Rogadinae), 246–7, 249–50
Alexeter (Ich.: Ctenopelmatinae), 481
Allobracon (Brac.: placement uncertain), 231, 237–8
Allomacrus (Ich.: Cylloceriinae), 360
Alloplitis (Brac.: Microgastrinae), 334
Allurus (Brac.: Euphorinae), 299–302
Allorhogas (Brac.: Doryctinae), 110–11, 185, 234–5
Alomya (Ich.: Alomyinae), 173–4, 387–8, 395, 397, 544
Alomyinae (Ich.: *see also* Phaeogenini), 345, 383, 387, 395, 397–8
Alophophion (Ich.: Ophioninae), 444
Alophosternum (Ich.: Pimplinae), 370
Alysia (Brac.: Alysiinae), 81, 109, 123, 177, 202, 251, 253–5, 463
Alysiasta (Brac.: Alysiinae), 251, 253
Alysiinae (Brac.), 12, 20, 46, 59, 62, 64, 69, 73, 81, 83, 85, 91–2, 97, 99, 108, 109, 114, 116, 122, 123, 134, 139, 140, 143, 144, 145, 15, 157, 159, 172, 177, 178, 179, 181, 203–4, 207, 229, 244, 250–255, 414, 439, 457, 458, 463, 478, 479
Amblyteles (for species in this book *see* *Ctenichneumon* or *Eutanyacra*)

The Braconid and Ichneumonid Parasitoid Wasps: Biology, Systematics, Evolution and Ecology, First Edition. Donald L. J. Quicke.

Amicrocentrinae (Brac.), 47, 169, 203, 207, 279, 287–9, 494
Amicrocentrum (Brac.: Amicrocentrinae), 47, 177, 288–9
Amphirhachis (Ich.: Banchinae), 403–4
Amplicella† (Ich.: Tanychorinae†), 194, 347
Amputoearinus (Brac.: Agathidinae), 275
Amydraulax (Ich.: Cryptinae), 393
Anaprixia (*see Paroligoneurus*), 321
Andesipolis (Brac.: Mesostoinae), 212, 226, 228–9, 237, 504
Aneurobracon (Brac.: Agathidinae), 273–5
Aneurobraconinae (*see* Mesocoelini)
Angitia (*see Diadegma*),
Ankylophon (Ich.: Tryphoninae), 411, 413
Ankylophonini (Ich.: Tryphoninae), 411, 413
Anomalinae (*see* Anomaloninae)
Anomalon (Ich.: Anomaloninae), 17, 20, 26, 66, 73, 85, 105, 174, 183–4, 196, 345, 400–1, 418, 430–1, 441–2, 445, 461, 491–2, 534
Anomaloninae (Ich.), 17, 20, 26, 73, 85, 174, 183–4, 196, 345, 400–1, 418, 430–1, 441–2, 445, 461, 491–2, 534
Anomalonini (Ich.: Anomaloninae), 430
Antestrigini (Brac.: Orgilinae), 293
Antestrix (Brac.: Orgilinae), 292–4
Apanteliini (Brac.: Microgastrinae), 324
Apanteles s.s. (Brac.: Microgastrinae), 83, 115–16, 147, 322, 324–5, 332, 334, 512
Apanteles (*see also Cotesia, Protapanteles*), 457
Apechoneura (Ich.: Labeninae), 27, 124, 170, 181, 354, 356
Apechthis (Ich.: Pimplinae), 98, 346, 374, 375, 456, 481, 498
Aphaereta (Brac.: Alysiinae), 59, 83, 92, 99, 253–4, 463
Aphidiinae (Brac.), 14–15, 30, 32, 38, 52, 55, 59, 62–4, 69, 73, 75, 78–80, 90–2, 115–16, 125, 134, 138, 156, 158, 160, 172, 192, 195–8, 202–4, 207, 210, 212–24, 229, 254, 294, 303, 454, 460, 480, 494, 496, 503–4, 535, 543
Aphidiini (Brac.: Aphidiinae), 75, 214, 216–17, 223–4
Aphidius (Brac.: Aphidiinae), 13–14, 26, 62, 72, 75, 93–4, 97, 115–16, 122, 133–5, 142, 156–60, 213–22, 457–8, 466, 471, 475–6, 487, 526
Aphrastobracon (*see Megalommum*),
Aphrastobraconini (Brac.: Braconinae), 268
Apoblacus (Brac.: Blacinae), 283
Aplomerus (Ich.: Xoridinae), 171, 349, 351
Apolophus (*see Scolomus*),
Apophua (Ich.: Banchinae), 147, 149, 151, 401, 404
Apozyginae (Brac.), 27, 49, 190, 192, 195, 197, 203, 207, 209–11, 278, 494, 504, 534
Apozyx (Brac.: Apozyginae), 49, 197, 210
Apsilops (Ich.: Cryptinae), 390
Aptesini (Ich.: Cryptinae), 389–91
Aptesis (Ich.: Cryptinae), 390–1
Arenetra (Ich.: Banchinae), 403
Argamania (Brac.: Braconinae), 269
Argamaniini (Brac.: Braconinae), 269
Aritranis (Ich.: Cryptinae), 391
Arotes (Ich.: Acaenitinae), 40, 79, 358
Arthula (Ich.: Cryptinae), 391
Ascogaster (Brac.: Cheloninae), 67, 68, 83, 93, 123, 147, 157, 160, 315–8, 465
Asiacentistes (Brac.: Euphorinae), 295
Asobara (Brac.: Alysiinae), 62, 64, 83, 96–7, 99, 108–9, 114, 116, 134, 139–45, 157, 159, 181, 253–4, 457, 458, 478, 479
Asperellus (*see Certonotus*),
Aspidobraconina (Brac.: Braconinae), 91, 179, 263, 457
Aspilodemon (Brac.: Mesostoinae), 212, 228
Aspilota (Brac.: Alysiinae), 172, 179, 251, 253–4
Astiphromma (Ich.: Mesochorinae), 422
Atanycolus (Brac.: Braconinae), 231, 260–2, 266, 268, 530
Atopandrium (Brac.: Alysiinae), 172
Atopotypus (*see Agriotypus*),
Atractodes (Ich.: Cryptinae), 393
Atrophini (Ich.: Banchinae), 403–5
Aulacocentrum (Brac.: Macrocentrinae), 177
Aulosaphes (Brac.: Lysiterminae), 178
Australoglypta (Ich.: Banchinae), 403
Austrohelcon (Brac.: Helconinae), 279
Austropion (Ich.: Ctenopelmatinae), 419
Austrozele (Brac.: Macrocentrinae), 40, 175, 284
Autriquella (Brac.: Aphidiinae), 221
Avga (Brac.: Mesostoinae), 229
Avgini (Brac.: Mesostoinae), 226, 228–9

B

Bactreodes (Brac.: Euphorinae), 309
Banchinae (Ich.), 3, 32, 40, 63, 75, 89, 92, 120, 138, 146–7, 149, 150, 157, 168, 178–9, 345, 392, 401–5, 406, 408, 426, 435, 444–5, 447, 461, 464, 482, 491–2, 502, 535, 540
Banchini (Ich.: Banchinae), 184, 403, 405
Banchogastra (Ich.: Ophioninae), 506
Banchus (Ich.: Banchinae), 39, 52, 75–7, 120, 387, 405, 424, 464, 472, 481
Barronia (Ich.: Eucerotinae), 447
Barycerotina (Ich.: Cryptinae), 391–2
Barycnemis (Ich.: Tersilochinae), 173, 198, 391, 429
Barytarbes (Ich.: Ctenopelmatinae), 148
Bassi (*see also* Agathidinae), 202
Bassinae (*see also* Agathidinae), 207–8
Bassus (Brac.: Agathidinae), 116, 208, 272–5
Bathyaulax (Brac.: Braconinae), 182, 269
Bathyplectes (Ich.: Campopleginae), 60, 62, 70, 78, 85, 105, 116, 121–2, 145, 148, 159, 392, 422, 432–5, 449, 457, 463, 468, 476, 485
Bathythrix (Ich.: Cryptinae), 102, 389, 393
Bathytrichina (Ich.: Cryptinae), 393
Bathyzonus (Ich.: Cryptinae), 448
Batothecoides (Brac.: Rogadinae), 246
Belesica (Ich.: *Cremastinae incertae sedis*), 105, 439
Bellopius (subgenus of *Opius* q.v.), 256–7
Betelgeuse (Brac.: Euphorinae), 303–4
Betylobracon (Brac.: Betylobraconinae), 173, 243
Betylobraconinae (Brac.), 173, 203–4, 207, 230, 243, 295, 309, 394, 535
Bicentra (Brac.: Braconinae), 17, 269
Bicurta (Ich.: Collyriinae), 359
Biolysia (*see Bathyplectes*)
Biosteres (Brac.: Opiinae), 31, 54, 62, 94, 123, 256–60
Binodoxys (Brac.: Aphidiinae), 143, 213–14, 216–17, 221, 458, 480
Biroia (Brac.: Agathidinae), 274
Bitomus (Brac.: Opiinae), 173, 178, 339
Blapsidotes (Ich.: Cryptinae), 389, 393
Brachistini (Brac.: Helconinae), 76, 177, 178, 182, 202, 279–81, 302
Brachycyrtinae (Ich.), 12, 183, 345, 353, 383, 398–400, 447, 491
Brachycyrtus (Ich.: Brachycyrtinae), 12, 398–9, 535

Brachyscleroma (Ich.: Sisyrostolinae), 345, 407, 428
Brachyscleromatinae (*see* Sisyrostolinae)
Bracon (Brac.: Braconinae), 2–4, 6, 9, 12–13, 15–17, 19, 21–22, 25, 27–33, 35, 39, 43, 46–48, 51–2, 54, 57, 59, 61–6, 68–9, 71–2, 75, 78- 80, 85, 87, 88, 90–2, 97, 99, 103–04, 107–10, 115–7, 120, 123–7, 130, 137–8, 145, 147, 156, 163–4, 167, 168–4, 176–86, 189–90, 192–3, 195–99, 260–1, 263–6, 268, 341, 344, 413, 430, 435, 439–40, 453–5, 459, 462, 464–6, 468, 470, 474, 479, 48–2, 484, 489–90, 492, 49–7, 500–1, 503–4, 509, 513, 533–5, 539–41
Braconella (Brac.: Braconinae), 174
Braconinae (Brac.), 15, 17, 31, 39, 46, 54, 59, 61, 63–4, 66, 68–9, 79, 91, 103–4, 115–7, 120, 123–4, 164, 167–70, 172, 174, 178–9, 181, 183, 193, 199, 260–269, 454–5, 459, 464–5, 482, 494–6, 503, 509
Braconini (Brac.: Braconinae), 268
Braunsia (Brac.: Agathidinae), 48
Bremiella (Ich.: placement uncertain), 344, 424, 534
Brulleia (Brac.: Helconinae), 280–1
Brulleini (Brac.: Helconinae), 279, 281, 290
Buathra (Ich.: Cryptinae), 166, 390–1
Bulborogas (Brac.: Rogadinae), 173, 249
Buluka (Brac.: Microgastrinae), 55, 177–8

C

Callojoppa (Ich.: Ichneumoninae), 397
Campocraspedon (Ich.: Diplazontinae), 179, 364
Campodorus (Ich.: Ctenopelmatinae), 418
Campoletis (Ich.: Campopleginae), 58, 74, 95–7, 122, 129, 145–6, 148–9, 152, 154, 156, 182, 405, 432–5, 474, 478
Campopleginae (Ich.), 40, 54, 61–2, 68, 72, 75, 80–1, 83, 88, 92, 94, 105, 109, 116, 138, 146, 148, 150–1, 154, 157, 159, 198–9, 344, 345–6, 353, 359, 400–2, 404, 432, 434–6, 439, 442, 445, 461, 466, 471, 475, 482, 497, 541
Campoplex (Ich.: Campopleginae), 40, 76, 344, 346, 433, 435, 482
Canalicephalus (Brac.: Acampsohelconinae), 284
Canberria (Brac.: Mesostoinae), 229
Canberriini (Brac.: Mesostoinae), 207, 212, 228
Cantharoctonus (Brac.: Rhysipolinae), 237, 242
Capitonius (Brac.: Cenocoeliinae), 171, 310
Cardiochiles (*see also Toxoneuron*), 312, 314
Cardiochilinae (Brac.), 138, 146–7, 158, 168, 183, 207, 312–4, 322, 472, 494, 509
Carinodes (Ich.: Ichneumoninae), 503
Catachora† (Ich.: Labenopimplinae†), 349, 413
Cedilla (Brac.: Braconinae), 265
Cedria (Brac.: Lysiterminae), 236, 245
Cedriini (Brac.: Lysiterminae), 193, 245
Cenocoeliinae (Brac.), 40, 43, 138, 170, 173, 203, 207, 279, 295, 310–11, 494
Cenocoelius (Brac.: Cenocoeliinae), 40
Centistes (Brac.: Euphorinae), 116, 299–302
Centistina (Brac.: Euphorinae), 299, 303–4
Centistini (Brac.: Euphorinae), 282, 295, 309
Cercobarcon (Brac.: Trachypetinae), 208–9, 413
Cercobarconinae (*see* Trachypetinae), 203, 207–9
Cephaloglypta (Ich.: Banchinae), 403
Ceratojoppini (Ich.: Ichneumoninae), 397
Certonotus (Ich.: Labeninae), 120, 170, 180, 355
Cervellus (Brac.: Braconinae), 169
Chaenusa (Brac.: Alysiinae), 252, 255
Chaoilta (*see Odontoscapus*),
Charmon (Brac.: Charmontinae), 47, 89, 126, 475, 535
Charmontia (Brac.: Charmontinae), 287
Charmontinae (Brac.), 89, 125, 180, 208, 278, 284, 287, 494
Chartobracon (Brac.: Braconinae), 171
Chasmias (Ich.: Ichneumoninae), 98, 394
Chasmodon (Brac.: Alysiinae), 172
Cheloninae (Brac.), 11, 39, 76, 77, 79, 116, 125, 138, 146, 147, 153, 157, 160, 168, 173, 177–8, 182, 305, 312–13, 315–9, 482, 496, 521
Chelonodoryctes (Brac.: Doryctinae), 177
Chelonogastra (Brac.: Braconinae), 179, 265
Chelonus (Brac.: Cheloninae), 77, 83–4, 116, 142–3, 147, 149, 152–3, 160, 315–7, 473
Chineator (Ich.: Mesochorinae), 422
Chorebus (Brac.: Alysiinae), 179, 253, 255
Chorinaeus (Ich.: Metopiinae), 424, 454
Chriodes (Ich.: Nesomesochorinae), 435, 442
Chrionota (Ich.: Ctenopelmatinae), 417
Chrionotini (Ich.: Ctenopelmatinae), 47, 416–8
Chrysocryptus (Ich.: Cryptinae), 393
Cidaphus (Ich.: Mesochorinae), 421–2
Claseinae (Ich.), 345, 398–9
Claseini (Ich.: Claseinae), 399
Clasis (Ich.: Claseinae), 398–400
Clinocentrini (Brac.: Rogadinae), 246–7, 249
Clinocentrus (Brac.: Rogadinae), 92, 101, 111–12, 160, 237, 247
Clistopyga (Ich.: Pimplinae), 103, 373
Clotildea (Brac.: Orgilinae), 293
Clydonium (Ich.: Pimplinae), 29
Clypeodromini (Ich.: Ichneumoninae), 397
Cnastis (Ich.: Poemeniinae), 378
Coccygomimus (*see Pimpla*),
Coelalysia (Brac.: Alysiinae), 254
Coelinidea (Brac.: Alysiinae), 255
Coeloides (Brac.: Braconinae), 13, 79, 100, 113, 128, 129, 260–3, 268, 464, 477
Coeloidini (Brac.: Braconinae), 268
Colastes (Brac.: Exothecinae), 250, 269–70, 279
Colastomion (Brac.: Rogadinae), 15, 92, 248
Coleocentrini (Ich.: Acaenitinae), 59, 358, 359
Coleocentrus (Ich.: Acaenitinae), 59, 358, 359
Coleopius (Brac.: Opiinae), 178
Collyria (Ich.: Collyriinae), 359
Collyriinae (Ich.), 182–3, 357, 359–60, 448, 492
Colpognathus (Ich.: Ichneumoninae), 388, 395, 397
Colpotrochia (Ich.: Metopiinae), 69, 70, 423
Compsobracon (Brac.: Braconinae), 266–7, 481
Compsobraconoides (Brac.: Braconinae), 263

Conoblasta (*see Glypta*)
Cosmophorini (Brac.: Euphorinae), 167, 302–5, 309
Cosmophorus (Brac.: Euphorinae), 12, 117, 119, 123, 172, 299, 302, 303, 309
Cotesia (Brac.: Microgastrinae), 13, 18, 54, 58, 61–2, 64–5, 67, 72, 77, 91–2, 94–7, 99, 109–10, 113–5, 120, 122–3, 128–9, 131, 132–4, 139, 142, 144–50, 154, 157–9, 161, 182, 205, 221, 322, 324–5, 327–35, 393, 421–2, 457, 459, 463, 469, 472, 477, 479–80, 483, 485, 498–9
Cotesiini (Brac.: Microgastrinae), 324
Crassomicrodus (Brac.: Agathidinae), 168, 247
Cratodactyla (Brac.: Rogadinae), 174
Cremastinae (Ich.), 26, 46–7, 105, 116, 174–5, 345, 398, 400–2, 430, 438–9, 444, 461, 491–2
Cremastus (Ich.: Cremastinae), 76, 438, 439
Cremnops (Brac.: Agathidinae), 176, 272, 339
Cremnoptini (Brac.: Agathidinae)
Cressonia (*see Daschyana*)
Cryptinae (Ich.)
Cryptophion (Ich.: Campopleginae),
Cryptopimpla (Ich.: Banchinae),
Cryptus (Ich.: Cryptinae),
Ctenichneumon (Ich.: Ichneumoninae), 61
Ctenocalini (Ich.: Ichneumoninae), 397
Ctenopelma (Ich.: Ctenopelmatinae), 182, 346, 418
Ctenopelmatinae (Ich.), 14, 39, 41, 47, 73, 75, 89, 100, 117, 121, 148, 174, 182–3, 197, 342, 345–6, 349, 400–2, 416–17, 420–1, 424–5, 428, 435, 439, 445, 447, 454–5, 461, 464, 466, 491–2, 534
Ctenopelmatini (Ich.: Ctenopelmatinae), 418
Cuniculobracon (Brac.: Helconinae), 280
Curriea (*see Megalommum*),
Cyanopterus (Brac.: Braconinae), 260
Cylloceria (Ich.: Cylloceriinae), 360–1
Cylloceriinae (Ich.), 14, 104, 345, 360–1, 425, 492

D

Dacnusa (Brac.: Alysiinae), 253, 255
Dacnusini (Brac.: Alysiinae), 203, 251–6, 439
Dapsilarthra (Brac.: Alysiinae), 251, 254
Daschyana (Ich.: Diacritinae), 360
Declotila (Brac.: Orgilinae), 293
Delomerista (Ich.: Pimplinae), 369
Delomeristini (Ich.: Pimplinae), 367, 369, 448
Deltaphyma (Brac.: Braconinae), 169
Dentichasmias (Ich.: Ichneumoninae), 98, 394
Dentifemura (Ich.: Acaenitinae), 174
Deuteroxorides (Ich.: Poemeniinae), 378
Devorgilla (*see Venturia*)
Diachasma (Brac.: Opiinae), 93, 256, 258,
Diachasmimorpha (Brac.: Opiinae), 31, 54, 94, 123, 155, 257–60, 439
Diacritinae (Ich.)
Diacritus (Ich.: Diacritinae), 360, 362, 367
Diadegma (Ich.: Campopleginae), 59, 60, 61, 64, 70, 74–6, 94, 96, 109, 120, 121, 122, 135, 148, 149, 157, 432, 433, 435, 437, 458, 469
Diadromus (Ich.: Ichneumoninae), 13, 55, 58, 61, 68–70, 82, 95, 97, 109, 116, 122, 130, 132, 145, 152, 158, 394, 395, 398
Diaparsis (Ich.: Tersilochinae), 428–9
Diapetimorpha (Ich.: Cryptinae), 13, 393, 474
Diaretiella (Brac.: Aphidiinae), 78, 134, 215–16
Dicamptus (Ich.: Ophioninae), 442
Digonogastra (Brac.: Braconinae), 39, 46, 68, 124, 170, 260, 265, 269, 525
Dimeris (Brac.: Pambolinae), 172
Dinocampini (Brac.: Euphorinae), 303–4
Dinocampus (Brac.: Euphorinae), 158, 299–300, 303–5, 330, 366, 475, 479
Diolcogaster (Brac.: Microgastrinae), 421
Diospilinae (*see* Diospilini)
Diospilini (Brac.: Helconinae), 203, 208, 224, 279, 281, 339
Diplazon (Ich.: Diplazontinae), 12, 14, 78, 81–2, 95, 104, 116, 345, 359–65, 398, 446, 448, 455, 457, 461, 491–2, 508
Diplazontinae (Ich.), 12, 14, 81–2, 104, 116, 345, 359–61, 363, 365, 398, 446, 448, 455, 461, 491–2, 508
Diradops (Ich.: Banchinae), 92, 403
Dirrhope (Brac.: Dirrhopinae), 319
Dirrhopinae (Brac.), 203, 207, 312, 319, 494, 535
Disophrys (Brac.: Agathidinae), 273
Distilirella (Brac.: Xiphozelinae), 288, 290
Dolichogenidea (Brac.: Microgastrinae), 120, 460
Dolichomitus (Ich.: Pimplinae), 33, 36, 42, 44, 59, 368–9, 379, 481
Dolichozele (Brac.: Macrocentrinae), 175, 284
Dolophron (Ich.: Campopleginae), 432
Dolopsidea (Brac.: Rhyssalinae), 238–40, 242
Doryctinae (Brac.), 38–40, 42, 48–9, 52, 54, 59, 63–4, 85, 103, 168, 170, 172, 174, 177–8, 181, 186, 199, 203, 207, 212, 229, 231–5, 459, 494, 496
Doryctini (Brac.: Doryctinae), 203, 235
Doryctobracon (Brac.: Opiinae), 46, 97, 114, 257, 260, 482
Doryctomorpha (Brac.: Mesostoinae *incertae sedis*), 174, 212, 226, 229–30
Dusona (Ich.: Campopleginae), 148, 432, 435
Dyscoletes (Brac.: Blacinae), 282–3
Dyscritulus (Brac.: Aphidiinae), 216–17, 221

E

Eadya (Brac.: Helconinae *incertae sedis*), 36, 78, 170, 281, 309–10
Earinini (Brac.: Agathidinae), 26, 272
Earinus (Brac.: Agathidinae), 176, 274–5, 265, 278
Earobia (Ich.: Tersilochinae), 349, 428, 455
Ecclitura (Brac.: Euphorinae), 303–4
Echthrini (*see* Aptesini), 389, 391
Echthromorpha (Ich.: Pimplinae), 505, 507
Echthrus (Ich.: Cryptinae), 166, 186, 390–1
Eclytini (Ich.: Tryphoninae), 411, 413
Eclytus (Ich.: Tryphoninae), 411, 414
Ecnomiinae (*see* Euphorinae: Ecnomiini)
Ecnomiini (Brac. Euphorinae), 158, 208, 282, 305–6
Ecnomios (Brac.: Euphorinae), 304, 306
Ecphysis (Ich.: Claseinae), 399
Eiphosoma (Ich.: Cremastinae), 174, 438, 458, 479
Elasmosoma (Brac.: Euphorinae), 307–8
Embobracon (Brac.: Doryctinae), 236
Endasys (Ich.: Cryptinae), 97
Endromopoda (Ich.: Pimplinae), 369
Enicospilus (Ich.: Ophioninae), 85, 161, 176, 414, 442, 444–5, 457, 499, 501, 504–5

Enizemum (Ich.: Diplazontinae), 363–5
Enytus (Ich.: Campopleginae), 432–3
Eobracon†(*see Eobraconus*†), 193–4, 199
Eobraconus† (Brac.: *incertae sedis*), 194
Eoichneumon† (Brac.: Eoichneumoninae†), 192, 194
Eoichneumoninae† (Brac.), 17, 190, 192–4, 346, 534
Ephedrini (Brac.: Aphidiinae), 75, 214, 216–17, 223–4
Ephedrus (Brac.: Aphidiinae), 53, 72, 75, 92, 116, 212, 214–6, 219–22, 462, 466
Ephialtes (*see also Pimpla* sensu Townes), 42, 346, 369–70, 480
Ephialtini (Ich.: Pimplinae), 49, 83, 145, 168, 346, 367–70, 373, 376, 379, 523
Epimicta (Brac.: Alysiinae), 252
Epsilogaster (Brac.: Mendesellinae), 322–3
Eremotylus (Ich.: Ophioninae), 445
Eriborus (Ich.: Campopleginae), 13, 54, 58, 67, 148, 432, 435, 437
Eridolius (Ich.: Tryphoninae), 413
Erromenus (Ich.: Tryphoninae), 14, 411
Erythrodolius (Ich.: Sisyrostolinae), 407
Eutanyacra (Ich.: Ichneumoninae), 394
Eubaeus† (Ich.: Labenopimplinae†), 349, 413
Eubazus (Brac.: Helconinae), 50, 279
Euceros (Ich.; Eucerotinae), 445–6, 534
Eucerotinae (Ich.), 73, 99, 105, 342–4, 398, 400, 445–8, 476, 492, 504
Eucoelinidea (Brac.: Alysiinae), 179
Eugalta (Ich.: Poemeniinae), 359, 378, 448
Eulimneria (*see Sinophorus*),
Euphorinae (Brac.), 3, 46, 50, 60, 62, 86, 88–9, 91–2, 94, 115–16, 157–8, 172–3, 178, 180, 182–5, 203, 208, 240, 243, 281–2, 290, 294–310, 446, 464–5, 471, 480, 496, 501, 506
Eurygenys (Ich.: Cremastinae *incertae sedis*), 439
Eurylabini (Ich.: Ichneumoninae), 397
Euryproctini (Ich.: Ctenopelmatinae), 416–8, 420
Euryproctus (Ich.: Ctenopelmatinae), 41, 418
Eurypterna (Ich.: Hybrizontinae), 174, 440–1
Euscelinus (Brac.: Doryctinae), 174
Eusterinx (Ich.: Orthocentrinae), 14, 366
Euurobracon (Brac.: Braconinae), 36, 44, 59, 261
Evaniodes (Brac.: Doryctinae), 232, 310
Excavarus (Ich.: Tryphoninae), 412
Exenterini (Ich.: Tryphoninae, 17, 411–3
Exenterus (Ich.: Tryphoninae, 411, 413, 418
Exephanes (Ich.: Ichneumoninae), 31
Exeristes (Ich.: Pimplinae), 61, 78, 83, 370, 376, 456, 474, 480
Exetastes (Ich.: Banchinae), 184, 405, 444
Exochus (Ich.: Metopiinae), 423, 481
Exodontiella (Brac.: Gnamptodontinae), 251, 255, 260, 271, 321

F

Facitorini (Brac.: Rogadinae), 243
Falcosyntretus (Brac.: Euphorinae), 309
Fopius (Brac.: Opiinae), 31, 94, 111, 120, 123, 257–8, 260, 525
Fornicia (Brac.: Microgastrinae), 177–8, 334
Forniciini (Brac.: Microgastrinae), 324
Fredegundia (Ich.: Pimplinae), 370
Furcadesha (Brac.: Braconinae), 179

G

Gabuniina (Ich.: Cryptinae), 392
Gahus (Ich.: Cremastinae), 439
Gastralysia (Brac.: Alysiinae), 177, 251
Gastrotheca (*see Physaraia*),
Gelanes (Ich.: Tersilochinae), 428
Gelinae (*see* Cryptinae)
Gelini (*see* Phygadeuontini), 389, 395
Gelis (Ich.: Cryptinae), 61, 93, 99, 102, 116, 170, 172, 184–5, 346, 388–90, 392–3, 483
Goedartiini (Ich.: Ichneumoninae), 397
Gilaromma (Ich.: Hybrizontinae), 440
Glabridorsum (Ich.: Cryptinae), 102, 389, 393
Glyphogaster (Brac.: Alysiinae), 251
Glypta (Ich.: Banchinae), 40, 47, 67, 122, 147, 149, 151, 403–4, 473, 482
Glyptapanteles (Brac.: Microgastrinae), 67, 147, 149, 328, 331, 334, 473
Glyptini (Ich.: Banchinae), 403–4, 406
Glyptomorpha (Brac.: Braconinae), 167, 260–1
Glyptorhaestus (Ich.: Ctenopelmatinae), 76, 419
Gnamptodon (Brac.: Gnamptodontinae), 270–1
Gnamptodontinae (Brac.), 266, 270–1
Gnaptodontinae (*see* Gnamptodoninae)
Gnaptogaster (Brac.: Gnamptodontinae), 271
Gnathobracon (Brac.: Braconinae), 167, 207, 269
Gnathobraconinae (Brac.: syn. Braconinae), 203, 207, 269
Gnathopleura (Brac.: Alysiinae), 254
Gnypetomorpha (Ich.: Cryptinae), 349
Gonotypus (Ich.: Campopleginae), 435
Goryphus (Ich.: Cryptinae), 389, 393
Grandia (Brac.: Alysiinae), 254
Gravenhorstiini (Ich.: Anomaloninae), 430, 431
Gregopimpla (Ich.: Pimplinae), 370
Grotea (Ich.: Labeninae), 81, 355
Groteini (Ich.: Labeninae), 186, 353, 355, 448
Grypocentrus (Ich.: Tryphoninae), 349, 411, 416, 419, 512
Grypokeros (Brac.: Blacinae), 283
Gyroneuron (Bra.: Rogadinae), 177

H

Habrobracon (Bra.:Braconinae), 31, 41, 54, 59, 61–2, 64, 67, 69, 72, 74, 75, 82, 83–4, 88, 108–10, 112, 115–7, 120, 122, 125, 129, 143, 156, 161, 260, 263, 265–8, 355, 458, 460, 462, 466, 468, 474, 499, 525
Habronyx (Ich.: Anomaloninae), 431, 458, 467
Hadrodactylus (Ich.: Ctenopelmatinae), 121, 417, 418
Hadrostethus (Ich.: Banchinae), 403
Hansonia (Brac.: Cardiochilinae), 85, 315
Helconidea (Brac.: Helconinae), 174
Hecabolini (Brac.: Doryctinae), 203, 235
Helconinae (Brac.), 16, 50–1, 76, 138, 168, 172, 174, 177–8, 180, 182, 197, 278–82, 284, 288, 291, 309–11, 337, 339, 496, 535
Helconini (Brac.: Helconinae), 52, 104, 170, 178, 203, 279–82
Helcostizus (Ich.: Cryptinae), 393
Helictes (Ich.: Orthocentrinae), 367
Helictinae (*see* Orthocentrinae)
Hellwigia (Ich.: Ophioninae), 12, 435, 437, 445
Hellwigiella (Ich.: Ophioninae), 29
Hellwigiellinae (*see* Ophioninae)
Hemigaster (Ich.: Cryptinae), 166, 186, 345, 389, 391
Hemigasterini auct. (*see* Aptesini)
Hemigyroneuron (subgen. of *Aleiodes*: Brac.: Rogadinae), 177, 249, 321
Hemiteles (Ich.: Cryptinae), 61, 177, 389, 481
Hemitelinae (*see* Cryptinae)

Hemitelini Authors (*see* Ich.: Cryptinae: Aptesini)
Heresiarchini (Ich.: Ichneumoninae), 346, 397
Heterogamus (Brac.: Rogadinae), 15, 123, 250
Heteropelma (Ich.: Anomaloninae), 174, 430, 432, 458
Heteropteron (Brac.: Cardiochilinae), 315
Heterospilini (Brac.: Doryctinae), 235
Heterospilus (Brac.: Doryctinae), 48, 78, 85, 109, 124, 172, 233–6, 459
Hidryta (Ich.: Cryptinae), 82, 389, 392
Himerta (Ich.: Ctenopelmatinae), 418
Histeromerinae (*see* Rhyssalinae.)
Histeromeroides (Brac.: Doryctinae), 232
Histeromerus (Brac.: Rhyssalinae), 32, 39, 53, 59, 109–10, 174, 238–40, 242–3, 302, 339, 372
Hodostates (Ich.: Ctenopelmatinae), 419
Homaspis (Ich.: Ctenopelmatinae), 139, 179, 418
Homolobinae (Brac.), 45, 50–1, 125, 290–2, 494
Homolobus (Brac.: Homolobinae), 39, 44, 45, 50–3, 63, 125, 290–1
Homotherus (Ich.: Ichneumoninae), 396
Hoplocryptus (Ich.: Cryptinae), 100, 116
Hormiinae (Brac.), 172, 198, 199, 234, 236–8, 242–5, 269, 454, 534
Hormiini (Brac.: Hormiinae), 203, 228,
Hormisca (Brac.: Hormiinae), 244
Hormius (Brac.: Hormiinae), 39, 172, 178, 243–4
Horogenes (*see Diadegma*),
Hyboteles (Brac.: Braconinae), 263
Hybrizon (Ich.: Hybrizontinae), 85, 169, 174, 190, 195–7, 342, 344–5, 439–1, 445, 492
Hybrizontinae (Ich.), 85, 169, 174, 190, 195–7, 342, 344–5, 439–1, 445, 492
Hydrangeocola (Brac.: Mesostoinae), 69, 212, 218, 228
Hydrangeocolinae (*see* Mesostoinae)
Hylesicida (Ich.: Banchinae), 403
Hymenoepimecis (Ich.: Pimplinae), 161, 372, 482
Hypamblis (Ich.: Ctenopelmatinae), 418
Hyperacmus (Ich.: Cylloceriinae), 173, 360–1
Hyposoter (Ich.: Campopleginae), 68, 83, 94, 97, 99, 109, 121, 142, 148–9, 151, 154–5, 182, 377, 432–5, 485

I

Iare (Brac.: Doryctinae), 232
Icaromimus (Ich.: Sisyrostolinae), 407
Ichneumon (Ich.: Ichneumoninae), 31, 394–6
Ichneumonini (Ich.: Ichneumoninae), 346, 397–8
Ichneumonini *sensu* Townes (*see* Heresiarchini)
Ichneumoninae (Ich.), 14, 31, 32–3, 61, 76, 91, 97, 104, 116, 126, 180, 184, 199, 344–5, 353, 383, 387–8, 394, 395, 397, 425, 448, 455, 461, 466, 482, 497, 502–3, 539, 542, 544
Ichneutes (Brac.: Ichneutinae), 320–1, 466
Ichneutinae (Brac.), 4, 15, 29, 123, 182, 312–13, 320–1, 339, 466, 490, 494, 535
Idiasta (Brac.: Alyssinae), 253
Idiogramma (Ich.: Tryphoninae), 345, 349, 411, 414
Idiogrammatini (Ich.: Tryphoninae), 414
Idiolispa (Ich.: Cryptinae), 389, 392, 448
Iphiaulacini (Brac.: Braconinae), 268
Iphiaulax (Brac.: Braconinae), 32, 36, 38, 260, 265, 268
Ischnocerus (Ich.: Xoridinae), 351
Ischnojoppini (Ich.: Ichneumoninae), 397
Ischyrocnemus (Ich.: placement uncertain), 424, 447, 534
Iseropus (Ich.: Pimplinae), 370
Isurgus (*see Tersilochus*),
Itamoplex (*see Cryptus*),
Itamuton (Ich.: Cryptinae), 393
Itoplectis (Ich.: Pimplinae), 13, 58, 68, 83, 93, 98, 100, 132, 358, 369, 374, 375–7, 433, 455–6, 462, 471–2, 474–6

J

Jarra (Brac.: Doryctinae), 231, 233, 482
Jezarotes (Ich.: Acaenitinae), 174
Joppini (*see* Ichneumonini), 346, 397
Joppocryptini (Ich.: Ichneumoninae), 397

K

Katytermus (Brac.: Lysiterminae), 91, 245
Kimavu (Brac.: Braconinae), 191
Khoikhoiinae (Brac.), 207, 312–13, 322
Khasurtella† (Ich.: Tanychorinae†), 190, 194, 346–7
Klutiana (Ich.: Nesomesochorinae), 437, 443
Korecnomios (Brac.: Helconinae *incertae sedis*), 282, 310

L

Labena (Ich.: Labeninae), 15, 167, 176, 180, 355–6
Labeninae (Ich.), 15, 18, 32, 38, 41–3, 49, 54, 62, 90, 165, 167–8, 174, 176, 179–80, 184, 345, 353–5, 383, 387, 398–9, 447, 449, 491
Labiini (Ich.: Labeninae), 448
Labenopimpla† (Ich.: Labenopimplinae†), 348
Labenopimplinae† (Ich.), 348, 356, 413
Labium (Ich.: Labeninae), 355, 398
Lagarotis (Ich.: Ctenopelmatinae), 418
Lamachus (Ich.: Ctenopelmatinae), 413, 418, 419
Lamnatibia (Ich.: Pimplinae), 368
Lapton (Ich.: placement uncertain), 344–5, 424, 354
Lasiophorus (Brac.: Braconinae), 169
Lathrolestes (Ich.: Ctenopelmatinae), 349, 391, 418–19, 463
Lathroplex (Ich.: Campopleginae), 432
Lathrostizus (Ich.: Campopleginae), 432–3, 466
Latibulus (Ich.: Cryptinae), 381–2
Laxiareola (Ich.: Sisyrostolinae), 407
Leiophroninae (*see* Euphorinae)
Leiophron (Brac.: Euphorinae), 299, 305–6, 482
Lemophagus (Ich.: Campopleginae), 131, 432–3, 435
Leptobaptopsis (Ich.: Banchinae), 403
Lepton (Brac.: Alysiinae), 255
Leptorhaconotus (Brac.: Doryctinae), 232, 309
Leptotrema (Brac.: Alysiinae), 174
Limerodops (Ich.: Ichneumoninae), 395
Liotryphon (Ich.: Pimplinae), 98
Liparophleps (*see Semirhytus*), 271
Lipolexis (Brac.: Aphidiinae), 214–15, 217, 220, 224
Lissonota (Ich.: Banchinae), 47, 92, 95, 147, 150, 157, 181, 184, 403–4
Lissonotini (*see* Atrophini)
Lissopimpla (Ich.: Pimplinae), 174, 376, 377
Listrodromini (Ich.: Ichneumoninae), 397
Lochetica (Ich.: Cryptinae), 393
Lodbrokia (Brac.: Alysiinae), 172
Lophyroplectus (Ich.: Ctenopelmatinae), 413, 418–9, 464
Loxocephalini (*see* Syntretini)

Lusius (Ich.: Ichneumoninae), 398
Lycorina (Ich.: Lycorininae), 40, 406–7, 491
Lycorinae (*see* Lycorininae)
Lycorininae (Ich.), 40, 47, 63, 73, 101, 105, 406, 427, 447, 492, 534
Lygurus (Ich.: Sisyrostolinae), 407
Lymeon (Ich.: Cryptinae), 69, 392
Lysibia (Ich.: Cryptinae), 99, 349, 389, 393
Lysiognathinae (*see* Tryphoninae)
Lysiphlebus (Brac.: Aphidiinae), 61, 66, 67, 69, 75, 114, 116, 134, 215–16, 218, 222, 474, 477
Lysiterminae (Brac.), 92, 172, 177, 178, 230, 236, 243–6, 534
Lysitermus (Brac.: Lysiterminae), 178, 245
Lytarmes (Ich.: Rhyssinae), 381

M

Macrocentrinae (Brac.), 40, 46–7, 54, 75–6, 91–2, 125, 139, 177, 180, 183–4, 203, 208, 278, 284–7, 478, 494, 501
Macrocentrus (Brac.: Macrocentrinae), 46–7, 54, 67, 69, 75, 86, 92, 114, 120, 139, 175, 184, 279–80, 284, 286–7, 471, 478, 525
Macrostomion (Brac.: Rogadinae), 15, 91, 110, 248
Malagopsis (Brac.: Braconinae), 169
Malasigalphus (Brac.: Sigalphinae), 276
Mannokeraia (Brac.: Euphorinae), 172, 295, 309
Mannokeraiini (Brac.: Euphorinae), 295, 336
Marjorietta† (Ich.: Townesitinae†), 350
Marshiella (Brac.: Euphorinae), 299, 309
Masona (Brac.: Masoninae), 172, 336–7
Masonbeckia (Brac.: Ichneutinae), 321
Masoninae (Brac.), 10, 172–3, 335–7, 494, 518, 534, 535
Mastrina (Ich.: Cryptinae), 393
Mastrus (Ich.: Cryptinae), 97, 98, 389, 391
Maxfischeria (Brac.: Maxfischeriinae), 212, 224–5
Maxfischeriinae (Brac.), 73, 190, 212, 224–5, 503, 535
Medophron (Ich.: Cryptinae), 390
Megaceria (Ich.: Ctenopelmatinae), 416, 418, 420, 449
Megachora†(Ich.: ?Tanychorinae†), 194
Megacoeloides (Brac.: Braconinae), 268
Megalohelcon (Brac.: Trachypetinae), 204, 208–9
Megalochus (Ich.: Tersilochinae), 428
Megalommum (Brac.: Braconinae), 39, 46, 177, 179, 263, 265
Megarhyssa (Ich.: Rhyssinae), 32, 41, 43–4, 53, 68, 120, 171, 379, 380–3, 461, 477, 481–2
Melalophacharops (Ich.: Campopleginae), 182, 432
Melanodolius (Ich.: Sisyrostolinae), 407, 427, 448
Melcha (*see Goryphus*),
Menaka (Ich.: Campopleginae), 435
Mendesellinae (Brac.), 44, 46, 162, 207, 312–13, 322–3
Mendesella (Brac.: Mendesellinae), 46, 322
Meringops (Ich.: Cryptinae), 391
Merinotus (Brac.: Braconinae), 135, 269
Mesocentrus (Brac.: Rogadinae), 243, 339, 535
Mesochorinae (Ich.), 32, 50, 99, 116–17, 184, 197, 345, 421–2, 492, 517
Mesochorus (Ich.: Mesochorinae), 32, 75, 116–17, 184, 421–2, 485
Mesocoelini (Brac.: Agathidinae), 272, 275, 339
Mesocoelus (Brac.: Agathidinae), 275
Mesocrina (Brac.: Alysiinae), 179
Mesoleiini (Ich.: Ctenopelmatinae), 416, 418, 420
Mesoleius (Ich.: Ctenopelmatinae), 75, 156, 416, 418, 481
Mesoleptus (*see* Cryptini), 393–4
Mesostenidea (*see* Cryptini), 181
Mesostenini (*see* Cryptini), 389, 392
Mesostoa (Brac.: Mesostoinae), 81, 172, 186, 212, 225–7, 229, 356, 504
Mesostoinae (Brac.), 12, 15, 20, 69, 81, 90, 172, 174, 181, 197, 212, 225–9, 503, 504, 535
Messatoporus (Ich.: Cryptinae), 391
Meteoridea (Brac.: Meteorideinae), 92, 337–8
Meteorideinae (Brac.), 27, 91, 92,335, 337–8, 494
Meteorinae (*see* Euphorinae)
Meteorini (Brac.: Euphorinae), 83, 86, 88, 115, 158, 182–3, 203, 281, 290, 294–99, 433, 471, 501
Meteorus (Brac.: Euphorinae), 50, 60, 86, 88–9, 92, 93–4, 97, 115–16, 145, 166, 172, 182, 184–5, 286, 295–9, 446, 464, 467, 470, 472, 475, 504, 530
Metopiinae (Ich.), 17, 59, 70, 73, 85, 101, 116, 168, 183–4, 344–5, 401, 421–4, 430, 447–8, 454–5, 461, 466, 491–2, 516
Metopius (Ich.: Metopiinae), 17, 59, 422–3
Microbracon (*see Bracon, Habrobracon*),
Microchelonus (*see Chelonus*),
Microctonus (Brac.: Euphorinae), 92, 116, 142, 157, 299–301, 308, 470
Microgaster (Brac.: Microgastrinae), 86, 147, 326–7, 487
Microgastrinae (Brac.), 12–4, 54–5, 61–2, 76, 77, 88, 91–2, 94, 100, 105, 109, 115–16, 134, 138, 146–7, 150, 157–8, 161, 168, 178, 180, 182, 193, 203, 207, 254, 256, 305, 311–3, 319, 321–335, 459, 471, 494, 495, 496, 501, 503, 539
Microgastrini (Brac.: Microgastrinae), 324
Microplitini (Brac.: Microgastrinae), 324
Microplitis (Brac.: Microgastrinae), 13, 15, 18, 32, 58, 61, 62, 64, 77, 85, 95, 97, 109, 130–2, 145, 147, 152, 154, 158–9, 322, 324–5, 327–8, 331–2, 334, 457, 459, 464, 472, 474, 475, 479
Microleptinae (Ich.), 342–3, 345, 360, 366, 425, 447–8
Microleptes (Ich.: Microleptinae), 344, 400, 447–8, 534
Microtypinae (Brac.), 208, 279,282, 292, 494
Microtypus (Brac.: Microtypinae), 292
Minanga (Brac.: Sigalphinae), 173, 178, 276, 278, 339
Minangini (Brac.: Sigalphinae), 277
Miracinae (Brac.), 12, 146, 147, 312–13, 317, 319, 321, 324, 335
Mirax (Brac.: Miracinae), 147, 336
Mithotynia (*see Histeromerus*),
Monganella (Ich.: Pedunculinae), 398
Monilobracon (Brac.: Braconinae), 100, 181, 261, 266, 381, 482
Monitoriella (Brac.: Doryctinae), 59, 186, 207, 231, 234–5
Monocoila (Brac.: Braconinae), 179
Monoctonia (Brac.: Aphidiinae), 215, 217
Monoctonus (Brac.: Aphidiinae), 139, 212, 216, 220–1
Muesebeckiini (Brac.: Ichneutinae), 320–1
Myiocephalini (*see* Syntretini)
Myllenyxis (Ich.: Rhyssinae), 46, 48, 383
Myrmeleonostenus (Ich.: Cryptinae), 393
Napo (Brac.: Euphorinae), 117, 149, 303
Nematopodius (Ich.: Cryptinae), 391

Nemeritis s.s. (Ich.: Campopleginae), 432, 435
Nemeritis canescens (*see Venturia*),
Neognamptodon (Brac.: Gnamptodontinae), 271

N

Neogreeneia (*see Anomalon*),
Neomicrotypus (Brac.: Microtypinae), 292
Neoneurinae (*see* Euphorinae: Neoneurini)
Neoneurini (Brac.: Euphorinae), 294, 306–7
Neoneurus (Brac.: Euphorinae), 306–7
Neorhacodes (Tersilochinae), 195, 198, 426–7, 442
Neorhacodini (Tersilochinae), 195, 197–8, 426, 427
Neotypus (Ich.: Ichneumoninae), 395
Nepiesta (Ich.: Campopleginae), 432
Neoxoridinae (*see* Poemeniinae)
Nesomesochorinae (Ich.), 345, 437, 439, 442
Netelia (Ich.: Tryphoninae), 47, 61, 74, 95–6, 125, 143, 161, 177, 180, 265, 377, 411, 413–6, 446, 448, 454, 455, 492, 504, 521
Neurateles (Ich.: Orthocentrinae), 179, 366
Nipponaetes (Ich.: Cryptinae), 389
Nipporicnus (Ich.: Cryptinae), 391
Nonninae (*see* Nesomesochorinae)
Notiopambolus (Brac.: Pambolinae), 179, 237
Notophrudus (Ich.: Tersilochinae), 174, 407, 428
Notopygus (Ich.: Ctenopelmatinae), 418
Notosemini (*see* Oedicephalini), 397
Notostilbops (Ich.: Stilbopinae), 406, 408, 411

O

Obisiphaga (Ich.: Cryptinae), 389
Odontocolon (Ich.: Xoridinae), 174–5, 351–2
Odontopimpla (Ich.: Pimplinae), 174
Odontoscapus (Brac.: Braconinae), 171
Oedemopsini (Ich.: Tryphoninae), 411–4, 448
Oedemopsis (Ich.: Tryphoninae), 421
Oedicephalini (Ich.: Ichneumoninae), 397
Oenonogastra (Brac.: Alysiinae), 251
Oetophorus (Ich.: Ctenopelmatinae), 417, 419
Olesicampe (Ich.: Campopleginae), 91–92, 121, 148, 432–3
Olethrodotini (Ich.: Ctenopelmatinae), 418
Oligoneurus (Brac.: Ichneutinae), 321
Oncometeorini† (Brac.: Euphorinae), 308
Oncophanes (Brac.: Rhyssalinae), 238–40
Opheltes (Ich.: Ctenopelmatinae), 419
Ophionellus (Ich.: Anomaloninae), 20, 197, 431
Ophioninae (Ich.), 12, 26, 28, 58–9, 70, 105, 161, 168, 175–6, 180, 183, 345, 400–1, 405, 426, 431, 437, 441–2, 445, 446, 447, 461, 492, 502, 505, 534, 536
Opiinae (Brac.), 10, 17, 46, 54, 62, 64, 69, 73, 76, 85, 91, 104, 123, 138, 173, 178, 182, 229, 241, 244, 250, 256–260, 439, 494, 496
Opiopterus (Brac.: Mesostoinae), 228–9
Opius (Brac.: Opiinae), 13, 64, 111, 123, 256–60
Orgilinae (Brac.), 48, 54, 55, 62, 76, 157, 172, 174, 494, 496, 535, 539
Orgilus (Brac.: Orgilinae), 48, 54, 55, 62, 77, 99, 100, 120, 157, 172, 174, 292–3, 468
Orthopelmatinae (Ich.), 62, 196, 342–3, 345, 383, 400, 492
Orthostigma (Alysiinae), 172
Osprynchotina (Ich.: Cryptinae), 345, 389
Osprynchotus (Ich.: Cryptinae), 391
Oxytorinae (Ich.), 174, 345, 360, 366, 424–5, 447, 448
Oxytorus (Ich.: Oxytorinae), 400, 424–5, 447
Ozlabium (Ich.: Labeninae), 355

P

Pachysomoides (Ich.: Cryptinae), 69, 391
Palabracon (Brac.: Braconinae), 169
Palaeocharmon† (Brac.: Charmontinae), 287
Palaeocharmontini† (Brac.: Charmontinae), 287
Palaeoichneumon† (Ich.: Palaeoichneumoninae†), 348
Palaeoichneumoninae† (Ich.), 26, 346, 348
Pambolinae (Brac.), 12, 172, 181, 183, 207, 230–1, 236–7, 244–5, 302, 494
Pambolus (Brac.: Pambolinae), 172, 236–7
Paniscus (*see Netelia*),
Panteles (Ich.: Stilbopinae), 47, 73, 95, 100, 406, 408–10, 418, 431
Paradelius (Brac.: Cheloninae), 482
Parabrulleia (Brac.: Helconinae), 281
Paralipsis (Brac.: Aphidiinae), 67, 218
Parallorhogas (Brac.: Doryctinae), 110–11, 231, 254
Parapanteles (Brac.: Microgastrinae), 331, 335
Paratanychora† (Ich.: Tanychorinae†), 193
Paroligoneurus (Brac.: Ichneutinae), 15, 251, 320–1
Parorthostigma (Brac.: Alysiinae), 251
Pauesia (Brac.: Aphidiinae), 213, 216, 218–19, 465
Paxylomma (*see Hybrizon*),
Paxylommatinae (*see* Hybrizontinae)
Pedunculinae (Ich.), 398–400
Pedunculus (Ich.: Pedunculinae), 398–400
Pelecystoma (*see Triraphis* and *Rogas*),
Peakelestes (Ich.: Ctenopelmatinae), 417
Pentatermus (Brac.: Lysiterminae), 244–5
Pergaphaga (Ich.: Ctenopelmatinae), 416, 449
Perilissus (Ich.: Ctenopelmatinae), 419
Perilitus (Brac.: Euphorinae), 92, 122, 158, 295, 299–300, 303, 308–9
Periope (Ich.: Metopiinae), 17
Perithoini (Ich.: Pimplinae), 367, 369
Perithous (Ich.: Pimplinae), 369
Peucobius (Tersilochinae), 407
Phaenocarpa (Brac.: Alysiinae), 181, 253–4, 256
Phaenodus (Brac.: Pambolinae), 172
Phaeogenes s.s. (Ich.: Ichneumoninae), 120, 394, 395, 397, 481
Phaeogenes (*see also Tycherus*),
Phaeogenini (Ich.: Ichneumoninae), 387, 388, 395, 397, 398, 503, 544
Phanerotoma (Brac.: Cheloninae), 11, 79, 147, 180, 205, 315–8
Phanerotomini (Brac.: Cheloninae), 317
Phanomeris (*see Shawiana*)
Pheloura (Brac.: Braconinae), 36, 265, 495
Pherombinae† (Ich.), 349
Pherombus† (Ich.: Pherombinae†), 349
Philomacroploea (Brac.: Braconinae), 263
Phobocampe (Ich.: Campopleginae), 433
Pholetesor (Brac.: Microgastrinae), 116, 147, 184, 332, 335
Pholichora see Hemigyroneuron
Phradis (Ich.: Tersilochinae), 429
Phrudinae (*see* Tersilochinae: Phrudini)
Phrudini (Ich.: Tersilochinae), 174, 407, 427–8

Phrudus (Ich.: Tersilochinae), 426–8
Phthorima (Ich.: Diplazontinae), 113, 179, 364, 438, 525
Phygadeuon (Ich.: Cryptinae), 166, 345, 388–91, 393–5
Phygadeuontinae (*see* Cryptinae)
Phygadeuontini (Ich.: Cryptinae), 166, 389–91, 393, 395
Physaraia (Brac.: Braconinae), 177–9
Phytodietini (Ich.: Tryphoninae), 411–12, 414, 427, 455
Phytodietus (Ich.: Tryphoninae), 47, 61, 411–12, 414, 421, 455, 464
Pimpla (Ich.: Pimplinae), 12–4, 54, 58, 61–2, 64, 72–5, 83, 88, 98, 111, 118, 128, 139, 142, 144–5, 165, 346, 358, 367–8, 370, 374–7, 392, 456, 459, 463, 471, 474, 479, 481, 483, 504
Pimplinae (Ich.), 14, 18, 31, 33, 49, 58–9, 61–2, 73, 75, 83, 90–1, 100–1, 103, 111, 116, 145, 165, 168, 174–5, 183, 344–6, 353, 359–61, 367–70, 377–9, 398, 406, 429, 455, 461, 466, 471, 474, 476, 486, 491–3, 502, 504–6, J, 508, 510, 514, 534, 536
Pimplomorpha (Ich.: Cremastinae), 26, 444
Pimplopterus (*see Lissonota*),
Pion (Ich.: Ctenopelmatinae), 52, 182, 359, 416, 419, 421
Pionini (Ich.: Ctenopelmatinae), 182, 359, 416, 419, 421
Planitorini (Brac.: Euphorinae), 243, 309
Planitorus (Brac.: Euphorinae), 243, 295, 309
Platylabini (Ich.: Ichneumoninae), 397–8
Plaxopsis (Brac.: Braconinae), 169
Plectochorus (*see Mesochorus*), 421
Pleolophus (Ich.: Cryptinae), 95, 97, 389
Plesiocoelus (Brac.: Agathidinae), 275
Plesiotypus (Brac.: Helconinae *incertae sedis*), 282, 292
Plynops (Brac.: Euphorinae), 299, 302
Podorgilus (Brac.: Orgilinae), 174, 292–3
Podoschistus (Ich.: Poemeniinae), 14, 170–1, 378
Poecilocryptini (Ich.: Labeninae), 353, 355–6
Poecilocryptus (Ich.: Labeninae), 356, 398
Poecilostictus (Ich.: Ichneumoninae), 394
Poemeniinae (Ich.), 14, 62, 168, 170–1, 345, 367, 378–9, 492, 516
Polemochartus (Brac.: Alysiinae), 15, 255
Polyaulon (Ich.: Cryptinae), 172, 389, 393
Polysphincta group (Ich.: Pimplinae), 90–1, 101, 103, 244, 367–9, 371–3, 504, 523
Porizontinae (*see* Campopleginae)
Praeichneumon† (Praeichneumonidae†), 191, 194
Praeichneumonidae†, 190–4, 196
Praini (Brac.: Aphidiinae), 75, 214, 217, 223
Praon (Brac.: Aphidiinae), 15, 69, 75, 94, 169, 213, 215–7, 220–1
Praonopterinae (Brac.: syn. Mesostoinae), 203, 207, 225
Praonopterus (Brac.: Mesostoinae), 169, 225, 229
Priopoda (Ich.: Ctenopelmatinae), 418–19, 454
Priotomis (Ich.: Cryptinae), 392
Pristomerus (Ich.: Cremastinae), 46, 50, 99, 174–5, 383, 438–9
Probles (Ich.: Tersilochinae), 46, 429
Probolus (Ich.: Ichneumoninae), 398
Procestus (Ich.: Banchinae), 174
Procinetus (Ich.: Acaenitinae), 357, 358, 359
Pronkia (Brac.: Meteorideinae), 337–9
Protaphidius (Brac.: Aphidiinae), 218, 220
Proteropini (Brac.: Ichneutinae), 320
Proterops (Brac.: Ichneutinae), 29, 320–1
Protohellwigia (*see Hellwigia*), 445
Protorhyssalinae† (Brac.), 27, 190, 192, 194, 196, 209–11
Protorhyssalodes†, 192–4, 209–11
Protorhyssalopsis† (Brac.: Protorhyssalinae†), 194
Protorhyssalus† (Brac.: Protorhyssalinae†), 192–4, 209–10
Pselaphaninae (*see* Sigalphinae), 208, 271
Pselaphanus (Brac.: Sigalphinae), 276, 278
Psenobolus (Brac.: Doryctinae), 173, 235
Pseudalomya (Ich.: Ichneumoninae), 387–8
Pseudodicrogeniinae (Brac.: Braconinae), 207, 269
Pseeudodicrogenium (Brac.: Braconinae), 167, 269
Pseudognamptodon (Brac.: Gnamptodontinae), 271
Pseudopauesia (Brac.: Aphidiinae), 217
Pseudopezomachus (Brac.: Alysiinae), 172
Pseudopimpla (Ich.: Pimplinae), 367, 370
Pseudopraon (Brac.: Aphidiinae), 212, 217
Pseudorhyssa (Ich.: Poemeniinae), 100, 378
Pseudorhyssini (Ich.: Poemeniinae), 378
Pseudospinaria (Brac.: Braconinae), 179
Pseudoyelicones (Brac.: Rogadinae), 177, 181
Psittacibracon (Brac.: Braconinae), 169
Psyttalia (Brac.: Opiinae), 13, 31, 117, 123, 256–60
Pycnophion (also *see Enicospilus*), 444
Pygmaeolus (Ich.: Tersilochinae), 12, 349
Pyracmon (Ich.: Campopleginae), 105, 342

Q

Quillonota (Ich.: Banchinae), 116

R

Reclinervellus (Ich.: Pimplinae), 371
Retusigaster (Brac.: Cardiochilinae), 315
Rhacodopteron (Ich.: Cryptinae), 389, 442
Rhadinobracon (Brac.: Braconinae), 231, 266
Rhamnura (Brac.: Braconinae), 17, 269, 511
Rhamnurini (Brac.: Braconinae), 269
Rhembobius (Ich.: Cryptinae), 394
Rhimphoctona (Ich.: Campopleginae), 78, 105, 432
Rhogadinae (*see* Rogadinae)
Rhogas (*see Rogas, see also Aleiodes*), *Trirhaphis*
Rhopalophorus (Brac.: Euphorinae), 302, 304
Rhorus (Ich.: Ctenopelmatinae), 419
Rhynchobanchus (Ich.: Banchinae), 3, 405
Rhynchophion (Ich.: Ophioninae), 444
Rhysipolinae (Brac.), 90, 101, 123, 180, 207, 229–31, 237–8, 242, 244, 455, 494, 534
Rhysipolis (Brac.: Rhysipolinae), 49, 101, 102, 160, 242, 372, 383
Rhyssalinae (Brac.), 11, 15, 40, 53, 59, 62–3, 109–10, 172, 183, 197, 204, 207, 210, 225, 228–31, 238–42, 260, 279, 302, 339, 455, 494, 506
Rhyssa (Ich.: Rhyssinae), 11, 15, 40, 43, 53, 59, 62–3, 100, 109–10, 172, 197, 378, 379–83, 455, 477, 506, 534
Rhyssalus (Brac.: Rhyssalinae), 53, 172, 192–4, 209–10, 238–9, 241
Rhyssella (Ich.: Rhyssinae), 378–9

Rhyssinae (Ich.), 31–2, 38, 42–3, 46, 49, 62, 112, 117–18, 124, 168, 170–1, 345, 355, 367, 376, 378–9, 382, 477, 482, 491–2, 502, 506, 516
Rhytimorpha (Brac.: Braconinae), 179
Rogadinae (Brac.), 18, 40, 53–4, 60, 63–4, 66, 73, 85, 88, 91, 92, 102, 104–5, 112, 116, 157, 160, 172–4,
Rogadinae (Brac.) (*continued*) 176–8, 180–1, 183, 184, 198, 199, 203–4, 207, 228–30, 236, 238, 243, 246–50, 454–5, 494, 496, 514, 521, 535
Rogadini (Brac.: Rogadinae), 246–7, 249–50
Rogas (Brac.: Rogadinae), 116, 249
Rothneyia (Ich.: Cryptinae), 178, 391
Rudimentifera† (Ich.: Palaeoichneumoninae†), 348

S

Sachtlebenia (Ich.: Banchinae), 405
Sania (Brac.: Khoikhoinae), 323
Sarops (Brac.: Alysiinae), 179
Saotis (Ich.: Ctenopelmatinae), 418
Sathropterus (Ich.: Tersilochinae), 429
Scambus (Ich.: Pimplinae), 369–70, 457–8, 466, 477, 486, 512
Scenocharops (Ich.: Campopleginae), 433–4
Schizoprymnus (Brac.: Helconinae), 177–8
Schizopyga (Ich.: Pimplinae), 371
Schiztobracon (Brac.: Braconinae), 179
Schlettereriella (Brac.: Doryctinae), 232
Scirtetes (Ich.: Campopleginae), 198, 434
Scolobatini (Ich.: Ctenopelmatinae), 417, 419–20
Scolobatinae (syn. Ctenopelmatinae; *see also* Scolobatini)
Scolomus (Ich: placement uncertain), 424
Seleucini (Ich.: Ctenopelmatinae), 416, 420
Seleucus (Ich.: Ctenopelmatinae), 179, 420, 428
Semirhytus (Brac.: Doryctinae), 271
Sericobracon (Brac.: Doryctinae *incertae sedis*), 91, 231, 236
Serrundabracon (Brac.: Braconinae), 46, 179, 439
Sesioctonus (Brac.: Agathidinae), 275
Shawiana (Brac.: Euphorinae), 250, 270
Sicophion (Ich.: Ophioninae), 445
Sigalphinae (Brac.), 27, 78, 91–2, 178, 202–3, 208–9, 271–2, 274–8, 284, 339, 466, 494
Sigalphus (Brac.: Sigalphinae), 92, 184, 202, 276–8
Sinophorus (Ich.: Campopleginae), 76, 99, 158, 432–3
Sinuatophorus (Brac.: Euphorinae), 46, 302–3
Sisyrostolinae (Ich.), 47, 345, 402, 407, 411, 428, 448, 535
Skiapus (Ich.: Ophioninae *incertae sedis*), 430, 435, 437, 441, 444–5, 534
Snellenius (Brac.: Microgastrinae), 324
Soter (Brac.: Braconinae), 179
Sozites (Ich.: Cryptinae), 389
Spasskia (Brac.: Helconinae), 178
Spathicopsis, 300–1, 308
Spathiinae (*see* Doryctinae), 178, 203, 235
Spathiini (Brac.: Doryctinae), 207, 231
Spathius (Brac.: Doryctinae), 42, 54, 61, 84, 88, 124, 172, 178, 231, 233, 235–6
Sphecoctonus (Ich.: Cryptinae), 391
Sphecophaga (Ich.: Cryptinae), 85, 116, 171, 389, 391–2, 480
Spilichneumon (Ich.: Ichneumoninae), 31
Sphinctini (Ich.: Tryphoninae), 411–12, 415, 447
Sphinctus (Ich.: Tryphoninae), 17, 398, 412–13, 415–16, 447
Spinadesha (Brac.: Braconinae), 179
Stantonia (Brac.: Orgilinae), 292–4, 504
Stenarella (Ich.: Cryptinae), 390–1
Stenomacrus (Ich.: Orthocentrinae), 170, 172, 366
Stephaniscini (Brac.: Doryctinae), 178
Stethantyx (Tersilochinae), 116, 428, 429
Sternaulopius (Brac.: Opiinae), 17
Stictopisthus (*see Mesochorus*),
Stilpnina (Ich.: Cryptinae), 104, 105, 345, 393
Stilpnus (Ich.: Cryptinae), 389, 393
Stiropiini (Brac.: Rogadinae), 246–7, 249
Streblocera (Brac.: Euphorinae), 12, 50, 299, 309–10
Surculus (Ich.: Cryptinae), 393
Sussaba (Ich.: Diplazontinae), 365
Sustenus (Ich.: Cremastinae), 26, 444
Sweaterella (Ich.: Cylloceriinae), 360
Symphya (*see Trachionus*)
Syndipnus (Ich.: Ctenopelmatinae), 68, 466
Syngaster (Brac.: Doryctinae), 110, 231, 461, 482
Synomelix (Ich.: Ctenopelmatinae), 418
Syntretellus (Brac.: Euphorinae), 309
Syntretini (Brac.: Euphorinae), 309
Syntretomorpha (Brac.: Euphorinae), 309
Syntretus (Brac.: Euphorinae), 309
Syrphoctonus (Ich.: Diplazontinae), 95, 362, 364, 365
Syrrhizus (*see Centistes*),
Syzeuctus (Ich.: Banchinae), 403–4

T

Tainitermini (Brac.: Euphorinae), 309
Tanychora† (Ich.: Tanychorinae†), 192–3, 344
Tanychorella† (Ich.: Tanychorinae†), 193–4
Tanychorinae† (Ich.), 192
Tatogastrinae (Ich.), 174, 345, 425–6
Tatogaster (Ich.: Tatogastrinae), 425–6
Telengaia (Brac.: Telengainae), 271
Telengaiinae (Brac.), 203, 207, 250, 271
Temelucha (Ich.: Cremastinae), 99–100, 438, 468
Terebrebus (Brac.: Alysiinae), 439
Termitobracon (Brac.: Doryctinae), 172, 236
Tersilochinae (Ich.), 12, 46–7, 73, 116, 168, 178, 195, 197–8, 344, 345, 349, 400–1, 407, 426–7, 439, 455, 497, 502
Tersilochus (Ich.: Tersilochinae), 116, 428, 429
Testudobracon (Brac.: Braconinae), 179
Tetrambon (Ich.: Ctenopelmatinae), 417
Tetrasphaeropyx see Aleiodes
Tetratermus (Brac.: Lysiterminae), 178, 245
Thaumatogelis (Ich.: Cryptinae), 172, 393
Theriinae (*see* Anomaloninae)
Therion (Ich.: Anomaloninae), 174, 345, 430
Therionini (*see* Gravenhorstiini), 430
Theronia (Ich.: Pimplinae), 38, 122, 359, 367, 373–4, 377, 378, 429, 448, 481, 498, 506
Therophilus (Brac.: Agathidinae), 273
Theroscopus (Ich.: Cryptinae), 389
Thoracoplites (Brac.: Rhyssalinae), 40, 53, 239, 241
Thrybius (Ich.: Cryptinae), 392
Thyreodon (Ich.: Ophioninae), 442, 444, 445
Toechorychus (Ich.: Cryptinae), 391, 448
Townesia (Ich.: Pimplinae), 369
Townesion (Ich.: Banchinae), 177–9,
Townesionini (Ich.: Banchinae), 179, 403, 405–6

Townesioninae (*see* Banchinae), 345
Townesites† (Ich.: Townesitinae†), 350
Townesitinae† (Ich.), 309
Toxares (Brac.: Aphidiinae), 214–15
Toxoneuron (Brac.: Cardiochilinae), 58, 64, 68, 72, 83, 112, 145–7, 154, 156, 158, 159, 160, 296, 312, 314–15, 321, 474–5, 478
Toxophoroides (*see Lycorina*),
Trachionus (Brac.: Alysiinae), 178, 255
Trachypetinae (Brac.), 27, 69, 90, 178, 190, 192, 195–8, 203, 205, 207–10, 271, 413, 494, 505, 534
Trachypetus (Brac.: Trachypetinae), 178, 205, 208
Trafana (Ich.: Cremastinae), 448
Tranosema (Ich.: Campopleginae), 148, 149, 154, 401, 432, 435
Trathala (Ich.: Cremastinae), 116, 438
Triaspidinae (*see* Helconinae: Brachistini), 208
Triaspis (Brac.: Helconinae), 177–8, 202, 279–80
Trichelyon† (Brac.: Rhysipolinae), 237
Trichomma (Ich.: Anomaloninae), 432
Triclistus (Ich.: Metopiinae), 76, 116, 423, 424
Tricyphus (Ich.: Ichneumoninae), 85, 394–5
Trieces (Ich.: Metopiinae), 424
Trigastrotheca (Brac.: Braconinae), 179, 263
Trioxys (Brac.: Aphidiinae), 62, 67, 75, 172, 213–7, 219, 221, 469, 550
Trirhaphis (Brac.: Rogadinae), 92, 104, 247–50
Trispinaria (Brac.: Braconinae), 179
Tritermus (Brac.: Lysiterminae), 178, 245
Trogus (Ich.: Ichneumoninae), 131, 395, 397, 463, 465, 499, 501
Tromatobia (Ich.: Pimplinae), 369, 373
Troticus (Brac.: Agathidinae), 273
Tryphoninae (Ich.), 6, 14, 17, 47, 49, 73, 76, 90–1, 96, 101, 161, 177, 180, 183, 197, 342, 343, 345, 349, 372, 377, 389, 398, 402, 408, 411–6, 429, 446–8, 454–5, 461, 491–2, 502, 521, 534
Tryphonini (Ich.: Tryphoninae), 411–12, 416
Tuberidelus (Brac.: Euphorinae), 302
Tycherus (Ich.: Ichneumoninae), 180, 394

U

Undabracon (Brac.: Braconinae), 265, 439
Urosigalphus (Brac.: Acampsohelconinae), 178, 284–5
Urotryphon† (Ich.: Labenopimplinae†), 349, 413
Ussurohelcon (Brac.: Helconinae), 279, 311
Utetes (Brac.: Opiinae), 97, 257–8, 482

V

Vaepellinae (Brac.: syn. Braconinae), 207, 269
Vaepellis (Brac.: Braconinae), 269
Venturia (Ich.: Campopleginae), 15, 54, 64–5, 72, 75–6, 82, 86, 94, 109, 114, 116–8, 122, 125, 135, 139, 143, 145–6, 432, 435–6, 449, 466, 474, 479, 483
Vervoortihelcon (Brac.: Helconinae), 279
Vervoortihelconina (Brac.: Helconinae), 279
Victorovia (Ich.: Cryptinae), 442
Victoroviella (Brac.: Braconinae), 169
Vipio (Brac.: Braconinae), 124
Vipiinae (*see* Braconinae), 207
Viponinae (*see* Braconinae), 207, 269
Vulgichneumon (Ich.: Ichneumoninae), 31, 396

W

Wahlamia (Ich.: Banchinae), 38, 392, 403
Westwoodiella (Brac.: Homolobinae), 291–2
Westwoodiellini (Brac.: Homolobinae), 291
Westwoodia (Ich.: Ctenopelmatinae), 417, 418, 420
Westwoodiini (Ich.: Ctenopelmatinae), 73, 416, 419–20, 449
Wroughtonia (Brac.: Helconinae), 51, 174, 280
Xanthopimpla (Ich.: Pimplinae), 98, 166, 367–8, 376–7, 509
Xenarcha (Brac.: Exothecinae), 270

X

Xenoschesis (Ich.: Ctenopelmatinae), 41, 121, 418
Xenothyrini (Ich.: Labeninae), 174, 353–5
Xiphozele (Brac.: Xiphozelinae), 288, 290
Xiphozelinae (Brac.), 47, 203, 208, 278, 288–90, 494
Xorides (Ich.: Xoridinae), 165–7, 186, 349–53
Xoridescopus (Ich.: Cryptinae), 392
Xoridinae (Ich.), 32, 49, 62, 64, 90, 165–7, 171, 174, 192, 345, 349, 351–2, 383, 492
Xyeloblacus (Brac.: Blacinae), 282–3
Xylonomus (*see Xorides*),
Xylophrurus (Ich.: Cryptinae), 391
Yelicones (Brac.: Rogadinae), 173, 176, 249, 504

Y

Yeliconini (Brac.: Rogadinae), 173, 243, 246–7, 249
Ypsistocerinae (*see* Doryctinae), 203, 207, 236
Ypsistocerini (Brac.: Doryctinae), 231, 535

Z

Zaglyptogastra (Brac.: Braconinae), 46, 48, 182, 265, 383, 439, 477
Zaglyptus (Ich.: Pimplinae), 39, 103, 373
Zanzopsis (Brac.: Braconinae), 169
Zatypota (Ich.: Pimplinae), 371–3
Zele (Brac.: Euphorinae), 290, 291, 295–9
Zelinae (*see* Homolobinae, also *see* Charmontiinae)
Zimmeriini (Ich.: Ichneumoninae), 397–8
Zurquilla (*see Nipponaetes*)

ICHNEUMONOIDEA SPECIES INDEX

Names in parentheses indicate other common but superceded generic combinations.

A

abruptorius, Exenterus, 413
aciculatus, Diacritus, 362
aclydis, Aleiodes, 248
acronycta, Diadegma, 148
aethiopoides, See aethiops
aethiops, Microctonus (*Perilitus*), 142
affinis, Exenterus, 413
agilis, Agathis, 32, 184, 185, 483, 485
albicoxa, Zatypota, 371, 372
albiditarsus, Zele, 298
albipes, Grypocentrus, 411–16
albitarsus, Cryptus, 393
albitarsus, Euceros, 445
albofascialis, Alophosternum, 370
albospina, Pambolus, 237
alebion, Microgaster, 487
alloeum, Diachasma, 93, 259
alloeus, See alloeum
alpestris, Pseudorhyssa, 378
alternans, Itoplectis, 374
alternipes, Acampsis, 139, 274, 277–8
amictorius, Exenterus, 413
amoenus, Arotes, 79
anastrephae, Utetes, 97, 257, 482
ancylivorus, Macrocentrus, 86, 120, 525
andina, Lamnatibia, 368
angaleti, Cotesia, 325
angelicae, Trioxys, 67
anglica, Agathis, 169
annulatus, Scambus, 369
annulicornis, Xoridescopus, 392
annulipes, Chelonus, 148, 315
antefurcalis, Pronkia, 338
antennalis, Pherombus†, 349
anthreni, Lathroplex, 432
antipoda, Doryctomorpha, 230
anurus, Bathyplectes, 62, 70, 121, 148, 392, 432–5, 468, 469
aotea, Aphaereta, 254
apiarus, Excavarus, 412
apicifer, Syzeuctus, 404
approximator, Rhyssella, 378–9
aprilis, Campoletis, 148
aptera, Autriquella, 221
apteris, Gelis, 393
arbitrium, Aleiodes, 348
arcanus, Campocraspedon, 364
argenteopilosus, Eriborus, 342
argiolatus, Latibulus, 392
argyraphaga, Hymenoepimecis, 161, 372, 482
arisanus, Fopius (*Biosteres*), 62, 94, 120, 123, 257–9, 525
armatus, Agriotypus, 385
armillatum, Diadegma (*Horogenes*), 61
arnaudi, Protorhyssalodes†, 211
asiaticus, Trioxys, 215
astrarches, Cotesia, 330
aterrima, Itoplectis, 377
atrata, Megarhyssa, 41, 44, 54, 379–80, 467
atricornis, See tuberculator
auctus, Neoneurus, 306
autographae, Meteorus, 295

B

basizonus, Pleolophus, 486
behriae, Phanerotoma, 11
benefactor, Olesicampe, 148, 432
berolinense, Elasmosoma, 307
bicincta, Neotheronia, 125
bicolor, Epsilogaster, 322
bicolor, Pholetesor, 116
bicolor, Sigalphus, 92, 184, 277–8
bicoloratus, Microplitis, 147
bifurcatus, Thoracoplites, 241
bignellii, Cotesia (*Apanteles*), 325
blackburni, Chelonus, 116, 147, 315, 317
blediator, Barycnemis, 173
boops, Loxocephalus, 306
boops, Polysphincta, 372
bostrichorum, Coeloides, 129, 261, 477
braconius, Colastes, 270
brevicaudis, Spathius, 231
brevicollis, Microctonus, 116, 300
brevicornis, See hebetor (*Habrobracon*)
brevicornis, Scambus, 370
brevicornis, Stenomacrus, 366
brooksi, Bassus, 272
browni, Metopius, 423
brunneri, Coeloides, 128, 465
buolianae, Scambus, 370, 458
buriatica, Khasurtella, 347
burra, Sphecophaga, See vesparum
busseolae, Dentichasmias, 98, 394
butcheri, Aleiodes, 177–8
buzurae, Aleiodes, 184, 248–9, 297

C

caeruliventris, Leurus, 120
calcitrator, See coxator
californica, Grotea, 81, 355
californicus, Ephedrus, 92, 214, 220, 462

The Braconid and Ichneumonid Parasitoid Wasps: Biology, Systematics, Evolution and Ecology, First Edition. Donald L. J. Quicke.

canescens, Venturia (*Nemeritis, Devorgilla*), 15, 52, 54, 64–5, 72, 75–6, 82, 86, 94, 109, 114–15, 117–18, 122, 125–6, 135, 139, 143, 145–6, 432, 434–6, 466, 474, 479
capdoliensis, Aenigmobracon†, 211
capeki, Cosmophorus, 302
capito, Diospilus, 281
cappaerti, Atanycolus, 262
cardui, Lysiphlebus, 66, 116, 218
caricivorae, Opius, 64, 257–8
carpatus, Apanteles, 328, 332
caudatus, Fopius, 257
caudatus, Liotryphon, 98
caudatus, Microctonus, 300
caudatus, Triaspis, 280
cembrae, Cosmophorus, 119, 302–3
cephi, Bracon, 251, 268, 466
cephi, Heterospilus, 78, 233
cerasicola, Ephedrus, 215, 221, 462, 466
chavarriai, Hansonia, 85, 315
chlorideae, Campoletis, 95, 97, 122, 148, 154, 432
chlorophthalma, Zele, 76, 297
chrysostictos, Diadegma (*Horogenes*), 94, 109, 433
cincta, Colpotrochia, 423
circumflexum, Therion, 430
circumscriptus, Pholetesor (*Apanteles*), 116, 322
citri, Asobara, 109, 140, 157, 254
clandestina, Olesicampe, 91–2
clausus, Hyposoter, 377
clavicornis, Itoplectis, 433
cnaphalocrosis, Trichomma, 432
coccinellae, Dinocampus (*Perilitus*), 116, 158, 300, 303–5, 330, 364, 475, 479
colemani, Aphidius, 115–16, 215, 466, 471
colesi, Microctonus, 300
collaris, Diadromus, 58, 64, 97, 116, 394–5
communis, Binodoxys, 213–14, 217, 458, 480
compressa, Eucoelinidea, 179
compressa, Phthorima, 364
compressiventris, Xiphozele, 290
compressor, Aleiodes, 179
comstockii, Exeristes, 83
concolor, Psyttalia (*Opius*), 123, 257–60
congregata, Cotesia (*Apanteles*), 65, 147, 149–50, 154–5, 157, 325, 332, 334, 498
coniferae, Dolichogenidea (*Apanteles*), 120
conquisitor, Itoplectis, 83, 98, 132, 259, 374, 376, 474–5
corax, Meteorus, 166, 296–7
corniger, Oxytorus, 425
coronarius, Aleiodes, 15, 181
corruptor, Gelis, 170
corvina, *See megacephalus*
costalimai, Monitoriella, 235
coxalis, Aleiodes, 70
coxator, Collyria, 359–60
crassifemur, Sinophorus (*Eulimneria*), 433
cremieri, Eurypterna, 441
croceipes, Microplitis, 13, 15, 62, 64, 77, 109, 130–2, 145, 147, 154, 158–9, 322, 325, 332, 334, 457, 474, 479
cruentatus, Charmon, 126, 287
cubiceps, Adelognathus, 384
cultratus, Chorebus, 179
cunctator, Clinocentrus, 112
cuneiformis, Seleucus, 420
curculionis, Bathyplectes, 60, 62, 85, 121–2, 159, 432–4, 457, 463, 468–9, 476, 485
curtus, Lemophagus, 432–3
curvimaculatus, Chelonus, 142–3, 147, 316–7
curvinervis, Amicrocentrum, 288–9
curvulus, Mesochorus, 422
cypris, Apanteles, 334

D

darlingtonia, Aleiodes, 177, 178
debellator, Alomya, 388
deceptor, Zele, 298
deesae, Stenobracon, 10, 58, 261, 263, 266
delusorius, Centistes, 301
demolitor, Microplitis, 147, 149, 152, 158–9, 334, 464
dendroctoni, Coeloides, 113, 464
dentipes, Odontocolon, 352
didymator, Hyposoter, 68, 94, 109, 142, 148, 151, 154, 433, 435
dimidiatus, Meteorus, 296
dimidiatus, Triclistus, 423
discitergus, Mesochorus, 421–2
disparsis, Pimpla, 118, 374, 471
domator, Stenarella, 390
dubitator, Acaenitus, 357, 359
dubius, Perilitis, 300, 308
dytiscivorus, Medophron, 390
dzhidensis, Praeichneumon†, 191

E

eadyi, Aphidius, 220
edwardsi, Proclitus, 367
eikoae, Paralipsis, 67
elasmopalpi, Orgilus, 293
elator, Earinus, 274
eliodis, Perilitus, 308
elongata, Limerodops, 395
elongata, Monitoriella, 234
enervis, Paralipsis, 218–19
ensator, Lathrolestes, 391, 463
enslini, Neorhacodes, 427
eques, Lamachus, 413
errebundus, Lemophagus, 432
ervi, Aphidius, 14, 72, 93, 115, 122, 133, 142, 156–60, 214–16, 218, 220–2, 457–8, 475–6, 487, 510, 526
eugenii, Triaspis, 279
eumerus, Ichneumon, 395
euphydryidis, Cotesia (*Apanteles*), 161
examinator, *See turionellae*
exarator, Spathius, 42
excelsa, Lissopimpla, 376
excilis, Amicrocentrum, 288
excitator, Coleocentrus, 358
exigua, Hyposoter, 83, 148, 155, 433
extensor, Charmon, 287–8, 535
extensor, Dolichomitus, 42

F

fabarum, Lysiphlebus, 114–16, 218, 222, 474
facetosa, Diolcogaster, 147, 334, 421
falcatus, Pygostolus, 309
femoralis, Banchus, 52, 405
fenestrale, Diadegma (*Angitia*), 435
filiformis, Xorides, 351
fissorius, Syrphoctonus (*Diplazon*), 95, 362–3
flavescens, Banchus, 405, 464, 472
flavicincta, Campoletis, 58, 148, 433
flavicoxis, Glyptapanteles (*Apanteles*), 147, 149, 334
flavicoxis, Pimpla, 165
flavidus, Ophion, 442, 444
flavipennis, Megalommum (*Aphrastobracon*), 263, 267
flavipes, Cotesia (*Apanteles*), 62, 77, 109, 122, 147, 322, 325, 477, 479
flavitestacea, *See leucobasis* (*Phanerotoma*)
flavoorbitalis, Cremastus, 116, 439
fletcheri, Psyttalia (*Opius*), 31, 260
folkertsorum, Maxfischeria, 225
formicoides, Aptenobracon, 235
fraternus, Protapanteles (*Apanteles*), 78
freja, Palaeoichneumon†, 348
frigidus, Euceros, 445–6
frontinalis, Hercus (*Eclytus*), 414
fugitivus, Hyposoter, 121, 148–9, 434
fulvescens, Theronia, 506
fulvicornis, Xorides, 167

fulvipes, Xenoschesis, 121
fumiferanae, Apanteles, 147, 322
fumiferanae, Glypta, 122, 147, 149, 151, 403–4
funebris, Chorinaeus, 424

G
galleriae, Apanteles, 83
gasseni, Centistes, 301
gayi, Grotea, 355
gelechiae, Habrobracon, 161, 458–60, 462
genevensis, Aphaereta, 254
geniculatae, Olesicampe, 138, 432
gibbosa, Agathis, 272
glabra, Chaenusa, 252
glomerata, Cotesia, 67, 72, 74, 95, 109–10, 113, 128, 131–3, 154, 157, 159, 221, 328, 331–4, 457, 459, 469, 472, 479, 483
glossinophaga, Coelalysia, 254
gonepterygis, Cotesia, 328–31
gracilis, Agriotypus, 385, 386
gracilis, Lipolexis, 215
graeffei, Aritranis, 391
grallator, Labena, 176
grandii, See cingulum
grandis, See cingulum (*Macrocentrus*)
grandis, Megalochus, 428
greenei, Megarhyssa, 380
griffithsi, Epimicta, 252
griseisetosus, Afrocampsis, 285
gutfreundi, Polysphincta, 161
gyrator, Meteorus, 296

H
hadrogaster, Chelonus, 317
haesitator, Glypta, 403
hebetor, Habrobracon (*Bracon, Microbracon*)[1], 31, 54, 59, 61, 64–5, 67, 69, 72, 74–5, 82–4, 108–10, 112–13, 115–17, 119, 120, 122–3, 125, 129, 143, 156, 260, 263–8, 462, 466, 468, 499, 525
heliophilus, Aritranis, 391
heterocerus, Tersilochus (*Isurgus*), 429
hirticornis, Lysiphlebus, 218, 477
hirtus, Opius, 256–7
horticola, Hyposoter, 94, 97, 432, 485
hospes, Agrothereutes, 462
hutsoni, Meteoridea, 337, 338
hymeniae, Cardiochiles, 314
hypermeces, Dolichomitus, 36, 369
hyperodae, Microctonus, 62, 116, 142, 308
hyphantriae, Cotesia, 325
hypochondriaca, See rufipes

I
ictericus, Meteorus, 296–7
inanitus, Chelonus, 77, 83–4, 142–3, 147, 149, 153, 160, 315–17, 473
indagator, Dolopsidea, 242
indicus, Trioxys, 216
indiensis, Glyptapanteles, 147, 149, 334
indistinctus, Pleolophus, 486
initiator, Atanycolus, 231
insidiator, Habronyx, 431
instigator, See rufipes
insularis, Chelonus, 147, 153
interrupta, Tranosemella, 148
introita, Diapetimorpha, 13, 393, 474
invisor, Phaeogenes, 120, 309, 397
iridescens, Macrocentrus, 286
irrorator, Sigalphus, 276
iwatensis, Mesochorus (*Plectochorus*), 421

J
jacunda, Diaparsis, 428
japonica, Asobara, 109, 114, 116, 143–5, 216
japonicus, Lysiphlebus, 215
japonicus, Meteorus, See pulchricornis
japonicus, Stilbops, 409
japonicus, Townesion, 406
jaragua, Cantharoctonus, 242
juglandis, Diachasmimorpha, 123, 257
juglandis, See hebetor (*Habrobracon*)

K
kariyai, Cotesia (*Apanteles*), 139, 147, 154, 157–9, 327–30, 332, 335
kerri, Mesostoa, 226–7
kimballi, Digonogastra, 124, 525
konowi, Xanthopimpla, 368

L
laborator, Buathra, 390
laetatorius, Diplazon, 82, 116, 141, 363–4, 457
lancifer, Dyscoletes, 283
laphygmae, Meteorus, 295, 470
larvicida, Microctonus, 300
latrobei, Australoglypta, 403
lebophagus, Enicospilus, 85
lepidus, Orgilus, 54, 55, 293
lepidus, Syngaster, 110, 231, 233, 461, 482
leucobasis, Phanerotoma, 147, 316–17, 319
leucobasis, Xyeloblacus, 283
leucotrochus, Adelognathus, 384
leviventris, Meteorus, 472
limnophilus, Spilichneumon, 31
lipari, Polemochartus, 255
liparidis, Glyptapanteles, 147, 473
longicauda, Apechoneura, 354, 356
longicauda, Phytodietus, 412
longicaudata, Diachasmimorpha (*Biosteres*), 31, 54, 94, 123, 257–9
longiceps, Sinuatophorus, 302
longifemoralis, Atanycolus, 530
loretosa, Stethantyx, 426
lucidator, Centistes, 302
lunator, Megarhyssa, 32, 68
luteator, Lophyroplectus, 97, 413, 418, 464
luteus, Erythrodolius, 408
luteus, Ophion, 74
luxemburgense, Elasmosoma, 307
lyctaea, Syntretus, 309
lymantrae, Hyposoter, 148

M
macrosiphagum, Toxares, 215
macrura, Megarhyssa, 380
mactator, Trogus, 395, 463
maculator, Itoplectis, 374
maculicoxis, Pseudorhyssa, 378
malacostomatos, Aleiodes, 249
malvaceorum, Agathis, 124
mandibularis, Townesites†, 350
manducator, Alysia, 81, 109, 253, 255, 463
manifestator, Ephialtes, 369
marginiventris, Cotesia (*Apanteles*), 13, 96, 129, 147, 322, 325, 328, 332, 335, 421, 422
masneri, Sania, 323
masumotoi, Reclinervellus, 371
mediator, Microplitis, 13, 32, 97, 159, 322, 325, 334, 457, 459, 472
mediator, Orthopelma, 400
medicaginis, Apanteles, 120, 325
megacephalus, Rhimphoctona, 78
melanocephala, Itoplectis, 80, 84, 374
melanoscela, Cotesia, 99, 147, 155, 322
melitaearum, Cotesia, 94, 97, 480, 483, 485, 498

[1] Currently treated as senior synonym of *brevicornis* and *serinopae*, but there is considerable morphological variation within this species, and at least one cryptic species has been discovered.

mellitor, Bracon, 85, 130, 156, 261, 265, 474, 479
mesocentrus, Dolichomitus, 368
mininae, Amplicella†, 347
minor, Marjorietta†, 350
minutus, Phradis, 429
mirabilis, Palaeoichneumon†, 348
moiwanus, Arotes, 358
mollipla, Diadegma, 469
molokaiensis, Enicospilus (Pycnophion), 444
monticola, Olesicampe, 121
mora, Rudimentifera†, 348
morabinarum, Perilitus, 308
murinanae, Cephaloglypta, 403
mutator, Lissonota, 92, 184, 403
mycetophila, Mesocrina, 179
mystacinus, Histeromerus ,32, 109, 239–40

N
napo, Yelicones, 176
naranyae, Itoplectis, 93, 98, 358, 456, 462, 471, 472
narayani, Chelonus, 112, 316
nawaii, Brachycyrtus, 399
nemorensis, Podorgilus, 292
nielseni, Reclinervellus, 371
niger, Echthrus, 166
niger, Oxytorus, 425
nigra, Amphirhachis, 404
nigriceps, Toxoneuron (Cardiochiles), 58, 64, 68, 72, 83, 112, 145–7, 154, 156, 158, 160, 296, 314–15, 321, 474, 478
nigricollis, Priopoda, 418
nigricornis, Blacus, 283
nigridens, Tycherus, 180, 394
nigridus, Phaeogenes, 395
nigrifrons, Exenterus, 413
nigripes, Aphidius, 214–17, 222
nigripes, Mesochorus, 116–17, 421–2, 485
nigrithorax, Sphinctus, 412
nigritulus, Phradis, 429
nipponica, Pimpla, 73, 98, 358, 370, 376, 392, 456
nitidulidis, Microctonus, 116, 308
nixalebion, Microgaster, 326, 487
nonagriae, Cotesia, 322
nortoni, Megarhyssa, 380–1
nursei, Goryphus (Melcha), 393

O
oberiae, Brulleia, 280
obscurellus, Triaspis, 280
obscurus, Phytodoetus, 61, 414
occupator, Exephanes, 31
ocularis, Phanerotoma, 317
onkotorebus, Histeromeroides, 232
oregmae, Lipolexis, 214
orientalis, Hadrodactylus, 417
ornatum, Enizemum, 363–5
osculator, Tycherus (Phaeogenes), 394
ovumancora, Maxfischeria, 225

P
palitans, Praon, 215
pallida, Stantonia, 293
pallidator, Aleiodes, 246–7, 249
pallidipes, Microplitis, 334
pallidipes, Triaspis, 279
pallidus, Trioxys, 214, 469, 530
pallipes, Aphaereta, 59, 83, 92, 253–4
pallipes, Peristenus (Leiophron), 306, 512
pallipes, Polysphincta, 116
pallipes, Triaspis, 280
paludator, Chasmias, 31
paludum, Monoctonus, 139, 220–1
pammi, Habronyx, 458
papyraceus, Neurateles, 366
paradoxa, Earobia, 455
paropsidis, Eadya, 281
parorgyae, Hyposoter, 433
pectinophorae, Hbrobracon, 115
pectoralis, Glyptomorpha, 261
pectoratorius, Diplazon, 120, 363
pedias, Pholetesor, 322
peleopodae, Carinichremylus, 245
pequodorum, Praon, 214–15, 220
perfecta, Netelia, 95
persicae, Ephedrus, 116
persuasoria, Rhyssa, 100, 379–83
petiolata, Tanychora†, 344, 347
petiolatus, Trachypetus, 208
phyllotomae, Shawiana, 270
phytophagus, Bracon, 186, 264
pictus, Homotrophus, 364
pictus, Pauesia, 219
pistaciaecola, Monoctonia, 217
plagiator, Ephedrus, 72
planiceps, Dyscritulus, 221
platensis, Temeluca, 438
pleuralis, Asobara, 109
pleuralis, Oetophorus, 417
plutellae, Cotesia (Apanteles), 62, 64, 96, 154, 157, 334, 463–4
podagricus, Triclistus, 424
pomorum, Scambus, 370
pedator, Xanthopimpla, 368
producta, Netelia, 96, 414
promissorius, Ichneumon, 394
prosopidis, Heterospilus, 48, 109, 124, 231, 459
pruinosa, Euceros, 445
pseudopallipes, Peristenus (Leiophron), 306, 512
pulchella, Sussaba, 365
pulchellus, Diadromus, 13, 61, 68–70, 82, 95, 97, 109, 122, 130, 133, 152, 155, 158, 394–5
pulcher, See nigripes
pulchricornis, Meteorus, 93, 94, 115, 116, 184, 185, 467, 472
pumila, Agathis, 272, 274, 480
punctus, Phradis, 429

Q
quadridentata, Ascogaster, 67, 68, 147, 465

R
rapae, Diaeretiella, 78, 111, 128, 131, 147, 154, 469, 472, 476
rastafari, Pambolus, 237
respinozae, Tricyphus, 85, 394–5
reticulatus, Ascogaster, 68, 83, 93, 160
rhopalosiphi, Aphidius, 62, 133, 213, 217, 222, 476
ridibundus, Mastrus, 98, 389, 391
rivalis, Hyposoter, 148
rixator, Megarhyssa, 382
roborator, Exeristes, 78, 83, 370, 474
robusta, Scirtetes, 434
romeroi, Sigalphus, 277–8
rostrale, Tranosema, 148, 149, 154, 401
rubecula, Cotesia (Apanteles), 67, 109, 131, 132, 134, 139, 142, 144, 145, 147, 332–4, 469
rufescens, Asobara, 457
ruficollis, Exeristes (Ephialtes), 61, 370, 371, 480
rufipes, Bassus, 116
rufipes, Pimpla, 12, 88, 144–5, 164, 375, 463
rufipes, Xorides, 351
rufiscutellaris, Glypta, 403
rufiventris, Homolobus, 291
rufiventris, Microplitis, 61, 95, 147, 325, 327
rufofemoratus, Austroearinus, 176

S
sarcitorius, Ichneumon, 31, 180, 396
satuatorius, Vulgichneumon, 31, 396
scaposum, Heteropelma, 430, 458
schuetzeanus, Panteles, 410
scrobiculatus, Chelonus, 317

seductor, Acroricnus, 54, 166, 391
semiclausum, Diadegma, 59, 60, 64, 70, 96, 120, 122, 135, 149, 157, 432, 433, 435, 437
semiflava, Alomya, 387
semiothesae, Euceros, 447
sensibus, Euceros, 446
sericata, Apsilops, 390
serinopae, See hebetor (Habrobracon)
serotinus, Sphinctus, 416
sesamiae, Cotesia (Apanteles), 54, 95, 109, 114, 131, 147, 322
seyrigi, Leptorhaconotus, 232
siculus, See concolor
simile, Orthopelma, 400
simplicipes, Apophua, 147, 149, 151
simulans, Praon, 94, 215
sinuata, Stethantyx, 429
sonorensis, Campoletis, 74, 96, 129, 146, 148, 149, 152, 154, 156, 432–4, 474, 478
spathiiformis, Evaniodes, 232
splendidulus, Microleptes, 448
splendidus, Syntretus, 309
stelleri, Microctonus, 300
stemmator, Xanthopimpla, 98, 166, 376
sternata, See maculicoxis
stigma, Agathis, 272
stigmaterus, See stigma
stigmator, Aleiodes, 248
stokesii, Glabriodorsum, 102, 393
subordinaria, Cotesia, 330
sulcatus, Chelonus, 317
svalbardicum, Diaretellus, 221
sylvaticae, Aspilota, 179

T

tabida, Asobara, 62, 83, 96, 99, 108–9, 114, 134, 140–2, 144–5, 160, 254, 457–8, 478–9
tachardiae, Bracon, 263
tarsatorius, Syrphoctonus (Diplazon), 78, 365
tarsoleuci, Buathra, 391, 393
tasmanica, Dolichogenidea, 460
tedellae, Choeras (Apanteles), 157
telengai, Cotesia, 480
tenthredinis, Mesoleius, 75, 156, 416, 449
tenuicaudus, Trioxys, 214
tenuifemur, Tranosema, 148
terebrans Eriborus (Diadegma), 148, 432
terebrator, Phradis, 429
testacea, Meteoridea, 337
testaceipes, Lysiphlebus, 116, 218
testaceus, Aleiodes, 157, 247
testaceus, Scolobates, 417
texanus, Alabagrus, 118–19, 275
tibiatorius, Diplazon, 364
tomostethae, Glyptorhaestus, 419
toreutae, Phanerotoma, 316
triangulifera, Lycorina, 407
tricarinatus, Trieces, 424
tricolor, Exenterus, 413
tristis, Blathyplectes, 116, 432–3
trogoides, Pselaphanus, 276
tuberculator, Eubazus (Brachistes), 279
tuberculifera, Microplitis, 327
tunetanusi, Agrothereutes, 392
turionellae, Pimpla (Coccygomimus), 13, 14, 61–4, 72, 74, 75, 83, 111, 139, 142, 144, 166, 374, 375, 474, 479

U

uchidai, Glypta, 404
ulmicola, Atanycolus, 268
unicolor, Agathis, 274
unicolor, Aleiodes, See pallidator
unifasciatus, Euceros, 445
unilachni, Pauesia, 213, 465
utilis, Trioxys, 217

V

vagatoria, Megarhyssa, 380, 382
vandenboschi, Fopius (Biosteres), 31, 62, 257
variegata, Schlettereriella, 232
varipes, Homotherus, 396
verdui, Cuniculobracon, 280
vesicarii, Monoctonia, 215
vesicarius, Scambus, 466
vesparum, Sphecophaga, 116, 171–2, 389, 392, 480
vestalis, Cotesia, 109, 147, 205, 325, 459, 479
vestigator, Certonotus, 120
vinulae, Netelia, 377, 415
vipioniformes, Procinetus, 358
viridescens, Pimpla, 368
vittatae, Microctonus, 116, 300
vitticole, Eiphosoma, 438, 458
volucre, Praon, 213, 217–18
vulgaris, Eriborus, 437
vulgaris, Phytodietus, 464
vulnerator, Pristomeris, 438
vulsus, Virgulibracon, 267

W

wissmannii, Protaphidius, 218

Y

yakushimensis, Batothecoides, 246
yakutatensis, Cotesia, 325

Z

zealandicus, Doryctopsis, 235
zealandicus, Microctonus, 300
zerus, Ophion, 175

www.ingramcontent.com/pod-product-compliance
Ingram Content Group UK Ltd.
Pitfield, Milton Keynes, MK11 3LW, UK
UKHW051039091026
15624UKWH00011B/578